PORPHYRINS AND METALLOPORPHYRINS

PORPHYRINS AND METALLOPORPHYRINS

A new edition based on the original volume by J. E. Falk

Edited by

KEVIN M. SMITH

University of Liverpool

 1975

ELSEVIER SCIENTIFIC PUBLISHING COMPANY

AMSTERDAM − OXFORD − NEW YORK

ELSEVIER SCIENTIFIC PUBLISHING COMPANY
335 Jan van Galenstraat
P.O. Box 211, Amsterdam, The Netherlands

AMERICAN ELSEVIER PUBLISHING COMPANY, INC.
52 Vanderbilt Avenue
New York, New York 10017

Library of Congress Cataloging in Publication Data
Main entry under title:

Porphyrins and metalloporphyrins.

    Includes bibliographical references and index.
    1. Porphyrins and porphyrin compounds.
I. Falk, J. E. Porphyrins and metalloporphyrins.
II. Smith, Kevin M.
QD401.P7 1975   547'.593   75-26551
ISBN 0-444-41375-8

ISBN 0-444-41375-8

Printed in The Netherlands

PREFACE

In 1964, J.E. Falk was able to give one man's view of the porphyrin and metalloporphyrin field at possibly the last time that such a major task was possible. Since then the area has mushroomed outwards and blossomed in a quite remarkable manner, and this may in no small way be due to the stimulus provided by the appearance of *Porphyrins and Metalloporphyrins*. Around the time of his death, Falk was addressing himself to the task of updating and revizing his highly successful book, and realizing the magnitude of the undertaking, had begun to gather about him various colleagues who might be willing to contribute to a multi-authored Second Edition. Alas, all of this came to nothing, but the demand for a new and expanded edition of *Falk* remained.

The present book represents an attempt by some of the leading authorities in the field to fill the gap left by the progress of porphyrin chemistry past the account written by Falk. It differs considerably from the original, mainly in size, but also in the organization of the chapters into eight sections. A detailed description of the content of each section would be out of place here, but it is worth commenting that sections dealing with synthetic and biological aspects have been added, as well as chapters dealing in depth with many spectroscopic methods which were only in their infancy in 1964. In the whole book Falk's 'systematic and rational exposition which would have been quite impossible some years ago' has been carried to the boundaries of present research.

An attempt has been made to retain the idea of a 'Laboratory Handbook' which was the underlying concept in *Falk*. Though the book has grown dramatically in size, a substantial section dealing with laboratory methods has been included. Many of the procedures have not been noticeably improved since 1964, and in these cases there are few changes from the account written by Falk; however, in other respects, the Laboratory Methods Section has been expanded and revized.

The manuscript deadline for the present book was 1st January 1975, and with the exception of a few chapters which arrived in late March 1975, literature published after December 1974 has not been considered. However, as always when accounts are written by active research workers in the field, the chapters contain abundant references to unpublished work, or work in press, as well as personal communications from other researchers. In order to preserve the timeliness of the contributions in this book, it was necessary to proceed to publication without two manuscripts which were estimated, by the authors, to be one month or more from completion on 1st June 1975.

It is a pleasure to record my thanks to Professor G.W. Kenner, F.R.S., for his advice and encouragement during the past two years. In contributed volumes of this type with about twenty chapters, crises arise at fairly regular intervals; I would like to thank Professor Dr J.W. Buchler (Aachen) and Dr J.-H. Fuhrhop (Stöckheim über Braunschweig) for unhesitating assistance during these difficult times, and for providing the encouragement to go ahead in the formative days of this project.

Liverpool, June 1975   Kevin M. Smith

CONTENTS

Chapter 2   *Synthesis and preparation of porphyrin compounds*
            Kevin M. Smith

SECTION B: BIOLOGICAL ASPECTS

Chapter 3   *Biosynthesis of porphyrins, chlorins, and corrins*
            A.R. Battersby and E. McDonald

Chapter 4   Heme cleavage: biological systems and chemical analogs
Pádraig O'Carra

SECTION C: COORDINATION CHEMISTRY OF METALLOPORPHYRINS

Chapter 5   Static coordination chemistry of metalloporphyrins
J.W. Buchler

Chapter 6   *Dynamic coordination chemistry of metalloporphyrins*
Peter Hambright

Chapter 7   *Metalloporphyrins with unusual geometry*
Minoru Tsutsui and Glenn A. Taylor

SECTION D: DETERMINATION OF MOLECULAR STRUCTURE

Chapter 8   Stereochemistry of porphyrins and metalloporphyrins
J.L. Hoard

Chapter 9   Mass spectrometry of porphyrins and metalloporphyrins
Kevin M. Smith

Chapter 10 *Nuclear magnetic resonance spectroscopy of porphyrins and metalloporphyrins*
Hugo Scheer and Joseph J. Katz

Chapter 11   *Vibrational spectroscopy of porphyrins and metalloporphyrins*
Hans Bürger

SECTION E: ELUCIDATION OF ELECTRONIC STRUCTURE

Chapter 12   *Mössbauer spectroscopy and magnetochemistry of metalloporphyrins*
Peter Hambright and Alan J. Bearden

Chapter 13   *Electron paramagnetic resonance spectroscopy of porphyrins and metalloporphyrins*
J. Subramanian

SECTION F: CHEMICAL REACTIVITY

Chapter 14 *Reversible reactions of porphyrins and metalloporphyrins and electrochemistry*
Jürgen-Hinrich Fuhrhop

Chapter 15   Irreversible reactions at the porphyrin periphery (excluding photochemistry)
Jürgen-Hinrich Fuhrhop

Chapter 16   Photochemistry of porphyrins and metalloporphyrins
Frederick R. Hopf and David G. Whitten

Chapter 17   Photochemistry of porphyrins in membranes and photosynthesis
David Mauzerall and Felix T. Hong

SECTION G: STRUCTURAL ANALOGS OF PORPHYRINS

Chapter 18   Structural analogs of porphyrins
A.W. Johnson

SECTION H: LABORATORY METHODS

Chapter 19   *Laboratory methods*
Jürgen-Hinrich Fuhrhop and Kevin M. Smith

APPENDIX: ELECTRONIC ABSORPTION SPECTRA   871

Section A

GENERAL AND SYNTHETIC ASPECTS

1

GENERAL FEATURES OF THE STRUCTURE
AND CHEMISTRY OF PORPHYRIN COMPOUNDS

KEVIN M. SMITH

The Robert Robinson Laboratories, University of Liverpool, U.K.

Massive contributions to our knowledge of the structure and chemistry of porphyrins, metalloporphyrins, and related compounds were accumulated in the first half of this century. Much of this has been summarized in the works of Willstätter and Stoll[1], Fischer and Orth[2], Fischer and Stern[3], and Lemberg and Legge[4]. An abbreviated description of the life and work of Hans Fischer[5] has enabled the contributions of the Munich School to be put into perspective and one can only marvel at the way this classical work was carried through without the aid of the sophisticated gadgetry and spectroscopic techniques that are today considered to be essential.

In 1964, Falk's *Porphyrins and Metalloporphyrins* was published; this rather short book pointed the way that contemporary porphyrin chemistry was to proceed and has since become a key work in the field. More recently, the more physical aspects of porphyrin research have been summarized in the report[6] of a Symposium held in New York by the New York Academy of Sciences.

It is the intention in this chapter to provide an overall view of the chemistry of porphyrins, metalloporphyrins, and related compounds. Its purview is a 'thumb-nail' sketch, which will be expanded in the subsequent chapters of this work.

1.1. Structures and Nomenclature

The porphin nucleus (1) consists of four 'pyrrole-type' rings joined by four methine bridges to give a macrocycle. This cyclic tetrapyrrole structure (1) was first suggested by Küster[7] in 1912; at that time it was thought that such a large ring would be unstable, and this structure was not accepted by Fischer, the father of contemporary porphyrin chemistry, until much later. However, Fischer did have a hand in the final proof, which was provided by his total synthesis of protoheme[8] in 1929.

Porphyrins and Metalloporphyrins, ed. Kevin M. Smith

TABLE 1

Trivial names and structures of common porphyrins

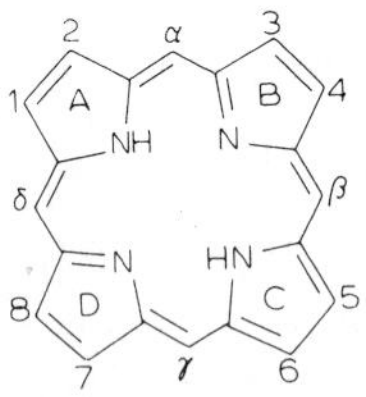

(1)

Porphyrin	1	2	3	4	5	6	γ	7	8
Etioporphyrin-I	Me	Et	Me	Et	Me	Et	H	Me	Et
Octaethylporphyrin	Et	Et	Et	Et	Et	Et	H	Et	Et
Deuteroporphyrin-IX	Me	H	Me	H	Me	P^H	H	P^H	Me
Mesoporphyrin-IX	Me	Et	Me	Et	Me	P^H	H	P^H	Me
Hematoporphyrin-IX	Me	CH(OH).Me	Me	CH(OH).Me	Me	P^H	H	P^H	Me
Protoporphyrin-IX	Me	V	Me	V	Me	P^H	H	P^H	Me
Coproporphyrin-I	Me	P^H	Me	P^H	Me	P^H	H	Me	P^H
Coproporphyrin-III	Me	P^H	Me	P^H	Me	P^H	H	P^H	Me
Uroporphyrin-I	A^H	P^H	A^H	P^H	A^H	P^H	H	A^H	P^H
Uroporphyrin-III	A^H	P^H	A^H	P^H	A^H	P^H	H	P^H	A^H
Chlorocruoroporphyrin	Me	CHO	Me	V	Me	P^H	H	P^H	Me
Pemptoporphyrin	Me	H	Me	V	Me	P^H	H	P^H	Me
Deuteroporphyrin-IX 2,4-di-acrylic acid	Me	Acr^H	Me	Acr^H	Me	P^H	H	P^H	Me
2,4-Diformyldeuteroporphyrin-IX	Me	CHO	Me	CHO	Me	P^H	H	P^H	Me
2,4-Diacetyldeuteroporphyrin-IX	Me	Ac	Me	Ac	Me	P^H	H	P^H	Me
Deuteroporphyrin-IX 2,4-disulfonic acid	Me	SO_3H	Me	SO_3H	Me	P^H	H	P^H	Me
Phylloporphyrin-XV	Me	Et	Me	Et	Me	H	Me	P^H	Me
Pyrroporphyrin-XV	Me	Et	Me	Et	Me	H	H	P^H	Me
Rhodoporphyrin-XV	Me	Et	Me	Et	Me	CO_2H	H	P^H	Me
Phylloerythrin	Me	Et	Me	Et	Me	CO———	CH_2	P^H	Me
Desoxophylloerythrin	Me	Et	Me	Et	Me	CH_2———	CH_2	P^H	Me
Pheoporphyrin-a_5	Me	Et	Me	Et	Me	CO———	CH | CO_2Me	P^H	Me

Side-chain abbreviations: Me = Methyl; Et = Ethyl; V = Vinyl; P^R = $CH_2CH_2CO_2R$; A^R = CH_2CO_2R; Ac = CO.Me; Acr^H = $CH=CH.CO_2H$

1.1.1. *Nomenclature*

Porphyrins are formally derived from porphin (1) by substitution of some or all of the peripheral positions with various side-chains. In the classical system of nomenclature, which will be used throughout this book, the peripheral positions are numbered from 1 to 8 (cf. 1) and the 'interpyrrolic' methine positions, usually termed '*meso*', are designated α, β, γ, and δ. The rings are usually lettered A, B, C, and D, though Roman numerals were preferred in some earlier texts.

Thus, the etioporphyrins, which carry four methyl and four ethyl groups in the eight peripheral positions (with one ethyl and one methyl on each 'pyrrole' subunit) exist as four isomers. These are given the 'type isomer' numbers I, (2); II (3); III (4); and IV (5). The correct systematic name for,

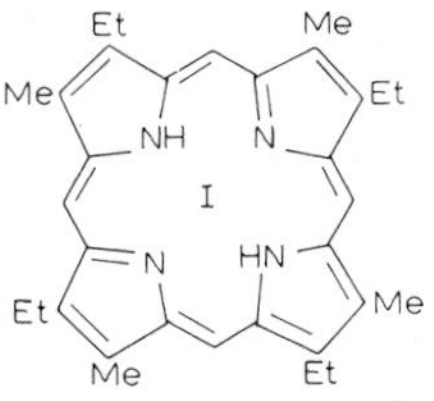

(2)

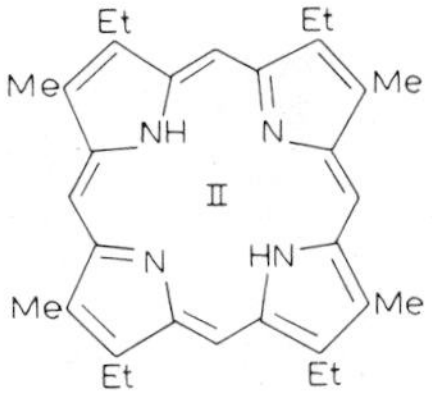

(3)

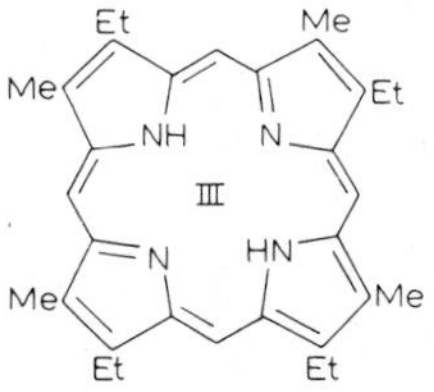

(4)

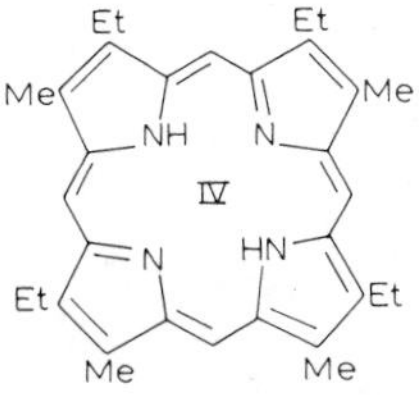

(5)

for example, etioporphyrin-III (the isomer related to natural products), is 2,4,6,7-tetraethyl-1,3,5,8-tetramethylporphin; it is perhaps unfortunate that trivial names have been attached to many porphyrins of biological significance, but this is because the precise structures of many of these compounds were only worked out some time after their discovery. Table 1 lists the trivial names and structures of the better known examples.

Nomenclature begins to get complicated when one considers the numeration of peripheral side-chains (an essential, for example, in ^{13}C spectra and X-ray investigations) or when complex metal-axial ligand systems are introduced. These fine details will be dealt with in the chapters concerned.

The common side-chain abbreviations are collected together at the foot of Table 1, and these will be used consistently throughout this book.

Open-chained tetrapyrroles (6) are numbered according to the current usage in bile pigment chemistry. Thus, the 'β-pyrrolic' positions are num-

References, p. 27

bered 1—8, the 'α-pyrrole' positions 1'—8', and the interpyrrolic carbons *a*, *b* and *c*. The tetrapyrrole shown in (6) is termed a *bilane* (the bile pigment analog bearing terminal lactam rings is *bilinogen*); introduction of unsaturation at position *a*, *b*, or *c* affords *bilenes* (e.g. (7), *b*-bilene) and one can proceed from there to *biladienes*, e.g. (8) and *bilatrienes*, e.g. (9). The bile pigment analogs of (8) and (9) (i.e. those possessing lactam oxygen functions at 1' and 8') are called *bilidienes* and *bilitrienes* respectively. Open-chained tetrapyrroles will be drawn in the cyclic (rather than linear) form throughout this book, in order to show them as they almost certainly exist. Space filling molecular models show that, due to steric repulsion of the 1—8 substituents, these molecules are forced into a coiled conformation, and this has had a profound influence on procedures for the synthesis of porphyrins from open-chained tetrapyrrolic intermediates (Chapter 2).

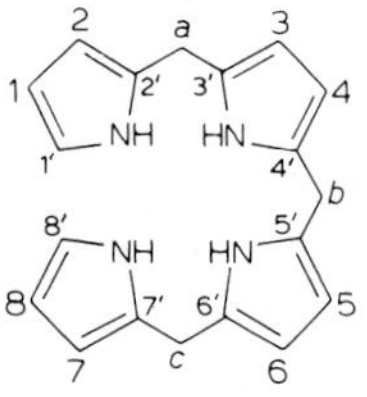

(6) 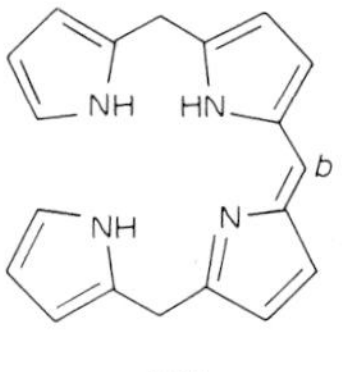   (7)

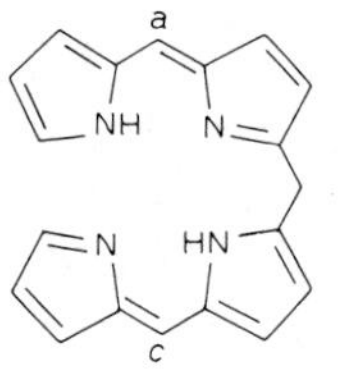

a,c-Biladiene   (9)

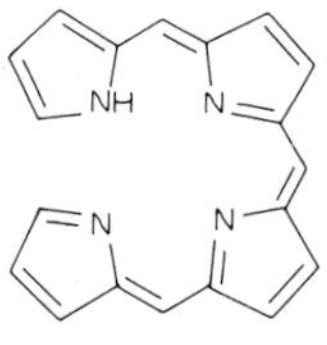

(8)

1.1.2. Related structures
1.1.2.1. Reduced porphyrin macrocycles

Chlorins are 7,8-dihydroporphyrins (10) in which the saturation is traditionally shown in ring D; magnesium complexes of chlorins are the various *chlorophylls*, which are not discussed in detail in this work, but are mentioned. Phlorins (11) are also dihydroporphyrins, and several routes to these have been published (Chapter 14). The macrocyclic conjugation in the chlorins is still maintained, but in phlorins, one hydrogen is added to nitrogen and the second to a *meso*-carbon. The resultant interruption in the macrocyclic conjugation of the latter is reflected in the visible absorption spectra.

Two types of tetrahydroporphyrin have been observed; these are the so-called a- and b-tetrahydroporphyrins. The a-compound (12) has two hydro-

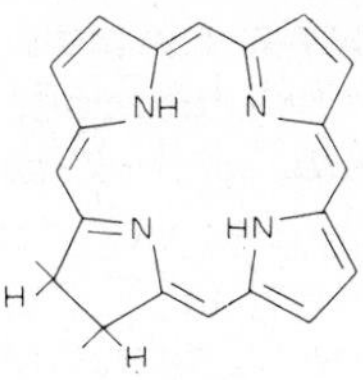

(10)

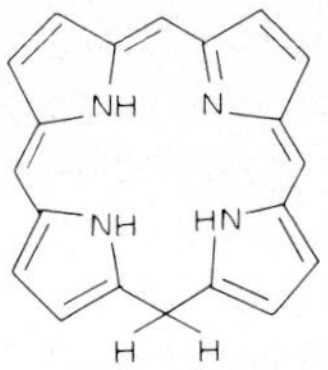

(11)

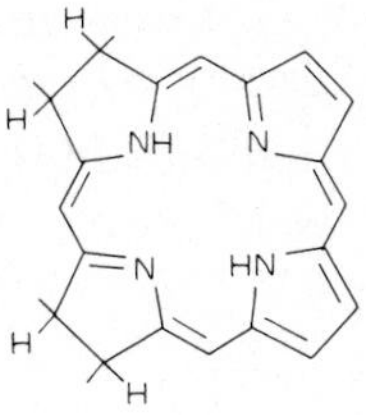

(12)

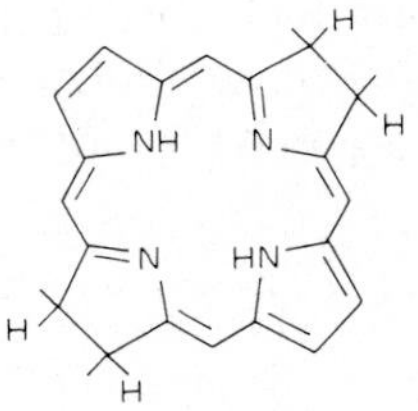

(13)

gens added in each of two adjacent rings, whereas the b-derivative (13) has opposite subunits reduced. The chromophore in (13) is found in bacterio-chlorophyll-*a* and in a slightly modified form in bacteriochlorophyll-*b* (which is now known[9] to be a dihydroporphyrin). Both of the structures (12) and (13) still retain full conjugation and aromaticity, though less resonance forms can be written for them. Tetrahydroporphyrin 'porphomethenes' are also known[10].

Examples of more highly reduced macrocycles are known. Corphins, e.g. (14) (Chapter 18) are at the hexahydroporphyrin state and have been synthesized[11] in the hope that they might represent a biosynthetic connection between the porphyrin and corrin (15) ring systems. The most important hexahydroporphyrins are the porphyrinogens (16), which are colorless and

(14)

(15)

(16)

References, p. 27

contain four pyrrole rings linked in a macrocycle by four methylene groups; the importance of these materials lies in the fact that they are the immediate biosynthetic precursors of the porphyrin and plant pigments, as well as vitamin B_{12} (Chapter 3).

1.1.2.2. Oxidized porphyrin macrocycles

Macrocycles with oxygen functions at one (17), two (18), three (19), and four (20) *meso*-positions have been prepared (Chapter 15). The oxophlorins (17) are of particular importance because iron complexes of these are thought to be intermediates in the catabolism of porphyrins leading to bile pigments (Chapter 4). Macrocycles with carbonyl groups on peripheral posi-

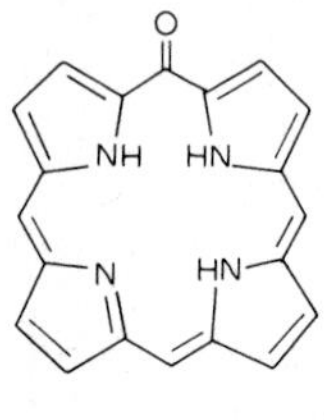

(17)

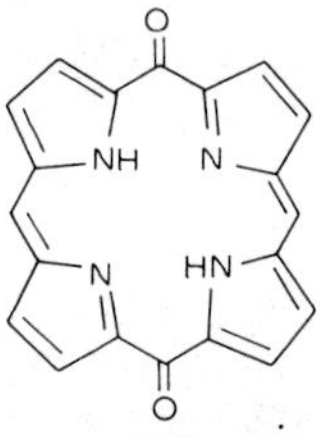

(18)

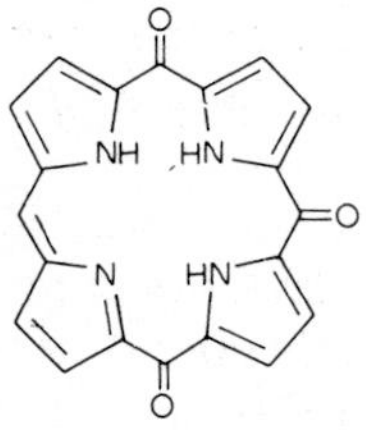

(19)

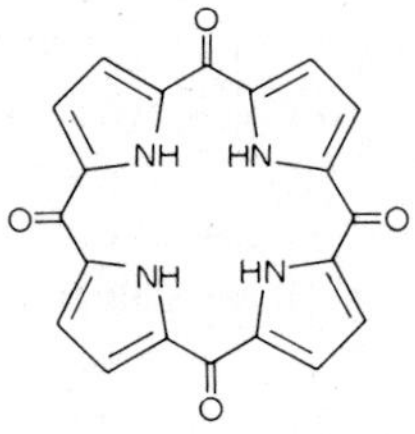

(20)

tions have also been prepared[1,2] by pinacol rearrangement of the corresponding glycol.

1.1.2.3. Porphyrin analogs

A great deal of work has been carried out on corroles (21), tetradehydrocorrins (22), and their derivatives, which have a carbon skeleton reminiscent

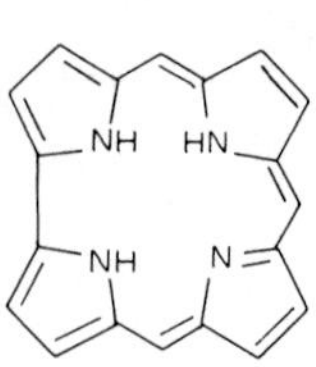

(21)

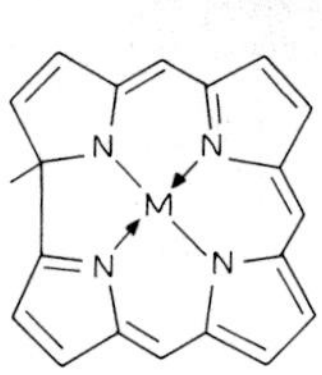

(22)

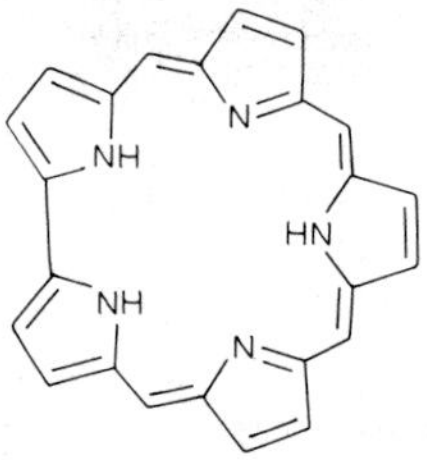

(23)

of corrin (15). Details of these, as well as sapphyrins (23) and porphyrin analogs containing mixed heteroatoms, can be found in Chapter 18.

1.1.3. Isomerism in porphyrins

With the assumption that like side-chains are not substituted on the same pyrrole nucleus, Fischer[2] pointed out that in a peripherally octasubstituted porphyrin with only two types of side-chain there are four possible type-isomers. These are illustrated in the etioporphyrin case in structures (2) to (5). Related isomers are found in the uroporphyrins (A^H, P^H side-chains) and coproporphyrins (Me, P^H). When there are three different side-chains, as in mesoporphyrin and protoporphyrin, fifteen type-isomers are possible. Protoporphyrin-IX (Table 1) is systematically related to etioporphyrin-III, as are uroporphyrin-III and coproporphyrin-III and their respective porphyrinogens, which are the biosynthetic precursors of protoporphyrin-IX and protoheme. Every porphyrin and metalloporphyrin known to have a metabolic function is modelled on this general pattern of side-chain substitution, including the prosthetic groups of all the known heme proteins, chlorophylls, and vitamin B_{12} (Chapter 3). Uroporphyrin-I and coproporphyrin-I, both related to etioporphyrin-I as described above for the type-III isomer, occur under certain pathological conditions, and are formed in some in vitro experiments on porphyrin biosynthesis.

The analytical identification of isomers, particularly of uro- and coproporphyrins, is of great practical importance. All of the chromatographic methods available (p. 839) have limitations, though high-speed high-pressure liquid chromatography will almost certainly prove to be successful in the long run. Melting points and X-ray powder diagrams have some limitations for the identification of porphyrin type-isomers, principally because porphyrins have the troublesome ability to crystallize in a variety of forms[13]. Infrared spectroscopy has been used[14] to differentiate between coproporphyrins I and III, but in recent times, nuclear magnetic resonance spectroscopy (Chapter 10) has been used with success for identification of a variety of type-isomers of porphyrins. A recently described method[15] involving n.m.r. spectroscopy of mercury(II) porphyrins has enabled differentiation of all four etio- and coproporphyrin type-isomers; the problem of detection of small quantities of one type-isomer in another, however, still remains, and it is here that one must look to improvement in chromatographic techniques.

1.2. General Chemistry
1.2.1. Aromaticity of the macrocycle

The porphin macrocycle is highly conjugated and a number of resonance forms can be written. There are nominally 22 π-electrons, but only 18 of these are included in any one delocalization pathway; this conforms with

Hückel's 4n + 2 rule for aromaticity. It also explains why chlorins and bacteriochlorins are aromatic compounds. Porphyrins are highly colored, their main absorption bands have very high extinction coefficients, and the intense 'Soret' band, found around 400 nm is characteristic of the macrocyclic conjugation; as might be expected, rupture of the macrocycle (for example, to give biliverdin-type bile pigments) results in disappearance of this band.

Aromatic character in porphyrin compounds has been confirmed by measurements of their heats of combustion[16]. Recent X-ray investigations (Chapter 8) of both metal-free porphyrins and metalloporphyrins have shown, in startling fashion, the basic planarity of the nucleus; this is a basic requirement for aromatic character.

Another criterion for aromaticity makes use of n.m.r. spectroscopy (Chapter 10). The shielded NH protons in porphyrins appear at very high field (approx. 14–15 τ) whereas the outer, *meso*-protons, being deshielded by the aromatic ring current, appear around zero τ. These shifts contrast well with those in bile pigments (e.g. methine protons between 3 and 4.5 τ in a,b,c-bilitrienes) which possess no macrocyclic ring current.

A number of electrophilic substitution reactions (Chapter 15) have been shown to occur on the unsubstituted positions of both porphyrins and metalloporphyrins; typical examples are nitration, halogenation, sulfonation, formylation, acylation, and deuteration. All of these reactions would be expected of an aromatic heterocycle, as would the basic stability of the nucleus to many reagents.

The startling abundance of doubly-charged ions and the stability of the molecular ion towards fragmentation in the mass spectra of porphyrins and their derivatives (Chapter 9) also attest to the aromatic nature of the nucleus.

1.2.2. Tautomerism in the macrocycle

The question of existence of different NH tautomers of porphyrins has long been argued. Infrared evidence has suggested that the NH groups in porphyrins are involved in intramolecular hydrogen bonds[17]. Some X-ray evidence[18], theoretical arguments, and [13]C n.m.r. studies[19] (which have been challenged[20]) appeared to favor the 16-membered 18π electron system (24) for the delocalization pathway in porphyrins. However, more recent X-ray studies of metal-free porphyrins have revealed (Chapter 8) that the bond lengths are more consistent with the structure (e.g. 25) in which there are two like-opposite nitrogens. Infrared data and calculations of orbital overlap[21] have led to the conclusion that the most stable form of porphin (1) is that in which the two inner hydrogens are bonded to opposite nitrogens. The less symmetrical form (26) with hydrogens bonded to adjacent nitrogens is considerably less stable because of penetration of each hydrogen into the Van der Waals sphere of the other.

A low-temperature n.m.r. study[22] has allowed the observation of the two tautomers (25) and (26); the rate of tautomerism has been measured[23] at three different temperatures for both *meso*-tetraphenylporphyrin and its N,N'-dideutero derivative, and the kinetic parameters, together with the kinetic isotope effect, are consistent with a two-step process (25) ⇌ (26) ⇌ (27) rather than a concerted two-proton shift (25) ⇌ (27). Line broadening

(24) 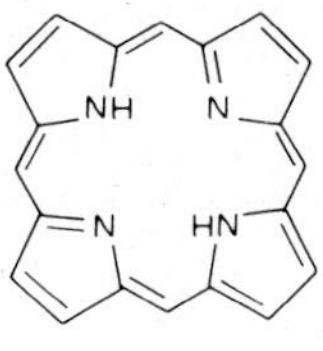 (25)

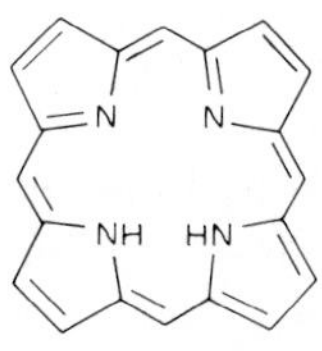

(26) 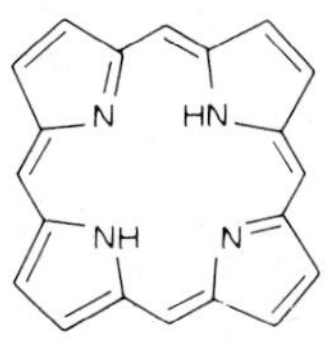 (27)

of some resonances in the ^{13}C n.m.r. spectra of metal-free porphyrins in deuterochloroform solution has been shown to be due to NH tautomerism rather than ^{14}N quadrupole effects[20].

1.2.3. Ionization of porphyrins
1.2.3.1. Nitrogen atoms

The porphyrin macrocycle can be regarded as an ampholyte with two pyrrolenine nitrogen atoms capable of accepting protons and two NH groups capable of losing protons. Full discussions of acid-base properties (Chapter 6) and replacement of the 'inner hydrogens' with metal ions (Chapters 5, 6) can be found elsewhere in this monograph.

The most useful scheme for assigning pK values to the porphyrin nucleus is that of Phillips[24]:

$$pK_1 = pH - \log[P^{2-}]/[PH^-]$$
$$pK_2 = pH - \log[PH^-]/[PH_2]$$
$$pK_3 = pH - \log[PH_2]/[PH_3^+]$$
$$pK_4 = pH - \log[PH_3^+]/[PH_4^{2+}]$$

Thus, pK_1 and pK_2 refer to the acidic equilibria involved in the dissociation of protons from the pyrrole-type nitrogens, and pK_3 and pK_4 to the basic equilibria associated with addition of protons to the pyrrolenine nitrogens.

References, p. 27

Porphyrins behave as very weak acids, sodium alkoxides being required to allow spectroscopic observation of the dianion (P^{2-}). Both pK_1 and pK_2 for etioporphyrin have been estimated[25] to be of the order of +16. The basicity of the porphyrins is evidenced by their solubility in dilute mineral acids, in which the dication (PH_4^{2+}) is usually observed spectroscopically. Measurements of basicity in aqueous solution are, of course, limited by solubility considerations, which have led to the use of techniques such as potentiomet-

TABLE 2

Basicity of substituted deuteroporphyrin-IX dimethyl esters

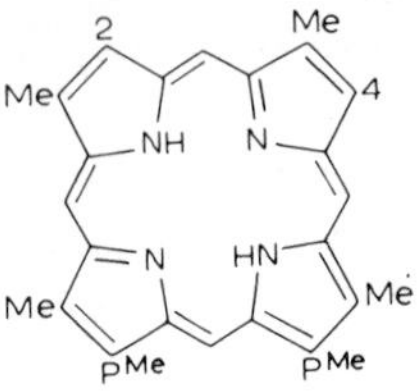

Porphyrin	Substituents		pK_3 [a] ± 0.1 unit	Band I [b] (nm)	Soret [a] (nm)
	2	4			
Meso	Et	Et	5.8	620	399
Deutero	H	H	5.5	618	398
Copro	p^{Me}	p^{Me}	5.5 [c]	620	399
Proto	V	V	4.8	630	408
2,4-Diacetyldeuterodiox-ime	C(Me)=NOH	C(Me)=NOH	4.5	625	403
2-Formyl-4-vinyl-deutero-oxime	CH=NOH	V	4.4	635	415
2,4-Diformyldeuterodi-oxime	CH=NOH	CH=NOH	4.3	639	414
4-Propionyldeutero	H	CO-Et	4.2		409
4-Formyldeutero	H	CHO	3.8[c]	640	413
2-Formyl-4-vinyldeutero (chlorocruoro)	CHO	V	3.7	644	
2,4-Diacetyldeutero	CO-Me	CO-Me	3.3	639	424
2,4-Dipropionyldeutero	CO-Et	CO-Et	3.2		423
4-Nitrodeutero	H	NO_2	3.2		401
2-Vinyl-4-cyanodeutero	V	CN	3.0 [d]		413
2,4-Di-methoxycarbonyl-deutero	CO_2Me	CO_2Me	3.0 [d]		423
2,4-Dibromodeutero	Br	Br	3.0[d]		
2,4-Diformyldeutero	CHO	CHO	3.0 [c,d]	651	436

[a] Measured at $25°C$ in 2.5% sodium dodecyl sulfate.
[b] In dioxan.
[c] By extrapolation from values at $20°C$.
[d] ± 0.2 units, inaccuracies introduced by ionic strength variations.

ric titration in glacial acetic acid[26] and spectroscopic titration in nitroben-zene[27] and aqueous detergent solutions[28].

The values of pK_3 (formation of the monocation) for a number of related porphyrin esters, determined in an aqueous sodium dodecyl sulfate, are given in Table 2; the base weakening (electron-withdrawing) effect of some side-chains is readily seen. The monocationic species of many porphyrins can be observed in anionic detergent systems, in which the three species PH_2, PH_3^+, and PH_4^{2+} are all solubilized. Alternatively, the monocations can be ob-served[29] in toluene—acetic acid solutions. The constants observed in deter-gent solutions are parallel to the true water values, and are about 1 pK unit more positive than the aqueous values for the porphyrin esters (Table 2), and would be about 1 pK unit more negative than the porphyrin free carboxylic acids.

pK_4 for a number of porphyrins has been determined in a similar way to the values in Table 2. The two values (pK_3 and pK_4) differ by only 2 or 3 pK units. It is this small difference which makes observation of the mono-cation difficult, and it is no doubt due to increased stabilization by reso-nance of the symmetrical dication. On the other hand, in anionic detergent systems, where monocations are fairly readily observed, they are presumably stabilized by interaction with the negative field of the detergent micelle. Values of $pK_3 + pK_4$ for some porphyrins have also been calculated[24] from phase-distribution data, and these are consistent with values from detergent systems.

The transition from porphyrin to chlorin (10), and also the presence of the isocyclic ring in chlorophylls, tend to decrease the basicity of the pyrrole nitrogens; the high HCl-numbers (p. 14) of chlorins, however, reflect in-creased lipid solubility more than the weaker basicity.

Oxophlorins (17) are relatively strong bases, accepting one proton on nitrogen to form isolable monocations (28); a second proton can be added to oxygen, giving the dication (29) derived from the tautomeric hydroxy-

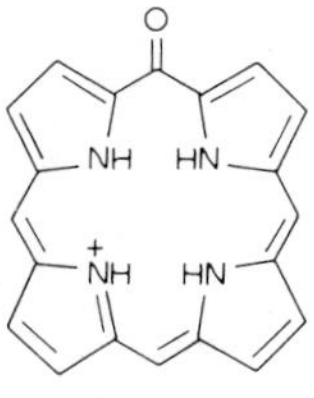

(28) 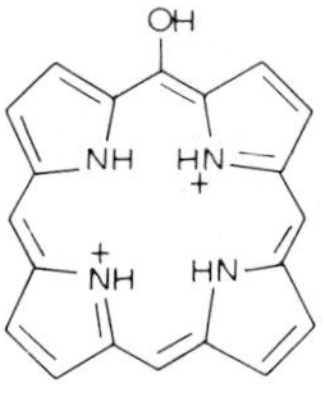   (29)

form[30]. Azaporphyrins add a third proton in solutions containing high con-centrations of sulfuric acid[29].

N.m.r. spectroscopy has been used to observe the slow (on the n.m.r. time scale) protonation of *meso*-tetraphenylporphyrin[31] and *N*-methyl porphy-rins[32]. This phenomenon could not be followed using the coproporphyrin esters and the higher barrier in *meso*-tetraphenylporphyrin is a result of the severe deformations from planarity of the porphyrin ring and the significant change[33] in the mean angle between the macrocycle and the phenyl rings induced by diprotonation[31]; there was no evidence of monoprotonated species.

1.2.3.2. Carboxylic acid side-chains

Naturally occurring porphyrins possess carboxylic acid side-chains which ionize simultaneously with the nuclear nitrogen atoms; the electrostatic field due to ionized carboxyl groups leads to an apparent increase in porphyrin basicity[34]. This complication is usually avoided by working with porphyrin esters, but, though the intrinsic pK_3 for coproporphyrin free acid in water can be estimated to be about 5, the effective pK_3 in aqueous solution is 7.2; in the biological environment, about 50% of this porphyrin would therefore be in the monocationic form[34].

TABLE 3

HCl numbers of porphyrins and chlorins

Compound	HCl Numbers	
	Porphyrins and chlorins	Corresponding methyl esters
Uroporphyrin-III		5.0
Coproporphyrin-III	0.09	1.7
Hematoporphyrin-IX	0.1	
Deuteroporphyrin-IX	0.3	2.0
Mesoporphyrin-IX	0.5	2.5
Etioporphyrin-I	3.0	
Protoporphyrin-IX	2.5	5.5
Chlorocruoroporphyrin	4.6	
Rhodoporphyrin-XV	4.0	7.5
Pyrroporphyrin-XV	1.3	2.5
Phylloporphyrin-XV	0.35	0.9
Pheoporphyrin-a_5	9.0	
Phylloerythrin	7.5	
Pheophytin-a		28.5*
Pheophorbide-a	15.0	
Pheophorbide-b		21.0
Mesopheophorbide-a	14.5	16.5
Chlorin-e_6	3.0	8.0
Mesochlorin-e_6	3.5	5.5
Rhodin-g_7	9.0	12.5

* Phytyl ester.

The common carboxylic acid side-chains found in natural porphyrins are acetic and propionic groups (e.g. in uroporphyrins); the pK's of corresponding pyrrole carboxylic acids are about 4.8. Direct measurements of the pK's of these acid groups on porphyrins have not been made, but calculations have shown[24] that if for a porphyrin monocarboxylic acid the pK is 4.8, the mean limiting pK (carboxylic groups all ionized) for a dicarboxylic acid would be approx. 5.7, for a tetracarboxylic acid 6.5, and for an octacarboxylic acid 7.3, the dissociation being modified by the electrostatic effect arising from other dissociated groups on the same molecule or in its immediate vicinity.

1.2.3.3. Acid (HCl) numbers

The HCl number (Willstätter number) of a porphyrin is defined[35] as that concentration of HCl in percent (w/v) which, from an equal volume of an ether solution of the porphyrin, extracts two thirds of the porphyrin. The HCl numbers for many porphyrins have been recorded[2] and typical examples are given in Table 3. These numbers depend jointly upon the dissociation of the porphyrin as a base, and upon its ether—water partition coefficient. In Table 3 the effects of hydrophilic, lipophilic, and electron-withdrawing substituents can all be seen.

Chlorins have greatly increased ether solubility and the increased lipid solubility of chlorophylls and their derivatives is partly due to this reduction, and partly to the esterification of their carboxylic groups with phytol and methanol. The isocyclic ring of chlorophylls and electrophilic side-chains such as acetyl (bacteriochlorophyll) and formyl (chlorophyll-*b*) also decrease the basicity of the ring nitrogens. The principal use of HCl numbers is in countercurrent distribution, a method which has found a great deal of application for the separation of mixtures of porphyrins (Chapter 19).

1.2.3.4. Complexation with metal ions

Porphyrins, chlorins, and many of their derivatives readily form complexes (in which one, or usually two, of the NH protons are lost) with a whole variety of metals. This is the subject of several of the subsequent chapters in this work and for that reason warrants no further amplification here.

1.2.4. Stability of porphyrin compounds

1.2.4.1. General features

Porphyrins with carboxylic acid side-chains do not have melting points, but their esters, and some metal complexes of these, melt in the region 200—300°C often without decomposition. Sublimation has occasionally been used for purification of porphyrins. It is historically significant that 'the whole classical edifice of porphyrin chemistry erected in Munich'[36]

References, p. 27

depended upon melting point comparisons, and these are still of great value, though they can be misleading[13,37,64].

The porphyrin nucleus is stable to concentrated sulfuric acid and neat trifluoroacetic acid, both of which are often used (Chapters 5, 6) to remove coordinated metals. It may be destroyed for example by perchloric acid, chromic acid, permanganate, or hydriodic acid, and therein lie the various methods for controlled degradation of the macrocycle for purposes of structure determination (Chapters 3, 15). Porphyrinogens are considerably less stable than their aromatized analogs and are readily randomized in acids[38].

1.2.4.2. Instability to light (photo-oxidation and photoreduction)

Solutions of porphyrins are relatively unstable to light (Chapters 16, 17). For example, vinyl containing porphyrins such as protoporphyrin-IX are rapidly photo-oxidized to afford photoprotoporphyrin and isophotoprotoporphyrin[39,40]. These green photo-products can often be seen as polar bands on columns used for chromatography of mixtures of porphyrins containing protoporphyrin-IX. Strongly electrophilic groups such as formyl have an inhibitory effect upon the oxidation of vinyl groups[41]. Radiochemically labeled porphyrins need to be bleached before their activity can be measured in a liquid scintillation counter; this is often achieved by allowing the porphyrin solution to stand in daylight in the presence of benzoyl peroxide. Magnesium porphyrins and zinc chlorins can be photo-oxidized to afford bile pigments bearing 1′-formyl substituents[42].

Porphyrins are rapidly photoreduced in the presence of compounds such as ascorbic acid, glutathione, or tertiary amines, to give phlorins (11). These can be quantitatively reconverted into porphyrins by mild oxidation with air or iodine. For a thorough description of photochemical reactions of porphyrin compounds, see Chapters 16, and 17.

1.2.4.3. Other examples of instability

Iron complexes of porphyrins can be ring-opened (to give bile pigments) by coupled oxidation with hydrogen peroxide and a reducing agent (Chapter 4). Thallium(III) trifluoroacetate and oxygen together accomplish the ring-opening of zinc and thallium(III) chlorins to give[43] 1,2-dihydro-a,b,c-bilitrienes.

1.3. Occurrence of Porphyrin Compounds
1.3.1. Metal complexes

The pyrrole pigments (porphyrins, chlorophylls, vitamin B_{12}, bile pigments, prodigiosins) constitute the most abundant coloring matters in natural systems. Protoheme (30), the iron(II) complex of protoporphyrin-IX, is the prosthetic group of both hemoglobin and myoglobin, and either this or a peripherally modified form is contained in the cytochromes[44] and the enzymes peroxidase and catalase. An impressive variety of chlorophylls

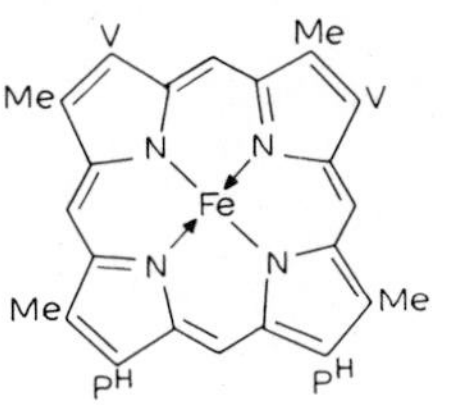

(30)

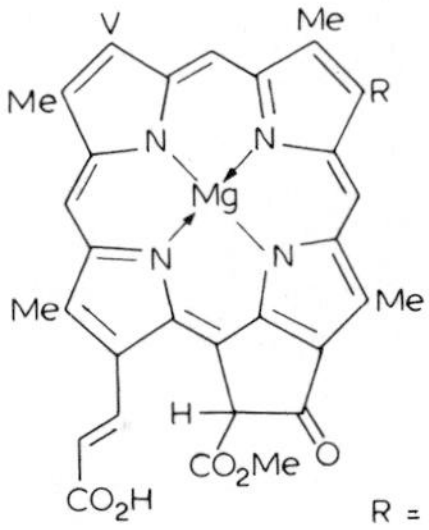

(31)   R = Et or V

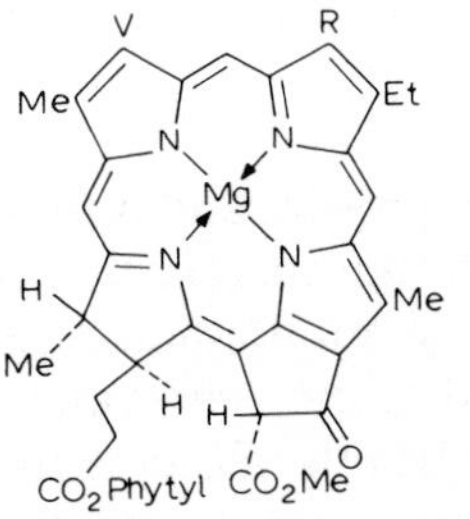

(32)

Chlorophyll – a   R = Me
Chlorophyll – b   R = CHO

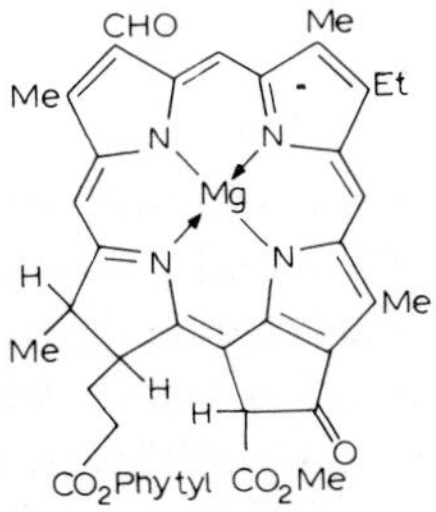

(33)

exists; with the exception of the chlorophylls-c (31), which are photosynthetic pigments in marine sources, they are all chlorins. The most abundant of the chlorophylls are chlorophyll-a and -b (32), the major photosynthetic pigments of the plant kingdom, which normally co-occur in a ratio of 3 : 1. Chlorophyll-d (33) is found, together with chlorophyll-a, in some species of *Rhodophyceae*, and the green sulfur bacteria *Chlorobium thiosulfatophilum*, *C. limicola*, and *Chloropseudomonas ethylicum* produce complex mixtures of polymethylated pigments called *Chlorobium* chlorophylls. The purple photosynthetic pigment of the *Thiorhodaceae* is bacteriochlorophyll-a (34), and bacteriochlorophyll-b [recently defined as (35)][9] is found in *Athiorhodaceae*.

The copper complex of uroporphyrin-III occurs in quite high concentrations in the wing feathers of *Turacus indicus*[45]; this is the source of most commercial samples.

Some types of marine worm possess a green oxygen-carrying pigment known as chlorocruorin; this consists of a protein combined with *Spirographis* hemin, which can be demetalated to afford *Spirographis* porphyrin (36).

References, p. 27

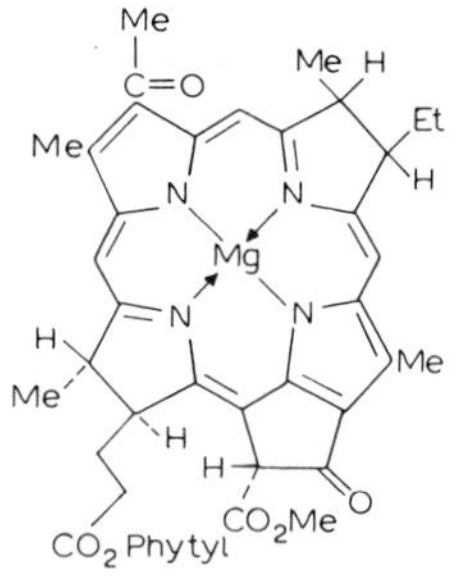

(34) 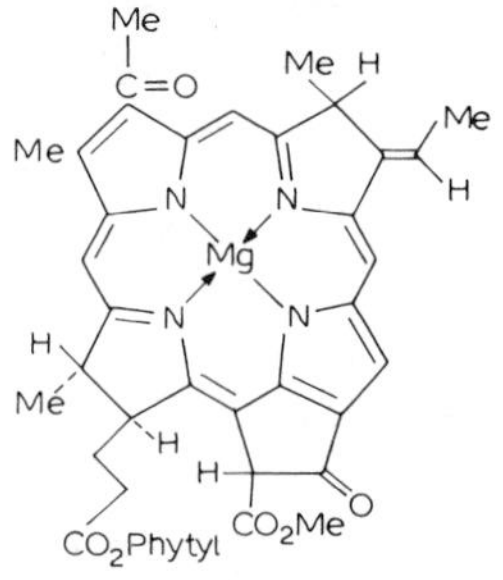   (35)

Chlorophyll degradation products are often found (as nickel or vanadium complexes) in oil shales and crude petroleum oils, in bituminous sands and asphaltic materials; the nature and origin of the petroporphyrins is still a topic of much speculation and investigation[46].

1.3.2. Metal-free porphyrins

The type I and III isomers of coproporphyrin and uroporphyrin, and protoporphyrin-IX are often found in the metal-free state in natural materials. No other isomer types are known to occur naturally. Relatively minute amounts of these porphyrins occur normally in tissue fluids and urine, and much larger amounts are found in certain pathological conditions, along with smaller amounts of porphyrins with 7, 6, 5, and 3 carboxyl groups[47]. Proto- and deutero-porphyrins occur in the feces of carnivores, and they result from bacterial decomposition of heme compounds ingested in meat or present due to intestinal hemorrhage. Phylloerythrin (37) arises similarly from ingested chlorophyll in herbivores.

Protoporphyrin-IX and coproporphyrins occur in high concentrations in the root nodules of leguminous plants[48] and in trace amounts in some higher plants; small amounts of uroporphyrin have been reported[49] in leaves. Protoporphyrin-IX, harderoporphyrin (a tricarboxylic monovinylporphyrin)[50] (38), and coproporphyrin occur[51] in the Harderian gland of many rodents. Protoporphyrin-IX is also found in the dorsal integument of worms, in a mutant of the yeast *Saccharomycopsis lipolytica*[52], and occurs ('ooporphyrin') in the pigmented areas of shells of many species of birds. Pempto-

(36) 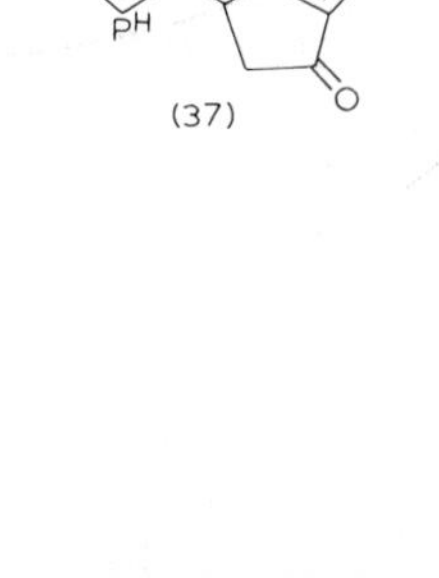   (37)

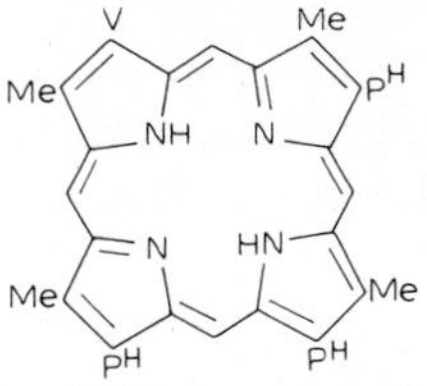

(38)

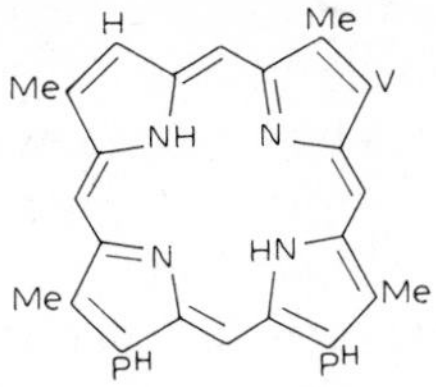

(39)

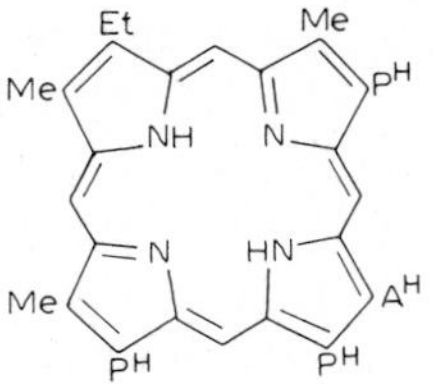

(40)

porphyrin, a fecal metabolite, has been assigned[53] structure (39). Uroporphyrin is found in the shells of molluscs[54]. Porphyrins, usually coproporphyrin, are excreted into the growth medium of many micro-organisms; isocoproporphyrin (40)[55] is formed in symptomatic cutaneous hepatic porphyria and in experimental porphyria due to hexachlorobenzene poisoning, and the suggestion has been put forward that it is formed by hydrogenation (by intestinal micro-organisms) of a vinyl-substituted precursor[56].

1.4. Chromophores of Porphyrin Systems

A pleasant bonus from carrying out research in the porphyrin field is the fact that one has the opportunity to observe the beautiful range of colors possessed by many of the compounds. The theoretical basis of the electronic absorption spectra of porphyrins and their derivatives is not within the scope of this brief Section; the intention here is rather to give a qualitative insight into the way in which the electronic absorption spectra vary with the nature of the chromophoric system and with the various substituents upon it. (See Appendix for Tables of spectra.)

In 1883, an intense absorption band at about 400 nm was discovered in hemoglobin by Soret[57]; this was later observed in porphyrins by Gamgee[58]. This 'Soret' band is the most intense band in the porphyrins and their derivatives, molar extinction coefficients around 400,000 often being recorded. It is found in all tetrapyrroles in which the nucleus is fully conjugated and can therefore be regarded as a characteristic of this macrocyclic conjugation. The intensity is weaker in chlorins and metallochlorins and, as might be expected, it is totally absent in the porphyrinogens. It is also lacking in bile pigments and is of only low intensity in metal-free oxophlo-rins, the metal complexes of which do possess Soret bands. The band is also present in vitamin B_{12} and in the metal complexes of bile pigments; both of these types of compound have interrupted conjugation in the ligand, but the pathway is maintained through the metal atom.

The Soret band is the band of choice for spectrophotometric determina-tions; it can be measured easily in cells of 10 mm path length at concentra-tions around 10^{-6} M. Commercial samples of porphyrins often have their purity expressed in terms of the extinction coefficient of the Soret band.

Another band in the ultraviolet (approx. 280 nm) is also present in por-phyrins, and is termed the γ band.

1.4.1. Porphyrins
1.4.1.1. In neutral solvents

Typical spectra are illustrated in Fig. 1; there are four satellite bands, numbered I to IV as shown, and in some porphyrins a further small band (Ia) between bands I and II can be discerned. All naturally occurring porphyrins have the etio-type spectrum (IV > III > II > I) (Fig. 1a), but several spectro-scopic types are found among the porphyrins derived from the heme pros-thetic groups of the natural heme proteins; for example, chlorocruorin and cytochromes-a and a_3 yield porphyrins with rhodo-type (Fig. 1b) and oxo-rhodotype (Fig. 1c) spectra. The extinction coefficients of porphyrin iso-mers do not differ significantly in the Soret and visible regions. Electrophilic side-chains tend to decrease the extinction coefficient.

Correlations occur between the nature of the side-chains on porphyrins and the positions and intensities of their visible absorption bands. Electro-philic side-chains such as vinyl and formyl cause shifts to longer wavelength of the Soret and satellite bands of porphyrins (and of metalloporphyrins and further complexes with additional ligands), (see Chapter 5). The dissocia-tion constants of the pyrrole nitrogen atoms serve as a measure of the degree of electron attraction in side-chains and a reasonable correlation between pK_3 and the wavelengths of the visible bands of different porphyrins can be shown (Table 2).

1.4.1.1.1. Etio-type spectra

The etio spectrum (Fig. 1a) is found in all porphyrins in which six or more peripheral positions carry side-chains such as methyl, ethyl, acetic acid, or propionic acid groups, the remaining positions being unsubstituted. The commonly occurring uro-, copro-, proto-, and deutero-porphyrins all have etio-type spectra.

The insulated carboxyl groups of acetic and propionic acid side-chains have only minimal effect upon the wavelength of absorption bands, and esterification of a uroporphyrin or a coproporphyrin hardly changes its spec-trum. Nevertheless, the absorption bands in coproporphyrin are some 5 nm to shorter wavelength than the corresponding bands in uroporphyrin; interestingly, a stepwise decrease of about 1 nm as each of the four acetic side-chains in uroporphyrin is convereted into methyl has been noted[59].

1.4.1.1.2. Rhodo-type spectra

A single carbonyl (aldehyde, ketone), a carboxylic acid, ester, or an acrylic acid side-chain causes band III to become more intense than band IV, affording a rhodo-spectrum (Fig. 1b; also Table 4) (III > IV > II > I), named after rhodoporphyrin-XV (41). These strongly electron-withdrawing groups which also cause shifts of absorption to longer wavelengths, are termed 'rhodofying' groups. An ionized carboxyl group is not electron-withdrawing, thus, the absorption spectra of rhodoporphyrin free acids in alkaline solution tend to be etio in type. The vinyl group is only weakly rhodofying, as are aldoxime and ketoxime groups. Oxime formation reverses rhodo-type spectra more or less completely to etio-type, and it also largely reverses the shifts to longer wavelength caused by the carbonyl group (Table 4).

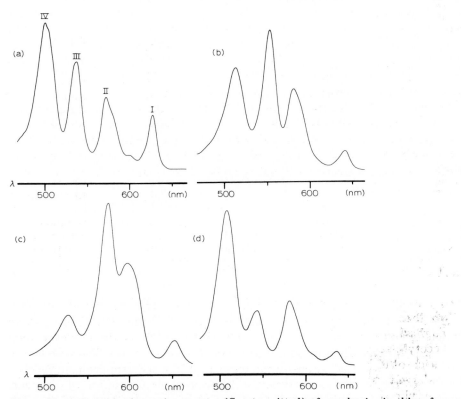

Fig. 1. Typical visible absorption spectra (Soret omitted) of porphyrins in chloroform: (a) etio-type, (b) rhodo-type, (c) oxorhodo-type, (d) phyllo-type.

References, p. 27

TABLE 4

Effect of substituent on wavelength and type of visible absorption spectrum of porphyrin esters

Deuteroporphyrin-IX derivatives	Substituent 2	Substituent 4	Band I (nm; dioxan)	Spectral type
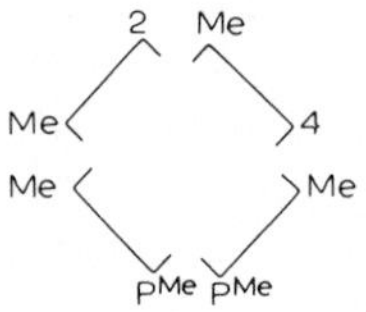	H	H	618	Etio
	Et	Et	620	Etio
	V	V	630	Etio
	H	CHO	640	Rhodo
	H	CH=NOH	630	Etio
	CHO	V	641	Rhodo
	CH=NOH	V	635	Etio
	CHO	CHO	650	Etio
	CH=NOH	CH=NOH	639	Etio
	CO.Me	CO.Me	639	Etio
	C(Me)=NOH	C(Me)=NOH	625	Etio

Pyrroporphyrin-XV derivatives	6		Band I	Spectral type
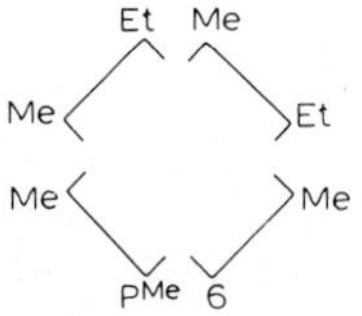	H		620	Etio
	V		624	Etio
	CHO		635	Rhodo
	CO.Me		632	Rhodo
	CO.Ph		627	Rhodo
	CO_2Me		632	Rhodo
	$CH=CH.CO_2Me$		624	Rhodo

2-Desethylpyrro-porphyrin-XV derivatives	2	6	Band I	Spectral type
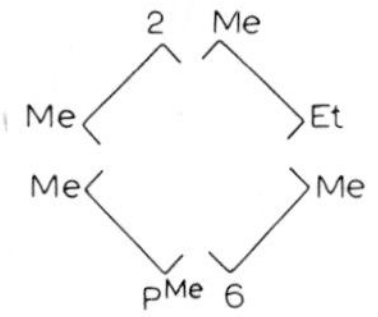	Et	CO_2Me	632	Rhodo
	V	CO_2Me	638	Rhodo-oxorhodo
	CO.Me	CO_2Me	637	Oxorhodo
	C(Me)=NOH	CO_2Me	631	Rhodo
	CO.Me	CO.Me	637	Oxorhodo

Two rhodofying groups on adjacent 'pyrrole' subunits cancel out each other's rhodofying effect, an etio-spectrum resulting (e.g. 2,4-diformyl- and 2,4-diacetyl-deuteroporphyrins-IX, Table 4); these groups do, however, reinforce each other's effect upon the wavelength of absorption. The wavelength shift is largely reversed upon oxime formation, the spectrum remaining etiotype. Acrylic acid groups are strongly rhodofying; condensation of diazoacetic ester with the double bond of these and vinyl groups (Chapter 15) cancels the rhodofying effect.

1.4.1.1.3. Oxorhodo-type spectra

Two rhodofying groups on diagonally opposite rings enhance each other's effect and the oxorhodo spectrum results (Fig. 1c) (III > II > IV > I), named after 'oxorhodo'- or 2-acetylrhodo-porphyrin (42). The oxime of this porphyrin has a rhodo-type spectrum, due to the carboxylic acid group remaining in position 6. Porphyrin β-keto-esters also possess[60] oxorhodo spectra, and these become etio in alkaline solution owing to ionization to the enolate anion.

1.4.1.1.4. Phyllo-type spectra

When four or more peripheral positions are unsubstituted the spectrum is not necessarily etio-type. Whereas symmetrical tetraethyl- and tetramethyl-porphins, and the symmetrical porphin-1,3,5,7-tetrapropionic acid have etio-spectra, 1,4-diethyl-2,3-dimethylporphin and porphin-1,4-dipropionic acid[61] have phyllo-type spectra (Fig. 1d) (IV > II > III > I), named after phylloporphyrin-XV (43). This is no doubt due to the development of asymmetry in the π-electron cloud, due to the electron-donating effect of the alkyl groups.

Apart from such part-substituted porphyrins, the phyllo-type spectrum is characteristic of substitution at a single meso-carbon, as in the chlorophyll derivatives phylloporphyrin-XV (43), desoxophylloerythrin (44), and various meso-acetoxy, meso-benzoyloxy-, and meso-alkoxy-porphyrins[30] (see p. 878).

1.4.1.1.5. Spectra of porphyrins containing isocyclic rings

The methyl esters of the chlorophyll derivatives phylloerythrin (37) and pheoporphyrin-a_5 (45) have oxorhodo-rhodo and oxorhodo-type spectra re-

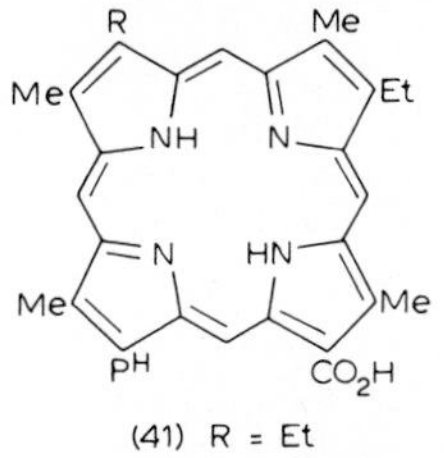

(41) R = Et
(42) R = CO·CH₃

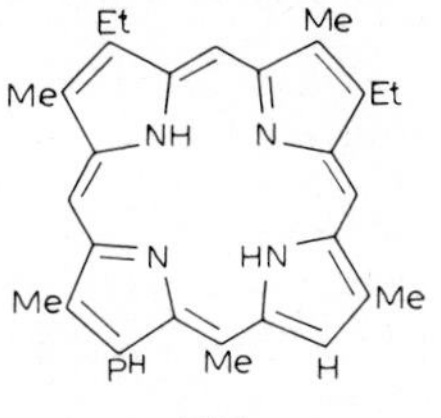

(43)

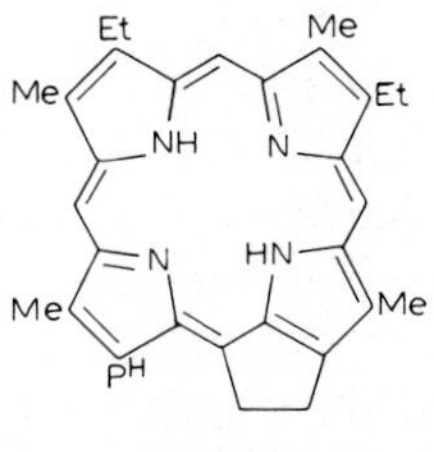

(44)

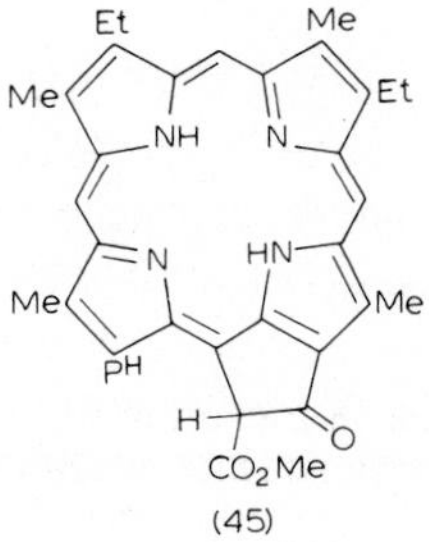

(45)

spectively. The carbomethoxy-substituted ring in pheoporphyrin-a_5 thus has an extremely strong rhodofying effect, and the isocyclic ring carbonyl group of phylloerythrin (which has lost the carbomethoxy group) has a much stronger effect than do open-chain keto groups or aldehydes (Table 4). As mentioned above, desoxophylloerythrin (44), which has lost the ring carbonyl, has a phyllo-type spectrum. The oxime of pheoporphyrin-a_5 has a rhodo-type spectrum, confirming the rhodofying effect of the carboxyl group. The oxime of phylloerythrin has a spectrum verging upon rhodo-type.

1.4.1.2. In acidic solvents

The Soret bands in the dicationic absorption spectra of porphyrins are displaced to longer wavelength compared with the neutral species. Dication spectra show two major satellite bands (Fig. 2a); weaker bands, which appear as shoulders to the main peaks, are also present. The spectral simplification is presumably a result of the approach towards square symmetry in the dication, and this is also manifest in the spectra of dianions and metalloporphyrins.

Porphyrin monocations have a Soret band (which is moved to shorter wavelength than the corresponding neutral porphyrin) and three visible bands (Fig. 2b).

1.4.1.3. In alkaline solvents

The spectra of porphyrin dianions closely resemble those of the corresponding dications (Fig. 2a); the Soret band is displaced to longer wavelength than in the corresponding neutral spectrum. The way in which alkali removes the rhodofying effect of porphyrins bearing carboxylic acid and keto-ester functions through ionization has already been mentioned.

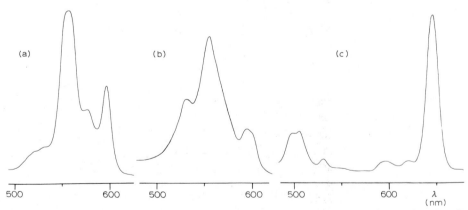

Fig. 2. Typical visible absorption spectra (Soret omitted) of: (a) porphyrin dication (in chloroform containing 0.5% TFA), (b) porphyrin monocation (in chloroform—acetic acid), (c) a chlorin in chloroform.

1.4.2 Chlorins

Thorough discussions of the variation of chlorin and chlorophyll visible absorption spectra with structural features have been given[3,5,62].

In the chlorins, band I in the visible region is very prominent (Fig. 2c), and is about 25 nm to longer wavelength than in porphyrins. The Soret : band I ratio is only about 5 (as against about 50 in porphyrins). The extinction coefficients of band IV and the Soret band, in neutral solvents, are comparable with those of the related bands in analogous porphyrins. Chlorin mono- and di-cations have spectra similar to those of the neutral compounds, band I and the Soret absorption being displaced to shorter wavelengths in the dications (and in metallochlorins). The Soret band in chlorins has a tendency to be split, this being more noticeable where there is distortion of the resonance pathway, as in pheophytin-d and bacteriopheophytin.

1.4.3. Metalloporphyrins

Simple square-planar chelates of porphyrins with divalent metal ions have a Soret and two visible bands, usually called the α and β bands. The wavelength of the α band appears to be related to bands I and III in the free porphyrin spectrum and the β band in metalloporphyrins to porphyrin bands II and IV. However, intensity changes are more difficult to correlate.

A very qualitative stability order for metalloporphyrins (see, however, Chapter 5) can be discerned by consideration of the relative intensities of the α and β bands; stable chelates have $\alpha > \beta$ while in less stable examples, the ratio is decreased or even reversed (Fig. 3). Further coordination with extra ligands causes a general shift of absorption bands to higher wavelength.

Though different metals in the same porphyrin nucleus exert their own, quite considerable, effects upon the wavelength of absorption, the effect of substituting vinyl for ethyl shows up independently of the nature of the metal ion.

A more thorough discussion of the spectra of metalloporphyrins can be found in Chapter 4.

1.4.4. Phlorins and oxophlorins

Phlorins (Chapter 14) have characteristic visible absorption spectra; the free base (11) possesses[63] two major bands, one at approx. 620 nm and the other centred at 387 nm. The shorter wavelength band is reminiscent of the Soret absorption found in porphyrins but is of considerably lower extinction, owing to the interrupted conjugation pathway. The 'imperial blue'[63] free base is converted into an olive-green monocation (46) with acid, and this features a broad absorption band at 723 nm with a sharper and more intense absorption to shorter wavelength, at 431 nm; additional less intense bands are also present. In strong acid, the dication (47) is produced and this has a typical porphodimethene absorption spectrum.

References, p. 27

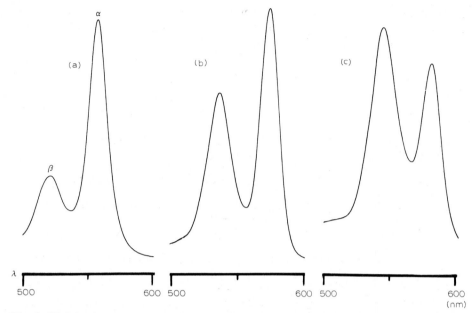

Fig. 3. Visible absorption spectra (Soret omitted) in chloroform of typical square planar metalloporphyrins: (a) nickel(II) octaethylporphyrin, (b) zinc(II) octaethylporphyrin, (c) cadmium(II) octaethylporphyrin.

Oxophlorins also have electronic absorption spectra which indicate interrupted conjugation in the free base and monocation[30]. The free base (17) spectrum closely resembles that of the phlorin free base, having bands at 400, (588), and 635 nm; the green monocation (48) absorbs at 407 and 706 nm, with additional fine structure between these main bands. The diprotonated species (49) has an absorption spectrum similar to that of a normal diprotonated porphyrin, and quite unlike that of the true phlorin.

References

1. R. Willstätter and A. Stoll, 'Investigations on Chlorophyll', Science Press, Lancaster, Ohio (1928).
2. H. Fischer and H. Orth, 'Die Chemie des Pyrrols', Akademische Verlagsgesellschaft, Leipzig Vols. I and IIi (1934 and 1937). (Now reprinted by Johnson Reprint Corp., New York.)
3. H. Fischer and A. Stern, 'Die Chemie des Pyrrols', Akademische Verlagsgesellschaft, Leipzig Vol. IIii (1940). (Also reprinted, see Ref. 2.)
4. R. Lemberg and J.W. Legge, 'Haematin Compounds and Bile Pigments', Interscience, New York (1949).
5. A. Treibs, 'Das Leben und Wirken von Hans Fischer', Hans Fischer Gesellschaft e. V., München (1971).
6. 'The Chemical and Physical Behavior of Porphyrin Compounds and Related Structures', Ann. N.Y. Acad. Sci., **206**, 1—761 (1973).
7. W. Küster, Hoppe-Seyler's Z. Physiol. Chem., **82**, 463 (1912).
8. H. Fischer and K. Zeile, Ann. Chem., **468**, 98 (1929).
9. H. Scheer, W.A. Svec, B.T. Cope, M.H. Studier, R.G. Scott, and J.J. Katz, J. Am. Chem. Soc., **96**, 3714 (1974).
10. D. Mauzerall, J. Am. Chem. Soc., **82**, 1832, 2601 (1960).
11. E.g. A. Eschenmoser, A.P. Johnson, P. Wehrli, and R. Fletcher, Angew. Chem., **80**, 622 (1968).
12. E.g. H.H. Inhoffen, Pure Appl. Chem., **17**, 443 (1968).
13. See, for example, F. Morsingh and S.F. MacDonald, J. Am. Chem. Soc., **82**, 4377 (1960).
14. C.H. Gray, A. Neuberger, and P.H.A. Sneath, Biochem. J., **47**, 87 (1950).
15. M.F. Hudson and K.M. Smith, J. Chem. Soc., Chem. Commun., 515 (1973).
16. A. Stern and G. Klebs, Ann. Chem., **500**, 91 (1932); **504**, 287 (1933); **505**, 295 (1933).
17. J.E. Falk and J.B. Willis, Aust. J. Sci. Res., Ser. **A**, **4**, 579 (1951).
18. L.E. Webb and E.B. Fleischer, J. Am. Chem. Soc., **87**, 667 (1965).
19. D. Doddrell and W.S. Caughey, J. Am. Chem. Soc., **94**, 2510 (1972).
20. R.J. Abraham, G.E. Hawkes, and K.M. Smith, J. Chem. Soc., Chem. Commun., 401 (1973); J. Chem. Soc., Perkin Trans. 2, 627 (1974).
21. G.M. Badger, R.L.N. Harris, R.A. Jones, and J.M. Sasse, J. Chem. Soc., 4329 (1962); S.F. Mason, ibid., 976 (1958).
22. C.B. Storm and Y. Teklu, J. Am. Chem. Soc., **94**, 1745 (1972); C.B. Storm, Y. Teklu, and E.A. Sokoloski, Ann. N.Y. Acad. Sci., **206**, 631 (1973).
23. R.J. Abraham, G.E. Hawkes, and K.M. Smith, Tetrahedron Lett., 1483 (1974).
24. J.N. Phillips, Rev. Pure Appl. Chem., **10**, 35 (1960).
25. W.K. McEwen, J. Am. Chem. Soc., **58**, 1124 (1936).
26. J.B. Conant, B.F. Chow, and E.M. Dietz, J. Am. Chem. Soc., **56**, 2185 (1934).
27. S. Aronoff, J. Phys. Chem., **62**, 428 (1958).
28. B. Dempsey, M.B. Lowe, and J.N. Phillips, in 'Haematin Enzymes', Eds. J.E. Falk, R. Lemberg, and R.K. Morton, Pergamon, London (1961), p. 29.

29. R. Grigg, R.J. Hamilton, M.L. Jozefowicz, C.H. Rochester, R.J. Terrell, and H. Wickwar, J. Chem. Soc., Perkin Trans. 2, 407 (1973).
30. A.H. Jackson, G.W. Kenner, and K.M. Smith, J. Chem. Soc. C, 302 (1968).
31. R.J. Abraham, G.E. Hawkes, and K.M. Smith, Tetrahedron Lett., 71 (1974).
32. A.H. Jackson and G.R. Dearden, Ann. N.Y. Acad. Sci., **206**, 151 (1973).
33. A. Stone and E.B. Fleischer, J. Am. Chem. Soc., **90**, 2735 (1968); S.J. Silvers and A. Tulinsky, ibid., **89**, 3331 (1967).
34. A. Neuberger and J.J. Scott, Proc. R. Soc. London, Ser. A, **213**, 307 (1952).
35. R. Willstätter and W. Mieg, Ann. Chem., **350**, 1 (1906).
36. M.T. Cox, A.H. Jackson, and G.W. Kenner, J. Chem. Soc. C, 1974 (1971).
37. J.L. Archibald, D.M. Walker, K.B. Shaw, A. Markovac, and S.F. MacDonald, Can. J. Chem., **44**, 345 (1966).
38. D. Mauzerall, J. Am. Chem. Soc., **82**, 2601 (1960).
39. H. Fischer and H. Bock, Hoppe-Seyler's Z. Physiol. Chem., **255**, 1 (1938).
40. J. Barrett, Nature London, 183, 1185 (1959) H.H. Inhoffen, H. Brockmann jr., and K.-M. Bliesener, Ann. Chem., **730**, 173 (1969).
41. J. Barrett and P.S. Clezy, Nature London, **184**, 1988 (1959).
42. J.-H. Fuhrhop and D. Mauzerall, Photochem. Photobiol., **13**, 453 (1971); P.K.W. Wasser and J.-H. Fuhrhop, Ann. N.Y. Acad. Sci., **206**, 533 (1973).
43. J.A.S. Cavaleiro and K.M. Smith, J. Chem. Soc., Perkins Trans. 1, 2149 (1973).
44. R. Lemberg and J. Barrett, 'Cytochromes', Academic Press, London (1973).
45. R.E.H. Nicholas and C. Rimington, Biochem. J., **50**, 194 (1951).
46. E.g. G.W. Hodgson, Ann. N.Y. Acad. Sci., **206**, 670 (1973); R. Bonnett, ibid., p. 722.
47. J.E. Falk, in 'Porphyrin Biosynthesis and Metabolism', Eds. G.E. Wolstenholme and E.C.P. Millar, Churchill, London (1955), p. 63.
48. H. Klüver, J. Psychol., **25**, 331 (1948); J.E. Falk, C.A. Appleby, and R.J. Porra, Symp. Soc. Exp. Biol., **13**, 73 (1959).
49. R.H. Goodwin, V.M. Koski, and O.v.H. Owens, Am. J. Botan., **38**, 629 (1951).
50. G.Y. Kennedy, A.H. Jackson, G.W. Kenner, and C.J. Suckling, FEBS Lett., **6**, 9 (1970); **7**, 205 (1970).
51. C. Rimington and G.Y. Kennedy, in 'Comparative Biochemistry', Eds. M. Florkin and H.S. Mason, Academic Press, London, Vol. 4, (1962), p. 557. See also Ref. 50.
52. J. Bassel, P. Hambright, and R. Mortimer, J. Bacteriol., in press.
53. S. Sano, T. Shingu, J.M. French, and E. Thonger, Biochem. J., **97**, 250 (1965); A.H. Jackson, G.W. Kenner, and J. Wass, Chem. Commun., 1027 (1967); J. Chem. Soc., Perkin Trans. 1, 480 (1974); P. Bamfield, R. Grigg, R.W. Kenyon, and A.W. Johnson, Chem. Commun., 1029 (1967); J. Chem. Soc. C, 1259 (1968).
54. R.E.H. Nicholas and A. Comfort, Biochem. J., **45**, 208 (1949).
55. M.S. Stoll, G.H. Elder, D.E. Games, P. O'Hanlon, D.S. Millington, and A.H. Jackson, Biochem. J., **131**, 429 (1973).
56. G.H. Elder, Biochem. J., **126**, 877 (1972).
57. J.L. Soret, Compt. Rend.. **97**, 1267 (1883).
58. A. Gamgee, Z. Biol. Munich, **34**, 505 (1897).
59. T.C. Chu and E.J. Chu, J. Biol. Chem., **234**, 2741 (1959).
60. M.T. Cox, T.T. Howarth, A.H. Jackson, and G.W. Kenner, J. Chem. Soc., Perkin Trans. 1, 512 (1974).
61. S. Aronoff and C.A. Weast, J. Org. Chem., **6**, 550 (1941).
62. A. Treibs, Ann. N.Y. Acad. Sci., **206**, 97 (1973).
63. R.B. Woodward, Ind. Chim. Belge, **27**, 1293 (1962).
64. A.R. Battersby, E. Hunt, M. Ihara, E. McDonald, J.B. Paine III, F. Satoh and J. Saunders, J. Chem. Soc., Chem. Commun., 994 (1974).

SYNTHESIS AND PREPARATION OF PORPHYRIN COMPOUNDS

KEVIN M. SMITH

The Robert Robinson Laboratories, University of Liverpool, U.K.

2.1. Scope

This chapter will review the methods available for the total synthesis of porphyrins and related compounds. A blow by blow account of the development of the art of porphyrin synthesis is not intended; rather an attempt will be made to outline the strategic considerations involved in the synthesis of a particular type of porphyrin. For example, it would be futile to attempt the synthesis of octaethylporphyrin (1) by laborious construction of an open-chain tetrapyrrolic intermediate; such symmetrically substituted compounds are best synthesized by polymerization of a suitable monopyrrole. On the other hand, there is no way that protoporphyrin-IX (2) can be synthesized by monopyrrole self-condensation, and far more sophisticated approaches are required in this case.

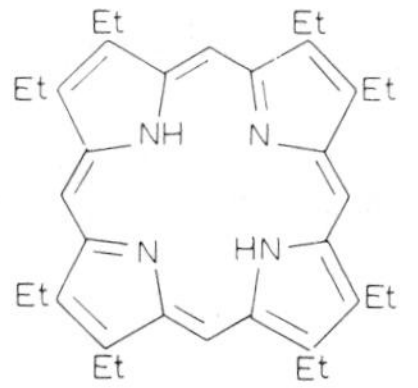

(1)

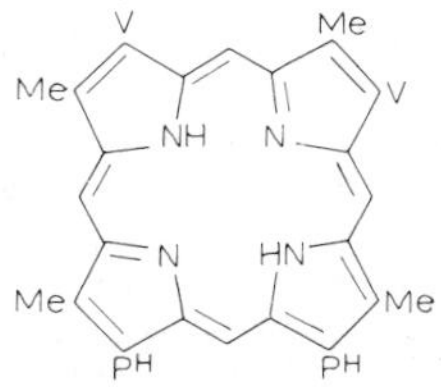

(2)

$$V = CH{=}CH_2$$
$$P^R = CH_2CH_2CO_2R$$

A section concerning methods for construction of dipyrrolic compounds is included; it should be borne in mind that many of the concepts and procedures for coupling monopyrroles have been applied to the construction of open-chain tetrapyrroles and to porphyrins. The wealth of monopyrrole syntheses which are available are not discussed; the reader is advized to consult recent reviews dealing with these[1,2].

Porphyrins and Metalloporphyrins, ed. Kevin M. Smith
© 1975, Elsevier Scientific Publishing Company, Amsterdam, The Netherlands

Porphyrin compounds occur widely in nature, and an abundant source of porphyrins is available through degradation of protoheme and chlorophyll. The final part of this chapter is devoted to the preparation of useful porphyrins from these natural sources.

2.2. Synthesis of Dipyrrolic Intermediates
2.2.1. Pyrromethenes

These compounds (3) have been used extensively[3] as intermediates in porphyrin syntheses. The most efficient and commonly used methods for synthesis of unsymmetrically substituted pyrromethenes is the acid catalyzed condensation of a 2-formylpyrrole (4) with a 2-unsubstituted pyrrole

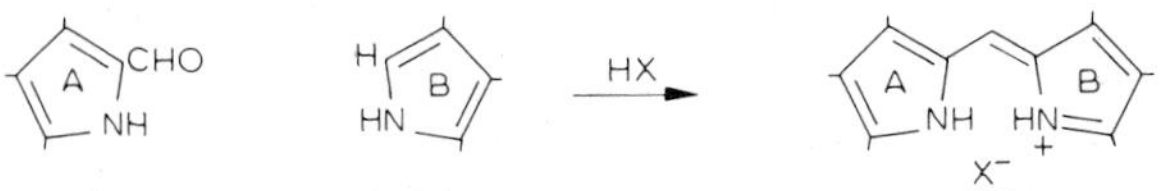

(4)   (5)   (3)

(5); the pyrromethene salt (3) is often produced in virtually quantitative yield[4]. Many successful applications of this route have tended to discourage the development of alternatives. However, heating of 2-bromomethylpyrroles with 2-bromopyrroles in solvents has received some attention; the best known example of this route is the preparation of a mixture of kryptopyrromethenes (6) and (7) by treatment of kryptopyrrole (8) with bromine in hot acetic acid[5]:

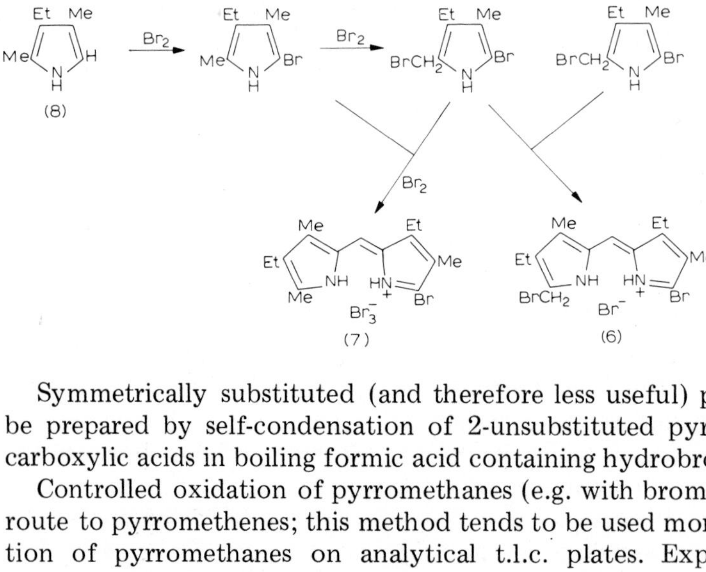

(8)

(7)   (6)

Symmetrically substituted (and therefore less useful) pyrromethenes can be prepared by self-condensation of 2-unsubstituted pyrroles or pyrrole-2-carboxylic acids in boiling formic acid containing hydrobromic acid[6].

Controlled oxidation of pyrromethanes (e.g. with bromine) is also a viable route to pyrromethenes; this method tends to be used more in the identification of pyrromethanes on analytical t.l.c. plates. Exposure to bromine vapor causes colorless pyrromethanes to turn red-pink on t.l.c. plates.

2.2.2. Pyrromethanes

Unsymmetrically substituted pyrromethanes (9) can be prepared by condensation of 2-acetoxymethylpyrroles (10) with 2-unsubstituted pyrroles (5) in methanol or acetic acid containing a catalytic amount (<0.1 equiv.) of

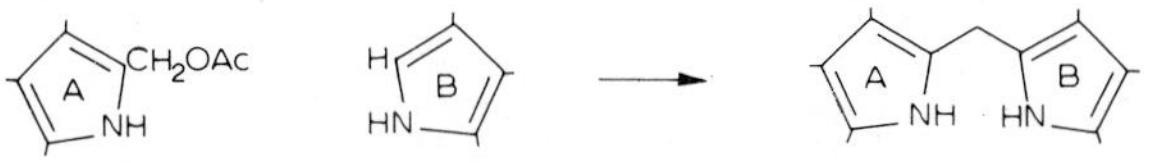

(10)   (5)   (9)

toluene *p*-sulfonic acid[7], in hot glacial acetic acid[8], or in refluxing glacial acetic acid containing sodium acetate[9]. Pyridinium salts of 2-bromomethylpyrroles react[10] with lithium salts of pyrrole-2-carboxylic acids in polar solvents at approx. 100°C to give good yields of unsymmetrical pyrromethanes provided that there is no additional strongly electron-withdrawing substituent on the pyrrole ring of the lithium carboxylate.

Symmetrically substituted pyrromethanes are readily prepared by self-condensation of bromomethylpyrroles in hot methanol[11], or by heating 2-acetoxymethylpyrroles in methanol—hydrochloric acid[12]. Alternatively, 2-unsubstituted pyrroles can be treated with formaldehyde to afford[13] pyrromethanes.

Reduction of pyrromethenes with sodium borohydride furnishes pyrromethanes.

2.2.3. Pyrroketones

The method of choice for the synthesis of unsymmetrical pyrroketones (11) is condensation of the phosphoryl chloride complex of a pyrrole-*N,N*-dimethylcarboxamide (12) with a 2-unsubstituted pyrrole[14]. After hydroly-

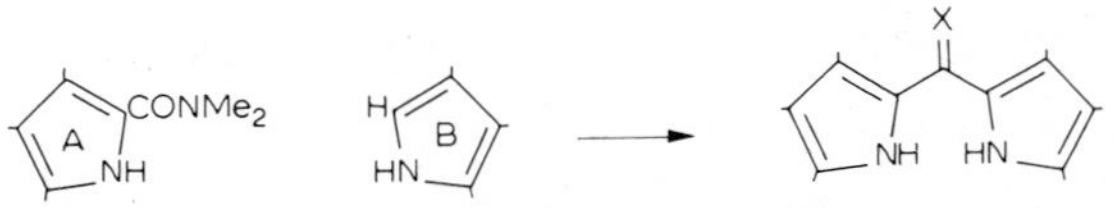

(12)   (5)   (11) X = O
                                    (13) X = $\overset{+}{N}Me_2$ Cl$^-$

sis of the intermediate imine salt (13) the pyrroketone is usually obtained in good yield. Alternatively, treatment of a pyrrole acid chloride with a pyrrole Grignard derivative gives pyrroketones[15].

Symmetrically substituted pyrroketones are accessible by treatment of pyrrole Grignard reagents with phosgene[16].

Oxidation of pyrromethanes with lead dioxide or lead tetra-acetate[17], or with bromine—sulfuryl chloride (or sulfuryl chloride alone)[18] also gives good yields of pyrroketones.

References, p. 55

2.3. Synthesis of Porphyrins
2.3.1. By polymerization of monopyrroles

This approach can only be used satisfactorily if the pyrrole 3 and 4 substituents are identical, that is, when $X = Y$ in (14). If $X \neq Y$, then poly-

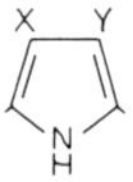

(14)

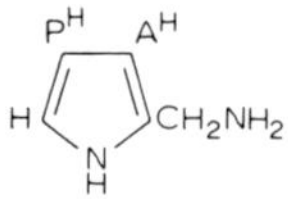

(15)

$$A^R = CH_2CO_2R$$

merization leads to a statistical mixture of all four possible type-isomers. (The only exception to this rule is the enzyme mediated tetramerization of porphobilinogen (15) to give uroporphyrinogen-III; see Chapter 3.)

Two major routes have been developed. The first general method involves polymerization of 2,5-di-unsubstituted pyrroles in the presence of agents which will provide the four carbons required for the *meso* methine groups of the product. The second approach is the tetramerization of pyrroles bearing 2-CH_2R substituents, the methylene carbon of which will provide the *meso*-carbons of the desired porphyrin.

Polymerization of 2,5-di-unsubstituted pyrroles with formic acid[19] is a viable method, but the most used procedure is that of Rothemund[20], involving the heating of pyrrole with aldehydes. This is the method by which *meso*-tetraphenylporphyrin (16) is obtained. The optimum conditions for

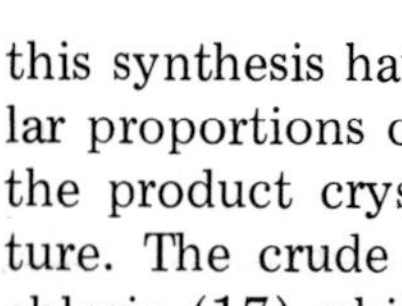

(16)

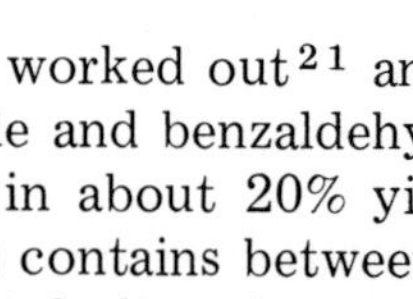

(17)

this synthesis have been worked out[21] and involve brief heating of equimolar proportions of pyrrole and benzaldehyde in propionic acid (Chapter 19); the product crystallizes in about 20% yield from the cooled reaction mixture. The crude product contains between 5 and 10% of *meso*-tetraphenylchlorin (17) which is best dealt with by brief treatment of the crude product with 2,3-dichloro-5,6-dicyanobenzoquinone[22]. Other methods are either time consuming or else unacceptably sacrificial. Because of overlapping absorptions in the visible spectra it is difficult to estimate the degree of contamination of the porphyrin with chlorin, though a calculation based on the extinction of the longest wavelength band is available[23]. However, if the zinc chelate of the crude product is made, the contaminant is clearly visible

owing to its long wavelength metallochlorin band, and the quinone dehydrogenation can be monitored as the reaction proceeds[24]. The Rothemund
method has been applied to a wide range of tetra-aryl-porphyrins.

Self-condensation of monopyrroles of the type (18), followed by aerial
oxidation, often affords good yields of symmetrical porphyrins. This approach is best exemplified by consideration of two different approaches to
octaethylporphyrin (1).

1) Mannich reaction of pyrrole (19) with dimethylamine and formaldehyde gives the 2-N,N-dimethylaminomethylpyrrole (20) which is heated in
refluxing acetic acid to give a 52% yield of (1)[25].

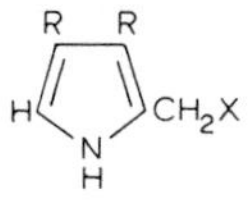

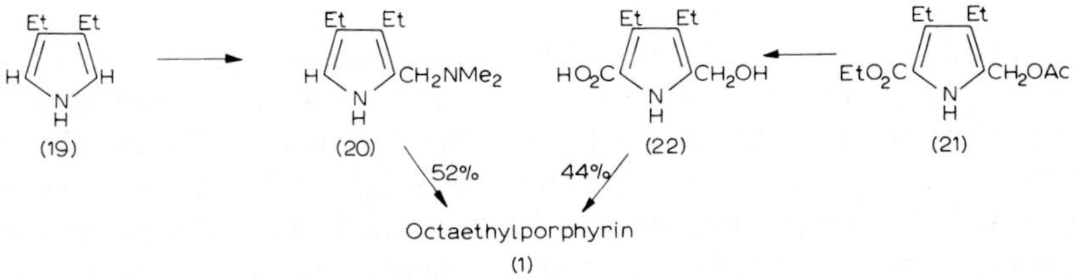

2) Hydrolysis of the pyrrole (21) gives the pyrrole (22) which is converted
into octaethylporphyrin in 44% yield by heating in acetic acid containing
potassium ferricyanide[26]. Fuller details of both of the above syntheses are
given in Chapter 19.

2.3.2. From dipyrrolic intermediates

All methods using the condensation of two dipyrrolic intermediates have
inherent symmetry restrictions; thus, this route is limited to the synthesis of
porphyrins which are centrosymmetrically substituted (produced by self-
condensation of a dipyrrolic compound) or porphyrins which possess symmetry in one or both halves of the molecule. The latter restriction is not
particularly limiting from the point of view of synthesis of naturally occurring porphyrins because rings C and D of uroporphyrin-III, coproporphyrin-
III, and protoporphyrin-IX are symmetrically substituted about the γ-meso-
carbon atom, and these molecules can be synthesized provided that the
strategy involves condensation of an A-B unit with a C-D unit.

2.3.2.1. From pyrromethenes

Almost all of the porphyrin syntheses from pyrromethenes were developed in Munich by H. Fischer and his collaborators. 5-Bromo-5′-methylpyr-

romethenes, e.g. (23), can be self-condensed in organic acid melts (succinic, tartaric, etc.) at temperatures up to 200°C, and good yields of porphyrins are obtained[27]. Alternatively, 5-bromo-5'-bromomethylpyrromethene hydrobromides, e.g. (6)[28], 5-bromo-5'-methylpyrromethene perbromides, e.g. (7)[29], or a mixture of both[30] (p. 30) can be heated in formic acid to give yields as high as 50% of centrosymmetrically substituted porphyrins such as

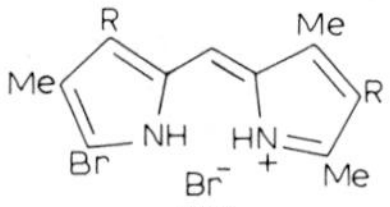

(23)

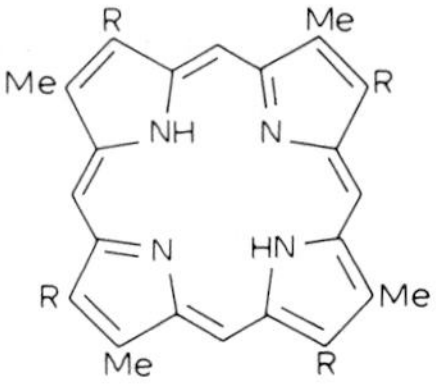

(24) R = Et
(25) R = P^H

etioporphyrin-I (24), coproporphyrin-I (25), or octaethylporphyrin (1). This type of synthesis can be modified to allow the synthesis of more diversely substituted porphyrins (e.g. those with the type-III side-chain pattern) by condensation of 5,5'-dibromopyrromethenes (26) with 5,5'-dimethyl- (27) or 5,5'-di(bromomethyl)pyrromethenes (28) in organic acid melts. In the example shown, namely Fischer's renowned synthesis of deuteroporphyrin-IX (29) which was an intermediate in the total synthesis of hemin[31], the formation of a mixture of porphyrins is avoided by use of one pyrromethene (26) which is symmetrically substituted about the methine carbon.

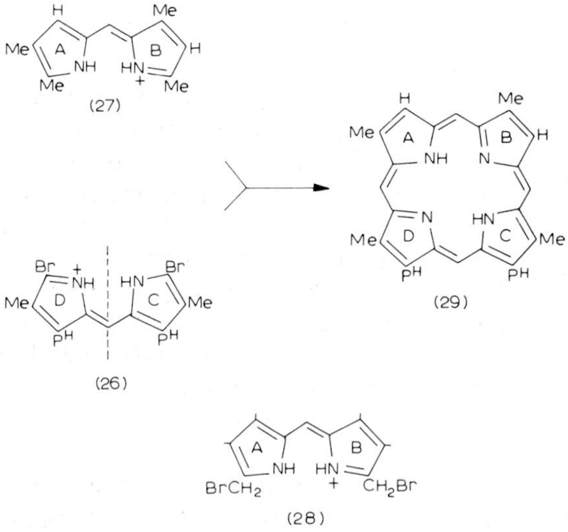

(27)

(29)

(26)

(28)

2.3.2.2. From pyrromethanes

Fischer's success with pyrromethenes tended to overshadow the development of porphyrin syntheses using pyrromethanes as intermediates; indeed, for a long time it was believed that pyrromethanes were inherently too unstable towards acidic reagents to be useful as porphyrin precursors. This is certainly the case in one of Fischer's procedures[32]; it has been shown[33] that the self-condensation of pyrromethane-5,5'-dicarboxylic acids, e.g. (30), in formic acid gives a mixture of type-isomers rather than pure etioporphyrin-II (31) or coproporphyrin-II (32).

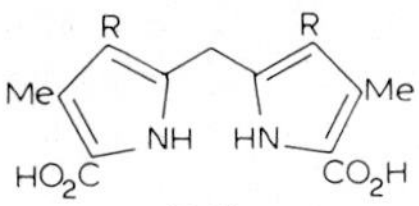

(30)

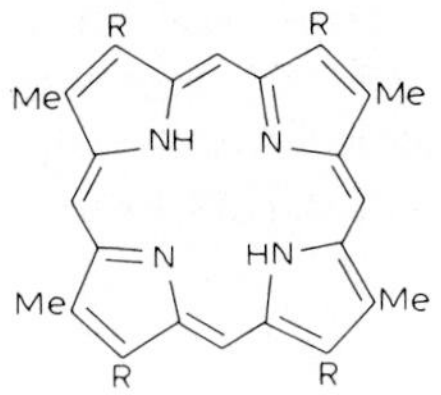

(31) R = Et
(32) R = pH

The breakthrough in porphyrin synthesis from pyrromethanes came with MacDonald's method[34] (and in a less general way, with Woodward's[35] synthesis of chlorophyll-a), in which a 5,5'-diformylpyrromethane, e.g. (33) is condensed with a 5,5'-di-unsubstituted pyrromethane (34) in the presence of an acid catalyst (usually hydriodic acid); yields of porphyrin are often as high as 60% based on pyrromethanes. Toluene p-sulfonic acid is a useful alternative[36] to hydriodic acid as the catalyst. The approach can also be

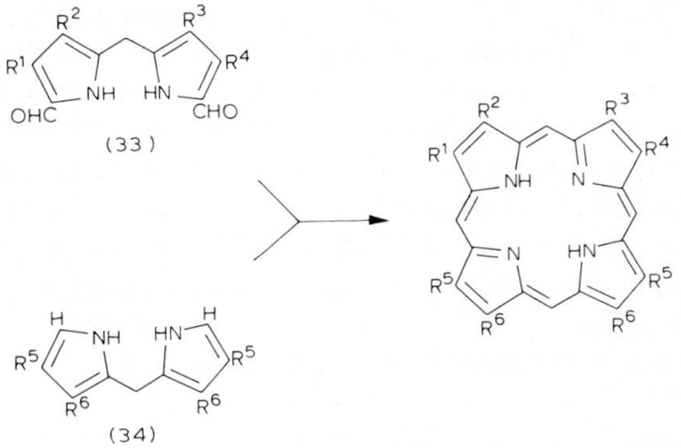

(33)

(34)

modified to include the preparation of centrosymmetrically substituted porphyrins by self-condensation of a 5-formyl-5'-unsubstituted pyrromethane.

The MacDonald procedure, therefore, fully complements the Fischer routes through pyrromethenes, each method having its advantages. The major disadvantage of the Fischer route is the vigorous thermal conditions

References, p. 55

required. Moreover, generally, pyrromethanes with complex substituents tend to be more readily prepared than the corresponding pyrromethene analogs and this has tended to encourage more extensive use of MacDonald's approach.

One example of the condensation of a 5,5'-bromomethylpyrromethene [of the type (28)] with a 5,5'-di-unsubstituted pyrromethane [of the type (34)] has been reported[34] and this gave a good yield of the isomerically pure porphyrin.

2.3.2.3. From pyrroketones

The pyrroketones most often encountered in porphyrin synthesis are the *a*-oxobilanes and *b*-oxobilanes (p. 38, 40). However, oxophlorins (35) can be prepared in good yield[37] by MacDonald-type condensation of 5,5'-diformyl-pyrroketones (36) with 5,5'-di-unsubstituted pyrromethanes (34) or the cor-

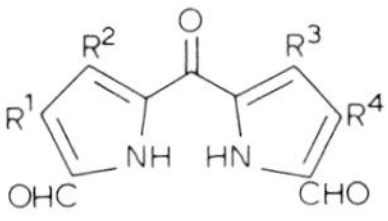

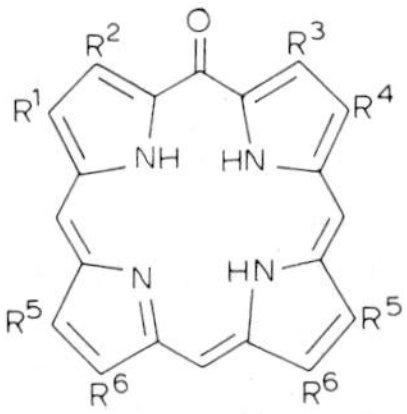

responding 5,5'-dicarboxylic acids. 5-Formyl-5'-hydroxymethylpyrroketones have been used[38] in place of (36). The diformylpyrroketones (36) are obtained[18] by direct oxidation of 5,5'-diformylpyrromethanes, which are readily available[39]. The same symmetry limitations apply as for the MacDonald synthesis from pyrromethanes; an added constraint is that the formyl groups which form the bridging carbon atoms between the two dipyrrolic halves must be sited on the pyrroketone moiety. 5,5'-Di-unsubstituted pyrroketones are not sufficiently nucleophilic to undergo condensation with 5,5'-diformylpyrromethanes, e.g. (33).

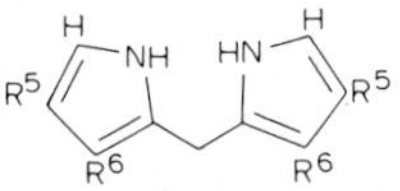

The conversion of oxophlorins into porphyrins is discussed on p. 41.

2.3.3. From open-chain tetrapyrrolic intermediates

The only *truly* general porphyrin syntheses are those which progress in a stepwise manner through discrete open-chain tetrapyrrolic intermediates. There can be no doubt that such syntheses are also the most esthetically pleasing.

Open-chain tetrapyrroles, as a general group, can exist in a variety of different forms, bilanes, bilenes, biladienes, and bilatrienes (Nomenclature, p. 6); a considerable number of synthetic approaches to these compounds have been reported in recent times. Once constructed, it is possible to check that the correct array of substituents is present, and mass spectrometry has been used extensively with this intention. Certain dangers are inherent in the use of open-chain tetrapyrroles for porphyrin synthesis; this is particularly so with the more saturated species such as bilanes and bilenes because methods for cyclization usually employ acidic reagents and the pyrromethane link, particularly when it is devoid of electron-withdrawing substituents on adjacent pyrrole rings, is notoriously unstable towards 'jumbling' or 'redistribution' reactions:

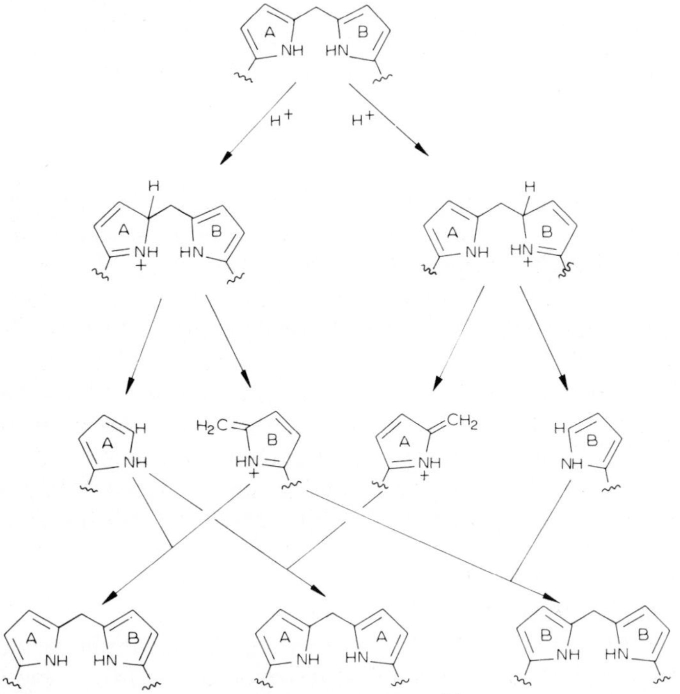

All arrows are reversible

Hence, the danger often exists that unless conditions for cyclization of the tetrapyrrole are extremely mild, or unless electron-withdrawing groups have been strategically placed within the molecule, then the synthesis may yield a mixture of several different porphyrins from a single, pure open-chain tetrapyrrole.

2.3.3.1. From bilanes and oxobilanes
2.3.3.1.1. Using bilane intermediates

In 1952, Corwin and Coolidge[40] reported the first attempt to synthesize a porphyrin using a discrete open-chain tetrapyrrole as an intermediate. This work was almost certainly many years ahead of its time and as a result some doubts have since been cast upon the isomeric purity of the product. Etio-porphyrin-II (31) was the target molecule and this was approached by treatment of the 5-unsubstituted pyrromethane (37) with formic acid (to give initially a b-bilene); catalytic hydrogenation of the crude product gave the bilane (38). After alkaline hydrolysis of the 1′ and 8′ esters the resultant dicarboxylic acid (39) was cyclized in formic acid to give a 36% yield of

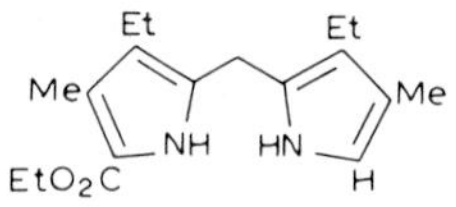

(37)

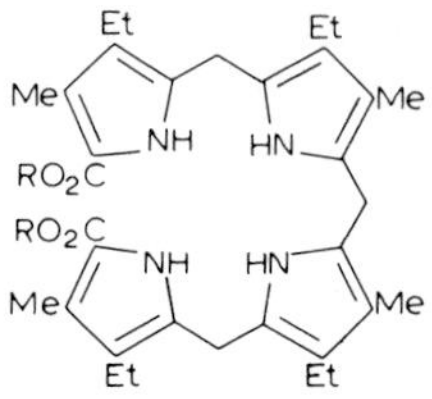

(38) R = Et
(39) R = H

porphyrin, claimed to be etioporphyrin-II. It was unfortunate that these workers chose an etioporphyrin as their synthetic objective because, at that time, there existed no way of differentiating one etioporphyrin from the three other type-isomers, or indeed from a mixture of all four. Since the publication of this synthesis the lability of bilanes[41] and porphyrinogens[42] towards redistribution reactions in acidic solution has been demonstrated and one is forced to conclude that the cyclization of (39) in formic acid would yield a mixture of etioporphyrins. It also seems likely that the conditions used in the synthesis of the bilane diester (38) were not conducive to the formation of only one isomer. However, Corwin and Coolidge's short paper[40] had a profound influence on the direction of research into porphyrin synthesis in the 1960s.

2.3.3.1.2. Using a-oxobilane intermediates

Condensation of lithium salts (40) of pyrromethane-5-carboxylic acids with pyridinium salts (41) of 5-chloromethylpyrroketones affords good yields of a-oxobilane-1′,8′-dibenzyl esters (42)[41]. These compounds (42) are highly crystalline, easily characterized, and mass spectrometry enables the sequence of pyrrole rings to be confirmed, though the precise order of the two side-chains on each pyrrole subunit, of course, cannot be verified. The oxo-function at the a position exerts a stabilizing influence, but the strongly electron-withdrawing nature of this moiety drastically reduces the nucleophilicity of ring A such that 1′,8′-di-unsubstituted a-oxobilanes (or the 1′,8′-dicarboxylic acid (43)) cannot be cyclized in the presence of an electrophilic one-carbon unit. It is therefore necessary to remove the oxo-function; hydrogenation of (42) gives (43) which is reduced with diborane to give the bilane-1′,8′-dicarboxylic acid (44). Acid promoted cyclization in the pres-

Scheme 1

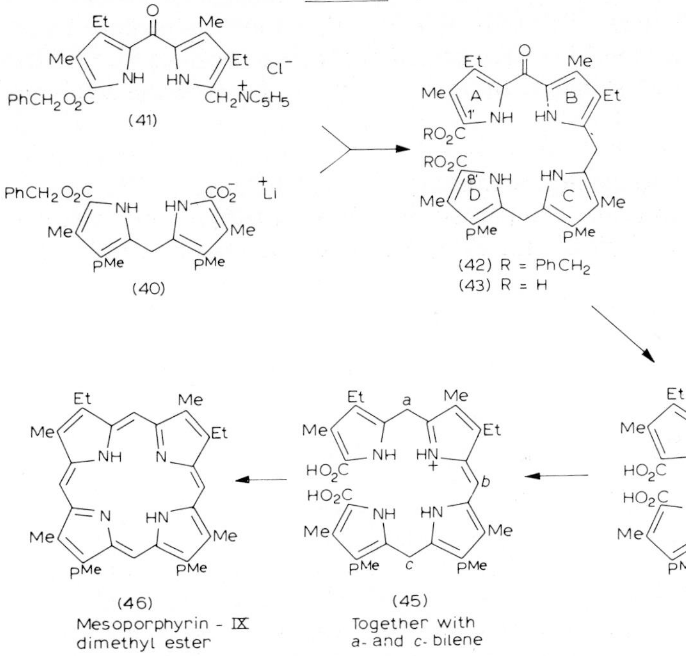

(41)

(40)

(42) R = PhCH$_2$
(43) R = H

(46)
Mesoporphyrin - IX
dimethyl ester

(45)
Together with
a- and c- bilene

(44)

ence of a one-carbon unit produces a complex mixture of porphyrins (cf. p. 37) and so the bilane is stabilized as the *b*-bilene (45) by oxidation with *t*-butyl hypochlorite. Cyclization in methylene chloride containing trichloroacetic acid and trimethyl orthoformate (as a linking one-carbon unit) gives the required isomerically pure porphyrin (46) in about 25% yield from (42).

This approach has been used for several synthetically demanding porphyrins; Scheme 1 shows the relatively routine synthesis of mesoporphyrin-IX dimethyl ester (46). In testing the generality and efficiency of a new synthesis it is essential that the product should have been well characterized by earlier workers so that comparisons can be made with literature data and with an authentic sample. More importantly, however, the target porphyrin should have pyrrole ring subunits which are differentially substituted, for example, with methyl—ethyl and methyl—propionate pairs of substituents. Then, if acid catalyzed redistribution of the open-chain tetrapyrrole occurs, the product porphyrin will consist of a mixture containing from one to four propionic ester groups; this is readily observed simply by thin-layer chromatography. If, on the other hand, the target molecule contains rings with only methyl—ethyl (or with only methyl—propionate) substituent pairs, as in the etioporphyrins (or coproporphyrins), then more refined techniques are required to check isomeric purity; it is only recently that it has been possible to identify individual etioporphyrins, for example[33].

The a-oxobilane porphyrin synthesis is quite general, and the intermediates (40) and (41) are readily obtainable; it suffers the disadvantage that it is rather complex and lengthy. After an initial flurry of activity[41,42] it has received few applications, being superseded by the b-oxobilane approach outlined below.

2.3.3.1.3. Using b-oxobilane intermediates

The phosphoryl chloride complexes of 5-*N*,*N*-dimethylamido-pyrromethanes (47) react in good yield with 5-unsubstituted pyrromethanes (48) by a

Scheme 2

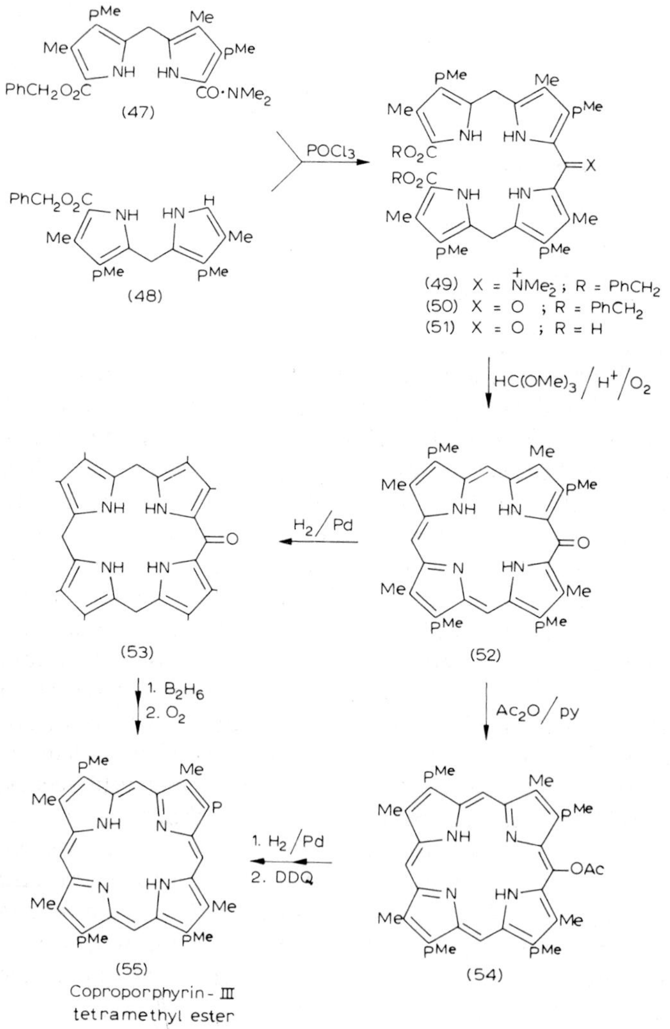
pMe
Me
PhCH2O2C
CO·NMe2
(47)
PhCH2O2C
(48)
POCl3
RO2C
X
(49) X = N+Me2 ; R = PhCH2
(50) X = O ; R = PhCH2
(51) X = O ; R = H
HC(OMe)3 / H+ / O2
H2 / Pd
(53)
(52)
1. B2H6
2. O2
Ac2O / py
1. H2 / Pd
2. DDQ
OAc
(55)
Coproporphyrin - III
tetramethyl ester
(54)

Vilsmeier—Haack type reaction, to give tetrapyrrolic imine salts (49)[43], which after chromatographic purification are readily hydrolyzed to give b-oxobilane-1',8'-dibenzyl esters (50). The b-oxo function affords protection against randomization in the B—C portion of the molecule, with the A and D rings being protected by the benzyl esters. However, the oxo-function is sufficiently remote from the terminal rings that it has no effect on the nucleophilicity of these rings and so b-oxobilanes can be cyclized directly, without the necessity to remove the oxo-group. Thus, catalytic hydrogenation of (50) furnishes the dicarboxylic acid (51) which cyclizes in methylene chloride containing trichloroacetic acid and trimethyl orthoformate to give the oxophlorin (52) after aerial oxidation and chromatographic purification; overall yields from (50) to (52) are usually high. The oxophlorin oxo-group can be removed in the following ways: (i) sodium amalgam reduction followed by re-oxidation of the resulting porphyrinogen, (ii) catalytic hydrogenation of the oxophlorin to give the macrocyclic pyrroketone (53) which is reduced with diborane to the corresponding porphyrinogen and the re-oxidized, or (iii) treatment with acetic anhydride in pyridine, giving the meso-acetoxyporphyrin (54) which is hydrogenated to give the porphyrinogen and then re-oxidized with 2,3-dichloro-5,6-dicyanobenzoquinone to give the meso-unsubstituted porphyrin (55). Of these three methods, the one most often used is the last, and yields of porphyrin (55) from the meso-acetoxy derivative are usually between 70 and 85%.

Scheme 2 outlines the synthesis of coproporphyrin-III tetramethyl ester[43]; the procedure has been used in the synthesis of a large number of complex porphyrins[44]. Though the method is lengthy and has a large number of steps, it has the advantages that it can be applied to a wide variety of porphyrins, affords the biologically interesting oxophlorins (see Chapter 4) as intermediates, and the starting materials, e.g. (47,48), are very readily available in quite large amounts.

2.3.3.2. From bilenes

Cyclization of a-bilenes gives[45] mixtures of porphyrins; all productive syntheses using bilenes have therefore made use of b-bilenes.

2.3.3.2.1. Using 1',8'-dimethyl-b-bilenes

1',8'-Dimethyl-b-bilenes can be prepared by two general methods. The first is particularly suited to synthesis of fully alkylated bilenes, but gives low yields of bilenes bearing electron-withdrawing substituents. Thus, condensation of two moles of a 2,3,4-trialkylpyrrole (56) with a 5,5'-di(methoxymethyl)pyrromethane (57) affords a b-bilene (58) with obvious symmetry restrictions[46−48]. A more useful synthesis is that involving condensation of a 5-formylpyrromethane (59) with a 5-unsubstituted pyrromethane or the corresponding 5-carboxylic acid (60)[49]; unsymmetrically substituted b-bilenes (61) result from this approach, but it is advisable[50] to have an electron-withdrawing group in the terminal rings of the b-bilene in order to avoid side-reactions.

b-Bilenes bearing 1'- and 8'-methyl groups are readily cyclized to give copper porphyrins (which can be demetalated with concentrated sulfuric acid) with copper salts in pyridine or dimethylformamide[47,48] and this procedure has been used on many occasions[49,51] for the cyclization of b-bilenes bearing electron-withdrawing groups.

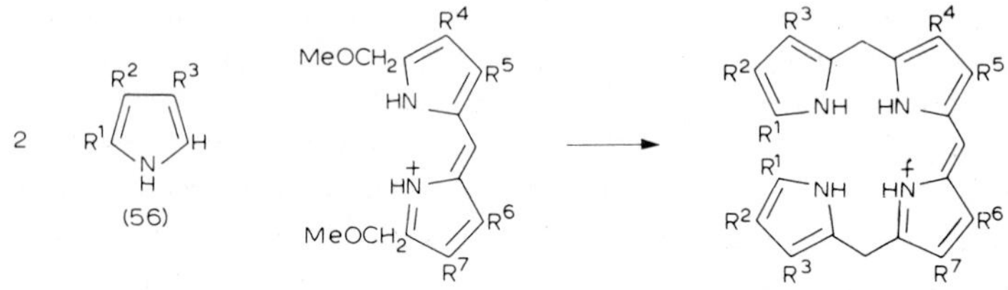

The mechanism by which one of the two terminal methyl groups is lost is still unknown (see also p. 45). This general method for porphyrin synthesis has many advantages, not the least of which are its simplicity and the way in which it can be used to prepare relatively large quantities of porphyrin. Its disadvantages concern symmetry and substituent limitations associated with the bilene intermediates.

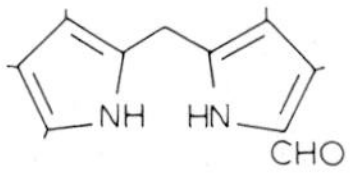

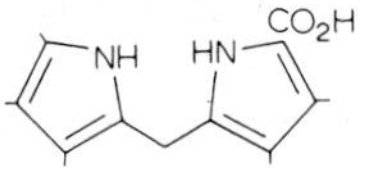

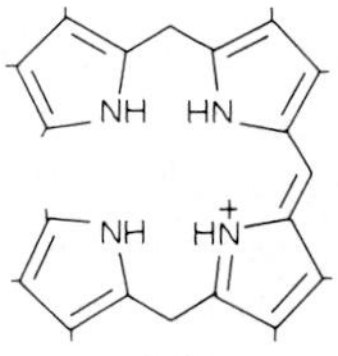

2.3.3.2.2. Using b-bilene-1',8'-diesters

b-Bilene-1',8'-dicarboxylic acids are intermediates in the synthesis of porphyrins from a-oxobilanes (p. 38). A more efficient approach to b-bilene-1',8'-dicarboxylates utilizes[52] the condensation of pyrromethanes (62) and (63) and furnishes b-bilene-1',8'-di-tert-butyl esters (64) in high yield as crystalline intermediates. Treatment with trifluoroacetic acid, followed by cyclization in methylene chloride containing trichloroacetic acid and trimethyl orthoformate (as a one-carbon linking unit) and then aeration gives good yields of porphyrin. Scheme 3 shows the synthesis[52b] of a phylloporphyrin (65) derived from the Chlorobium chlorophylls 660.

Scheme 3

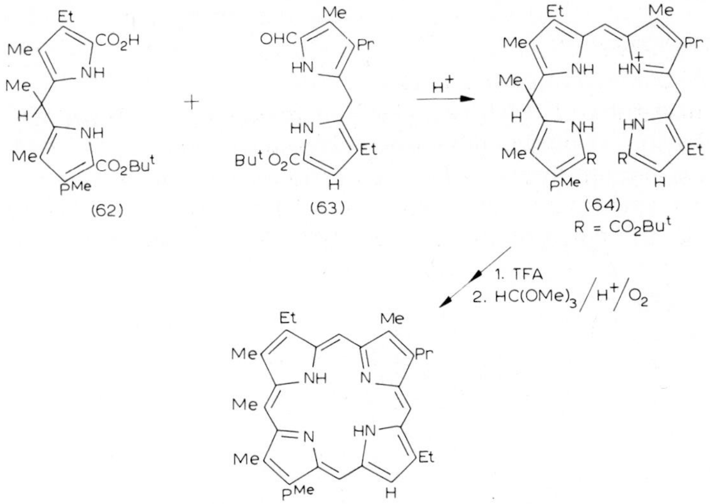

(62)   (63)   (64)
R = CO₂Buᵗ

1. TFA
2. HC(OMe)₃/H⁺/O₂

(65)

The method is simple, direct, and has crystalline intermediates at all stages. However, it tends to be erratic, particularly when electron-withdrawing substituents are sited on the B—C portion of the *b*-bilene, and this results in the production of porphyrin mixtures.

2.3.3.2.3. *Using other b-bilenes*

Deoxophylloerythroetioporphyrin (67) has been synthesized[53] from a *b*-bilene (66); a *b*-bilene (68) was also a short-lived intermediate in Woodward's

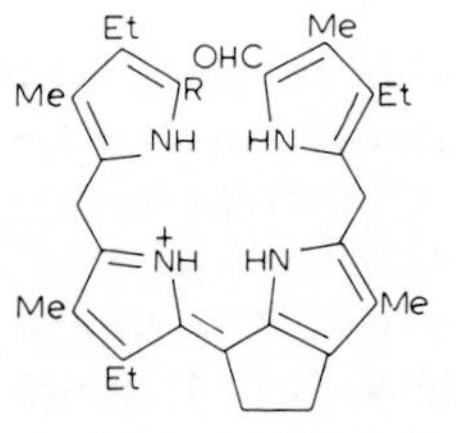

(66)
R = CO₂CH₂C₆H₄OMe

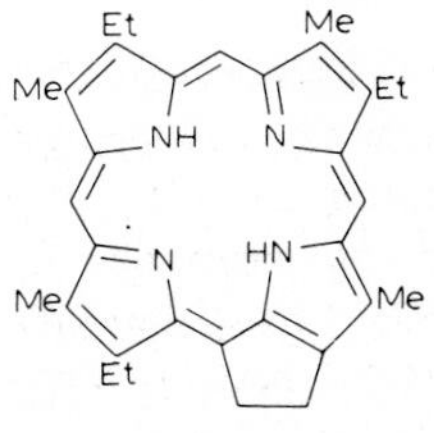

(67) 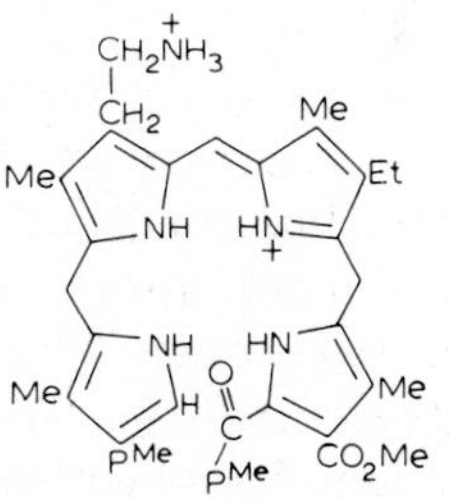   (68)

synthesis of chlorin-e_6 in his formal total synthesis of chlorophyll-a[35]. 1′-*tert*-Butoxycarbonyl-8′-methyl-*b*-bilenes have been prepared[54] but these are relatively difficult to cyclize to porphyrin and afford poor yields. Condensation of 5-formyl-5′-methylpyrromethanes with 5-unsubstituted-5′-

References, p. 55

(*N*,*N*-dimethylformimino)pyrromethane salts in methanol—acetic acid, to give[51] porphyrins, proceeds via a *b*-bilene intermediate.

2.3.3.3. From a,c-biladienes
2.3.3.3.1. Using 1′,8′-dimethyl-a,c-biladienes

Symmetrically substituted 1′,8′-dimethyl-*a,c*-biladienes (71) can be synthesized by condensation of pyrromethane-5,5′-dicarboxylic acids (69) with 2 moles of a 2-formyl-5-methylpyrrole (70), and the cyclization of these to give porphyrins using the copper salt method has been extensively investigated in parallel with the corresponding *b*-bilenes[47−49,55]. Scheme 4 out-

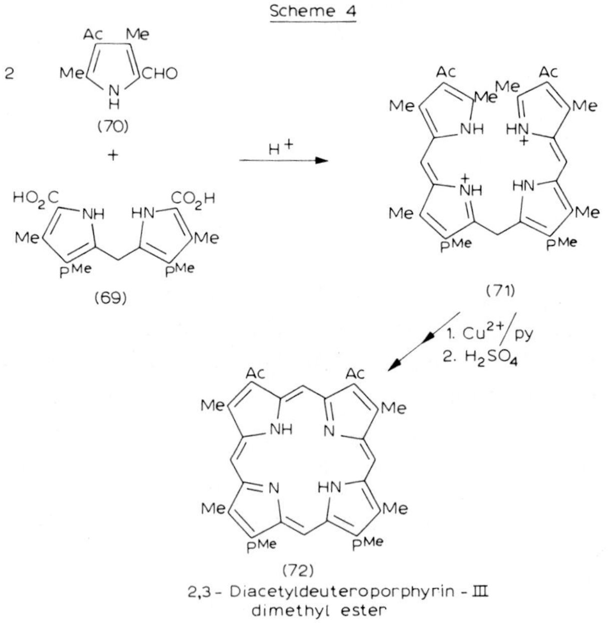

Scheme 4

(70)

(69)

(71)

1. Cu^{2+}/py
2. H$_2$SO$_4$

(72)
2,3 – Diacetyldeuteroporphyrin – III
dimethyl ester

lines the synthesis of 2,3-diacetyldeuteroporphyrin-III (72) using this method[49]. High yields of porphyrins are often obtained after removal of the chelating copper atom. All of the symmetry limitations inherent in this approach have been eliminated[56] by stepwise construction of the 1′,8′-dimethyl-*a,c*-biladiene. Thus, catalytic hydrogenation of a benzyl *tert*-butyl pyrromethane-5,5′-dicarboxylate (73) gives the carboxylic acid (74) which condenses under mild conditions[52a] with a 2-formyl-5-methylpyrrole (75) to afford a high yield of a stable, readily characterized, tripyrrene (76). After treatment with trifluoroacetic acid and a second 2-formylpyrrole (77), a high yield of 1′,8′-dimethyl-*a,c*-biladiene (78) with an unsymmetrically disposed array of substituents is obtained; cyclization by the normal copper salt method, followed by demetalation with sulfuric acid in trifluoroacetic acid gives a good yield of porphyrin (79). Several porphyrins have now been

Scheme 5

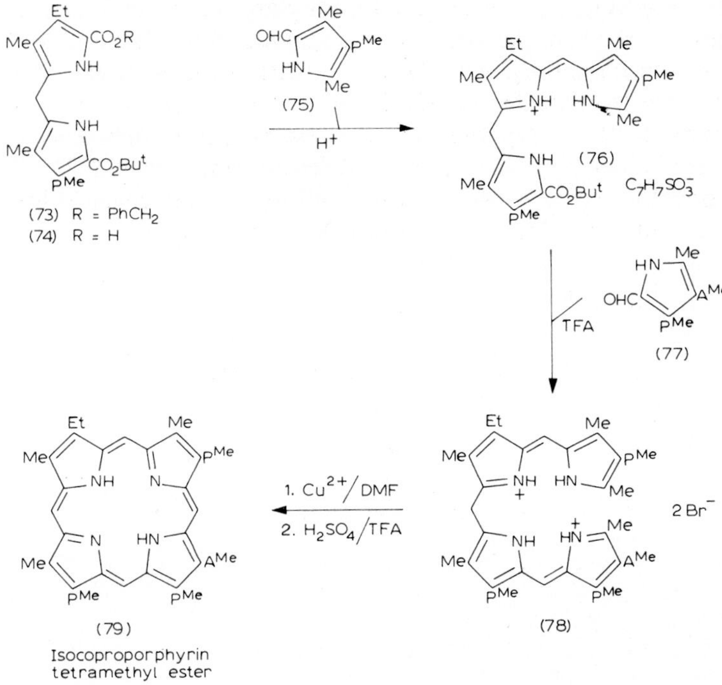

(73) R = PhCH₂
(74) R = H

(75)

(76)

(77)

TFA

(78)

(79)

Isocoproporphyrin
tetramethyl ester

1. Cu²⁺/DMF
2. H₂SO₄/TFA

synthesized by this method[56]; Scheme 5 outlines the approach to isocopro-porphyrin (79), a compound found in the feces of patients suffering from symptomatic cutaneous hepatic porphyria, or in rats with porphyria induced by hexachlorobenzene poisoning.

1′,8′-Dimethyl-*a,c*-biladienes are undoubtedly highly effective interme-diates for porphyrin synthesis; good yields of porphyrin are obtained and there are no obvious symmetry or substituent difficulties.

In the cyclization to porphyrin, one of the two terminal methyl groups in the *a,c*-biladiene is lost while the other presumably forms the bridging methine carbon in the product. There has been little, if any, comment on the precise mechanism of this cyclization, but it has been shown[57] that cycliza-tion of a 1,8-di-unsubstituted *a,c*-biladiene gives a porphyrinic product bear-ing an extra formyl substituent which presumably arises from the 'extra' methyl group in the *a,c*-biladiene:

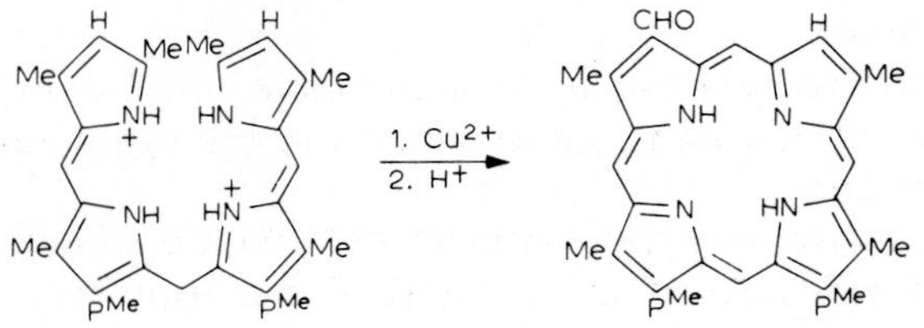

1. Cu²⁺
2. H⁺

References, p. 55

2.3.3.3.2. Using 1'-bromo-8'-methyl-a,c-biladienes

5-Bromo-5'-bromomethylpyrromethenes (80) alkylate 5-methyl-5'-unsub-stituted pyrromethenes (81) in the presence of stannic chloride to give[58] high yields of 1'-bromo-8'-methyl-a,c-biladienes (82) after removal of a chelating tin atom with hydrobromic acid. These tetrapyrroles can then be cyclized to porphyrin (83) in very high yield by heating in o-dichloroben-zene[58] or at room temperature in dimethylsulfoxide and pyridine[59]. A whole variety of porphyrins have been synthesized by this approach[60,61]; Scheme 6 shows the synthesis of rhodoporphyrin-XV ester (83).

Scheme 6

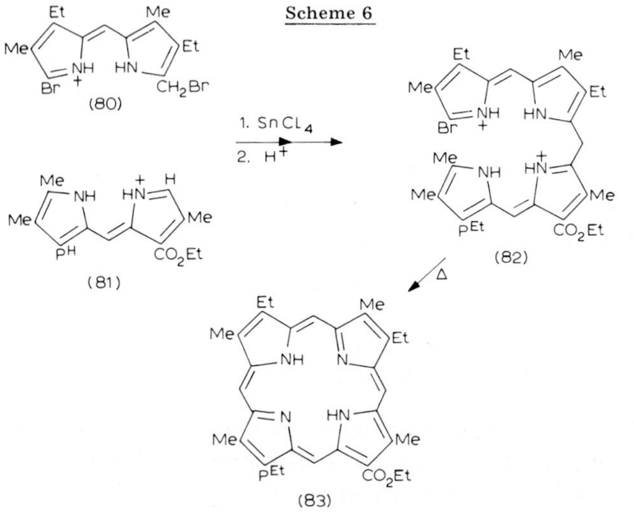

(80)

1. SnCl$_4$
2. H$^+$

(81)

(82)

Δ

(83)

Rhodoporphyrin - XV
diethyl ester

The 1'-bromo-8'-methyl-a,c-biladiene route affords good yields in all of its stages (rhodoporphyrin syntheses almost invariably give somewhat lower yields), of which there are few. It is easily applied to large-scale preparations and is applicable to a large variety of unsymmetrical porphyrins. Its main disadvantage is that the pyrromethene starting materials (80) and (81) are difficult to obtain in more complex cases.

1'-Bromo-8'-methyl-a,c-biladienes can also be cyclized to porphyrins by treatment with cupric salts and other one-electron oxidizing agents[62].

2.3.3.3.3. Using other a,c-biladienes

Heating of 1'-unsubstituted 8'-methyl-a,c-biladienes furnishes porphyrins, but these compounds are more often used as intermediates in the synthesis of tetradehydrocorrins (Chapter 18).

Condensation of 5,5'-di-unsubstituted pyrromethanes (84) with one mole of a 2-formyl-5-methylpyrrole (85) affords high yields of the tripyrrenes

(86); these are apparently uncontaminated with by-products due to reaction of the formylpyrrole at both ends of (84). The tripyrrene intermediates can then be reacted with a different formylpyrrole (87) to give high yields of the appropriate *a,c*-biladiene (88). Several of these biladienes have been cyclized to give a variety of porphyrins (89)[63]. This new modified procedure still

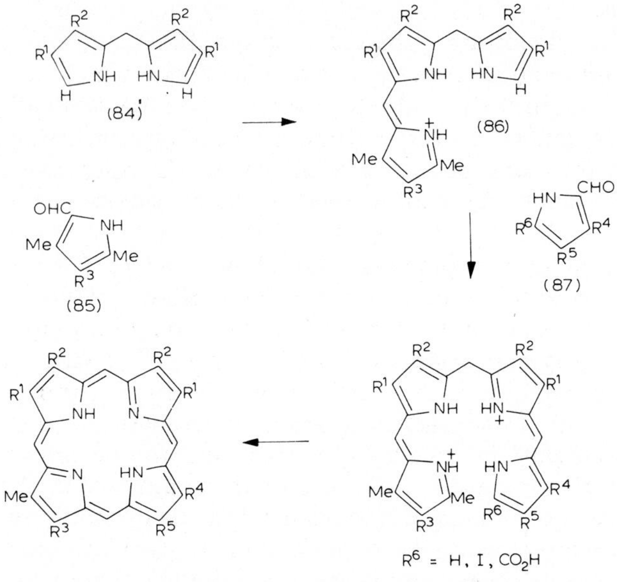

retains the limitation that the pyrromethane (84) must be symmetrically substituted about the interpyrrolic carbon atom.

In a potentially very useful modification of the *a,c*-biladiene approach, *a,c*-biladiene-1′,8′-dicarboxylic acids have been efficiently cyclized with formaldehyde[61].

2.3.4. General conclusions

Synthetic approaches to porphyrin compounds have developed dramatically within the past ten years[1,64,65]. In the 1920s and 1930s Fischer's school in Munich introduced several highly efficient methods, usually involving condensations of pyrromethenes. Though symmetry limitations are inherent in this concept of porphyrin synthesis, the Munich school produced literally dozens of now classical syntheses of protoheme and chlorophyll degradation products; when syntheses were ambiguous and produced mixtures of porphyrins, methods were developed to enable the separation of these mixtures, and even the most minor by-product was investigated and characterized. Intelligent use of substituent combinations permitted these separations; for example, if the target porphyrin possessed two ester functions then the synthesis was devised such that both of these esters were on one of the two pyrromethene halves. Then, the by-products produced from self-condensation of each half would have zero and four ester groups (or carboxylic acid functions) and these would be readily separable from the required porphyrin. However, Fischer's methods tended to rely upon rather drastic reaction conditions and the recent developments have preferred milder synthetic operations.

Since Fischer's time, and following a hiatus between 1940 and about 1960, it has become unfashionable to use porphyrin syntheses which produce mixtures of porphyrins. Instead, with the exception of MacDonald's endeavors[34], workers have tended to concentrate on porphyrin synthesis involving isolable open-chain tetrapyrrolic intermediates, devoting much effort to achieve conditions which would allow relatively labile side-chains to be carried through intact.

However, these sophisticated developments should be used only when necessary; it would be nonsensical to use them for the synthesis of simple model porphyrins. meso-Tetraphenylporphyrin (16) and octaethylporphyrin (1) are best prepared by polymerization of a monopyrrole. Regularly or centrosymmetrically substituted porphyrins such as etioporphyrin-I should be prepared by self-condensation of a readily available pyrromethene, such as (6) or (7); such pyrromethenes are so accessible that octaethylporphyrin has also been prepared by this method[30]. Because of the symmetry in the lower half of many porphyrins of biological importance [e.g. deuteroporphyrin-IX, (29)] these can be approached by condensation of two pyrromethenes or two pyrromethanes because the lower (C—D) dipyrrolic unit can condense in either of the two possible ways and yet still afford only one product. Porphyrins, which because of their complex array of substituents, possess no symmetry characteristics, must be approached synthetically from open-chain tetrapyrroles.

2.4. Preparation of Porphyrins by Degradation of Natural Pigments*
2.4.1. From hemoglobin
2.4.1.1. Protohemin

Several methods for the isolation of protohemin (90) from blood have been reported. The most common method[66] is to pour strained citrated, heparinized, or defibrinated blood into hot acetic acid containing sodium chloride; the hemin separates on cooling. Alternatively[67], the protein can be precipitated with a solution of strontium chloride and concentration of the filtrate affords crystalline protohemin. A simple method for the isolation of protoheme (the iron(II) analog of (90)) on a small scale has been reported[68].

---

* Fuller details of many of these procedures are given in Chapter 19.

Hemins should not be esterified with diazomethane; the normal mineral acid—alcohol mixture is preferable.

2.4.1.2. Hematoporphyrin-IX (91)

Hematoporphyrin-IX was the first porphyrin to be isolated; it was prepared by the action of concentrated sulfuric acid on blood in 1867 by Thudichum[69]. This treatment obviously involves removal of the iron from protohemin (90) with concomitant Markownikoff hydration of the 2- and 4-vinyl groups. It is now prepared by treatment of protohemin with hydrogen bromide in acetic acid[70], followed by decomposition of the HBr adduct with water; if methanol is used in this last stage then the dimethyl ether is obtained. The propionic acid side-chains of hematoporphyrin-IX and its dimethyl ether are best esterified with diazomethane. Mineral acid—alcohol esterification furnishes[71] a mixture of hematoporphyrin-IX dimethyl ester, protoporphyrin-IX dimethyl ester, and the corresponding monovinyl—mono-hydroxyethyl-porphyrins.

Hematoporphyrin-IX has two chiral carbon atoms, one in each of the 2- and 4-side-chains; an optically active form has been obtained[72] by degradation of cytochrome-*c*. However, due to the hydration—dehydration equilibrium, synthetic hematoporphyrin-IX is optically inactive.

2.4.1.3. Protoporphyrin-IX (92)

The most common method for preparation of protoporphyrin-IX is to treat protohemin with ferrous sulfate in hydrochloric acid[73]; the metalloporphyrin is reduced and iron(II) is readily removed by the acid. Commercial samples of hematoporphyrin-IX are often of high purity (unlike those of protoporphyrin-IX); hence, a method for the preparation of protoporphyrin-IX from hematoporphyrin-IX, involving brief heating in *ortho*-dichlorobenzene containing toluene *p*-sulfonic acid, has been reported[74].

The dimethyl ester can be obtained by esterification with either diazomethane or methanol—sulfuric acid (p. 835). A method for the direct preparation of protoporphyrin-IX dimethyl ester from protohemin has been described[75].

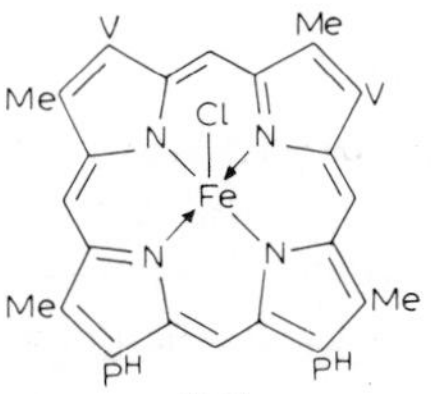

$$V = CH{=}CH_2$$
$$P^R = CH_2CH_2\,CO_2R$$

(90)

(91) R = $-\overset{\underset{|}{OH}}{CH}-CH_3$

(92) R = V

(93) R = Et

(94) R = $-H$

(95) R = $-COCH_3$

(96) R = $-CHO$

(97) R = $-\overset{\underset{|}{OH}}{CH}-CH_2OH$

(98) R = $-CH=CHCO_2H$

(99) R = P^H

(100) R^1 = R^2 = $-CH_2CH(OMe)_2$

(101) R^1 = R^2 = $-CH_2CH_2OH$

(102) R^1 = R^2 = $-CH_2CH_2Cl$

(103) R^1 = R^2 = $-CH_2CH_2Br$

(104) R^1 = R^2 = $-CH_2CH_2CN$

(105) R^1 = $-\overset{\underset{|}{OH}}{CH}-CH_2OH$; R^2 = V

(106) R^1 = V ; R^2 = $-\overset{\underset{|}{OH}}{CH}-CH_2OH$

(107) R^1 = $-CHO$; R^2 = V

(108) R^1 = V ; R^2 = $-CHO$

(109) R^1 = V ; R^2 = $-CH_2CH(OMe)_2$

(110) R^1 = $-CH_2CH(OMe)_2$; R^2 = V

(111) R^1 = V ; R^2 = $-CH_2CH_2OH$

(112) R^1 = $-CH_2CH_2OH$; R^2 = V

(113) R^1 = V ; R^2 = P^{Me}

(114) R^1 = P^{Me} ; R^2 = V

(115) R^1 = $-H$; R^2 = $-CH_2CH_2OH$

(116) R^1 = $-CH_2CH_2OH$; R^2 = $-H$

(117) R^1 = $-H$; R^2 = V

(118) R^1 = V ; R^2 = H

2.4.1.4. Mesoporphyrin-IX (93)

This compound was a key derivative in the establishment of the structure of both protohemin and chlorophyll-a[3]. Mesoporphyrin-IX is considerably more stable than protoporphyrin-IX, principally because the latter is highly activated towards photo-decomposition. Hence, biosynthetic investigations of the origin of protoporphyrin-IX have often concentrated upon mesoporphyrin-IX which is readily obtained by reduction of the 2- and 4-vinyl groups. The classical method for preparation of (93) involves[76] treatment of protohemin with hydriodic acid in acetic acid. However, the method of choice[3] is catalytic hydrogenation over palladium in formic acid. Either protoporphyrin, its ester, or protohemin may be used, iron being removed from the last during the reaction. Mesohemin is best prepared by introduction of iron into (93). Esterification of mesoporphyrin-IX can be achieved with either diazomethane or alcohol and mineral acid.

2.4.1.5. Deuteroporphyrin-IX (94) and its derivatives

Deuteroporphyrin-IX itself has a place in history because it was the first porphyrin formed (p. 34) in Fischer's classical total synthesis of protohemin[3,31]. Deuterohemin can be prepared from protohemin by brief heating in a resorcinol melt[77], the vinyl groups being replaced by hydrogens; intermediate resorcinol adducts have been isolated[78].

Various 2,4-disubstitution products can be prepared, usually by aromatic electrophilic substitution of the hemin. Diacetyldeuteroporphyrin-IX (95) was an intermediate in the protohemin total synthesis and was obtained by Friedel—Crafts acetylation of deuterohemin. Compound (95) can also be obtained by Jones oxidation of hematoporphyrin-IX. The full range of 2,4-disubstituted deuteroporphyrins are listed in Table 4 (p. 22); in addition, 2,4-dibromo, dicyano-, and others have been prepared[3].

2,4-Diformyldeuteroporphyrin-IX (96) can be obtained by treatment of protoporphyrin-IX with osmium tetroxide (to give the bis-glycol (97)) followed by periodate[79,80]. Knoevenagel condensation with malonic acid gives[79—81] the bis-acrylic acid (98), hydrogenation of which produces[81] coproporphyrin-III (99) in good overall yield.

2.4.1.6. Coproporphyrin-III (99)

The concentration of coproporphyrinogen-III present in natural systems is usually minute, though with expert biochemical knowledge it is possible to culture mutant organisms which produce large quantities of biosynthetic intermediates. Thus, any synthesis of a biosynthetic intermediate from a terminal product such as protoporphyrin-IX or protohemin is of great potential importance. Coproporphyrin-III (99) prepared from 2,4-diformyldeuteroporphyrin-IX as described above was used[81] in radiochemically labeled form to prove that coproporphyrinogen-III is indeed a biosynthetic precursor of protoporphyrin-IX (Chapter 3). A second synthesis of coproporphyrin-III (99) from protoporphyrin-IX (92) has been devised[74a]; treatment of protoporphyrin-IX dimethyl ester with three equivalents of thallium(III) nitrate in methanol gives the bis-acetal (100) which after hydrolysis and reduction with sodium borohydride affords the bis-anti-Markownikoff hydrated product (101). (Anti-Markownikoff hydration of porphyrin vinyl groups can also be accomplished in lower yield using diborane[82]). This sequence is particularly useful in that it accomplishes the protection of the highly reactive porphyrin vinyl groups; the vinyl can be regenerated by treatment with thionyl chloride in presence of dimethylformamide (to give (102)) followed by dehydrohalogenation carried out on the zinc porphyrin with *tert*-butoxide[42,83]. However, treatment of (101) with thionyl bromide gives the bis-bromoethyl derivative (103) which with cyanide furnishes the bis-cyanoethyl compound (104). Methanolysis then produces coproporphyrin-III tetramethyl ester in 37% yield from protoporphyrin-IX dimethyl ester.

2.4.1.7. Other porphyrins of biological significance

With the development of sophisticated chromatographic techniques (p. 839) a much wider range of natural porphyrins have become accessible from protoporphyrin-IX. Modification of Mauzerall and Sparatore's method[80] involving treatment of protoporphyrin-IX with osmium tetroxide gives[84] a mixture of the two mono-vinyl—mono-glycols (105) and (106) after column chromatography. These two compounds can be separated preparatively by conventional thick-layer chromatography on Kieselgel silica and treatment of the individual bands affords *Spirographis* porphyrin dimethyl ester (107) and its isomer (108). Treatment of protoporphyrin-IX dimethyl ester with only two equivalents of thallium(III) nitrate in methanol (one equivalent merely accomplishes metalation of the macrocycle) gives a mixture of mono-acetal—mono-vinyl compounds, (109) and (110), which are separable from starting material and bis-acetal (100) by column chromatography. Hydrolysis and reduction of the mono-acetals affords[84] the mono-hydroxyethyl—mono-vinyl compounds (111) and (112) which are just separable by preparative thick-layer chromatography. Treatment of each of these compounds as described earlier (p. 51) in the analogous synthesis of coproporphyrin-III furnishes harderoporphyrin trimethyl ester (113) and its isomer (114). Alternatively resorcinol fusion of the hemins of the individual mono-vinyl derivatives (111) and (112) gives[84] the mono-hydroxyethyl deuteroporphyrins (115) and (116) respectively, after demetalation. These compounds have already been transformed[85] into pemptoporphyrin dimethyl ester (117) and its isomer (118).

2.4.2. *From the plant chlorophylls**

Fischer and Stern[86] devoted a whole volume to the constitution and degradative chemistry of the chlorophylls. It would be unrealistic to attempt to summarize this in the minimal space available here. In view of the fact that chlorophylls are not dealt with in great depth in this book, only those degradations which furnish porphyrins in reasonable and useful quantities will be discussed.

Chlorophyll nomenclature uses italic letters and subscript numbers; the latter indicate the number of oxygen atoms in the molecule, and therefore, pheoporphyrin-a_5 has five oxygen atoms and is a porphyrin from the chlorophyll-*b* series of degradation products. In the chlorophyll-*b* series, a formyl group replaces the methyl at position 3, and so the number of oxygen atoms increases to 6. In the chlorins and rhodins, where the isocyclic ring carbonyl group of chlorophyll-*a* and -*b* respectively has become a carboxyl group, there are 6 and 7 oxygen atoms. The letters *a* and *e* are interchangeable, and are used in the chlorophyll-*a* series only (e.g. in chlorin-e_6), while *g* is used for chlorophyll-*b* derivatives (e.g. in rhodin-g_7). Much of the remaining chlorophyll nomenclature defies logical interpretation.

Chlorophylls-*a* and -*b*, (119) and (120) respectively, can be extracted in large quantities from leaf tissue[87,88]. These were first separated by a classical application[89] of chromatography, and subsequently by solvent partition[90]. The chlorophyll-*a* to chlorophyll-*b* ratio is usually about 3 : 1.

---

* For a schematic summary of the full degradative chemistry of the chlorophylls, see ref. 86, p. 40.

Mild acid treatment of the chlorophylls affords the metal-free pheophy-
tins-a and b, (121) and (122) respectively; this is usually the form in which
the pigments are stored prior to further degradation, and at one time, pheophytin mixture was commercially available. The mixture is conveniently

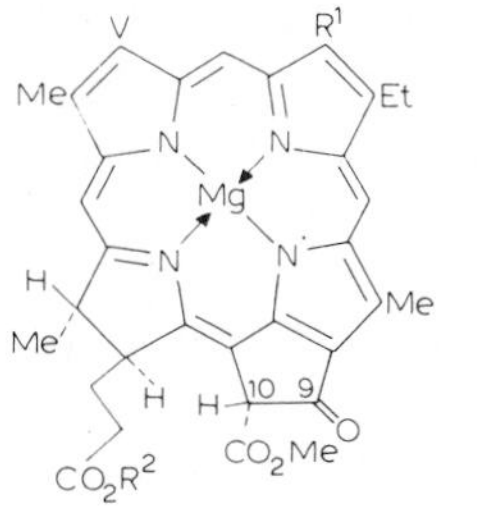

(119) R¹ = Me ; R² = Phytyl
(120) R¹ = CHO ; R² = Phytyl
(125) R¹ = R² = Me
(126) R¹ = CHO ; R² = Me

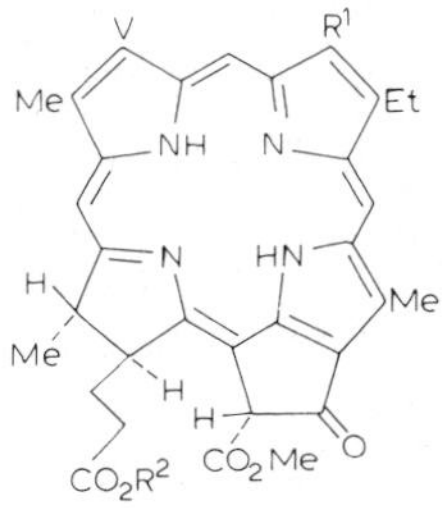

(121) R¹ = Me ; R² = Phytyl
(122) R¹ = CHO ; R² = Phytyl
(123) R¹ = R² = Me
(124) R¹ = CHO ; R² = Me

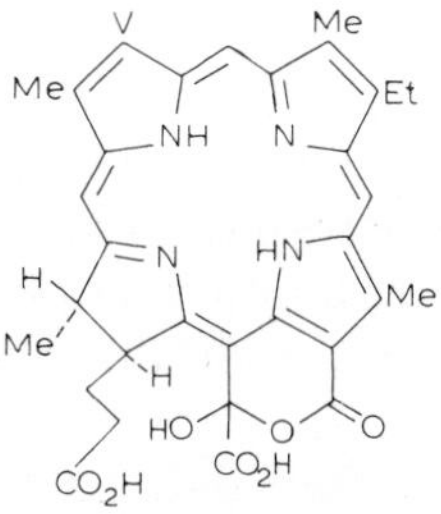

(127)

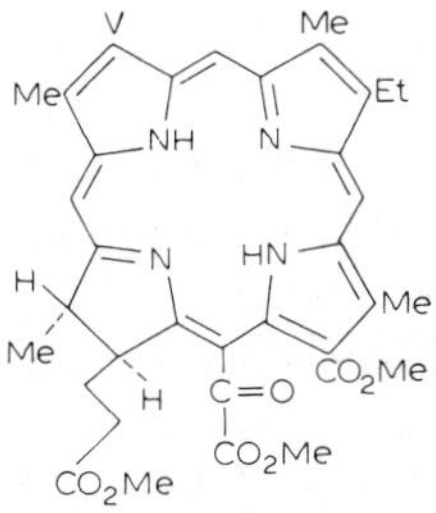

(128)

separated on a large scale by making use of the reaction[91,92] of the formyl
group in the 'b' series with Girard's reagent T, followed by chromatographic
separation[92].

Methanolysis of pheophytin-a or -b produces the corresponding methyl
pheophorbides-a (123) and -b (124). It is possible[93] to transesterify the
phytyl residue without removal of the magnesium atom; with methanol, the
methyl chlorophyllides (125) and (126) result.

Aerial oxidation of pheophytin-a (121) in alkaline solution cleaves the
C_9—C_{10} bond with concomitant loss of phytol, giving the 'unstable chlorin'
(127)[92,94]. Evaporation of the solution produces purpurin-18, whereas
esterification with diazomethane furnishes purpurin-7 trimethyl ester
(128)[92,94]. If the latter compound is heated in collidine, a 76% yield of
2-vinylrhodoporphyrin-XV dimethyl ester (129) is obtained[92]. In an analogous fashion, the meso (dihydro, with the 2-vinyl reduced to ethyl) compound affords an 81% yield of rhodoporphyrin-XV dimethyl ester (130).

References, p. 55

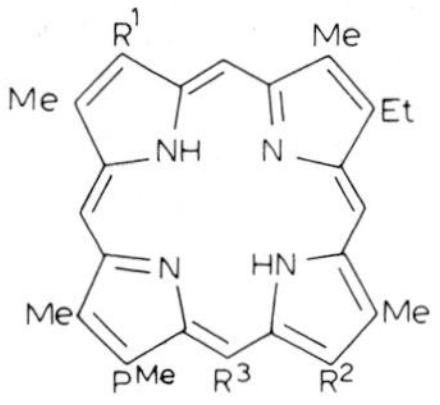

(129) R^1 = V; R^2 = CO_2Me ; R^3 = H
(130) R^1 = Et; R^2 = CO_2Me ; R^3 = H
(140) R^1 = Et ; R^2 = H ; R^3 = Me
(141) R = Et ; R^2 = R^3 = H

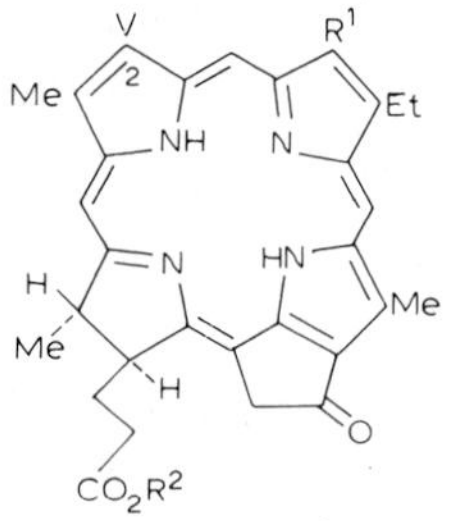

(131) R^1 = Me ; R^2 = H
(132) R^1 = CHO ; R^2 = H
(134) R^1 = R^2 = Me
      Et replaces V in position 2

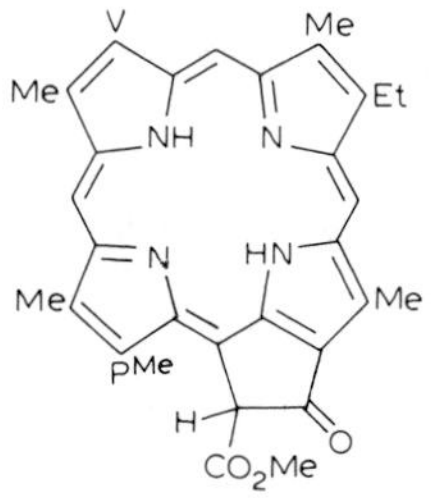

(133)

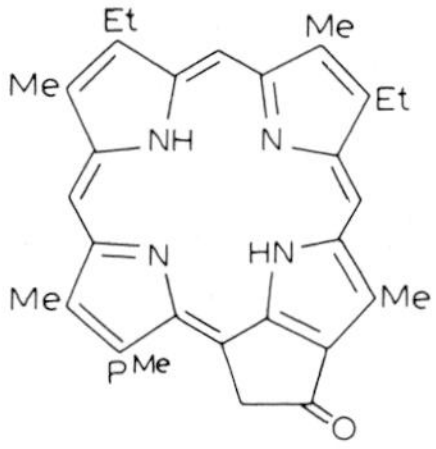

(135)

If methyl pheophorbide-*a* (or -*b*) is heated in collidine[92] the corresponding pyropheophorbide-*a* (131) (or -*b*, 132) is produced in virtually quantitative yield; heating in pyridine affords a lower yield[95]. Similar reactions take place with the meso-series of compounds.

Treatment of chlorins with high-potential quinones, such as 2,3-dichloro-5,6-dicyanobenzoquinone gives the corresponding porphyrins. For example[92], methyl pheophorbide-*a* (123) gives 2-vinylpheoporphyrin-a_5 dimethyl ester (133), and methyl mesopyropheophorbide-*a* (134) yields phylloerythrin methyl ester (135).

Pheophorbide-*a* and pheoporphyrin-a_5 can be cleaved[86] with alkali to give chlorin-e_6 (136) and chloroporphyrin-e_6 (137) respectively; methanolysis of

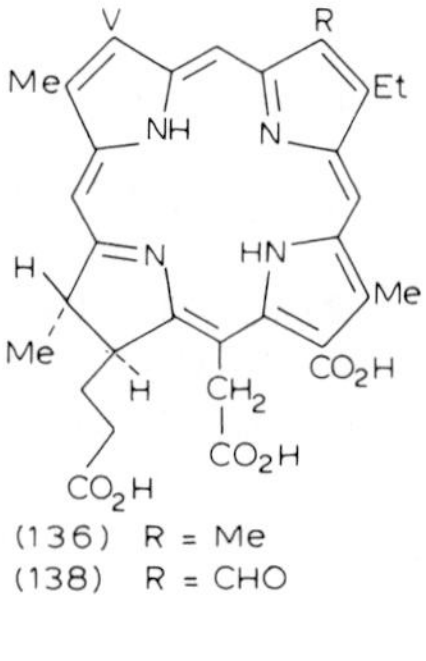

(136) R = Me
(138) R = CHO

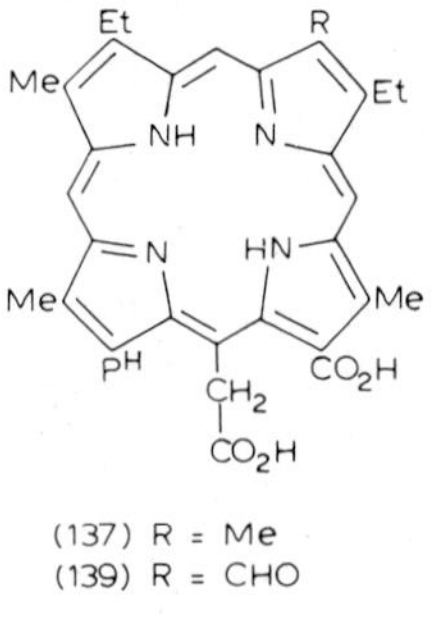

(137) R = Me
(139) R = CHO

pheophorbide-a with methanolic diazomethane affords chlorin-e_6 trimethyl ester. The isocyclic ring is easily reformed in these molecules.

Completely analogous reactions occur in the b series of compounds; the analogue of chlorin-e_6 in the b series is rhodin-g_7 (138) and that of chloroporphyrin-e_6 is rhodinporphyrin-g_7 (139).

Vigorous alkaline treatment of chlorophyll, followed by methanolysis affords[96] phylloporphyrin-XV methyl ester (140) and pyrroporphyrin-XV methyl ester (141); pyrroporphyrin-XV is also obtained from phylloporphyrin-XV by treatment with alkali.

References

1. A.H. Jackson and K.M. Smith in 'Total Synthesis of Natural Products', Vol. 1, Ed. J.W. ApSimon, Wiley, New York (1973), p. 143.
2. E. Baltazzi and L.I. Krimen, Chem. Rev., **63**, 511 (1963).
3. H. Fischer and H. Orth, 'Die Chemie des Pyrrols', Akademische Verlagsgesellschaft, Leipzig Vol. IIi (1937).
4. Ref. 3, p. 2.
5. Ref. 3, pp. 73, 106.
6. Ref. 3, p. 3.
7. J.A.S. Cavaleiro, A.M.d'A. Rocha Gonsalves, G.W. Kenner, and K.M. Smith, J. Chem. Soc., Perkin Trans. 1, 2471 (1973).
8. P.S. Clezy and A.J. Liepa, Aust. J. Chem., **23**, 2443 (1970).
9. E.J. Tarlton, S.F. MacDonald, and E. Baltazzi, J. Am. Chem. Soc., **82**, 4389 (1960).
10. A. Hayes, G.W. Kenner, and N.R. Williams, J. Chem. Soc., 3779 (1958); A.H. Jackson, G.W. Kenner, and D. Warburton, ibid., 1328 (1965).
11. H. Fischer and H. Orth, 'Die Chemie des Pyrrols', Akademische Verlagsgesellschaft, Leipzig, Vol. I, (1934), p. 333.
12. A.F. Mironov, T.R. Ovsepyan, R.P. Evstigneeva, and N.A. Preobrazenskii, Zh. Obshch. Khim., **35**, 324 (1965).
13. Ref. 11, p. 332.
14. J.A. Ballantine, A.H. Jackson, G.W. Kenner, and G. McGillivray, Tetrahedron, **22**, 241 (Suppl. 1) (1966).
15. Ref. 11, p. 362.
16. Ref. 11, p. 361.
17. J.M. Osgerby and S.F. MacDonald, Can. J. Chem., **40**, 1585 (1962).
18. P.S. Clezy, A.J. Liepa, A.W. Nichol, and G.A. Smythe, Aust. J. Chem., **23**, 589 (1970).
19. Ref. 3, p. 175.
20. P. Rothemund, J. Am. Chem. Soc., **57**, 2010 (1935); P. Rothemund and A.R. Menotti, ibid., **63**, 267 (1941).
21. A.D. Adler, F.R. Longo, J.D. Finarelli, J. Goldmacher, J. Assour, and L. Korsakoff, J. Org. Chem., **32**, 476 (1967).
22. G.H. Barnett, M.F. Hudson, and K.M. Smith, Tetrahedron Lett., 2887 (1973).
23. G.M. Badger, R.A. Jones, and R.L. Laslett, Aust. J. Chem., **17**, 1028 (1964).
24. G.H. Barnett, M.F. Hudson, and K.M. Smith, J. Chem. Soc., Perkin Trans. 1 (1975), in press.
25. H.W. Whitlock and R. Hanauer, J. Org. Chem., **33**, 2169 (1968); cf. U. Eisner, A. Lichtarowicz, and R.P. Linstead, J. Chem. Soc., 733 (1957).

26. H.H. Inhoffen, J.-H. Fuhrhop, H. Voigt, and H. Brockmann jr., Ann. Chem., **695**, 133 (1966); cf. W. Siedel and F. Winkler, ibid., **554**, 162 (1943).
27. H. Fischer and J. Klarer, Ann. Chem., **448**, 178 (1926); H. Fischer, H. Friedrich, W Lamatsch, and K. Morgenroth, ibid., **466**, 147 (1928); ref. 3, p. 169.
28. Ref. 3, p. 193.
29. K.M. Smith, J. Chem. Soc., Perkin Trans. 1, 1471 (1972).
30. H. Fischer and R. Bäumler, Ann. Chem., **468**, 58 (1929); ref. 3, p. 188.
31. H. Fischer and K. Zeile, Ann. Chem., **468**, 98 (1929).
32. Ref. 3, p. 165.
33. M.F. Hudson and K.M. Smith, J. Chem. Soc., Chem. Commun, 515 (1973)
34. G.P. Arsenault, E. Bullock, and S.F. MacDonald, J. Am. Chem. Soc., **82**, 4384 (1960).
35. R.B. Woodward, Angew. Chem., **72**, 651 (1960); Pure Appl. Chem., **2**, 383 (1961).
36. J.A.S. Cavaleiro, A.M.d'A. Rocha Gonsalves, G.W. Kenner, and K.M. Smith, J. Chem. Soc., Perkin Trans. 1, 1771 (1974); J.A.S. Cavaleiro, G.W. Kenner, and K.M. Smith, ibid., 1188 (1974).
37. P.S. Clezy, A.J. Liepa, and G.A. Smythe, Aust. J. Chem., **23**, 603 (1970).
38. R. Bonnett, M.J. Dimsdale, and G.F. Stephenson, J. Chem. Soc. C, 564 (1969).
39. R. Chong, P.S. Clezy, A.J. Liepa, and A.W. Nichol, Aust. J. Chem., **22**, 229 (1969).
40. A.H. Corwin and E.C. Coolidge, J. Am. Chem. Soc., **74**, 5196 (1952).
41. A.H. Jackson, G.W. Kenner, and G.S. Sach, J. Chem. Soc. C, 2045 (1967).
42. A.H. Jackson, G.W. Kenner, and J. Wass, J. Chem. Soc., Perkin Trans. 1, 1475 (1972); 480 (1974); R.P. Carr, A.H. Jackson, G.W. Kenner, and G.S. Sach, J. Chem. Soc. C, 487 (1971); T.T. Howarth, A.H. Jackson, and G.W. Kenner, J. Chem. Soc., Perkin Trans. 1, 502 (1974).
43. A.H. Jackson, G.W. Kenner, G. McGillivray, and K.M. Smith, J. Chem. Soc. C, 294 (1968).
44. R.P. Carr, A.H. Jackson, G.W. Kenner, and G.S. Sach, J. Chem. Soc. C, 487 (1971); G.Y. Kennedy, A.H. Jackson, G.W. Kenner, and C.J. Suckling, FEBS Lett., **6**, 9 (1970); **7**, 205 (1970); P.J. Crook, A.H. Jackson, and G.W. Kenner, Ann. Chem., **748**, 26 (1971); A.H. Jackson, G.W. Kenner, and J. Wass, J. Chem. Soc., Perkin Trans. 1, 480 (1974); T.T. Howarth, A.H. Jackson, and G.W. Kenner, ibid., 502 (1974); M.T. Cox, A.H. Jackson, G.W. Kenner, S.W. McCombie, and K.M. Smith, ibid., 516 (1974); J.A.S. Cavaleiro, A.M.d'A. Rocha Gonsalves, G.W. Kenner, and K.M. Smith, ibid., 1771 (1974); G.W. Kenner and K.M. Smith, Ann. N.Y. Acad. Sci., **206**, 138 (1973).
45. J. Ellis, A.H. Jackson, A.C. Jain, and G.W. Kenner, J. Chem. Soc., 1935 (1964).
46. H. Fischer and A. Kurzinger, Hoppe-Seyler's Z. Physiol. Chem., **196**, 213 (1931).
47. A.W. Johnson and I.T. Kay, J. Chem. Soc., 2418 (1961); I.D. Dicker, R. Grigg, A.W. Johnson, H. Pinnock, K. Richardson, and P. van den Broek, J. Chem. Soc. C, 536 (1971).
48. G.M. Badger, R.L.N. Harris, and R.A. Jones, Aust. J. Chem., **17**, 1013 (1964).
49. P.S. Clezy and A.J. Liepa, Aust. J. Chem., **24**, 1027 (1971).
50. J.M. Conlon, J.A. Elix, G.I. Feutrill, A.W. Johnson, M.W. Roomi, and J. Whelan, J. Chem. Soc., Perkin Trans. 1, 713 (1974).
51. P.S. Clezy, V. Diakiw, and N.W. Webb, J. Chem. Soc., Chem. Commun., 413 (1972); P.S. Clezy and V. Diakiw, ibid., 453 (1973); Aust. J. Chem., **26**, 2697 (1973); P.S. Clezy, A.J. Liepa, and N.W. Webb, ibid., **25**, 1991 (1972).
52. (a) A.H. Jackson, G.W. Kenner, and K.M. Smith, J. Chem. Soc. C, 502 (1971); (b) M.T. Cox, A.H. Jackson, and G.W. Kenner, ibid., 1974 (1971).
53. M.E. Flaugh and H. Rapoport, J. Am. Chem. Soc., **90**, 6877 (1968).
54. P.S. Clezy and C.J.R. Fookes, Aust. J. Chem., **27**, 371 (1974).

55. S.M. Nasr-ala, A.G. Bubnova, G.V. Ponomarev, and R.P. Evstigneeva, Uch. Zap. Mosk. Inst. Tonkoi Khim. Tekhnol., 1, 81 (1971) (Chem. Abstr. 76, 99634p (1972); A.F. Mironov, V.D. Rumyantseva, M.A. Kulish, T.V. Kondukova, B.V. Rozynov, and R.P. Evstigneeva, Zh. Obshch. Khim., 41, 1114 (1971); V.D. Rumyantseva, A.F. Mironov, and R.P. Evstigneeva, Zh. Org. Khim., 7, 828 (1971).
56. J.A.B. Almeida, G.W. Kenner, K.M. Smith, and M.J. Sutton, J. Chem. Soc., Chem. Commun., 111 (1975).
57. M.A. Kulish, A.F. Mironov, B.V. Rozynov, and R.P. Evstigneeva, Zh. Obshch. Khim., 41, 2743 (1971).
58. R.L.N. Harris, A.W. Johnson, and I.T. Kay, J. Chem. Soc. C, 22 (1966).
59. D. Dolphin, A.W. Johnson, J. Leng, and P. van den Broek, J. Chem. Soc. C, 880 (1966).
60. P. Bamfield, R.L.N. Harris, A.W. Johnson, I.T. Kay, and K.W. Shelton, J. Chem. Soc. C, 1436 (1966); A.W. Johnson, Chem. Brit., 3, 253 (1967); P. Bamfield, R. Grigg, R.W. Kenyon, and A.W. Johnson, J. Chem. Soc. C, 1259 (1968); R. Grigg, A.W. Johnson, and M. Roche, ibid., 1928 (1970); R.V.H. Jones, G.W. Kenner, and K.M. Smith, J. Chem. Soc., Perkin Trans. 1, 531 (1974).
61. A.R. Battersby, G.L. Hodgson, M. Ihara, E. McDonald, and J. Saunders, J. Chem. Soc., Perkin Trans. 1, 2923 (1973).
62. R. Grigg, A.W. Johnson, R. Kenyon, V.B. Math, and K. Richardson, J. Chem. Soc. C, 176 (1969).
63. R.P. Evstigneeva, A.F. Mironov, and L.I. Fleiderman, Dokl. Akad. Nauk SSSR, 210, 1090 (1973).
64. R.L.N. Harris, A.W. Johnson, and I.T. Kay, Quart. Rev. (London), 20, 211 (1966).
65. K.M. Smith, Quart. Rev. (London), 25, 31 (1971).
66. H. Fischer, Org. Synth., 3, 442 (1955).
67. T.C. Chu and E.J. Chu, J. Biol. Chem., 212, 1 (1955); R.F. Labbe and G. Nishida, Biochim. Biophys. Acta, 26, 437 (1957).
68. A. Egyed, Experientia, 28, 1396 (1972).
69. J.L.W. Thudichum, Report Med. Off. Privy Council, Appendix 7, 10, 152 (1867).
70. M. Nencki, Arch. Exp. Pathol. Pharmakol., 24, 430 (1888).
71. V.N. Luzgina, E.I. Filippovich, and R.P. Evstigneeva, Zh. Obshch. Khim., 43, 2762 (1973).
72. K.-G. Paul, Acta Chem. Scand., 5, 389 (1951).
73. R. Lemberg, B. Bloomfield, P. Caiger, and W. Lockwood, Aust. J. Exp. Biol. Med. Sci., 33, 435 (1955); D.B. Morell and M. Stewart, ibid., 34, 211 (1956); D.B. Morell, J. Barrett, and P.S. Clezy, Biochem. J., 78, 793 (1961).
74. (a) G.W. Kenner, S.W. McCombie, and K.M. Smith, Ann. Chem., 1329 (1973); (b) R.P. Carr, A.H. Jackson, G.W. Kenner, and G.S. Sach, J. Chem. Soc. C, 487 (1971).
75. M. Grinstein, J. Biol. Chem., 167, 515 (1947).
76. H. Fischer and R. Müller, Hoppe-Seyler's Z. Physiol. Chem., 142, 120, 155 (1925).
77. O. Schumm, Hoppe-Seyler's Z. Physiol. Chem., 178, 1 (1928); see also ref. 3.
78. P.A. Burbidge, G.L. Collier, A.H. Jackson, and G.W. Kenner, J. Chem. Soc. B, 930 (1967).
79. H. Fischer and K. Deilmann, Hoppe-Seyler's Z. Physiol. Chem., 280, 186 (1944).
80. F. Sparatore and D. Mauzerall, J. Org. Chem., 25, 1073 (1960).
81. A.R. Battersby, J. Staunton, and R.H. Wightman, J. Chem. Soc., Chem. Commun., 1118 (1972).
82. R. Thomas, Dissertation, Technische Hochschule, Braunschweig (1967).
83. E.g. J.A.S. Cavaleiro, G.W. Kenner, and K.M. Smith, J. Chem. Soc., Perkin Trans. 1, 2478 (1973).
84. G.W. Kenner, J.M.E. Quirke, and K.M. Smith, unpublished results.
85. A.H. Jackson, G.W. Kenner, and J. Wass, J. Chem. Soc., Perkin Trans. 1, 480 (1974).

86. H. Fischer and A. Stern, 'Die Chemie des Pyrrols', Akademische Verlagsgesellschaft, Leipzig Vol. IIii, (1940).
87. Ref. 86, p. 48.
88. H.H. Strain and W.A. Svec in 'The Chlorophylls', Eds. L.P. Vernon and G.R. Seely, Academic Press, New York (1966), p. 21.
89. M. Tswett, Ber. Dtsch. Bot. Ges., **24**, 384 (1906).
90. R. Willstätter and M. Isler, Ann. Chem., **390**, 269 (1912); A. Stoll and E. Wiedemann, Helv. Chim. Acta, **16**, 739 (1933).
91. H.R. Wetherell and M.J. Hendrickson, J. Org. Chem., **24**, 710 (1959).
92. G.W. Kenner, S.W. McCombie, and K.M. Smith, J. Chem. Soc., Perkin Trans. 1, 2517 (1973).
93. Ref. 86, p. 52.
94. Ref. 86, p. 108.
95. Ref. 86, p. 73.
96. Ref. 3, p. 358.

Section B

BIOLOGICAL ASPECTS

BIOSYNTHESIS OF PORPHYRINS, CHLORINS AND CORRINS

A.R. BATTERSBY and E. McDONALD

University Chemical Laboratory, Lensfield Road, Cambridge, CB2 1EW, U.K.

3.1.Introduction

The many fascinating biosynthetic problems posed by the complex macro-cycles of the natural porphyrins, chlorins, and corrins have steadily attracted the interest of biochemists and chemists and remarkable progress has been made over the past thirty years. Scheme 1 shows simply in broad outline, the

Scheme 1

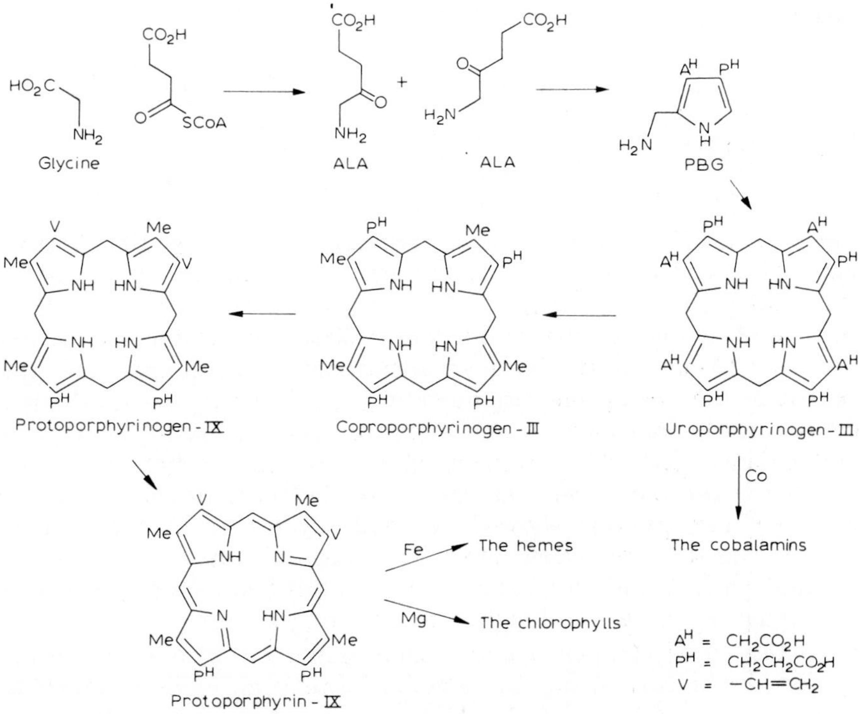

Porphyrins and Metalloporphyrins, ed. Kevin M. Smith

biosynthetic pathway which has emerged from these researches. Our detailed account will be divided into four main sections covering:

1. The biosynthesis of porphobilinogen (PBG)
2. The formation of the macrocyclic rings of uroporphyrinogen-I and uroporphyrinogen-III
3. Side-chain modification and aromatization leading to protoporphyrin-IX
4. The iron, magnesium, and cobalt branches

In each section the key experimental evidence for the pathway will be summarized[1] and the more recent mechanistic and enzymatic studies will be described in detail. Consistent mechanistic schemes will be presented in those cases where the experimental evidence justifies such rationalizations.

3.2. The biosynthesis of porphobilinogen (PBG)

3.2.1. Historical background and fundamental studies

The simplest amino acid glycine (1) is an essential starting material for protein synthesis, and succinic acid (2) plays an important part in the tricarboxylic acid cycle. It might be argued however that the most important function of these molecules is to react together to form 5-aminolevulinic acid (ALA) (3) at the commencement of porphyrin biosynthesis. The natural porphyrins (suitably modified) provide the key materials for photosynthesis[2], respiration[3], and electron transport[4], and so play a crucial role in all living systems.

Scheme 1 (a)

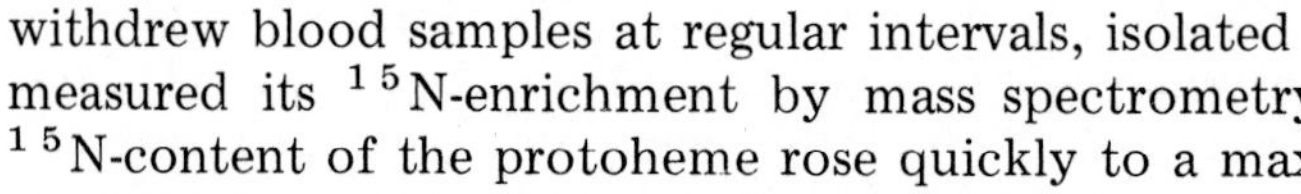

Glycine (1) was first shown to be a precursor of the porphyrins in 1945 by Shemin who submitted his own body for the biosynthetic experiment by swallowing 66 grams of [^{15}N]glycine over a three day period. He then withdrew blood samples at regular intervals, isolated the protoheme (4) and measured its ^{15}N-enrichment by mass spectrometry. He found[5] that the ^{15}N-content of the protoheme rose quickly to a maximum, remained fairly constant for nearly four months, and then dropped steadily to the level of natural abundance. The ^{15}N-level in globin, the protein to which protoheme is bound in the red blood cell, followed a similar series of changes but in contrast the plasma protein showed no tendency to retain ^{15}N after an initial rise. It was therefore calculated that hemoglobin was being protected from metabolism for an average of 127 days, and this figure corresponds well with other reliable estimates of the lifetime of a red blood cell.

This spectacular experiment heralded the beginning of a classical series of investigations to be carried out by Shemin, Rittenberg, and co-workers at

Columbia University during the next ten years, and the first task was to degrade the protoheme from the experiment described above. Chromic acid oxidation of the bis-methanol adduct from (4) gave[6] the maleimides (5) and (6) each with the same ^{15}N-enrichment, thus demonstrating that glycine provides the nitrogen atom for both types of pyrrole residues which make up the macrocycle despite their different patterns of substitution.

Scheme 2

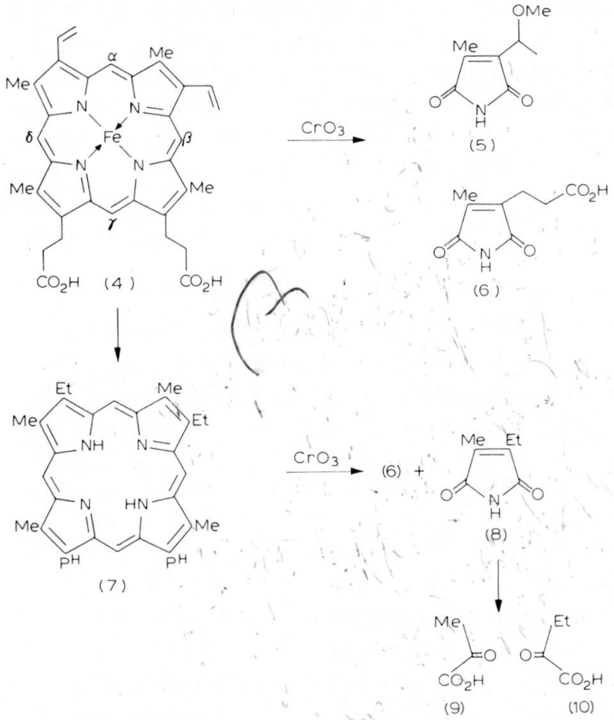

Incubation of ^{15}N-glycine in vitro with blood from various animals led not only to the development of a convenient experimental system for subsequent biosynthetic experiments but also gave some insight into the exact location of protoheme biosynthesis. Thus protoheme enriched with ^{15}N was produced most efficiently by whole duck blood[7a] which has nucleated red cells. The erythrocytes (red cells) of normal human[7b] or rabbit blood[7c] are formed in the bone marrow and they lose their internal structure before emerging into circulation. Such peripheral blood with nonnucleated erythrocytes is unable to use glycine for protoheme synthesis. An intermediate situation is found for humans with sickle-cell anemia[7b] and for immature rabbits[7c]. Their blood contains some immature (but non-nucleated) erythrocytes known as reticulocytes, which allow glycine to be incorporated into protoheme to a limited extent.

References, p. 116

Duck blood was obviously ideal for further biosynthetic experiments, and after optimizing the incubation conditions[7d] it was shown that [2-^{14}C]glycine, but not [1-^{14}C]glycine, gave rise to radioactive protoheme. Degradation gave[8] maleimides (5) and (6) with equal molar activities but they carried only one half of the total radioactivity present in the protoheme. Evidently the remaining half was present at the four *meso*-carbon atoms (labeled $\alpha,\beta,\gamma,\delta$) and assay of the carbon dioxide formed in the chromic acid oxidation later showed this to be so[9]. Doubly labeled [2-^{14}C, ^{15}N]glycine was also incorporated into protoheme, and the resultant dilution of ^{15}N was double that for the ^{14}C. Taken together, these results show that eight glycine molecules provide eight carbon atoms (including the four *meso* atoms) together with the four nitrogen atoms of protoheme. A similar and equally thorough study of the incorporation of [^{15}N]glycine[9a] and doubly labeled glycine[9b] into protoheme by chicken erythrocytes was carried out in Neuberger's laboratory. The protoheme was degraded via mesoporphyrin-IX (7) to the maleimides (6) and (8) and the same conclusions summarized above were drawn.

Both research groups now used sodium [1-^{14}C]acetate and [2-^{14}C]acetate as precursors and they degraded their radioactive protoheme to the maleimides (6) and (8)[9b,10b]. The first step of further degradation of the maleimides[11] was decarboxylation of the hematinic acid (6) to (8) by heating at 175°C in ammoniacal ethanol. Methylethylmaleimide (8) was therefore available from all four rings of protoheme, and it was oxidized by $OsO_4/KClO_3$ followed by periodate to yield the keto-acids (9) and (10) isolated as the corresponding phenylhydrazones. Permanganate oxidation then afforded acetic acid and propionic acid respectively from the two hydrazones and finally the acids were decarboxylated by the Schmidt procedure. This powerful and experimentally demanding sequence was applied to the radioactive protoheme from both acetate feedings, and the precise labeling pattern was elucidated. The rather complex results[11] could be explained if acetate were first converted into an unsymmetrical C-4 unit and in 1952[12] this was identified as succinyl coenzyme A (11). Acetate is converted into succinate in the course of the tricarboxylic acid cycle[13] (Scheme 3) and it was subsequently shown[14] that activity from the intermediates in this conversion, viz. citrate (12) and 2-oxoglutarate (13) (^{14}C-labeled) is incorporated into protoheme in the expected manner by hemolyzed duck blood.

With the identity of the two primary precursors well established, and the knowledge that glycine must lose its carboxyl group en route to protoheme, several research groups deduced that 5-aminolevulinic acid (ALA) (3) might be a key intermediate. Numerous syntheses were reported and several routes were developed (e.g. Ref. 15, 16) to allow specific labeling of ALA. It was then demonstrated that [^{15}N]ALA was 40 times more effective[15] than glycine for protoheme synthesis, and that [5-^{14}C]ALA yielded protoheme with a labeling pattern[16] identical to that from [2-^{14}C]glycine. Protoheme was

Scheme 3

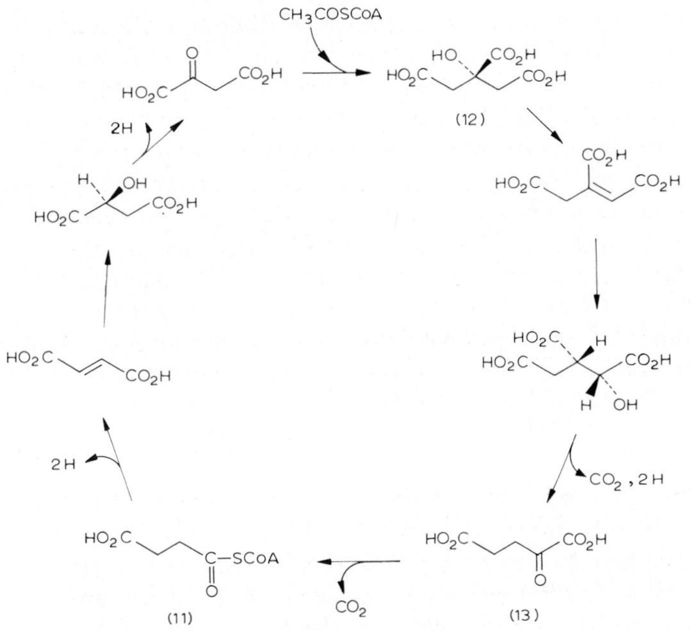

Scheme 3 (a)

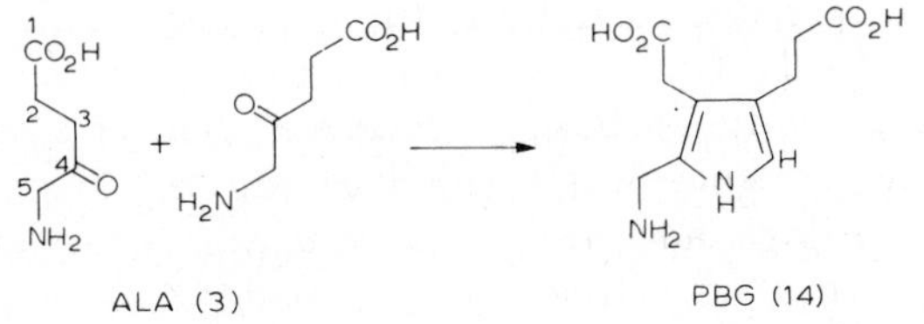

ALA (3)   PBG (14)

also shown to be synthesized from $[^{15}N]$ ALA by lysed chicken erythrocytes[17].

It was at this stage in the early 1950s that the results from researches in apparently different areas combined synergistically. A substance, named porphobilinogen (PBG), had been known for many years in the urine of patients with porphyria but its instability prevented isolation in a pure state until 1952[18]. Its structure (14) was established in the following year[19]. Parallel to this development, the knowledge summarized above about the incorporation of ALA into protoheme leads on chemical grounds to the idea of combining two ALA molecules to form structure (14). The coalescence was thus perfect and the biosynthetic importance of PBG was almost immediately demonstrated. Thus ALA and PBG were equally well utilized[17] by hemolyzed chicken erythrocytes for protoheme synthesis, and if an ALA incubation was allowed to run for a short time, an α-free pyrrole appeared[20] having chromatographic behavior identical to that of PBG. Subsequently a highly purified preparation[21] capable of converting ALA to PBG on a 200 mg scale was isolated from duck blood. Using this enzyme, [5-^{14}C]ALA was converted into PBG which had twice the molar radioactivity of the ALA. This PBG and the original radioactive ALA were incubated with whole duck blood and gave two samples of protoheme with identical molar activities[21].

With the conclusion of these fundamental studies, attention turned towards the enzymes which catalyze the early stages of porphyrin biosynthesis. Aminolevulinic acid synthetase[22] (ALA synthetase) carries out the conversion of glycine and succinic acid into ALA whilst ALA dehydratase[23] catalyzes the formation of PBG from ALA. These enzymes have recently been reviewed in detail[22,23] but there have been further developments since that time. These new results will be incorporated into our summary of the chemically significant aspects of these enzymes.

3.2.2. ALA Synthetase

This enzyme has been purified from bacterial[24-27] and mammalian[28-30] sources. The bacterium used most frequently is *Rhodopseudomonas spheroides*, and the mammalian tissue preferred is the liver. In normal animals only low levels of ALA synthetase can be detected but the enzyme can apparently be induced by certain drugs which cause porphyria. The bacterial and mammalian enzymes are both difficult to handle but progress has been made recently by the discovery[25] that in the presence of glycerol, 2-mercaptoethanol, and moderate buffer concentrations, the bacterial enzyme can be stored for long periods at $-15°C$.

The optimum pH for both enzymes is close to pH 7.5. Recent estimates of molecular weight range from 57,000 to 100,000 for the bacterial enzyme[24-27] and from 113,000 to 300,000 for the mammalian case[22,29,30] (Table 1).

ALA synthetase activity has only occasionally been detected in plants[31,32] and the low levels and instability of the enzyme have precluded any detailed study. There is therefore a possibility that ALA can be formed in plants from precursors other than glycine and succinyl coenzyme A and there are strong indications that this is so[33].

3.2.2.1. Mechanistic studies
3.2.2.1.1. Cofactor requirement

Whatever the source of enzyme, pyridoxal phosphate (15) is required as a cofactor, and kinetic studies have shown that this is always bound more tightly than succinyl coenzyme A, which in turn is more strongly bound than glycine [see Table 1 for K_m (Michaelis constant) values]. The holoenzyme (i.e. enzyme plus coenzyme) absorbs[27] at 330 and 415 nm, the ratio of the peaks varying with pH. Removal of pyridoxal phosphate yields the

TABLE 1

Properties of ALA synthetase

Source	Rhodopseudomonas spheroides			Rat	Rat	Rabbit	Rat
Molecular weight	60,000–80,000	57,000	61,000	100,000	—	200,000	300,000
pH optimum	7.4–7.8	7.3	7.2	—	7.0–8.5	7.6	7.5
K_m (glycine) (M)	5×10^{-3}	1×10^{-2}	5×10^{-3}	—	2.5×10^{-3}	1×10^{-2}	1×10^{-2}
K_m (succinyl CoA) (M)	5×10^{-6}	2.5×10^{-5}	10×10^{-5}	—	2×10^{-4}	6×10^{-5}	7×10^{-5}
K_m (pyridoxal P) (M)	—	—	5×10^{-6}	—	—	—	3×10^{-6}
Reference	24	25	26	27	28	30	29

References, p. 116

apoenzyme which has no long wavelength absorption and will not catalyze the formation of ALA unless the pyridoxal phosphate is replaced (approx. one mole of cofactor per 100,000 molecular weight units). Sodium borohydride has little effect on the holoenzyme under neutral conditions but in a slightly acidic medium the 415 nm absorption disappears[27] and the enzyme is irreversibly inactivated; the product absorbs at 330 nm. These facts can be rationalized by Scheme 4 involving binding of the coenzyme to the enzyme as a Schiff's base (16) (absorbing at 415 nm) which is accessible to a neigh-

Scheme 4

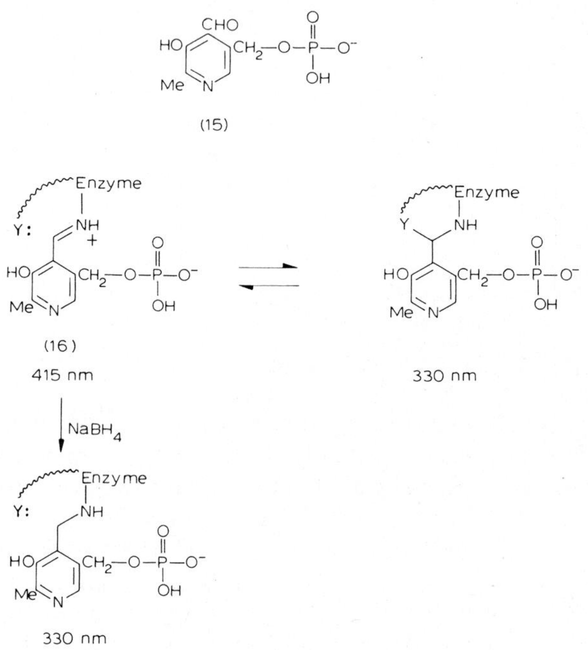

boring nucleophile Y. This scheme would also account for the powerful inhibitory effect of cyanide.

It is likely that the phosphate group of the coenzyme also plays some part in binding, and this could explain the inhibitory effect[27] of ATP and pyrophosphate and perhaps also the effect of M^{2+} ions, which act as promoters at low concentration, but inhibit at higher concentration[26,29].

3.2.2.1.2. Binding and activation of glycine

Glycine is not decarboxylated by the holoenzyme in the absence of succinyl coenzyme A. $(2RS)$-$[2$-$^3H_1]$Glycine loses 52% of its radioactivity during the enzymic formation of ALA. These facts[22] and the nature of the cofactor strongly support the mechanism outlined in Scheme 5 in which an anion generated from the complex (17) undergoes acylation by succinyl coenzyme A.

Scheme 5

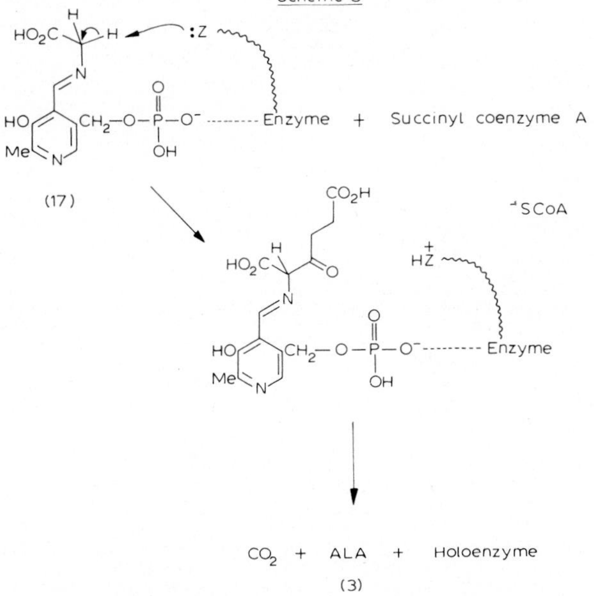

The above loss of approx. 50% of the tritium from $(2RS)$-$[2\text{-}^3H_1]$glycine is in keeping with hydrogen removal from C-2 being stereospecific; a random process would be expected to take place with a much higher retention of tritium[34] as a result of the kinetic isotope effect. The sense of the stereospecificity has been determined as follows for the ALA synthetase from *Rhodopseudomonas spheroides*. $(2R)$-, and $(2S)$-$[2\text{-}^3H_1]$Glycine (Scheme 6) were prepared using the enzyme serine hydroxymethylase; in Scheme 6, H_A of glycine is the *pro*-S hydrogen atom. Each enantiomer was mixed with $[2\text{-}^{14}C]$glycine and incubated with purified ALA synthetase and succinyl coenzyme A. ALA (3) was not isolated (since any tritium at C-5 might be lost during handling by chemical exchange via the enol) but was reduced with borohydride in situ. The amino alcohol (18) was then cleaved by periodate and C-5 was isolated and counted as the dimedone derivative of formaldehyde. The ALA formed from $(2R)$-$[2\text{-}^3H_1]$glycine was found[35] to have lost all the original tritium, whereas that from the $(2S)$-enantiomer had retained 97%. Evidence for the illustrated configuration of H_A in ALA will be covered later.

No amino acid has yet been found which can replace glycine as a substrate for ALA synthetase, but L-alanine and L-threonine, act as inhibitors, and L-cysteine is a much more effective inhibitor than the D-enantiomer. Amino malonate is a powerful inhibitor of the enzyme. This information[22] suggests

Scheme 6

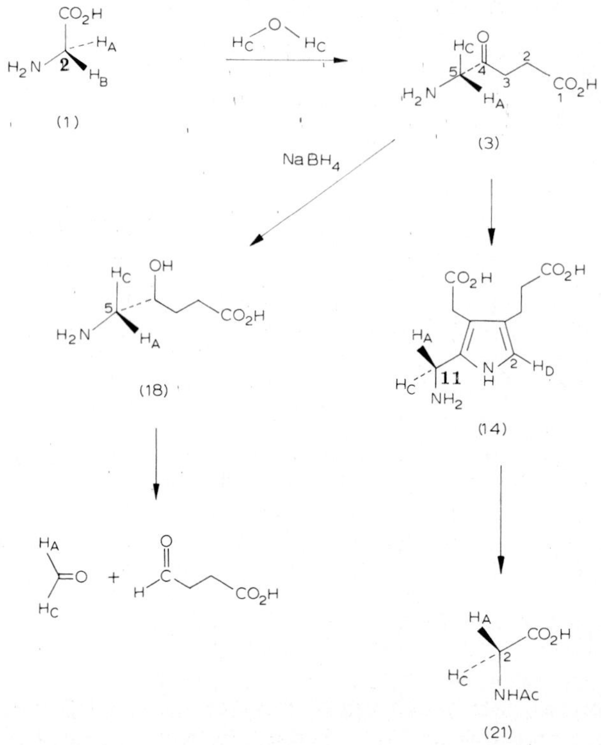

that the space occupied by H_A of glycine can also be occupied by a CH_3, $CH_2 CH_2 OH$, $CO_2 H$ or $CH_2 SH$ group. D-Amino acids could only bind in the same sense as glycine if the space occupied by H_B of glycine could accommodate a substituent group; this seems not to be the case.

It would be useful to know whether ALA synthetase holoenzyme can catalyze the exchange of hydrogen at C-2 in those L-amino acids which act as inhibitors.

3.2.2.1.3. Binding of succinyl coenzyme A and the acylation step

It is not known whether succinyl coenzyme A acylates a functional group on the enzyme which then transfers the acyl group or whether the thiol ester is bound in a noncovalent manner. The methyl ester of succinyl coenzyme A is a reasonably good substrate for the enzyme, and other coenzyme A esters will react though much more slowly[22].

3.2.2.1.4. The decarboxylation step

One possibility is that the initial product of the enzymic reaction considered in the previous section is 2-amino-3-oxoadipic acid (19) since this compound is known[36] to decarboxylate spontaneously at pH 7 with a half-life of 10 min. However, the alternative is that decarboxylation is enzyme catalyzed and the nature of the enzyme product complex (20) (see later for justification of the illustrated stereochemistry) is perfect for such assistance. A distinction between these possibilities has been made[37] by analyzing the stereochemistry of the $[5\text{-}^3H_1]$ ALA [(3) in Scheme 6, $H_A = {}^3H$] produced enzymatically from $2S[2\text{-}^3H_1]$ glycine.

Scheme 6 (a)

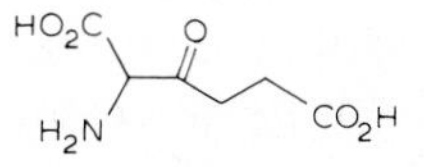

The incubation was run in the presence of ALA dehydratase (see next section), so that the $[5\text{-}^3H_1]$ ALA (3) produced was converted in situ into $[11\text{-}^3H_1]$ PBG (14). Acetylation followed by oxidation gave N-acetyl-$[2\text{-}^3H_1]$ glycine (21) which was hydrolyzed. The absolute configuration of the tritium in the resultant glycine was determined by incubating it with serine hydroxymethylase when 87% of the tritium was lost by exchange with the medium. Since this enzyme is known to catalyze specific exchange of the *pro-S* hydrogen of glycine, it is evident that the PBG, and hence the ALA are stereospecifically tritiated as illustrated*. The labeled ALA has largely the (5S)-configuration. In contrast, non-enzymic decarboxylation of 2-amino-3-oxo$[2\text{-}^3H_1]$ adipic acid would have given equal amounts of the (5R) and (5S)-enantiomers.

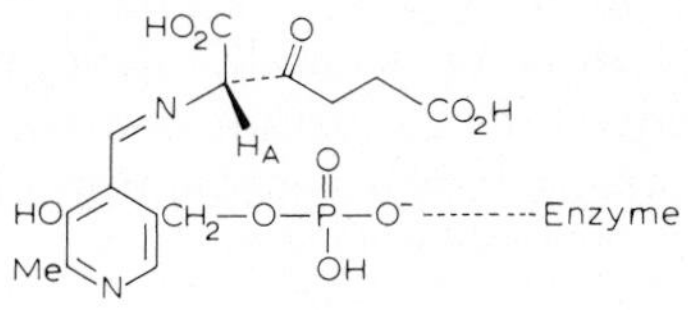

These results also lead to the important conclusion that the conversion of glycine into PBG occurs by an inversion-retention mechanism and this is quite secure. That is, either the acylation involves inversion, with retention in the subsequent decarboxylation step, or vice versa. The single example of enzymic decarboxylation of natural α-amino acids whose stereochemistry has been determined occurred with retention of configuration. On this slender evidence (one swallow does not make a summer) and knowledge of related C-acylations, inversion followed by retention seems the more likely.

The foregoing experiments should also have revealed whether it is the *pro-R* or *pro-S* hydrogen of glycine which survives at C-2 of PBG (14, H_D) but this point did not receive comment.

---

* The same conclusion was also reached simultaneously by Professor D. Arigoni and coworkers (Zurich): personal communication.

References, p. 116

3.2.2.2. Inhibition of ALA synthetase in vivo

ALA synthetase is strongly inhibited[22] by protoheme (Fe^{2+}) and proto-hemin (Fe^{3+}) and this may allow an organism to control porphyrin synthesis and avoid generating unnecessary amounts of the biosynthetic intermediates. There is also some indication that protoheme can influence the biosynthesis of ALA synthetase itself.

The mechanism of this feedback inhibition is not known but it has recently been suggested[38] that a vital functional group at the active site of the enzyme is made unavailable by reaction at the iron atom of protoheme as an axial ligand. The lack of inhibition by hemoglobin, myoglobin, and the bisimidazole complex of protoheme is cited[38] in support of this mechanism.

If ALA synthetase is indeed an important control enzyme for porphyrin biosynthesis, it is not surprising that it is present at only low levels, and that it has a high turnover rate.

3.2.3. ALA Dehydratase

Preparations capable of converting ALA (3) into PBG (14) have been isolated from many bacterial[23,43,44], plant[23] and animal[23,39−42] sources. In several cases the enzymes have been partially purified and their properties have been tabulated[23]. Estimates of molecular weight fall in the range 240,000—280,000 and the bovine liver enzyme[39] has a single zinc atom associated with a unit of molecular weight 275,000. When the enzyme is treated with 1 M-urea it dissociates[23] into two subunits of molecular weight 140,000 and it is interesting that this figure was reported[40] some years ago for the molecular weight of bovine liver enzyme. Furthermore treatment of the enzyme with detergent (sodium dodecyl sulfate) causes dissociation into smaller subunits of molecular weight ~40,000. These subunits are apparently identical with one another and each can bind a single ALA molecule[41]. Each subunit consists of 270 amino acid residues which have been identified[41] but not sequenced. In the bovine liver case, treatment of the approx. 40,000 molecular weight subunits with guanidine hydrochloride and dithiothreitol yields[42] a 'monomer' of molecular weight ~18,000, which has four cysteine residues. Thus it would seem that the intact enzyme consists of 12—14 subunits in two groups of 6—7 associated with a zinc atom, and that it can bind 6—7 ALA molecules.

A valuable practical contribution has come[45] from binding bacterial ALA dehydratase covalently to Sepharose; a column of the resultant gel has been

Scheme 6 (b)

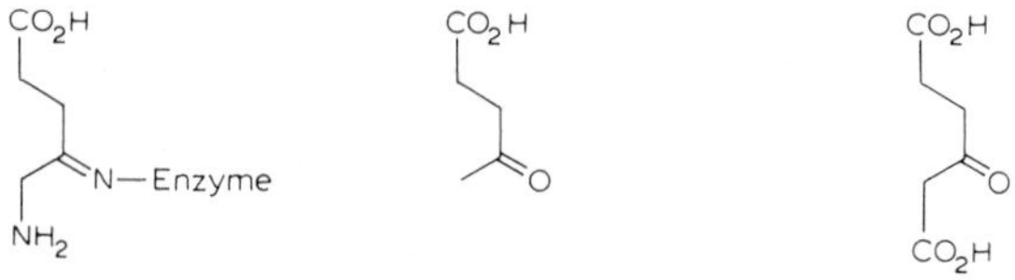

(22)   (23)   (24)

used to synthesize PBG continuously at a rate of 200 mg per day for one month.

3.2.3.1. Binding of ALA

The dehydratase alone is unaffected by sodium borohydride but in the presence of [^{14}C] ALA and borohydride, it is irreversibly deactivated and the radioactivity becomes bound to protein[23]. This points to the formation of a Schiff's base (22) presumably with a lysine residue; a similar reaction can also be demonstrated with levulinic acid (23) and its ester, but not with 3-oxoglutaric acid (24). These findings[23] show that the site occupied by the NH$_2$ group of ALA is unable to accommodate a CO$_2$H group. The carboxyl group of ALA is considered to be important for binding since levulinic acid (23) is a competitive inhibitor of the dehydratase but its ester is not[23]. The mechanism proposed[23] for the condensation reaction reasonably uses Schiff's base formation to enable C-3 of the bound ALA (22) to act as a nucleophile towards the carbonyl group of a second ALA molecule (Scheme 7). The

Scheme 7

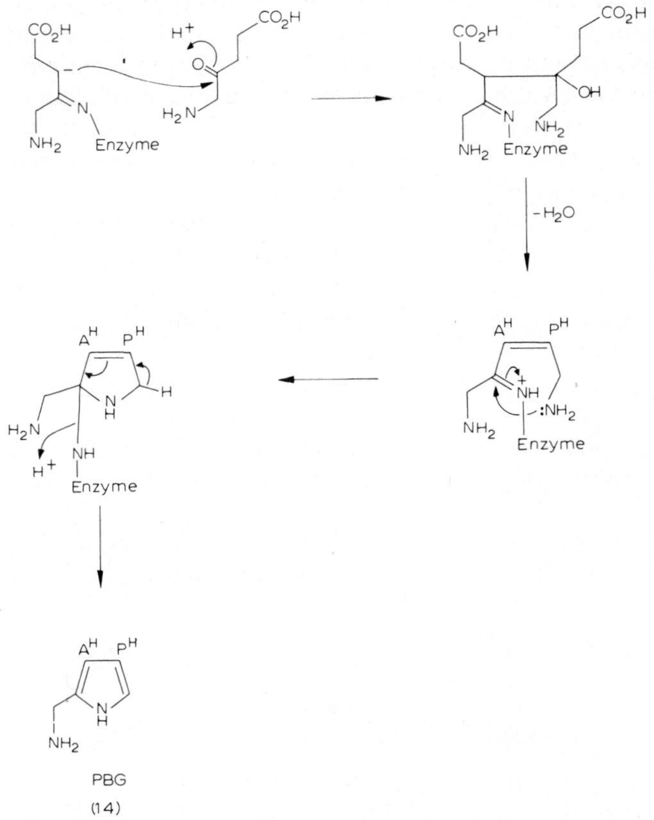

PBG

(14)

subsequent steps are supported not by factual evidence but by good chemical analogy. However, it will be evident that there is nothing sacrosanct about the *sequence* in Scheme 7 and there are many possible variations including the use of *both* ALA units as bound imines.

Although the only significant function of porphobilinogen is as a porphyrin precursor, an enzyme (pyrrolo-oxygenase) has recently been discovered[46−48] which oxidizes PBG to a lactam; this enzyme may have a control function.

3.3. Conversion of porphobilinogen into uroporphyrinogen-III
3.3.1. Porphyrinogens as biosynthetic intermediates

Once PBG had been identified as the unique monopyrrolic precursor of the porphyrins, it seemed probable that uroporphyrin-III (25) would precede coproporphyrin-III (26) en route to protoheme (4). It was therefore rather a shock[49,50] when [14C] coproporphyrin-III (26) was not incorporated into protoheme (4). However, in 1956, uroporphyrin-III (25) isolated from *Turaco* feathers, was reduced to the hexahydroporphyrin (27), which is at the same oxidation level as PBG. When the product, uroporphyrinogen-III (27), was incubated aerobically[51] with hemolyzed duck blood, an increase was observed in the levels present of uroporphyrin-III (25), coproporphyrin-III (26), and protoporphyrin-IX (28). Furthermore, the addition of uropor-

Scheme 8

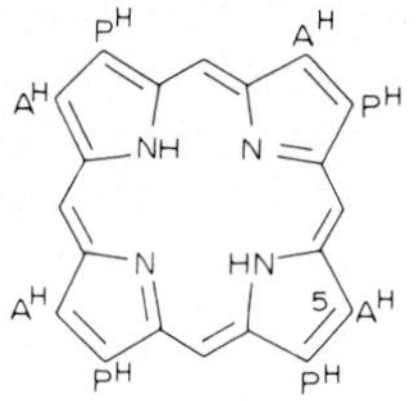

Uroporphyrin-III
(25)

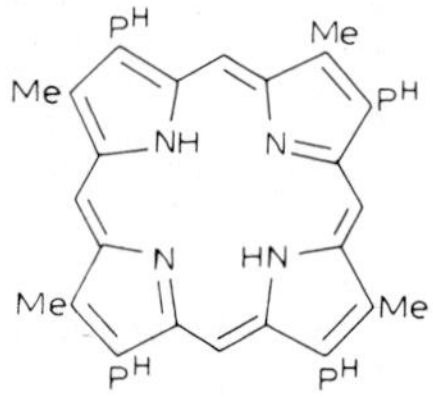

Coproporphyrin-III
(26)

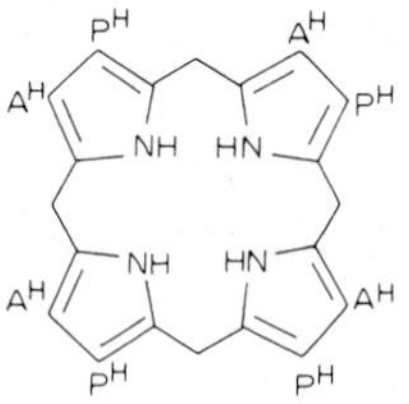

Uroporphyrinogen-III
(27)

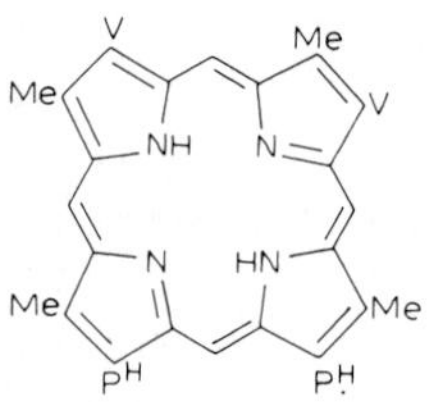

Protoporphyrin-IX
(28)

phyrinogen-III to the enzyme system enhanced the incorporation of ^{59}Fe into protoheme and reduced the specific activity of protoheme biosynthesized from [2-^{14}C] glycine.

Further support for the intermediacy of the porphyrinogens will appear later, and it now seems certain that uroporphyrin-III and coproporphyrin-III are artefacts formed by autoxidation of the porphyrinogens during incubation and work up. Early reports that uroporphyrin-III was incorporated into protoheme[52] and protoporphyrin-IX[50,53] by rabbit bone marrow and chicken blood hemolyzate respectively are unlikely to be accurate[54].

3.3.2. Outline of the problem

It was brought out in Scheme 1 and above that uroporphyrinogen-III (27) is the first macrocyclic substance formed on the biosynthetic pathway to the porphyrins. The chemical process by which PBG (14) is converted specifically into this particular isomer is one of the most fascinating stages in a remarkable series of enzymic conversions and one which has led to intense speculation. The interest arises from the fact that there are four possible uroporphyrin isomers* designated types I, II, III, and IV as shown in Scheme 9. Yet the vitally important porphyrin derivatives such as protoheme, the chlorophylls, and the cytochromes are all based on the type-III isomer**. Further, we should consider the most straightforward way in which four molecules of PBG (14) might be expected to form a macrocycle (Scheme 9). The basic nitrogen of PBG could be protonated as illustrated (or alternatively converted into a good leaving group by some Lewis acid) and the resultant methylenepyrrolenine (29) could attack the α-free position of a second molecule of PBG to form a pyrromethane (30). Two more PBG units could be added stepwise in a similar way to form the bilane (32) which on ring closure to the free α-position of the terminal unit would generate uroporphyrinogen-I (33). So by forming uroporphyrinogen-III (27), Nature *specifically forms the unexpected isomer* and the present section will cover available knowledge of this process which is of central importance in living things. The specificity which has been emphasized above contrasts with the acid catalyzed conversion of PBG into uroporphyrinogen which gives rise to all four isomers[57] in the proportions expected on statistical grounds[58], viz. one half type-III, one quarter type-IV, one eighth type-II and one eighth type-I.

---

* This holds generally for porphyrins and porphyrinogens when the eight peripheral sites are filled by four groups A and four groups B and each pyrrole residue carries one group A and one group B.

** Uroporphyrin-I (34) is formed in the shell of an oyster *Pinctada vulgaris*[55] and in the urine of patients suffering from congenital porphyria[56]. Uroporphyrin-II (35) and uroporphyrin-IV (36) have not been found in living systems.

References, p. 116

Scheme 9

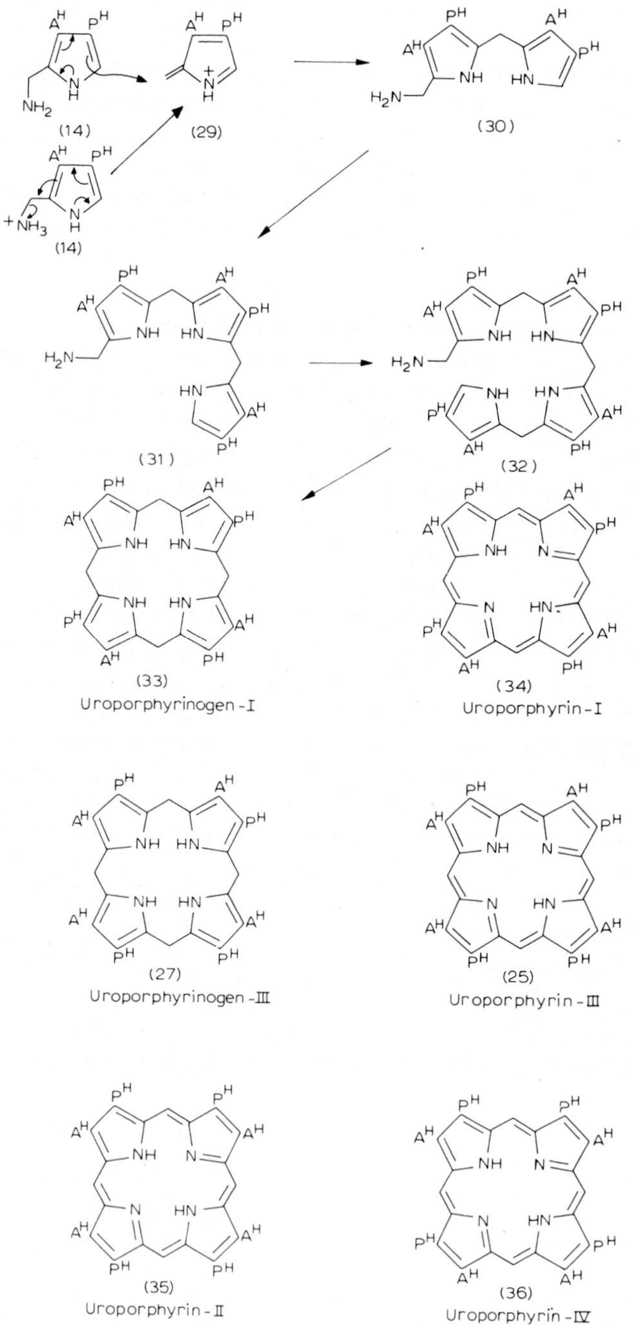

(14)   (29)   (30)

(14)

(31)   (32)

(33)
Uroporphyrinogen-I

(34)
Uroporphyrin-I

(27)
Uroporphyrinogen-III

(25)
Uroporphyrin-III

(35)
Uroporphyrin-II

(36)
Uroporphyrin-IV

$A^H = CH_2CO_2H$ $P^H = CH_2CH_2CO_2H$

It was found early in the 1950s that the biological formation of uro-porphyrinogen-III (27) requires the cooperative effort of two proteins which differ in their stability to heat[59]. Porphobilinogen deaminase (also called uroporphyrinogen-I synthetase; we will refer to it as *deaminase*) is relatively stable and it smoothly converts PBG into uroporphyrinogen-I (33). The other protein, uroporphyrinogen-III cosynthetase (referred to here as *cosynthetase*) is inactivated by heating at 55°C for 30 min. When PBG is treated with deaminase in the presence of an excess of cosynthetase, the product is uroporphyrinogen-III (27)[60]. These two proteins will be discussed more fully in turn.

3.3.3. Porphobilinogen deaminase (Uroporphyrinogen-I synthetase)

This enzyme has been prepared[61−63] in partially purified form from plant, bacterial, algal, avian, and mammalian sources. In addition, highly purified samples have been isolated and the data obtained for these prepara-tions are collected in Table 2. The findings generally agree well including the values for the Michaelis constant (K_m) if one bears in mind the slight spread of pH values used and the normal difficulty in determining K_m. What is clear from these K_m studies is that the enzyme has a high affinity for PBG. It is generally accepted that deaminase as isolated is one protein of molecular weight somewhat under 40,000. Many inhibitors of deaminase have been

TABLE 2

Properties of porphobilinogen deaminase

Source	Molecular weight	Isoelec-tric point	Optimum pH	K_m (PBG)	Refer-ence
Wheat germ	—	—	8.2	50 μM	64
Rhodopseudomonas spheroides	35,000—36,000	4.46	7.8—8.0	40 μM at pH 7.8—8.0	65
Rhodopseudomonas spheroides	36,000—39,000	—	7.6	13—20 μM at pH 7.6	66
Chicken erythrocytes	—	<4.6	7.4—8.2	18 μM at pH 8.0	67
Euglena gracilis	38,000—40,000	—	7.0—8.1	30 μM at pH 8.0 phos-phate buffer	68
Spinach leaves	38,000—40,000	4.2—4.5	—	72 μM at pH 8.2	69, 70
Human erythrocytes	20,000—30,000	—	7.4—8.2	130 μM at pH 7.4	71

References, p. 116

examined and the results are collected in Refs. 60, 64—66 and 71. Mention should be made of the strong inhibition by sulfhydryl reagents such as p-chloromercuribenzoate, mercuric chloride, and N-ethylmaleimide, an effect which is reversed by 0.1 M-cysteine. Metal-ion chelating agents including o-phenanthroline and 2,2'-bipyridyl have no effect on the activity of deaminase from R. spheroides[66] but deaminase is inactivated by pyrrolo-oxygenase[72]. The interaction of deaminase with certain pyrroles and pyrro-methanes will be considered later as well as inhibition by ammonia and other compounds of structure RNH_2.

3.3.4. Uroporphyrinogen-III cosynthetase

Fewer detailed studies have been made of cosynthetase than of deaminase largely because no simple assay is available. Preparations have to be tested for cosynthetase activity by determining how much protein is required to form a chosen percentage (say 50%) of uroporphyrinogen-III (27) when incubated with a standard mixture of deaminase and PBG[73]. Since it is by no means easy to determine accurately the amounts of the derived uropor-phyrin-I (34) and uroporphyrin-III (25) in a mixture of the two, the whole procedure is rather slow. However, a somewhat more rapid assay (see below) has recently been described.

That cosynthetase is a protein is now beyond doubt (e.g. Ref. 73) and prep-arations containing it have been obtained from chicken, rabbit and human erythrocytes[71,74,75,76], Chlorella[59], R. spheroides[61], Euglena gracilis[79,80] and from several other plant and animal sources[60,63,77,81,82]. More highly purified cosynthetase has been isolated from spinach[70], mouse spleen[73], and wheat germ[69]. The isoelectric point of this last preparation[69] was 6.1—6.7 and its molecular weight was estimated to be about 60,000—65,000. The optimum pH for cosynthetase from mouse spleen was 7.7—7.9 and it was strongly inhibited[73] by 20 μM uroporphyrin-III (25). Cosynthetase from human erythrocytes was inhibited by sulfhydryl reagents[71].

Deaminase and cosynthetase which have been separated can be recom-bined to restore the original activity in producing uroporphyrinogen-III[70,81].

3.3.5. Enzymic formation of uroporphyrinogen-I (33)
3.3.5.1. Stoichiometry

The major problem of how uroporphyrinogen-III (27) is biosynthesized is best approached here by looking first at the formation of uroporphyrino-gen-I (33) from PBG (14) by deaminase. Four moles of PBG are converted by this enzyme into uroporphyrinogen-I and the yield has been shown[60,64] to be at least 91%; the difference from 100% could be due to pyrrolo-oxy-genase[72] and/or to less than quantitative yield in the conversion of uropor-phyrinogen-I (33) chemically into uroporphyrin-I (34) for spectroscopic determination. This view that four moles of PBG are quantitatively converted is supported by the finding that exactly four moles of ammonia (101 ± 2%) are released into the medium during the formation of one mole of uroporphyrinogen-I[60]. There are indications that the elimination of ammonia and consumption of PBG follow roughly parallel time courses[60].

3.3.5.2. Intermediates

When deaminase converts PBG into uroporphyrinogen-I under normal conditions, it does so without the liberation of any detectable intermediates (see Refs. 83, 84 and Refs. therein). However, a stepwise building process on the enzyme surface is chemically attractive and could involve the intermediate formation of a pyrromethane (30), a tripyrrole (31), and a bilane (32) or modified forms of these substances (see later). Interest in the synthesis of di-, tri-, and tetrapyrrolic systems of this sort has been stimulated not only by these chemical considerations but also by the results of inhibition experiments. Thus it was found that the action of deaminase on PBG in the presence of ammonium ions or of hydroxylamine (which is more effective) led to the accumulation of substances having properties in keeping with oligopyrrolic structures[83,84]. These products will be discussed more fully later.

Scheme 10

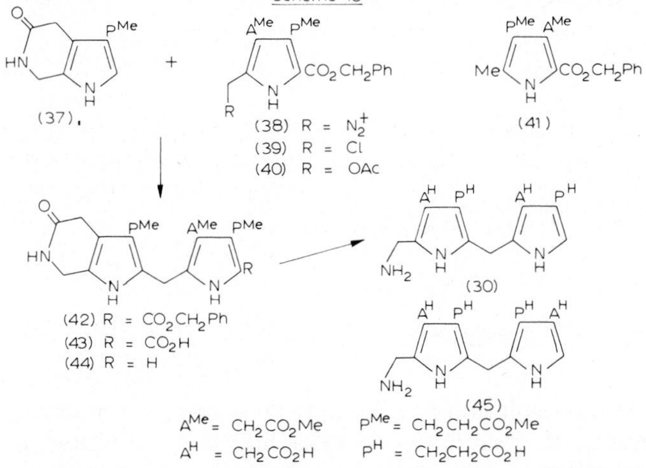

The synthetic strategy for the pyrromethanes (e.g. 30) was governed by the expected instability of these substances, and all three research groups[86−91] used internal protection of the aminomethyl group. Scheme 10 shows the synthesis[86−90] of two aminomethylpyrromethanes which we will refer to as AP · AP (30) and AP · PA (45); the side-chains are 'read' from the aminomethyl side. Briefly, PBG-lactam (37) was coupled with one of the reactive pyrroles (38—40) to afford the pyrromethane (42) which was debenzylated and the acid (43) decarboxylated. Alkaline hydrolysis of the resultant α-free pyrromethane (44) then uncovered the aminomethyl group to yield the tetracarboxylic acid (30). The AP · PA isomer (45) was obtained from the lactam (37) and the pyrrole (41) by an analogous sequence[86,90]. Preparation[91] of isoPBG-lactam (46) allowed synthesis[91,92] of the two remaining pyrromethane isomers, PA · AP (53), and PA · PA (54) as in Scheme 11. Proton n.m.r. has been used to characterize the various pyrromethanes[94,86−91] and by comparison of the spectra of the four isomers (30), (45), (53), and (54) it was evident[91] that discrimination between AP · AP (30) and PA · AP (53) was difficult and the same held true for the pair AP · PA (45) and PA · PA (54). However, all four could be distinguished[91] when the spectra were sharp and of good quality. This is of importance for current research on the isolation of biosynthetic intermediates.

Scheme 11

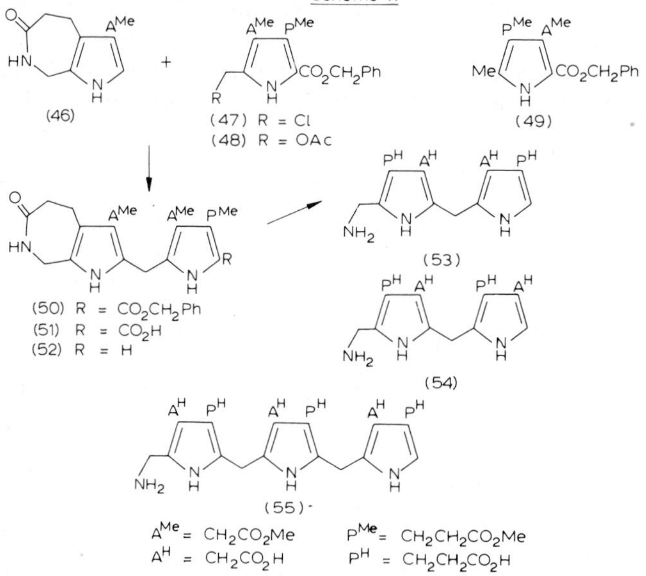

Recent work has led to the isolation of oligopyrrole systems from experiments in which the action of deaminase on PBG had been inhibited by ammonium ions or hydroxylamine. One product from the work with hydroxylamine[84] behaved in a similar way on electrophoresis to the synthetic AP · AP pyrromethane (30). Also, the n.m.r. spectra of the isolated and synthetic samples were closely similar. Though this evidence is in keeping with the suggested aminomethyl structure (30) for the enzymic product, it is also consistent with the (at least) equally probable hydroxyamino structure (cf. Ref. 65) (56).

Several tracer experiments have shown that activity from synthetic ^{14}C-AP·AP pyrromethane (30) is incorporated into uroporphyrinogen-I

(33) when (30) is incubated with PBG and deaminase. Apparently AP · AP competed more successfully with PBG for the active site of the spinach enzyme[84] than it did when wheat germ is the source[87]. It was further found[84] that the foregoing enzymically formed pyrromethane (30) or (56) behaved similarly with PBG and deaminase in yielding radioactive uroporphyrinogen-I. Neither the synthetic AP · AP pyrromethane nor the enzymically produced material acted as precursor of uroporphyrinogen-I in the absence of PBG[84,87]. There was no evidence for the incorporation of AP · PA pyrromethane (45) into uroporphyrinogen-I[87]. Further, the results obtained when the radio-inactive synthetic tripyrrole (55) was incubated with ^{14}C-PBG and deaminase are interpreted as showing incorporation of the tripyrrole into uroporphyrinogen-I to a very small extent[95].

A second family of substances isolated from the inhibition experiments with amines gave further insight into the enzymic process. When inhibition was brought about by ammonium ions, a product was formed to which the bilane structure (32) was assigned[85,65] on the following grounds (a) the product is formed from PBG, it contains an α-free pyrrole residue and is converted non-enzymically into uroporphyrinogen-I (33); (b) the latter conversion follows first-order kinetics; (c) the formation of uroporphyrinogen-I establishes the order of the side-chains in the isolated product; (d) ammonia is released as the product is converted into uroporphyrinogen-I and the effect of pH on the rate of the conversion is consistent with a mechanistic requirement for protonation of an amino group.

Scheme 12

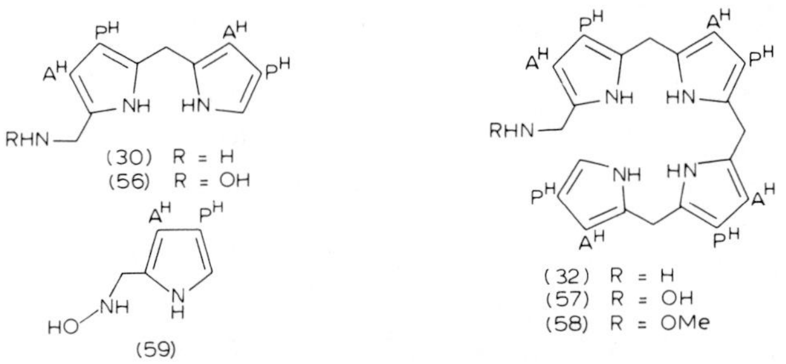

(30) R = H
(56) R = OH

(59)

(32) R = H
(57) R = OH
(58) R = OMe

It is particularly interesting in relation to our later discussion of the enzymic formation of uroporphyrinogen-III (p. 92) that the rate of formation of uroporphyrinogen-I from the aminomethylbilane (32) was not increased by added deaminase. Two further products resulted when deaminase acted on PBG in the presence of hydroxylamine and O-methylhydroxylamine and the structures (57) and (58), respectively, were assigned to them[65]. Again these products were converted non-enzymically into uroporphyrinogen-I (33) with release of the base used for inhibition. Clear evidence was given that the product from the experiment with hydroxylamine does not contain a primary amino group and this is relevant to the earlier comment about the structure of the pyrromethane formed enzymically in the presence of hydroxylamine.

The foregoing evidence in support of structure (32) for the product of enzymic inhibition by ammonium ions is certainly strong; however, in the longer term, the interest of this substance and of the analogs (57) and (58) would justify the additional difficult work needed to establish their constitutions unambiguously.

It should be mentioned finally that the chemistry of PBG and of simple analogues has been extensively studied in vitro, especially their reactions with acid and their conversion into dipyrrolic systems and uroporphyrinogen macrocycles[57,58,96-98]

3.3.5.3. Possible mechanistic scheme

One can now consider a scheme for the formation of uroporphyrinogen-I from PBG by deaminase which fits all the data summarized so far. This considers a nucleophilic group, R-XH, on the enzyme displacing the protonated amino group of PBG by S_N2 attack as illustrated (Scheme 13) (or replacing it by an S_N1 process) to form intermediate (60). Stepwise build-up via (61) would then give the bilane (62) attached to the enzyme. A ring-

Scheme 13

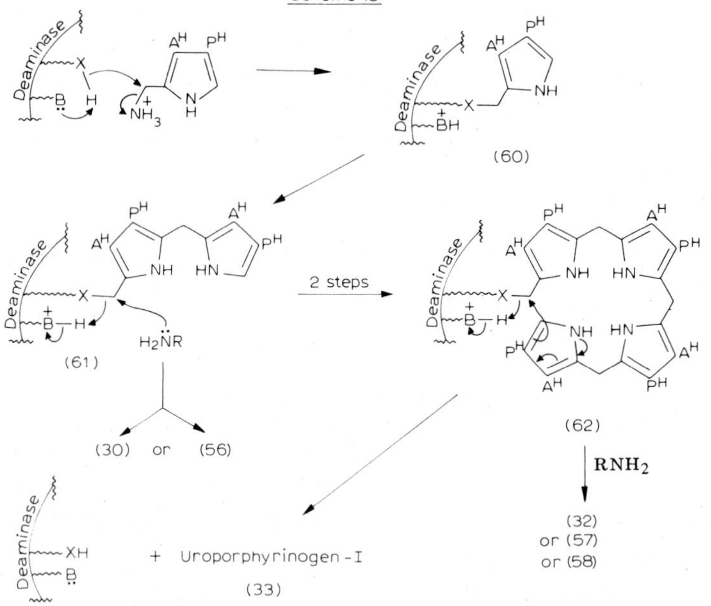

Hypothetical enzymic formation of uroporphyrinogen-I and products from inhibited preparations

closure step, perhaps slower than the building steps, could then regenerate the original nucleophile*.

The relatively vast concentrations of nucleophilic bases used for the inhibition work already described could bring about displacement of any of the intermediates in Scheme 13. Indeed, there is evidence that the hydroxy-analog of PBG (59) is formed when hydroxylamine is the inhibitor and importantly, this modified product is only formed when enzyme is present[65].

3.3.6. Enzymic formation of uroporphyrinogen-III (27)

There has been intense interest for many years in the conversion of PBG specifically into uroporphyrinogen-III (27) by the cooperative action of deaminase and cosynthetase and a great variety of experimental studies has been made. These have provided, up to about 1970, the following information about the enzymic process:

(i)   Four molecules of PBG are converted into one of uroporphyrinogen-III[60,73,75,76,54,99] and no pyrroles detectable by Ehrlich's reagent remain in solution; this matches the findings already reported for the enzymic formation of uroporphyrinogen-I.

(ii) No free formaldehyde can be detected as PBG is converted enzymically into uroporphyrinogen-III at pH 8.2 nor is added ^{14}C-formaldehyde incorporated by the combination of deaminase and cosynthetase into uroporphyrinogen-III at pH 8.2—8.5[76,100]. An earlier report[74] that formaldehyde is produced in this enzymic reaction may well have been caused by some (enzymic?) reaction unrelated to uroporphyrinogen formation taking place in the mixture of enzymes making up the cell-free preparation.

(iii) Uroporphyrinogen-I (33) is not converted into the type-III isomer (27) by cosynthetase alone or by deaminase and cosynthetase together; thus the isomerization occurs at some earlier stage[60,75,101].

(iv) The purest samples of cosynthetase (i.e. free of deaminase) did not combine PBG molecules one with another. Also when PBG was incubated first with cosynthetase which was then destroyed by heat and the resultant mixture was treated with deaminase, only uroporphyrinogen-I (33) was formed[60]. It is clear that cosynthetase is not transforming PBG into a substrate for deaminase.

---

* This Scheme illustrates one reasonable view without attempting to be too detailed; so we do not intend to imply that other types of binding interactions (e.g. through the carboxyl groups) are not important. Indeed, evidence will be given later that the carboxyl groups are involved in binding. Also, the suggested nucleophilic attack though chemically attractive is not an *essential* step; further studies are necessary to provide more detailed information.

(v) Whilst pure deaminase converts PBG into uroporphyrinogen-I (33), addition of increasing amounts of cosynthetase causes a gradually increasing proportion of the type-III isomer (27) to be formed. The product is essentially pure uroporphyrinogen-III when PBG is treated with deaminase in the presence of a considerable excess of cosynthetase[60,73,81].

(vi) The rate of consumption of PBG by deaminase is increased by added cosynthetase provided the experiment is run either with low substrate concentration[60] or at pH 9[73].

(vii) Opsopyrroledicarboxylic acid (59A) Scheme 14, which featured in speculations on the mechanism of formation of the type-III macrocycle, is not involved in the enzymic production of uroporphyrinogen-III from PBG (or of uroporphyrinogen-I either); indeed, the dicarboxylic acid (59A) acts as a competitive inhibitor of the consumption of PBG[102,1a]. In contrast, PBG α-carboxylic acid (60A) Scheme 14, is without effect on the enzymic reaction[61]. Neither the ethyl analog (61A) of PBG nor the non-basic pyrrole (62A) acted as substrates for deaminase but they did act as inhibitors of porphyrin formation from PBG. In contrast, the dimethyl ester of (62A) had no inhibitory effect, pointing to the involvement of one or both carboxyl groups in binding[71].

Scheme 14

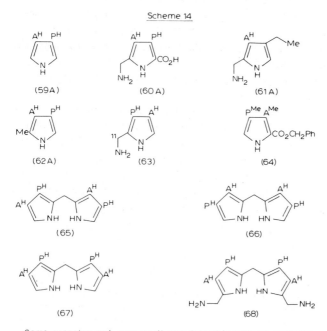

Some pyrroles and pyrromethanes tested in enzyme systems

(viii) IsoPBG (63) alone is not a substrate for the deaminase-cosynthetase combination and when PBG (14) is included in the medium, isoPBG acts as a weaker competitive inhibitor than was opsopyrroledicarboxylic acid (59A) above[103]. These studies left open some possibility that isoPBG might have been incorporated with PBG into uroporphyrinogen-III. Unambiguous information on this point is essential and therefore [11-^{14}C]isoPBG (63) was synthesized[104] from the pyrrole (64) by a modification of the published route[91]. When this labeled isoPBG together with unlabeled PBG were treated with the deaminase-cosynthetase system from *Euglena gracilis*, the incorporation into uroporphyrinogen-III (27) [as measured by the isolated protoporphyrin-IX (69) enzymically formed from it] was less than 0.01%[104]. IsoPBG (63) need therefore be considered no further as a precursor of uroporphyrinogen-III (27).

(ix) The three pyrromethanes (65), (66), and (68) were not incorporated alone or with PBG into uroporphyrinogen-III by deaminase-cosynthetase from spinach[45] and these three substances together with pyrromethane (67) were tested without added PBG as precursors of uroporphyrinogens in the deaminase-cosynthetase preparation from *R. spheroides*; they were all essentially ineffective as precursors of the macrocycles[105].

3.3.7. Analysis of hypotheses regarding formation of the type-III isomer

By 1970, more than twenty hypothetical schemes had been published[106] aimed at explaining the specific formation of the type-III macrocycle in Nature and several additional ones could be suggested (e.g. Ref. 88). Some are eliminated by the data summarized above but the majority remained. Recently, a series of researches based upon carbon-13 labeling has led to the discovery of the nature of the rearrangement process and as a result, the number of viable mechanisms is now strictly limited. The approach depends on following the origin and fate of the four carbon atoms which build the *meso*-bridges of uroporphyrinogen-III, that is, carbons α, β, γ and δ of structure (27) (see Scheme 15). The importance of these bridges can be brought out by comparing two of the hypothetical pathways.

Scheme 15 shows a slight modification of one proposal[108] wherein the bilane (62) is built in the way already discussed (Schemes 9 and 13), that is, each PBG unit is joined head-to-tail without rearrangement as far as the putative intermediate (62). Ring-closure to the free α-position would account for what is found using deaminase alone whereas the presence of the cosynthetase protein could modify the process so that cyclization is favored at the substituted α-position of the terminal pyrrole residue. The spiro-system (71) so generated could fragment as shown to allow final ring-closure of the methylenepyrrolenine (72), the overall effect being reversal of ring-D. This proposal involves only one rearrangement, which is intramolecular with

Scheme 15

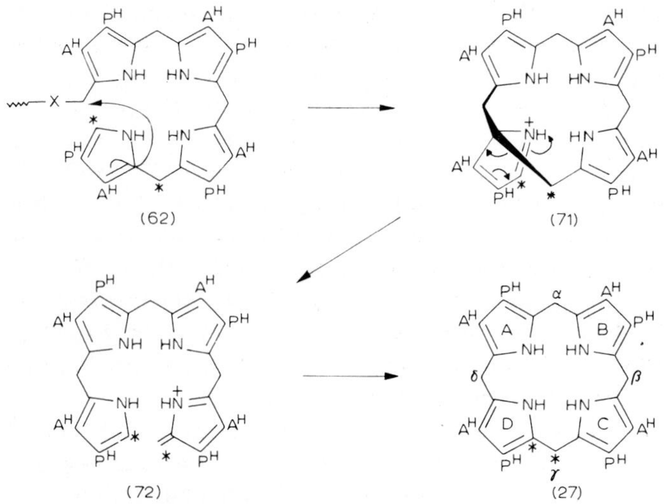

(62)   (71)

(72)   (27)

The "spiro" hypothesis for formation of
type-III porphyrins

respect to the PBG unit forming ring-D as can be seen by following the
asterisks in Scheme 15.

Small changes from the original proposal[109] have also been made in the
second sequence shown in Scheme 16. Here it is suggested that the first three
PBG units are all joined head-to-head with subsequent intramolecular rearrangement as illustrated. The fourth PBG unit is inserted by a head-to-tail
step and final ring-closure produces uroporphyrinogen-III (27). In this proposal, the α, β and γ-meso carbons are all involved in intramolecular rearrangements and so in principle, it should be possible to distinguish it from the

Scheme 15 (a)

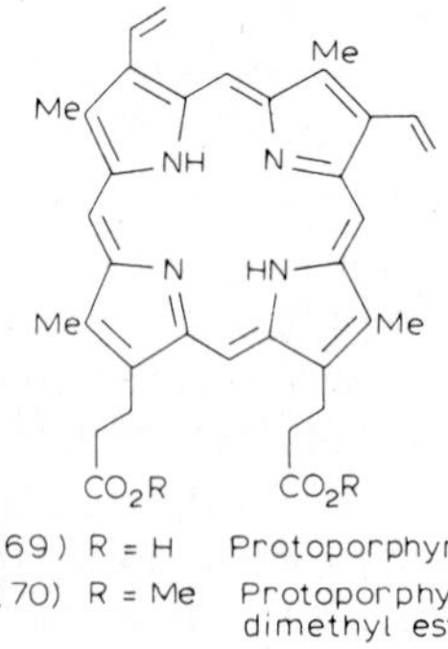

(69) R = H   Protoporphyrin - IX
(70) R = Me   Protoporphyrin - IX
                dimethyl ester

first proposal. *Indeed, such an analysis of all possible schemes showed that they generally differed in the origin and fate of the* meso-*bridge atoms.*

The approach used to track the individual *meso*-carbons through the biosynthetic process was based on carbon-13 n.m.r. spectroscopy and the porphyrin to be studied must be so selected that carbon-13 atoms at these four key positions give signals with distinguishable chemical shifts. This would not be expected for uroporphyrin-III (25) (which would be the form in which enzymically produced uroporphyrinogen-III would be isolated) because each pyrrole residue carries an identical pair of substituents. However, differences were expected in the case of protoporphyrin-IX (69). Success depended on this being so and on the conversion of enzymically formed uroporphyrinogen-III (27) into protoporphyrin-IX (69) by the enzymes acting further along the pathway.

Scheme 16

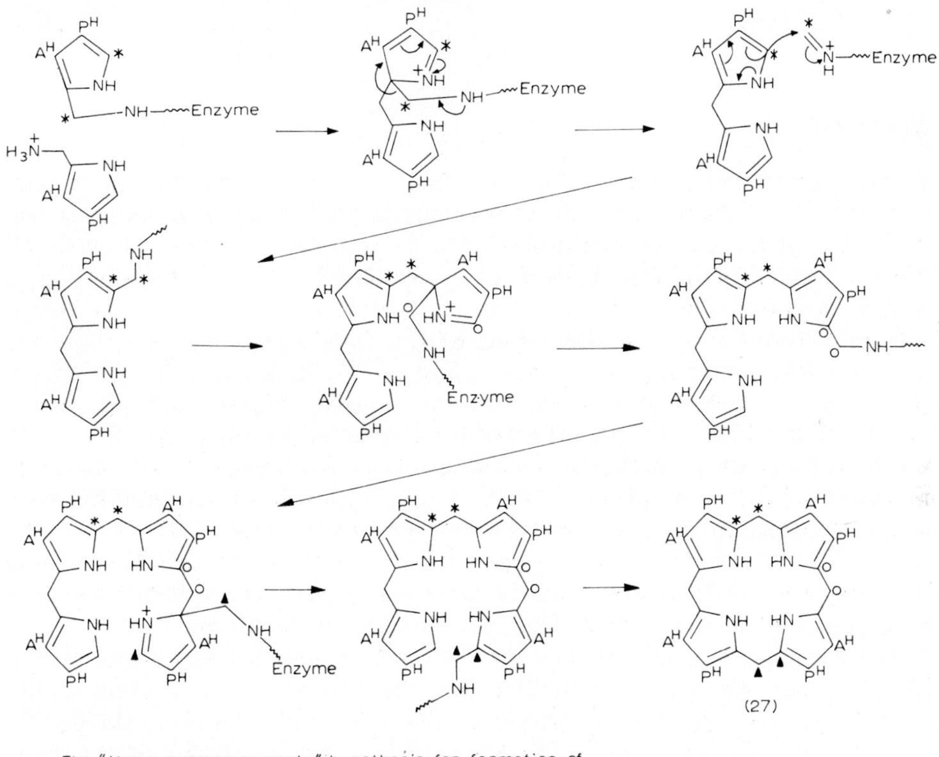

The "three rearrangements" hypothesis for formation of type-III porphyrins. (Note that —NH~~~Enzyme could also be —X~~~Enzyme, see scheme 13)

3.3.8. Basic studies with carbon-13 n.m.r. and integrity of the type-III macrocycle

The natural abundance proton-decoupled ^{13}C-spectrum of protoporphyrin-IX dimethyl ester (70) showed four distinguishable signals in the 96—98 p.p.m. region which were well separated from other signals[110]. These were proved to arise from the four *meso*-carbons by preparing labeled protoporphyrin-IX biosynthetically from specifically labeled [11-^{13}C] PBG (Scheme 17); the ^{13}C-spectrum of the resultant protoporphyrin-IX dimethyl ester

Scheme 17

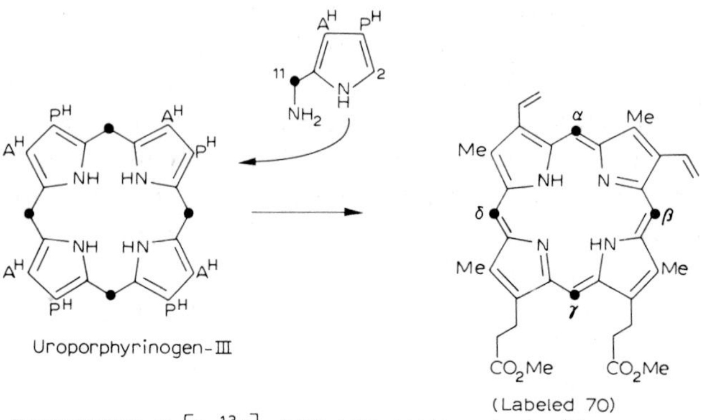

Uroporphyrinogen-III

(Labeled 70)

Incorporation of [11-^{13}C] –PBG into protoporphyrin-IX

showed great enhancement of the four signals in the 96—98 p.p.m. region[110,111]. Assignment of the signals to individual *meso*-carbons was made by unambiguous synthesis of three samples of protoporphyrin-IX dimethyl ester (70), one labeled with carbon-13 at the β-*meso* carbon, a second at the γ-*meso*, and a third at the δ-*meso* position. Their ^{13}C-n.m.r. spectra allowed rigorous assignment of the four signals and the sequence reading from low to high field was α, β, δ, γ[112]. In addition, protoporphyrin-IX was converted chemically into the 2,4-diacetyldeuteroporphyrin[113] dimethyl ester (73)* and the effect of the change of peripheral groups was to spread the signals from the *meso*-carbons. They were assigned by converting the three synthetic samples of ^{13}C-labeled protoporphyrin-IX dimethyl ester into the corresponding 2,4-diacetyldeuteroporphyrin esters (73)*. Again, the signals from the *meso*-carbons appeared in the same order, α, β, δ, γ reading from low to high field; importantly there was a chemical shift difference of 6.9 p.p.m. between the signals from the α and γ-carbon atoms[112].

These studies opened the way for ^{13}C-n.m.r. research on protoporphyrin-IX (69). But we have seen earlier that the rearrangement leading to the type-III macrocycle occurs between PBG (14) and uroporphyrinogen-III (27). Clearly, protoporphyrin-IX can only be used in studies of this rearrangement if it is established that the basic ring structure of the type-III macrocycle (e.g. 27) once formed remains unchanged. This was done by converting specifically labeled uroporphyrinogen-III (27) and, separately, coproporphyrinogen-III (74) into protoporphyrin-IX (69) by the mixture of

---

* 3,8-Diacetyldeuteroporphyrin using the 1—24 numeration.

Scheme 17 (a)

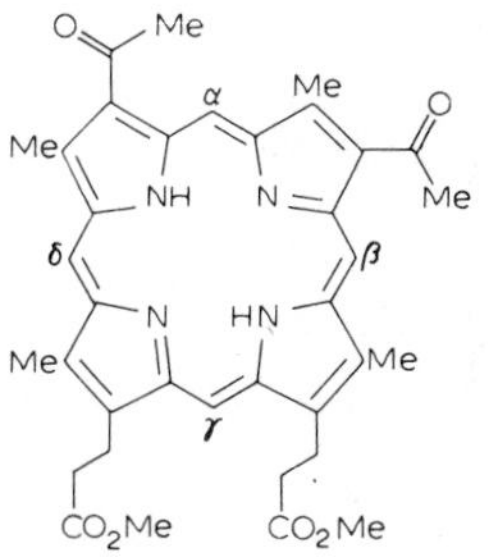

(73) 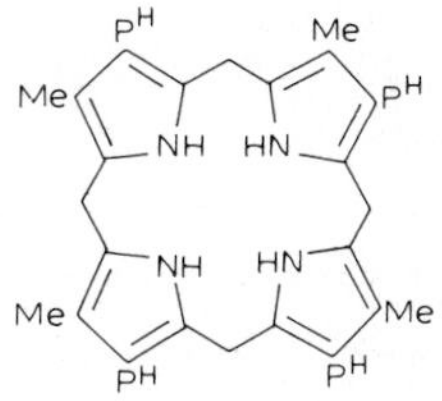   (74)

enzymes operating late in the sequence. Degradation showed that no change of labeling position occurred along this part of the biosynthetic pathway[114,115]. Accordingly, data on labeling patterns and $^{13}C-^{13}C$ couplings obtained for enzymically formed protoporphyrin-IX (69) hold true for uroporphyrinogen-III (27) and so are directly informative about the rearrangement process.

3.3.9. Nature of the rearrangement process from experiments with $^{13}C_2$-PBG

[5-^{13}C] Aminolevulinic acid (3) available from a new synthesis[111] was converted by aminolevulinic acid dehydratase into doubly ^{13}C-labeled PBG; each labeled site carried 90 atom % ^{13}C so that approx. 81% of the PBG molecules contained two ^{13}C-atoms. Deaminase transformed this product into uroporphyrinogen-I (33, Scheme 18) which was aromatized to form the corresponding porphyrin (75). Its n.m.r. spectrum showed that the directly bonded ^{13}C-atoms give rise to a 72 Hz coupling[116]. An additional 5.5 Hz coupling shown by this spectrum was assigned to the longer range coupling indicated by the asterisks on structure (75). This interpretation was proved to be correct by synthesis of the esters of uroporphyrin-III (25) and proto-porphyrin-IX (70) with 81% of the molecules carrying two ^{13}C-atoms[117] (see Scheme 19). Both samples (25) and (70) showed the 5.5 Hz coupling; this assignment is of great importance because, as will be seen below, *intact incorporation of PBG can be revealed by the presence of the 5.5 Hz coupling.*

A further decisive step in this study was dilution of the doubly ^{13}C-labeled PBG with unlabeled PBG (Scheme 18) before conversion by avian erythrocyte enzymes into uroporphyrinogen-III. The result of this dilution is that the majority of single molecules in the uroporphyrinogen-III so formed contain only one labeled PBG unit and carry only two ^{13}C-labels. This uroporphyrinogen-III was converted enzymically into protoporphyrin-IX, isolated as its dimethyl ester. The proton decoupled n.m.r. spectrum of this product, and of the corresponding 2,4-diacetyldeuteroporphyrin-IX (as 73)*

---

* 3,8-Diacetyldeuteroporphyrin using the 1—24 numeration.

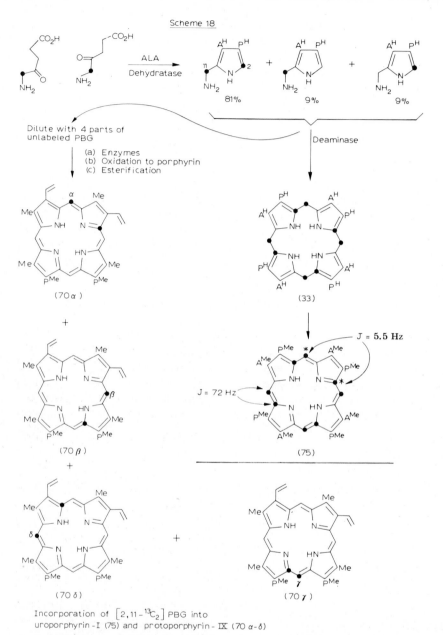

Incorporation of $[2,11-^{13}C_2]$ PBG into
uroporphyrin-I (75) and protoporphyrin-IX (70 α-δ)

which was chemically derived from it, showed signals from the α, β and
δ-*meso* carbons which were all 5.5 Hz doublets whereas that from the
γ-*meso* carbon appeared as a 72 Hz doublet. Thus the protoporphyrin-IX
dimethyl ester is composed largely of the species (70α), (70β), (70γ), and
(70δ).

Scheme 19

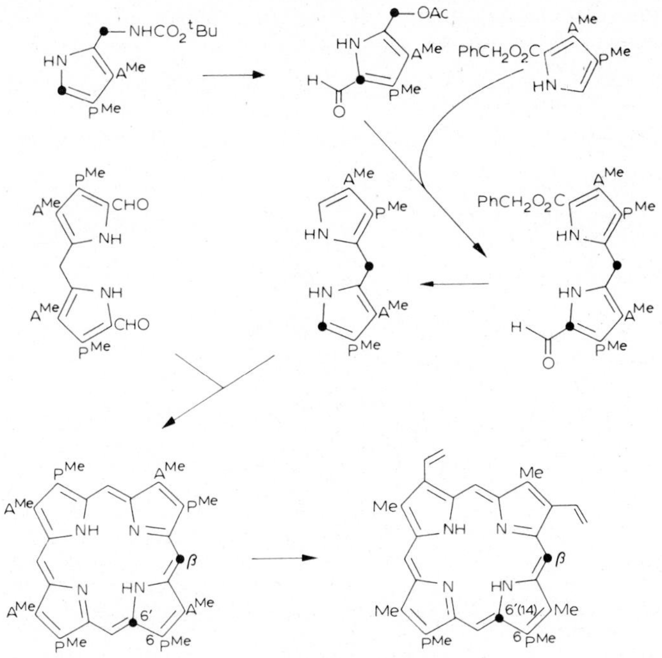

Synthesis of $[\beta, 6'-^{13}C_2]$ uroporphyrin-III and
$[\beta, 6'-^{13}C_2]$ protoporphyrin-IX methyl esters,
$[10,14-^{13}C_2]$ using the 1-24 numeration

It follows that the rearrangement process by which the type-III porphyrins are biosynthesized is characterized by these features:

(i)   The three PBG units which form ring A and its attached δ-*meso* bridge, ring B and the α-*meso* bridge and ring C with its β-*meso* bridge are all incorporated *intact without rearrangement.*

(ii)   The PBG unit forming ring D is built in *with rearrangement which is intramolecular with respect to that PBG unit.*

(iii) The rearranged carbon atom becomes the γ-*meso* bridge.

These features of the rearrangement process place very strict limitations on the mechanisms which can now be considered for formation of the type-III porphyrins.

An identical pattern of signals was observed[118] when the above work (using avian blood) was repeated using the deaminase-cosynthetase enzyme system from *Euglena gracilis* showing that the same fundamental mechanism operates in organisms of quite different evolutionary development.

Most of the speculative mechanisms for the rearrangement process are eliminated by these findings. That in Scheme 15 remains for consideration together with a sequence such as Scheme 20 in which the intramolecular nature of the rearrangement is preserved by carrying the migrating carbon atom on some enzymic functional group. It will be obvious that several minor sequential variations of Scheme 20 are possible.

Scheme 20

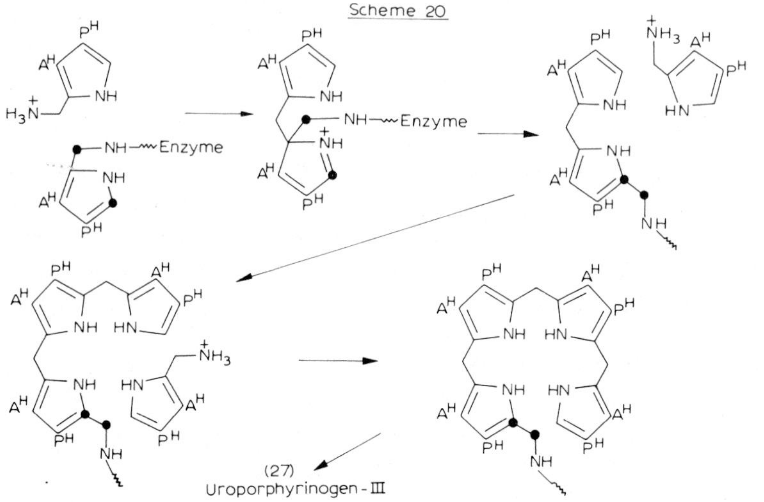

(27)
Uroporphyrinogen - III

Hypothetical sequence for biosynthesis of uroporphyrinogen - III. (Note that —NH⁓Enzyme could also be —X⁓Enzyme, see scheme 13)

3.3.10. Enzymic studies with pyrromethanes

We can now turn to the considerable efforts which have been made, also since about 1970, to discover intermediates between PBG and uroporphyrinogen-III since these too can provide crucially important information about the nature of the rearrangement process. The synthesis of the four isomeric aminomethylpyrromethanes (30), (45), (53), and (54) [Scheme 10 and 11] has already been described together with the enzymic incorporation of AP · AP (30) into uroporphyrinogen-I. It should be emphasized at the outset that researches on the enzymic incorporation of aminomethylpyrromethanes into porphyrinogens are bedeviled by concomitant *non-enzymic* conversion of the substrates into porphyrinogens which produces a troublesome blank. This is a major difficulty for all workers in the area and undoubtedly it has contributed to some reports of somewhat dissimilar results from different researchers. We will take each of the four pyrromethanes in turn, starting with the ones where agreement is complete, and outline the present position.

3.3.10.1. The AP · PA pyrromethane (45)

Both in Buenos Aires[86,87] and Cambridge[119] it was found that this pyrromethane (45) was not a substrate alone or in admixture with PBG for deaminase-cosynthetase enzyme systems active in producing the uroporphyrinogen-III macrocycle (27). Enzymes from wheat germ[87], duck erythro-cytes[119], and *Euglena gracilis*[119] were used in these experiments.

The pyrromethane (45) underwent non-enzymic conversion[87,119] in 10—12% yield (after aromatization) into uroporphyrin-II (35) which was homogeneous within the limits of the chromatographic analytical methods. This is the expected product from straightforward dimerization of (45) with-out rearrangement.

3.3.10.2. The PA · PA pyrromethane (54)

The Cambridge group[120] using ^{14}C-labeled (54) and the deaminase-cosynthetase enzyme system from *Euglena gracilis* found that the amount of radioactivity built into the biosynthesized type-III macrocycle did not differ significantly from that found in blank experiments with totally inactivated enzymes. Carbon-14 labeling of the PA · PA 'precursor' was specifically at the methylene of the aminomethyl group.

Bearing in mind the findings using carbon-13 described in section 3.3.9, it would have been remarkable if PA · PA (54) *had* been incorporated; we can see no way of generating (54) which does not contravene the requirements set out in section 3.3.9; so the two studies interlock.

3.3.10.3. The AP · AP pyrromethane (30)

To assess the researches on this pyrromethane (30) and on the PA · AP isomer below, it is essential to summarize how the experiments have been carried out by different workers. One way, which we will call the *dilution method*, involves treatment with the enzyme of a mixture of ^{14}C-labeled PBG (of known molar activity) together with *radio-inactive* pyrromethane. If the resultant uroporphyrin-III (25) has a lower molar radioactivity than that isolated from control experiments, then this result is interpreted as showing incorporation of the pyrromethane (and conversely, if a lower molar radio-activity is not observed, then the interpretation is one of no incorporation).

Success with this method demands (a) extreme accuracy in determinations of molar activities (since the result usually depends on a relatively small difference between two larger values) and (b) high chemical purities of all products (impurities could either lower the molar activity by adding non-porphyrin material to the weight, if weighing is used for assay; if spectro-scopy is the method of assay, impurity could give a false value for porphyrin present by absorbing at the selected wavelength).

The second way, which we will call the *direct method* involves incubation of *radioactive* pyrromethane and radioinactive PBG with the enzyme system. This method requires rigorous radiochemical as well as chemical purity of the product (entrapped radio-labeled impurities would clearly cause errors).

Both approaches have, in addition, to contend with the problem of the blank discussed at the outset of this section.

The Buenos Aires group[86,87] using mainly the dilution method and two experiments by the direct one, came to the conclusion that pyrromethane AP · AP (30) is not incorporated with PBG into uroporphyrinogen-III (27). Further, it was found that the formation of (27) is inhibited by the AP · AP pyrromethane, especially when in considerable excess over PBG, and the enzymic reaction is diverted to form more uroporphyrinogen-I (33). Their deaminase-cosynthetase enzyme was from wheat germ.

Duck erythrocytes and *Euglena gracilis* were used as sources of deaminase-cosynthetase in separate experiments in Cambridge[88,119,120] by the direct method. Pyrromethane AP · AP (30) was used labeled at the bridge methylene group with carbon-14 in one set of experiments and with carbon-13 in another. The uroporphyrinogen-III formed initially was further transformed in situ and the isolated protoporphyrin-IX dimethyl ester (70) was shown to be labeled heavily at the α and γ *meso*-carbons by chemical degradation (for ^{14}C-labeled material) and by n.m.r. spectroscopy (for the ^{13}C labeled sample). The incorporations of activity* with the duck blood enzymes were 1.6—9.0% and with *Euglena* 0.2—0.5%.

Current experiments with highly active enzyme preparations by the direct method will determine the extent of the chemical and enzymic contributions to the above non-random incorporations.

Both studies of non-enzymic conversion of pyrromethane (30) into porphyrins largely agree in finding conversion at pH 7.4—8.2, after aromatization, into [results (a)[87]] 4.5% uroporphyrin-I (34) and 4.5% uroporphyrin-III (25) and/or uroporphyrin-IV (36); compare [results (b)[119]] 9% uroporphyrin-I (34), 2% uroporphyrin-III (25) and 1% uroporphyrin-III and/or uroporphyrin-IV (36).

3.3.10.4. The PA · AP pyrromethane (53)

This pyrromethane is converted non-enzymically via uroporphyrinogen-II into pure uroporphyrin-II (35) in 8—10% yield at pH 7.4; when PBG is included in the reaction mixture, approx. 2% of uroporphyrinogen-III (27) is formed in addition (estimated as the corresponding porphyrin)[87].

Using deaminase-cosynthetase from wheat germ and human erythrocytes, the Buenos Aires group[93] have reported results interpreted as showing that the PA · AP pyrromethane (53) is enzymically incorporated into uroporphyrinogen-III (27) and not into uroporphyrinogen-I (33). Ten or twenty nanomoles of the ^{14}C-pyrromethanes were used to give a considerable excess over the radio-inactive PBG (6 nmol) and the incorporations* of radioactivity by the direct method were 0.6—0.9%. If calculated another way, the results correspond to 10—26% of the 0.53—0.77 nanomoles of isolated uro-

---

* Incorporation values for all pyrromethanes discussed here are calculated as percentages thus:

$$\text{Incorporation} = \frac{\text{total activity in isolated porphyrin}}{\text{total activity of precursor used}} \times 100$$

porphyrin-III having been formed from the 10—20 nanomoles of PA · AP pyrromethane which were present.

The Cambridge group[119,120] used deaminase-cosynthetase from both duck erythrocytes and *Euglena gracilis* with ^{14}C-PA · AP and radio-inactive PBG. Two conditions were used:

(a) with ^{14}C-PA · AP at high specific activity and an excess of radio-inactive PBG,

(b) with the ^{14}C-PA · AP at somewhat lower specific activity but now in excess over the PBG.

In these experiments, the incorporation of radioactivity into uroporphyrinogen-III (27), as measured by the crystalline protoporphyrin-IX isolated was in the range 0.01—0.015% for *Euglena* and 0.06—0.12% for duck erythrocytes.

3.3.10.5. Conclusions

All the foregoing researches clearly show that it is difficult to insert an externally added pyrromethane into the biosynthetic pathway to uroporphyrinogen-III. This may be a reflection of some chemical modification of the first PBG unit bound in the normal process along the lines suggested in Schemes 13 and 15 but this is not the only possibility.

The results so far strongly indicate that AP · PA (45) and PA · PA (54) pyrromethanes are not biosynthetic intermediates on the pathway to uroporphyrinogen-III nor do they appear to be closely related to the true intermediates.

Current and future work on the other two pyrromethanes AP · AP (30) and PA · AP (53) to settle rigorously their status for the biosynthesis of the type-III macrocycle is awaited with interest. Though it will be extremely difficult to achieve, the following requirements ideally should be met in these experiments. Proof will be required (a) that the isolated porphyrin isomers, especially type-I and type-III, are chemically (i.e. also isomerically) pure, (b) that in the ^{14}C-series, they are radiochemically pure, (c) that incorporations observed are specific and the site(s) of labeling are determined.

If, in addition, conditions can be found for the enzymic runs which allow the observed conversion of pyrromethane and PBG into uroporphyrinogen-III to be greater (preferably considerably greater) than that in the blank, then the present difficulties of interpreting hard-won results would be vastly diminished. At the time of writing, the Buenos Aires group interpret their results so far as eliminating AP · PA (45) as a precursor, with PA · AP (53) being incorporated into the type-III system whereas AP · AP (30) is not. The Cambridge group interpret their findings so far as eliminating AP · PA (45) and PA · PA (54) but feel that a rigorous distinction between AP · AP (30) and PA · AP (53) as precursors of the type-III system requires further research.

References, p. 116

3.3.11. Enzymic studies with a tripyrrole and a bilane

The tripyrrole (55) did not act as a precursor of uroporphyrinogen-III (27) either with PBG and cosynthetase or with PBG and deaminase-cosynthetase. Also, addition of large quantities (relative to the PBG used) of the tripyrrole to an enzyme system making both uroporphyrinogen-I (33) and uroporphyrinogen-III (27) caused reduction in total porphyrinogen formed and an increase in the proportion of uroporphyrinogen-I (33)[95].

The proposed ^{14}C-aminomethylbilane (32), isolated from the inhibition experiments with ammonia described earlier, was not incorporated[85] into uroporphyrinogen-III when it was incubated with approx. 40 molar excess of unlabeled PBG in the deaminase-cosynthetase enzyme preparation. In considering this result, the difficulty of inserting isolated bilane into the enzymic sequence should be borne in mind particularly when the formation of uroporphyrinogen-I from the isolated bilane was not accelerated either by added deaminase enzyme.

It has been mentioned several times that intermediates do not accumulate during the normal conversion of PBG into uroporphyrinogen-I or uroporphyrinogen-III. However, reports have appeared[121,122] that the enzyme from soybean callus tissue releases pyrrolic intermediates and that a partially purified fraction on incubation with PBG and cosynthetase alone yields uroporphyrinogen-III.

3.3.12. Conclusions

The current position on this fascinating problem is that we now know exactly *what happens* as PBG is converted into uroporphyrinogen-III. Future interest centres on finding out *how it happens* and also on the role of cosynthetase which at present is incompletely understood. There are indications though from sedimentation studies[69] and by experiments with deaminase bound to Sepharose[71] that the two proteins are physically associated during the synthesis of uroporphyrinogen-III. It has been further suggested[87] that cosynthetase acts as a 'specifier protein' in the way lactoalbumin* works[71] during the biosynthesis of lactose.

3.4. Side-chain modifications and aromatization to protoporphyrin-IX
3.4.1. Decarboxylation of uroporphyrinogen-III (27)

Enzymic decarboxylation of the four acetic acid side-chains of uroporphyrinogen-III (27) (8 CO_2H groups) affords coproporphyrinogen-III (74) (4 CO_2H groups) and the reduced porphyrin can be isolated in >90% yield[123] if the incubation is carried out anaerobically. The pyrrole units of

---

* This possible relationship of the function of cosynthetase to that of lactoalbumin was independently suggested to us by Dr. B. Middleton (Cambridge) in 1972.

Scheme 20(a)

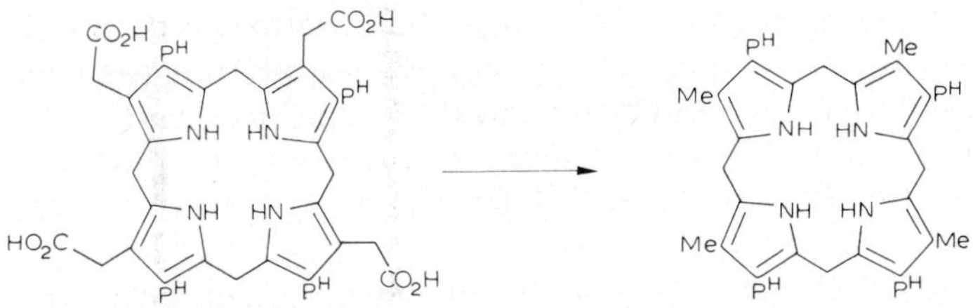

(27)

(74)

$pH = CH_2CH_2CO_2H$

the porphyrinogen macrocycle remain in their original locations as this transformation occurs[114], that is no scrambling occurs.

Intermediates were observed by several research groups[46,59,99a,123], and the chromatographic behavior of the corresponding porphyrins indicated that they posessed seven, six, or five carboxyl groups. It seems, however, that a single 'enzyme' is able to effect the four successive decarboxylations[124], and that it has a rather low substrate specificity[46,61b,123]. Thus the enzyme converts all four isomers of uroporphyrinogen into coproporphyrinogen with the natural isomer giving the greatest yield (III > IV > II > I). However, the rate of decarboxylation of uroporphyrinogen-III is only slightly greater than that of the type-I isomer[123,125,126]. Uroporphyrin-III (25) is decarboxylated one hundred times more slowly than uroporphyrinogen-III (27), whilst ALA and PBG are not affected by the enzyme[123].

3.4.1.1. The partially decarboxylated intermediates

At first it was assumed[123] that the heptacarboxylic acid fraction observed chromatographically above would be a mixture of all four possible isomers, but there is now clear evidence (especially from ^{1}H-n.m.r.[128]) that the material from chicken erythrocytes[126-128] is a single compound. It gives a methyl ester m.p. 213–215°C and the same m.p. is observed for the porphyrin isolated from the urine of porphyria patients[129]. Though there is no direct evidence, it seems very probable that these porphyrins are all identical and they have been variously named porphyrin-208, pseudouroporphyrin, and phyriaporphyrin.

The heptacarboxylic porphyrin is decarboxylated when heated with hydrochloric acid at 200°C, and the product is coproporphyrin-III[126-128] (26). Radioactive 'phyriaporphyrin' was isolated after incubating hemolyzed chicken blood with [^{14}C] ALA. This ^{14}C-labeled porphyrin was converted by chemical reduction into the porphyrinogen which was reincubated with fresh enzyme system and yielded[130] radioactive protoporphyrin-IX (28). Furthermore, the specific activity of protoporphyrin-IX biosynthesized from [2-^{14}C] glycine was lowered[130] in the presence of radio-inactive uroporphyrinogen-III (27), 'phyriaporphyrinogen', or coproporphyrinogen-III (74) but not by uroporphyrinogen-I (33) or 'phyriaporphyrin'. It is clear therefore

References, p. 116

that 'phyriaporphyrin' (a) belongs to the type-III series of porphyrins, and (b) is a true intermediate on the porphyrin biosynthetic pathway. By means of similar experiments the hexa- and pentacarboxylic porphyrinogens have been shown to be true biosynthetic intermediates[131]. The hexa- and penta-carboxylic porphyrins from porphyric rats have recently been identified[224a] as (76; $R^1 = R^4 = Me$, $R^2 = R^3 = A^H$) and (76; $R^1 = R^2 = R^4 = Me$, $R^3 = A^H$) respectively.

Further information on the structure of the heptacarboxylic porphyrin was obtained by incubating hemolyzed chicken blood with highly enriched $[2,11-^{13}C_2]$porphobilinogen (Scheme 21). From the investigation described earlier (p. 87) it is possible largely to assign the *meso*-carbon ^{13}C-reso-nances of the isolated heptamethyl ester (76) by their multiplicities: the signals from the α and β *meso*-carbons must appear as double doublets, $J = 72, 5$ Hz, that from the γ-*meso* carbon as a triplet $J = 72$ Hz, and that from the δ-*meso* position as a triplet $J = 5$ Hz. In practice, the four *meso*-carbon atoms have virtually identical chemical shift values, but addition of small portions of $Pr(fod-d_9)_3$ to a chloroformic solution of the methyl ester causes the α, β and γ resonances to move upfield approximately twice as far as the δ resonance. The shift reagent would be expected to complex more readily

Scheme 21

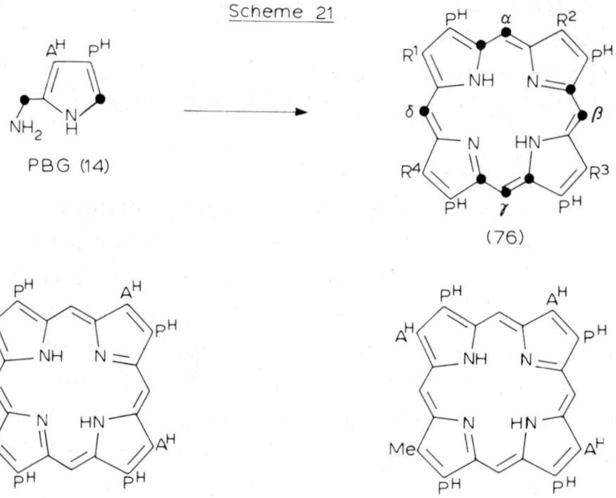

PBG (14)

(76)

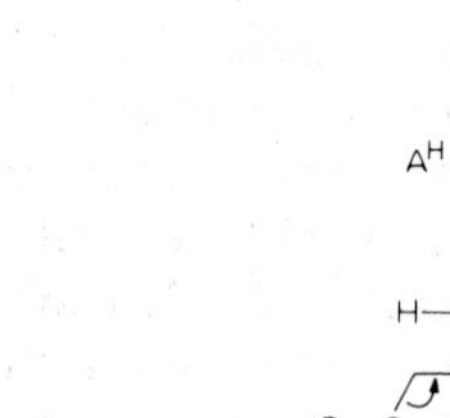

(77)

(78)

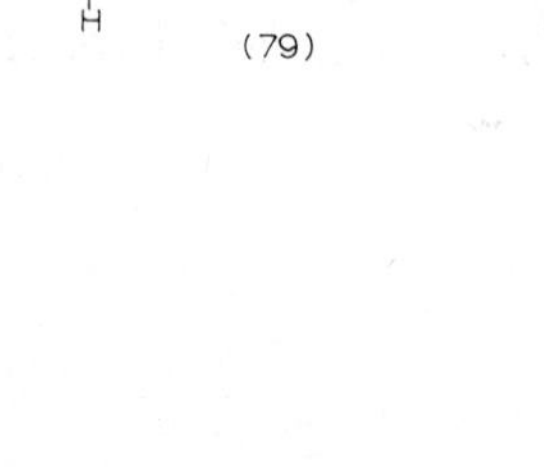

(79)

near to a *meso*-position flanked by two ester side-chains than it does near to one flanked by an ester chain and a methyl substituent; this is certainly true for proton spectroscopy[132]. The foregoing result therefore strongly suggests that the heptacarboxylic porphyrin should be represented by formula (76) with $R_2 = R_3 = CH_2CO_2H$. Thus, two possible structures (77) and (78) must be considered.

Porphyrin (77) has been unambiguously synthesized[128] as have the two isomers with the methyl substituent in rings B and in ring C (rather than in ring A). These three synthetic porphyrins could be distinguished by their ^{1}H-n.m.r. spectra (always taken at comparable concentrations), and none was identical to the natural material. Specific decarboxylation of uroporphyrinogen-III (27) must therefore take place in ring D to give the heptacarboxylic porphyrinogen with the substitution pattern of porphyrin (78). Enzyme catalyzed protonation of uroporphyrinogen-III (27) could give rise to the hypothetical intermediate (79), which is suitably activated for the decarboxylation to occur.

3.4.2. Oxidative decarboxylation of coproporphyrinogen-III (74)

Scheme 21 (a)

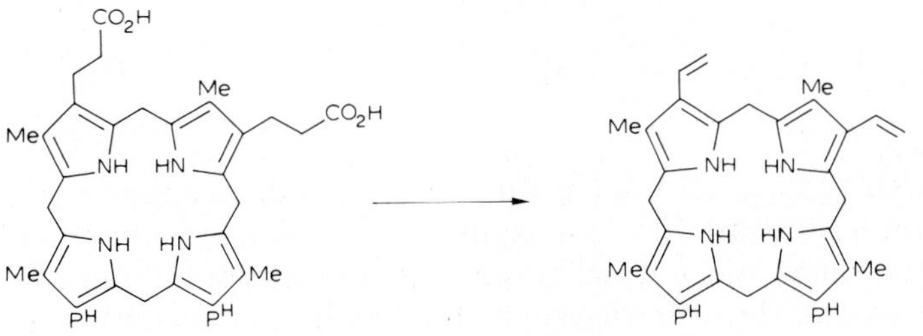

(74) $P^H = CH_2CH_2CO_2H$ (80)

The next stage in the biosynthesis of protoporphyrin-IX (28) is the conversion of the propionate side-chains on rings A and B of coproporphyrinogen-III (74) into vinyl groups. Formally this is an oxidative decarboxylation, and up to 1968 every biological system studied had been shown to have an absolute requirement for oxygen in this step. This seemed paradoxical because there are several bacteria which live under anaerobic conditions, and yet have normal heme proteins based on protoporphyrin-IX (28). Recently, it has been possible to observe the oxidative decarboxylation of coproporphyrinogen-III under anaerobic conditions, with cell free preparations from *Pseudomonas*[133], *Rhodopseudomonas spheroides*[134], and *Chromatium*[134]. The incubation media contained S-adenosylmethionine, Mg^{2+}, ATP, NADH, $NADP^+$, and succinic acid, and the experimental conditions (including the method of reduction of coproporphyrin-III to the porphyrinogen) seem to be critical[134b], because this anaerobic process has not been observed by other research groups working with *Chromatium*[135], *Pseudomonas*[136a], and *Micrococcus*[136b].

3.4.2.1. Enzymic studies

The enzyme involved in the normal oxygen-requiring process is called *coproporphyrinogenase*. It has been obtained in purified form from bacterial[134b], plant[137], and mammalian sources[138−139]. The properties of the enzyme are summarized in Table 3.

Coproporphyrinogen-III is preferred as substrate over the type-IV isomer, and coproporphyrinogen-I and II remain unchanged[140]. The porphyrinogens prepared from uroporphyrin-III (25), mesoporphyrin-IX (7), and hematoporphyrin-IX (81) also remain unchanged, but the porphyrinogen (82) is

Scheme 21(b)

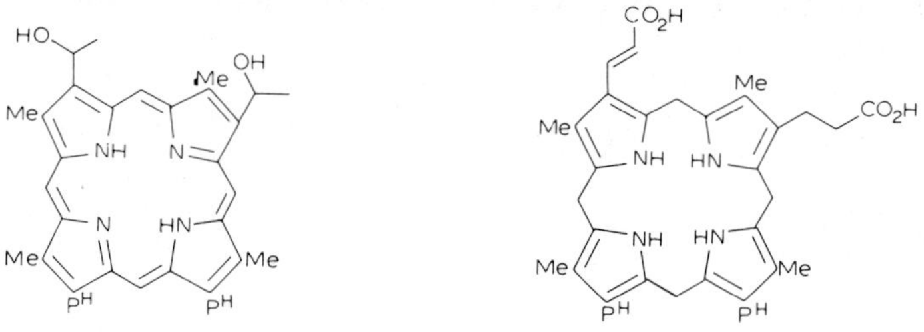

(81)   (82)

$P^H = CH_2CH_2CO_2H$

transformed by the enzyme to an unidentified tricarboxylic porphyrin[138]. Coproporphyrin is not a substrate for the enzyme[138,140]. When incubations are carried out in the dark, the normal product is protoporphyrinogen-IX (80), but when the oxygen level is raised, further oxidation occurs[140] to give a compound absorbing at 500 nm, proposed to be a tetrahydroproto-

TABLE 3

Properties of coproporphyrinogenase (aerobic form)

Source	Beef liver mitochondria	Rat liver mitochondria	Tobacco mitochondria	*Rhodopseudomonas spheroides*	*Chromatium*
Molecular weight	—	80,000	—	44,000	—
pH optimum	7.7	7.4	—	—	6.4
K_m	20—30 μM	30 μM	36 μM	30—50 μM	35 μM
Reference	138	139	137	134b	135b

porphyrin[141]. Some crude preparations catalyze the formation of protoporphyrin-IX in the dark[140], and are presumably contaminated with en-
zyme(s) involved in the aromatization step (see later).

3.4.2.2. Mechanistic studies

Coproporphyrinogen-III (74) has been synthesized in specifically labeled
form and incubated with a cell-free system from *Euglena gracilis*. The isolated protoporphyrin-IX (28) was degraded and the results[115] showed that
rings A and B of the precursor had retained their identity in the product; no
scrambling of the pyrrole rings occurs at the porphyrinogen oxidation level.
Synthetic [8-^{2}H$_2$] and [9-^{2}H$_2$]PBG (Scheme 22) incubated separately in
the same system afforded[142], after aromatization, coproporphyrin-III (26)
and protoporphyrin-IX (28). Mass and ^{1}H-n.m.r. spectroscopy on the crystalline methyl esters of these porphyrins showed that coproporphyrin-III (26)
had been formed from both labeled PBGs without loss or rearrangement of
deuterium, while the protoporphyrin-IX (28) had lost only the obligatory
one deuterium atom from each propionate residue during the formation of
the vinyl groups. Mechanisms involving acrylic acids[143] or ketones as intermediates were thus eliminated and the evidence was consistent with only
two possibilities in which the oxidative step is either C-hydroxylation or
hydride removal. These will be considered more fully below.

Scheme 22

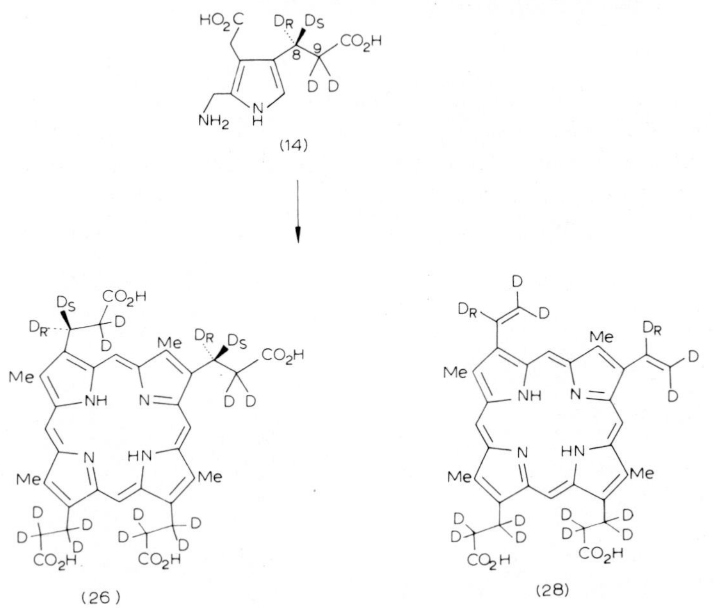

Further experiments[142] with (8RS)-[8-^{3}H$_1$, 8-^{14}C]PBG (Scheme 22)
gave protoporphyrin-IX in which the vinyl side-chains of rings A and B had

exactly half the tritium activity of the propionate groups on rings C and D. Provided that a normal kinetic isotope effect is experienced in the C—H bond-breaking step, then this result shows that the oxidation is a stereo-specific process. A series of enzymic steps were used to prepare PBG labeled stereospecifically in the propionic acid side-chain[144]. The results obtained when this product was converted enzymically into protoporphyrin-IX sup-ported the removal (for each propionate residue) of the *pro-S* hydrogen atom from the methylene adjacent to the macrocycle (Scheme 22).

3.4.2.3. Intermediates

Whatever the mechanism of the foregoing oxidative decarboxylation may be, it would be surprising if the process operated simultaneously on both propionate side-chains of ring A and ring B. In agreement with this view, a small amount (6%) of a tricarboxylic porphyrin was transiently formed[138] during the conversion of coproporphyrinogen-III into protoporphyrin-IX. Further, the unnatural tricarboxylic porphyrinogen (83) was found to act as a substrate[140] for the enzyme, and also the 2-vinyl tripropionic acid (84)* was isolated[145] from Harderian glands of the rat. The porphyrinogen corre-sponding to (84) is incorporated by a cell-free extract from *Euglena gracilis* into protoporphyrin-IX six times better than that from the 4-vinyl isomer[146] (85)†. These results show that the preferred biosynthetic route involves formation of the vinyl side-chain at position 2* before modification of the propionate chain at position 4†.

Scheme 22 (a)

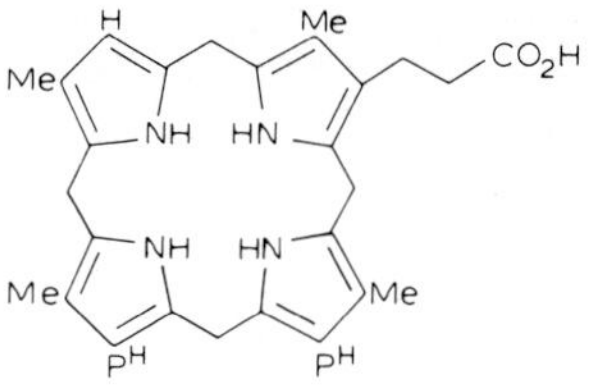

(83) 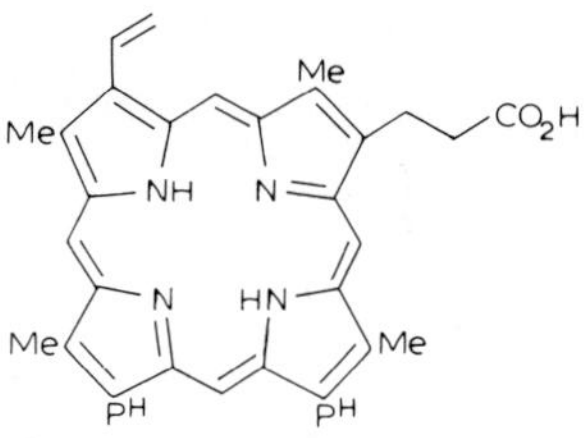   (84)

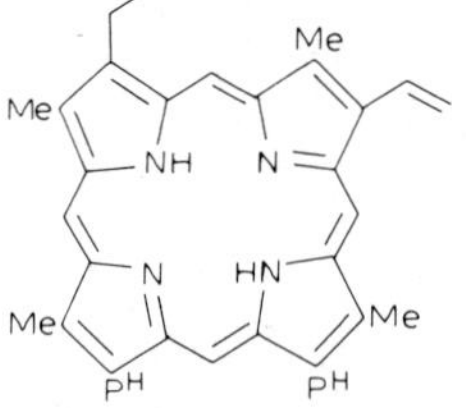

(85)

$$p^H = CH_2CH_2CO_2H$$

---

* 3-Side-chain using the 1—24 numeration.
† 8-Side-chain using the 1—24 numeration.

Intermediates in the oxidative decarboxylation itself have also attracted considerable interest. The stereospecificity of hydrogen removal from the methylene adjacent to the macrocycle and the involvement of oxygen suggest, though by no means prove, that the side-chains may first undergo enzymic hydroxylation. The porphyrin (86) having two β-hydroxypropionate side-chains was therefore prepared, and the corresponding porphyrinogen was reported to be converted into protoporphyrin-IX by coproporphyrinogenase from bovine liver[146]. This report is not in agreement with the foregoing results with the ring-A monovinyl derivative (84) nor with the following ones. The diol (86) and two isomeric porphyrins (87) and (88) each having a single β-hydroxypropionate side-chain were prepared, all in labeled form[147]. The corresponding porphyrinogens were tested with enzyme systems from *Euglena gracilis* and (for the diol) also from beef liver mitochondria. The diol was not significantly incorporated into protoporphyrin-IX under any conditions whereas the isomer hydroxylated at the 2-side-chain* (porphyrinogen from (87)) was somewhat more efficiently incorporated than was the 4-isomer† (porphyrinogen from 88) into protoporphyrin-IX; parallel incubations were run in each case with labeled coproporphyrinogen-III (74) to act as a reference substance.

Scheme 22 (b)

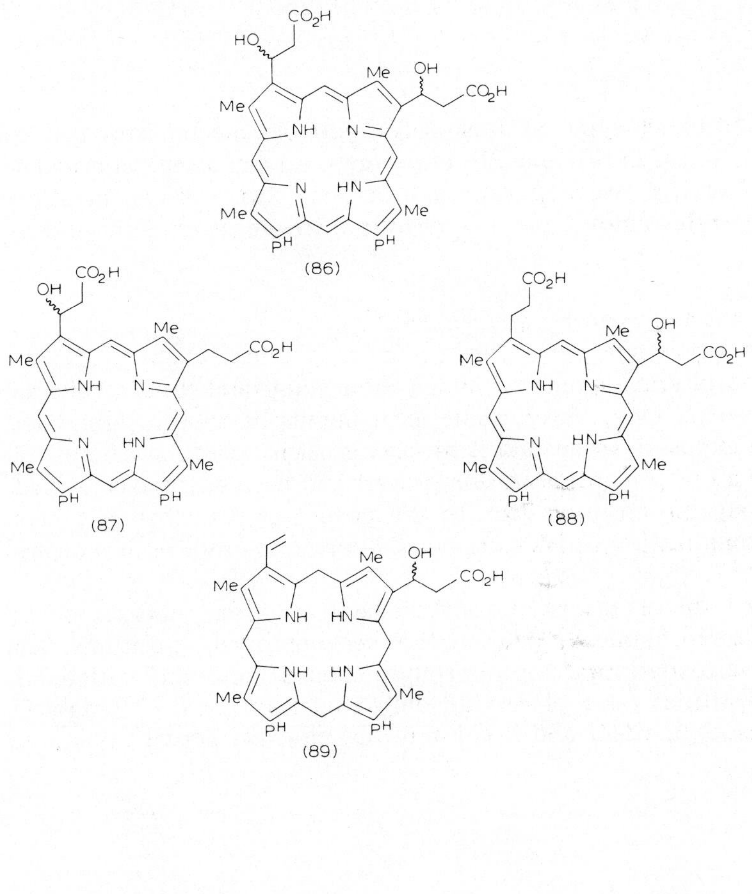

(86)

(87)

(88)

(89)

A porphyrin recently isolated from the feces of a porphyria patient is reported[148] to have either structure (87) or (88).

To sum up, it seems probable that the sequence of events catalyzed by the oxygen-requiring coproporphyrinogenase is coproporphyrinogen-III (74) → 2-β-hydroxypropionate* (as 87) → 2-vinylporphyrinogen* (as 84) → 2-vinyl-4-β-hydroxypropionate ((89), not yet investigated) → protoporphyrinogen-IX (80). Further research is needed in this area and is in progress.

Scheme 23

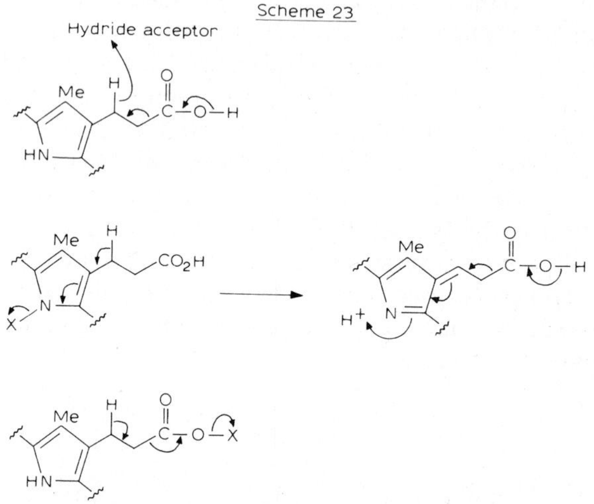

For the anaerobic enzyme at least three possible mechanisms can be envisaged (A, B, and C in Scheme 23), but there is no experimental evidence in this area. Clearly a two-step variation of A is also possible, involving assistance of hydride removal by the pyrrolic nitrogen atom, followed by fragmentation.

3.4.3. Aromatization of protoporphyrinogen-IX (80)

The autoxidation of porphyrinogens occurs readily in daylight to give the corresponding porphyrins, which are much more stable and easier to characterize. Consequently, most biosynthetic experiments have been terminated by a deliberate exposure to air and light, and in such cases one cannot tell whether any or all of the aromatization process had been enzyme-catalyzed. No enzyme catalyzing aromatization to the porphyrin level has ever been purified, but aromatization activity has been observed in crude coproporphyrinogenase[138,140].

Clear evidence for or against the involvement of an enzyme should be available from stereochemical studies at the meso-methylene positions. The photochemical autoxidation of coproporphyrinogen-III randomly tritiated at the four meso-positions takes place with only 4% tritium loss[149] as a result of the kinetic isotope effect and it is interesting that the actual value is so low. In contrast, an enzyme would be expected to oxidize each *meso*-position in a stereospecific manner resulting in exactly 50% loss of tritium[34]. This was found to be the case[150] for the protoporphyrin-IX formed when *meso*-tritiated ^{14}C-coproporphyrinogen-III (as (74)) was incubated with the enzyme from chicken erythrocytes, so supporting enzymic aromatization.

Scheme 24

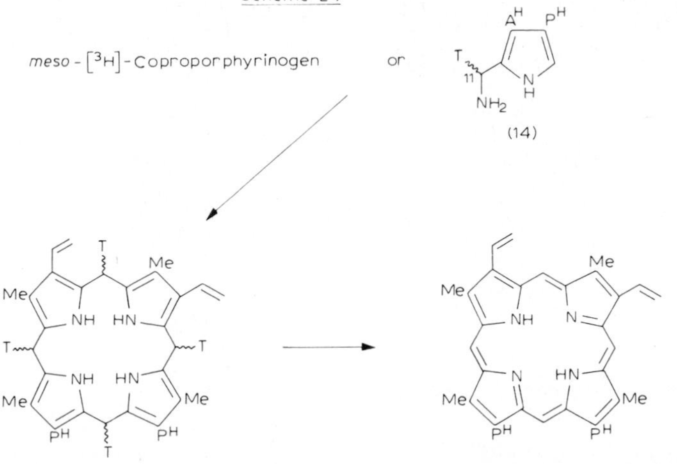

meso - [³H] - Coproporphyrinogen   or

(14)

(80)   (28)

A different approach[151] used (11-*RS*)-[11-^{14}C, 11-³H₁]PBG (Scheme 24) for incubation aerobically *in the dark* with a cell-free system from *Euglena gracilis*. Half of the reaction mixture was examined after 45 min, when approx. 40% of the resultant protoporphyrinogen-IX (80) had been converted into the corresponding porphyrin. After aerial oxidation, protoporphyrin-IX (28) was isolated and purified to constant activity by recrystallization of the methyl ester. The ³H-retention was 77%. The remaining half was treated in a similar way after the incubation had proceeded for 5 h. All the protoporphyrinogen-IX had been aromatized at this stage and the ³H-retention was 69%. The difference in tritium retention between the two experiments and from the high 'chemical' value shows that enzymic aromatization is taking place but that non-enzymic air oxidation is competing in this isolated system.

3.4.3.1. Intermediates

Protoporphyrinogen-IX (80) is a hexahydroporphyrin, and the aromatization can be considered to proceed via a tetrahydroporphyrin and a dihydroporphyrin[141]. It is therefore of interest that an absorption appears at 503 nm during the early stages of normal aerobic growth of *E. coli*[152] and of yeast[153]. This chromophore disappears after disruption of the cells, but the substance responsible for it has recently been isolated[153] from yeast grown anaerobically. In all its properties it resembles[152,154,155] tetrahydroprotoporphyrin-IX (90), and its aromatization is catalyzed by the broken cells of yeast or *E. coli* grown aerobically[153,155]. This catalysis is specific for tetrahydroprotoporphyrin-IX (90); under the same conditions tetrahydrocoproporphyrin-III (91) remains unchanged and this surely points to the existence of at least one enzyme involved in the aromatization process leading to protoporphyrin-IX (28).

Scheme 24 (a)

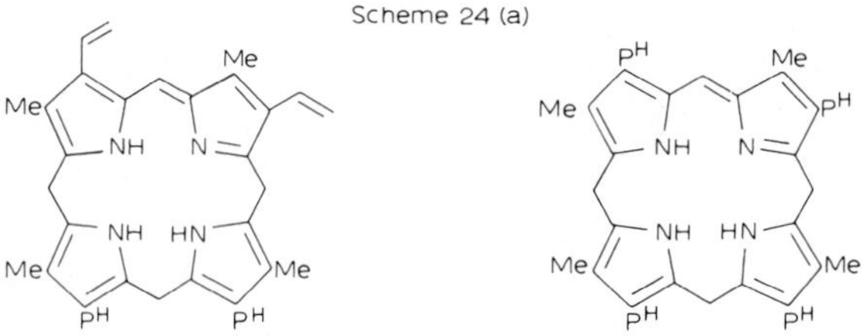

(90)   (91)

For these two structures the location of the sp2 *meso* carbon is unknown. The α-isomer has been arbitrarily illustrated.

3.5. The iron, magnesium and cobalt branches
3.5.1. The iron branch
3.5.1.1. Formation of protoheme (92)

There are numerous reports that the chelation of Fe^{2+} by protoporphyrin-IX can be catalyzed with crude enzyme preparations from bacteria[156], avian erythrocytes[157], and mammalian liver[158]. The Fe^{2+} insertion also takes place chemically under specially defined, mild conditions[158,159], and the enzymic and chemical methods both exhibit the following features:
   (i)   pH optimum near pH 8.0,
   (ii) glutathione inhibits the reaction when Fe^{2+} is supplied but enhances the rate at which Fe^{3+} is built in,
   (iii) the porphyrin must have two free carboxyl groups[160]: the corresponding esters are not converted into hemes.

An observation which strongly supports the existence of an enzyme, *ferrochelatase*, is that the boiled preparations lose the ability to catalyze heme formation. Recently, 80% of the ferrochelatase activity of broken *Spirillum itersonii* cells has been found to reside in the membrane fraction. A mutant which requires protoheme for normal growth yields a membrane fraction devoid of ferrochelatase activity[156].

3.5.1.2. The hemoproteins

Protoheme (92) itself is the prosthetic group of hemoglobin, myoglobin, and the cytochromes-*b*. Protoheme (92) is also the precursor[161] of the cytochromes-*c*, in which the protein is covalently bound to the heme by the addition of an —SH residue to each vinyl side-chain. The cytochromes-*a* have heme-*a* as their prosthetic group, which is a considerably modified form of protoheme carrying a long hydrocarbon side-chain (93); its structure has yet to be rigorously established. Heme-*a* (93) has been shown to be derived from protoheme (92) in *Staphylococcus*[162].

Scheme 24 (b)

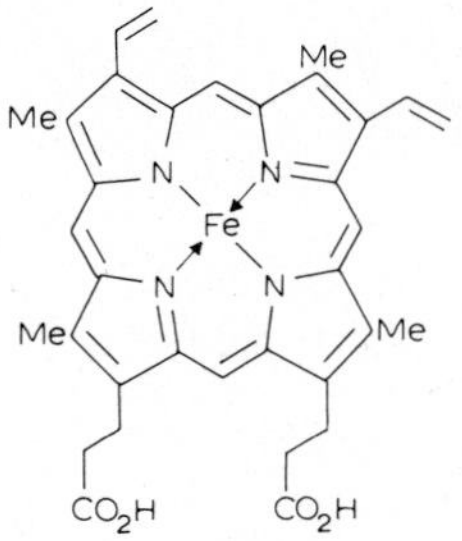

(92)

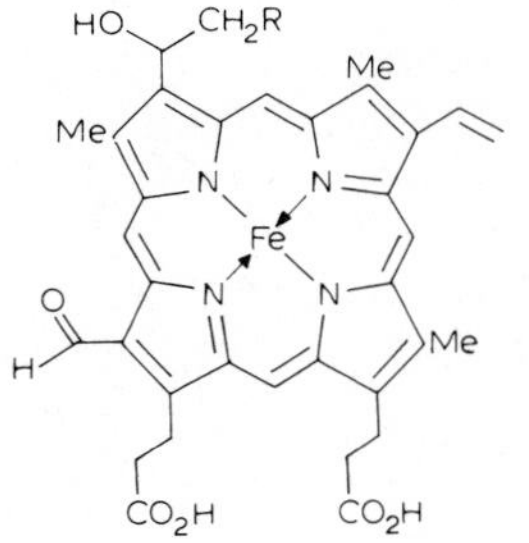

(93) R = hydrocarbon chain

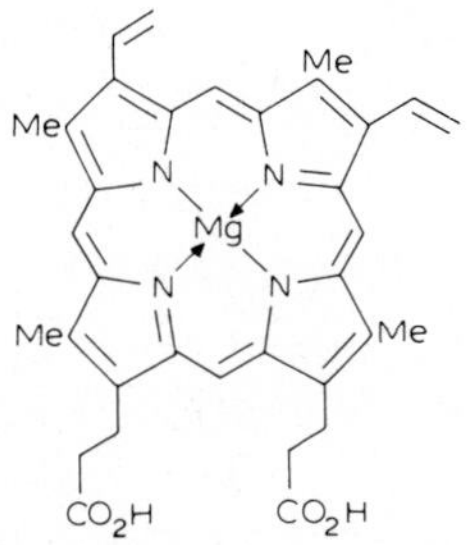

(94)

3.5.2. The magnesium branch

A full investigation of the late stages of biosynthesis of the chlorophylls has been hindered by the insolubility of the intermediates and the relevant enzymes. The pathway was first outlined by Granick[163] on the basis of the intermediates which accumulate in mutants of *Chlorella vulgaris* which are unable to make chlorophyll itself. The evidence obtained during the twenty intervening years has generally supported these early suggestions for chlorophyll biosynthesis.

Protoporphyrin-IX (28) is considered to be the last metal-free precursor of chlorophyll-*a* and bacteriochlorophyll for the following reasons:

(a) It accumulates[164] in the brown *Chlorella* mutant W_5.

(b) It is produced by cell-free preparations[165] of normal photosynthetic organisms, e.g. *Euglena gracilis*.

(c) The established precursors of protoporphyrin-IX (28), viz. $[2-^{14}C]$ glycine[166,167], $[2,3-^{14}C_2]$ succinate[167], and $[^{14}C]$ ALA[168,169] are incorporated by *Chlorella vulgaris*[166], by wheat[167,169], and by cucumber[168] into chlorophyll derivatives.

References, p. 116

(d) Labelled protoporphyrin-IX has been incorporated into chlorophyll-*a*
by preparations from bean leaves[170] and from tobacco leaves[171]; and
into bacteriochlorophyll by *R. spheroides*[172].

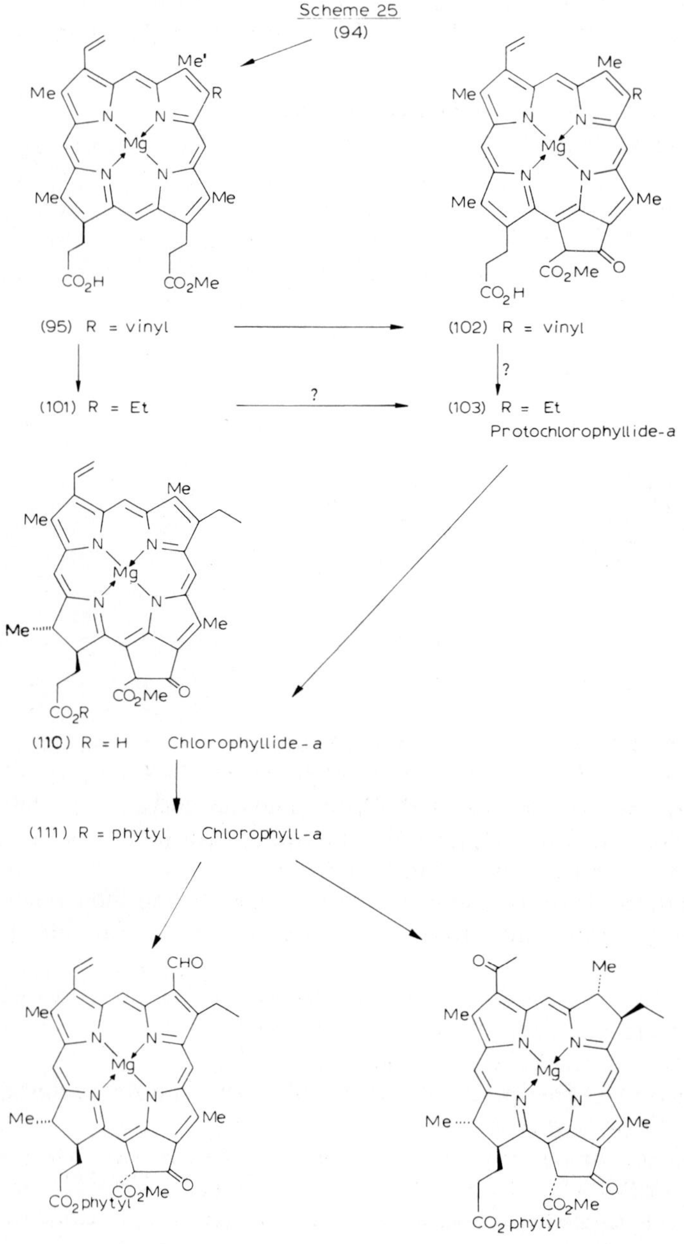

Scheme 25
(94)

(95) R = vinyl ⟶ (102) R = vinyl

(101) R = Et —?→ (103) R = Et
Protochlorophyllide-*a*

(110) R = H   Chlorophyllide-*a*

(111) R = phytyl   Chlorophyll-*a*

(113) Chlorophyll-*b*   (114) Bacteriochlorophyll-*a*

he likely seque..ce of .vent. beyond protoporphyrin-IX is summarized i .
Scheme 25, and the evidence for each step is given below.

3.5.2.1. Chelation of magnesium

Magnesium protoporphyrin-IX (94) has been isolated[173] from an orange
Chlorella mutant 60. The chelation step in *R. spheroides* is inhibited by
oxygen[174].

3.5.2.2. Esterification of magnesium protoporphyrin-IX

A different *Chlorella* mutant 60A produces the monomethyl ester of magnesium protoporphyrin-IX[175] (95) Scheme 25 and this porphyrin has also
been isolated from etiolated barley[175] and from *R. spheroides*[176]. The methyl donor is *S*-adenosyl-methionine (96), and when this is supplied in radioactive form ([14]C-*methyl*) to a culture of *R. spheroides*, radioactive bacterio-
chlorophyll-*a* (114) (Scheme 25) can be isolated, labeled exclusively in the
ester methyl group[177].

Scheme 25 (a)

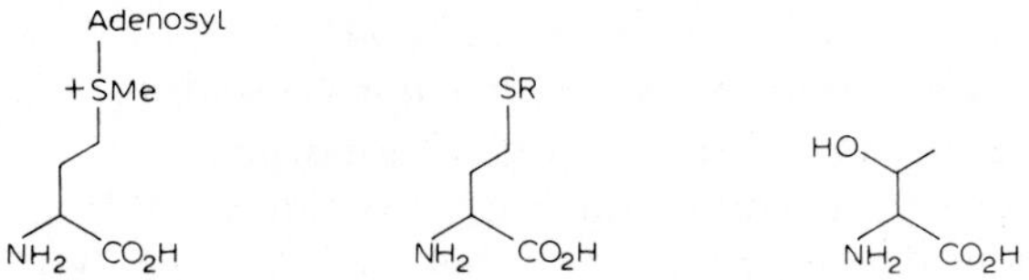

The enzyme catalyzing the esterification is located in the chromatophores
of *R. spheroides*[178] and in the chloroplasts of *Zea mays*[179]. In both systems magnesium protoporphyrin-IX ·is a much better substrate than the
metal free porphyrin[178,179] (or porphyrinogen), but the zinc and calcium
analogues, and the magnesium complexes of mesoporphyrin-IX (7) and deuteroporphyrin-IX (97) become esterified[180]. The wheat seedling enzyme[181]
has a pH optimum of 7.7 and K_m values of 36 and 48 μM for the porphyrin
and S-adenosylmethionine, respectively.

In *R. spheroides*, bacteriochlorophyll production is inhibited by ethionine
(98) and by threonine (99) resulting in the excretion of coproporphyrin-III
(26) into the medium[177,182]. Both inhibitory compounds are known to
impede the synthesis of S-adenosylmethionine (96) and the effect can be
reversed by supplying methionine (100).

It is interesting that coproporphyrin-III accumulates in these inhibition
experiments rather than protoporphyrin-IX; the reason for this is not clear.

3.5.2.3. Reduction of the vinyl group at position 4*

The point at which reduction of the vinyl group occurs in the multi-step

---

* Position 8 using the 1—24 numeration.

References, p. 116

sequence is not yet known with certainty and the evidence below is some-what conflicting.

A cell-free extract from wheat seedlings is reported to catalyze[183] the reduction of magnesium protoporphyrin-IX monomethyl ester (95) by NAD^3H to afford tritiated (101). The fact that the divinyl derivative with the isocyclic ring present (102) is a poor substrate for the enzyme is cited as evidence[183] that reduction of one vinyl group occurs before the formation of carbocyclic ring. On the other hand, the divinyl ester (102) has been isolated[184] from *R. spheroides* grown in the presence of 8-hydroxyquinoline, and material with closely similar spectral properties has been observed in a *R. spheroides* mutant[185], in marine flagellates[186], and in the seed coats of *Cucurbita pepo*[187].

3.5.2.4. Oxidative formation of the carbocyclic ring

The carbocyclic (isocyclic) ring could reasonably be formed by oxidative cyclization of one or other of the keto-esters (104) or (105) thereby gener-ating magnesium vinylpheoporphyrin-a_5 (103) (protochlorophyllide-*a*) or the divinyl compound (102). The acyclic β-keto-esters (104) and (105) are reported to be present in a *Chlorella* mutant, as are the corresponding alco-hols (106) and (107)[188]. The isolation of the divinyl pheoporphyrin (102) was referred to in the previous section, and protochlorophyllide-*a* (103) has been found in the yellow *Chlorella* mutant 31[189] and in etiolated seed-lings[168].

Scheme 25 (b)

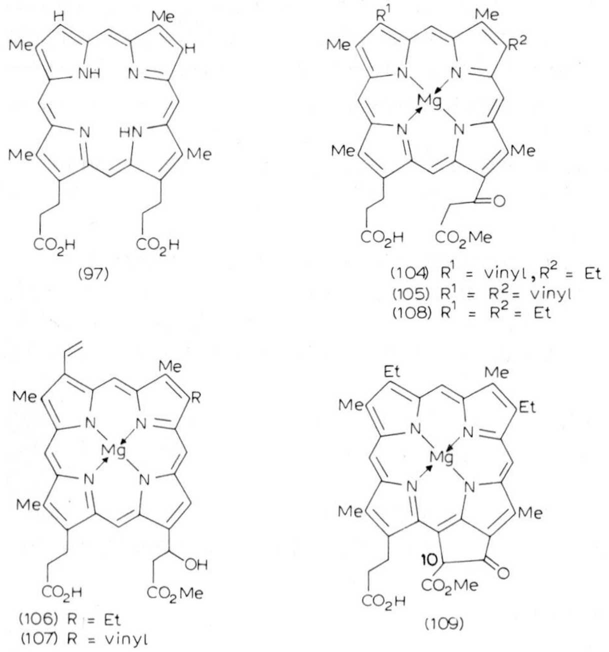

(97)

(104) R^1 = vinyl, R^2 = Et
(105) R^1 = R^2 = vinyl
(108) R^1 = R^2 = Et

(106) R = Et
(107) R = vinyl

(109)

The carbocyclic ring has been successfully closed[190] *in vitro* by iodine oxidation of (108), but the product (7% yield) had incorporated a methoxy group at C-10 from the solvent. More recently, the metal-free porphyrins corresponding to (104) and (108) have been oxidized by thallium(III) tris-(trifluoroacetate) to afford[191] the corresponding pheoporphyrins (103) and (109) in 37% and 69% yield. Biosynthetic studies with these keto-esters are in progress.

3.5.2.5. *Reduction of ring D*

In green plants, chlorins are generated only in light. Etiolated seedlings accumulate the porphyrin protochlorophyllide (103), and so the formal *trans*-hydrogenation of ring D could be a photochemical reduction[192] or one switched on by light. The reaction has been extensively studied[192] and it is found that the protochlophyllide is bound to protein forming a so-called holochrome.

It is significant that in *Chlorella*, chlorins are generated in the dark[163].

3.5.2.6. *Formation of Chlorophyll-a*

All that remains for the formation of chlorophyll-*a* (111) from chlorophyllide-*a* (110) is esterification of the propionate carboxyl group with the C_{20} alcohol phytol (112).

The enzyme chlorophyllase[193] catalyzes the reverse reaction (i.e. the hydrolysis of the phytyl ester in chlorophyll-*a*, chlorophyll-*b*, and bacteriochlorophyll) and it has been suggested that the same enzyme may catalyze the forward esterification if the reaction takes place in a lipid environment within the cell.

Scheme 25 (c)

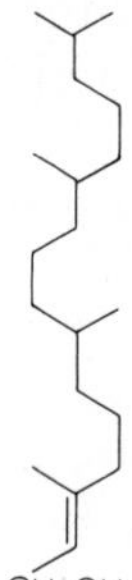

(112) 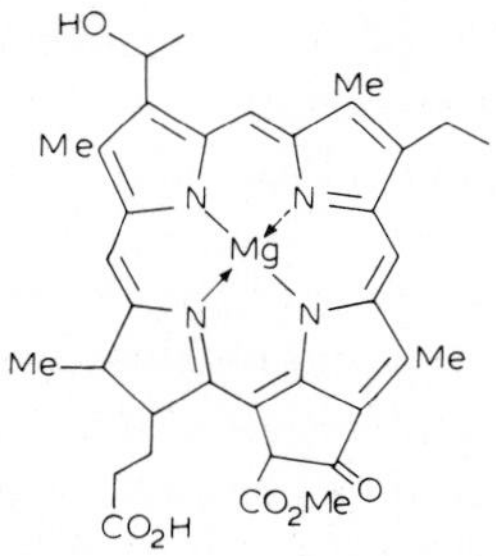   (115)

3.5.2.7. *Chlorophyll-b*

Chlorophyll-*b* (113) differs from chlorophyll-*a* (111) in having a formyl rather than a methyl substituent at C-3*. The biosynthetic relationship be-

---

* Position 7 using the 1—24 numeration.

tween these two chlorophylls has long been debated, but a recent review of the evidence led to the conclusion[194] that chlorophyll-*a* is probably the direct precursor of chlorophyll-*b*.

3.5.2.8. *Bacteriochlorophyll*

It is likely that the biosynthetic pathway to bacteriochlorophyll (114) is the same as that for the plant chlorophylls at least as far as chlorophyllide-*a* (110). Support for this suggestion comes from the isolation of the corresponding metal-free chlorin (as 110, no Mg) from a mutant of *R. spheroides* and from the wild strain grown in a medium containing 8-hydroxyquinoline[184]. (The lability of magnesium chlorins might account for the absence of metal).

If chlorophyllide-*a* (110) is indeed an intermediate en route to bacteriochlorophyll (114), then the subsequent structural changes are:

(a) Oxidation of the 2*-vinyl substituent to acetyl

(b) Overall *trans* hydrogenation of ring B

(c) Esterification with phytol

The first step may well be hydration of the vinyl group, since the 1-hydroxyethyl chlorin (115) has been isolated from *R. spheroides*[195,196]. The sequence of the other required changes is unknown.

3.5.3. *The cobalt branch*
3.5.3.1. *Studies using precursors labeled with carbon-14 and carbon-13*

Coenzyme B_{12} (116) is the biologically-active form of vitamin B_{12} (117) and this complex molecule presents a series of formidable challenges to chemists. Determination of the detailed structure[197] and the total synthesis[198] of the vitamin stand in the select group of great achievements in organic chemistry. Two further challenges viz. elucidation of the biosynthesis of vitamin B_{12} and determination of its role and mechanism of action in biology[199] are now under active investigation.

The origin of the complex side-chain attached to the propionate group at C-17 has been studied[200], but this review will be concerned only with the biosynthesis of the natural corrin nucleus (118). In common with the porphyrin nucleus this consists of four pyrrole residues built into a macrocyclic assembly though with the important difference of a direct C-1/C-19 link in the corrins. Furthermore, the acetate and propionate chains of vitamin B_{12} are clearly arranged in the type-III pattern familiar in the natural porphyrins, with the characteristic 'inversion' in ring D. The existence of a shared biosynthetic pathway to porphyrins and corrins was first demonstrated by the incorporation of [1,4-^{14}C] ALA[201,202] and [2,11-^{14}C] PBG[203] into vitamin B_{12} by the actinomycete ATCC 11072. Radioactivity from [^{14}C] me-

---

* 3-Vinyl using 1—24 numeration.

Scheme 26

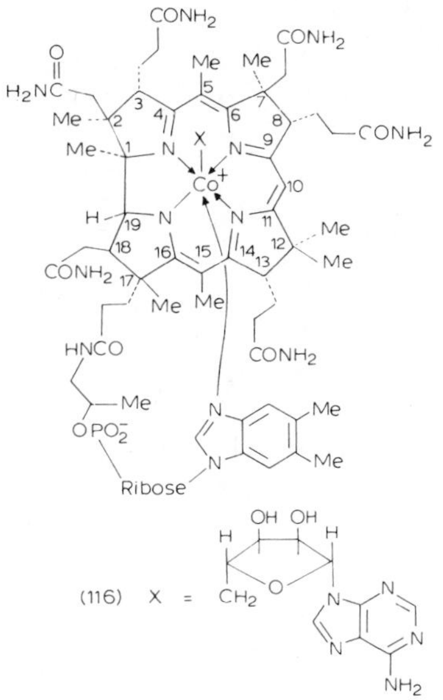

(116) X = CH₂

(117) X = CN

thionine was incorporated into the peripheral methyl groups and the results showed that at least six were methionine-derived[202,204]. The advent of ^{13}C-n.m.r. has now permitted a detailed study of the mode of incorporation of these primary precursors.

Three groups have studied the incorporation of [5-^{13}C] ALA (3) into vitamin B_{12} by *Propionibacterium shermanii*. Two of them recorded the ^{13}C-n.m.r. spectrum of the resultant vitamin B_{12} in D_2O and both concluded[205,206] that none of the enriched signals was due to a methyl group. The third team confirmed[207] that no methyl resonance showed enhancement and their spectra were measured in C_6D_6 after methanolysis[208] of the vitamin B_{12} to give the relatively simpler system for spectroscopy, heptamethyl cobyrinate (119). The stereochemistry is unaffected by this treatment. Clearly the angular methyl substituent at C-1 does *not* arise from rearrangement of the δ-*meso* carbon atom of a porphyrinogen. In all three studies, *seven* carbons were found to be ^{13}C-enriched after incorporation of [5-^{13}C] ALA so only *three* of the four C-11 methylene groups of the four original PBG units (14) appear in vitamin B_{12}; one is lost at some stage.

When [2-^{13}C] ALA was fed to the organism a single methyl resonance was enhanced[206] thus confirming the conclusion of earlier ^{14}C-studies[202] that one of the C-12 methyl substituents arises by decarboxylation of an acetate side-chain. It was found[206] and confirmed[207,210] that the remaining seven methyl resonances were all enhanced in corrins biosynthesized from

Scheme 27

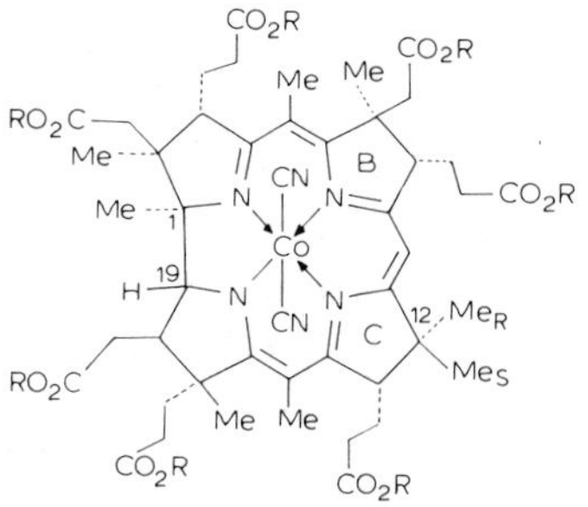

(118) R = H
(119) R = Me

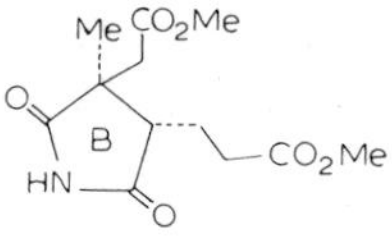

(121)

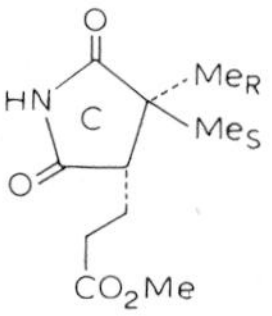

(120)

[S-$^{13}CH_3$]methionine, and so the methyl substituent at C-1 arises from S-adenosylmethionine.

At this stage it was possible to identify which of the C-12 methyl groups originates from ALA and which from methionine[207]. Ozonolysis[211] of the heptamethyl cobyrinate (119) from the [S-$^{13}CH_3$]methionine feeding yielded the optically active ring C imide (120). The high field methyl resonance in its ^{1}H-n.m.r. spectrum was of diminished intensity relative to unlabeled material and ^{13}C-satellite signals (J ^{13}C-^{1}H = 128 Hz) were clearly visible. This high-field signal has been unambiguously assigned[211] to the C-12 α-methyl group. This α-methyl group (12-pro-R) is thus the one which comes from methionine and, as for the other three rings, the elements of methane have formally been added in a $trans$-fashion. This result was confirmed[209] by assigning the C-12α and C-12β ^{13}C-resonances of the corrin on the basis of the γ-effect. In a third study[210], each of the ^{13}C-methyl resonances was carefully correlated with the corresponding signals in the ^{1}H-n.m.r. spectrum but the conclusion which was reached indicates that some of the ^{1}H assignments[212] must now be regarded with suspicion.

In order to check whether or not the methyl groups of methionine are transferred intact, $P.$ $shermanii$ was incubated with [S-C^{2}H$_3$]methionine[213]. The heptamethyl cobyrinate (119) derived from the isolated vitamin B$_{12}$ was ozonized to give ring B imide (121) and ring-C imide (120). In each case the mass spectrum showed that intact transfer of a CD_3 group had occurred.

3.5.3.2. Intermediates in corrin biosynthesis

The foregoing results establish that the corrin skeleton of vitamin B_{12}, like uroporphyrinogen-III, is built up from four molecules of PBG (14) and the possibility has long been attractive[214,215] that uroporphyrinogen-III (27) may itself be a direct corrin precursor. The recent experimental tests of this idea have led to conflicting results. Incubation of [^{14}C]-uroporphyrinogen-III with *P. shermanii* cells led in two studies[114,216] to no incorporation but in a third[217], using a large quantity, relative to *P. shermanii* cells, of mixed ^{14}C-labeled uroporphyrinogen isomers (either I and III or all four together), activity was incorporated into vitamin B_{12} (0.4—3.0% incorporation). A similar result was found when the four mixed ^{14}C-labeled uroporphyrinogen isomers (I-IV) were incubated with a cell-free system[218,*] and [*methyl*-^{3}H]methionine to give heptamethyl [^{3}H,^{14}C]cobyrinate after esterification of the products. There was no incorporation of activity from uroporphyrinogen-I into the vitamin[217] in the whole cell system but when this system was incubated with the uroporphyrinogen I-IV mixture labeled with ^{13}C at the same carbon of each PBG residue, the isolated vitamin B_{12} was labeled in the same way.

There are a number of possible explanations of the above differing results but it suffices to say here that all workers in this area recognise that the situation will remain uncertain until a sample of uroporphyrinogen-III labeled at one or two known carbons has been specifically incorporated *without randomization* into a corrin[†].

The conversion of uroporphyrinogen-III (27), or of a closely related precursor into cobyrinic acid (118) requires, in a sequence yet to be determined, the following changes:

(i)   introduction of seven C-methyl groups,
(ii) decarboxylation of the acetic acid side-chain on ring-C,
(iii) extrusion of the δ-*meso* carbon atom and new ring closure between C-1 and C-19,
(iv) introduction of cobalt,
(v)   possible redox changes which quantitatively are dependent on the oxidation level of the introduced cobalt, and of the extruded carbon atom.

Cobalt-free corrins have been isolated[219] from *Chromatium*, and an en-

---

* A.R. Battersby, E. McDonald, M. Ihara, and B. Middleton described at the European Symposium on Bio-organic Chemistry, May, 1973, a similar cell free preparation from *P. shermanii*, differing mainly in the range of additives used, which biosynthesizes cobyrinic acid (118) from PBG.
† The intact incorporation of specifically labeled uroporphyrinogen-III into vitamin B_{12} has recently been reported[224b,c].

References, p. 116

zyme preparation from the corrin producing bacterium *C. tetanomorphum* catalyzes the chelation of cobalt by porphyrins[220]. The relevance of these observations is still uncertain and there is as yet no experimental evidence concerning the *sequence* of the many steps which have to occur after the four PBG units have been joined one to another. So without knowing rigorously yet whether uroporphyrinogen-III (27) is transformed into the corrin nucleus, it seems premature to give here detailed speculative schemes. Everyone in the field has several working hypotheses (and some favourites) on which current research is based and some features of those which have been published[221-223] resemble the writers' schemes. The future promises exciting research on the biosynthesis of corrins for many years.

Epilog

The aim of this survey has been not only to give a bird's eye view of the field as it stands late in 1974 but also, to bring out the current intense interest in biosynthetic research on porphyrins, chlorins and corrins and to show how many marvellous and highly important problems still remain to be solved. The co-operative interaction of chemists, biochemists, and X-ray crystallographers could fill in many of the gaps in knowledge over the next decade.

References

1. More detailed accounts of the historical developments are available in the following reviews:
   (a) L. Bogorad, in "Comparative Biochemistry of Photoreactive Systems', Ed. M.B. Allen, Academic Press, New York (1960), pp. 226—56.
   (b) S. Granick and D. Mauzerall, in 'Metabolic Pathways', Ed. D.M. Greenberg, Academic Press, New York (1961), pp. 525—616.
   (c) L. Bogorad, in 'The Chlorophylls', Eds. L.P. Vernon and G.R. Seely, Academic Press, New York (1966), p. 481.
   (d) B.F. Burnham, in "Metabolic Pathways", Ed. D.M. Greenberg, Academic Press, New York (1969), Vol. III, pp. 403—537.
   (e) G.S. Marks, "Heme and Chlorophyll", van Nostrand, London (1969), pp. 121—162.
2. E.g. E. Rabinowitch and Govindjee, "Photosynthesis", Wiley, New York (1969).
3. M.F. Perutz, Nature London, **228**, 726 (1970); ibid., **237**, 495 (1972).
4. R. Lemberg and J. Barrett, "Cytochromes", Academic Press, London (1973).
5 D. Shemin and D. Rittenberg, J. Biol. Chem., **159**, 567 (1945); ibid., **166**, 627 (1946).
6. J. Wittenberg and D. Shemin, J. Biol. Chem., **178**, 47 (1949).
7. (a) I.M. London, D. Shemin, and D. Rittenberg, J. Biol. Chem., **173**, 799 (1948);
   (b) ibid., p. 797;
   (c) ibid., **183**, 749 (1950);
   (d) ibid., p. 757.
8. J. Wittenberg and D. Shemin, J. Biol. Chem., **185**, 103 (1950).

9. (a) H.M. Muir and A. Neuberger, Biochem. J., **45**, 163 (1949);
   (b) ibid., **47**, 97 (1950).
10. (a) N.S. Radin, D. Rittenberg, and D. Shemin, J. Biol. Chem., **184**, 745 (1950);
    (b) ibid., p. 755.
11. D. Shemin and J. Wittenberg, J. Biol. Chem., **192**, 315 (1952).
12. D. Shemin and S. Kumin, J. Biol. Chem., **198**, 827 (1952).
13. J.M. Lowenstein, in "Metabolic Pathways", Ed. D.M. Greenberg, Academic Press, New York (1967), Vol. I, 3rd ed., p. 146.
14. J.C. Wriston, Jr., L. Lack, and D. Shemin, J. Biol. Chem., **215**, 603 (1955).
15. D. Shemin and C.S. Russell, J. Am. Chem. Soc., **75**, 4873 (1953).
16. (a) D. Shemin, T. Abramsky and C.S. Russell, J. Am. Chem. Soc., **76**, 1204 (1954);
    (b) D. Shemin, C.S. Russell, and T. Abramsky, J. Biol. Chem., **215**, 613 (1955).
17. A. Neuberger and J.J. Scott, Nature London, **172**, 1093 (1953).
18. R.G. Westall, Nature London, **170**, 614 (1952).
19. G.H. Cookson and C. Rimington, Nature London, **171**, 875 (1953); Biochem. J., **57**, 476 (1954).
20. E.I.B. Dresel and J.E. Falk, Nature London, **172**, 1185 (1953).
21. R. Schmid and D. Shemin, J. Am. Chem. Soc., **77**, 506 (1955).
22. P.M. Jordan and D. Shemin, in "The Enzymes", Ed. P.D. Boyer, Vol. VII, Academic Press, New York (1972), p. 339.
23. D. Shemin, ibid p. 323.
24. S. Tuboi, H.J. Kim, and G. Kikuchi, Arch. Biochem. Biophys., **138**, 147 (1970).
25. G.W. Warnick and B.F. Burnham, J. Biol. Chem., **246**, 6880 (1971).
26. T. Yubisui and Y. Yoneyama, Arch. Biochem. Biophys., **150**, 77 (1972).
27. M. Fanica-Gaignier and J. Clement-Metral, Eur. J. Biochem., **40**, 13, 19 (1973).
28. B.H. Kaplan, Biochim. Biophys. Acta, **235**, 381 (1971).
29. P.L. Scholnick, L.E. Hammaker, and H.S. Marver, J. Biol. Chem., **247**, 4126 (1972).
30. Y. Aoki, O. Wada, G. Urata, F. Takaku, and K. Naka, Biochem. Biophys. Res. Commun., **42**, 568 (1971).
31. E.A.W. De Xifra, A.M. Del C. Batlle and H.A. Tigier, Biochim. Biophys. Acta., **235**, 511 (1971).
32. R.J. Porra, R. Barnes, and O.T.G. Jones, Hoppe-Seyler's Z. Physiol. Chem., **353**, 1365 (1972).
33. S.I. Beale and P.A. Castelfranco, Biochem. Biophys. Res. Commun., **52**, 143 (1973).
34. A.R. Battersby, Acc. Chem. Res., **5**, 148 (1972).
35. Z. Zaman, P.M. Jordan, and M. Akhtar, Biochem. J., **135**, 257 (1973).
36. W.G. Laver, A. Neuberger, and J.J. Scott, J. Chem. Soc., 1483 (1959).
37. M.M. Abboud, P.M. Jordan, and M. Akhtar, J. Chem. Soc., Chem. Commun., 643 (1974).
38. R.J. Porra, E.A. Irving, and A.M. Tennick, Arch. Biochem. Biophys., **148**, 37 (1972).
39. P.E. Gurba, R.E. Sennett, and R.D. Kobes, Arch. Biochem. Biophys., **150**, 130 (1972).
40. A.M. Del C. Batlle, A.M. Ferramola, and M. Grinstein, Biochem. J., **104**, 244 (1967).
41. D. Doyle, J. Biol. Chem., **246**, 4965 (1971).
42. E.L. Wilson, P.E. Burger, and E.B. Dowdle, Eur. J. Biochem., **29**, 563 (1972).
43. D.L. Nandi and D. Shemin, Arch. Biochem. Biophys., **158**, 305 (1973).
44. Y.K. Ho and J. Lascelles, Arch. Biochem. Biophys., **144**, 734 (1971).
45. D. Gurne and D. Shemin, Science, **180**, 1188 (1973).
46. R.B. Frydman, M.L. Tomaro, A. Wanschelbaum, and B. Frydman, FEBS Lett., **26**, 203 (1972).
47. R.B. Frydman, M.L. Tomaro, A. Wanschelbaum, E.M. Anderson, J. Awruch, and B. Frydman, Biochemistry, **12**, 5253 (1973).

48. M.L. Tomaro, R.B. Frydman, and B. Frydman, Biochemistry, **12**, 5263 (1973).
49. Footnote 7 to ref. 16(b).
50. E.I.B. Dresel and J.E. Falk, Biochem. J., **63**, 388 (1956).
51. R.A. Neve, R.F. Labbe, and R.A. Aldrich, J. Am. Chem. Soc., **78**, 693 (1956).
52. K. Salomon, J.E. Richmond, and K.I. Altman, J. Biol. Chem., **196**, 463 (1952).
53. J.E. Falk, E.I.B. Dresel, and C. Rimington, Nature London, **172**, 292 (1953).
54. cf. L. Bogorad, J. Biol. Chem., **233**, 516 (1958).
55. A. Comfort, Science, **112**, 279 (1950).
56. G.S. Marks, "Heme and Chlorophyll", van Nostrand, Princeton (1969), p. 165.
57. D. Mauzerall, J. Am. Chem. Soc., **82**, 2601 (1960).
58. D. Mauzerall, J. Am. Chem. Soc., **82**, 2605 (1960).
59. L. Bogorad and S. Granick, Proc. Nat. Acad. Sci. U.S.A., **39**, 1176 (1953).
60. L. Bogorad, Science, **121**, 878 (1955); J. Biol. Chem., **233**, 501, 510 (1958).
61. H. Heath and D.S. Hoare, Biochem. J., **72**, 14 (1959); D.S. Hoare and H. Heath, Biochem. J., **73**, 679 (1959).
62. E.Y. Levin and D.L. Coleman, J. Biol. Chem., **242**, 4248 (1967).
63. H.A. Sancovich, A.M.C. Batlle, and M. Grinstein, Biochim. Biophys. Acta., **191**, 130 (1969); E.B.C. Llambias and A.M.C. Batlle, Biochim. Biophys. Acta., **220**, 552 (1970); Biochim. Biophys. Acta, **227**, 180 (1971); Biochem. J., **121**, 327 (1971).
64. R.B. Frydman and B. Frydman, Arch. Biochem. Biophys., **136**, 193 (1970).
65. R.C. Davies and A. Neuberger, Biochem. J., **133**, 471 (1973).
66. P.M. Jordan and D. Shemin, J. Biol. Chem., **248**, 1019 (1973).
67. Dr. B. Middleton, unpublished work, 1972—74, University of Cambridge.
68. Dr. D.C. Williams, unpublished work, 1973—74, University of Cambridge.
69. M. Higuchi and L. Bogorad, Ann. N.Y. Acad. Sci., in press.
70. L. Bogorad, Methods Enzymol., **5**, 885 (1962).
71. R.B. Frydman and G. Feinstein, Biochim. Biophys. Acta., **350**, 358 (1974).
72. R.B. Frydman and B. Frydman, Biochim. Biophys. Acta., **293**, 506 (1973).
73. E.Y. Levin, Biochemistry, **7**, 3781 (1968); **10**, 4669 (1971).
74. D. Shemin, T. Abramsky, and C.S. Russell, J. Am. Chem. Soc., **76**, 1204 (1954) and refs. therein; D. Shemin, C.S. Russell, and T. Abramsky, J. Biol. Chem., **215**, 613 (1955); R. Schmid and D. Shemin, J. Am. Chem. Soc., **77**, 506 (1955).
75. S. Granick and D. Mauzerall, J. Biol. Chem., **232**, 1119 (1958).
76. W.H. Lockwood and A. Benson, Biochem. J., **75**, 372 (1960).
77. K.D. Gibson, A. Neuberger, and J.J. Scott, Biochem. J., **58**, XLI (1954).
78. P. Cornford, Biochem. J., **91**, 64 (1964).
79. E.F. Carell and J.S. Kahn, Arch. Biochem. Biophys., **108**, 1 (1964).
80. A.R. Battersby, J. Baldas, J. Collins, D.H. Grayson, K.J. James, and E. McDonald, J. Chem. Soc., Chem. Commun., 1265 (1972).
81. E. Stevens and B. Frydman, Biochim. Biophys. Acta., **151**, 429 (1967).
82. E. Stevens, B. Frydman, and R.B. Frydman, Biochim. Biophys. Acta, **158**, 496 (1968).
83. L. Bogorad, Ann. N.Y. Acad. Sci., **104**, 676 (1963).
84. J. Plusec and L. Bogorad, Biochemistry, **9**, 4736 (1970).
85. R. Radmer and L. Bogorad, Biochemistry, **11**, 904 (1972).
86. B. Frydman, S. Reil, A. Valasinas, R.B. Frydman, and H. Rapoport, J. Am. Chem. Soc., **93**, 2738 (1971).
87. R.B. Frydman, A. Valsinas, and B. Frydman, Biochemistry, **12**, 80 (1973).
88. A.R. Battersby, 23rd International Congress of Pure and Applied Chemistry, Special Lectures, **5**, 1 (1971).
89. J.M. Osgerby, J. Pluscec, Y.C. Kim, F. Boyer, N. Stojanac, H.D. Mah, and S.F. MacDonald, Can. J. Chem., **50**, 2652 (1972).

90. A.R. Battersby, D.A. Evans, K.H. Gibson, E. McDonald, and L. Nixon, J. Chem. Soc., Perkin Trans. 1, 1546 (1973).

91. A.R. Battersby, J.F. Beck, and E. McDonald, J. Chem. Soc., Perkin Trans. 1, 160 (1974).

92. Details are not available at the moment of the synthesis used by Frydman et al. for PA · AP (53); see ref. 93.

93. R.B. Frydman, A. Valasinas, H. Rapoport, and B. Frydman, FEBS Lett., 25, 309 (1972).

94. Y.C. Kim, Can. J. Chem., 47, 3259 (1969).

95. R.B. Frydman, A. Valasinas, S. Levy, and B. Frydman, FEBS Lett., 38, 134 (1974).

96. G.H. Cookson and C. Rimington, Biochem. J., 57, 467 (1954).

97. R.G. Westall, Nature London, 170, 614 (1952).

98. R.B. Frydman, S. Reil, and B. Frydman, Biochemistry, 10, 1154 (1971).

99. E.I.B. Dresel and J.E. Falk, Biochem. J., 63, 80, 388 (1956) and refs. therein.

100. L. Bogorad and G.S. Marks, J. Biol. Chem., 235, 2127 (1960).

101. L. Bogorad and G.S. Marks, Biochim. Biophys. Acta, 41, 356 (1960).

102. A.T. Carpenter and J.J. Scott, Biochem. J., 71, 325 (1959).

103. A.T. Carpenter and J.J. Scott, Biochim. Biophys. Acta, 52, 195 (1961).

104. A.R. Battersby, E. McDonald, and R.E. Markwell, unpublished work.

105. D.S. Hoare and H. Heath, Biochim. Biophys. Acta., 39, 167 (1960).

106. For leading refs. see E. Margoliash, Annu. Rev. Biochem., 30, 551 (1961), P. Maitland, Q. Rev., 4, 45 (1950) and refs. 85, 107—109.

107. E. Bullock, Nature London, 205, 70 (1965).

108. J.H. Mathewson and A.H. Corwin, J. Am. Chem. Soc., 83, 135 (1961).

109. R. Robinson, "The Structural Relations of Natural Products", Clarendon Press, Oxford (1955).

110. A.R. Battersby, J. Moron, E. McDonald, and J. Feeney, J. Chem. Soc., Chem. Commun., 920 (1972).

111. A.R. Battersby, E. Hunt, E. McDonald, and J. Moron, J. Chem. Soc., Perkin Trans. 1, 2917 (1973).

112. A.R. Battersby, G.L. Hodgson, M. Ihara, E. McDonald, and J. Saunders, J. Chem. Soc., Chem. Commun., 441 (1973); J. Chem. Soc., Perkin Trans. 1, 2923 (1973).

113. See D. Doddrell and W.S. Caughey, J. Am. Chem. Soc., 94, 2510 (1972) for ^{13}C-spectrum of this substance at natural abundance.

114. B. Franck, D. Gantz, F.-P. Montforts, and F. Schmidtchen, Angew. Chem., Int. Edn. Engl., 11, 421 (1972).

115. A.R. Battersby, J. Staunton, and R.H. Wightman, J. Chem. Soc., Chem. Commun., 1118 (1972).

116. A.R. Battersby, E. Hunt, and E. McDonald, J. Chem. Soc., Chem. Commun., 442 (1973).

117. A.R. Battersby, E. McDonald, and J. Saunders, in preparation.

118. A.R. Battersby, G.L. Hodgson, E. Hunt, E. McDonald, and J. Saunders, in preparation.

119. A.R. Battersby, K.H. Gibson, E. McDonald, L.N. Mander, J. Moron, and L.N. Nixon, J. Chem. Soc., Chem. Commun., 768 (1973).

120. A.R. Battersby, J.F. Beck, G.L. Hodgson, E. McDonald, R.E. Markwell, and J. Saunders, in preparation.

121. E.B.C. Llambias and A.M.C. Batlle, FEBS Lett., 6, 285 (1970).

122. A.M. Stella, V.E. Parera, E.B.C. Llambias, and A.M.C. Batlle, Biochim. Biophys. Acta., 252, 481 (1971).

123. D. Mauzerall and S. Granick, J. Biol. Chem., 232, 1141 (1958).

124. (a) J.M. Tomio, R.C. Garcia, L.C. San Martin de Viale, and M. Grinstein, Biochim. Biophys. Acta., 198, 353 (1970);
(b) ibid., 309, 203 (1973).

125. G. Romeo and E.Y. Levin, Biochim. Biophys. Acta., **230**, 330 (1971).

126. J.E. Falk, E.I.B. Dresel, A. Benson, and B.C. Knight, Biochem. J., **63**, 87 (1956).

127. A.M. Del C. Batlle and M. Grinstein, Biochim. Biophys. Acta., **57**, 191 (1962); ibid., **82**, 1 (1964).

128. A.R. Battersby, E. Hunt, M. Ihara, E. McDonald, J.B. Paine III, F. Satoh, and J. Saunders, J. Chem. Soc., Chem. Commun., 994 (1974).

129. J. Canivet and C. Rimington, Biochem. J., **57**, 476 (1954).

130. A.M. Del C. Batlle and M. Grinstein, Biochim. Biophys. Acta., **62**, 197 (1962); ibid., **82**, 13 (1964).

131. L.C. San Martin de Viale and M. Grinstein, Biochim. Biophys. Acta, **158**, 79 (1968).

132. M.S. Stoll, G.H. Elder, D.E. James, P. O'Hanlon, D.S. Millington, and A.H. Jackson, Biochem. J., **131**, 429 (1973).

133. A.F.M. Enteshamuddin, Biochem. J., **107**, 446 (1968).

134. (a) G.H. Tait, Biochem. Biophys. Res. Commun., **37**, 116 (1969).
      (b) G.H. Tait, Biochem. J., **128**, 1159 (1972).

135. (a) M. Mori and S. Sano, Biochem. Biophys. Res. Commun., **32**, 610 (1968).
      (b) S. Sano and M. Mori, Biochim. Biophys. Acta, **264**, 252 (1972).

136. (a) N.J. Jacobs, J.M. Jacobs, and P. Brent, J. Bacteriol., **102**, 398 (1970);
      (b) ibid., **107**, 203 (1971).

137. W.P. Hsu and G.W. Miller, Biochem. J., **117**, 215 (1970).

138. S. Sano and S. Granick, J. Biol. Chem., **236**, 1173 (1961).

139. A.M. Del C. Batlle, A. Benson, and C. Rimington, Biochem. J., **97**, 731 (1965).

140. R.J. Porra and J.E. Falk, Biochem. J., **90**, 69 (1964).

141. D. Mauzerall, J. Am. Chem. Soc., **84**, 2437 (1962).

142. A.R. Battersby, J. Baldas, J. Collins, D.H. Grayson, K.J. James, and E. McDonald, J. Chem. Soc., Chem. Commun., 1265 (1972).

143. J. French, D.C. Nicholson, and C. Rimington, Biochem. J., **120**, 393 (1970).

144. Z. Zaman, M.M. Abboud, and M. Akhtar, J. Chem. Soc., Chem. Commun., 1263 (1972).

145. G.Y. Kennedy, A.H. Jackson, G.W. Kenner, and C.J. Suckling, FEBS Lett., **6**, 9 (1970).

146. S. Sano, J. Biol. Chem., **241**, 5276 (1966).

147. A.R. Battersby, J. Staunton, and R.H. Wightman, in preparation.

148. G.H. Elder and J.R. Chapman, Biochim. Biophys. Acta, **208**, 535 (1970).

149. J.A.S. Cavaleiro, G.W. Kenner, and K.M. Smith, J. Chem. Soc., Chem. Commun., 183 (1973), J. Chem. Soc. Perkin Trans. I, 1188 (1974).

150. A.H. Jackson, D.E. Games, P. Couch, J.R. Jackson, R.B. Belcher, and S.G. Smith, Enzyme, **17**, 81 (1974).

151. A.R. Battersby, E. McDonald, and J.R. Stephenson, in preparation.

152. K. Olden and W.P. Hempfling, J. Bacteriol., **113**, 914 (1973).

153. R. Poulson, J. Boon, and W.J. Polglase, Can. J. Chem., **52**, 21 (1974).

154. P. Labbe, C. Volland, and P. Chaix, Biochim. Biophys. Acta, **143**, 70 (1967).

155. R. Poulson and W.J. Polglase, Biochim. Biophys. Acta, **329**, 256 (1973).

156. H.A. Dailey Jr. and J. Lascelles, Arch. Biochem. Biophys., **160**, 523 (1974) and refs. therein.

157. Y. Yoneyama, H. Ohyama, Y. Sugita, and H. Yoshikawa, Biochim. Biophys. Acta, **62**, 261 (1962).

158. R. Tokunaga and S. Sano, Biochim. Biophys. Acta, **264**, 263 (1972) and refs. therein.

159. R.J. Kassner and H. Walchak, Biochim. Biophys. Acta, **304**, 294 (1973).

160. R.J. Porra and O.T.G. Jones, Biochem. J., **87**, 181, 186 (1963).

161. E.M. Colleran and O.T.G. Jones, Biochem. J., **134**, 89 (1973).

162. P. Sinclair, D.C. White, and J. Barrett, Biochim. Biophys. Acta., **143**, 427 (1967).

163. S. Granick, Annu. Rev. Plant Physiol. 2, 115 (1951).
164. S. Granick, J. Biol. Chem., 172, 717 (1948).
165. E.F. Carell and J.S. Kahn, Arch. Biochem. Biophys., 108, 1 (1964).
166. R.J. Della Rosa, K.I. Altman, and K. Salomon, J. Biol. Chem., 202, 771 (1953).
167. D.W.A. Roberts and H.J. Perkins, Biochim. Biophys. Acta, 58, 486, 499 (1962).
168. C.A. Rebeiz, M. Yaghi, M. Abou-Haidier, and P.A. Castelfranco, Plant Physiol., 46, 57, 543 (1970).
169. R.K. Ellsworth and C.A. Nowak, Anal. Biochem., 51, 656 (1973).
170. G.W. Kenner, R.P. Carr, and M.T. Cox, footnote (1) in ref. 190(a).
171. J. Shien, G.W. Miller, and M. Psenak, Biochem. Physiol. Pflanz., 165, 100 (1974).
172. A. Gorchein, Biochem. J., 127, 97 (1972).
173. S. Granick, J. Biol. Chem., 175, 333 (1948).
174. A. Gorchein, Biochem. J., 134, 833 (1973).
175. S. Granick, J. Biol. Chem., 236, 1168 (1961).
176. O.T.G. Jones, Biochem. J., 86, 429 (1963).
177. K.D. Gibson, A. Neuberger, and G.H. Tait, Biochem. J., 83, 550 (1962).
178. G.H. Tait and K.D. Gibson, Biochim. Biophys. Acta, 52, 614 (1961).
179. R.J. Radmer and L. Bogorad, Plant Physiol., 42, 463 (1967).
180. K.D. Gibson, A. Neuberger, and G.H. Tait, Biochem. J., 88, 325 (1963).
181. R.K. Ellsworth and J.P. Dullaghan, Biochim. Biophys. Acta, 268, 327 (1972).
182. J. Lascelles, Biochem. J., 100, 184 (1966).
183. R.K. Ellsworth and A.S. Hsing, Biochim. Biophys. Acta, 313, 119 (1973).
184. O.T.G. Jones, Biochem. J., 88, 335 (1963); ibid., 89, 182.
185. J. Lascelles, Biochem. J., 100, 175 (1966).
186. T.R. Ricketts, Phytochemistry, 5, 223 (1966).
187. O.T.G. Jones, Biochem. J., 101, 153 (1966).
188. R.K. Ellsworth and S. Aronoff, Arch. Biochem. Biophys., 130, 374 (1969).
189. S. Granick, J. Biol. Chem., 183, 713 (1950).
190. (a) M.T. Cox, T.T. Howarth, A.H. Jackson, and G.W. Kenner, J. Am. Chem. Soc., 91, 1232 (1969); ibid., J. Chem. Soc., Perkin Trans. 1, 512 (1974).
191. (a) G.W. Kenner, S.W. McCombie, and K.M. Smith, J. Chem. Soc., Chem. Commun., 844 (1972);
     (b) ibid., J. Chem. Soc., Perkin Trans. 1, 527 (1974).
192. J.H.C. Smith, in "Comparative Biochemistry of Photoreactive Systems", Ed. M.B. Allen, Academic Press, New York (1960), p. 257.
193. M. Holden, Biochem. J., 78, 359 (1961).
194. A.A. Schlyk, Ann. Rev. Plant Physiol., 22, 169 (1971).
195. O.T.G. Jones, Biochem. J., 91, 572 (1964).
196. J. Lascelles, Biochem. J., 100, 175 (1966).
197. (a) D.C. Hodgkin, J. Pickworth, J.H. Robertson, K.N. Trueblood, R.J. Prosen, and J.G. White, Nature London, 176, 325 (1955),
     (b) R. Bonnett, J.R. Cannon, A.W. Johnson, I. Sutherland, A.R. Todd, and E.L. Smith, ibid., 176, 328 (1955).
198. (a) R.B. Woodward, Pure Appl. Chem., 33, 145 (1973).
     (b) A. Eschenmoser, XXIIIrd IUPAC Congress, Vol. 2, 69 (1971).
199. T.C. Stadtman, Science, 171, 859 (1971).
200. P. Renz and R. Weyhenmeyer, FEBS Lett., 22, 124 (1972); S.H. Lu and W.L. Alworth, Biochemistry, 11, 608 (1972) and refs. therein.
201. (a) D. Shemin, J.W. Corcoran, C. Rosenblum, and I.W. Miller, Science, 124, 272 (1956);
     (b) D. Shemin and J.W. Corcoran, Biochim. Biophys. Acta, 25, 661 (1957).
202. R.C. Bray and D. Shemin, J. Biol. Chem., 238, 1501 (1963).

203. S. Schwartz, K. Ikeda, I.M. Miller, and C.J. Watson, Science, **129**, 40 (1959).
204. R. Bray and D. Shemin, Biochim. Biophys. Acta, **30**, 647 (1958).
205. C.E. Brown, J.J. Katz, and D. Shemin, Proc. Nat. Acad. Sci. U.S.A., **69**, 2585 (1972).
206. A.I. Scott, C.A. Townsend, K. Okada, M. Kajiwara, P.J. Whitman, and R.J. Cushley, J. Am. Chem. Soc., **94**, 8267 (1972).
207. A.R. Battersby, M. Ihara, E. McDonald, J.R. Stephenson, and B.T. Golding, J. Chem. Soc., Chem. Commun., 404 (1973).
208. A. Eschenmoser and L. Werthemann, Diss. No. 4097, Eidgenössische Technische Hochschule Zürich.
209. A.I. Scott, C.A. Townsend, and R.J. Cushley, J. Am. Chem. Soc., **95**, 5759 (1973).
210. C.E. Brown, D. Shemin, and J.J. Katz, J. Biol. Chem., **248**, 8015 (1973).
211. A. Eschenmoser and P. Dubs, Diss. No. 4297, Eidgenössische Technische Hochschule Zürich.
212. J.D. Brodie and M. Poe, Biochemistry, **10**, 914 (1971); ibid., **11**, 2534 (1972).
213. A.R. Battersby, M. Ihara, E. McDonald, J.R. Stephenson, and B.T. Golding, J. Chem. Soc., Chem. Commun., 458 (1974).
214. R.J. Porra, Biochim. Biophys. Acta, **107**, 176 (1965).
215. B.F. Burnham and R.A. Plane, Biochem. J., **98**, 13C (1966).
216. G. Müller and W. Dieterle, Hoppe-Seyler's Z. Physiol. Chem., **352**, 143 (1971).
217. A.I. Scott, C.A. Townsend, K. Okada, M. Kajiwara, and R.J. Cushley, J. Am. Chem. Soc., **94**, 8269 (1972).
218. A.I. Scott, B. Yagen, and E. Lee, J. Am. Chem. Soc., **95**, 5761 (1973).
219. J.I. Toohey, Proc. Nat. Acad. Sci. U.S.A., **54**, 934 (1965).
220. R.J. Porra and B.D. Ross, Biochem. J., **94**, 557 (1965).
221. H.C. Friedman and L.M. Cagan, Annu. Rev. Microbiol., **24**, 159 (1970); J.H. Mathewson and A.H. Corwin, J. Am. Chem. Soc., **83**, 135 (1961).
222. D. Dolphin, Bioorg. Chem., **2**, 155 (1973).
223. A.I. Scott, E. Lee, and C.A. Townsend, Bioorg. Chem., **3**, 229 (1974).
224. Royal Society Discussion Meeting on "Biosynthesis of Porphyrins, Chlorophyll, and Vitamin B$_{12}$", London, February (1975). Papers presented by: (a) A.H. Jackson, (b) A.I. Scott, (c) A.R. Battersby and E. McDonald. To be published in Phil. Trans. Roy. Soc. (London).

HEME-CLEAVAGE: BIOLOGICAL SYSTEMS AND CHEMICAL ANALOGS

PÁDRAIG O'CARRA

Department of Biochemistry, University College, Galway, Ireland

4.1. Introduction
4.1.1. Chemical and biological derivation of bilins from hemes

Heme compounds are particularly susceptible to oxidative attack at the *meso* positions, i.e., at the carbon bridges linking the pyrrole rings. Such oxidation readily leads to rupture or elimination of the carbon bridge, resulting in cleavage of the porphyrin macrocycle and formation of an open chain tetrapyrrolic structure (Scheme 1).

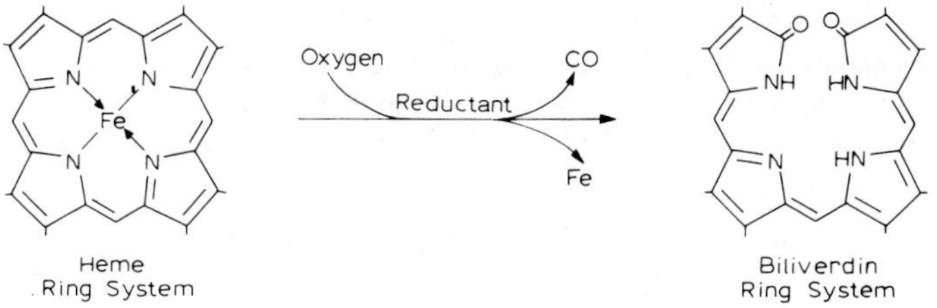

Scheme 1. Oxidative cleavage of the heme macrocycle to an open chain tetrapyrrole (bilin) structure.

This chapter is particularly concerned with a type of oxidative heme cleavage that occurs in the presence of certain reductants, such as ascorbate. The reaction proceeds by concurrent oxidation of the reductant and the carbon bridge of the heme molecule by molecular oxygen. This is usually, if somewhat confusingly, referred to as coupled oxidation of heme (i.e., co-oxidation of heme and reductant) and is by far the most extensively studied mode of chemical heme-cleavage (other types of cleavage reaction undergone by metalloporphyrins are reviewed in Chapter 15). The particular importance of this coupled oxidation-type of cleavage stems from the fact that it has long been regarded as a likely chemical model for the formation of

Porphyrins and Metalloporphyrins, ed. Kevin M. Smith
© 1975, Elsevier Scientific Publishing Company, Amsterdam, The Netherlands

naturally occurring open chain tetrapyrroles by an analogous biological process.

Open chain tetrapyrroles are often referred to collectively as bile pigments, a term deriving from the circumstance that the earliest such compounds to be characterized were the pigments of animal bile — the green biliverdin and its yellow reduced product bilirubin (Scheme 2). The related, but less incongruous terms *bilins* or *bilinoids* are preferable as general terms for open chain tetrapyrroles collectively, since some of these compounds are colorless, and open chain tetrapyrroles have been isolated from a wide variety of natural sources having no connection with bile.

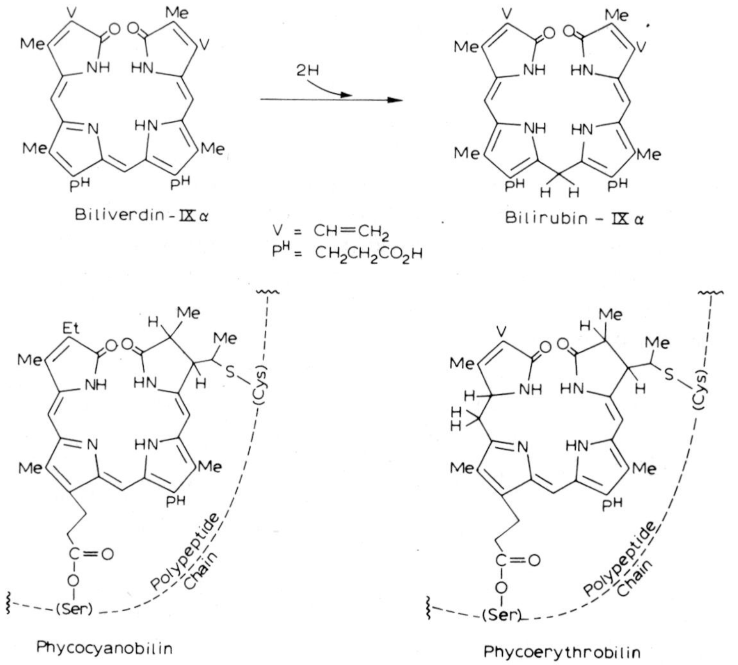

Biliverdin - IX α

Bilirubin - IX α

$$V = CH{=}CH_2$$
$$P^H = CH_2CH_2CO_2H$$

Phycocyanobilin

Phycoerythrobilin

Scheme 2. Some naturally occurring bilins (see Refs. 1,2).

Besides occurring in the bile of certain animal groups, biliverdin also occurs as the green skin pigment of a variety of reptile, amphibian, and insect species and is also responsible for the blue-green coloration of the egg-shells of many birds. In the plant kingdom, bilins play important functional roles as photoreceptors and photosynthetic light harvesters. In certain types of algae the bilins phycocyanobilin and phycoerythrobilin, covalently attached to apoproteins (Scheme 2), act as important photosynthetic chromoproteins called phycoerythrins and phycocyanins. A related bilin constitutes the prosthetic group of phytochrome, a widely distributed photoreceptor of plants which governs photomorphogenic responses such as germination and flowering. (See Refs. 1 and 2 for recent detailed accounts of naturally occurring bilins).

The probable derivation of the natural bilins by oxidative cleavage of porphyrin or heme precursors was deducible from the structural relationships which were revealed in the 1930s by the work of Fischer's school and that of Lemberg and co-workers (see, e.g., Ref. 3). The contemperaneous discovery of coupled oxidation of protoheme[4,5] demonstrated the chemical feasibility of such a pathway and suggested protoheme rather than protoporphyrin-IX as the more likely immediate precursor, since porphyrins, unlike hemes, do not readily undergo cleavage by coupled oxidation. Isotopic tracer studies later confirmed the derivation of animal bilins from protoheme (e.g., Refs. 6—12). Permeability problems have interfered with efforts to carry out similar tracer studies with algae and other plants, but the results of such efforts, although not clear-cut, support the view that the phycocyanobilin of blue-green algae (Scheme 2) is also derived by cleavage of protoheme[13,14].

The fate of the oxidatively eliminated carbon bridge was established in the 1950s by Sjöstrand[15,16,16a] who showed that heme catabolism in animals results in stoichiometric formation of carbon monoxide, a very unusual metabolite which appears to be uniquely associated with heme cleavage, at least in animals. Subsequent studies of the chemical cleavage of heme, by coupled oxidations in vitro, demonstrated a similar formation of carbon monoxide from the eliminated carbon bridge[17,18], a fact that supported the view that the in vitro systems represent close chemical analogs of the in vivo cleavage process. Recently, Troxler and co-workers[19,20] demonstrated that formation of phycocyanobilin in blue-green algae is also accompanied by a stoichiometric release of carbon monoxide, a finding that strongly supports the other evidence indicating that the formation of the algal bilins involves a heme-cleavage mechanism analogous to that operating in mammals.

The initial bilin, produced by oxidative cleavage of protoheme and loss of the iron atom, is a biliverdin (Scheme 1). In mammals the biliverdin formed is immediately converted to bilirubin (Scheme 2) by enzymic reduction. It is not known at present whether the prototropic and reductive process required to produce the characteristic structural features of phycocyanobilin and phycoerythrobilin in algae (Scheme 2) take place prior to, or subsequent to, the heme-cleavage step.

The unsymmetrical arrangement of side-chains in protoheme-IX, the naturally occurring isomer, ensures that the four carbon bridges (α,β,γ and δ) are non-equivalent (Scheme 3). Four different isomeric biliverdins may therefore arise by cleavage of protoheme-IX at these different bridge positions. The four possible isomers are designated by the suffixes-IXα, -IXβ, -IXγ, and -IXδ which indicate the parent heme and the eliminated bridge (Scheme 3). The same system of suffixes is used to denote the isomeric nature of bilins derived, formally or actually, from these isomeric biliverdins

References, p. 150

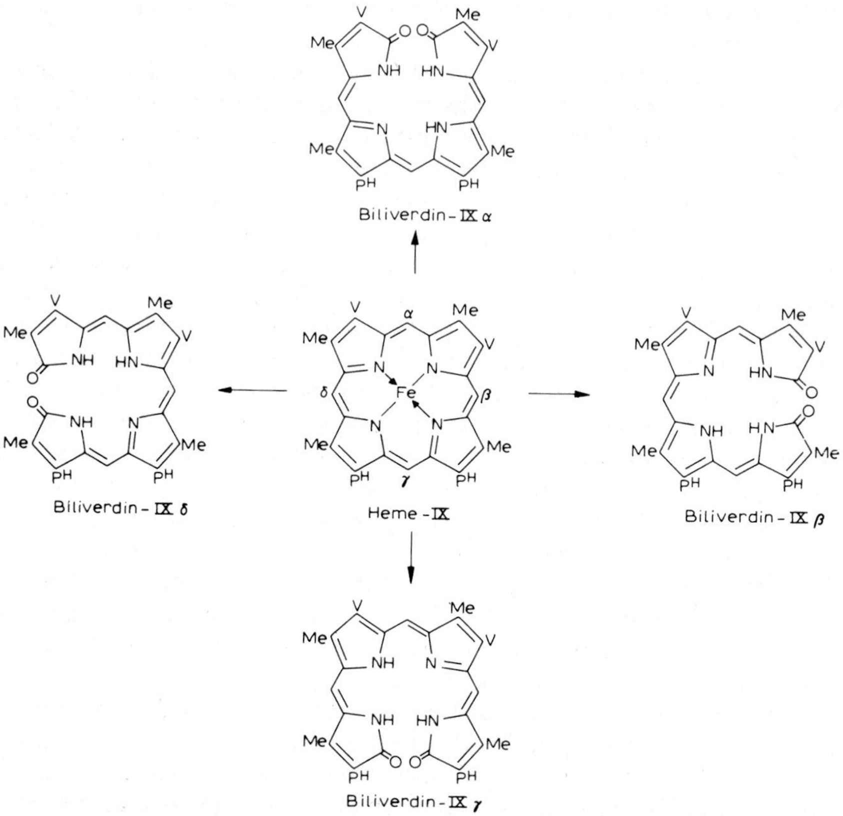

Scheme 3. Derivation of four isomeric biliverdins by cleavage of protoheme-IX at the four methine bridge positions.

by processes not involving any shuffling of the order of side-chains (e.g., bilirubin-IXα, deuterobiliverdin-IXα etc.). It was originally thought that all natural bilins were of the IXα isomeric type, but some exceptions to this 'IXα rule' have recently been demonstrated, the most outstanding being the biliverdin imparting the green pigmentation to the caterpillars of the cabbage white butterfly (*Pieris brassicae*). This has been shown to be biliverdin-IXγ[21]. Further, the biliverdin found in hepatic catalase is an equimolar mixture of the IXα and IXβ isomers[22], and minor traces of the IXβ and IXδ isomers are detectable in the bilirubin of mammalian bile[23]. However, the overwhelming bulk of such biliary bilirubin is the IXα isomer, and most other natural bilins, including the algal bilins and the biliverdin from a variety of species other than *Pieris*, are of the IXα type also.

It is clear therefore that the formation of most natural bilins involves cleavage of protoheme exclusively, or almost exclusively, at the α carbon bridge. This α-specificity has figured prominently in the recurrent controversies regarding the mechanism and causative factors of mammalian heme catabolism. It is a phenomenon that must be accounted for in any mechanism proposed for biological heme cleavage. Indeed, many workers have regarded the α-specificity as the key phenomenon whose satisfactory explanation would automatically solve the broader question of the cleavage mechanism.

Broadly speaking, two opposed theories regarding the causative factors in biological heme cleavage and the origin of the α-specificity have been repeatedly advocated over the years. The first type of theory was originated by Lemberg (see, e.g., Refs. 24,25) who proposed that the cleavage mechanism in vivo is a purely chemical process very similar to the coupled oxidation of heme achieved in vitro; on this basis the α-specificity was originally attributed to an intrinsic lability of the α-carbon bridge of the heme molecule making it much more susceptible than the other bridges to oxidative cleavage. The other type of theory, advocated most prominently by Wise and Drabkin[26-28], Nakajima et al.,[29-32] and Tenhunen, Schmid and co-workers[33-37] invokes a conventional, enzyme-catalyzed metabolic process in which the α-specificity is imposed by the specificity of enzyme(s) whose functional role is heme cleavage.

A third point of view, which could be regarded as intermediate between the preceeding two, has been put forward by O'Carra and Colleran[38-40] who suggest that biological heme cleavage may result from chemical events, closely analogous to coupled oxidation in vitro, leading to incidental or accidental cleavage of heme groups within the heme-binding sites of certain oxygen-activating hemoproteins. According to this viewpoint, the heme binding sites of these hemoproteins impose the α-specificity and certain other formal attributes of enzymatic processes on the cleavage, although the true functions of the hemoproteins are probably totally unrelated to heme cleavage.

4.1.2. Mammalian heme catabolism

Until very recently the study of biological heme cleavage has been concentrated almost entirely on heme breakdown in mammals, particularly in man and the laboratory animals used in medical research.

This concentration of attention stems partly from the technical difficulties that have been encountered in studies of heme cleavage in other organisms and partly from the intense medical interest in mammalian heme catabolism generated by the medical conditions causing jaundice, and in particular by jaundice in infants which can lead to the complication known as kernicterus. The most characteristic symptoms of these conditions are caused by bilirubin, the yellow catabolite produced as a result of mammalian heme catabolism (see Schemes 2 and 4). Bilirubin is normally excreted rapidly in the bile, to which it gives the characteristic yellow color, but where such excretion is impaired, or when bilirubin formation is excessive owing to excessive heme breakdown, the excess bilirubin accumulates in the skin and tissues, giving rise to the yellow coloration characteristic of jaundice. In infants, excessive levels of bilirubin in the circulation can result in accumulation of the pigment at the base of the brain resulting in serious brain damage, this condition being known as kernicterus.

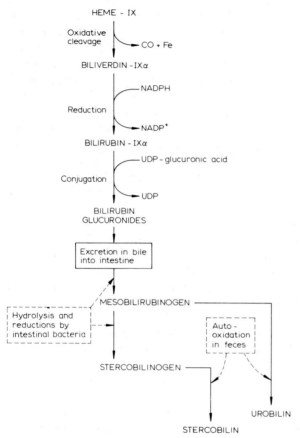

Scheme 4. Metabolic transformations of bilins in mammals.

The pathway of bilirubin-IXα formation from protoheme, and the subsequent transformations which it undergoes, are indicated in Scheme 4. It seems rather paradoxical that while the basic biochemistry and physiology of most of the transformations subsequent to the heme cleavage step are now reasonably well understood, the initial and decisive cleavage step itself still remains the subject of argument and controversy. A more fundamental paradox, to which attention has recently been drawn by O'Carra and co-workers[41] concerns the biological rationale governing mammalian heme catabolism. None of the bilin metabolites in mammals has ever been found to be of any functional importance (unlike the various bilin products elaborated by certain invertebrates and particularly by plants — see above). The mammalian bilins seemingly represent merely catabolites destined for rapid excretion. Most non-mammalian vertebrates excrete the initial bilin product, biliverdin, directly and without further transformation. For example, the pigment in the bile of birds, reptiles and amphibia (which is green) is free, unconjugated biliverdin. Only mammals and certain fish appear to transform the seemingly harmless and easily disposed of biliverdin into toxic and highly insoluble bilirubin, thereby necessitating the complex sequence of transport and conjugation steps entailed in its disposal. It has been suggested by O'Carra and Colleran[38−40] that the initial heme-cleavage step may possibly be a metabolic accident rather than a 'purposeful' metabolic process (see above, and pp. 144−150). However, the reductive conversion of biliverdin-IXα into bilirubin-IXα involves a well-defined and seemingly functionally-specific enzyme system, biliverdin reductase[42−45]. It seems unlikely, therefore, that this step could be a 'metabolic accident'.

Since this chapter is concerned specifically with the heme-cleavage step itself, the subsequent bilin transformation outlined in scheme 4 will not be further considered here.

A very large proportion of the body heme of vertebrates is contained in the hemoglobin of the circulating red blood cells. It has been known for over a century that a major proportion of biliary bilirubin is derived by catabolism of hemoglobin released when the red blood cell becomes senescent and is scavenged and destroyed by the reticuloendothelial system (after a circulating life-span of about 120 days in man or about 60 days in the rat). Whether such catabolism of hemoglobin takes place mainly in the spleen or in the liver has been the subject of much investigation and controversy which has not yet been fully resolved. It seems likely that both organs play a role.

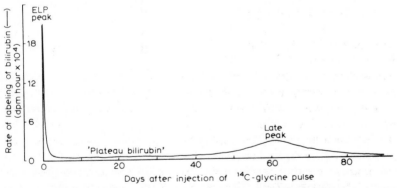

Fig. 1. Time course of appearance of ^{14}C labeled bilirubin after injection of a pulse of ^{14}C-glycine into a Gunn rat. [Based on results of Robinson et al.[46]]. The Gunn rat is a defective mutant that cannot conjugate or readily excrete bilirubin, therefore the rate of labeling of bilirubin in this experiment was conveniently monitored in samples of blood serum rather than bile.

References, p. 150

The general heme pool in experimental animals can be radioisotopically labeled by injecting ^{14}C-glycine intravenously into the animals. This allows the time pattern of heme catabolism to be followed by monitoring the flux of label in bilirubin. The results of a typical experiment of this type are illustrated in Fig. 1.

Three phases are distinguishable in the time-course of ^{14}C-bilirubin formation. A rapid initial phase, characterized by a sharp peak of labeled bilirubin, appears within hours of the administration of ^{14}C-glycine and tails off rapidly. This is termed the early-labeled pigment or ELP fraction (Fig. 1) and it is known to be attributable, at least partly, to a rapid turnover of the heme of certain hemoproteins in the liver, particularly that of the microsomal cytochrome P-450. (see, e.g., Refs. 46—49). The ELP peak is followed by a so-called 'plateau' which is usually attributed to catabolism of certain hemoproteins with a slow rate of turnover (e.g. myoglobin). This 'plateau' is followed by a broad 'late peak' representing catabolism of erythrocyte hemoglobin. The time lag which elapses before the appearance of this late peak is due to the fact that hemoglobin, which is labeled many weeks or months previously during the original pulse of labeled glycine, is at that time imprisoned in immature red blood cells and is not released and catabolized until these red cells become senescent and rupture after the elapse of their normal life-span. This life span varies with species, and the time of appearance of the late peak of ^{14}C-bilirubin varies accordingly. In the rat it peaks after about 60 days (Fig. 1).

When ^{14}C-labeled δ-aminolevulinic acid is injected instead of ^{14}C-glycine, the label is taken up almost entirely by the liver and very little labeling of hemoglobin or of other extra-hepatic hemoproteins takes place. The time course of formation of labeled bilirubin then shows only the ELP peak attributed to the catabolism of hepatic heme[46].

4.2. Coupled oxidation of hemes and hemoproteins
4.2.1. General characteristics and requirements

As mentioned above, coupled oxidation of the methine bridges is strongly dependent on the presence of the central iron atom of heme, and uncomplexed porphyrins show no significant tendency to undergo such coupled oxidation.

Oxidative cleavage of other metal complexes has been described but, although the mechanism of such cleavage bears some resemblance to that of the coupled oxidation of heme compounds, the conditions and requirements of the reactions differ considerably. Thus, for example, some reported examples of oxidative cleavage of zinc and magnesium porphyrin complexes required a photochemical step[50,51]. (See Chapter 15, for a general treatment of such cleavage mechanisms).

In the classical studies of Warburg and Negelein[4] in 1930 coupled oxidation of heme was performed in pyridine with hydrazine as co-oxidized reductant. Elevated temperatures and a high oxygen tension are required to achieve satisfactory reaction rates with this system. In subsequent studies Lemberg and co-workers[5,25,52−55] replaced hydrazine with ascorbate. This reductant promotes rapid coupled oxidation of hemes in pyridine—water mixtures at temperatures in the physiological range and at atmospheric oxygen tension. To date, ascorbate remains the reductant most widely used to promote coupled oxidation of hemes.

Lemberg (cf. Refs. 5,25) considered that protoheme underwent negligible coupled oxidation in the absence of pyridine but found that certain proteins, when complexed with protoheme, promoted coupled oxidation in a manner reminiscent of the effect of pyridine. Hemoproteins or heme-protein complexes which undergo such coupled oxidation under near-physiological conditions include hemoglobin, myoglobin, and heme-albumin[55−59]. Catalase undergoes similar coupled oxidation but requires much higher concentrations of ascorbate[60]. Since the apoproteins, like pyridine, ligate to the iron atom of the heme molecule at one or both of the free axial positions, it was concluded that the binding of such axial ligands activates the heme group in some way which renders the bridge positions susceptible to oxidative cleavage[24,25]. Kench later showed that protoheme, in aqueous solution and uncomplexed with any nitrogenous ligand, undergoes very slow coupled oxidation in the presence of high ascorbate concentrations and under vigorous aeration[57,59]. The requirement for the axial ligands is therefore not absolute, but the activating effect of such ligands is nonetheless very considerable.

Coupled oxidation of heme or its pyridine hemochrome does not give rise directly to biliverdin but to a green product that may be converted to biliverdin by treatment with acid. This green product, originally termed 'green hemin' by Warburg and Negelein[4] is now usually referred to as verdoheme (or verdohemochrome when it is complexed with pyridine).

4.2.2. Role of axial ligands and effect of heme environment

The activating effect of pyridine has usually been assumed to be attributable to electronic effects associated with the formation of pyridine hemochrome by binding of pyridine to the iron atom of the heme molecule. However, when the pyridine concentration is decreased below about 25% (v/v) — e.g., to about 10% (v/v) in water — the rate of coupled oxidation decreases drastically, although the heme remains almost entirely in the complexed pyridine hemochrome form[39,61]. If the pyridine concentration is maintained at about 10% while a proportion of the water is replaced by an organic solvent, such as ethanol, the rate of coupled oxidation increases in proportion to the ratio of organic solvent added. Very rapid coupled oxidation takes place in predominantly hydrophobic solvent mixtures containing concentrations of pyridine as low as 5%[39,61,62].

References, p. 150

It therefore seems that part of the effect of high pyridine concentrations may be a general solvent effect reproducible with other organic solvents. However, most other organic solvents do not promote coupled oxidation unless some pyridine is present. Pyridine must therefore promote some specific activating effect, presumably dependent on its role as an axial ligand.

In myoglobin and hemoglobin the heme groups are ligated to an imidazolyl residue and embedded in a largely hydrophobic environment (cf. Refs. 63,64). The same activating effects that operate in the case of pyridine may therefore operate also to promote coupled oxidation of the heme groups of these hemoproteins[38]. Peculiarly, imidazole itself in free solution does not promote coupled oxidation and, indeed, it inhibits the coupled oxidation of heme in pyridine[61]. This may be because imidazole in free solution can bind very strongly to both axial positions and this may prevent any interaction of oxygen with the heme group. As discussed below, it is possible that binding of oxygen to the heme group represents one of the necessary preliminary steps in the mechanism of the coupled oxidation process.

It has been shown by Brown and co-workers[65] that the slow rate of coupled oxidation of uncomplexed heme in aqueous solution is probably partly attributable to the fact that the heme molecules exist largely in the dimerized state under such conditions. The monomeric form seems to undergo coupled oxidation far more readily than the heme dimer, and Brown suggests that the activating effect of axial ligands may be due simply to a dispersing or 'dimer-breaking' effect. However, against this argument must be counted the fact, mentioned above, that the activating effect of pyridine in pyridine—water mixtures falls off drastically at pyridine concentrations in the region of 10% although the heme remains almost entirely in the pyridine-complexed monomeric form. Therefore, while a dimer-breaking effect may well account for part of the activating effect of pyridine and of hemoproteins[65] it seems unlikely to account for a very substantial part of it.

Brown et al.[65] also make the interesting observation that protoheme readily undergoes coupled oxidation in dimethyl sulfoxide in the absence of pyridine. These authors relate this phenomenon again to a dimer-breaking effect of this solvent. However, as they point out, dimethyl sulfoxide also acts as an axial ligand. It may thus activate both by complexing with the heme and by providing a non-aqueous environment, as suggested above in relation to the effects of pyridine and hemoproteins.

4.2.3. Chemical pathway and intermediates

The studies of Lemberg and co-workers[5,52—56] and of Fischer, Libowitzky, and Stier[66—69] in the 1930s and 1940s established a sequence of chemical events[24,25] which, with comparatively minor modifications, has formed the basis of most mechanisms since proposed for oxidative heme cleavage. Scheme 5 represents a version of this sequence of events that seems most consistent with all currently available evidence. In this illustration axial

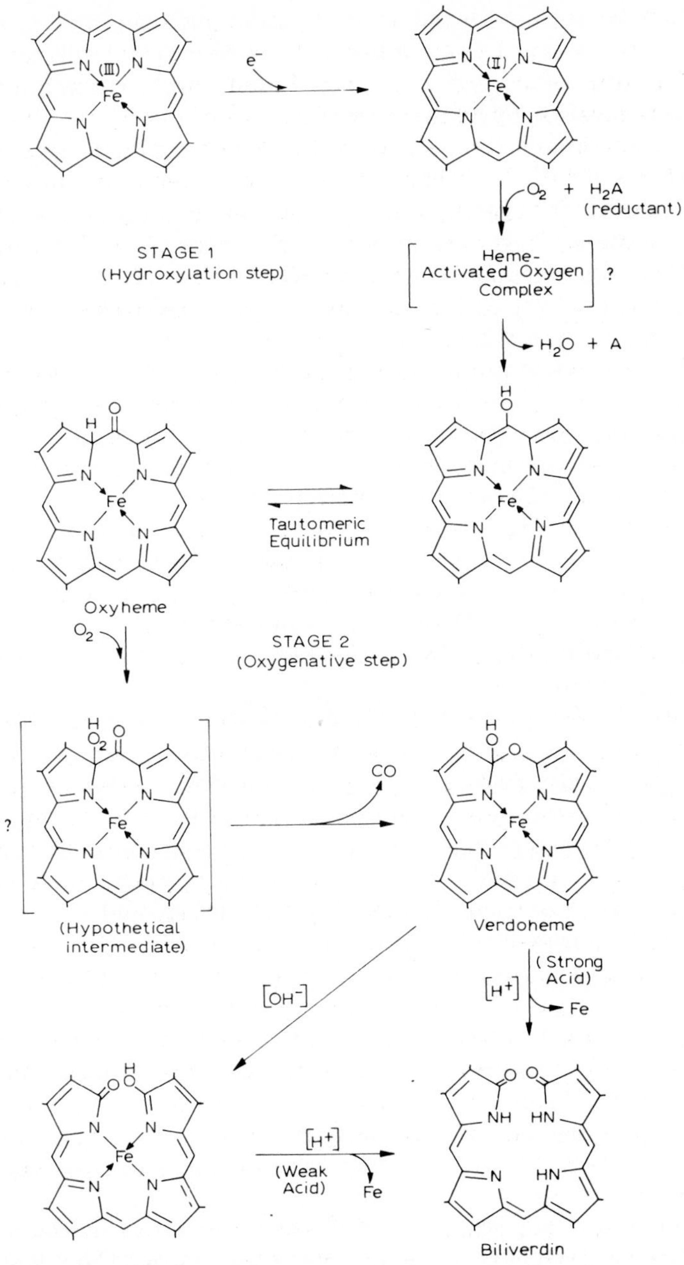

Scheme 5. Probable pathway of coupled oxidation of heme and hemochromes. As indicated, current evidence suggests that the iron must be reduced to the ferrous form to initiate the sequence of reactions, but the oxidation state of the iron in the subsequent steps is uncertain. It seems probable that it is converted to the ferric form at some point during Stage 2.

ligands are omitted, since such evidence as is available indicates that the oxidative cleavage of uncomplexed heme follows the same pattern as that of hemochromes. As discussed below, it is not clear to what degree this scheme applies also to coupled oxidation of hemoproteins.

Two stages in the oxidation of the carbon bridge can be distinguished, as indicated in Scheme 5. The first stage involves a hydroxylation of the methine bridge by 'activated oxygen', while the second stage involves an auto-oxidative or oxygenative reaction in which molecular oxygen reacts with the hydroxylated bridge leading to its elimination as carbon monoxide and formation of verdoheme. The latter can then be converted to biliverdin by treatment with strong acids or alkalis.

The hydroxylated intermediate produced in stage 1 can exist in two tautomeric forms. The keto form (Scheme 5) predominates under the normal conditions of coupled oxidation[5]. In the literature pertaining to heme-cleavage such structures have been termed *oxyhemes* and this term will be used here in preference to the more cumbersome synonym ferro-oxophlorin*.

The role of the reductant (e.g. ascorbate) in the coupled oxidation is confined to stage 1 (Scheme 5) leading to the formation of the oxyheme intermediate. Coupled oxidation seems to proceed only when the central iron atom is in the ferrous form. Therefore, if the starting material is a ferriheme derivative, a preliminary role of the reductant seems to be to reduce this to the ferroheme form.

The 'activated oxygen', which now reacts with a methine bridge and hydroxylates it, is produced by reduction of molecular oxygen by the reductant. The precise nature of this 'activated oxygen' remains unclear. If the central iron atom is pre-reduced (i.e., if the starting material is a ferroheme derivative), the hydroxylation step may be effected by hydrogen peroxide in the absence of reductant or oxygen[68]. This was for many years taken as indicating that the 'activated oxygen' formed from reductant and oxygen consisted simply of hydrogen peroxide. However, addition of catalase only marginally affects coupled oxidations of pyridine hemochromes or of hemoproteins with ascorbate[39,61], seeming to rule out free hydrogen peroxide as the hydroxylating agent in such systems. Superoxide dismutase also fails to significantly inhibit coupled oxidations[61], seeming to similarly rule out the superoxide radical in the free form.

However, it seems possible that the heme derivative itself participates catalytically in the formation of the 'activated oxygen' from oxygen and

---

* The oxyheme intermediate in slightly alkaline solution, and particularly in pyridine, is very similar to verdoheme in color (green) and the two compounds may have been confused occasionally in the past. They can be most readily distinguished by acidifying the solution with HCl. Verdoheme remains green, while an oxyheme turns a brown-red color very similar to that of heme itself. This color change is attributable[70] to the fact that alkaline pH favors the green keto tautomer (the oxyheme form) while acid pH causes conversion to the red enol tautomer (the hydroxyheme form) (Scheme 5).

reducing equivalents, and that this 'activated oxygen' remains ligated to the iron atom of the heme in the interval before it reacts with the methine bridge. In such a bound form either peroxide or superoxide would probably be unaffected by catalase and superoxide dismutase respectively. The oxyheme intermediate may be readily detected during coupled oxidation of pyridine hemochromes in aqueous pyridine and may be isolated from the reaction mixture in trace amounts[5,66,67]. The accelerating effect of predominately non-aqueous media on the rate of coupled oxidation seems to operate more on the hydroxylation stage than on the subsequent reactions, so that the oxyheme intermediate accumulates to a much greater extent and may be isolated in better yield[61].

Although verdoheme is generally regarded as the product of coupled oxidation of heme with reductants, the reductant appears to play no essential role beyond the oxyheme stage. Therefore, the end product of the coupled oxidation is, strictly speaking, the oxyheme. The subsequent oxidation of the oxyheme, resulting in cleavage of the carbon bridge and its elimination as carbon monoxide, involves only molecular oxygen. This oxygenative cleavage presumably involves some intermediate adduct(s) formed by addition of the oxygen molecule at a position adjacent to the keto bridge. Such intermediates must be transitory, however, and have never been isolated or otherwise directly demonstrated. The intermediate indicated in Scheme 5 was suggested originally by Lemberg[25]. More recently it has been suggested by Jackson and Kenner[71] that an intermediate of the sort illustrated in Scheme 6 would be more consistent with the rather unusual expulsion of the keto bridge as carbon monoxide. The suggested one-step formation of such an intermediate would require the oxygen molecule to add across the bridge position in the singlet rather than the triplet state (Scheme 6). Subsequent evidence[72] has failed to support an involvement of singlet oxygen, however, and the intermediate suggested by Lemberg, and indicated in Scheme 5, must again be regarded as a more likely, though still very hypothetical, possibility. It seems possible that the hydroperoxide grouping of such an adduct might subsequently react across the keto bridge forming the suggested alternative intermediate, illustrated in Scheme 6, before elimination of the keto bridge as carbon monoxide.

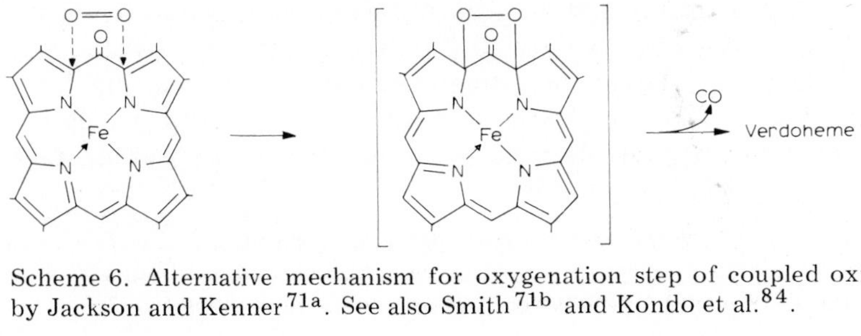

Scheme 6. Alternative mechanism for oxygenation step of coupled oxidation suggested by Jackson and Kenner[71a]. See also Smith[71b] and Kondo et al.[84].

References, p. 150

The product remaining after expulsion of carbon monoxide is verdoheme. This is a green pigment in which final opening of the tetrapyrrole macrocycle has not yet been achieved. An oxygen bridge still acts as a link across the position from which the carbon bridge has been expelled. The verdoheme structure indicated in Scheme 5 is that favored by Lemberg[5]. More recent work of Levin[73] seems to confirm this structural assignment and argues against alternative structures, such as that illustrates in Scheme 7, which have been proposed from time to time*.

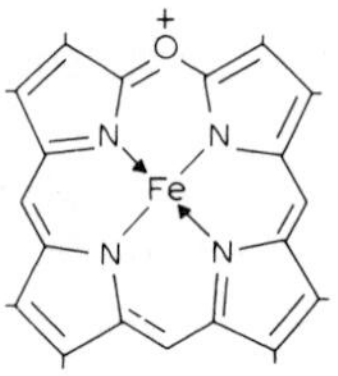

Scheme 7.

Verdoheme is non-oxidatively converted into biliverdin by treatment with strong acids such as HCl and H_2SO_4 which cleave the oxygen bridge and displace the iron atom (Scheme 5). This acid-promoted conversion has often been regarded as hydrolytic, but the cleavage of the oxygen bridge involves rearrangement rather than hydrolysis, and the iron atom is displaced by protons. In keeping with this, the conversion into biliverdin-IXα takes place readily in non-aqueous acid media (e.g. in methanol or glacial acetic acid containing HCl or H_2SO_4). Note that the alternative suggested structure in Scheme 7 would require a hydrolytic step to achieve conversion into biliverdin. In strong alkali, also, the bridge oxygen grouping of verdoheme rearranges to yield the open chain tetrapyrrole but the iron atom is retained. The resulting iron—biliverdin complex is readily dissociated by weak acids such as acetic acid (Scheme 5).

Coupled oxidation of the heme groups of hemoproteins such as hemoglobin and myoglobin has generally been considered to proceed by a mechanism similar to that followed by pyridine hemochromes and free heme. There are certainly many points of close resemblance. The hemoprotein must first be reduced to the ferrous form and, in general, requirements for efficient coupled oxidation are similar. However, during coupled oxidation of hemoproteins the cleavage proceeds rapidly to the open chain tetrapyrrole stage and no detectable accumulation of either oxyheme or verdoheme intermediate is observed. After elimination of the carbon bridge as carbon monoxide the cleaved tetrapyrrole remains bound to the apoprotein, from

---

* There is good evidence, however, that compounds of the type indicated in Scheme 7 may be formed during oxidative cleavage of oxyhemes under certain conditions[70]. Such conditions seem to include a largely anhydrous or non-aqueous environment.

which it may be dissociated by weak acids such as acetic acid. The tetrapyrrole product released thus is biliverdin itself. Since acetic acid at room temperature dissociates the biliverdin—iron complex but does not split verdoheme to biliverdin, it must be concluded that the cleavage process in hemoproteins proceeds all the way to the biliverdin—iron complex, rather than halting at the verdoheme stage.

The biliverdin—iron—apoprotein complexes formed from the hemoproteins by coupled oxidation are usually referred to as choleglobins, a term coined by Lemberg. While the choleglobin formed from myoglobin is relatively stable and well defined, that formed from hemoglobin undergoes complicating side-reactions seemingly caused by reactive thiol groups of the globin (which are absent from the myoglobin apoprotein). The thiol groups in hemoglobin tend to add across the vinyl side-chains of the biliverdin and thereby link the pigment covalently to the apoprotein. This is accompanied by denaturation and precipitation of the protein. Owing to such side-reactions, the proportion of the biliverdin product released by acetic acid decreases progressively if the coupled oxidation is prolonged unduly. Blockage of the thiol groups with thiol reagents, such as ethylenimine or p-hydroxymercuribenzoate, minimizes this complication[13,61].

4.2.4. Specificity of bridge cleavage

As explained in the introduction (p. 125), cleavage of protoheme-IX is potentially capable of giving rise to four different isomeric forms of biliverdin owing to the non-equivalence of the α, β, γ, and δ bridge positions (Scheme 3). It was originally thought, however, that coupled oxidation in vitro, like heme-cleavage in vivo, was specific for the α bridge, and that the biliverdin so produced was entirely the -IXα isomer. This α-specificity was supposed to arise as a result of intrinsic electronic features of the heme molecule itself. Lemberg and co-workers originally proposed that the propionic acid side-chains exerted an inductive effect which made the α bridge opposite them more susceptible to oxidative cleavage (see Ref. 25), and a somewhat similar effect was proposed more recently by Woodward[74]. Pullman and Perault[75] in 1959 published molecular orbital calculations that were also generally taken as implying preferential attack at the α bridge. In this case, inclusion of the effects of the vinyl side-chains in the calculations seems to have been the key factor leading to a rather tentative conclusion of a differential activation of the α methine bridge.

Doubt was cast on the validity of these ideas by Petryka et al. who in 1962 reported that the biliverdin produced by coupled oxidation of pyridine hemochrome was not solely the -IXα isomer but a mixture of isomers[76]. The technique used in these studies — the Nicolaus degradative method[77] — did not reveal which isomers were present, nor did the more sensitive degradative technique later introduced by Rüdiger for examination of isomeric bilins[78]. However, more informative methods of isomer analysis which have recently

References, p. 150

TABLE 1

Isomer Composition of the Biliverdin products Formed by Coupled Oxidation of Pyridine Hemochromes of proto-, meso- and deutero- hemes with ascorbate*

Heme/Bili-verdin type	Substituents in 2 and 4 positions	% Isomer composition of biliverdin product			
		IXα	IXβ	IXγ	IXδ
Proto	Vinyl	32	25	23	20
Meso	Ethyl	33	26	17	24
Deutero	H	10	24	23	43

* Based on Refs. 13, 23, 61, 79, 81.

been developed show that coupled oxidation of pyridine hemochrome produces all four possible biliverdin isomers.

These newer analytical methods developed independently by Rüdiger[79] O'Carra and Colleran[23,38,80] and Bonnett and McDonagh[81,82] all depend on thin layer chromatographic separation of the biliverdin isomers, allowing both their identification and quantitation. These methods showed conclusively that oxidative cleavage of pyridine hemochrome is almost entirely random as regards the four carbon bridges*. The isomer composition of a typical biliverdin sample so obtained and analyzed is shown in Table 1. A slight preponderance of the α isomer is indeed observed, which might suggest a very marginal differential activation of the α bridge by either the vinyl groups or propionic acid groups, although not on the scale originally envisaged (see above). However, this must be doubted on the basis of some further results of O'Carra and Colleran[23,61] who showed that the slight preponderance of α-cleavage persists when the propionic acid groups are esterified before the heme is subjected to coupled oxidation and also when the vinyl groups are reduced to ethyl groups as in mesoheme (see Table 1).

Analysis of the deuterobiliverdin produced by coupled oxidation of the pyridine complex of deuteroheme-IX shows again the production of all four isomeric deuterobiliverdins (Table 1) but in this case there is a rather striking

---

* The anomalous conclusion of Nichol and Morell[83], that coupled oxidation of pyridine hemochrome produces only the β or the δ isomer, may be attributable to mis-interpretation of the complex physical data on which this identification was based. Alternatively, the β or δ isomer may have been preferentially isolated during the preparation of the crystalline material used by these authors for the identification, leaving the other isomers in the mother liquor. Such preferential crystallization of isomers has been noted in the case of the deuterobiliverdins[13] and it is clearly not advisable to subject biliverdin preparations to crystallization or other potential fractionation steps before isomer analysis[23,82].

preponderance of cleavage at the δ bridge[61,79]. The reason for this effect is not clear, but it is presumably related to some change of electronic configuration caused by the absence of side-chains in positions 2 and 4 of deuteroheme-IX. Coupled oxidation of pyridine hemochrome (proto or meso type) under a variety of conditions (varied pH, solvent composition, ascorbate concentration, oxygen tension) produced no change in the isomer composition of the biliverdin product[13,61]. Coupled oxidation of uncomplexed heme and of heme—albumin complex[38] also yielded an essentially random mixture of biliverdin isomers (Table 2). Thus the lack of intrinsic bridge specificity remains remarkably unaffected by environmental effects of a generalized nature. Although such changes in reaction conditions cause considerable changes in the rate of coupled oxidation, the effects of such changes seem to apply to all four bridge positions equally.

However, a dramatic change in the bridge specificity of cleavage is brought about by the specific heme-binding sites of those hemoproteins that undergo coupled oxidation (excluding heme—albumin which is not regarded as being a particularly well-defined or functional hemoprotein). When isomer analysis was extended to the biliverdin produced by coupled oxidation of such hemoproteins, a considerable degree of cleavage specificity was revealed[38,61]. As shown in Table 2 and Fig. 2, coupled oxidation of myoglobin yields only the α biliverdin isomer, while coupled oxidation of hemoglobin yields the α and β isomers but no traces of the γ or δ. Catalase also yields the α and β isomers. Thus the heme binding sites of these hemoproteins impose a strict bridge specificity on the cleavage process. This specificity was found to be retained when the native protoheme of these hemoproteins was replaced by mesoheme or deuteroheme[61]. Coupled oxidation of these artificial hemoproteins yielded, once again, only the α isomer from the myoglobins and the α and β isomers from the hemoglobins. This confirms the conclusion that the imposition of bridge specificity is a property of the heme-binding sites of these proteins and does not derive in any significant way from the heme itself. In further confirmation of this, the imposed bridge specificity is abolished when the structure of the binding site is disrupted by denaturation of the apoprotein[38]. Thus, treatment of myoglobin and hemoglobin with denaturants such as 8M urea causes the bridge cleavage to revert to an essentially random pattern (see Table 2).

The specificity of the cleavage for both the α and β bridges in hemoglobin, as opposed to the purely α-specific cleavage in myoglobin, has been studied in some detail by Colleran and O'Carra[61,62]. The proportion of the β isomer produced varies considerably for hemoglobins from different species, being low, for example, for hemoglobin from guinea-pig (80%α + 20%β) and rat (76%α + 24%β). While the proportion of the β isomer formed by human hemoglobin is comparatively high (Table 2), that produced by hemoglobin from perch (a freshwater fish) is even higher, more β isomer than α isomer being produced in this case (40%α + 60%β). This two-way specificity of

TABLE 2

Coupled oxidation (with ascorbate) of heme, pyridine hemochrome, heme-albumin and various functional hemoproteins*

Heme derivative**	% Isomeric composition of biliverdin product			
	IXα	IXβ	IXγ	IXδ
Heme	31	24	24	21
Pyridine Hemochrome	32	25	23	20
Heme-Albumin (human)	28	21	31	20
Myoglobin (horse or whale)	100	0	0.	0
Hemoglobin (human)	60	40	0	0
Catalase (beef liver)	48	52	0	0
Myoglobin in 8M Urea	32	21	25	22
Hemoglobin in 8M Urea	33	21	25	21

 * Taken from Refs. 38 and 61.
** Starting material in each case was in the ferriheme form but this was reduced to the ferroheme form during the initial lag in the time-course of the coupled oxidations (see, e.g., Fig. 3, p. 148).

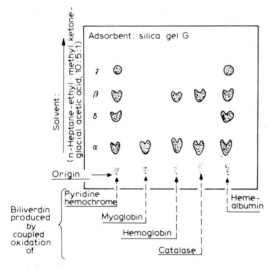

Fig. 2. Thin layer chromatographic analysis of the isomeric biliverdins produced by coupled oxidation of various heme complexes and hemoproteins with ascorbate[61]. The chromatographic system is that of O'Carra and Colleran[23].

hemoglobin is retained when the tetrameric subunit structure is dissociated to yield the monomeric subunits and it is also retained by both types of subunit of hemoglobin when these are separated chromatographically[13,61,62].

2,3-Diphosphoglycerate, a metabolite which acts as an allosteric effector of hemoglobin and considerably modifies its oxygen-binding behaviour, does not noticeably affect the ratio of α to β cleavage, but haptoglobin when bound to hemoglobin causes a considerable drop in proportion of β cleavage[13,61,62]. For example, the proportions of the two isomers produced by human hemoglobin change from 60%α + 40%β to about 90%α + 10%β when haptoglobin is added.

The structural basis of the imposed specificity of heme cleavage in myoglobin and hemoglobin is not at all clear, despite the fact that the classic X-ray diffraction work of Perutz, Kendrew and their associates[63,64] has revealed in great detail the complete three-dimensional structures of the heme-binding sites of both these proteins, together with the exact positioning of the heme groups therein. It can be concluded however, that the bridge specificity effect is not due simply to a steric hindrance effect, i.e., a differential masking of the bridges that are not cleaved. In both myoglobin and hemoglobin (in which the heme-binding sites seem very similar structurally) the α bridge appears to be the least accessible, being buried at the bottom of the heme binding crevice. The γ, on the other hand, is at the opening of the crevice and would seem to be the most accessible of the bridges from a steric point of view. The β and δ bridges, one at each side of the crevice, both seem problematical as regards accessibility.

The failure of such detailed structural knowledge to provide a better insight into the mechanism of the bridge-specificity effect presumably reflects the fact that knowledge of the covalent structures, even when it encompasses full conformational detail, still leaves us largely in ignorance of the subtler non-covalent effects which are probably more important in imposing biological-type specificity than the covalent chemistry itself. In this connection it may be significant that the environment around the α bridge created by the neighboring amino acid side-chains is noticeably more hydrophobic than that around the other bridges in both myoglobin and hemoglobin, while the only noticeable difference between the environment of the β bridge in hemoglobin compared to myoglobin seems to be a somewhat closer packing of hydrophobic amino acid side-chains around the β-methine bridge in hemoglobin. The suggestion that this differential hydrophobicity may possibly be responsible for the specificity of the attack on the α bridge in myoglobin and on both the α and β bridges in hemoglobin[38] derives some support from the evidence that non-aqueous solvents tend to accelerate coupled oxidation of pyridine hemochrome, though in free solution situations this acceleration applies equally to all four bridges (see above).

On the basis of evidence, some of which has been mentioned above, it seems likely that the 'activated oxygen' which initiates the oxidative attack on the bridge position may be derived by reduction of an oxygen molecule complexed to the central iron atom of the heme group in myoglobin or hemoglobin. Such an 'activated oxygen' group might then be differentially attracted by non-covalent effects towards the α bridge in myoglobin or towards both the α and β bridges in hemoglobin.

4.3. Biological heme cleavage
4.3.1. Pathway in vivo

It has long been assumed that heme cleavage in biological systems proceeds by a chemical pathway similar to that of coupled oxidations in vitro. One outstanding similarity is the expulsion of the cleaved carbon bridge as carbon monoxide in both the in vivo and in vitro systems (see above, p. 125). Coupled oxidation of hemoproteins might be presumed to represent a closer model for in vivo heme cleavage than the much-studied coupled oxidation of pyridine hemochrome, on whose characteristics the mechanism outlined in Scheme 5 is largely based. However, studies of such hemoprotein systems have so far provided little additional insight as regards the chemical intermediates of the pathway; as mentioned above (p. 136) the cleavage seems to proceed rapidly to the open tetrapyrrole stage and no intermediates have been reliably characterized. Presumably these intermediates exist transiently, but no attempts seem to have been made, thus far, to apply techniques suitable for the study of such transient states. The hemoprotein systems, however provide very useful models for the study of other aspects of the in vivo cleavage process, as outlined below.

Little direct evidence for the involvement of intermediates of the type indicated in Scheme 5 in the in vivo processes was available until the recent work of Kondo and co-workers[84] who injected synthetically-prepared ^{3}H-labeled α-oxymesohemin[70] into rats and observed a rapid conversion to mesobilirubin, a result taken to confirm the intermediacy of an oxyheme-type structure in the in vivo pathway in mammals. The scheme proposed by Kondo et al. for heme cleavage in vivo also takes note of the conclusions drawn by Tenhunen and co-workers[35] from ^{18}O tracer experiments designed to reveal the origin and fate of the oxygen groups participating in the cleavage process. These experiments refer to heme cleavage promoted in vitro by microsomal preparations, rather than true in vivo heme cleavage, but they are generally accepted as providing a reliable reflection of the process in vivo. The results of Tenhunen et al., summarized in Scheme 8, indicated that the two oxygen atoms introduced into the bilin product, and the atom in carbon monoxide, are all derived ultimately from molecular oxygen and that no oxygen is incorporated from water, ruling out a hydrolytic step in the cleavage mechanism. This would also be the result expected from a mechanism of the type indicated in Scheme 5. Tenhunen et al.[35] and Kondo et

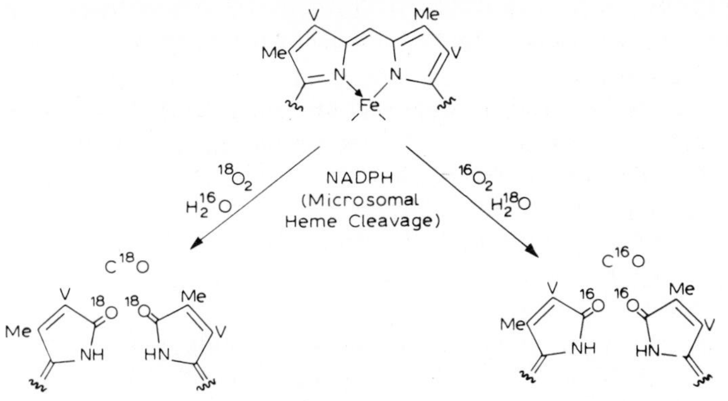

Scheme 8. Summary of the results of ^{18}O experiments reported by Tenhunen et al.[35] revealing the origin of the oxygen atoms incorporated into the products of NADPH-dependent heme-cleavage promoted by microsomal preparations.

al.[84] considered the ^{18}O results excluded the intermediacy of verdoheme-type compounds, but this conclusion was based on the assumption of a structure for verdoheme similar to that shown in Scheme 7. This basically represents an anhydro form of the verdoheme structure rather than that favored by Lemberg and by Levin (p. 136) and indicated in Scheme 5. As discussed above, it seems likely that the structure shown in Scheme 5 is the intermediate formed during coupled oxidations under aqueous conditions in vitro. Since this latter structure undergoes conversion to biliverdin non-hydrolytically (see p. 136), its intermediacy would not necessarily lead to incorporation of oxygen from water. The results of Tenhunen et al. do, however, seem to cast doubt on the biological relevance of the type of mechanism recently proposed Besecke and Fuhrhop[51], in which an anhydro-verdoheme type compound, of the sort shown in Scheme 7, was indicated as an intermediate. This mechanism[51] differs in a number of other respects from that indicated in Scheme 5. However, the model chemical experiments on which these authors based their proposals involved photochemical oxidative cleavage of nickel and zinc metallohydroxyporphyrins, rather than heme, and were performed under non-aqueous conditions. All of these differences could alter significantly the mechanism of oxidative bridge cleavage (see p. 130). For example, the initial hydroxylation step in such cleavages of nonferrous metalloporphyrins is usually photochemically-promoted and dependent on the non-aqueous environment. Some years ago, Barrett[50] demonstrated a similar photochemically-promoted cleavage of a magnesium—porphyrin complex, the implication being that although such a photochemical process could hardly be of any direct relevance to mammalian heme cleavage, it might be of relevance to the formation of plant bilins in view of the fact that the dominant metalloporphyrin of most plants is chlorophyll, a magnesium—dihydroporphyrin complex which resides in a hydrophobic membrane environment and participates prominently in photo-chemical processes (see Chapter 17).

The work of Nichols and Bogorad[85] suggested the involvement of a pho-tochemical event in the formation of the algal bilin, phycocyanobilin (Scheme 2), but current opinion seems to favour the view that this repre-sents a control or triggering mechanism, rather than a step in the actual cleavage pathway. Nichols and Bogorad[85] determined the action spectrum of this photochemical response and interpreted it in terms of the involve-ment of a heme-type photoreceptor rather than of chlorophyll. Most other available evidence now seems to indicate that the algal bilins, like those of animals, are probably derived from heme precursor(s) rather than chloro-phyll (see Section 4.1), but much work has yet to be done.

4.3.2. Causative factors

As mentioned in the introduction, much debate has centred around the nature of the factors promoting biological heme cleavage.

The striking acceleration of in vitro coupled oxidations by axial ligands, in-cluding complexed proteins, suggested to many early workers that heme cleav-age in vivo might involve similar causative factors, i.e., coupled oxidation, pro-moted relatively nonspecifically by a variety of heme-ligating substances and probably involving ascorbate as reductant. Associated with this viewpoint was the idea that the α-specificity of the cleavage process was an intrinsic property of the protoheme group and arose automatically without the inter-vention of a specific enzymic system. This latter view has now, of course, been disproved by the demonstration that coupled oxidations of heme, sim-ple hemochromes, or relatively non-specific heme—protein complexes such as heme-albumin, all involve essentially random methine-bridge cleavage (Table 2). However, the demonstration that the apoproteins of certain func-tional hemoproteins impose bridge specificity on the coupled oxidation of their heme groups seemed to provide a possible alternative explanation for the in vivo α-specificity on the basis of a non-enzymic coupled oxidation of heme while still bound specifically to such apoproteins. There are stumbling blocks to such an interpretation, however. While totally α-specific cleavage by such a mechanism has been demonstrated in the case of myoglobin in vitro[38], the catabolism of this particular hemoprotein contributes only a minor proportion of bilin output. In vertebrates, a large proportion of the bilin output results from the catabolism of hemoglobin and the bridge specificity imposed by the apoprotein in this case is not entirely α-specific, a relatively large proportion of the β-isomer being produced in addition to the α. The secondary bilin transformation steps in mammals (Scheme 4) (in particular the enzymic reduction of biliverdin-IXα to bilirubin-IXα) seem likely to impose a barrier to the conversion of any β-isomer into bilirubin. The formation of a proportion of β-isomer in mammals might thus have escaped detection[38,44]. However, no such secondary metabolic transformation steps complicate interpretation of the bridge specificity of the in vivo cleavage process in non-mammalian vertebrates, since in most of such species the biliverdin is excreted directly without further modification. In such animals (e.g. chicken, frog) this biliverdin is almost entirely -IXα in composition ($>$99%)[41], seeming to rule out direct, non-enzymic coupled oxidation of hemoglobin as a significant mechanism of heme-cleavage in vivo.

The observation that the coupled oxidation of hemoglobin is rendered more α-specific by the binding of haptoglobin[61,62] might seem to suggest a modified scheme involving this hemoglobin-binding plasma protein, which has also been implicated in heme catabolism in other contexts. Haptoglobin merely diminishes, rather than abolishes, the production of the β-isomer, but a more serious impediment to its involvement is the fact that many species seem to lack this protein completely (e.g. ox, sheep, and probably other ruminants)[41]. These and other considerations suggest that the heme groups of hemoglobin are probably cleaved α-specifically by some external agency after being detached from the globin apoprotein. A number of workers have postulated enzyme systems with this function.

For many years the most prominent example was the 'heme α-methenyl oxygenase' system proposed by Nakajima and coworkers[29-32]. This was described as a soluble hepatic enzyme acting on hemoglobin or pyridine hemochrome. However, critical investigation subsequently revealed[39,86,87] that this activity was in reality attributable to ascorbate, possibly loosely bound to proteins in the tissue extracts. This acted on the hemochrome 'substrates' by promoting a purely chemical coupled oxidation identical to that already well known. The rather unusual 'formylbiliverdin' structure assigned to the product of this system by Nakajima[30] was not confirmed; the product was shown, rather to be simply the verdoheme—pyridine complex[39,86,87]. Contrary to the claims made for it[32], this system was also shown[39] to promote essentially random rather than α-specific cleavage, as in the case of the coupled oxidation of pyridine hemochrome. Wise and Drabkin[26,27] have briefly described heme-cleaving activity in homogenates of the hemophagous organ of the dog, but the relevance of their findings to the problem of hemoglobin catabolism is uncertain. The organ referred to is a specialized area of the placenta. Its function is obscure and this particular activity has not been reported in any other species but the dog.

Currently the most convincing evidence in favor of an enzymic system relevant to hemoglobin catabolism stems from the work of Tenhunen, Schmid and co-workers who observed and characterized heme-cleaving activity in microsomal preparations from liver and other tissues which they attribute to a microsomal heme cleaving enzyme, termed heme oxygenase[33-37]. Their evidence indicated an involvement of cytochrome *P*-450 in this heme-cleaving system. This involvement was interpreted in terms of a function similar to that which *P*-450 is known to perform in the hydroxyla-tion of metabolites and drugs, that is, it is envisaged that the *P*-450 activates oxygen which is then transferred to an external active site where it is em-ployed to hydroxylate a heme substrate molecule. This mechanism, as en-visaged by Tenhunen, Schmid et al.,[36,88,89], is illustrated in Scheme 9.

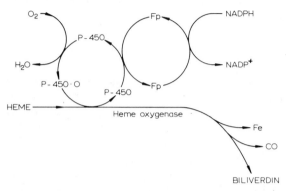

Scheme 9. Microsomal heme-cleavage as visualized by Tenhunen, Schmid and co-workers in terms of a heme oxygenase enzyme whose specific function is heme-cleavage. (Based on the scheme proposed by Tenhunen, Marver and Schmid in Refs. 36, 88 and 89).

Yoshida and co-workers[90] and Maines and Kappas[90a] have recently reported NADPH-dependent microsomal heme cleavage activity that does not seem to be associated with cytochrome *P*-450. Nichol[91] previously re-ported heme-cleavage activity in chicken macrophages which he interpreted similarly to be unassociated with a component formally recognizable as *P*-450. These activities, however, are dependent on the flavoprotein NADPH-dehydrogenase which transfers electrons from NADPH. This micro-somal flavoprotein 'normally' feeds electrons from NADPH to cytochrome *P*-450. The heme-cleavage end of these systems described by Yoshida et al., and by Nichol might thus be imagined as taking the place of *P*-450 in the microsomal electron transport system, rather than being linked to *P*-450 as in the model envisaged by Tenhunen, Schmid et al., (Scheme 9). Such a situation bears some resemblance to the model proposed by O'Carra and Colleran[38–40] for microsomal heme cleavage based on a coupled oxidation mechanism. This model was derived from studies of the non-enzymic coupled oxidation of hemoproteins, but it incorporates many of the formal features attributed to enzymic heme-cleaving systems (see below). The features distinguishing this model from the purely enzymic ones may be formulated as follows. The heme cleavage is still regarded as being caused basically by a type of coupled oxidation of heme in situ in specific heme-binding sites. It is postulated that biological heme cleavage can arise as a result of a fortuitous or 'accidental' suicide reaction of the heme groups of certain hemoproteins which are exposed to this 'risk' by their oxygen-binding and oxygen-activating function. On the basis of current evidence it seems probable that at least some biological heme cleavage is 'accidental' in this sense, but that much of the cleavage may arise from the action of enzymic systems whose specific purpose is heme cleavage. However, the proposed microsomal enzymic systems mentioned above are so similar in many essential respects to the 'accidental' ones described below, that it is tempting to speculate that any 'purposeful' heme cleaving system may have evolved from an 'accidental' system in which the heme group originally functioned as an 'at risk' component in an oxygen-activating capacity, rather than as a substrate subjected 'purposely' to heme cleavage.

The proposals of O'Carra and Colleran[38−40] arise basically because the striking α-specificity imposed by the apoprotein of myoglobin on the coupled oxidation of its heme group (and a similar specificity revealed in the case of a microsomal hemoprotein thought to be cytochrome P-450), seems to provide rather too characteristic and relevant a connection with in vivo heme cleavage to be dismissed as merely coincidental.

Myoglobin was also shown to promote α-specific cleavage of added or exogenous heme under conditions promoting only very slow and random coupled oxidation of the exogenous heme itself (added as ferriheme). A typical experiment demonstrating α-specific catalysis of heme-cleavage by myoglobin is illustrated in Fig. 3, and the interpretation of the mechanism of the effect is outlined schematically in Scheme 10. It is envisaged that when the endogenous heme group of a myoglobin molecule is cleaved by coupled oxidation, the biliverdin—iron complex is quickly displaced by an exogenous heme molecule which in turn may undergo coupled oxidation to biliverdin and be displaced in the same way. In promoting such repeated cycles of heme cleavage, the apoprotein is acting with all the formal characteristics of an α-specific heme-cleaving enzyme, the heme-binding crevice being equivalent to the enzyme active site. Some results obtained by the same workers by incubating microsomal preparations with NADPH in the presence and absence of added or exogenous heme were interpreted in terms of a similar quasi-enzymic process taking place in the heme-binding site of a microsomal hemoprotein, suggested to be cytochrome P-450. That is, the observed NADPH-promoted formation of biliverdin from endogenous microsomal heme was interpreted as a type of coupled oxidation of the heme group of P-450 with NADPH acting as the reductant in place of the ascorbate used in the case of myoglobin. Conversion of added heme to biliverdin was considered to take place by the same cyclic type of repeated cleavage and replacement as was demonstrated in the case of myoglobin. The suggested mechanism for this heme-cleavage effect of cytochrome P-450 is summarized in Scheme 11. The biliverdin produced in these experiments was also identified as the IXα isomer.

Certain other analogies between cytochrome P-450 and myoglobin lend credence to the view that the mechanism of heme cleavage in microsomes is

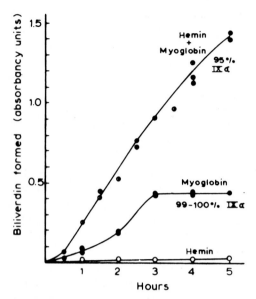

Fig. 3. Demonstration of the phenomenon of (apo)myoglobin acting as an α-methine-specific heme-cleaving enzyme. Hemin alone, metmyoglobin alone, or hemin + metmyo-globin were incubated aerobically with ascorbate in 0.1 M phosphate buffer, pH 7 for the indicated periods of time. The biliverdin product was isolated, assayed spectrophoto-metrically and the yield of biliverdin was plotted *versus* time of incubation. The bili-verdin samples were also analyzed chromatographically to determine their isomer com-positions with the results indicated on the curves. The lag phase at the beginning seems to be caused by the necessary reduction of the metmyoglobin to myoglobin. The leveling-out of the curve for myoglobin alone after 3 to 4 hours corresponds with essentially complete cleavage of its endogenous heme component to biliverdin. [Reproduced from P. O'Carra and E. Colleran, FEBS Letters, 5, 295 (1969).]

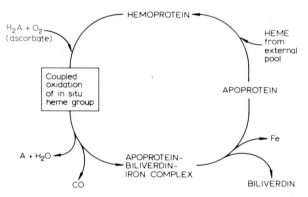

Scheme 10. Schematic representation of the mechanism envisaged by O'Carra and Col-leran[38] for the quasi-enzymic process by which the coupled oxidation of free hemin is accelerated and made α-specific by the presence of catalytic amounts of myoglobin (see Fig. 3).

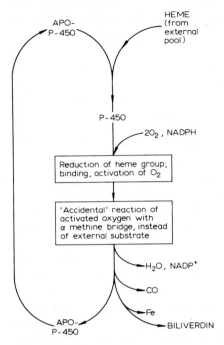

Scheme 11. Mechanism suggested by O'Carra and Colleran[38,39,40] for heme-cleaving activity of microsomal preparations.

analogous to that occurring in the better-defined myoglobin system. Both hemoproteins bind molecular oxygen when the iron is in the reduced form. In the presence of ascorbate, the oxymyoglobin seems to accept further reducing equivalents which serve to activate the oxygen, leading to hydroxylation and subsequent cleavage of the α-methine bridge. Oxy-P-450 similiarly accepts a further reducing equivalent from NADPH (probably *via* a flavoprotein) and this serves to activate the bound oxygen which is then 'normally' released to hydroxylate exogenous substrates, such as drugs, steroids etc. (see, for example, Ref. 92). It is not difficult to imagine the activated oxygen periodically and 'accidentally' hydroxylating the methine bridge of the heme group instead.

Such an 'accidental' or quasi-enzymic mechanism could account for a large part of the ELP fraction of bilirubin (see p. 130), which is thought to be derived by turnover of hepatic P-450. It is known that the heme group turns over at a far more rapid rate than the apoprotein[93-95] a fact in keeping with this hypothesis.

If the heme of catabolized hemoglobin finds its way into the general pool of hepatic heme it might very well be used to replace the rapidly turning-over heme of P-450 and thus could be cleaved 'at second hand' in P-450. Alternatively a functionally-specific activity may have evolved to dispose of excess heme accumulating from hemoglobin catabolism. It seems possible that such an activity could arise most readily as a particularly labile or 'accident prone' form of P-450.

The view of Yoshida et al. that microsomal heme-cleaving activity is independent of P-450 (see above) is based largely on a lack of correlation between the level of the activity and the level of spectrally identifiable P-450 in microsomal preparations. However, the evidence does not exclude the possibility that the heme-cleaving activity may be caused by a spectrally-altered form of P-450. Such alterations of P-450 are well known, and indeed some data of Maines and Kappas[90a] might be interpreted as supporting such an interpretation. These authors found that treatment of microsomal preparations, in vitro, with 4 M urea led to considerable increases in the heme-cleaving activity while concomitantly decreasing the level of spectrally-identifiable P-450, presumably owing to conversion to one of the modified forms of P-450.

Hepatic catalase, as isolated, contains biliverdin in a 1 : 3 ratio with respect to heme. The observation of Morell and O'Carra[22], that this biliverdin is an equimolar mixture of the α and β isomers indicates that it is probably formed in situ in catalase itself, since this isomeric composition is identical with that produced by coupled oxidation of catalase in vitro (Table 2). This presumably represents an example of 'accidental' or fortuitous heme cleavage; catalase, in performing its function of disproportionating hydrogen peroxide, binds the peroxide via its heme groups. This again represents a complex of heme with activated oxygen and presumably puts the heme 'at risk'.

Acknowledgement

I thank the Medical Research Council of Ireland for their support of the studies on heme-cleavage carried out at Galway.

References

1. P. O'Carra and C. O'hEocha in 'Chemistry and Biochemistry of Photosynthetic Pigments', 2nd ed., Ed. T.W. Goodwin, Academic Press, London, in press.
2. W. Rüdiger, in 'Fortschritte der Chemie organischer Naturstoffe', Eds. W. Herz, H. Grisebach, and G.W. Kirby, Vol. XXIX, Springer-Verlag, Vienna (1971) p. 60.
3. H. Fischer and H. Orth, 'Die Chemie des Pyrrols', Vol. II, Parts 1 and 2, Akademische Verlagsgesellschaft, Leipzig (1937, 1940).
4. O. Warburg and E. Negelein, Chem. Ber., **63**, 1816 (1930).
5. R. Lemberg, Biochem. J., **29**, 1322 (1935).
6. I.M. London, R. West, D. Shemin, and D. Rittenberg, J. Biol. Chem., **184**, 351, 365 (1950).
7. C.H. Gray, A. Neuberger, and P.H.A. Sneath, Biochem. J., **47**, 87 (1950).

8. C.H. Gray, H.M. Muir, and A. Neuberger, Biochem. J., **47**, 542 (1950).
9. I.M. London, J. Biol. Chem., **184**, 373 (1950).
10. J.D. Ostrow, J.H. Jandl, and R. Schmid, J. Clin. Invest., **41**, 1628 (1962).
11. G.W. Ibrahim, G.W. Schwartz, and C.J. Watson, Metab. Clin. Expt., **15**, 1129 (1966).
12. S.H. Robinson, C.A. Owen Jr., E.V. Flock, and R. Schmid, Blood, **26**, 823 (1965).
13. E. Colleran, Ph.D. Thesis, National University of Ireland, Galway (1971).
14. P. O'Carra, Biochem. J., **119**, 2P (1970).
15. T. Sjöstrand, Nature London, **168**, 729, 1118 (1951).
16. T. Sjöstrand, Acta Physiol. Scand., **26**, 334, 338 (1952).
16a.T. Sjöstrand, Ann. N.Y. Acad. Sci., **174**, 5 (1970).
17. G.D. Ludwig, W.S. Blakemore, and D.L. Drabkin, Biochem. J., **66**, 38 P (1957).
18. F.K. Anan and H.S. Mason, J. Biochem. Tokyo, **49**, 765 (1961).
19. R.F. Troxler, A. Brown, R. Lester, and P. White, Science, **167**, 192 (1970).
20. R.F. Troxler, Biochemistry, **11**, 4235 (1972).
21. W. Rüdiger, W. Klose, M. Vuillaume, and M. Barbier, Experientia, **24**, 1000 (1968).
22. D.B. Morell and P. O'Carra, Ir. J. Med. Sci., **143**, 181 (1974).
23. P. O'Carra and E. Colleran, J. Chromatogr., **50**, 458 (1970).
24. R. Lemberg and J.W. Legge, 'Hematin Compounds and Bile Pigments', Interscience Publishers Inc., New York (1949).
25. R. Lemberg, Rev. Pure Appl. Chem., **6**, 1 (1956).
26. C.D. Wise and D.L. Drabkin, Fed. Proc., Fed. Am. Soc. Exp. Biol., **23**, 323 (1964).
27. C.D. Wise and D.L. Drabkin Fed. Proc., Fed. Am. Soc. Exp. Biol., **24**, 222 (1965).
28. D.L. Drabkin, Ann. N.Y. Acad. Sci., **174**, 49 (1970).
29. H. Nakajima, T. Takemura, O. Nakajima, and K. Yamaoka, J. Biol. Chem., **238**, 3784 (1963).
30. H. Nakajima, J. Biol. Chem., **238**, 3797 (1963).
31. O. Nakajima in 'Bilirubin Metabolism', Eds, I.A.D. Bouchier and B.H. Billing, Oxford: Blackwell (1967) p. 55.
32. O. Nakajima and C.H. Gray, Biochem. J., **104**, 20 (1967).
33. R. Tenhunen, H.S. Marver, and R. Schmid, Proc. Nat. Acad. Sci. U.S.A., **61**, 748 (1968).
34. R. Tenhunen, H.S. Marver, and R. Schmid, J. Biol. Chem., **244**, 6388 (1969).
35. R. Tenhunen, H. Marver, N.R. Pimstone, W.F. Trager, D.Y. Cooper, and R. Schmid, Biochemistry, **11**, 1716 (1972).
36. R. Tenhunen, Semin. Hematol., **9**, 19 (1972).
37. N. Pimstone, R. Tenhunen, P.T. Seitz, H.S. Marver, and R. Schmid, J. Exp. Med., **133**, 1264 (1971).
38. P. O'Carra and E. Colleran, FEBS Lett., **5**, 295 (1969).
39. E. Colleran and P. O'Carra, Biochem. J., **119**, 905 (1970).
40. P. O'Carra and E. Colleran, Proceedings of Bilirubin Meeting, Hemsedal, Norway (1974).
41. P. O'Carra, E. Colleran, and M. Delaney, Proceedings of Bilirubin Meeting, Hemsedal, Norway (1974).
42. J.W. Singleton and L. Laster, J. Biol. Chem., **240**, 4780 (1965).
43. R. Tenhunen, M.E. Ross, H.S. Marver, and R. Schmid, Biochemistry, **9**, 298 (1970).
44. E. Colleran and P. O'Carra, Biochem. J., **119**, 16P (1970).
45. P. O'Carra and E. Colleran, Biochem. J., **125**, 110P (1971).
46. S.H. Robinson, M. Tsong, B.W. Brown, and R. Schmid, J. Clin. Invest., **45**, 1569 (1966).
47. T. Yamamoto, J. Skandenberg, A. Zipursky, and L.O.G. Israels, J. Clin. Invest., **44**, 31 (1965).
48. R. Schmid, H.S. Marver, and L. Hammaker, Biochem. Biophys. Res. Commun., **24**, 319 (1966).

49. S.A. Landau, Ann. N.Y. Acad. Sci., **174**, 32 (1970).
50. J. Barrett, Nature, London, **215**, 733 (1967).
51. S. Besecke and J.H. Fuhrhop, Angew. Chem., Internat. Edn. Engl., **13**, 150 (1974).
52. R. Lemberg, B. Cortis-Jones, and M. Norrie, Biochem. J., **32**, 149 (1938).
53. R. Lemberg, B. Cortis-Jones, and M. Norrie, Biochem. J., **32**, 171 (1938).
54. R. Lemberg, M. Norrie, and J.W. Legge, Nature London, **114**, 551 (1939).
55. R. Lemberg, W.H. Lockwood, and J.W. Legge, Biochem. J., **35**, 363 (1941).
56. R. Lemberg, J.W. Legge, and W.H. Lockwood, Nature London, **142**, 148 (1938).
57. J.E. Kench, C. Gardikas, and J.F. Wilkinson, Biochem. J., **47**, 129 (1950).
58. J.E. Kench and C. Gardikas, Biochem. J., **52**, XV (1952).
59. J.E. Kench, Biochem. J., **56**, 669 (1954).
60. J.E. Kench, Biochem. J., **52**, XXVII (1952).
61. P. O'Carra and E. Colleran, submitted for publication
62. P. O'Carra and E. Colleran, Biochem. J., **119**, 42P (1970).
63. J.C. Kendrew, Science, **139**, 1259 (1963).
64. M.F. Perutz, Proc. Roy. Soc. London, Ser. B, **173**, 113 (1969).
65. N.A. Brown, R.F.G.J. King, M.E. Shillcock, and S.B. Brown, Biochem. J., **137**, 135 (1974).
66. H. Fischer and H. Libowitzky, Hoppe-Seyler's Z. Physiol. Chem., **251**, 198 (1938).
67. H. Libowitzky, Hoppe-Seyler's Z. Physiol. Chem., **265**, 191 (1940).
68. E. Stier, Hoppe-Seyler's Z. Physiol. Chem., **272**, 239 (1942).
69. E. Stier, Hoppe-Seyler's Z. Physiol. Chem., **273**, 47 (1942).
70. A.H. Jackson, G.W. Kenner, and K.M. Smith, J. Chem. Soc. C, 302 (1968).
71. (a) A.H. Jackson and G.W. Kenner in 'Porphyrins and Related Compounds', (Biochem. Soc. Symposium, No. 28), Ed. T.W. Goodwin, Academic Press, London (1968), p. 3.
    (b) K.M. Smith, Q. Rev. (London), **25**, 31 (1971).
72. A.H. Jackson, Proceedings of Bilirubin Meeting, Hemsedal, Norway (1974).
73. E.Y. Levin, Biochemistry, **5**, 2845 (1966).
74. R.B. Woodward, Ind. Chim. Belge, **27**, 1293 (1962).
75. B. Pullman and A.M. Perault, Proc. Nat. Acad. Sci. U.S.A., **45**, 1476 (1959).
76. Z.J. Petryka, D.C. Nicholson, and C.H. Gray, Nature London, **194**, 1047 (1962).
77. R.A. Nicolaus, Rass. Med. Sper., **7**, Suppl. 2, p. 1 (1960).
78. W. Rüdiger, Hoppe-Seyler's Z. Physiol. Chem., **350**, 291 (1969).
79. W. Rüdiger in 'Porphyrins and Related Compounds' (Biochem. Soc. Symposium, No. 28), Ed. T.W. Goodwin, Academic Press, London (1968) p. 121.
80. (a) E. Colleran and P. O'Carra, Biochem. J., **115**, 13P (1969);
    (b) P. O'Carra and E. Colleran, J. Chromatogr., in press (1975).
81. R. Bonnett and A.F. McDonagh, Chem. Commun., 237 (1970).
82. R. Bonnett and A.F. McDonagh, J. Chem. Soc., Perkin Trans. 1, 881 (1973).
83. A.W. Nichol and D.B. Morell, Biochim. Biophys. Acta, **184**, 173 (1969).
84. T. Kondo, D.C. Nicholson, A.H. Jackson, and G.W. Kenner, Biochem. J., **121**, 601 (1971).
85. (a) K.E. Nichols and L. Bogorad, Bot. Gaz. Chicago, **124**, 85 (1962);
    (b) L. Bogorad, Rec. Chem. Prog., **26**, 1 (1965).
86. E.Y. Levin, Biochim. Biophys. Acta, **136**, 155 (1967).
87. R.F. Murphy, C. O'hEocha, and P. O'Carra, Biochem. J., **104**, 6C (1967).
88. R. Schmid, in 'Metabolic Basis of Inherited Disease', 3rd edn., Eds. J.B. Stanbury, J.B. Wyngaarden, and D.S. Fredrickson, McGraw Hill, New York (1972) p. 1141.
89. R. Tenhunen, H.S. Marver, and R. Schmid, Trans. Assoc. Am. Physicians, **82**, 363 (1969).
90. T. Yoshida, S. Takahashi, and G. Kikuchi, J. Biochem. Tokyo, **75**, 1187 (1974).
90a. M.D. Maines and A. Kappas, Proc. Nat. Acad. Sci. U.S.A., **71**, 4293 (1974).
91. A.W. Nichol, Biochim. Biophys. Acta, **222**, 28 (1970).

92. R.W. Estabrook, J. Baron, J. Peterson, and Y. Ishimura, in 'Biological Hydroxylation Mechanisms', (Biochem. Soc. Symposia, No. 34), Eds. G.S. Boyd and R.M.S. Smellie, Academic Press, London (1972) p. 159.
93. W. Levin and R. Kunzman, J. Biol. Chem. **244**, 3671 (1969).
94. R.C. Garner and A.E.M. McLean, Biochim. Biophys. Acta, **37**, 883 (1969).
95. W. Levin, M. Jacobson, E. Sernatenger, and R. Kunzman, Drug Metab. Dispos., **1**, 275 (1973).

COORDINATION CHEMISTRY OF METALLOPORPHYRINS

STATIC COORDINATION CHEMISTRY OF METALLOPORPHYRINS

J.W. BUCHLER

Institut für Anorganische Chemie, Technische Hochschule Aachen, D-51 Aachen, Federal Republic of Germany

5.1. Introduction

Metalloporphyrins are porphin derivatives in which at least one of the lone electron pairs residing on the central nitrogen atoms of porphin [(1); $H_2(P)$*] is shared with a metal ion acting as a Lewis acid[1-12]. In the normal case, a metal ion not only coordinates to the two lone pairs on the pyrrolenine nitrogen atoms of a free acid porphyrin**, but furthermore replaces the two hydrogen atoms on the pyrrole nitrogen atoms; thus, two more lone pairs are donated from the porphinato ligand to the valence shell of the metal ion which now nearly or completely occupies the center of the porphin hole as depicted in (2), Fig. 1. The bar graph (A) represents (2) as viewed sideways along the straight line penetrating C_δ, M, and C_β and defines the schematical representations of the coordination types used here-after, see e.g. Section 5.2. The formation of a metalloporphyrin M(P), (2), may be arbitrarily regarded as the combination of a dinegative porphin anion $(P)^{2-}$ [resulting from deprotonation of $H_2(P)$] with a dipositive metal ion assuming the square planar geometry (A) [Fig. 1, eq. (A)]. The negative charge of $(P)^{2-}$ is delocalized through the π-electron system over the four nitrogen atoms. The system $(P)^{2-}$ found in most of the metalloporphyrins represents a rigid, closed, macrocyclic tetradentate dinegative chelating ligand. The formation of a metalloporphyrin [eq. (A)] is called 'metalation', the removal of the metal [reverse of eq. (A)] 'demetalation'[1].

$$H_2(P) \quad \underset{\underset{2\,H^+}{\underset{(A1)}{\longleftarrow}}}{\overset{-2\,H^+}{\longrightarrow}} \quad (P)^{2+} \quad \underset{\underset{-M^{2+}}{\underset{(A3)}{\longleftarrow}}}{\overset{M^{2+}}{\longrightarrow}} \quad M(P) \qquad (A)$$

---

* See Table 1 for the abbreviations used in this article.
** The free acid of porphin, $H_2(P)$, is often called 'free base porphin' because the term 'free acid porphyrin' applies also to species containing carboxylic acid functions like $H_2(\text{Proto-IX})$. In this context, the free acid $H_2(P)$ corresponds to the porphinato dianion $(P)^{2-}$

Porphyrins and Metalloporphyrins, ed. Kevin M. Smith
© 1975, Elsevier Scientific Publishing Company, Amsterdam, The Netherlands

TABLE 1
Abbreviations used in this article

a) Porphyrin free acids

H$_2$(P)	= Porphin
H$_2$(Pc)	= Phthalocyanine
H$_2$(OEP)	= Octaethylporphyrin
H$_2$(TPP)	= *meso*-Tetraphenylporphyrin
H$_2$(TTP)	= *meso*-Tetra(*p*-tolyl)porphyrin
H$_2$(TPyP)	= *meso*-Tetra(4-pyridyl)porphyrin
H$_2$(TAP)	= *meso*-Tetra(4-methoxyphenyl)porphyrin
H$_2$(Etio-II)	= Etioporphyrin-II
H$_2$(Proto-IX)	= Protoporphyrin-IX
H$_2$(Proto-IX-DME)	= Protoporphyrin-IX Dimethyl ester
H$_2$(Meso-IX)	= Mesoporphyrin-IX
H$_2$(Meso-IX-DME)	= Mesoporphyrin-IX Dimethyl ester
H$_2$(Deut-IX)	= Deuteroporphyrin-IX Dimethyl ester
H$_2$(2,4-AcDeut-IX)	= 2,4-Diacetyldeuteroporphyrin-IX
H$_2$(OEPMe$_2$)	= α,γ-Dimethyl-α,γ-dihydro-octaethylporphyrin
H$_2$(OEC)	= Octaethylchlorin
H$_2$(TPC)	= *meso*-Tetraphenylchlorin
H$_2$(Phy-*a*)	= Pheophytin-*a*
H$_2$(Bphy-*a*)	= Bacteriopheophytin-*a*
H$_2$(Chlorin-*e*$_6$-TME)	= Chlorin-*e*$_6$ trimethyl ester

b) Metalloporphyrins

Mg(OEP)	Octaethylporphinatomagnesium
Fe(Etio-I)Cl	Chloro(etioporphinato-I)iron(III)
Sn(TPP)Cl$_2$	Dichloro(tetraphenylporphinato)tin(IV)

c) Neutral donor ligands L (or solvents)

Et$_2$O	Diethyl Ether
THF	Tetrahydrofuran
MeOH	Methanol
EtOH	Ethanol
DMSO	Dimethylsulfoxide
DMF	Dimethylformamide
Me$_2$NH	Dimethylamine
Et$_3$N	Triethylamine
Pip	Piperidine
Py	Pyridine
Col	Collidine
Lut	Lutidine
t-BuPy	4(tert.Butyl)pyridine
ImH	Imidazole
1-MeIm	1-Methylimidazole
(2-MeIm)H	2-Methylimidazole
1,2-MeIm	1,2-Dimethylimidazole
PPh$_3$	Triphenylphosphine
P(OMe)$_3$	Trimethylphosphite
AsPh$_3$	Triphenylarsine

TABLE 1 (continued)

d) Uninegative, unidentate organic anionic ligands X

X^-	Name
OAc^-	Acetate
$(acac)^-$	Acetylacetonate
OEt^-	Ethoxide
$(Im)^-$	Imidazolide
OMe^-	Methoxide
OR^-	Alkoxide
OPh^-	Phenoxide
R^-	General aliphatic anion

e) Bidentate and polydentate ligands

General type	Abbreviation	Ligand name
XL (anionic bidentate)	OAc^-	Acetate
	$(acac)^-$	Acetylacetonate
	$(dbm)^-$	Dibenzoylmethanate
	N_3^-	Azide
$(XL)_2$ (dianionic tetra-dentate)	$(pim)^{2-}$	Pimelate, $(CH_2)_5(CO_2)_2^{2-}$
	$(phthal)^{2-}$	o-Phthalate
	$(suc)^{2-}$	Succinate, $(CH_2)_2(CO_2)_2^{2-}$

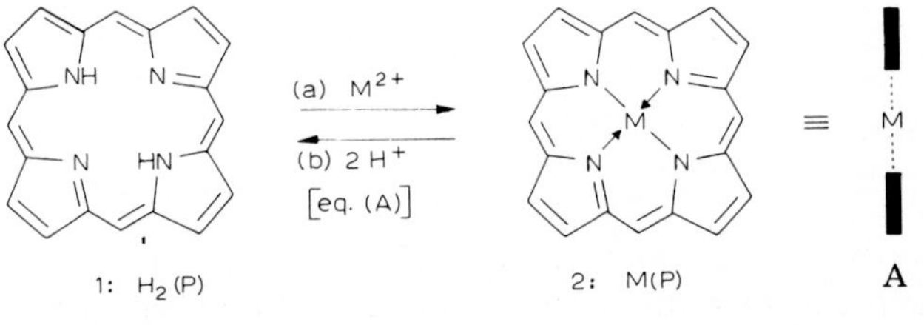

Fig. 1. Formation of the equatorial coordination group of a monometallic metalloporphy-rin M(P) (2). (Aa): metalation, (Ab): demetalation of a porphyrin $H_2(P)$ (1).

Equation (A), however, gives only a very simplified picture of the various chemical processes involved. They are explained below and listed formally in Scheme 1. In fact, the chemical processes which are logically separated into types (A1) to (A5) in the scheme probably occur in all sorts of interchanged sequences.

I. *Protonation/deprotonation equilibria* [(A1), left half of eq. (A)]. Porphins may be protonated and deprotonated at the central nitrogen atoms according the quations (A1) listed in Scheme 1.

References, p. 224

Scheme 1

Formal steps of metalloporphyrin synthesis

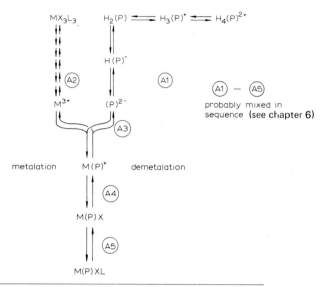

I	Protonation - deprotonation equilibria	A1
II	Deconvolution of the metal ion from the metal carrier	A2
III	Formation of the equatorial metal nitrogen plane	A3
IV	Adjustment of the charge balance	A4
V	Completion of the coordination sphere	A5

M: metal atom
(P): porphin Ligand
X: uninegative anion
L: neutral electron pair donor

II. Deconvolution of the metal ion from the 'metal carrier'. The compound which brings the metal into the reacting system will be called 'metal carrier' subsequently. The right half of eq. (A), $(A3)$, implies that a bare $(P)^{2-}$ ion combines with a 'naked' M^{2+} ion to yield M(P). This would be an ideal situation for the formation of the four ligand-to-metal bonds. But under normal reaction conditions, these bare species will not be present, and other activated species have to be generated which successively allow the formation of the 4 M—N bonds. In the course of the metalation, the metal ion must be 'unwrapped' of the ligands surrounding it in the metal carrying compound. This may be a very decisive step in a metalation reaction and explains why sometimes quite unique chemicals have to be used as metal carriers.

III. Formation of the equatorial MN_4 plane. The formation of the equatorial coordination group is straightforward [eq. $(A3)$, Scheme 1], but compli-

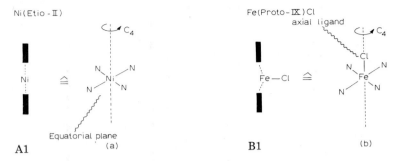

Fig. 2. Coordination types of nickel etioporphyrin-II [short formula Ni(Etio-II); (A1)] and hemin [Fe(Proto-IX)Cl; (B1)]. The bold bar represents the projection of the porphyrin plane as in Fig. 1. The bar graphs are shown together with the usual representations of the square planar coordination groups (a) of the NiN$_4$-system in Ni(Etio-I) (roughly of local symmetry D$_{4h}$) and the square pyramidal coordination group (b) of the FeN$_4$Cl-system in Fe(Proto-IX)Cl (local symmetry C$_{4v}$). These drawings serve to illustrate the terms 'coordination type' (Section 5.2), 'equatorial plane' ('equatorial coordination chemistry', Section 5.3), 'axial ligand' ('axial coordination chemistry', Section 5.4).

cated by the fact that all the other reactions may, and mostly do, interfere with the consecutive addition of the 4 pyrrole nitrogen atoms to the metal atom. A good example is Ni(Etio-II), (A1) (Fig. 2), where the reaction stops with the formation of the planar array. The local symmetry of the porphinato-nickel core roughly is D$_{4h}$, a 4-fold rotational axis C$_4$ penetrating the nickel atom.

IV. Adjustment of the charge balance. As most metalloporphyrins represent a rather nonpolar and large moiety which is therefore soluble only in organic solvents, they will tend to strongly associate with ions X of negative charge, if the metal has a positive charge $Z > 2$. [Scheme 1, eq. (A4)]. Usually, an uncharged metalloporphyrin is formed by coordination of the anion to the central metal if it has enough electron acceptor sites left after the formation of the equatorial MN$_4$ plane. This is the case in hemin Fe(Proto-IX)Cl where $Z = 3$, the chloride ion neutralizing the excess charge by coordinating to the iron ion forming the square pyramid (B1) (Fig. 2). The local symmetry of the FeN$_4$Cl-coordination group is C$_{4v}$. The iron-chlorine bond is coaxial with the main rotational (C$_4$) axis. From this picture, the name 'axial ligand' is derived for all extraplanar ligands, although in many examples they are not coaxial with rotational axes of the molecules. The adjustment of electroneutrality is then one of the driving forces to build up the axial coordination groups further discussed in Sections 5.2 and 5.4.1.

V. Completion of the coordination sphere. The four coordinate, square planar geometry (A) shown in Fig. 1, which is a consequence of the steric conditions of the $(P)^{2-}$ ligand, can be realized with only a few metals, most of which already prefer the coordination number CN = 4 under square planar

Fe(Proto-IX)Py$_2$ Rh(Etio-I)Cl·H$_2$O

C1   (a)   C2   (b)

Fig. 3. Short formula, bar graphs and central coordination groups of the distorted-octahe-
dral coordination type (C) (Section 5.2) in bis/pyridine)hemochrome [Fe(Proto-IX)Py$_2$-
(C1)] and in aquochlorooctaethylporphinatorhodium(III) [Rh(OEP)Cl · H$_2$O, (C2)].

geometry with unidentate ligands, e.g. Ni, Pd, Pt. Under normal conditions,
the metal ions once embedded into the equatorial N$_4$ plane tend to complete
their coordination sphere by further complexing with neutral donor molecules L, e.g. water or pyridine [Scheme 1, eq. ($A5$)]. A typical example is
the distorted octahedron of the pyridine hemochrome Fe(Proto-IX)Py$_2$ (C1)
(Fig. 3). Likewise, this process may happen with metalloporphyrins in which
the metal has an oxidation number Z > 2 and where the desired coordination number CN of the metal is not yet reached by uptake of neutralizing
anions according eq. ($A4$). This may be the case in Rh(Etio-I)Cl · H$_2$O (C2)
(Fig. 3).

The completion of the coordination sphere in the general case where CN
> Z + 2 is then the other main driving force in determining the various
coordination groups (Sections 5.2 and 5.4).

It is obvious from eqs. ($A1$)—($A5$) that strong acids will reverse the reactions thus causing 'demetalation' if equilibria ($A1$)—($A5$) are established.
Summarizing the situation by inspection of Scheme 1 tells us that the formation of a general metalloporphyrin M(P)XL under the conditions of the
scheme requires the fission of 8 bonds and the formation of 6 new bonds.
Consequently, the kinetics and mechanism of metalloporphyrin formation
according eqs. ($A1$)—($A5$) are a field of its own which is treated separately in
Chapter 6 concerning 'dynamic coordination chemistry of metalloporphyrins'. It must be stated that the reactions ($A1$)—($A5$) may occur in all sorts
of combined sequences which are very difficult to resolve experimentally.
The research groups working in this field have a very difficult task.

Moreover, the Scheme 1 in itself is still simplified as it neglects any change
in oxidation state of the metal ion involved and the formation of bimetallic
metalloporphyrins which may in some cases occur during the formation of
the monometallic species mentioned so far. [Monometallic means a 1 : 1
molar ratio of metal per porphyrin, bimetallic a 2 : 1 molar ratio. This is
evident from the general formulae: M(P) versus M$_2$(P)]. The two main examples of bimetallic metalloporphyrins are shown in Fig. 4. The coordination

types of bimetallic metalloporphyrins are presented in Section 5.2. The chemistry of these 'unusual' metalloporphyrins is treated in Chapter 7.

This article will be confined to the static aspects of the following topics:

1) *Classification of metalloporphyrins* (Section 5.2.)

A classification of metalloporphyrins seems necessary because the knowledge on metalloporphyrins has enormously grown since 1964, when Falk's monograph[1] appeared.

2) *Equatorial coordination chemistry* (Section 5.3)

Here, the synthetic aspects of the reactions (*A1*), (*A2*), and (*A3*), i.e. the formation and dissociation of the four metal-to-nitrogen bonds, will be discussed irrespective of all axial ligand chemistry. Some generalizations will be presented.

3) *Axial coordination chemistry* (Section 5.4)

Under this heading, the synthesis and characterization of metalloporphyrins with specific axial ligands will be described. The clear separation of 'equatorial' and 'axial' chemistry in metalloporphyrins has proven to be very useful[1,12].

5.2. Classification of metalloporphyrins

In order to obtain a general view of all the various metalloporphyrins known, the chemical and physical properties of which will be discussed in subsequent chapters, a three-dimensional classification of the various metalloporphyrins which have entered discussion in the literature will be presented here. This undertaking will give the quickest answer to the question of whether or not a certain kind of metalloporphyrin exists. The three dimensions will be: 1) 'Equatorial' stoichiometry; 2) Coordination type; 3) 'Axial' stoichiometry.

1) The 'equatorial' stoichiometry defines the metal-to-porphyrin ratio thus providing a means to devide the set of all possible metalloporphyrins $M_h(P)_k X_m L_n$ into classes (see Table 2): I. Monometallic mononuclear species [M(P)]. II. Semimetallic species $[M(P)_2] = [M_{1/2}(P)]$ in which two porphyrin disks are held together by one metal ion; III. Bimetallic species $[M_2(P)]$ which are monocuclear as regards the porphin moiety (inorganic chemists may still call these species binuclear complexes the porphinato ligand acting as a bridging unit). IV. Monometallic binuclear species $[M(P)]_2$ where two monometallic mononuclear species are combined by metal—metal bonds or bridging ligands. V. Polymetallic mononuclear species $[M_h(P)]$ where h = 3 or 4. VI. Polynuclear Species $[M_h(P)_k]$ where h, k > 2.

2) Each of the classes generated by an analysis of the equatorial stoichiometry may then be divided into various coordination types according stereochemistry (Table 2). Thus, the monometallic mononuclear species which are presently discussed may be split into the coordination types (A)—(H) (Figs.

TABLE 2

Classification of general metalloporphyrins $[M_h(P)_k X_m L_n]$ according equatorial or axial stoichiometry and stereochemistry

Classes generated by equatorial stoichiometry	Character of the $[M_h(P)_k]$-moiety	Types generated by stereochemistry	Number of sub-classes generated by axial stoichiometry
I M(P)	Monometallic Mononuclear h, k = 1	(A) (B) (C) (D), (E), (F), (G), (H) [Fig. 5—7]	(A): 2 (B): 4 (C): 6
II M(P)$_2$	Semimetallic h = 1, k = 2	(K) [Fig. 8]	
III M$_2$(P)	Bimetallic h = 2, k = 1	(L), (M), (N), (P), (Q) [Fig. 9]	
IV [M(P)]$_2$	Monometallic Binuclear h, k = 2	(AA) (BB) (GG) (CC) [Fig. 10]	(CC): 3
V M$_3$(P) M$_4$(P)	Polymetallic h = 3,4; k = 1	(R), (S), (T) [Fig. 11]	
VI M$_h$(P)$_k$	Polynuclear h > 2 k ⩾ 2	(DKD), (BCB), (CCC) Poly (C), Poly (G) [Fig. 12, 13]	

M: metal (P): porphyrin ligand, X: anionic donor site, L: neutral donor site

5—7) by introducing a variable number of axial donor atoms Z belonging to any kind of neutral or negatively charged axial ligands completing the coordination sphere. The coordination types designate the configuration of the donor atoms Z by giving their orientation with respect to the plane of the porphin system. Without or just with a single donor atom [types (A),

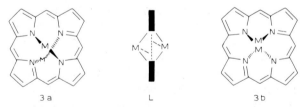

3 a   L   3 b

Fig. 4. Two principal constitutional isomers of bimetallic metalloporphyrins M_2(P). (The metal ions may form additional coordinate bonds to a third or fourth nitrogen atom; in the latter case, the two species are identical. This topic is extensively treated in Chapter 7.)

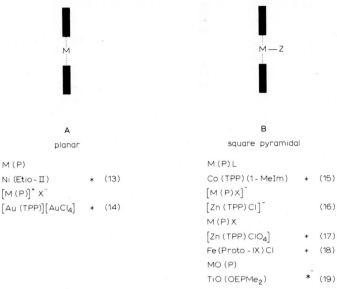

A

planar

M (P)

Ni (Etio - II)   *   (13)

[M (P)]⁺ X⁻

[Au (TPP)][AuCl₄]   *   (14)

B

square pyramidal

M (P) L

Co (TPP) (1 - MeIm)   *   (15)

[M (P) X]⁻

[Zn (TPP) Cl]⁻   (16)

M (P) X

[Zn (TPP) ClO₄]   *   (17)

Fe (Proto - IX) Cl   *   (18)

MO (P)

TiO (OEPMe₂)   *   (19)

Fig. 5. Bar graphs (A) and (B) of square planar and square pyramidal monometallic mononuclear metalloporphyrins (Figs. 5—13 also show the general formulae and examples of subclasses; references to examples in brackets. For meaning of * and ✕, see text.)

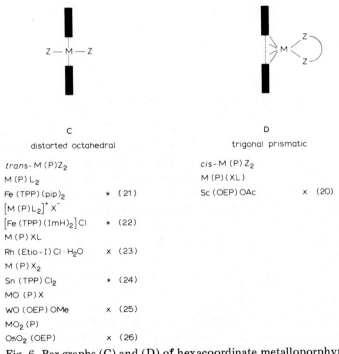

C

distorted octahedral

trans- M (P) Z₂

M (P) L₂

Fe (TPP) (pip)₂   *   (21)

[M (P) L₂]⁺ X⁻

[Fe (TPP) (ImH)₂] Cl   *   (22)

M (P) X L

Rh (Etio - I) Cl · H₂O   ✕   (23)

M (P) X₂

Sn (TPP) Cl₂   *   (24)

MO (P) X

WO (OEP) OMe   ✕   (25)

MO₂ (P)

OsO₂ (OEP)   ✕   (26)

D

trigonal prismatic

cis- M (P) Z₂

M (P) (XL)

Sc (OEP) OAc   ✕   (20)

Fig. 6. Bar graphs (C) and (D) of hexacoordinate metalloporphyrins.

References, p. 224

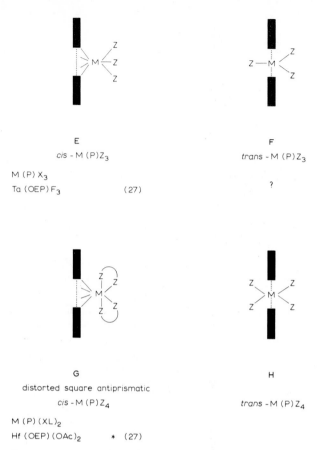

E
cis - M (P)Z₃

M (P) X₃
Ta (OEP) F₃   (27)

F
trans - M (P) Z₃

?

G
distorted square antiprismatic
cis - M (P) Z₄

M (P) (XL)₂
Hf (OEP) (OAc)₂   *   (27)

H
trans - M (P) Z₄

Fig. 7. Bar graphs (E)—(H) of heptacoordinate and octacoordinate metalloporphyrins.

(B)] no configurational problem arises. With two or more axial donor atoms Z, *cis* and *trans* configurations (C)—(H) are introduced. Arbitrarily, the set of configurations is closed with CN = 8; however, with very big ions, CN = 9 might occur. When two donor atoms are combined in a bidentate chelating ligand, this is expressed by Z—Z or (ZZ). In those types where the metal is drawn out of the porphyrin plane, the lines connecting the metal to the porphyrin bar each indicate a metal-to-nitrogen bond.

Each general formula given below a coordination type is associated with the formula of a specific example and a reference describing its existence (Ref. 13—45 in brackets). An asterisk is added when the coordination type has been established by X-ray crystallography (see Chapter 8), a cross indicates sufficient assignments by vibrational, nuclear magnetic resonance, or mass spectrometry.

The semimetallic class has so far only a single type, (K) (Fig. 8). The five

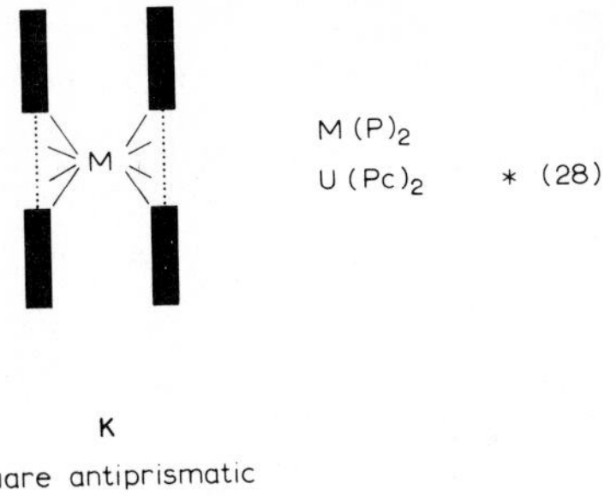

M (P)$_2$
U (Pc)$_2$ * (28)

K
square antiprismatic

Fig. 8. Bar graph (K) of a semimetallic metalloporphyrin.

bimetallic types (L)—(Q) (Fig. 9) are composed of two univalent metal ions
M$^+$ (or M$'^+$) or univalent metal-ligand residues MZ$_2^+$ [e.g. Rh(CO)$_2^+$] and (MZ$_3$)
[e.g. Re(CO)$_3^+$]. The *cis*-type (P) is thought to originate from B(OR)$_2^+$-
groups which have been condensed to a B$_2$O(OR)$_2$-moiety. Neither the
types (L) and (N) (Fig. 9) nor the types (R), (S), and (T) (Fig. 11) have so
far been isolated in their fully metalated forms; the species are therefore

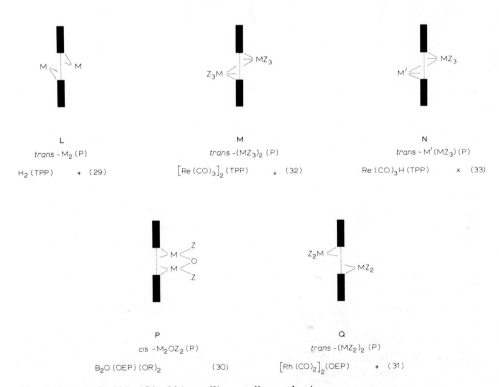

L
trans - M$_2$ (P)

H$_2$ (TPP)   * (29)

M
trans -(MZ$_3$)$_2$ (P)

[Re (CO)$_3$]$_2$ (TPP)   * (32)

N
trans - M$'$ (MZ$_3$) (P)

Re (CO)$_3$H (TPP)   × (33)

P
cis - M$_2$OZ$_2$ (P)

B$_2$O (OEP) (OR)$_2$ (30)

Q
trans - (MZ$_2$)$_2$ (P)

[Rh (CO)$_2$]$_2$ (OEP)   * (31)

Fig. 9. Bar graphs (L)—(Q) of bimetallic metalloporphyrins.

References, p. 224

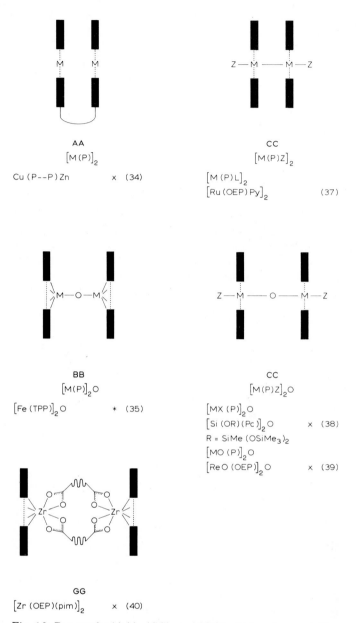

Fig. 10. Bar graphs (AA)⊢(CC), and (GG) of binuclear metalloporphyrins.

represented by their proton analogs (if the proton may be taken as the simplest possible univalent metal ion). Figures 10, 12 and 13 show species which are termed binuclear or polynuclear because two or more porphyrin nuclei are connected by two or more metal ions. They may be derived from

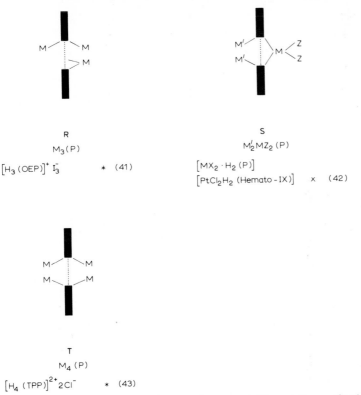

R

$M_3(P)$

$\left[H_3 (OEP) \right]^+ I_3^-$ *   (41)

S

$M'_2 M Z_2 (P)$

$\left[MX_2 \cdot H_2 (P) \right]$

$\left[PtCl_2 H_2 (Hemato - IX) \right]$ x   (42)

T

$M_4 (P)$

$\left[H_4 (TPP) \right]^{2+} 2Cl^-$ *   (43)

Fig. 11. Bar graphs (R)—(T) of tri- and tetrametallic metalloporphyrins.

the combination of the symbols classifying the stereochemistry of each monometallic subunit.

The association of two square planar metalloporphyrins has so far only been realised by linking two porphyrin moieties together at their periphery in (AA) (Fig. 10). The resulting compounds are useful for the investigation of radiative energy transfer between metalloporphyrin moieties at a fixed distance[34]. The species $[M(P)Z]_2$ (CC) (Fig. 10) may be prepared, but the material with this designation in the literature[36] is in fact $Ru(OEP)Py_2$[37]. While these compounds are extensively treated in Chapter 7, we shall have to mention some of them here in conjunction with condensation or oligomerisation reactions of the plain monometallic metalloporphyrins to which the rest of this chapter is devoted.

3) Axial stoichiometry is invoked when the sites Z are specified and offered by the following kinds of ligands: neutral unidentate donors L, uninegative anions X^-, dinegative oxide (O^{2-}) or sulfide (S^{2-}) ions which are either doubly bound to a single metal or act as bridging groups between two metal

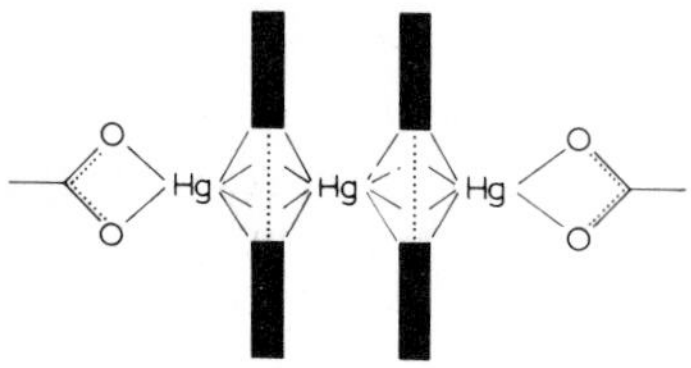

DKD

Hg$_3$(OAc)$_2$(Etio-I)$_2$ x   (44)

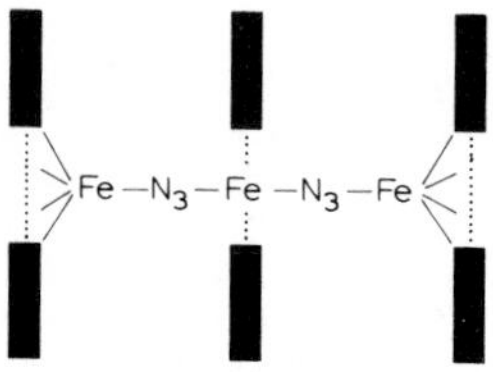

BCB

Fe$_3$(TPP)$_3$(N$_3$)$_2$ x   (45)

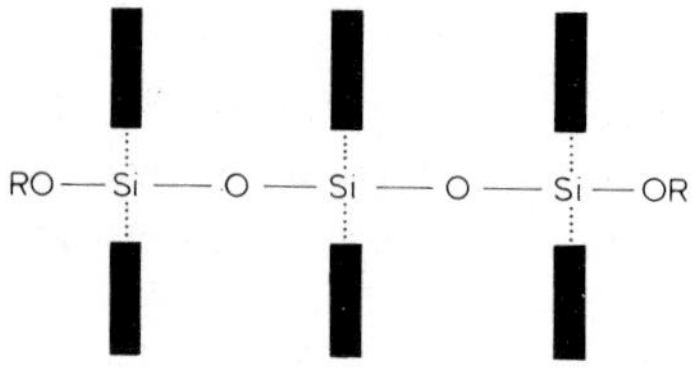

CCC

[Si$_3$O$_2$(OR)$_2$(Pc)$_3$]   (R = SiMe(OSiMe$_3$)$_2$) x (38)

Fig. 12. Bar graphs (DKD), (BCB), and (CCC) of some trinuclear metalloporphyrins.

ions, or combinations of anionic and neutral donor sites as found in bidentate chelating or bridging ligands (XL) or (XL)$_2$, e.g. acetate and dicarboxylates. Hence, the stoichiometry devides the coordination types into subclasses which are noted in the general terms below the corresponding coordination types in Figs. 5, 6, and 10. As a sufficient number of examples is only known with the types (A), (B) (Fig. 5), (C) (Fig. 6), and (CC) (Fig. 10), only these four types can at the moment show stoichiometric subclasses, and the importance of the basic types (A), (B), (C) becomes quite obvious.

Table 2 and Figs. 5—13 thus provide a synoptical representation of the kind of compounds encountered in the broad field of substantial knowledge

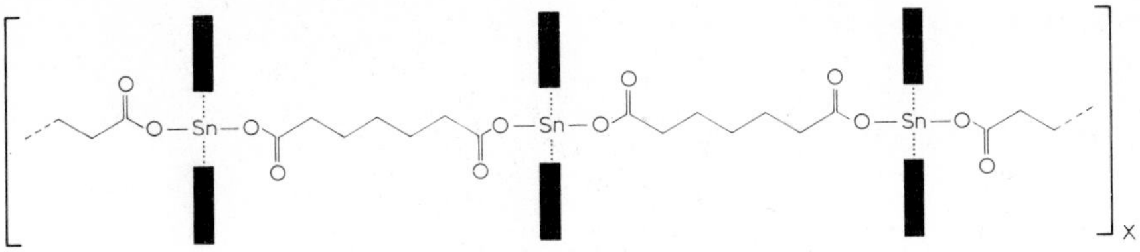

Poly - C

$[Sn (OEP) (pim)]_x$ x (40)

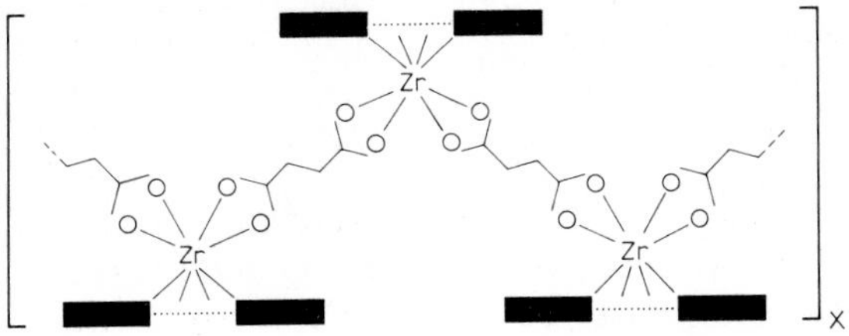

Poly - G

$[Zr (OEP) (suc)]_x$ x (40)

Fig. 13. Bar graphs poly-(C) and poly-(D) of polynuclear metalloporphyrins.

pertinent to the coordination chemistry of metalloporphyrins. Certainly the manifold of species depicted will grow considerably in the near future.

The following sections intend to animate the skeleton designed above. First, the preparation and characterization of various respresentatives of monometallic metalloporphyrins will be described (Sections 5.3, 5.4).

5.3. Equatorial coordination chemistry

This chapter will deal with the synthetic aspects of the equatorial MN_4 coordination group of metalloporphyrins.

5.3.1. Stages in the development of the 'Periodic table of metalloporphyrins'.

The periodic table of metalloporphyrins is obtained when, in any periodic table of the elements, those electropositive elements are indicated which are known to form adducts with a porphyrin to give one of the various coordination compounds classified in Section 5.2. Forming these adducts the metal may just be associated to yield a rather loose complex or be 'inserted' into the hole between the four nitrogen atoms, becoming more or less coplanar with the porphyrin ring. Figure 14 shows the periodic table of metallopor-

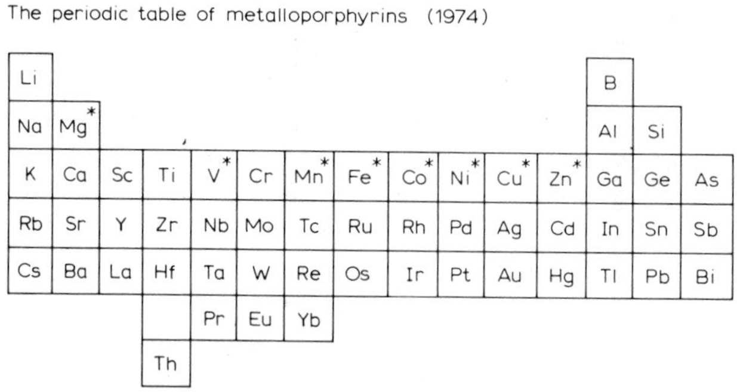

J.W. BUCHLER

The periodic table of metalloporphyrins (1974)

Li												B		
Na	Mg*											Al	Si	
K	Ca	Sc	Ti	V*	Cr	Mn*	Fe*	Co*	Ni*	Cu*	Zn*	Ga	Ge	As
Rb	Sr	Y	Zr	Nb	Mo	Tc	Ru	Rh	Pd	Ag	Cd	In	Sn	Sb
Cs	Ba	La	Hf	Ta	W	Re	Os	Ir	Pt	Au	Hg	Tl	Pb	Bi
				Pr	Eu	Yb								
			Th											

Fig. 14. The periodic table of metalloporphyrins. (Metals which are inserted by Nature are marked with an asterisk.)

phyrins in its present state; it is obvious nowadays that there is no principal limitation in the scope. Almost every metal has been forced to form some kind of complex with a porphyrin or a phthalocyanine; phthalocyanine complexes will only be mentioned when the appropriate porphyrin complex is missing. In fact, some of the metal ions, e.g. Pa(IV), U(IV), have so far only been combined with phthalocyanines. This illustrates the principal difference in the procedures to obtain the complexes. Whereas metals are normally 'inserted' into a preformed porphyrin to yield metalloporphyrins, metallophthalocyanines are obtained by 'wrapping' the phthalocyanine stepwise about the metal ion which acts as a template in the condensation of the four phthalodinitrile units[46]. This process is easier than the insertion: in the closed, planar, rigid porphyrin ligand the free electron pairs on the nitrogen atoms point inwards; to start complexation, either the pyrrole rings must be tilted so that the electron pair can reach the acceptor site at the metal ion, or the metal ion must have a very loosely held ligand shell surrounding it in order to offer more than one free acceptor site at once to slip easily into the porphyrin hole. X-ray analyses show that the pyrrole rings can be tipped considerably outwards of the porphyrin plane (see Chapter 8), but nevertheless, resonance energy is consumed to bring this movement about in such a rigid, aromatic molecule. Template syntheses unfortunately do not work so efficiently in the porphyrin series as compared with the phthalocyanines.

The development of the periodic table of metalloporphyrins has started with those species which are provided by nature[1]. Iron porphyrins form the prosthetic groups of the various heomoproteins (see also Chapter 1), magnesium chlorins collect the light energy for photosynthesis, a copper porphyrin is found in the wing feathers of *Turacus indicus*. Manganese porphyrins seem to occur in blood[1] and have been discussed as participants in the photosynthetic system[47]. Zinc porphyrins have been found in yeast mutants which accumulate large amounts of Zn(Proto-IX)[48]. Nickel and vanadyl porphyrins have been extracted from oil shales, crude petroleum oils, bituminous sands and asphaltic materials[1,49,50]. Fe, Co, Cu, and Zn are incorporated into porphyrins by chromatophore preparations from purple bacteria as well as preparations from liver mitochondria[51]. Cobalt porphyrins may be furthermore regarded as naturally occurring because vitamin B_{12} is a closely related cobalt 'tetrapyrrole' complex[9]. The natural occurrence of just these metalloporphyrins will be discussed at the end of Section 5.3. This basic stage of metalloporphyrins where nature has priority is the first stage of development indicated in Fig. 14.

The second stage was the recognition of hemin and chlorophyll as species containing the $-N_4 FeCl-$ and $-N_4 Mg-$ coordination groups, respectively, by H. Willstätter in 1913 in his remarkable book on chlorophyll[4]. In 1904, Laidlaw[52] had observed the first artificial re-insertion of iron into protoporphyrin. He also had described cobalt hematoporphyrin. Zaleski[53] had investigated the incorporation of iron, manganese, copper, and zinc. In 1909, Milroy[54] had synthesized the first nickel and stannic porphyrins. His compound, presumably $Sn(Hemato-IX)(OAc)_2$, was the first metalloporphyrin with an unnatural metal. Finally, Willstätter and Forsen (1913) re-inserted Mg into chlorophyll derivatives[4]. Optical spectroscopy had already played a very important role at that time.

The third stage is indicated by the culmination of Hans Fischer's work in 1940 which was reviewed recently by Treibs[6]. In Fischer's famous monograph on pyrrole pigments[2,3], metalloporphyrins containing Na, K, Mg, V, Mn, Fe, Co, Ni, Cu, Zn, Cd, Ag, Sn, and Tl are documented. In 1925, Hill[55] had investigated the spectra of hematoporphyrin complexes containing Na, K, Ag, Mg, Zn, Al, Mn, Fe(II), Fe(III), Co(II), Co(III), Ni, Cu, As, Sn, and Pb. Treibs[56] had contributed to the knowledge of complexes with Cd, Pb, Al, Ga, In, V, Ge, Sb. Stern[57] had reported the spectra of MX_n(Meso-IX-DME) where MX_n was VO, Co, CoCl, Ni, Cu, Zn, GaCl, $GeCl_2$, Pd, Ag, InCl, $SnCl_2$, Hg, TlCl, Pb. All these compounds had been prepared by Treibs. Haurowitz, Kittel, and Clar[48] had discussed the spectra of tetramethyl—hematoporphyrin-IX complexes with Mg, Zn, Cd, Cu, Ag, Ni, Co, Fe, FeCl, MnCl, Pb, and Tl as well as selected determinations of magnetic susceptibilities. Theorell[59] had made Pd(TPP) and Pt(TPP). In the phthalocyanine series, even $Li_2(Pc)$, Be(Pc), and Cr(Pc)Cl had been obtained by Linstead[46,60].

The fourth stage was reached with Falk's monograph[1] about 1964. Not very many newly introduced metals joined the party: metalloporphyrins with Li, Rb, Cs, B, Ca, Ba, and $Si(Pc)(OH)_2$[61] are quoted there, $Sr(Meso-IX-DME)$[62] and $[Au(TPP)][AuCl_4]$[4] elsewhere; a compound which may have been $Be_2(Meso-IX-DME)X_2$ has been characterized only by its fluoresence[62]. However, the 'further coordination' — i.e. the axial coordination chemistry in the terms used here — received a thorough evaluation in the monograph[1] (see Section 5.4).

TABLE 3

Novel metals combined with porphyrins since 1964

Metals (references in brackets)	Senior author
As, Sb, Bi (65)	Treibs
Ru (66), Rh (66, 67), Ir (68), Mo (69)	Fleischer
Ti (70), Cr (71), Tc (72), Re (73)	Tsutsui
Sc, Zr (20), Hf, Nb, Ta, W (27), Os (74)	Buchler
Lanthanides: Eu, Pr, Yb (75)	Horrocks
Actinides: (Th, Ac, U) (75—77)	Horrocks, Lux

Since 1964, the field has grown enormously. Especially inorganic chemists directed their interest on to metalloporphyrins because the very important role of metals in biological systems was increasingly recognized by chemists and their sponsors. The metalloporphyrins obtained since 1964 have been termed 'novel' or 'unusual'[63] or even 'exotic'[64], as the lower left and upper right corner of the periodic table seemed rather remote from the natural iron porphyrins on paper as well as in a chemical sense: sometimes quite elaborate methods for metal insertion had to be used. The novel metal ions introduced since 1964, the leaders of the responsible research groups, and references are given in Table 3. Metal hydrides, alkyls, carbonyls, acetylacetonates, and phenoxides, compounds which have become readily available only in recent years, served as metal carriers. Thus, the extension to the last stage and the present state of the periodic table of metalloporphyrins is partly due to the rapid development of organometallic chemistry in these years.

5.3.2. Problems associated with metal insertion

No general method for the insertion of all metal ions is available, because the various ions behave too differently. Let us recall the logically separated steps $(A1)$ to $(A5)$ in Scheme 1 (p. 160) to outline the difficulties encountered in metalation reactions. There are: a) solubility problems, b) problems associated with the acidity or basisity (donor strength) of the solvent, c) problems caused by the intrinsic stability of the metal carrier, e.g. MX_3L_3 (Scheme 1), d) problems arising from lability of the metalloporphyrin moiety under the reaction conditions.

a) Solubility problems

The metallic salts normally available are hydrophilic, whereas the usual porphyrins H_2P are lipophilic. To favor the central association step, e.g. (A3),

$$M^{3+} + (P)^{2-} \quad \leftrightharpoons \quad M(P)^+ \qquad (A3)$$

high concentrations of the reactants must be achieved. Thus either the metal salt must be modified to obtain a metal carrier which is soluble in organic solvents, or hydrophilic groups must be attached to the porphyrin to make it water soluble. The latter approach has been realized, e.g. with the following porphyrins: $\alpha,\beta,\gamma,\delta$-tetra(4-hydroxysulfonylphenyl)porphin [$H_2(TH_4SP)$][78], $\alpha,\beta,\gamma,\delta$-tetra($N$-methyl-4-pyridyl)porphinium tosylate [$H_2(TMPyP)$] [C_7-H_7-$SO_3]_4$[79], and with 2,4-bis(3,6-diazahexyl)deuteroporphyrin-IX-bis-(ethylenediamide)[80]. The sulfonic acid-, quaternary ammonium- and amino groups present in these porphyrins render them water soluble.

Far more generally applicable is the choice of a metal salt with lipophilic anions making it sufficiently soluble in organic solvents; such metal derivates are metal acetates[1], metal hydrides[81], metal alkyls[71,81,82], metal carbonyls[71], metal acetylacetonates[20,25,39,65,75,83,84], and metal phenoxides[20,25,39,40]. Purely inorganic metal carriers, e.g. metal halides of various kinds, may be solubilized by further coordinating these halides to organic solvent molecules with a weak electron donor site, e.g. dimethylformamide[85] (Me_2NCHO, oxygen donor), sulfolane (C_4H_8SO, oxygen donor)[85], or benzonitrile (C_6H_5CN, nitrogen donor)[27,39,86,87]; with its high boiling point the latter is particularly suitable for dissolving and cracking polymeric or clustered metal halides.

b) *Problems associated with the acidity or basicity of the solvent*

Inspection of Scheme 1 at first glance suggests that basic solvents should be the best solvents because they drive the deprotonation equilibria (A1) to the $(P)^{2-}$ level, thus favoring the metalation step (A3). Basic solvent molecules, however, are not only Brønsted bases, but Lewis-bases which may more or less strongly bind as neutral donor ligands L to the metal ions behaving as Lewis acids. Hence, in the presence of strongly basic solvents, the set of equilibria indicated by (A2) will be shifted so as to impede the deconvolution of the metal carrier. A more thorough insight into the Scheme 1 (see p. 160) therefore tells us that a compromise has to be made between the conflicting requirements of equilibria (A1) and (A4).

c) *Problems caused by the intrinsic stability of the metal carrier*

It is obvious on inspection of Scheme 1 that a very labile metal carrier MX_3L_3 will favor the metalation. Those metal compounds in which the coordination sphere is held together very tightly will not serve as good metal carriers.

The higher the charge of the metal ion in the educt of metalloporphyrin formation, the more sluggish it will behave as a metal carrier for simple electrostatic reasons: the negatively charged anions X and the negatively polarized ligands L are more strongly attracted by the metal ion if it carries a high charge; high charges impede deconvolution of the metal carrier.

A consequence of this is the advice that in any metalation reaction the metal carrier should have the metal in its lowest conveniently accessible oxidation state (see Section 5.3.6).

Metal carbonyls are therefore a good choice as metal carriers unless they become themselves too unreactive because of cluster formation. [This may be the reason why $Fe(CO)_5$ and $Ru_3(CO)_{12}$ are good metal carriers[63] while osmium carbonyls are not[88]; the latter are now known to form rather stable cluster aggregates, e.g. $Os_6(CO)_{18}$ [89].]

Because of the high charge of their central metals and the additional stabilization by the chelate effect of the 2,4-pentanedionato ligands, the metal(III) and metal(IV) acetylacetonates $M(acac)_3$ and $M(acac)_4$ are not the best metal carriers, but because of their other qualities they have proved very useful (see Section 5.3.3.3). Aquation, especially of highly charged ions, and consecutive formation of polynuclear ions or oxo complexes may also decrease reactivity towards porphyrins. Therefore, nonaqueous media are generally preferred in metalation reactions.

d) *Problems arising from lability of the metalloporphyrin moiety under the reaction conditions*

A variety of metalloporphyrins are labile towards protic or Lewis acids [reversal of eq. (*A1*) or (6)]. These have then to be prepared under conditions where even the protons produced on metalation according eq. (6) can be removed from the equilibrium. Basic conditions are advisable in these cases,

$$ML_m X_n + H_2(P) \quad \rightleftharpoons \quad M(P)L_m X_{n-2} + 2\,HX \tag{6}$$

e.g. pyridine for the preparation of the acid-labile Mg-porphyrins according to eq. (7)[20,90,91], of Cd-, Hg-, and Pb-porphyrins[14,91-94].

$$[Mg(Py)_6][ClO_4]_2 + H_2(OEP) \xrightarrow{py} Mg(OEP)Py_2 + 2[PyH][ClO_4] + 2\,Py$$

$$\tag{7}$$

Metal alkoxides have been used (eq. 8) to obtain the labile species Mg(Proto-IX-DME)[4,95], and in solution $Li_2(TPP)$, $Na_2(TPP)$, and $K_2(TPP)$ may be prepared[14] by using the alkali hydroxides in absolute methanol. Hydrides or alkyls are even more effective (see next paragraph), as any protic species present is consumed by excess hydride or alkyl.

$$Mg(OC_3H_7)_2 + H_2(\text{Proto-IX-DME}) \rightarrow Mg(\text{Proto-IX-DME}) + 2\,C_3H_7OH \tag{8}$$

5.3.3. Examples and scopes of selected successful insertion procedures*

In this section, eight generalized metalating systems will be described. They are listed in Table 4 and suffice to introduce all the metals shown in the periodic table of monometallic metalloporphyrins presented in Table 5. The preparation of an alleged Be(Meso-IX-DME)[62] could not be reproduced[85], and like the boron complexes, Be should rather form bimetallic metalloporphyrins; Be(P) or BX(P) should not exist because the coordination spheres of Be and B are already saturated with 4 ligands in a tetrahedral array. Actinides have been introduced into the phthalocyanine system by condensation of phthalodinitrile in the presence of suitable metal halides; thus, $U(Pc)_2$ and $Th(Pc)_2$ $(K)^{28,76}$ are obtained from UI_4 and ThI_4, respectively. Even sapphyrin derivatives containing a linear UO_2-moiety are now known[77].

The reaction times — if given at all — of the procedures presented may vary considerably. This is, however, no serious problem, as all metal insertions are normally monitored by spectrophotometry (see Section 5.3.4).

TABLE 4

Selected metalating systems serving to prepare monometallic metalloporphyrins

No.	Metalating system [a]	Temp. ($^\circ$C)	Scope of metals inserted [b]
1	$MX_m L_n$/HOAc	100	Zn, Cu, Ni, Co, Fe, Mn, Ag, In, V, Hg, Tl, Sn, Pt, Rh, Ir
2	MX_m/Py	115—185	Mg, Ca, Sr, Ba, Zn, Cd, Hg, Si, Ge, Sn, Pb, Ag, Au, Tl, As, Sb, Bi, Sc
3	$M(acac)_n$/solvent	180—240	Mn, Fe, Co, Ni, Cu, Zn, Al, Sc, Ga, In, Cr, Mo, Ti, V, Zr, Hf, Eu, Pr, Yb, Y, Th
4	MX_m/PhOH	180—240	Ta, Mo, W, Re, Os (X = O, OPh, acac, Cl)
5	MCl_m/PhCN	191	Nb, Cr, Mo, W, Pd, Pt, Zr, In
6	MCl_m/DMF	153	Zn, Cu, Ni, Co, Fe, Mn, V, Hg, Cd, Pb, Sn, Mg, Ba, Ca, Pd, Ag, Rh, In, As, Sb, Tl, Bi, Cr
7	MR_m/solvent	25	Mg, Al, Ti
8	$MX_m(CO)_n$/solvent	80—200	Cr, Mo, Mn, Tc, Re, Fe, Ru, Co, Rh, Ir, Ni, Os

[a] Metalating systems are classified either by solvent or by metal carrier; the choice depending on the more essential component of the metalating system. (For details, see text.)
[b] Magnesium is inserted into pheophytin and other chlorophyll derivatives of similar lability using magnesium alkoxides as metal carriers[1,4,7,8,95] (see Section 5.3.2).

---

* For further examples of metal insertion, see Chapter 19.

TABLE 5

The periodic table of monometallic metalloporphyrins indicating the respective metalating systems defined in Table 4 (the necessary specifications are given in text)

— Main group metals

Mg 2															

Early transition metals | **Iron metals** | | | — Copper and zinc groups | | **Main group metals**

				Mn 1,3,6,8	Fe 1,3,6,8	Co 1,3,6,8	Ni 1,3,6,8			Al 2,3,7	Si 2		

Element				
Ca 2				
Sc 2,3	Ti 3,7	V 1,3,6	Cr 3,5,6,8	
Sr 2				
Y 3	Zr 3	Nb 5	Mo 5,4,8	
Ba 2				
La 3	Hf 3	Ta 4	W 4,5	

Iron metals

Mn 1,3,6,8	Fe 1,3,6,8	Co 1,3,6,8	Ni 1,3,6,8
Tc 8	Ru 8	Rh 8	Pd 5
Re 4,8	Os 4	Ir 8	Pt 5

— Platinum Metals

Copper and zinc groups

Cu 1,3,6	Zn 1,3,6
Ag 2,6	Cd 2,6
Au 1	Hg 1,2,6

Main group metals

Al 2,3,7	Si 2	
Ga 3	Ge 2	As 2
In 1,3,6	Sn 2	Sb 2
Tl 1	Pb 2,6	Bi 2

Lanthanides ... Eu, Pr, Yb 3

Th 3

Work-up should not be commenced until the reaction has either gone to completion or else come to a point where the amount of products does not increase any more.

At this point it must be mentioned that, with all the following methods, only those porphyrins may be metalated the side-chains of which can withstand the metalation conditions. Protoporphyrin-IX and its esters must be metalated using special precautions because the vinyl groups are very easily attacked by various chemicals (see Chapter 19). A safe procedure for the insertion of iron is mentioned in paragraph 5.3.3.

5.3.3.1. The acetate method

Under the term 'acetate method' those metalation reactions will be discussed where the protons of the porphyrins to be metalated are transferred to acetate ions or propionate ions (eq. 9, 11)[1,2,3,14,39,91−93,96−114]. In the examples marked with an asterisk in eq. (9), the original metal oxidation

$$H_2(P) + MX_m L_n + 2\ NaOAc \xrightarrow[100°C,\ 2h]{HOAc} M(P)X_{(n-2)} + 2\ HOAc + 2\ NaX + nL$$

$$(9)$$

$MX_m L_n$: $Mn(OAc)_2 \cdot 4\ H_2O^*$, $FeSO_4 \cdot 7\ H_2O^*$, $FeCl_3$, $Co(OAc)_2 \cdot 4\ H_2O$, $Ni(OAc)_2 \cdot 4\ H_2O$, $Cu_2(OAc)_4 \cdot 2\ H_2O$, $Zn(OAc)_2 \cdot 2H_2O$, $InCl_3$, $VOSO_4 \cdot 5\ H_2O$, $SnCl_2 \cdot 2\ H_2O^*$, $AgNO_3^*$, $[PtCl_4]^{2-}$, $[Rh(CO)_2 Cl]_2^*$, $[Ir(CO)_3 Cl]^*$

states are not stable within the porphyrin under aerobic conditions, and autoxidation occurs [eq. (10); further discussion of this topic in Section 5.3.6]. This is the case with Mn, Fe, Ag, and Sn; the autoxidation of Sn(II) is practically inevitable[94], while the other three by exclusion of oxygen can be kept in the Mn(II), Fe(II) or Ag(I) states. In the last case a bimetallic complex $Ag_2(P)$ is formed[92] which collapses on prolonged heating to Ag(P) and metallic silver. Examples are given in eq. (10a) and 10b)[99,101].

$$4\ Fe(Deut\text{-}IX\text{-}DME) + O_2 + 4\ HOAc \rightarrow 4\ Fe(Deut\text{-}IX\text{-}DME)OAc + 2\ H_2O$$

$$(10a)$$

$$2\ Sn(Etio\text{-}I) + O_2 + 2\ C_2H_5COOH \rightarrow 2\ Sn(Etio\text{-}I)(C_2H_5COO)_2 + H_2O$$

$$(10b)$$

As is seen with examples (10a) and (10b), an acetate or propionate anion

References, p. 224

combines with the metalloporphyrin moiety if the metal exceeds the divalent state and no other anion is present.

A variant of the method is given in eq. (11). The solvent mixtures employed allow the dissolution of both reactants, $CHCl_3$ favoring the porphy-

$$M(OAc)_n + H_2(P) \xrightarrow[\substack{or\ CH_2Cl_2/THF \\ 30-60°C}]{CHCl_3/MeOH} M(P)OAc_{(n-2)} + 2\ HOAc \qquad (11)$$

$$M(OAc)_n = Hg(OAc)_2,\ Tl(OAc)_3\ [or\ Tl(CF_3COO)_3]$$

rin, MeOH the metal acetate. As many metalloporphyrins containing divalent metal ions are completely insoluble in MeOH, addition of excess MeOH after the reaction leads to crystallization of the product while the excess metal carrier either remains in solution or can be washed away with water. The low temperatures allowed Smith[109] to isolate intermediates in the metalation step: e.g. $H_2(TPP)$ and excess $Hg(OAc)_2$ in THF/CH_2Cl_2 produce the bimetallic species $Hg_2(TPP)(OAc)_2 \cdot 2H_2O$ (L) which collapses to yield 2 moles of Hg(TPP) in the presence of excess $H_2(TPP)$. Also, treatment of $H_2(OEP)$ with the oxidizing metal carriers $Tl(OAc)_3$ and $Tl(CF_3COO)_3$ in THF/CH_2Cl_2 at 25°C and subsequent chromatography furnished $Tl(OEP)OH \cdot H_2O$, presumably of type (B)[110].

With the rather sluggishly reacting palladium complexes, another intermediate of a metalation has recently been isolated (eq. 12)[42].

$$H_2(Hemato\text{-}IX) \xrightarrow[\substack{HOAc\ 24h,60°C}]{K_2PtCl_4} H_2Pt(Hemato\text{-}IX)Cl_2 \xrightarrow[\substack{24h,85°C}]{HOAc}$$

$$(S)$$

$$Pt(Hemato\text{-}IX)$$

$$(A) \qquad (12)$$

The metalation of porphyrins with $[RhCl(CO)_2]_2$ [63,66,108] or $[Ru(CO)_3\text{-}Cl_2]_2$[66] can be realized in HOAc, but is treated in Section 5.3.3.8.

The acetate method can be performed very conveniently by extracting gram quantities of porphyrins from a thimble into the boiling solution of the metal carrier in glacial acetic acid containing sodium acetate. Thus, it is assured that no crystalline starting material is left when the reaction is finished. On cooling, the metalloporphyrins frequently crystallize directly from the solution; alternatively addition of hot water or methanol accomplishes the precipitation. The product then may be purified by chromatography on alumina, grade III, with CH_2Cl_2 or $CHCl_3$ as eluent.

This procedure has often served to prepare the chlorohemins Fe(TPP)Cl and Fe(OEP)Cl directly from $FeCl_3$[2,4,102-106]: 1.0 g of $H_2(OEP)$ is extracted during 8–12 h from a soxhlet-thimble into a boiling solution of

1.0 g $FeCl_3$ in 0.2 l of glacial HOAc containing 0.5 g of NaOAc. On cooling, the chlorohemin crystallizes in a practically pure state. It is astonishing that with the OAc^- present, no acetate is formed.

The hemin preparation shows that this method of metal incorporation is not restricted to divalent ions; $InCl_3$ and $VOSO_4 \cdot 5H_2O$ have been used as metal carriers in syntheses of In(TPP)Cl (B)[107] and VO(OEP) (B)[100]; in fact, Fe(III), In(III), and V(IV) are rather easy to insert by this method.

For the insertion of iron into protoporphyrin-IX via the acetate method, glacial acetic acid which has been carefully purified from oxidizing impurities is used at temperatures not exceeding $80°C$. Iron is administered either as $FeSO_4 \cdot 7H_2O$ or as a $Fe(OAc)_2$ solution in HOAc which has been freshly prepared by dissolving pure iron in HOAc under nitrogen. It is important that no metallic iron be present when the porphyrin reacts with the $Fe(OAc)_2$ solution. The porphyrin is added as a pyridine solution. The procedure should be run under nitrogen for 10—20 min. Autoxidation afterwards provides Fe(Proto-IX)OAc which is taken into ether. The ether solution is freed from unreacted H_2(Proto-IX) with 25% (w/v) HCl and then further processed as required[1,2,99]. A recent example given by Kenner[111] uses $FeSO_4 \cdot 7H_2O$/NaCl/HOAc/NaOAc at $80°C$ for 15 min. If the whole procedure is run strictly under nitrogen, and in the absence of nitrogenous bases, the bare ferrous porphyrin of type (A) may be directly obtained in the solid state. Thus, Fe(Etio-III) (A) has been made by Fischer with $Fe(OAc)_2$ in HOAc[112], and Fe(TPP) (A)[113] with $Fe(C_2H_5CO_2)_2$ in propionic acid. These compounds are stable in air for a short period, but autoxidize instantaneously in solution.

5.3.3.2. The pyridine method

Pyridine is a very good solvent for porphyrins and metal salts when the latter are capable of forming pyridine complexes. Its basicity promotes eq. (A1) (Scheme 1) according to eq. (13); the metalloporphyrin formed may further fill up its coordination sphere either furnishing mixed complexes and

$$H_2(P) + MX_m + 2 Py \rightarrow M(P)X_{(m-2)} + 2[PyH]^+X^- \tag{13}$$

$$M(P)X + Py \rightarrow M(P)XPy + Py \rightarrow [M(P)Py_2]^+X^- \tag{14}$$

$$M(P) + 2 Py \rightarrow M(P)Py + Py \rightarrow M(P)Py_2 \tag{15}$$

$$MX_m + nPy \rightarrow MX_m Py_n \tag{16}$$

salts [eq. (14)] or mono- and bispyridine metalloporphyrins (e.g. eq. 15), respectively. However, the metal carrier, e.g. MX_m, may itself react with pyridine and thus be reduced in activity (eq. 16). MX_n may be either a metal perchlorate [Mg(II)[90,91], Ca(II), Sr(II), Ba(II), Fe(II)[91]], a metal hydroxide in methanol [Mg(II), Ca(II), Sr(II), Ba(II)[3,4,14,62]], an acetate [Zn(II), Cd(II), Hg(II), Pb(II), Cu(II), Ag(I)[14,92,114]], a halide [Fe(II), Fe(III), Co(II)[114], Al(III)[55], As(III), Sb(III)[65,115], Sc(III)[64], Si(IV)[25,96,116], Ge(IV)[25,96,117], Sn(II)[14,25], Au(III)[14,118], Tl(I)(?)[14]], or a nitrate [Bi(NO$_3$)$_3$[65,115]]. The higher the oxidation state of the metal ion indicated before in brackets, the more drastic are the reaction conditions for the reasons pointed out in Section 5.3.2. SiCl$_4$ has to be reacted in sealed tubes at 180°C; for Si and Ge, the reactions do not go to completion.

For the preparation of Cd(OEP)Py, Pb(OEP), Ag(OEP), and Hg(OEP) from the respective metal acetates, this method was recently checked[93]. The acetates were boiled several hours in pyridine with H$_2$(OEP), the largest part of the pyridine removed in vacuo, the remaining few milliliters taken up in benzene or CH$_2$Cl$_2$ and then extracted with 2 N aqueous NH$_3$ to remove the residual pyridine and inorganic salts. The remaining organic phase was then taken to dryness and the remaining purple solid recrystallized from CH$_2$Cl$_2$/MeOH [Ag, Hg, Pb] or Py [Cd(OEP)Py]. Chromatography on alumina or silica cannot be performed with Hg(II), Pb(II) and Cd(II) porphyrins because of their ready cleavage in acidic media.

5.3.3.3. The acetylacetonate method

Metal acetylacetonates have been used only very recently and have provided routes to novel metalloporphyrins[20,25,39,40,65,75,83,84,119−122]. Their great advantage is their solubility in organic solvents, the weakness of the acid liberated (eq. 17), and their ready availability and easy manipulation (see Table 4); many of them are commercially accessible.

$$H_2(P) + M(acac)_n \rightarrow M(P)(acac)_{(n-2)} + 2\,H(acac) \tag{17}$$

With M(acac)$_2$ (M = divalent metal), metalations can be carried out in benzene[83,120] or CHCl$_3$[20] and the solution after concentration may be placed directly on the chromatography column, the excess M(acac)$_2$ being retained at the top of the column. The main importance of the acetylacetonate method lies in its applicability to metal carriers containing trivalent, tetravalent and even hexavalent ions, especially 'early transition metal' ions[20,25]. The more labile acetylacetonates, e.g. TiO(acac)$_2$, VO(acac)$_2$, V(acac)$_3$, Mn(acac)$_3$, Fe(acac)$_3$, Co(acac)$_3$ and In(acac)$_3$, may even react without solvent just by mixing them with the porphyrin and heating to 200−250°C. However, to achieve better mixing, a high-boiling solvent should be added. Phenol is especially useful, because it may catalyze the metalation by partly or totally replacing the acetylacetonate ligands (e.g. eq. 18) which deactivate

$$Zr(acac)_4 + 4\,PhOH \rightarrow Zr(acac)_3OPh + 3\,PhOH + H(acac) \rightarrow \rightarrow \rightarrow$$

$$Zr(OPh)_4 + 4\,H(acac) \tag{18}$$

TABLE 6

Preparation of metalloporphyrins via the acetylacetonate method

Metal carrier	amount (g)	Ref.	$(OEP)H_2$ amount (g)	Solvent[a] amount (g)	Temp. (°C)	Time (min)	Product isolated	%	Type
$Zn(acac)_2$	3.0	(122)	2.0	10.0	200	10	$Zn(OEP)$	87	(A)
$Fe(acac)_2$	2.0	(122)	0.40	5.0	200	10	$[Fe(OEP)]_2O$	70 [b]	(BB)
$Al(acac)_3$	1.08	(20)	0.30	1.8	230	120	$Al(OEP)OPh$	62 [b]	(B)
$Sc(acac)_3$	0.10	(20)	0.07	3.0	170	120	$Sc(OEP)(acac)$	55 [b]	(D)
$Ga(acac)_3$	0.48	(25)	0.39	2.5	220	20	$Ga(OEP)OPh \cdot C_6H_6$	92	(B)
$In(acac)_3$	2.05	(20)	1.03	3.0	230	5	$In(OEP)OPh$	84	(B)
$TiO(acac)_2$	0.30	(123)	0.10	none	290	10	$TiO(OEP)$	83	(B)
$VO(acac)_2$	0.30	(20)	0.10	none	290	10	$VO(OEP)$	75	(B)
$Zr(acac)_4$	0.91	(39)	0.20	1.0	200—220	50 [c]	$Zr(OEP)(acac)OPh$	82	(E)
$Hf(acac)_4$	0.32	(27)	0.20	0.7	190—210	600 [c]	$Hf(OEP)(OAc)_2$	99	(G)

[a] Solvent: phenol in all cases except $Sc(acac)_3$ where imidazole was used.
[b] Product isolated by sublimation in vacuo (10^{-3} Torr).
[c] After half of the duration of the procedure, the temperature is raised, and the reaction mixture recharged with approx. 1/3 of the initial amount of $M(acac)_4$.

References, p. 224

the metal carrier by their chelate effect. The high boiling point serves to remove the acetylacetone from the equilibria (17,18) by distillation; the high solubility of both reactants in phenol allows use of very high concentrations, so that gram quantities of $H_2(P)$ and $M(acac)_n$ can be heated together in an open test tube containing a few grams of phenol. The resulting slurry should be stirred to prevent occlusion of unreacted porphyrin, and in difficult cases, e.g. with $Cr(acac)_3$, $Mo(acac)_3$, or $MoO_2(acac)_2$, it should be ascertained that the distilling liquids are allowed to escape, thus gradually concentrating the melt. Finally, it should be noted that commercially available acetylacetonates often contain chloride impurities. When metalloporphyrins of the type $(P)MX_m$ are to be prepared, a chloride-free acetylacetonate must be applied in order to avoid contamination of X^- with Cl^- in the product. The intrinsic stability of the acetylacetonate has an observable influence on the reaction rate. The inertness of octahedral Cr(III) and Mo(III) complexes as compared with analogous Sc(III) and Al(III) systems necessitates more drastic conditions for the reactions of $Cr(acac)_3$ and $Mo(acac)_3$ ($240°C$, 5 h) and octaethylporphyrin (in phenol) as compared with $Al(acac)_3$ ($230°C$, 2 h, phenol) or $Sc(acac)_3$ ($170°C$, 2 h, imidazole) to achieve completion of the reaction. $Mn(acac)_3$, $Fe(acac)_3$, and $Co(acac)_3$, however, react much faster ($230°C$, 10 min) due to a prior reduction of the metals, namely Fe(III) → Fe(II) and Co(III) → Co(II)[20]. Some useful examples and data are compiled in Table 6. Of general interest may be the introduction of iron into $H_2(P)$ with $Fe(acac)_2$ in phenol because of its applicability to a variety of alkyl- and arylporphyrins which are currently investigated as heme models.

For the preparation of $[Fe(OEP)]_2O$, 0.4 g of $H_2(OEP)$ are heated with 2 g $Fe(acac)_2$ in 5 g PhOH to $200°C$ for 10 min with stirring. After cooling, the product is dissolved in CH_2Cl_2 and stirred with 70 ml of 25% aqueous NaOH for 30 min; the organic layer is separated, washed with water until neutral and taken to dryness in vacuo. The residue is chromatographed on neutral alumina, grade III, in CH_2Cl_2. The main fraction is taken to dryness and recrystallized from CH_2Cl_2/n-hexane[122]. μ-Oxobis[porphinatoiron(III)] complexes serve as useful starting materials for a variety of other complexes (P)FeX (see Section 5.4).

Very recently, Eu(TPP)(acac) has been prepared by reacting $H_2(TPP)$ with $Eu(acac)_3 \cdot H_2O$ in 1,2,4-trichlorobenzene at $240°C$ for 4 h[75]. The method seems applicable to the entire series of porphinatolanthanides. $Th(TPP)(acac)_2$ (G) is also accessible[75].

5.3.3.4. The phenoxide method

As is suggested by eq. (18) (p. 182), $Zr(OPh)_4$ might be a good metal carrier eq. (19). Indeed, the reaction of a 1 : 4 molar mixture of $H_2(OEP)$ and commercially available $Zr(OPh)_4$ produces a quantitative yield of $(OEP)Zr(OPh)_2$ at $220°C$ in 15 min[122]. $Al(OPh)_3$ is incorporated at $240°C$. As few phenoxides are easily accessible, this method has only found limited use with the following metal compounds which seem to be converted to metal phenoxides: $TaCl_5$[27,39,64], MoO_3[123], H_2WO_4[25], Re_2O_7[25], and OsO_4[74]. Only the use of $Zr(OPh)_4$, $Al(OPh)_3$, $TaCl_5$, and Re_2O_7 can be recommended to insert the respective metal ions.

$$Zr(OPh)_4 + H_2(OEP) \xrightarrow{220°C} Zr(OEP)(OPh)_2 + 2\,PhOH \qquad (19)$$

5.3.3.5. The benzonitrile method

Benzonitrile $(PhCN)$[39] is a good solvent for porphyrins at its boiling point; it is a weak base and may form complexes with metal halides[86], e.g. $[PdCl_2(PhCN)_2]$, by boiling it with those anhydrous metal chlorides which are known to exist as dimers, polymers, or clusters in the solid state and therefore are otherwise unreactive or sparingly soluble. $PdCl_2$[86,87], $PtCl_2$[87], $MoCl_2$[25,27], $CrCl_3$[27] have been used as well as the more reactive $ZrCl_4$[122], $NbCl_5$[27], and WF_5[27]. An anionic tungsten complex, administered as $K_3[W_2Cl_9]$, required very long reaction times. This salt has, however, been frequently used[25,123] because it is rather easily prepared. Care must be taken to obtain pure metal carriers, because metal impurities are often incorporated quicker than the tungsten ions. The metalations in PhCN should be performed in a stream of dry nitrogen which drives off the HCl liberated from the metal halides and $H_2(OEP)$, eq. (20).

$$CrCl_3 + H_2(OEP) \xrightarrow[195°C,30min]{PhCN} Cr(OEP)Cl + 2\,HCl \qquad (20)$$

5.3.3.6. The dimethylformamide method

Adler[85,124] has prepared 68 different metalloporphyrins using oxygen donor solvents such as dimethylformamide (DMF), tetramethylurea, and sulfolan which are good solvents for the metal carriers and the porphyrins. Best results are obtained with anhydrous metal chlorides, the high boiling point driving off the HCl evolved during the reaction eq. (21)[85].

$$4\,CrCl_2 + 4\,H_2(TPP) + O_2 \xrightarrow[153°C,10min]{DMF} 4\,Cr(TPP)Cl + 2\,H_2O + 4\,HCl \quad (21)$$

By this method, 10 g of $H_2(TPP)$ were reacted in 1 l of DMF. The crude product crystallizes simply by adding 1 l of H_2O and cooling. The method may cause surprising results with catalytically active transition metal compounds which can decarbonylate DMF. Thus, incorporation of dimethylamine Me_2NH may occur eq. (22)[125]. The solvent itself is sufficiently basic to allow the preparation of labile metalloporphyrins such as $Pb(OEP)$[96].

$$H_2(Etio\text{-}I) \xrightarrow[-2CO,-2HCl]{RhCl_3,2DMF} [Rh(Etio\text{-}I)(Me_2NH)_2]^+Cl^- \qquad (22)$$

References, p. 224

5.3.3.7. The metal organyl method

Grignard reagents have been used to introduce Mg into pheophytin as early as 1913[4]. Later, it was shown that modifications of the side-chains occur, as would nowadays be expected[3]. The organometallic method therefore, is restricted to the preparation of metalloporphyrins with unreactive side-chains. Nevertheless, metal alkyls are the ideal metal carriers. They are themselves Lewis acids in most cases, thereby promoting association with the porphyrin nitrogen atoms, and provide at the same time the strongest known bases to accept the protons from $H_2(P)$, e.g. according to eq. (23)[101].

$$H_2(\text{Etio-I}) + 2\,\text{MeMgI} \xrightarrow[35°C, 5h]{Et_2O} \text{Mg(Etio-I)} + MgI_2 + 2\,\text{MeH} \qquad (23)$$

$AlEt_3$ and the closely related AlH_3 also react at low temperatures[81,82] to form ultimately Al(OEP)OH. Higher temperatures are required with $TiPh_2$[63,70,126] and $(C_5H_5)_2TiCl_2$[127]; $TiPh_2$ is probably highly associated, and $(C_5H_5)_2TiCl_2$ is a coordinatively saturated molecule. As $AlEt_3$ is commercially available, it is the suggested metal carrier for the introduction of Al into porphyrins.

5.3.3.8. The metal carbonyl method

Novel metalloporphyrins[63] have been obtained for the first time from the following metal carbonyls: $Cr(CO)_6$[126], $Mo(CO)_6$[69], $Tc_2(CO)_{10}$[72], $Re_2(CO)_{10}$[73], $[RuCl_2(CO)_3]_2$[66], $[Rh(CO)_2Cl]_2$[67], and $[Ir(CO)_2Cl]_2$[68]. The carbonyls can be reacted in inert organic solvents, e.g. decalin, benzene, or toluene. The metal carbonyl chlorides may be obtained in situ from the corresponding metal trichlorides and reduction in ethanol in the presence of CO, and reacted in HOAc. The method is indispensable for the preparation of the bimetallic Tc(I)- and Re(I)-porphyrins (see Chapter 7) and for the incorporation of Ru(II)[128−133], Rh(II)[108], Rh(III)[23,31,134], and Ir(III)[68]. For the insertion of Ru(II), $Ru_3(CO)_{12}$ in boiling benzene or decalin is very suitable. A stoichiometric reaction can only be formulated under the unproven assumption[63] that H_2 is evolved according to e.g. eq. (24):

$$Cr(CO)_6 + H_2(\text{Meso-IX-DME}) \rightarrow Cr(\text{Meso-IX-DME}) + H_2 + 6\,CO \qquad (24)$$

The insertion of Rh is realized in various steps, e.g. eq. (25):

$$H_2(\text{OEP}) \xrightarrow[C_6H_6]{[Rh(CO)_2Cl]_2} [Rh(CO)_2]_2(\text{OEP}) \xrightarrow[reflux]{CHCl_3} [Rh(\text{OEP})Cl \cdot H_2O]$$

$$\qquad\qquad\qquad\qquad (Q) \qquad\qquad\qquad\qquad\qquad (C) \qquad\qquad (25)$$

References, p. 224

The method has further been used to prepare iron porphodimethene complexes, e.g., eq. (26)[122]:

$$H_2(OEPMe_2) \xrightarrow[\text{2. Al}_2O_3 \text{ chromatography}]{1. Fe(CO)_5/I_2} [Fe(OEPMe_2)]_2O \qquad (BB) \qquad (26)$$

As the reaction of OsO_4 with $H_2(OEP)$ in PhOH only gave poor yields of an Os-porphyrin[74] and $Os_3(CO)_{12}$ failed completely[88], a special method has been developed. A cold solution of OsO_4 in diethyleneglycol monomethyl ether is added dropwise to a boiling solution of $H_2(OEP)$ in the same solvent. Evaporation of the solvent, chromatography, and crystallization in the presence of a suitable donor L (see Section 5.4) furnishes $Os(OEP)(CO)L$ in very good yields[74]. Labile intermediates of the type $OsH_x(CO)_y(OR)_z$ are thought to be involved. They are formed by hydride abstraction and decarbonylation of the solvent.

5.3.4. Criteria of successful metal insertion: absorption spectroscopy in the visible region and other methods

A successful metal insertion can be established by various methods which allow one to state either the absence of the porphyrin educt or the presence of the metalloporphyrin product. They are: visible absorption spectroscopy, observation of fluorescence, infrared spectroscopy, and thin layer chromatography (t.l.c.). The infrared or the t.l.c. analysis require taking a small sample of the reaction mixture to dryness. For the infrared assay, even the metal carrier should be removed, e.g. by washing with water. Absence of either the NH-bands in the infrared spectrum or the typical spot of the porphyrin on the t.l.c. plate then indicate the end of the reaction. The silica t.l.c. plate normally shows the desired metalloporphyrins in the type $M(P)(A)$ as spots migrating faster than $H_2(P)$ [although $M(P)$ may be less soluble than $H_2(P)$], whereas those of the type $P(M)X_n$ are usually more strongly adsorbed.

The most versatile procedure is to take a small volume from the reaction mixture with a syringe, dilute, and measure the electronic absorption spectrum in the visible region. The end of the reaction is indicated by disappearance of the porphyrin band I (see Chapter 1) which is located at about 620 nm in all octa-alkylporphyrins [e.g. $H_2(Meso-IX-DME)$ and $H_2(OEP)$] and about 645 nm in the tetra-arylporphyrins (see Fig. 15, 16). A skilful experimentalist may even recognize this change in light absorption by means of a hand spectroscope. This procedure works because most metalloporphyrins show their longest wavelength absorptions (α-band, see below) hypsochromic with respect to band I of the porphyrin.

A more positive proof is the identification of the metalloporphyrin spectrum itself. As reported below (pp. 189—190), metalloporphyrins can be classified by their spectral absorption and emission characteristics ('optical

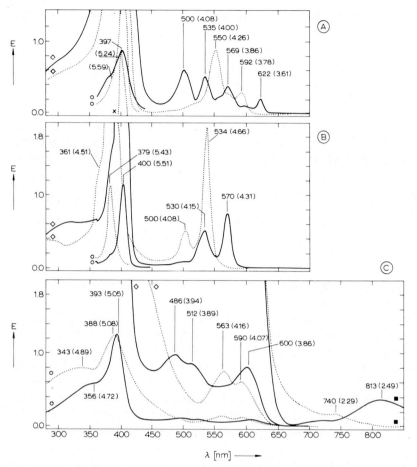

Fig. 15. Visible absorption spectra of some derivatives of octaethylporphyrin. [Unicam SP 800 B, CH_2Cl_2 as solvent, E = absorbance, λ in nm; digits give λ_{max} (log ϵ in brackets); ($\diamond$) d = 1 cm, ($\circ$) d = 0.1 cm, ($\blacksquare$) d = 5 cm, ($\times$) d = 0.1 cm and 1.0 to be added to absorbance, λ-values are accurate with 1%, above 700 nm readings are 1% too high.][122]

(A) ——— $H_2(OEP)$; . . . . . . $H_4(OEP)^{2+}$ (10% CF_3COOH).

(B) ——— $Zn(OEP)$; . . . . . . $Pt(OEP)$.

(C) ——— $Fe(OEP)F$; . . . . . . $[Fe(OEP)]_2O$ (the latter in benzene).

taxonomy of metalloporphyrins', M. Gouterman[135]). In this context, the three absorption types are sufficient to allow the recognition of the metalloporphyrin formed in a reaction. They are briefly presented below and compared in Fig. 15 for the $H_2(OEP)$ series and in Fig. 16 for the $H_2(TTP)$ series. Fig. 15 serves to illustrate the kind of spectra one might encounter in natural and octa-alkylporphyrins. Fig. 16 gives examples pertinent to the widely used tetra-arylporphyrins. These few examples may suffice here, as

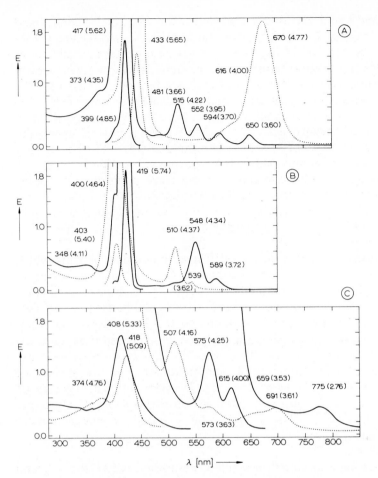

Fig. 16. Visible absorption spectra of some derivatives of *meso*-tetra(*p*-tolyl)-porphin (experimental details see Fig. 15)[122].

(A) ——— $H_2(TTP)$; . . . . . . $H_4(TTP)^{2+}$ (10% CF_3COOH).

(B) ——— $Zn(TTP)$; . . . . . . $Pt(TTP)$.

(C) ——— $[Fe(TTP)]_2O$; . . . . . . $Fe(TTP)Cl$ (both in benzene).

the spectral shifts caused by substitution of the porphyrin periphery are normally preserved on metalation, i.e. in most cases the peripheral substituents and the metal act as separated perturbations to the porphyrin chromophore.

1) *The 'normal spectrum'*[64,135]

On metalation, the 4-banded spectrum, e.g. of $H_2(OEP)$ (Fig. 15) or of $H_2(TTP)$ (Fig. 16), is altered to give a spectrum showing 2 bands in the visible region; the Soret band is retained. The two visible bands are labeled α and β

which parallel band I and III in the free acids. The position of the α-band varies within the range 570—610 nm for the $M(OEP)X_n$ series, and within 590 and 630 nm for the $M(TPP)X_n$ series; the position of the β-band varies accordingly. Increasing atomic number of the metal within a given coordination type and subclass (see Section 5.2) causes a bathochromic shift. The Soret band varies accordingly within a smaller wavelength range[62,92]. The normal spectrum is observed with all d^0 or d^{10} ions. The VO^{2+} ion (d^1) and Eu(III)(f^6) behave practically as a d^0 ion and have a normal spectrum, whereas Sb(III), Bi(III), and Pb(II) behave abnormally[62,65,115]. The normal light absorption gives these substances their purple color. The d^0 and d^{10} ions allow fluorescence to be observed with the corresponding metalloporphyrins. Practically all transition metal ions with open shells produce abnormal spectral types for the porphyrin. The d^1—d^5 configurations give rise to the 'hyper type', and the d^6—d^9 configurations cause the 'hypso' type.

2) *The 'hyper spectrum'*[64,135,136]

Treibs[65] had introduced the term 'allo' for all metalloporphyrins which are not purple, but brown or green in solution. All 'allo' species belong to the 'hyper' class which is now defined as a 'shifted metalloporphyrin spectrum with the α, β, Soret, and one or more extra bands being present' (see Ref. 135). Most of the metal ions belonging to this hyper class can easily realize lower oxidation states within the porphin which cause then a normal light absorption (Cr, Mo, Mn, Fe, Os). The Mn(III)- and Fe(III)-porphyrins are the most notable examples in the field, see e.g. the spectra of Fe(OEP)F and $[Fe(TTP)_2]O$ (Fig. 15, 16). Occasionally, the spectra of Mn(III)- and Fe(III)-porphyrins are taken as prototypes of all abnormal metalloporphyrin spectra[137]. The 'hyper' type is caused by charge transfer and other metal-ligand interactions.

3) *The 'hypso spectrum'*[64,135]

The d^6—d^9 ions have in common a filled d—subshell where the d_{xz} and d_{yz} orbitals are apt to overlap with the empty porphyrin π^* levels (see Ref. 135) which are thereby pushed up in energy thus causing a hypsochromic shift of the porphyrin π—π^* transitions with respect to the normal metalloporphyrins. This effect increases with increasing atomic number, e.g. in the series Ni(II) < Pd(II) < Pt(II)[138,139], and is pronounced with the Pt(II) porphyrins, see Pt(OEP) and Pt(TTP) (Fig. 15, 16); the spectrum shows just a hypsochromically shifted normal shape. Therefore, the 'hypso' spectrum is sometimes not recognized as a special case[137].

It is still under investigation, whether a fourth type ('batho spectrum') might be introduced to sort out some difficulties in the interpretations[135]. The brief introduction presented here may enable the synthetic chemist to identify the product he looks for. As none of the metalloporphyrins containing open shell metal ions (d^1—d^9) shows a fluorescence which is sufficiently intense to be observed by the naked eye, the end of a metalation producing these complexes may be observed by the disappearence of the strong fluorescence of the porphyrin educt which is excited by irradiating the Pyrex reaction vessel with a small u.v. lamp.

Finally, it may be mentioned that the recording of a fluorescence excitation spectrum provides a very sensitive test for purity of a metalloporphyrin prepared. The emission intensity must follow the absorption spectrum. Very small amounts of impurities can be detected when their fluorescence is more intense than that of the main product.

5.3.5. Oxidation states of the central metal ions

As has been pointed out before, the oxidation state of the metal ion in a metalloporphyrin often does not correspond to the oxidation state of the ion in the metal carrier. Therefore, it is worthwhile to discuss the oxidation states so far encountered in metalloporphyrin chemistry. They have been compiled in Table 7; those oxidation states which are adopted under aerobic conditions, i.e. in presence of water and oxygen, are given as underlined figures; henceforth they will be called the 'stable' states. The topic has been reviewed extensively by Fuhrhop[137].

It is obvious that the porphinato ligand neither stabilizes high nor low oxidation states particularly. Most oxidation states are well known within metal chemistry[140]. One may ask the question: why does the metal adopt a certain oxidation state as 'stable' under aerobic conditions when there is a choice? A choice is presented, of course, in the transition metal series and with the 5th- and 6th-row heavy metals where the 'inert electron pair'[140] becomes important, and as is evident from Table 7, the metals really take their choice. (The table does not show all the possible metal oxidation states because only monometallic species are considered in this chapter.) In the opinion of the author, the balance of electron donor-acceptor power and size of the respective ion is the essential feature.

To obtain a general view considering every metal in the periodic table, a rather primitive model must be used. A metalloporphyrin may be regarded as a 'zero-dimensional' crystal composed of a rigid, square N_4^{2-} anion and a metal ion M^{z+} which are held together by mainly electrostatic forces. Co-valent bonding becomes essential only with the metals that show a rather high electronegativity (E_N) and then gives an additional contribution to the bond strength. Accepting the dominance of ionic bonding, the ionic radius r_i of the central metal gives an appropriate measure of its size. Arbitrarily, the values for octahedral coordination within metal oxides are taken as standard radii r_i, if not otherwise stated[141].

A comparison of the 'stable' metal ion oxidation numbers, Z_{ox}, and the ionic radii of various oxidation states is displayed in Table 8. First, we take a mean ionic radius of 64 pm (picometer) as best-fit for the hole in the N_4^{2-}-moiety of the porphinato ligand. This choice is justified by the planar structure (A) and the M—N bond lengths, 201 pm, found in Pd(TPP) (see Chapter 8); according to Hoard[143] a value of approx. 201 pm for the

TABLE 7

Periodic table of monometallic metalloporphyrins showing metal oxidation numbers (Z_{ox}) and spectral absorption types

Mg 2 n													
Ca 2 n	Sc 3 n	Ti 4 n	V 4 n	Cr 2* 3* 4°	Mn 2* 3* 4°	Fe 1† 2† 3* 4°	Co 1* 2† 3†	Ni 2† 3°	Cu 2†	Zn 2 n	Al 3 n	Si 4 n	
Sr 2 n	Y 3 n	Zr 4 n	Nb 5 n	Mo 4 n 5*		Ru 2† 3†	Rh 1† 2† 3†	Pd 2† 4°	Ag 2† 3†	Cd 2 n	Ga 3 n	Ge 4 n	As 5 n
Ba 2 n	La 3 n	Hf 4 n	Ta 5 n	W 5*	Re 5*	Os 2† 4* 6*	Ir 3†	Pt 2†	Au 3†	Hg 2 n	In 3 n	Sn 4 n	Sb 3* 5 n
		Th 4 n	...Eu... 3 n								Tl 3 n	Pb 2* 4 n	Bi 3*

Underlined figures indicate 'stable oxidation states' (see text). n: normal; *: hyper; †: hypso type of absorption spectrum; o: indicates spectrum not reported.

Note added in proof: Recent evidence[122,269] indicates that the original assignments of oxidation states of the group V metal ions[115] should be reversed; the present state of knowledge is presented in Tables 7, 8, and 10.

TABLE 8

Comparison of stable metal oxidation numbers (Z_{ox}; stable within metalloporphyrins under aerobic conditions) and ionic radii (r_i).

M	Stable Z_{ox}	Effective ionic radii (given in pm) for Z_{ox} =							
		+2		+3		+4	+5	+6	+7
		low spin	high spin	low spin	high spin				
Ti	4		86		67	60††			
V	4		79		64	59	54		
Cr	3		82		62	55	(~50)	(<50)	
Mn	3		82		65	54			
Fe	2,3	61	77	55	64				
Co	2,3	65	73	52	61				
Ni	2	[60]*	70		60				
Cu	2	[62]	73						
Nb	5				70	69	64		
Ta	5				67	66	64		
Mo	5				67	65	63	60	
W	5					65	(62)	58	
Re	5					63	(61)	52	57
Ru	2,3	(78)†		68†		62			
Os	2,4,6	(79)†				63		?	
Rh	2,3	(78)†		[66]†		62			
Ir	3	(79)†		[(71)]†		63			
Pd	2	[64]				62			
Pt	2	(78)†							
Ag	2	[(84)]†		[65]†					
Au	3			[70]†					
Sn	4		(90)			69			
Pb	2		118			77			
As	5				(63)		50		
Sb	3,5				74		61		
Bi	3				102		79		

r_i are effective ionic radii for CN = 6 taken from Ref. 141. Values in square brackets refer to square planar coordination (CN = 4). Values in bowed brackets are interpolated or calculated from values tabulated in Ref. 141 for radii systems other than the effective ionic radii. Radii corresponding to oxidation states found in metalloporphyrins are underlined.
* Taken from Ref. 142.
† Backbonding reduces size indicated.
†† r_i (CN = 5):53 pm.

center-nitrogen distance corresponds to a minimization of radial strain in the porphinato core of a metalloporphyrin. Upper and lower limits for coplanar M—N bond lengths are then substantiated by $Sn(TPP)Cl_2$, 210 pm, and

Ni(OEP), 193 pm; the respective ionic radii amount to 69 pm for Sn(IV) and 60 pm[142] for Ni(II). Hence, a metal ion M^{n+} of ionic radius between 69 and 60 pm may produce a coplanar MN_4-moiety in a metalloporphyrin; the small ions, however, produce considerable radial strain (see Chapter 8).

The subsequent discussion of Table 8 will show that the mean ionic radius, 64 pm, may be used to understand the 'stable' oxidation numbers if some corrections are applied which are induced by additional phenomena, e.g. $d_\pi - p_\pi$-bonding, redox potentials, and 'softness' of the ions under consideration. We shall not predict, however, that the '64 pm-ions' must sit right in the MN_4-plane, because any asymmetric axial coordination will pull the central metal out of plane towards that ligand which is more strongly bound (see Chapter 8); in this context, center-metal distances up to 50 pm will be tolerated as 'fitting into the porphyrin N_4-cavity'.

Ti(IV) is favored over Ti(III) because the latter is too strongly reducing. Although at first glance titanyl porphyrins have CN = 5 and thus should have r_i = 53 which may cause 'rattling' of the ion within the N_4-hole, multiple dative bonding of the oxo-ligand effectively 'inflates' the Ti(IV) ion to make it comfortable about the N_4 plane. The same argument holds for V(IV); the VO^{2+} ion is very stable even in aqueous media and its occurrence in porphyrins is no surprise; V(III) is a strong reductant. Progressing through the first transition series, Cr(III) and Mn(III) are definitely favored. All the high-spin M(II) ions are too big. Fe(II) is only stabilized with axial π-acceptor ligands (e.g. Py, CO, O_2, NO, 1-MeIm). Co(II), Ni(II), and Cu(II) shrink sufficiently by π-donation to the porphyrin ligand alone. High-spin Co(III) is too strongly oxidizing; Co(III) therefore needs high field ligands converting it to a 'normal' low-spin Co(III). The radial strain caused by Ni(II) in square planar geometry may be relieved by raising CN to 6 by addition of two axial imidazole ligands[144]. Nb(V) and Ta(V) are obviously quite stable. As Mo, W, and Re usually appear in their higher oxidation states combined with oxide ligands attached via multiply dative bonding, their 'stable' M(V) state seems reasonable, the rather suitable r_i-values for the M(IV) states becoming effectively larger by π-bonding and the M(V) values thus more favorable. One might argue that the M(VI) state should also be stable because of the 'double bonding inflation'. However, with MoO_2^{2+} and WO_2^{2+}, the trans-dioxo-array has never been found because of the severe trans-effect of the multiply bound oxide ions[25] which repel each other. With ReO_2^+, trans-species are known[140], but the single positive charge of this linear unit is not compatible with the porphinato anion. Many derivatives of the 'osmyl ion', trans-OsO_2^{2+}, have been prepared[140]. Consequently, the existence of OsO_2(OEP) is not too surprising[26], although this is the only metalloporphyrin in which the central metal ion is hexavalent. This is the highest oxidation state yet observed within metalloporphyrins.

Unfortunately, the radii presented for the noble metal series Ru . . . Au (Table 8) are mostly rough estimates because of the limited data available[141]. On first glance, one would expect that the M(IV) state is 'stable', but this is only found with Os(OEP)(OMe)$_2$ [26]. [The finding that OsO$_2$(OEP) exists is then a consequence of the 'double bonding inflation'.] All the other metals adopt the M(III) or M(II) states, although the values of the radii given are rather too large. Here, an opposite bonding scheme operates: π-back donation from the metal to the porphyrin ligand and the axial ligands reduces the electron density at the metal thus causing an effective shrinkage of the ion. Hence, the big Os(II) fits into the porphyrin in the presence of CO as an axial ligand which is known as a strong d^2-electron acceptor yielding compounds such as Os(CO)Py(OEP)[74]. Osmium porphyrins therefore are not only unique because of the Os(VI) state, but furthermore because of the large span of 'stable' oxidation states: Os(II), Os(IV), Os(VI).

Summarizing, the noble metals exist in low oxidation states although the r_i-values are unfavorable: back-bonding and the generally observed 'soft-acid' behavior of these metal ions is not incorporated in these values; 'soft acids' are polarizable acceptors which may be come 'caved in' when embedded in the porphyrin cavity[39].

The radii r_i for square planar Ag(III) and Au(III) indeed indicate good fit of these ions. Au(III) is therefore a 'stable', and Ag(III) a very easily accessible oxidation state[145]. Pd(IV) porphyrins have recently been observed[146]. Cr(II)[124,126], Mn(II)[124,147], Fe(I)[148], Co(I)[137,149], and Mo(IV)[20,150] are some 'unstable' but spectrally well-defined oxidation states that have been generated within metalloporphyrins; very 'unstable' Cr(IV), Mn(IV), Fe(IV), Ni(III), and Sn(II) are under discussion[137].

Turning now to the remaining five main group metals (Table 8), the effect of the 'inert pair'[140] increases from Sn to Pb. Sn(IV) is exceedingly stabilized relative to Sn(II) as seen from the radii. With Pb, both Pb(II) and Pb(IV) are uncomfortable in size. Still, the Pb(II) state is stabilized by the inertness of its 6s^2-pair, Pb(IV) is strongly oxidizing, but 'unstable' Pb(OEP)Cl$_2$[96] has been isolated. The pattern of radii helps to understand that As(V), Sb(V) as well as Sb(III), and Bi(III) are the 'stable' states of the respective metals[65,115,122].

One might desire to start the same kind of arguments using covalent radii, but the available data are too scanty to allow a general survey.

5.3.6. Demetalation, transmetalation, and stability orders

As pointed out before (p. 157), the removal of a metal ion from a porphyrin [schematically according to eq. (27)] is called 'demetalation'

$$M(P)XL + 2\,HX + 2\,L \rightarrow H_2(P) + MX_3L_3 \tag{27}$$

In some cases, metals may be replaced by other metals. This process is named 'transmetalation' and is formally depicted in eq. (28):

References, p. 224

$$M(P)XL + M'X_3L_3 \rightarrow M'(P)XL + MX_3L_3 \tag{28}$$

The higher the excess of HX or $M'X_3L_3$ in a given set of demetalations or transmetalations, the more stable is the corresponding metalloporphyrin. Stability of a metalloporphyrin is therefore defined as 'stability towards acids' or 'stability in the presence of another metal compound'. Since Falk's monograph[1], only little experimental progress pertinent to a quantitative understanding of stability phenomena has been made. The stability constant (eq. 29) of Zn(Meso-IX), determined by Phillips[151], is still the only figure known. The high value obtained agrees well with values observed with other

$$Zn^{2+} + (Meso-IX)^{2-} \leftrightharpoons Zn(Meso-IX) \tag{29}$$

$$K_{ass} = \frac{[Zn(Meso-IX)]}{[Zn^{2+}][(Meso-IX)^{2-}]} 10^{29}$$

macrocyclic tetramine chelates[152] and illustrates the enormous stability of metalloporphyrins. Their high values, the limited solubility of the porphyrins, and the interference of other equilibria, e.g. protonation/deprotonation and ligand exchange (see eq. $(A1)$, $(A2)$, $(A4)$, $(A5)$ in Scheme 1, p. 160), render the measurement of stability constants very difficult[12].

However, qualitative statements have been made[1,151−156]. The details of the various investigations will not be reiterated, but rather a schematic treatment will be given. This is based on empirical stability classes (Table 9) and an invented quality S_i which is called the 'stability index'[39,157] (see Section 5.3.6.2). Demetalations and transmetalations will then be discussed inasmuch as they do not proceed in accord with expectations. Finally, correlations of stability with spectroscopic data are examined.

5.3.6.1. Stability classes

According to Falk[1] and Phillips[151], one may discern five kinds of reagents showing decreasing protic acidity and reactivity in demetalation

TABLE 9

Assignment of stability classes to metalloporphyrins towards acids

Stability class	Reagent (25°C, 2 h)	Behavior
I	100% H_2SO_4	Not completely demetalated
II	100% H_2SO_4	Completely demetalated
III	$HCl/H_2O-CH_2Cl_2$	Demetalated
IV	100% HOAc	Demetalated
V	$H_2O-CH_2Cl_2$	Demetalated

TABLE 10

Stability index S_i and stability classes of metallo-octa-alkylporphyrins with metal ions in a given oxidation number Z

Z	M	E_N*	r_i(pm)	E_N/r_i $\times 100$	S_i	Stability class Good fit	Bad fit	Remarks	Ref.
2	Pd	2.20	[64]	3.44	6.88	I			86
	Ni	1.91	[60]	3.18	6.37	II			55
	Cu	1.90	[62]	3.06	6.12	II			55
	Pt	2.28	(78)	2.92	5.85		I	r_i improper	·86
	Co	1.88	65	2.89	5.78	II		Low-spin Co(II)	55
	Fe	1.83	77	2.38	4.76	III			55
	Ag	1.93	(84)	2.30	4.60		II	r_i improper	152
	Zn	1.65	74	2.23	4.46	III			55
	Sn	1.96	93	2.11	4.22	—		Not investigated	—
	Cr	1.66	82	2.02	4.05	—		Not investigated	—
	Pb	2.33	118	1.97	3.95	IV			122
	Hg	2.00	102	1.96	3.92	IV		V in CH_2Cl_2/H_2O	122
	Mn	1.55	82	1.89	3.78	IV			65
	Mg	1.31	72	1.82	3.64	IV			153
	Cd	1.69	95	1.78	3.56	IV			122
	Ca	1.00	100	1.00	2.00	V			62
	Sr	0.95	116	0.82	1.64	V			62
	Ba	0.89	136	0.65	1.31	V			62
3	Au	2.54	[70]	3.63	10.88	I			154
	Ir	2.20	[(71)]	3.10	10.38	I			68
	Rh	2.28	[66]	3.45	10.36	I			68
	Sb	2.05	74	3.38	10.14		III		115,122
	Ru	2.2	68	3.24	9.71	I			122
	Co	1.88	61	3.08	9.25		II	Co(III) → Co(II)	55
	Al	1.61	53	3.04	9.11	I			55
	Ga	1.81	62	2.92	8.76	II			122
	Fe	1.83	64	2.86	8.58	II			55
	Cr	1.66	62	2.68	8.03		I	Cr(III) very inert	123
	Mn	1.55	65	2.38	7.15	II			122
	Tl	2.04	88	2.32	6.95	II			110
	In	1.78	79	2.25	6.76	III			122
	Bi	2.02	102	1.98	5.94	III			115,122
	Sc	1.36	73	1.86	5.59	IV			84
	Y	1.22	89	1.37	4.11	<II			75
	La	1.10	106	1.03	3.11	—		Not investigated	—
4	Si	1.90	40	4.75	19.00	I			122
	Ge	2.01	54	3.72	14.89	I			122
	Os	2.2	63	3.49	13.97	I			123
	Pb	2.33	78	2.99	11.95	—		Not investigated	—
	Sn	1.96	69	2.84	11.36	I			54
	V	1.63	59	2.76	11.05	I		II when reduced	65

TABLE 10 (continued)

Z	M	E_N*	r_i(pm)	E_N/r_i × 100	S_i	Stability class Good[b] fit	Bad fit	Remarks	Ref.
4	Ti	1.54	60	2.57	10.27	II			127
	Zr	1.33	72	1.85	7.39	II			123
	Hf	1.3	71	1.83	7.32	II			123
	Ce	1.12	80	1.40	5.60	—		Complex not known	—
	Th	1.3	100	1.30	5.20	<II			75
5	As	2.18	50	4.36	21.80	I			115,122
	W	2.36	(62)	3.81	19.03	I			123
	Mo	2.16	63	3.43	17.14		II	Mo(V) → Mo(IV)	123
	Sb	2.05	61	3.36	16.80	I		Reduction?	115,122
	Re	1.9	(61)	3.11	15.57	I			123
	Nb	1.6	64	2.50	12.50		III	Reduction?	123
	Ta	1.5	64	2.34	11.72		III	Reduction?	123

Ionic radii r_i for CN = 6; radii in square brackets for CN = 4; radii given in bowed brackets are estimated only, see Table 8.
* Pauling-electronegativities; values with 2 decimals taken from Ref. 140, with 1 decimal from Ref. 161.

reactions. The metalloporphyrins are then divided into five 'stability classes' which indicate their resistance towards protic acids of different strength[39,157], see Table 9. They are determined for a given metalloporphyrin by subsequently treating it with the reagents indicated in Table 9, starting with the mildest conditions, and observing the behavior of the metalloporphyrin. Concentrated sulfuric acid and glacial acetic acid are reacted directly with the solid metalloporphyrins with stirring for 2 h at room temperature and then poured onto a 10-fold excess of ice, thereby precipitating the unreacted metalloporphyrin or the porphyrin free acid. In the case of aqueous HCl and water, CH_2Cl_2 is added and the two-phase system vigorously stirred for 2 h, the products then being recovered from the organic phase. Divalent ions may be removed from metalloporphyrins of class II also by action of trifluoroacetic acid[23]. Stability classes may also be taken from metalation studies. Fe(II) porphyrins must at least belong to class IV as they can be made in acetic acid. The removal of iron from hemes is best performed with hydrochloric acid after reduction of Fe(III) to Fe(II), thus indicating class III for Fe(II)[1,2]. Useful preparative methods for the isolation of natural porphyrins from hemes have been described by Morell[158] and Grinstein[159] and have been discussed by Falk[1], see also Chapter 19.

Stability classes of most of the known monometallic metalloporphyrins and references are compiled in Table 10.

5.3.6.2. Appraisal of a stability index

The extension of the periodic table of metalloporphyrins provides so many species containing different metals in varying oxidation states that one may look for the metal parameters determining the stability of the corresponding metalloporphyrins. The utility of ionization potentials, electronegativities (E_N), the charge-to-radius ratio (Z/r_i or Z^2/r_i) and the 'Van Uitert—Fernelius function' $Z \cdot E_N$ have been critically examined by Rosotti[160]. The expression $Z \cdot E_N$, the product of the charge number of the ion, Z, and the electronegativity of the metal, serves to predict the thermodynamic stability of open chelates $M(acac)_Z$ and $M(dbm)_Z$ with metal ions in various oxidation states. All the expressions mentioned before do not produce a satisfactory correlation with experimental stability orders of metalloporphyrins. The author has then tested the quantity $S_i = 100 \cdot Z \cdot E_N/r_i$, called 'stability index', and found reasonable agreement with stability orders[39,157]. For simplicity, the electronegativity values of the elements irrespective of their oxidation state are taken. Only the Pauling values[140,161] work. The 'effective ionic radii' are taken from the recent compilation by Shannon and Prewitt[141] for coordination numbers (CN) and spin states indicated in Table 10, which shows all the figures used for the evaluation of S_i. The inverse relationship of stability and ionic radius is meant to express the destabilizing influence of r_i once it exceeds a value of about 70 pm.

The approach is highly arbitrary and intuitive. Nevertheless, it challenges further research, especially to look for a finer subdivision of the stability classes. The postulated increase of stability with increasing atomic number within some groups of the periodic table is a consequence of the Pauling E_N values, and its relevance remains to be tested. Table 10 will now be discussed for demetalation reactions.

5.3.6.3. Demetalation

It has been known for some time[1,2,3] that demetalation is much more easily achieved with lower oxidation states of a given metal. This feature emerges clearly from Table 10. However the correlation between stability class and stability index S_i is not consistent within the series of ions differing in Z.

With the divalent ions, the decrease in S_i is correctly associated with a decrease in stability class, with two exceptions: Pt(II) and Ag(II). This is due to their r_i values being improperly used; the required square planar effective ionic radii are so far unknown[141]. In fact, Pt(II) porphyrins are probably the most stable M(II) porphyrins (see Section 5.3.6.5), Pt(OEP) being converted into a black, insoluble material on heating with H_2SO_4 at $100°C$ for 1 h[123]. The hitherto reported stability order[1,12,151] for divalent ions Pt > Pd > Ni > Co > Ag > Cu > Zn > Mg > Cd > Sn > Ba is, indeed, misleading since the order of Ni, Co, Ag, and Cu has been taken from spectroscopic and kinetic rather than thermodynamic measurements[86,139,151,153,156,162], and the Sn(II)-porphyrins investigated were in fact Sn(IV)-porphyrins. As Cu(Etio-I) is only partially demetalated by CF_3COOH, whereas Ni(Etio-I) and Co(Etio-I) completely lose their metal ion in this reagent[23], the relative position of these metals must be further secured. In the next chapter it will become clear that the rates of metalation and demetalation reactions are dependent not only upon the acid strength, but also upon the nature of anions and neutral donor ligands present in the system, and this may likewise be true for the stability constants. In this situation, one may only speculate on the reasons why the metallopoprhyrin stability order deviates from the stability orders of complexes with other ligands, e.g. the Mellor–Maley and Irving–Williams series[160].

The position of Pb(II) and Hg(II) in class IV is difficult to prove. In CH_2Cl_2/H_2O, Hg(OEP) is demetalated, in benzene/H_2O, it is stable for 2 h[122]. Demetalation of Hg(TPP) in CH_2Cl_2 containing traces of acids has also been observed[163]. In CH_2Cl_2/H_2O, Pb(OEP) is only slightly demetalated. The low electronegativity and the large radius render Ca-, Sr-, and Ba-porphyrins so unstable that they have not been characterized by elemental analysis of a solid sample.

In the series of tervalent metal ions, the stability of the complexes Sb(OEP)Cl and Co(OEP)XL is overestimated by S_i. This may be due to an intrinsic thermal lability: As-, Sb-, and Bi-porphyrins are decomposed by heating in solvents above $60°C$ with formation of $H_2(OEP)$. [This is not the case with Hg(OEP) and Pb(OEP).] Co(III) in porphyrins is easily reduced to Co(II) and then expelled by H_2SO_4; reduction probably occurs because the axial ligands stabilizing the Co(III) state are removed by protonation. Cr(III) is enormously inert to reduction within the porphyrin system[136] and may be as inert to equatorial substitution as it is labile to axial substitution in porphyrins (see Chapter 6). It is expected from the lability of Sc(III)-porphyrins and the S_i values that monometallic Y(III)- and La(III)-porphyrins will be even more labile. Nevertheless, Eu(TPP)(acac), Pr(TTP)(acac), and Y(TTP)(acac) have been prepared[75].

No serious deviation of stability classes and S_i is seen in the field of tetravalent ions. However, VO(OEP) is easily demetalated by H_2SO_4 containing reducing impurities. Fresh H_2SO_4, or H_2SO_4 to which a trace of H_2O_2 has been added, do not demetalate VO(OEP)[123].

From the values of S_i, all M(V)-porphyrins should belong to class I. The aberration of Mo is certainly due to the easily accessible Mo(IV) state[20]. The same argument may hold for Sb(III) and Nb(V), which are of class III contrary to their high value of S_i. Nb(V) forms lower oxidation states more readily than Ta(V)[140]. It is remarkable that some ions of group V, Nb(V), Ta(V), Sb(III), and Bi(III), are expelled by alkali. These ions are known to from anionic oxo and hydroxo complexes very easily, and one can imagine that addition of several hydroxide ions pulls the cation out of the porphyrin ring. The concomitant increase in coordination number then inflates the ion, thereby reducing the stability of the metalloporphyrin. As Zr(IV) and Hf(IV) have CN = 8 within metalloporphyrins (see Chapter 8), they should actually have a lower S_i value; however, it is not clear whether CN = 8 is retained in acidic media in these complexes.

Clearly, univalent metals are rather loosely bound to porphyrins, and they show low stability indices. Indeed, the metalloporphyrins $M_2(P)$ (M = Li, Na, K, Rb, Cs, Ag) belong to class V; they are considered in Chapter 6. For the synthetic chemist it may be helpful to note that metalloporphyrins showing stability classes IV cannot be purified by chromatography on alumina or silica gel because of partial or complete demetalation on the columns[20,84,122,153]. In these cases gel permeation chromatography on Sephadex LH 20 and Merckogel 2000[84], or column chromatography on magnesol—cellulose mixtures[154] may help. Metalloporphyrins in stability class V do not seem to have been isolated in the crystalline state because of their easy hydrolysis.

At this point — knowing the principal features of metalation and demetalation — one may speculate why VO-, Ni-, and Cu-porphyrins are found in mineral oil although Al, Ca, Fe, K, Mg, Na, Si, and Ti are two or three orders of magnitude more abundant than these three metals, and Ba, Cr, Mn, Sr, Zn, and Zr are roughly of equal abundance on earth. (The less abundant metals, e.g. Sn, are not considered.) V, Ni, and Cu are favored over Ca, Fe, K, Mg, Na, Ti, Ba, Mn, Sr, Zn, and Zr by their greater adherence to the porphyrin ligand. The remaining three metals Al, Cr, and Si yield very stable metalloporphyrins; however, their formation requires far more rigorous conditions than the formation of Cu-, Ni-, and VO-porphyrins. In fact V(IV) is the easiest tetravalent metal ion to be inserted into a porphyrin. These three metals then are introduced under rather mild conditions and are bound so firmly that they 'survive' within the porphyrin.

5.3.6.4. Transmetalation

Stability orders can also be derived from transmetalation reactions, see eq. (28) (p. 196), since it is reasonable that the more 'adhesive' metal ion replaces more weakly bound ions. The following pairs have been studied (the arrow indicates the direction of the metalation):

A) $Na_2 \rightarrow K_2$, $Li_2 \rightarrow Na_2$, $Li_2 \rightarrow Hg$, $Li_2 \rightarrow Zn$, $Pb \rightarrow Hg$, $Hg \rightarrow Cu$ and $Hg \rightarrow Zn$[164], $Ba \rightarrow Hg$, $Ba \rightarrow Pb$[92] (all TPP complexes, pyridine, 20°C); $Pb \rightarrow Cu$, $Pb \rightarrow Zn$, $Zn \rightarrow Cu$ (all TPP complexes, pyridine, 114°C)[164]. As expected, neither Ag(II), Sn(IV), Co(II), and Cu(II) are replaced by Zn(II)[164]. It is remarkable that Zn(II) and Cu(II) do not replace Mg(II) in boiling pyridine; this may be due to formation of $Mg(TTP)Py_2$ (C) which may be attacked more slowly than the less crowded $Zn(TPP)Py$ (B).

B) $Cd \rightarrow Zn$, $Pb \rightarrow Zn$, $Hg \rightarrow Zn$, $Zn \rightarrow Cu$ (in pyridine, 25—45°C, with TPP complexes)[165].

C) $Zn \rightarrow Cu$ (much more slowly with etioporphyrin than with tetra-aryl-porphyrins)[156,166].

D) $Cd \rightarrow Cu$, $Pb \rightarrow Cu$, $Zn(II) \rightarrow Au(III)$ [with (Proto-IX—DME) in pyridine][153].

(E) $Sn(IV) \rightarrow Li_2$, $Fe(III) \rightarrow Li_2$, $Cu(II) \rightarrow Li_2$, $Co(II) \rightarrow Li_2$, $Co(III) \rightarrow Li_2$, $Ni(II) \rightarrow Li_2$, $V(IV) \rightarrow Li_2$, $Pd(II) \rightarrow Li_2$ $Pt(II) \rightarrow Li_2$. These transmetalations have been carried out with lithium metal in ethylenediamine at 116°C. The stability of the corresponding metalloporphyrins is thought to increase in the given sequence (Sn $\rightarrow$ Pt), the apparently low stability of Sn(IV)-, Fe(III)-, and Co(III)-porphyrins being caused by prior reduction $Sn(IV) \rightarrow Sn(II)$, $Fe(III) \rightarrow Fe(II)$, and $Co(III) \rightarrow Co(II)$. The stability order has been derived from an increasing amount of metallic Li necessary to effect transmetalation. Experiments have been performed with $M(Etio-I)X_n$, $M(TPP)X_n$, and Fe(OEP)Cl. Excessive reduction of the porphyrins had accompanied trans-metalation which could be reverted with quinones of high oxidation potential. As the di-lithium porphyrins are easily demetalated by water, this procedure serves to demetalate the otherwise stable Sn(IV)- and Pd(II)-porphyrins. Lithium ions in ethylenediamine also cause the replacement of Cu, Co, Ni, and Pd. With the catalytically active transition metal derivatives M(TPP) (M = Co, Pd, Pt) destruction of the porphyrin ligand invalidates the method. The mechanism does not seem to involve a simple electrophilic attack of Li^+ ions[86].

F) $Cu(II) \rightarrow Al(III)$, $Fe(III) \rightarrow Al(III)$, $Zn(II) \rightarrow Al(III)$, $Sn(IV) \rightarrow Al(III)$. These reactions have been observed with $M(OEP)X_n$ and di-isobutyl aluminum hydride in THF. Cu(OEP) gives good yields of Al(OEP)OH at 25°C, whereas Fe, Zn, and Sn are only replaced by less than 20% Al at 25°C in 2 h. The apparently greater lability of Cu(II) relative to Zn(II) is due to a reduction to Cu(I).

Under normal transmetalation conditions, Mg(Phy-a), Mg(Phy-b), Fe(TPP)Cl, Fe(Proto-IX), Cu(Phy-a), and $Zn(TPP)$[12,165,166] do not ex-change with the respective identical but radioactive metal ions. This is not surprising since the ions mentioned are reluctant to transmetalation anyway. That radioactive sodium is incorporated may be expected from the lability of the alkali metal porphyrins[164]. As a whole, the course of transmetalations is in accord with the knowledge gained from demetalation reactions. Reduction of a central metal ion facilitates demetalation and transmetalation. Metals with a low charge, a big diameter, and a low covalent bonding contribution will be very easy to remove from a porphyrin, as expressed by the stability index S_i.

5.3.6.5. Correlations of spectral data and stability

Taking into account the development of the periodic table of metalloporphyrins as achieved now, a correlation of spectral data with stability becomes less obvious than two decades ago. Any correlation of fluorescence intensity and stability towards acids as suggested earlier[153] must be discarded (see Ref. 135). The often-cited[1,12,139,151,156,162] stability order derived from the relative position of the α-band in the optical spectrum obtains for a selection of divalent metal ions only, and the origin of the spectral effects as presented by Williams[162] may be interpreted also from a further viewpoint as follows.

Although varying in their stability class from I—IV, 18 metallooctaethylporphyrins containing the ions Mg(II), Al(III), Si(IV), Sc(III), Ti(IV), V(IV), Zn(II), Ga(III), Ge(IV), As(V), Zr(IV), Nb(V), Mo(IV), In(III), Sn(IV), Sb(III), Hf(IV), and Ta(V) have normal absorption spectra showing the α-band between 567 and 582 nm, the β-band between 528 and 545 nm, and the Soret band between 399 and 412 nm[20,25,27,115]. The mean values (in CH_2Cl_2 or C_6H_6) are α : 573, β : 536, Soret : 405 nm. These spectra are found irrespective of coordination type or deviation of the metal from coplanarity with the porphyrin ligand (Ti, V, Zr, and Hf protruding considerably, see Chapter 8). Thus the mean values given above should be the reference values for metallooctaalkylporphyrins, irrespective of stability.

Considering now the metalloporphyrins M(OEP) (A), the following positions of the α-band are observed in benzene for the divalent metals indicated:

M	Pb	Zn	Hg	Ag	Cu	Co	Ni	Pd	Pt
λ_{max}(nm)	582	570	566	563	554	554	554	548	536
Ref.	93	20	122	122	20	20	20	87	87

The order Pd > Ni > Co > Cu > Zn > Mg has been determined in $CHCl_3$ [167]. With the octahedral species $M(OEP)Py_2$ (C) in pyridine, the following series exists:

M	Mn	Mg	Fe	Co	Ru	Os
λ_{max}(nm)	584	582	548	548	520	510
Ref.	105	168	122	122	169	169

With M(OEP) (A) or M(OEP)Py (B) the values in pyridine are:

M	Pb	Zn	Cd	Hg
λ_{max} (nm)	578	580	586	589
Ref.	94	122	93	122

Clearly, the measurements do not strictly prove a correlation between stability class and position of the α-band. In benzene, Hg(II) is given the wrong place. In pyridine, Fe(II) and Pb(II) are estimated too stable.

In the author's opinion, the increasing deviations from the 'normal' α-band position at ~ 573 nm reflect either increasing stabilization by $d_\pi - p_\pi$ backbonding in the case of hypsochromic shifts (see Section 5.3.4. and Ref. 135) or increasing destabilization by steric distortion of the porphinato ligand when bathochromic shifts are observed. The bimetallic species all show bathochromically shifted spectra irrespective of stability (see Chapter 7)! As has been shown by Brunings and Corwin, a torsion at the methene bridges about a double bond causes a bathochromic shift of the spectrum of pyrromethenes[170]. With porphyrins and chlorins these effects are, of course, not so pronounced, but well established[104,171].

An asymmetric coordination with a metal ion which is pulled out of the porphyrin plane by a single anionic axial ligand, e.g. in VO(OEP) or TiO(OEP), or which has just too high a coordination number to fit into the porphyrin cavity, e.g. in Zr(OEP)(OAc)$_2$ (see Chapter 8), does not necessarily imply such bathochromic shifts. The high degree of covalent and thus directional bonding in Cd-, Pb-, and Hg-porphyrins may, however, allow a stronger steric effect on the porphyrin ligand. If this argument holds, Sr- and Ba-porphyrins also should not have bathochromic shifts, like the Zr- and Hf-porphyrins, because their bonding is more ionic. Nevertheless, they do show such effects[62]! But has it ever been excluded that the Sr- and Ba-porphyrins are bimetallic complexes? This might well be, as they have never been characterized in the solid state and are formed in presence of an excess of metal ion.

Another suggested correlation of stability with spectroscopic data uses the intensity ratio of the α- and β-bands, a high ratio I_α / I_β indicating low stability, and vice versa, with metallotetraphenylporphyrins[92,151,164]. Roughly speaking, the opposite holds for the complexes M(OEP) (A) and M(OEP)Py$_2$ (C) mentioned before, the ratio increasing with increasing stability[167], with some reversions occurring when the series is extended[122]. Furthermore, considerable variations of the intensity ratios occur when axial ligands are changed at the same metal[93]. Thus, this correlation does not help much at present.

Finally, correlations have been proposed between the positions of certain bands in the infrared spectra (see Chapter 11) and the stability with M(TPP)

(M = Cu, Co, Ni, Pd, Pt)[172] and with M(OEP) (M = Mg, Zn, Cu, Co, Ni, Pd)[167,173]. The increase of π-bonding between the metal and the porphinato ligand may very well be reflected in the series given. However, a discussion of the more extensive data on a variety of other $M(OEP)X_m L_n$ complexes which have been collected[174] seems premature at present.

Generally, the properties of a metalloporphyrin are influenced by the geometry, charge, electronegativity, and the degree of σ- and π-bonding of not only the metal-porphyrin system but also the metal-axial ligand system. Thus, any conclusion drawn from investigations of only a few metalloporphyrins will be of limited use[11].

5.3.7. The influence of structurally modified porphyrin ligands on the stability of metalloporphyrins

Structural modifications of the porphyrin ligand have a definite influence on the stability of the corresponding metalloporphyrins. The following modifications will be discussed: 1) ring opening, 2) ring contraction, 3) hydrogenation, 4) central and peripheral substitution.

1) Ring opening

Pyrromethenes, a,c-biladienes, and b-bilenes are open bidentate or tetradentate chelating ligands which are closely related to porphyrins and form a variety of zinc, copper, cobalt, and nickel complexes[175-177] (see Chapter 18). The 'macrocyclic effect', i.e. the extra stabilization of a closed chelate versus an open, similar derivative[152], is obvious. Copper pyrromethene complexes transfer the metal to H_2(Etio-II)[178], and the a,c-biladiene nickel complexes are instantaneously demetalated by $CF_3 COOH$[176], whereas the electronically closely related porphodimethene complex $Ni(OEPMe_2)$ (A)(Table 1) withstands 100% $H_2 SO_4$ for 2 hr[171], (see Chapters 10,15 for porphodimethene structures).

2) Ring contraction

The corrins (see Chapter 18) may be regarded as ring-contracted porphyrins. Cobalt corrins belong to stability class I[10], in contrast to the cobalt porphyrins of class II. Thus, ring contraction may stabilize the M—N bonds when the central ion is small enough to allow such a contraction.

3) Hydrogenation

The metal complexes of chlorins are less stable than the corresponding porphyrin derivatives. The dissociation constants of various magnesium porphyrins and chlorins according to eq. (30) have been measured in phenol at $100°$ C by Corwin and Wei[154]. They have been found to be definitely greater

$$Mg(P) + 2\ PhOH \leftrightharpoons Mg(OPh)_2 + H_2(P) \tag{30}$$

for the chlorin complexes in the three series Mg(Etiochlorin-II)/Mg(Etio-II), Mg(pyropheophorbide-a-ME)/Mg(vinylphylloerythrin-ME), and Mg(chlorin-e_6-TME)/Mg(vinylchloroporphyrin-e_6-TME). Closure of the isocyclic ring raises the dissociation constant. The reduced tendency of forming chlorin complexes has been utilized for the otherwise difficult separation of chlorins and porphyrins by selective complexing of the latter; thus, Zn(TPP) is formed much faster than Zn(TPC) from Zn(OAc)$_2$ in trichloroethylene[179], and treatment of a mixture of H$_2$(OEP) and H$_2$(OEC) with VOSO$_4$ · 5H$_2$O in HOAc furnishes a mixture of VO(OEP) and H$_2$(OEC)[180], the resulting mixtures being more easily separated by chromatography than the original mixtures.

Principally, the same stability order may be expected for porphyrin and chlorin complexes. Berezin[181] has studied the relative dissociation rates (which frequently are taken as parallel to the dissociation constants[156]) of a variety of metallopheophytins. They follow an order which justifies the stability order Fe(III) > Pd > Cu > Ni > Co > Zn > Mg > Cd > Hg for these chlorophyll complexes in ethanol/acetic acid. (The mildness of this demetalating agent again exemplifies the lower stability of metallochlorins as compared with metalloporphyrins.)

Although the chlorin ligands are difficult to prepare[11,102,104,106,179], metal derivatives M(OEC)X$_n$ [MX$_n$ = Mg[121], Zn[182], Cu[102], Ni[121], Pd[86,121], Pt[86], TlOH(H$_2$O)[182], AlOH[82], FeCl[106] and SnCl$_2$[94]] as well as M(TPC)X$_n$ [MX$_n$ = Na$_2$, Mg, Zn, Cd, Cu, Ag, Co, Sn(OAc)$_2$][179] are known. The Sn- and Al-complexes are stable towards concentrated HCl[82,183]; in contrast to Pd(OEP), Pd(OEC) has been demetalated in reasonable yield after treatment with lithium in ethylenediamine[86], and Sn(OEC)(OAc)$_2$ has been demetalated with NaBH$_4$/HCl[184]. Cu(II)-, Zn(II)-, Pd(II)-, Pt(II)-, Al(III)-, and Sn(IV)-complexes of tetrahydroporphyrins and a Cu(II)-complex of a hexahydroporphyrin have so far been characterized only by optical absorption spectra[82,86,94,102]. As far as can be presently estimated, a progressive hydrogenation of the methine bridges will normally labilize the metal complexes of the corresponding hydroporphyrins. Their study is complicated by their extreme sensitivity towards oxygen[39]. Therefore, the discussion will be limited to two air-stable alkyl derivatives, namely α,γ-dimethylporphodimethene [α,γ-dimethyl-α,γ-dihydro-octaethylporphyrin, H$_2$(OEPMe$_2$)[39,82,87,171], see Chapter 14] and acetonepyrrole ($\alpha,\beta,\gamma,\delta$-octamethyl-porphyrinogen)[185]. The latter does not form metal derivatives at all, perhaps because of an unfavorable orientation of the four pyrrole rings which may be tilted about the methine bridges in this hexahydroporphyrin derivative.

The porphodimethene complexes are less stable towards acids than their porphyrin analogs, with the exception of Ni(OEPMe$_2$)[171], which is not demetalated by 100% H$_2$SO$_4$, in contrast to Ni(OEP). This increased stability is caused by a rooflike-folding of the porphodimethene skeleton which offers shorter and thus more favorable Ni—N bond lengths to the rather small Ni(II) ion than the porphyrin cavity[19,39,87,171] (see Chapter 8).

4) *Central and peripheral substitution*

Various porphyrin derivatives are obtained when the hydrogen atoms at the central N-atoms, at the methene C- or the peripheral C-atoms are replaced by other substituents. Clearly, the formation of metal complexes is impeded by *N*-alkylation: in 0.005 M NH_4OH/DMF *N*-methyltetraphenylporphyrin H(N-MeTPP) is only in an equilibrium with its copper complex according to eq. (30), the stability constant being 3.75 ± 0.25.

$$CuCl_2 + H(N\text{-}MeTPP) \rightleftharpoons CuCl(N\text{-}MeTPP) + HCl \tag{30}$$

The metalation and demetalation rates are also enhanced; dilution of the DMF-solutions of CuCl(N-MeTPP) instantaneously decomposes the complex[186]. The demethylation of these porphyrins is greatly facilitated by insertion of Zn, Cu, Ni, and Pd[23,186], steric strain being relieved by expulsion of the methyl group.

It is tempting to discuss the effects of *meso*- and peripheral substituents in conjunction with the nitrogen basicities of the respective porphyrins (see Chapter 6), but the available data do not allow an unambiguous interpretation[86]. Increasing substitution of the porphyrin periphery with electron-withdrawing substituents definitely stabilizes the Mg(II)-porphyrins in phenol[154]; in H_2SO_4/HOAc, the stability increases in the order M(Etio-I) < M(TPP) < M(2,4-AcDeut-IX-DME) for M = Cu, Ni[86]. These substituents reduce the basicity of the central nitrogen atoms[1,151] and apparently impede attack of protons upon the porphyrin. The converse is found in transmetalations[86,156,166] with Li^+ and Cu^{2+} as the intruding ions. Here, the true M—N_4-bond strength of the leaving metal may be decisive. Although the chlorins are presumably less basic than porphyrins[121], they are demetalated more readily than the corresponding porphyrins both in acid media and via transmetalation with lithium[86,154]. Therefore, an intrinsic understanding of the substituent effects on the metalloporphyrin stability remains to be achieved.

5.4. *Axial coordination chemistry*

5.4.1 *General remarks*

Both the important classes of natural metalloporphyrins, the iron porphyrins and the magnesium chlorins, have their function associated with axial ligation phenomena: the former bind the dioxygen molecule to a Fe(II) ion initiating its transport or reduction, the latter aggregate via the basic 9-keto-oxygen of one chlorophyll molecule occupying a vacant axial coordination site above the Mg ion of a second chlorophyll molecule (see Chapters 10, 16, 17). Therefore, a detailed treatment of axial coordination chemistry of metalloporphyrins seems necessary. Over and above, the synthetic chemists dealing with metalloporphyrins of all kinds necessarily run into problems associated with separation, purification, and analysis of metalloporphyrins with specific axial ligands, because the manifold of metals (see Fig. 14) can in principle be multiplied with the manifold of axial donor ligands produced by a combination of nonmetallic elements.

The knowledge gained in this field has grown enormously within the last 10 years, and the material is scattered over a vast amount of publications. In order to facilitate a systematic treatise of the field which seems premature at the moment, only a compilation of monometallic metalloporphyrins with specific axial ligands is presented. Tables 11—18 (with approx. 60 additional references) encompass a representative selection of monometallic species nearly all of which have been isolated in a crystalline state. The complexes are distributed between the respective tables according to their stereochemical types and their subclass generated by axial stoichiometry (see Section 5.2, Table 2, and Figs. 5—7, 10). For the details of the corresponding preparations and the spectral details, the reader is referred to the original papers. The space available in this context allows only a brief delineation of general trends (Section 5.4.2) For visible absorption spectra, see p. 884.

The metalloporphyrins containing iron or cobalt group metals deserve a

TABLE 11

Porphinatometal complexes M(P) (A)

Complex	Central metals
M(P)	Mg, Cu (2)
M(Proto-IX)	Fe, Ni, Cu, Zn (2)
M(Proto-IX-DME)	Mg (95), Cu, Zn (2), Co, Ni, Cu, Zn, Ag, Cd, Pb (1,153)
M(Meso-IX-DME)	Cr (126), Fe (2,99,112) Co (58), Cu (2), Ag (2,58), Zn (1,58)
M(Deut-IX-DME)	Co, Pd, Ag, Zn (223), Ni (98)
M(Etio-I)	Mg, Co, Ni, Cu, Zn, Ag (2,101)
M(Etio-II)	Co, Ni, Cu, Zn (156)
M(Etio-III)	Mg, Cu (2), Fe (2,112)
M(Hemato-IX)	Co (52), Ni (54), Co, Ni, Cu, Zn, Ag, Pb (55), Pt (42)
M(TM-Hemato-IX)	Mg, Fe, Co, Ni, Cu, Zn, Ag, Cd, Pb, (58), Zn, Cu, Mg (2)
M(OEP)	Mg (91), Co, Ni, Cu, Zn (20,105,167), Ag, Hg, Cd (122), Pd (87, 167), Pt (87), Pb (93,96), Os (169)
M(OEPMe$_2$)	Zn, Cu, Ni, Pd, Pt (19,39,87,171)
M(OEC)	Mg (121), Zn (182), Cu (102), Ni (121), Pd (86,121), Pt (86)
M(Pheophytin-a)	Mg, Co, Ni, Cu, Zn, Cd, Pd, Hg (181)
M(TPP)	Fe (113,187), Co, Ni, Cu, Zn, Ag, Cd, Hg (14,92), Rh (108), Pd, Pt (59,139,172)
M(TPC)	Mg, Zn, Cd, Cu, Ag, Co (179)

Ionic species: Au(TPP)$^+$ (14,118), Ag(OEP)$^+$ (145)
(References in brackets)

TABLE 12

Porphinatometal complexes M(P)L, (B)

Complexes	Metals	Axial ligands L
M(Etio-III)L	Mg	MeOH (101)
M(OEP)L	Co	1-MeIm (188)
	Cd	Py (122)
M(TPP)L	Mg	H_2O (189)
	Fe	NO (190), H (2-MeIm) (187) [a]
	Co	NO (191,254), 1,2-MeIm (192), (1-MeIm) (15)
	Pb, Hg	Py (14,193) [b]
M(TPyP)L	Zn	Py (143)

[a] EtOH-solvate.
[b] Existence proved in solution only.

special treatise (Section 5.4.3) because the knowledge of their axial coordination chemistry is necessary to understand the catalytic function of iron and cobalt porphyrins and of corrins.

The main problem is the identification of the axial ligands which represent only a small part of a large molecule. This topic is briefly touched in Section 5.4.4. Many special topics will be dealt with in the other chapters of this book.

TABLE 13

Porphinatometal complexes $M(P)L_2$ and $[M(P)L_2]^+$ (C)

Complexes	Axial ligands	Metals
$M(Proto-IX-DME)L_2$	Py	Fe (99)
$M(Meso-IX)L_2$	ImH	Fe (194)
$M(Meso-IX-DME)L_2$	Py	Fe (2,99,112)
$M(Deut-IX-DME)L_2$	Py	Fe (99)
$M(OEP)L_2$	Py	Mg (91), Mn, Fe (105), Co (122, 195), Ru (37), Os (26)
	3-Pic	Co (195)
	$P(OMe)_3$	Os (74)
$M(TPP)L_2$	Py	Fe (196)
	Pip	Fe (21,197), Co (198)
	PPh_3	Ru (129)
$[M(Etio-I)L_2]^+$	$NHMe_2$	Rh (125)
$[M(TPP)L_2]^+$	ImH	Fe (197), Co (199)
	Pip	Co (200)
$[M(P)L_2]^{4+}$ [a]	Py	Ni (144)

[a] (P) = $\alpha, \beta, \gamma, \delta$-Tetra (N-methyl-4-pyridyl)porphinium ion.

TABLE 14

Porphinatometal complexes M(P)X (B) and M(P)(XL) (D)

Complexes	Axial ligands X	Metals
M(P)X	Cl	Fe (2)
M(Proto-IX)X	Cl, Br, I, SCN, OAc	Fe (2,4,7)
M(Proto-IX-DME)X	Cl, OMe	Fe (2,99)
M(Meso-IX-DME)X	F, Cl, Br, I, OMe	Fe (2,99,201)
	Cl	Mn, Al, Ga, In, Sb (65)
M(Deut-IX-DME)X	F, Cl, Br, I, OPh, N_3	Fe (2, 202)
	OCN, NCS, NCSe, SPh	Fe (203)
M(Etio-III)X	Cl, OAc, OMe	Fe (2)
M(OEP)X	F	Al, Ga, In, Mn (204),
		Fe (204, 205)
	Cl	Sb (115), Al, In (122)
	Cl, Br, I, N_3, NCS	Fe (2, 84, 105, 106, 205)
	OMe, OPh	Al (25), Ga (122), Fe (84)
	OPh	Al (20,25), Ga, In, Mn (25)
	OAc	Ga (122), Fe (84, 205)
	OAc, $OSiMe_3$, $OSiPh_3$	Al (81, 82)
	OH	Al (20,25,81,82), Ga (122)
		Cr (123)
	Me	Rh (134)
	OH^a, Cl, I, CN	Tl (110)
M(OEPMe₂)X	Cl, OMe, OAc, OPh	Fe (204), Al, Ga (82,87,122)
M(TPP)X	Cl	Cr (85), Mn (78, 206)
		In (107)
	Cl, Br, I, NCS	Fe (207)
	COOEt, Ac	Rh (133,208)
	SnR_3	Fe (148)
M(TTP)X	Cl	Fe, Co (114)
M(TAP)X	Cl, Br, I, OAc, CF_3CO_2, NCS, N_3	Fe (209)
M(OEP)(XL)	OAc	Sc (20), In (122)
	(acac)	Sc (20)

a Also contains a molecule of water, presumably as a solvate.

5.4.2. Preparation of metalloporphyrins with specific axial ligands

Axial ligands of metalloporphyrins are normally subject to various kinds of exchange equilibria. Neutral ligands L are involved in addition-elimination-sequences (eqs. 31, 32) the possible products of which are listed in Tables 11—13, 17, and 18.

$$M(P) + 2 L \rightleftharpoons M(P)L + L \rightleftharpoons (P)ML_2 \tag{31}$$

$$M(P)X + L \rightleftharpoons M(P)XL \tag{32}$$

TABLE 15

Porphinatometal complexes [M(P)]$_2$O and MO(P) (μ-oxo- and terminal oxo ligands)

Complexes	Porphinato ligands	Metals
[M(P)]$_2$O	(Proto-IX)	Fe (210)
	(Proto-IX-DME)	Fe (203,211)
(BB)	(Deut-IX-DME)	Fe (202)
	(OEP)	Fe (84), Sc (64, 84), Al (81)
	(TPP)	Fe (212, 213), Sc (64), Mn (78)
	(TAP) and 7 others	Fe (209)
	(OEPMe$_2$)	Fe (122)
MO(P)	(Meso-IX-DME)	Ti (126), V (2,49)
	(Etio-I)	V (100)
(B)	(OEP)	Ti (20,127), V, Mo (20)
	(OEPMe$_2$)	Ti, V (39,87)
	(TPP)	Mo (150)

Anionic ligands show substitution equilibria such as (eq. 33) which frequently are disturbed or suppressed by hydrolysis (eq. 34) and consecutive con-

$$M(P)X_n + nX' \leftrightharpoons M(P)X'_n + nX \qquad (33)$$

densation (eq. 35) yielding binuclear μ-oxo species or elimination of water (eq. 36) furnishing a metalloporphyrin where the metal carries a multiply bound terminal oxide ligand.

$$M(P)X_n + nH_2O \leftrightharpoons M(P)(OH)_n + nHX \qquad (34)$$

$$2\,M(P)OH \leftrightharpoons (P)M\text{-}O\text{-}M(P) + H_2O \qquad (35)$$

$$M(P)(OH)_2 \leftrightharpoons (P)M{=}O + H_2O \qquad (36)$$

The hydroxo complexes emerging from reaction (eq. 34) may be isolated most easily for M = Al(III), furthermore for M = Ga(III), Cr(III), Si(IV), Ge(IV), and Sn(IV) (Tables 14, 16). Their isolation for M = Sc(III), Fe(III), and In(III) has never been achieved; obviously these metals protrude from the porphyrin plane and condensation becomes very easy[84]; Mn(III) should therefore not give a hydroxide either. The ease of condensation observed with Nb(V), Mo(V), W(V), and Re(V) is ascribed to the *trans*-influence of the M = O system found in these complexes[25]. As a whole μ-oxo bridges are typically formed with M = Fe(III), Sc(III), Si(IV), Ge(IV), Sn(IV), Nb(V), Mo(V), W(V), Re(V), and sometimes with M = Al(III), Mn(III) (Tables 15, 16; types (BB) and (CC), see p. 168). Terminal oxide ligands are observed

TABLE 16

Porphinatometal complexes $M(P)X_2$, $MO(P)X$, $[MO(P)]_2O$, $[M(P)X_2]^-$, and $M(P)(XL)_2$

Complexes, type	Axial ligands	Metals
M(Meso-IX-DME)X$_2$ (C)	OH, OMe, OPh, OAc	Ge (216)
	Cl	Ge (65,216), Sn (65)
M(Etio-I)X$_2$ (C)	OH	Si (116), Sn (96)
	OSiMe$_3$, OEt, OSi(OR)$_3$	Si (116)
	O$_2$CPh, F, Cl, Br, I	Sn (96)
	O$_2$CEt	Sn (101)
M(OEP)X$_2$ (C)	OH	Ge (93), Si, Sn (122)
	OMe	Si, Ge, Sn (25), Os (26)
	OPh	Si, Ge, Sn (25)
	OAc	Si (122), Ge (93), Sn (93,215)
	F	Si, Ge, Sn (204), Ti, Zr, Hf (122)
	Cl	Si, Ge, Sn, Pb (96)
	O (doubly bound)	Os (26)
M(TPP)X$_2$ (C)	Cl	Sn (14,24)
MO(OEP)X (C)	F	Nb, Mo, W, Re (27,123)
	OMe, OAc, Cl	Mo, W (25, 123)
	OPh	Mo, W, Re (25)
	Br, I, I$_3$, OCD$_3$	W (25, 123)
	I, I$_3$	Nb, Ta (64,123)
[MO(OEP)]$_2$O (CC)	μ-Oxo bridges, terminal oxides doubly bound	Mo, W, Re (27)
[MO(TPP)]$_2$O (CC)		Mo (150)
		(Alleged MoO (TPP)OH)
[M(Etio-I)X$_2$]$^-$ (C)	CN	Co (217)
[M(OEP)X$_2$]$^-$	CN	Fe (218)
[M(TPP)X$_2$]$^-$	CN	Ru (129)
M(OEP)(XL)$_2$ (G)	OAc, (acac), (dbm), (pim)	Zr, Hf (20,27,39,40)
[M(OEP)(XL)$_2$]$_2$ (GG)	(phthal), (pim)	Zr, Hf (40)

with M = Ti(IV), V(IV), Nb(V), Mo(V), W(V), Re(V), and Os(VI); the multiply bound M = O moiety is very reluctant to replacement (Tables 15, 16).

As a consequence of the above mentioned properties, complexes with specific axial ligands L and (or) X are generally obtained by adding an excess of the ligands L and (or) X during the preparation, purification (in most cases by chromatography) or crystallization of the products[39]. Water should be excluded as it can compete with L or induce hydrolysis and its secondary reactions. The presence of water cannot be avoided in the course of chromatography on alumina or silica. Normally, chromatography is necessary to remove unreacted porphyrin free acid after metal insertion. Thus, specific axial ligands are supplied mostly by crystallization in the presence of an

TABLE 17

Porphinatometal complexes M(P)XL (C) with mixed neutral and anionic ligands

Porphinato ligand (P)	Axial ligands X	L	Metals M
(Proto-IX-DME)	F, Cl, Br, I, N_3, OCN, NCS	H_2O	Mn (97)
(Meso-IX-DEE)	Cl	H_2O [a]	Rh (68)
(Hemato-IX-DME)	Cl	H_2O	Mn (206, 219)
(Hemato-IX-DEE)	Cl	CO	Ir (68)
(Etio-I)	Me, Et, Ph, Ac	H_2O	Fe, Co (220)
	OAc	Py	Mn (221)
	Br	Py, NH_3	Co (217)
	F, Cl, Br, I, OAc, N_3, NCO, NCS	H_2O	Mn (147, 206, 222)
	Cl, OAc	H_2O [a]	Co (20,105) Mn, Cr (20)
(OEP)	Br	Py	Co (20,105) Mn, Cr (20)
	OAc	H_2O	Mn (105,122)
	Cl	H_2O [a]	Rh (134)
	OPh	PhOH	Cr (25)
	OMe	NO	Ru, Os (169)
(TPP)	Cl	CO	Rh (66,133)
	H	H_2O [a]	Rh (208)
	NO_2	Lut	Co (224)
	N_3	MeOH [a]	Mn (119)

[a] These M(P)XL are formulated with a second L presumably acting as a solvate.

excess of these ligands. Axial ligands and the solvents used should have the highest possible purity.

The most suitable solvents, methylene chloride and chloroform, must be freed from alcohols which are used as stabilizers and hydrogen chloride originating from the decomposition of these solvents in order to avoid a spurious formation of alkoxo and chloro complexes. If methanol is added to the solvents during chromatography and crystallization of metalloporphyrins containing metals M = Al(III), Ga(III), Fe(III), Si(IV), Ge(IV), Sn(IV), Os(IV), Mo(V), and W(V), the corresponding methoxides are formed (Tables 14, 16, 17). They are also obtained by methanolysis of μ-oxo complexes (M = Fe, Mo, W; Tables 15, 16).

Substituted pyridines are the most widely used neutral axial ligands. Despite early reports claiming that ortho-substituted pyridines bind to metalloporphyrins, it has recently been shown that 2,4,6-collidine and similar ligands do not bind to iron or cobalt porphyrins[248,258]; the axial ligation as deduced from optical and e.s.r. spectra was actually caused by minute impurities consisting of sterically unhindered pyridines. This observation illustrates the necessity of using definitely pure axial ligands.

TABLE 18

Porphinatometal complexes M(P)LL' (C) with mixed neutral ligands L and L'

Porphinato ligand (P)	Axial ligands L	L'	Metals M
(Proto-IX-DME)	N_2	Py	Fe (203,225)
	CO	Py	Fe (99,203,226)
	CO	1-MeIm, MeNC, Me_2NNH_2	Fe (203,226)
	MeNC	Py	Fe (203,226)
(Meso-IX-DME)	NO	NO	Ru (227)
	CO	Py	Fe (99,203)
	CO	ImH, Py	Ru (228)
(Deut-IX-DME)	CO	Py	Fe (99,203,226)
(2,4-AcDeut-IX)	CO	Py	Fe (99,203,226)
(Etio-I)	CO	Py	Ru (36)
(OEP)	N_2	THF	Os (169, 251)
	CO	THF, NEt_3, Pip, ImH	Ru (36)
	CO	THF, MeOH, EtOH, NEt_3, Py, $AsPh_3$, PPh_3	Os (74)
(OEPMe$_2$)	CO	Py	Os (169)
(TPP)	NO	1-MeIm	Fe (229)
	NO	N-MePip	Mn (229)
	CO	Py, 1-MeIm, Pip	Fe (196)
	CO	Py	Ru (129,227,230)
	CO	EtOH, Py, t-BuPy	Ru (130,131)
(picket-fence-P) [a]	O_2, CO	1-MeIm, THF	Fe (231,232)
(crown-P) [a]	O_2, CO	1-MeIm	Fe (233)

[a] For a description of these porphyrins, see Fig. 17.

Generally, neutral ligands L are administered as such in gaseous or liquid phases. Anionic ligands X^- are offered as salts, e.g. NaX- if necessary, by shaking CH_2Cl_2-solutions of the metalloporphyrin with H_2O-solutions of NaX. More efficient is admission of the corresponding acid HX, if the metalloporphyrin is not demetalated by it. HX also splits μ-oxo complexes, while these are formed frequently by treating M(P)X with aqueous NaOH or KOH. A very good system for preparing acetates and other derivatives of carboxylic acids is heating the metalloporphyrin with the acid in pyridine and slowly adding water until crystals appear.

Most difficult to prepare and to identify are those metalloporphyrins which contain either mixed anionic (Table 16), mixed anionic and neutral (Table 17), or mixed neutral axial ligands (Table 18). The neutral ligands L of the species M(P)XL (Table 17) are easily lost and in some cases not unambiguously identified. In the case of the species M(P)LL' (Table 18) it is obvious that these mixed complexes are only isolable because L and L' strongly differ in their σ-donor-π-acceptor balance; the strong π-acceptors L exert a *trans* effect preventing the binding of a second ligand of the same type. As these compounds are very important with respect to the natural hemes, they are further discussed in Section 5.4.3. The dinitrosyl mentioned in Table 18 is regarded as a mixed species because there is evidence that such species contain NO^+ and NO^- which have different π-acceptor capacity[266]. A whole host of mixed complexes Co(P)LL' has been studied at low temperatures by ESR spectroscopy[252–258]. Their isolation may be difficult due to lability caused by population of the d_{z^2}-orbital in these d^7-complexes.

5.4.3. Axial coordination chemistry of metalloporphyrins containing group VIII metals

While the most stable coordination type of Ni(II)-, Pd(II)-, and Pt(II)-porphyrins is the 'bare' species M(P) (A), the square planar d^8 configuration with the filled d_{z^2}-orbital allowing addition of axial ligands only to a limited extent (see Table 13), the bare d^6 species represented by Fe(II)-, Ru(II)-, and Os(II)-porphyrins are obviously the most reactive ones, only the bare Fe(II)-porphyrins being known at present (Table 11); the Co(II)- and Rh(II)-porphyrins are intermediate in reactivity. The essential chemistry which can be formally derived by ligation and redox reactions of these bare species M(P) is depicted in Schemes 2—6 presenting the behavior of Fe-,

Scheme 2. The chemistry of iron porphyrins [L = Py, ImH, 1-MeIm, Pip, etc.; Ref. 1,2,7,9,12,84,99,112,113,187,190, 194,196,197,201—203,205, 207,209—214,218, 225, 226,229,231--248,254,259—266].

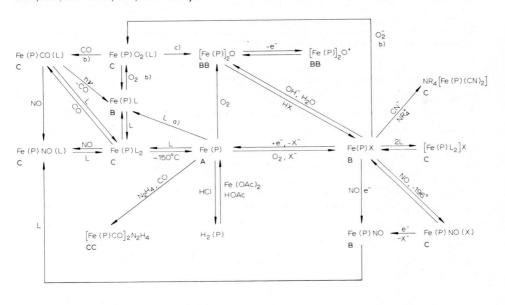

a) L = H (2 - MeIm)   b) At -50°C   c) Above - 50°C

<u>Scheme 3</u>. The chemistry of ruthenium porphyrins [Ref. 36,66, 129—132,169,227,228, 230; { }: not sufficiently characterized in the solid state].

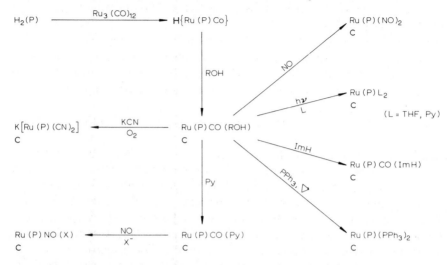

Ru-, Os-, Co-, and Rh-porphyrins; Ir-, Ni-, Pd-, and Pt-porphyrins only show a few interesting reactions. The references given within the schemes can be traced using Tables 11—18 if the compound under consideration has been isolated as a solid. Pioneering work pertinent to axial ligation of iron porphyrins has been done by Caughey and his collaborators in the last 10 years (see

<u>Scheme 4</u>. The chemistry of osmium porphyrins [(P) = (OEP); Ref. 26,74,169,251; the complexes are octahedral (C) if not otherwise stated].

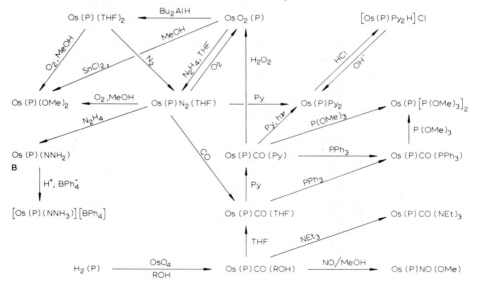

Scheme 5. The chemistry of cobalt porphyrins[Ref. 15,20,105,114,191,192,195, 198—200,217,220,224,253—258,266]. a) Compounds not yet sufficiently characterized in the solid state. b) A = NO, O_2 for L = Py, 4-Pic, 3,4-Lut, 1-MeIm, $P(OMe)_3$. c) L = Py, t-BuPy, Pip, 3-Pic, PR_3 (R = Me, Et, Bu), PMe_2Ph. d) A = O_2. e) A = CO, CNMe, NO, SO_2, O_2, PF_3, $P(OMe)_3$.

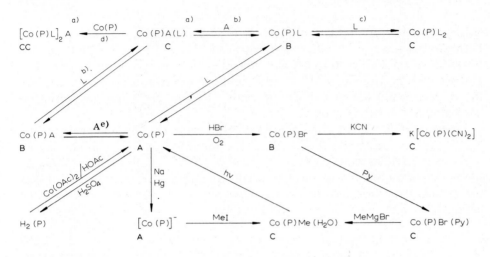

Scheme 6. The chemistry of rhodium porphyrins [Ref. 23,31,66,68,108,125,133,134, 208].

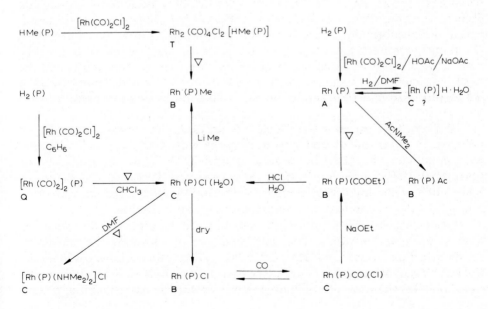

References, p. 224

Tables 11–18, and also Section 5.4.4). In the absence of π-acceptor ligands, Fe(P) is irreversibly oxidized to $[Fe(P)]_2O$ in the solid state and in solution (eq. 37),

$$4 \, Fe(P) + O_2 \rightarrow 2 \, [Fe(P)]_2 O \qquad\qquad (37)$$

although there are some reports of reversible binding of O_2 to solid Fe(Meso-IX)(ImH)$_2$[194], solid Fe(Proto-IX)(ImH)$_2$[234], solid Fe(Proto-IX-DME)[263], and Fe(Proto-IX-DEE)L generated in a polystyrene matrix [L = 1-(2′-phenylethyl)imidazole][235]; the products, however, seem to contain the oxygen molecule in a bonding mode differing from that occurring in hemoglobin[263] (see Section 5.4.4.).

Below $-50°C$, many hemochromes [i.e. iron porphyrins of the type Fe(P)Py$_2$] are oxygenated according the simplified eq. (38) which may be split up in various steps given in Scheme 2[237–244]. The binding of O_2 to the hemes is monitored by electronic spectra although it must be stated that the typical near infrared bands occurring at about 900 nm in oxygenated hemoglobin have not yet been traced in these quite recent experiments.

$$Fe(P)L_2 + O_2 \rightarrow Fe(P)O_2(L) + L \qquad\qquad (38)$$

The kinetic stability of the products Fe(P)O$_2$(L) increases in the order Py $<$ Pip $<$ (1-MeIm) for L[244]. It is interesting that for L = HIm a rapid irreversible oxidation occurs even at $-78°C$[242]. This observation lends support to the old assumption that the globin pocket should preserve the oxygenated heme from protic attack[267]. The oxygenations may be achieved in solvents such as toluene, ether, methylene chloride, dimethylformamide, 1-butanol, and dimethylsulfoxide; increasing polarity seems to favor adduct formation. (It should be noted that CH_2Cl_2 does react with Fe(P) at $25°C$, but not at $-50°C$.) To effect O_2 binding, it is neither necessary to covalently link an axial base molecule to the porphyrin nucleus nor to provide an empty axial coordination site. The O_2 molecule can indeed replace an axial base according eq. (38).

The 'bare' Fe(P)-species (see Table 11) necessary for these studies can be prepared by insertion of iron into porphyrins with Fe(OAc)$_2$ under N_2 (see Section 5.3.3.1, p. 179), by reduction of Fe(P)X with Cr(acac)$_2$[187] or CaH$_2$/Pd/H$_2$O[241], or by heating hemochromes Fe(P)Py$_2$ in vacuo to $140-160°C$ whereby Py is removed[263]. During reductions in situ using Na$_2$S$_2$O$_4$ or N$_2$H$_4$ an excess of reductant may cause trouble.

Nevertheless, the globin pocket is necessary to stabilize the dioxygen adduct at room temperature. This can be shown by imitating the pocket by firmly attaching suitable substituents to the porphyrin periphery of a heme. Collman has synthesized a heme carrying four pivaloylamide residues whose steric bulk covers one side of the porphyrin plane; he has called it a 'picket-

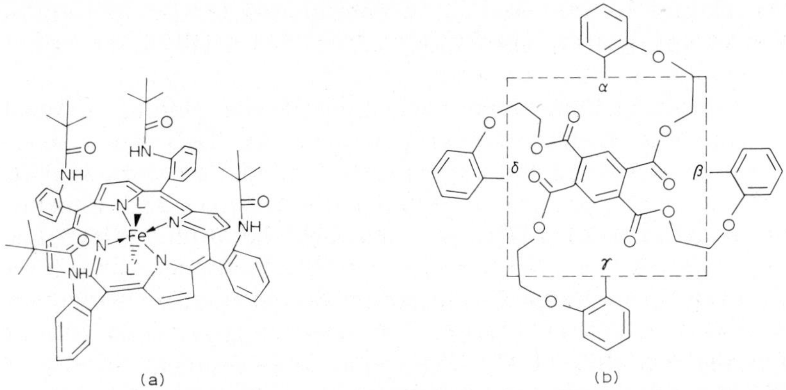

(a)   (b)

Fig. 17. Synthetic porphyrins designed to mimic myoglobin: a) Configuration of the 'picket fence' heme [Collman et al.[231,232,264,265]; Fe(picket-fence-P)LL'; see Table 18]. b) Schematic drawing of the 'crown-porphyrin' [Baldwin et al.[233]; H$_2$(crown-P); see Table 18; the dotted line represents a porphin unsubstituted at positions 1—8 and carrying a crown-ether-ester array coupled at positions α-δ above one side of the porphin plane.

fence' heme (Fig. 17a)[231,232,264,265]. Baldwin has prepared a porphyrin where four side-chains are locked together thus providing a rather rigid fused-ring cavity (Fig. 17b; the author suggests the name 'crown' heme for this molecule as it partly resembles the crown ethers)[233]. Both hemes reversibly bind dioxygen at room temperature in dilute solutions as long as an excess of the axial base (preferredly 1-MeIm) is present to avoid attack of oxygen from the unprotected side of the heme thus effecting autoxidation according eq. (37), and the oxygen adducts can be isolated and characterized in the solid state (Section 5.4.4). In order to really mimic the active site of hemoglobin, the tetraphenylporphyrin moiety present in these two hemes has to be replaced by a protoporphyrin or at least an octa-alkylporphyrin moiety because the four *meso*-phenyl groups have a definitively different electronic influence on the metal than eight peripheral alkyl groups; but the achieved isolation of solid O$_2$-hemes has enormously stimulated research in this field.

The chemistry of Fe- and Ru-porphyrins is governed by the +2- and +3-oxidation states, higher oxidation states at present not being easily accessible presumably because strong oxidizing agents preferentially attack the porphyrin π-system[169]. However with Os-porphyrins, Os(II), Os(IV), and Os(VI) are easily reached and interconverted. Thus, Os-porphyrins are very versatile for fundamental axial ligation studies. In the absence of strong π-acceptor ligands, Fe(II) is autoxidized to Fe(III), while Os(II) is autoxidized to Os(VI) (Schemes 2—4). The Co- and Rh-porphyrins are especially noteworthy as they provide several examples of metal to carbon bonds which may be relevant to vitamin B_{12}. A coordinated CO group is transformed into an axially ligated carboethoxy group, and axially bound alkyl groups are generated from various starting materials (Schemes 5, 6).

Besides the interest in hemes and heme models, the binding of small π-acceptor molecules to metalloporphyrins deserves by itself attention of inorganic chemists. The species and reactions encountered on working with Fe-, Ru-, Os-, Co-, and Rh-porphyrins are depicted in Schemes 2—6. Generally, the stability of adducts M(P)A(L) for a given metalloporphyrin increases in the order $N_2 < O_2 < CO < NO$ for $A^{9,251,254,266}$. Their stability for a given $A = N_2$, CO or NO increases with increasing atomic numbers for homologous metals. Thus, Os(OEP)N_2(THF)[251] is slowly formed at 25°C from solutions containing Os(OEP)(THF)$_2$ (Scheme 4) while a similar reaction is not known with Fe(II)-porphyrins at least at room temperature. The reported Fe(Proto-IX-DME)N_2(Py) (see Table 18) seems dubious in view of the fact that Py and O_2 replace N_2 in the Os-system (Scheme 4). As Py replaces O_2 in Fe(picket-fence-P)O_2(1-MeIm)[231], there is no reason why it should leave N_2 untouched in an iron porphyrin. The thermal stability of porphinatometal carbonyls M(P)CO(L) also increases with increasing atomic number of the metal. While Fe(P)CO(Py) is labile at 25°C, Ru(P)CO(Py) is stable and Os(OEP)CO(Py) can be vaporized at 200°C to give a molecular ion in a mass spectrometer albeit with low relative intensity, the ions Os(OEP)CO$^+$ and Os(OEP)$^+$ being the predominant species. The increasing stability is due to increased M—CO backbonding, the CO-stretching frequency progressively decreasing in the order Fe (1950) > Ru (1930) > Os (1900 cm^{-1}) for M^{74}. Therefore, Ru and Os cannot replace Fe in hemoglobin; even traces of CO would accumulate and irreverisbly poison the active site at the metal.

The same kind of *cis*- and *trans*-effects which are found with Fe-porphyrins may also be studied with Ru- and Os-porphyrins[74,169]. N.m.r. experiments show that in complexes Os(OEP)LL' the *meso*-protons are progressively shielded, e.g. in the series CO/MeOH < CO/Py < P(OMe)$_3$/P(OMe)$_3$ < Py/Py for L/L', as the σ-donor strength of L and L' increases (*cis*-effect). In complexes Os(OEP)CO(L), the infrared stretching frequency of the CO ligand rises with increasing π-acceptor-capacity of the *trans*-ligand L, indicating diminishing back-bonding to the CO group in the order THF (1897) > MeOH (1898) > Py (1902) > AsPh$_3$ (1914) > PPh$_3$ (1926 cm^{-1}) for L (*trans*-effect). For L = P(OMe)$_3$, the back-bonding to the CO ligand is obviously so poor that the latter is replaced by another P(OMe)$_3$ molecule[74]. With Co(II)-porphyrins, no CO binding occurs in the presence of nitrogenous bases, and therefore not in coboglobin[254]. Generally, there is a delicate balance between the π-acceptor- and σ-donor properties of two ligands in the *trans*-positions; two ligands will repel each other from the *trans*-positions when they are held mainly by back-donation from the metal. Therefore, bis(N_2)- or bis(CO)-complexes will be labile and bis(NO)-complexes disporportionate to derivatives of NO^+ and NO^-, as has been found [169,266].

Another interesting aspect is back-bonding from the metal to the porphyrin ligand, which has been invoked as the cause for the 'hypso'-type absorption spectrum (Section 5.3.4, p. 190). Because the porphinato ligand competes in π-acceptance with the axial ligands, it is expected that the hypsochromic shift as compared with the 'normal' absorption (p. 189) diminishes with increasing π-acceptor capacity of an axial ligand. Indeed, the wavelength of the α-band shows a progressive bathochromic shift in the series $Os(OEP)Py_2$ (512) < $Os(OEP)N_2(THF)$ (525) < $Os(OEP)CO(Py)$ (540) < $Os(OEP)NO(OMe)$ (568 nm), providing another example of a *cis*-effect, this time more sensitive to a variation in π-bonding.

As a whole, the rigid equatorial probe provided by the porphinato ligand is well suited to try a separation of the cooperating or antagonistic electronic effects. From the few examples selected from numerous studies it becomes clear that the further investigation of the group VIII metalloporphyrins presently under way in various laboratories throughout the world will be of benefit not only to bioinorganic chemistry, but also to organometallic and coordination chemistry.

5.4.4. Identification of axial ligands

In the age of X-ray crystallography and spectroscopy it may be old-fashioned to stress the necessity of elemental analyses. The author, however, has the experience that a) good commercial analytical laboratories can provide sufficiently precise analytical data on metalloporphyrins for C, H, N, O, halogens, and nonvolatile metal oxides (despite the remarks of other research groups, see footnote in Ref. 125), and b) that these figures are suitable to identify axial ligands. Of course, about 20 mg of a sample has to be sacrificed to obtain C, H, N, and additionally, O or halogen values, but apart from a metal analysis which may be omitted because the presence of the metal is evident from electronic or mass spectra, a total elemental analysis should be ascertained. A high N : C and N : H ratio points to the presence of nitrogenous bases. In those cases where the metal is sufficiently electronegative to allow an oxygen determination in its presence and the porphyrin contains little or no oxygen, the oxygen figures are very helpful. Thus, mononuclear and binuclear metalloporphyrins, e.g. $Fe(P)OMe$ and $[Fe(P)]_2O$, can be very easily discerned: the latter have roughly half the oxygen percentage of the former.

Other chapters in this book give detailed accounts of the applications of modern physical and spectroscopic techniques to porphyrins and their derivatives; most of these methods are sensitive to axial ligands and therefore may be used to their identification. Here only a few hints will be listed; the references of Tables 11—18 may lead the reader to the solution of any specific problem.

References, p. 224

The identification of neutral axial ligands can be accomplished by thermo-gravimetry[251,268]. If a sensitive apparatus for these measurements is at hand, this method may be routinely used also for the detection of solvate or included solvent molecules.

The same — though not in a quantitative manner — can be accomplished in a mass spectrometer. Most neutral ligands are partially or totally elimi-nated on heating before the metalloporphyrin starts to sublime and are then detectable in the vapor phase.

Mass spectrometry (see Chapter 9) is in general most helpful for the detection of axial ligands in molecules subliming without decomposition. In some cases, however, due to thermal degradation, either no porphyrin mass spectrum [$OsO_2(OEP)^{26}$; $Os(OEP)N_2(THF)^{251}$], or artefacts are ob-served. The latter may show molecular ions of apparently lower molecular weight, when thermal reduction accompanied with elimination of anions occurs, e.g. with Mo-, W-[20,25], Mn-, and Fe-porphyrins[122,124], or mole-cular ions of higher mass than expected. This phenomenon may be due to some thermal aggregation[37,130] or to thermal condensation of hydroxo complexes according eq. (35)[81,82]. The thermal lability of As-, Sb-, and Bi-porphyrins causes some unexpected ions (e.g. As_4^+, Sb^+, Bi^+) to be ob-served[115]. With Hg- and Tl-porphyrins, ion-molecule reactions may occur (see Chapter 9). Even if an intense molecular ion of the expected composi-tion is found, one cannot be sure that the solid sample does not contain dimeric, nonvolatile species of the same composition. This has been found with Zr- and Hf-porphyrins bearing dicarboxylates as ligands. While the vapor only contains the species [M(OEP)(pim)], a $CHCl_3$-solution consists of a mixture of [M(OEP)(pim)] and [M(OEP)(pim)]$_2$ (see types (G), Fig. 7, and (GG), Fig. 10; Table 16) as was deduced from careful molecular weight determinations by vapor pressure osmometry. As a consequence, one should rely on mass spectra alone only when the coordination type is well established.

Both neutral and anionic ligands give rise to specific infrared (Table 6, Chapter 11) and n.m.r. signals, the latter being easily observable only with diamagnetic species and showing characteristic high-field shifts due to ring current effects (see Chapter 10). Methoxide causes a specific CH stretch about 2780 cm^{-1} [25,82,84,99], hydroxide a sharp OH stretch between 3580 and 3640 cm^{-1} [39], and acetate an antisymmetric COO stretch about 1560—1665 cm^{-1} and a symmetric COO stretch about 1320—1450 cm^{-1}. If the wavenumber difference between these two is large (300 cm^{-1}), the acetate is bound as a monodentate ligand [e.g. M(P)OAc (B) (Table 14), M(P)(OAc)$_2$ (C) (Table 16), and MO(OEP)OAc (C) (Table 16)]. If, on the contrary, this frequency difference is small (150 cm^{-1}), bidentate bonding occurs, producing an abnormal coordination type, e.g. in Sc(OEP)OAc (D) (Table 14), Zr(OEP)(OAc)$_2$ (G) (Table 16 and Chapters 7 and 8) and Tl(OEP)OAc · H$_2$O^{110} [possibly representing type (E)].

Of course, all axial ligands produce characteristic metal-ligand vibrations (Table 5, Chapter 11). Terminal MO systems absorb between 820 and 1100 cm^{-1}, low values being produced when a second oxygen donor is located *trans* to the MO group, as in MO(OEP)OMe (M = Mo, W)[25] or in OsO_2(OEP)[26]. This is a consequence of the lone pairs at the oxygen atoms competing in dative π-bonding to the vacant d_{xz}, d_{yz}-orbitals at the metal[25] (*trans*-influence). The μ-oxo complexes $[MO(P)]_2O$ (CC) show the same phenomenon: while in complexes $[Fe(P)]_2O$ (BB) characteristic Fe—O—Fe absorptions occur at 800—900 cm^{-1}, the O=M—O—M=O system has very intense bands at about 630—690 cm^{-1} in $[MO(OEP)]_2O$ (CC)[27]. Furthermore, the μ-oxo bridges can be indirectly identified by an upfield shift of some resonances in the porphyrin n.m.r. spectra as compared with mononuclear species[39,64,84] (see also Chapter 10).

The light absorption of metalloporphyrins showing a 'normal absorption spectrum' is not very sensitive to a variation in axial ligands. However, remarkable heavy-atom effects of the ligands are found in their light-emission (see Ref. 135). The metalloporphyrins showing the 'hyper' or 'hypso' type absorption show notable ligand dependences (see Fig. 15, 16; Section 5.4.3., and Ref. 135) which may be used for the assignment of coordination types[246,247].

Finally, some remarks on the identification of small π-acceptor molecules should be made. N_2, CO, NO, NO_2, and isonitriles (Table 17, 18) as well as O_2[260-264] can be detected by their respective infrared stretching frequencies. The old dispute on the mode of bonding of the oxygen molecule in hemoglobin or myoglobin[236,249,250,267] now seems to have come to a decision in favor of Pauling's formulation which takes type (C) for the oxygenated heme group carrying a monodentate kinked FeO_2 grouping; Griffith's model with a bidentate O_2 molecule mainly fixed by π-bonding would represent the hitherto unknown and improbable coordination type (F) (Fig. 7). This decision has been made on the grounds of an X-ray analysis of Fe(picket-fence-P)O_2(1-MeIm)[265] (see Chapter 8) and by infrared investigations[260-262]. Monodentate O_2 ligands in kinked MO_2 systems are known to absorb at about 1100 cm^{-1}, and such bands have been detected in hemoglobin[260], myoglobin[261], and coboglobin[262]. At variance with these findings is the observation of an O_2 stretch at 1385 cm^{-1} in Fe(picket-fence-P)O_2(1-NMeIm)[264] in KBr at −175°C. It remains to be clarified whether this difference arises from the different conditions of the measurements (H_2O, room temperature on one hand, KBr, −175°C on the other hand), or from different π-acceptor capacities of the porphinato ligands used, or from steric repulsions being different in the cavities of the globin or the model porphyrin.

A more subtle problem is the distribution of electrons between the iron and the O_2 molecule: is it better represented by [Fe(II)] $\cdot$ O_2 (dioxygen complex) or by [Fe(III)] $\cdot$ O_2^- (superoxide complex)? That these formulations indeed represent the extremes that have to be discussed follows from the possible formation by combination of Fe(II)-porphyrins with dioxygen and Fe(III)-porphyrins with electrochemically generated superoxide ion to yield the same product (Scheme 2)[246]. The occurrence of near infrared bands which are also typical for Fe(III)-porphyrins[246,247,259] points to an important contribution of the Fe(III) case[250]. In the Fe—CO system, they are absent, therefore only the view [Fe(II)] CO is realistic. These assignments are further supported by Mössbauer spectra (see Chapter 12)[259].

References

1. J.E. Falk, "Porphyrins and Metalloporphyrins", Elsevier, Amsterdam (1964).
2. H. Fischer and Orth, "Die Chemie des Pyrrols", II. Band, Pyrrolfarbstoffe, 1. Hälfte, Akademische Verlagsgesellschaft, Leipzig (1937).
3. H. Fischer and A. Stern, "Die Chemie des Pyrrols", II. Band, Pyrrolfarbstoffe, 2. Hälfte, Akademische Verlagsgesellschaft, Leipzig (1940).
4   R. Willstätter and A. Stoll, "Untersuchungen über Chlorophyll — Methoden und Ergebnisse", Springer, Berlin (1913).
5. L.P. Vernon and G.R. Seely (Eds.), "The Chlorophylls", Academic Press, New York and London (1966).
6. A. Treibs, "Das Leben und Wirken von Hans Fischer", Hans Fischer-Gesellschaft, München (1971).
7. H. Fischer, Synthese des Hämins, Naturwissenschaften, 17, 611 (1929).
8. R.B. Woodward, Totalsynthese des Chlorophylls, Angew. Chem., 72, 651 (1960).
9. E. Antonini and M. Brunori, "Hemoglobin and myoglobin in their reactions with ligands", North-Holland Publishing Co., Amsterdam and London (1971).
10. J.M. Pratt, "Inorganic chemistry of vitamin B_{12}", Academic Press, London and New York (1972).
11. H.H. Inhoffen, J.W. Buchler, and P. Jäger, Fortschr. Chem. Org. Naturst. [Wien], 26, 284 (1968).
12. P. Hambright, Coord. Chem. Rev., 6, 247 (1971).
13. M.B. Crute, Acta Crystallogr., 12, 24 (1959).
14. R. Rothemund and A.R. Menotti, J. Am. Chem. Soc., 70, 1808 (1948).
15. W.R. Scheidt, J. Amer. Chem. Soc., 96, 90 (1974).
16. A. Forman, D.C. Borg, R.H. Felton, and J. Fajer, J. Am. Chem. Soc., 93, 2790 (1971).
17. L.D. Spaulding, P.G. Eller, J.A. Bertrand, and R.H. Felton, J. Am. Chem. Soc., 96, 982 (1974).
18. D.F. Koenig, Acta Crystallogr., 18, 663 (1965).
19. F.P. Dwyer, J.W. Buchler, and W.R. Scheidt, J. Am. Chem. Soc., 96, 2789 (1974).
20. J.W. Buchler, G. Eikelmann, L. Puppe, K. Rohbock, H.H. Schneehage, and D. Weck, Ann. Chem., 745, 135 (1971).
21. L.J. Radonovich, A. Bloom, and J.L. Hoard, J. Am. Chem. Soc., 94, 2073 (1972).
22. D.M. Collins, R. Countryman, and J.L. Hoard, J. Am. Chem. Soc., 94, 2066 (1972).
23. R. Grigg, G. Shelton, A. Sweeney, and A.W. Johnson, J. Chem. Soc., Perkin Trans. 1, 1789 (1972).
24. D.M. Collins, W.R. Scheidt, and J.L. Hoard, J. Am. Chem. Soc., 94, 6689 (1972).
25. J.W. Buchler, L. Puppe, K. Rohbock, and H.H. Schneehage, Chem. Ber., 106, 2710 (1973).

26. J.W. Buchler and P.D. Smith, Angew. Chem., **86**, 378 (1974); Angew. Chem., Int. Ed. Eng., **13**, 341 (1974).
27. J.W. Buchler and K. Rohbock, Inorg. Nucl. Chem. Lett., **8**, 1073 (1972).
28. A. Gieren and W. Hoppe, Chem. Commun., 413 (1971).
29. S.J. Silvers and A. Tulinsky, J. Am. Chem. Soc., **89**, 3331 (1967).
30. R. Thomas, Diplomarbeit, Technische Hochschule Braunschweig (1964).
31. A. Takenaka, Y. Sasada, T. Omura, H. Ogoshi, and Z.I. Yoshida, J. Chem. Soc., Chem. Commun., 792 (1973).
32. D. Cullen, E. Meyer, T.S. Srivastava, and M. Tsutsui, J. Am. Chem. Soc., **94**, 7603 (1972).
33. T.S. Srivastava, C.-P. Hrung, and M. Tsutsui, J. Chem. Soc., Chem. Commun., 447 (1974).
34. F.P. Schwarz, M. Gouterman, Z. Muljiani, and D. Dolphin, Bioinorg. Chem., **2**, 1 (1972).
35. A.B. Hoffmann, D.M. Collins, V.W. Day, E.B. Fleischer, T.S. Srivastava, and J.L. Hoard, J. Am. Chem. Soc., **94**, 3620 (1972).
36. G.W. Sovocool, F.R. Hopf, and D.G. Whitten, J. Am. Chem. Soc., **94**, 4350 (1972).
37. D.G. Whitten, personal communication.
38. T.R. Janson, A.R. Kane, J.F. Sullivan, K. Knox, and M.E. Kenney, J. Am. Chem. Soc., **91**, 5210 (1969).
39. J.W. Buchler, L. Puppe, K. Rohbock, and H.H. Schneehage, Ann. N.Y. Acad. Sci., **206**, 116 (1973).
40. J.W. Buchler, H. Habets, J. van Kaam, and K. Rohbock, unpublished; Diplomarbeit H. Habets, Technische Hochschule Aachen (1974); Diplomarbeit J. van Kaam, Aachen (1972).
41. N. Hirayama, A. Takenaka, Y. Sasada, E.-I. Watanabe, H. Ogoshi, and Z.-I. Yoshida, J. Chem. Soc., Chem. Commun., 330 (1974).
42. J.P. Macquet and T. Theophanides, Can. J. Chem., **51**, 219 (1973).
43. E.B. Fleischer and A.L. Stone, Chem. Commun., 332 (1967).
44. M.F. Hudson and K.M. Smith, J. Chem. Soc., Chem. Commun., 515 (1973).
45. I.A. Cohen, Ann. N.Y. Acad. Sci., **206**, 453 (1973).
46. F.H. Moser and A.L. Thomas, "Phthalocyanine compounds", Reinhold, New York (1963).
47. M. Calvin, Rev. Pure Appl. Chem., **15**, 1 (1965).
48. T.P. Pretlow and F. Sherman, Biochim. Biophys. Acta, **148**, 629 (1967).
49. A. Treibs, Ann. Chem., **517**, 172 (1935).
50. G. Hodgson, Ann. N.Y. Acad. Sci., **206**, 670 (1973).
51. O.T.G. Jones, in "Porphyrins and Related Compounds", Ed. T.W. Goodwin, Academic Press, London and New York (1968), p. 131.
52. P.P. Laidlaw, J. Physiol. London, **31**, 464 (1904).
53. J. Zaleski, Hoppe-Seyler's Z. Physiol. Chem., **37**, 54 (1902), Hoppe-Seyler's Z. Physiol. Chem., **43**, 11 (1904).
54. J.A. Milroy, J. Physiol. London, **38**, 384 (1909).
55. R. Hill, Biochem. J., **19**, 341 (1925).
56. A. Treibs, Angew. Chem., **49**, 31 (1936).
57. A. Stern and M. Dezelic, Z. Phys. Chem., Abt. A, **180**, 131 (1937).
58. F. Haurowitz, Chem. Ber., **68**, 1795 (1935).
59. H. Theorell, Enzymologia, **4**, 192 (1937).
60. P.A. Barrett, C.E. Dent, and R.P. Linstead, J. Chem. Soc., 1719 (1936).
61. R.D. Joyner and M.E. Kenney, Inorg. Chem., **1**, 717 (1962).
62. R.S. Becker and J.B. Allison, J. Phys. Chem., **67**, 2669 (1963).
63. D. Ostfeld and M. Tsutsui, Acc. Chem. Res., **7**, 52 (1974).

64. M. Gouterman, L. Karle Hanson, G.-E. Khalil, J.W. Buchler, K. Rohbock, and D. Dolphin, J. Am. Chem. Soc., in press.
65. A. Treibs, Ann. Chem., **728**, 115 (1969).
66. E.B. Fleischer, R. Thorp, and D. Venerable, Chem. Commun., 475 (1969).
67. E.B. Fleischer and D. Lavallee, J. Am. Chem. Soc., **89**, 1132 (1967).
68. N. Sadasivan and E.B. Fleischer, J. Inorg. Nucl. Chem., **30**, 591 (1968).
69. T.S. Srivastava and E.B. Fleischer, J. Am. Chem. Soc., **92**, 5518 (1970).
70. M. Tsutsui, R.A. Velapoldi, K. Suzuki, and T. Koyano, Angew. Chem., **80**, 914 (1968).
71. M. Tsutsui, M. Ichakawa, F. Vohwinkel, and K. Suzuki, J. Am. Chem. Soc., **88**, 854 (1966).
72. M. Tsutsui and C.P. Hrung, J. Am. Chem. Soc., **95**, 5777 (1973).
73. D. Ostfeld, M. Tsutsui, C.P. Hrung, and D.C. Conway, J. Am. Chem. Soc., **93**, 2548 (1971).
74. J.W. Buchler and K. Rohbock, J. Organometal. Chem., **65**, 223 (1974).
75. W.DeW. Horrocks Jr., D.-P. Wong, and R. Venteicher, J. Am. Chem. Soc., **96**, 7149 (1974), and personal communications.
76. F. Lux, D. Dempf, and D. Graw, Angew. Chem., **80**, 792 (1968).
77. T.J. Marks, Nachr. Chem. Tech., **22**, 346 (1974).
78. E.B. Fleischer, J.M. Palmer, T.S. Srivastava, and A. Chatterjee, J. Am. Chem. Soc., **93**, 3162 (1971).
79. P. Hambright and E.B. Fleischer, Inorg. Chem., **9**, 1757 (1970).
80. R.R. Das, R.F. Pasternack, and R.A. Plane, J. Am. Chem. Soc., **92**, 3312 (1970).
81. H.H. Inhoffen and J.W. Buchler, Tetrahedron Lett., 2057 (1968).
82. J.W. Buchler, L. Puppe, and H.H. Schneehage, Ann. Chem., **749**, 134 (1971).
83. W.P. Hambright, A.N. Thorpe, and C.C. Alexander, J. Inorg. Nucl. Chem., **30**, 3139 (1968).
84. J.W. Buchler and H.H. Schneehage, Z. Naturforsch., Teil B, **28**, 433 (1973).
85. A.D. Adler, F.R. Longo, F. Kampas, and J. Kim, J. Inorg. Nucl. Chem. **32**, 2443 (1970).
86. U. Eisner and M.J.C. Harding, J. Chem. Soc., 4089 (1964).
87. J.W. Buchler and L. Puppe, Ann. Chem., 1046 (1974).
88. M. Tsutsui, personal communication.
89. J. Lewis, Pure Appl. Chem., in press.
90. S.J. Baum, B.F. Burnham, and R.A. Plane, Proc. Nat. Acad. Sci. U.S.A., **52**, 1439 (1964).
91. J.-H. Fuhrhop and D. Mauzerall, J. Am. Chem. Soc., **91**, 4174 (1969).
92. G.D. Dorough, J.R. Miller, and F.M. Huennekens, J. Am. Chem. Soc., **73**, 4315 (1951).
93. J.W. Buchler and L. Puppe, unpublished experiments; L. Puppe, Dissertation, Technische Hochschule Aachen, (1972).
94. D.G. Whitten, J.C.N. Yau, and F.A. Carrol, J. Am. Chem. Soc., **93**, 2291 (1971).
95. S. Granick, J. Biol. Chem., **175**, 333 (1948).
96. M. Gouterman, F.P. Schwarz, P.D. Smith, and D. Dolphin, J. Chem. Phys., **59**, 676 (1973).
97. L.J. Boucher, J. Am. Chem. Soc., **90**, 6640 (1968).
98. B.D. McLees and W.S. Caughey, Biochemistry, **7**, 642 (1968).
99. J.O. Alben, W.H. Fuchsman, C.A. Beaudreau, and W.S. Caughey, Biochemistry, **7**, 624 (1968).
100. J.G. Erdman, V.G. Ramsey, N.W. Kalenda, and W.E. Hanson, J. Am. Chem. Soc., **78**, 5844 (1956).
101. H. Fischer and W. Neumann, Ann. Chem., **494**, 225 (1932).

102. U. Eisner, J. Chem. Soc., 3461 (1957).
103. U. Eisner, A. Lichtarowitz, and R.P. Linstead, J. Chem. Soc., 733 (1957).
104. H.H. Inhoffen, J.W. Buchler, and R. Thomas, Tetrahedron Lett., 1141 (1969).
105. R. Bonnett and M.J. Dimsdale, J. Chem. Soc., Perkin Trans. 1, 2540 (1972).
106. H.W. Whitlock Jr., R. Hanauer, M.Y. Oester, and B.K. Bower, J. Am. Chem. Soc., 91, 7485 (1969).
107. M. Bhatti, W. Bhatti, and E. Mast, Inorg. Nucl. Chem. Lett., 8, 133 (1972).
108. B.R. James and D.V. Stynes, J. Am. Chem. Soc., 94, 6225 (1972).
109. M.F. Hudson and K.M. Smith, Tetrahedron Lett. 2223 (1974).
110. R.J. Abraham, G.H. Barnett, and K.M. Smith, J. Chem. Soc., Perkin Trans. 1, 2142 (1973).
111. P.J. Crook, A.H. Jackson, and G.W. Kenner, Ann. Chem., 748, 26 (1971).
112. H. Fischer, A. Treibs, and K. Zeile, Hoppe-Seyler's Z. Physiol. Chem., 195, 20 (1931).
113. S.M. Husain and J.G. Jones, Inorg. Nucl. Chem. Lett., 10, 105 (1974).
114. N. Datta-Gupta and T.J. Bardos, J. Pharm. Sci., 57, 300 (1968).
115. J.W. Buchler and K.L. Lay, Inorg. Nucl. Chem. Lett., 10, 297 (1974).
116. D.B. Boylan and M. Calvin, J. Am. Chem. Soc., 89, 5472 (1967).
117. J.E. Maskasky and M.E. Kenney, J. Am. Chem. Soc., 95, 1443 (1973).
118. E.B. Fleischer and A. Laszlo, Inorg. Nucl. Chem. Lett., 5, 373 (1969).
119. V.W. Day, B.R. Stults, E.L. Tasset, R.O. Day, and R.S. Marianelli, J. Am. Chem. Soc., 96, 2650 (1974).
120. A. Bluestein and J.M. Sugihara, Inorg. Chem., 12, 690 (1973).
121. J.-H. Fuhrhop, Z. Naturforsch., Teil B, 25, 255 (1970).
122. J.W. Buchler, K.L. Lay, L. Puppe, and H. Stoppa, unpublished experiments.
123. J.W. Buchler and K. Rohbock, unpublished experiments; K. Rohbock, Dissertation, Technische Hochschule Aachen (1972).
124. L. Edwards, D.H. Dolphin, M. Gouterman, and A.D. Adler, J. Mol. Spectrosc., 38, 16 (1971).
125. L. Karle Hanson, M. Gouterman, and J.C. Hanson, J. Am. Chem. Soc., 95, 4822 (1973).
126. M. Tsutsui, R.A. Velapoldi, K. Suzuki, F. Vohwinkel, M. Ichakawa, and T. Koyano, J. Am. Chem. Soc., 91, 6262 (1969).
127. J.-H. Fuhrhop, Tetrahedron Lett., 3205 (1969).
128. G.M. Brown, F.R. Hopf, J.A. Ferguson, T.J. Meyer, and D.G. Whitten, J. Am. Chem. Soc., 95, 5939 (1973).
129. B.C. Chow, and I.A. Cohen, Bioinorg. Chem., 1, 57 (1971).
130. J.J. Bonnet, S.S. Eaton, G.R. Eaton, R.H. Holm, and J.A. Ibers, J. Am. Chem. Soc., 95, 2141 (1973).
131. S.S. Eaton, G.R. Eaton, and R.H. Holm, J. Organometal. Chem., 39, 179 (1972).
132. M. Tsutsui, D. Ostfeld, and L.M. Hoffman, J. Am. Chem. Soc., 93, 1820 (1971).
133. I.A. Cohen and B.C. Chow, Inorg. Chem., 13, 488 (1974).
134. H. Ogoshi, T. Omura, and Z. Yoshida, J. Am. Chem. Soc., 95, 1666 (1973).
135. J.W. Buchler and M. Gouterman, manuscript in preparation.
136. M. Gouterman, L. Karle Hanson, G.-E. Khalil, W.R. Leenstra, and J.W. Buchler, J. Chem. Phys., 62, 2343 (1975).
137. J.-H. Fuhrhop, Struct. Bonding Berlin, 18, 1 (1974).
138. R.S. Becker and J.B. Allison, J. Phys. Chem., 67, 2662 (1963).
139. D.W. Thomas and A.E. Martell, Arch. Biochem. Biophys., 76, 286 (1958).
140. F.A. Cotton and G. Wilkinson, "Advanced Inorganic Chemistry", 3rd ed., Interscience, New York (1972).
141. R.D. Shannon and C.T. Prewitt, Acta Crystallogr., B 25, 925 (1969).

142. H. Rieck and R. Hoppe, Z. Anorg. Allg. Chem., **392**, 139 (1972).
143. J.L. Hoard, Science, **174**, 1295 (1971).
144. J.F. Kirner, J. Garofalo, Jr., and W.R. Scheidt, Inorg. Nucl. Chem. Lett., **11**, 107 (1975).
145. K. Kadish, D.G. Davis, and J.-H. Fuhrhop, Angew. Chem., **84**, 1072 (1972).
146. D. Dolphin, personal communication.
147. L.J. Boucher and H.K. Garber, Inorg. Chem., **9**, 2644 (1970).
148. I.A. Cohen, D. Ostfeld, and B. Lichtenstein, J. Am. Chem. Soc., **94**, 4522 (1972).
149. H.W. Whitlock Jr. and B.K. Bower, Tetrahedron Lett., 4827 (1965).
150. E.B. Fleischer and T.S. Srivastava, Inorg. Chim. Acta, **5**, 151 (1971).
151. J.N. Phillips, Rev. Pure Appl. Chem., **10**, 35 (1960).
152. D.K. Cabbiness and D.W. Margerum, J. Am. Chem. Soc., **91**, 6540 (1969).
153. J.E. Falk and R.S. Nyholm, in "Current Trends in Heterocyclic Chemistry", Eds. A. Albert, G.M. Badger, and C.W. Shoppee, Butterworths, London (1958), p. 130.
154. A.H. Corwin and P.E. Wei, J. Org. Chem., **27**, 4285 (1957).
155. A. Mac Cragh and W.S. Koski, J. Am. Chem. Soc., **87**, 2496 (1965).
156. W.S. Caughey and A.H. Corwin, J. Am. Chem. Soc., **77**, 1509 (1955).
157. J.W. Buchler, Habilitationsschrift, Technische Hochschule Aachen (1971).
158. D.B. Morrell, J. Barrett, and P.S. Clezy, Biochem. J., **78**, 793 (1961).
159. M. Grinstein, J. Biol. Chem., **167**, 515 (1947).
160. F.J.C. Rossotti, in "Modern Coordination Chemistry", Eds. J. Lewis and R.G. Wilkins, Interscience, New York (1960) p. 1.
161. L. Pauling, "The Nature of the Chemical Bond", 3rd ed., Cornell University Press, Ithaca (1960).
162. R.J.P. Williams, Chem. Rev., **56**, 299 (1956).
163. M.F. Hudson and K.M. Smith, Tetrahedron Lett. 2227 (1974).
164. J.W. Barnes and G.D. Dorough, J. Am. Chem. Soc., **72**, 4045 (1950).
165. C. Grant and P. Hambright, J. Am. Chem. Soc., **91**, 4195 (1969).
166. H. Baker, P. Hambright, L. Wagner, and L. Ross, Inorg. Chem., **12**, 2200 (1973).
167. H. Ogoshi, N. Masai, Z. Yoshida, J. Takemoto, and K. Nakamoto, Bull. Chem. Soc. Jpn., **44**, 49 (1971).
168. J.W. Buchler and H.H. Schneehage, unpublished results; H.H. Schneehage, Dissertation, Technische Hochschule Aachen (1971).
169. J.W. Buchler and P.D. Smith, unpublished results.
170. K.J. Brunings and A.H. Corwin, J. Am. Chem. Soc., **64**, 72 (1942).
171. J.W. Buchler and L. Puppe, Ann. Chem., **740**, 142 (1970).
172. D.W. Thomas and A.E. Martell, J. Am. Chem. Soc., **81**, 5111 (1959).
173. H. Ogoshi and Z. Yoshida, Bull. Chem. Soc. Jpn., **44**, 1722 (1971).
174. H. Bürger and J.W. Buchler, unpublished observations.
175. I.D. Dicker, R. Grigg, A.W. Johnson, H. Pinnock, K. Richardson, and P. van den Broek, J. Chem. Soc., C, 536 (1971), and preceding papers.
176. H.H. Inhoffen, J.W. Buchler, L. Puppe, and K. Rohbock, Ann. Chem., **747**, 133 (1971).
177. Y. Murakami, Y. Matsuda, and S.-I. Kobayashi, J. Chem. Soc., Dalton Trans., 1734 (1973).
178. A.H. Corwin and M.H. Melville, J. Am. Chem. Soc., **77**, 2755 (1955).
179. G.D. Dorough and F.M. Huennekens, J. Am. Chem. Soc., **74**, 3974 (1952).
180. J.W. Buchler and T. Kals, unpublished experiments; T. Kals, Diplomarbeit, Technische Hochschule Aachen (1972).
181. B.D. Berezin and A.N. Dobrysheva, Zh. Fiz. Khim., **44**, 2804 (1970).
182. J.A.S. Cavaleiro and K.M. Smith, J. Chem. Soc., Perkin Trans. 1, 2149 (1973).
183. J.W. Buchler and H.H. Schneehage, Angew. Chem., **81**, 912 (1969).

184. J.-H. Fuhrhop and T. Lumbantobing, Tetrahedron Lett. 2815 (1970).
185. A.H. Corwin, A.B. Chivvis, and C.B. Storm, J. Org. Chem., 29, 3702 (1964).
186. D.K. Lavallee and A.E. Gebala, Inorg. Chem., 13, 2004 (1974).
187. J.P. Collman and C.A. Reed, J. Am. Chem. Soc., 95, 2048 (1973).
188. R.G. Little and J.A. Ibers, J. Am. Chem. Soc., 96, 4452 (1974).
189. R. Timkovich and A. Tulinsky, J. Am. Chem. Soc., 91, 4430 (1969).
190. W.R. Scheidt and M.E. Frisse, J. Am. Chem. Soc., 97, 17 (1975).
191. W.R. Scheidt and J.L. Hoard, J. Am. Chem. Soc., 95, 8281 (1973).
192. P.D. Dwyer, P. Madura, and W.R. Scheidt, J. Am. Chem. Soc., 96, 4815 (1974).
193. J.R. Miller and G.D. Dorough, J. Am. Chem. Soc., 74, 3977 (1952).
194. A.H. Corwin and S.D. Bruck, J. Am. Chem. Soc., 80, 4736 (1958).
195. R.G. Little and J.A. Ibers, J. Am. Chem. Soc., 96, 4440 (1974).
196. D.L. Anderson, C.J. Weschler, and F. Basolo, J. Am. Chem. Soc., 96, 5599 (1974).
197. L.M. Epstein, D.K. Straub, and C. Maricondi, Inorg. Chem., 6, 1720 (1967).
198. W.R. Scheidt, J. Am. Chem. Soc., 96, 84 (1974).
199. J.W. Lauher and J.A. Ibers, J. Am. Chem. Soc., 96, 4447 (1974).
200. W.R. Scheidt, J.A. Cunningham, and J.L. Hoard, J. Am. Chem. Soc., 95, 8289 (1973).
201. J.G. Erdman and A.H. Corwin, J. Am. Chem. Soc., 69, 750 (1947).
202. N. Sadasivan, H.I. Eberspaecher, W.H. Fuchsman, and W.S. Caughey, Biochemistry, 8, 534 (1969).
203. W.S. Caughey, in "Inorganic Biochemistry", Ed., G.I. Eichhorn, Vol. 2, Elsevier, Amsterdam (1973) p. 797.
204. J.W. Buchler and K.L. Lay, unpublished results; K.L. Lay, Diplomarbeit, Technische Hochschule Aachen (1972).
205. H. Ogoshi, E. Watanabe, Z. Yoshida, J. Kincaid, and K. Nakamoto, J. Am. Chem. Soc., 95, 2845 (1973).
206. L.J. Boucher, Ann. N.Y. Acad. Sci., 206, 409 (1973).
207. C. Maricondi, W. Swift, and D.K. Straub, J. Am. Chem. Soc., 91, 5205 (1969).
208. B.R. James and D.V. Stynes, J. Chem. Soc., Chem. Commun., 1261 (1972).
209. M.A. Torrens, D.K. Straub, and L.M. Epstein, J. Am. Chem. Soc., 94, 4160, 4162 (1972).
210. S.B. Brown, P. Jones, and I.R. Lantzke, Nature London, 223, 960 (1969).
211. M. Wicholas, R. Mustacich, and D. Jayne, J. Am. Chem. Soc., 94, 4518 (1972).
212. E.B. Fleischer and T.S. Srivastava, J. Am. Chem. Soc., 91, 2403 (1969).
213. I.A. Cohen, J. Am. Chem. Soc., 91, 1980 (1969).
214. R.H. Felton, G.S. Owen, D. Dolphin, A. Forman, D.C. Borg and F. Fajer, Ann. N.Y. Acad. Sci., 206, 504 (1973).
215. H.H. Inhoffen, H. Parnemann, and R.G. Foster, Illinois Inst. Techn. NMR Newsletter, 65, 45 (1964); H. Parnemann, Dissertation, Technische Hochschule Braunschweig (1964).
216. L.J. Velenyi and M.E. Kenney, Abstr. of Papers, 6th Internat. Conf. on Organometallic Chemistry, Amherst (Mass.), Aug. 13—17 (1973).
217. A.W. Johnson and I.T. Kay, J. Chem. Soc., 2979 (1960).
218. H.A.O. Hill, personal communication.
219. P.A. Loach and M. Calvin, Biochemistry, 2, 361 (1963).
220. D.A. Clarke, D. Dolphin, R. Grigg, A.W. Johnson, and H.A. Pinnock, J. Chem. Soc. C, 881 (1968).
221. A. Yamamoto, L.K. Phillips, and M. Calvin, Inorg. Chem., 7, 847 (1968).
222. G. Engelsma, A. Yamamoto, E. Markham, and M. Calvin, J. Phys. Chem., 66, 2517 (1962).
223. W.S. Caughey, R.M. Deal, C. Weiss, and M. Gouterman, J. Mol. Spectrosc., 16, 451 (1963).
224. J.A. Kaduk and W.R. Scheidt, Inorg. Chem., 13, 1875 (1974).

225. S. McCoy and W.S. Caughey, Biochemistry, **9**, 2387 (1970).
226. W.S. Caughey, C.H. Barlow, D.H. O'Keefe, and M.C. O'Toole, Ann. N.Y. Acad. Sci., **206**, 296 (1973).
227. T.S. Srivastava, L. Hoffman, and M. Tsutsui, J. Am. Chem. Soc., **94**, 1385 (1972).
228. M. Tsutsui, D. Ostfeld, J.N. Francis, and L.M. Hoffman, J. Coord. Chem., **1**, 115 (1971).
229. P. Piciulo, G. Rupprecht, and W.R. Scheidt, J. Am. Chem. Soc., **96**, 5293 (1974).
230. R.G. Little and J.A. Ibers, J. Am. Chem. Soc., **95**, 8583 (1973).
231. J.P. Collman, R.R. Gagné, T.R. Halbert, J.-C. Marchon, and C.A. Reed, J. Am. Chem. Soc., **95**, 7868 (1973).
232. J.P. Collman, R.R. Gagné, and C.A. Reed, J. Am. Chem. Soc., **96**, 2629 (1974).
233. J.E. Baldwin, J. Huff, J. Almog, C. Wilkerson, and R. Dyer, Section Lecture, XVIth Internat. Conf. on Coordination Chemistry, Dublin, Aug. 19—24 (1974).
234. A.H. Corwin and Z. Reyes, J. Am. Chem. Soc., **78**, 2437 (1956).
235. J.H. Wang, J. Am. Chem. Soc., **80**, 3168 (1958).
236. L. Pauling, Nature London, **203**, 182 (1964).
237. C.K. Chang and T.G. Traylor, J. Am. Chem. Soc., **95**, 5810 (1973).
238. C.K. Chang and T.G. Traylor, J. Am. Chem. Soc., **95**, 8475 (1973).
239. C.K. Chang and T.G. Traylor, J. Am. Chem. Soc., **95**, 8477 (1973).
240. W.S. Brinigar, C.K. Chang, J. Geibel, and T.G. Traylor, J. Am. Chem. Soc., **96**, 5597 (1974).
241. W.S. Brinigar and C.K. Chang, J. Am. Chem. Soc., **96**, 5595 (1974).
242. J. Almog, J.E. Baldwin, R.L. Dyer, J. Huff, and C.J. Wilkerson, J. Am. Chem. Soc., **96**, 5600 (1974).
243. G.C. Wagner and R.J. Kassner, J. Am. Chem. Soc., **96**, 5593 (1974).
244. C.J. Weschler, D.L. Anderson, and F. Basolo, J. Chem. Soc., Chem. Commun., 757 (1974).
245. D.V. Stynes, H.C. Stynes, B.R. James, and J.A. Ibers, J. Am. Chem. Soc., **95**, 4087 (1973).
246. H.A.O. Hill, D.R. Turner, and G. Pellizer, Biochem. Biophys. Res. Commun., **56**, 739 (1974).
247. D.W. Smith and R.J.P. Williams, Struct. Bonding Berlin, **7**, 1 (1970).
248. P. Mohr and W. Scheler, Eur. J. Biochem., **8**, 444 (1969).
249. J.J. Weiss, Nature London, **202**, 83 (1964).
250. J.B. Wittenberg, A.B. Wittenberg, J. Peisach, and W.E. Blumberg, Proc. Nat. Acad. Sci. U.S.A., **67**, 1846 (1970).
251. J.W. Buchler and P.D. Smith, Angew. Chem., **86**, 820 (1974), Angew. Chem., Int. Ed. Eng., **13**, 745 (1974).
252. S.A. Cockle, H.A.O. Hill, and R.J.P. Williams, Inorg. Nucl. Chem. Lett., **6**, 131 (1970).
253. F.A. Walker, J. Am. Chem. Soc., **92**, 4235 (1970); ibid. **95**, 1150, 1154 (1973).
254. B.B. Wayland and M.E. Abd-Elmageed, J. Am. Chem. Soc., **96**, 4809 (1974), and preceding papers.
255. M.J. Carter, D.P. Rillema, and F. Basolo, J. Am. Chem. Soc., **96**, 392 (1974).
256. D.V. Stynes, H.C. Stynes, J.A. Ibers, and B.R. James, J. Am. Chem. Soc., **95**, 1142 (1973).
257. J.A. Ibers, D.V. Stynes, H.C. Stynes, and B.R. James, J. Am. Chem. Soc., **96**, 1358 (1974), and preceding papers.
258. R.G. Little, B.M. Hoffman, and J.A. Ibers, Bioinorg. Chem., **3**, 207 (1974).
259. M. Weissbluth, "Hemoglobin: Cooperativity and Electronic Properties", Springer, Heidelberg (1974).
260. C.H. Barlow, J.C. Maxwell, W.H. Wallace, and W.S. Caughey, Biochem. Biophys. Res. Commun., **55**, 91 (1973).

261. J.C. Maxwell, J.A. Volpe, C.H. Barlow, and W.S. Caughey, Biochem. Biophys. Res. Commun., **58**, 166 (1974).

262. J.C. Maxwell and W.S. Caughey, Biochem. Biophys. Res. Commun., **60**, 1309 (1974).

263. W.H. Fuchsman, C.H. Barlow, W.J. Wallace, and W.S. Caughey, Biochem. Biophys. Res. Commun., **61**, 635 (1974).

264. J.P. Collman, R.R. Gagné, H.B. Gray, and J.W. Hare, J. Am. Chem. Soc., **96**, 6522 (1974).

265. J.P. Collman, R.R. Gagné, C.A. Reed, W.T. Robinson, and G.A. Rodley, Proc. Nat. Acad. Sci. U.S.A., **71**, 1326 (1974).

266. B.B. Wayland and L.W. Olson, J. Am. Chem. Soc., **96**, 6037 (1974).

267. E. Bayer and P. Schretzmann, Struct. Bonding Berlin, **2**, 181 (1967).

268. R.Y. Lee and P. Hambright, J. Inorg. Nucl. Chem., **32**, 477 (1970).

269. M. Gouterman, personal communication.

DYNAMIC COORDINATION CHEMISTRY OF METALLOPORPHYRINS

PETER HAMBRIGHT

Chemistry Department, Howard University, Washington, D.C. 20059, USA

6.1. Introduction

This chapter will review recent developments in the dynamic coordination chemistry of metalloporphyrins. The field is outlined in Fig. 1. The literature is extensive, and earlier monographs are still of fundamental importance[1-5].

METAL-PORPHYRIN INTERACTIONS

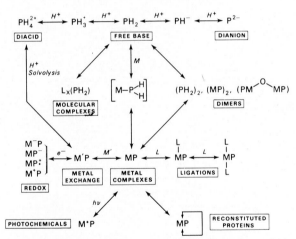

Fig. 1. A general outline of the interactions of metal ions with porphyrin molecules.

The main porphyrin types that have been used in mechanistic studies are shown in Fig. 2. To enable a concise treatment, the following abbreviations are used:

H_2(Proto-IX)	Protoporphyrin-IX
H_2(Proto-IX-DME)	Protoporphyrin-IX dimethyl ester[5]
H_2(Deut-IX)	Deuteroporphyrin-IX

Porphyrins and Metalloporphyrins, ed. Kevin M. Smith
© 1975, Elsevier Scientific Publishing Company, Amsterdam, The Netherlands

H$_2$(S-Deut-IX-DME)	A series of 2,4-disubstituted deuteroporphy-rin-IX dimethyl esters[6]
H$_2$(DaDeut-IX)	2,4-Diacetyldeuteroporphyrin-IX[6]
H$_2$(Deut-IX-DSA)	Deuteroporphyrin-IX dimethyl ester 2,4-disulfonic acid
H$_2$(Meso-IX)	Mesoporphyrin-IX
H$_2$(Uro-III)	Uroporphyrin-III
H$_2$(Copro-III-TME)	Coproporphyrin-III tetramethyl ester
H$_2$(OEP)	Octaethylporphyrin[7]
H$_2$(Etio-I)	Etioporphyrin-I
H$_2$(TPP)	meso-Tetraphenylporphyrin[8−10]
H$_2$(TPPS$_3$), H$_2$(TPPS$_4$)	tri-[11,12] or tetra-[13] p-sulfonated H$_2$(TPP)
H$_2$(TPPC$_4$)	meso-(p-CO$_2$H)tetraphenylporphyrin[10]
H$_2$(p-X · TPP)	meso-(p-X)tetraphenylporphyrin
H$_2$(TPyP)	meso-Tetrapyridylporphyrin[14]
H$_2$(TMPyP)	meso-tetrakis-(4-N-methylpyridyl)porphyrin[15]
H$_2$(EnProto-IX-DME)	Ethylenediamine substituted[16] H$_2$(Proto-IX-DME)
H(NMeOEP)	N-Methyloctaethylporphyrin
H$_2$(Hemato-IX-DME)	Hematoporphyrin-IX dimethyl ester

X = X' = —⟨O⟩ , (TPP)

X = X' = —⟨O⟩N , (TP$_y$P)

X = X' = —⟨O⟩N$^+$—CH$_3$, (TMP$_y$P)

X = X' = —⟨O⟩—SO$_3^-$, (TPPS$_4$)

X = X' = —⟨O⟩—COOH , (TPPC$_4$)

X = —⟨O⟩—SO$_3^-$, X' = —⟨O⟩ , (TPPS$_3$)

S = —H , (Deut-IX-DME)

S = —C$_2$H$_5$, (Meso-IX-DME)

S = —CH=CH$_2$, (Proto-IX-DME)

S = —SO$_3^-$, (Deut-IX-DSA)

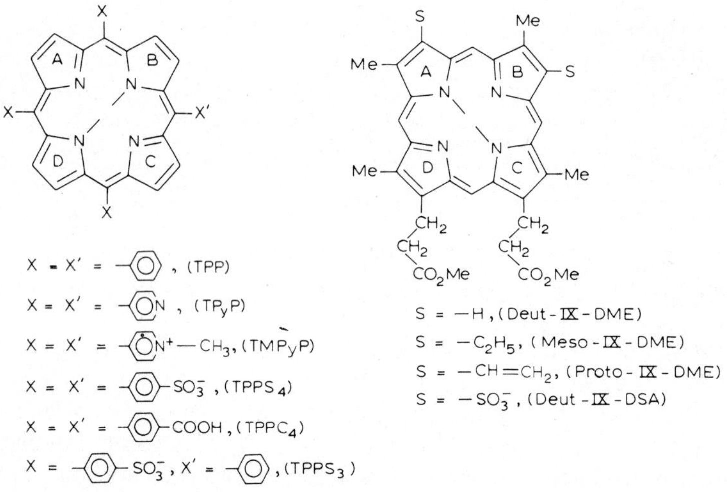

Fig. 2. Types of porphyrins that have been most widely used in kinetic studies.

6.2. Acid-base properties

A knowledge of the relative basicity of metal free porphyrins towards protons is important in understanding electronic effects on metal-porphyrin reactivity. The free base porphyrin[4] (PH_2) can add two protons to its imine type nitrogen atoms to form mono- (PH_3^+) or di-cations (PH_4^{2+}), or lose two pyrrole type protons to produce the mono- (PH^-) or di-anions (P^{2-}). The proton dissociation equilibria are described by pK values as follows:

$$PH_4^{2+} \underset{H^+}{\overset{pK_4}{\rightleftharpoons}} PH_3^+ \underset{H^+}{\overset{pK_3}{\rightleftharpoons}} PH_2 \underset{H^+}{\overset{pK_2}{\rightleftharpoons}} PH^- \underset{H^+}{\overset{pK_1}{\rightleftharpoons}} P^{2-} \tag{1}$$

The imine-pyrrole proton distinction is a formalism since ^{13}C and ^1H NMR work has demonstrated[17,18] the rapid N—H tautomerism in PH_2. The NMR results were more in line[17] with the porphyrin being represented as an 18 membered conjugated ring with two isolated peripheral pyrrole positions rather than as a sixteen membered inner ring with four such positions, favored by crystal structure data[19].

The protonation constants are usually determined from the changes in the characteristic absorption spectra of the porphyrin species upon titration with acids or bases[5], and problems arise when only the Soret bands are monitored. For example $H_2(TPyP)$[20] and $H_2(TMPyP)$[15] apparently had a single isosbestic point in the Soret, which indicated that only PH_2 and PH_4^{2+} were present in appreciable concentrations. An examination of the corresponding visible region, however, clearly showed the lack of pH independent isosbestic points, and the monocation was demonstrated[21]. In the Soret region, it is always difficult to distinguish between a single isosbestic point, and a group of curves that overlap over a finite wavelength range. The mathematical analysis of such Soret data is often misleading, since any nonintegral n (number of protons or ligands added) values obtained may be easily attributed to specific artifacts rather than the presence of additional species. Therefore it is best to use the visible region, with its multiplicity of bands, for equilibrium calculations. The relative concentrations of the acid forms of porphyrins are strongly dependent on the ionic strength and temperature[21]. Most porphyrins are dimers or polymers in solution, with formation constants large enough such that in the usual spectrophotometric range (10^{-4} to 10^{-6} M), appreciable association occurs[12]. Thus, a fair amount of work is needed to establish the nature of the porphyrin types present.

Detergents[22-26] have been used to monomerize and bring into aqueous solution the otherwise insoluble porphyrin esters. With anionic (sodium lauryl sulfate, SLS) and neutral (Tween) detergents, both pK_3 and pK_4 can be obtained. Cationic detergents (cetyltrimethylammonium bromide CTAB) destabilize PH_3^+, and only PH_2 and PH_4^{2+} are observed. The results found by several groups are shown in Table 1.

Monomerization of the porphyrin occurs above a critical micelle level[27], and the apparent pK values change somewhat with detergent concentration. The pK is a function both of the intrinsic porphyrin basicity, and the ratio of the partition coefficients of the various species between the aqueous and

TABLE 1

Basicity of 2,4-substituted deuteroporphyrin dimethyl esters in detergents

Substituents	pK_3 [a] ± 0.1	pK_3 [b] ± 0.1	pK_4 [c] ± 0.2	pK_3+pK_4 [d] ± 0.2
2,4-C_2H_5	5.8	5.85	2.1	2.1
2,4-$(CH_2)_2$—CO_2CH_3		5.58	1.80	2.24
2,4-H	5.5	5.5		
2,4-(2'-Ethoxycarbonyl cyclopropyl)	4.8			
2,4-CH=CH$_2$	4.8	4.80	1.84	<0.5
2,4-C(CH$_3$)=NOH	4.5	4.5		
2-CH=NOH, 4-CH=CH$_2$		4.45		
2,4-CH=NOH	4.3	4.3		
2 (and 4) H, COC$_2$H$_5$	4.2	4.2		
2 (and 4) H, —CHO	3.8	3.8		
2-CHO, 4-CH=CH$_2$		3.75		
2,4-COCH$_3$	3.3	3.35		
2,4-COC$_2$H$_5$	3.2	3.2		
α (and β) H, —NO$_2$	3.2	3.20		
2,4-CO$_2$CH$_3$	3.0	~3.0		
2,4-CHO		~3.0		
2,4-Br	3.0	~3.0		
2 (and 4)-CH=CH$_2$, —CN		~3.0		
2,4-CN	<3.0			
α,β-NO$_2$	≪3.0			

[a] 25°C in 2.5% SLS, Ref. 26.
[b] 25°C in 2.5% SLS, Ref. 5.
[c] 20°C in 2.5% SLS, Ref. 25.
[d] 20°C in CTAB, Ref. 25 only PH$_2$ and PH$_4^{2+}$ are observed.

micelle phase[22]. The hydrocarbon porphyrin region may be solubilized by the detergent, with the central porphyrin nitrogen atoms free to equilibrate with the aqueous environment[28]. While no claim can be made for absolute pK values, the relative basicities form a self-consistent series. The fact that only PH$_2$ and PH$_4^{2+}$ are found in CHCl$_3$—HCl, whereas all three forms are present[29] in EtOH—HCl, shows the complicated solvent dependence of the acid types.

pK_3 is more sensitive to substituent effects than pK_4, and Table 1 shows that electron-donating substituents on the porphyrin periphery increase the basicity of the free base porphyrin (higher pK_3) towards proton acceptance. A linear Hammett relationship was found between $\log(K_3/K_3$ [H$_2$(Deut-IX-DME)]) versus $\Sigma \sigma_m$ for the 2,4-substituents (s) in the H$_2$(S-Deut-IX-DME) series[30]. The lack of a good correlation with σ_p or σ_p^-, and the high ρ of -3.50 indicates that protonation is very sensitive to substituent effects, and that the mode of electronic transmission is inductive in character, rather than a direct resonance interaction between the central nitrogen atoms and peripheral substituents. No relationship[25] was found between the basicity of the porphyrin, and the positions or intensities of the porphyrin absorption bands[26].

The insolubility of many porphyrins in aqueous solution has led to basicity measurements in nonaqueous solvents[5]. In toluene—acetic acid[31], the volume of HOAc needed to produce 1 : 1 mixtures of PH_2 and PH_3^+ were 56 ml [$H_2(OEP)$], 54 ml [$H_2(Etio-I)$], 0.9 ml [$H_2(NMeOEP)$], and 1.1 ml for $H_2[(N-(3-Br-n-propyl)OEP)]$. No evidence for the potential[5] PH_5^{3+} or PH_6^{4+} types has been reported.

The majority of porphyrins are in the PH_2 form up to pH 14. One exception is the tetrapositively charged $H_2(TMPyP)$, with $pK_2 = 12.9$. This compound and its metal complexes slowly decompose in base[15,32]. Dimethylsulfoxide increases the basicity of aqueous solutions of NaOH, and by this method[33,34] the N-alkyl octaethylporphyrin derivatives, which initially have a single central proton, are converted into mono-anions with a relative dissociation constant K_2' of 54. In contrast, both central protons of $H_2(OEP)$ and $H_2(Etio-I)$ dissociate, giving $K_1', K_2' = 870$ and 1259, respectively. No evidence of the mono-anion is found. Earlier estimates[35] of $pK_A = 14-15$ for $H(NMeEtio)$ were probably orders of magnitude low[33].

Table 2 shows the results of basicity measurements on water soluble porphyrins and miscellaneous species[36−40]. The tetrapositively charged free base form of $H_2(TMPyP)$ (4+) is a weaker base[21] than the negative $H_2(TPPS_4)$ (4−)[37], $H_2(TPPS_3)$ (3−) or $H_2(TPPC_4)$ (4−) porphyrins[12]. The latter three compounds have an apparent $(pK_3 + pK_4) \simeq 5$, with nonintegral n values and the pK depends on wavelength. These porphyrins, along with $H_2(EnProto-IX-DME)$[38] and $H_2(Deut-IX-DSA)$[37] are definitely dimeric in moderately concentrated solutions, and further work will probably reveal distinct pK_3 and pK_4 values for both the monomer and dimer forms[38b]. Such dimerization of $H_2(Deut-IX-DSA)$ could account for certain controversies in the literature[39,40]. While the peripheral field effects cannot as yet be satisfactorily separated from the inductive and resonance contributions of these substituents[36], it appears that positively charged substituents decrease, while negative substituents increase the apparent basicity of the free base porphyrin towards protons. The proton reaction is favored if the product has a smaller net positive charge.

With the exception of $H_2(TPP)$[41], N-methyl derivatives are stronger bases by 3—4 pK units than non-N-alkylated species, and they are also stronger free base acids. In fact, $trans-N_A N_B$-dimethyl-octaethylporphyrin is hard to obtain in an unprotonated form[42]. Such results arise because the bulky central N—Me groups force the porphyrin nucleus to distort from planarity. Consequently several of the central nitrogen atoms have more sp^3 (more basic) than sp^2 character[42].

Meso-substituted porphyrins have a somewhat smaller difference between

References, p. 271

TABLE 2

Basicity of water soluble and related porphyrins

Type (charge) [a]	pK_3 ±0.1	pK_4 ±0.1	ΔpK	μ, Temp.	Ref.
H_2(TMPyP)(4+)	1.5	0.4	1.1	0.2, 22°C	21
	2.0	0.7	1.3	0.7	
	2.5	1.0	1.5	2.0	
	1.5	0.6	0.9		
H_2(EnProto-IX-DME)(7+)	5.0	2.8	2.2	0.1, 28°C	38
H_2(Copro-III)(−4)	7.2	4.2	3.0	0.005—0.1, 20°C	36
H(NMeCopro-III)(−4)	11.3	0.7	10.6		36
H(NMeCopro-III-TME)(O)	8.3	0.7	7.6		36
H_2(Deut-IX-DSA)(−2)	4.7	0.3	4.3		36
H_2(TPP)(O)	4.4	3.9	0.5	Nitrobenzene-HClO$_4$	41
H(NMeTPP)(O)	5.6	3.9	1.6		41

[a] Sum of the formal charges on the porphyrin periphery.

pK_3 and pK_4 (~1.5 units) than found in the peripherally substituted Deu-tero-Copro-Etio series (~2—3 units). Models indicate that three or four pro-tons cannot occupy the central cavity without strong Van der Waals and coulombic repulsions. Crystal structures show that the di-protonated forms of H_2(TPP)[43], H_2(TPyP)[43] and the monocation of H_2(OEP)[44] are rather distorted porphyrins. One theory[43] is that upon adding a proton to the relatively planar PH_2 (with its lone pairs directed inwards), the non-planar PH_3^+ that forms has its pyrrole rings tilted alternately upwards and down-wards with respect to the mean porphyrin plane. Such deformation facili-tates protonation of the remaining lone pair, thus limiting the existence of the mono-cation. In the *meso*-substituted series, this deformation allows the phenyl rings to rotate more nearly coplanar with the porphyrin ring facili-tating resonance interaction, and delocalizing the added charge in PH_4^{2+}. Consequently PH_3^+ may be more stable in the peripherally-substituted series. The proton exchange reaction between PH_2 and PH_4^{2+} with H_2(TPP) is much slower than for H_2(Copro-I-TME)[18,45]. Calculations indicate that the phenyl rings are probably 19° more coplanar [40° more for M(P)] with the porphyrin in solution as opposed to the solid state[46]. PMR results suggest that metallo-TPP is more flexible than metallo-Etio compounds[47].

Di-cations and mono-cations are strongly associated through hydrogen bonding with anions in non-aqueous media. The strength of the bond depends on the nucleophilicity of the anion[48].

6.3. Isotopic exchange of the central metal ion

Isotopic exchange reactions between coordinated and free metal ions (M) (eq. 2) have received some attention.

$$M_A + M_A^*(P) \rightleftharpoons M_A^* + M_A(P) \tag{2}$$

Under various conditions, negligible exchange was found in the following systems: Fe^{3+} with $Fe^{3+}(TPP)$[49], $(Proto-IX)$[49], or $(pheophytin)$[49]; Co^{2+} with $Co(Meso-IX-DME)$[50], $(phthalocyanine)$[51], $(tetrasulfonated\ phthalo-cyanine)$[51] or vitamin B_{12}[52]; Cu^{2+} with $Cu(pheophytin)$[49], Mg with chlo-rophyll[53,54] a or b, Sn with a $Sn(porphyrin)$[55]; or Zn with $Zn(TPP)$[56], $(phthalocyanine)$[57] or $(TMPyP)$[56]. For example no exchange was found for $Zn/^{65}Zn(TPP)$ in refluxing pyridine $(114°C)$ or dimethylformamide $(153°C)$, or in the $Zn/^{65}Zn(TMPyP)$ system in water $(pH = 6, 56°C)$ for periods of up to three days[56], with ratios of M_A to $M_A^*(P)$ of 10^4 to 1. This inertness to isotopic exchange is a feature of the fused ring systems found in porphyrins, phthalocyanines, pheophytins, and corrins. Open-chain ligands with four coordinate metals are much more labile[57]. In general, the breaking of four M—P bonds and concurrent formation of four new bonds in such fused systems is not compensated by a large net free energy change nor aided by relatively accessible vacant coordination positions for the entering metal ion.

In pyridine, the $Na^+/Na_2(TPP)$ exchange at room temperature was complete within the time of separation, approx. 6 min[55]. The weakness of the ionic sodium—porphyrin bonds is indicated by their rapid hydrolysis with traces of water. Under more vigorous conditions or with longer reaction times, isotopic exchange can no doubt be demonstrated. Thus the $Fe^{2+}/Fe^{3+}(Deut-IX)$ exchange occurs in melted resorcinol[5].

6.4. Reactions of metalloporphyrins with free base porphyrins

The reaction of $M(P_A)$ with a different free base $P_B H_2$ (eq. 3) could provide relative metalloporphyrin stability data.

$$M(P_A) + P_B H_2 \rightleftharpoons M(P_B) + P_A H_2 \tag{3}$$

If the equilibrium favors $M(P_B)$, it would be more 'stable' than $M(P_A)$ under the experimental conditions. Stability conclusions cannot be reached if the reaction is not observed to occur, since the attainment of equilibrium could be a slow process. There is no necessary connection between the thermo-dynamic position of the equilibrium and the kinetically determined time it takes to reach such a position.

References, p. 271

H_2(Etio-II) displaced Mg from Mg(Etiochlorin) in boiling acetic acid[58]. In a competition experiment Mg(Etiochlorin) formed first, and was replaced in longer times by (Etio). Since chlorins are less basic than porphyrins; the conclusion was that metalloporphyrin stability parallels porphyrin basicity.

In phenol[59] at $100°C$, Mg^{2+} will partially metalate PH_2, and the solvent will partially demetalate Mg(P), i.e., Mg(P) + $2C_6H_5OH$ = PH_2 + $Mg(OC_6H_5)_2$. The relative dissociation constants for Mg(Etio) and Mg(Etiochlorin) were 17.4 and 355, respectively. The ratio of these dissociation constants is the equilibrium constant for the $P_B H_2/M(P_A)$ reaction (eq. 3). Thus [Mg(Etio)/Mg(Etiochlorin)] = 20.4, in agreement with the results above.

6.5. Electrophilic substitution reactions

In contrast to the isotopic exchange process, electrophilic substitution reactions of one metal ion for another coordinated to a porphyrin (eq. 4) involve a larger net free energy change.

$$M_A + M_B(P) \rightleftharpoons M_A(P) + M_B \qquad (4)$$

Qualitative observations and detailed mechanistic work has been done on such systems[60-64]. M_A/M_B(TPP) reactions in pyridine show no equilibrium behavior[55]. Either the reactions went completely or not at all under the

TABLE 3

Rate data for M*—M(P) substitution reactions

Reaction	Rate law	Rate constants [a]	Half life [b]	Solvent, Temp. (°C)	Ref.
Zn/Cd(TPP)	k(Zn)(CdP)	1.9×10^{-2}	53 sec	Py, 25	62
Zn/Hg(TPP)	k(Zn)(HgP)	27×10^{-2}	4 sec	Py, 25	62
Zn/Pb(TPP)	k(Zn)^2(PbP)	4×10^{-4}	1 h	Py, 44	62
Cu/Zn(TPP) [c]	k_1(Cu)(ZnP)/- $(1+k_2$(Zn))	k_1 = 1.1×10^{-4}	2.5 h	Py, 114	62
Cu(NO$_3$)$_2$/Zn(TPPS$_4$) [d]	k(Cu)(ZnP)	5×10^{-3}	3.3 min	H_2O, 30	63
Cu(NO$_3$)$_2$/Zn(TMPyP)	k(Cu)(ZnP) [e]	2.5×10^{-3}	6.7 min	H_2O, 49	56

[a] Rate constants in appropriate $M^{-x} sec^{-1}$ units, acetate salts unless noted.
[b] Half life calculated for 1 M concentration of each reactant.
[c] k_2 term probably an artifact, and is not included in $t_{1/2}$ calculation.
[d] (k_{H_2O}/k_{D_2O}) = 1, ΔH^* = 14.3 cal/mol, ΔS^* = −21 eu, μ = 0.5.
[e] Probably k(Cu)(ZnP)(NO$_3^-$), E_a^* = 15.2 kcal/mol, μ = 2.0.

stated conditions. Assuming that replacement occurs only if a more stable metalloporphyrin forms, the following room temperature sequence was found: [Cu,Zn] > [Hg > Pd] > [Li > Na > K]. In the later two categories, the smaller ion gives the more stable metalloporphyrin. This order was correlated with bond strength. Kinetic studies have suggested possible mechanisms for the substitution process, and rate data are listed in Table 3.

The Zn/Cd(TPP) and Zn/Hg(TPP) reactions in pyridine[62] at $25°$C were first order in each reactant, with Hg^{2+} being displaced fourteen times faster than Cd^{2+}. The absorption spectra of Zn(TPP) and Hg(TPP) cross at seven wavelengths in the visible region, and seven excellent isosbestic points were found as the Zn/Hg(TPP) substitution proceeds. With 10^{-2} M Zn the half-life was 4 min while the Zn/H_2(TPP) reaction took several hours. The addition of 2 M H_2O (to react with any P^{2-}) did not change the rate. Such evidence argues strongly against the free base being an intermediate in electrophilic substitution.

Zn/Pb(TPP) in pyridine was first order in porphyrin and second order in zinc[62]. With 5×10^{-2} M Zn at room temperature, $\sim 10\%$ Zn(TPP) was produced in 17 min. By the addition of 10^{-2} M $Hg(OAc)_2$, complete Zn(TPP) formation occurs within this time. The Hg catalysis of the Zn/Pb(TPP) reaction was explained by the following sequence:

$$Hg(II) + Pb(TPP) \rightarrow Hg(TPP) + Pb(II) \tag{5}$$

$$Zn(II) + Hg(TPP) \rightarrow Zn(TPP) + Hg(II) \tag{6}$$

The order for Zn/M(TPP) is Hg > Cd > [Cu, Zn] where the weaker the M—P bond, the faster the rate. A relationship between porphyrin basicity and rate was shown in the Cu/Zn(S-Deut-IX-DME) series in refluxing pyridine. The displacement rates increased in the order of decreasing pK_3 values[56], H_2(Br-Deut-IX-DME) > H_2(Proto-IX-DME) > H_2(Meso-IX-DME) > H_2-(Etio). The more basic the porphyrin, the more firmly the coordinated zinc is held, and the slower it is to be replaced.

In aqueous solution, the Cu/Zn(P) rates are much faster than in pyridine, due to the greater solvating power of H_2O and the weaker M—H_2O versus M—N bonding. For Cu/Zn(TMPyP)[56] and Cu/Zn(TPPS$_4$)[63], the rates were first order in each reactant, and no inhibition by added zinc was noted. The Cu/Zn(TMPyP) rate law is probably first order in NO_3^- (overall third order), which is usual for H_2(TMPyP). No Zn/Cu(TMPyP) reaction occurred, nor was solvolysis found in the absence of Cu. High concentrations of Co^{2+} and Ni^{2+} did not replace Zn for periods of up to 15 h. With Cu/Zn(TPPS$_4$), $(k_{H_2O}/k_{D_2O}) = 1$. The Cu/Zn(EnProto-IX-DME) reaction rate[64] was first order in Zn(EnProto-IX-DME), and became independent of copper at high copper concentrations. The rate decreased with an increase in pH, and no analytical rate law could be fitted to the data.

The rate laws for most of the Cu/Zn(P) reactions define an activated

References, p. 271

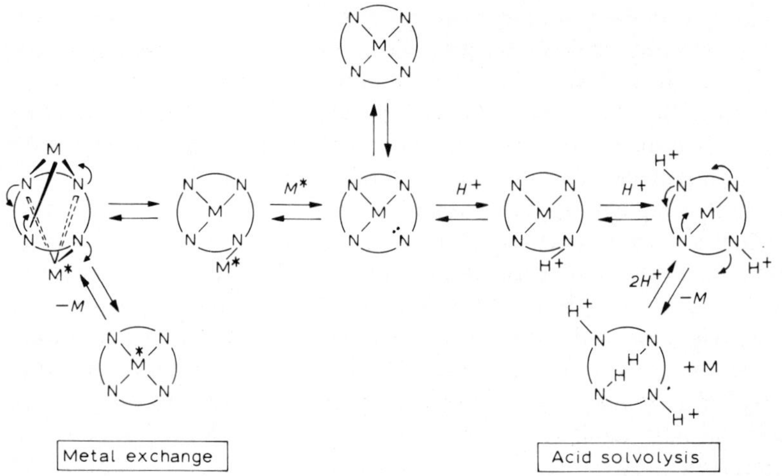

Fig. 3. Representations of possible pathways for acid solvolysis and metal ion exchange reactions. Porphyrin deformation is indicated by the curved arrow directions.

complex of the composition [Cu—P—Zn]*. Free base intermediates do not occur, and the lack of zinc inhibition indicates that no zinc depleted forms are present before the rate determining step. Figure 3 shows a pathway consistent with the results. Without invoking attack of the entering metal ion on the porphyrin pi cloud or coordinated nitrogen atoms (which could occur), intermediates are considered which have one of the Zn—N bonds broken, to provide a lone pair for the entering metal ion. The crystal structure of μ-[meso-tetraphenylporphyrinato]-bis-[tricarbonylrhenium(I)] and the similar [$(CO)_3$Tc-P-Re$(CO)_3$] show that two metals on opposite sides of the porphyrin plane can each bond to three adjacent nitrogen atoms[65]. A similar structure may be an early intermediate in these reactions. The greater strength of the Cu—N bonds over those of Zn—N provide the driving force where bond breaking is concerted with bond making. Porphyrin deformation further aids this process, and the opposite pyrrole rings may tilt upwards (or downwards) to facilitate overlap between the lone pairs and the metal center.

The fact that Ni^{2+} and Co^{2+} do not displace Zn from Zn(TMPyP) as rapidly as does Cu indicates that the aquo ion water exchange rates are important. The relative rates[32] of Co^{2+} and Cu^{2+} reacting with H_2(TMPyP) are in the order 1 : 1060, and the aquo ion solvent exchange rates are 1 : 1250, respectively. Following the rapid formation of an outer sphere complex [$(H_2O)_3M(H_2O)_3$...(P)Zn(H_2O)], the rate should depend on how fast M can desolvate and simultaneously be in a position to bond with the porphyrin nitrogen atoms. Thus the more labile the metal, the weaker the P-departing metal bond and the stronger the incoming M—P bond, the faster the reaction. The rate determining step could be the breaking of the third Zn—N bond, which destroys the stabilizing chelate effect. Substitution reactions of metal ions between flexible open-chain ligands are more facile, because the ligand can partially unwrap from about the coordinated metal, providing lone pair positions for the entering metal to occupy[66]. The relative rigidity of the porphyrin makes such an associative process difficult.

6.6. Acid solvolysis reactions

Acid catalyzed solvolysis reactions involve the replacement of a coordinated metal ion by protons (eq. 7), and most studies have been done under conditions that drive the reaction to the diprotonated form.

$$M(P) + 4H^+ \rightleftharpoons PH_4^{2+} + M^{2+} \tag{7}$$

Metals can be placed in classes[5,67a] depending upon whether they are or are not demetalated by acids of various strengths (H_2SO_4 to H_2O). In general, trivalent metals (Fe^{3+}, Mn^{3+}) are more inert than divalent forms (Fe^{2+}, Mn^{2+}) and a typical stability order[5] is Pt(II) > Pd(II) > Ni(II) > Co(II) > Ag(II) > Cu(II) > Fe(II) > Zn(II) > Mg(II) > Cd(II) > Li_2 > Na_2 > Ba(II) > K_2 > $[Ag(I)]_2$. Similar results are found for M(TPP) and M(OEP) complexes in refluxing ethylenediamine, which are ranked by the number of equivalents of Li required to produce demetalation[68]. The metal ion order fairly well follows[69] the 'stability index' of a metal ion, S_i, where $S_i = (Z E_N / r_i)$. The quantity (Z/r_i) is the charge to radius ratio of the ion, and is related to ionic stabilization effects. E_N is the electronegativity of the cation, and is a covalent parameter. A full treatment of S_i is found in Chapter 5.

The acid labile[70] Mg, Co, Ag, Pb and Sn phthalocyanines, M(Pc), show only the azo-protonated free base $H_2(Pc)H^+$ spectra when dissolved in 16 M H_2SO_4, and their solvolysis rates are similar to the rate of hydrolysis breakdown of phthalocyanine $H_2(Pc)$ (Fig. 4) itself ($t_{1/2}$ = 5 h in 17.6 M H_2SO_4, 25°C). In contrast, Cu, Co, Ni, Ag^{2+}, Zn, VO^{2+} and Sn(IV) are rather inert, and the coordinated metals stabilize the ring towards acid decomposition. For $Cl_2Sn(Pc)H^+$, the solvolysis rate law is first order in metal complex and second order in H^+.

The kinetics of degradation[71] of the Co, Ni, Cu, Zn and Cd macrocycles shown in Fig. 4 were studied in H_2SO_4. The rates were first order in metal complex, and the different metal ion stability orders noted for each species were ascribed to differences in M—N bond strengths. Metallophthalocyanines are more stable than the macrocycles. In 2.6 M H_2SO_4 at 25°C, the half life for Cu(Pc)H$^+$ was 1.4×10^8 years, while for the copper complex of species II, $t_{1/2}$ = 9.5 min. The stability of the cobalt derivatives increased with the number of metal—nitrogen bonds, with relative $t_{1/2}$ values of 17 h for

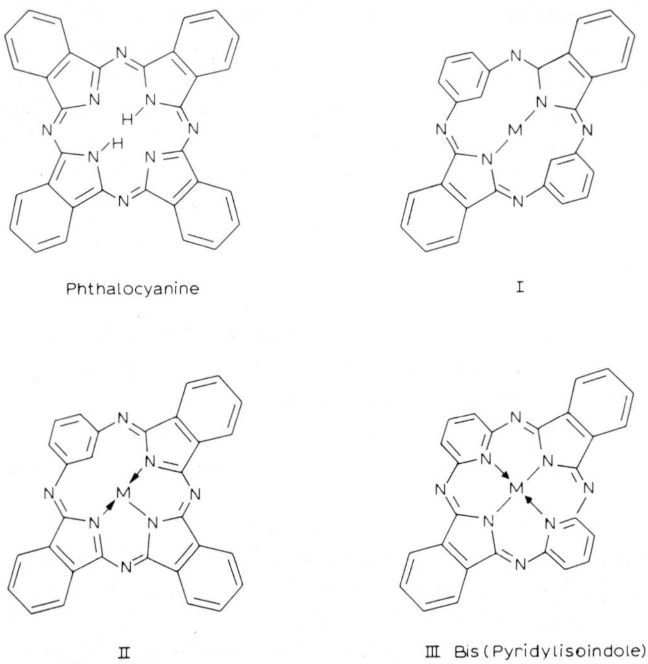

Phthalocyanine   I

II   III Bis(Pyridylisoindole)

Fig. 4. Macrocyclic ligands related to the porphyrin molecule.

four bonds (III), 57 min for three bonds (II), and 15 min for the two coordinate (I). The polyphthalocyanine field has been reviewed[72].

Mg(Deut-IX-DME) solvolysis[73] in alcohol—pyridine—water—$HClO_4$ mixtures at 42°C, showed a three term rate law with Py and $(H_2O)(Py)$ catalysis. The rate increased with the coordinating power of the solvent, EtOH (1): MeOH (260): H_2O (>8000), and rate data are given in Table 4. The pyridine and water catalysis could arise by these molecules occupying coordination positions on the Mg and thereby weakening its attachment to the porphyrin. It was noted that Zn is 0.2 Å further from the porphyrin plane when bonded to pyridine as opposed to water. Absolute rate theory was used to give the magnitudes and the probable sequence of the reaction events. Magnesium porphyrins are thermodynamically unstable in water[6 7b].

In MeOH—HCl, the Zn(Etio) solvolysis[74] showed $R = k(ZnP)(HCl)^3/(\rho + (HCl))$ while Zn(NMeEtio) had an $(HCl)^2$ numerator term. The more basic Zn(Etio) has a displacement half-life of 0.11 sec while the less basic Zn(Br-Deut-IX-DME) was 5000 times slower. A linear relationship between $\log k/k_{(Deut-IX-DME)}$ and pK_3 was demonstrated, indicating that the rates parallel porphyrin basicity, with bond making more important than bond breaking. The general mechanism is of the following form with S as solvent molecules or added bases:

$$Zn(P) + 2H^+ + S \underset{k_{-1}}{\overset{k_1 K_N}{\rightleftharpoons}} [Zn(P) \cdot 2H^+ \cdot S] \qquad (8)$$

$$[Zn(P) \cdot 2H^+ \cdot S] + H^+ \overset{k_2}{\rightarrow} Products \qquad (9)$$

The protonated $Zn(P)$ is a steady state intermediate formed in various pre-equilibrium steps ($k_1 K_N$). k_2 is rate limiting at low acidities giving rise to the $(H^+)^3$ term. $(H^+) > \rho$ at high acidities ($\rho = k_{-1}/k_2$) producing an $(H^+)^2$ dependence. As shown in Fig. 3, a proton may bond to a vacant N position on $Zn(P)$, causing porphyrin deformation and facilitating the attachment of two more protons. This tri-protonation breaks the chelate effect, making the zinc one-coordinate with respect to the porphyrin, and blocks the reformation of Zn—N bonds[74,75]. For $Zn(NMeEtio)$ (which may be initially three coordinate in zinc), only two protons are required to produce the one-coordinate species.

In $HOAc$—H_2SO_4, $R = k(CuP)^2(H^+)^4$ was found for $Cu(Etio)$ solvolysis[61]. The pre-equilibrium formation of $[Cu(P) \cdot 2H^+]$ followed by two of these types reacting together in the rate determining step was the suggested mechanism. Alternatively, $Cu(Etio)$ could be slightly dimerized, and the dimer could react in steps with four protons. For $Mn(II)(Hemato-IX)$ or the dimethyl ester, the rate law was first order in porphyrin[76], first order in (H^+) at the intermediate pH values, and higher than first order outside of these limits ($4.6 > pH > 6$). The half-life was 19.2 hours at pH 7, in 45% EtOH, 25°C.

$Zn(TPPS_4)$ in aqueous solution[63] follows $R = k(ZnP)(H^+)^2$. The rate increased somewhat with ionic strength and $(k_{H_2O}/k_{D_2O}) = 0.23$. While the ground state energy of $Zn(TPPS_4)$ would be fairly independent of solvent, the transition state will have protons partially attached to the central nitrogen atoms. This would result in a lower energy activated complex for N—D than N—H, and thus a faster rate. The acid solvolysis of $[(en)_2CoF_2]$ also shows this behavior[77]. With $Zn(TMPyP)$ in water[15,75], the relative solvolysis rates in 1 M acid were 675 (HCl), 186 (HBr), 85 (HNO_3), 8 (H_2SO_4) and 1 (HI). The rate law was $R = k(ZnP)(H^+)^2(X^-)^n$ with $n = 2$ for the hydrogen halides and $n = 1$ for NO_3^-. The anion term may involve both ion-pairing with this tetrapositive porphyrin, and some Zn—X interaction. The $(H^+)^2$ term is common in aqueous media for porphyrin and phthalocyanines, and could be a limiting form of $[(H^+)^3/\rho + (H^+)]$ noted in eq. 9 under less acidic conditions.

In summary, N—H bond making is important as shown by the D_2O isotope effects and rate studies of Zn porphyrins as a function of porphyrin basicity. Bond breaking is also involved, as a rationalization of the relative solvolysis orders of different metal ions. The overall process can be viewed as a stepwise displacement reaction, where metal ligation and porphyrin defor-

References, p. 271

TABLE 4

Rate data for acid catalyzed solvolysis reactions

Complex	Rate law	Rate constants [a]	Solvent, Temp. (°C)	Activation [b] parameters	Ref.
Zn(Etio)	$k(ZnP)(H^+)^3/[\rho + (H^+)]$	$k = 2 \times 10^5$	MeOH–HCl, 25		74
Zn(NMeEtio)	$k(ZnP)(H^+)^2/[\rho + (H^+)]$	$\rho = 6.6 \times 10^{-3}$ $k = 2.5 \times 10^2$ $\rho = 2.5 \times 10^{-1}$	MeOH–HCl, 25		74
Mg(Deut-IX-DME)	$k(MgP)(H^+)^3/[\rho + (H^+)]$	$k = 3.1 \times 10^5$			
		$\rho = 5 \times 10^{-5}$	MeOH–HClO$_4$, 42	$k_1(\Delta H^* = 2.2,$ $\Delta S^* = -27)$ $k_1/\rho_1(\Delta H^* = -3.7,$ $\Delta S^* = 0)$	73
	$k(MgP)(Py)(H^+)^3/[\rho + (H^+)]$	$k = 1.2 \times 10^3$ $\rho = 4 \times 10^{-5}$ $k = 7.2 \times 10^8$	EtOH–HClO$_4$, 42		73
	$k(MgP)(Py)(H_2O)(H^+)^2$	$\rho = 5 \times 10^{-6}$ 2.7×10^9	MeOH–Py–HClO$_4$, 42 MeOH–Py–HClO$_4$–H$_2$O, 42	$k_2(\Delta H^* \sim 5, \Delta S^* \sim -1)$	73 73
Zn(TMPyP)	$k(ZnP)(H^+)^2(X^-)^2$	$Cl^-\,(6.8 \times 10^{-2})$ $Br^-\,(1.9 \times 10^{-2})$ $I^-\,(\sim 1 \times 10^{-4})$	H$_2$O, –HX, 25		75 15
Zn(TMPyP)	$k(ZnP)(H^+)^2(X^-)$	$NO_3^-\,(8.5 \times 10^{-3})$	H$_2$O, –HNO$_3$, 25		75
Zn(TPPS$_4$) [c]	$k(ZnP)(H^+)^2$	10.9	H$_2$O, $\mu = 0.1$, 30	$\Delta H^* = 8.8, \Delta S^* = -24$	63

[a] Rate constants in appropriate M^{-x} sec^{-1} units.

[b] Unit of ΔH^*, kcal/mol, ΔS^*, eu.

[c] $(k_{H_2O}/k_{D_2O}) = 0.23$ (30°C).

mation act in a concerted fashion, without the necessity of high energy $[Zn^{2+} \cdots P^{2-}]$ intermediates. Models for this process are found with H_2(Hemato-IX) and $PtCl_4$, in which an intermediate of the composition $[Cl_2Pt \cdot \cdot PH_2]$ has been isolated[78]. In addition, the [monohydrogen-(Meso-IX-DME)—(tricarbonylrhenium)(I)] complex may have a $Re(CO)_3$ bonded to 3 nitrogen atoms on one side of the porphyrin plane and a proton on the other[79]. Fluxional behavior is found, indicating movement of the metal carbonyl and proton around the ring[80].

The acid demetalation[81] of Hg(TPP) in methylene chloride proceeds through a bimetallic intermediate: $2Hg(P) + 2HCl \rightarrow PH_2 + [Cl-Hg-P-Hg-Cl]$ followed by further reaction with HCl to produce PH_2 and $HgCl_2$. No evidence is found for binuclear species in basic solvents. An interesting parallel can be drawn when considering the disproportionation of the monohydrogen technetium complex[82], which reacts as $2[H-P-Tc(CO)_3] \rightarrow PH_2 + [(CO)_3Tc-P-Tc(CO)_3]$. Trinuclear double sandwich complexes [X—Hg—P—Hg—P—Hg—X] are known in the (Copro) and (Etio) series[83].

In a novel approach, Fe(III)P was reduced to acid labile Fe(II)P forms by Hg^0 in HOAc containing tetramethylammonium chloride[84]. An equilibrium constant of the form $K_0 = (PH_4^{2+})(FeCl_2)(Hg_2Cl_2)^{\frac{1}{2}}/([Fe(III)PCl][Cl^-]^2 \cdot [Hg]^0[H^+]^4)$ could be measured. The equilibrium constants for (Copro-III-TME) and (Meso-IX-DME) were 3.6×10^{-3} and 150×10^{-3}, respectively. Thus the weaker base (Copro-III-TME) was less demetalated under the reaction conditions.

6.7. Metal ion incorporation reactions

A number of kinetic studies have been done to elucidate the mechanisms of metal ion incorporation into porphyrin molecules (eq. 10). As will be shown, the difficulties have been both technical and conceptual.

$$M^{2+} + PH_2 \underset{k_r}{\overset{k_f}{\rightleftharpoons}} M(P) + 2H^+ \qquad\qquad K_e \qquad\qquad\qquad (10)$$

6.7.1. Equilibrium constants

Several equilibrium constants, K_e, for eq. (10) have been measured. $K_e = 6.0 \times 10^{-3}$ M at 80°C and 2.5×10^{-3} M at 60°C for Zn/H_2(Proto-IX-DME) in 0.25% $CTAB^{25}$. The conventional formation constant, $K_f = (MP)/(M)(P^{2-})$ is equal to K_e/K_1K_2, where K_1 and K_2 are the acid ionization constants of PH_2; $K_1K_2 \geqslant 10^{-32}$ for H_2(Etio). Thus for Zn(Proto-IX-DME), $\log K_f \geqslant 29$, an indication of the high metalloporphyrin stability. In water at 25°C with H_2(Hemato-IX)[85a] and $Zn(OAc)_2$, $K_e \simeq 1.3 \times 10^{-7}$, and $\log K_f \geqslant 25$. In HOAc at constant pH, the Cu/PH_4^{2+} reaction had $K_f = 1.8 \times$

10^5 $[H^+]^4$. K_e is equal to the (k_f/k_r) in eq. (10). With Zn/H_2 (EnProto-IX-DME)[16] at pH 6.23, k_f = 14.2 M^{-1} min^{-1} and the reverse rate was 1.4 × 10^{-7} min^{-1}. Assuming a second order term in (H^+), $K_e \simeq 3 \times 10^{-8}$, and log $K_f \geqslant 24$. Macrocycles may be less solvated than open-chain ligands, and this enthalpy effect favors the reaction[85b]

6.7.2. Sitting-atop complexes

Several groups have put forward evidence for intermediates in the incorporation process termed 'sitting-atop' complexes (SAT). For example[86] H_2 (Proto-IX-DME) and acetone solutions of Cu or Ni react to directly form a M(P) at ambient temperatures whereas Fe^{2+}, Fe^{3+} and Co^{2+} first produce a spectrally distinct species, the SAT. Upon heating, the SAT is transformed into the M(P). The visible bands of the Fe^{3+}-SAT resembled PH_4^{2+} and an initial suggestion was that the SAT was, in fact, a dication[4]. Depending on MeOH : $CHCl_3$ composition, equilibrium constants were calculated for 1 : 1 $[FeCl_3-H_2(Proto-IX-DME)]SAT$ or 2 : 2 $[Fe_2Cl_6 ... \{H_2(Proto-IX-DME)\}_2]$-SAT complexes, where the N—H bonds were intact. It was later proposed[87] that EtOH in chloroform solvolyzes Fe_2Cl_6, forming H^+, Cl^-, and $[Fe_2Cl_5-(OEt)_x]^-$. The released (H^+) will protonate PH_2 to PH_4^{2+}, which is strongly associated with the negative iron anion. A complex grown under nonaqueous siting atop conditions was shown by X-ray diffraction[43] to be $[PH_4^{2+} \cdots Cl^- \cdots (FeCl_4)^-]$. Thus the non-aqueous SAT complexes appear to be associated dications.

Further SAT evidence was found with H_2 (TPyP) in aqueous solution[88]. At pH = 2, spectrophotometric titrations of the red H_2 (TPyP) with almost any metal salt produced SAT green species, and equilibrium calculations, along with an apparent isosbestic point in the Soret indicated 1 : 1 complexation, with formation constants ranging from 1.5 M^{-1} for K^+ to 87 M^{-1} for Cu^{2+}. The incorporation kinetics were second order in metal, first order in PH_2, and inverse first order in (H^+). The demonstration[89] of a $[(Li^+)(Cu^{2+})]$ term indicated that the sitting-atop metal was not the ion that inserted. The interpretation was that the first metal formed a SAT, which deformed the porphyrin and facilitated proton dissociation (from the porphyrin or coordinated metal). The second metal incorporates from the distal side to produce the MP. This deformation hypothesis was used to explain the metal ion incorporation, substitution, and exchange facts known at that time[90].

With the fully water soluble H_2 (TMPyP)[15], however, the incorporation rate law was found to be first order in free base, metal, and anion[21,32]. The mono and di-cation forms were unreactive, and their apparent pK values depended strongly on ionic strength. Thus the green H_2 (TPyP)-SAT complexes[88] were mixtures of PH_4^{2+} and PH_3^+, produced by ionic strength changes at constant pH, transforming the red PH_2 into the green acid forms by a shift in pK values. The second order dependence on metal ion for

$H_2(TPyP)$ was actually first order in M^{2+} and first order in X^-. The inverse proton term arose from the fact that the unsuspected PH_3^+ form is unreactive. This story is not concluded since other workers[91] are finding complicated interactions between alkali and alkaline earth salts and the *meso*-tetrapyridyl-porphyrin types.

A number of porphyrin adducts with inert metal ions show SAT attributes. With $H_2(Hemato-IX)$[78], the ion pair $[PtCl_4^{2-}\cdots PH_4^{2+}]$ forms at $25°C$, and is converted to $[Cl_2Pt = PH_2]$ at $60°C$. Further heating at $80°C$ produces Pt(P). The use of such platinum complexes in chemotherapy has been discussed[92]. The $[H-TPP-Re(CO)_3]$ adducts[65] are SAT models. In the reaction of $H_2(TPP)$ or $H_2(OEP)$ with $[Rh(CO)_2Cl]_2$, an anionic porphyrin species bonded to $[(CO)_2-Rh-Cl-Rh-(CO)_2]^+$ forms[93]. Each Rh atom could be bonded to two adjacent nitrogen atoms on the same side of the ring. The initial product of the rhodium carbonyl dimer reaction with $H_2(NMeOEP)$ is a neutral N—Me porphyrin, bonded to $[(CO)_2-Rh-Cl_2-Rh-(CO)_2]$, where each Rh could be attached to one adjacent N-atom[94]. Upon heating, an oxidative addition reaction forms $P-Rh(III)-CH_3$, with the Rh—C bond the result of an alkyl group migration.

6.7.3. Non-aqueous, mixed solvent and detergent kinetics

The rate laws in acetic acid of Cu, Co and Mn incorporation into $H_2(TPyP)$[95] and $H_2(Hemato-IX)$[96,97] which were shown to be first order in metal and porphyrin were later found[98,99] to be one-half order in total Cu, Co and Ni, and between half and first order in Mn and Zn. The half order dependence could be due to a monomer-dimer equilibrium with the low concentration monomer as the reactant. An unexplained inverse proton dependence[95] found for $Cu/H_2(TPyP)$ in $HOAc-H_2O$ mixtures can now be attributed to the presence of the unreactive PH_3^+, which must dissociate into PH_2 before reaction. Studies of metal incorporation as a function of the mole fractions of $HOAc-H_2O$[95] or $HOAc-EtOH$[100] show pronounced rate maxima and minima, corresponding to different metal ion solvation forms and solvent properties. A method for the estimation of transition metal ions based on their reactivities with porphyrins has been proposed[101].

Incorporation studies in a common solvent have been done using a series of porphyrin types. With $Cu(OAc)_2$ in unbuffered EtOH, the rate law was first order in metal and porphyrin for a number of porphyrins, chlorins and pheophorbides[102]. A plot of ΔH^* versus ΔS^* is fairly linear (Fig. 5), indicating a common mechanism. The slope of such a graph[103] is the isokinetic temperature T_i, the absolute temperature at which all members of the series that conform to the line react at the same rate. A comparison of the properties of molecules when derived from kinetic measurements in the range of T_i must be made with caution. For the copper study[102] in EtOH, $T_i \simeq 25°C$, and since the reactions were studied in this range $(15-45°C)$, no

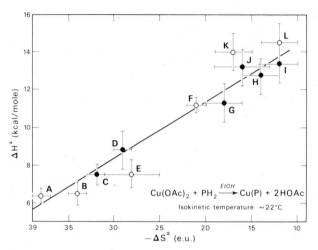

Fig. 5. A graph of $\Delta H*$ versus $\Delta S*$ for copper incorporation into porphyrins (solid dots) and related ligands (open circles) in ethanol. A. Rhodin-g_7 trimethyl ester; B. Chlorin-e_6 trimethylester; C. Meso-IX-DME; D. Proto-IX-DME; E. Pheophorbide-a; F. The chlorin of porphin; G. Etiochlorin-II; H. Etio-I; I. Etio-II; J. Porphin; K. Pheophytin-b; L. Pheophytin-a. See Ref. 102.

conclusions can be drawn as to how the peripheral substituents affect the porphyrins' reactivity.

A similar study[104a] with $CuCl_2$ and fifteen *ortho* and *para* substituted $H_2(TPP)$ ligands was done in dimethylformamide, a superior solvent for metal incorporation[105]. The rate laws were first order in metal and porphyrin down to 1 : 1 concentrations of the reactants, and the linear $\Delta H*—\Delta S*$ plot in Fig. 6 supports a common mechanism. $T_i \simeq 57°C$, which was in the same range as the measurements. The reactant in DMF was possibly $CuCl^+$, and the rates increased with the donicity of the solvent[104b]. The order was

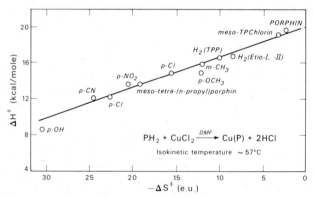

Fig. 6. A graph of $\Delta H*$ versus $\Delta S*$ for copper incorporation into porphyrins in DMF. See Ref. 104a.

Cu > Cd > Zn > Pd > Co > Ni ≫ Mg, and with the exception of Mg, ΔG^* at 70.2°C was in the narrow range of 20.4 to 23.2 kcal/mol.

Zn and Cd incorporation[106] into H_2(Etio) and H_2(NMeEtio) in DMF was first order in each reactant, with the N—Me rate 10^5 times that of H_2(Etio). The higher basicity of the former might favor an internal conjugate base mechanism[107]. The N—Me-porphyrin has one less proton to lose, which is more acidic than those in H_2(Etio), and the metal should sit well above the porphyrin plane. All of these factors contribute to an increased N—Me-porphyrin rate, as less porphyrin nuclear rearrangement need occur. The shifts in the Zn(Etio) absorption bands upon pyridine ligation are more pronounced than with Zn(NMeEtio)[108]. Such shifts may arise from interactions between the porphyrin pi cloud and the ligand (a stereoelectronic effect)[109], and the further the metal from the plane, the less the shift. The Cu complex of H(NMe TPP) dissociates upon dilution[41] while Cu(TPP) is stable. In this context, the N—Me group is lost more readily (Cu > Zn > Cd)[108] from M(NMeEtio) than from H_2(Etio) itself in refluxing pyridine. This is a reflection of Lewis acidity and a driving force towards planarity[110,111]. With N—Me Ni and Pt corroles, the alkyl group migrates to the peripheral positions[112].

The reactions of twelve porphyrins with $ZnCl_2$ were studied in pyridine—water solutions. H_2(Meso-IX-DME)[113] had the complicated form $R = (PH_2)[Zn]_T/(1 + 981[Cl]_T + 94[Zn]_T)$. The rate was half order in total Zn in the absence of added chloride. Irrespective of porphyrin basicity, the incorporation rates at 45°C were very similar[98], both for the H_2(p-X · TPP) and H_2(S-Deut-IX-DME) forms.

In 10% pyridine—water[114] at pH 8.2, the rates of Fe(II) incorporation into H_2(Hemato-IX) (2—), H_2(Copro-III) (4—) and H_2(Uro-III) (8—) increased with the number of carboxylic acid groups. Their respective activation energies, 23.7, 23.9 and 24.6 kcal/mol, were about the same, and the rate increase (0.22, 8.2 and 88 $M^{-1}sec^{-1}$) was an activation entropy effect. The greater the number of negative groups about the porphyrin, the more an $[Fe^{2+} \cdots PH_2]$ adduct would be stabilized. The idea was advanced that several ligands must be removed from the iron before metal incorporation, in order to explain the slow porphyrin rates as compared with other complexes.

The terminal step in heme biosynthesis may be the incorporation of Fe(II) into H_2(Proto-IX), and a number of groups have studied the enzyme (ferrochelatase, heme synthetase) responsible. Under optimum conditions[115], in the absence of cellular material, the Fe^{2+}/PH_2 reaction was just as fast, or faster than found with enzyme preparations. The relative amounts of Fe(P) formed under standard conditions were 200 nmoles H_2(Uro-III), 163 H_2(Copro-III), 46 H_2(Hemato-IX) and 19 for both H_2(Meso-IX) and H_2(Proto-IX). The rates appeared to be independent of porphyrin basicity and dependent on the number of negatively charged substituents, as noted above.

References, p. 271

In contrast, earlier work[5] on Cu/PH$_2$ in 2.5% SLS showed that the rate increases with porphyrin basicity, in the order H$_2$(Meso-IX-DME) > H$_2$(Proto-IX-DME) > H$_2$(DaDeut-IX-DME) as 7.0 : 2.5 : 1. At 30°C, Cu(Meso-IX-DME) forms 20,000 times faster in anionic 2.5% SLS than in the cationic detergent 0.25% CTAB. The rate change was ascribed[28,116,117] to an attraction between the anionic micelle containing the porphyrin and Cu^{2+}, and involved an E_A difference of 7 kcal/mol. Certain 1 : 1 complexes (not 1 : 2 or 1 : 3) like sodium diethyldithiocarbamate catalyzed metal ion incorporation by factors of up to 200, and these particular 1 : 1 copper chelates were effective anti-fungal and anti-bacterial agents, which might act by inhibiting heme synthesis[117].

H$_2$(Etio) was metalated in benzonitrile[118] by a series of Ni(II)-bis-1,3-diketones. Several ligands showed rates first order in metal and porphyrin, while others were zero order in metal. The reactions were inhibited by free ligand. In the former case, NiL$_2$ could react to form L$^-$ + [LNi $\cdots$ PH$_2$], which goes to product in a slow step. In the latter, either depolymerization, or a rapid and complete initial formation of [PH$_2$—NiL$_2$], followed by a subsequent loss of L$^-$ was suggested. Many CuL$_2$ reactions show marked induction periods[108]. The irradiation of H$_2$(Etio) with light in the 400 nm region[119] catalyzed the incorporation of CuL$_2$, but not Ni or ZnL$_2$. Following illumination, the reactions proceed in the dark with extents proportional to the initial light period. The mechanism involved a photoexcited PH$_2$ reducing Cu(II)L$_2$ to Cu(I)L, followed by reaction with ground state PH$_2$, and a terminal [Cu(I)PH] oxidation by L.

Without special conditions[120,121], Mg is difficult to insert into porphyrins. With H$_2$(Deut-IX-DME)[122] in buffered MeOH at 100°C, $R = k_1$ (PH$_2$)/(H$^+$) + k_2(Mg)(PH$_2$)(Py)/(H$^+$). k_1 could be the formation of (PH$^-$), which rearranges slowly prior to rapid Mg incorporation. The dissociation of a proton from the tertiary complex to form (Mg—Py—PH$^-$) could explain k_2. It was noted that the results could equally well be accounted for by the reaction BH $\rightleftharpoons$ B$^-$ + H$^+$ giving rise to the inverse proton term. B$^-$ would react with PH$_2$ in k_1 and with the tertiary complex in k_2. BH could be traces of H$_2$O, MeOH, etc. While the latter explanation might be favored because of the high pK_2 value normally exhibited by H$_2$(Deut-IX-DME), the former proposal cannot be excluded on kinetic grounds alone.

6.7.4. Aqueous solutions

The Cu/H$_2$(Deut-IX-DSA) reaction[123] was first order in total copper and free base. The inverse proton dependence was due to PH$_2$—PH$_3^+$, with the monocation unreactive. While pK_3 was in agreement from both kinetics and independent equilibrium studies, no account was taken of H$_2$(Deut-IX-DSA) dimerization[37] or the differential reactivities of copper complexes in the acetic acid buffer.

With Cu/H$_2$(TPPS$_4$) in acetic acid buffers[124], the rate was first order in

$[Cu]_T$, and PH_2, with Cu^{2+} and $Cu(OAc)^+$ showing equal reactivities. The observed $[(CuOAc^+)(PH_4^{2+})]$ term is better formulated as $[(CuOAc^+)(PH_3^+)]$. At high copper concentrations the rate was independent of copper, but it was never clear that precipitation was not the rate limiting factor[108]. An isotope effect (k_{H_2O}/k_{D_2O}) of 1.37 was found, and a value of 2.22 was noted[63] with $Zn/H_2(TPPS_4)$. This system had a rate law first order in each reactant, with a specific rate constant of 4.8×10^{-1} $M^{-1}sec^{-1}$ at $30°C$. The rate was a composite form, with $Zn(OH)^+$ fifty times more reactive[32] than Zn^{2+}. While $H_2(TPPS_4)$ is dimerized[37], reactions at high $(10^{-5} M)$[63] and low $(10^{-7} M)$[32] concentrations showed no dependence on total porphyrin. In general dimers are five to ten times less reactive towards cations than monomers[125].

For $Cu/H_2(EnProto-IX-DME)$ in acetate buffers[64], the rate was first order in porphyrin, and reached a limiting value at high copper concentrations. The rate was different for Cu^{2+} and $CuOAc^+$, and increased with an increase in pH. The $Zn/H_2(EnProto-IX-DME)$ reaction[16] was first order in each reactant at constant pH, and an additional term $k(Zn)(PH_2)(B)$ was found where B was a nitrogenous base such as pyridine or imidazole. The base catalysis was explained by $Zn(Py)_x^{2+}$ species as reactants[32].

Tetrasulfonated phthalocyanines[126] are strongly dimerized in aqueous solution, and the reaction goes through the monomer. This monomer reacts with a base (B^-) to form a low concentration monoanion ($pK_2 \sim 9.2$), which complexes with the metal in a slow step. The reactivity order was Cu $\gg$ Zn > Co > Mn > Ni.

With NO_3^- anions[32], the $M/H_2(TMPyP)$ reactions were first order in metal, free base porphyrin and NO_3^-. PH_3^+ and PH_4^{2+} were unreactive. The ratio of the observed rate constants (Cu, Zn, Mn, Co, Ni as 46,000 : 1000 : 50 : 43 : 1)[45] to the corresponding water exchange rates for the aquo ions were fairly constant; the incorporation rates spanned at 46,000 fold range, while the ratio varied by less than a factor of five. The number of nitrogen donors about a metal in general increase the lability of the remaining coordinated water molecules, and with $Zn/H_2(TMPyP)$, the rate order was $Zn(Py)_3^{2+}$ > $Zn(Py)_2^{2+}$ > $Zn(Py)^{2+}$ and $Zn(NH_3)_3^{2+}$ > $Zn(NH_3)_4^{2+}$. For monosubstituted $[(H_2O)_5NiX]$ complexes, the rate of water exchange increases with the sigma donor power of X, and is relatively independent of the charge. The order of [ZnX] complexes with $H_2(TMPyP)$ was OH^- (E_A = 1.65) > pyridine (1.20) > acetate (0.95) > H_2O (0.0). The numbers in parenthesis are proportional to donor power[127]. $H_2(Uro-III)$ (8−) reacts with cationic zinc forms much faster than $H_2(TPPS_4)$ (4−), and $H_2(TMPyP)$ (4+) $\gg$ $H_2(Hemato-IX)$ (2−) with anionic $Cu(OH)_3^-$.

6.7.5. Mechanisms of incorporation

Kinetic studies can demonstrate that certain pathways are improbable, but give no information on which of the mathematically consistent mechanisms actually occur. There is no evidence for a two proton dissociation mechanism, where the dianion reacts with the metal: $PH_2 \rightarrow P^{2-} + 2H^+$ then $P^{2-} + M^{2+} \rightarrow M(P)$.

The initial monoanion formation, $PH_2 \rightarrow PH^- + H^+$, could be followed by $M^{2+} + PH^- \rightarrow M(P) + H^+$. The limiting form $R\alpha(PH_2)(M)/(H^+)$ has probably been observed with $Zn(NH_3)_3^{2+}/H_2(TMPyP)$, whose rate increases[32] with pH. The specific rate constant is $> 10^5$ M^{-1} sec^{-1}, and this porphyrin has a $pK_2 \sim 12$. The evidence for this form with $Mg/H_2(Deut\text{-}IX\text{-}DME)$ in $MeOH$[122] can arise by other means; the water soluble phthalocyanine with a low pK_2 may use this pathway[126].

This leaves the associative mechanism $M + PH_2 \rightleftharpoons (M\cdot\cdot PH_2)$, then $(MPH_2) \rightarrow M(P) + 2H^+$, which was considered probable on the basis of correlations of rate with porphyrin basicity[5]. The recent evidence is that such correlations do not exist. This simple associative rate law is found in the majority of incorporation studies, but in itself gives no detailed information. The rate law defines the composition of the activated process as $[M\text{-}P\text{-}2H]^*$, and does not indicate the solvent composition nor the configuration. The moderate deuterium isotope effects show that the protons are somewhat dissociated. Assuming no coordination shell expansion, the rate law[32] first order in $Zn(NH_3)_4^{2+}$ shows that the metal can have its full complement of ligands in the activated complex.

The order of divalent ions $Cu > Zn > Co, Mn, Fe > Ni$ is found for both porphyrins and phthalocyanines. This may be related to how fast the aquo ion can vacate a coordination position. With substituted metal ions, their rates of incorporation correlate with their ability to labilize a coordinated water[32]. The increased rate of cation incorporation as the number of negative peripheral ligands[32,114] increases could be due to increased outer sphere association constants, and adduct stabilization.

The comparative rigidity of the porphyrin nucleus makes their lone pairs relatively unavailable to cations, and probably some porphyrin nucleus deformation must occur. With 14-membered macrocyclic ligands[128], the copper incorporation rates are 10^3-10^4 slower than with a corresponding open-chain ligand, which is 10^9 times more reactive than $H_2(Deut\text{-}IX\text{-}DSA)$. The comparisons between porphyrins and their N-alkyl counterparts shows that deformation[106] may be critical. Open-chain ligands favor stepwise ligand replacement, while some degree of multiple metal desolvation[114,128] has been suggested for the slow macrocycle and porphyrin reactions. The dissociation of the central protons may not be rate limiting[114] in H_2O, but could be in DMF[104], where the rate is fairly independent of the nature of the metal ion. The high reactivity of PH^- is attributable to the three lone pair nitrogen atoms, its distorted structure (similar to PH_3^+), and the absence of the two proton N—H tautomerism.

Metalloporphyrin formation may be an extreme example of a sterically controlled substitution process[129]. An outer sphere complex forms between the metal ion and porphyrin. A ligand dissociates from the metal providing a vacant position for a porphyrin nitrogen atom to occupy. Due to the steric constraints of the porphyrin nucleus, before a second (or third) M—P bond has a chance to form, the first (or second) bond dissociates. The low probability of 'effective' collisions thus slows down the rate of metal ion incorporation.

6.8. Electron transfer, exchange and catalysis

Although the field appears to be developing in the opposite manner, information on the mechanisms of electron transfer reactions involving metalloporphyrins (eq. 11) may aid in understanding the redox behavior of heme proteins[130].

$$M(III)P + \text{Reducing agent} \rightleftharpoons M(II)P + \text{Oxidizing agent} \qquad (11)$$

Recent work has been done[131–134] on ferricytochrome-c and Cr^{2+}, $Fe(EDTA)^{2+}$, $Ru(NH_3)_6^{2+}$ and $S_2O_4^{2-}$. Reduction can occur either by a remote pathway with reaction at the porphyrin edge ($k \sim 10^6$ M^{-1} sec^{-1}), or by an adjacent attack at the buried iron atom, whose rate ($k \sim 30–60$ sec^{-1}) may be limited by the opening of a protein crevice or slow substitution at the iron center. Cr^{2+} may reduce a tyrosine residue which in turn reduces the ferriporphyrin[135].

The first porphyrin mechanistic studies[136] involved the oxidation of Cr^{2+}, V^{2+}, Eu^{2+} and $Cr(bipy)_3^{2+}$ by Mn(III) or Fe(III)TPyP. The $k_0 = (k_1 + k_2(X^-))$ [Red] where the iron reactions were more strongly halide catalyzed than those of manganese, in the order $SCN^- > Cl^- > Br^- > I^-$. Chloride and thiocyanate were incorporated into the product chromic ion which excludes activated complexes of the form [Cl—M(P)—Cr]* or (A) [(P)M—H_2O—X—Cr]* but does not distinguish between (B) [M(P)—Cl—Cr]*, or (C) [M(P)—Cr—Cl]*, since both Fe(II)P and Fe(III)P are substitution labile. The demonstration[137] of 96% chloride ion exchange from ^{36}Cl—Fe(III)TPP into the oligomeric $Cr(acac)_2$ in benzene-1% pyridine during electron transfer (where Cl—Fe(P) is inert) implies an inner sphere mechanism (B) under these conditions. Approximately 9% ^{36}Cl$^-$ from Bu_4N^{36}Cl was found in the Cr(III) product with cold Cl—Fe(III)P, which could be due to a minor outer sphere pathway. Alternatively, PMR measurements under similar conditions[138] show that chloride will exchange by an associative S_N2 mechanism, where the iron or indium[139] atom can 'invert' through the porphyrin plane. The Cr^{2+}/Co(III)TMPyP reaction[140] and a reinvestigation of the Fe(III)TPyP work showed $k_0 = (k_1 + k_2/[H^+])$- [Cr^{2+}]. For Co(III), k_1 and k_2 were first order in chloride, and both first and second order in SCN^-. The inner sphere complex [(P)Co—OH—Cr]*

References, p. 271

was found for k_2, and the relative substitution inertness of the cobaltic porphyrin allowed demonstration that the anion catalysis was of the outer sphere variety (Type A). The much larger anion effects found with the substitution labile ferric porphyrin[136] suggests that the halide is bonded to the metal center before electron transfer, and promotes the reaction irrespective of whether the reductant is bridging $(Cr(H_2O)_5^{2+})$ or non-bridging $(V(H_2O)_6^{2+})$. Electron transfer probably did not take place through the porphyrin ring.

The oxidation of dithionite $(S_2O_4^{2-})$ by Mn(III)Hemato-IX was studied in 25% MeOH–H_2O to prevent porphyrin aggregation[141]. The reaction went through (SO_2^-), and the rates for (OH)–Mn(III)P (7.4×10^4 M^{-1} sec^{-1}) and $(OH)_2$–Mn(III)P (2.1×10^4 M^{-1} sec^{-1}) were similar. With $S_2O_4^{2-}$ and a series of Mn(III) and Co(III) substituted deuteroporphyrins[142] in 4.0 M pyridine–water, the higher the potential needed to reduce Cl–Mn(III)P/Mn-(II)P, the slower was the reduction rate. The driving force thus contributes to the reaction rate. No strong mechanistic conclusions could be drawn due to the unfavorable reaction conditions. Low spin bis-pyridine Fe(II) and Fe(III) porphyrins rapidly exchange electrons, and the addition of CO to produce the mixed carbonyl–pyridine Fe(II) complex blocks this pathway. No exchange is found between the low spin iron porphyrins and high spin Fe(III) forms[144a,b].

When Fe(III)OEP was reduced[145] with Fe^0 in N-methylpyrrolidone-CH_3CO_2T, and reoxidized with O_2 or H_2O_2, five percent tritium was incorporated at the $meso$-positions. H · or H-donor attack could occur at the porphyrin edge, with subsequent radical reduction of the Fe(III). While this could account for hemoprotein reduction when the irons' axial positions are ligated or buried in the protein, other workers noted[146] that the small effect could also be explained by electrophilic attack on the Fe(II) complex. Fe(II) is isoelectronic (d^6) with Co(III), which had a half-life for $meso$-proton exchange in refluxing deuterotrifluoroacetic acid of ~ 11 min. Co(III) and Co(II)Etio, H_2(Etio), H_2(OEP) and Fe(III)OEP ($t_{1/2}$ - 2300 min) did not differ in $meso$ exchange rates by more than a factor of 200, in contrast to observations that metalloporphyrins[147] undergo electrophilic attack faster than metal free compounds.

The reduction of O_2 to H_2O required four moles of $(Py)_2$Fe(II)P, and it was shown to proceed by both outer sphere $[O_2 \cdots (Py)_2 Fe(P)]$ and inner sphere $[O_2–Fe(P) \cdot Py]$ pathways in aqueous solution[148]. The fact that the rates decreased with solvent polarity in EtOH-benzene showed that charge separation $[(P)Fe^{3+}\cdots O_2^-]$ was important; oxygen 'carrying' rather than reduction would be favored by the hydrophobic, low dielectric cavity that iron atoms occupy in hemoglobin and myoglobin. In benzene-Py, however, two ferrous porphyrins were implicated in O_2 reduction[149a]. The mechanism involved addition of O_2 to a vacant coordination position on [Fe(II)P · Py], formed by dissociation of the ferrous dipyridinate. A second unsaturated porphyrin reacts with the dioxygen adduct to produce a binuclear $[Py-(P)Fe-O_2-Fe(P)-Py]$. This could decompose[150] in the slow step to $[Py-(P)Fe(IV)-O]$, which combines with $(Py)_2 Fe(II)P$ to form the oxy-bridged dimer $[P-Fe-O-FeP]$. The mechanism required that O_2 occupy a coordination position on the iron (which rules out a normal outer sphere reduction)[149b], and that two fairly simultaneous reducing equivalents are necessary, as shown in O_2 reactions[130] with $Co(CN)_5^{3-}$, Fe^{2+} and $Cr(NH_3)_6^{2+}$. The $Fe(P)-O_2$ reaction is slower the lower the porphyrin basicity.

In toluene[151] at $23°C$, O_2 reacts with mono-amine cobalt complexes $[(P)Co(II)-L]$ to produce either binuclear peroxy adducts $[L-(P)Co-O_2-Co(P)-L]$ or hydroxy monomers. $[(P)Co(II)L]$ combines with O_2 to form $[L-(P)Co-O_2]$, which rearranges in a slow step, followed by a faster reaction with $[(P)Co(II)L]$ leading to the product. Benzimidazole and imidazole (Im) reacted a hundred times faster than 1-methylimidazole (1-MeIm) and this was attributed to hydrogen bonding of the former two ligands in solution with the coordinated dioxygen, which promotes the peroxy species.

Nitro and nitroso compounds, quinones, alkyl halides, oxygen and peroxides are found[152] to oxidize ferrous porphyrins. The kinetics[153] of the coupling of alkyl halides ($RX \rightarrow R_2$), hydrogenolysis ($RX \rightarrow RH$), and the elimination of vicinal halides have been studied with Fe(II)Deut-IX. The reactions are consistent with the homolytic breaking of the $R-X$ bond by an oxidative-addition mechanism: $(P)Fe(II) + RX \rightarrow [(P)Fe-X-R] \rightarrow (P)Fe(III)X + R\cdot$. Related work[154] was done with Fe(II) hemoglobin or myoglobin. It was noted that many of the in vitro dehalogenations also occur in vivo. For example dogs breathe in CCl_4 and exhale $CHCl_3$, and the heme center could be involved in this process. The reverse reaction[155], where Fe(II)TPP can be formed by photolysis of $Cl-Fe(TPP)$ in azobenzene or cyclohexene has been demonstrated.

Solvents with dielectric constants above 12 (MeOH, Py) cause the slow reversible oxidation of Co(II) to Co(III)Meso-IX-DME[156a]. Stripping the solvent from the (P)Co(III) product reforms Co(II). While similar observations on the solvent, temperature and ligand dependence of this system have been made[157], others feel that what appears spectrophotometrically to be a redox process[158] is simply the change in coordination number of the cobalt species. Other induced redox reactions with $Py_2 Fe(II)P$ are known[156b]. The reported induced oxidation[157] of Co(II) and Fe(II) (followed by degradation) porphyrins by olefins and acetylenes were ascribed to peroxide impurities[152], since thirty such clean substrates under N_2 showed no Fe(II)P oxidation. This apparent contradiction was resolved[155] when it was noted that the concurrent oxidation of olefins and Fe(II)P occurs only if O_2 is present, and while hydroperoxides will decompose Fe(II)P, they will not lead to olefin oxidation. The autoxidation of olefins and active methylene compounds (cyclohexene to cyclohex-2-enol, cyclohex-2-enone and cyclohexane oxide) by Cl—Fe(III)TPP occurs by a free radical chain mechanism in O_2 at $25°C$. Co(II) and Rh(III) porphyrins are also active. Vinyl monomers[159] are photopolymerized with Co(II) and Fe(III)TPP in the presence of amines and CCl_4, and other types of metalloporphyrin catalysis have been explored[160].

As a model for natural nitrogenase enzymes ($N_2 \rightarrow NH_3$), Co(III)TPPS$_4$ and borohydride in aqueous solution reduce typical nitrogenase substrates[161] such as acetylene, azide, cyanide and acetonitrile. The suggested mechanism involved the reduction of Co(III) to Co(I)P with BH_4^-, pi-complex formation with acetylene, then a two electron transfer to produce a pi bonded $[(P)Co(III) \cdot H—C\equiv C—H]$. Two protons are added to form a pi$[(P)Co(III) \cdot H_2C = CH_2]$, which splits off ethylene (cis $D_2H_2C_2$ in D_2O) to reform the original porphyrin. Some ethane is produced and the (P)Co(III)—ethylene adduct was isolated[162]. The initial turnover number was 60 moles acetylene/min/mole catalyst at $25°C$. Other metalloporphyrins capable of two electron reduction such as $O = Mo(V)$ and Rh(III) showed catalytic activity. The cobalt system produced ~ 0.9 mol NH_3/mol of catalyst, and the indication was that the N in NH_3 did not come from the porphyrin[162]. Other workers[163] found that the reductive degradation by BH_4^- of various cobalt complexes containing nitrogen ligands often produces NH_3. Using electrochemical reduction to mimic BH_4^-, numerous porphyrin products were obtained from Co(TPPS$_4$). With $4 : 1$ ($^{15}N_2$): O_2, and Co(TPPS$_4$), ammonia unenriched in ^{15}N was formed, and in the same amounts as when the reactions were run under argon or argon-O_2. The conclusion was that the previously observed NH_3 came either from the porphyrin, and/or from difficult to remove nitrogenous impurities in the air.

Catalase is involved in the disproportionation of H_2O_2 to O_2 and H_2O, and iron[164] and cobalt porphyrins have been studied as model systems. With Co(III)Hemato-IX[165], H_2O_2 dissociates a proton which allows the formation of $[(P)Co(III)—O_2H]$. Another H_2O_2 adds trans forming $[H_2O_2—Co(P)—O_2H]$, which decomposes in a slow step to (P)Co(III), H_2O, OH^- and O_2. Co(TPPS$_4$) behaves in a similar fashion[166]. Two electrons are transferred from one coordinated peroxide through the cobalt into the other peroxide, and thus no 'free' radicals are formed. Certain nitrogenous bases[167] inhibit the reaction by complexing with the cobalt, while imidazole shows a peroxidase-like effect. When coordinated to $[(P)Co—O_2H]$, the imidazole may be oxidized by electron transfer through the cobalt into the peroxide. In this connection, Fe(II)Proto-IX combined with 4-4'-bipyridyl (L) and stabilized with poly-L-lysine catalyzes the aerobic oxidation of cytochrome-c, similar to cytochrome oxidase[168,169]. The [...Fe(II)-L-Fe(II)... O_2] system forms a delocalized pathway favoring cooperative multiple electron transfer into coordinated O_2. Bidentate[170] nonconjugated ligands such as 1,2-di(4-pyridyl)ethane were non-catalytic.

6.9. Stability constants — thermodynamics

Lewis bases can coordinate at one or both axial positions on the metalloporphyrin, and stability constants (eq. 12) have been measured for various metal and ligand types.

$$M(P) \underset{-L}{\overset{K_1}{\rightleftharpoons}} L - M(P) \underset{-L}{\overset{K_2}{\rightleftharpoons}} (L)_2 - M(P) \qquad (12)$$

The primary focus of this section is on thermodynamic, rather than kinetic information.

6.9.1. Cu, Zn, Cd, Hg and VO

These metalloporhyrins form $1:1$ complexes in non-aqueous solution, with the room temperature stability order Zn > Cd > Hg > Cu > VO. For substituted pyridines and Zn, Cd and Hg(TPP), the formation[171,172] constants increase with ligand basicity, a linear log K_f–pK relationship was found, and the Hammett equation was followed, with ρ values of 1.5, 2.2 and 3.1, respectively. If Hg accepted more of the electron pair than zinc, the pyridine nitrogen atom adjacent to the mercury would have a higher formal positive charge, and be more sensitive to substituent donor effects. The class b character of Hg and class a behavior of zinc is in line with this notion, and the ρ values are thus a measure of the sigma polarizability of the metal ion[172]. The enthalpies of adduct formation[173] for Zn(TPP) increase in the order N > P > O > S, showing that the Zn—L bond is primarily electrostatic.

For pyridine (Py) or piperidine (Pip) with VO, Cu, Zn and Ni substituted deuteroporphyrins[173-177] and Pip with VO and Ni (p-X · TPP) and (m-X · TPP)[178], K_f increases with a decrease in porphyrin basicity. Electron withdrawing groups place a higher formal charge on the metal, making it more attractive to nucleophiles. The stronger the metal to ligand interaction, the weaker the metal to porphyrin interaction[179]. The Hammett relationship[30] is found with σ_p^- and σ_p^- (not σ_m), indicating that a resonance (not inductive) interaction occurs between the $d\pi$ metal orbitals and the π orbitals on the porphyrin. For the H_2(S-Deut-IX-DME)[30] series with Pip, + ρ = 1.6, 0.66 and 0.24 for Ni, Cu and Zn;·1.32 and 0.45 for Ni and VO(p-X · TPP) respectively[178], and 1.65 for Ni(m-X · TPP)[178]. The resonance and inductive contributions to the substituent effects are similar in magnitude for both of the (p-X · TPP) series, while inductive effects are about twice as important as resonance effects for Ni(m-X · TPP)[178]. The order was rationalized in terms of the effective nuclear charges of the metals, the need for filled $d\pi$ orbitals[30], and the necessity for the phenyl ring to rotate more coplanar for increased resonance possibilities[178]. The solvent effects on ΔH and ΔS have been discussed[180]. Thermal analysis data show that the temperatures required for ligand removal in the solid state parallel the formation constants in solution[181]. A linear relationship[169] was found between the heats of formation of the 3 and 4-substituted pyridines with $B(CH_3)_3$ and the formation constants of these ligands with Zn, Cd and Hg(TPP)[172].

6.9.2. Mg, Ni and Fe(II)

Mg porphyrins are invariably five coordinate[182] and intermolecularly associated in non-associating solvents[183] where a basic group on one molecule coordinates to a Mg on another. Py—Mg(P) is the exclusive form at 1 : 1 molar ratios[182] and while $Mg(P)Py_2$ solids can be isolated, the second Py is removed 130° above the first[181]. The association constants for the (Py—Mg(P) + Py) reaction are small (~ 0.2 M^{-1}), and show little dependence on porphyrin or ligand basicity. One hypothesis is that peripheral effects are transmitted to the metal only through d orbitals[30]. Another is that electron-withdrawing substituents strengthen the bonding of both the fifth and sixth ligands, and consequently no basicity effect is shown for the $5 \rightarrow 6$ process[179].

Diamagnetic four coordinate Ni(II) porphyrins can add one or two ligands and become paramagnetic. With certain exceptions, the initial indication[179−184] was that only the $4 \rightarrow 6$ process could be observed. However $H_2(TPP)$[180], $H_2(Meso-IX-DME)$[185] and $H_2(Deut-IX-DME)$[180] are now reported to show L—Ni(P) forms. Most of the $4 \rightarrow 6$ data was taken in the Soret, and as noted earlier (p. 235), this often leads to complications. For example, amine addition to Co(P) shows no Soret band shifts[158], but K_1 can be readily determined in the visible. 2 : 1 Amino—Co(P) complexes do shift the Soret. The large amount of total ligand needed to measure K_2 for Ni or Co(P) makes interpretation difficult, because solvent—ligand, ligand—porphyrin and ligand—ligand interactions both decrease the amount of 'free' ligand present[185], and significantly change the nature of the medium. It has been suggested[158] that the $4 \rightarrow 5$ data reported for Ni(P) is actually the $5 \rightarrow 6$ process, with the readily formed diamagnetic L—Ni(P) going to paramagnetic $[(L)_2-Ni(P)]$ types, and Pip—Ni(P) was later shown to be diamagnetic, with a spectrum similar to Ni(P)[178]. In general, high ligand and low porphyrin basicity favor 2 : 1 Ni(P) complexation.

The anomalous absorption spectra[14] of Ni(TPyP) has been explained by studies on Ni(TMPyP)[186a] where a diamagnetic four coordinate Ni(P) is in equilibrium with a six coordinated water ligated form in water. Acetone promotes four coordination while Py and imidazole (Im) favor hexacoordination. $K_0 = [Ni(P)(H_2O)_2/Ni(P)] = 1.23$ at 25°C. For Im $K_1 = 8.39$ M^{-1} (3.44 M^{-1} for Py) and $K_2 = 0.19$ M^{-1} at 25°C. The kinetics of the K_0 pathway have been studied by laser T-jump methods[186b].

With substituted pyridines, Fe(II)S-Deut-IX-DME derivatives in non-aqueous media show only the $4 \rightarrow 6$ ligation[187,188]. The point that $\Delta G°$ alone is not a reliable indication of the behavior of a system was well documented. As expected for a low spin d^6 configuration, the ΔH values depend

TABLE 5

Formation constants for metalloporphyrins and bases in nonaqueous solutions

Complex	Ligand [a]	Log K (25°C)	$-\Delta H$ (kcal/mol)	$-\Delta S$ (eu)	Solvent [b]	Ref.
Mg(TPP)Py	Py	−0.75	2.37	11.4	L	182
Zn(TPP)	PiP	5.05			B	171
Zn(TPP)	Py	3.68	9.2	14	B	174
Zn(TPP)	4-CNPy	2.9			B	171
Zn(TPP)	TMTU[S]	1.08	5.6	13.8	B	173
Zn(TPP)	TMU[O]	1.86	6.5	13.3	B	173
Zn(TPP)	TPPP[P]	1.42	7.7	19.4	B	173
Cd(TPP)	Py	3.53	8.7	13	B	174
Hg(TPP)	Py	1.19	4.9	11	B	174
Ni(TPP)	Pip	2.0			T	178
Ni(TPP)	2 Pip	−0.46	5.6	21	T	178
Ni(Deut-IX-DME)Py	Py	−0.74	0.8	6	B	180
Ni(Meso-IX-DME)	2 Pip	−1.96	4.8	25	THF	178
Cu(Meso-IX-DME)	Pip	−0.85	2.3	11.6	THF	178
VO(Meso-IX-DME)	Py	−1.28			B	176
VO(p-Me · TPP)	PiP	−0.42	5.6	21	T	178
Fe(II)(Proto-IX-DME)	2 Py	0.44	16	50	C	188
Fe(II)(Proto-IX-DME)	2 Py	0.22	10.8	35	CH	188
Fe(II)(Proto-IX-DME)	2 Py	1.32	3.3	5	B	188
Co(II)(p-OMe · TPP)	Py	2.68	8.5	16	T	225
Co(II)(p-OMe · TPP)	Pip	3.39	6.8	7	T	225
Co(II)(p-OMe · TPP)Pip	Pip	−0.07	1.7	6	T	225
Co(II)(p-OMe · TPP)Py	O_2	−0.18	9.3	32	T	225

[a] TMTU, 1,1,3,3-tetramethylthiourea; TMU, 1,1,3,3-tetramethylurea; TPPP, triphenylphosphine; Pip, piperidine; Py, pyridine; 4-CNPy, 4-cyanopyridine.
[b] B, benzene; THF, tetrahydrofuran; C, carbon tetrachloride; CH, chloroform; T, toluene; L, 2,6-dimethylpyridine.

on the sigma-pi characteristics of both the porphyrin and the ligand. The $-\Delta S$ and $-\Delta H$ values were high in CCl_4, and decreased in cases by up to 50 eu. and 17 kcal/mol in going to $CHCl_3$ or C_6H_6. If the solvent—porphyrin interactions are minimal in CCl_4, the large negative entropy change (−70 eu) could be associated with the loss of rotational and translational freedom as two ligands bind with Fe(II)P. This unfavorable ligand addition entropy may be compensated by the simultaneous desolvation of solvent molecules bound to the porphyrins in $CHCl_3$. Chloroform could hydrogen bond to the porphyrin pi cloud and/or its nitrogen atoms before adduct formation. Representative data are shown in Table 5. 2-Methylthioethanol combines with Fe(II)Proto-IX in successive overlapping steps[189]. The earlier work in this area has been reviewed[5,190].

References, p. 271

Fe(II)P forms many mixed complexes[191] such as Py—Fe(P)—CO. The more basic the porphyrin and the more sigma bonding is the ligand, the stronger (more double bond character) the Fe—CO bond[150,192]. The Fe(II) can reduce its excess negative charge donated by the porphyrin and ligands by delocalizing it in a pi fashion into the acceptor orbitals of CO. It does so more readily the higher the porphyrin basicity (a *cis* effect), and ligand (a *trans* effect) basicity.

Fifty percent formation of [(NO)—Fe(P)—Pip] from Pip$_2$—Fe(P) in piperidine at 23°C was at $P_{1/2}$ = 0.18 and 0.4 Torr for H$_2$(Proto-IX) and H$_2$(TPP), while for the [CO—Fe(P)—Pip] complex, $P_{1/2}$ = 5.8 and 75 Torr, respectively[193]. The higher basicity of H$_2$(Proto-IX) favored mixed ligand stability, in terms of [(P)Fe(III)—NO$^-$] structures for NO, and CO was found to be the stronger acceptor. At equilibrium, (K_{NO}/K_{CO}) = 30, while for a sheep hemoglobin, the ratio is ~1500. The unfavorable entropy term in solution as compared with the rigid protein environment could explain this behavior. The spontaneous reduction of Fe(III) to Fe(II) in piperidine was used to advantage in this system. The kinetics of the CO adduct formation are treated on p. 271.

Fe(II) porphyrins that duplicate the myoglobin function of reversibly binding molecular oxygen have been synthesized. Pyrroporphyrin-N-[3-(imidazoyl)propyl]-amide has an imidazole intra-molecularly[194] attached to Fe(II)P, and the distal CO can be removed by heating in the solid state and replaced by O$_2$. Such adducts are also formed at −45°C in methylene chloride[195a]. In comparison with the usual bis-pyridine or imidazole Fe(II) porphyrins, intramolecularly bonded Py or Im groups enhance both CO and Py affinity by a factor of 10, which may be due to a chelate entropy effect[195b]. While the internal Im and Py adducts show about the same affinity for CO, the imidazole complex[195c] is 3800 times more effective in binding O$_2$. The pi orbitals of Im may transfer charge to the $d\pi$ orbitals of Fe(II), making the iron a better pi donor to O$_2$. Oxygenation is favored in polar, aprotic solvents[195d]. Ferromesoporphyrin-bis-3(1-imidazoyl)propyl amide[195e] and other simple bis-amine ferrous porphyrins[195f−h] form oxygen adducts at low temperature.

Fe(II)-*meso*-tetra($\alpha,\alpha,\alpha,\alpha$,-O-pivanamidephenyl)porphyrin[196a] can be isolated as the diamagnetic bis(1-MeIm) or (1-MeIm)—Fe(P)—O$_2$ complex at 25°C. The Fe—O—O bond is end-on and bent[196b]. Other bases (Im, Py, Pip...) also form O$_2$ adducts in solution. With an excess of ligand, O$_2$ addition is reversible, whereas in 1 : 1 ligand to porphyrin ratios, the Fe(II) dimer slowly forms. In the solid state[197], 1 mole of O$_2$ can be pumped off of the (1-MeIm—O$_2$) complex and paramagnetic [THF—Fe(P)—O$_2$] adducts are formed by solid-gas reactions. This 'picket fence' porphyrin places dioxygen in a hydrophobic pocket, and as a consequence (1), protonation of O$_2^-$ to H$_2$O$_2$ is inhibited and (2) the reduction of O$_2$ by the 2 Fe(P)/O$_2$ mechanism is avoided. With similar ligands, Co(II) forms dioxygen adducts[198a].

The nitrogenous base prevents O_2 coordination on the exposed side of the ring. The Fe(II) complex of 9,10 bridged 9-10-dihydroanthracene[198] has a 5 Å cavity, and O_2 addition is found at low temperatures with pyridine on the exposed face. 'Capped' ferrous porphyrins also carry O_2[198b].

It is known that solid [(Im)$_2$—Fe(II)Proto-IX] or (Meso-IX) can reversibly bind molecular oxygen, without the loss of imidazole[199]. Thermomagnetic microbalance techniques[108] and chemical analysis[200] show that solid [(Py)$_2$—Fe(II)P] can form 2 : 1 peroxy type complexes with oxygen, by loss of two pyridine groups.

6.9.3. Fe(III), Mn(III) and Co(III)

Ferric porphyrins form oxybridged dimers in basic solution[201—203], and higher aggregates are known[204]. The earlier ligand addition studies have been reviewed[3,5,22,205]. In DMF[206], acetone[207], CH_2Cl_2[208] and $CHCl_3$[209] (not C_6H_6), the equilibrium 2 Im + (P)FeCl → [(Im)$_2$—Fe(III)-P]$^+$Cl$^-$ was demonstrated, with no evidence for a monoimidazole adduct. In the latter three solvents, $-\Delta S°$ was approximately constant (~45 eu), with $-\Delta H$ ~7 kcal/mol. ΔG increased with solvent polarity[209]. High or mixed spin mono-Py or Im adducts can be formed in the solid state[210], and in solution[211]. For Fe^{3+}(Proto-IX) monomerized in aqueous—ethanol buffers[212], the reaction 2 Im + H$^+$ + Fe(P)—OH → (Im)$_2$—Fe^{3+}(P) has K ~ $10^{14}M^{-3}$ (25°C). Phenanthroline[213] makes a 1 : 1 pi complex with Fe^{3+}(Deut-IX-DME), and such complexation enhances the porphyrins' affinity for imidazole. A number of molecular complexes[114,214—216] with nitroarenes, steroids, purines and other species are known.

High spin mono- and difluoro (but not dichloro) species are formed in DMF with Fe(III)(Deut-IX-DME)[217]. Starting with bisimidazole hemichromogens, EPR studies indicate that 1 and 2 fluorides can interact with the coordinated N—H groups, and a third (monofluoro, F—Im) type is present. The bis(β-imidazoylethyl) amides of iron (Deut-IX) and (Meso-IX) and related internal complexes have been synthesized[218], and are found to be substitution inert at high pH in the Fe(III) form, but labile in the Fe(II) state. No dioxygen complexes were noted at room temperature.

The solid-gas reaction [Cl—Fe(TPP)]$_s$ + NO (g) → [Cl—Fe(TPP)—NO]$_s$ had K_p = 3.45 atm^{-1} (25°C). The NO group is readily lost upon dissolution in organic solvents, and magnetic data supported NO as a reductant[219], forming high spin [(TPP)Fe(II)—NO$^+$]. Fe(III)Proto-IX-DME and the superoxide ion form a 1 : 1 adduct [O$_2$—Fe(P)], which could be a superoxide complex of Fe^{3+} or a dioxygen—Fe(II) species[220].

The coordination chemistry of manganese porphyrins has been fully reviewed[221]. The ratios of the stability constants for nitrogenous base complexation to the oxidized and reduced forms of Co(III) and Mn(III)Hemato-IX have been reported[222—223a]. Cobalt complexes of bis(hystidine methylester)Deut-IX have been prepared[223b].

6.9.4. Co(II)

Co(II) porphyrins and amines (L) undergo the following types of reactions[224–226]: $Co(P) + L \rightleftharpoons Co(P)L$ (K_1); $Co(P)L + L \rightleftharpoons Co(P)L_2$ (K_2); $Co(P)L + O_2 \rightarrow Co(P)L—O_2$ (K_3); $Co(P) + O_2 \rightleftharpoons Co(P)—O_2$ (K_4), $Co(P)L_2 + O_2 \rightarrow Co(P)L—O_2 + L$ (K_5) and $Co(P)L—O_2 + Co(P)L \rightarrow$ dimers (K_6). In general, with 0.1 molar ligand concentrations at 25°C, k_6 is slow, K_2, K_3 and K_4 are small, and Co(P) and Co(P)L are the main forms. For [(p-OMe · TPP)Co], the isoequilibrium temperature for K_1 was 87°C in toluene[225], and linear log K_1—pK_L plots were found at 25°C for various imidazoles, aromatic and aliphatic amines (see Table 5). The ΔH values for the aromatic ligands were greater than for the aliphatic cases, indicating M → L pi donation is important.

With Co(Proto-IX-DME)[158], a $\sigma\rho$ plot for K_1 gave $\rho = 0.6$, similar to that found for Zn(TPP). Apparently[227] 2,6-dimethylpyridine does not bind to Co(P), and such complexing with this ligand as has been found may be due to trace impurities, as noted in iron porphyrin systems[228].

K_2 and K_3 can be determined from low temperature EPR[229] or spectrophotometric measurements. At 208°K (70°K above T_i) there was no general relationship between amine basicity and K_3 with [(p-OMe · TPP)Co]. —ΔH was between 8.5—9.5 kcal/mol and —ΔS = ~30 eu for all bases. K_3 and K_4 were similar for Co(Proto-IX-DME)[158], and no strong correlation was found between ligand basicity and ΔG, ΔH or ΔS. Other workers used a different method to analyze the K_3 data for Co(Proto-IX-DME), and concluded[230] the system was poorly defined and that, in fact, no numerical data (K_3...) could be obtained. The initial workers[231] then showed that their original limiting spectra analysis was in agreement with a non-linear least-squares procedure, and they reaffirmed their previous contentions.

The greater pi donor character of imidazole as compared with pyridine allows more pi electron transfer to the cobalt atom, which can be delocalized into the oxygen pi-antibonding orbitals, favoring stronger bonding[158]. Even the weak base DMF, which is a good pi donor, allows dioxygen formation. The amines activate Co(P) for oxygen addition[229], since K_3 is a factor of 100 larger than K_4, for Co(P)—O_2 formation. Polar aprotic solvents stabilize [L-(P)Co(III)—O_2^-], leading to increased K_3 values[232]. EPR data show that Co(II)TPP complexes with CO, CH_3CN and R_3P, and such 1 : 1 adducts form dioxygen species. For the oxygen complexes[232], the odd electron occupies a localized π^* orbital on oxygen, while for closed shell ligands, the electron is in a $(d_{z^2})^1$ orbital. Co(TPP) also reacts with NO and tetracyanoethylene.

When reconstituted with hemoglobin or myoglobin, Co(II) porphyrins (coboglobins) carry molecular oxygen slightly less efficiently than their Fe(II) counterparts[234]. An extensive amount of work is being done along these lines[235].

6.10. Dimerization and hydrolysis

Evidence from a variety of sources[1,5,12] indicate that many porphyrins and metalloporphyrins dimerize in solution:

$$2\ PH_2 \underset{k_r}{\overset{k_f}{\rightleftharpoons}} (PH_2)_2 \quad K_D \tag{13}$$

K_D can be obtained from spectrophotometric dilution studies[12]. Dimers usually have lower extinction coefficients and wider band widths than monomers, and dimerization is often missed if a large ($10^{-4}-10^{-7}$ M) concentration range is not investigated. Apparently only monomers fluoresce, and K_D can also be measured in this manner[38]. With water soluble porphyrins, dimerization is usually not found (a) in detergents[27], (b) at low ionic strengths[12], (c) in ethyleneglycol—H_2O[12] or EtOH—H_2O^2[12]. Exceptions do occur. Dimerization is promoted by 1 : 1 electrolytes, where the association depends on the concentration and not the type of salt added[12].

The water soluble tetrasulfonated phthalocyanines are extensively dimerized[236], with log K_D values at 60°C of 7.0 H_2(TSPc), 5.3 (Co), 7.3 (Cu) and 6.0 Zn(TSPc). Ferrous phthalocyanine in DMSO dissociates very slowly[237], and kinetic data, mainly from temperature-jump relaxation studies, is given in Table 6.

TABLE 6

Dimerization rate and equilibrium results for porphyrin types and metalloporphyrins

Complex, Temp. (°K)	$K_D(M^{-1})$	$k_f(M^{-1}sec^{-1})$	$k(sec^{-1})$	Ref.
H_2(EnProto-IX-DME), 298 [a]	4.3×10^6	2.0×10^8	50	38
H_2(TPPS$_3$), 298	4.8×10^4	2.2×10^8	4.6×10^3	12
H_2(TPPS$_4$), 292	9.6×10^4	1.3×10^8	1.4×10^3	239
Cu(TPPS$_4$), 292	6.7×10^4	6.2×10^7	9.2×10^2	239
Pd(TPPS$_4$), 292	2.1×10^5	7.5×10^7	3.6×10^2	239
Ag(TPPS$_4$), 292	6.9×10^3	4.8×10^6	7.0×10^2	239
H_2(TPPC$_4$), 298	4.6×10^4	6.4×10^7	1.4×10^3	12
Cu(TPPC$_4$), 298	1.7×10^5	4.5×10^7	2.7×10^2	238
Ni(TPPC$_4$), 298	1.6×10^5	6.2×10^7	4.0×10^2	238
Zn(OEP), 300 [b]	5.5×10^3			242
Zn(Etio), 300	3.4×10^6			242
Zn(OEP)$^{+\cdot}$, 293	2.5×10^4	1.3×10^8	5.0×10^3	241
VO(TSPc), 301 [c]	5.0×10^6	4.5×10^6	9.0×10^{-1}	60a
Co(TSPc), 331	2.1×10^5	5.4×10^2	2.8×10^{-3}	60b

[a] k_f, E_A = 4.1, k_r, E_A = 12.4 kcal/mol.
[b] In methylcyclohexane, Zn(OEP)($-\Delta H$ = 8.6, $-\Delta S$ = 21.5 eu), Zn(Etio)(9.5, -17.8).
[c] In CHCl$_3$—MeOH, $-\Delta H$ = 17.5 kcal/mol, $-\Delta S$ = 38 eu.

References, p. 271

The tetrapositive $H_2(TPyP)$, $H_2(TMPyP)$ and its Cu, Ni and Zn complexes do not associate ($K_D < 100$) in water[12,238]. The free base and monocations of the positive $H_2(EnProto-IX-DME)$ and related derivatives aggregate[38]. While $Zn(TPPC_4)$ does not dimerize, its tetranegative Cu and Ni complexes do associate[238]. The unsymmetrical trinegative[238] $H_2(TPPS_3)$ and symmetrical tetranegative[239] $H_2(TPPS_4)$ free bases associate, as well as the Cu, Ni, Pd and Ag(II)TPPS$_4$ forms. However VO^{2+}, Zn, Fe^{3+}, Fe—O—Fe, Cr^{3+}, Mn^{3+} and $Co^{3+}(TPPS_4)$ do not dimerize[239]. Free base $H_2(S-Deut-IX-DME)$ form dimers[240], as does $H_2(Deut-IX-DSA)$[239]. The cation radicals $Zn(Hemato-IX)^{+\cdot}$ in water and $Zn(OEP)^{+\cdot}$ in 10 : 1 MeOH—CHCl$_3$ [but not $Zn(OEChlorin)^{+\cdot}$] form dimers[241], where the extent increases with the solvent dielectric constant. $H_2(Etio)$, $H_2(OEP)$ and their zinc derivatives dimerize at low temperatures (220—140° K) in methylcyclohexane solutions[242], but $H_2(TPP)$ and $Zn(TPP)$ (due to phenyl-steric factors) do not associate under these conditions. Dicyano-ferric (Proto-IX) dimerizes by intermolecular stacking of rings A and D[278].

The trend in water is that metalloporphyrins with five coordinate metal ions do not associate while four coordinate metals (Ni, Pd, Cu) dimerize. While the immediate explanation is that such five coordination does not allow the porphyrins to come close enough to interact, vanadyl porphyrins in nonaqueous media are associated[243]. Some positive porphyrin types associate while others do not. The dimerization constants in Table 6 are similar in magnitude, and the association rate constants are not far below the diffusion controlled limit. The indication is that solvent structural rearrangements, steric factors for effective collision, and some electrostatic repulsion lowers the observed rates[238]. One hypothesis is that the positive charges in $H_2(TMPyP)$ delocalize the pi electron density across the surface of the molecule, and thus inhibit π—π interactions, which are favored by negative substituents which localize electron density, leading to stacking forces of the Van der Waals type. The electrolytes create an ionic atmosphere about the porphyrins which favor coulombic interaction, since dimerization does not occur for some porphyrins in the absence of electrolytes[12,38b].

Trivalent metalloporphyrins can hydrolyze by the loss of one or two protons (eq. 14,15), which is formally equivalent to the removal of a proton from a coordinated water at low acidities, and the addition of a OH^- group at higher pH. Before hydrolysis, it is seldom clear whether the sixth axial position is vacant, or occupied by a weakly bonded solvent molecule.

$$MP(H_2O)_2 \rightleftharpoons MP(H_2O)(OH)^- + H^+ \quad K_{A1} \tag{14}$$

$$MP(H_2O)(OH)^- \rightleftharpoons MP(OH)_2^{2-} + H^+ \quad K_{A2} \tag{15}$$

Oxybridge dimerization can be formulated as follows:

$$2 MP(OH)-X \rightleftharpoons O-(MP-X)_2 \quad K_3 \tag{16}$$

$$2 MP(H_2O)_2 \rightleftharpoons O-(MP-X)_2 + nH^+ \quad K_4 \tag{17}$$

The iron porphyrins differ in the types of species observed in solution. For $Fe^{3+}(TPPS_4)$, only K_4 (X = H_2O, n = 2) is shown[37,244]. For the dipropionic acid of Fe^{3+}(Deut-IX-DSA), K_{A1} and K_3 (X = H_2O) can be obtained[245], while for Fe^{3+}(EnProto-IX-DME) K_{A1}, K_{A2} and K_3 (X = OH^-) are known[246]. By combination of K_{A1}, K_{A2} and K_3, all of the dimer results can be formulated in terms of K_4. Table 7 shows both hydrolysis and dimerization data for various metal types.

In comparison with the aquo ion, the porphyrin nucleus itself makes coordinated water molecules less acidic, and more so the higher the porphyrin basicity. In general, the lower the formal charge of an ion, the less it hydrolyzes. Thus ferrous porphyrins do not hydrolyze[5] up to pH 12, while ferric porphyrins have pK_{A1} values from 4 to 7, and Fe^{3+} itself has pK_{A1} of 2.8. The ferric complexes of $(TPPS_4)$ and (Deut-IX-DSA) have the same tetranegative peripheral charge, and dimerize to the same extent. Water molecules coordinated to positive ferric porphyrins are more acidic than with negative porphyrins, and it appears that when comparing porphyrin charge types, hydrolysis is favored if it leads to a product having a lower net charge[245].

In the H_2(S-Deut-IX-DME) series[150], NMR studies show that more basic porphyrins are less dimerized and Fe^{3+}(Proto-IX) is more associated than Fe^{3+}(Deut-IX). For the ferric(Deut-IX)[240] reaction (2 Fe(P) ⇌ Fe$_D$ + H^+),

TABLE 7

Hydrolysis data for trivalent metalloporphyrins

Ion	Porphyrin [a]	pK_{A1}	pK_{A2}	Log K_3	pK_4	Ref.
Fe(III)		2.8		2.7	2.9	248
	(Meso-IX-DME)(0) [b]	6.2				249
	(Meso-IX)(2−) [c]	7.0				212
	(Deut-IX-DME)(0) [b]	6.0				249
	(Proto-IX-DME)(0) [b]	5.4				249
	(Proto-IX)(2−) [c,d]	6.5	~13			212
	(En Proto-IX-DME)(7+)	4.5	6.0	5.0	16.0	246
	$(TPPS_4)$(4−)				8.6	244
	(Deut-IX-DSA)(4−)	7.7		6.5	8.9	245
	(TMPyP)(4+)	~4.9				250
Co(III)	(TMPyP)(4+)	6.0	10.0			262
Cr(III)	$(TPPS_4)$(4−)	4.8	7.9			260
Rh(III)	$(TPPS_4)$(4−)	~7.5				261
Mn(III)	(Hemato-IX)	~6	11.9			221

[a] Charge on the porphyrin periphery, reactions in water at ~27°C unless noted.
[b] 2.5% SLS, 25°C.
[c] EtOH−H_2O
[d] pK_{A2} in water.

References, p. 271

$K_D = 3.4 \times 10^{-2}$ (25°C). $k_f = 2.6 \times 10^8$ M^{-1} sec^{-1} and $k_r = 1.1 \times 10^3$ sec^{-1} at pH 6.8. It was suggested that Fe(P) and Fe(P)OH were the reactants. With Fe^{3+}(TPPS$_4$), $pK_4 = 8.6$ at 20°C, and the kinetically consistent mechanism was found[37] to be $(P)Fe(H_2O) \rightleftharpoons [(P)Fe-OH] + H^+$ (k_1, k_{-1}), then $2[(P)Fe-OH] \rightleftharpoons (P-Fe-O-Fe-P)$ (k_2, k_{-2}). With the unobserved hydrolyzed form as a steady state intermediate, $k_1 = 3.4$ sec^{-1}, and $k_{-2} = 0.9$ sec^{-1}. k_{-2} is similar to that found for other iron complexes[248,249] as EDTA (1.2 sec^{-1}) and H_2O (0.35 sec^{-1}). H_2O may be the attacking nucleophile, forming a dihydroxy-bridged intermediate. The dissociation of this dimer studied by stopped-flow techniques[244] followed $R = [k_3 + k_4(H^+)][Fe_D]$, with $k_3 = 41$ sec^{-1} and $k_4 = 842$ M^{-1} sec^{-1} (25°C). k_3 is much higher than the k_{-2} found above, and k_4 is a reflection of proton attack at the oxybridge facilitating dimer dissociation[249,250].

6,11. Solvent exchange

The solvent exchange rates between free and coordinated molecules have been measured by PMR techniques[251-253]. With respect to the solvated metal ion, the metalloporphyrins' axial ligand is labilized by a factor of 10^3 in MeOH and 10^5 in DMF. The water exchange rate may be 10 to 100 times faster than found for MeOH, and the rate enhancement shows up as more positive ΔS^* values for metalloporphyrins. In pyridine—H_2O solutions at 25°C, Fe^{3+}(Proto-IX) is 21% in the high spin form[254]. For $[H_2O-Fe(P)-Py]$, exchange rates (τ^{-1}) of 2.8×10^4 sec (H_2O) and 5×10^3 (Py) were observed. These lifetimes were longer than found for pure high spin ferric porphyrins[251].

6.12. Ligand substitution mechanisms

The synthesis of a wide range of water soluble porphyrin types had led to studies on the mechanisms of axial ligand substitution in metalloporphyrins. The kinetics of cyanide addition[249] to the ferric dimer of (TPPS$_4$) show $R = 3.8 \times 10^4$ (dimer) (HCN)/[1 + 3×10^{10} (H$^+$)] at 25°C. HCN was the reactant, forming H$^+$ and CN—Fe—O—Fe, which rearranges in an intramolecular process to NC—Fe—O—Fe. This subsequently dissociates to the dicyano monomer by two pathways, 9.5×10^{-6} M^{-1} sec^{-1} (H$^+$) and 24 M^{-2} sec^{-1} (HCN)(CN$^-$). The (H$^+$) or (HCN) could protonate the bridge oxygen, leading to OH$^-$ loss and cyanide addition.

While the Fe^{3+}(TPPS$_4$) dimers are not split by detergents[249], substituted deuterohemins[27,255a,b] are monomerized by SLS, CTAB or Tween. CN$^-$ reacts with HO—Fe(P) to form HO—Fe(P)—CN, which then combines with HCN to give H_2O and CN—Fe(P)—CN. This rearranges in an observable (not in SLS) intramolecular step to $NC-Fe(P)-CN$. The equilibrium constants for cyanide addition[249,255a,b] to (Meso-IX), (Deut-IX) and (Proto-IX) (4.4, 69 and 12.6×10^{-5} M^{-2}, respectively) in CTAB do not parallel the pK_3 values, showing that the low spin Fe^{3+}(Deut-IX) is a better sigma acceptor from CN^- than (Meso-IX), and a better pi donor than (Proto-IX). This peculiar order is also found for myoglobins reconstituted with these hemin types[235] in their reactions with O_2 and CO.

The unhydrolyzed Fe^{3+}(Deut-IX-DSA) and (EnProto-IX-DME) form only monoimidazole complexes at low pH, and the rates were similar irrespective of total porphyrin charge[245,246]. For (EnProto-IX-DME), $K = 3.8 \times 10^3$ M^{-1}, $k_f = 8.5 \times 10^6$ M^{-1} sec^{-1}, and $k_r = 1.8 \times 10^3$ sec^{-1}. The imidazolium ion was unreactive. An ion pair interchange mechanism was suggested:

$$(P)Fe(H_2O) + Im \underset{}{\overset{K_{OS}}{\rightleftharpoons}} [(P)Fe(H_2O) \cdot \cdot Im] \underset{k_{-d}}{\overset{k_d}{\rightleftharpoons}} (P)Fe-Im + H_2O \qquad (18)$$

Since $k_f = K_{OS} k_d$, with K_{OS} as high as 100 M^{-1}, the water exchange rate ($\sim 8.5 \times 10^4$ sec^{-1}) is about a thousand times that of aquo iron(III).

Two imidazole ligands complex with Fe^{3+}(Proto-IX) in aqueous ethanol[256]. The imidazolium ion was about 150 times more reactive than imidazole, and an ion pair interchange process was advanced.

At constant pH, the rates of formation of $[X_2-Co(III)Hemato-IX]$ approached a limiting value as the concentration of CN^- or SCN^- increased[257]. This was consistent with either the ion pair mechanism, or a limiting S_{N1} dissociative mechanism:

$$(H_2O)_2 M(P) \underset{k_{-1}}{\overset{k_1}{\rightleftharpoons}} (H_2O)-M(P) + H_2O \qquad (19)$$

$$(H_2O)-M(P) + X \underset{k_{-2}}{\overset{k_2}{\rightleftharpoons}} X-(H_2O)M(P) \qquad (20)$$

The presence of the first ligand makes the second anion incorporation rapid. The (P)Fe(III)/CN reaction showed no rate limiting behavior at high CN^- concentration. This was the first study[257] to point out that the porphyrin ligand labilizes the first coordination sphere of a classically inert metal ion, by a factor of 10^6 in the Co(III) case. Both cobalamin[258] and the vitamin B_{12} model methylaquocobaloxime complexes[259] show similar behavior. The Cr(III)TPPS$_4$ substitution reactions with Py, F^- and CN^- were also interpreted in a D mechanistic framework[260], where the porphyrin labilization factor was $\sim 10^3$, and approx. 10^2 for Rh(III)TPPS$_4$[261]. Since Cr(III)P does not have an accessible Cr(II) state, the indication was that the

References, p. 271

(P)Co(III) $\rightleftharpoons$ (P)Co(II)$^+$ equilibrium does not account for the enhanced porphyrin reactivity.

The Co(III)TMPyP/SCN$^-$ reaction has been studied by several groups[262,263] with similar results. For mono-SCN formation[262] at pH 2, 25°C, k_1 = 2.1 M^{-1} sec^{-1} and k_{-1} = 3.1 × 10^{-4} sec^{-1}. For bis-SCN production from the mono-SCN, k_2 = 2.8 × 10^4 and k_{-2} = 3.0 × 10^3. The observations that the rate of mono-SCN formation approached a limiting value at high SCN, and that ΔS^* is close to zero indicated a D mechanism. The bound SCN$^-$ group activates the *trans* water molecule by a factor of $\sim 10^4$, and also decreases by a thousand its acidity. The stability constants[263] for mono-ligation of Co(III)TMPyP were SCN > I > Br > Cl, in accord with the porphyrin acting as a soft acid, delocalizing charge into the coordinated Co(III). This mixing of metal and porphyrin orbitals makes a formal d^6 classification meaningless for the cobaltic ion, and may explain its enhanced lability as compared with classical Werner complexes[260,262,263]. Co(III) TPPC$_4$ and anions have been studied[186b].

A dissociative mechanism[264] was advanced for the exchange of free 1-Me Imidazole with [(1-MeIm)$_2$—Fe(III)TPP] in CDCl$_3$. At 35°C, $K_{eq} \simeq 10^3$, and the dissociation rate was ~ 60 sec^{-1} at 298°C, with E_A ~ 17 kcal/mol and ΔS^* ~ 7 eu. The exchange rate was about ten times faster with the more basic (OEP), which might stabilize the five-coordinate intermediate.

The fact that the ferric ion sits above the porphyrin plane, and the restricted rotation of the aryl groups[265] of [(p-Me · TPP)FeCl] places the *meso*-aryl protons in two different magnetic environments characterized by a m-H doublet separated by ~ 110 Hz at 35°C in CDCl$_3$. This doublet broadens into a single peak at a ratio of Bu$_4$N$^+$Cl$^-$/Fe(P) of ~ 6.5, indicating that the iron atom is 'inverting' through the porphyrin plane by the S$_{N2}$, associative process[132]:

$$Cl^* + (P)FeCl \rightarrow [Cl^*{-}(P)Fe{-}Cl] \rightarrow Cl^*{-}(P)Fe + Cl \qquad (21)$$

with an inversion rate of ~ 450 sec^{-1}. This inversion was much slower with the more basic (OEP), which could destabilize the six-coordinate intermediate. Chloride can simultaneously exchange by other pathways not sensed by this NMR technique.

Nitrogenous bases coordinated to [(p-isopropyl · TPP)RuCO] can exchange by two kinetic processes[169]. Intermolecular exchange can occur where the rate limiting step is Ru—N bond breaking. The rates range from 10^{-1} to 10^4 seconds at 25°C (ΔS^* ~ 2.3 to 7.1 eu), with α-methyl substituted pyridine or pyridazine ligands reacting fastest. Intramolecular exchange occurs with molecules such as 3,6-dimethylpyridazine, and involves two equivalent donor sites on the same ligand, i.e. [(P)—Ru(L—L*)] $\rightarrow$ [(P)—Ru(L*—L)]. The intramolecular exchange was found to be 20—80 times faster than the intermolecular process[266]. The initial observations of intramolecular exchange with imidazole[267] was later shown to actually be an intermolecular substitution[268,269]. The exchange of Py with Zn^{171}, Mg^{182} and $Co(III)^{117}$ porphyrins followed by NMR has been used to determine the position of the metal with respect to the porphyrin plane.

The formation of [CO—Fe(P)—Pip] from the low-spin $(Pip)_2Fe(P)$ involves a dissociative pathway with a five coordinate monopiperidine intermediate[270]. The same D mechanism is found for bis-ligated low spin Fe(II) phthalocyanines[271] (L = Im, Py, Pip, 2-MeIm) which form [L-Fe(Pc)—CO] complexes. The ferrous porphyrins are about as labile as the phthalocyanines, both of which are faster than the glyoximes. It was proposed that the five coordinate intermediate in the porphyrin system is in a high spin d^6 state, while the corresponding phthalocyanine has a low or mixed spin configuration. This is supported by the synthesis of the high spin five coordinate $[Im—Fe(III)P^+]Cl^-$, which is produced by heating the low spin bis-imidazole form[210].

The rate law for the formation of $(Im)_2Fe(Pc)$ from $[(DMSO)_2—Fe(Pc)]$ is first order in complex and imidazole[272,273] up to 1.5 molar Im. The DMSO (solvent) could compete more effectively than imidazole for the D intermediate (k_{-1} (DMSO) $\gg k_2$ (Im)). Thus it is not clear if the reaction is associative or dissociative in nature, since mechanistic decisions can be made neither by authority nor analogy[274].

Metalloporphyrins will play a useful role in Nuclear Medicine. Deuterium labeled $H_2(TPPS_3)$ concentrates[11] to a higher extent in tumor tissues than metalloderivatives. ^{109}Pd porphyrins show high tumor/organ ratios[275]. Depending on the metal ion, metalloporphyrins are absorbed on lymph nodes[276]. With ^{109}Pd porphyrins, the selective radiation destruction of lymphatic tissue responsible for homograph rejection has been demonstrated, by successful skin transplants without the use of antibiotics[277].

Acknowledgement

This work was performed under the auspices of the United States Atomic Energy Commission.

References

1. P. Hambright, Coord. Chem. Rev., **6**, 247 (1971).
2. H. Fischer and H. Orth, "Die Chemie des Pyrrols" Akademische Verlagsgesellschaft, Leipzig (1937), Vol. 2, Part 1.
3. R. Lemberg and J.W. Legge, 'Haematin Compounds and Bile Pigments', Interscience, New York (1949).
4. J.E. Falk, R. Lemberg, and R.K. Morton (Eds.), 'Haematin Enzymes', Pergamon, London (1961).

5. J.E. Falk, 'Porphyrins and Metalloporphyrins', Elsevier, Amsterdam (1964).
6. W.S. Caughey, J.O. Alben, W.Y. Fujimoto, J.L. York, J. Org. Chem., **31**, 2631 (1966).
7. H.W. Whitlock and R. Hanauer, J. Org. Chem., **33**, 2169 (1968).
8. A. Adler, F.R. Longo, J.D. Finarelli, J. Goldmacher, J. Assour, and L. Korsakoff, J. Org. Chem., **32**, 476 (1967).
9. N. Datta-Gupta and T.J. Bardos, J. Heterocycl. Chem., **3**, 495 (1966).
10. A. Adler, L. Sklar, F.R. Longo, J.D. Finarelli, M. Finarelli, J. Heterocycl. Chem., **5**, 669 (1968).
11. J. Winkelman, G. Slater, and J. Grossman, Cancer Res., **27**, 2060 (1969).
12. R.F. Pasternack, P.R. Huber, P. Boyd, G. Engasser, L. Francesconi, E. Gibbs, P. Fasella, G.C. Venturo, and L. Hinds, J. Am. Chem. Soc., **94**, 4511 (1972).
13. E.B. Fleischer, J.M. Palmer, T.S. Srivastava, and A. Chatterjee, J. Am. Chem. Soc., **93**, 3162 (1971).
14. E.B. Fleischer, Inorg. Chem., **1**, 493 (1962).
15. P. Hambright and E.B. Fleischer, Inorg. Chem., **9**, 1757 (1970).
16. T.P. Stein and R.A. Plane, J. Am. Chem. Soc., **91**, 607 (1969).
17. C.B. Storm, Y. Teklu, and E.A. Sokoloski, Ann. N.Y. Acad. Sci., **206**, 631 (1973).
18. R.J. Abraham, G.E. Hawkes, and K.M. Smith, Tetrahedron Lett., 71 (1974).
19. E.B. Fleischer, Acc. Chem. Res., **3**, 105 (1970).
20. E.B. Fleischer and L. Webb, J. Phys. Chem., **67**, 1131 (1963).
21. H. Baker, P. Hambright, and L. Wagner, J. Am. Chem. Soc., **95**, 5942 (1973).
22. J.N. Phillips, Rev. Pure Appl. Chem., **10**, 35 (1960).
23. J.N. Phillips in "Current Trends in Heterocyclic Chemistry, Eds. A. Albert, G.M. Badger, and C.W. Shoppe, Butterworths, London (1958), p. 30.
24. J.E. Falk and J.N. Phillips in "Chelating Agents and Chelate Compounds", Eds. D.P. Mellor and F.P. Dwyer, Academic Press, New York (1964).
25. B. Dempsey, M.B. Lowe, and J.N. Phillips, Ref. 4, p. 31.
26. W.S. Caughey, W.Y. Fujimoto, and B.P. Johnson, Biochemistry, **5**, 3830 (1966).
27. J. Simplicio, Biochemistry, **11**, 2525, 2529 (1972).
28. M.B. Lowe and J.N. Phillips, Nature London, **190**, 262 (1961).
29. B.F. Burnham and J.J. Zuckerman, J. Am. Chem. Soc., **90**, 1547 (1970).
30. E.W. Baker, C.B. Storm, G.T. McGrew, and A.H. Corwin, Bioinorg. Chem., **3**, 49 (1973).
31. R. Grigg, R.J. Hamilton, M.L. Jozefowicz, C.H. Rochester, R.J. Terrell, and H. Wickwar, J. Chem. Soc., Perkin Trans. 2, 407 (1973).
32. P. Hambright and P.B. Chock, J. Am. Chem. Soc., **96**, 3123 (1974).
33. J.A. Clarke, P.J. Dawson, R. Grigg, and C.H. Rochester, J. Chem. Soc., Perkin Trans. 2, 414 (1973).
34. T.I. Strelkova and G.P. Gurinovich, Biophysics (USSR), 13, 1164 (1968).
35. W.K. McEwen, J. Am. Chem. Soc., **58**, 1124 (1936).
36. A. Neuberger and J.J. Scott, Proc. R. Soc. London, Ser. A, **213**, 307 (1952).
37. J. Sutter, P. Hambright, M. Krishnamurthy, and P.B. Chock, Inorg. Chem., **13**, 2764 (1974).
38. (a) R.R. Das, R.F. Pasternack, and R.A. Plane, J. Am. Chem. Soc., **92**, 3312 (1970);
    (b) W.E. White and R.A. Plane, Bioinorg. Chem., **4**, 21 (1974).
    (c) R.R. Das and R.A. Plane, J. Inorg. Nucl. Chem., **37**, 147 (1975).
    (d) R.R. Das, J. Inorg. Nucl. Chem., **37**, 153 (1975).
39. R.E. Walter, J. Am. Chem. Soc., **75**, 3860 (1953).
40. J.J. Scott, J. Am. Chem. Soc., **77**, 325 (1955).
41. D.K. Lavallee and A.E. Gebala, Inorg. Chem., **13**, 2004 (1974).
42. A.H. Jackson and G.R. Dearden, Ann. N.Y. Acad. Sci., **206**, 151 (1973).
43. E.B. Fleischer and A. Stone, J. Am. Chem. Soc., **90**, 2735 (1968).

44. N. Hirayama, A. Takenaka, Y. Sasada, E. Watanabe, H. Ogoshi, and Z. Yoshida, J. Chem. Soc., Chem. Commun., 330 (1974).
45. R.J. Abraham, G.E. Hawkes, M.F. Hudson, and K.M. Smith, J. Chem. Soc., Perkin Trans. 2, 204 (1975).
46. A. Wolberg, J. Mol. Struct., 21, 61 (1974).
47. C.B. Storm, J. Am. Chem. Soc., 92, 1423 (1970).
48. H. Ogoshi, E. Watanabe, and Z. Yoshida, Tetrahedron, 29, 3241 (1973).
49. S. Ruben, M.D. Kamen, M.B. Allen, and P. Nahinsky, J. Am. Chem. Soc., 64, 2297 (1942).
50. N. Ashelford and D. Meller, Aust. J. Sci. Res., 5A 784 (1952).
51. B. West, J. Chem. Soc., 3115 (1952).
52. R.N. Boos, C. Rosenblum, and D.T. Woodbury, J. Am. Chem. Soc., 73, 5446 (1951).
53. R.S. Becker and R.K. Sheline, Arch. Biochem. Biophys., 54, 259 (1955).
54. S. Aronoff, Biochim. Biophys. Acta, 60, 193 (1962).
55. J.W. Barnes and G.D. Dorough, J. Am. Chem. Soc., 72, 4045 (1950).
56. H. Baker, P. Hambright, and L. Ross, Inorg. Chem., 12, 2200 (1973).
57. D.C. Atkins and C.S. Garner, J. Am. Chem. Soc., 74, 3528 (1952).
58. A.H. Corwin and M.H. Melville, J. Am. Chem. Soc., 77, 2755 (1955).
59. A.H. Corwin and P.E. Wei, J. Org. Chem., 27, 4285 (1962).
60. (a) R.D. Farina, D. Halko, and J.H. Swinehart, J. Phys. Chem. 76, 2343 (1972); (b) Z. Schelly, R.D. Farina, and E.M. Eyring, ibid., 74, 617 (1970).
61. W.S. Caughey and A.H. Corwin, J. Am. Chem. Soc., 77, 1509 (1955).
62. C. Grant and P. Hambright, J. Am. Chem. Soc., 91, 4195 (1969).
63. S.K. Cheung, L.F. Dixon, E.B. Fleischer, D.Y. Jeter, and M. Krishnamurthy, Bio-inorg. Chem., 2, 281 (1973).
64. R.R. Das, J. Inorg. Nucl. Chem., 34, 1263 (1972).
65. D. Ostfeld and M. Tsutsui, Acc. Chem. Res., 7, 52 (1974).
66. D.W. Margerum, Rec. Chem. Prog., 24, 237 (1963).
67. (a) R. Hill, Biochem. J. 19, 341 (1925); (b) R.J. Kassner and P.S. Facuna, Bioinorg. Chem., 1, 165 (1972).
68. U. Eisner and M.J.C. Harding, J. Chem. Soc., 4089 (1964).
69. J.W. Buchler, L. Puppe, K. Rohbock, and H.H. Schneehage, Ref. 42, p. 116.
70. B.D. Berzin, Russ. J. Phys. Chem., 36, 258 (1962).
71. R.P. Smirnov and B.D. Berezin, Russ. J. Phys. Chem., 43, 1398 (1969).
72. A.A. Berlin and A.T. Sherle, Inorg. Macromol. Rev., 1, 235 (1971).
73. R. Snellgrove and R.A. Plane, J. Am. Chem. Soc., 90, 3185 (1968).
74. B. Shah, B. Shears, and P. Hambright, J. Am. Chem. Soc., 93, 776 (1971).
75. B. Shah and P. Hambright, J. Inorg. Nucl. Chem., 32, 3420 (1970).
76. D.G. Davis and J.G. Montalvo Jr., Anal. Chem., 41, 1195 (1969).
77. F. Basolo, W.R. Matousch, and R.G. Pearson, J. Am. Chem. Soc., 78, 4833 (1956).
78. J.P. Macquet and T. Theophanides, Can. J. Chem., 51, 219 (1973).
79. D. Ostfeld, M. Tsutsui, C.P. Hrung, and D.C. Conway, J. Am. Chem. Soc., 93, 2549 (1971).
80. M. Tsutsui and C.P. Hrung, J. Am. Chem. Soc., 96, 2638 (1974).
81. M.F. Hudson and K.M. Smith, Tetrahedron Lett., 2223, 2227 (1974).
82. M. Tsutsui and C.P. Hrung, J. Coord. Chem., 3, 193 (1973).
83. M.F. Hudson and K.M. Smith, J. Chem. Soc., Chem. Commun., 515 (1973).
84. A.H. Corwin and R. Singh, J. Org. Chem., 28, 2476 (1963).
85. (a) D.A. Brisbin and R.J. Balahura, Can. J. Chem., 44, 2157 (1966); (b) F.P. Hinz and D.W. Margerum, J. Am. Chem. Soc., 96, 4993 (1974).
86. E.B. Fleischer and J.H. Wang, J. Am. Chem. Soc., 82, 3498 (1960).
87. B.F. Burnham and J.J. Zuckerman, J. Am. Chem. Soc., 90, 1547 (1970).

88. E.B. Fleischer, E.I. Choi, P. Hambright, and A. Stone, Inorg. Chem., **3**, 1284 (1964).
89. P. Hambright, J. Inorg. Nucl. Chem., **32**, 2449 (1970).
90. R. Khosropour and P. Hambright, J. Chem. Soc., Chem. Commun., 13 (1972).
91. Professor F. Longo, personal communication.
92. M.J. Cleare, Coord. Chem. Rev., **12**, 349 (1974).
93. M.Z. Yoshida, H. Ogoshi, T. Omura, E. Watanabe, and T. Kuorsaki, Tetrahedron Lett., 1077 (1972).
94. H. Ogoshi, T. Omura, and Z. Yoshida, J. Am. Chem. Soc., **95**, 1666 (1973).
95. E.I. Choi and E.B. Fleischer, Inorg. Chem., **2**, 94 (1963).
96. D.A. Brisbin and R.J. Balahura, Can. J. Chem., **46**, 3431 (1968).
97. D.J. Klingham and D.A. Brisbin, Inorg. Chem., **9**, 2034 (1970).
99. D.A. Brisbin and G.D. Richards, Inorg. Chem., **11**, 2849 (1972).
100. O. Berezin and L.V. Klopova, Russ. J. Phys. Chem., **45**, 1242 (1971).
101. D.A. Brisbin and J.O. Asgil, J. Chem. Educ., **51**, 211 (1974).
102. B.D. Berezin and O.I. Koifman, Russ. J. Phys. Chem.; **46**, 24 (1972) and **45**, 820 (1971).
103. J.E. Leffer, J. Org. Chem., **20**, 1202 (1955).
104. (a) F.R. Longo, E.M. Brown, D.J. Quimby, A.D. Adler, and M. Meot-ner, Ref. 42, p. 420;

     (b) V. Gutman and R. Schmid, Coord. Chem. Rev., **12**, 263 (1974).
105. A.D. Adler, F.R. Longo, F. Kampas, and J.B. Kim, J. Inorg. Nucl. Chem., **32**, 2443 (1970)
106. B. Shah, B. Shears, and P. Hambright, Inorg. Chem., **10**, 1818 (1971).
107. D.B. Rorabacher, Inorg. Chem., **5**, 1890 (1966).
108. P. Hambright, unpublished results.
109. A.H. Corwin, Ref. 42, p. 201.
110. B. Shears and P. Hambright, Inorg. Nucl. Chem. Lett., **6**, 679 (1970).
111. R.C. Ellington and A.H. Corwin, J. Am. Chem. Soc., **68**, 1112 (1946).
112. R. Grigg, A.W. Johnson, and G. Shelton, Ann. Chem., **746**, 32 (1971).
113. P. Hambright, Ref. 42, p. 443.
114. R.J. Kassner and J.H. Wang, J. Am. Chem. Soc., **88**, 5170 (1966).
115. R.J. Kassner and H. Walchak, Biochim. Biophys. Acta, **304**, 294 (1973).
116. J.N. Phillips, Enzymologia **32**, 13 (1968).
117. M.B. Lowe and J.N. Phillips, Nature London, **194**, 1058 (1962).
118. A. Bluestein and J.M. Sugihara, Inorg. Chem., **12**, 690 (1973).
119. A. Bluestein and J.M. Sugihara, J. Inorg. Nucl. Chem., **35**, 1048 (1973).
120. P. Wei, A.H. Corwin, and R. Arellano, J. Org. Chem., **27**, 3344 (1962).
121. S.J. Baum, B.F. Burnham, and R.A. Plane, Proc. Nat. Acad. Sci. U.S.A., **52**, 1439 (1964).
122. S.J. Baum and R.A. Plane, J. Am. Chem. Soc., **88**, 910 (1966).
123. J. Weaver and P. Hambright, Inorg. Chem., **8**, 167 (1969).
124. N. Johnson, R. Khosropour, and P. Hambright, Inorg. Nucl. Chem. Lett., **8**, 1063 (1972).
125. R. Kolski and R.A. Plane, private communication.
126. I. Schiller, K. Bernauer, and S. Fallab, Experientia, **17**, 540 (1961).
127. J.O. Edwards, J. Am. Chem. Soc., **76**, 1540 (1954).
128. D.K. Cabiness and D.W. Margerum, J. Am. Chem. Soc., **92**, 2151 (1970).
129. "Inorganic Reaction Mechanisms", Eds. J. Burgess, D. Hague, R.D.W. Kemmitt, and A. McAuley, The Chemical Society, London (1971), Vol. 1, Chap. 3.
130. L.E. Bennett, Prog. Inorg. Chem., **18**, 1 (1973).
131. J.K. Yandell, D.P. Fay, and N. Sutin, J. Am. Chem. Soc., **95**, 1131 (1973).
132. H.L. Hodges, R.A. Holwerda, and H.B. Gray, J. Am. Chem. Soc., **96**, 3132 (1974).
133. R.X. Ewall and L.E. Bennett, J. Am. Chem. Soc., **96**, 940 (1974).

134. C. Creutz and N. Sutin, Proc. Nat. Acad. Sci. U.S.A., **70**, 1701 (1973); Inorg. Chem. **13**, 2041 (1974).
135. C.J. Grimes, O. Piszkiewicz, and E.B. Fleischer, Proc. Nat. Acad. Sci. U.S.A., **71**, 1408 (1974).
136. P. Hambright and E.B. Fleischer, Inorg. Chem., **4**, 912 (1965).
137. I.A. Cohen, Ref. 42, p. 453; I.A. Cohen, C. Jung, and T. Governo, J. Am. Chem. Soc., **94**, 3003 (1974).
138. G.N. La Mar, J. Am. Chem. Soc., **95**, 1662 (1973).
139. S.S. Eaton and G.R. Eaton, J. Chem. Soc., Chem. Commun., 576 (1974).
140. R.F. Pasternack and N. Sutin, Inorg. Chem., **13**, 1956 (1974).
141. J. James and P. Hambright, J. Coord. Chem., **3**, 183 (1973).
142. P. Hambright and P.B. Chock, Inorg. Chem., **13**, 3029 (1974).
143. I. Tabushi and S. Kojo, Tetrahedron Lett., 1577 (1974).
144. (a) J.A. Weightman, N.J. Hoyle, and R.J.P. Williams, Biochim. Biophys. Acta, **244**, 567 (1971); (b) K. Yamamoto and T. Kwan, Bull. Chem. Soc. Jpn., **45**, 664 (1972).
145. C.E. Castro and H.F .Davis, J. Am. Chem. Soc., **91**, 5405 (1969).
146. J.B. Paine and D. Dolphin, J. Am. Chem. Soc., **93**, 4080 (1971).
147. A. Grigg, A. Sweeney, and A.W. Johnson, Chem. Commun., 1237 (1970).
148. O.W. Kao and J.H. Wang, Biochemistry **4**, 342 (1965).
149. (a) I.A. Cohen and W.S. Caughey, Biochemistry, **7**, 636 (1968); (b) K. Yamamoto and T. Kwan, Bull. Chem. Soc. Jpn. **45**, 664 (1972).
150. W.S. Caughey, in 'Inorganic Biochemistry', Ed. G.L. Eichorn, Elsevier, Amsterdam (1973), Chap. 24.
151. D V. Stynes, H.C. Stynes, J.A. Ibers, and B.R. James, J. Am. Chem. Soc., **95**, 1142 (1973).
152. R.S. Wade, R. Havlin, and C.E. Castro, J. Am. Chem. Soc., **91**, 7530 (1969).
153. R.S. Wade and C.E. Castro, J. Am. Chem. Soc., **95**, 216 (1973).
154. R.S. Wade and C.E. Castro, J. Am. Chem. Soc., **95**, 231 (1973).
155. D. Paulson, R. Ullman, R.D. Sloane, and G.L. Closs, J. Chem. Soc., Chem. Commun. 186 (1974).
156. (a) M. Tsutsui, R. Velapoldi, L. Hoffman, K. Suzuki, and A. Ferrari, J. Am. Chem. Soc., **91**, 3337 (1969); (b) M. Tsutsui and T.S. Srivastava, Ref. 42, p. 404.
157. D.G. Whitten, E.W. Baker, and A. Corwin, J. Org. Chem., **28**, 2363 (1963).
158. D.V. Stynes, H.C. Stynes, B.R. James, and J.A. Ibers, J. Am. Chem. Soc., **95**, 1796 (1973).
159. T. Okimoto, M. Takahashi, Y. Inaki, and K. Takemoto, Polymer Lett., 121 (1974).
160. J. Manassen, J. Catal., **33**, 133 (1974).
161. E.B. Fleischer and M. Krishnamurthy, J. Am. Chem. Soc., **94**, 1382 (1972).
162. E.B. Fleischer and M. Krishnamurthy, Ref. 42, p. 32.
163. J. Chatt, C.M. Elson, and G.L. Leigh, J. Am. Chem. Soc., **95**, 2408 (1973).
164. S.B. Brown, T.C. Dean, and P. Jones, Biochem. J., **117**, 741 (1970).
165. P. Waldmeier and H. Sigel, Inorg. Chem., **11**, 2174 (1972).
166. H. Ruesch and H. Sigel, Chimia, **27**, 533 (1973).
167. P. Waldmeier and H. Sigel, J. Inorg. Nucl. Chem., **35**, 1741 (1973).
168. J.H. Wang, J. Am. Chem. Soc., **77**, 822, 4715 (1955).
169. S.S. Eaton, G.R. Eaton, and R.H. Holm, J. Organomet. Chem., **39**, 179 (1972).
170. J.H. Wang, Acc. Chem. Res., **3**, 90 (1970).
171. C. Kirksey, P. Hambright, and C. Storm, Inorg. Chem., **8**, 2141 (1969).
172. C. Kirksey and P. Hambright, Inorg. Chem., **9**, 958 (1970).
173. G.C. Vogel and L.A. Searby, Inorg. Chem., **12**, 936 (1973).
174. C.J. Miller and G. Dorough, J. Am. Chem. Soc., **74**, 3977 (1972).
175. E.W. Baker, M.C. Brookhart, and A.H. Corwin, J. Am. Chem. Soc., **86**, 4587 (1964).

176. E. Higginbotham and P. Hambright, Inorg. Nucl. Chem. Lett., **8**, 747 (1972).
177. P. Hambright, Chem. Commun., 470 (1967).
178. F.A. Walker, E. Hui, and J.M. Walker, J. Am. Chem. Soc., **97**, 2390 (1975).
179. B. McLees and W.S. Caughey, Biochemistry, **1**, 642 (1968).
180. S.J. Cole, G.C Curthoys, E.A. Magnusson, and J.N. Phillips, Inorg. Chem., **11**, 1024 (1972).
181. R. Lee and P. Hambright, J. Inorg. Nucl. Chem., **32**, 447 (1970).
182. C. Storm, A.H. Corwin, R. Arellano, M. Martz, and R. Weintraub, J. Am. Chem. Soc., **88**, 2525 (1966).
183. J.J. Katz, Ref. 150, Chap. 29.
184. W.S. Caughey, R. Deal, B. McLees, and J.O. Alben, J. Am. Chem. Soc., **84**, 1735 (1962).
185. R.J. Abraham and P.F. Swinton, J. Chem. Soc. B, 903 (1969).
186. (a) R.F. Pasternack, E.G. Spiro, and M. Teach, J. Inorg. Nucl. Chem., **36**, 599 (1974); (b) R.F. Pasternack and N. Sutin, personal communication.
187. S.J. Cole, G.C. Corthoys, and E.A. Magnusson, J. Am. Chem. Soc., **92**, 2991 (1970).
188. S.J. Cole, G.C. Corthoys, and E.A. Magnusson, J. Am. Chem. Soc., **93**, 2153 (1971).
189. T.H. Davies, Biochim. Biophys. Acta, **286**, 84 (1972).
190. W.A. Gallagher and W.B. Elliott, Biochem. J. **97**, 189 (1965).
191. J.H. Wang, A. Nakahara, and E.B. Fleischer, J. Am. Chem. Soc., **80**, 1109 (1958).
192. J.O. Alben and W.S. Caughey, Biochemistry, **7**, 175 (1968).
193. D.V. Stynes, H.C. Stynes, B.R. James, and J.A. Ibers, J. Am. Chem. Soc., **95**, 4087 (1973).
194. C.K. Chang and T.G. Traylor, Proc. Nat. Acad. Sci. U.S.A., **70**, 2647 (1973).
195. (a) C.K. Chang and T.G. Traylor, J. Am. Chem. Soc., **95**, 5810 (1973); (b) ibid., **95**, 8475 (1973); (c) ibid., **95**, 8477 (1973); (d) W.S. Brinigar, C.K. Chang, J. Geibel, and T.G. Traylor, ibid., **96**, 5597 (1974); (e) W.S. Brinigar and C.K. Chang, ibid., **96**, 5595 (1974); (f) G.C. Wagner and R.J. Kassner, ibid., **96**, 5593 (1974); (g) D.L. Anderson, C.J. Weschler, and F. Basolo, ibid., **96**, 5599 (1974); (h) J. Almog, J.E. Baldwin, R. Dyer, J. Huff, and C.J. Wilkerson, ibid., **96**, 5600 (1974).
196. (a) J.P. Collman, R. Gagne, T.R. Halbert, J.C. Marchon, and C.A. Reed, J. Am. Chem. Soc., **95**, 7868 (1973); (b) J.P. Collman, R. Gagne, C.A. Reed, T.R. Halbert, G. Lang, and W. Robinson, J. Am. Chem. Soc., **97**, 1427 (1975).
197. J.P. Collmann, R.R. Gagne, and C.A. Reed, J. Am. Chem. Soc., **96**, 2629 (1974).
198. (a) J.E. Baldwin and J. Huff, J. Am. Chem. Soc., **95**, 5757 (1973); (b) J. Almog, J.E. Baldwin, and J. Huff, J. Am. Chem. Soc., **97**, 228 (1975).
199. A.H. Corwin and Z. Reyes, J. Am. Chem. Soc., **78**, 2437 (1956).
200. J.O. Alben, W.H. Fuchsman, C.A. Beaudreau, and W.S. Caughey, Biochemistry **7**, 624 (1968).
201. I.A. Cohen, J. Am. Chem. Soc., **91**, 1980 (1969).
202. E.B. Fleischer and T.S. Srivastava, J. Am. Chem. Soc., **91**, 2403 (1969).
203. S.B. Brown, T.C. Dean, and P. Jones, Biochem. J. **117**, 733 (1970).
204. J. Shack and W.M. Clark, J. Biol. Chem., **171**, 143 (1947).
205. R. Cowgill and W.M. Clark, J. Biol. Chem., **198**, 33 (1952).
206. M. Momenteau, Biochim. Biophys. Acta, **304**, 814 (1973).
207. J.M. Duclos, Bioinorg. Chem., **2**, 263 (1973).
208. C.L. Coyle, P.A. Rafson, and E.H. Abbott, Inorg. Chem., **12**, 2007 (1973).
209. J.M. Duclos, personal communication.
210. (a) L. Bullard, R. Panayappan, A. Thorpe, P. Hambright, and G. Ng, Bioinorg. Chem., **3**, 41 (1973); (b) H. Ogoshi, E. Watanabe, and Z. Yoshida, Chem. Lett., 989 (1973).

211. G.N. LaMar and F.A. Walker, J. Am. Chem. Soc., **94**, 8607 (1972).
212. T.H. Davies, Biochim. Biophys. Acta **329**, 108 (1973).
213. E.H. Abbott and P.A. Rafson, J. Am. Chem. Soc., **96**, 7378 (1974).
214. H.A.O. Hill, A.J. MacFarlane, and R.J.P. Williams, J. Chem. Soc. A, 1704 (1969).
215. C.D. Barry, H.A.O. Hill, J.P. Sadler, and R.J.P. Williams, Proc. R. Soc. London, Ser. A, **334**, 493 (1973).
216. D. Mauzerall, Biochemistry, **4**, 1801 (1965).
217. M. Momenteau, J. Mispelter, and D. Lexa, Biochim. Biophys. Acta, **320**, 652 (1973).
218. C.E. Castro, Bioinorg. Chem., **4**, 45 (1974).
219. L. Vaska and H. Nakai, J. Am. Chem. Soc., **95**, 5431 (1973).
220. H.A.O. Hill, D.R. Turner, and G. Pellizer, Biochem. Biophys. Res. Commun., **56**, 739 (1974).
221. L.J. Boucher, Coord. Chem. Rev., **7**, 289 (1972).
222. D.G. Davis and J. Montalvo, Anal. Lett., **1**, 641 (1968).
223. (a) D.G. Davis and L.A. Truxillo, Anal. Chim. Acta, **64**, 55 (1973); (b) M. Momenteau and D. Lexa, Biochem. Biophys. Res. Commun., **58**, 940 (1974).
224. F.A. Walker, J. Am. Chem. Soc., **92**, 4235 (1970).
225. F.A. Walker, J. Am. Chem. Soc., **95**, 1151 (1973).
226. H.C. Stynes and J.A. Ibers, J. Am. Chem. Soc., **94**, 1559 (1972).
227. R.G. Little, B.M. Hoffman, and J. Ibers, Bioinorg. Chem., **3**, 207 (1974).
228. P. Gallagher and W.B. Elliot, Ref. 42, p. 480.
229. F.A. Walker, J. Am. Chem. Soc., **95**, 1154 (1973).
230. R.M. Guidry and R.S. Drago, J. Am. Chem. Soc., **95**, 6645 (1973).
231. J.A. Ibers, D.V. Stynes, H.C. Stynes, and B.P. James, J. Am. Chem. Soc., **96**, 1358 (1974).
232. B.B. Wayland, J.V. Minkiewicz, and M.E. Abd-Elmageed, J. Am. Chem. Soc., **96**, 2795 (1974).
233. H.C. Stynes and J.A. Ibers, J. Am. Chem. Soc., **94**, 5125 (1972).
234. G.C. Hui, C.A. Spillburg, C. Bull, and B.M. Hoffman, Proc. Nat. Acad. Sci. U.S.A., **69**, 2122 (1972).
235. H. Yamamoto, F.J. Kayne, and T. Yonetani, J. Biol. Chem., **249**, 682 (1974).
236. I. Schiller and K. Bernauer, Helv. Chim. Acta, **46**, 3002 (1963).
237. J.G. Jones and M.V. Twigg, Inorg. Nucl. Chem. Lett., **8**, 307 (1972).
238. R.F. Pasternack, L. Francesconi, D. Raff, and E. Spiro, Inorg. Chem., **12**, 2606 (1973).
239. J. Sutter, M. Krishnamurthy, and P. Hambright, J. Chem. Soc., Chem. Commun., 13 (1975).
240. P. Jones, K. Prudhoe, and S.B. Brown, J. Chem. Soc., Dalton Trans., 912 (1974).
241. J.-H. Fuhrhop, P. Wasser, D. Riesner, and D. Mauzerall, J. Am. Chem. Soc., **94**, 7996 (1972).
242. K.A. Zachariasse and D.G. Whitten, Chem. Phys. Lett., **22**, 527 (1973).
243. F.E. Dickson and L. Petrakis, J. Phys. Chem., **74**, 2805 (1970).
244. E.B. Fleischer, J.M. Palmer, T.S. Srivastava, and A. Chatterjee, J. Am. Chem. Soc., **93**, 3162 (1971).
245. B.G. Kolski and R.A. Plane, Ref. 42, p. 604.
246. B.G. Kolski and R.A. Plane, J. Am. Chem. Soc., **94**, 3740 (1972).
247. Y. Inada and K. Shibata, Biochem. Biophys. Res. Commun., **9**, 323 (1962).
248. R.G. Wilkins and R.E. Yelin, Inorg. Chem., **8**, 1470 (1969).
249. P. Hambright and P.B. Chock, J. Inorg. Nucl. Chem. **37**, in press (1975).
250. G.S. Wilson and B.P. Neri, Ref. 42, p. 568.
251. J. Hodgkinson and R.B. Jordan, J. Am. Chem. Soc., **95**, 763 (1973).
252. L. Rusank and R.B. Jordan, Inorg. Chem., **11**, 196 (1972).

253. N.S. Angerman, B.B. Hasinoff, H.B. Dunford, and R.J. Jordan, Can. J. Chem., **47**, 3217 (1969).

254. H.A. Degani and D. Fiat, J. Am. Chem. Soc., **93**, 4281 (1971).

255. (a) J. Simplicio and K. Schwenzer, Biochemistry, **12**, 1923 (1973); (b) P. Hambright, M. Krishnamurthy, and P.B. Chock, J. Inorg. Nucl. Chem., **37**, 557 (1975).

256. B.B. Hasinoff, H.B. Dunford, and D.G. Horne, Can. J. Chem., **47**, 3217 (1969).

257. E.B. Fleischer, S. Jacobs, and L. Mestichelli, J. Am. Chem. Soc., **90**, 2527 (1968).

258. D. Thusius, J. Am. Chem. Soc., **93**, 2629 (1971).

259. H.A.O. Hill, Ref. 150, Chap. 30.

260. E.B. Fleischer and M. Krishnamurthy, J. Am. Chem. Soc., **93**, 3784 (1971).

261. M. Krishnamurthy and E.B. Fleischer, personal communication.

262. R.F. Pasternack and M.A. Cobb, J. Inorg. Nucl. Chem., **35**, 4327 (1973).

263. K.R. Ashley, M. Berggren, and M. Cheng, J. Am. Chem. Soc., **97**, 1422 (1975).

264. G.N. LaMar and F.A. Walker, J. Am. Chem. Soc., **94**, 8607 (1972).

265. F.A. Walker and G.N. LaMar, Ref. 42, p. 328.

266. G.R. Eaton and G.N. LaMar, Coord. Chem. Rev., In Press (1975).

267. M. Tsutsui, D. Ostfeld, J.N. Francis, and L.M. Hoffman, J. Coord. Chem., **1**, 115 (1971).

268. S.S. Eaton, G.R. Eaton, and R.H. Holm, J. Organometal. Chem., **32**, C52 (1971).

269. J.W. Faller and J.W. Sibert, J. Organometal. Chem., **31**, C5 (1971).

270. D.V. Stynes and B.R. James, J. Chem. Soc., Chem. Commun., 325 (1973).

271. D.V. Stynes and B.R. James, J. Am. Chem. Soc., **96**, 2733 (1974).

272. H.P. Bennetto, J.G. Jones, and M.V. Twigg, Inorg. Chim. Acta, **4**, 180 (1970).

273. J.G. Jones and M.V. Twigg, Inorg. Chem., **8**, 2120 (1969).

274. R.G. Yalman, in "Mechanisms of Inorganic Reactions", Advances in Chem. Series, No. 49, Amer. Chem. Soc., Washington, D.C. (1965), p. 47.

275. P. Hambright, R. Fawwaz, J. McRae, P. Valk, and A.J. Bearden, Bioinorg. Chem., **5**, in press (1975).

276. R. Fawwaz, W. Hemphill, and H.S. Winchell, J. Nucl. Med., **12**, 231 (1971).

277. R. Fawwaz, F. Frye, W. Loughman, and W. Hemphill, J. Nucl. Med., **15**, 997 (1974).

278. G.N. La Mar and D. Viscio, J. Am. Chem. Soc., **96**, 7354 (1974).

METALLOPORPHYRINS WITH UNUSUAL GEOMETRY

MINORU TSUTSUI

Texas A&M University, Department of Chemistry, College Station, Texas 77843, U.S.A.

and GLENN A. TAYLOR

Union Carbide Corp., Research and Development Department, Chemical and Plastic Division, Tarrytown, N.Y. 10591, U.S.A.

7.1. Introduction

There is a growing interest in metalloporphyrins among chemists because of the unique nature of the coordination chemistry, for both the porphyrin ligand as well as the metal ion, of these materials. Recent progress in the chemistry of synthetic metalloporphyrins has shown that the porphyrin moiety can act as a bi-, tri- or hexadentate ligand, as well as the usual tetradentate ligand[1-7]. In addition, the metal ion has been observed to possess 4-, 5,- 6-, or 8-coordination. Furthermore, they are of interest to chemists because of their obvious relevance as biological models. This high level of interest in porphyrins and metalloporphyrins is fully justified by their behavior as complex physicochemical systems and the biological circumstance that iron porphyrins serve as the heme or prosthetic groups in several classes of the hemeproteins[1]. For example, the varied chemical and physical properties of metalloporphyrins are reflected in the biological significance of porphyrinic compounds (such as chlorophyll[8], hemoglobin[9], cytochrome[10], and vitamin B_{12}[11,12]) in photosynthesis, gas transport, enzymatic catalyses, metabolic regulation and control, electron-transport, etc. Changes or modifications of general porphyrin metabolism have been associated with cancer, drug metabolism, and specific disease syndromes[13]. Chemists are also interested in the synthesis of bridged and metal—metal bonded complexes as models for studying the role of the heme group in our life processes, (mitochondrial electron transfer mechanism) through a better understanding of the electron distribution in these metalloporphyrin complexes. These complexes as well as the out-of-plane metalloporphyrins will

Porphyrins and Metalloporphyrins, ed. Kevin M. Smith

be discussed in this section. Clearly, the better understanding of the structure and chemical properties of metalloporphyrins are of great importance toward understanding their biological functions.

As we have already seen (Chapter 6), kinetic studies on the formation and reactivity of metalloporphyrins have suggested the existence of a 'sitting-a-top' (SAT) intermediate (Fig. 11) in these mechanisms. In this configuration the metal ion deforms the porphyrin macrocycle so as to facilitate the incorporation of the metal ion in the subsequent step. Such intermediates are also postulated in the oxygen transport mechanism of hemoglobin.

Besides the kinetic evidence, there are also structural arguments for the existence of these out-of-plane metalloporphyrins. We do not wish to give a detailed argument based on structural considerations for the existence of out-of-plane metalloporphyrins, since Hoard will deal with this in his contribution (Chapter 8). We would like, however, to give a general description of the main points. Porphyrins like other π-macrocycles have a central 'hole' or 'core' (Fig. 1) of fixed size which can be altered by the puckering of the porphyrin macrocycle, this consequently limits it to a narrow range of variation. The range of the variation of the 'core' has been observed to lie between 2.098 Å and 1.929 Å. It is easy to see that in certain complexes the metal ion is unable to fit into this hole and therefore lies out of the porphinato nitrogen plane. Numerous X-ray diffraction analyses have confirmed this hypothesis. It is of considerable interest to establish metal—nitrogen distances as these critically affect theoretical calculations through the overlap of metal—nitrogen orbitals. Thus, any regularities deducible from X-ray studies are extremely important, especially since it is not always possible to carry out structural determinations for each metalloporphyrin complex that would be studied theoretically. With this in mind, we might venture the following generalizations:

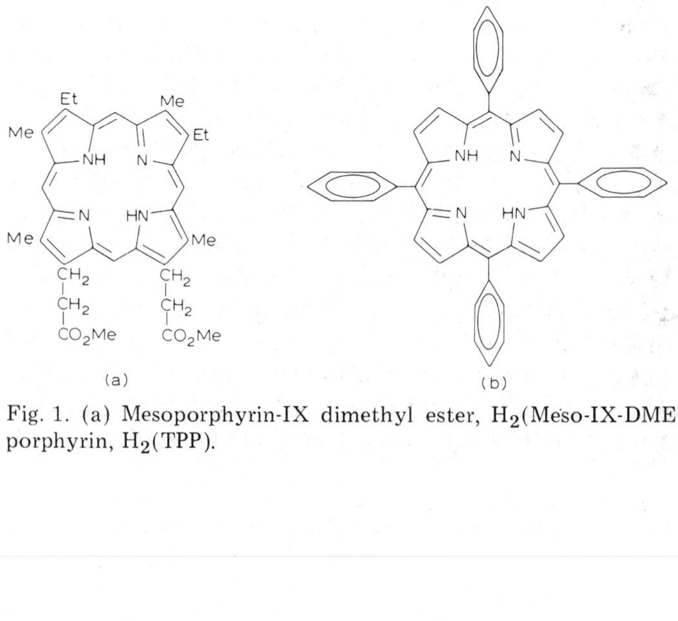

Fig. 1. (a) Mesoporphyrin-IX dimethyl ester, H_2(Meso-IX-DME). (b) *meso*-Tetraphenyl-porphyrin, H_2(TPP).

(a) The porphyrin central 'core' or 'hole' radius has been observed to lie between a minimum of 1.929 Å[14] for the highly ruffled tetragonal form of Ni(OEP) to a maximum of 2.098 Å[15] for the highly planar form of $Sn^{IV}(TPP)Cl_2$.

(b) As pointed out by Hoard[16], for structures with identical metal ions, but in different spin states, the high-spin structure has a larger metal—nitrogen distance and is nonplanar in all structures so far determined. An example of this phenomenon is the ferrous-ferric ions taking the difference between the empirical ionic radii of ferrous and ferric iron as 0.12 Å for a coordination number of six to the high-spin ferrous porphyrin gives 2.19 Å for the length of the complexing bonds to the nitrogen atoms. Taking the 'core' radius as 2.01 Å then yields a calculated out-of-plane displacement of 0.87 Å for the ferrous ion. Although these values are surely overestimated, the prediction that the ferric ion lies in the plane and the ferrous out-of-plane appears valid[17]. Interconversion between these two forms has been postulated[7] as part of the mechanism for the hemoglobin oxygen transport system. In hemoglobin, the iron atom in the oxygenated form lies roughly in-plane, while the metal lies out-of-plane in the deoxy form.

(c) The nitrogen to center distance for the first row transition elements with partly filled d shells is smaller than that found for elements such as Mg^{18}, Zn^{19} and $Sn^{15,20}$. It is the largest for $V = O^{21}$, at the beginning of the series, which is in agreement with the trend in atomic radii.

It must be noted that these are generalizations and are by no means intended to cover the subtleties of porphyrin conformation, such as the variation of ring structural parameters with the central core radius, the nonplanarity of closed shell metals such as Zn and Mg, unusual axial ligand properties, etc. These aspects will be covered later by Hoard (Chapter 8).

Certainly most of the research in metalloporphyrins stems from interest in the biological systems and by their potential medicinal value, and would be justification enough for a great deal of research interest. However, metalloporphyrins are studied for other reasons as well, such as the search for new semiconductors[22], superconductors[23], catalysts[24], and chemical shift reagents[25]. Some of these aspects will be discussed in other chapters. Several porphyrin-related compounds, particularly the phthalocyanines, have proved useful as dyes[26], such as the fast drying blue inks and the blue and green coloring in security notes and bonds. Even without their biological and industrial implications, the properties of metalloporphyrins would be studied for their purely theoretical importance.

While in the last few years we have seen a substantial growth in metalloporphyrin research[1-6], a rapid development in a unique area of chemistry, out-of-plane metalloporphyrins, has been particularly noticeable. Clearly, the importance of these, out-of-plane, complexes warrants the allocation of a separate section to their description. The theoretical arguments, kinetic and structural, have been described in detail by Buchler, Hambright and Hoard in their chapters, and briefly mentioned in the introduction. Therefore, they will not be discussed, except in the way of emphasizing a point, any further in this section. In addition, the physical properties, except for those unique to these complexes, will be discussed under the appropriate sections in this book. We wish to confine ourselves only to the detailed synthesis and structures of these out-of-plane complexes and metalloporphyrins with unusual geometries.

7.2. Syntheses
7.2.1. Traditional method

The major problem in the traditional method (prior to 1966) of organo-metalloporphyrin syntheses is that of the difficulty in dissolving both the free porphyrin and the metallic salt simultaneously in the same solution under reactive conditions. These syntheses may be classified into one of two general reaction types: 1) reaction of a porphyrin with a metallic salt in an acidic medium (e.g. acetic acid), 2) reaction of a porphyrin with a metallic salt in a basic medium (e.g. pyridine). While these reactions are successful and useful, they suffer from certain deficiencies as general or convenient methods of syntheses. This is due to the fact that good solvents for the porphyrins in their un-ionized forms are generally poor solvents for simple metallic ions and vice versa.

$$M^{II} + H_2(P) \rightarrow M(P) + 2 H^+ \tag{1}$$

M = metal
P = porphyrinic material

Acidic media typically require a large excess of the metallic salt, usually a thousand fold or greater, in order to force the reaction toward metal insertion. This is particularly true where the thermodynamic stability of the complex is low, such as in the formation of zinc complexes, or where the formation of the porphyrin acid cation, PH_4^{2+}, is facile. In the basic medium, the porphyrin must compete, often disadvantageously, with the solvent as a complexing agent for the metal ion. The yields are frequently low and the rates of reaction are generally slow for these methods. Thus, it is often inconvenient to prepare a relatively large amount of the material at one time, and the further workup of the product for purification is frequently troublesome or tedious. Recently, new synthetic techniques, which have been shown to be capable of circumventing this problem, have been developed.

7.2.2. Carbonyl method

The first of the new synthetic methods to be discovered was the insertion of a metal from a carbonyl complex[27,28]. Instead of simple ligand exchange with the porphyrin dianion, the porphyrin is oxidized by the reduction of the porphyrin pyrrole protons. For example:

$$Cr(CO)_6 + H_2(P) \xrightarrow[\text{decalin}]{185°C} Cr^{II}(P) + 6\ CO\uparrow + H_2\uparrow \qquad (2)$$

P = porphyrinic material

The mechanism for this class of reactions has never been studied, but it is believed that the evolution of hydrogen and carbon monoxide, together with the fact that metal carbonyls dissolve in the same organic solvents as do porphyrins, drives the reaction toward metal insertion. By inference from the rhenium complexes, *vide infra*, it can be surmised that the metal is oxidized in a series of one-electron steps, each of which is accompanied by the reduction of a pyrrole proton.

$$2\ M(O) + 2\ H_2(P) \xrightarrow[\Delta]{\text{Solvent}} 2\ H\text{-}M^I(P) + H_2\uparrow \qquad (3)$$

$$2\ H\text{-}M^I(P) \xrightarrow[\Delta]{\text{Solvent}} 2\ M^{II}(P) + H_2\uparrow \qquad (4)$$

Solvent = decalin, mesitylene, or toluene, benzene

The carbonyl method has been used to synthesize previously unknown complexes of chromium[27,28], molybdenum[30,31], ruthenium[32,33], rhodium[34-37,39], iridium[38,39], rhenium[40,41], and technetium[42,43]. Metalloporphyrins containing some of these metals (chromium[44,45], molybdenum[44], rhenium[46], and rhodium[47]) have subsequently been prepared by other methods, but the complexes formed from carbonyls are still unique. In some cases (ruthenium(II), iridium(I), rhenium(I), and technetium(I)) one or more carbonyl ligands are retained by the metal. Also, metalloporphyrin complexes with the metal ion in a low oxidation state are generally obtained from metal carbonyls. Furthermore, several novel M(I) porphyrin complexes which have been prepared using metal carbonyls, *vide infra*, are certainly unique to this method (Table 1).

7.2.3. Hydride method

A similar method involves the oxidation of hydride ions from a metal hydride complex. For example[48]:

$$AlH_3 + H_2(P) \xrightarrow[\text{THF}]{65°C} \xrightarrow{+H_2O} Al(P)OH + 3\ H_2\uparrow \qquad (5)$$

References, p. 310

TABLE 1

Spectral data for several metalloporphyrin complexes prepared via the carbonyl method

Compound	M.P. (°C)	UV, max (log ϵ) (in CH_2Cl_2)	IR, cm^{-1} (KBr pellet)	PMR (in $CDCl_3$)
H(Meso-IX-DME)Re(CO)$_3$	175–177	392 nm (Soret) 485 585	1740 (ester, CO) 1880 (νM—CO) 2020 (νM—CO) 3380 (νN—H)	−0.5 τ (s), =CH— −0.2 (s), =CH— 5.3–7.0 (m), alkyl 8.0–8.4 (t), alkyl 14.90 (s), NH
H(TPP)Re(CO)$_3$	302–304	402 nm (6.12) 473 (5.62) 670 (4.92)	1875 (νM—CO) 2010 (νM—CO) 3350 (νN—H)	0.89 τ (q), pyrrole 1.12 (d), pyrrole 1.28 (s), pyrrole 1.70 (m), phenyl 2.13 (m), phenyl 14.00 (s), NH
(Meso-IX-DME)[Re(CO)$_3$]$_2$	246–248	400 nm (Soret) 480 sh 520	1730 (ester, CO) 1900 (νM—CO) 2015 (νM—CO)	−0.30 τ (s), =C— 5.3–7.0 (m), alkyl 8.0–8.4 (t), alkyl
TPP[Re(CO)$_3$]$_2$	350, dec.	408 nm (Soret) 485 sh 513	1900 (νM—CO) 2025 (νM—CO)	0.80 τ (s), pyrrole 1.70 (m), phenyl 2.20 (m) phenyl
(OC)$_3$Re(Meso-IX-DME)Tc(CO)$_3$	238–240	398 nm (5.04) 480 sh (3.80) 513 (4.46)	1720 (ester, CO) 1740 (ester, CO) 1925 (νM—CO) 2030 (νM—CO) 2045 (νM—CO)	−0.60 τ (s), =CH— 5.3–7.0 (m), alkyl 8.0–8.5 (m), alkyl

H(Meso-IX-DME)Tc(CO)$_3$	181–182	388 nm (4.17) 473 (3.70) 580 (3.17)	1735 (ester, CO) 1900 (νM—CO) 1920 (νM—CO) 2025 (νM—CO) 3380 (νN—H)	−0.5 τ (s), =CH— −0.2 (s), =CH— 5.3—7.0 (m), alkyl 8.0—8.4 (t), alkyl 14.90 (s), NH
(Meso-IX-DME)[Tc(CO)$_3$]$_2$	227–229	396 nm (4.43) 480 sh (3.60) 507 (3.93)	1740 (ester, CO) 1925 (νM—CO) 2036 (νM—CO)	−0.40 τ (s), =CH— 5.3—7.0 (m), alkyl 8.0—8.3 (t), alkyl
(TPP)[Tc(CO)$_3$]$_2$	322–325	403 nm (Soret) 475 sh 504 670	1915 (νM—CO) 2030	

References, p. 310

The small number of metals for which hydrides are available has probably restricted the use of this method. One other example of a metal hydride reacting with a porphyrin is the reaction of sodium borohydride with *meso*-tetraphenylporphyrin. The product, which has never been fully characterized, may be a boron porphyrin[49].

7.2.4. Organometallic method

Another method developed to circumvent this solubility problem, using an organometallic compound as the metal source, was reported by Tsutsui and co-workers in 1966. Diphenyltitanium was found to react with mesoporphyrin-IX dimethyl ester, H_2(Meso-IX-DME), to give a titanyl porphyrin complex[50]. The reason for its success is that both the porphyrin and organometallic compound used here are soluble in a non-polar solvent, and, furthermore, the stable organic moiety formed in the reaction helps drive the equilibrium in the desired direction — metal insertion. Presumably, air oxidizes the titanium to the more stable +4 state. Although this method has not received much attention, except for the acetylacetone complexes, it seems reasonable that other organometallic compounds could be used in a similar manner.

$$(C_6H_5)_2Ti + H_2(\text{Meso-IX-DME}) \xrightarrow[\text{mesitylene}]{240°C} \xrightarrow{+O_2} O=Ti(\text{Meso-IX-DME})$$

$$+ 2C_6H_6 \tag{6}$$

7.2.5. Acetylacetone derivatives

Buchler and co-workers[44] found that metal acetylacetone derivatives were both a readily available source of metal and reasonably soluble in organic solvents. They studied the reaction of octaethylporphyrin[51], H_2(OEP), with a number of metal acetylacetones, using melts of phenol, quinoline and imidazole as solvents.

$$M(\text{acac})_n + H_2(\text{OEP}) \rightarrow M(\text{OEP})(\text{acac})_{n-2} + 2\,H\text{-(acac)} \tag{7}$$

In addition to a number of previously known metalloporphyrins[51−53], new complexes of scandium(III)[44], zirconium(IV)[44] and hafnium(IV)[46,51] were prepared. Of their reactions the only ones that failed were attempted syntheses of cerium and thorium porphyrins.

Indeed, these new synthetic techniques present a powerful tool to the chemists for synthesizing metalloporphyrins. They have, however, several serious disadvantages toward being a generalizable method: The inconvenience involved in the prior acquisition or preparation of the specific derivatives to be used as starting material; the instability of competitive reactions of some of these complexes; and the lack of generality for the various types, especially the hydride complexes.

7.2.6. Using solvents with high dielectric constants

Instead of supplying the metal in the form of a complex, Adler and co-workers solved the solubility problem by using N,N-dimethylformamide (DMF) as a solvent[45]. DMF, due to its high dielectric constant, can dissolve both porphyrins and metal salts. While one cannot exclude the possibility of another form of solvent assistance, such as a stabilized transition state, no evidence of this has been found. Adler has obtained improved yields for many of the previously known compounds by using DMF. However, this method will also prove to be useful for the synthesis of new compounds. Among the compounds thus synthesized was the once-elusive chromium porphyrin.

$$4\ H_2(P) + 4\ CrCl_2 + O_2 \xrightarrow[DMF]{153°C} 4\ Cr^{III}(P)Cl + 4\ HCl\uparrow + 2\ H_2O \tag{8}$$

Due to the simplicity of this method and the ready availability of starting materials, it is not surprising to see compounds which were once made by other methods now being prepared in DMF[47].

The use of polar reaction media had not been previously unknown. It is presumably due to their lower dielectric constants that solvents such as acetone, dioxane, and ethanol[54] were not so successful as DMF.

Buchler[51] further extended the synthesis of metalloporphyrins in a medium with a high dielectric constant. Using a phenol melt and appropriate metal halides or oxides, he prepared complexes of scandium, tantalum, tungsten, osmium, and rhenium. In benzonitrile, porphyrins and metal halides reacted to give an equally novel array of metalloporphyrins, including those of chromium, molybdenum, tungsten, and niobium. For example:

$$H_2(P) + H_2WO_4 \xrightarrow[C_6H_5OH]{220°C} O{=}W(P)(OC_6H_5) \tag{9}$$

$$H_2(P) + MoCl_2 \xrightarrow[C_6H_5CN]{195°C} [O{=}Mo(P)]_2O \tag{10}$$

The tungsten complex is particularly noteworthy, since an attempted synthesis using tungsten hexacarbonyl had proven unsuccessful[55].

7.2.7. Progress

These new synthetic methods have allowed the syntheses of metalloporphyrins with labile ligation. Such compounds are, of course, a great aid to our understanding of the kinetics and mechanisms of these molecules as both biological models and coordination compounds.

In 1964, Falk[1] listed porphyrin complexes for 28 different metals. Today, except for some of the lanthanide and actinide series, at least one porphyrin

References, p. 310

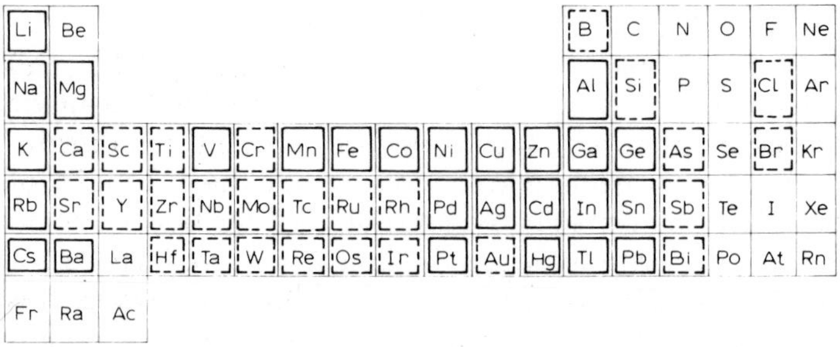

Fig. 2. Elements for which a metalloporphyrin has been made. Metals boxed with full lines were mentioned by Falk[1], but those boxed with broken lines have been prepared more recently.

complex has been made for virtually every metallic element (Fig. 2). However, due to the importance of the oxidation state and axial ligation, it would be incorrect to assume that all porphyrin complexes of potential interest are known. What might be stated is that any reasonable porphyrin complex desired by the modern chemist should be attainable through the synthetic methods now available.

7.3. Structures

7.3.1. Bridged and metal—metal bonded species

Among the reasons for interest in bridged and metal—metal bonded complexes has been the desire to prepare model compounds for studying the role of heme iron in mitochondrial electron transfer[56]. In mitochondria, the cellular bodies in which oxidation occurs, heme proteins called cytochromes are crucial links in the electron transfer chain. A typical chain, as found in beef heart, is[57]:

$$e^- \rightarrow \text{cyt-}b \rightarrow \text{cyt-}c_1 \rightarrow \text{cyt-}c \rightarrow \begin{bmatrix} \text{Cu} \\ \text{cyt-}a_3 \\ \text{cyt-}a_1 \end{bmatrix} \rightarrow O_2$$

The classes of cytochromes (a, b, c, etc.) differ in their spectra, oxidation potentials, porphyrin substitution patterns, and attached proteins. However all (except the d cytochromes) are iron porphyrins.

While electron transfer via the porphyrin ring is a possibility[58], work has concentrated on transfer through the axial sites[59]. Cytochrome-b is known to coordinate two imidazoles in its axial position[60], while the cytochrome-c binds to an imidazole and a methionine sulfur[61]. It has been theorized that an electron leaving cytochrome-c goes by way of the methionine sulfur[62]. However, imidazole, with its double bonds, could also be a suitable bridge for electron transfer.

By comparison with the b and c cytochromes, considerably less is known about the structure of cytochrome oxidase (the boxed species in the electron transfer scheme shown above). Even the order of and connection between the copper and the a cytochromes are unclear. But it is thought to proceed through an axial ligand bridge between the iron and copper ions; for this reason a variety of model compounds can be considered relevant to the cytochrome oxidase system[63].

7.3.1.1. Nitrogen bridged

Ostfeld and Cohen[64] have studied the polymer [−Fe(TPP)(imidazolate)−]$_n$ (Fig. 3). Magnetic studies have indicated the presence of spin coupling through the imidazolate ligands, thus demonstrating the ability of this ligand to serve as an electron bridge when complexed to an iron porphyrin.

Another recently prepared bridging system is a trimer involving two iron(III) porphyrin azides and an iron(II) porphyrin[64]. The magnetic moment of the ferric ion in this compound is 5.0 B.M., independent of temperature. This indicates the presence of four unpaired electrons instead of the five expected for high-spin iron(III). That pairing occurs between only two of the electrons can be explained by the occurrence of a single orbital available for electron occupancy across the length of the molecule. Three possible structures have been proposed for this (2 : 1) complex (Fig. 4). While all three agree with the magnetic data, the reactions of Cr(II) (acac)$_2$ and Fe(TPP)Cl, which proceed via an inner sphere process[65], indicate that either structure 1 or 2 is probably the correct one.

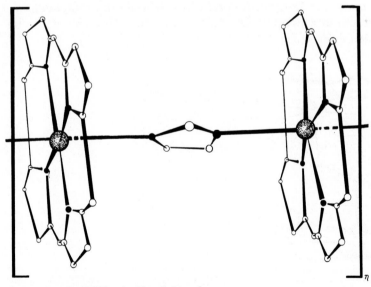

Fig. 3. The Fe(TPP)-μ-imidasolate polymer.

References, p. 310

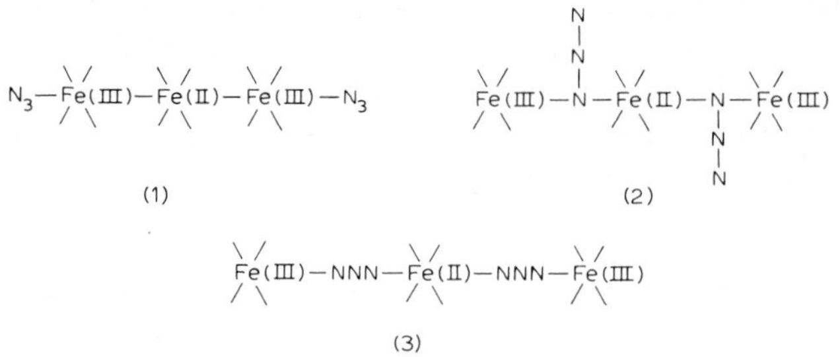

$$N_3—Fe(III)—Fe(II)—Fe(III)—N_3 \qquad Fe(III)—N—Fe(II)—N—Fe(III)$$

(1)

(2)

$$Fe(III)—NNN—Fe(II)—NNN—Fe(III)$$

(3)

Fig. 4. Proposed structures of an azide bridged iron porphyrin, taken from ref. 64.

A third nitrogen bridged organometalloporphyrin is the μ-hydrazidobis-(protoheme dimethyl ester carbonyl) (Fig. 5)[66]. This complex was prepared by the reduction of μ-oxo-bis(protohemin dimethyl ester) with hydrazine hydrate under a CO atmosphere.

7.3.1.2. Oxygen bridged

The best known example of metal—metal interaction in metalloporphyrins is the oxo-bridged 'hematin' dimer (Fig. 6), which has been only recently recognized to be a dinuclear complex and not the mononuclear hydroxide[67]. The strong anti-ferromagnetic coupling[68] between the two iron(III) atoms shows the ability of this bridge to conduct electrons. Caughey[63] has proposed the possibility of iron—iron or iron—copper oxobridges in cytochrome oxidase. Similar oxygen bridged dimers have also been prepared for porphyrin complexes of aluminum(III)[51] and scandium(III)[51]. Other examples are in the porphyrin complexes of niobium, molybdenum, tungsten, and rhenium in which two O = M(V) groups have been shown to be joined by an oxygen bridge (Fig. 7)[46,51].

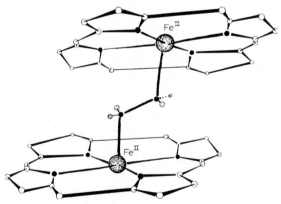

Fig. 5. Structure of the μ-hydrazido-bis(protoheme dimethyl ester carbonyl), taken from ref. 66.

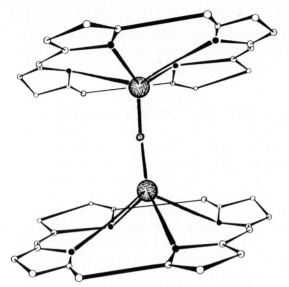

Fig. 6. Structure of the oxo-bridged 'hematin' dimer.

The scandium complex forms dimers, similar to the iron dimer, instead of the non-existent mononuclear hydroxide[44]. The porphyrin hydroxide of aluminum(II), however, does exist and is dimerized by heating under vacuum[48].

$$Al^{III}(P)OH \xrightarrow[\text{vacuum}]{\Delta} (P)Al^{III}\text{-}O\text{-}Al^{III}(P) \tag{11}$$

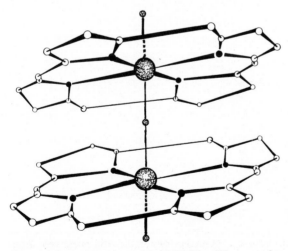

Fig. 7. Structure of the μ-oxo-$[(P) M = O]_2$ complex.

This leads to the possibility that other +3 metals (e.g., Co(III), Mn(III), Ga(III)) can form oxo-bridged porphyrin dinuclear complexes upon removal of water from the hydroxide. An oxo-bridged dinuclear complex of manganese(III) phthalocyanine is, in fact, already known[2].

7.3.1.3. Halide bridged

Although several halide bridging species have been identified spectroscopically as intermediates, none have been isolated and thoroughly characterized.

Cohen[65] has shown the reaction between Fe(TPP)Cl and $(acac)_2$ Cr to proceed via an inner sphere process with a bridging chlorine intermediate.

$$-[Fe \ldots \ldots Cl \ldots \ldots Cr]-$$

In addition, Fleischer and Wang[69], based upon their kinetic data, proposed the existence of a dinuclear structure, of the "sitting-atop' type of complex, as an intermediate species (Fig. 8), in the reaction of H_2(Proto-IX-DME) with anhydrous ferric chloride, $(FeCl_3)$.

7.3.1.4. Metal—metal bonds

A good case for the preparation of a metal—metal bonding species in these

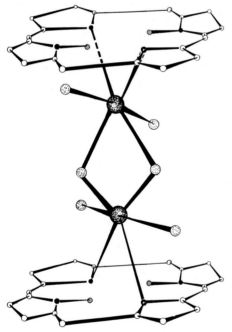

Fig. 8. Proposed 'sitting-atop' structure for the dinuclear dimetallic iron porphyrin intermediate.

complexes has been made by Whitten and co-workers[70] with the photo-chemical dimerization of a ruthenium porphyrin (Fig. 9). Upon irradiating a pyridine solution of ruthenium carbonyl octaethylporphyrin, two molecules of carbon monoxide are expelled:

$$2[Ru(OEP)(Py)(CO)] \xrightarrow{h\nu} Py(OEP)Ru\text{---}Ru(OEP)Py + 2\ CO\uparrow \qquad (12)$$

This reaction does not occur, however, when the porphyrin macrocycle is *meso*-tetraphenylporphyrin[70]. Presumably the TPP phenyl rings, which are known to lie almost perpendicular to the porphyrin plane[71], interact such that the two porphyrin macrocycles cannot approach close enough for a metal—metal bond to be formed. A study of this compound, $(Ru(OEP)Py)_2$, by single crystal X-ray diffraction would be desirable.

Metal—metal interaction has also been observed to occur in the solid chromium(II) mesoporphyrin-IX dimethyl ester complex. The observed mag-netic moment, 2.84 B.M.[27], for each chromium atom is substantially less than the expected[72] 4-spin value of 4.90 B.M. for a square planar chro-mium(II). However, solutions of the chromium complex do not show a reduced moment, and a value of 5.19 B.M. is observed[27]. Apparently the close approach of chromium atoms in the solid phase causes a sufficient rise in the energy of the d_{z^2} orbitals, thus causing electron pairing to occur. As in the case of ruthenium porphyrin dimerization, *meso*-tetraphenylporphyrin seems also to inhibit metal—metal bonding in the chromium complexes. For example, the magnetic moment of solid $Cr^{II}(TPP)$ has been observed to be 4.9 B.M., which is normal for a square planar complex of Cr(II). This is not true in all cases and in fact metal—metal interaction may be a more general phenomenon for paramagnetic metalloporphyrins containing no axial ligands than previously thought, since a somewhat low magnetic moment has also been found for $Rh^{II}(TPP)$[73].

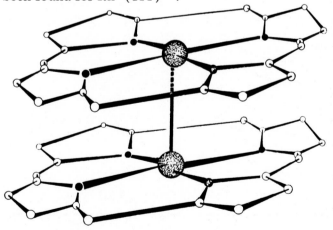

Fig. 9. Ruthenium porphyrin dimer.

References, p. 310

Two other excellent examples have recently been reported, namely $(Zn(Pc)Cl)_2$ [74] and $(Zn(OEP)Br)_2$ [75]. The reaction of zinc(OEP) or zinc(Pc) with reagents such as thionyl chloride, sulfuryl chloride, phosphorus oxychloride, phosphorus pentachloride, chlorine, bromine, and iodine[76,77] produces a radical dimer species. Oxidation titrations (dichromate method[78]) have shown the Zn(Pc)Cl complex to be one electron above Zn(Pc), inferring the alternate formulations $Zn^{III}(Pc)Cl$ or the $Zn^{II}(Pc)Cl$ radical species. The zinc phthalocyanine complex was observed to be diamagnetic over an extended temperature range from $-196°C$ to $+100°C$. Since the oxidative titration data show that the Zn(Pc)Cl complex contains an odd number of electrons, the diamagnetism must arise through spin pairing. The most simple procedure to accomodate this pairing would be dimerization, and a dimeric radical structure $(Zn^{II}(Pc)Cl)_2$ has been proposed for this species (Fig. 10). This observation parallels zinc porphyrin chemistry where diamagnetic dimeric zinc (OEP) complexes, $(Zn(OEP)Br)_2$, have been characterized. While the detailed structure of such a dimer is not known, it seems probable that two five coordinate $Zn^{II}(Pc)Cl$ radical units are stacked with their porphyrinic bases parallel to each other and subsequent electron coupling through the π clouds. It is relevant that, in porphyrin chemistry, a mononuclear paramagnetic radical species is obtained when the zinc porphyrin has bulky substituents which presumably prevent the π clouds from getting sufficiently close together[79]. The only metal—metal bond in which a metalloporphyrin bonds to a nonporphyrin metal that has been reported is an adduct of trimethyl tin and iron meso-tetraphenylporphyrin. Mössbauer and far infrared data indicate that this air-sensitive, diamagnetic compound contains a metal—metal bond between iron(III) and tin(IV).

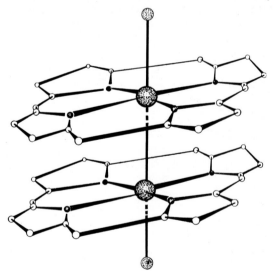

Fig. 10. Structure of $[Zn(P)Br]_2$ and $[Zn(Pc)Cl]_2$.

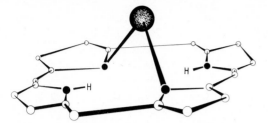

Fig. 11. Fleischer's 'sitting-atop' model.

7.3.2. Unusual geometries
7.3.2.1. Monometallic

Two models have been proposed for the mononuclear monometallic complexes. Fleischer's[80] 'sitting-atop' (SAT) complex (Fig. 11) and Hoard's[7] square—pyramidal complex (Fig. 12). In his kinetic studies, Fleischer proposed the 'sitting-atop' complex as the intermediate in the metal insertion mechanism. In this configuration the metal ion deforms the porphyrin so as to facilitate the incorporation of the metal ion. On the other hand, Hoard, based on structural arguments, proposed the square—pyramidal complex as the intermediate in the oxygen transport mechanism of hemoglobin.

Two examples of this type of complex occur with zirconium(IV)[44] (Fig. 13) and hafnium(IV)[46,51], which form stable, out-of-plane, mononuclear monometallic porphyrin complexes each containing two bidentate acetate ligands. A single crystal X-ray diffraction analysis[81] has confirmed that the metal atom is out of the porphyrin plane, and on the S_2 axis normal to the porphyrin plane with both acetate ligands on the same side of the porphyrin ring, in agreement with both proposed models.

Another type, dinuclear monometallic, out-of-plane, complex is the sandwich, 8-coordinate, compound of tin(IV)[82] phthalocyanine (Fig. 14). Again, an X-ray analysis which has shown this complex to have the tin atom out of both phthalocyanine planes and on the S_2 axis normal to the phthalocyanine

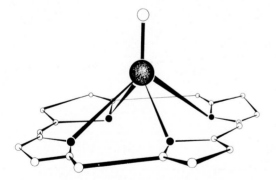

Fig. 12. Hoard's square—pyramidal complex.

References, p. 310

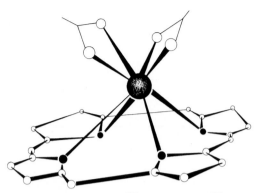

Fig. 13. Structure of $Zr^{IV}(OEP)$ and $Hf^{IV}(OEP)$.

plane in agreement with the proposed models.

$$Sn^{IV}(Pc)Cl_2 + Na_2(Pc) \xrightarrow[\substack{\Delta \\ 90\,min}]{Chloronaphthalene} Sn^{IV}(Pc)_2 + 2\,NaCl \qquad (13)$$

Pc = phthalocyanine

Several other examples of this type have been reported such as the X-ray diffraction analysis of bis(phthalocyanine)U(IV)[83] which showed it to have a similar structure as the tin(IV) complex. In addition, Misumi and Kasuga[84] proposed a similar structure for a series of lanthanide complexes, La, Ce, Nd, Eu, Er, and Yb. However, no X-ray analysis has been reported for these latter compounds.

Recently, a new type of mononuclear monometallic out-of-plane complex, was prepared. When H_2(Meso-IX-DME) and $Tc_2(CO)_{10}$ in a mole ratio of 1 to 0.6 were mixed in decalin and heated under argon at $150°C$ (oil bath)

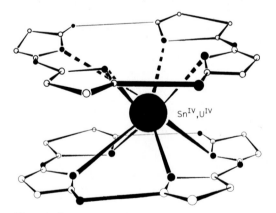

Fig. 14. Structure of Sn(IV) and U(IV) bis-(phthalocyanine).

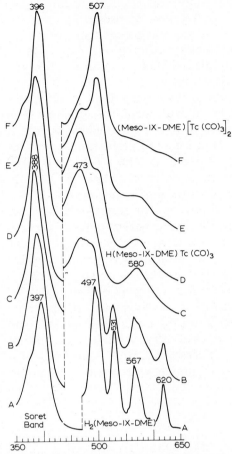

Fig. 15. Visible absorption spectra changes for the $Re_2(CO)_{10}$ and $Tc_2(CO)_{10}$ reaction.

for 2—3 h, an unusual technetium metalloporphyrin complex, monohydro-gen(mesoporphyrin-IX dimethyl esterato)tricarbonyltechnetium(I), H(Me-so-IX-DME)Tc(CO)$_3$, was formed[85]. The completion of the reaction was determined by visible spectroscopy; when absorptions at 388 (Soret) and 473 nm reached maxima, the reaction was stopped (Fig. 15).

The new complex was isolated from the cooled decalin solution by using a molecular still apparatus to remove the decalin at room temperature under vacuum. The remaining solid material was chromatographed on a talcum or Sephadex LH-20 column.

$$1.0\ H_2(P) + 0.6\ M_2(CO)_{10} \xrightarrow[200°C]{decalin} H(P)[M(CO)_3] \qquad (14)$$

M = Re, Tc

H(P) = H(Meso-IX-DME), H(TPP)

References, p. 310

Elution, of the column with a benzene/cyclohexane (1 : 1) mixture gave a pale green band for the technetium[85] and yellowish brown for the rhenium[86] complexes. Evaporation of the solvent and subsequent crystallization of the complex from absolute alcohol/chloroform gave a dark greenish brown solid for the rhenium complex and the technetium complex. Spectral data indicate that the rhenium and technetium atom lies both off the S_2 axis normal to the porphyrin plane and out of the porphyrin plane (Fig. 16).

The p.m.r. spectrum of H(TPP)Re(CO)$_3$ in deuterochloroform showed an upfield singlet at 14.0 τ for the N—H proton, two broadened multiplets at 2.13 and 1.70 τ for the phenyl protons, and three different type signals for the β-pyrrole protons. One AB quartet centered at 0.89 τ (J_{AB} = 5.0 Hz, νAB = 12.0 Hz, and the intensity ratio of the outer to inner lines are 1 : 2.2), a doublet centered at 1.12 τ (J = 2.0 Hz), and a singlet centered at 1.28 τ, with relative intensities of 2 : 1 : 1 respectively. The results can be interpreted if the rhenium atom coordinates with three adjacent pyrrole nitrogen atoms of the porphyrin and the proton binds with the uncoordinated pyrrole nitrogen atom (Fig. 16). The peripheral protons of the two pyrrole rings are expected to show an AB quartet pattern. The doublet centered at 1.12 τ is due to long-range coupling of β-protons with a nitrogen-bonded proton of the same pyrrole ring. The singlet at 1.28 τ can be assigned to the peripheral-protons of the remaining pyrrole ring. Thus, the p.m.r. spectral data lend support to the structure previously proposed. The variable temperature p.m.r. spectra of H(TPP)Re(CO)$_3$ and H[Meso-IX-DME] · Re(CO)$_3$ dissolved in 1,1,2,2-tetrachloroethane, showed completely reversible changes in

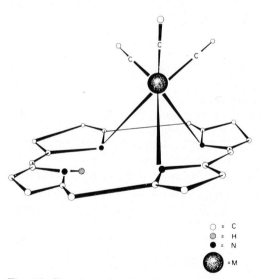

○ = C
◉ = H
● = N
⬤ = M

Fig. 16. Structure showing the coordination sphere around the rhenium and technetium ion [H(Meso-IX-DME)Re(CO)$_3$; H(TPP)Re(CO)$_3$; and H(Meso-IX-DME)Tc(CO)$_3$].

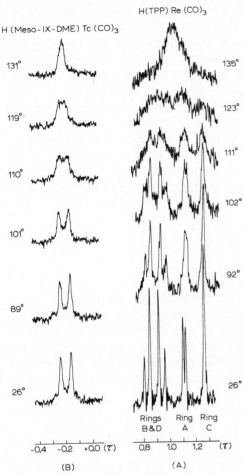

Fig. 17. HA-100 NMR variable temperature spectra of H(TPP)Re(CO)₃ and H(Meso-IX-DME)Tc(CO)₃ dissolved in 1,1,2,2,-tetrachloroethane.

the spectral peaks. As the temperature is raised, the peaks broaden, coalesce, and gradually sharpen (Fig. 17). The above fluxional phenomenon is best explained by an intramolecular rearrangement of the rhenium-carbonyl group among the four ring nitrogens of porphyrin and either synchronous or subsequent movement of the N—H proton. A solution containing H[Meso-IX-DME] · Re(CO)₃ and excess, free, H₂[Meso-IX-DME] showed no broadening of the free ligand bridged methine proton peaks in the fast exchange region. Furthermore, the coalescence temperatures were not shifted by changes in the concentration of complexes, within experimental error (approx. ±5°C). These results lend support to the fact that the thermal rearrangement is an intramolecular process, with a free energy of activation of

References, p. 310

~19 Kcal/mole. Thus, the temperature dependent p.m.r. spectra of H[Meso-IX-DME] $Re(CO)_3$, H[Meso-IX-DME] $Tc(CO)_3$ and H[TPP] $Re(CO)_3$[87], have shown these complexes to exhibit dynamic behavior. The rhenium and technetium atoms migrate about the face of the ring via an intramolecular (fluxional) mechanism[88]. Dynamic, fluxional, behavior by coordinated ligands has previously been observed, but this appears to be the first example of dynamic, fluxional behavior by the metal atom in a metalloporphyrin complex.

In addition, whereas the $H(P)[Re(CO)_3]$ is thermally stable toward decomposition, both the technetium and rhodium complexes, *vide infra*, are not. The exact mechanism for the rhodium complex is unclear, but the technetium complex disproportionates forming the dinuclear technetium complex and free porphyrin[85].

$$2\ H(P)[Tc(CO)_3] \xrightarrow[\text{decalin}]{190°C} H_2(P) + (P)[Tc(CO)_3]_2 \qquad (15)$$

7.3.2.2. Dimetallic

The second class of out-of-plane metalloporphyrin are the dimetallic complexes. Hambright[89] has proposed, from kinetic studies, a model for the mononuclear dimetallic complexes (Fig. 18), in which the metal atoms sit above and below the porphyrin plane, and on the S_2 axis normal to the porphyrin plane. Tsutsui and co-workers have synthesized the mononuclear dimetallic complexes of both rhenium[40,86] and technetium[90].

$$1.0\ H_2(P) + 0.6\ M_2(CO)_{10} \rightarrow H(P)[M(CO)_3] \qquad (16)$$

$$1.0\ H(P)[M(CO)_3] + 0.6\ M_2'(CO)_{10} \rightarrow [M'(CO)_3](P)[M(CO)_3] \qquad (17)$$

$$M \cong M' = Re,\ Tc$$

and

$$1.0\ H_2(P) + 1.0\ M_2(CO)_{10} \rightarrow (P)[M(CO)_3]_2 \qquad (18)$$

$$M = Re,\ Tc$$

These complexes were prepared in a similar manner to the mononuclear monometallic rhenium and technetium complexes previously discussed. Crystals for X-ray diffraction analysis were grown in a solution of dioxane. Roughly, the X-ray analysis[40,91] has shown that the metal atoms lie above and below the porphyrin plane in agreement with Hambright's model, but off the S_2 axis normal to the porphyrin plane (Fig. 19). The distance between metal atoms is 3.114 ± 0.013 Å, too long for a formal bond, but short enough to allow some metal—metal interaction.

In more detail, the single crystal X-ray diffraction analyses for both the

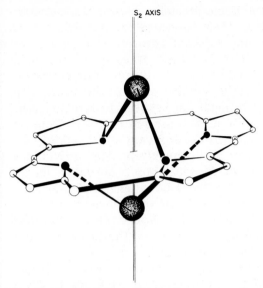

Fig. 18. Hambright's model for the mononuclear dimetallic porphyrin complexes.

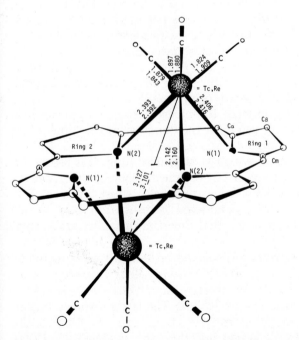

Fig. 19. Coordination sphere showing bond distances around the Re; Tc atoms of (TPP)[Re(CO)₃]₂ and (TTP)[Tc(CO)₃]₂. Inner values are for technetium; outer values are for rhenium.

tricarbonylrhenium and technetium homodimetallic *meso*-tetraphenylporphyrin complexes have shown both of them to be centrosymmetric, $P2_1/C$, having two metal atoms bonded to three adjacent nitrogen atoms, one above and one below the plane of the macrocyclic ligand. In each case the metal atoms lie 1.42 Å above the plane defined by the four pyrrole nitrogen atoms. The metal ions are not positioned directly over the center of the macrocycle, but are set to one side centering each metal ion between to three adjacent nitrogen atoms. N(2) is bonded to both metal ions. The distances from the metal ion to the uncoordinated fourth nitrogen atom are 3.217 and 3.212 Å for the rhenium and technetium complexes respectively. The M—N(1) distance is 2.16 Å in both cases. The two M—N(2) bonds are much longer, averaging 2.39 Å in both complexes. The M—C bond lengths appear to be normal. There is, however, a small but significant difference in the M—M distances, with the Tc—Tc distance being 0.026 Å shorter. This small difference in the covalent radii of congeners in the second and third row transition metal series can be attributed to the lanthanide contraction effect.

The porphyrin macrocycle is highly distorted. The distortion is such that if one considers the mean plane of the macrocycle, pyrrole ring 1, [N(1), C(1)-C(4)], is 'bent' towards the metal atom to which it is coordinated, while pyrrole ring 2, [N(2), C(6)-C(9)], lies almost in the mean plane of the macrocycle. The individual pyrrole rings are quite planar, but the angle between the least squares planes of the two adjacent rings is 17.3° in both complexes. Another measure of the bending is the perpendicular distance of an atom in one pyrrole ring from the least squares plane defined by the equivalent ring across the macrocycle. Thus, N(1)' lies 1.83 Å from the plane of pyrrole ring 1. N(2)' lies 0.36 Å from the plane of pyrrole ring 2. (The primes indicate an atom related by a center of symmetry.)

The distortion of the macrocycle leads to unusual distances opposite pyrrole nitrogen atoms. The optimum value for the diameter of the 'hole' of an undistorted metalloporphyrin complex has been estimated[2] to be 4.02 Å. The N(1)—N(1)' distance is 4.52 Å, which is unusually long. It is caused by the large deviation of the pyrrole ring from the mean plane of the macrocycle. The N(2)—N(2)' distance is 3.65 Å, an unusually short distance. This shortening apparently arises so that octahedral coordination may be attained. It is of interest to note that the two Cm—Cm distances are normal and approximately the same.

The involvement of N(2) in two coordination bonds might be expected to affect the double bond character of the Cα—N(2) bonds. This effect is indeed seen in both complexes where the average Cα—N(2) distance is 1.42 Å as compared to the Cα—N(1) distance of 1.37 Å. The latter distance is typical of that found in normal metalloporphyrin complexes. In addition, the [C(6)—N(2)—C(9)] angle is four degrees less than the [C(1)—N(1)—C(4)] angle. The standard deviations of the non-metal atoms in (TPP)[Re(CO)₃]₂ are fairly large due to the high X-ray scattering power of the rhenium atom, so in this case the significance of these differences is hard to judge. However, in $(TPP)[Tc(CO)_3]_2$ the standard deviations of the non-metal atoms are much lower and the differences in these bond lengths and angles can be judged significantly.

To a lesser extent the bond lengths and angles around $C\alpha$ atoms on pyrrole ring 2 are affected. The average $C\alpha-C\beta$ distance is slightly smaller than the analogous difference in pyrrole ring 1. In this case, however, the significance of these differences may only be classified as marginal. In normal metalloporphyrin complexes, the $C\alpha-C\beta$ distance has been found to be remarkably constant at approximately 1.44 Å. There is a 1.0° difference in the averaged $N-C\alpha-C\beta$ angle for the two independent rings, a difference also on the verge of significance. The $C\beta-C\alpha-Cm$ and $N-C\alpha-Cm$ angles are considerably different in the two rings, but this is probably a factor of the distortion of the macrocycle.

As is normal for metallo-*meso*-tetraphenylporphyrin complexes, the phenyl rings are rotated considerably out of the plane of the macrocycle. In $(TPP)[Tc(CO)_3]_2$, phenyl ring 1, $[C(11)-C(16)]$, is rotated 82.4° from the least squares plane of the four pyrrole nitrogen atoms, while phenyl ring 2, $[C(17)-C(22)]$, is rotated 53.8°. Phenyl ring 1 is rotated 116.6° and 100.4° with respect to pyrrole rings 1 and 2. The equivalent numbers for phenyl ring 2 are 108.3° and 116.7°. The angles for $(TPP)[Re(CO)_3]_2$ are very similar. The average carbon-carbon distance in the two phenyl rings is 1.372 and 1.375 Å for the rhenium and technetium complexes respectively.

The reaction of $Re(CO)_5X$, $[X = Cl, Br]$, with $H_2[OEP]$ and $H_2[Meso-IX-DME)$ in decalin at 150°C for 4 h gave a product with an elemental analysis in agreement with the formula $(P)[Re(CO)_2X]_2$ [92]. The product formed from the reaction of $H_2(TPP)$ with $Re(CO)_5X$ [92] could not be isolated, but is thought to have a structure similar to that obtained from the reaction of $H_2(TTP)$ and $Cr(CO)_6$ [93], *vide infra*. All the complexes are very moisture sensitive. Although a detailed X-ray analysis has not been carried out, preliminary data indicates that the products from the $H_2[Meso-IX-DME]$ and $H_2[OEP]$ reaction have structures similar to $(TPP)[Re(CO)_3]_2$ (Fig. 20).

Several derivatives of the monorhenium metalloporphyrin complex have recently been reported[86]. The reaction of $H(P)[Re(CO)_3]$ with excess $Re_2(CO)_{10}$ to form the dirhenium complex, $(P)[Re(CO)_3]_2$, as well as the reaction of $H(P)[Re(CO)_3]$ with $Tc_2(CO)_{10}$ to form the heterodimetallic $[Tc(CO)_3](P)[Re(CO)_3]$ complex (Fig. 21), suggested the possibility of making a series of complexes in which the porphyrin ring is bound to both a rhenium atom and an atom of another metal.

Although this could not be done with all metals, $H(Por)Re(CO)_3$ was reacted further with a few mono- and divalent cations of heavy metals (Ag^+, Hg^{2+}, and Pb^{2+}) in a basic solvent[86]. Evidence for these reactions came from changes in the visible absorption spectra of the solutions (Table 2). It is presumed that monovalent cations replace the remaining pyrrole proton,

References, p. 310

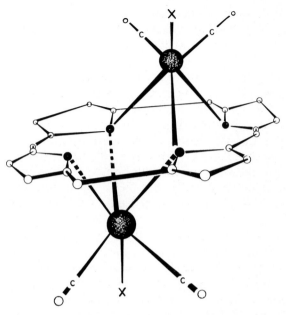

Fig. 20. Proposed structure for $(OEP) \cdot [Re(CO)_2X]_2$; X = Cl, Br.

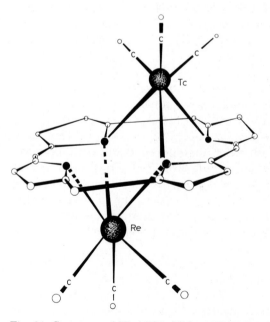

Fig. 21. Structure of $[Re(CO)_3](Meso\text{-}IX\text{-}DME)[Tc(CO)_3]$.

TABLE 2

Visible absorption spectra for H(Meso-IX-DME)Re(CO)$_3$, (I), and its unstable derivatives

Derivatives	λ_{max} (in acetone), nm
H(Meso-IX-DME)Re(CO)$_3$, (I)	400 (Soret); 480, 580
(I) + CH$_3$COOAg	387 (Soret); 480, 560, 580
(I) + (CH$_3$COO)$_2$Hg	385 (Soret); 480, 530, 570, 580
(I) + (CH$_3$COO)$_2$Pb $\cdot$ 3 H$_2$O	388 (Soret); 460, 580
(I) + (CH$_3$COO)$_2$Cu $\cdot$ H$_2$O	395 (Soret); 520, 560*

* This is the visible absorptions of Cu^{II}(Meso-IX-DME)

forming a 1 : 1 complex. Divalent cations are thought to similarly replace a pyrrole proton from each of two molecules, forming a sandwich-type complex with the heavy metal ion situated between two porphyrin rings, similar to the trimetallic mercury complex, *vide infra*. The metal-nitrogen bonds thus formed are weak and the products could not be purified due to decomposition in air. Partial decomposition is also shown by the molecular weight and analytical results. It is of interest that an attempt at the reaction of H(P)[Re(CO)$_3$] with neutral copper acetate, in 2 : 1 molar ratio, by stirring in a basic solvent at room temperature for 5 h results in expulsion of the rhenium-carbonyl moiety by the cupric ion to form a square planar copper(II) porphyrin complex.

Very recently, several rhodium complexes have been obtained from di-μ-chloro-tetracarbonyldirhodium(I), [RhCl(CO)$_2$]$_2$, and a porphyrin.

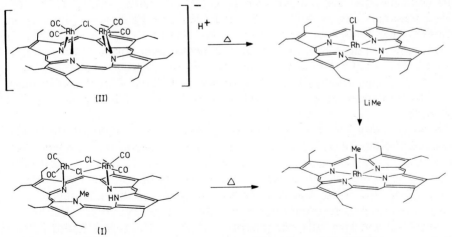

Fig. 22. Reaction sequence proposed for [Rh(CO)$_2$Cl]$_2$ and H$_2$(OEP); compounds I and II.

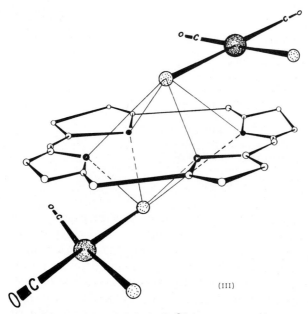

Fig. 23. Structure of $[H_4(OEP)]^{2+} \cdot 2[Rh(CO)_2Cl_2]^-$; compound III.

Thus, 1) mesoporphyrin-IX diethyl ester in hot glacial acid afforded $Rh^I(Meso-IX-DEE)$[39]; 2) (N-methyl)octaethylporphyrin in benzene at room temperature gave the mononuclear dimetallic $Me(OEP)[Rh(I)Cl(CO)_2]_2$ complex, formulated schematically as I, which upon heating yielded the mononuclear monometallic $MeRh^{III}(OEP)$ complex[35]; 3) octaethylporphyrin in benzene at room temperature gave $H^+[(OEP)Rh^I_2(CO)_4Cl]$[36], formulated schematically as II (Fig. 22), and $ClRh^{III}(OEP)2H_2O$, which upon treatment with methyl-lithium formed the $MeRh^{III}(OEP)$ complex[36]; 4) meso-tetraphenylporphyrin in hot glacial acetic acid forms $ClRh^{IV}(TPP)$-H_2O and $Rh^{III}(TPP)$, which could be reduced by hydrogen to $H(TPP)Rh \cdot 2H_2O$[73]; 5) and finally, octaethylporphyrin in hot chloroform forms a salt-like mononuclear dimetallic complex $[H_4(OEP)]^{2+} \cdot 2[Rh(CO)_2Cl_2]^-$[94], III (Fig. 23).

The latter complex is very interesting, especially in light of the X-ray analysis of the complex, reported by Yoshida and co-workers[37], which they proposed to have the structure II, after it liberated HCl in solution. The resulting compound IV was shown to have two rhodium atoms per porphyrin (Fig. 24), one above and one below the plane of the macrocycle. The rhodium ions possess four-coordination, square planar symmetry, bonded to two carbonyls and two adjacent pyrrole nitrogens. The formation of IV can be best envisioned by starting with a complex similar to III, than for a complex similar to II. However, this is not to say that structures I or II do

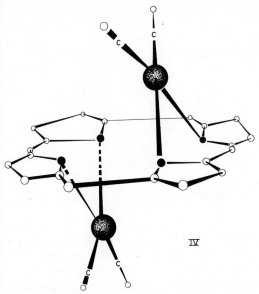

Fig. 24. Structure of (OEP)[Rh(CO)$_2$]$_2$, compound IV.

not exist, only that structure III allows an easier interpretation of all the experimental results.

It would appear that the metal carbonyl complexes prefer to coordinate to adjacent nitrogen atoms (Fig. 25), rather than alternate nitrogen atoms, as previously proposed (Figs. 11 and 18), causing the metal atoms to lie off the S$_2$ axis normal to the porphyrin plane.

It should also be noted that unlike most porphyrin complexes where the metal atom is coordinated to all four pyrrole nitrogens, the above complexes

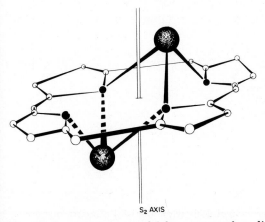

Fig. 25. Proposed new model for mononuclear dimetallic porphyrin complexes.

References, p. 310

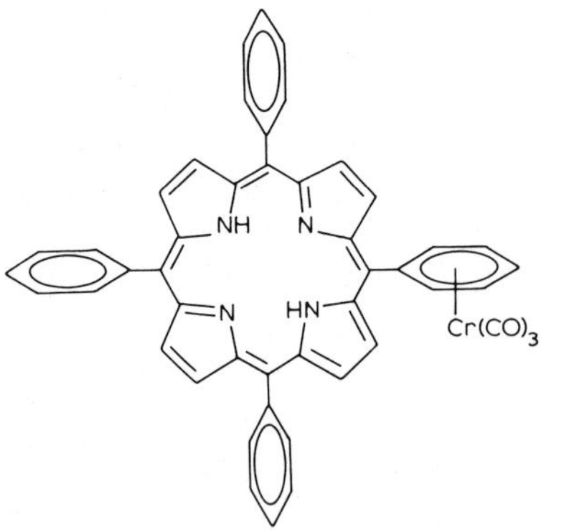

Fig. 26. Structure of π-(Cr(CO)$_3$)-(TPP). 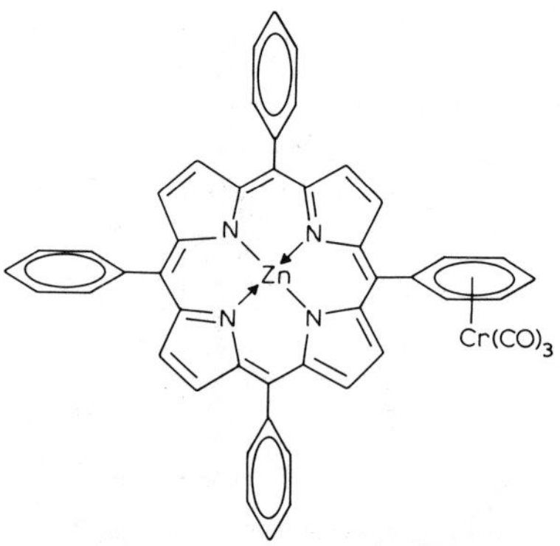   Fig. 27. Structure of π-(Cr(CO)$_3$)-Zn(TPP).

[Re, Tc, Rh] vary from 2 to 3. The coordination to 2 or 3 nitrogens is predicted by the 18-electron rule[95] with the metal in a low oxidation state (+1,
d^6, d^8). Magnetic susceptibility measurements for the rhenium complex lend
support to the metal being d^6, diamagnetic. Furthermore, porphyrins and

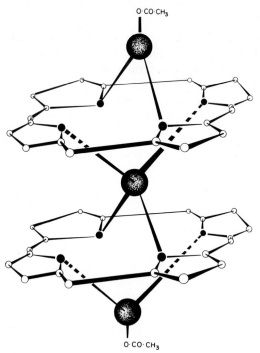

Fig. 28. Proposed structure of the dinuclear trimetallic [triple decker] mercury porphyrin
complex.

certain other unsaturated ligands (e.g. bipyridyl) are distinguished by their delocalized π systems. Overlap between the delocalized π-electron orbitals on the porphyrin and the metal orbitals of proper symmetry produces a moderately high ligand field strength[96]. Also, by back-accepting π-electron density from complex metals, the porphyrin facilitates the reduction of the complexes' metal to a low oxidation state[73,97,98]. It should be mentioned, however, that not all unusual porphyrin oxidation states are so easily explained. For example, the most stable silver porphyrin has the metal in a +2 and not a +1 state[99]. No explanation for this has yet been proposed. Whether this model (Fig. 25), is valid for metals possessing ligands other than carbonyls which are coordinated to less than 4 pyrrole nitrogen atoms in unknown at this time.

Another interesting type of mononuclear dimetallic complex is the metal chelates of tricarbonylchromium(O) π-complexes of *meso*-tetraphenylporphyrin recently reported by Gogan and Siddiqui[93]. They showed that the reaction of $Cr(CO)_6$ with $H_2[TPP]$ in the absence of air produces a π-bonded tricarbonylchromium(O) complex (Fig. 26), in which the chro-

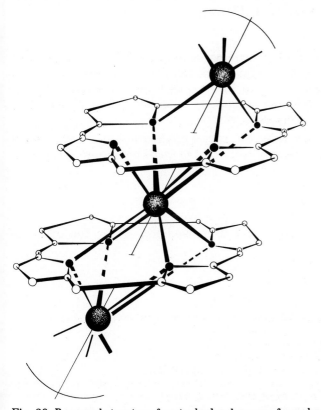

Fig. 29. Proposed structure for stacked polymers of porphyrins.

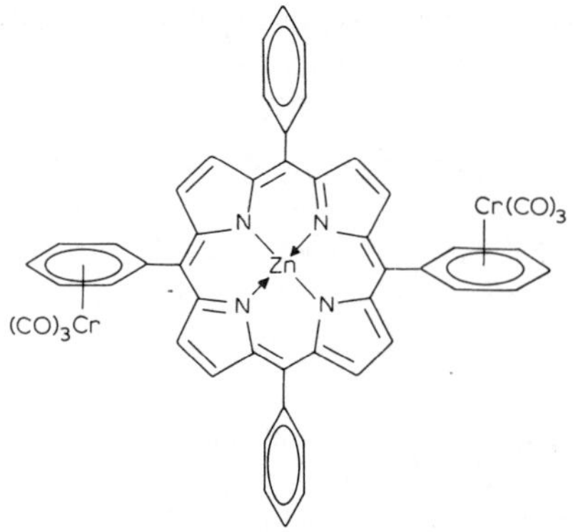

Fig. 30. Structure of bis-(π-Cr(CO)$_3$)-Zn(TPP).

mium(O) ion is π-bonded to one of the phenyl rings on the *meso*-tetraphenylporphyrin macrocycle. A similar reaction occurs for the reaction of Cr(CO)$_6$ with a M[TPP] complex [M = Mn, Co(II), Ni(II), Cu(II) and Zn(II)]. However, only the Zn^{II}(TPP)Cr(CO)$_3$ complex was stable enough to be isolated and characterized (Fig. 27).

7.3.2.3. Trimetallic

Recently, a dinuclear trimetallic mercury complex[100] (Fig. 28) has been prepared which illustrates a third class of out-of-plane complex. This compound also illustrates the possibility of forming more extensive stacked polymers (Fig. 29). Another type of trimetallic complex is the mononuclear trimetallic bis[π-tricarbonylchromium(O)]—zinc[*meso*-tetraphenylporphyrin] (Fig. 30)[93]. In this complex, the tricarbonylchromium(O) groups are each π-bonded to a different phenyl ring on the *meso*-tetraphenylporphyrin macrocycle.

Acknowledgement

The preparation of this article was supported in part by the office of Naval Research.

References

1. J.E. Falk, 'Porphyrins and Metalloporphyrins', Elsevier, Amsterdam (1964).
2. E.B. Fleischer, Acc. Chem. Res., **3**, 105 (1970).
3. P. Hambright, Coord. Chem. Rev., **6**, 247 (1971).
4 L.J. Boucher, Coord. Chem. Rev., **7**, 289 (1972).
5. D. Ostfeld and M. Tsutsui, Acc. Chem. Res., **7**, 52 (1974).
6. A.D. Adler, Ed., 'The Chemical and Physical Behavior of Porphyrin Compounds and Related Structures', Ann. N.Y. Acad. Sci., **206**, (1973).
7. J.L. Hoard, Science, **174**, 1295 (1971).

8. L.P. Vernon and G.R. Seely, Eds., 'The Chlorophylls: Physical, Chemical, and Biological Properties', Academic Press, New York and London (1966).
9. E. Antonini and M. Brunori, 'Hemoglobin and Myoglobin in their Reactions with Ligands', North-Holland/American Elsevier (1971).
10. R. Lemberg and J. Barrett, 'Cytochromes', Academic Press, London and New York (1973).
11. E.L. Smith, 'Vitamin B_{12}', 3rd ed., John Wiley & Sons, Inc., New York (1965).
12. G.N. Schrauzer, Acc. Chem. Res., 1, 97 (1968).
13. A. Goldberg and C. Rimington, "Diseases of Porphyrin Metabolism", Thomas, Springfield (1962).
14. E.F. Meyer, Acta Crystallogr., Sect. B, 28, 2162 (1972).
15. D.M. Collins, W.R. Scheidt, and J.L. Hoard, J. Am. Chem. Soc., 94, 6689 (1972).
16. J.L. Hoard, M.J. Hamor, T.A. Hamor, and W.S. Caughey, J. Am. Chem. Soc., 87, 2312 (1965); and J.L. Hoard, "Structural Chemistry and Molecular Biology", Eds., A. Rich and N. Davidson; W.H. Freeman, San Francisco (1968), p. 573.
17. M.F. Perutz, Nature London, 228, 726, 734 (1970) and ref. 7.
18. R. Timkovich and A. Tulinsky, J. Am. Chem. Soc., 91, 4430 (1969).
19. D.M. Collins and J.L. Hoard, J. Am. Chem. Soc., 92, 3761 (1970); M.D. Glick, G.H. Cohen, and J.L. Hoard, ibid., 89, 1996 (1967).
20. D.L. Cullen and E.F. Meyer, Jr., J. Chem. Soc. D, 629 (1971).
21. R.C. Pettersen, Acta Crystallogr., Sect. B, 25, 2527 (1969).
22. M. Tsutsui, work in progress.
23. A.D. Adler, J. Polym. Sci., Part C, 29, 73 (1970).
24. E.B. Fleischer and M. Krishnamurthy, J. Am. Chem. Soc., 94, 1382 (1972).
25. A.R. Kane, J.F. Sullivan, D.H. Kenny, and M.E. Kenney, Inorg. Chem., 9, 1445 (1970) and refs. therein.
26. D. Patterson, "Pigments, an Introduction to their Physical Chemistry", Elsevier, New York (1967) p. 44.
27. M. Tsutsui, M. Ichikawa, F. Vohwinkel, and K. Suzuki, J. Am. Chem. Soc., 88, 854 (1966).
28. M. Tsutsui, R.A. Velapoldi, K. Suzuki, F. Vohwinkel, J. Ichikawa, and T. Koyano, J. Am. Chem. Soc., 91 6262 (1969).
29. M. Tsutsui, M. Ichikawa, F. Vohwinkel, and K. Suzuki, J. Am. Chem. Soc., 88, 854 (1966).
30. E.B. Fleischer and T.S. Srivastava, Inorg. Chim. Acta, 5, 151 (1971).
31. T.S. Srivastava and E.B. Fleischer, J. Am. Chem. Soc., 92, 5518 (1970).
32. J.J. Bonnet, S.S. Eaton, G.R. Eaton, R.H. Holm, and J.A. Ibers, J. Am. Chem. Soc., 95, 2141 (1973).
33. B.C. Chow and I.A. Cohen, Bioinorg. Chem., 1, 57 (1971).
34. E.B. Fleischer and D. Lavalle, J. Am. Chem. Soc., 89, 7132 (1967).
35. H. Ogoshi, T. Omura, and Z. Yoshida, J. Am. Chem. Soc., 95, 1666 (1973)
36. Z. Yoshida, H. Ogoshi, T. Omura, E. Watanabe, and T. Kurosaki, Tetrahedron Lett., 11, 1077 (1972).
37. A. Takenaka, Y. Sasada, T. Omura, H. Ogoshi, and Z. Yoshida, J. Chem. Soc., Chem. Commun., 792 (1973).
38. E.B. Fleischer and N. Sadasivan, Chem. Commun., 159 (1967).
39. N. Sadasivan and E.B. Fleischer, J. Inorg. Nucl. Chem., 30, 591 (1968).
40. D. Ostfeld, M. Tsutsui, C.P. Hrung, and D.C. Conway, J. Am. Chem. Soc., 93, 2548 (1971); M. Tsutsui, D. Ostfeld, and L.M. Hoffman, J. Am. Chem. Soc., 93, 1820 (1971).
41. D. Cullen, E. Meyer, T.S. Srivastava, and M. Tsutsui, J. Am. Chem. Soc., 94, 7603 (1972); and T.S. Srivastava, C.P. Hrung, and M. Tsutsui, J. Chem. Soc., Chem. Commun., 447 (1974).

42. M. Tsutsui and C.P. Hrung, J. Am. Chem. Soc., **95**, 5777 (1973).
43. M. Tsutsui and C.P. Hrung, J. Am. Chem. Soc., **96**, 2638 (1974).
44. J.W. Buchler, G. Eikelmann, L. Puppe, K. Rohbock, H.H. Schneehage, and D. Weck, Ann. Chem., **745**, 135 (1971).
45. A.D. Adler, F.R. Longo, F. Kampas, and J. Kim, J. Inorg. Nucl. Chem., **32**, 2443 (1970).
46. J.W. Buchler and K. Rohbock, Inorg. Nucl. Chem. Lett., **8**, 1073 (1972).
47. L.K. Hanson, M. Gouterman, and J.C. Hanson, J. Am. Chem. Soc., **95**, 4822 (1972).
48. H.H. Inhoffen and J.W. Buchler, Tetrahedron Lett., 2057 (1968).
49. P.S. Clezy and J. Barrett, Biochem. J., **78**, 798 (1961).
50. M. Tsutsui, R.A. Velapoldi, K. Suzuki, and T. Koyano, Angew. Chem., Int. Ed. Engl., **7**, 891 (1966).
51. J.W. Buchler, L. Puppe, K. Rohbock, and H.H. Schneehage, Ann. N.Y. Acad. Sci., **206**, 116 (1973).
52. J.W. Buchler, L. Puppe, K. Rohbock, and H.H. Schneehage, Chem. Ber., **106**, 2710 (1973).
53. J.W. Buchler and K. Rohbock, J. Organometal. Chem., **65**, 223 (1974).
54. Ref. 1, p. 135.
55. T.S. Srivastava and M. Tsutsui, unpublished observations.
56. W.S. Caughey, J.L. Davies, W.H. Fuchsman, and S. McCoy in "Structure and Function of Cytochromes", Eds. K. Okunki, M.D. Kamen, and I. Sekuzu, University Park Press, Baltimore (1968) p. 20.
57. H.R. Mahler and E.H. Cordes, "Biological Chemistry", Harper and Row, New York (1971), p. 683.
58. C.E. Castro and H.F. Davis, J. Am. Chem. Soc., **91**, 5405 (1969).
59. I.A. Cohen, C. Jung, and T. Governo, J. Am. Chem. Soc., **94**, 3003 (1972).
60. F.S. Motthews, P. Argos, and M. Levine, Cold Stream Harbor Symp. Quant. Biol., **36**, 387 (1971).
61. R.E. Dickerson, T. Takano, D. Eisenberg, O.B. Kallai, L. Samson, A. Cooper, and E. Margoliash, J. Biol. Chem., **246**, 1511 (1971).
62. T. Tankano, R. Swanson, O.B. Kallai, and R.E. Dickerson, Cold Stream Harbor Symp. Quant. Biol., **36**, 397 (1971).
63. W.S. Caughey, Advan. Chem. Ser., No. 100, 248 (1971).
64. I.A. Cohen, Ann. N.Y. Acad. Sci., **206**, 453 (1973); J.P. Collman and C.A. Reed, J. Am. Chem. Soc., **95**, 2048 (1973).
65. I.A. Cohen, C. Jung and T. Governo, J. Am. Chem. Soc., **94**, 3003 (1972).
66. W.S. Caughey, C.H. Barlow, D.H. O'Keffee, and M.C. O'Toole, Ann. N.Y. Acad. Sci., **206**, 296 (1973).
67. I.A. Cohen, J. Am. Chem. Soc., **91**, 1980 (1969).
68. T.H. Moss, H.R. Lilienthal, G. Moleski, G.A. Smythe, M.C. McDaniel, and W.S. Caughey, J. Chem. Soc., Chem. Commun., 263 (1972).
69. E.B. Fleischer and J.H. Wang, J. Am. Chem. Soc., **82**, 3498 (1960).
70. W. Sovocol, F.R. Hopf, and D.G. Whitten, J. Am. Chem. Soc., **94**, 4350 (1972).
71. E.B. Fleischer, C.K. Miller, and L.E. Webb, J. Am. Chem. Soc., **86** 2342 (1964).
72. B.N. Figgs and J. Lewis, Prog. Inorg. Chem., **6**, 134 (1964).
73. B.R. James and D.V. Stynes, J. Am. Chem. Soc., **94**, 6225 (1972).
74. J.F. Meyers, G.W. Rayner Canham, and A.B.P. Lever, in press.
75. J.-H. Fuhrhop, P. Wasser, D. Riesner, and D. Mauzerall, J. Am. Chem. Soc., **94**, 7996 (1972).
76. G.W. Rayner Canham, J. Myers, and A.B.P. Lever, Chem. Commun., 483 (1973).
77. P. Mertens and H. Vollman, U.S. Patent 3,651,082, Mar. 21, 1972; and Farbenfabriken Bayer Aktiengesellschaft, French Patent 1,580,611, September 5, 1969.
78. J.A. Elvidge, J. Chem. Soc., 869 (1961).

79. L.D. Spaulding, P.G. Eller, J.A. Berstrand, and R.H. Felton, J. Am. Chem. Soc., **96**, 982 (1974).
80. S.K. Cheung, F.L. Dixon, E.B. Fleischer, D.Y. Jeter, and M. Krishnamurthy, Bioinorg. Chem., **2**, 281 (1973); and ref. 69.
81. J.L. Hoard, Los Angeles ACS meeting, (1973); and ref. 51.
82. W.E. Bennett, D.E. Broberg, and N.C. Baenzyiger, Inorg. Chem., **12**, 930 (1973).
83. A. Gieren and W. Hoppe, Chem. Commun., 413 (1971).
84. S. Misumi and K. Kasuga, Nippon Kagaku Zasshi [Japan], **92**, 335 (1971).
85. M. Tsutsui and C.P. Hrung, J. Coord. Chem., **3**, 193 (1973).
86. D. Ostfeld, M. Tsutsui, C.P. Hrung, and D.C. Conway, J. Coord. Chem., **2**, 101 (1972).
87. M. Tsutsui and C.P. Hrung, J. Am. Chem. Soc., **96**, 2638 (1974).
88. F.A. Cotton, Acc. Chem. Res., **1**, 257 (1968).
89. P. Hambright, J. Inorg. Nucl. Chem., **32**, 2449 (1970); H. Baker, P. Hambright, and L. Wagner, J. Am. Chem. Soc., **95**, 5942 (1973); E.B. Fleischer, E.I. Choi, P. Hambright, and A. Stone, Inorg. Chem., **3**, 1284 (1964).
90. M. Tsutsui and C.P. Hrung, Chem. Lett., 941 (1973); and ref. 85.
91. D. Cullen, E. Meyer, C.P. Hrung, and M. Tsutsui, in preparation.
92. M. Tsutsui and C.P. Hrung, unpublished results.
93. N.J. Gogan and Z.U. Siddiqui, Can. J. Chem., **50**, 720 (1972).
94. E. Cetinkaya, A.W. Johnson, M.F. Lappert, G.M. McLaughlin, and K.W. Muir, J. Chem. Soc., Dalton Trans., 1236 (1974).
95. P.R. Mitchell and R.V. Parish, J. Chem. Educ., **46**, 811 (1969).
96. M. Zerner, M. Gouterman, and H. Koboyaski, Theor. Chim. Acta., **6**, 363 (1966); H. Koboyaski and Y. Yanagawa, Bull. Chem. Soc. Jpn, **45**, 450 (1972).
97. D.A. Clarke, D. Dolphin, R. Grigg, A.W. Johnson, and H.A. Pinnock, J. Chem. Soc. C, 881 (1968).
98. I.A. Cohen, D. Ostfeld, and B. Lichtenstein, J. Am. Chem. Soc., **94**, 4522 (1972).
99. G.D. Dorough, J.R. Miller, and F.M. Huennekens, J. Am. Chem. Soc., **94**, 4315 (1951).
100. M.F. Hudson and K.M. Smith, J. Chem. Soc., Chem. Commun., 515 (1973).

DETERMINATION OF MOLECULAR STRUCTURE

STEREOCHEMISTRY OF PORPHYRINS AND METALLOPORPHYRINS

J .L. HOARD

Department of Chemistry, Cornell University, Ithaca, New York 14853, U.S.A.

8.1. General considerations

Authoritative determinations of crystalline structure and molecular stereo-chemistry for porphyrins and metalloporphyrins, achieved by the methods of X-ray diffraction analysis, date onward from 1962. Apart from an earlier praiseworthy, but inconclusive, study of a highly disordered crystal[1], all reported X-ray analyses of structure for the manifold derivatives of porphin have been based upon the three-dimensional {*hkl*} data afforded by single crystals. Counter measurement of the diffracted intensities — the intensities of the Bragg 'reflections' — and the lavish use of modern computers have contributed to the efficiency, selectivity, and precision of the structure analyses. Certain limitations of X-ray diffraction analysis, some of which can be circumvented in structure determinations for molecular crystals, are subsequently to be mentioned.

8.1.1. Nomenclature

An immediate concern is the choice of nomenclature that is to be used for the metalloporphyrins in this chapter. Given that the parent porphyrins are to be systematically described as peripherally substituted derivatives of porphin (itself the simplest porphyrin), it is readily seen that the systematic nomenclature of coordination chemistry is fully applicable to the metalloporphyrins. The unique feature of a metalloporphyrin is the requirement that the coordination group include a macrocyclic 'porphinato' ligand which derives from a parent porphyrin by the excision of the central pair of hydrogen nuclei from the parent species. The systematic nomenclature for this macrocyclic ligand, including naturally the pattern of peripheral substitution on the porphinato core (or porphin skeleton), is then derived from that already specified for the parent porphyrin by the substitution of 'porphinato' for 'porphin'. Rather than to refer repeatedly to a given metalloporphyrin by its systematic name — which is likely to be quite lengthy — it is

Porphyrins and Metalloporphyrins, ed. Kevin M. Smith

usually desirable to make use of a rationally condensed formula for the complexed species. The principal rule to be observed in writing this type of formula is that the symbol, formula, or abbreviation used for each ligand (but not the symbol of the metal atom) is to be enclosed in parentheses. Thus, for example, (TPP) is used for the tetraphenylporphinato ligand derived from $\alpha,\beta,\gamma,\delta$-tetraphenylporphin; the omission of the Greek letters is justified by the fact that rather more than half of the determined structures for crystalline metalloporphyrins are for the metal derivatives of this symmetrically substituted porphin, whereas none is in prospect for a metal derivative of any other tetraphenylporphin. Conservation of mass then requires that the condensed formula for the parent porphyrin be written as H_2(TPP).

Both the versatility and the relative simplicity of the specified combination of systematic nomenclature and rationally condensed formulae may be illustrated for a few metalloporphyrins of fully determined stereochemistry. The Fischer system is used to specify the pattern of substitution on the porphinato core; see Fig. 1. Two four-coordinate metalloporphyrins that are subsequently to receive quite detailed treatment are $1,2,\cdots 8$—octaethylporphinatonickel(II)[2,3], Ni(OEP), and $\alpha,\beta,\gamma,\delta$-tetraphenylporphinatoiron(II)[4], Fe(TPP). Chloro-$\alpha,\beta,\gamma,\delta$-tetraphenylporphinatoiron(III)[5], Fe(TPP)(Cl) or (Cl)Fe(TPP) [the writer's preference], is, of course, a five-coordinate species;

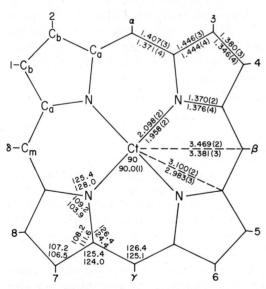

Fig. 1. Diagram of the carbon-nitrogen skeleton in the porphinato core of a metalloporphyrin which retains real or effective D_{4h} geometry. The notation, C_a, C_b, and C_m, for the three chemically distinctive classes of carbon atoms is employed throughout this chapter. Values of the principal radii (Å), bond lengths (Å), and angles (deg.) in (Cl)$_2$Sn-(TPP)[26] and in the planar form of Ni(OEP)[3] are entered on the diagram; the upper datum in each pairing is the value in (Cl)$_2$Sn(TPP).

both name and formula are correctly suggestive of a complex that has little or no salt-like character. Bis(imidazole)-$\alpha,\beta,\gamma,\delta$-tetraphenylporphinatoiron-(III) chloride[6], [(Im)$_2$Fe(TPP)]Cl or, more specifically, [(Im)$_2$Fe(TPP)]$^+$-Cl$^-$ is, by contrast, the chloride salt of the six-coordinate cationic complex. Diacetato-1,2$\cdots$8-octaethylporphinatozirconium(IV), (OAc)$_2$Zr(OEP), and the isostructural hafnium(IV) complex are eight-coordinate species, a fact that was strongly indicated, but not proven, prior to structure determination[7]. The abbreviated formula, Ni(Etio-I), is appropriate for 1,3,5,7-tetramethyl-2,4,6,8-tetraethylporphinatonickel(II)[8], the nickel derivative of etioporphyrin-I, H$_2$(Etio-I).

Determinations of crystalline structure are available for just five metalloporphyrins which are characterized by asymmetric patterns of peripheral substitution on the porphinato core. All of these are metal derivatives of deuteroporphyrin-IX, H$_2$(Deut-IX), which is systematically named as 1,3,5,8-tetramethyl-6,7-di(2'-carboxyethyl)porphin. Noting that protoporphyrin-IX, H$_2$(Proto-IX), and mesoporphyrin-IX, H$_2$(Meso-IX), are the respective 2,4-divinyl and 2,4-diethyl derivatives of deuteroporphyrin-IX, and that in the dimethyl ester of any member of the family the substituents in the 6,7-positions on the porphin ring become 2'-carbomethoxyethyl groups, the systematic naming of the metalloporphyrins of determined stereochemistry is straightforward. Thus μ-oxo-bis[1,3,5,8-tetramethyl-2,4-divinyl-6,7-di-(2'-carbomethoxyethyl)porphinatoiron(III)][9], O[Fe(Proto-IX-DME)]$_2$, is the oligomeric product of the spontaneous condensation of two molecules of the unstable hydroxoiron(III) derivative of protoporphyrin-IX dimethyl ester, (OH)Fe(Proto-IX-DME)*; 2(OH)Fe(Proto-IX-DME) $\rightarrow$ O[Fe(Proto-IX-DME)]$_2$ + H$_2$O.

---

* The specification of a metalloporphyrin as the metal derivative of a familiar or previously specified porphyrin, exemplified here for (OH)Fe(Proto-IX-DME) and above for Ni(Etio-I), is compatible with any systematic nomenclature which recognizes that the metalloporphyrin, as just one of the products of a substitutional or displacement reaction, necessarily has lost the central pair of hydrogen nuclei that were present in the parent porphyrin. Relative to the neutral components, metal atom and porphyrin molecule, the metalloporphyrin lacks two hydrogen atoms. In the applicable classic type of chemical nomenclature (which has seen little or no use in metalloporphyrin chemistry), a metalloporphyrin is a metal porphinate — usually, to be sure, a peripherally substituted porphinate. As a substitute for the more recently developed nomenclature of coordination chemistry, this classic type of nomenclature has little to commend it. But as a formally correct nomenclature which observes mass balance, it would always have been (and is still) a most acceptable substitute for the common practice of using 'metal porphyrin', the purely additive coupling of the names of the components from which the complexed species derives, for the complex itself — a practice which wholly ignores conservation of mass. The use of nickel tetraphenylporphine, nickel tetraphenylporphyrin, nickel etioporphyrin-II, etc., is fully analogous to the clearly indefensible use of methane chlorine or chlorine methane instead of methyl chloride or chloromethane.

The allure of defective terminology (which owes much to its brevity) would be much

8.1.2. Historical notes

Two series of investigations dating from the period, 1936—1940, are deserving of special mention. Prior to 1963, the best available model for the stereochemistry of the porphin skeleton in the porphyrins and metalloporphyrins was an indirect product of Robertson and Woodward's classic X-ray analyses of crystalline structure for phthalocyanine[10] and its nickel(II)[11] and platinum(II)[12] derivatives. Less generally appreciated was the special relevance to the stereochemistry of the porphinatoiron complexes (iron porphyrins) of Pauling and Coryell's[13,14] discovery that the numerous derivatives of hemoglobin divide naturally into magnetically distinctive classes which differ from one another in the electronic configuration of the ground state of the iron atoms in the protohemes; low-spin and high-spin porphinatoiron species containing either iron(II) or iron(III) correspond to the respective maximum and minimum number of spin pairings among the $3d$ electrons in the valence shell of the iron atom[13-16].

The bearing of these early studies on metalloporphyrin stereochemistry was clarified during an initial period (1962—1965) of sustained effort by four groups of investigators in which nine authoritative analyses of crystalline structure for porphyrins and metalloporphyrins were reported in the scientific literature. These structure determinations, conveniently grouped into four stereochemically related categories, were (A) for Ni(Etio-I)[8] and the nickel(II) derivative of 2,4-diacetyldeuteroporphyrin-IX dimethyl ester, Ni(2,4-DAcDeut-IX-DME)[17]; (B) for both the tetragonal[18] and the triclinic[19] modifications of H_2(TPP) and for porphin itself[20,21]; (C) for isostructural crystals of the copper(II) and palladium(II) derivatives of H_2(TPP), Cu(TPP) and Pd(TPP)[22]; (D) for chlorohemin, (Cl)Fe(Proto-IX)[23], and the methoxoiron(III) derivative of mesoporphyrin-IX dimethyl ester, (MeO)Fe(Meso-IX-DME)[24].

---

reduced in an obvious revision — a simplification — of the existing systematic nomenclature whereby 'porphin', 'porphinato', and (if used) 'porphinate' were consistently replaced by 'porphyrin', 'porphyrinato', and 'porphyrinate', respectively. All porphyrins and metalloporphyrins would still be named as derivatives of the simplest porphyrin which, for purposes of nomenclature, would be called 'porphyrin' instead of 'porphin'. A simple, but accurate, quasi-systematic nomenclature could be used for the metal derivatives of the numerous porphyrins which carry trivial names: thus, (etioporphyrinato-I)nickel(II), Ni(Etio-I), instead of the so-called 'nickel etioporphyrin-I'; chloro-(protoporphyrinato-IX)iron(III), (Cl)Fe(Proto-IX), for chlorohemin. The particularly important family of the 'iron porphyrins', correctly described as porphinatoiron complexes in the present systematic nomenclature, would become porphyrinatoiron complexes (also, perhaps, iron porphyrinates) in the revised system. Both precision and simplicity, in the writer's judgment, would be well-served by the suggested revision in nomenclature. Some attempt to formulate a chemically and semantically defensible nomenclature which will be generally used by porphyrin chemists is surely in order.

8.1.3. Considerations of symmetry

Presentation of the stereochemical conclusions that were derived from these early studies is complicated by a general limitation of X-ray structural analysis as a technique for determining molecular symmetry. It is almost invariably found that the symmetry required of a molecule in a crystalline metalloporphyrin — the symmetry of its environment — is that of a subgroup of the higher symmetry which, at least theoretically, the externally unconstrained species would be expected to display*. Since it is usually the higher symmetry which commands primary interest, criteria are needed for assaying the probability that the observed lower symmetry is attributable to packing constraints in the crystal.

Several points are to be noted in this connection. (1) The highest symmetry which may be displayed by a metal derivative of a symmetrically substituted porphyrin such as $H_2(TPP)$ or $H_2(OEP)$ is determined by the mutual interactions of the metal atom, porphinato core, and any axial ligand or ligands which may be present; full D_{4h} symmetry, which is illustrated by the diagram of the porphinato core in Fig. 1, is theoretically attainable only in four-coordinate derivatives, M(TPP) and M(OEP), and in six-coordinate species, $(L)_2 M(TPP)$ and $(L)_2 M(OEP)$, wherein each of the chemically identical axial ligands can observe C_{4v} symmetry. (2) It is not certain that all externally unconstrained molecules, M(TPP) and M(OEP) are stabilized by the planar conformation required in D_{4h} symmetry; for those molecules in which the M(II) atom is especially small, e.g., the diamagnetic d^8 nickel(II), a somewhat plausible alternative is a strongly ruffled conformation of D_{2d} symmetry[3,25]. (3) The more or less drastic reduction from the symmetry expected for the unconstrained species that is commonly required in the crystalline phase generally is most strongly displayed in the orientations of the peripheral substituents relative to the porphinato core; not infrequently, however, the experimentally determined geometry of the core approximates closely to a higher symmetry than that of the molecular environment. (4) Deformations of the core from planarity that are not rationally attributable to any intramolecular source may be discounted as the consequence of constraints imposed by the molecular packing; moderate deformations have rather little effect upon the bond angles and still less upon the bond dis-

---

* This phenomenon is merely a byproduct of the overriding need to fulfill two basic requirements. The arrangement of the molecules in the crystal must observe three-dimensional periodicity while achieving a packing density which assures thermodynamic stability of the phase in circumstances specified as follows. The equilibrium crystallization at fixed temperature and pressure of a metalloporphyrin from solution is thermodynamically the result of a favorable decrease in the partial molar enthalpy of this component that fully compensates for the accompanying unfavorable decrease in its partial molar entropy. The beneficial consequence of a reduction in the symmetry required of the molecule in the crystal is an increase in the number of parameters which may be adjusted to yield the most favorable packing of the molecules.

References, p. 376

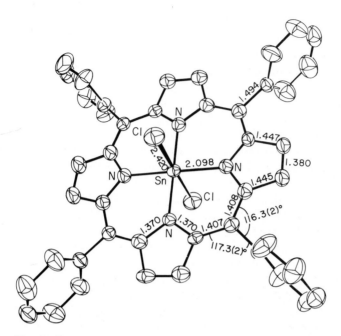

Fig. 2. Computer-drawn model in perspective of the $(Cl)_2Sn(TPP)$ molecule (hydrogen atoms omitted) as it exists with required C_{4h} symmetry in the crystalline phase. Each atom is represented by an ellipsoid having the shape and relative size concomitant with its thermal motion in the crystal. The only significant departure of the porphinato core from D_{4h} geometry is displayed in the inequality of the pair of external angles subtended at a methine carbon atom.

tances in the quasi-aromatic porphinato core[25]. (5) In just one fully ordered structural type, that observed for tetragonal crystals of dichloro-$\alpha,\beta,\gamma,\delta$-tetraphenylporphinatotin(IV), $(Cl)_2Sn(TPP)$[26], is planarity of the porphinato core required. The very precisely determined stereochemistry of this molecule, as it occurs with required C_{4h} symmetry in the crystal, is illustrated in Fig. 2.

There are two particularly illuminating structures for crystalline metalloporphyrins in which the observed conformations of the cores are adequately and most simply described by models of D_{4h} geometry even though the required molecular symmetry is just a center of inversion, C_i. These are the structures reported for monoclinic crystals of dichloro-1,2 ··· 8-octaethylporphinatotin(IV) nitromethane solvate[27], $(Cl)_2Sn(OEP) \cdot 2CH_3NO_2$, and for triclinic crystals of Ni(OEP)[3]; their special relevance derives from the fact that tin(IV) and nickel(II) afford, respectively, the largest and the smallest metal atoms that are known to be centered in the porphinato core (at Ct in Fig. 1). An averaged model of D_{4h} geometry for the core of the Ni(2,4-DAcDeut-IX-DME) molecule, as this species occurs with no required symmetry in triclinic crystals of the 2 : 1 benzene solvate, has been used in several earlier discussions[25,28-30] of metalloporphyrin stereochemistry. The difference between this model and that of recent date[3] for the flat core in the Ni(OEP) species are of a uniformly minor nature, as is demonstrated in Section 8.2.

Of most immediate interest are the lengths of the complexing Ni—N and and Sn—N bonds within porphinato cores of effectively D_{4h} geometry. The reported Ni—N distances of 1.957—1.960 Å in Ni(OEP)[3], Ni(2,4-DAcDeut-IX-DME)[17], and in the statistically disordered structure of Ni(Etio-I)[8], are fully 0.10 Å longer than the bonds that are rather commonly formed by the diamagnetic d^8 nickel(II) atom with the nitrogen atoms of four monodentate ligands. Sn—N distances of 2.082(5) Å in $(Cl)_2Sn(OEP)$[27] and 2.098(2) Å in $(Cl)_2SnTPP$[26] are by contrast, $\geqslant 0.08$ Å shorter than is expected with nitrogen atoms of monodentate ligands. These data are evidently in agreement with the expectation that the macrocyclic porphinato ligand in a metalloporphyrin must resist undue radial contraction or expansion; they are compatible, moreover, with the empirical concept[24] of a particular value of the M—N bond distance or, more generally, of the *radius of the central hole*[24] — the $Ct \cdots N$ radius (Fig. 1) — that, within a narrow range of variation, serves to minimize radial strain in the porphinato core of metalloporphyrins. The value first suggested[24] as achieving such minimization, ~ 2.01 Å, retains its utility for discussion of the stereochemistry of planar or *quasi*-planar cores in metalloporphyrins[25,28-30].

The $Ct \cdots N$ = Ni—N distance of 1.96 Å observed in Ni(OEP)[3], Ni(Etio-I)[8], and Ni(2,4-DAcDeut-IX-DME)[17] is 0.13 Å larger than the 1.83 Å reported for this parameter in phthalocyaninatonickel(II), Ni(Pc)[11]. Furthermore, the averaged values of 2.05—2.06 Å observed for the $Ct \cdots N$ radius in both crystalline modifications of $H_2(TPP)$[18,19] and in porphin[20] is 0.13—0.14 Å larger than the $Ct \cdots N$ radius of 1.92 Å reported for phthalocyanine, $H_2(Pc)$. It appears, consequently, that the value of $Ct \cdots N$ which minimizes radial strain in metallophthalocyanines must be < 1.90 Å. It is, of course, the replacement of the bridging $C_m H$ groups in a porphyrin by nitrogen (N_m) atoms in phthalocyanine that is primarily responsible for the substantial quantitative alterations in the $Ct \cdots N$ radii. The most direct and important stereochemical consequence of this substitution is to be seen in $C_a C_m C_a$ angles (Fig. 1) of 125.1, 123.9, and 125.2° in Ni(OEP), Ni(2,4-DAcDeut-IX-DME), and $H_2(TPP)$[18], respectively, that are replaced by $C_a N_m C_a$ angles of $\sim 117°$ in Ni(Pc) and $H_2(Pc)$. This contraction of the bond angle at the bridging atom is presumably attributable to the larger spatial demands of the lone pair of electrons on an N_m atom.

In the tetragonal structural type established in 1964 for crystalline Cu(TPP) and Pd(TPP)[22], the M(TPP) molecule exists in a strongly ruffled conformation of required S_4 symmetry and the geometry of the porphinato core approximates rather closely to the requirements of the point group,

D_{2d}. Bond lengths of Cu—N = 1.981(7) and Pd—N = 2.009(9) Å were seen to be ordered as expected relative to 1.960 Å for the Ni—N distance in Ni(2,4-DAcDeut-IX-DME)[17]. Given this background, the observation in 1972 that the tetragonal structural type[2] in which Ni(OEP) crystallizes also requires a molecular symmetry of S_4 and a strongly ruffled core of *quasi-D_{2d}* geometry was not especially surprising; only the shortness of the Ni—N bonds, 1.929 (3) Å was initially disconcerting. It was then shown[25] that a D_{2d} ruffling of the porphinato core, though somewhat inimical to the most favorable delocalization of π bonding, is geometrically conducive to a significant shortening of the complexing bonds relative to those expected with a planar core. This analysis made it evident that a generally superior pattern of σ bonding should be afforded by a marked D_{2d} ruffling of the core in those metalloporphyrins wherein the complexing M—N bonds are significantly stretched from the lengths preferred with nitrogen atoms of monodentate ligands. Subsequent determinations of crystalline structure for triclinic Ni(OEP) (*vide supra*) and monoclinic $\alpha,\beta,\gamma,\delta$-tetra($n$-propyl)porphinato-copper(II), Cu(TPrP)[31], have provided direct support for the purely geometrical aspects of the earlier analysis[25]; within the nearly planar cores observed in these crystals, the lengths of the complexing bonds are Ni—N = 1.958 (3) and Cu—N = 2.000 (5) Å, longer by ~0.03 and ~0.02 Å, respectively, than the values cited above for these parameters in the ruffled species.

In any general consideration of the molecular symmetries that are theoretically available to the metal derivatives of a symmetrically substituted porphyrin such as H_2(TPP) or H_2(OEP), the first point to be emphasized is that D_{4h}, D_4, C_{4v}, and D_{2d}, are the only point groups which require that structural equivalence be maintained among the atoms of each chemical class. Figure 1 is a multipurpose diagram of the porphinato core in which this special condition is observed; the three chemically distinctive types of carbon atoms are denoted by C_a, C_b, and C_m. Only in D_{4h} is planarity of the core required; for each of the other point groups, Fig. 1 becomes a diagram of a nonplanar core as projected along the unique 4-fold axis onto a mean equatorial plane. Inasmuch as the stabilizing contribution of delocalized π bonding to the energy of the porphinato core is maximized by a conformation of D_{4h} symmetry, it appears that several features of the D_{4h} core should be retained, nearly or exactly, in any competing nonplanar conformation. These are (1) planarity of the pyrrole rings; (2) planarity of the trigonal bond systems centered at the bridgehead carbon (C_a) atoms; and (3) planarity of the trigonal bond systems centered at the methine carbon (C_m) atoms. These specifications, indeed, seem never to be ignored in the more unsymmetrical cores that have been observed in crystalline metalloporphyrins. Of the three groups other than D_{4h} that require structural equivalence of the stereochemical parameters within each chemical class, C_{4v} and D_{2d} are each fully compatible with (1), (2), and (3); D_4, by contrast, is wholly incompatible with (2) and is of no further theoretical or experimental interest in metalloporphyrin stereochemistry.

The combination of specifications (1) and (2) leads to a convenient structural subunit for describing conformation in either D_{2d} or C_{4v}; this is the planar or *quasi*-planar entity which comprises a pyrrole ring and its attached pair of bridging C_a—C_m bonds — just one quarter of the porphinato core. The symmetry elements of D_{2d} include the unique S_4 (or $\bar{4}$) axis of rotary-reflection (or rotary-inversion) a pair of mutually perpendicular 2-fold axes which are also perpendicular to the S_4 axis, and a pair of mutually perpendicular mirror planes which intersect in the unique axis and are diagonally oriented at 45° from the 2-fold axes; see Fig. 3. In the D_{2d} conformation of the porphinato core that commands interest, the complexing M—N bonds lie on the 2-fold axes and the methine carbon (C_m) atoms lie in the diagonal mirror planes. Starting from the D_{4h} conformation, a rotation through the angle ϕ of a structural subunit around the 2-fold axis which bisects it requires the simultaneous rotation of the other subunits through either $+\phi$ or $-\phi$ as determined by the symmetry operations of S_4. Such rotations of the subunits produce foldings of the chelate rings along the (now slanting) lines defined by the $Ct \cdots C_m$ radii in the mirror planes. One pair of trans-core C_m carbon atoms are displaced above the equatorial plane of the core (or coordination group), the other an equal distance below. The S_4 ruffling of the core is, in effect, a circular standing wave in which the rotational spacing of crests alternating with troughs is 90° and the spacing of crests (or troughs) is 180°.

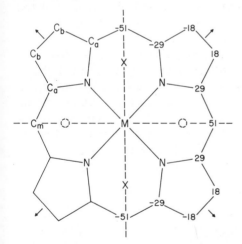

Fig. 3. Diagram in projection of the porphinato core in the ruffled form of the Ni(OEP)[2] molecule as slightly idealized to D_{2d} symmetry. The pair of vertical mirror planes are indicated by broken lines, the equatorial two-fold axes by arrows. In the right-hand half of the diagram, the symbol for each carbon atom is replaced by the displacement of the atom, in units of 0.01 Å, from the mean plane of the core (in which the nitrogen atoms necessarily lie). This diagram serves also to show that a D_{2d}-ruffled configuration is the sterically ideal choice for a bis(pyridine)metal derivative. The crosses and the broken circles represent the projected positions from above and below, respectively, of the 2,6-hydrogen atoms of the pyridine ligands.

References, p. 376

As is also indicated in the diagram of Fig. 3, the sterically ideal configuration for a bis(pyridine)metal derivative of H_2(TPP) or H_2(OEP) observes D_{2d} symmetry. Destabilizing steric interactions of the 2,6-hydrogen atoms of the pyridine ligands with porphinato nitrogen and carbon atoms are thereby minimized.

The symmetry elements of D_{4h} that are retained in C_{4v} are the 4-fold axis and the four mirror planes which intersect in this axis; each pair of diagonally positioned C_m atoms lie in a mirror plane as do also each pair of diagonally positioned nitrogen atoms (Fig. 1). Connecting four of the structural subunits at the methine carbon atoms in agreement with C_{4v} yields a 'domed', 'tented', or 'truncated pyramidal' conformation; the core atoms are sorted into descending parallel tiers, each of which is perpendicular to the 4-fold axis, in the order, 4N, $8C_a$, $4C_m$, and $8C_b$. This conformation, which has no symmetry-defined center, is evidently appropriate for an externally unconstrained five-coordinate metalloporphyrin, such as (Cl)Fe(TPP)[5] or isothiocyanato-$\alpha,\beta,\gamma,\delta$-tetraphenylporphinatoiron(III), (SCN)Fe(TPP)[32], in which the axial ligand can observe the required C_{4v} symmetry. The metal atom (M) lies on the 4-fold axis, but not at the center (Ct) of the configuration defined by the four nitrogen atoms; it is displaced above the plane (P_N) of the nitrogen atoms at a distance, $M \cdot\cdot P_N = M \cdot\cdot Ct$, which, depending on the choice of metal atom, may range from a few hundredths to several tenths Å. The coordinated atom (X) of the axial ligand is positioned at the apex of the square-pyramidal coordination group which is illustrated in the diagram of Fig. 4.

Although the mean plane (P_N) of the four nitrogen atoms and the mean plane (P_C) of the 24-atom porphinato core usually are not exactly parallel in a five-coordinate species as it exists with little or no required symmetry in the crystal, either the difference of the $M \cdot\cdot P_C$ and $M \cdot\cdot P_N$ distances or the $Ct \cdot\cdot P_C$ distance serves well enough as a measure of the *net* doming of the core; the difference between the two estimates usually is very small and of trivial interest. The observational statistics show that the net doming is more

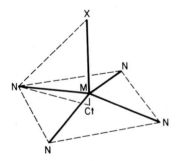

Fig. 4. The square-pyramidal (C_{4v}) coordination group which serves as a somewhat idealized model for the five-coordinate metal derivatives of the porphyrins.

likely to be <0.05 Å than >0.05 Å; especially, moreover, when the axial ligand is an aromatic base, the mean displacement (disregarding sign) of the 24 atoms in the core from P_C is quite likely to be larger than the net doming. These observations provide evidence not only that the detailed conformation of the core is dominated by packing constraints in the crystal, but also that the net doming in the externally unconstrained species is likely in general to be <0.05 Å.

The stereochemical data and the conclusions drawn therefrom[24] that were reported in 1965 for the biologically significant five-coordinate (Cl)Fe-(Proto-IX)[23] and (MeO)Fe(Meso-IX-DME)[24] species may now be summarized. Both crystalline phases are characterized by some packing disorder which primarily involves the 2,4-substituents on the porphinato cores. Averaging of the bond parameters within each chemical class leads to stereochemical descriptions of both the porphinato cores and the coordination groups that are notably similar for the two species; the only substantial difference is that required in the lengths of the axial complexing bonds: Fe—Cl = 2.218 (6) and Fe—OMe = 1.842 (4) Å. The net doming of the porphinato core is 0.065 Å in (Cl)Fe(Proto-IX) and 0.035 Å in (MeO)Fe-(Proto-IX-DME), respectively. Presumably because the porphinato cores of paired molecules in crystalline (MeO)Fe(Proto-IX-DME) are smoothly packed back to back, the core of each molecule does exhibit a fairly uniform quasi-C_{4v} doming toward the iron atom. Referring now to Fig. 4, Fe—N = 2.068 ± 0.006 Å, $Ct \cdots N = 2.015 \pm 0.008$ Å, and Fe $\cdots Ct$ = Fe $\cdots P_N = 0.465 \pm 0.010$ Å cover the ranges in these parameters that were reported for the high-spin (Cl)Fe(Proto-IX) and (MeO)Fe(Meso-IX-DME) species. These data were essential to the formulation[24] in 1965 of the overall pattern for the dependence of stereochemistry on the spin-state of the iron atom, whether ferrous or ferric iron, in the family of the porphinatoiron species — a pattern that has been uniformly supported in subsequent investigations of crystalline structure[4−6,32−36,115,116,118].

8.2. Stereochemistry of the porphinato core in metalloporphyrins

Averaged values of the stereochemical parameters in the porphinato cores of sixteen metalloporphyrins are listed in Tables 1 to 3. In the four- and six-coordinate metalloporphyrins listed in Tables 2 and 3, $Ct \cdots N$ is nearly or exactly equal to the complexing bond length, M—N; in the five- and the eight-coordinate species (the last four entries in the tables), M—N $> Ct \cdots N$. It is the latter parameter which is especially significant for the discussion of the quantitative features of the porphinato core.

In selecting metalloporphyrins for inclusion in Tables 1 to 3, preference has been given to structure determinations based upon X-ray intensity data that were measured to comparatively high $(\sin \theta)/\lambda = 1/2d$, wherein θ, λ, and

TABLE 1

Averaged parameters of the cores in nickel and tin metalloporphyrins[a].

Parameter	$(Cl)_2Sn(TPP)$[26]	$(Cl)_2Sn(OEP)$[27]	Planar $Ni(OEP)$[3]	Ruffled $Ni(OEP)$[2]
Distances in Å				
$M-N$	2.098(2,—)	2.082(5,2)	1.958(2,2)	1.929(3,—)
$N-C_a$	1.370(2,0)	1.379(7,5)	1.376(4,6)	1.387(4,2)
C_a-C_m	1.407(3,1)	1.386(8,10)	1.371(4,4)	1.373(5,2)
C_a-C_b	1.446(3,1)	1.437(8,12)	1.443(4,3)	1.449(5,5)
C_b-C_b (b)	1.380(3,—)	1.368(8,5)	1.346(4,2)	1.362(5,—)
C_b-C_e (b)	1.495(2,—)(c)	1.497(8,13)	1.495(5,7)	1.501(5,4)
$M\cdots C_m$	3.469(2,—)	3.424(6,10)	3.381(3,1)	3.355(4,—)
$M\cdots C_a$	3.100(2,1)	3.099(6,4)	3.006(3,4)	2.983(3,1)
$M-Cl$	2.420(1,—)	2.453(2,—)		
Angles in degrees				
NMN	90	90.0(2,1)	90.0(1,2)	90.0(0,0)
MNC_a	125.4(1,1)	125.9(3,4)	128.0(2,2)	127.4(2,1)
C_aNC_a	109.2(2,—)	108.2(5,2)	103.9(2,4)	105.1(3,—)
NC_aC_m	126.4(2,1)	124.4(5,4)	124.4(3,3)	124.0(2,0)
NC_aC_b	108.2(2,1)	108.3(5,7)	111.6(3,3)	110.6(2,1)
$C_bC_aC_m$	125.4(2,1)	127.3(6,4)	124.1(3,4)	125.0(2,1)
$C_aC_mC_a$	126.4(2,—)	129.5(6,4)	125.1(3,1)	124.1(2,—)
$C_aC_bC_b$	107.2(2,1)	107.6(5,3)	106.5(3,4)	106.8(3,4)

[a] The first figure in parentheses following each averaged value is the usual estimated standard deviation (esd) given by the structure analysis for each individual parameter of the chemical class. The second figure is the esd for an individual parameter, calculated from the observed deviations from the mean, on the assumption that the values averaged represent independent estimates of the same physical quantity.

[b] Bond connecting an ethyl carbon (C_e) atom to the core.

[c] Value for C_m-C_ϕ, the linkage between methine and phenyl carbon atoms.

d are the Bragg angle, the wave length of the X-radiation, and the effective interplanar spacing, respectively. Although intensity measurements were carried to, or somewhat beyond, the Cu K_α limit (where $(\sin\theta)/\lambda = 1/\lambda = 0.65$ $Å^{-1}$) in ten of the sixteen structure determinations from which the stereochemical data listed in the tables are drawn, the fraction of reflections having values of $(\sin\theta)/\lambda$ near 0.65 $Å^{-1}$ that were retained as observed in the structure analyses varied enormously — from <0.10 in $(3\text{-Pic})_2Co(OEP)$[40] to >0.98 in $(Cl)_2Sn(TPP)$[26]. It is noteworthy that the fraction of reflections from within the Cu K_α limiting sphere ($d \gg \lambda/2 = 0.77$ Å) that was retained for structure refinement was much higher for five of the listed metal derivatives of $H_2(TPP)$ and for $(Py)Zn(TPyP)$ than for any metal derivative of $H_2(OEP)$ except $(OAc)_2Hf(OEP)$. The pertinent fractions for $(3\text{-Pic})_2Co-(OEP)$, tetragonal $Ni(OEP)$, triclinic $Ni(OEP)$, and $(Cl)_2Sn(OEP)$, 0.75, 0.63,

TABLE 2

Bond lengths in the cores of selected metalloporphyrins[a].

Metalloporphyrin[b]	Distances in Å						
	$Ct\cdot\cdot N$	$N-C_a$	C_a-C_m	C_a-C_b	C_b-C_b	C_m-C_ϕ	$Ct\cdot\cdot C_m$
Ni(2,4-DAcDeut-IX-DME)[17]	1.960(7)	1.383(10)	1.375(10)	1.446(10)	1.350(10)	—	3.402(7)
(3,5-Lut)Co(NO$_2$)(TPP)[37]	1.954(2)	1.374(4)	1.400(5)	1.442(4)	1.358(5)	1.500(5)	3.386(4)
Fe(TPP)[4]	1.972(4)	1.382(6)	1.392(6)	1.436(6)	1.353(6)	1.509(6)	3.395(4)
[(Im)$_2$Fe(TPP)]Cl[6]	1.989(4)	1.378(7)	1.392(8)	1.437(8)	1.350(9)	1.498(8)	3.424(6)
(Pip)$_2$Fe(TPP)[33]	2.004(3)	1.384(4)	1.396(5)	1.444(5)	1.347(6)	1.499(5)	3.436(4)
[(Pip)$_2$Co(TPP)]NO$_3$[38]	1.979(3)	1.384(4)	1.389(5)	1.435(5)	1.356(5)	1.494(5)	3.434(3)
(Pip)$_2$Co(TPP)[39]	1.987(2)	1.381(2)	1.392(2)	1.444(3)	1.345(3)	1.498(3)	3.433(3)
(3-Pic)$_2$Co(OEP)[40]	1.992(1)	1.374(2)	1.381(2)	1.449(2)	1.355(3)	1.508(3)[c]	3.402(3)
(1-MeIm)Co(TPP)[41]	1.973(3)	1.387(5)	1.385(6)	1.441(6)	1.350(6)	1.506(6)	3.423(6)
(Py)Zn(TPyP)[42]	2.047(2)	1.369(4)	1.406(4)	1.447(4)	1.355(4)	1.491(4)	3.448(3)
(2-MeIm)Fe(TPP)[36]	2.044(4)	1.377(7)	1.403(7)	1.447(7)	1.347(7)	1.499(8)	3.454(6)
(OAc)$_2$Hf(OEP)[7]	2.016(3)	1.382(6)	1.385(7)	1.447(6)	1.353(8)	1.506(7)[c]	3.414(6)
Averaged bond lengths for sixteen metalloporphyrins[d]							
—	1.379[6]	1.390[11]	1.443[5]	1.354[10]	1.499[5]	—	

[a] The figure in parentheses following each datum is the usual estimated standard deviation (esd) given by the structure analysis for each individual parameter within the chemical class.

[b] Some abbreviations: 3,5-Lut = 3,5-lutidine; Pip = piperidine; Im = imidazole; 3-Pic = 3-picoline; MeIm = 3-methylpyridine; MeIm = methyl-imidazole; (Py)Zn(TPyP) = pyridine-α,β,γ,δ-tetra(4-pyridyl)porphinatozinc; OAc = acetato.

[c] C_b-C_e (see Table 1).

[d] Data from Table 1 are included. The [esd] following each datum is defined in the text.

TABLE 3

Bond angles in the cores of selected metalloporphyrins[a].

Metalloporphyrin[b]	Angles in degrees							
	NMN	C_aNC_a	MNC_a	NC_aC_m	NC_aC_b	$C_bC_aC_m$	$C_aC_mC_a$	$C_aC_bC_b$
Ni(2,4-DAcDeut-IX-DME)[17]	90.0(2)	104.4(5)	127.8(5)	125.3(5)	111.0(5)	123.7(5)	123.8(5)	106.8(5)
(3,5-Lut)Co(NO$_2$)(TPP)[37]	90.1(1)	106.4(2)	126.7(2)	125.4(3)	109.7(3)	124.5(3)	121.6(3)	106.9(3)
Fe(TPP)[4]	90.01(4)	105.4(3)	127.3(3)	127.3(4)	110.2(3)	124.5(3)	123.5(3)	107.1(4)
[(Im)$_2$Fe(TPP)]Cl[6]	90.0(2)	106.1(4)	126.9(2)	126.0(5)	109.7(5)	124.2(5)	123.1(5)	106.7(5)
(Pip)$_2$Fe(TPP)[33]	90.0(1)	105.2(3)	127.2(2)	125.6(3)	110.2(3)	124.1(3)	124.1(3)	107.2(3)
[(Pip)$_2$Co(TPP)]NO$_3$[38]	90.0(1)	104.9(3)	127.5(2)	125.9(3)	110.5(3)	123.6(3)	123.0(3)	107.0(3)
(Pip)$_2$Co(TPP)[39]	90.0(1)	104.8(1)	127.4(1)	125.8(2)	110.5(2)	123.6(2)	123.4(2)	107.0(2)
(3-Pic)$_2$Co(OEP)[40]	90.0(1)	105.0(1)	127.5(1)	124.5(2)	111.0(2)	124.5(2)	125.8(2)	106.5(2)
(1-MeIm)Co(TPP)[41]	89.98(7)[c]	104.6(3)	127.6(3)	125.4(4)	110.7(4)	123.8(4)	123.5(4)	107.0(4)
(Py)Zn(TPyP)[42]	88.8(1)[c]	106.6(2)	126.5(2)	125.7(2)	109.8(2)	124.5(3)	125.2(3)	106.9(3)
(2-MeIm)Fe(TPP)[36]	87.7(2)[c]	106.7(4)	126.4(4)	125.9(5)	109.3(4)	124.8(5)	124.8(5)	107.4(5)
(OAc)$_2$Hf(OEP)[7]	78.4(1)[d]	105.3(4)	124.8(3)	124.5(4)	110.4(4)	125.1(4)	126.6(4)	107.0(3)

[a] The figure in parentheses following each datum is the usual estimated standard deviation given by the structure analysis for each individual parameter within the chemical class.

[b] Some abbreviations: 3,5-Lut = 3,5-lutidine; Pip = piperidine; Im = imidazole; 3-Pic = 3-picoline; 3-methylpyridine; MeIm = methylimidazole; (Py)Zn(TPyP) = pyridine-α,β,γ,δ-tetra(4-pyridyl)porphinatozinc; OAc = acetato.

[c] Five-coordinate species in which the metal atom lies out-of-plane from the porphinato nitrogen atoms.

[d] Eight-coordinate species; Hf··Ct = 1.012(3) Å; Hf–N = 2.257(4) Å; the porphinato core is markedly domed.

0.61, and 0.53, respectively, are arranged in descending order to reflect the increasing dependence of the refinement on the data of lower resolution from within the inner half of the Cu K_α limiting sphere. That such enhanced dependence on the low-resolution data is definitely inimical to the evaluation by standard least-squares techniques of interatomic C—C and C—N distances which best approximate to true internuclear separations is theoretically and experimentally documented in Appendix I.

Whenever the accurate measurement of intensity data can be carried to sufficiently high $(\sin\theta)/\lambda$ — a condition more than amply fulfilled for $(Cl)_2Sn(TPP)$ — the low-resolution data which reflect the asymmetric scattering of X-rays by the valence shell electrons can profitably be excluded from a least-squares refinement of structure that is designed to give realistic positions for the carbon and nitrogen atoms. (The limiting condition on the use of this modified procedure is more precisely defined in Appendix I.) The refinements VI and VII specified in Appendix I were of this type; they lent their fullest support to the earlier assignment of atomic positions taken from the Fourier synthesis of the structure factors measured for the $(Cl)_2Sn(TPP)$ crystal. By taking the lengths obtained for a given chemical type of bond from the Fourier synthesis and from refinements VI and VII as independent determinations of the same quantity, there were obtained, N—C_a = 1.370 (1), C_a—C_m = 1.407 (2), C_a—C_b = 1.446 (2), C_b—C_b = 1.380 (1), and C_m—C_ϕ = 1.495 (2) Å, wherein the figure in parentheses following each mean length is the estimated standard deviation in the last significant figure as calculated from the deviations from the mean. These very low *esd* values are fully indicative of the simultaneous triviality of the differences in lengths given by the three techniques and of the deviations from D_{4h} geometry.

Depending, on the other hand, upon the weighting scheme employed, conventional least-squares refinements using all of the $(Cl)_2Sn(TPP)$ data gave C_b—C_b distances of 1.369 (3) and 1.366 (3) Å; the weighting which gave the lower value was of the type considered appropriate by most crystallographers. These data, other considerations given in Appendix I, and subsequent investigations in the Cornell Laboratory (in part by Professor Robert E. Hughes and coworkers) support the following conclusions regarding the use of all of the data in conventional least-squares refinement of structure. (1) Relative to the true internuclear separation, the length of a bond having substantial multiple bond character always should be underestimated. (2) The magnitude of the underestimate should increase with decreasing range in $(\sin\theta)/\lambda$ of the accurately measurable data. (3) Given data of unduly restricted scope and a structural model having a plethora of adjustable parameters, the purely mathematical minimization of the function employed in least-squares refinement may yield results that are, at best, only in qualitative agreement with (1) and (2).

In view of the qualifying conclusion (3), it would be unsafe to conclude that all of the C_b—C_b bond lengths listed in Tables 1 to 3, save that in

$(Cl)_2Sn(TPP)$, are underestimated by $\geqslant 0.010$ Å. The writer does suggest, however, that the averaged value, 1.354 Å, for $C_b—C_b$ in the sixteen species (Table 2) is certainly too low and that the internuclear $C_b—C_b$ distance does not fall below 1.350 Å in any metalloporphyrin. It is further to be noted that $C_b—C_b$ distances ranging from 1.333 to 1.337 Å, apparently quite as short as the length (1.337 Å) of the double bond in ethylene, are reported for chlorohemin[23], for $Cu(TPP)$[22], and for $(Pyridine)Ru(CO)(TPP)$[44]. Inasmuch as the structure determinations were based upon X-ray data of especially small range in $(\sin\theta)/\lambda$, underestimation of the $C_b—C_b$ bond length by $\geqslant 0.02$ Å is not unexpected and is in qualitative agreement with the definitive examples of this phenomenon that are cited in Appendix I. The still shorter length, 1.327 (12) Å, reported for the $C_b—C_b$ bond in $(EtOH)Ru(CO)(TPP)$[45] comes from a structure determination in which the combination of a statistically required molecular symmetry of $C_i—\bar{1}$ with data of limited scope means that the uncertainty in the bond length is only roughly suggested by the large, but formally calculated, *esd* of 0.012 Å.

From inspection of the stereochemical data listed in Tables 1 to 3 and entered in part on the diagrams of Figs. 1 to 3, it is seen that the *quasi*-aromatic character of the core in metalloporphyrins may be somewhat perturbed, but not largely modified, by some combination of the several factors discussed above. The bond lengths entered in the bottom line of Table 2 are mean values of these parameters for the entire group of sixteen metalloporphyrins; they are qualitatively consistent with the classical Kekulé-type formulae whereby the $N—C_a$, $C_a—C_b$, $C_b—C_b$, and $C_a—C_m$ linkages are assigned double-bond characters of 1/4, 1/4, 3/4, and 1/2, respectively. The estimated standard deviation associated with each mean value is formally calculated from the observed deviations from the mean. These deviations include components roughly classified as arising from (1) random variations in the determination of what may be a *quasi*-invariant quantity; (2) systematic errors in the measurement of the diffracted intensities; (3) deficiencies in the structural model used during least-squares refinement; and (4) real variations in this bond length from one metalloporphyrin to another. Some appraisal of the total input from components (1) to (3) is required in order to identify an objectively significant component (4).

The minimization of the extraneous components (1) to (3) during structure refinement for $(Cl)_2Sn(TPP)$ in the extraordinarily favorably circumstances detailed in Appendix I allows the derived stereochemical parameters to serve as a quantitative standard for a porphinato core of D_{4h} geometry. These parameters, unfortunately, apply directly only to the most expanded core yet observed in a metalloporphyrin rather than to a core in which the $Ct \cdots N$ radius, ~2.01 Å, corresponds to the minimization of equatorial strain. Thus the $Ct \cdots C_m$, $Ct \cdots C_a$, $C_a—C_m$, and $C_b—C_b$ distances (Fig. 1 and Tables 1 and 2) take maximum values in $(Cl)_2Sn(TPP)$ and, in agreement with a mechanical model of linked springs[25,26], the $N—C_a$ bond is compressed to the minimal length (1.370 (1) Å); the C_a—C_b distance of 1.446 (2) Å, however, differs little from the mean value of 1.443 (5) Å for the sixteen metalloporphyrins. Further evidence that the C_b—C_b bond length is consistently underestimated in conventional least-squares refinement is provided by noting (Table 3) that the *esd* of 0.010 Å associated with the mean value of C_b—C_b relative to the 0.005 Å associated with the mean C_a—C_b length is an outright reversal of expectations based upon the well-established relation between π bond order and interatomic distance in bonds connecting trigonally hybridized carbon atoms. Any specified alteration in the length of a C_b—C_b linkage having a π bond order of ~0.75 implies a significantly larger alteration (of opposite sign) in the length of the C_a—C_b bonds having a π bond order of ~0.25.

Cruickshank and Sparks[46] have shown that the length of a C—C bond, r_p, of π bond order, p, in a number of benzenoid hydrocarbons is given satisfactorily as a function of p by the relation

$$r_p = 1.477 - (1.477 - 1.337)(1.333p)/(1 + 0.333\,p) \tag{1}$$

with constants in (1) determined from data as follows: for the respective π bond orders of 1, 1/2, and 1/3, r_1 = 1.337 Å in ethylene, $r_{1/2}$ = 1.397 Å in benzene, and $r_{1/3}$ = 1.421 Å in graphite. The constant, r_0 = 1.477 Å, is then the *predicted* length of a pure σ bond ($p = 0$) between two trigonally hybridized carbon atoms. For C_a—C_b = 1.443 and C_b—C_b = 1.360 Å, equation (1) gives $p = 0.20$ and $p = 0.80$, respectively. A more rational value for r_0, as demonstrated below, is the experimentally definitive length, 1.495 (2) Å, of the C_m—C_ϕ bond in $(Cl)_2Sn(TPP)$. On an empirical basis it then appears that C_a—C_b = 1.445, C_b—C_b = 1.365—1.370, C_a—C_m = 1.397, and N—C_a = 1.375 Å correspond rather well to Kekulé formulae for a prototypic core in which the aromatic character is divided equally between the inner 16-membered and the outer 20-membered ring systems (Fig. 1).

Although the dihedral angle between the plane of the porphinato core and that of each phenyl group in crystalline $(Cl)_2Sn(TPP)$ is exactly 90° in the equilibrium configuration, a quite minor conjugation of the two ring systems through the C_m—C_ϕ bond is in principle allowed by the thermally excited libration of the phenyl group around this bond. The observed values of this dihedral angle in the eight crystalline derivatives of $H_2(TPP)$ listed in Table 2 and in $(Py)Zn(TPyP)$ range downward to ~60°, affording the possibility that enhanced conjugation might lead to an observable decrease in the C_m—C_ϕ distance below the 1.495 (2) Å in $(Cl)_2Sn(TPP)$. In fact, however, the observed C_m—C_ϕ bond lengths in seven of these species lie between 1.498 and 1.509 Å, and the two lower values, 1.494 (5) and 1.491 (4) Å, which are from structure determinations wherein the low-resolution data from within the chromium K_α sphere comprise <0.25 and <0.20, respectively, of the data employed in the refinements of structure, are objectively consistent with the 1.495 (2) Å observed in $(Cl)_2Sn(TPP)$. The statistical impact of these data is to suggest that underestimation of the $Ct \cdot\cdot C_m$ radius — an apparent shifting of the C_m atom toward the center of the core as the consequence of foreshortening of the aromatic C_a—C_m distances — more than outweighs any otherwise observable effect of conjugation on the C_m—C_ϕ bond length.

The greater dependence on low-resolution data for structure refinement along with enhanced thermal motions of the C_m atoms in crystals of the derivatives of $H_2(OEP)$ that are listed in the tables should lead to more significant underestimation of the Ct—C_m radii and the C_a—C_m bond lengths than in the derivatives of $H_2(TPP)$. A further uncertainty in determining the position of a C_m atom in a derivative of $H_2(OEP)$ stems from the use of a necessarily empirical model (in several variations) for the scattering of X-radiation by the electron contributed to the C—H bond by the bound hydrogen atom. The similar uncertainty which attends the determination of the position of a C_b carbon atom in derivatives of $H_2(TPP)$ is presumably responsible in part for the considerable variation in the reported values of C_b—C_b bond lengths (Table 2).

No foreshortening of the C_a—C_m bonds in Ni(OEP) as an experimental artifact can be large enough to invalidate the conclusion that this parameter, with experimentally determined values of 1.371 and 1.373 Å in the respective planar and ruffled conformations of the molecule (Fig. 1 and Table 1) is indeed substantially smaller than the 1.407 Å observed in $(Cl)_2Sn(TPP)$. Given the relatively constrained geometry of the five-membered pyrrole rings, the very substantial shortening of the M $\cdot\cdot$ C_a radius in the *quasi*-planar core of Ni(OEP) relative to the value observed in the most expanded core of $(Cl)_2Sn(TPP)$, ~0.094 Å, is the direct consequence of the still larger difference in the lengths of the complexing M—N bonds, ~0.140 Å. Near or exact invariance of the C_a—C_m distance would then demand highly unfavorable values for the $C_aC_mC_a$ and NC_aC_m angles in the chelate rings. Thus it is seen that the bridging pairs of C_a—C_m bonds become the principal repository for cumulative strain in the porphinato core.

It is clear that in chemically analogous, externally unconstrained, metal derivatives of $H_2(OEP)$ and $H_2(TPP)$ at room temperature, significantly longer C_a—C_m bonds are called for in the latter species: (1) because the libration of the phenyl groups around the C_m—C_ϕ bonds degenerates in effect to a rotation restricted by a moderate potential barrier and thus allows substantial conjugation of the phenyl groups with the core; (2) because the vibrations which involve stretch in the C_a—C_m bonds and thus cooperate in allowing this restricted rotation surely are so strongly excited as to bespeak anharmonicity. (The fairly slow racemization in solution at room temperature of the $\alpha,\alpha,\alpha,\alpha$-atropisomer of *meso*-tetra(*o*-aminophenyl)porphyrin[47] is suggestive in this connection.) Stretch in the C_a—C_m bonds requires, of course, alterations in the electronic structure of the porphinato core, most particularly in the delocalization of the π electron density. The n-propyl substituents in $H_2(TPrP)$ and its metal derivatives may also be expected to require stretch in the C_a-C_m bonds through a mechanism of thermally excited anharmonic vibrations analogous to (2) and, perhaps, through the postulated hyperconjugation[48] of the porphinato core with alkyl substituents. Thus it is scarcely surprising to find significant differences in the visible absorption spectra recorded for porphin, $H_2(TPrP)$, and $H_2(TPP)$ in benzene solution at room temperature[49,50].

The low-spin d^7 cobalt(II) atoms in crystalline $(Pip)_2Co(TPP)$ and $(3\text{-}Pic)_2Co(OEP)$ are centered in CoN_6 coordination groups of similar dimensions and, within the excellent internal precision attained in both structure refinements, the C_a atoms are assigned nearly identical positions in the two averaged structures. Consequently, the observed differences in the C_a-C_m bond length, 0.011 (3) Å, the $C_a C_m C_a$ angle, 2.4 (3)°, and the $Co \cdots C_m$ radius, 0.031 (4) Å, (Tables 2 and 3) are formally determined by the position assigned to the C_m atom. The C_a-C_m bonds in crystalline $(Pip)_2Co(TPP)$ may be slightly stretched as the consequence of the distinctively nonplanar conformation of the porphinato core; being also rather less dependent on low-resolution intensity data, the derived value of the C_a-C_m bond length is presumably less susceptible to significant underestimation than that in crystalline $(3\text{-}Pic)_2Co(OEP)$.

Comparison of the stereochemical parameters listed in Table 1 for the highly expanded cores of the dichlorotin(IV) derivatives of $H_2(TPP)$ and $H_2(OEP)$ show that the rather large differences reported for the C_a-C_m distance, 0.021 (10) Å, the $Sn \cdots C_m$ radius, 0.045 (10) Å, and the $C_a C_m C_a$ angle, 3.1 (6)°, carry especially large estimated standard deviations which come almost wholly from the structure determination for crystalline $(Cl)_2\text{-}Sn(OEP)$ nitromethane solvate[3] — a determination necessarily based upon X-ray data of quite limited scope. Indeed, the value obtained for the $C_a C_m\text{-}C_a$ angle, 129.5 (6) Å, in the 6-membered chelate rings in $(Cl)_2Sn(OEP)$ is $\sim 3.0°$ larger than that reported for any other metalloporphyrin (Tables 1 and 3). Noting that the positions assigned to the C_a atoms in the averaged structure of $(Cl)_2Sn(OEP)$ are ostensibly the same (within 0.002 Å) as those observed in $(Cl)_2Sn(TPP)$, it is geometrically evident that the effects of any underestimation of the C_a-C_m distance must be magnified in both the $C_a C_m C_a$ angle and the $Sn \cdots C_m$ radius. The probability that the C_a-C_m distance of 1.386 (10) Å in $(Cl)_2Sn(OEP)$ may be significantly underestimated is otherwise supported by (1) the conclusion that the C_a-C_m distances in $(3\text{-}Pic)_2Co(OEP)$ and $(Pip)_2Co(TPP)$ differ by no more than 0.010 Å, and (2) the observation that structure determination for $(OAc)_2\text{-}Hf(OEP)$, based upon X-ray data extending beyond the Cu K_α limit, yields $C_a-C_m = 1.385$ (7) Å along with the comparatively small $Ct \cdots N$ radius of 2.016 (3) Å.

References, p. 376

8.3. The free base porphyrins and related metal-free species
8.3.1. Free bases

Fully ordered structures (apart from thermal motion) have been reported for crystals of porphin[21,50], *meso*-tetraphenylporphyrin, $H_2(TPP)$[19,50], *meso*-tetra(n-propyl)porphyrin, $H_2(TPrP)$[49,50], octaethylporphyrin, $H_2(OEP)$[48], and mesoporphyrin-IX dimethyl ester, $H_2(Meso-IX-DME)$[51]. The positions assigned to the central pair of hydrogen atoms in the porphyrin molecules follow the same pattern in all of the five structures; one pair of diagonally situated pyrrole rings carries N—H bonds, the other pair does not. The molecule is thereby limited to some class of 2-fold symmetry. No symmetry is required in the crystals for either the porphin or the H_2-(Meso-IX-DME) molecule; a center of inversion is required in each of the three other species. Averaged bond parameters within the two chemically distinctive types of pyrrole rings are listed for each of the five porphyrins in Table 4; also listed are the comparative data for two metalloporphyrins, $Cu(TPrP)$[31] and $Ag(TPP)$[52], which crystallize isomorphously with their respective parent porphyrins. The parameters of the bridging C_a—C_m bonds given in the last two lines of Table 4 are dependent upon factors which are generally applicable to both porphyrins and metalloporphyrins.

There is cogent evidence, the writer judges, that the precision of the structural parameters listed for the five crystalline porphyrins decreases in the order, $H_2(OEP)$, $H_2(Meso-IX-DME)$, $H_2(TPrP)$, $H_2(TPP)$, and porphin; this is the order of increasing dependence upon low-resolution X-ray data and, saving only the interchange of $H_2(TPP)$ and porphin in the sequence, is also that based upon the estimated standard deviations reported by the authors. The dependence on low-resolution data was minimized in the structure analyses of $H_2(OEP)$ and $H_2(Meso-IX-DME)$ by including the largest practicable fractions of the reflections recordable with Cu K_α radiation. But in the other three analyses, particularly in that of porphin, the measurement of diffracted intensities was restricted, as a matter of experimental convenience, to unduly small ranges in $(\sin\theta)/\lambda$. The data for porphin included only reflections from within a limiting sphere having just 0.45 of the volume of the Cu K_α limiting sphere and the data/parameter ratio employed during least-squares refinement of the structural parameters, including the 56 pliable parameters assigned to the 14 hydrogen atoms, was only 4.42. Reported lengths of the 12 C—H bonds range from 0.92 to 1.17 Å, and the averaged value, 1.06 Å, is ~0.08 Å longer than the expected separation of the centroids of electron density — which is all that can be determined by X-ray analysis. The scatter of the individual N—C and C—C bond lengths around the mean values entered in Table 4 is notably large for at least four types of bonds in porphin, but is as notably small for *all* of the structural parameters reported for $H_2(OEP)$. It may be concluded that the structure analyses provide no objectively convincing evidence for a difference in the bond parameters of the porphin skeleton as it exists in porphin and in $H_2(OEP)$.

TABLE 4

Averaged parameters in porphyrins and two relevant metalloporphyrins[a].

Parameter	Porphin	H_2(TPrP)	Cu(TPrP)	H_2(TPP)	Ag(TPP)	H_2(OEP)	H_2(Meso-IX-DME)
	A. Pyrrole rings carrying N—H bonds in the porphyrin						
$Ct\cdot\cdot N$	2.056	2.080	1.994	2.099	2.101	2.098	2.039
$N—C_a$	1.380	1.376	1.378	1.374	1.365	1.367	1.365
$C_a—C_b$	1.431	1.437	1.448	1.428	1.445	1.438	1.438
$C_b—C_b$	1.365	1.352	1.344	1.355	1.365	1.373	1.368
C_aNC_a	108.6	110.0	106.0	109.2	108.6	109.6	109.8
NC_aC_b	107.9	106.7	109.8	107.3	108.6	107.7	107.5
$C_aC_bC_b$	107.9	108.3	107.2	108.1	107.1	107.4	107.6
NC_aC_m	125.2	127.1	126.7	126.0	126.4	125.0	125.3
	B. Pyrrole rings lacking N—H bonds in the porphyrin						
$Ct\cdot\cdot N$	2.029	2.040	2.005	2.026	2.082	2.026	1.971
$N—C_a$	1.377	1.372	1.383	1.364	1.368	1.364	1.366
$C_a—C_b$	1.452	1.443	1.449	1.455	1.447	1.462	1.457
$C_b—C_b$	1.345	1.341	1.345	1.347	1.350	1.353	1.359
C_aNC_a	106.1	106.6	106.5	106.2	108.7	105.7	105.7
NC_aC_b	109.8	109.5	109.3	110.3	108.2	110.8	110.9
$C_aC_bC_b$	107.1	107.3	107.4	106.8	107.5	106.3	106.3
NC_aC_m	125.0	126.4	126.6	126.3	126.5	125.1	124.7
$C_a—C_m$	1.382	1.396	1.389	1.400	1.410	1.392	1.388
$C_aC_mC_a$	127.1	125.0	122.8	125.6	125.6	127.6	127.7

[a] Distances in Å, angles in degrees. Ranges in the esd for individual distances and angles given by the authors are: porphin[21], 0.004—0.007 Å and 0.3—0.5°; H_2(TPrP)[49], 0.004 Å and 0.2—0.4°; Cu(TPrP)[31], 0.008 Å and 0.55°; H_2(TPP)[19], 0.005—0.010Å and 0.5—0.8°; Ag(TPP)[52], 0.004—0.005 Å and 0.3°; H_2(OEP)[48], 0.002 Å and 0.1°; H_2(Meso-IX-DME)[51], 0.003 Å and 0.2—0.3°.

Comparison of the bond parameters (Table 4) in H_2(Meso-IX-DME) and H_2(OEP) shows that in no parameter is the observed difference large enough to have objectively possible significance. It is rather surprising, consequently, to note that the two $Ct\cdot\cdot N$ radii in H_2(Meso-IX-DME) are shorter by ~0.059 and ~0.055 Å than the corresponding radii in H_2(OEP). Since this reduction in the 'size of the central hole' shortens the distance between the central pair of hydrogen atoms[51] to an estimated 2.05 (10) Å as compared with 2.36 (4) Å in H_2(OEP), it must be achieved by way of an appropriately nonplanar conformation of the porphin skeleton that is required by the packing of the bulky molecules in the crystal. Three of the pyrrole rings are tilted from coplanarity with the mean plane of the porphin skeleton by 1.4—1.6°, the fourth ring by 2.5°' the mean and maximum deviations of the atoms in the skeleton from the mean plane are 0.025 and 0.066 Å, respec-

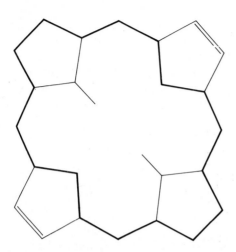

Fig. 5. Representation of the bonding pattern for the porphin skeleton in a porphyrin that corresponds to the pair of Kekulé formulae wherein each of the nitrogen atoms which carry N—H bonds is restricted to a pair of single bonds with its carbon neighbors. This pattern calls for full resonance within the heavily outlined, 18-membered, carbon-nitrogen ring, and to the exclusion of resonance from all other bonds in the skeleton.

tively. Thus rather modest deviations from planarity that are distributed throughout the skeleton give it a conformation that does not approximate closely to any recognizable symmetry[51].

The bonding pattern for the porphin skeleton in a porphyrin that is displayed in Fig. 5 corresponds formally to the pair of Kekulé formulae wherein each of the nitrogen atoms which carry N—H bonds is restricted to a pair of single bonds with its carbon neighbors. This pattern corresponds to full resonance (π bond order of $1/2$) within the heavily outlined, 18-membered, carbon-nitrogen ring and to the exclusion of resonance from all other bonds in the skeleton; C_a—C_b, C_b—C_b, and, of course, the N—C_a bonds are then divided each into two classes accordingly as they occur within or without the 18-membered ring. Using the schedule of expected bond lengths for π bond order of 0, 1/2, and 1 given earlier (p. 333), the data listed in Table 4 may be used to test the conclusion that this bonding pattern represents the dominant resonance structures in the porphin skeleton of the porphyrins[19,21,49,50].

If the pattern of Fig. 5 were to be a precise representation of the bonding in the skeleton, the following differences in the lengths of bonds lying, respectively, within and without the 18-membered ring would be expected: in C_b—C_b bonds, 0.060 Å; in C_a—C_b bonds, —0.095 Å; and in N—C_a bonds, —0.08 Å. The respective observed differences (Table 4) are <0.020, $-|<0.025|$, and $-|<0.005|$Å, of which the last is especially striking because it suggests that the presence or absence of the N—H bonds has little or no direct effect on the length and, presumably, the strength of the N—C_a bonds.

The conservative conclusion is that the substitution of a pair of protons for the metal cation in a metalloporphyrin does perturb, but does not largely alter, the 4-fold pattern of delocalized π bonding that characterizes the porphinato core therein. The effects of this perturbation are more strongly displayed in the bond angles (Table 4) than in the bond lengths; this is evidently the geometrical consequence of the mutual repulsion of the central pair of hydrogen atoms for one another.

Data that are listed in Table 4 for Cu(TPrP)[31] and Ag(TPP)[52] are compatible with porphinato cores which approximate to the 4-fold symmetry expected for the externally unconstrained species. The approximation to D_{4h} geometry is rather better for Cu(TPrP) as the consequence of less restrictive packing constraints in the crystals. Near planarity of the porphin skeleton is observed in the monoclinic crystals of H_2(TPrP) and Cu(TPrP), whereas a distinctively nonplanar conformation of the skeleton, in which two oppositely situated pyrrole rings are each tilted $6.6°$ from the mean plane, is maintained in the triclinic crystals of H_2(TPP) and Ag(TPP). Since the longer Ag—N bonds (2.101 (3) Å) extend to these misaligned pyrrole rings, it is probable that the shorter Ag—N distance of 2.082 (3) Å is appropriate within a strictly planar conformation of D_{4h} symmetry. As a predictable consequence of the difference in the bond radii of the Cu(II) and Ag(II) atoms, the stereochemistry of the pyrrole rings in Cu(TPrP) is generally quite similar to that of the *type B* rings in H_2(TPrP) while in Ag(TPP), by contrast, the stereochemical similarity is with the *type A* rings in the parent porphyrin.

The earlier observation[53] that Ag(TPP) and H_2(TPP) from a complete series of triclinic solid solutions at room temperature is quite in keeping with the comparative data listed for the pure components in Table 4. An especially intriguing feature of the study[53] was the simultaneous appearance, in relatively small amount, of tetragonal crystals having the composition, Ag-$(TPP)_x \cdot H_2(TPP)_{1-x}$, for $0.08 < x < 0.54$. The subsequent determination of structure for a tetragonal crystal in which $x = 0.54$ led to an extremely simple molecular packing in which the required molecular symmetry of C_{4h} is satisfied exactly by the Ag(TPP) species, but only statistically by molecules of the parent porphyrin[54]; indeed, the packing of the molecules in the crystal is extraordinarily similar to that observed in crystalline $(Cl)_2$Sn-(TPP)[26], wherein the axial chloro ligands of the Sn(IV) atom take a relatively minor part in determining the intermolecular contacts. These observations lead to a number of probable conclusions which receive further support from other studies that are to be cited.

(1) Crystallization of the tetragonal phase is induced by the marked preference of the Ag(TPP) species for a planar conformation; although the preference of the Ag(II) atom for square-planar bonds is significant in this connection, the dominant factor is the large Ag—N bond distance (>2.08 Å) which requires a highly expanded porphinato core that is not well-served by

References, p. 376

any appreciable departure from planarity[25]. (2) The fact that the mean molecular volume of the tetragonal phase, 870 Å[3], exceeds that of the triclinic phase by ~8.5% suggests that the latter is enthalpically favored at room temperature; in agreement with this tentative conclusion is the observation that the tetragonal phase appears only for compositions rich in H_2-(TPP) that gain entropic stabilization from the configurational disorder contributed by the molecules of this component. It follows that stability of the tetragonal phase over a wider range of compositions should be favored at higher temperatures. (3) The configurational disorder contributed by the H_2(TPP) molecules may be static or dynamic: the first possibility[54] assigns equal probability to two orientations of the H_2(TPP) molecule that differ by a rotation of 90°; the alternative is N—H tautomerism whereby the central pair of protons are continually redistributed among the four nitrogen atoms with a frequency that, at room temperature, is fast on the *nmr* time scale[55].

Published *nmr* studies[56-58] provide convincing evidence for rapid N—H tautomerism in several porphyrins in solution, quite specifically in H_2-(TPP)[58]. The provision in a crystalline phase of sites which require or permit the H_2(TPP) molecules to observe statistically effective 4-fold symmetry should be particularly conducive, both thermodynamically and mechanistically, to the dynamic type of disorder associated with N—H tautomerism. It may then be emphasized that the tetragonal structural type observed for crystalline Cu(TPP)[22], Pd(TPP)[22], Fe(TPP)[4], Co(TPP)[59], Ni(TPP)[60], and, most particularly, for crystals of pure H_2(TPP)[18] that were grown by deposition from the vapor phase at low pressure and elevated temperature[61] is particularly well-suited to this end. Within the precision of the measured lattice constants[18,19], the molar volume of tetragonal H_2(TPP) is comparable with, perhaps slightly smaller than, that of triclinic H_2(TPP). Thus the relative thermodynamic stabilities of the tetragonal and triclinic phases of pure H_2(TPP) at room temperature remain in doubt. It is highly probable, however, that the presence in solution of a very small concentration of Cu(TPP) or other appropriate M(TPP) species will ensure crystallization of the tetragonal phase of H_2(TPP). A study of this problem is in progress[62].

A variation of the phenomenon just described, which did not require phase differentiation, was apparently operative during crystallization of the porphin used in the determination of structure first reported[20,63]. The crystal was found to contain a significant concentration of an unidentified metal derivative in solid solution; the mol fraction of this impurity, evaluated during structure refinement on the assumption that the metal atoms were copper(II), was estimated to be ~0.079. Correction of the amplitude data for the scattering of the metal impurity then led to a quantitative description of the porphin molecule that displayed no objectively significant departure from D_{4h} geometry consistent with the low estimated standard deviations of 0.003—0.004 Å in the N—C and C—C bond lengths and 0.2° in the bond angles. The average of the twelve C—H distances, 1.00 (6) Å, and of the four N—H distances involving 'half-atoms' of hydrogen, 0.85 (5) Å corresponded satisfactorily to the expected separations of the centroids of electron density. The authors[20,63] suggested, moreover, that the observed disorder in the configuration of the porphin molecule was probably indicative of N—H tautomerism. It should be emphasized that the crystals of porphin and, indeed, those of tetragonal $H_2(TPP)$ and $Ag(TPP)_x \cdot H_2(TPP)_{1-x}$ as well, all provided useful X-ray intensity data to maximum values of $(\sin \theta)/\lambda$ which fully matched or exceeded that attained in the precise structure analysis of $H_2(OEP)$. Consequently, the configurational disorder in the porphyrin molecules did not seriously limit the determination of physically rational values for the bond parameters.

In the briefly reported structure of α-benzoyloctaethylporphyrin[64], the pyrrole rings contiguous to the single *meso* substituent are inclined at angles of 6.4 and 3.9° from the mean plane of the porphin nitrogen atoms and at an angle of 9.9° to one another. Each nitrogen atom, nonetheless, carries statistically a half-atom of hydrogen.

8.3.2. Related free bases

Structures have been reported for crystals of phyllochlorin methyl ester[65] and methyl pheophorbide a[66,67], free bases which are related to chlorophyll a; both the chlorin moiety and the fifth isocyclic ring of chlorophyll a are present in methyl pheophorbide a. The structural data for this base[67] are suggestive of N—H tautomerism, but with some preference for localization of the hydrogen atoms on the pair of nitrogen atoms that is predicted by approximate theory[68]. An *nmr* study of methyl pheophorbide in solution is consistent with a unique pair of localized N—H bonds[69].

8.3.3. Porphyrin acids

The porphyrin dications, $H_4(TPP)^{2+}$ and $H_4(TPyP)^{2+}$, wherein all four of the porphinato nitrogen atoms in *meso*-tetraphenylporphyrin and *meso*-tetra(4-pyridyl)porphyrin are protonated, occur in crystals of determined structures[63,70]. Inasmuch as the precision attained in both analyses was necessarily rather low, as indicated by standard deviations which averaged ~0.019 and ~0.032 Å in the respective $H_4(TPyP)^{2+}$ and $H_4(TPP)^{2+}$ ions, it is the striking qualitative features of the conformational geometries that command primary interest. The conformations overall of the two cations are notably similar; the S_4-$\overline{4}$ geometry which is required of the $H_4(TPP)^{2+}$ ion is closely approached in the observed C_2-2 conformation of the $H_4(TPyP)^{2+}$ species. This S_4 conformation, which is quite unlike that described earlier for molecules of M(TPP), M = Cu, Pd, Fe, Co, and Ni, may be stereochemically characterized as follows.

Starting from a planar D_{4h} conformation of the porphin skeleton (Fig. 1) and using the four C_m carbon atoms as pivots, the pyrrole rings with their attached C_a—C_m bonds are tilted alternately up and down with respect to the unique $\bar{4}$ axis. The resulting disphenoidal (S_4 or $\bar{4}$) configuration of the four positively charged N—H groups in the $H_4(TPP)^{2+}$ ion is well-designed to minimize the destabilizing effects of the Coulombic interactions and of closed-shell overlap of the hydrogen atoms. Concomitant with the tilted pyrrole rings is a nonplanar D_{2d} conformation of the porphin skeleton that, for any substantial angle of tilt, is more than usually inimical to the delocalization of the π bonding. A more smoothly varying conformation for the inner 16-membered ring is achieved by limiting the symmetry to S_4.

In the observed S_4 conformation of $H_4(TPP)^{2+}$, the pyrrole rings are tilted from coplanarity with the mean plane of the porphin skeleton by no less than 33.0°, and in the quasi-S_4 $H_4(TPyP)^{2+}$ species, the corresponding pair of quasi-equivalent angles are 27.3 and 28.2°. As the consequence of the markedly disphenoidal configuration of the pyrrole rings, the aromatic substituent, phenyl or 4-pyridyl, on a C_m carbon atom approaches much more nearly to coplanarity with the contiguous pair of C_a—C_m bonds than is allowed in the neutral porphyrins. It then follows, the authors suggest[70], that the enhanced conjugation of these aromatic substituents with the porphin ring systems is responsible for the observed shifts to the red in the ultraviolet-visible absorption spectra of the $H_4(TPP)^{2+}$ and $H_4(TPyP)^{2+}$ ions relative to the spectra of the parent free bases. In a more recent communication[71] devoted to trans-$N_a N_b$-dimethyltetraphenylporphyrin, it is pointed out that a bright green color of a tetraphenylporphyrin system does not provide a unique test for the presence of the $H_4(TPP)^{2+}$ species; the green color is presumably associated with any conformation of the porphin skeleton that leads to a sufficient enhancement of conjugation with the phenyl groups.

In marked contrast with the disphenoidal geometry of the $H_4(TPP)^{2+}$ ion, a nearly planar conformation of the octaethylporphyrin dication, $H_4(OEP)^{2+}$, is observed[72] in a crystalline salt formulated as $[H_4(TPP)^{2+}]$ $[Rh(Cl)_2(CO)_2^-]_2$. Two chlorine atoms, one each from a pair of contiguous anions, and the four porphin nitrogen atoms define an octahedral grouping of required C_i-$\bar{1}$ symmetry within which the distribution of the hydrogen atoms is not determined by the structure analysis. The four independent Cl $\cdot\cdot$ N separations, which range from 3.26 to 3.42 Å and average to 3.34 Å, could correspond to localized, if rather weak, hydrogen bonds provided that the N—H bonds were allowed to point more or less directly toward the chlorine atoms — e.g., by reason of sp^3 hybridization of the nitrogen bond orbitals. Noting, however, that the unimportance of any contribution from sp^3 hybridization is attested by the averaged N—C bond distance of 1.38 Å, it appears rather that negative charge carried by the chlorine atoms provides electrostatic stabilization of the quasi-D_{4h} geometry of the octahedral grouping and the nearly planar conformation of the porphin skeleton. Disorder in the configuration of the hydrogen atoms is certainly not precluded by the X-ray data.

The undetected configuration of the hydrogen atoms again presents a problem in the brief report[73] on the stereochemistry of the octaethylporphyrin monocation, $H_3(OEP)^+$, as observed in a crystalline tri-iodide salt. Three of the pyrrole rings, which are coplanar within ±0.03 Å, define a mean plane to which the fourth ring is inclined by 14°. The ring which carries no N—H bond is convincingly identified by its especially small CNC angle of 102°; this ring, which is included in the nearly coplanar set of three, is diagonally situated with respect to the ring which is tilted out-of-plane. The tilt of the latter ring is assigned by the authors to the mutual repulsions of the hydrogen atoms; they suggest that the unduly tight H ·· H contacts may need to be relieved through partial sp^3 hybridization of the pertinent nitrogen bonding orbitals, but they cite no structural parameters which provide support for this phenomenon.

The stereochemical parameters of the N-carboethoxymethyloctaethylporphyrin monocation, $H_2(N\text{-}ROEP)^+$ [R = carboethoxymethyl], as this species occurs in crystals of the iodide salt, do provide evidence for the partial sp^3 hybridization of the bonding orbitals of one nitrogen atom — the atom which carries the R substituent[74]. The pyrrole ring which carries the N—R or, more specifically, the N—C_R bond is tilted from the plane defined by the nitrogen atoms of the other three rings by ~19.1°. Bond angles subtended at the nitrogen atom are C_aNC_a = 107° and C_aNC_R (average) = 115°. The observed N—C_R distance, 1.51 Å, is ostensibly longer than the standard value, 1.475 Å, for an aliphatic N—C bond, and the N—C_a distance, 1.405 Å, is apparently ~0.02 Å shorter than the estimated value for a bond between aliphatic nitrogen and trigonally hybridized carbon atoms. The bond lengths, C_a—C_b = 1.41 and C_b—C_b = 1.38 Å, are simultaneously indicative of a compensating enhancement of resonance. It is the internal consistency of the overall pattern that carries conviction; the esd associated with an individual N—C or C—C bond length is, at best, ~0.020 Å. Quantitatively definitive determinations of structure for the porphyrin acids would seem to require the use of the fully adequate intensity data that are afforded by crystals at low temperature.

8.4. Stereochemistry of the coordination groups in metalloporphyrins

An especially illuminating correlation of stereochemistry with the d electron configurations in the valence shells of the metal(II) atoms in four-coordinate metalloporphyrins is exhibited in the isostructural series of M(TPP) species, M = Fe, Co, Ni, Cu, Pd. Thus the pattern of M—N bond distances in the Fe(TPP)[4], Co(TPP)[59], Ni(TPP)[60], and Cu(TPP)[22] structures, 1.972 (4), 1.949 (4), 1.928 (3), and 1.981 (7) Å, respectively, is fully correlated with the required presence of an odd electron in the $3d_{x^2-y^2}$ orbital of the paramagnetic (S = ½) d^9 Cu(II) atom and the absence of such an electron from the $3d_{x^2-y^2}$ orbitals of the respective intermediate-spin (S = 1) d^6

Fe(II), low-spin $(S = \frac{1}{2})$ d^7 Co(II), and low-spin $(S = 0)$ d^8 Ni(II) atoms[4]. The marked S_4 ruffling of the M(TPP) molecule in this structural type requires only a quite trivial departure from planarity in the square-planar pattern of complexing M—N bonds that is best-suited to Cu(II), Ni(II), and, most particularly, Pd(II) complexes; indeed, the Pd—N distance[22] of 2.009 (9) Å is suggestive of little or no strain in the complexing bonds. It is scarcely an exaggeration, moreover, to conclude that the unusual electronic state of the iron atom in the Fe(TPP) molecule is dictated by the complexing properties of the macrocyclic porphinato ligand[4,24].

The five-coordinate species, (Cl)Fe(TPP)[5], $(H_2O)Zn(TPP)$[75], $(H_2O)Mg$-(TPP)[76], (ON)Co(TPP)[77], and (ON)Fe(TPP)[78], crystallize in a disordered variant of the $(Cl)_2Sn(TPP)$ structural type[26]; the required molecular symmetry of C_{4h} is satisfied statistically by allowing the polar axis of the five-coordinate molecule to take, with equal probability, either the parallel or the antiparallel orientation with respect to the fourfold axis of the tetragonal crystal[5]. Although this disordered structural type appears to be structurally and thermodynamically available to any pure (X)M(TPP) species wherein the axial ligand (X) is not unduly large, the stereochemical parameters of the coordination group, M—N, M—X, and M ·· Ct in Fig. 4, retain their marked dependence upon the d electron configuration of the metal atom. The precision with which the axial parameters can be specified is primarily dependent upon the precision with which the separation, $2z_M$, of the two 'half-atoms' of the metal at $\pm z_M$ is determined[5]. Resolution of these half-atoms theoretically requires structure amplitudes which extend to reflections having Bragg spacings at least as small as $2z_M/0.61 = 3.3z_M$[79]. This minimum condition could not be satisfied for crystalline (ON)Co(TPP), wherein $3.3z_M$ is estimated to be only ~0.32 A, and it was not satisfied by the very limited data recorded for $(H_2O)Zn(TPP)$, wherein the required minimum spacing is ~1.0 Å. Underestimation of the Zn ·· Ct distance and overestimation of the Zn—OH bond length, each by ~0.15 Å, is indicated by comparative data from pertinent ordered structures (vide infra).

For the systematic consideration of the coordination geometries in the metalloporphyrins, each complex is assigned to a distinctive group in which the geometry is directly related to the d electron configurations in the valence shells of the metal atoms. The order in which these groups are to be discussed, though seemingly capricious, is chosen so as ultimately to focus attention on the d^5 and d^6 iron complexes — the metalloporphyrins in which the spin state of the metal atom becomes a usual and customary variable in the determination of the stereochemistry. For quite specific choices of axial ligands, the low-spin d^7 cobalt(II) species are of comparable interest and importance.

8.4.1. The d^{10} and d^0 metalloporphyrins

Bond parameters in the core of the five-coordinate d^{10} (Py)Zn(TPyP) molecule[42] are listed in Tables 2 and 3. Using N_P and N_X to denote the respective porphinato and pyridine nitrogen atoms, the principal parameters of the coordination group (Fig. 4) are $(Zn-N_P)_{av}$ = 2.073 (3), $Zn-N_X$ = 2.143 (4), $(Ct \cdots N_P)_{av}$ = 2.047 (3), and $Zn \cdots Ct$ = 0.33 Å. Since steric interactions of the pyridine ligand with the porphinato core are altogether trivial[42], the absence of stretch from this source in the $Zn \cdots Ct$, $Zn-N_X$, and $Zn-N_P$ distances is assured. Long, comparatively weak, $Zn-N_X$ bonds in (Py)Zn(TPyP) and (Py)Zn(TPP) are consistent with the observed equilibrium constant, $\sim6 \times 10^3$ in benzene at room temperature, for the complexing reaction, pyridine + Zn(TPP) → (Py)Zn(TPP)[80,81]. The evidence, furthermore, for the complexing of a second pyridine (or substituted pyridine) ligand is wholly negative[80,81].

Structure determination for crystalline (Py)Zn(OEP)[82] yields $(Zn-N_P)_{av}$ = 2.068 (3), $Zn-N_X$ = 2.200 (3), $(Ct \cdots N_P)_{av}$ = 2.043 (3), and $Zn \cdots Ct$ = 0.31 Å. The sterically unfavorable orientation of the pyridine ligand, which leads to moderately tight contacts of 2.65 and 2.71 Å between pyridine hydrogen and porphinato nitrogen atoms, seems to provide the only plausible basis[82] for the observed stretch of nearly 0.06 Å in the $Zn-N_X$ bond relative to the length in the (Py)Zn(TPyP) molecule; so large a strain associated with so little stress is further indicative of a weak $Zn-N_X$ bond.

In the radical perchlorato-$\alpha,\beta,\gamma,\delta$-tetraphenylporphinatozinc(II) species, $(ClO_4)Zn(TPP)[83]$, charge compensation for the loss of an electron from the highest filled molecular orbital of the porphinato core is provided by the coordination of the perchlorate ion as the axial ligand of the zinc(II) atom. The Zn–O distance of 2.079 (8) Å, which represents the shortest metal to perchlorate-oxygen bond yet reported[83], is quite in line with the known strong affinity of the d^{10} zinc ion for oxo, hydroxo, and aquo ligands. Other structural parameters in $(ClO_4)Zn(TPP)$, including $(Zn-N)_{av}$ = 2.076 (9), $(Ct \cdots N)_{av}$ = 2.046 (9), $Zn \cdots Ct$ = 0.35 Å, and the bond distances in the core, differ insignificantly from those in (Py)Zn(TPyP).

The quite inadequate X-ray data employed for refinement of the disordered $(H_2O)Zn(TPP)$ structure[75] (p. 344) led to apparent stereochemical parameters of Zn–N = 2.05, $Zn-OH_2$ = 2.20, $Ct \cdots Zn$ = 0.19, $Ct \cdots N$ = 2.043, and $Ct \cdots OH_2$ = 2.39 Å. Noting that it is the z-parameter of the zinc atom that could not be reliably determined, and that a $Zn-OH_2$ distance $\leqslant 2.05$ Å is expected from other structure analyses[83,84], it is likely that a better approximation to the actual parameters in the $(H_2O)Zn(TPP)$ molecule is obtained by taking $Zn-OH_2$ to be 2.05 Å along with unaltered values of $Ct \cdots N$ and $Ct \cdots OH_2$ to give Zn–N = 2.075 and $Ct \cdots Zn$ = 0.34 Å[42].

It has long been recognized that chemically analogous six-coordinate species containing the d^{10} zinc(II) and d^0 magnesium(II) atoms are generally isostructural and, indeed, very nearly isodimensional. The five-coordinate isostructural $(H_2O)Zn(TPP)[75]$ and $(H_2O)Mg(TPP)[76]$ molecules are also expected to be, and probably are, nearly isodimensional. Reported values of the parameters in the statistically disordered structure of $(H_2O)Mg(TPP)$ are $Mg-N = 2.072$ (7), $Mg-OH_2 = 2.10$, $Ct \cdots Mg = 0.027$, $Ct \cdots N = 2.055$ (6), and $Ct \cdots OH_2 = 2.37$ (2) Å. Underestimation of the out-of-plane displacement $(Ct \cdots Mg)$ of the magnesium atom with concomitant overestimation of the $Mg-OH_2$ bond length, each by $\geqslant 0.05$ Å, is compatible with the limited resolution achieved in the structure determination; it is, moreover, rendered plausible by relevant stereochemical data from determinations of structure for the pair of isomorphous crystals which are now to be discussed.

Inasmuch as the structural differences between ethyl chlorophyllide $a \cdot 2H_2O$[85] and ethyl chlorophyllide $b \cdot 2H_2O$[86] are of a quite minor nature, attention may be focused on the stereochemical data obtained from the variant a. One of the water molecules is in fact tightly bonded to the magnesium ion at 2.033 (5) Å in the five-coordination group. The four $Mg-N$ bond lengths range from 2.018 (6) to 2.168 (6) Å and average to 2.084 Å; the $Ct \cdots Mg$ distance is 0.39 Å. A prominent role is taken by the uncoordinated water molecules in the 'one-dimensional polymers' which exist in the crystal. The authors' proposal that these polymers afford a rational basis for a structural model of chlorophyll aggregation in photosynthetic organisms is buttressed by other considerations advanced by them[85,86].

Five-coordination of the magnesium ion in chlorophyll was foreseen to be highly probable by several workers. The earlier observations[81,87] that equilibria exemplified by pyridine + $(Py)Mg(TPP) = (Py)_2Mg(TPP)$ are detectable in solution at $30°C$, but with equilibrium constants restricted to the range from 0.07 to 0.32 for nine choices of the parent porphyrin[87], were indicative of the improbability of six-coordinate magnesium(II) in chlorophyll. Very long axial bonds, ~ 2.34 Å, were observed in disordered crystals of the six-coordinate $(Py)_2Mg(Etio-II)$ species[30,88]. The pertinence of the five-coordinate stereochemistry reported for the d^1 vanadyl derivative of deoxophylloerythroetioporphyrin[89] became evident when the structural data were presented[30] as $(V-N)_{av} = 2.065$, $Ct \cdots V = 0.48$, and $V-O = 1.62$ Å for comparison with the analogous parameters in $(H_2O)Mg(TPP)$[76] and $(Cl)Fe(TPP)$[5].

The complexing properties of the closed-shell d^{10} Zn(II) and d^0 Mg(II) atoms derive primarily from their respective unoccupied $4sp^3$ and $3sp^3$ orbitals. In a porphinato complex, one p orbital is useful only for axial complexing; it then appears that the $M-N$ bonds in a four-coordinate complex should not be materially shorter than those in the five-coordinate species. The data cited above suggest that centering of the metal atom in the core with $M-N = 2.06-2.07$ Å would correspond to unstable equilibrium — a maximum in the potential energy. Suppose, for example, that the complexed water molecules could be pumped out of a tetragonal $(H_2O)M(TPP)$ phase with retention of the basic crystalline arrangement; a sharply defined dependence of the molecular conformation on temperature would then be expected. Below a transition temperature at which the heat capacity (but neither the enthalpy nor the entropy) of the crystal is discontinuous, the metal atoms would be distributed equally between two, symmetry equivalent, out-of-plane positions. Above the transition temperature, by contrast, thermally excited vibrations would allow the metal atom to pass freely over the potential barrier and thus give rise to a rapid inversion of conformation. The metal atom and, in lesser degree, the complexed nitrogen atoms would then exhibit extraordinarily large *rms* amplitudes for vibration about their mean positions parallel to the 4-fold axis. This analysis exemplifies the expected behavior of a Zn(TPP) or Mg(TPP) molecule in an effectively centro-symmetric environment. A molecular conformation in which a unique out-of-plane position is taken by the metal atom could be stabilized in a polar environment.

Stereochemical parameters of the coordination groups in the $(Cl)_2 Sn(TPP)$ and $(Cl)_2 Sn(OEP)$ molecules are included in the more comprehensive data listed for these species in Table 1. The one other example of a determined structure for a d^{10} (or d^0) metalloporphyrin wherein the metal atom is centered among the nitrogen atoms is that for the dimethoxogermanium(IV) derivative of porphin[90]. Averaged values of Ge—N = 2.015 and Ge—O = 1.822 Å are obtained for the pair of structurally nonequivalent molecules in the crystal.

In the chlorothallium(III) derivative of octaethylporphyrin[91], the parameters of the five-coordination group (Fig. 4) are $(Tl—N)_{av}$ = 2.20, Tl—Cl = 2.45, $Ct \cdots Tl$ = 0.69, and $(Ct \cdots N)_{av}$ = 2.09 Å. Strong complexing by the very large d^{10} Tl(III) atom and, less significantly, the observed 'doming' or 'tenting' of the porphinato core toward the thallium atom contribute to the large expansion of the $Ct \cdots N$ radius from the ~2.01 Å expected for the minimization of strain in a planar core to the observed 2.09 Å; this $Ct \cdots N$ radius, indeed, is seen to be comparable with the Sn—N = $Ct \cdots N$ distance of 2.098 Å in the most expanded planar core of $(Cl)_2 Sn(TPP)$.

Structure determinations for monoclinic crystals of $(OAc)_2 Hf(OEP)$[7] and orthorhombic crystals of $(OAc)_2 Zr(OEP)$[7] lead, as expected, to nearly iso-dimensional configurations for the two eight-coordinate molecules. The slightly idealized version of the observed coordination polyhedra that is defined by the four acetato-oxygen and the four porphinato-nitrogen atoms is a C_{2v} adaptation of the D_{4d} square antiprism; see Fig. 6. The rectangular shape of the upper base is required by the small 'bite' of the bidentate acetato ligands. The resulting separation of the O $\cdots$ N packing contacts into two classes leads to a nearly equilateral, but slightly disphenoidal, conformation of the lower base, and to the separation of the M—N bond lengths into two classes. Parameters in the hafnium complex of C_{2v} symmetry are Hf—O = 2.278 (3), $(Hf—N)_1$ = 2.266 (5), $(Hf—N)_2$ = 2.248 (3), $Ct \cdots N$ = 2.016 (3), Hf $\cdots Ct$ = Hf $\cdots P_N$ = 1.012 (3), and Hf $\cdots P_O$ = 1.480 (3) Å (P_O and P_N are the planes of the respective upper and lower bases of the coordination polyhedron). Parameters in the core of the $(OAc)_2 Hf(OEP)$ molecule are entered

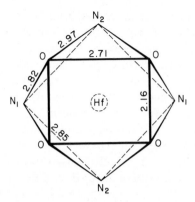

Fig. 6. Diagram illustrating the C_{2v} geometry of the coordination group in the (OAc)$_2$Hf-
(OEP)[7] molecule. The hafnium atom lies 1.012 (3) Å above the mean plane of the
nitrogen atoms, 1.480 (3) Å below the plane of the oxygen atoms. (Hf—N)$_{av}$ = 2.257 (3),
and Hf—O = 2.278 (3) Å.

in Tables 2 and 3. Also to be noted is the marked doming of the cores
toward the M(IV) atoms in both the hafnium and the zirconium complexes.

In the oxotitanium(IV) derivative of α,γ-dimethyl-α,γ-dihydro-octaethyl-
porphyrin[92], (O)Ti(Me$_2$OEP), the macrocyclic ligand is strongly folded in
near conformity with C_{2v} symmetry along the line joining the reduced
(methene) C-α and C-γ carbon atoms. The titanyl (TiO^{2+}) ion and the *syn*-
axially oriented α-methyl and γ-methyl groups occupy 'chimney' positions
along the 'ridge' of the 'roof' formed by the folding of the macrocycle. The
dihedral angle at the ridge between the two halves of the roof is 141.8°; and
in each pyrromethene half, the pair of pyrrole rings which are bridged by an
unaltered (β or δ) methine carbon atom depart from coplanarity by only
4.6°. Although the four pyrrole rings retain structural equivalence in C_{2v}
symmetry, each ring carries nonequivalent pairs of C$_a$—N, C$_a$—C$_b$, and
C$_a$—C$_m$ bonds (Fig. 1). Thus the lengths of the bridging bonds to methene
(C$_{m1}$) and methine (C$_{m2}$) carbon atoms are C$_{a1}$—C$_{m1}$ = 1.504 (4) and
C$_{a2}$—C$_{m2}$ = 1.386 (4) Å, of which the latter is still typical for a metalloporphyrin (see Table 2). Bond lengths in the pyrrole rings are N—C$_{a1}$ = 1.347
(4), N—C$_{a2}$ = 1.401 (4), C$_{a1}$—C$_{b1}$ = 1.416 (4), C$_{a2}$—C$_{b2}$ = 1.429 (4), and
C$_{b1}$—C$_{b2}$ = 1.386 (5) Å. The qualitative correlation of the similar bonding
pattern observed in Ni(Me$_2$OEP)[93] [the nickel(II) derivative of the same
dihydroporphyrin] with the applicable Kekulé formulae is discussed in the
earlier paper[93].

Parameters in the coordination group (Fig. 4) of the d^0 (O)Ti(Me$_2$OEP)
complex are (Ti—N)$_{av}$ = 2.110 (3), Ti—O = 1.619 (4), Ti $\cdots$ Ct = 0.58, and
(Ct $\cdots$ N)$_{av}$ = 2.031 (3) Å. The authors point out that the Ti—N bond length,
long though it be, is 0.06—0.13 Å shorter than the values observed in other
complexes containing Ti—N bonds[93].

8.4.2. The d^9 and d^8 metalloporphyrins

Excepting only one six-coordinate paramagnetic d^8 nickel(II) complex[94], all of the d^9 and d^8 metalloporphyrins of determined structures are four-coordinate with planar or nearly planar coordination groups. As pointed out earlier (p. 323), however, the length of the complexing bonds may depend upon the conformation of the porphinato core. Thus within the *quasi*-planar cores of Cu(TPrP)[31] and Ag(TPP)[52], $(Cu-N)_{av}$ = 2.000 (5) and $(Ag-N)_{av}$ = 2.091 (3) Å (Table 4), but in the strongly S_4-ruffled Cu(TPP)[22] species, $Cu-N$ = 1.981 (7) Å; and within the nearly planar cores of Ni(OEP)[3] and Ni(2,4-DAcDeut-IX-DME)[17], $(Ni-N)_{av}$ = 1.958 (2) and 1.960 (7) Å, respectively (Tables 1 and 2), whereas in the S_4-ruffled Ni(OEP)[2] and Ni(TPP)[60] species, $Ni-N$ = 1.929 (3) and 1.928 (3) Å, respectively. The displacement of the methine carbon atoms from the equatorial plane in Ni(TPP), alternately 0.45 Å up and down in the 'circular standing wave', is a maximum for the structural type, and is comparable with the analogous ruffling parameter of 0.51 Å in tetragonal Ni(OEP).

It should be noted that the (statistically satisfied) D_{2d} ruffling which is required of the Ni(Etio-I)[8] molecules in the disordered tetragonal structure is of the general type described above (p. 341) for the porphyrin dication, $H_4(TPP)^{2+}$; the methine carbon atoms lie on the 2-fold axes in the equatorial plane and serve as pivots for the tipping out of plane, alternately up and down, of the pyrrole rings. Such ruffling, in contrast with that in Ni(TPP) and Ni(OEP), may lead to somewhat longer M—N bonds than those in a D_{4h} conformation. The observed value of Ni—N in the modestly ruffled conformation of Ni(Etio-I)[8] is 1.957 (13) Å.

In the nickel(II) derivative of deoxophylloerythrin methyl ester[95], the averaged value of the four observed Ni—N bond lengths (1.89, 1.94, 1.97, and 2.00 Å) is 1.95 Å. The shortest bond is to the pyrrole ring which is fused along one edge with the fifth isocyclic ring.

An extraordinarily large *quasi*-S_4 ruffling of the core in the α,γ-dimethyl-α,γ-dihydro-octaethylporphinatonickel(II) molecule, Ni(Me$_2$OEP)[93], is conducive to an averaged Ni—N bond length of only 1.908 (5) Å, a value which exceeds that considered most favorable with monodentate ligands by $\leqslant 0.05$ Å. The reduced (methene) C-α and C-γ atoms lie 1.02 Å above the mean plane of the core, the unaltered (methine) C-β and C-δ atoms lie 0.89 Å below. An illuminating discussion of the stereochemistry, including the correlation of the observed bond parameters in the core with the appropriate Kekulé formulae, is provided by the authors.

The stereochemistry of the six-coordinate paramagnetic (S = 1) bis(imidazole-$\alpha,\beta,\gamma,\delta$-tetra(4-$N$-methylpyridyl)porphinatonickel(II) cation, (Im)$_2$Ni(TMePyP)$^{4+}$, as it occurs in a crystalline acetone solvate of the perchlorate salt, [(Im)$_2$Ni(TMePyP)][ClO$_4$]$_4$ · 2C$_3$H$_6$O, has been reported[94]. *Quasi*-planarity of the porphinato core is preserved in this centrosymmetric octahedral complex. Equatorial and axial Ni—N bond lengths are $(Ni-N_P)_{av}$ =

2.038 (4) and Ni—N_X = 2.160 (4) Å, respectively. In the usually observed, diamagnetic, four-coordinate, nickel(II) complexes, the pairing of electrons in the $3d_{z^2}$ orbital of the metal atoms is conducive to the full utilization of the $3d_{x^2-y^2}$ orbital in the characteristic pattern of square-planar bonds. Formation of the paramagnetic, six-coordinate, nickel(II) complex requires that one of these electrons be transferred to the $3d_{x^2-y^2}$ orbital with a concomitant weakening and lengthening of the complexing bonds relative to those in the diamagnetic species. Relative to an expected length of ~2.08 Å for the complexing bonds to monodentate ligands in a paramagnetic NiN_6 coordination group of O_h symmetry, the shorter Ni—N_P bonds, 2.038 Å, in the $(Im)_2Ni(TMePyP)^{4+}$ ion provide another example of the constraints imposed by the porphinato ligand.

The axial Ni—N_X bond length of 2.160 Å is apparently in qualitative agreement with the low stability constant[96] observed for the $(Im)_2Ni$-$(TMePyP)^{4+}$ ion in 30% acetone—water solution at rcom temperature; it is, on the other hand, shorter by >0.23 Å than the axial Co—N_X bond lengths in the $(Pip)_2Co(TPP)^{39}$ and $(3\text{-Pic})_2Co(OEP)^{40}$ molecules wherein the $3d_{z^2}$ orbitals of the d^7 low-spin cobalt(II) atoms also are singly occupied. Noting that the equatorial Co—N_P bond lengths in these Co(II) complexes are longer by only ~0.03 Å than the Ni—N_P bonds in the Ni(2,4-DAcDeut-IX-DME)[17] and (planar) Ni(OEP)[3] molecules, it follows that the difference of >0.23 Å in the axial bond lengths is much too large to be attributed solely to the difference of one protonic unit in the nuclear charges carried by the cobalt and nickel atoms. It appears rather that the positively-charged pyridyl-nitrogen atoms must draw electron density from the central region of the cation and thus enhance the complexing power and the effective oxidation state of the nickel atom.

The Pd—N bond length of 2.009 (9) Å in the observed S_4-ruffled conformation of the molecule[22] corresponds simultaneously to the preferred value of this parameter with monodentate ligands and to the minimization of equatorial strain in a *planar* porphinato core. In view of the earlier discussion of the rather extreme conditions in which planarity of the core in the Ag-(TPP) molecule can be preserved in a crystalline phase (p. 339), there is little reason to doubt that the observed ruffling of the Pd(TPP) molecule in the crystal is solely the consequence of the superior structural and thermodynamic properties of the structural type. The analogous platinum complex, Pt(TPP), may be expected to crystallize isostructurally and nearly isodimensionally with Pd(TPP).

A structure determination for $\alpha,\beta,\gamma,\delta$-tetraphenylporphinatogold(III) chloride[97] leads to a nearly square-planar coordination group in which $(Au\text{—}N)_{av}$ = 2.00 (3) Å. The Au—Cl distance of 3.01 (1) Å is suggestive of electrostatic stabilization of the crystalline arrangement.

8.4.3. The d^1, d^2, and d^3 metalloporphyrins

Apart from the d^1 vanadyl derivative of deoxophylloerythroetioporphyrin[89] (p. 346), no other metalloporphyrin containing a d^1, d^2, or d^3 metal atom is represented among the complexes of determined stereochemistry.

8.4.4. The $4d^5$ and $4d^6$ metalloporphyrins

The relatively orthodox stereochemistry displayed in six-coordinate low-spin metalloporphyrins formed by ruthenium and rhodium provides background for the much richer stereochemistry of manganese, iron, and cobalt complexes. A notable feature of the $(C_6H_5)Rh(Cl)(TPP)$ molecule[98] is the phenyl-rhodium(IV) σ bond. Reported bond lengths in this single example of a $4d^5$ metalloporphyrin of determined structure are $(Rh-N)_{av} = 2.04$, $Rh-C = 2.05$, and $Rh-Cl = 2.35$ Å.

Stereochemical parameters have been reported for three d^6 ruthenium(II) complexes, $(Py)_2Ru(OEP)$[99], $(Py)Ru(CO)(TPP)$[44], and $(C_2H_5OH)Ru(CO)$-(TPP)[45]. In the molecule with two pyridine ligands[99], the averaged equatorial and axial $Ru-N$ bond lengths are 2.041 (6) and 2.095 (6) Å, respectively. In the $(Py)Ru(CO)(TPP)$ molecule, $(Ru-N_P)_{av} = 2.052$ (9), $Ru-N_{Py} = 2.193$ (4), $Ru-C = 1.838$ (9), and $Ru \cdots Ct$, the out-of-plane displacement of the metal atom toward the CO ligand, is 0.068 Å. The RuCO bond angle is 178.4 (7)°. On the basis of structural data observed for a number of other complexes, the authors conclude that the $Ru-N_{Py}$ bond is lengthened by ~ 0.1 Å as the consequence of interaction with the trans $Ru-CO$ linkage[44]. It seems probable, however, that some part of the observed lengthening is ascribable to the rather tight contacts between pyridine hydrogen and porphinato nitrogen atoms that further characterize the molecular configuration. In the statistically disordered $(C_2H_5OH)Ru(CO)(TPP)$ molecule of crystallographically required $C_i\text{-}\overline{1}$ symmetry[45], $Ru-Ct$ is not determinable, but $(Ru-N)_{av} = 2.049$ (5), $Ru-CO = 1.77$ (2), and $Ru-OEt = 2.21$ (2) Å; the RuCO angle is given as 175.8 (1.9)°.

Determinations of crystalline structure for $[(Me_2NH)_2Rh(Etio-I)]Cl \cdot 2H_2O$[100] and $[(Me_2NH)_2Rh(TPP)]Cl$[101] provide stereochemical parameters for the d^6 cationic species. In the centrosymmetric $(Me_2NH)_2Rh(Etio-I)^+$ ion, $(Rh-N_P)_{av} = 2.038$ (8) and $Rh-N(amine) = 2.090$ (8) Å. The analogous parameters in the $(Me_2NH)_2Rh(TPP)^+$ ion are 2.03 and 2.11 Å, respectively.

8.4.5. The manganese, iron, and cobalt metalloporphyrins

Referring only, of course, to metalloporphyrins of determined structures, the following classification applies: (1) all of the d^4 metalloporphyrins are high-spin ($S = 2$) manganese(III) complexes; (2) the $3d^5$ metalloporphyrins include three manganese(II) complexes, but are otherwise iron(III) derivatives occurring as both high-spin ($S = 5/2$) and low-spin ($S = \frac{1}{2}$) species; (3) the $3d^6$ metalloporphyrins include the iron(II) complexes in three spin states

$(S = 0, 1, 2)$ and the low-spin $(S = 0)$ cobalt(III) species; (4) all of the d^7 metalloporphyrins are low-spin $(S = \frac{1}{2})$ cobalt(II) derivatives.

In a low-spin iron(II), iron(III), or cobalt(III) metalloporphyrin, the pairing of electron spins in the d_{xy}, d_{yz}, and d_{xz} valence-shell orbitals of the metal atom allows the full utilization of the unoccupied $d_{x^2-y^2}$ and d_{z^2} orbitals for complexing the quadridentate porphinato macrocycle and two axial ligands. In any high-spin iron or manganese(II) metalloporphyrin, by contrast, the $3d_{x^2-y^2}$ and $3d_{z^2}$ orbitals in the metal atom retain each one unpaired electron. The presence of the electron in the $d_{x^2-y^2}$ orbital is responsible for the substantial displacement of the iron atom from the plane (P_N) of the nitrogen atoms that is observed in every high-spin porphinato-iron species: thus $Fe-P_N$ (or $Fe \cdot \cdot Ct$) usually is >0.40 Å. Although this large displacement is altogether favorable to the sterically unhindered complexing of a single axial ligand, it is highly inimical to the formation of a stable high-spin six-coordinate species[24]. In either a high-spin manganese(III) or a low-spin cobalt(II) metalloporphyrin, the unoccupied $3d_{x^2-y^2}$ orbital in the metal atom is fully available for complexing with the porphinato macrocycle, and the singly occupied $3d_{z^2}$ orbital is adaptable to the somewhat weaker complexing of one axial ligand; the stability of the six-coordinate species is limited by the presence of the unpaired electron in the d_{z^2} orbital of the metal atom, but not, as in a high-spin iron complex, by the additional requirement of a large out-of-plane displacement of the metal atom.

Approximate theoretical considerations suggest that the axial bonds in all of the foregoing five-coordinate complexes should be rather easily stretched when subjected to moderate stresses. On either an electrostatic (i.e., crystal field) or molecular orbital basis, the charge associated with the odd electron in the $3d_{z^2}$ orbital of the metal atom is somewhat concentrated in, though not limited to, the sixth coordination position as a 'phantom ligand'. Overlap, positive or negative, of the $3d_{z^2}$ orbital with the σ orbital of the axial ligand is mostly confined to the region of the complexing bond; the antibonding molecular orbital, wherein the odd electron is accommodated, is predominantly of $3d_{z^2}$ character. Concomitant with decreasing overlap of metal and ligand orbitals — with, that is, stretch in the axial bond — are movements, opposite in sign, of the energy levels associated with the bonding and antibonding orbitals toward a common intermediate level (for infinite M—X). Consequently the net stabilizing energy must decrease rather slowly with increasing M—X and thus give rise to a relatively small force constant. Comparative data for the high-spin iron(III) and manganese(III) metalloporphyrins are illuminating in this connection.

8.4.5.1. The high-spin iron and manganese metalloporphyrins

The generally observed five-coordinate stereochemistry of high-spin porphinatoiron(III) complexes is exemplified in $(Cl)Fe(TPP)$[5], chlorohemin[23],

(MeO)Fe(Meso-IX-DME)[24], (SCN)Fe(TPP)[32], (N_3)Fe(TPP)[102], $(O_2NC_6-H_4S)$Fe(Proto-IX-DME)[103], and the oligomeric $O[Fe(Proto\text{-}IX\text{-}DME)]_2$[9] and $O[Fe(TPP)]_2$[35] species wherein the iron(III) atoms are antiferromagnetically coupled[104]. Representative parameters in the coordination group (Fig. 4) of the monomeric species are $Fe-N_P$ = 2.065, $Ct \cdots N_P$ = 2.015, and $Ct \cdots Fe$ = $Fe \cdots P_N$ = 0.45 Å. It is to be noted that the axial ligands in all of these complexes are anionic species, and that the bond lengths, $Fe-Cl$ = 2.192 (12)[5] and 2.218 (6)[23], $Fe\text{-}OCH_3$ = 1.842 (4)[24], $Fe-NCS$ = 1.957 (5)[32], and $Fe-N_3$ = 1.91 Å[102] are quite generally as short as, or shorter than, the values expected for low-spin octahedral species; especially short $Fe-O$ bonds, 1.763 (1)[35] and ~1.73 Å, are observed in the oligomeric species. Relative, however, to the averaged $Fe-Cl$ bond distance of 2.205 Å, the $Fe-SR$ bond distance of 2.32 (1) Å in the p-nitrothiophenoxoiron(III) derivative of protoporphyrin-IX dimethyl ester[103] is ~0.065 Å longer than would be expected from the difference (0.05 Å) in the covalent radii[105] of the liganded sulfur and chlorine atoms.

The methoxo-, azido-, and, presumably, the isothiocyanatoiron(III) complexes readily add pyridine to give the low-spin six-coordinate species. McLees[106] found, on the other hand, that chlorodeuterohemin, (Cl)Fe-(Deut-IX), in chloroform and pyridine solutions has effective magnetic moments of 5.94 and 5.70 *BM*, respectively. He suggested, furthermore, that the reported[107] downward drift, extending over periods of several days, of the effective moment of chlorohemin in pyridine to values between 2 and 3 *BM* might be ascribed to chemically reducing impurities in the pyridine. (That the principal product of slow overall reaction might be low-spin $[(Py)_2Fe(Proto\text{-}IX)]Cl$, with the displaced chloride ions hydrogen-bonded to the propionic acid substituents, is a rather interesting conjecture.) Quite pertinent to this problem are the stereochemical data from structure determinations for both the five- and the six-coordinate high-spin manganese(III) derivatives of $H_2(TPP)$.

Parameters in the coordination group of the (Cl)Mn(TPP) molecule from structure determinations for crystals of the respective acetone[108] and chloroform[109] solvates are in excellent agreement. Averaged values are $Mn-N_P$ = 2.008, $Ct \cdots N_P$ = 1.990, $Mn \cdots Ct$ = $Mn \cdots P_N$ = 0.265, $Mn-Cl$ = 2.373, $Ct \cdots Cl$ = 2.638, and $Cl \cdots N_P$ = 3.30 Å (see Fig. 4). Several points are to be emphasized: the $Mn-Cl$ bond is ~0.16 Å longer than the $Fe-Cl$ linkage in chlorohemin, although an increase of only ~0.03 Å can be justified by the difference of one protonic unit in the nuclear charges carried by the metal atoms. In view of the preference of the relatively strong $Mn-N_P$ bonds for a planar conformation, the out-of-plane displacement of the Mn(II) atom, $Mn \cdots Ct$ = 0.265 Å, is quite surprisingly large. It is the $Cl \cdots N_P$ packing contacts which primarily determine the overall stretch in the axial $Ct \cdots Cl$ distance, though not the division of this stretch between the additive components, $Ct \cdots Mn$ and $Mn-Cl$. The $Cl \cdots N_P$ distance of 3.30 Å, which is only

~0.20 Å shorter than the sum of the van der Waals radii of chlorine (1.80 Å) and aromatic nitrogen (1.70 Å) atoms, corresponds to the moderate stress which, according to approximate theory, should lead to substantial stretch in the axial bond system.

The stereochemical data provided by structure determination for the benzene solvate of $\alpha,\beta,\gamma,\delta$-tetraphenylporphinatoazidomanganese(III)[110] are quite in agreement with the preceding analysis. Parameters in the coordination group of the $(N_3)Mn(TPP)$ molecule are $(Mn-N_P)_{av} = 2.005$ (3), $Mn-N_{az} = 2.045$ (4), $Mn \cdot\cdot Ct = 0.23$, $(Ct \cdot\cdot N_P)_{av} = 1.992$, and $(N_{az} \cdot\cdot N_P)_{av} = 3.025$ Å. The $N_{az} \cdot\cdot N_P$ contacts correspond again to a moderate intramolecular stress which requires substantial stretch in the axial bond system. Thus the $Mn-N_{az}$ bond is longer by ~0.135 Å than the $Fe-N_{az}$ bond in $(N_3)Fe(TPP)$ and by ~0.09 Å than the $Fe-NCS$ linkage in (SCN)-$Fe(TPP)$. The out-of-plane displacement of the $Mn(III)$ atom, 0.23 Å, is again impressively large.

Stereochemical data from determinations of crystalline structure are available for two six-coordinate, high-spin, manganese(III) derivatives of $H_2(TPP)$. Parameters in the octahedral coordination group of $(Py)Mn(Cl)$-(TPP)[111] are $(Mn-N_P)_{av} = 2.009$ (3), $Mn-Cl = 2.468$ (1), $Mn-N_{Py} = 2.444$ (4), $(Ct \cdot\cdot N_P)_{av} = 2.005$ (3), $Ct \cdot\cdot Mn = 0.12$, $Ct \cdot\cdot Cl = 2.588$, $Ct \cdot\cdot N_{Py} = 2.324$, $(Cl \cdot\cdot N_P)_{av} = 3.275$, and $(N_{Py} \cdot\cdot N_P)_{av} = 3.06$ Å. Alterations in the parameters of the $(Cl)Mn(TPP)$ molecule that are required by the complexing of the pyridine ligand are a reduction of 0.145 Å in the out-of-plane displacement of the $Mn(III)$ atom along with a largely compensating stretch of 0.095 Å in the $Mn-Cl$ bond that leaves the critical $Cl \cdot\cdot N_P$ contact distance at 3.275 Å, an ostensible reduction of ~0.025 Å. Of the two weak axial bonds, the $Mn-N_{Py}$ linkage is evidently the weaker. This difference should be accentuated in a high-spin $(Py)Fe(Cl)(TPP)$ molecule wherein a $Ct \cdot\cdot Fe$ displacement <0.30 Å is improbable. Although the presence of the overlarge chloro ligand is prejudicial to the stability of any six-coordinate porphinatoiron(III) species, it is clear that the use of a 1-alkylimidazole instead of pyridine as the sixth ligand is more likely to yield a stable low-spin complex.

Parameters in the octahedral coordination group of the $(CH_3OH)Mn(N_3)$-(TPP) molecule[112] are $(Mn-N_P) = 2.031$ (7), $Mn-N_{az} = 2.176$ (9), $Mn-O = 2.329$ (8), $(Ct-N_P)_{av} = 2.029$, $Ct \cdot\cdot Mn = 0.085$, $(N_{az} \cdot\cdot N_P)_{av} = 3.035$, and $(N_P \cdot\cdot O)_{av} = 3.025$ Å. The principal alterations in the parameters of the $(N_3)Fe(TPP)$ molecule that attend the evidently weak complexing of the methanol ligand are a reduction of 0.145 Å in $Ct \cdot\cdot Mn$ and an increase of 0.131 Å in the $Mn-N_{az}$ bond length; the critical $N_{az} \cdot\cdot N_P$ packing distance is almost unaffected. It is further to be noted that the observed $quasi$-S_4 rufflings of the porphinato cores in the $(Cl)Mn(TPP)$, $(N_3)Mn(TPP)$, and $(Py)Mn(Cl)(TPP)$ species are conducive to slightly shortened $Mn-N_P$ bond lengths (2.005–2.009 Å) whereas the asymmetrically nonplanar core in

$(CH_3 OH)Mn(N_3)(TPP)$ certainly is not.

The geometry of the binuclear $N_4 FeOFeN_4$ coordination group in the oligomeric $O[Fe(TPP)]_2$ molecule[35] approximates to the noncrystallographic symmetry of $D_{4d} - \bar{8}2m$; see Fig. 7. Each square-pyramidal half of the group (Fig. 4) has $(Fe—N)_{av}$ = 2.087 (3), Fe—O = 1.763 (1), Fe $\cdot\cdot$ Ct = 0.50, and $(Ct—N)_{av}$ = 2.027 Å. The FeOFe bond angle at the central oxygen atom is 174.5 (1)° and the distance between the mean planes of the two porphinato cores is ~4.40 Å. These parameters carry over with little alteration to the $O[Fe(Proto-IX-DME)]_2$ oligomer[9]. Serious positional disorder in the crystal together with nearly prohibitive requirements for computing make it unprofitable to attempt highly detailed refinement of the structure.

Parameters in the coordination groups of the high-spin five-coordinate 2-methylimidazoleiron(II)[36] and 1-methylimidazolemanganese(II)[113] deriv-

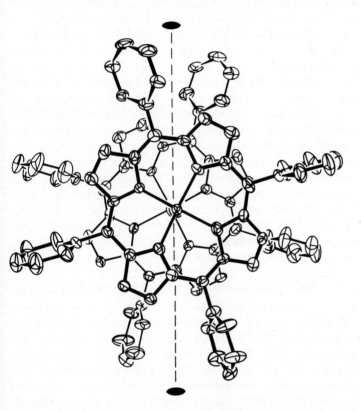

Fig. 7. Computer-drawn diagram in perspective of the $O[Fe(TPP)]_2$ molecule as viewed nearly along the *quasi*-$\bar{8}$ axis. Each atom is represented by an ellipsoid having the shape and the relative size concomitant with its thermal motion in the crystal. The twofold axis passes through the bridging oxo oxygen atom, thus requiring structural equivalence of the upper and lower halves of the oligomer. (Reprinted with permission from J. Amer. Chem. Soc., **94**, 3620 (1972). Copyright by the American Chemical Society.)

References, p. 376

atives of H_2(TPP), (2-MeIm)Fe(TPP) and (1-MeIm)Mn(TPP), respectively, are now available for comparison with the data cited above for the high-spin iron(III) and manganese(III) complexes and, indeed, with stereochemical expectations of long standing[24,28-30]. In the (2-MeIm)Fe(TPP) molecule (for which bond parameters in the porphinato core are listed in Tables 2 and 3), (Fe—N_P)$_{av}$ = 2.086 (4), Fe—N_X = 2.161 (5), ($Ct \cdot\cdot N_P$)$_{av}$ = 2.044 (4), and $Ct \cdot\cdot$ Fe = 0.42 Å (Fig. 4). The corresponding parameters in the (1-MeIm)Mn(TPP) molecule, (Mn—N_P)$_{av}$ = 2.128 (2), Mn—N_X = 2.192 (2), ($Ct \cdot\cdot N_P$)$_{av}$ = 2.065 (2), and $Ct \cdot\cdot$ Mn = 0.515 Å, are, without exception, significantly larger. The increase of $\sim$0.03 Å in the length of the axial M—N_X bond is about that expected from the decrease of one protonic unit in the nuclear charge carried by the metal atom; the slightly larger increase, $\sim$0.04 Å, in the M—N_P bond distance includes the further correlation with the accompanying small, but energetically demanding, increase of $\sim$0.02 Å in the $Ct \cdot\cdot N_P$ radius of the already expanded core. Relative to parameters in the (SCN)Fe(TPP) molecule of (Fe—N_P)$_{av}$ = 2.065 (5), Fe—N_X = 1.957 (5), $Ct \cdot\cdot N_P$ = 2.01, and $Ct \cdot\cdot$ Fe = 0.485 Å, the axial M—N_X bonds in (2-MeIm)-Fe(TPP) and (1-MeIm)Mn(TPP) are seen to be much longer, ostensibly by 0.204 and 0.235 Å, respectively, than the analogous Fe—NCS bond in the iron(III) complex. The square-pyramidal geometry together with the large value (0.485 Å) of $Ct \cdot\cdot$ Fe in the coordination group of the (SCN)Fe(TPP) molecule encourages the sterically uninhibited Coulombic interaction of the SCN^- anion with the high-spin iron cation. At least 0.15 Å of the observed shortening of the Fe—NCS bond relative to the axial bond formed between electrically neutral species in the (2-MeIm)Fe(TPP) molecule is directly ascribable to this source. As a further consequence of this intimate ion-pairing, the strength of the bonding interactions of the d^5 iron atom with the porphinato core in (SCN)Fe(TPP) is reduced toward that of the d^6 iron atom with the core in (2-MeIm)Fe(TPP). In an (imidazole)Fe(TPP)$^+$ ion, somewhat shortened Fe—N_P bonds, a smaller $Ct \cdot\cdot$ Fe displacement, and an Fe—N_X bond length in the range 2.10—2.15 Å may be expected.

The preparation[43] and the structure[36] of (2-MeIm)Fe(TPP) provide an especially vivid example of thermodynamic behavior that is dependent upon the relative structural merits of competing crystalline phases. In (strictly anaerobic) benzene solution, the equilibrium constant (K_1) for the formation of high-spin (2-MeIm)Fe(TPP) from intermediate-spin ($S = 1$) Fe(TPP) and 2-methylimidazole is $\sim$2 $\times$ 10^4 m^{-1}. It was found[43], nonetheless, that crystallization from either benzene or benzene—methanol solution afforded only the structurally elegant Fe(TPP) phase[4] (see Tables 2 and 3; also p. ●●●). In the centrosymmetric crystals of (2-MeIm)Fe(TPP) · C_2H_5OH, obtained from benzene—ethanol solution[43], the required molecular symmetry of C_2 (a 2-fold axis) is satisfied statistically by two equally probable orientations of the (2-MeIm)Fe(TPP) molecule[36]. The structural framework in the crystal is primarily determined by the packing of the porphinato entities, but the stabilization of this framework relative to that in the Fe(TPP) phase requires the presence, as loosely packed filler, of both the complexed 2-methylimidazole and the solvating ethanol molecules[36]. The rather grotesquely nonplanar conformation of the porphinato core that is required in the (2-MeIm)Fe(TPP) · C_2H_5OH structure may lead to a slight expansion of $Ct \cdots N_P$ and a concomitant larger contraction of $Ct \cdots Fe$ relative to the values of these parameters in an externally unconstrained (2-MeIm)Fe(TPP) molecule. It is seen, nevertheless, that the parameters in the coordination groups of the high-spin (SCN)Fe(TPP), (2-MeIm)Fe(TPP), and (1-MeIm)Mn-(TPP) molecules fit satisfyingly into a physically rational sequence.

A most striking difference between the high-spin (L)Mn(TPP) and (L)Fe-(TPP) species, wherein L is a 1-alkylimidazole or other strongly complexing ligand that is not sterically inhibitory to six-coordination, is their markedly different behavior toward conversion into low-spin $(L)_2M(TPP)$ species. The presence of 1-methylimidazole or pyridine in large excess notwithstanding, only the high-spin (L)Mn(TPP) species appear to be crystallizable from toluene solution[113]. From benzene solutions of Fe(TPP) and imidazole or a 1-alkylimidazole, by contrast, the crystallization of the high-spin (L)Fe(TPP) appears to be precluded by thermodynamic constraints[36,43]. Thus, for example, the equilibrium constant (K_2) for the complexing of a second 1-MeIm ligand by the (1-MeIm)Fe(TPP) molecule to give the low-spin $(1\text{-MeIm})_2Fe(TPP)$ species is approximately an order of magnitude larger than the constant (K_1) for the complexing of the first 1-MeIm ligand by the intermediate-spin Fe(TPP) species[36]. It follows that crystallization of the five-coordinate complex is allowed only if its solubility under optimum conditions be smaller by a factor of at least $(K_1/K_2)^{1/2}$, i.e., $(1/10)^{1/2} = 0.32$, than the solubilities of both the four- and the six-coordinate species[36]. A K_2/K_1 ratio of ~10 holds also for L chosen as imidazole or 4-t-butylimidazole and is probable with pyridine or piperidine. These thermodynamic criteria for the preferential crystallization of a five-coordinate (L)Fe(TPP) seem not to be met with any of the cited choices of L[43]. The successful preparation of crystalline (2-MeIm)Fe(TPP) · C_2H_5OH derives from the fact that 2-methylimidazole is sterically inhibitory $(K_2 \ll K_1)$ in $(2\text{-MeIm})_2Fe(TPP)$, though not in the least so in (2-MeIm)Fe(TPP)[36].

The stereochemistry of $\alpha,\beta,\gamma,\delta$-tetraphenylporphinatomanganese(II), Mn-(TPP), as this high-spin four-coordinate species is observed with a required center of inversion in crystals of the toluene solvate[113], presents a problem of interpretation. Orthodox structure analysis leads to $(Mn-N_P)_{av} = (Ct \cdots N_P)_{av} = 2.082$ (2) Å. It leads also, however, to an extraordinarily large thermal parameter of ~9.65 Å^2 with a concomitant root-mean-square (rms) amplitude of 0.35 Å for vibration of the metal atom perpendicular to the mean plane of the porphinato core; the values, for comparison, of the analogous thermal parameters in crystals of Fe(TPP) and Ni(TPP) are only ~3.2 Å^2. The authors point out that a disordered model in which the manganese atom is equally distributed between two, symmetry-equivalent, out-of-plane positions with $Ct \cdots Mn = \pm 0.19$ Å, an *rms* amplitude of ~ 0.28 Å (still very high), $(Mn-N_P)_{av} = 2.090$ (9) Å, and $(Ct \cdots N_P)_{av}$ unchanged at 2.082 Å gives equally good agreement with the X-ray data. The writer judges that centering of the manganese atom in the plane does correspond to unstable equilibrium — a subsidiary maximum in the potential energy — but that, at room temperature, the thermally excited vibrations allow the metal atom to pass freely over the potential barrier. Only below a transition temperature at which the heat capacity (but neither the enthalpy nor the entropy) of the crystal is discontinuous, should the metal atoms be distributed between two, symmetry-equivalent, out-of-plane, equilibrium positions.

The 'picket fence' porphyrin, *meso*-tetra($\alpha,\alpha,\alpha,\alpha$-*o*-pivalamidophenyl)porphyrin, $H_2(TpivalPP)$, represents a major synthetic achievement of Collman et al.[47,114]. The semiquantitative geometries of the nearly isostructural low-spin $(O_2)Fe(1-MeIm)$[115] and $(CO)Fe(1-MeIm)$[116] derivatives of H_2-(TpivalPP), illustrated in Fig. 8 for the slightly simpler carbonyl complex, are prime subjects for subsequent discussion. Of immediate interest is the observation that the Mössbauer spectra given by the four-coordinate iron(II) derivative[114] of $H_2(TpivalPP)$ are fully characteristic of a high-spin porphinato-iron(II) complex. It may be, but it does not follow, that the stereochemistry of the coordination group follows the pattern observed in crystalline high-spin $Mn(TPP)$[113]. Given the highly polar geometry of the flexible molecule (Fig. 8), a pyramidal coordination group in which the iron atom lies substantially out-of-plane from the nitrogen atoms is a rational probability.

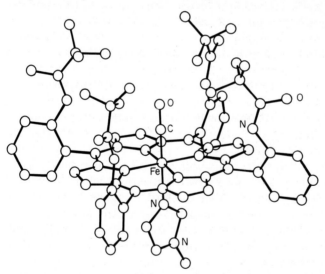

Fig. 8. Diagram in perspective of the $(OC)Fe(1-MeIm)(TpivalPP)$ molecule — the carbonyl-1-methylimidazoleiron(II) derivative of the 'picket fence' porphyrin. Just one of the two equivalent orientations of the 1-MeIm ligand is shown.

8.4.5.2. The low-spin iron and cobalt porphyrins

Comparison of the centro-symmetric $quasi\text{-}D_{4h}$ geometries of the coordination groups in three chemically analogous low-spin complexes, $(Pip)_2$ Fe-(TPP)[33], $[(Pip)_2 Co(TPP)^+]$[38], and $(Pip)_2 Co(TPP)$[39], illuminates the correlation of stereochemistry with the d electron configurations in the valence shells of the respective iron(II), cobalt(III), and cobalt(II) atoms. (See Tables 2 and 3 for bond parameters in the porphinato cores of these species.) The respective equatorial and axial M—N bond lengths in $(Pip)_2$ Fe(TPP) are $(Fe-N_P)_{av}$ = 2.004 (3), and $Fe-N_{Pip}$ = 2.127 (3) Å; steric interactions of piperidine hydrogen atoms (five on each ligand) with, for the most part, porphinato nitrogen are responsible for substantial stretch in the axial bonds[33]. Substitution of the isoelectronic cobalt(III) atom for iron(II) gives the $(Pip)_2 Co(TPP)^+$ cation with shorter equatorial and axial Co—N bond lengths of 1.978 (3) and 2.060 (3) Å, respectively[38]. The reduction of the cobalt(III) atom to cobalt(II) by the insertion of an electron in its $3d_{z^2}$ orbital yields the neutral $(Pip)_2 Co(TPP)$ molecule in which the equatorial Co—N distance is only slightly increased to 1.987 (2) Å, whereas the axial Co—N bond length is grossly stretched to 2.436 (2) Å[39]. A subsequent determination of structure for the low-spin bis(3-methylpyridine)cobalt(II) derivative of octaethylporphyrin, $(3\text{-Pic})_2 Co(OEP)$, provides concordant results: namely, equatorial and axial bond lengths of 1.992 (1) and 2.386 (2) Å, respectively[40].

For systematic discussion of the six-coordinate stereochemistry of the low-spin iron and cobalt(III) metalloporphyrins whenever one or both of the axial ligands is an aromatic nitrogen bse, the dihedral angle between the plane of the aromatic ring and an xz coordinate plane — which may or may not be completely definable by symmetry — is a fundamental *orientation* parameter of the coordination group. To the precision required for discussion of the experimentally determined configurations, it suffices to take the unique z axis as the normal to the mean plane (P_N) of the four porphinato nitrogen atoms with the origin at Ct and to fix the orientation of the xz reference plane by requiring it to pass through one of these nitrogen atoms. The z axis thus defined may, however, become a 2-fold axis as in the diagram shown in Fig. 9 for a pyridine ligand. Given this symmetry, the dihedral angle of interest, ϕ, is reproduced in the angle made by the trace of the pyridine ring on the xy plane with the x axis. This same relation holds with any aromatic ligand provided the plane of the aromatic ring be not tipped away from the z axis — from, that is, the normal to P_N. It is usually most illuminating, if not quite precise, to regard the dihedral angle, ϕ, as the angle through which the plane of the ligand is rotated around the z axis from coincidence with the xz reference plane.

With a fixed M—N_{Py} bond distance, steric interactions between pyridine hydrogen and porphinato nitrogen atoms are maximized for $\phi = 0$, minimized for $\phi = \pm45°$. It is noteworthy, moreover, that the sterically ideal

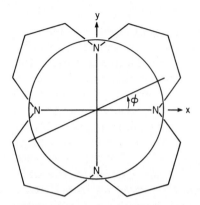

Fig. 9. Diagram of the inner 16-membered ring in a metalloporphyrin whereon the circle represents the projected positions of the 2,6-hydrogen atoms of a pyridine ligand when this ligand is rotated around a real or effective 2-fold axis of symmetry (the z axis). A general orientation of the pyridine ligand is represented by its trace on the equatorial xy plane. In this simplest case, the parameter, ϕ, is merely the angle by which the ligand is rotated around the unique (z) axis from coincidence with the xz coordinate plane; it is also, of course, the dihedral angle between the ligand plane and the xz coordinate plane; see text.

configuration for a bis(pyridine) metal derivative of $H_2(TPP)$ or $H_2(OEP)$ observes D_{2d} symmetry as indicated in Fig. 3; with a moderate ruffling of the porphinato core, all steric interactions of pyridine hydrogen with porphinato carbon atoms are reduced to triviality and only the $H_{Py} \cdots N_P$ contacts need to be considered.

The porphinato core in the nitro-3,5-dimethylpyridinecobalt(III) derivative of $H_2(TPP)$, $(3,5\text{-Lut})Co(NO_2)(TPP)$[37], as this molecule exists in the crystal with a required 2-fold axis of symmetry, is strongly ruffled in approximate agreement with D_{2d} geometry; methine carbon atoms are displaced by $+0.56$ and -0.64 Å from the mean plane of the core. As pointed out by the authors[37], the magnitude of this ruffling is much larger than is needed to allow the equatorial $Co-N_P$ bonds to take the value (1.95–1.96 Å) which is generally preferred with nitrogen of monodentate ligands and, consequently, that the ruffling is primarily attributable to packing relations in the crystals. Bond parameters in the coordination group are $(Co-N_P)_{av} = 1.954$ (2), $Co-NO_2 = 1.948$ (4), and $Co-N_{Lut} = 2.036$ (4) Å. The departure of the dihedral angle, $\phi = 35.4°$, from the sterically most favorable value of $45°$ (Fig. 9) is presumably a further consequence of packing constraints.

Averaging the parameters derived for the coordination groups of two structurally non-equivalent, but centro-symmetric, $(Im)_2Co(TPP)^+$ cations which were found to be present in a crystal of rather poor quality for a precise refinement of structure[117] yields $(Co-N_P)_{av} = 1.98$ (2) and $(Co-N_{Im})_{av} = 1.93$ (2) Å. The two imidazole rings are coplanar with $\phi =$

43°, an orientation which approaches perfection (45°) for minimizing the steric interactions of the 2,4-hydrogen atoms on each ligand with the porphinato nitrogen atoms, though not with the porphinato carbon atoms. But since the steric requirements of imidazole are substantially less demanding than are those of pyridine — the consequence of the differing geometries of five- and six-membered rings — the axial $Co-N_{Im}$ bonds are surely very short in the $(Im)_2 Co(TPP)^+$ cation relative to the $Co-N_{Lut}$ distance of 2.036 (4) Å in $(3,5-Lut)Co(NO_2)(TPP)$.

Stereochemical data from X-ray structural analyses are available for four low-spin porphinatoiron(III) complexes. In the $(Im)_2 Fe(TPP)^+$ cation[6], as it exists in the crystalline salt, $[(Im)_2 Fe(TPP)^+][Cl^-] \cdot CH_3 OH$, $(Fe-N_P)_{av}$ = 1.989 (4), $Fe-N(Im_a)$ = 1.957 (4), $Fe-N(Im_b)$ = 1.991 (5), and $Ct \cdots Fe$ = 0.009 Å. The dihedral angle between the planes of the two imidazole ligands is 57°, and the values of the orientation parameter, ϕ in Fig. 9, are 39° and $-18°$ for the ligands Im_a and Im_b, respectively. A marked *quasi-S_4* ruffling of the core eases all contacts of the 2,4-hydrogen atoms in both ligands with porphinato carbon atoms, but the pair of unduly tight $H \cdots N_P$ contacts that persist for Im_b provide evidence for a significant stretch, ostensibly 0.034 Å, in the $Fe-N$ bond to this ligand. Further support for this interpretation comes from the observation that the out-of-plane displacement of the iron atom, though only ~0.009 Å, is toward Im_b[6]. The averaged $Fe-N_P$ bond length of 1.989 (4) Å is seen to be a rational value between those, 2.004 (3) and 1.978 (3) Å in the $(Pip)_2 Fe(TPP)$ molecule[33] and the $(Pip)_2 Co(TPP)^+$ ion[38], respectively.

In the $(Im)_2 Fe(OEP)^+$ ion, as it occurs with a required center of inversion in a chloroform solvate of the perchlorate salt[34], the coplanar imidazole ligands are rotated only ~7° from coincidence with the reference plane (Fig. 9). Individual values for the three independent $Fe-N$ distances, of which two are equatorial, are not reported; it is stated[34] that the three $Fe-N$ distances average to 2.01 Å and, within the standard deviations, are in agreement with those in the $(Im)_2 Fe(TPP)^+$ ion. If, as is certainly probable, $(Fe-N_P)_{av} \ll 2.00$ Å, then $Fe-N_{Im} > 2.03$ Å is required to give the averaged value overall of 2.01 Å. Calculated distances of 2.34 and 2.45 Å for the tight $H_{Im} \cdots N_P$ contacts also are given.

Structure determination[118] for crystals of the composition, $(1-MeIm)_2-Fe(Proto-IX) \cdot CH_3OH \cdot H_2O$, provides convincing evidence that the complex is not an iron(II) derivative of $H_2(Proto-IX)$, but is rather an iron(III) derivative of $H_2(Proto-IX^-)$, thus written to indicate the loss of a hydrogen ion from one of the carboxylic acid groups of the porphyrin. Otherwise put, the electrically neutral complex is a porphinatoiron(III) species in which the counter ion is a carboxylate group of the porphyrin. One-dimensional polymeric ribbons of the complex are stabilized in the crystal by the presence of a very strong hydrogen-bond between the propionate groups of every pairing of adjacent molecules; $O \cdots H \cdots O = 2.41$ Å. Parameters in the coordination

References, p. 376

group are $(Fe-N_P)_{av} = 1.990$ (5), $Fe-N(1-MeIm_a) = 1.966$ (5) and $Fe-N-(1-MeIm_b) = 1.988$ (5) Å. The observed values of the orientation parameters (ϕ in Fig. 9) are $+16°$ for 1-MeIm$_a$ and $-3°$ for 1-MeIm$_b$. Each axial ligand is then involved in a pair of close $H_{Im} \cdots N_P$ contacts that are calculated[118] to be 2.5 (1) and 2.1 (1) Å, of which the latter is astonishingly, if not quite incredibly, short. Although the coordination group cannot retain the center of inversion which characterizes the otherwise related geometry in the $(Im)_2$-Fe(OEP)$^+$ ion, the magnitude of the difference in the two axial bond lengths, ostensibly 0.022 Å, is not guaranteed by the precision attained in the structure determination. As a further consequence of the limited precision, the quite varied pattern of bond lengths in the porphinato ligand does not lend itself to unambiguous interpretation[118]. This pattern, indeed, is remarkably similar to that reported in 1965 for the Ni(2,4-DAcDeut-IX-DME)[17] species.

It is highly probable that externally unconstrained bis(imidazole)iron(III) and bis(imidazole)cobalt(III) derivatives of the various porphyrins utilize a common, qualitatively defined, coordination geometry with only quantitative alterations thereto. That determinations of crystalline structure do not suffice to identify this qualitatively unique geometry is made abundantly clear by the foregoing discussion.

Parameters in the coordination group of the pyridineazidoiron(III) derivative of $H_2(TPP)$, $(Py)Fe(N_3)(TPP)$[119], are $(Fe-N_P)_{av} = 1.990$ (6), $Fe-N_{Py} = 2.085$ (6), $Fe-N_{az} = 1.926$ (6), and $Ct \cdots Fe - 0.031$ Å. The linear azido ligand is coordinated end-on to give the usual bent, but planar, configuration; the FeNN angle is $125°$. The orientation parameters, ϕ in Fig. 9, of the planes of the pyridine and the Fe-N-N-N bond system are $50°$ and $40°$, respectively. Values of ϕ near $45°$ or $-45°$ for the plane of the Fe-N-N-N bond system require the central nitrogen atom of the azido group to make contact with two porphinato nitrogen atoms at 3.00—3.04 Å; this is sterically much preferable to a ϕ near $0°$ or $90°$ which would require a single, but substantially tighter, contact of ~2.82 Å. To what degree the orientation of the Fe-N$_3$ bond system is influenced by the possibilities for a π component in the complexing bond is, as is usual, an unanswered question. It is not improbable that the choice of the observed molecular configuration, with ϕ near $45°$ for both axial bond systems in preference to the orthogonal combination of $+45°$ and $-45°$, is determined by packing constraints in the crystalline phase. Comparison of the Fe-N$_{Py}$ distance (2.085 Å) with the Co-N$_{Py}$ bond length (2.036 Å) in $(3,5-Lut)Co(NO_2)(TPP)$[37] suggests that the Fe-N$_{Py}$ bond may be somewhat weakened and slightly stretched through interaction with the stronger *trans* bond to the azido ligand; the iron atom is drawn out-of-plane toward the azido ligand by ~0.031 Å.

Of four determinations of crystalline structure for five-coordinate low-spin cobalt(II) metalloporphyrins wherein the axial ligand is an aromatic base, the structural analyses of the 1-methylimidazolecobalt(II) and the 1,2-

dimethylimidazolecobalt(II) derivatives of $H_2(TPP)$, (1-MeIm)Co(TPP)[41] and (1,2-Me$_2$Im)Co(TPP)[120], respectively, are particularly noteworthy. Parameters in the square-pyramidal coordination group (Fig. 4) of the (1-MeIm)Co(TPP) molecule[41] are $(Co-N_P)_{av}$ = 1.977 (3), $Co-N_{Im}$ = 2.157 (3), $Co \cdot \cdot P_N$ = 0.13, and $Co \cdot \cdot P_C$ = 0.14 Å. Inasmuch as P_N and P_C are the mean planes of the porphinato nitrogen atoms and of the 24-atom porphinato core, respectively, the difference of $Co \cdot \cdot P_C$ and $Co \cdot \cdot P_N$, 0.01 Å, is a measure of the *net* doming of the core toward the cobalt atom. The less precisely determined parameters which were subsequently obtained for the analogous derivative of $H_2(OEP)$, (1-MeIm)Co(OEP)[121], differ from those given above in just one minor respect: the $Co \cdot \cdot P_C$ distance of 0.16 Å leads to a net doming of the core of 0.03 Å. Such net doming, as observed in crystalline five-coordinate metalloporphyrins, rarely exceeds 0.05 Å, and is typically smaller than the mean deviation of the 24 atoms in the core from its mean plane.

Structure determination for the 3,5-dimethylpyridinecobalt(II) derivative of $H_2(TPP)$, (3,5-Lut)Co(TPP)[122], leads to $Co \cdot \cdot P_N$ = 0.14 and $Co-N_{Py}$ = 2.161 (5) Å, as compared with 0.13 and 2.157 (3) Å, respectively, in the (1-MeIm)Co(TPP) molecule. It appears, consequently, that in the absence of significant steric constraints, pyridine and imidazole display nearly identical behavior as complexing agents.

The preparation and structural analysis of crystalline (1,2-Me$_2$Im)Co-(TPP)[120] were undertaken to determine the stretch in the axial bond system that should attend the substitution of 1,2-dimethylimidazole, with its sterically active 2-methyl substituent, for 1-methylimidazole as the axial ligand. Parameters obtained for (1,2-Me$_2$Im)Co(TPP) are $(Co-N_P)_{av}$ = 1.985 (2), $Co-N_{Im}$ = 2.216 (2), $Co \cdot \cdot P_N$ = 0.15, and $Co \cdot \cdot P_C$ = 0.18 Å. The perpendicular distance ($N_{Im} \cdot \cdot P_N$) of the axial nitrogen atom (N_{Im}) from the mean plane (P_N) is 2.35 Å, $\sim$0.016 Å less than the sum of $Co-N_{Im}$ and $Co \cdot \cdot P_N$ because the $Co-N_{Im}$ bond is tipped from the normal to P_N by nearly 7°. With this orientation of the axial bond, the hydrogen atoms of the librating 2-methyl substituent on the imidazole ring are relieved from unduly short contacts with atoms of the porphinato core: $H_{Me} \cdot \cdot N_P$ and $H_{Me} \cdot \cdot C_P$ are $\geqslant$2.83 Å in the equilibrium orientation of the group. Simultaneously, however, the 4-hydrogen substituent on the ligand is brought into fairly tight contacts with two atoms of the core: $H_{Im} \cdot \cdot N_P$ = 2.65 and $H_{Im} \cdot \cdot C_P$ = 2.57 Å. In crystalline (1-MeIm)Co(TPP)[41], moreover, the unfavorable orientation of the (1-MeIm) ligand leads to one short $H_{Im} \cdot \cdot N_P$ contact distance of 2.62 Å and, presumably, a small concomitant stretch in the $N_{Im} \cdot \cdot P_N$ distance.

The much shorter ligand—core contacts, $H_{Im} \cdot \cdot N_P$ = 2.1 and 2.5 Å, involving each of the axial ligands in the (1-MeIm)$_2$Fe(III) derivative of H_2-(Proto-IX$^-$) are apparently responsible for, at most, a stretch of $\sim$0.03 Å in the $Fe-N_{Im}$ bonds[118]. It is the more striking, consequently, to see that the

References, p. 376

easing of the destabilizing, but far less demanding, ligand—core steric interactions which accompany the substitution of 1,2-dimethylimidazole for 1-methylimidazole in the five-coordinate (L)Co(TPP) species requires a stretch of 0.059 (4) Å in the Co—N_{Im} bond and an increase from 0.13 to 0.15 Å in the Co $\cdot\cdot$ P_N displacement. That only a very small stress is associated with the ligand—core interactions is strongly suggested by the comparatively small value of Co $\cdot\cdot$ P_N. The substantially larger Mn $\cdot\cdot$ P_N displacements of 0.265 Å in (Cl)Mn(TPP)[108,109] and 0.23 Å in (N_3)Mn(TPP)[110] are required by not particularly tight Cl $\cdot\cdot$ N_P and N_{az} $\cdot\cdot$ N_P packing contacts in the respective coordination groups of these species (p. 353). The specific conclusion of subsequent interest is that the axial Co—N_{Im} bond in any five-coordinate porphinatocobalt(II) species is readily stretched when subjected to a quite modest tension, whether directly or indirectly applied.

8.4.5.3. The nitrosylmetal(II) metalloporphyrins

Preparative and structural studies carried out by Scheidt and coworkers on the (ON)Mn(II), (ON)Fe(II), and (ON)Co(II) derivatives of H_2(TPP) afford a detailed description of the stepwise qualitative and quantitative alterations in the stereochemistry of the axial M—N—O bond system that are concomitant with the stepwise addition or removal of a $3d$ electron. On the basis of earlier experimental and theoretical studies, the *quasi*-linearity of the Mn—N—O bond system observed in (ON)Mn(4-MePip)(TPP)[124] [4-MePip = 4-methylpiperidine] and the strongly bent Co—N—O bond system observed in (ON)Co(TPP)[77] were expected. Of prime interest, consequently, are the stereochemistries of the low-spin six-coordinate (ON)Fe(1-MeIm)(TPP)[124] and five-coordinate (ON)Fe(TPP)[78] species.

Bond lengths in the coordination group of the (ON)Fe(1-MeIm)(TPP) molecule are $(Fe—N_P)_{av}$ = 2.008 (4), Fe—NO = 1.743 (4), and Fe—N_{Im} = 2.180 (4) Å; the out-of-plane displacement of the iron atom, Fe $\cdot\cdot$ P_N = 0.07 Å, is toward the nitrosyl ligand. The authors conclude that the bond to the 1-methylimidazole ligand is ~0.20 Å longer than the normally expected value by reason of the extraordinary character of the bonding interactions in the *trans* Fe—N—O linkage. Of subsequent interest is the FeNO angle of 142.1 (6)°.

In the (ON)Mn(4-MePip)(TPP) molecule[124], the Mn—N—O linkage is nearly linear; the MnNO angle is 176.2 (5)°. Bond lengths in the octahedral coordination group are $(Mn—N_P)_{av}$ = 2.028 (4), Mn—NO = 1.644 (5), and Mn—N_{Pip} = 2.206 (5) Å. Although these data suggest the probability that the long Mn—N_{Pip} bond is subjected to a structural *trans* effect similar to that borne by the Fe—N_{Im} bond in the (ON)Fe(1-MeIm)(TPP) molecule, the authors point out that the substantial steric requirements of the piperidine ligand also should lead to a very long axial bond. The displacement, Mn $\cdot\cdot$ P_N = 0.10 Å, of the manganese atom toward the nitrosyl ligand leaves the N_{Pip} $\cdot\cdot$ P_N distance at ~2.11 Å, and this is already smaller than the

$N_{Pip} \cdots P_N = Fe-N_{Pip}$ distance of 2.127 (3) Å in the centrosymmetric $(Pip)_2 Fe(TPP)$ species[33].

The $(ON)Co(TPP)$[77] and $(ON)Fe(TPP)$[78] species crystallize in the tetragonal structural type first observed for $(Cl)Fe(TPP)$[5] in which statistical conformity with C_{4h} symmetry is required of the five-coordinate molecules. But whereas two orientations (parallel and antiparallel to the tetragonal axis) of a $(Cl)Fe(TPP)$ molecule suffice to this end, the bent M—N—O linkage in either $(ON)M(TPP)$ species requires no less than eight orientations of the molecule for satisfaction of the C_{4h} symmetry. The plane of the M—N—O bond system takes four equivalent values of the orientation parameter (Fig. 9), $(\phi + 90n)^{\circ}$, $n = 0, 1, 2$, and 3.

Complications of statistical disorder notwithstanding, the structural data obtained for crystalline $(ON)Fe(TPP)$ lead to an illuminating description of the molecular stereochemistry. Parameters of the coordination group (Fig. 4) are $Fe-N_P = 2.001$ (3), $Fe-N_{NO} = 1.717$ (7), $Ct \cdots Fe = Fe \cdots P_N = 0.21$, $Ct \cdots N_{NO} = 1.928$ (6), $N_P \cdots N_{NO} = 2.770$, $O \cdots N_P = 3.30$ and 3.34 Å, and $N-O = 1.12$ (1) Å; the FeNO angle is 149.2 (6)$^{\circ}$ and the orientation angle, ϕ, is 40.6°. The most precisely determined axial parameter is $Ct \cdots N_{NO}$, but the rather high resolution of the data[78] provides insurance against substantial underestimation of $Ct \cdots Fe$ and concomitant overestimation of $Fe-N_{NO}$. The fairly large out-of-plane displacement (0.21 Å) of the low-spin iron atom is the joint consequence of the short $Fe-N_{NO}$ bond and tight (2.770 Å) $N_P \cdots N_{NO}$ packing distance. Several points regarding the observed values of the FeNO and orientation (ϕ) angles are to be emphasized. (1) A linear Fe—N—O linkage would be *fully* compatible with the observed packing of the molecules in the crystal. (2) All intermolecular and intramolecular contacts in the crystal are compatible with the conclusion that the FeNO angle is allowed to take very nearly its preferred value. (3) With the observed FeNO angle of 149.2°, but with $\phi = 0$ (instead of 40.6°), the single $O \cdots N_P$ contact within the molecule is calculated to be 3.22 Å, larger still by 0.12 Å than the sum of the Van der Waals packing radii for oxygen and aromatic nitrogen atoms. It is quite probable, consequently, that the axial bonding in the externally unconstrained molecule is better served with $\phi = 45^{\circ}$ than with $\phi = 0$ (Fig. 9). Two slightly short intermolecular contacts involving the oxygen atom are apparently responsible for the observed deviation (4.4°) of ϕ from the ideal 45°.

Although formally isostructural with $(ON)Fe(TPP)$, two special features of the otherwise comparable structure of crystalline $(ON)Co(TPP)$[77] seriously limited the precision with which some of the stereochemical parameters of the molecule could be determined. Unambiguously determined parameters include the orientation angle of the Co—N—O plane, $\phi = 39.7^{\circ}$, and the following distances: $Ct \cdots N_P = 1.976$ (3), $Co-N_P = 1.978$ (4), $Ct \cdots N_{NO} = 1.927$ (10), $N_P \cdots N_{NO} = 2.760$ (10), and $O \cdots N_P = 3.04$ and 3.09 Å. Because the separation of the 'half-atoms' of cobalt in the disordered structure is

References, p. 376

much too small to be resolved (see p. 344), the objectively derived values of the dependent parameters, $Co \cdots P_N$ = 0.094 (52) and $Co-N_{NO}$ = 1.833 (53) Å, carry very large estimated deviations; they do, however, fit rationally into the stereochemical pattern observed for other nitrosylcobalt(II) complexes[77]. Positioning of the nitrosyl nitrogen atom as two 'half-atoms' on the statistically required 4-fold axis leads to an unexpectedly large CoNO angle of 135° and an impossibly short N—O bond length of 1.01 Å. Noting then that the rather tight $O \cdots N_P$ contacts (3.04 and 3.09 Å) bespeak a small displacement of the nitrogen atom from the normal to P_N in the individual molecule, the statistical model should allow for 8-fold disorder in the positioning of this atom. Although this displacement is much too small to be resolved by X-ray analysis, its existence is supported by the otherwise unduly large value of the apparent thermal parameter of the nitrogen atom and by its documented occurrence in other nitrosylcobalt(II) complexes[77]. To demonstrate that the matter is nontrivial, it may be exphasized that a displacement of the nitrogen atom off-axis by >0.10 Å corresponds to a tipping of the $Co-N_{NO}$ bond from the normal by >3.1°, to a CoNO angle <128.5° (as compared with 135.2°), and to a C—O bond length >1.10 Å (as compared with 1.01 Å). The same phenomenon may be expected to be present in the similar geometry displayed by the isoelectronic Fe—O—O bond system in the dioxygen-1-methylimidazoleiron(II) derivative of the 'picket fence' porphyrin (vide infra).

In summary, the nitrosylmetal(II) metalloporphyrins of determined structures display two remarkable trends in their stereochemical parameters that are correlated with the number of electrons which must be accommodated in the axial M—N—O bond systems. Both the observed ordering in the $M-N_{NO}$ bond lengths, $Co-N_{NO} > Fe-N_{NO} > Mn-N_{NO}$, and the accompanying stepwise transition from a strongly bent to a linear M—N—O geometry are suggestive of the increasing importance of π interactions[78,124]. Of many theoretical discussions of nitrosyl complexes, that of Hoffmann et al.[125] appears to be the most illuminating.

8.4.5.4. The iron(II) derivatives of the 'picket fence' porphyrin

Independent determinations of structure for the monoclinic crystalline solvates of the dioxygen-1-methylimidazoleiron(II) and the carbonyl-1-methylimidazoleiron(II) derivatives of meso-tetra($\alpha,\alpha,\alpha,\alpha$-o-pivalamidophenyl)porphyrin, $(O_2)Fe(1-MeIm)(TpivalPP)$[115] and $(OC)Fe(1-MeIm)(TpivalPP)$[116], respectively, lead to the conclusion that, apart from the configurations of the Fe—C—O and Fe—O—O bond systems, the two complexes share a common stereochemistry; see Fig. 8. For reasons too numerous to detail here, the useful X-ray data are much too limited in respect to resolution, number, and mean intensity to permit more than a semiquantitative evaluation of the structural parameters of the molecules. In the structure as based on the centrosymmetric space group, C_2/c, the required molecular symmetry of C_2 (a 2-fold axis) can only be satisfied statistically by either of the iron(II) complexes; two orientations of the 1-MeIm molecule that differ by a rotation of 180° around the 2-fold axis must contribute equally to the representation of this axial ligand. Referring again to the diagram in Fig. 9, the orientation of the composite 1-MeIm ligand is specified by an angle, ϕ, of ~24°. The Fe—C—O linkage in (OC)Fe(1-MeIm)(TpivalPP) appears to be strictly linear, an expected, but nevertheless important, result because FeCO angles of ~135° and 145 ± 15° in carbonylhemoglobins have been reported[126,127]. In (O_2)Fe(1-MeIm)(TpivalPP), by contrast, the strongly bent Fe—O—O linkages occur in four orientations that approximate closely to C_{4v} geometry while observing in pairs the exact symmetry of C_2; the values of the orientation parameter, ϕ in Fig. 9, are 45° and 225° for one pair of ligands, 135° and 315° for the other. Thus the orientational disorder in the Fe—O—O linkages is quite like that described above for one-half of the isoelectronic Co—N—O linkages in (ON)Co(TPP)[77].

Although subject to possible further refinement[115], only the published values of the stereochemical parameters in the coordination group of the dioxygen complex may be cited at this time. These are: Fe—N_{Im} = 2.07 (2), Fe—O = 1.75 (2), O—O = 1.23 (8) and 1.26 (8) Å, with FeOO angles of 135 (4) and 137 (4)°. With the iron centered in the P_N plane, these data give calculated contacts between the terminal oxygen and porphinato nitrogen atoms of 3.02 and 3.05 Å, slightly smaller than those in the (ON)Co(TPP) molecule. It is probable, consequently, that the complexed oxygen atom lies somewhat off the normal to the P_N plane, as does the complexed nitrogen atom in (ON)Co(TPP) and other nitrosylcobalt(II) complexes, and that the FeOO angle is less than 130°. It is probable also that the O—O distance lies in the range between the superoxide and peroxide values of about 1.28 and 1.48 Å, respectively; it is surely larger than the 1.21 Å in molecular oxygen.

A partial recasting in molecular orbital terms of Pauling's theoretical bonding pattern[128] for an Fe—O—O linkage having the foregoing geometry shows that it corresponds formally to a double bond ($\sigma + \pi$) between an Fe(III) and a trigonally hybridized, essentially neutral, oxygen atom and a single bond joining the latter to the negatively charged, terminal, oxygen atom[77]; the π bond involves charge transfer from the d_{xz} and d_{yz} orbitals of the iron atom to the stable π orbital of the oxygen atom. The perhaps more usual iron(III)-superoxide formulation has the resultant charge of the ligand distributed more or less equally over the two oxygen atoms; the π^* orbital of the ligand accepts this charge (~1 electronic unit) from the d_{xz} and d_{yz} orbitals of the iron atom. Both formulations attribute a pronounced ferric character to the iron atom as is seemingly required by other experimental observations[129], and both require double-bond character of the Fe—O linkage; they differ materially in respect to the nature and the length of the O—O bond. Neither of these approximate bonding patterns is excluded by the semiquantitative data cited above.

References, p. 376

The possibility that both atoms of the dioxygen ligand are symmetrically coordinated to the iron atom in the protoheme has received a good deal of attention. The extraordinary steric disability of the seven-coordinate geometry which would then be required was first emphasized in 1968[29], and was further documented in a later paper[77]. With the structural results for the $(O_2)Fe(1-MeIm)(TpivalPP)$ molecule at hand, it becomes most unlikely that symmetrical coordination of the oxygen molecule will be found in any hemoprotein.

Presumably because they are based upon a substantially larger number of usable amplitude data (though still confined to the chromium K_α sphere), the bond parameters of each chemical type in the porphinato core of the $(OC)Fe(1-MeIm)(TpivalPP)$ molecule display less scatter and better approximations to the expected averaged values than do those in the dioxygen complex; the two independent $Fe-N_P$ distances, 1.97 (2) and 2.03 (2), average to 2.00 Å. The axial bond parameters are subject to further refinement using improved models for the disordered 1-MeIm ligand and, especially, for the troublesome toluene solvate. Final parameters are now expected to fall within the ranges, 2.04—2.08 Å for $Fe-N_{Im}$, 1.59—1.63 Å for $Fe-C$, 1.24—1.30 Å for $C-O$, and 2.87—2.89 Å for $O\cdots Fe$. Relative to precisely determined parameters for $Fe-C-O$ linkages in other complexes, the overall $O\cdots Fe$ distance is short by 0.03—0.06 Å, $Fe-C$ is short by 0.13—18 Å, and $C-O$ is long by 0.08—0.14 Å; the accurate determination of the position of the comparatively light carbon atom in this multiple bond system demands X-ray data of much higher resolution than those which are now available.

8.4.6. Stereochemistry of the protoheme in hemoglobin

Each of the four possible combinations of ferrous or ferric iron in a high- or a low-spin ground state is realized in one or more of the hemes as these occur in the several families of the hemoproteins. A brief discussion of the stereochemistry of the protoheme or ferroheme — the iron(II) derivative of protoporphyrin-IX — as this porphinatoiron complex exists in the oxygen carrier, hemoglobin[130,131], is in order. The hemoglobin (Hb) molecule consists of four subunits, two α chains and two β chains; each subunit carries a protoheme in a hydrophobic pocket in the globin with, however, the carboxylic acid groups of the 6,7-substituents on the porphinato core exposed to the surrounding medium. The protoheme is directly attached to the globin framework through an axial complexing bond from the iron atom to an imidazole nitrogen atom of the proximal histidine residue. Molecular oxygen occupies the sixth position in the coordination group of the iron atom in the low-spin oxyhemoglobin (oxy-Hb) molecule; there is, of course, no sixth ligand in the high-spin deoxyhemoglobin (deoxy-Hb) species. The equilibrium uptake of oxygen by deoxy-Hb is described approximately by the Hill equation[132]

$$y/(1-y) = KP^n \qquad (2)$$

wherein y is the fractional saturation, K is a constant, P is the partial pressure of oxygen, and n is an empirical measure of the degree to which the uptake is modulated by the cooperative interactions of the four subunits in the molecule — the classic 'heme—heme' interaction[132]. For uptake of oxygen by a monomeric species such as myoglobin, $n = 1$, but for normal human or horse hemoglobin, n is 2.7 to 2.9. Reversible oxygenation gives rise also to the Bohr effect[132] and to distinctively different quaternary structures for the globin frameworks in deoxy-Hb and oxy-Hb[130]. Direct heme—heme interactions are an implausible source of the cooperative phenomena because the minimum spacing of any pairing of iron atoms in either deoxy-Hb or oxy-Hb is >25 Å[130]. The decrease in the Gibbs energy attributable solely to cooperative interactions during oxygenation is predominantly or wholly of entropic origin and, consequently, is rationally identified with an entropy-dominated conformational change[132]. It was then noted[28] that a movement >0.5 Å of the iron atom *relative* to the porphinato core, a movement concomitant with oxygenation of the heme and the accompanying change in spin state[24], could be expected to require cooperative movements (translations and rotations of groups) in the globin framework that would provide the starting point for a stereochemical mechanism of cooperative interaction.

In Perutz's subsequently developed mechanism[130,131] for the reversible oxygenation of hemoglobin, the primary trigger for initiating cooperative interaction of the subunits is the shrinkage of perhaps as much as 0.9 Å[123] in the distance ($N_\epsilon \cdots P_\mu$) separating the coordinated nitrogen atom (N_ϵ) of the histidine residue from the mean plane (P_μ) of the protoheme that accompanies the transition in spin state. Because the determination of the position of the P_μ plane involves the viewing edge-on of the protoheme as a rather thick band of unresolved electron density to which the peripheral substituents contribute in part, P_μ is to be distinguished from the mean plane (P_C) of the porphinato core and, of course, from the mean plane (P_N) of the porphinato nitrogen atoms. The $Fe \cdots P_\mu$ distances in the nonequivalent α and β subunits, both estimated to be ~ 0.75 Å, are the most accurately determined parameters in the deoxy-Hb molecule. Comparison of this datum with the $Fe \cdots P_C$ distance of 0.47 Å in the model[123] for an externally unconstrained five-coordinate imidazoleporphinatoiron(II) molecule provides direct support for Perutz's conclusion derived from spectral studies[133] that the axial connection in deoxy-Hb is in tension. The $Fe \cdots P_C$ distance, 0.47 Å, in the reference molecule is the sum of the $Fe \cdots P_N$ displacement, 0.42 Å, observed in (2-MeIm)Fe(TPP) $\cdot$ C_2H_5OH and an allowance of 0.05 Å for the difference between $Fe \cdots P_C$ and $Fe \cdots P_N$ as fully representative of the unconstrained species. Thus the averaged value of this doming parameter is 0.044 Å in the iron(III) complexes, (Cl)Fe(Proto-IX)[23], (MeO)Fe(Meso-IX-DME)[24], and O[Fe(TPP)]$_2$[35]; the unrepresentative values of ostensibly zero in (Cl)Fe(TPP)[5] and of 0.13 Å in (2-MeIm)Fe(TPP) $\cdot$ C_2H_5OH[36] are clearly dictated by packing constraints in the crystals. Most illuminating is the high-spin (1-MeIm)Mn(TPP) species in which Mn $\cdots$ P_N is 0.515 Å and the doming parameter is still only 0.04 Å[113]. Unless the Fe $\cdots$ P_μ distance in deoxy-Hb is overestimated by at least 0.25 Å, it appears that the axial connection must be in tension.

The axial Fe—N_ϵ bond provides the mechanism whereby the reaction of the globin to tension is assured at one terminus of the axial system. At the other terminus, the required support must come from packing contacts between the globin and, for the most part, peripheral substituents on the core of the protoheme. This restraint must be exerted on the side of the protoheme from which the M—N_ϵ bond emerges; the presence on this side of the histidine ligand together with the generally protective distribution of the substituents must limit, perhaps preclude, the close approach of the *restraining* groups in the globin to any part of the porphinato core. In this model, consequently, a doming of the entire porphinato core toward the metal atom is permitted[123].

Of the estimated Fe $\cdots$ P_μ distance of 0.75 Å in deoxy-Hb, perhaps as much as 0.55 Å may be attributable to the Fe $\cdots$ P_N displacement, leaving only ~0.20 Å to be assigned to doming of the core. The Fe—N_P bond length of 2.086 Å in the (2-MeIm)Fe(TPP) molecule corresponds, of course, to the observed Fe $\cdots$ P_N displacement of 0.42 Å and to $Ct \cdots N_P$ = 2.044 Å, but equally well to Fe $\cdots$ P_N = 0.55 Å and $Ct \cdots N_P$ = 2.010 Å. Since equatorial strain in a planar or *quasi*-planar core is minimized for a $Ct \cdots N_P$ radius of ~2.01 Å, it follows that the difference in energy of the two configurations is certainly small. That a *quasi*-C_{4v} doming of the core of ~0.20 Å is quite readily attainable is perhaps best exemplified in the conformations observed in the $(OAc)_2Hf(OEP)$ and $(OAc)_2Zr(OEP)$ molecules as these occur as three structurally nonequivalent species within monoclinic and orthorhombic crystals[7]. Only a very small tension in the axial connection within deoxy-Hb is required by a doming of this magnitude[123].

Little and Ibers[121] strongly question — indeed, seemingly reject — the tension model for the axial connection in deoxy-Hb on the joint grounds (1) that the Fe $\cdots$ P_μ distance of 0.75 Å is very likely overestimated by 0.20 Å and (2) that observations using other physical methods that have been interpreted[133] as supporting the tension model are probably better ascribed to other sources. They present, on the other hand, a model for the axial connection that, by implication, is a model without tension, in which the Fe $\cdots$ P_C distance of 0.55 Å observed in the disordered (2-MeIm)Fe(TPP) $\cdot$ C_2H_5OH structure is combined with an Fe—N_ϵ distance of 2.27 Å from a hexapyridineiron(II) ion to give an overall $N_\epsilon \cdots P_C$ distance of 2.82 Å just 0.08 Å short of the 2.90 Å estimated for the $N_\epsilon \cdots P_\mu$ distance in deoxy-Hb[130,131]. No reason for rejecting the Fe—N_ϵ distance of 2.161 (5) Å observed in the (2-MeIm)Fe(TPP) complex in favor of the 2.27 (3) Å observed in a complex of quite different geometrical and electronic structure is given, but the rejection certainly is injudicious: the Mn—N_ϵ (=Mn—N_{Im}) distance of 2.192

(2) Å in the ordered structure taken by the high-spin (1-MeIm)Mn(TPP) molecule[113] is fully supportive of an Fe—N_ϵ distance of 2.16 ± 0.01 Å in an unstretched axial bond system. And in taking the Fe $\cdots$ P_C distance to be 0.55 Å, the evidence[123] (vide supra) that this displacement is larger by ~0.08 Å than is rationally expected in an externally unconstrained (2-MeIm)Fe(TPP) molecule or other imidazoleporphinatoiron(II) species is quite ignored. The stretch overall in their model for the axial connection is then ~0.19 Å. A sufficient condition, the writer judges, for the existence of some stretch in the overall axial connection is that Fe $\cdots$ P_N be > 0.45 Å; and even though the interpretation of *nmr* and other spectral data may not be unambiguous, it is clear that such data cannot be used to establish a purely negative conclusion — the *absence* of tension in the connection.

8.4.7. The axial connection in deoxycobaltohemoglobin

Hoffman et al.[134−136] have shown that cobaltohemoglobin, CoHb, the species obtained through replacement of the protohemes in hemoglobin by molecules of the low-spin ($S = \frac{1}{2}$) cobalt(II) derivative of protoporphyrin-IX, Co(Proto-IX), displays a reversible *and cooperative* uptake of oxygen that is qualitatively similar to the uptake of oxygen by hemoglobin. Thus the low-spin d^7 cobalt(II) atom serves as a qualitatively acceptable substitute for the d^6 iron atom both in the high-spin five-coordinate hemes of deoxy-Hb and in the low-spin oxygenated six-coordinate hemes of oxy-Hb. In each of the unoxygenated metalloporphyrins, the unpaired electron in the $3d_{z^2}$ orbital of the metal atom stabilizes five-coordinate geometry. The studies of Hoffman et al. suggest that the geometry of the $(O_2)Co(Hist)(Proto-IX)$ complex is nearly identical with that of the oxygenated protoheme by reason of the transfer from the cobalt atom of most of the charge associated with the odd electron to the antibonding $2p\pi^*$ orbital of the dioxygen ligand; the interactions of an effectively d^6 cobalt(III) atom with the N_ϵ and the four N_P atoms can then simulate the interactions of the low-spin iron atom with these ligands. It is presumed, though not as yet confirmed by X-ray structural analysis, that deoxy-CoHb and deoxy-Hb share essentially the same, relatively compact, quaternary structure designated as type T[130], whereas oxy-CoHb and oxy-Hb share the more open quaternary structure of type R which is considered by Perutz to be adequately represented by the most intensively investigated structure of methemoglobin, met-Hb[130,131].

The primary problem which arises in adapting the tension model for the axial connection in deoxy-Hb to the connection in deoxy-CoHb stems from the fact that the unpaired electron in the $3d_{x^2-y^2}$ orbital of the high-spin iron atom guarantees an Fe $\cdots$ P_N displacement $\geqslant 0.42$ Å whereas the absence of such an electron in the low-spin cobalt atom allows the Co $\cdots$ P_N displacement to be only 0.13 Å in the (1-MeIm)Co(TPP) molecule[41]. More stretch and enhanced tension in the axial connection and, to be sure, greater strain in the supporting globin are required in deoxy-CoHb than in deoxy-

Hb. The last condition requires the overall $N_\epsilon \cdots P_\mu$ distance in deoxy-CoHb to be significantly shorter than in deoxy-Hb. Two limiting cases are conveniently distinguished. (1) Taking $Fe \cdots P_\mu = 0.55$ Å as the lower limit in deoxy-Hb, the required inequality in deoxy-CoHb becomes $N_\epsilon \cdots P_\mu <$ 2.70 Å. (2) Taking Perutz's estimate of 0.75 Å for $Fe \cdots P_\mu$ as the upper limit (but noting that it implies stretch in the $Fe-N_\epsilon$ bond) requires $N_\epsilon \cdots P_\mu <$ 2.95 Å in deoxy-CoHb[123]. Some attempt to set rational limits on the stretch in the $N_\epsilon \cdots P_N$ component of $N_\epsilon \cdots P_\mu$ and on the doming of the core implied thereby may be in order.

It is clear from the earlier analysis (p. 363) that the $Co-N_\epsilon$ (or $Co-N_{Im}$) bond length of 2.216 Å and the $Co \cdots P_N$ displacement of 0.15 Å observed in the $(1,2-Me_2Im)Co(TPP)$ structure[120] correspond to a very small tension in the axial connection. Combining these parameters with an allowance of 0.25 Å for the doming of the porphinato core gives a model for case (1) above in which $N_\epsilon \cdots P_\mu$ is ostensibly 2.62 Å; how nearly this model (or any other) meets the requirement of uniform tension throughout the axial connection is, at best, a matter for qualitative judgment.

The axial $Co-N$ bond lengths of 2.436 (2) and 2.386 (2) Å in the respective six-coordinate $(Pip)_2Co(TPP)$[39] and $(3-Pic)_2Co(OEP)$[40] species correspond to weak bonds. But as a stretched bond in a five-coordinate species, an axial $Co-N$ length of 2.42 Å corresponds to a very much stronger linkage because most of the charge associated with the odd electron in the $3d_{z^2}$ orbital of the cobalt atom is then concentrated as a 'phantom ligand' in the *trans* axial position (p. 352). A $Co-P_\mu$ displacement of perhaps as much as 0.20 Å would be expected to accompany such a highly stretched $Co-N_\epsilon$ bond to give $N_\epsilon \cdots P_N > 2.60$ Å. In deoxy-CoHb, however, a smaller $N_\epsilon \cdots P_N$ of 2.45 to 2.50 Å along with a net doming of the porphinato core of 0.30–0.35 Å would seem to fit the roughly specified requirements of case (2) above.

It is evident that neither the overall $N_\epsilon \cdots P_\mu$ separation nor, of course, the division of this quantity between the $N_\epsilon \cdots P_N$ and the doming components in the tension model for deoxy-CoHb can be satisfactorily delimited on the basis of the experimental data that are currently available. It is evident also that the magnitude and the importance of these quantitative ambiguities in the model are minimized for case (1) above wherein $N_\epsilon \cdots P_\mu < 2.70$ Å — the consequence of taking $Fe-P_\mu$ in deoxy-Hb to be only 0.55 Å as suggested by Little and Ibers[121]. Their proposed model for the axial connection in deoxy-CoHb has $N_\epsilon \cdots P_\mu = 2.31$ Å, the observed value of $N_{Im} \cdots P_C$ in crystalline $(1-MeIm)Co(OEP)$[121]. It is by design a tensionless model which serves to accentuate the conceivable differences, rather than the probable similarities, between the axial connections in deoxy-Hb and deoxy-CoHb. For the detailed inferences drawn from this unconstrained model, the original papers should be consulted[121,137].

8.4.8. Porphinato complexes of unconventional coordination geometry

In the centrosymmetric $[(OC)_2 Rh]_2 (OEP)$ molecule[138], one rhodium atom lies 1.32 Å above, the other an equal distance below, the plane (P_N) of the porphinato nitrogen atoms; the Rh $\cdot\cdot$ Rh separation is 3.094 Å. By forming bonds with just two, contiguously situated, nitrogen atoms and with two molecules of carbon monoxide, each rhodium(I) atom achieves planar four-coordination. The porphinato ligand displays bidentate behavior toward each of the rhodium atoms. Averaged bond lengths are Rh—N = 2.084 (7) and Rh—C = 1.85 Å. The Rh $\cdot\cdot$ N distance separating unbonded atoms is 3.07 Å.

In the centrosymmetric $[(OC)_3 Re]_2 (TPP)$ molecule[139], one rhenium atom lies 1.42 Å above, the other 1.42 Å below, the P_N plane; the Re $\cdot\cdot$ Re distance is 3.127 Å. One pair of diagonally situated nitrogen atoms are coordinated to both metal atoms at an averaged distance of 2.39 Å, whereas each of the other two nitrogen atoms is coordinated to just one rhenium atom at the shorter distance, 2.16 Å; the remaining Re $\cdot\cdot$ N distance between unbonded atoms is ~3.2 Å. Three carbon monoxide ligands complete the octahedral coordination group around each rhenium atom. Apart from insignificant alterations in dimensions, the foregoing description applies equally to the analogous technetium species[140].

Acknowledgments

The substantial contributions of the writer's coworkers, past and present, to studies of metalloporphyrin stereochemistry carried out at Cornell University are gratefully acknowledged. These studies have received financial support from the National Institutes of Health, the National Science Foundation, and the Materials Science Center at Cornell University. Special thanks are due Dr. Alice Ann Sayler for assistance rendered during the preparation of this chapter.

Appendix I

The atomic form factor of the carbon atom, f_C, which is plotted as a function of $(\sin \phi)/\lambda$ in Fig. 10, provides an example of the *spherically symmetric approximations* to the scattering factors for X-rays of the various atoms in a structure that are employed in standard least-squares refinements of the structural parameters. Since no atom in a metalloporphyrin carries the spherically symmetric distribution of electron density which is concomitant with a spherically symmetric form factor, it follows that the position assigned to an atom by the usual least-squares refinement represents an effective centroid of electron density that may not approximate closely to the position of the atomic nucleus (excepting, of course, a position for the metal atom that may be dictated by symmetry). From inspection of the graph of

References, p. 376

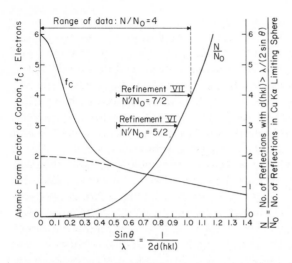

Fig. 10. Graphs (1) of the spherically symmetric form factor of the carbon atom as a function of $(\sin\theta)/\lambda$ and (2) of the dependence, for a given crystal, of the number of theoretically recordable reflections on $(\sin\theta)/\lambda$.

f_C vs $(\sin\theta)/\lambda$, it is seen that coherent scattering of X-radiation by the valence-shell electrons is of consequence only for $(\sin\theta)/\lambda$ less than ~ 0.5 Å^{-1}; for all larger values of $(\sin\theta)/\lambda$, the coherent scattering of X-rays by the carbon atom is almost solely attributable to the pair of electrons in the closed K shell. It is also evident from the graph that the electron density associated with this pair of electrons must be largely concentrated within a very small volume surrounding the nucleus of the atom. It follows that a structural model in which the atoms are assumed to be spherically symmetric scatterers of X-rays becomes a physically realistic model for the accurate determination of nuclear positions for the carbon and all heavier atoms by least-squares procedures provided that the low-resolution data which are dominated by the valence-shell scattering be excluded from the final refinement of the structural parameters.

Exclusion of the low-resolution data is practicable, however, only if higher-resolution data which meet minimum specifications in respect to number and quality are available for the refinement. The coherent scattering factor for any atom in the crystal takes the form $f_i \exp[-B_{hkl}(\sin\theta)^2/\lambda^2]$, wherein f_i is the form factor for the atom at rest (as in Fig. 10) and the thermal parameter, B_{hkl}, is directly proportional to the mean-squared amplitude of vibration of the atom normal to the Bragg plane, (hkl). Higher-resolution data which observe the minimum specifications are obtainable only if these thermally excited vibrations are kept within bounds which allow the coherent scattering of the X-radiation by the K-shell electrons in the carbon (and nitrogen) atoms to hold up strongly for $(\sin\phi)/\lambda$ *at least as large* as the maximum value (0.648 Å^{-1}) theoretically attainable with copper K_α radiation. For many, perhaps most, porphyrins and metalloporphyrins, fully adequate higher-resolution data are obtainable only at low temperature. The outstanding exception to this general rule is provided by the data recorded at room temperature from a tetragonal crystal of $(Cl)_2 Sn(TPP)$[26].

The graph of N/N_O vs $(\sin \theta)/\lambda$ in Fig. 10 shows that, for any given crystal, the number of theoretically recordable reflections within a limiting sphere of radius, $[(\sin \theta)/\lambda]_{max}$, is directly proportional to the third power of this radius or, otherwise put, to the volume of the sphere. Since N_O is taken to be the number of reflections in the Cu K_α limiting sphere of radius, $1/\lambda = 1/1.54 = 0.648 \text{ Å}^{-1}$, the number of reflections, N, in the chromium limiting sphere of radius, $1/2.29 = 0.436 \text{ Å}^{-1}$, is only $0.305 N_O$; these reflections, moreover, are seen to be strongly dependent upon the scattering from the valence-shell electrons. Given a data set which includes a dominant proportion of the higher-resolution reflections, it is quite straightforward to exclude, as a minimum, the chromium-sphere data from the final least-squares refinement. Since the contributions to the observed intensity data from the hydrogen atoms and from concentrations of electron density that are associated with multiple bond character (or any other nonspherically symmetric source) are thereby eliminated, the purely mathematical minimization of the function employed in least-squares refinement is constrained by a realistic structural model having the minimum number of variable parameters. Exclusion of the low-resolution data may proceed in steps of increasing N/N_O until, in particular, there is no significant increase in the length of any bond having multiple bond character.

In a metalloporphyrin, it is, of course, the apparent length of the *quasi*-double bond connecting a pair of C_b atoms that is most sensitive to the refinement procedure. This fact is clearly demonstrated by the structure determination for $(Cl)_2 Sn(TPP)$[26] and it receives consistent support of a less definitive semiquantitative nature from the structure analyses for $(Py)Zn$-$(TPyP)$[42], $Ag(TPP)$[52], $(OAc)_2 Hf(OEP)$[7], and $Ni(TPP)$[60].

Some 3909 of the 4450 independent reflections included within the limiting sphere for which $(\sin \theta)/\lambda = 1.03 \text{ Å}^{-1}$ and $N/N_O = 4$ (Fig. 10) were recorded as observed and used in the analysis of structure for $(Cl)_2 Sn(TPP)$. Although appreciable contributions to the intensities from the carbon atoms were presumably confined to the smaller sphere for which $N/N_O = 2$, the extension of the measurements to include virtually all observable reflections served two highly desirable ends. (1) It led to a well-defined Fourier synthesis from which accurate positions for the carbon and nitrogen atoms could be determined. (2) It led also to unusually well-defined thermal parameters for the strongly scattering tin and chlorine atoms and, consequently, to nearly invariant contributions from these atoms to the seven variants of least-squares refinement that were carried to convergence. Data from the inner half of the Cu K_α sphere ($N/N_O = \frac{1}{2}$) were excluded from refinements

References, p. 376

VI and VII as indicated in Fig. 10. All of the data were included in the other five refinements. Hydrogen contributions were ignored in refinement V, but were included in all of the others (pro forma in VI and VII). Refinements I to IV differed from one another in the use of two weighting schemes and, additionally, in II and IV by the neglect of the small absorption corrections to the data.

The eight distinctive procedures for evaluating atomic positions in the porphinato core led to averaged values for six of the seven structurally nonequivalent C—C and C—N bond lengths — the C_b—C_b distance excepted — that are most aptly characterized by the following statement: the maximum deviation from the averaged length exceeded 0.002 Å in just one instance, a C_a—C_b distance overlong by 0.004 Å from refinement V. The C_b—C_b bond lengths given by refinements I to IV ranged from 1.366 to 1.372 Å; refinements I and III, based on data corrected for absorption, gave values of 1.366 (3) and 1.369 (3) Å, respectively. The C_b—C_b distances given by the Fourier synthesis and by refinements V, VI, and VII ranged from 1.379 (3) to 1.381 (3) Å, with a mean deviation from 1.380 Å of 0.0005 Å. It is clear that the electron contributed to each C_b—H bond by a hydrogen atom, but ignored in the model used in refinement V, just compensates for half of the π bonding density in the C_b—C_b bond while slightly overcompensating for half of the smaller π density in the C_a—C_b bond.

Refinements of structure (by Dr. A.A. Sayler) for $(Cl)_2 Sn(TPP)$ and $Ni(TPP)^{60}$ in which only the data from within the chromium sphere were used yield C_b—C_b distances which are underestimated by 0.024 and 0.019 Å, respectively. Less readily interpretable variations of 0.010—0.015 Å in several other bond lengths in the porphinato cores also are observed.

References

1. M.B. Crute, Acta Crystallogr., 12, 24 (1959).
2. E.F. Meyer, Jr., Acta Crystallogr., B28, 2162 (1972).
3. D.L. Cullen and E.F. Meyer, Jr., J. Amer. Chem. Soc., 96, 2095 (1974).
4. J.P. Collman, J.L. Hoard, G. Lang, L.J. Radonovich and C.A. Reed, J. Amer. Chem. Soc., 97, 2676 (1975).
5. J.L. Hoard, G.H. Cohen and M.D. Glick, J. Amer. Chem. Soc., 89, 1992 (1967).
6. R. Countryman, D.M. Collins and J.L. Hoard, J. Amer. Chem. Soc., 91, 5166 (1969); D.M. Collins, R. Countryman and J.L. Hoard, ibid., 94, 2066 (1972).
7. N. Kim. J.L. Hoard, J.W. Buchler and K. Rohbock, J. Amer. Chem. Soc., to be submitted.
8. E.B. Fleischer, J. Amer. Chem. Soc., 85, 146 (1963).
9. L.J. Radonovich, W.S. Caughey and J.L. Hoard, unpublished determination of structure.
10. J.M. Roberston, J. Chem. Soc., 1195 (1936).
11. J.M. Robertson and I. Woodward, J. Chem. Soc., 219 (1937).
12. J.M. Robertson and I. Woodward, J. Chem. Soc., 36 (1940).
13. L. Pauling and C.D. Coryell, Proc. Natl. Acad. Sci. USA, 22, 159 (1936); ibid., 22, 210 (1936).

14. C.D. Coryell, F. Stitt and L. Pauling, J. Amer. Chem. Soc., **59**, 633 (1937).
15. C.D. Russell and L. Pauling, Proc. Natl. Acad. Sci. USA, **25**, 517 (1939).
16. F. Stitt and C.D. Coryell, J. Amer. Chem. Soc., **61**, 1263 (1939).
17. T.A. Hamor, W.S. Caughey and J.L. Hoard, J. Amer. Chem. Soc., **87**, 2305 (1965).
18. J.L. Hoard, M.J. Hamor and T.A. Hamor, J. Amer. Chem. Soc., **85**, 2334 (1963); M.J. Hamor, T.A. Hamor, and J.L. Hoard, ibid., **86**, 1938 (1964).
19. S.J. Silvers and A. Tulinsky, J. Amer. Chem. Soc., **86**, 927 (1964); ibid., **89**, 3331 (1967).
20. E.B. Fleischer and L.E. Webb, J. Amer. Chem. Soc., **87**, 667 (1965); L.E. Webb and E.B. Fleischer, J. Chem. Phys., **43**, 3100 (1965). See also ref. 21.
21. B.M.L. Chen and A. Tulinsky, J. Amer. Chem. Soc., **94**, 4144 (1972).
22. E.B. Fleischer, J. Amer. Chem. Soc., **85**, 1353 (1963); E.B. Fleischer, C.K. Miller, and L.E. Webb, ibid., **86**, 2342 (1964).
23. D.F. Koenig, Thesis, The Johns Hopkins University, Baltimore, Md., U.S.A. (1962); Acta Crystallogr., **18**, 663 (1965).
24. J.L. Hoard, M.J. Hamor, T.A. Hamor, and W.S. Caughey, J. Amer. Chem. Soc., **87**, 2312 (1965).
25. J.L. Hoard, Ann. New York Acad. Sci., **206**, 18 (1973).
26. D.M. Collins, W.R. Scheidt, and J.L. Hoard, J. Amer. Chem. Soc., **94**, 6689 (1972).
27. D.L. Cullen and E.F. Meyer, Jr., Acta Crystallogr., B29, 2507 (1973).
28. J.L. Hoard, in 'Hemes and Hemoproteins', B. Chance, R.W. Estabrook, and T. Yonetani, Eds., Academic Press, New York, 1966, pp. 9—24.
29. J.L. Hoard, in 'Structural Chemistry and Molecular Biology', A. Rich and N. Davidson, Eds., W.H. Freeman, San Francisco, 1968, pp. 573—594.
30. J.L. Hoard, Science, **174**, 1295 (1971).
31. I. Moustakali and A. Tulinsky, J. Amer. Chem. Soc., **95**, 6811 (1973).
32. A. Bloom and J.L. Hoard, to be published.
33. L.J. Radonovich, A. Bloom and J.L. Hoard, J. Amer. Chem. Soc., **94**, 2066 (1972).
34. A. Takenaka and Y. Sasada, Chem. Soc. Japan, Chem. Letters, 1235 (1972).
35. E.B. Fleischer and T.S. Srivastava, J. Amer. Chem. Soc., **91**, 2403 (1969); A.B. Hoffman, D.M. Collins, V.W. Day, E.B. Fleischer, T.S. Srivastava and J.L. Hoard, ibid., **94**, 3620 (1972).
36. D.A. Buckingham, J.P. Collman, J. L. Hoard, G. Lang, L.J. Radonovich, C.A. Reed and W.T. Robinson, to be published; see also Ref. 123.
37. J.A. Kaduk and W.R. Scheidt, Inorg. Chem., **13**, 1875 (1974).
38. W.R. Scheidt, J.A. Cunningham and J.L. Hoard, J. Amer. Chem. Soc., **95**, 8289 (1973).
39. W.R. Scheidt, J. Amer. Chem. Soc., **96**, 84 (1974).
40. R.G. Little and J.A. Ibers, J. Amer. Chem. Soc., **96**, 4440 (1974).
41. W.R. Scheidt, J. Amer. Chem. Soc., **96**, 90 (1974).
42. D.M. Collins and J.L. Hoard, J. Amer. Chem. Soc., **92**, 3761 (1970).
43. J.P. Collman and C.A. Reed, J. Amer. Chem. Soc., **95**, 2048 (1973).
44. R.G. Little and J.A. Ibers, J. Amer. Chem. Soc., **95**, 8583 (1973).
45. J.J. Bonnet, S.S. Eaton, G.R. Eaton, R.H. Holm and J.A. Ibers, J. Amer. Chem. Soc., **95**, 2141 (1973).
46. D.W.J. Bruickshank and R.A. Sparks, Proc. Roy. Soc. (London), A258, 270 (1960); D.W.J. Bruickshank, Tetrahedron, **17**, 155 (1962).
47. J.P. Collman, R.R. Gagne, T.R. Halbert, J.-C. Marchon and C.A. Reed, J. Amer. Chem. Soc., **95**, 7868 (1973).
48. J.W. Lauher and J.A. Ibers, J. Amer. Chem. Soc., **95**, 5148 (1973).
49. P.W. Codding and A. Tulinsky, J. Amer. Chem. Soc., **94**, 4151 (1972).
50. A. Tulinsky, Ann. New York Acad. Sci., **206**, 47 (1973).
51. R.G. Little and J.A. Ibers, J. Amer. Chem. Soc., **97**, in press (1975).

52. L.J. Radonovich, A.D. Adler and J.L. Hoard, J. Amer. Chem. Soc., to be submitted.
53. G. Donnay and C.B. Storm, Mol. Cryst., 2, 287 (1967).
54. M.L. Schneider and G. Donnay, Amer. Crystallogr. Assoc., 1971 Summer Meeting, Abstr. Kl1; M.L. Schneider, J. Chem. Soc., Dalton Trans., 1972, 1093.
55. See the appended Discussion to Ref. 50, pp. 67—69 therein.
56. E.D. Becker, R.B. Bradley and C.J. Watson, J. Amer. Chem. Soc., 83, 3743 (1961).
57. C.B. Storm, Y. Teklu and E.A. Sokolski, Ann. New York. Acad. Sci., 206, 631 (1973).
58. R.J. Abraham, G.E. Hawkes and K.M. Smith, Tetrahedron Letters, 1974, 1483.
59. P. Madura and W.R. Scheidt, to be published.
60. A.A. Sayler and J.L. Hoard, to be published.
61. A.D. Adler and W. Shergalis, Private Communication (1964) to the writer.
62. In Professor W.R. Scheidt's laboratory.
63. E.B. Fleischer, Accounts Chem. Res., 3, 105 (1970).
64. M.B. Hursthouse and S. Neidle, Chem. Comm., 1972, 449.
65. W. Hoppe, G. Will, J. Gassmann and H. Weichselgartner, Z. Kristallogr., 128, 18 (1969).
66. J. Gassmann, I. Strell, F. Brandl, M. Sturm and W. Hoppe, Tetrahedron Letters, 1971, 4609.
67. M.S. Fischer, D.H. Templeton, A. Zalkin and M. Calvin, J. Amer. Chem. Soc., 94, 3613 (1972).
68. C. Weiss, H. Kobayashi and M. Gouterman, J. Mol. Spectrosc., 16, 415 (1965).
69. G.L. Closs, J.J. Katz, F.C. Pennington, M.R. Thomas and H.H. Strain, J. Amer. Chem. Soc., 83, 3743 (1961).
70. E.B. Fleischer and A.L. Stone, J. Chem. Soc., Chem. Comm., 1967, 332; Allen Stone and E.B. Fleischer, J. Amer. Chem. Soc., 90, 2735 (1968).
71. E.B. Fleischer and M. White, personal communication.
72. E. Cetinkaya, A.W. Johnson, M.F. Lappert, G.M. McLaughlin and K.W. Muir, J. Chem. Soc., Dalton Trans., 1974, 1236.
73. N. Hirayama, A. Takenaka, Y. Sasada, E.-I. Watanabe, H. Ogoshi and Z.-I. Yoshida, J. Chem. Soc., Chem. Comm., 1974, 330.
74. G.M. McLaughlin, J. Chem. Soc., Perkin II, 1974, 136.
75. M.D. Glick, G.H. Cohen and J.L. Hoard, J. Amer. Chem. Soc., 89, 1996 (1967).
76. R. Timkovich and A. Tulinsky, J. Amer. Chem. Soc., 91, 4430 (1969).
77. W.R. Scheidt and J.L. Hoard, J. Amer. Chem. Soc., 95, 8281 (1973).
78. W.R. Scheidt and M.E. Frisse, J. Amer. Chem. Soc., 97, 17 (1975).
79. R.W. James, 'The Optical Principles of the Diffraction of X-Rays', Cornell University Press, Ithaca, New York, 1965, p. 400.
80. C.H. Kirksey, P. Hambright and C.B. Storm, Inorg. Chem., 8, 2141 (1969).
81. J.R. Miller and G.R. Dorough, J. Amer. Chem. Soc., 74, 3927 (1952).
82. D.L. Cullen and E.F. Meyer, Jr., Amer. Crystallogr. Assoc., Summer Meeting, Aug., 1971, Abstract K15; private communication.
83. L.D. Spaulding, P.G. Eller, J.A. Bertrand and R.H. Felton, J. Amer. Chem. Soc., 96, 982 (1974).
84. H. Montgomery and E.C. Lingafelter, Acta Crystallogr., 16, 748 (1963).
85. C.E. Strouse, Proc. Natl. Acad. Sci. U.S.A., 71, 325 (1974); H.-C. Chow, R. Serlin and C.E. Strouse, personal communication (1975).
86. R. Serlin, H.-C. Chow and C.E. Strouse, personal communication (1975).
87. C.B. Storm, A.H. Corwin, R.A. Arellano, M. Martz and R. Weintraub, J. Amer. Chem. Soc., 88, 2525 (1966).
88. D.M. Collins, R. Countryman and J.L. Hoard, unpublished data.
89. R.C. Pettersen, Acta Crystallogr., B25, 2527 (1969); R.C. Pettersen and L.E. Alexander, J. Amer. Chem. Soc., 90, 3873 (1968).

90. A. Mavridis and A. Tulinsky, to be published.
91. D.L. Cullen and E.F. Meyer, Jr., Amer. Crystallogr. Assoc., Spring Meeting, March, 1974, Abstract Q6; private communication.
92. P.N. Dwyer, L. Puppe, J.W. Buchler and W.R. Scheidt, Inorg. Chem., 14, in press (1975).
93. P.N. Dwyer, J.W. Buchler and W.R. Scheidt, J. Amer. Chem. Soc., 96, 2789 (1974).
94. J.F. Kirner, J. Garofolo, Jr. and W.R. Scheidt, Inorg. Nucl. Chem. Letters, 11, 107 (1975).
95. R.C. Pettersen, J. Amer. Chem. Soc., 93, 5629 (1971).
96. R.F. Pasternack, E.G. Spiro and M. Teach, J. Inorg. Nucl. Chem., 36, 599 (1974).
97. R. Timkovich and A. Tulinsky, private communication.
98. E.B. Fleischer and D. Lavallee, J. Amer. Chem. Soc., 89, 7132 (1967).
99. F.R. Hopf, T.P. O'Brien, W.R. Scheidt and D.G. Whitten, J. Amer. Chem. Soc., 97, 277 (1975).
100. L.K. Hanson, M. Gouterman and J.C. Hanson, J. Amer. Chem. Soc., 95, 4822 (1973).
101. E.B. Fleischer, F.L. Dixon and R. Florian, Inorg. Nucl. Chem. Letters, 9, 1303 (1973).
102. W.R. Scheidt and J.L. Hoard, unpublished results.
103. S. Koch, S.C. Tang, R.H. Holm, R.B. Frankel and J.A. Ibers, J. Amer. Chem. Soc., 97, 916 (1975).
104. D.H. O'Keeffe, C.H. Barlow, G.A. Smythe, W.H. Fuchsman, T.H. Moss, H.R. Lilienthal and W.S. Caughey, Bioinorg. Chem., in press.
105. L. Pauling, 'The Nature of the Chemical Bond', Cornell University Press, Ithaca, New York, U.S.A., 1960, pp. 224 and 246.
106. B.D. McLees, Ph.D. Dissertation, The Johns Hopkins University, Baltimore, Maryland, U.S.A., 1964. See Hoard et al.[24] for a summary of McLees' conclusions.
107. R. Havemann, W. Haberditzl and K.-H. Mader, Z. physik. Chem., 218, 71 (1962); R.A. Rawlinson and P.B. Scutt, Australian J. Sci. Res., 45, 173 (1952).
108. B.M.L. Chen and A. Tulinsky, private communication.
109. L.J. Radonovich, A.A. Sayler and J.L. Hoard, unpublished results.
110. V.W. Day, private communication.
111. J.F. Kirner and W.R. Scheidt, to be published.
112. V.W. Day, B.R. Stults, E.L. Tasset, R.O. Day and R.S. Marianelli, J. Amer. Chem. Soc., 96, 2650 (1974).
113. B. Gonzales, J. Kouba, S. Yee, C.A.-Reed, J.F. Kirner and W.R. Scheidt, J. Amer. Chem. Soc., 97, in press (1975).
114. J.P. Collman, R.R. Gagne, C.A. Reed, T.R. Halbert, G. Lang and W.T. Robinson, J. Amer. Chem. Soc., 97, 1427 (1975).
115. J.P. Collman, R.R. Gagne, C.A. Reed, W.T. Robinson and G.A. Rodley, Proc. Natl. Acad. Sci. U.S.A., 71, 1326 (1974).
116. J.P. Collman, R.R. Gagne, T.R. Halbert, J.L. Hoard, C.A. Reed and A.A. Sayler, unpublished study.
117. J.W. Lauher and J.A. Ibers, J. Amer. Chem. Soc., 96, 4447 (1974).
118. R.G. Little, K.R. Dymock and J.A. Ibers, J. Amer. Chem. Soc., 97, in press (1975).
119. K. Adams, P.G. Rasmussen and W.R. Scheidt, to be published.
120. P.N. Dwyer, P. Madura and W.R. Scheidt, J. Amer. Chem. Soc., 96, 4815 (1974).
121. R.G. Little and J.A. Ibers, J. Amer. Chem. Soc., 96, 4452 (1974).
122. W.R. Scheidt and J. Ramanuja, unpublished results.
123. J.L. Hoard and W.R. Scheidt, Proc. Natl. Acad. Sci. U.S.A., 70, 3919 (1973); for a quantitative emendation, ibid., 71, 1578 (1974).
124. P.L. Piciulo, G. Rupprecht and W.R. Scheidt, J. Amer. Chem. Soc., 96, 5293 (1974).

125. R. Hoffmann, M.M.L. Chen, M. Elian, A.R. Rossi and D.M.P. Mingos, Inorg. Chem., 13, 2666 (1974), and numerous references cited therein.
126. E.A. Padlan and W.E. Love, J. Biol. Chem., 219, 4067 (1974).
127. R. Huber, O. Epp and H. Formanek, J. Mol. Biol., 52, 349 (1970).
128. L. Pauling, Nature (London), 203, 182 (1964); L. Pauling in 'Haemoglobins', F.J.W. Roughton and J.C. Kendrew, Eds., Butterworths, London, 1949, p. 57.
129. J.J. Weiss, Nature (London), 202, 83 (1964); J.B. Wittenberg, B.A. Wittenberg, J. Peisach and W.E. Blumberg, Proc. Natl. Acad. Sci. USA, 67, 1846 (1970); A.S. Koster, J. Chem. Phys., 56, 3161 (1972).
130. M.F. Perutz, Nature (London), 228, 726 (1970).
131. M.F. Perutz and L.F. TenEyck, Cold Spring Harbor Symp. Quant. Biol., 36, 295 (1972).
132. J. Wyman, Advan. Protein Chem., 19, 223 (1964).
133. M.F. Perutz, Nature (London), 237, 495 (1972), and references cited therein.
134. B.M. Hoffman and D.H. Petering, Proc. Natl. Acad. Sci. USA, 67, 637 (1970).
135. B.M. Hoffman, C.A. Spilburg and D.H. Petering, Cold Spring Harbor Symp. Quant. Biol., 36, 343 (1972).
136. G.C. Hsu, C.A. Spilburg, C. Bull and B.M. Hoffman, Proc. Natl. Acad. Sci. USA, 69, 2122 (1972).
137. J.A. Ibers, J.W. Lauher and R.G. Little, Acta Crystallogr., B30, 268 (1974).
138. A. Takenaka, Y. Sasada, T. Omura, H. Ogoshi and Z.-I. Yoshida, J. Chem. Soc., Chem. Commun., 792 (1973).
139. D. Cullen, E. Meyer, T.S. Srivastava and M. Tsutsui, J. Amer. Chem. Soc., 94, 7603 (1972).
140. D. Cullen and E. Meyer, private communication.

MASS SPECTROMETRY OF PORPHYRINS AND METALLOPORPHYRINS

KEVIN M. SMITH

The Robert Robinson Laboratories, University of Liverpool, U.K.

9.1. Introduction

Quantities of materials available in natural product chemistry are often minute, and the technique of mass spectrometry has the advantage that, using only diminutive samples, it can provide accurate information, not only on molecular weights and elemental compositions of compounds, but also details of the nature of some of the functions within complex molecules. Both of these factors are of obvious utility in structural investigations of porphyrins and metalloporphyrins.

Electronic absorption spectroscopy can be used to elucidate the gross structure of porphyrins and their derivatives, such as whether the nucleus is reduced (as in chlorins, phlorins, bacteriochlorins, etc.) or whether certain types of group are conjugated with the macrocycle. On the other hand, n.m.r. spectroscopy can provide precise information on the nature of porphyrin side-chains; in some cases even type-isomers can be identified[1]. N.m.r. spectroscopy does, however, suffer the disadvantages that relatively large samples of good solubility are required. Nonetheless, a combination of n.m.r. and mass spectrometry has been used in several definitive structural investigations in the porphyrin area.

The major breakthrough in porphyrin mass spectrometry came (approx. 1964) with the introduction of 'direct' insertion probes; before that time it had been virtually impossible to measure the spectra of involatile substances, though, using extreme measures, some macrocycles had been examined[2].

Even with direct inlet systems, the mass spectra of porphyrins and metalloporphyrins are usually measured using the esters (if applicable) rather than carboxylic acids. Volatility can also be enhanced by the preparation of silicon complexes[3] or trimethylsilyl ethers[4].

Fragmentation characteristics of pyrroles[5], and linear di-, tri-, and tetrapyrrolic compounds[6,7] have been recorded. These investigations had been

Porphyrins and Metalloporphyrins, ed. Kevin M. Smith
© 1975, Elsevier Scientific Publishing Company, Amsterdam, The Netherlands

intended as preliminaries to assist in the interpretation of porphyrin mass spectra; in the event, these hopes were never realised because simple pyrrolic compounds possessed fragmentation pathways which were far more complex than those of the corresponding macrocycles.

Complete porphyrin mass spectra are rarely recorded. Two definitive papers[8,9] which outlined general trends found in mass spectra have been published, and readers are advised to consult these. Contemporary papers tend to report only molecular ions, major fragmentations, and unusual features of individual mass spectra. Much of what follows is drawn from the two major papers[8,9]; reference is given only to information derived from other sources. A brief, but informative review[10] of tetrapyrrole mass spectra has appeared.

9.2. Porphyrins and chlorins
9.2.1. The molecular ion

In the absence of labile or certain heavily functionalized side-chains, the base (100%) peak is usually the molecular ion. Figure 1 shows the typical mass spectrum of octaethylporphyrin (1). The predictable dominance of the molecular ion is particularly useful for molecular weight determination. Mass spectra of porphyrins and their derivatives are best recorded at the lowest possible temperature (usually approx. 200—250°C); if considerably higher temperatures are used, more opportunity for pyrolysis and for fragmentation of the molecular ion is afforded. At very high temperatures (400—800°C), thermal degradation of the nucleus into monopyrroles is known to occur[11].

Using conditions (beam energy 12 eV) such that only molecular ions were observed, the petroporphyrins[12] from a variety of asphaltenes have been shown[13] to consist of two major and one minor homologous series of poly-alkylporphyrins containing no oxygen functions. Evidence for transalkylation during diagenesis has been gained[14] from a mass spectrometric investigation of the products from heating vanadyl and nickel octa-alkylporphyrins in presence of certain minerals.

The intensity of molecular ions in the mass spectra of porphyrins has also been used[15] in studies of deuteration (as a general example of electrophilic substitution) of porphyrins, metalloporphyrins, and some reduced derivatives. It is a simple mathematical operation to derive an accurate estimate of the extent of meso-deuteration of porphyrins under standardized exchange conditions, using the intensities of the molecular ion (M) and the (M + 1) to (M + 4) ions.

As in normal mass spectrometry[16], the physical appearance of the molecular ion enables one to ascertain the presence of halogens, metals, etc. in compounds. This is of great help in the identification of unknown metalloporphyrins because of the general tendency for metal ions not to be lost in fragmentation processes and because the precise isotopic compositions of

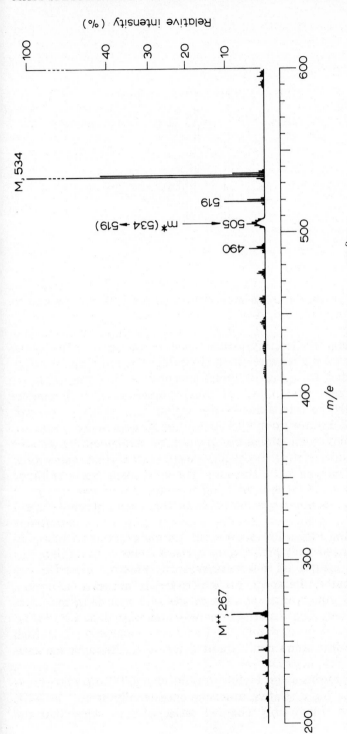

Fig. 1. Mass spectrum of octaethylporphyrin (1) (A.E.I. MS 12; source temp. approx. 220°C).

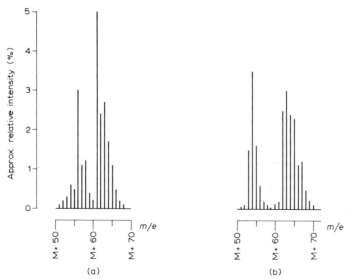

Fig. 2. High mass cluster of ions due to metal impurities: (a) A.E.I. MS 9; (b) A.E.I. MS 12 (both at Liverpool).

metals are known. Metal-free porphyrin mass spectra almost invariably possess a cluster of peaks to higher mass than the molecular ion (Fig. 2); these peaks can be attributed to traces of metalloporphyrins and the origin of these impurities has been the subject of some discussion. The anomalous peaks usually have relative intensities in the region of 1 to 5%, even with porphyrin samples which show no involatile residues in semi-micro elemental analysis. Their relatively high intensities cannot be explained by greater volatility of metalloporphyrins, because, except for certain cases (e.g. silicon), this does not appear to be the case. The most likely explanation of the high mass peaks appears to be scavenging of metal ions by the porphyrin in the source of the spectrometer; it may even be that each instrument has a 'fingerprint' of metal ions which is unique, depending upon the parameters and construction of the ionization source. A striking example of this phenomenon was experienced with preliminary trials of a now defunct Series of mass spectrometer. A sample of mesoporphyrin-IX dimethyl ester (2) was sent to the manufacturer; the spectrum which was returned to Liverpool possessed a base peak which corresponded to the lead chelate of mesoporphyrin-IX dimethyl ester. The manufacturer was informed that the new instrument must have an inappropriately sited lead seal somewhere in its high vacuum system, and after some denials and delay, the offending seal was tracked-down.

In simple alkylporphyrins (Fig. 1) the doubly charged molecular ion is invariably observed, having a relative intensity occasionally as high as 50%. Metastable ions within the doubly charged series of ions show that the

TABLE 1

Structural formulae of compounds discussed in the text of this chapter

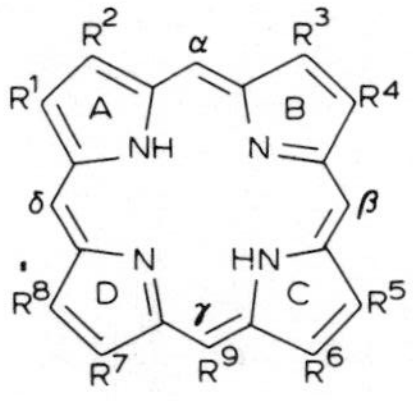

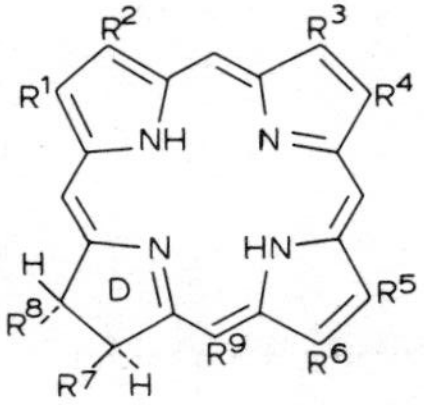

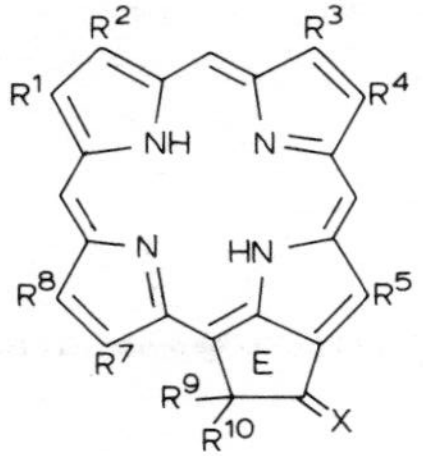

Compound	Nucleus	Substituents										
		R^1	R^2	R^3	R^4	R^5	R^6	R^7	R^8	R^9	R^{10}	Others
(1)	A	Et	Et	Et	Et	Et	Et	Et	Et	H		δ-Et
(2)	A	Me	Et	Me	Et	Me	p^{Me}	p^{Me}	Me	H		α,β,δ Ph
(4)	A	Me	p^{Me}	Me	p^{Me}	Me	p^{Me}	p^{Me}	Me	H		
(5)	A	Me	CH(OH)Me	Me	CH(OH)Me	Me	p^{Me}	p^{Me}	Me	H		
(6)	A	Me	Et	Me	Et	Me	Et	Me	$CO(CH_2)_7Me$	H		
(8)	A	Me	H	Me	H	Me	p^{H}	p^{H}	Me	H		
(9)	B	Et	Et	Et	Et	Et	Et	Et	Et	H		
(10)	A	Me	Et	Me	Et	Me	H	p^{Me}	Me	Me		
(11)	A	Me	Et	Me	Et	Me	H	p^{Me}	Me	Et		
(12)	A	Me	p^{Et}	Me	Et	Me	CO_2Et	p^{Et}	Me	H		
(13)	A	Me	Et	Me	Pr	Et	H	p^{Me}	Me	Ph		
(14)	A	H	H	H	H	H	H	H	H	H		
(15)	C	Me	Et	Me	Et	Me		Et	Me	H	H	X = H_2
(16)	D	Me	V	Me	Et	Me		p^{Me}	Me	H	CO_2Me	
(17)	C	Me	Et	Me	Et	Me		p^{Me}	Me	H	CO_2Me	X = O
(18)	D	Me	V	Me	Et	Me		p^{Me}	Me	H	H	
(19)	C	Me	Et	Me	Et	Me		Acr^{Me}	Me	H	CO_2Me	X = O

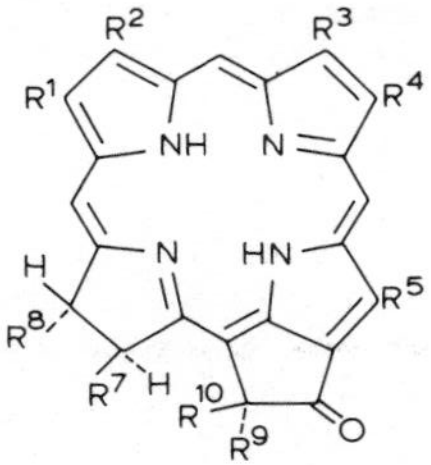

TABLE 1 (continued)

Compound	Nucleus	Substituents										Others
		R^1	R^2	R^3	R^4	R^5	R^6	R^7	R^8	R^9	R^{10}	
(20)	D	Me	V	CHO	Et	Me		P^{Me}	Me	H	CO_2Me	
(23)	B	Me	V	Me	Et	Me	CO_2Me	P^{Me}	Me	CH_2CO_2Me		
(24)	B	Me	Et	Me	Et	Me	CO_2Me	P^{Me}	Me	CH_2CO_2Me		
(25)	B	Me	V	CHO	Et	Me	CO_2Me	P^{Me}	Me	CH_2CO_2Me		

$P^R = CH_2CH_2CO_2R$; $V = CH=CH_2$; $Acr^{Me} = CH=CHCO_2Me$; $Pr = CH_2CH_2CH_3$

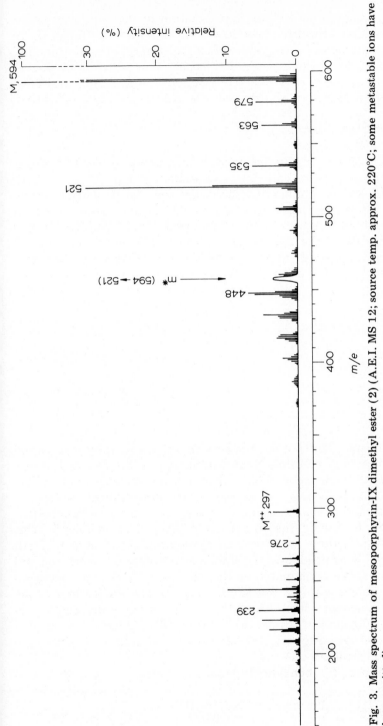

Fig. 3. Mass spectrum of mesoporphyrin-IX dimethyl ester (2) (A.E.I. MS 12; source temp. approx. 220°C; some metastable ions have been omitted).

doubly charged parent is subject to fragmentation in the same way as the singly charged analog, though there are some important differences. In more highly functionalized compounds [e.g. mesoporphyrin-IX dimethyl ester, (2)] (Fig. 3), the doubly charged molecular ion tends to be smaller than many of the other doubly charged ions, no doubt as a result of the extra fragmentation pathways afforded by the additional functionalization.

9.2.2. Side-chain fragmentations
9.2.2.1. Peripheral side-chains

A characteristic feature of the mass spectra of porphyrinic compounds is the way in which the ions are split into two separate groups (Figs. 1 and 3). The highest mass group contains the molecular ion and its fragmentation products. After a relatively bare region the doubly charged series of ions is observed. Below about m/e 200 there are few peaks, indicating that there is no extensive cleavage of the macrocyclic nucleus. This situation does not obtain in the mass spectra of molecules such as the xanthoporphyrinogens (3) which possess[17] quite intense ions corresponding to mono-, di-, and

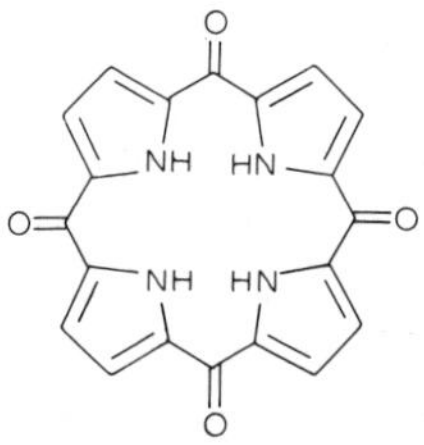

(3)

tri-pyrrolic fragments. The porphyrinogens (hexahydroporphyrins) would also be expected to show multiple fragmentations, possibly similar to those of pyrromethanes, but on account of their chemical instability and handling difficulties, these compounds do not appear to have been studied.

In organic mass spectrometry, a major driving force and stabilizing effect for fragmentation is usually the formation of even-electron ions[16]. This principle holds firm in the mass spectra of porphyrinic compounds for both the singly and doubly charged ions; the stability difference between even and odd electron ions is even accentuated by the macro-ring[18]. Several successive cleavages of side-chains are observed (Figs. 1 and 3), and the features of the spectra are well in accord with the stability of the aromatic nucleus, which allows wide delocalization of the positive charges. By and large, the porphyrin ring acts as an inert support and allows detailed studies of side-chain fragmentations to be carried out.

Major cleavages produce 'benzylic' type ions:

$$[\text{Porphyrin—CH}_2\text{-R}]^+ \rightarrow [\text{Porphyrin—CH}_2]^+ + \text{R}^\bullet$$

For example, the major cleavages of mesoporphyrin-IX dimethyl ester (2) (Fig. 3) are:

(i)   loss of CH_3^{+} (m/e 579) due to benzylic fission of an ethyl group,

(ii) cleavage of $CH_3O^{\cdot}$ (m/e 563) from a propionate group,

(iii) cleavage of $CH_3O_2C^{\cdot}$ (m/e 535) from a propionate group, and

(iv) benzylic fission of $CH_3O_2CCH_2^{+}$ (m/e 521) from a propionate group.

Other peaks in the singly charged series merely represent similar successive cleavages of the other ethyl and propionate groups, together with ions arising from hydrogen transfers. In the mass spectrum of coproporphyrin-III tetramethyl ester (4), successive cleavages of all four propionate residues (together with appropriate metastable ions) are observed. Nuclear carbomethoxy groups are cleaved completely (M—59) and with transfer of one hydrogen (M—58). Porphyrins bearing peripheral β-keto-ester substituents afford[19] a peak corresponding to acetylporphyrin (M—58) as the base peak, and only on rare occasions has the molecular ion been observed[20].

Other cases in which the base peak is not the molecular ion have been reported. This is usually limited to porphyrins and chlorins having very labile side-chains; a prominent example is hematoporphyrin-IX dimethyl ester (5), which has two (1-hydroxyethyl) functions. The base peak here appears at m/e 590 (corresponding to protoporphyrin-IX dimethyl ester) owing to double dehydration, a process which is temperature dependent. Several natural products feature (1-hydroxyethyl) side-chains (e.g. the *Chlorobium* chlorophylls) and the lability of these groupings should be kept in mind in structural investigations. The chemical difference between (1-hydroxyethyl) and (2-hydroxyethyl) substituents is also borne out in the masss spectrometry of compounds containing these groups. (2-Hydroxyethyl)porphyrins do not undergo thermal dehydration in the source of the mass spectrometer, but tend to cleave with loss of 31 mass units ($HOCH_2^{\cdot}$) affording a benzylic ion[21].

After comparison of a large number of porphyrin mass spectra, a qualitative stability order for substituents has been assembled, and more recently, expanded[10]:

$$-H > -CH=CH_2 \geqslant -CHO > -CH_2CH_3 > -CH_2CH_2CO_2Me > -CH_2CH_2CH_3$$

$$> -CH_2CO_2Me > -CH_2CH_2NHCOMe > -CO_2Me > -OCOMe > -CO-$$

$$CH_2CO_2Me.$$

Though the mass spectra of porphyrin 'type' isomers (e.g. coproporphyrins) do show some differences in the relative intensities of some of their fragment ions, these differences are no greater than those due to running the spectra of the same compound on different days, and even with the most careful standardization of operating parameters, no headway with type isomer differentiation by mass spectrometry has been made.

References, p. 397

Longer peripheral side-chains follow the usual fragmentation pathways. For example, the porphyrinyl n-octyl ketone (6) cleaves[22] by normal α-cleavage of the C_8H_{17} side chain (m/e 477; 25%) as well as by McLafferty rearrangement to give the ion (7):

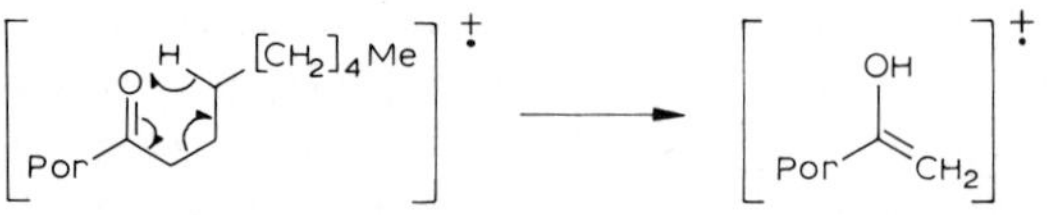

(6)   (7)
m/e 590 (100%) m/e 492 (20%)

Trimethylsilyl ethers of six hydroxyporphyrins (some of which would be expected to undergo thermal dehydration before fragmentation) have been prepared[4] and the mass spectra of these showed major fragmentations involving the side-chains bearing the ether groupings to afford benzylic ions.

Owing to lack of volatility, porphyrin free carboxylic acids have not been extensively examined by mass spectrometry. However, the mass spectrum of deuteroporphyrin-IX (8) shows principal singly charged ions at m/e 510 (100%), 492 (100), 466 (33), 464 (33), 451 (35), 422 (80), and 405 (27). The major fragmentation appears to be loss of water, and other large peaks correspond to losses of CO_2, $2CO_2$, HCO_2H, as well as normal benzylic cleavage of $\cdot CH_2 CO_2 H$.

Some chlorins bearing vinyl groups have been shown[23] to give rise to intense (M + 2) ions, owing to hydrogenation in the source of the spectrometer. This phenomenon is strongly temperature dependent and since the corresponding mesochlorins (ethyl in place of vinyl) do not show this process, the vinyl group has been implicated in the hydrogenation; this anomaly also occurs in some vinylporphyrins.

Most chlorin mass spectra are broadly similar to those of their porphyrin counterparts, benzylic cleavages predominating. Thus, the whole substituent is usually lost from the reduced ring. For example, the most significant fragmentation in the spectrum of $trans$-octaethylchlorin (9) is loss of an ethyl radical, whereas in octaethylporphyrin (1) (Fig. 1) the major fission involves loss of a methyl radical (by benzylic cleavage of an ethyl group). Hence, the notable fission in the chlorin must be from the reduced ring:

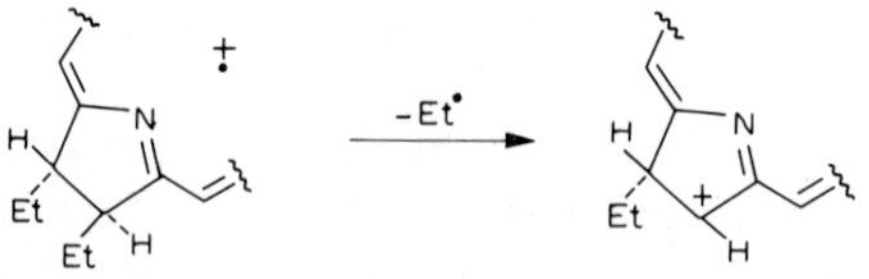

Similar considerations apply to the loss of 87 mass units ($CH_2 CH_2 CO_2 Me$) from the methyl esters of many chlorophyll degradation products containing the reduced ring D. Special features associated with $meso$-substituents and carbocyclic rings in many of these derivatives are discussed in Section 9.2.2.2.

Chlorophylls and bacteriochlorins have proven to be highly resistant to mass spectrometric analysis, although pheophytin-*a* has given a poor spectrum. Apparently, coordination to the magnesium atom has an adverse effect upon the volatility of many chlorophyll derivatives.

9.2.2.2. meso-Substituents and carbocyclic rings

meso-Methyl and *meso*-dimethyl-porphyrins do not appear to show any evidence of expulsion of the *meso*-groups; for example, apart from the molecular weight difference of 15 mass units, the spectrum of pyrroporphyrin-XV methyl ester (10) is very similar to that of γ-phylloporphyrin-XV methyl ester (11). However, with bulkier groups, fission at the porphyrin − *meso*-substituent bond is common. The major fragmentation in the mass spectra of *meso*-ethyl-porphyrins [e.g. (12)[9], (13)[24]] is loss of 27 mass units, with smaller peaks corresponding to loss of 28 and 29. This is due to loss of the ethyl substituent, with transfer of hydrogens to the macrocycle. Hoffman[9] claimed that this phenomenon, with transfer of hydrogens, when available, from the expelled *meso*-substituent to the macrocycle, was common. More recently, labeling studies have shown[24] that the hydrogens are transferred to the nucleus from the environment within the source (e.g. from water) and not from the substituent, at least in the case of *meso*-ethylporphyrins. Cleavage at the porphyrin − *meso*-substituent bond is also predominant with CHO and CO_2Me substituents, and more surprisingly with CH_2CO_2Me and CH_2CH_2-CO_2Me which might have been expected to afford benzylic ions. It is clear that the enhanced fragmentation tendency for whole *meso*-substituents is due to stability gained by loss of sterically demanding *meso*-groups (cf. Ref. 25). Fission between the nucleus and *meso*-substituent is not the predominant fragmentation when the *meso*-function has a double or triple bond in conjugation with the macrocyclic π-system (except in the case of *meso*-tetraphenylporphyrin).

The differences between fragmentation characteristics of *meso*- and peripheral substituents (e.g. CH_2CO_2Me, $CH_2CH_2CO_2Me$) can be of use in structural assignments. For example, *meso*-formyl groups are lost very readily, with transfer of one hydrogen to the nucleus. On the other hand, peripheral formyl groups are much more resistant to fragmentation under electron impact.

The most intense ion in the spectra of *meso*-acetoxyporphyrins is usually that due to cleavage of ketene from the molecular ion; a large (M − CO · Me) peak is also observed. In such cases, the tendency is for fission to occur such that the oxygen atom is retained by the macrocycle, and this is presumably a consequence of the extra stabilization afforded to the positive charge by the hetero-atom[26]:

References, p. 397

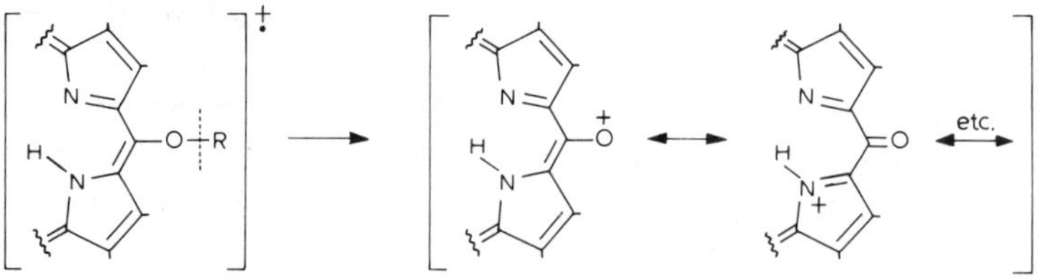

R = CO·Me, CO·Ph, CO·CF$_3$, Me , Et

Detailed fragmentation pathways of oxophlorins and a variety of other *meso*-substituted porphyrins have been reported[41].

As might be expected from the preceding discussion, *meso*-tetraphenyl-porphyrin (14) and its derivatives show extremely intense molecular ions, with peaks corresponding to successive losses of aryl groups, accompanied by additional hydrogen transfers[27].

The most intense fragment ions in the spectra of compounds with carbocyclic ('E') rings are usually associated with fragmentation from and loss of the 10-CO$_2$Me function, and in view of the naming of derivatives lacking the 10-function as 'pyro-' compounds, it is hardly surprising that these cleavages are temperature dependent. In the case of desoxophylloerythroetioporphyrin (15), which lacks the 10-substituent, the fragments observed[28] are quite normal, without any noticeable influence by the carbocyclic ring. Methyl pheophorbide-*a* (16) (Figure 4), its meso-derivative, and pheoporphyrin-*a*$_5$ methyl ester (17) feature a characteristic loss of 32 mass units (MeOH), this cleavage being absent in the 'pyro-' compounds [e.g. methyl pyropheophorbide-*a* (18)]. Methanol is not cleaved from the propionate residue because ions corresponding to loss of this *and* methanol are observed. Further work, using labeled compounds, has confirmed[29] the nature of this characteristic cleavage, and this fragmentation was used[30] as evidence for the presence of a normal carbocyclic ring grouping in a degradation product from chlorophyll-*c*; on the other hand, the *Chlorobium* chlorophylls have a 'pyro-' carbocyclic ring system.

Methyl pheophorbide-*a* (16) (Fig. 4), methyl pheophorbide-*b* (20), and their meso-derivatives show the direct loss of 147 mass units, attested by a strong metastable ion. This involves[8,31] fission of the 10-CO$_2$Me group together with the 7-CH$_2$CH$_2$CO$_2$Me and one hydrogen atom, to give ions such as (21) or (22). Possibly more remarkable is the loss of 159 mass units

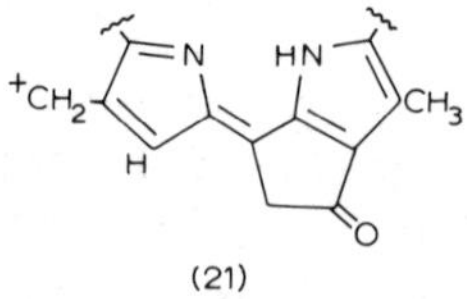

(21) 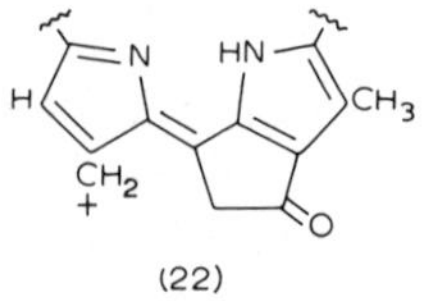   (22)

(again confirmed by a metastable ion) from the trimethyl esters [(23), (24), and (25) respectively] of chlorin-*e*$_6$, mesochlorin-*e*$_6$, and rhodin-*g*$_7$; these compounds also showed the loss of 147 mass units mentioned earlier. This loss of 159 involves[8,31] the 7-CH$_2$CH$_2$CO$_2$Me and γ-CH$_2$CO$_2$Me moieties, with

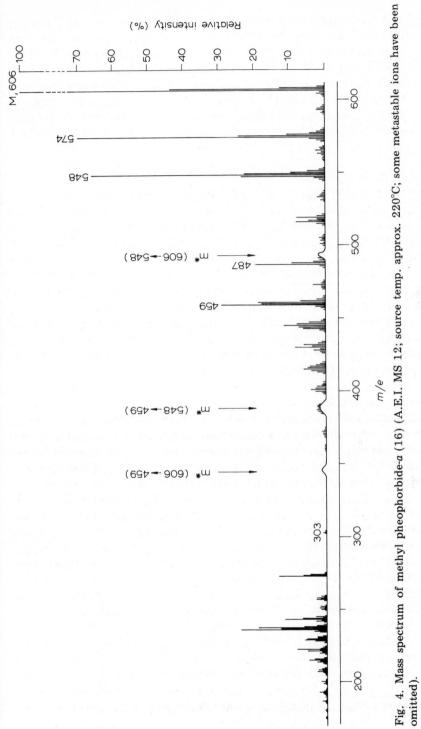

Fig. 4. Mass spectrum of methyl pheophorbide-*a* (16) (A.E.I. MS 12; source temp. approx. 220°C; some metastable ions have been omitted).

transfer of one hydrogen to the macrocycle. The fact that these major cleavages are confirmed by large metastable ions resulted in perhaps the first realization that metastable ions need not arise by a single one-step fragmentation, but could be due to two or three individual cleavages succeeding each other so rapidly as to be virtually simultaneous. Nickel mesopyropheophorbide-a methyl ester also exhibits[31] a metastable ion attesting the cleavage of the 7-$CH_2CH_2CO_2Me$ as well as carbon monoxide from the carbocyclic ring; again a one-step concerted fragmentation cannot be considered for this process.

9.2.2.3. N-Methylporphyrins

The molecular ion is always the base peak in the mass spectra of N-monomethylporphyrins, and at 260°C there is little evidence of expulsion of the N-methyl group[32]. In the spectra of N,N'-dimethylporphyrins, the base peak is two mass units higher than the expected molecular ion, and the compounds with opposite N-methyl groups also shows intense (M + 4) ions; moreover, cleavages corresponding to loss of CH_3 [or CH_5 from (M + 2)] are very pronounced. N,N',N''-Trimethylporphyrins show a base peak one mass unit higher than expected, with a further intense ion at (M + 3). These unusual features are believed to be associated with the high basicity of N-methylporphyrins, and presumably arise in the mass spectrometer by intermolecular transfer of hydrogens, either from water vapor or from one molecule of porphyrin to another.

9.2.3. Doubly charged ions

The abundance and intensity of doubly charged ions in the mass spectra of porphyrinic compounds are a direct indication of the aromaticity and stability of these macrocycles; indeed, even triply charged ions can be observed in some spectra. The composition of the doubly charged ions has been investigated by high resolution mass measurement, and most ions arise by fragmentations which are entirely different from their counterparts in the singly charged series (i.e. not merely by loss of a second electron from a singly charged ion). The even-electron ion rule is usually obeyed, as discussed earlier. Some fragments (F) correspond to those from the singly charged series, i.e. $F^{2+} = F^+/2$, but this is a relatively rare occurrence. More often, the correlation $F^{2+} = (F^+ - 1)/2$ is encountered.

Consideration of Figs. 1, 3, and 4 shows how the relative intensity of the doubly charged molecular ion can decrease with functionalization of the molecule; simple porphyrins and chlorins have impressive M^{2+} ions, but highly substituted compounds such as methyl pheophorbide-a have diminutive M^{2+} ions, presumably as a consequence of the numerous fragmentation pathways afforded by the presence of several oxygen atoms in the sidechains.

A noteworthy feature in the spectra of all compounds containing propionate side-chains is the novel cleavage of ketene (CH_2CO) from this function in the doubly charged series of ions. This is illustrated by the ions at m/e 276 and 239 in the spectrum of mesoporphyrin-IX dimethyl ester (Fig. 3); the analogous cleavages are not observed in the singly charged fragments, suggesting an electronic state specificity for this cleavage:

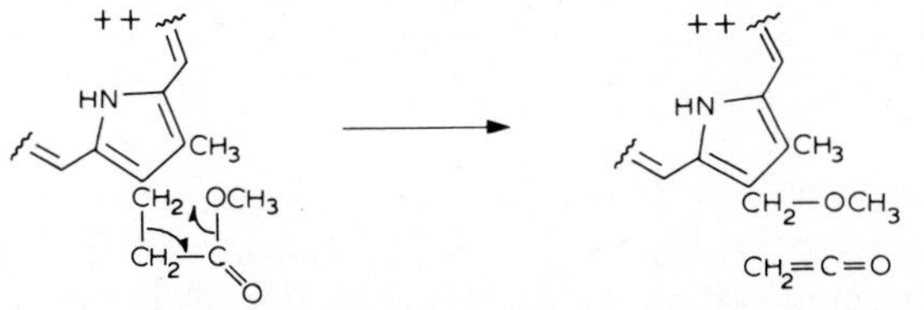

The fragmentation is remarkably diagnostic, and does not occur with acrylic side-chains; this was used as evidence against the presence of a propionate side-chain in chlorophyll-c and its derivatives[30].

Some features of the doubly-charged region in the mass spectrum of cobalt(II) deuteroporphyrin-IX dimethyl ester have been discussed[40].

9.3. Metalloporphyrins

Metal complexes of porphyrins undergo fragmentation in a similar manner to the free bases, the only differences being in the physical appearance of the ions owing to the isotopic compositions of the metals. Except in very unusual cases (see later), the metal atom is not lost in any fragmentation process, and this might be expected because of the stability of the macrocyclic nucleus towards cleavage.

Conflicting reports of less, similar, and more fragmentation of metalloporphyrins relative to their free bases are probably due to temperature effects. Depending upon source temperature, octaethylporphyrinatoaluminum(III) hydroxide gives[33] either a normal molecular ion or else one corresponding to the dehydrated dimer. Silicon complexes of porphyrins are so volatile that they can be subjected to gas chromatographic separation[3]. However, a systematic study of silicon complexes has not been attempted.

Certain differences between fragmentation characteristics of meso-tetraphenylporphyrin (14) and its metal chelates have been noted[27]. In addition, the extent of fragmentation

$$(E) = \left[\frac{\Sigma I_{\text{fragments}}}{I_M} \times 100\% \right]$$

in the singly charged region is greater for the metal derivatives than for the free ligand. For example[27], (E) = 5.5% for (14) whereas it is 17.6% for the iron(II) chelate. Likewise, (E) for meso-tetra(p-chlorophenyl)porphyrin is

References, p. 397

13.5% whereas that for the magnesium chelate is 28.2%. The increased peripheral fragmentation in the cases of these metal complexes is in accord with charge withdrawal from the periphery by charge localization on the metal atom. In contradistinction, metal chelates of *meso*-tetrapropylporphyrin afford mass spectra which are very similar to that of the metal-free ligand.

9.4. Anomalies*

From the foregoing discussion it will be seen that porphyrins and metalloporphyrins undergo fragmentations in the mass spectrometer which are easily rationalized in terms of normal organic mass spectrometry. A series of 'rules' could almost be compiled in view of the reproducible behavior of most porphyrinic compounds. In such circumstances, it is advisable to be aware of any anomalies which might cause misinterpretation of data. Some examples have already been mentioned, such as thermal degradation of molecules (e.g. hematoporphyrin-IX dimethyl ester) in the source of the mass spectrometer so that the molecular ion is not the base peak, or is not even observed in extreme cases. More often, however, it is the presence of peaks above the expected molecular ion that can cause confusion. The most common example of this is the cluster of peaks 50—70 mass units above the molecular ion (Fig. 2) which are due to metal ions scavenged from the instrument. Hydrogenation of some vinyl groups, and (M + 2) and (M + 4) ions in *N*-methylporphyrins are also troublesome; (M + 2) peaks have also been encountered in the mass spectra of $\alpha\gamma$-dioxoporphodimethenes[34], and in acrylic porphyrins[40].

Ion-molecule reactions in the source of the mass spectrometer have been observed[27] in *meso*-tetra-arylporphyrins, where ions as large as (M + 124) are produced; similarly, a ruthenium porphyrin has been shown to undergo ion-molecule reactions with perfluorokerosene and perfluorotributylamine mass standards, giving[35] anomalous high-mass ions with the composition (M + $C_n F_{2n}$) (after loss of an axial ligand), where $n = 1$—4.

Metals are not usually cleaved from metalloporphyrins in their mass spectra and so it is normally safe to assume that peaks corresponding to free base porphyrin in the mass spectrum of a metalloporphyrin are due to inefficient metalation. However, thallium(III)[36] and mercury(II)[37] porphyrins do tend to lose the metal in the mass spectrometer, and in the thallium case, the metal-free porphyrin ligand is often the base peak. As a result of the high pressure of thallium ions present in the source, ion-molecule reactions occur with the thallium porphyrin (after loss of its axial ligand) to give ions with the composition $(Tl)_2$ Porphyrin. A metal-metal bond seems likely, because

---

* For further discussion of anomalous mass spectra of metalloporphyrins, see Chapter 5.

this anomaly only occurs with ions which have lost their axial ligand and a ruthenium porphyrin dimer having a metal-metal bond has been shown to be stable under electron impact[38].

meso-Tetraphenylchlorin gives[39] a mass spectrum which corresponds to that of the porphyrin analog; this novel dehydrogenation is due to electron-impact excitation and not thermal effects in the source of the spectrometer.

References

1. R.J. Abraham, P.A. Burbidge, A.H. Jackson, and D.B. Macdonald, J. Chem. Soc. B, 620 (1966); R.J. Abraham, G.H. Barnett, E.S. Bretschneider, and K.M. Smith, Tetrahedron, **29**, 553 (1973); M.F. Hudson and K.M. Smith, J. Chem. Soc., Chem. Commun., 515 (1973).
2. A. Hood, E.G. Carlson, and M.J. O'Neal in "The Encyclopedia of Spectroscopy", Ed. G.L. Clark, Reinhold, New York (1960), p. 616; W.L. Mead and A.J. Wilde, Chem. Ind. London, 1315 (1961); H.C. Hill and R.I. Reed, Tetrahedron, **20**, 1359 (1964).
3. D.B. Boyland and M. Calvin, J. Am. Chem. Soc., **89**, 5472 (1967); D.B. Boyland, Y.I. Alturki, and G. Eglinton in "Advances in Organic Geochemistry", Ed. F. Vieweg, Brunswick (1969).
4. J.R. Chapman and G.H. Elder, Org. Mass Spectrom., **6**, 991 (1972).
5. H. Budzikiewicz, C. Djerassi, A.H. Jackson, G.W. Kenner, D.J. Newman, and J.M. Wilson, J. Chem. Soc., 1949 (1964).
6. A.H. Jackson, G.W. Kenner, H. Budzikiewicz, C. Djerassi, and J.M. Wilson, Tetrahedron, **23**, 603 (1967); A.H. Jackson, K.M. Smith, C.H. Gray, and D.C. Nicholson, Nature London, **209**, 581 (1966).
7. W.J. Cole, D.J. Chapman, and H.W. Siegelman, J. Am. Chem. Soc., **89**, 3643, 5976 (1967); Biochemistry, **7**, 2929 (1968); C.J. Watson, D.A. Lightner, Z.J. Petryka, E. David, and M. Weimer, Proc. Nat. Acad. Sci. U.S.A., **58**, 1957 (1967); H.L. Crespi, L.J. Boucher, G.D. Norman, J.J. Katz, and R.C. Dougherty, J. Am. Chem. Soc., **89**, 3642 (1967); H.L. Crespi, U. Smith, and J.J. Katz, Biochemistry, **7**, 2232 (1968); A.W. Nichol and D.B. Morrel, Biochim. Biophys. Acta, **184**, 173 (1969); R. Bonnett and A.F. McDonagh, J. Chem. Soc., Perkin Trans. 1, 881 (1973).
8. A.H. Jackson, G.W. Kenner, K.M. Smith, R.T. Aplin, H. Budzikiewicz, and C. Djerassi, Tetrahedron, **21**, 2913 (1965).
9. D.R. Hoffman, J. Org. Chem., **30**, 3512 (1965).
10. R.C. Dougherty in "Biochemical Applications of Mass Spectrometry", Ed. G.R. Waller, Wiley, New York (1972), p. 591.
11. R.L. Levy, H. Gesser, E.A. Halevi, and S. Saidman, J. Gas Chromatogr., **2**, 254 (1964); D.G. Whitten, K.E. Bentley, and D. Kuwada, J. Org. Chem., **31**, 322 (1966).
12. For reviews of the geochemistry of porphyrins, see G.W. Hodgson, Ann. N.Y. Acad. Sci., **206**, 670 (1973); R. Bonnett, ibid., p. 726.
13. E.W. Baker, J. Am. Chem. Soc., **88**, 2311 (1966); E.W. Baker, T.F. Yen, J.P. Dickie, R.E. Rhodes, and L.F. Clark, ibid., **89**, 3631 (1967).
14. R. Bonnett, P. Brewer, K. Noro, and T. Noro, J. Chem. Soc., Chem. Commun., 562 (1972).
15. R. Grigg, G. Shelton, A. Sweeney, and A.W. Johnson, J. Chem. Soc., Perkin Trans. 1, 1789 (1972); J.B. Paine III and D. Dolphin, J. Am. Chem. Soc., **93**, 4080 (1971); G.W. Kenner, K.M. Smith, and M.J. Sutton, Tetrahedron Lett., 1303 (1973).

16. F.W. McLafferty, "Mass Spectrometry of Organic Ions", Academic Press, New York (1963); K. Biemann, "Mass Spectrometry, Organic Chemical Applications", McGraw-Hill, New York (1962); J.H. Beynon, "Mass Spectrometry and its Applications to Organic Chemistry", Elsevier, Amsterdam (1960).

17. H.H. Inhoffen, J.-H. Fuhrhop, and F. v.d. Haar, Ann. Chem., **700**, 92 (1966).

18. H. Budzikiewicz and F. v.d. Haar, Org. Mass Spectrom., **1**, 323 (1968); H. Budzikiewicz, Adv. Mass Spectrom., **4**, 313 (1968).

19. M.T. Cox, T.T. Howarth, A.H. Jackson, and G.W. Kenner, J. Chem. Soc. Perkin Trans. 1, 512 (1974).

20. R.K. Ellsworth and S. Aronoff, Arch. Biochem. Biophys., **125**, 269 (1968); **130**, 374 (1969).

21. A.H. Jackson, G.W. Kenner, and J. Wass, J. Chem. Soc., Perkin Trans. 1, 480 (1974).

22. R.V.H. Jones, G.W. Kenner, and K.M. Smith, J. Chem. Soc., Perkin Trans. 1, 531 (1974).

23. H. Budzikiewicz and S.E. Drewes, Ann. Chem., **716**, 222 (1968).

24. M.T. Cox, A.H. Jackson, and G.W. Kenner, J. Chem. Soc. C, 1974 (1971).

25. R.B. Woodward, Angew. Chem., **72**, 652 (1960); Ind. Chim. Belge, **27**, 1293 (1962).

26. G.H. Barnett, M.F. Hudson, S.W. McCombie, and K.M. Smith, J. Chem. Soc., Perkin Trans. 1, 691 (1973).

27. A.D. Adler, J.H. Green, and M. Mautner, Org. Mass Spectrom., **3**, 955 (1970); M. Meot-Ner, J.H. Green, and A.D. Adler, Ann. N.Y. Acad. Sci., **206**, 641 (1973).

28. E.W. Baker, A.H. Corwin, E. Klesper, and P.E. Wei, J. Org. Chem., **33**, 3144 (1968).

29. H. Budzikiewicz, F. v.d. Haar, and H.H. Inhoffen, Ann. Chem., **701**, 23 (1967).

30. R.C. Dougherty, H.H. Strain, W.A. Svec, R.A. Uphaus, and J.J. Katz, J. Am. Chem. Soc., **92**, 2826 (1970).

31. H.H. Inhoffen, P. Jäger, R. Mahlhop, and C.-D. Mengler, Ann. Chem., **704**, 188 (1967). See also refs. 18 and 29.

32. A.H. Jackson and G.R. Dearden, Ann. N.Y. Acad. Sci., **206**, 151 (1973).

33. H.H. Inhoffen and J.W. Buchler, Tetrahedron Lett., 2057 (1968).

34. J.-H. Fuhrhop, Chem. Commun., 781 (1970); K.M. Smith, ibid., 540 (1971).

35. D. Rosenthal, F.R. Hopf, D.G. Whitten, and M.M. Bursey, Org. Mass Spectrom., **7**, 497 (1973).

36. K.M. Smith, Org. Mass Spectrom., **6**, 1401 (1972); R.J. Abraham, G.H. Barnett, and K.M. Smith, J. Chem. Soc., Perkin Trans. 1, 2142 (1973).

37. M.F. Hudson and K.M. Smith, unpublished results.

38. G.W. Sovocool, F.R. Hopf, and D.G. Whitten, J. Am. Chem. Soc., **94**, 4350 (1972).

39. M. Meot-Ner, A.D. Adler, and J.H. Green, Org. Mass Spectrom., **7**, 1395 (1973).

40. P.S. Clezy, C.L. Lim, and J.S. Shannon, Aust. J. Chem., **27**, 2431 (1974).

41. P.S. Clezy, C.L. Lim, and J.S. Shannon, Aust. J. Chem., **27**, 1103 (1974).

NUCLEAR MAGNETIC RESONANCE SPECTROSCOPY
OF PORPHYRINS AND METALLOPORPHYRINS*

HUGO SCHEER** and JOSEPH J. KATZ

Chemistry Division, Argonne National Laboratory, Argonne, Illinois, 60439, U.S.A.

10.1. Introduction

The rapid development of proton nuclear magnetic resonance (n.m.r.) spectroscopy since about 1960 has had a strong influence on the study of almost all classes of organic compounds[1]. There are, however, few categories of compounds for which such a wealth of information can be obtained by n.m.r. as for porphyrins. This circumstance arises for the most part from the large magnetic anisotropy (ring current) of the aromatic macrocycle of these compounds[2,3]. The ring current functions as a built-in chemical shift reagent, and spreads the proton magnetic resonance ([1]Hmr) spectrum of porphyrins over the unusually large range of more than 15 p.p.m. This in consequence generally renders the [1]Hmr spectra first order, simplifying interpretation and assignment, and makes [1]Hmr a very sensitive probe of structural modifications. The ring current effects, in addition, allow detailed studies of molecular interactions of porphyrins in solution.

In the early applications of n.m.r. to porphyrins, [1]Hmr was the most widely used as an analytical tool, and the new structural insights that resulted were a major reason for the revival of interest in porphyrin chemistry. A few examples of important pioneering work may be cited here. The first [1]Hmr spectra of porphyrins were reported by Becker and Bradley[2a], and by Ellis et al.[2b], and an early survey on a variety of porphyrin structures was carried out by Caughey and Koski[4]. Based on the extensive synthetic work of Jackson, Kenner, and Smith[5], a series of researches was carried out by Abraham on a number of special aspects of porphyrin behavior, especially

---

* Work performed under the auspices of the U.S. Energy Research and Development Administration.
** Stipendiate of the Deutsche Forschungsgemeinschaft, Bonn—Bad Godesberg, W. Germany; present address: Institut für Botanik der Universität, D 8000 München 19, W. Germany.

Porphyrins and Metalloporphyrins, ed. Kevin M. Smith
© 1975, Elsevier Scientific Publishing Company, Amsterdam, The Netherlands

the effects of substitution[8a,8b], the self-aggregation[6−8] in solution, iso-merism of porphyrins[6,8,9], and interactions of nucleophiles with the central metal atom of metalloporphyrins[10,11]. Inhoffen et al.[12] investigated a great variety of chlorophyll derivatives, and widely applied octaethylporphyrin H_2(OEP) as a powerful model compound for the naturally-occurring porphy-rins. Closs et al.[13] studied molecular interactions in chlorophylls and chloro-phyll derivatives, and were able to delineate many of the salient features of the self-aggregation of the chlorophylls in the convenient form of aggrega-tion maps. In addition to studies undertaken on porphyrins themselves, ligand molecules bound to them have been the subject of investigation by n.m.r. The ring current induced shifts (RIS) by the porphyrin macrocycle on the chemical shifts of axial ligands serve as an alternative probe and thus supplement and complement the pseudocontact shift produced by lantha-nide shift reagents (LIS) so important for conformational and stereochemical studies[14−19].

Kowalsky's[20] early report of sharp proton resonance lines in the porphy-rin moiety of cytochrome-c that lie far outside the usual chemical shift range for protons led to an extensive study of paramagnetic metal complex-es[21−29]. The extremely large proton chemical shifts observed in these com-pounds are produced by hyperfine nuclear interactions with the unpaired electrons of the central metal atom. As these hyperfine shifts are dependent on oxidation state, spin state, and axial ligands coordinated to the central metal ion, n.m.r. has been used as a probe in structural and functional studies of heme and hemoproteins[21,22].

In recent years, nuclei other than protons, especially ^{13}C, have become important in n.m.r. spectroscopy[30]. Although of the same absolute magni-tude, the ring current effect in [13]Cmr is small relative to the magnitude of the intrinsic chemical shifts and the ring current plays only a minor role[31], whereas paramagnetic contributions from low-lying excited states make a decisive contribution to the ^{13}C chemical shifts[30]. The influence of metala-tion on the electronic structure of porphyrins has been studied in some detail[32,33], and two n.m.r. publications focus on the conjugation pathway in porphyrins[32,34].

In contrast to the unusual chemical shifts often observed, coupling con-stants in porphyrins are quite normal. The [1]Hmr subspectra of various sub-stituents are in most cases first order, and long range coupling constants are usually only observed in porphyrins with unsubstituted peripheral (β)-posi-tions. Recently, some data on ^{1}H coupling constants with [13]C[35−37], [15]N[33], and [205]Tl[8,9,11,32] have been published and have been given straight-forward explanations.

10.1.1. The chemical shift
The magnetic resonance frequency ν of a nucleus is given by

$$\nu = \frac{\gamma}{2\pi}.$$ (1)

The gyromagnetic ratio γ is a natural constant for a particular nucleus, and H is the magnetic field experienced by it. Although the latter is usually very close to the external magnetic field, H_0, applied in the experiment, the field at a particular nucleus is modified by its chemical environment. The additional local magnetic field produced by neighboring nuclei with magnetic properties are proportional to H_0, and eq. (1) can then be rewritten as:

$$\nu = \frac{\gamma}{2\pi} H_0 (1-\sigma),$$ (2)

where the shielding constant σ is a measure of the modification of the external magnetic field H_0 by the chemical environment.

Shielding of a particular proton from the external magnetic field results from currents induced within the electron system of the atom and its surroundings (Larmor precession)[1,38], and the overall shielding is usually divided into several contributions to the chemical shift* experienced by a particular nucleus[1]. Local magnetic effects arise from changes (with respect to the free atoms) in the density and the shape of the electron cloud surrounding a particular proton, and long-range magnetic effects occur from magnetically anisotropic groups in the neighborhood of a particular proton or group of protons. Both of these effects contain diamagnetic contributions, which reflect changes in the magnitude of the electron density, and paramagnetic contributions that originate from changes in the shape of the electron cloud**. In an alternative and equivalent representation, paramagnetic shifts arise from distortions of the ground state orbitals from the mixing of the wave functions of the ground state and low-lying excited states. In compounds containing unpaired electrons, hyperfine interactions result from contact shifts (non-zero spin density at the nucleus) whose effects are transmitted through the chemical bonds in the molecule, and pseudocontact shifts transmitted through space. Proton chemical shifts arising from the presence of unpaired spins can be orders of magnitude larger than those observed in diamagnetic molecules. In addition to all of these internal ef-

---

* Due to the small magnitude of the shielding constant σ ($\sim 10^{-5}$) and the difficulties in measuring its absolute value, it is usually expressed as the *chemical shift* relative to that of a reference compound. Throughout this chapter, the chemical shift is given in δ units (parts per million, p.p.m.) where $\delta = -10^6$ $(\sigma - c)$, and c is the shielding constant for the protons in the usual internal standard tetramethylsilane (TMS). Another commonly employed internal standard is hexamethyldisiloxane (HMS), whose protons come into resonance at slightly higher field than TMS. In some publications, τ is used instead of δ as a measure of chemical shift. These two quantities are related by $\tau = 10 - \delta$.

** The terms paramagnetic and diamagnetic shifts are sometimes used in a different sense designating low-field and high-field shifts, respectively. On the other hand, paramagnetic as well as diamagnetic contributions (as defined above) describe shielding mechanisms which can be both positive (= shielding, high-field shift) and negative (= deshielding, low-field shift).

References, p. 514

fects, solvent-induced proton chemical shifts may occur from more or less specific interactions of solute molecules with a solvent that possesses magnetic anisotropy*.

10.1.2. The aromatic ring current

In the [1]Hmr spectra of diamagnetic porphyrins, the long-range diamagnetic contribution of the aromatic macrocyclic system to the chemical shift is the most important single factor that distinguishes porphyrins from similar non-aromatic structures. Consequently, we shall describe this ring current term in somewhat more detail.

If a closed loop of electrons is subjected to an external magnetic field, a Larmor precession[1] of the entire π-cloud is induced. The circulation of the electrons (ring current) gives rise to a secondary magnetic field that is shown in Fig. 1. This effect is strongly anisotropic, it does not average out to zero by random tumbling of the molecule, and thus the ring current gives rise to an anisotropic shielding effect on protons within the range of the ring cur-

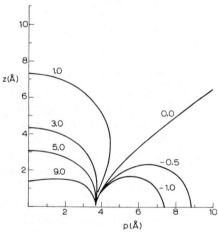

Fig. 1. The magnetic anisotropy of the porphyrin ring system (from Ref. 40). The iso-shielding lines (incremental shift Δ[p.p.m.]) were obtained by a classical ring current calculation (*vide infra*) with two circular loops above and below the macrocyclic plane. The radius '*r*' and spacing '*s*' (see Fig. 2) and π-electron number were adjusted to fit the [1]Hmr data observed for the stacked system (10) (see text). The calculation included four additional loop-pairs for the peripheral benzene rings of the phthalocyanine, but only the contribution of the inner tetra-azaporphyrin system is shown here. The abcissa gives the radial distance from the center of the macrocycle, and the ordinate, the z (out-of-plane) coordinate. The three dimensional picture is obtained by rotating this cross-section around the z-axis.

---

* Bulk magnetic susceptibility changes due to the geometry of the sample and the magnetic properties of the solvent system employed are usually dealt with by use of internal standards, or, where practical, by appropriate susceptibility corrections[1].

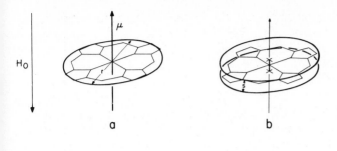

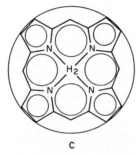

Fig. 2. Schematic drawings of the classical ring current models for porphyrins. (a) The single-loop approach with the radius 'r' and the π-electron number as variables, approximated by the magnetic point dipole μ; (b) the loop-pair approach with the spacing 's' as additional variable; (c) the network or multi-loop pair approach, viewed from the top, with each circle representing a pair of loops above and below the macrocycle. For discussion see text.

rent. This classical 'ring current' model of Pauling[39] has been widely used as a criterion for aromaticity or anti-aromaticity, depending on the sign and magnitude of the related shielding constant. (For a detailed discussion, see Ref. 41.) For an aromatic system such as the porphyrins, the magnetic shielding resulting from the ring current is positive for nuclei on the outside of the loop, and negative for nuclei positioned within the loop (Fig. 1). The first approach to the calculation of the shielding of aromatic nuclei by this classical picture was made by Pople[42]. He assumed a single loop in the plane of the aromatic system, the magnetic field of which can be treated approximately (for protons at the periphery) by a point-dipole in the center of the loop (see Fig. 2a). This treatment has been refined by Waugh and Fessenden[43] and by Johnson and Bovey[44], who used instead two separate loops situated symmetrically at 0.45 Å and 0.65 Å, respectively, above and below the plane of the aromatic ring (Fig. 2b).

Based on the work of London[45], a molecular orbital treatment was developed by Pople[46], McWeeney[47], and Hall and Hardisson[48a] that essentially gives a similar picture as the classical ring current approach, but is more

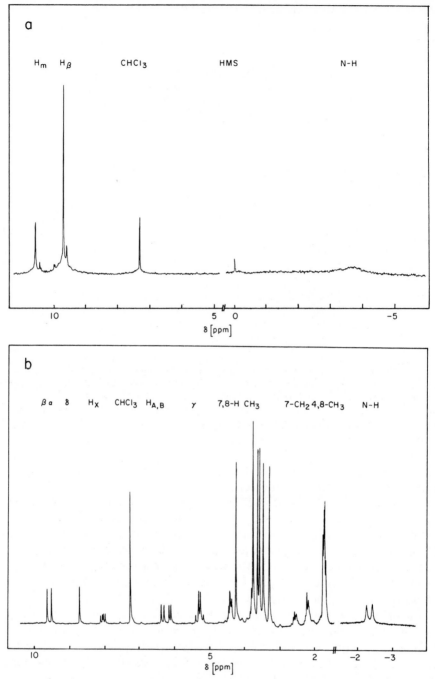

Fig. 3. ^{1}Hmr spectra (220 MHz pulse FT) of (a) porphin (1), saturated solution ($\sim 5 \times 10^{-5}$ M) in C^{2}HCl$_3$, 2000 pulses, repetition rate 0.49 sec, spectrum width 4500 Hz; and (b) chlorin-e_6 trimethyl ester (14), 5×10^{-3} M in C^{2}HCl$_3$, 1000 pulses, repetition rate 2.45 sec, spectrum width 3500 Hz.

versatile in complex systems. Although the ring current approach neglects other contributions (i.e., paramagnetic or σ-system terms) and gives results that are often only in qualitative agreement with experiment, this approximation has proved most useful for the qualitative and even semi-quantitative interpretation of chemical shifts in various aromatic and anti-aromatic systems. (For a recent review on a discussion of the limitations of the ring current approach, see Ref. 41.)

The sizeable ring current effect associated with the π-system of the aromatic porphyrin macrocycle was recognized as a dominant feature of the first [1]Hmr studies of porphyrins[2−4]. In the [1]Hmr spectrum of porphin (1) (Fig. 3a), the resonances of the peripheral protons are shifted about 5 p.p.m. to lower field as compared to those of pyrrole[1], whereas the resonances of the inner N protons are shifted about 11 p.p.m. to higher field. Becker et al.[2a] used both the single loop (point dipole) and double loop model, assuming a radius of 3.3 Å for the 18 π-electron loop, and a variable spacing for the loop pair in the latter (Fig. 2a,b). Ellis et al.[2b] used a double loop pair model (Fig. 2b) to provide a semi-quantitative description of the ring current effect. These authors obtained a self-consistent model with loops of a radius of 3.7 Å, a spacing of 1.28 Å and an effective ring current of 18.8 π-electrons*. In contrast to these studies, which consider only a single pair of loops, Abraham[3] used a network approach[39], in which auxiliary pyrrole loops and chelate hexagon loops as well are all explicitly taken into account in addition to the main macrocycle loop itself (Fig. 2c). The calculated chemical shifts obtained by this procedure are too large by the same factor of 1.5 that was observed earlier in the application of this model to other polycyclic compounds. A similar approach has been used more recently by Mamaev[48b] for the study of some principal porphyrins and of substituent effects. While earlier investigations were focussed on the ring current effect on protons within the plane of the macrocycle, its effect on protons above (and below) the macrocycle plane was studied experimentally by Storm et al.[16], Katz et al.[51] and Janson et al.[52] From the chemical shifts for N-substituents in porphyrins and axial ligands in metalloporphyrins, a semi-empirical formula was deduced that gives with fair accuracy the chemical shift of protons within the ring current loop but above or below its plane[16,17].

10.1.2.1. Ring current in related macrocycles

The porphyrin macrocycle can formally be regarded as a bridged diaza[18]annulene (2) with two isolated peripheral double bonds, or as a tetraaza[16]annulene dianion (3) with four isolated double bonds (for leading references, see Refs. 12, 18, 19 and 32).

---

* The 'best self consistent fit' obtained in this semi-empirical approach characteristically gives fictitious or unreal values for the number of electrons in the ring current (Section 10.1.2.2).

TABLE 1

¹Hmr chemical shifts (δ[p.p.m.] from TMS) of 18 π-aromatic compounds related to the porphyrins.

Compd.				Ref.
(4) 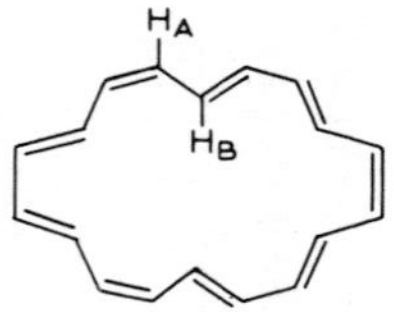		H_A (quadruplet)	H_B (quintuplet)	
	obsd.	9.03	-2.26	53,54
	calcd.	11.06	-12.6	50
			THF$-^2H_8$, 0°C	
(5) 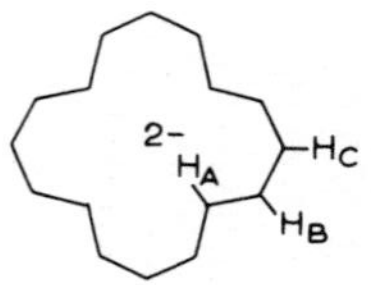		H_A	H_B	H_C
	obsd.	-8.07	8.77	7.40
			(J_{AB} = 13 Hz, J_{BC} = 9.5 Hz)	
	calcd.	-10.9	10.2	10.2
			THF$-^2H_8$, -30°C, Li⁺ as gegenion	50
(6) 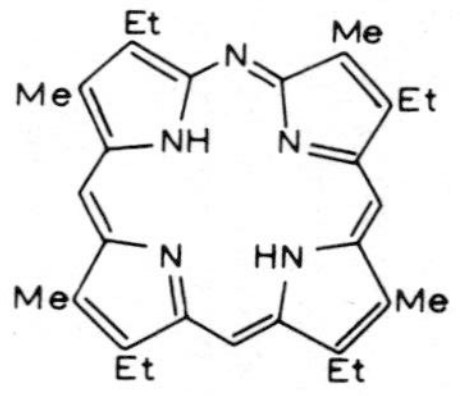	Methine—H:	γ: 10.02		
		β,δ: 9.90		
	Ring—CH₃:	3.58, 3.50		
	CH₂—CH₃:	3.91, 1.83	C²HCl₃	55
(7) 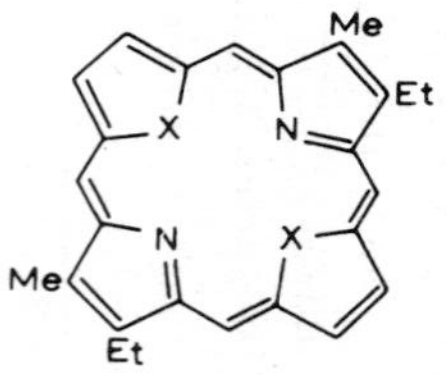	Methine—H		β-furane or β-thiophene—H	
	a) X = O, Y = O: 11.71		10.98	56
	b) X = O, Y = S: 10.59, 10.05		10.01, 9.69	56
	c) X = S, Y = S: 10.68, 10.71			57
			Dihydrobromide in TFA	

(8)

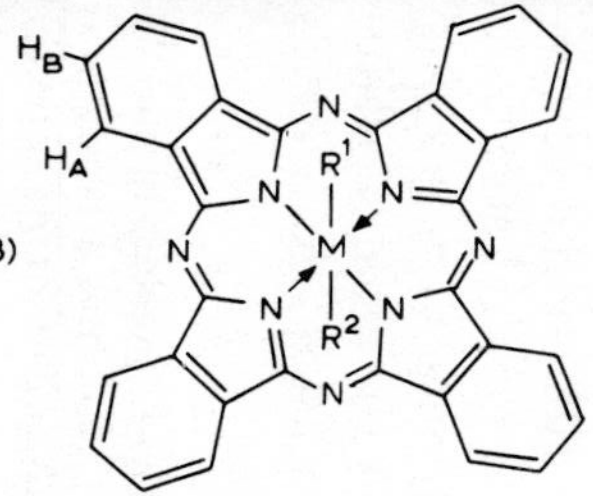

	H_A	H_B	C^2HCl_3	52
M=Si R^1=CH$_3$ R^2=OSi CH$_3$[OSi(CH$_3$)$_3$]$_2$ −2.99 −1.26 −6.31	9.60	8.27		
M=Si R^1=R^2=OSiCH$_3$[OSi(CH$_3$)$_3$]$_2$ −2.90 −1.22	9.63	8.30		
M=Si R^1=R^2=OSi(CH$_2$CH$_3$)$_3$ −2.48 −1.25	9.68	8.36		
M=Ge R^1=R^2=OSi(CH$_2$CH$_3$) −2.42 −1.24	9.70	8.38		

(9)

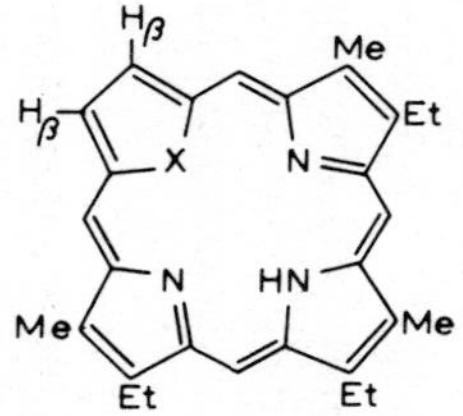

	Methine—H	β-thiophene or β-furane	N—H	C^2HCl_3 56
X=O	10.12, 9.98	9.69	—	
X=S	10.06, 10.00	9.98	−4.98	

References, p. 514

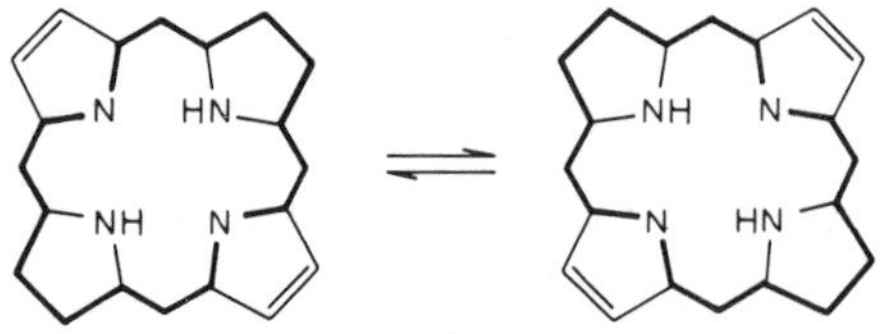

Indeed, striking similarities in electronic transition and magnetic resonance spectroscopy are evident between porphyrins and the [16]-annulene di-anion (5)*[49,50].

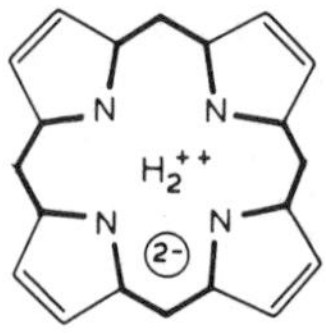

As far as electron states are concerned, porphyrins may be regarded as only a special case of bridged 18 π-heteroannulenes. The n.m.r. properties of several related structures are listed in Table 1. All compounds show strongly shielded inner protons and strongly deshielded outer protons (relative to similar structures with interrupted macrocyclic conjugation, Section 10.2.6) in a manner characteristic of protons of aromatic systems. Replacement of one N—H group in a pyrrole ring of porphyrins by O or S (see Chapter 18) leaves the methine proton resonances of the macrocycle almost unchanged (see Table 1), while in the disubstituted porphyrins the methine protons are more strongly deshielded. As substitution in these compounds is accompanied by local structural changes implicit in the synthesis (i.e., removal of alkyl side chains, change in the entire geometry of the molecule), the chemical shift effects resulting contain contributions other than ring current effects.

The [16]-annulene di-anion (5) and the [18]-annulene (4) have been thoroughly investigated by Oth et al.[50] (Table 1). Based on the 'normal' value of $\delta = 5.8$ p.p.m. for olefinic protons[50], the chemical shifts for both systems were calculated assuming contributions only from the diamagnetic ring current (based on London's treatment[45]) and the negative charge in the case of the di-anion. The agreement with the experimental data is good for the [16]-annulene di-anion and fair for the [18]-annulene, although again in both cases the influence of the ring current is exaggerated.

The ring current effect in silicon and germanium phthalocyanines was studied by Janson et al.[52] These authors circumvented the difficulties involved in defining a reference system in an elegant way. Silicon and germanium phthalocyanines can form stacked complexes of the general structure (10), where X can vary from 0 to 4. In this way, the influence of a new ring added to the stack at one end on the phenyl protons and the axial substituent R of the successive porphyrin macrocycles could be measured for various

---

* It should be noted that in solutions of the di-anion, structure (5) is present in the 8S configuration characteristic for porphyrins, while the annulene (4) assumes a configuration different from the one present in porphyrins (Table 1) and thus cannot be directly compared.

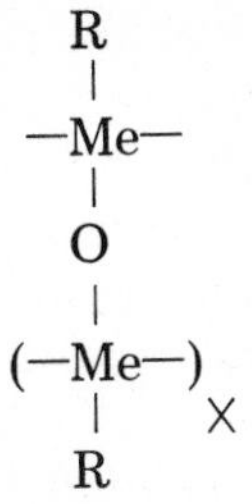

R
|
—Me—   R = CSiCH$_3$[OSi(CH$_3$)$_3$]$_2$
| M = Si, Ge
O   X = 0—4
|
(—Me—)
| X
R

(10)

stack heights. Semi-empirical calculations of the ring current (according to the treatment of Johnson and Bovey[44]) led to a self-consistent set of parameters for a five double loop ring current model. In the model of Janson et al.[52] four benzenoid loops[44] were added to the central macrocycle loop, which has a diameter of 3.90 Å, a separation of 0.64 Å, and an effective ring of 8.43 electrons*. The ^{1}Hmr spectrum of tetrabenzoporphin, the parent compound of the phthalocyanines, was reported recently[60]. Due to the common presence of paramagnetic impurities[58], no systematic study has been done on this class of porphyrins.

10.1.2.2. Ring current and structure

Almost any structural modification to the macrocyclic system changes its ring current, as indicated by changes in the chemical shifts of protons remote from the point of structural change. In spite of its success in describing the general features and in providing good estimates of the proton shifts observed in some porphyrins, the ring current model has to be used carefully, however, in attempts to make quantitative predictions for the consequences of structural modifications in porphyrins. Several reasons can be adduced for this situation. First, although the ring current is a major contributor to the chemical shift, it is not the only source for the unusual shifts observed in aromatic compounds (see above). Second, the ring current is not localized as in a wire, but in orbitals that are subject to hybridization changes. It is thus characteristically found that the 'best self-consistent fit' in ring current calculations is usually obtained with fictitious or unreal values for π-electron number as well as the ring current radius and the distance of π-cloud separation. Third, porphyrins are polycyclic systems, and therefore the true ring current may be affected not only in intensity but also in position by changes in the relative contributions within the various loops into which the total ring current is decomposed[3]. In principle, a much better insight into the relative distribution of the ring current can be provided by ^{13}C spectra.

---

* See footnote on p. 405.

However, the relatively small ring current effects on ^{13}C chemical shifts make separation of the ring current contributions from other operative factors very difficult[31], and the two publications on the subject arrive at opposite conclusions[32,34]. Fourth, local changes can lead to conformational changes in the macrocycle as a whole (Section 10.4.3) thereby changing the magnetic environment of a proton remote from the site of structural modification.

In spite of these limitations, the simple double loop ring current model (Fig. 3b) has been extremely useful from a practical point of view in the interpretation of the ^{1}Hmr spectra of various classes of porphyrins. To evaluate the contribution of the ring current to the ^{1}Hmr of structurally altered porphyrins the following criteria are usually helpful: (a) only well-assigned signals of protons close to the aromatic systems, but far from the locus of modification should be used; (b) resonances of protons inside versus outside the aromatic macrocycle ought to experience opposite shifts; (c) side effects from conformational changes of the ring system have to be taken into account.

Some general aspects of the ring current model as applied to porphyrins may also be summarized here: (a) the ring current is larger in both metal complexes (for some exceptions, see Section 10.2.8.1) and di-cations of porphyrins (Section 10.2.7). This effect is explained by Abraham[3] as a result of increased resonance stabilization in these classes of compounds. (For an alternative explanation, see Ref. 41) (b) Steric hindrance, that is, effects that reduce π—π overlap by distorting the planar macrocycle structure causes a decrease in ring current. This point is further elaborated below (see N-substitution, (Section 10.2.4), meso-substitution (Section 10.2.3), and stereochemistry (Section 10.4.3)). (c) Decrease in the electron density of the π-system diminishes the ring current and may thus cause up-field proton shifts, even though a decrease in electron density generally leads to a reduced shielding and down-field shifts. The latter behavior, a down-field shift upon introduction of electron-withdrawing groups, is usually observed in benzene derivatives, where, for example, the ortho, meta, and para proton signals in benzaldehyde are shifted to lower field by 0.58, 0.21 and 0.27 p.p.m., respectively, as compared to the protons in benzene itself[1]. In porphyrins, however, the deshielding effect of electron withdrawal is usually overcompensated by the simultaneous decrease in ring current that results from lowering of the electron density. Thus, the peripheral α, β, and δ methine proton signals in 9-keto pheophorbides are shifted by 0.43, 0.26, and 0.41 p.p.m., respectively, to lower fields as compared to the respective 9-desoxo compound (Tables 5, 11, 12). This effect was first investigated by Caughey et al.[4] and is quite general in porphyrin ^{1}Hmr. As expected from these considerations, the interior N—H proton signals move in just the opposite direction when the ring current is lowered, and are deshielded by 1.5 p.p.m. in the above cited case.

10.1.3. Practical considerations for ^{1}Hmr of porphyrins

The ^{1}Hmr spectra of porphyrins, especially some of the metalloporphyrins, are strongly solvent, concentration and temperature dependent. This is due to the tendency of porphyrins to experience self-aggregation, and this, in combination with the strong magnetic anisotropy of the porphyrins has major consequences for the ^{1}Hmr spectra (Section 10.4.1). In the free porphyrin bases, aggregation is weak, and parallels the π—π aggregation behavior generally observed in aromatic molecules[6]. In metalloporphyrins, self-aggregation or ligation to donor (Lewis base) molecules is usually much stronger and more specific, and occurs by interaction of polar side chains or donor groups in one molecule with axial interaction sites on the central metal ions of another[13,59]. Aggregates of both types have been very useful in the study of molecular interactions with porphyrins (see Section 10.4.1). The formation of self-aggregates on the one hand, or coordination interaction products with extraneous nucleophiles on the other, however, presents a serious problem in making structural deductions from n.m.r. data. Under aggregating conditions the accurate determination and assignment of chemical shifts becomes especially important, as aggregation shifts of more than 2 p.p.m. may occur for the resonances of particular protons as a result of close proximity to the ring current of another macrocycle. A rigorous approach to the problems proposed by aggregation requires mapping the concentration-dependence of the chemical shifts and extrapolation to infinite dilution, but this procedure is really practical only for certain important compounds. The aggregation problem in the assignment of chemical shifts can in general be circumvented by recording the spectra in trifluoroacetic acid (TFA), in which both π—π and coordination-aggregates are broken down by dication formation or by preferential ligation of the metal axial coordination sites with TFA. For sufficiently stable compounds this is a very useful approach, particularly because TFA is an excellent solvent even for otherwise only poorly soluble free base porphyrins.

For compounds unstable in TFA, however, other solvent systems must be used. In metalloporphyrins, strong self-aggregation caused by donor—acceptor interactions involving the central metal ion can be avoided by addition of small amounts of bases (tetrahydrofuran, methanol) to chloroform (or other nonpolar solvents) to compete for the metal coordination site. For free base porphyrins, the concentration should be maintained as low as possible and constant for a series of compounds. Several laboratories use standard concentrations for recording porphyrin spectra whenever possible. For example, the Braunschweig group uses 0.05 M in C^2HCl_3 as their standard, a concentration that is usually on the monomer side for free porphyrin bases, and gives a reasonable signal-to-noise (S/N) ratio in single scan continuous-wave (cw) n.m.r. experiments. However, this requires about 10 mg of material dissolved in 0.3 ml of solvent in a 5 mm n.m.r. sample tube, a concentra-

TABLE 2

Incremental shifts (Δ[p.p.m.$_x$]) in C^2HCl_3 due to the aromatic ring current in porphyrins

N	1	2	3	4	5
Methine	−9.5	−3.8	−1.2	−0.6	n.e.
β-pyrrole	−8.5	−3.3	−1	−0.3	n.e.
N	+4	+5.9	+3.7	+2.8	+1.9
Chlorin	−4	−0.9	−0.2	−	−

The expected chemical shift for a proton or a substituent in various positions of the porphyrin can be estimated from its chemical shift in the related aliphatic compound, R—CH$_3$, and the listed increments. N is the number of bonds between the respective proton and the indicated C or N atom, respectively. Upon reduction of the ring current the increments are reduced proportionally.

tion sometimes not accessible because of insolubility or unavailability of material.

As lower concentrations, down to less than 5×10^{-4} M, are sufficient for routinely recording ^{1}Hmr spectra in modern n.m.r. equipment by pulse Fourier transform spectroscopy, a lower standard concentration is probably useful where a broad range of compounds are to be investigated. In any case, the solvent system should always be quoted together with the concentrations at which the ^{1}Hmr spectra have been recorded.

In ^{13}Cmr, ring current effects are relatively less important, and weak self-aggregation of free porphyrin bases plays a lesser role in determining the relative magnitude of chemical shifts. Consequently, solubility limitations and the sizeable amount of material needed for ^{13}Cmr spectra at natural abundance are the major problems for a broad application of this technique. Concentrations of 0.1 M are desirable for ^{13}C natural abundance work, more if a high S/N ratio is desired.

10.2. ^{1}HMR spectra
10.2.1. The ^{1}Hmr spectra of diamagnetic porphyrins
The ^{1}H chemical shifts of six important porphyrins selected to illustrate the n.m.r. behavior of typical porphyrins are given in Table 3, and two typical spectra are shown in Fig. 3.

10.2.1.1. Porphin
The spectrum of free base porphin (1), the parent compound of the porphyrins, was obtained only recently because of its poor solubility[60]. Porphin shows three signals (Fig. 3a), two at low field assigned to the methine and the β-protons, and one at high field assigned to the NH protons. At room temperature, the N—H exchange between the possible tautomeric

TABLE 3

^{1}Hmr chemical shifts (δ[p.p.m.] from TMS) of some principal porphyrins

Structure	Chemical shifts	Remarks	Ref.
(1) 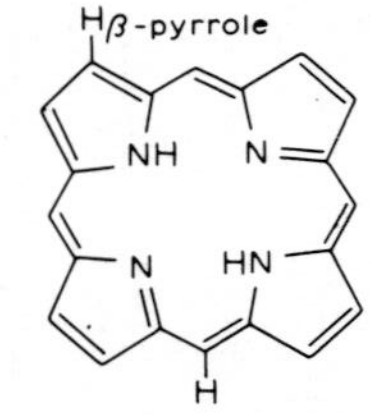	$H_{Methine}$: 10.58 $H_{\beta\text{-pyrrole}}$: 9.74 NH: −3.76	Satd. solution in C^{2}HCl$_3$	60,67
(11) 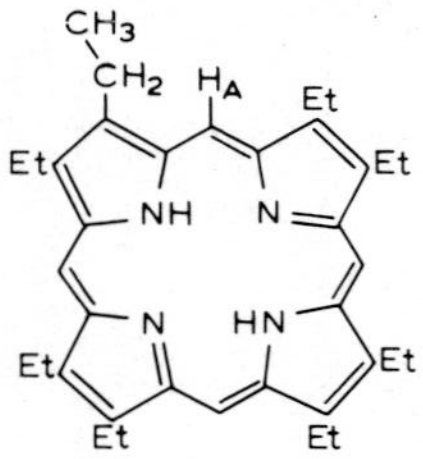	$H_{Methine}$: 10.18 CH_2: 4.14 CH_3: 1.95 NH: −3.74	RT, C^{2}HCl$_3$	71
(12) 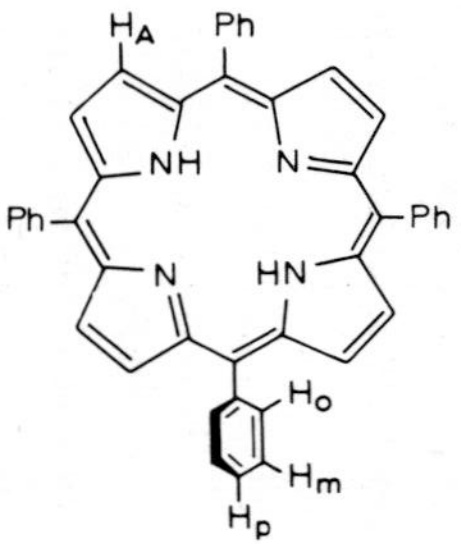	H_A: 8.75 H_o: 8.30 (m) H_m: 7.80 (m) H_p: 7.80 (m) NH: —	30°, C^{2}HCl$_3$/CS$_2$	74

References, p. 514

TABLE 3 (continued)

		Remarks	Ref.
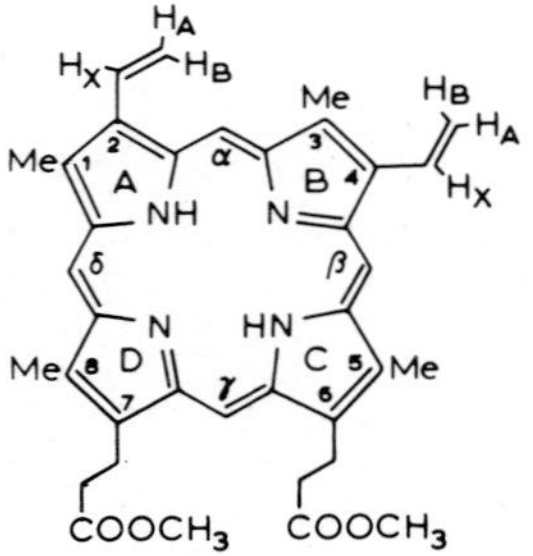 Methine—H: $\alpha, \beta, \gamma, \delta$ (s):	9.96, 9.95, 9.83, 9.82	220 Mz, C^2HCl_3	68
2-vinyl: H_X, H_A, H_B:	8.10, 6.09, 6.26		
4-vinyl: H_A, H_A, H_B:	8.14, 6.09, 6.26		
	ABX, $J_{AB} = 2$		
	$J_{AX} = 12$		
	$J_{BX} = 19$		
6,7-CH_2—CH_2—$COOCH_3$:	3.59, 3.16, 3.59		
Arom. CH_3; 1, 3, 5, 8:	3.48, 3.48, 3.54, 3.55		
NH:	-4.08		
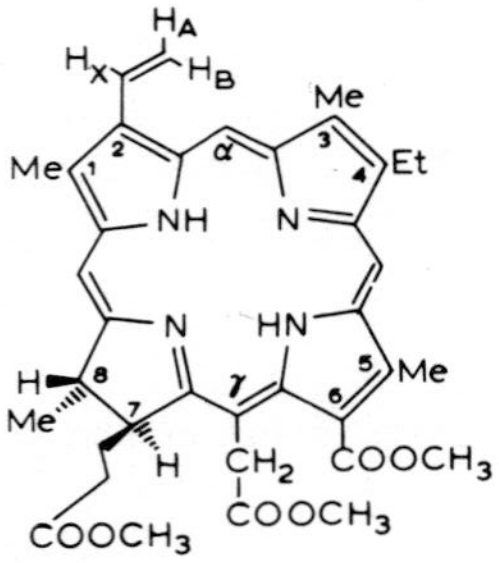 β, α, δ—H:	9.65, 9.50, 8.70 (s)	0.05 m in C^2HCl_3	72,73
H_A, H_B, H_X:	6.07, 6.28, 8.00 (ABX)		
	$J_{AB} = 2, J_{AX} = 12, J_{BX} = 18$ Hz		
γ-CH_2—$COOCH_3$:	5.27 (AB, $J = 17$ Hz), 3.73 (s)		
7,8-H:	4.38 (m), 4.40 (q, 7Hz)		
4-CH_2—CH_3:	3.73 (q, 7Hz), 1.69 (t, 7Hz)		
1,3,5-CH_3:	3.43, 3.24, 3.55 (s)		
6,7′′-$COOCH_3$:	4.23, 3.60 (s)		
8-CH_3:	1.73 (d, 7Hz)		

(13)

(14)

13

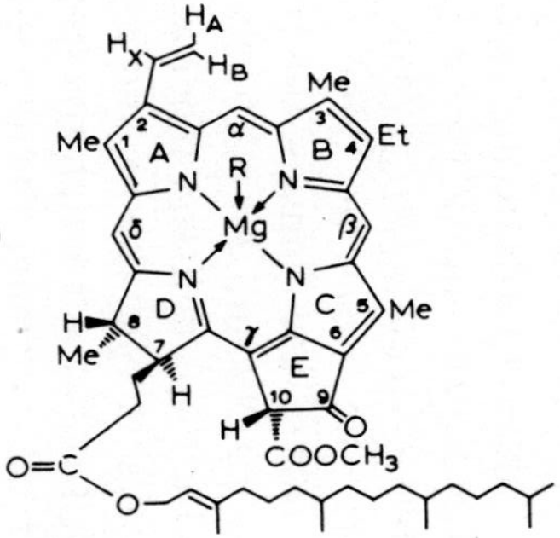

$C^2HCl_3/$
$C^2H_3O^2H$

β, α, γ—H:	9.50, 9.23, 8.28
H_A, H_B, H_X:	5.97, 6.13, 7.92 (ABX, $J_{AB} = 1.7$, $J_{AX} = 11.2$, $J_{BX} = 18.1$)
10-H:	6.22 (s)
7,8-H:	4.14 (m), 4.27 (q), $J_{7H,8H} \leqslant 2$ Hz
4-CH_2—CH_3:	3.75 (q), 1.72 (t)
10 α-CH_3:	3.97 (s)
1,3,5-CH_3:	3.28 (s), 3.25 (s), 3.60 (s)
8-CH_3:	1.78 (q)
P-1 H_2:	4.30 (d)
P-2-H:	5.10 (t)
P-3-CH_3:.	1.52 (s)
P-7, 11, 15-CH_3:	0.71—0.75 (s)

(15)

For details see Section 10.2.1; for chlorophyll-a (15) see Section 10.2.8. The numbering system for the substituents indicated in structures (13—15) is used throughout this chapter.

forms is fast on the n.m.r. time scale and only one resonance is observed for each group of protons. At low temperatures the β-pyrrole proton signal is split, however, and the exchange kinetics can be studied by ^{1}Hmr[61].

The β-protons in porphin are magnetically equivalent, but in asymmetrically substituted porphyrins the following vicinal coupling constants of the β-pyrrole protons ($^3J_{H\beta-H\beta}$) were obtained: $J = 4.5$ Hz in a partially alkyl-substituted porphyrin[62], $J = 5$ Hz in a Re(I) complex of *meso*-tetraphenyl-porphyrin H_2(TPP)[63], and $J_1 = 4.70$ Hz, $J_2 = 4.83$ Hz in an isoporphyrin[64], and $J_1 = 4.5$ Hz[65] and $J_2 = 4.7$–5.3 Hz[66] in some *meso*-tetraphenylchlo-rins. The long-range coupling constant of the β-pyrrole H with the N—H ($^4J_{H-NH} = 1$ Hz) was observed in a Re(I) TPP complex[63], and the splitting of 2 Hz which was observed[65] for the β-pyrrole protons in tetrahydro-*meso*-tetraphenylporphyrin is probably also due to coupling with N—H.

10.2.1.2. Octaethylporphyrin

Octaethylporphyrin [H_2(OEP) (11)] is today the most widely used model compound for structural studies related to the naturally occurring porphyrins. The assignment of the H_2(OEP) n.m.r. spectrum proceeds directly from ring current considerations and the multiplet structure of the resonances[71]. The decrease of the ring current induced low-field shift with increasing distance of the peripheral protons from the macrocycle is clearly evident in the methine, methylene, and the methyl protons, which appear at increasingly high field in that order (Table 3).

One of the major reasons for the use of H_2(OEP) as a model compound is its high symmetry. Because of fast N—H exchange, the parent compound has four-fold symmetry and only one signal is observed for each group of protons. Although this four-fold symmetry is often reduced by chemical modifications, many derivatives retain two-fold symmetry, and in these compounds the ^{1}Hmr spectra can often be interpreted by inspection on the basis of multiplet structure, relative intensities, and symmetry arguments.

10.2.1.3. meso-Tetraphenylporphyrin

meso-Tetraphenylporphyrin[H_2(TPP) (12)][70] is the parent of a variety of compounds not related structurally to the naturally-occurring porphyrins. The ^{1}Hmr spectrum of *meso*-tetraphenylporphyrin shows two resonances (β-pyrrole H, N—H) for the macrocyclic protons, and two signals for the three phenyl protons well separated from the first two. Due to steric hindrance, the phenyl rings in H_2(TPP) are out of the plane of the macrocy-cle, they do not rotate freely (see Section 10.2.3 and 10.4.2.), and meso-meric interactions between the four phenyl groups and the macrocycle are efficiently reduced. The very similar chemical shifts for the m- and p-protons of the phenyl groups can be explained on this basis. Although the m-protons are closer to the macrocycle, they are out of its plane, and are thus posi-tioned in a less deshielded region. As in porphin (1) the N—H tautomerism is again rapid at ambient temperature on the [1]Hmr time scale, but can be studied at low temperatures[69,74-76].

10.2.1.4. Protoporphyrin-IX dimethyl ester

Protoporphyrin-IX dimethyl ester [H_2(Proto-IX-DME), (13)] is the principal porphyrin from which most of the naturally-occurring tetrapyrrole pigments are derived. Except for the porphyrin plane, the compound lacks a symmetry element, and under suitable conditions[68], all of the expected resonances are resolved. Although the assignment to a certain group of substituents (i.e., β-pyrrole CH_3 groups) is straightforward, precise assignment within these groups is a difficult task. In the case of (9) it was accomplished by a careful aggregation study[68]. Again, only one set of N—H signals is observed in all instances at ambient temperature because of fast exchange in the N—H tautomers.

10.2.1.5. Chlorin-e_6 trimethyl ester

Chlorin-e_6 trimethyl ester (14) is a key compound in chlorophyll chemistry and was the ultimate goal of Woodward's chlorophyll synthesis[77]. Chlorin-e_6 trimethyl ester serves to some extent as the prototype of chlorin-type molecules in which one of the pyrrole rings is reduced (see Section 10.2.5). Assignment of part of the [1]Hmr spectrum of chlorin-e_6 trimethyl ester (Fig. 3b) was carried out by Woodward[78] and later by Caughey[4]. To clarify some contradictory assignments, the [1]Hmr spectrum of chlorin-e_6 trimethyl ester was reassigned by Jeckel et al.[72,73], whose data are included in Table 3. The order β, α, δ for the three methine proton resonances, in the order of increasing field, is characteristic of a great variety of chlorophyll-a derivatives, which have no deshielding substituents in the 2-position. The vinyl group at position 2 in chlorin-e_6 trimethyl ester gives rise to a characteristic ABX subspectrum[1]. All aromatic and ester methyl groups in this compound occur as well-resolved singlets in a narrow range between $\delta = 3$ and 4 p.p.m. The protons in the 4-ethyl group gives rise to a quadruplet and a triplet, both of which are assignable by inspection. The γ-CH_2 protons are magnetically anisotropic because of the neighboring ring asymmetric center. These protons give rise to an AB subspectrum, which is generally observed for γ-CH_2 substituents in the phorbin series (Section 10.2.3). The most complex part of the [1]Hmr spectrum arises from the substituents at the reduced pyrroline ('chlorin') ring D. Due to the sp^3 hybridization of the 7 and 8 positions, all protons in the alkyl side-chains are one bond more remote from the aromatic system than in the true porphyrins, and their signals are thus less deshielded by the ring current. The signals are further complicated by the asymmetric centers at C-7 and C-8 and the resulting complex spin systems. The 8-CH_3 group gives rise to a doublet ($J = 7$ Hz), and the neighboring 8-H proton shows the expected quadruplet structure. The small additional coupling of the 8 to the 7 proton, from which the 7,8 *trans*-configuration was inferred[14], is rarely resolved. The 7-proton is coupled to three nonequivalent protons (8-H, 7'-H_A, 7'-H_B), but its resonance peak shows up very often as a characteristic pattern of two broad signals separated by 7 Hz[79]. All four protons of the propionic acid side-chain are magnetically nonequivalent due to the neighboring asymmetric centers at C-7 and C-8. Although potentially useful for conformational studies, no complete assignment for the chemical shifts of these protons has been reported as yet. Recently, the 7b-methylene resonances have been observed as an AB double doublet (δ = 2.50, 2.18 p.p.m., J = 16 Hz) in a selectively deuterated pyropheophorbide-a[80]. In the 100 MHz spectrum, only one N—H signal is observed at high-field for (14), but upon cooling[75] or in the 220 MHz spectrum at low concentration (5×10^{-3} M)[80] two N—H signals are well resolved (Fig. 3b). This splitting is typical for chlorins, and is in particular very pronounced in the phorbins (Section 11.2.3). It is indicative of a more pronounced localization of the N—H protons in chlorins than in porphyrins[72,73] (Section 10.4.2).

10.2.1.6. Chlorophyll-a

For a discussion of the n.m.r. spectrum of the chlorophyll-a (Chl-a, (15)) and the other chlorophylls, the reader is referred to Section 10.2.8.2.

10.2.2. β-Substitution

Porphyrins derived from natural pigments usually have substituents at all of the eight β-pyrrole (peripheral) positions, and n.m.r. spectra recorded on β-substituted porphyrins are so numerous that we can only try to show here some general trends in chemical shifts as observed in some characteristic examples.

Substitution of all β-pyrrole positions of porphyrin with alkyl groups shields the methine protons by 0.20—0.24 p.p.m., and the N—H protons by 0.36 to 0.46 p.p.m., a shielding that is identical within experimental error for different alkyl groups (methyl, ethyl, n-propyl)[8b]. The shielding effect may be discussed in terms of a long range (dipole) effect of the alkyl groups, but in addition a ring current change (via an inductive effect) is possible[8b]. For compounds substituted only by alkyl groups or by acetic or propionic acid side-chains, the effects on the methine positions are additive with respect to the next neighbors, but even in these cases a polarization of the entire macrocycle that results in long-range effects can already be observed. Incremental shifts of the methine proton resonance (in TFA) of +0.11, +0.02, and +0.11 p.p.m. for a neighboring alkyl, 2-carbomethoxyethyl and vinyl group, respectively, are reported by Abraham et al.[8b].

Both nearest neighbor and ring current effects are much more pronounced with substituents other than the ones cited above, and for such compounds a simple nearest-neighbor incremental treatment is no longer possible. In an early review on the n.m.r. of various β-substituted porphyrins, Caughey[4] detected a decrease in ring currents with increasingly electron withdrawing

TABLE 4

[1]Hmr chemical shifts (δ[p.p.m.] from TMS) of β-pyrrole substituted oligomethylporphyrins, selected from the work of Clezy et al.[81—35]

Parent Compound, Substituents	Methine—H	Substituent—H	Propionic Ester Protons [CH$_2$CH$_2$-(COOR)]	Ester, Ring-CH$_3$	Ref.
Porphin	11.22				8a
Octamethylporphyrin	10.98			3.77	8a
Hexamethyl-Porphin					
6,7-diP^{Me}	11.20(1)		4.72, 3.32	3.79	81
	11.06(3)			3.84	
2,3-diAc	11.87(1)			4.02	81
	11.05(2)			3.69	
	10.91(1)				
2-Ac, 5P^{Me}	11.51(1)	3.58 (Ac)	4.65, 3.35	4.13(1)	81
	11.15(1)			3.80(1)	
	10.95(2)			3.81(2)	
				3.72(3)	
2-COOEt, 5-P^{Me}	11.88(1)	5.12, 1.95	4.65, 3.35	4.13(1)	81
	11.21(1)	(COOC$_2$H$_5$)		3.82(2)	
	11.01(2)			3.80(1)	
				3.77(3)	
2-CHOH—CH$_3$, 3-COOEt	12.26	6.90 (*q*)		4.09(1)	81
	11.13	3.30 (*d*)		3.80(1)	
	10.92	(CHOHCH$_3$)		3.72(4)	
	10.90	5.10 (*q*)			
		1.87 (*t*)			
		(COOC$_2$H$_5$)			
2-Ac, 3-COOEt	12.22(1)	5.07 (*q*)		4.05(2)	81
	11.11(2)	1.86 (*t*)		3.70(4)	
	10.98(1)	(COOC$_2$H$_5$)			
		3.57			
		(COCH$_3$)			
1-OCH$_3$, 3-H	11.07(1)	9.62 (β_{pyrr} H)		3.78(2)	82,83
	10.88(3)			3.70(4)	
4-NHCOCH$_3$, 7-P^{Me}	11.08	3.93 (NHCOCH$_3$)	4.67, 3.30	3.82(1)	84
	11.05			3.77(2)	
	11.02			3.80(2)	
	10.97			3.72(2)	
Pentamethyl-Porphin					
5,8-diP^{Me}, 2-CN	11.21(1)		4.65, 3.21	4.01(1)	81
	11.01(3)			3.82(2)	
				3.80(4)	
2-Ac, 3-COOEt, 8-Br	12.17(1)	5.06, 7.88		4.05(2)	81
	11.12(1)	(OC$_2$H$_5$)		3.69(2)	
	11.04(1)	3.55 (Ac)		3.74(1)	
	10.91(1)				

TABLE 4 (continued)

Parent Compound, Substituents	Methine—H	Substituent—H	Propionic Ester Protons [CH_2CH_2 (COOR)]	Ester, Ring CH_3	R
2,3-Ac, 5-Et	11.83(1) 11.02(2) 10.90(1)	3.54 (Ac) 4.20 (q) 1.78 (t) (C_2H_5)		4.02(2) 3.69(3)	8
1-OCH_3, 3-Ac, 7-P^{Me}	11.28(1) 10.98(1) 10.90(1) 10.82(1)	4.98 (OCH_3) 4.23, 1.26 (OC_2H_5) 3.52 (Ac)	4.58, 3.26	4.05(1) 3.74(1) 3.69(3)	8
4-$NHCOCH_3$, 6,7-P^{Et}	11.23 11.12 11.08 11.04	3.88 ($NHCOCH_3$)	4.62, 3.18	3.68(5)	8
2-$NHCOCH_3$, 4-COOEt, 7-P^{Et}	11.76(1) 11.15(1) 10.98(2)	2.91 ($NHCOCH_3$) 5.09, 1.90 ($COOC_2H_5$)	4.67, 3.28	4.02(1) 3.76(2) 3.72(2) 3.61(1)	8
1-COOEt, 6,7-C_2H_5	11.75 11.12 10.92 10.88 11.42	5.10, 1.85 ($COOC_2H_5$) 4.25, 1.85 (C_2H_5) 3.52($COCH_3$)		4.09(1) 3.72(4) 4.05(1)	8 8
1-$COCH_3$, 6,7-C_2H_5	11.09 10.89 10.84	4.20, 1.79 (CH_2CH_3)		3.70(4)	
1-CHOH—CH_3; 6,7-C_2H_5	11.28(1) 10.99(3)	7.00, 3.30 ($CHOHCH_3$) 4.25, 1.82 (CH_2CH_3)		3.82(1) 3.75(4)	8
5-vinyl; 2,3-C_2H_5	11.03(2) 10.98(2)	8.30, 6.50 ($CHCH_2$) 5.25, 1.80 (CH_2CH_3)		3.80(1) 3.71(4)	8
4-Ac; 6,7-P^{Et}	11.10(1) 11.03(1) 10.88(2)	3.52 ($COCH_3$*)	4.63, 3.22	4.07(1) 3.71(4)	8
Tetramethyl-Porphin 6,7-diP^{Me}, 2,3-COOEt	12.57(1) 11.18(1) 11.12(2)	5.09 (q) 1.79 (t) (OC_2H_5)	4.68, 3.39	4.08(2) 3.78(4)	8
6,7-diP^{Me}, 2,3-diCN	11.30(1) 11.08(3)		4.70, 3.30	4.08(2) 3.72(4)	8

TABLE 4 (continued)

Parent Compound, Substituents	Methine—H	Substituent—H	Propionic Ester Protons $[CH_2CH_2(COOR)]$	Ester, Ring CH_3	Ref.
2,3,6,7-tetraP^{Me}	11.22(2) 11.08(2)		4.75, 3.35	3.84(4) 3.78(4)	81
6,7-diP^{Me}, 2,3-diAc	11.88(1) 11.16(1)	3.55(OAc)	4.60, 3.30	4.05(2) 3.78(4)	81
5,8-diP^{Me}, 2-COOEt, 1-H	11.91(1) 11.03(1) 11.35(1) 10.99(1)	10.48 (β_{pyrrol} H) 5.13 1.93 (OC_2H_5)	4.70, 3.35	3.83(6)	81
2,3-diNHAc, 6,7-diP^{Me}	11.28(1) 11.07(3)	2.95 (NHCOCH$_3$)	9.70, 3.35	3.82(2) 3.75(2) 3.71(2)	84
2,3-diNH$_2$, 6,7-diP^{Me}	11.13(1) 10.79(1) 10.72(2)		4.60, 3.30	3.78(2) 3.73(2) 3.62(2)	84
2,3-diNHCOOEt, 6,7-diP^{Et}	11.22(1) 11.18(1) 11.04(2)	4.65, 1.50 (COOC$_2$H$_5$)	4.65, 3.31	3.80(2) 3.75(4)	84
2-H, 4-Ac, 6,7-diP^{Et}	11.50(1) 11.10(2) 10.95(1)	9.61 (H) 3.58 (COCH$_3$*)	4.68, 3.29	4.10(1) 3.80(2) 3.78(1)	85
2-H, 4-C$_2$H$_3$, 6,7-diP^{Et}**	9.81(1) 9.72(2) 9.66(1)	8.51, 6.10 (CHCH$_2$) 8.75(H)	4.16, 3.09	3.52(1) 3.44(1) 3.39(2)	85

* Tentative assignment
** In C^2HCl_3

For the numbering system, see structure (13). If not otherwise indicated, all spectra were recorded with TFA as solvent. Abbreviations: Et = C_2H_5, P^{Me} = $CH_2CH_2CO_2Me$, P^{Et} = $CH_2CH_2CO_2Et$, Ac = $COCH_3$.

References, p. 514

substituents at the pyrrole β-positions. This general trend is clearly visible in Table 4, which lists a selection of chemical shift data of synthetic porphyrins collected from the work of Clezy et al. Although in most cases the basic features of the spectra can be discerned, it is clear from these data that a detailed interpretation of the spectra is not possible without a complete assignment of all signals. This is especially true for the methine and aromatic methyl proton signals, which appear in the spectra as singlets, and which cannot be assigned from their multiplicity.

As the effects of substitution are at best difficult to estimate per se, a careful choice of a completely assigned reference compound from which the substituent effects can be deduced by stepwise structural modifications is necessary in any particular investigation. Inhoffen et al.[72] completely assigned the spectrum of chlorin-e_6 trimethyl ester (14), which is a suitable reference compound for chlorins and pheophorbides, by a systematic variation of certain substituents (for details, see Ref. 73). As an example of a different approach, the careful study of aggregation shifts made possible the chemical shift assignment of chlorophyll-a (15)[13] and of a series of three type IX isomer porphyrins[68]. These can now serve as reference compounds for other chlorophylls and porphyrins.

The [1]Hmr data of Inhoffen et al.[86] was used to deduce the substituent effects of —CHO, —COOCH$_3$, —COCH$_3$ and —CHOHCH$_3$ in positions 2 and 4 in mono- and disubstituted deuteroporphyrins-IX*. The reference compound used was deuteroporphyrin-IX DME (16), whose [1]Hmr spectrum was completely assigned by Janson et al.[68] As the latter data were obtained in C^{2}HCl$_3$, the first step was to assign the signals observed for the reference compound in TFA. The same order with respect to field for the methine and methyl resonances of H$_2$(Deut-IX-DME) was assumed for dilute solutions in both CDCl$_3$[68] and in TFA[86]. For the assignment of the protons in the substituted compounds, an incremental shift similar in size for the two most remote resonances was assumed, a slightly different increment for the closer group, and the most strongly deviating increment for the nearest neighbor. Although some of the assignments so obtained are still not completely unambiguous, a self-consistent set of data was obtained by this procedure (Table 5).

The β-substitution effects for —CHO, —COOCH$_3$, COCH$_3$ and —CHOHCH$_3$ can be summarized as follows: 1) All of these substituents decrease the ring current substantially, and the incremental shift for remote proton signals is of comparable size for all four substituents; 2) different chemical shift increments are observed for the same substituent in position 2 or 4; 3) the substituent effects are not additive, as shown by the chemical shifts in the disubstituted compounds; 4) different relative increments for the signals of the propionic acid side-chain indicate changed conformations of the latter and/or the ring system for different substituents; 5) the most pronounced effects, which vary characteristically for the substituents, are the neighboring group effects due to long-range shielding. The adjacent methine proton signal is strongly deshielded by the substituents cited, and this effect increases in the order —CHOHCH$_3$ $\leqslant$ —COCH$_3$ < —CHO < —COOCH$_3$. The effect on the nearest methyl group is similar, with one

---

* The original assignment of Fischer et al.[87] for the 2- and 4-monoformyldeuteroporphyrin-IX dimethyl ester was revised[88] after the publication of the n.m.r. data[86], hence, the interchanged substituents in Table 5.

TABLE 5

¹Hmr chemical shifts (δ[p.p.m.] from TMS in TFA) of H_2(Deut-IX-DME) (16) and its mono- and di-substituted derivatives; and incremental shifts (Δ[p.p.m.]) as compared to the respective signals in the parent compound (16).

(16)

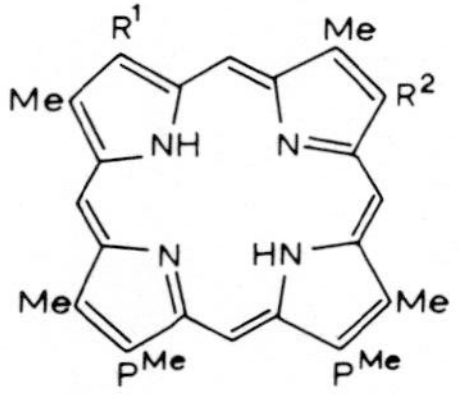

| | R¹ | H | H | CHO | COOCH₃ | COCH₃ | CHOHCH₃ | H | H | H | H | CHO | COCH₃ |
	R²	H*	H	H	H	H	H	CHO	COOCH₃	COCH₃	CHOHCH₃	CHO	COCH₃
Methine—H,	α	10.06	11.09	11.61	11.73	11.43	11.42	11.02	11.08	10.96	10.95	11.51	11.63
	β	10.03	11.02	10.82	10.79	10.74	10.88	11.43	11.73	11.37	11.41	11.51	11.34
	γ	10.10	11.21	11.10	11.03	11.00	11.12	11.02	11.08	10.96	11.03	11.01	10.97
	δ	9.99	10.98	11.15	11.09	11.00	10.95	10.89	10.94	10.79	10.91	11.01	10.96
Δ	α			−0.52	−0.64	−0.34	−0.33	+0.07	+0.01	+0.13	+0.14	−0.42	−0.54
	β			+0.20	+0.23	+0.28	+0.14	−0.41	−0.71	−0.35	−0.39	−0.47	−0.32
	γ			+0.11	+0.18	+0.21	+0.09	+0.19	+0.13	+0.25	+0.18	+0.20	+0.24
	δ			−0.17	−0.11	−0.02	+0.03	+0.09	+0.04	+0.19	+0.07	−0.03	+0.02
Arom. CH₃,	1	3.60	3.83	4.16	4.61	4.06	3.72	3.77	3.78	3.71	3.72	4.10	4.01
	3	3.57	3.80	3.77	3.74	3.75	3.66	4.13	4.61	4.06	3.84	4.08	4.01
	5	3.68	3.86	3.77	3.77	3.75	3.66	3.77	3.78	3.74	3.79	3.78	3.71
	8	3.70	3.87	3.77	3.77	3.75	3.66	3.77	3.78	3.74	3.79	3.78	3.66
Δ	1			−0.33	−0.78	−0.23	+0.11	+0.06	+0.05	+0.12	+0.11	−0.28	−0.18
	3			+0.03	+0.06	+0.05	+0.14	−0.33	−0.81	−0.26	−0.04	−0.27	−0.21
	5			+0.09	+0.09	+0.11	+0.20	+0.09	+0.08	+0.12	+0.07	+0.08	+0.15
	8			+0.10	+0.10	+0.12	+0.21	+0.10	+0.09	+0.13	+0.08	+0.09	+0.21
6',7'-CH₂		4.38	4.72	4.61	4.62	4.58	4.60	4.61	4.61	4.59	4.65	4.58	4.56
Δ				+0.11	+0.10	+0.14	+0.12	+0.11	+0.06	+0.13	+0.07	+0.14	+0.16

TABLE 5 (continued)

| R^1 | H | H | CHO | $COOCH_3$ | $COCH_3$ | $CHOHCH_3$ | H | H | H | H | CHO | $COCH_3$ |
R^2	H*	H	H	H	H	H	CHO	$COOCH_3$	$COCH_3$	$CHOHCH_3$	CHO	$COCH_3$
6″,7″-CH_2 Δ	3.24	3.30	3.29 / +0.01	3.25 / +0.05	3.24 / +0.06	3.17 / +0.13	3.27 / +0.03	3.26 / +0.04	3.23 / +0.07	3.26 / +0.04	3.23 / +0.07	3.20 / +0.10
β-pyrrolic—H Δ	9.06 9.04	9.66	9.53 / +0.13	9.50 / +0.16	9.45 / +0.21	9.60 / +0.06	9.58 / +0.08	9.60 / +0.06	9.57 / +0.09	9.57 / +0.09		
Substituent			11.66	3.83	3.54	6.86/ 2.19	11.57	3.84	3.52	6.84/ 2.30	11.54/ 11.79	3.48/ 3.50

* 0.004 m in C^2HCl_3
For details see text.

TABLE 6

Expectation ranges (δ [p.p.m.] from TMS in C^2HCl_3) of β-pyrrole substituents in positions 1, 3, and 5 of porphyrins

(17)

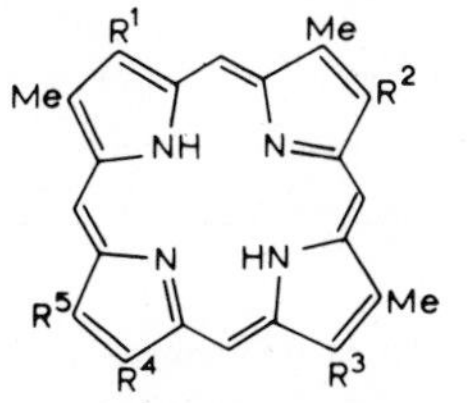

R^4 = Me, Et, P^{Me}
R^5 = Me, Et

R^1			
H	8.78—8.87		
CHCH$_2$	7.8—8.2 (8.2—8.3 in TFA) (H$_X$) 5.9—6.3 (6.4—6.5 in TFA) (H$_{A,B}$) ABX, $J_{AB} \sim 2$, $J_{AX} \sim 17$, $J_{BX} = 12$		
COOCH$_3$	4.3 (4.6 in TFA) (s)		
CHO	10.4—11.1 (s)		
CH$_2$CH$_2$OCOCH$_3$	3.8—4.4	4.5—4.8	1.9—2.1
CH$_2$CH$_2$OH	3.8—4.5 (4.6 in TFA), 3.20		
CH$_2$CH$_2$CN		3.21	
CH$_2$CH$_2$Cl	4.1—4.6	3.20	
CH$_2$CH$_2$Br	4.3—4.5		
CO—CH$_2$—(CH$_2$)$_6$CH$_3$	2.0	1.5—0.8	
CH$_2$CHO	4.77 (d)	10.12 (t)	J = 3 Hz
CH$_2$CH(OCH$_3$)$_2$	4.2—4.3	5.0—5.1	3.4—3.5
CH$_2$CH$_2$OTS	4.75 (m)	4.90 (m)	7.64 (d), 7.29 (d) 2.38 (s)

R^2			
H	8.83—8.93		
CHCH$_2$	7.8—8.2 (8.2 in TFA)	6.0—6.4 (6.4—6.5 in TFA)	
COOCH$_3$	4.3		
CH$_2$CH$_2$OCOCH$_3$	4.0—4.4 (3.08 for β-OAc)	4.6—4.8	2—2.1
CH$_2$CH$_2$OH	3.8—4.5 (4.6 in TFA)	3.20	
CH$_2$CH$_2$Cl	3.9—4.4	3.20	
CH$_2$CH$_2$Br	4.35		
CHO	11.08		
CH$_2$CH(OCH$_3$)$_2$	4.32	5.13	3.4—3.5

R^3			
COOCH$_3$	4.4—4.6		
COCH$_2$COOCH$_3$	4.4—4.6	3.5—3.6	
C(OH)=CH—COOCH$_3$	3.3	6.1—6.2	
C(OCH$_3$)=CH—COOCH$_3$	3.88*	5.81	3.91*
C(OAc)=CH—COOCH$_3$	2.46	6.56	3.93

References, p. 514

TABLE 6 (continued)

R^3			
CO—O—CO—Bu^t	1.6—1.7		
CO—O—CO—Et	4.76	1.66	
CO—N—CH=CH—N=CH	8.3—8.4*	7.8—7.9	7.2—7.3*

* Signals assigned to either one of the indicated positions.

Selected from the work of Jackson, Kenner, Smith et al.[91—99]. If not otherwise indicated, the chemical shifts are listed according to the proton sequence in the substituent formula (from left to right, i.e., —CH$_2$ before —CH$_3$).

exception: the —CHOHCH$_3$ group has a strong neighbor group effect on the nearest methine proton resonance, and a negligible effect on the nearest methyl resonance. As both the methine and the methyl protons are at comparable distances to the substituents, the discrepancy is probably due to a preferred conformation of the substituent (or its solvate) (see Section 10.4.2.2).

The chemical shifts of β-pyrrole substituents can be estimated from the incremental shifts listed in Table 2. The expectation ranges for a variety of β-pyrrole substituents in porphyrins of the general structure (13), are listed in Table 6, and some further examples can be found in Table 8 and in the appropriate column in Table 4. Conjugated β-dicarbonyl substituents (Table 7) are usually present both in the keto and the enol form[89,90]. As the total ring current is affected by this (generally slow) tautomerism, the n.m.r. spectrum shows two sets of lines characteristic of a slowly equilibrating mixture. The assignment of the tautomers is possible by temperature dependent studies and by the relative intensities of the signals.

Some n.m.r. data of H$_2$(TPP) derivatives substituted at β-pyrrole positions (Table 8) have been reported by Callot[100,101]. The shielding effect of the phenyl rings (Section 10.2.3.) shifts the signals of the substituents to considerably higher field as compared to the respective substituents in *meso*-unsubstituted porphyrins (Tables 6 and 7). A further noteworthy feature is the nonequivalence of β-pyrrole methylene protons, which is indicated by the multiplet rather than a triplet structure in the methylene resonances of the ω-bromo-octyl derivatives. This nonequivalence, as well as the centrosymmetric structure discussed for the tetra-substituted products, are judged to be strong indications for a nonplanar structure of the macrocycle.

10.2.3. Meso substitution

Meso-substitution changes the [^1]Hmr spectrum of porphyrins in three important ways: 1) The ring current is reduced, and within broad limits the extent of the reduction is independent of the nature of the substituents; 2)

TABLE 7

^{1}Hmr chemical shifts (δ[p.p.m.] from TMS) of porphyrins with conjugated β-dicarbonyl substituents.

Compd.		Keto-form		Enol-form	Solvent	Ref.
(18) a) R=CO—CH(OCH$_3$)COOCH$_3$ b) R=C(OH)=C(OCH$_3$)COOCH$_3$	Methine—H	10.52 (2) 9.96 9.72		10.34 (2) 10.12 9.88	C^{2}HCl$_3$	90
	R: OH CH	— 5.83		12.2 —		
(19) a) R=CO—CH$_2$—COOCH$_3$ b) R=C(OH)=CH—COOCH$_3$	Methine—H	10.33 9.46 9.39 9.30		10.20 9.62 9.56 9.39	C^{2}HCl$_3$	89
	Vinyl—H$_{ABX}$		7.93 6.08	6.08		
	R: CH$_{(2)}$ CH$_3$ OH 5-CH$_3$	4.57 — 3.92	3.61	13.30 4.03		
(20) a) R=CO—CH$_2$—COOCH$_3$ b) R=C(OH)=CH—COOCH$_3$	Methine—H	10.68 9.82 (2) 9.80		10.48 10.01 9.91 (2)	C^{2}HCl$_3$	89
	R: CH$_{(2)}$ CH$_3$ OCH$_3$	4.59 3.88 3.57 (2) 3.50 3.42		6.15 3.99 3.71 3.67 3.54 3.46		

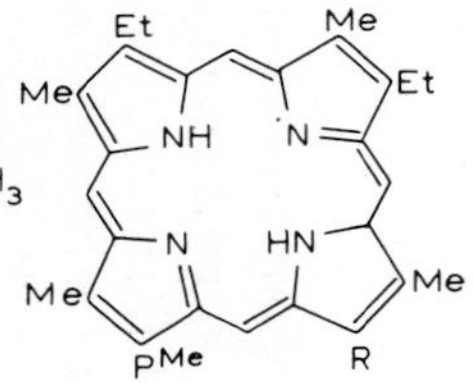

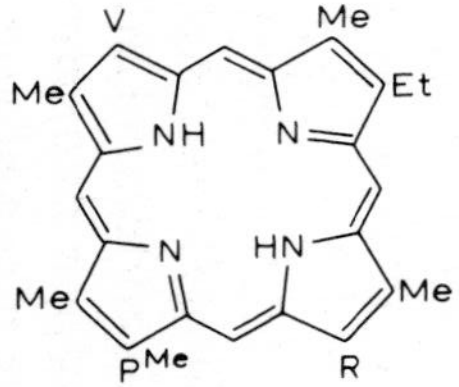

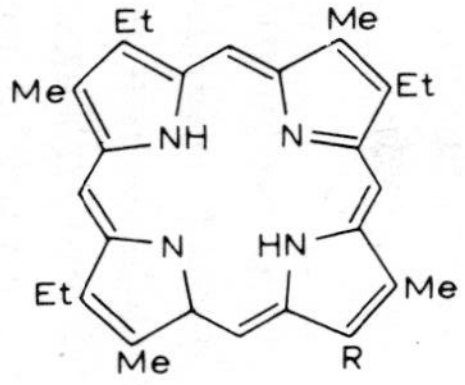

See also Sections 10.2.8.2. and 10.2.8.4.

TABLE 8

^{1}Hmr chemical shifts (δ[p.p.m.] from TMS in C^2HCl_3) of β-pyrrole substituents in H_2(TPP) derivatives substituted at β-pyrrole positions

Substituent(s)	Substituent Resonances	Remarks References
1-CHCH$_2$	H$_X$: 6.35, H$_A$: 5.68, H$_B$ = 5.17 J_{AB} = 2.3, J_{AX} = 17.3, J_{BX} = 11.8	100
1-CHCH$_2$	H$_X$: 6.35, H$_A$: 5.86, H$_B$ = 4.97 J_{AB} = 2.2, J_{AX} = 16.6, J_{BX} = 10.6	100 (Ni-complex)
1-CHO	9.20, 9.33; for 1-CHO and 2-H	100
1,3-di-Br	β-pyrrole—H: 8.80(1); 8.82(1); 8.70(4)	101
1,5-di-Br	β-pyrrole—H: 8.63(2); 8.50, 8.70 (d, J = 5.1)	101
1,3,5,8-tetra-Br	β-pyrrole—H: 8.50	101
1-(CH$_2$)$_8$-Br	CH$_2$—(CH$_2$)$_6$—CH$_2$Br: 2.8 (m), 1.2—2 (m), 3.40 (t) Phenyl-H: 7.7—8.2 β-pyrrole-H: 8.6—8.85 (m)	101
1,3-di-(CH$_2$)$_8$-Br	CH$_2$—(CH$_2$)$_6$—CH$_2$Br: 2.8 (m), 1.2—2 (m), 3.40 (t) Phenyl—H: 7.75—8.15 β-pyrrole—H: 8.5—8.8 (m)	101
1-OC$_2$H$_5$	CH$_2$CH$_3$: 4.20 (q), 1.08 (t) Phenyl—H: 7.7—8.2 β-pyrrole—H: 8.75	101
1-O-(CH$_2$)$_6$-Br	O—CH$_2$—(CH$_2$)$_4$—CH$_2$Br: 4.17 (t), 1.25—1.80, 3.43 Phenyl—H: 7.7—8.2 β-pyrrole—H: 8.75	101
1,3,5,7-tetra-CN	β-pyrrole—H: 8.69	101 (Ni-complex)

If not otherwise indicated, chemical shifts are listed according to the proton sequence in the substituent formula (from left to right).

the methine proton opposite to the *meso* substituent is more strongly shifted to higher field than the neighboring methines; 3) protons in the vicinity of the substituent experience additional shielding effects.

The overall reduction of the ring current can be rationalized[3] in terms of the network theory[39], because a barrier to conjugation at the *meso* position affects the full ring current rather than only one branch of it[3]. The principal reason for the reduced ring current in *meso*-substituted porphyrins appears to be steric hindrance between the *meso*- and β-pyrrole substituents, an explanation which is well supported by the decrease in the effect on the ring current with decrease in the size of the β-substituent[102]. If the neighboring β-pyrrole position is unsubstituted, only a minor decrease ($\sim$3%) in ring current is observed on *meso*-substitution[8a], which is somewhat more than the decrease observed for the introduction of a β-substituent.

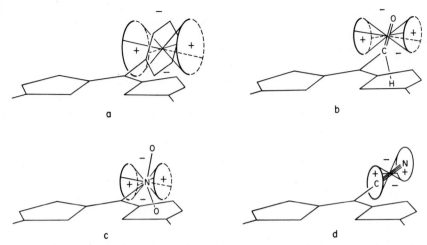

Fig. 4. Magnetic anisotropy of *meso*-substituents. Schematic representation of the porphyrin macrocycle and the zero shielding surface. (a) Phenyl; (b) CHO; (c) NO$_2$; and (d) CN.

For similar steric reasons, non-linear substituents such as -phenyl, -NO$_2$ or -CHO are not coplanar with the macrocycle ring, which efficiently reduces mesomeric interactions (Fig. 4). Thus, the shielding effect of a nitro[103,104] or a carbonyl group[71] on the resonances of the remaining methine protons is comparable to that of a methyl[8a] or hydroxymethyl[71] group. There are, however, two outstanding exceptions to this rule, and these are the amino[103,104] and the hydroxyl groups. Upon introduction of a *meso*-NH$_2$ substituent, the methine proton resonances are more shielded by about 1 p.p.m. than by other *meso*-substituents. This is likely due to contributions from imino-phlorin-like tautomeric structures (21a), in which the ring current is interrupted (see Section 10.2.7). The presence of these mesomeric imino-phlorin structures, which are protonated at the opposite methine position, is evidenced by the ready ^{1}H—^{2}H exchange in *meso*-amino porphyrins[103]. It is interesting to compare the ^{1}Hmr spectrum of derivatives with *N*-pyrrole substituents in a *meso*-position for which cross conjugated phlorin-like structures have also been discussed[85]. These compounds exhibit a normal n.m.r.

(21)

a R = N—H
b R = O

(22)

References, p. 514

spectrum, however, indicating that the bulky pyrrole substituent cannot assume the coplanar conformation.

The second exception is the *meso*-hydroxy group. In acidic solutions the di-cation of the *meso*-hydroxyporphyrin is observed, while in neutral solutions[93,105,106] the tautomeric oxophlorin free base or monocation (see Section 10.2.7). can be observed. (For leading references, see Ref. 49.) Hydroxyporphyrin-cations present in TFA exhibit well-resolved [1]Hmr spectra with the characteristics of *meso*-substitution, and various *meso*-hydroxy porphyrins[93,106a,107−109] and similar structures with thiophen and furan rings[106] have been studied in detail in acidic media. In contrast, the [1]Hmr spectra of the free base oxophlorins present in neutral solutions are generally[105,106] difficult to observe[93,109,110] and exhibit line-broadening because of the presence of small amounts of the oxophlorin π-radical[111] (see Section 10.2.7).

The strong shielding of the methine resonance opposite to the *meso*-substituent can be ascribed to a conformational change in which the entire macrocycle is folded across the two opposite *meso* positions[8a]. This would be expected to reduce the deshielding effects most efficiently at the substituted *meso* position and at the one opposite to it. An indication for the presence of this preferred conformation is the stronger deshielding of the methine resonance in α,γ-disubstituted porphyrins as compared to the α,β-isomers[103,104]. This interpretation of the *meso* effect is further supported by the stability of α,γ-porphodimethenes relative to other systems with interrupted ring current (see Ref. 112 and Section 10.4.3 for X-ray results).

Protons in the vicinity of the *meso*-substituent experience additional shielding effects. These are partly steric in origin, because these groups are forced out of the plane of the macrocycle, but magnetic anisotropies of the *meso*-substituent appear to play the predominant role (Fig. 4). With $-CHO$[71], phenyl[8a] or $-NO_2$ groups[104], the neighboring β-substituents are in a region of positive shielding because of the preferred out-of-plane conformation of the *meso*-substituent. Just the opposite is true for the *meso*-cyanoporphyrins, in which the neighboring β-substituent are in a region of strong negative shielding[71].

The most prominent signals for some selected *meso*-substituted porphyrins are listed in Table 9, with references to further examples listed for every substituent. Some δ-alkyl substituted porphyrins related to structures postulated for *Chlorobium* chlorophylls have been reported by Cox et al.[118], and various β-alkoxy porphyrins have been characterized by Clezy et al.[62,93,107,109].

The influence of *meso*-substitution in chlorins is similar to that in porphyrins, but only if the substitution is at a position distant from the reduced ring (see Clezy, Ref. 62). If the *meso*-substituent is adjacent to the pyrroline ring, the effects are less pronounced and more complex. (See, for example,

TABLE 9

^{1}Hmr chemical shifts (δ[p.p.m.] from TMS) of *meso*-substituted derivatives of principal porphyrins

Parent Compound	Meso Substituent (Position)	Methine—H ^{a}	NH	Substituent Resonances	Cond.	Ref. ^{b}
H$_2$(OEP)	H	10.18	-3.74	—	C^{2}HCl$_3$	113
H$_2$(OEP)	H	10.98	-4.65	—	CF$_3$COOH	114
H$_2$(OEP)	Cl	10.71, 10.58	-2.96, -3.74	—	CF$_3$COOH	114
H$_2$(OEP)	Cl$_2$(α,γ)	10.47	-2.62		CF$_3$COOH	114
	Cl$_4$	—	-0.92		C^{2}HCl$_3$	114
H$_2$(OEP)	Br	10.43, 10.38	-2.68, -3.63	—	CF$_3$COOH	114
H$_2$(OEP)	NO$_2$	10.85, 10.76	-3.17, -3.66	—	CF$_3$COOH	104
	(NO$_2$)$_2$,α,β	10.49	-2.07	—	CF$_3$COOH	(103)
	(NO$_2$)$_2$,α,γ	10.75	-2.31	—	CF$_3$COOH	
	(NO$_2$)$_3$	10.51	-1.37	—	CF$_3$COOH	
H$_2$(OEP)	NH$_2$	9.40, 8.95	1.01, -0.35	—	CF$_3$COOH	104 (103)
H$_2$(OEP)	NHCOOEt	10.2, 10.0	-2.95		CCl$_4$	115
H$_2$(OEP)	CHO	9.98, 9.87	-3.30	12.74	C^{2}HCl$_3$, 0.05 m	71
H$_2$(OEP)	CN	9.98, 9.89		—	C^{2}HCl$_3$, 0.05 m	71
H$_2$(OEP)	CH$_2$OH	10.07, 9.88		6.79	C^{2}HCl$_3$, 0.05 m	71
H$_2$(OEP)	CH$_2$OSO$_2$—CH$_3$	10.18, 9.98	-3.00	6.45 (α-CH$_2$) 3.92 (CH$_3$)	C^{2}HCl$_3$	122
H$_2$(OEP)	CH$_2$OC$_2$H$_5$	10.14, 10.05	-3.00	6.43 (α-CH$_2$) 3.95 (q), 1.61 (t)	C^{2}HCl$_3$	122
H$_2$(OEP)	CH$_3$	10.07, 9.87	-2.86	4.63	C^{2}HCl$_3$	122
H$_2$(OEP)	OH	10.36, 10.09	-2.20, -3.03	—	CF$_3$COOH	106a (116)
H$_2$(OEP)	OCOCH$_3$	10.04, 9.88	-3.4	2.83	C^{2}HCl$_3$	113 (96, 62 89)
H$_2$(OEP)	OCOCF$_3$	10.4, 9.84	-3.56	8.56 p.p.m. from CF$_3$CCl$_3$	C^{2}HCl$_3$	116
H$_2$(OEP)	OCOPh	10.17, 9.99	-3.36	8.92—8.70 (o) 7.79—7.63 (m, p)	C^{2}HCl$_3$	106a (117)

TABLE 9 (continued)

Parent Compound	Meso Substituent (Position)	Methine—H [a]	NH	Substituent Resonances	Cond.	Ref. [b]
H$_2$(OMP)	H	10.98	−4.82	—	CF$_3$COOH	8a
H$_2$(OMP)	CH$_3$	10.62, 10.48	−3.57, −4.33	4.83	CF$_3$COOH	8a
H$_2$(OMP)	(CH$_3$)$_2$ α, γ	10.39	−3.66	4.66	CF$_3$COOH	(118, 102, 119)
	Ph$_4$	—	−0.76	8.32, 7.90 (Ph), 1.84 (CH$_3$)	C^{2}HCl$_3$/TFA	120
Porphin	H	11.22	−4.40	9.92 (β-H)	CF$_3$COOH	8a
Porphin	(CH$_3$)$_4$	—	−3.01	4.73 (meso-CH$_3$), 9.55 (β-H)	CF$_3$COOH	8a
Porphin	(C$_6$H$_5$)$_4$	—	−2.07	8.59 (o), 8.08 (m, p), 8.85 (β-H)	CF$_3$COOH	8a (120)
H$_2$(Etio-I)	NHCOCH$_3$	10.68, 10.59	−3.3, −3.48, −4.3, −4.4		CF$_3$COOH	103
H$_2$(Copro-II-TME)	SCN (β)	10.94, 10.64		—	CF$_3$COO^{2}H	121
H$_2$(Copro-II-TME)	SH (β)	10.08, 8.45		—	C^{2}HCl$_3$	121
A Hexamethyl-P	S-COCH$_3$	9.65	−4.0	2.06	C^{2}HCl$_3$	123
A Hexamethyl-P	A substituted pyrrole	10.81		—	CF$_3$COO^{2}H	85

[a] The low field resonance is always due to the proximate, the high field one for the opposite methine protons.
[b] References in brackets ( ) indicates additional information regarding the same substituent.

Ref. 114 for halogen-substituted chlorins.) One reason for this is certainly the diminished steric interaction[77] of this *meso*-substituent with the neighboring group when the β-carbon atoms in the pyrroline group change hybridization from sp^2 to sp^3. More indirect effects involve conformational changes of the more mobile reduced (pyrroline) ring, which are exemplified in the detailed analysis of δ-chloro-chlorins[124] and of peripheral complexes of pheophorbides[80]. The ^{1}Hmr spectra of three isomeric mono-*meso* acetoxy derivatives of the H$_2$(OEP) mono-geminiketone (35) were studied by Inhoffen et al.[105]. An incremental shift of Δ = 0.16 to 0.18 p.p.m. for the sets of opposed methine protons were observed; the magnitude of the shifts was independent of the position of the *meso*-acetoxy group. The ^{1}Hmr spectra of the *meso* hydroxy isomers are also reported in the same publication[105]. While two of the isomers are present in neutral solution (CDCl$_3$) as oxophlorins, the isomers in which the gemini ketone carbonyl groups and the hydroxy substituent are adjacent has a spectrum typical for a *meso*-substituted porphyrin. To our knowledge, this is the first instance of a metal-free neutral *meso*-hydroxyporphyrin, the stability of which is attributable to hydrogen-bond formation between the two neighboring oxygen functions[105].

Most of the chlorins derived from the chlorophylls have *meso*-substituents. The spectrum of chlorin-e_6 trimethyl ester (14) is discussed in detail at the beginning of this section (10.2.1.5), and the spectra of some selected compounds of similar structure are listed in Table 10. Some examples in which the γ- and the 6-substituents are linked together by ring formation are also listed in Table 10. The isocyclic five-membered ring so formed is the principal characteristic of the phorbin structure common to all chlorophylls. (For a discussion of the n.m.r. spectra of the chlorophylls, see Section 10.2.8.2.) As a consequence of the new bond, which is formed between carbon atom 6 and the γ methine carbon, the steric interaction of the substituents at C-γ and C-6 is efficiently reduced[77] and they become essentially co-planar[128]. This is principally expected to enhance the macrocycle ring current. At the same time, however, considerable strain is introduced into pyrrole ring C[128c,d,129] and the conformation of its substituents at position 6 (i.e., C-9) is changed in such a way that conjugation with the aromatic system of the macrocycle is facilitated. In the compounds listed in Table 10, the first effect is clearly overcompensated by the latter, and the methine protons are shielded by an additional 0.2—0.3 p.p.m. relative to compounds lacking a ring E.

As evidenced by the marked increase in ring current upon reduction of the C=O function to a CH$_2$ group (Δ methine ≃ 0.45 p.p.m.), a major contribution to the shielding comes from the keto C=O group at position 9. This general principle is clearly visible from the data in Tables 11 and 12, in which some changes in the n.m.r. spectra of phorbins (Table 11) and pheoporphyrins (Table 12) upon substitution of the isocyclic ring E are listed. These

References, p. 514

TABLE 10

[1]Hmr chemical shifts (δ[p.p.m.] from TMS in C^2HCl_3) of γ-substituted 7,8-chlorins

(23)

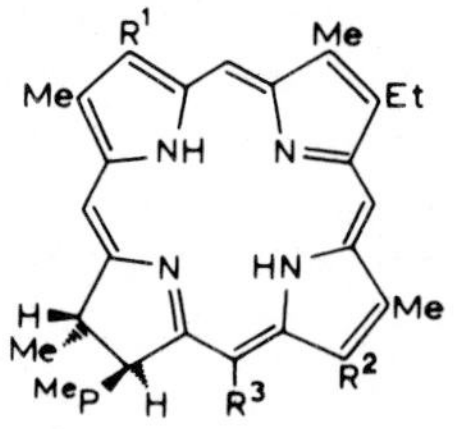

R^2	R^3	Methine—H (β, α, δ)	N—H	γ-CH$_2{}^a$	2'-H$_x$	Re
		R^1 = Vinyl				
CH$_2$COOCH$_3$	H	9.70, 9.70, 8.81	−2.02	5.35	8.10	7
CH$_2$COOCH$_3$	COOCH$_3$	9.65, 9.50, 8.70	−1.04	5.27	8.00	7
CH———————C=O | COOCH$_3$		9.36, 9.24, 8.50	−1.80	4.93	7.86	12
		R^1 = H			2-H	
CH$_2$COOCH$_3$	H	9.73, 9.54, 8.81	−2.20, −2.34	5.39	8.81	7
CH$_2$COOCH$_3$	COOCH$_3$	9.69, 9.36, 8.72	−1.58	5.31	8.72	7
		R^1 = Ethyl			2'-CH$_2$	
CH$_2$COOCH$_3$	H	9.70, 9.53, 8.74	−2.08	5.35	3.90	7.
CH$_2$COOCH$_3$	COCH$_3$	9.65, 9.37, 8.65	−1.50	5.20	3.85	7.
CH$_2$COOCH$_3$	COOCH$_3$	9.64, 9.32, 8.61	−1.36	5.25	3.83	7
H$_3$COOC—CH——C=O		9.39, 9.14, 8.43	−1.83	5.13	3.64	12
		R^1 = Acetyl			2''-CH$_3$	
CH$_2$COOCH$_3$	H	9.62, 10.20, 8.92	−1.84, −2.16	5.34	3.27	7.
CH$_2$COOCH$_3$	COCH$_3$	9.64, 10.11, 8.90	−1.42, −1.64	5.22	3.26	7.
CH$_2$COOCH$_3$	COOCH$_3$	9.65, 10.09, 8.88	−1.60	5.30	3.23	7
H$_3$COOC—CH——C=O		9.42, 9.86, 8.68	—	5.12	3.19	12

^{a} γ-CH$_2$: AB, J = 17 Hz, Δv = 0.08—0.13 p.p.m.

changes can generally be interpreted in terms of the electron-withdrawing effect of the substituents. Thus, introduction of (electron-withdrawing) carbonyl groups in the 9 or in the 9 and 10 positions leads to a pronounced decrease in the ring current[125,127,130]. These changes are about twice as large as for a conjugated carbonyl substituent in position 2, 4 or 6 (see Section 10.2.2), which is not a part of an additional ring[72,131,132]. This clearly reflects the coplanar conformation of the 6 and γ-substituents in the

TABLE 11

[1]Hmr chemical shifts (δ[p.p.m.] from TMS in C^2HCl_3) of pheophorbides with various substituents at the isocyclic ring E

(24)

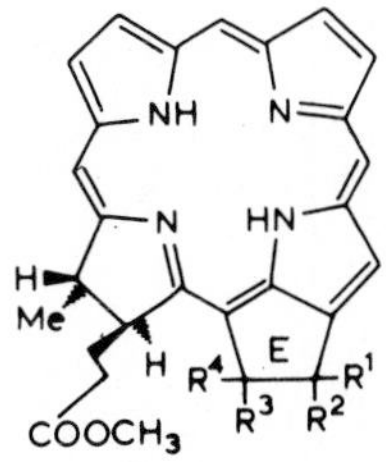

R^1	R^2	R^3	R^4	Methine—H	Vinyl—$H_x{}^a$	NH	Remarks	Ref.
H	H	H	H	9.77, 9.58, 8.87	—	−1.68 −3.53	2-Ethyl	127
	=O	H	H	9.32, 9.23, 8.47	7.90	−1.80	—	79
H	OH	H	H	9.83, 9.56, 8.86	8.15	—	3-CH_2OH	133
	=O	H	$COOCH_3$	9.36, 9.24, 8.50	7.87	−1.72	10(R)	79
H	OH	H	$COOCH_3$	9.86, 9.86, 8.93	8.20	−3.2	9(R), 10(R)	125
	=O	OCH_3	$COOCH_3$	9.53, 9.36, 8.58	7.92	−1.80	10(S)	79, 134
H	OH	OCH_3	$COOCH_3$	9.89, 9.70, 8.94	8.18	−3.0	9(S), 10(S)	130
	=O	H	OCH_3	9.51, 9.39, 8.65	7.94	−2.08	10(S)	79
C====C								
H_3CO—C=O- -Mg—O				9.01, 8.83, 8.00	7.77	—	Mg-chelate	135
H	O—$C(CH_3)_2$—O H			9.69, 9.69, 8.83	—	—	2-Ethyl	136
H	OH	H	CH_2OH	9.79, 9.53, 8.85	8.07	—	3,7''-CH_2OH	136

^{a} See Formula (13).

pheophorbides imposed by the additional ring E. In the absence of X-ray structural information on 9-desoxo-pheophorbides[128d], the steric effect on the geometry of rings C and E that accompanies the change from sp^3 to sp^2 configuration of C-9 is much more difficult to estimate, but can probably not be neglected. The isocyclic five-membered ring E has two potentially asymmetric centers. Although stereoisomers at C-9 and C-10 influence the resonances of neighboring substituents in a characteristic way (Section 10.4.3), long-range effects via the ring current are generally negligible.

Some systems with a fused ring between a *meso* and a β-pyrrole position other than the isocyclic five-membered ring of the phorbins have been investigated by n.m.r. The relief in steric strain resulting from enlargement of ring E in phorbins is clearly demonstrated by the increased ring current in pheophorbide lactones[129]. The [1]Hmr spectra of isopheoporphyrins are

References, p. 514

TABLE 12

[1]Hmr chemical shifts (δ[p.p.m.] from TMS) of pheoporphyrins with various substituents at the isocyclic ring E.

(25)

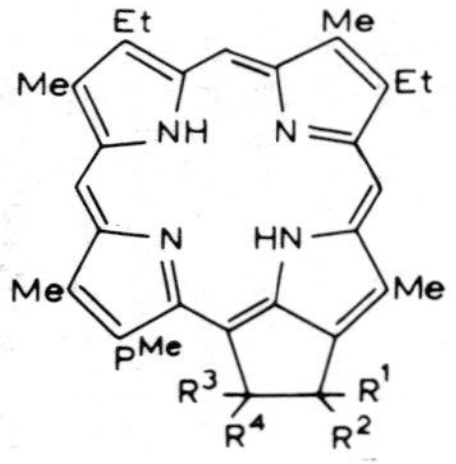

R¹	R²	R³	R⁴	Methine—H	NH	R³-Resonance	Remarks	Ref.
H	H	H	H	9.88, 9.79, 9.72	—	—	0.05 m, C²HCl₃	127
keto C=O	keto C=O	H	H	9.72, 9.67, 9.26	—	—	0.05 m, C²HCl₃	132
keto C=O	keto C=O	keto C=O		9.70, 8.95, 8.89	—	—	0.1 m, C²HCl₃	137
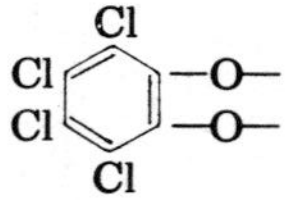	[dichlorophenyl structure]	keto C=O	COOCH₃	11.28(1), 11.12(2)	—	—	CF₃COO²H	137
keto C=O		OCH₃	COOCH₃	10.15, 10.05, 9.81	−2.75	—	0.03 m, C²HCl₃	126
keto C=O		OC₂H₅	H	9.97, 9.82, 9.80	−3.2/−4.1	~3.85/1.25	0.03 m, C²HCl₃	135
keto C=O		OC₃H₇	H	9.96, 9.84, 9.77		~4.20/1.40/0.72	0.03 m, C²HCl₃	135
keto C=O		O-i-prop	COOCH₃	9.96, 9.79, 9.50	−3.7	5.10/1.32/0.71	0.05 m, C²HCl₃, 2-vinyl	138

reported by Dougherty et al.[139] and establish a structure in which the γ-substituent is linked to C-7 rather than to C-6[77]. A thiacyclic structure in rapid tautomeric exchange with its mirror image is proposed by Clezy and Smythe[123] to account for the unusual chlorin-like [1]Hmr spectrum of the product obtained by hydrolysis of a *meso*-thioacetoxy-porphyrin.

10.2.4. N-Substitution

The n.m.r. spectra of N-alkylated porphyrins are generally interpreted in terms of the steric hindrance imposed on the macrocycle by central substituent(s) that do not fit into the inner cavity of the macrocycle. The chemical shift changes observed appear to arise from three different effects which can all be attributed to steric distortions. These include decrease in the ring current, changed local shielding patterns due to conformational changes, and an increased sp^3 hybridization at the substituted N-atom(s). These effects were first analyzed by Caughey et al.[140] for N-methyl- and N-ethyl-etioporphyrin-II (26).

Caughey et al.[140] discuss a conformation of the macrocycle in which the N-substituted ring and the neighboring rings are twisted out of the macrocycle plane, the latter to a smaller extent and in the opposite direction, while the opposite ring remains in the plane of the macrocycle. Confirmation of this interpretation was recently obtained by X-ray analysis[141], although the X-ray structural data showed a similar tilt of the opposite ring C as well. The resonances of the protons in the side chains of the (opposite) ring C are slightly shifted to higher field because of the reduced ring current, and the shielding is even more pronounced in the resonance signals of the protons of ring A, which are considerably tilted out-of-plane. The methine resonances under these circumstances are expected to move to higher field, which is indeed observed for the β and γ protons. This effect is partially compensated, however, by the less effective shielding by the out-of-plane alkyl substituents on the α and δ protons.

A series of N-mono-, di- and tri-substituted porphyrins (27—29) was investigated by Jackson et al.[142]. While the interpretation of the spectra for the N-monosubstituted porphyrins is similar to the one given above[140], it should be noted that the methine resonances were assigned the opposite way. The ring current of N-alkyl porphyrins is increased stepwise by formation of the mono- and di-cation (27a,b,c) although in the latter the conformation is changed as well[142]. The free bases do not aggregate, and the spectra are concentration independent. The mono-cation, however, forms a complex with the free base, the kinetics of which were studied by [1]Hmr[142]. In addition to the mono-substituted compounds, three N,N-dimethylporphyrin isomers (28a,29a,b), as well as their dications, were investigated and their spectra were again interpreted in terms of conformational and ring current changes arising from steric hindrance[142].

As the N-alkyl groups are in the center of the macrocycle, their proton

TABLE 13

¹Hmr chemical shifts (δ[p.p.m.] from TMS) of N-alkylporphyrins and their cations

Compound (26)

a) R = H
b) R = CH₃ (R = CH3)
c) R = C₂H₅ (R = C2H5)

Formula	a	b	c	Ref.
N–CH₃	—	−4.89	−2.37	140
N–CH₂	—	—	−5.16	
N–H	−3.79	−3.12	—	
1-CH₃	3.62	3.20	3.22	
4,8-CH₃		3.50	3.52	
5-CH₃		2.66	3.65	
2-CH₂		3.96	3.94	
3,6,7-CH₂	4.11	4.14	4.12	
2'-CH₃		1.42	1.39	
3',6',7'-CH₃	1.87	1.85	1.86	
α,δ-H	10.11	10.01	10.08	
β,γ-H		9.97	9.96	

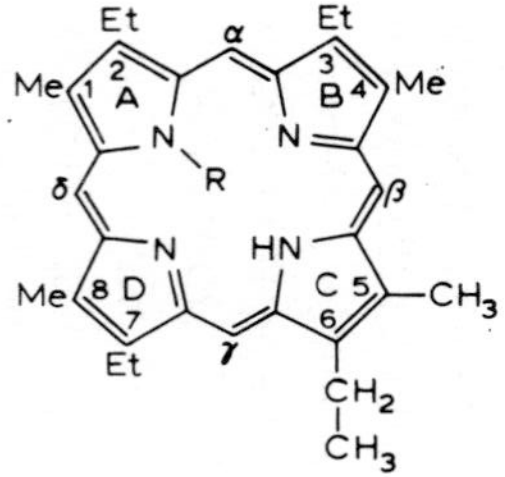

Compound (27)

a) Free base
b) Mono cation (ClO₄⁻)
c) Dication
d) Zn-Complex (Cl⁻)
e) Zn-Complex (I⁻)

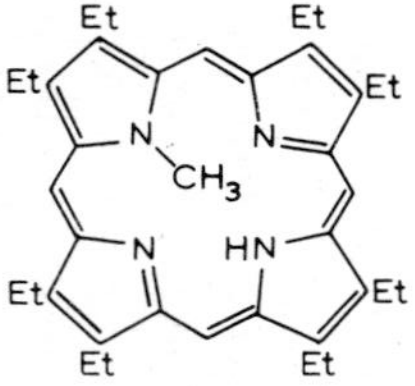

Formula	a	b	c	d	e	Ref.
N–CH₃	−4.76	−5.18	−5.20	−4.61	−6.05	142
N–H	—	—	−3.96	—	—	
1',2'-CH₃	1.48	1.44	1.54	1.75	0.90	
3',4',7',8'-CH₃	1.90	1.91	1.94	1.94	1—1.5	
5',6'-CH₃	1.91	2.00	2.00			
1,2-CH₂	3.72	4.84	3.94	3.98	3.08	
3,4,7,8-CH₂	3.96	to	4.40	4.04	3.7—4.2	
5,6-CH₂	4.00	3.85		4.08		
α,δ-H	9.89	10.55	11.00	10.22	9.65	
β,γ-H	9.94	10.64	11.12	10.31	8.65	

(28)
a) R=H
b) R=CH₃ (up); Cl⁻

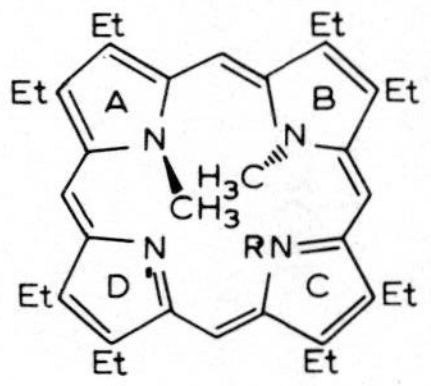

(29)
a) CH₃/CH₃ trans
b) CH₃/CH₃ cis

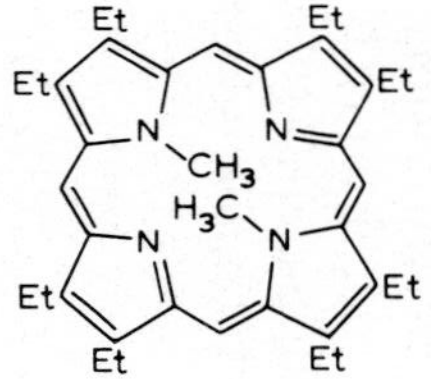

	28a	28b	29a	29b	142
$N_{A,C}$—CH₃	−5.8	−3.92	−5.30	−3.52	
N_B—CH₃		−7.08			
CH₃	1.55	1.27	1.71	1.28	
	1.71	1.50	1.94	1.84	
	1.98	1.63			
	1.99	1.89			
CH₂		3.3—4.1	3.8—4.2	3.39—3.98	
Meso—H	10.20	9.96	9.80	9.68	
	10.39	9.92			

References, p. 514

TABLE 14

[1]Hmr chemical shifts (δ [p.p.m.] from TMS) of some N-substituted derivatives of (OEP).

				Ref.
29c	a b c CH$_2$CH$_2$CH$_2$–Br N—N | | N — N	a: b: c:	−4.9 (t) −1.5 (m) 1.5 (t)	143a
29d	COOEt | CH N—N |Co| N — N	CH:	−2.25 (s)	143b
29e	COOEt | CH N—N | | N — N	CH:	−5.78 (s)	143b
29f	CH$_2$ N—N | | N — N	CH$_2$:	−8 (broad)	143b

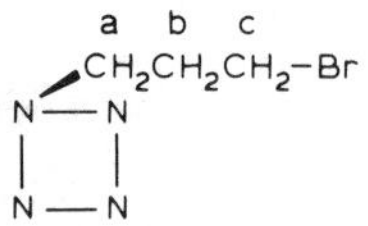

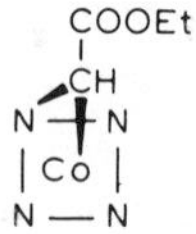

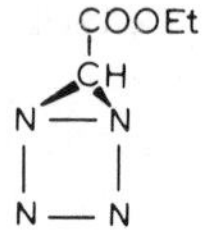

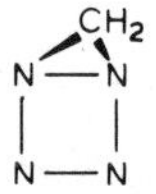

Schematic representation of the porphyrin system.

resonance occurs at extremely high field due to the strong shielding provided by the aromatic system in this region. The N-alkyl resonances of some principle N-substituted porphyrins are listed in Table 13. The N-alkyl incremental shifts were used by Storm and Corwin[16] as a probe from which an empirical formula was derived for the ring current shifts of central, out-of-plane substituents[17]. The decreasing shielding effect with increasing distance from the macrocyclic plane (Fig. 1) is clearly visible in the N(ω-bromopropyl)octaethylporphyrin and alkyl-Co-porphyrins[143a,b,199a], which are listed in Table 14 together with some unusual N-substituted H(OEP) derivatives. The spectrum of N-methyl-$meso$-tetraphenylporphyrin was recently reported[143c].

10.2.5. Chlorins and related structures

In chlorins (30) and bacteriochlorins (31,32) one or two of the macrocycle peripheral double bonds are reduced without loss of the macrocyclic ring current. In most natural products, both carbon atoms in the reduced pyrrole ring(s) become sp^3 hybridized, but bacteriochlorophyll-b (41)[231] and several synthetic compounds (54,56) contain an exocyclic double bond at one of the chlorin positions.

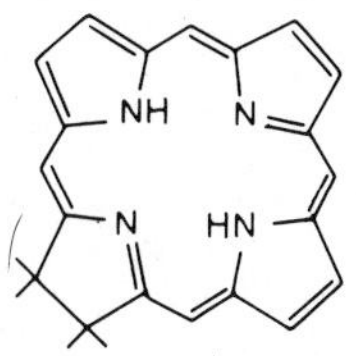

(30)

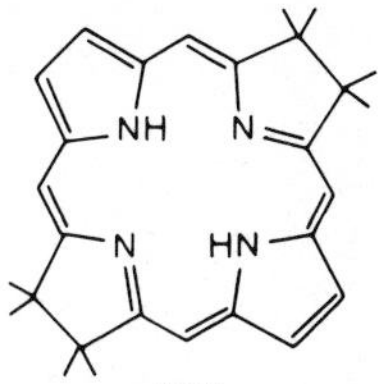

(31)

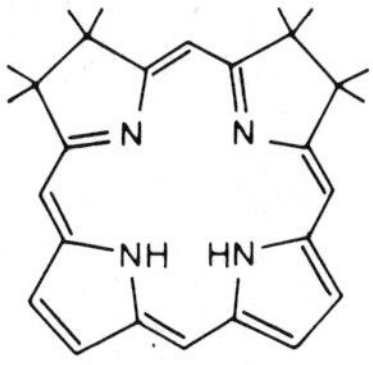

(32)

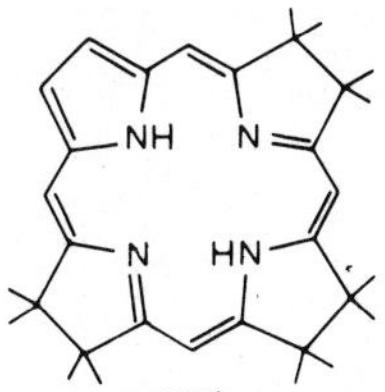

(33)

Removal of one of the peripheral double bonds leads to a decrease in the ring current, as indicated by the upfield shift of peripheral proton signals and a down field shift of the N—H signals (Table 15). The decrease is moderate in chlorins and bacteriochlorins, but very pronounced in the isobacteriochlorins (32). In the latter compounds, the two N—H protons are for the most part located at the two neighboring (non-reduced) pyrrole rings, a structure which is unfavorable for a large ring current for both steric and electronic reasons. A similar trend is observed in the gemini-porphin-diketones[143] with neighboring pyrroline rings (isomers 35a—c) vs. the 'opposite' isomers (35d) and (35e) (Table 17). A strongly decreased ring current is also observed in the gemini-porphin-triketones[143], which are examples of the dihydro-bacteriochlorin structure (33).

The [1]Hmr spectra of the basic chlorins and bacteriochlorins are listed in Table 15, and some selected examples of chemical shifts in chlorins are compared in Table 16 with those of the corresponding porphyrins. While the fully unsaturated porphyrins generally exhibit one set of closely spaced methine resonances, chlorins (30) show two sets about 0.8—1.0 p.p.m. apart even in the absence of highly anisotropic substituents. The high field set is assigned to the methine protons next to the reduced ring, the low field set to the remote methine protons. The low-field set reflects a moderate decrease in the macrocyclic ring current (Δ = +0.38 p.p.m. for octaethylchlorin vs. octaethylporphyrin (11)), and corresponding increments in chemical shift are observed for the resonances of substituents not attached to the reduced ring. The unusual shift of the neighboring methine resonances is best explained by a picture of the macrocycle introduced by Woodward[78,151]. In this model, the four pyrrole rings are considered to remain to some extent autonomous aromatic subunits that borrow electron density from the methine positions. Removal of a peripheral double bond (as by addition of 2H to ring D) results in the loss of an aromatic subunit, and an increase in the electron density at

TABLE 15

¹Hmr chemical shifts (δ [p.p.m.] from TMS) of some principal chlorins, bacteriochlorins, and isobacteriochlorins (see structures (30—33)).

	Methine—H	β_pyrrole—H	N—H	Chlorin—H	Remarks	Ref.
Porphin	10.58	9.74	−3.76	—	C²HCl₃	60
Chlorin (7,8)	9.62 (α,β), 8.92 (γ,δ)	9.03 (3,4), 8.63 (2,5), 8.52 (1,6)	−2.75	4.25	C²HCl₃	60

Octaethyl-	Methine—H	CH₂	CH₃	N—H	Chlorin—H	Remarks	Ref.
Porphyrin	10.18	4.14	1.95	−3.74	—	C²HCl	71(144)
Chlorin (7,8-H) (*trans*)	9.80 (α,β), 8.95 (γ,δ)	3.88 (2 × CH₂), 3.91 (2 × CH₂), 4.01 (2 × CH₂), 2.20 (7,8-CH₂)	1.81 (2 × CH₃), 1.77 (2 × CH₃), 1.75 (2 × CH₃), 1.05 (7′,8′-CH₃)	−2.46	4.54*	0.05 m in C²HCl₃	126(145, 146,65,147, 127,144)
Bacteriochlorin (3,4,7,8-H)	8.82	1.9—2.3(M, 3,4,7,8-CH₂), 3.73(1,2,5, 6-CH₂)	0.97 (3′,4′,7′, 8′-CH₃), 1.70 (1′,2′,5′, 6′-CH₃)	−1.88	4.2—4.35	0.05 m in C₆²H₆	111,147
Iso-Bacteriochlorin (1,2,7,8-H)	6.80, 6.82 (δ), 7.40, 7.42 (α, γ), 8.46, 8.48 (β)	1.6—2.3 (M,1, 2,7,8—CH₂), 3.25 (3,6-CH₂), 3.37 (4,5-CH₂)	0.98(1,8-CH₃), 1.03 (2′,7′-CH₃), 1.46 (3′,6′-CH₃), 1.50 (4′,5′-CH₃)	+2.96	3.4—3.8	0.05 m in C²HCl₃	147(127,144 146,148)

Tetraphenyl-	β_pyrrole—H	Phenyl—H (o, m, p)	N—H	Chlorin—H	Remarks	Ref.
Porphyrin	8.75	8.30	7.80	—	C²HCl₃	74
Chlorin (7,8-Dihydro)	8.34 (s,3,-4-H), 8.10, 8.49	7.6—8.5	7.80	1.3	4.10	C²HCl₃	65

				C^2HCl_3	65
				3.92	
				1.3	
		7.52			

| Bacteriochlorin (3,4,7,8-Tetrahydro) | (AB,1,2, 5,6-H, $J=4.5$) 7.85 (d, $J=2,1,2$, 5,6-H) |

* X-part of an ABX spectrum, $J_{AX} = 4$, $J_{BX} = 6.8$.

TABLE 16

[1]Hmr chemical shifts (δ[p.p.m.] from TMS) of some selected chlorins as compared to the respective porphyrins.

Selected Chlorins		Porphyrin	Chlorin	Remarks	Ref.
(37) 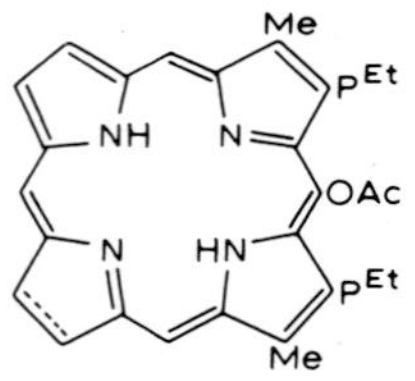	Methine—H	11.08, 10.98	9.80, 8.53	TFA	62
	β_{pyrrole}—H	9.83	9.02, 8.55		
	Aromatic CH_3	3.69	3.30, 3.22		
	Chlorin—H	—	4.70		
	N—H	—	−3.0		
(38) 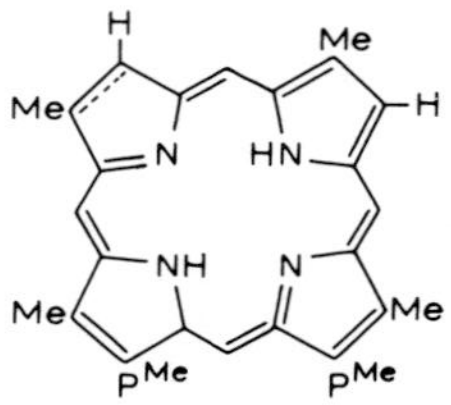	Methine—H	9.46 (3), 9.37 (1)	9.50, 9.52	C^2HCl_3	149,150
	β_{pyrrole}—H	8.61, 8.54	8.41 (2, α, δ) 8.60 (4-H) 3.77—4.77 (2-H)		
	Ar., Ester CH_3	3.39, 3.30 3.25, 3.14	3.29 (3) 1.75 (1—CH_3) (d, $J = 7$)		
	Chlorin—H	—	3.77—4.77		
	N—H	−4.70	−2.75		
(39) 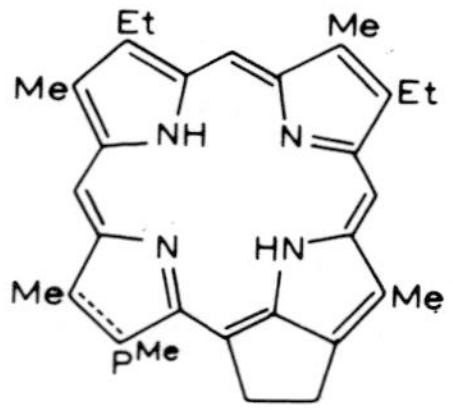	Methine—H	9.88, 9.79	9.77, 9.58	0.05 m, C^2HCl_3	127
	9-CH_2	9.72	8.87 (δ)		
		4.00	3.99		
	Ar., Ester CH_3	3.67, 3.59 3.47, 3.41 3.32	3.43 (2) 3.53 (2) 1.80 (8-CH_3) (d, $J = 7$)		
	Chlorin—H	—	4.3—4.6		
	N—H	—	−1.68, −3.53		

TABLE 17

[1]Hmr chemical shifts (δ[p.p.m.] from TMS in C^2HCl_3) of the geminiporphin mono-, di-, and tri-ketones of $H_2(OEP)$

(34), 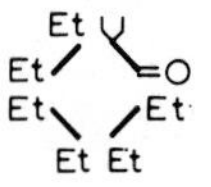 (35a), (35b), (35c), (35d), (35e), (36a), (36b) — structural diagrams

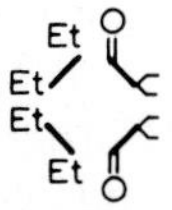

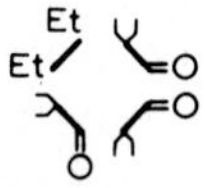

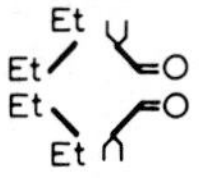

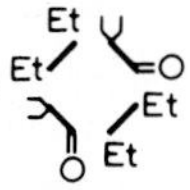

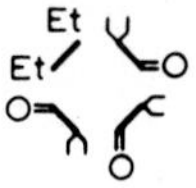

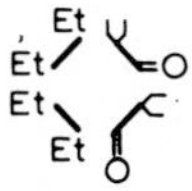

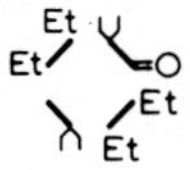

Compound	N—H	Methine—H		Nuclear		Geminal	
				$CH_2(q)$	$CH_3(t)$	$CH_2(q)$	$CH_3(t)$
34	−2.8	9.13		4.0 (m)	1.9 (m)	2.76	0.46
		9.86					
		9.88					
		9.96					
35a	−1.58	8.83	2X	3.86	1.74	2.60	0.48
		9.59		3.95	1.80		
		9.79					
35b		8.39		3.7 (m)	1.67	2.56	0.42
		8.58			1.70		0.54
		9.24			1.72		
		9.37					
35c		7.42	2X	3.52	1.61	2.5 (m)	0.49
		8.81		3.57	1.62		
		9.05					
35d	−1.84	8.78	2X	3.84	1.76	2.63	0.44
		9.59	2X	3.89	1.78		
35e	−2.63	9.05	2X	3.94	1.80	2.70	0.44
		9.71	2X	3.98	1.83		
36a		8.10		3.62	1.63	2.5 (m)	0.49
		8.36					
		8.45					
		8.86					
36b		7.78		3.51	1.57	2.36	0.45
		8.01		3.56	1.60		0.55
		8.08					0.58
		8.90					

Schematic representation of the porphyrin ring system. The heavy bars correspond to the peripheral bonds between adjacent β-pyrrole carbons. From Ref. 143.

References, p. 514

the neighboring methine positions, which accounts for the high field shift of the signals for the neighboring methine protons[152].

In the bacteriochlorins (31) all four methine protons are adjacent to a (reduced) pyrroline ring. The methine resonances thus occur in the same chemical shift range as the high field set in chlorins, and the additional (small) high field shift reflects an additional decrease in ring current. In the isobacteriochlorins (32) one methine proton is between two reduced pyrroline rings, two protons are next to one reduced ring each, and one proton is situated between two pyrrole rings. Thus, three sets of methine resonances are observed, each about 0.8—1.0 p.p.m. apart from the next set for the reasons set forth above.

In the gemini-porphin-ketones (34—36) one of the chlorin positions in each pyrroline ring is part of a carbonyl group. All known isomeric gemini-porphin-mono- (34), di- (35) and tri-ketones (36) are listed in Table 17. In a detailed [1]Hmr investigation, Inhoffen and Nolte[143] showed that the position of a methine proton resonance is mainly determined by its nearest neighbors. For the methine protons next to one or two geminal diethyl substituents, incremental shifts of δ = 0.8 and 1.6 p.p.m., respectively, could be deduced[143]. (For *meso*-substituted gemini-porphin-ketones, see Section 10.2.3.)

In contrast to the generally straightforward interpretation of the substituent subspectra in porphyrins, the subspectra of substituents of the (reduced) pyrroline ring in chlorins are often very complex. One reason is the reduced ring current shifts. All protons of substituents attached to the pyrroline ring are one C—C bond further removed from the aromatic system than in the corresponding porphyrins. The subspectra are, therefore, less spread out and are often not first order. The spectra are further complicated by the magnetic non-equivalence of the methylene protons of alkyl side-chains because of the asymmetric quaternary β carbon atoms of the pyrroline ring[13]. An additional complicating factor in these compounds is the possibility of spin—spin coupling between the substituents. Finally, structural isomers can be encountered. (The stereochemical aspects of chlorins are dealt with in Section 10.4.3, and the [1]Hmr spectra of compounds related to chlorophylls are treated in Section 10.2.8.2.)

Some further examples of chemical shifts associated with reduced pyrroline ring(s) are listed in Table 18. Chlorins in nature generally have one alkyl substituent and one proton at either one of the quaternary chlorin positions, and coupling between the two protons is observed[13]. Corresponding dioxychlorins (40), in which the chlorin protons are replaced by hydroxy and alkoxy-substituents, have been investigated by Inhoffen et al.[153]. Bacteriochlorophyll-*b* (41) has an ethylidene group[155] attached to a pyrroline ring (see also Section 10.2.8.2), and synthetic chlorins with a methylene substituent have been reported by Jackson et al.[91]. These authors[91] also studied some compounds (44,45) in which a methylpyrroline ring is spiroannelated

TABLE 18

Chemical shifts (δ [p.p.m.] from TMS in C^2HCl_3) of substituents of the (reduced) pyrroline ring in chlorins and bacteriochlorins

		Ref.
(40[a])	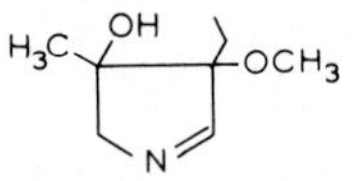 H₃C–OH, –OCH₃, N=	2.95 (3-OCH₃) 1.58 (3-CH₃) 4.50 (4-OH) 1.48 (4'-CH₃)	153
(41[b])	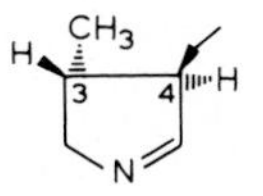 H, CH₃ 3 4 H, N=	1.77 (3-CH₃) 4.24 (3-H) 4.20 (4-H) 1.09 (4'-CH₃)	154
(42[c])	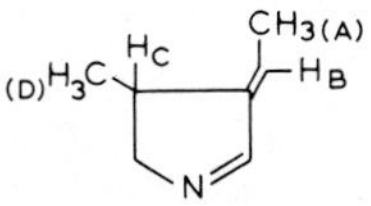 (D)H₃C, H_C, CH₃(A), H_B, N=	2.01 (CH₃(A)), d, $J_1 = 7$ 6.84 (H_B), dq, $J_2 = 2$ 4.93 (H_C), dq, $J_3 = 7$ 1.66 (CH₃,D), d	155
(43[d])	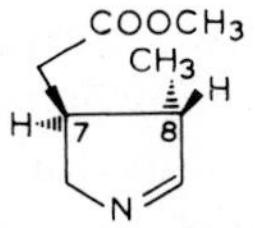 COOCH₃, CH₃, H 7 8 H, N=	1.82 (d, 8-CH₃) $J_{8,8'} = 7$ 4.40 (q, 8-H) $J_{7,8} \leqslant 2$ 4.13 (m, 7-H) 2.1–2.8 (m, 7a–CH₂), (m, 7b–CH₂) $J_{7b_A–7b_B} = 19$	156
(44[e])	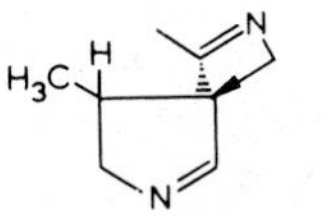 H₃C, H, N, N=	1.67 (s, CH₃) 1.39 (m, CH₂CH₃) 3.05 (d, CH₃) 4.20 (m, CH)	91
(45[e])	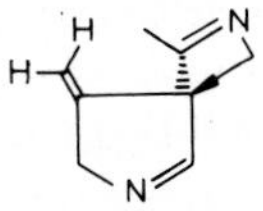 H, N, N=	6.73, 5.67 (CH₂) 1.67 (s, CH₃) 1.6 (m, CH₂CH₂)	91

(46[f])

a) $R_B = H_B, R_C = H_C$
b) $R_B = H_B, R_C = COOCH_3$
c) $R_B = COOCH_3$, $R_C = H_C$
d) $R_B = R_C = COOCH_{3(B,C)}$

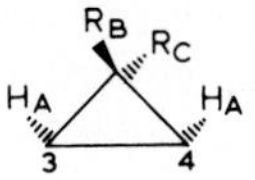

a) 3.88 (H_A), 1.65 (H_B), 0.60 (H_C)   66
$J_{AB} = 8.2$, $J_{AC} = 3.3$, $J_{BC} = 3.0$
b) 4.34 (H_A), 2.68 (H_B), $J_{AB} = 8.1$
2.74 (COOCH₃)
c) 4.47 (H_A), 1.58 (H_C), $J_{AC} = 2.6$
3.76 (COOCH₃)
d) 4.79 (H_A), 3.82 (CH₃,B), 2.34 (CH₃,C)

(47[g])
a) Isomer I
b) Isomer II

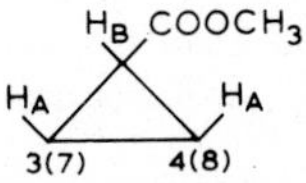

a) 3.70 (H_A), 1.42 (H_C), 3.70 (CH₃)   66
b) 3.72 (H_A), 1.40 (H_C), 3.63 (CH₃)

Only a partial structure is shown; the compounds are derived from: (a) methyl bacteriopheophorbide-*a*; (b) Bchl-*a*; (c) Bchl-*b*; (d) pheophytin-*a*; (e) an Etio-type chlorin; (f) *meso*-tetraphenylchlorin; and (g) *meso*-tetraphenylbacteriochlorin.

References, p. 514

to one of the chlorin positions. A group of chlorins with condensed cyclo-propane rings (46,47) was investigated in some detail by Callot et al.[66,157,158], who showed that their stereochemistry can be deduced both by chemical shift and spin—spin coupling arguments (cf. Section 10.4.3). A substituted octaethylchlorin, in which one 'extra' hydrogen is replaced by a methyl group, is reported by Fuhrhop[158a], and recently the spectrum of 7,8-diethyl-octamethylchlorin has been studied[158b].

10.2.6. Systems with interrupted conjugation

Tetrapyrrole structures related to porphyrins, in which the ring current is more or less reduced or abolished, are currently of great interest because of their biochemical importance. We confine our discussion in this section to structures in which the porphyrin skeleton is retained*. In the non-aro-matic** structures which contain the porphyrin skeleton, but not the con-jugation, the conjugation is usually interrupted at one or more of the bridg-ing methine positions. The resulting subunits have either four (one interrup-tion) or two (two interruptions at opposite positions) pyrrole rings in con-jugation, or the pyrrole subunits are isolated. While the system with four conjugated pyrrole rings may show indications of a residual ring current across the interrupted methine bridge, the ^{1}Hmr spectra of the other com-pounds are very similar to those of pyrromethenes (two conjugated pyrrole subunits) and pyrroles, which are representative non-macrocycle systems. Although the appropriate linear oligopyrrole systems are unlikely to be in the same cyclic conformations as the porphyrins, the chemical shifts are similar to those of porphyrin-derivatives with interrupted conjugation.

For systems without strongly anisotropic groups, the following ranges for the proton chemical shifts are observed: methine-H: δ = 5.8—6.8 p.p.m.; β-pyrrole H: δ = 6—7 p.p.m.; peripheral ethyl groups: δ = 2.5—2.9 p.p.m. (CH_2), and δ = 1.2—1.3 p.p.m. (CH_3); peripheral CH_3: δ = 1.8—2 p.p.m. Significant deviations from these chemical shift values indicate either an aromatic (diamagnetic) or an anti-aromatic (paramagnetic) ring current.

A characteristic feature of these conjugation-interrupted systems is a very large (up to 12.6 p.p.m.) shift to lower field frequently observed for the N—H protons. In contrast to the porphyrins with an aromatic macrocycle, the N—H protons in the interrupted systems occupy peripheral positions with respect to the aromatic subunits, and are subjected to their combined

---

* For leading references to the ^{1}Hmr spectroscopy of corrins and related structures, in which two of the pyrrole rings are directly linked together, see Ref. 159. For a discussion of bile pigments, see Ref. 160. Illustrative of other porphyrin-like systems with inter-rupted conjugation, a homoazaporphyrin has been reported by Grigg[161], hemiporphy-razines have been studied in detail by Kenney et al.[162a], and a series of related structures is reviewed by Haddon et al.[41].

** The ring current criterion is generally used to determine whether or not a macrocycle aromatic system is present.

deshielding influence. Although no systematic [1]Hmr data are available, the strong variation of the N—H resonance within a given system indicates an additional structural dependence of the N—H shift, probably involving a change in hydrogen bonding.

One hetero porphyrin, a *meso*-thia-porphyrin[57] (54) in which the ring current is interrupted by substitution at a *meso* carbon, is described in the literature. The resonances of (54) occur at considerably lower field than in linear tetrapyrroles[160], and the chemical shifts are indicative of a residual ring current, suggesting that the non-bonding sulfur electrons participate to a certain extent in the aromatic π-system. A porphyrin-like character for (54) is also consistent with the methine chemical shift pattern, which shows two resonances at lower field, and one originating in the C—H group opposite to the sulfur bridge at higher field. This pattern is characteristic of *meso*-substituted porphyrins (Section 10.2.3) and is opposite to that observed in biliverdins[160a]. The possibility of participation by the non-bonding electrons of heteroatoms is clearly demonstrated by an oxaporphyrin reported by Besecke and Fuhrhop[162], the cation of which shows a typical aromatic [1]Hmr spectrum.

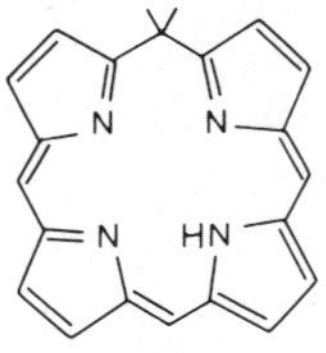

(48) 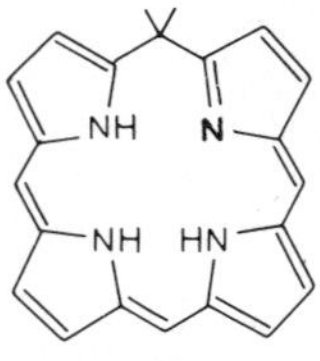 (49) 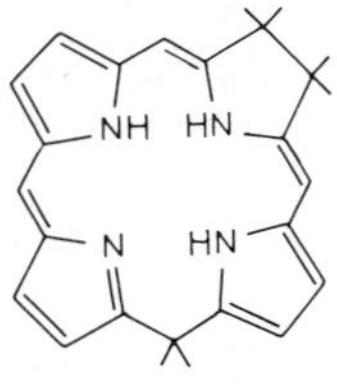 (50)

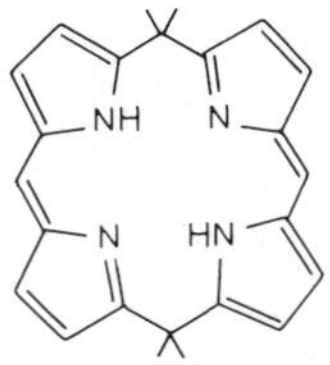

(51) 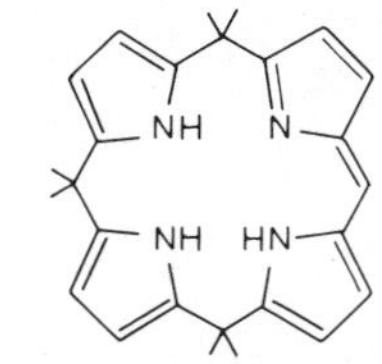 (52) 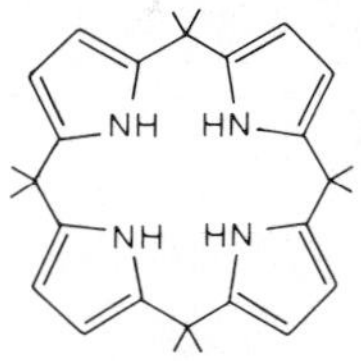 (53)

Isoporphyrins (48)[64,163], phlorins (49)[151] and chlorin-phlorins (50)[153] are isomers of the porphyrins, chlorins (30) and bacteriochlorins (31), respectively, in which the conjugation in the ring is interrupted by quaternization of one methine bridge. Very often the more stable cross conjugated oxo- or imino-derivatives are studied instead of the corresponding structure with a quaternized *meso*-carbon (Table 19). The loss of the aromatic ring current in Zn(iso-TPP) (55) is clearly indicated[64] by the high field shift of the β-pyrrole protons and the presence of only one high field multiplet for the aromatic phenyl protons, whose chemical shifts cover a considerable range in H_2(TPP)[8a,120]. Similar shifts are reported for an *iso*-OEP Pd com-

References, p. 514

TABLE 19

^{1}Hmr chemical shifts (δ[p.p.m.] from TMS) of porphyrin derivatives with interrupted conjugation.

	Chemical shifts, δ	Solvent	Ref.
(54) a) R = H b) R = CH$_3$ c) Dication of a 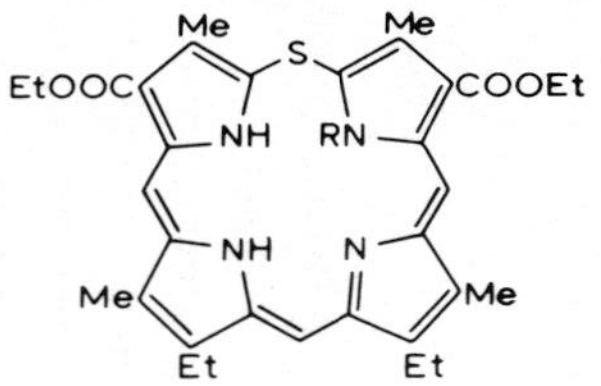	a) Methine—H: 7.57 (β, δ), 6.30 (γ) N—H: 4.56 b) Methine—H: 8.83, 7.84, 6.06 N—CH$_3$: 2.92 c) Methine—H: 8.30, 7.58 (γ, δ) 5.05 (2H, β)	a and b: C^{2}HCl$_3$ c: TFA	57
(55) 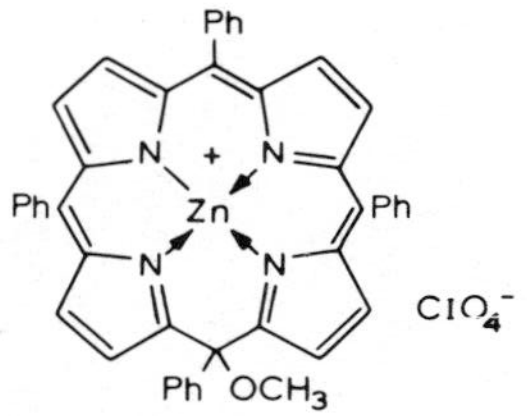	Phenyl—H: 7.04 (m) β-H: 6.25, 6.52/6.51, 6.57 2 × AB, J_1 = 4.70, J_2 = 4.85	C$_6$^{2}H$_6$	64
(56)	$\beta_{pyrrole}$—H: 6.97–6.77	C$_6$^{2}H$_6$	165

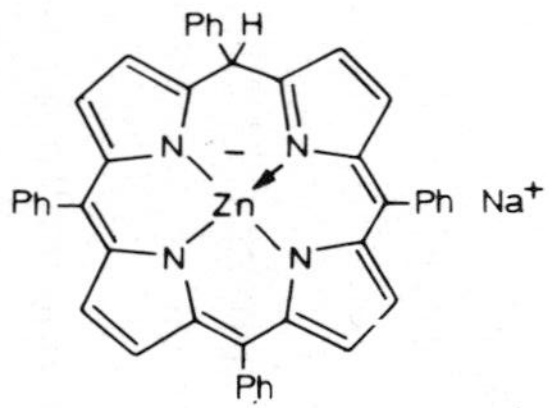

(57)

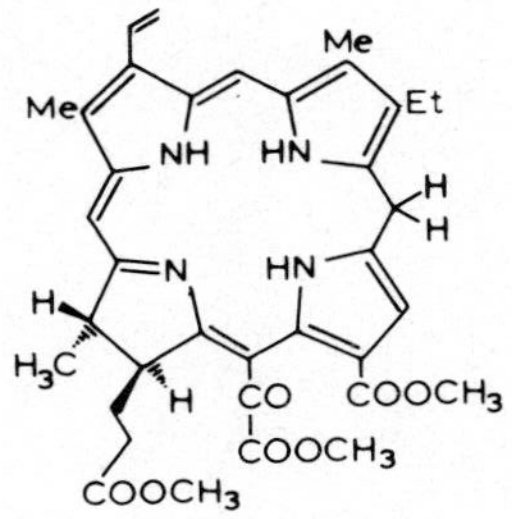

Assignment	Shift	Solvent	Ref.
Methine-H:	6.73, 5.45	C²HCl₃	166
β-CH₂:	3.80		
Vinyl (H_X,A,B):	6.70, 5.40, 5.35		
OCH₃:	3.58(1X), 3.56 (2X),		
CH₃ (ring):	2.17 (1X), 2.08 (2X),		
8,4 -CH₃:	1.28, 1.16		
NH:	8.60 (1), 5.9 (2)		

(58)

Assignment	Shift	Solvent	Ref.
Methine-H:	7.85 (2), 6.98 (1)	C²HCl₃	93
CH₂:	3.31, 2.98		
CH₃:	1.38		

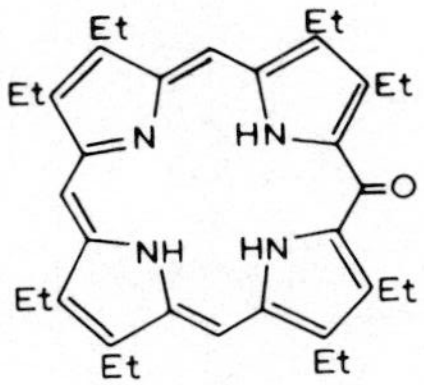

(59)

Assignment	Shift	Solvent	Ref.
Methine-H:	7.3—8	C²HCl₃ (aromatic in TFA)	106
Ring—CH₃:	2.6		
NH:	1.6		

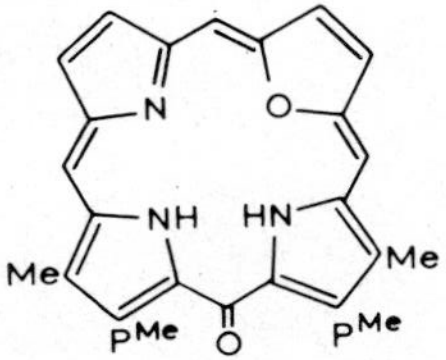

(60)

Assignment	Shift	Solvent	Ref.
Methine-H:	7.87, 7.50 (β, δ), 6.65 (α)	C²HCl₃ (aromatic in TFA)	105
CH₂CH₃(a):	3.60—2.94, 1.54—1.14		
CH₂CH₃(b):	2.07, 0.67		

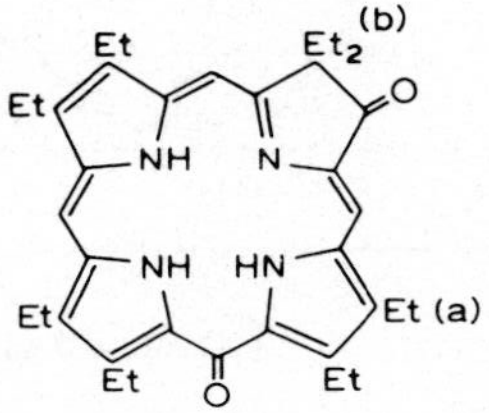

References, p. 514

TABLE 19 (Continued)

	Chemical shifts, δ	Solvent	Ref.
(61)	Methine—H: 5.52 (s) Phenyl: 7.40, 7.21, 7.01 (m) CH₃: 1.90, 1.77 (s)	C^2HCl_3	120
(62)	Methine—H: 5.52 (s) Phenyl: 7.40, 7.21, 7.01 (m) CH₃: 1.90, 1.77 (s)	C^2HCl_3	120
(63)	Methine—H: 6.58 (β, δ), 4.08 (q, α, γ) CH₃: 1.69 (d) CH₂CH₃ (a): 2.51, 1.13 CH₂CH₃ (b): 2.41, 1.10 NH: 12.58	$C_6{}^2H_6$	119
(64) a) R¹ = R² = O, 　M = H₂ b) R¹ = O, 　R² = NH, 　M = Zn	a) Methine—H: 6.68 　CH₂: 2.71, 2.50 　CH₃: 1.14, 1.12 b) Methine—H: 6.47, 6.81 　NH: 9.72	C^2HCl_3 C^2HCl_3	174 175

(65)
a) R=H$_2$
b) R=O

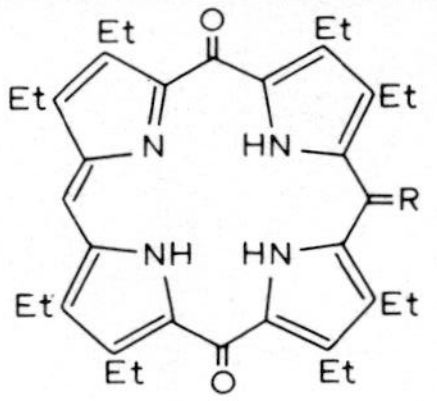

		Shift	Solvent	Ref.
a)	Methine—H:	6.86	C^{2}HCl$_3$	167
	Methylene—H:	3.83		
	CH$_2$:	2.77, 2.59, 2.45		
	CH$_3$:	1.28, 1.09		
	N—H:	10.33		
b)	Methine—H:	6.96	C^{2}HCl$_3$	174
	CH$_2$CH$_3$:	2.4—3.0, 1.0—1.4		
	N—H:	10.33		

(66)
a) R=H
b) R=CH$_3$

		Shift	Solvent	Ref.
a)	Methine—H:	5.38 (s)	C^{2}HCl$_3$	120
	Phenyl:	7.24 (m)		
	$\beta_{pyrrole}$—H:	5.68, 5.78 (2 × d, J = 6)		
	N—H:	8.2 (b, m)		
b)	Methine—H:	5.34 (s)		
	Phenyl:	7.0 (m)		
	CH$_3$:	1.77 (s)		
	N—H:	6.41 (b, s)		

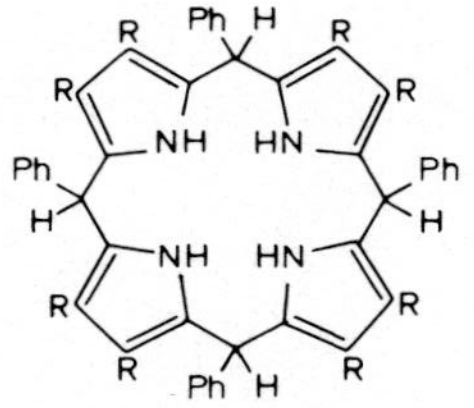

(67)
a) R=O
b) R=H$_2$

		Shift	Solvent	Ref.
a)	CH$_2$:	2.98	C$_5$^{2}H$_5$N	167
	CH$_3$:	1.26		
	N—H:	11.9		
b)	Methylene—H:	3.83 (s)		
	CH$_2$CH$_3$:	2.71, 2.42/1.17, 1.06		
	N—H:	8.72		

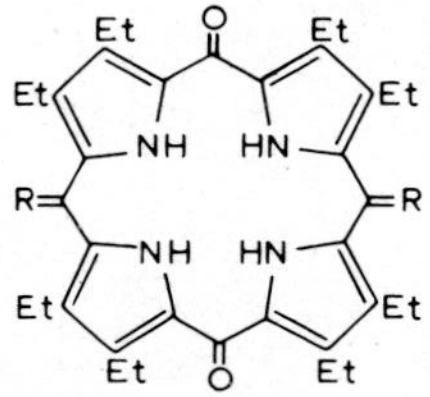

(70)
a) R=R^1=Et, OH
b) R=O,R^1=Et$_2$

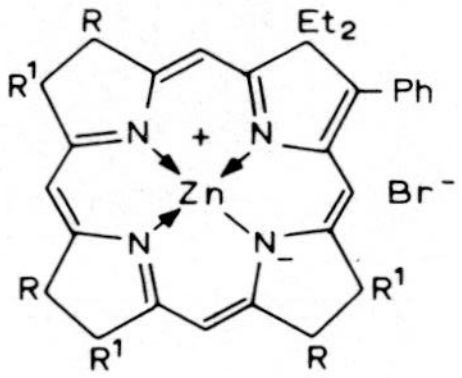

		Shift	Solvent	Ref.
a)	Methine—H:	5.97, 6.07, 6.12, 6.18	C^{2}HCl$_3$	177
b)	Methine—H:	6.47, 6.63, 6.80, 7.14		

References, p. 514

TABLE 19 (continued)

	Chemical shifts, δ	Solvent	Ref.
	CH₃: 1.37, 1.47, 1.48, 1.50 (s, each 2 × CH₃)	C^2HCl_3	178
	CH₂: 3.06, 3.17, 3.27 (s, each 1 × CH₂)		
	CH: 5.86, 5.90, 6.10, 6.24, 6.29 (s, each 1H)		
	N—H: 10.1, br.		
	CH₃: 1.34	TFA	178
	CH₂: 2.79		
	CH: 5.47		

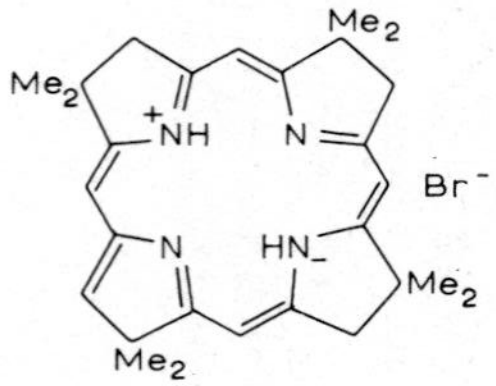

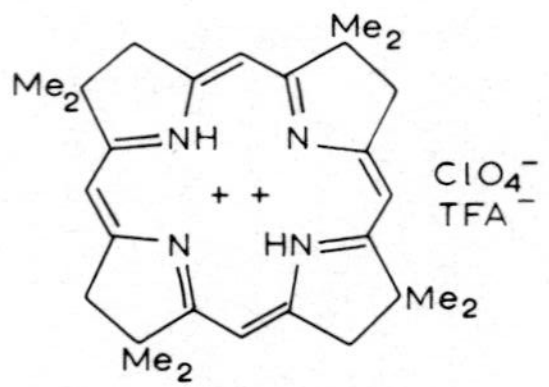

(71)

(72)

plex[163]. Phlorins (49) were first characterized by Woodward[151]. They are in acid-base equilibrium with chlorins (30)[164] and are unstable against oxidants[111]. Only marginal [1]Hmr information is available on phlorins[165], but a series of 7,8-chlorin-β-phlorins ((50), see Table 19)[153,166] as well as a 7,8-chlorin-γ-phlorin have been studied in detail[166a]. Both phlorins and chlorin-phlorins show strongly deshielded N—H protons and shielded peripheral protons characteristic of a ring current that has all but vanished. The hydroxyporphyrin ⇌ oxophlorin tautomeric systems (21b ⇌ 22b) have been extensively studied[93,106,106a,108,109,167]: The equilibrium is shifted at higher pH towards the isomer (21b) with interrupted conjugation. The 'olefinic' [1]Hmr spectrum in neutral solution indicates an oxophlorin structure. Upon addition of acid, the ring current gradually increases as the monocation is formed and converted to the fully aromatic hydroxyporphyrin dication[69,106a] (see also Section 10.2.3). The n.m.r. spectra of the oxophlorins are generally difficult to observe[109−111,117], because oxophlorins are extremely easily oxidized and the [1]Hmr lines are broadened by spin exchange with the oxophlorin radical[111] present in small amounts[110]. Both the position and the pattern of the methine proton resonances again indicate that a residual ring current is present in these cross-conjugated systems.

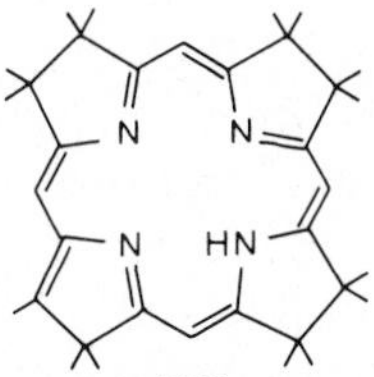

(68) 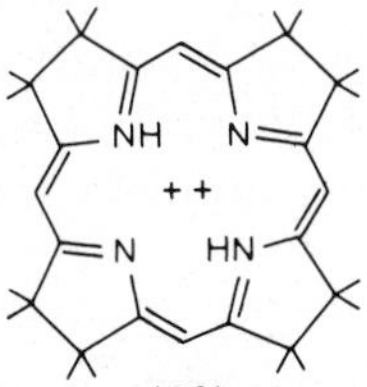   (69)

Porphodimethenes (51) are tautomers[168] of phlorins (49) in which the ring current is interrupted at two opposite methine bridges. This structure contains two isolated pyrromethene subunits, and in fact the [1]Hmr spectrum of the Zn complex (59) is very similar to that of the Zn-pyrromethene (60)[120]. Metal complexes of α,γ-dimethyl-octaethyl-β,δ-porphodimethenes (63) have been extensively investigated by Buchler et al.[169,170]. A structure[171] in which the macrocycle is folded like a roof and the meso-methyl and axial ligands occupy 'chimney' positions is deduced from steric and n.m.r. considerations[170]. Only one doublet and quadruplet, respectively, for the two methyl groups and the protons at the quaternized bridges are observed, and variations in the chemical shifts of these protons are related to the folding angle and long-range shielding effects of the axial ligands. The N—H signals in the free base occur at extremely low field (δ = 12.58 p.p.m.) indicating strong ring current shifts from the pyrromethene subunits and H-bonding. One of the isomers formed by photo-reduction of α,γ-dimethyl OMP-Zn was proved by [1]Hmr to be the corresponding β,δ-porphodimethene[172], and the structure of the Krasnovskii photoreduction product of chlorophyll-a (see Section 10.2.8.2) was also shown to be an α,γ-porphodimethene (reduced at the β,δ-positions)[173]. Dioxo- and iminooxo-porphodimethenes (64) were studied by Smith[174] and Fuhrhop[175], respectively. The chemical shifts reported for these compounds agree with those cited in the above examples. An interesting feature is the observation of two methine signals in the spectrum of the imino-oxophlorin (64b), indicating that the molecule has no σ_v plane because of the nonlinear C = NH substituent.

In the porphomethenes (52), the conjugation is interrupted at all but one bridge. [1]Hmr spectra of the dioxo- and trioxo-OE-porphomethenes (65) show signals for the peripheral protons that fall in the same range as those observed in oxo-porphodimethenes[167,174]. In contrast to the latter compounds reported only as metal complexes, the free bases were measured in the case of the porphomethene (65). The N—H resonances occur at very low field, and are about 2.6 p.p.m. more strongly deshielded than in pyrrole. The [1]Hmr spectrum of a true porphomethene has been reported by Shulga et al.[174a].

In the porphyrinogens (53) the ring current is interrupted at all four methine bridges, and as a result, the porphyrinogen spectra are very similar to those obtained from pyrroles. The most significant difference between the TPP-porphyrinogen (66a) and its octamethyl-derivative (66b), is the shift to higher field of the phenyl protons and the shift to lower field of the N—H proton upon methyl-substitution of the peripheral pyrrole positions[120]. As the N—H shift is especially dependent on molecular structure and intermolecular interactions, no structural conclusions or generalizations can be drawn from the few data available. However, the observation of two doublets ($J = 6$ Hz) for the β-pyrrole protons in (66a) clearly indicates a symmetry lower than C_4 for these porphyrinogens.

The main spectral feature of the oxoporphyrinogens (67) (xanthoporphyrinogens)[167,176] is the extreme low field shift of the N—H resonance, and to a lesser extent, of the —CH$_2$ quadruplets, a shift which increases with increasing oxo-substitution. The extreme nature of the shifts in the oxo-compounds has already been noted in the oxo-porphomethenes (65), and must somehow be related to the presence of *meso* carbonyl groups. While the shift of the resonances of the peripheral groups can be accounted for by the magnetic anisotropy of the C = O group, the N—H signals must be subject to additional shifts, presumably by (intermolecular) hydrogen-bonding.

In the corphins (68)[177,178] (see Chapter 18) the ring current is interrupted in an essentially different manner at one α-pyrrolic carbon atom and at one N atom. The former efficiently blocks the macrocyclic conjugation, and thus olefinic spectra are observed for (70) and (71). The considerable high-field shift for the methine protons ($\Delta \sim 0.5$—1.0 p.p.m.) in the oxo-substituted corphin (70)[177] is probably due to the long range deshielding effect of the β-pyrrole carbonyl groups.

The [1]Hmr spectra of the corphin[178] and metallocorphin[177,178] mono-cations are characteristic of a highly asymmetric structure and indicate localized double bonds rather than rapid tautomerism. On the other hand, the di-cation (72) exhibits only one signal each for all of its methyl, methylene, and methine protons, indicative of protonation at carbon to form the symmetric structure (69). The tetraaza[16]annulene conjugated system which results is anti-aromatic, and the [1]Hmr spectrum shows the expected high-field shift of all peripheral signals (Δ_{CH_3} = +0.4—0.8 p.p.m., Δ_{CH_2} = 0.3—0.5 p.p.m., Δ_{CH_3} = +0.03—0.16 p.p.m.) as compared to the free base mono-cation[178].

10.2.7. Porphyrin acids

Because of aggregation and the frequent low solubility of porphyrins in organic solvents, trifluoroacetic acid is widely used as a solvent for [1]Hmr measurements. In this strongly acidic solvent system, the porphyrins are usually present as N,N'-diprotonated di-cations, the resonances of which are considerably changed with respect to the free base. The C—H resonances are shifted to lower field by 0.8—1.0 p.p.m. in the protonated species, the N—H resonances to higher field by 0.4—1.0 p.p.m. This effect was first discussed by Abraham[3], who proposed an enhanced ring current from the larger resonance energy (and therefore higher aromaticity) associated with the D_{4h} symmetry of the di-cation as compared with the D_{2h} symmetry of the free base. This argument is identical to that advanced to account for the high basicity of porphyrins[179]. While the increased ring current in the di-cation is the principal contributor to the methine proton chemical shifts, the effect on the N—H protons is partly compensated for by the deshielding resulting from the positive charges at the nitrogen atoms. Expansion of the π-system to the periphery of the macrocycle may play an additional role[180]. An alternative explanation for the protonation effect was put forward by Haddon et al.[41]. In this view, the positive charges at nitrogen lead to a general deshielding that is more than compensated for in the case of the N—H protons by the abolition of hydrogen-bonding[1] and the (anisotropic) nitrogen lone-pairs. The effect of hydrogen-bonding was recently distinguished from other contributions by Ogoshi et al.[180], who reported the [1]Hmr spectra of $H_4(OEP)^{2+}$ diacids in chloroform-$^2H_1$. The [1]Hmr spectra of the di-cations in a neutral solvent are distinctly different from those of the di-cation in TFA, and the spectra also show marked variations that depend on the gegenions present. This is especially true for the resonances of the N—H signals, and the close correlation of the N—H stretching vibration to the magnitude of the chemical shifts suggested hydrogen-bonding to the gegenion as the main factor in this effect[180].

Judged from the examples cited in Table 20, the effect of protonation is quite general for porphyrins, and the spectra in TFA provide a valuable basis for correlations. The effects are more complicated, however, in porphyrins

TABLE 20

[1]Hmr chemical shifts (δ [p.p.m.] from TMS) of some porphyrins and their dications

	Methine	β-H		NH	Solvent	Ref.
$H_2(P)$	10.58	9.74		-3.76	C^2HCl_3	60
$H_4(P)^{2+}$	11.22	9.92		-4.40	TFA	8b

	β-H	o-H	m,p-H		Solvent	Ref.
$H_2(TPP)$	8.75	8.3	7.80		C^2HCl_3	74
$H_4(TPP)^{2+}$	8.67	8.67	8.01		C^2HCl_3/TFA	69

	Methine	CH_2 $(q, J = 7\,Hz)$	CH_3 $(t, J = 7\,Hz)$		Solvent	Ref.
$H_2(OEP)$	10.18	4.14	1.95	-3.74	C^2HCl_3	71
$H_4(OEP)^{2+}$	10.98	4.28	1.87	-4.65	TFA	113
$H_4(OEP)Cl_2$	10.49	4.04	2.04	-2.07	C^2HCl_3	180
$H_4(OEP)(ClO_4)_2$	10.58	4.10	1.80	-4.58	C^2HCl_3	180
$H_4(OEP)(BF_4)_2$	10.61	4.13	1.83	-4.92	C^2HCl_3	180
$H_2(OEC)$ [a]	9.68	3.88	1.80	-2.49	C^2HCl_3	144
$H_4(OEC)^{2+}$ [b]	8.84	2.22	1.06			
	9.95	3.78	1.60	-0.28	TFA	144
	8.80	2.33	1.17	-1.04		
$H_2(OEBC)$ [c]	8.49	3.42	1.53	$-$		
	7.47	3.30	1.50			
	6.86	1.91	1.05		C^2HCl_3	144
			1.01			
$H_4(OEBC)^{2+}$ [d]	8.98	3.41	1.44	$-$	TFA	144
	7.73		1.40			
	7.07 (d)	2.02	1.12			

	Methine	CH_3	$CH_2(P)$		Solvent	Ref.
H_2(Copro-I-TME)	9.96	3.55	4.32	-3.89	C^2HCl_3	3
H_4(Copro-I-TME)$^{2+}$			3.20			
	11.11	3.83	4.67	-4.26	TFA	8b
			3.32			

	Methine	CH_2	CH_3	$N-CH_3$	Solvent	Ref.
$H_2[(CH_3)_2OEP]$ [e]	9.80	3.8—4.2	1.94	-5.30	C^2HCl_3	169
			1.71			
$H_4[(CH_3)_2OEP]^{2+}$ [f]	11.40	4.37	1.97	-5.11	TFA	169
		4.56	2.20			

[a] *trans*-octaethylchlorin
[b] Dication of a
[c] Octaethyl-iso-bacteriochlorin (see structure (32)).
[d] Dication of c
[e] α, γ-dimethyl-octaethylporphyrin
[f] Dication of e

with reduced peripheral bonds or for those with N-substituents. The influence of protonation on the ^{1}Hmr of N-mono-, di- and tri-alkyl substituted porphyrins was studied by Jackson et al.[142]. Although the influence of N-substitution on the ring current is not very pronounced[140], the rigidity of the macrocyclic ring is changed, which in turn changes the ring behavior on protonation. The spectra of both mono- and di-cations can be interpreted on this basis. The ^{1}Hmr spectra of two peripheral reduced OEP derivatives, H_2(OEC) and a,b-H_2(OEBC)*, and their di-cations have been reported by Bonnet et al.[144] (Table 20). The methine protons remote from the reduced rings are again deshielded, but the other signals do not follow the usual precedents, and the N-protons for example are strongly deshielded in these instances.

The behavior of H_2(TPP) is unusual because the β-protons are shielded when the di-cation is formed[69], an effect which is probably to be attributed to conformational changes of the anisotropic *meso*-phenyl groups. H_2(TPP) shows an unusually slow exchange of the N—H protons in TFA, which has been explained[69] in terms of pronounced changes in the geometry of the macrocycle accompanying di-cation formation[181,182]. X-ray diffraction reveals the macrocycle in TPP free base to be fairly planar[183,184], while the di-cation shows extreme deviations from planarity[182]. It is these structural changes that probably account for the decrease in ring current implied by the low-field shift of the β-protons.

10.2.8. Metal-porphyrin complexes

The primary effect of metal complex formation in porphyrins is similar to that of di-cation formation in that the symmetry of the complex is enhanced and thus the strength of the ring current increased. However, metalation has an additional pronounced influence that depends on the ligand structure and type of axial ligation, the electronegativity of the metal, and the spin state of the central metal ion. The spin state of the metal is highly important, for the hyperfine shifts from interaction of protons with unpaired (electron) spins on the central metal may have consequences that outweigh any of the other contributions.

Metal complexes of porphyrins can occur in a variety of stoichiometric relations, and they can exist in a variety of structures. (For a characterization and classification of metal porphyrins, see Chapter 5 and Refs. 48, 169.) Factors important in determining the structure of metal porphyrins are: the stoichiometry, by which the common 1 : 1 metalloporphyrins are differentiated from bridged structures (μ-porphinato- or μ-metallo-complexes) or layered structures such as compounds (10) and (82); the size of the metal ion and the number of axial ligands, which largely determine whether the metal is in-plane or out-of-plane (Section 10.4.3); and the nature of the axial ligand(s), by which the ligand field is determined. N.m.r. has been widely

---

* See footnotes in Table 20 for nomenclature.

used, sometimes as the decisive tool, to characterize metal complexes over the entire range of compounds that can be prepared. Here, only some characteristic features of the n.m.r. spectroscopy of these compounds will be described.

10.2.8.1. Diamagnetic 1 : 1 metal complexes

In these compounds, one metal ion is chelated by the four central N-atoms of the one porphyrin macrocycle. The basic questions that arise for this type of metal complex are the identity and the number of the axial ligand(s) and whether or not the metal is in the plane of the macrocycle. The answer to both questions is considerably assisted by n.m.r. The axial ligand occupies a region strongly shielded by the ring current, and thus the proton signals of the ligand occur at unusually high field*. A quantitative study of this effect based on ring current data and comparison with nonporphyrin complexes, can yield, among other information, the extent of the out-of-plane displacement of the metal[17]. The magnetic anisotropy of side-chain methylene protons in alkyl-substituted porphyrins and of the phenyl ring protons in $H_2(TPP)$ can be used as additional criteria for the position of the metal ion (Sections 10.4.2.2 and 10.4.2.3). Such an anisotropy in magnetic environment is observed for several out-of-plane complexes and for asymmetric ligated structures[11,185], and has been studied in detail by Abraham et al.[186]. The arguments based on such magnetic anisotropy are valid even in cases where the ligand protons cannot be seen by n.m.r.

Complexes of octaethylporphyrin with various metals (1 : 1) are listed in Table 21. With a few exceptions, the chemical shifts of the resonances cover a narrow range that extends from $\delta = 10$ to 10.7 p.p.m. for the methine protons, $\delta = 4$—4.3 p.p.m. for the CH_2 quadruplet, and $\delta = 1.8$—2 p.p.m. for the CH_3 triplet. Although to our knowledge the ^1Hmr spectrum of low spin $Fe^{II}(OEP)$ is not known, several compounds related to Fe^{II} protoporphyrin-IX have been investigated by Caughey et al.[199,200]. While the chemical shift data recorded in pyridine solution for the methine protons fall in the above range[200], the spectra obtained in pyridine/water mixtures show a strong high-field shift for all proton signals influenced by the ring current[199]. With only a few exceptions, the chemical shifts of the peripheral protons are determined by the oxidation state of the central metal, and there is a well-marked trend toward increased chemical shift with increased oxidation state of the metal. This conclusion is consistent with Caughey's[4] original statement that complexation decreases the porphyrin ring current, because, in his pioneer investigation of metalloporphyrins, only complexes of divalent Zn^{II}, Ni^{II} and Pd^{II} were investigated, and so the effect of oxidation state of the metal could not be detected. According to the oxidation state of the central

---

* Although many ligands bear protons, it is sometimes difficult to detect the ligand signals by ^1Hmr, which is especially true for water[11,185]. In one case[11], an —$OCOCF_3$ ligand has been detected by ^{19}Fmr.

TABLE 21

[1]Hmr chemical shifts (δ[p.p.m.] from TMS) of 1:1 metal complexes of octaethylporphyrin $H_2(OEP)$, (11)

Central Metal	Methine—H	CH_2	CH_3	Ligand(s) (δ_H)	Solvent	Reference
H_2	10.18	4.14	1.95	—	C^2HCl_3	187
Mg^{II}	10.06	4.08	1.91	(quinoline)$_2$ (5.1, 5.7, 6.5, 7.02, 7.45)	C^2HCl_3	185 [a]
Al^{III}	10.31	4.14	1.86	OPh (5.60)	C^2HCl_3	189
	10.38	4.13	1.94	OMe	C^2HCl_3	189
	10.2	4.09	1.91	OH (−1.82)	C^2HCl_3	189
Ga^{III}	10.13	4.05	1.87	OPh (5.61)	C^2HCl_3	189
In^{III}	10.30	4.14	1.95	OPh (2.59, 5.80)	C^2HCl_3	189
Tl^{III}	10.32	4.23, 4.17	1.95	OH, H_2O	C^2HCl_3	11
	10.30	~4.1	1.92	OAc (0.05), H_2O		
	10.34	~4.1	1.94	$OCOCF_3$		
	10.32	~4.1	1.91	I		
	10.24	~4.1	1.93	CN		
Sc^{III}	10.39	4.16	1.90	Acac (0.04)	C^2HCl_3	185
Si^{IV}	9.85	4.14	1.99	(OMe)$_2$ (−2.95)	C^2HCl_3	189
	10.07	4.12	1.90, 1.93	(OPh)$_2$ (1.33, 5.49)	C^2HCl_3	189
Ge^{IV}	10.36	4.17	1.98	(OMe)$_2$ (−3.01)	C^2HCl_3	189
	10.30	4.15	1.93	(OPh)$_2$ (1.35, 5.43)	C^2HCl_3	189
Sn^{II}	10.48	4.11, 4.13	1.87	—	$C_5{}^2H_5N$	175a, 190, 191
Sn^{IV}	10.32			(OAc)$_2$	C^2HCl_3	187
	10.40	4.20	2.01	(OMe)$_2$(−2.57)	C^2HCl_3	189
	10.32	4.15	1.93	(OPh$_2$) (1.45, 5.40)	C^2HCl_3	189
	10.66	4.30	2.10	Cl_2	C^2HCl_3	190,191
Pb^{II}	10.44	4.14, 4.16	1.90	—	$C_5{}^2H_5N$	190,191
Zn^{II}	10.05				C^2HCl_3	187
Cd^{II}	9.99	4.05	1.86		Dioxan	122
Ti^{IV}	10.48	4.18	1.99	=O	C^2HCl_3	192
Re^{V}	10.55	4.19	1.95	=O,OPh, (1.35, 5.26)	C^2HCl_3	189
Mo^{IV}	10.58	4.20	1.98	=O	C^2HCl_3	185
Co^{III}	10.00	4.09	1.83	Br, Py (6.3—5.7 (2), 4.9—4.6 (3))	C^2HCl_3	117
Co^{III}	10.08	4.00	1.88	CH_3 (−5.20)	C^2HCl_3	198 (199,200)
Ni^{II}	9.77	3.93	1.83	—	C^2HCl_3	122
Pd^{II}	10.08	4.03	1.90	—	C^2HCl_3	122
Ru^{II}	9.75	3.88	1.82	CO, Py (1.07, 4.76, 5.65, 2:2:1)	C^2HCl_3	193,194
Rh^{III}	10.31	4.15	1.99	Cl	C^2HCl_3	195
Rh^{III}	9.96	4.01	1.90	CH_3(−6.47, J_{H-RH} =3 Hz)	C^2HCl_3	196
Os^{VI}	10.75	4.25	2.13	=O	C^2HCl_3	197

[a] See also Refs. 13, 16 and 188 and Section 10.2.8.2.

metal, the methine-resonances are observed in the following well-defined ranges: δ = 9.75—10.08 p.p.m. for divalent metals, δ = 10.13—10.39 p.p.m. for trivalent metals, δ = 10.30—10.58 p.p.m. for tetravalent metals, δ = 10.55 p.p.m. for pentavalent Re and δ = 10.75 for hexavalent Os. Exceptions are observed for the complexes with the large, out-of-plane ions Sn^{II} and Pb^{II} [86,175a], which show the methine resonances at usually low field, and for Co^{III} [140] and Rh^{III} [89] complexes, which show resonances at unusually high field. The latter complex (Rh^{III}) shows a pronounced dependence of the chemical shifts on the axial ligand[89,241], which is probably related to the kind of axial bond involved: Cl (ionic) lies within the regular range[241], CH_3 (covalent) lies in the range of divalent metals[89]. In first order, this chemical shift dependence can be explained by the same electrostatic model invoked by Fuhrhop[200a] for the redox potentials of porphyrins. More highly charged central metal ions will reduce the electron density on the porphyrin ligand and thus deshield the peripheral protons*.

The metal complexes of (OEP) are characteristic for the metal complexes of the naturally-occurring β-pyrrole substituted porphyrins. The properties of metal complexes of *meso*-tetraphenylporphyrin including a $Co^I(TPP)$[202] are reported by several groups[11,63,201]. The publications on some metal complexes of octaethylchlorin[187,190] and of Sn^{IV} octaethyl—isobacteriochlorin[190] should be mentioned for leading references on reduced porphyrins. (For chlorophylls, see next section.) The metal complexes of two cyclic systems with interrupted conjugation have been investigated; these are the Tl^{III}-dioxoporphodimethenes[174] and extensive series of porphodimethene complexes[169].

10.2.8.2. The chlorophylls

Chlorophylls are magnesium complexes of the phorbin system, which are characterized by an isocyclic five-membered ring attached to the γ-carbon and carbon 6. The chlorophylls are derived from true porphyrins, chlorins, or bacteriochlorins. Because of the essential role chlorophylls play as the primary photo-acceptors in plant and bacterial photosynthesis, a great deal of work has focused on their molecular structure, their interactions in solution, and their structure-function relationships. N.m.r. work to 1966 has been reviewed[188], and some basic features of the n.m.r. spectroscopy of the phorbin system are discussed in Sections 10.2.3 and 10.2.5. Here, we wish first to review ^{1}Hmr spectral data of some recently characterized important chlorophyll structures. Spectral parameters of these newly characterized

---

* It should be noted that in the metalloporphyrins, electron withdrawal has the opposite (deshielding) effect as compared to the shielding upon withdrawal of electrons by peripheral substituents. This discrepancy might be explained by a contraction and expansion of the main loop, respectively, but it shows the ambiguities that can arise from the ring current approach (see Section 10.1.2).

TABLE 22

¹Hmr chemical shifts (δ[p.p.m.] from TMS) of the chlorophylls

	Chl-a[e] (15)	Chl-b[e]	Chl-c_1[d] (74)	Chl-c_2[d] (74)	Bchl-a[f,i]	Bchl-b[f] (75)	Bchl-d[g] (76a)	Bchl-c[g] (76b)	Bchl-e[h] (76c)
Methine α	9.23	9.87	9.95*	10.10*	9.40	9.41	9.67	9.78	9.41*
β	9.50	9.55	9.90*	10.00*	8.52	8.93	9.34	9.41	10.63*
δ	8.28	8.18	9.80*	9.92*	8.38	8.39	—	9.20	—
3-CHO	—	10.92	—	—	—	—	—	—	~11.1
2-Vinyl[a] H_X	7.92	7.85	8.28	8.33	—	—	—	—	—
H_A	5.97	5.98	6.34	6.35	—	—	—	—	—
H_B	6.13	6.15	6.04	6.06	—	—	—	—	—
4-Vinyl[a] H_X	—	—	—	8.33	—	—	—	—	—
H_A	—	—	—	6.32	—	—	—	—	—
H_B	—	—	—	6.04	—	—	—	—	—
7-Acrylic[b] 7a	—	—	8.89	8.99	—	—	—	—	—
7b	—	—	6.61	6.67	—	—	—	—	—
10-H$_{(2)}$	6.22	6.10	6.72	6.84	6.44	6.43	4.92(s)	~5.18[h]	5.18(A,B)
3-H	—	—	—	—	4.10	4.93(dd)	n.r.	n.r.	n.r.
4-H	—	—	—	—	3.86	—	n.r.	n.r.	n.r.
7-H	4.14[j]	4.15[j]	—	—	4.10	4.10	n.r.	n.r.	n.r.
8-H	4.27[j]	4.45[j]	—	—	4.21	4.21	6.58	n.r.	6.54(q)
2a-H	—	—	—	—	—	6.84(dd)	—	—	—
4a-H	—	—	—	—	—	—	—	—	—
1-CH$_3$	3.28	3.22	3.5–4[c],*	3.5–4[c],*	3.33*	3.34*	3.33*	3.34*	3.51
2a-CH$_3$	—	—	—	—	3.00*	2.99*	n.r.	n.r.	1.49(d)
3-CH$_3$	3.25	n.r.	3.5–4[c],*	3.5–4[c],*	1.58(d)	1.66(d)	3.40*	3.50*	n.r.
4a-CH$_3$	1.72(d)	3.52	1.67	—	n.r.	2.01(d)	n.r.	—	n.r.
5-CH$_3$	3.60	n.r.	3.5–4[c],*	3.5–4[c],*	3.44*	3.45*	—	—	—
8-CH$_3$	1.78(d)	3.95	3.5–4[c],*	3.5–4[c],*	1.41(d)	1.41(d)	—	—	—
10a-CH$_3$	3.97	n.r.	3.5–4[c],*	3.5–4[c],*	3.66*	3.66*	—	3.70*	n.r.
δ-CH$_3$	—	—	—	—	—	—	3.75*	—	3.60
4-CH$_2$	3.75(d)	n.r.	4.26[c]	—	~2.5	—	n.r.	n.r.	3.84

References, p. 514

TABLE 22 (continued)

	Chl-a[e] (15)	Chl-b[e]	Chl-c_1[d] (74)	Chl-c_2[d] (74)	Bchl-a[f,i]	Bchl-b[f] (75)	Bchl-d[g] (76a)	Bchl-c[g] (76b)	Bchl-e[h] (76c)
5-CH$_2$	—		—	—	—	—	n.r.	n.r.	3.99(q)
7-CH$_2$	2.0—2.5	~2.35	n.r.	n.r.	~2.5	~2.4	n.r.	n.r.	n.r.
7a-CH$_2$	2.0—2.5	~2.35	n.r.	n.r.	~2.5	~2.4	n.r.	n.r.	n.r.

* Tentative assignment from intercomparison with other chlorophylls.

[a] ABX spectrum with $J_{AB} \sim 2$ Hz, $J_{AX} \sim 12$ Hz, $J_{BX} \sim 18$ Hz.

[b] AX spectrum with $J \sim 16$ Hz

[c] in TFA[139].

[d] in tetrahydrofuran-^{2}H$_8$.[139,203]

[e] in C^{2}HCl$_3$/C^{2}H$_3$O^{2}H[13].

[f] in pyridine—^{2}H$_5$.[80,155]

[g] Resonances of one of the pheophorbides of the homologues, in C^{2}HCl$_3$, Ref. 207.

[h] See Ref. 206.

[i] See Ref. 204 and footnote p. 466 for the esterifying alcohols in Bchl a.

[j] in pyridine.

All spectra are obtained in disaggregated (monomeric) solution.

chlorophylls are listed in Table 22 together with those of chlorophylls-a and
-b, and bacteriochlorophyll-a, the principal natural chlorophylls. The chlorophylls show very pronounced solvent and concentration dependence, which
result from chlorophyll self-aggregation (see Section 10.4.1.1) and chlorophyll—solvent interactions (Section 10.4.1.2). All chemical shift values referred to in this section were obtained on disaggregated monomeric chlorophyll
solutions, and are thus typical of chlorophyll · L_1 and for chlorophyll · L_2
species[59].

10.2.8.2.1. Chlorophylls-c_1 and -c_2

The chlorophylls-c (74) are minor accessory pigments in diatoms and
many marine micro-organisms and brown algae, and are closely related to the
chlorophylls found in other photosynthetic organisms. Early [1]Hmr results
were obtained on a mixture of the two pigments, which are difficult to
separate by the usual sugar column chromatography, and the results were
later confirmed on the fully separated compounds[139,203]. Both of these
closely related pigments are *porphyrin* free acids, which lack an esterifying
alcohol. The [1]Hmr spectra of the chlorophylls-c as compared to all other
chlorophylls derived from chlorins and bacteriochlorins show a low field
shift of the methine signals and the spectra are simple in the medium and high
field region. The broad unresolved high-field resonance associated with the
aliphatic protons of the long chain esterifying alcohol, as well as all signals
typical of reduced pyrroline rings are missing from the c_1 and c_2 spectra.
Apart from the CH_3 singlets, Chl-c_2 shows no resonances at high field below
δ = 6 p.p.m., and chlorophyll-c_1 shows only the ethyl proton resonances in
this region. The low field region in both compounds is dominated by the
complex patterns of the vinylic protons. Both compounds show an AX
pattern[1] for one *trans*-acrylic side-chain proton, and ABX patterns[1] for one
(Chl-c_1), or two vinyl groups (Chl-c_2). Although the vinyl signals of Chl-c_1
and -c_2 do overlap, a quantitative analysis of Chl-c mixtures is possible by
[1]Hmr.

10.2.8.2.2. Bacteriochlorophyll-b

Bchl-b (75) is the pigment responsible for the extreme long wavelength
absorption of *Rhodopseudomonas viridis* and some other photosynthetic
bacteria[211]. Bchl-b has an ethylidene side-chain in position 4 in place of the
ethyl side-chain present in Bchl-a[155]. Thus, the main difference in the [1]Hmr
spectrum of b as compared to that of Bchl-a are the resonances of ring B
protons. Both the 3- and the 4a-protons give rise to double doublets (J_1 = 2
Hz, J_2 = 7 Hz) at low field (δ = 4.93 and 6.84 p.p.m.). By double resonance
experiments these were proved to be coupled to each other (J = 2 Hz) and to
a high field methyl group each (J = 7 Hz) at higher field and assigned to
protons 3 and 4a, respectively. As a further consequence of the 4,4a double
bond, the β-proton resonance is shifted to lower field, while all other

References, p. 514

resonances are essentially identical to those of Bchl-*a* (Table 21)*. Obviously, the small shielding effect expected to result from the introduction of the conjugated double bond is compensated for by (possibly steric) effects.

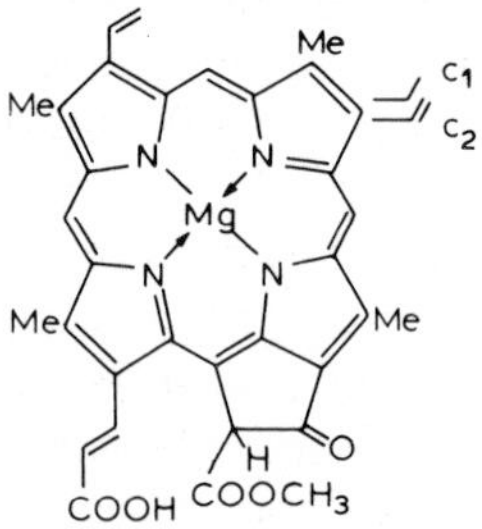

(74)

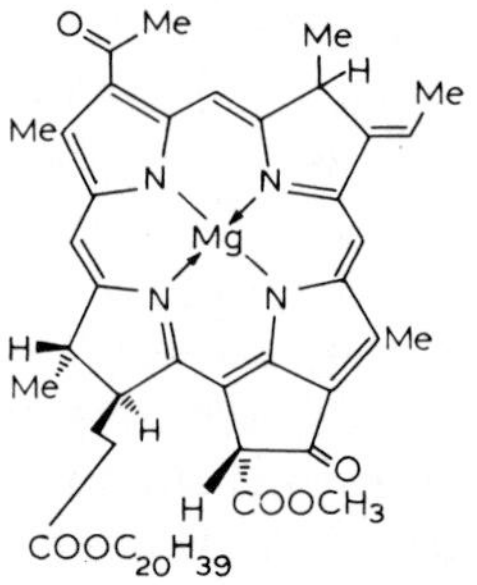

(75)

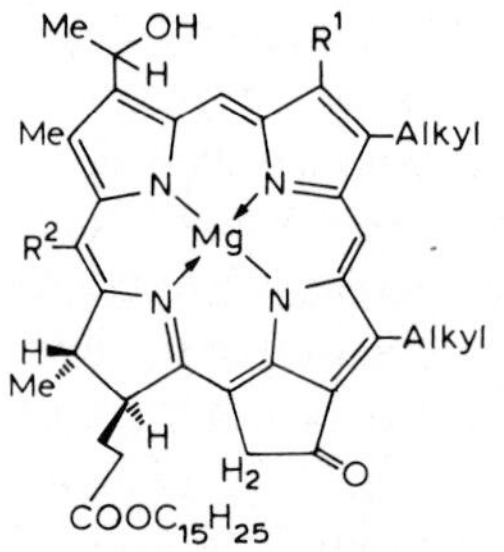

(76)

(77)

(78)

a; $R^1 = R^2 = Me$
b; $R^1 = Me, R^2 = H$
c; $R^1 = CHO, R^2 = Me$

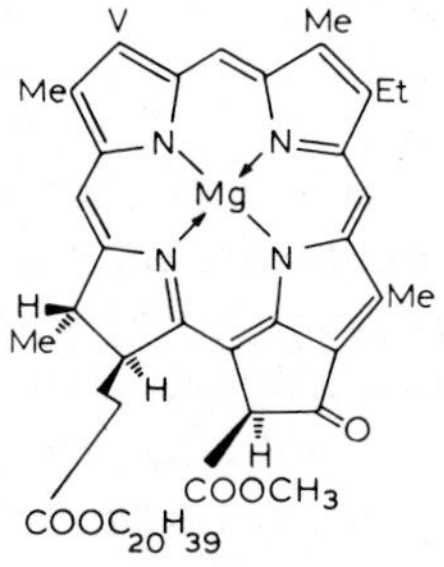

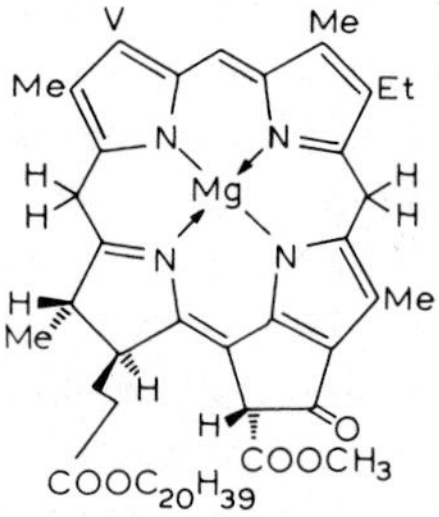

V = CH=CH₂

10.2.8.2.3. *Bacteriochlorophyll-c, -d, and -e*

The bacteriochlorophylls-*c*, -*d*, and -*e* present in the green photosynthetic bacteria (*Chlorobium* species) are unique among all natural chlorophylls in that they appear to be a mixture of various homologs[205]. All *Chorobium* chlorophylls have a 2(α-hydroxyethyl)-substituent characterized by a low-field quadruplet (2a-H) at 6.1—6.6 p.p.m. and a high field doublet. These chlorophylls lack a 10-COOCH₃ group, but the typical (Section 10.2.3) AB double doublet[206] expected for the 10-methylene protons is often only

---

* Bchl-*b* from *Rh. viridis* as well as Bchl-*a* from *Rhodopseudomonas* strains contain phytol as the esterifying alcohol. In contrast, Bchl-*a* from *Rhodospirillum rubrum* contains geranyl-geraniol instead[204]. The latter alcohol contains four double bonds. In the ¹Hmr spectrum, this is manifested by a general deshielding of all resonances from the esterifying alcohol (as compared to phytol), and additional olefinic resonances at δ = 4—5 p.p.m.

poorly resolved[207]. Finally, all *Chlorobium* chlorophylls have farnesol as the esterifying alcohol, which was characterized among other criteria by the olefinic CH_3 singlets at about 1.6 p.p.m. and the 1-methylene doublet at 3.96 p.p.m.[208].

Bchl-*d* (*Chlorobium* chlorophyll '650')* is a mixture of homologues of structure (76b). The [1]Hmr spectrum of one of the pheophorbides (Table 22) is reported by Mathewson et al.[207]. The spectrum shows only three low-field methyl singlets, the signal position indicating homologation at position C-5. The methine protons show a very unusual pattern as compared to 2-desvinyl-2-hydroxyethyl-pyromethylpheophorbide-*a*[135]. The α-proton is deshielded by 0.23 p.p.m., the β-proton by 0.18 p.p.m., and the γ-proton by 0.78 p.p.m. These differences are unexpected. They may be due, however, to the unusual aggregation behavior of 2-(α-hydroxyethyl)-pheophorbides[213].

Bacteriochlorophyll-*c* (*Chlorobium* chlorophyll '660' (76a)) is considered to have an alkyl substituent at one of the methine bridges, as deduced from the presence of only two methine signals in the pheophorbides. The position of this alkyl substituent was discussed[207,209,210] mainly on the basis of n.m.r. arguments. While the loss of the high-field methine-signal suggested a δ-substituent[210], the presence of one acid-exchangeable methine proton indicated the δ-proton was still present and indicated substitution in the α or β position[207,209]. (For unsuccessful attempts to correlate the structure of porphyrins derived from Bchl-*c* with synthetic δ- and β-alkylporphyrins, see Ref. 118.) Substitution of the δ position was shown recently to be the correct assignment, a conclusion supported by n.m.r. studies (aggregation, substituent induced shifts) of model compounds[206,212,213].

Very recently, still another series of at least three *Chlorobium* chlorophylls, designated Bchl-*e* (76c) was investigated and described by Brockmann[206]. This family of chlorophylls has the same relationship to Bchl-*c* as does Chl-*b* to Chl-*a*. The spectra have features similar to those of Bchl-*c*, but the presence of a CHO group is proven by the appropriate CHO resonance in both the [1]Hmr and [13]Cmr spectra.

10.2.8.2.4. *Chlorophyll related structures*

A variety of structures related to the chlorophylls have been characterized by [1]Hmr. In the chlorphyllides, the propionic ester side-chain is transesterified, usually with methanol or ethanol. Besides the loss of all signals related to the long chain alcohols in chlorophylls and the appearance of the signals related to the introduced alcohol, the [1]Hmr spectrum remains unchanged upon transesterification. The spectral changes observed upon substitution at C-10[129,212], removal of the 10-COOCH$_3$ group[214], the reduction of the 9-CO to a CH_2 group[80], cleavage of ring E[212], or the hydrogenation of the

---

* The number indicates the wavelength (nm) of the absorption band in the red, measured in ether solution.

2-vinyl group[80] are similar to those observed in the (metal free) pheophorbides (Section 10.2.3).

The structure of two long known chlorophyll derivatives has been proven recently by [1]Hmr; Chl-a' (and other 'prime' chlorophylls) were identified as C-10 epimers[215], and the product of the Krasnovskii photoreduction was shown to be β,δ-dihydro Chl-a[173]. Chlorophyll-a' (Chl-a', (77)) is the 10-epimer of Chl-a[215] which is present as an artifact in equilibrium amounts of about 15% in chlorophyll preparations[216]. Its presence is manifested in the [1]Hmr spectrum by small satellite peaks or shoulders at the high field side of the methine-H, and by a distinct satellite peak accompanying the 10-H resonances at about 0.12 p.p.m. towards higher field. Similar satellites are observed for Chl-b and Bchl-a, as well as for their pheophytins and some related structures. They are absent, however, in compounds without an asymmetric C-10. The 10-epimeric structure of Chl-a' was proved[215] by comparison of the distinctly different chemical shifts of the C-10 protons in Chl-a and -a' with the ones in pyropheophorbides[79], and by equilibration and aggregation experiments of the epimers. These experiments were substantially facilitated by carrying out the experiments with [2]H-chlorphyll-[10-^{1}H], in which the 10-H resonances can be studied without interference from other signals.

The Krasnovskii photoreduction was the first[217] and probably most widely studied photoreaction of the chlorophylls, and porphyrins in general[126,218]. The structure of the Krasnovskii reaction product of Chl-a was recently shown to be β,δ-dihydro Chl-a (78) by carrying out the reaction in dilute chlorophyll solution with hydrogen sulfide as reductant directly in a sealed n.m.r. tube[173]. The [1]Hmr spectrum of the porphodimethene that is formed shows the typical high field shifts observed for systems with interrupted ring current (see Section 10.2.6). The meso-methine and methylene signals and the protons at the reduced meso positions were assigned by using [2]H_2S as the reducing agent, and correlating signals in the reduction product with the signals in the regenerated Chl-a from the isotope content and position.

10.2.8.3. Unusual metalloporphyrins with central metal

Several porphyrin—metal complexes with a stoichiometry deviating from 1 : 1 have been characterized by n.m.r. recently (Table 23). The most useful aids in rationalizing the [1]Hmr spectra of these substances were again symmetry and ring current arguments. The appearance of four methine proton resonances testifies to the coordination of two neighboring N-atoms to the two metal atoms in the non-axial dirhodium complex (80), otherwise a spectrum consistent with higher symmetry would be observed[196]. The mirror plane in the Re complex (81) is established by the pattern of the β-proton signals, the inner hydrogen by a resonance at the very high field of −4.0 p.p.m. and by its spin coupling with the β-protons in the same ring[63]. Two

TABLE 23

¹Hmr chemical shifts (δ [p.p.m.] from TMS) of metalloporphyrins with stoichiometries other than 1:1.

Compound	Data	Solvent	Ref.
(80) (OEP)	Methine—H: 10.35, 10.50, 10.55, 10.58 Ethyl—CH₃: 1.72, 1.77, 1.80, 1.84, 1.87, 1.89, 1.96, 1.98 N—H: -3.74 N—CH₃: -5.90	C^2HCl_3	196
(81) (TPP)	β-pyrrole protons: Ring A: 8.88 (d, J_{H-NH} = 2Hz) Ring B, D: 9.11 (AB, J = 5Hz, Δ = 0.12 p.p.m.) Ring C: 8.72 (s) Phenyl—H: 7.87 (m, p); 8.30 (o) N—H: -4.0	C^2HCl_3	63
(82) (Etio I–IV; Copro I–IV TME)	Etio-I: meso-H: 8.97 (J_{H-Hg} = 10Hz) ring—CH₃: 3.48; 3.45 (s) OAc: -0.24 (s)	C^2HCl_3	219
	¹⁹⁹Hg (Indor): 17.8866 (2 : 1) 17.8854		223
(83) (OEP)	Methine—H: 9.55 (s) CH₂: 1.65 (t), 1.89 (t) CH₃: 3.94, 3.98		220

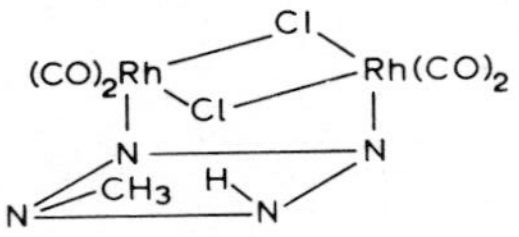

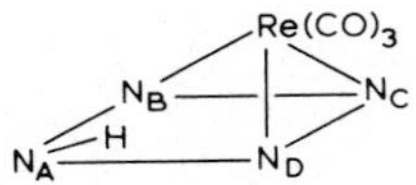

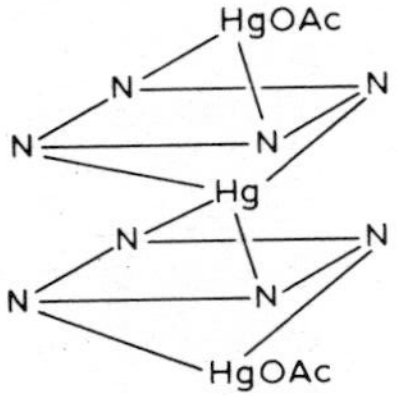

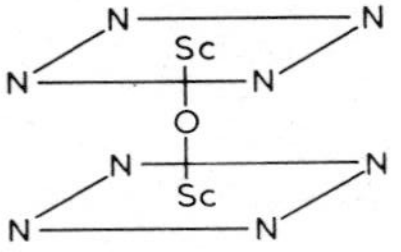

TABLE 23 (continued)

		Solvent	Ref.
(84) (OEP) 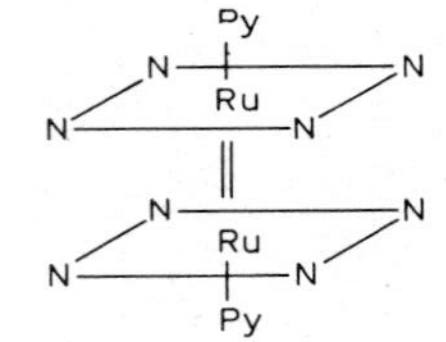	Methine—H: 7.61 (s) CH$_2$CH$_3$: 5.92, 1.45 (t) Pyridine—H: 2.6, 5.3 (2 : 2) (geometry not discussed)		193
(85) (OEP) 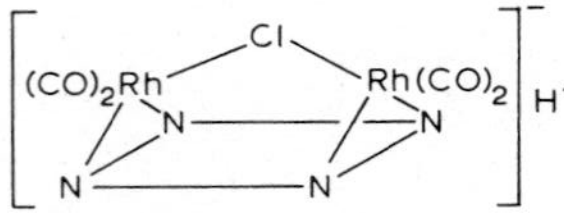	Methine—H: 10.04, 10.35 CH$_2$: 4.00, 4.10 CH$_3$: 1.89, 1.67	C^{2}HCl$_3$	195
(86) (OEP) 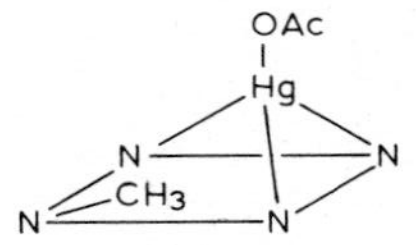	Methine—H: 10.03, 10.00 CH$_2$: 3.95 (m) CH$_3$: 1.95 (m), 1.72 (t) N—CH$_3$: −5.0 OAc: 0.1 ($J_{^1H—^{199}Hg}$ = 10 Hz)	C^{2}HCl$_3$	223

isomers (meso, racemic) account for the splitting of the ring-methyl group resonances in the stacked Hg_3P_2 complex (82); the high field acetate resonance is characteristic of inner protons, and the two ^{199}Hg indor* lines indicate the presence of one and two equivalent Hg atoms, respectively[219,223] (see Section 10.4.3).

A characteristic feature of the sandwich structures is the pronounced shielding effect of one ring on the other. This shielding increases the closer the rings are to each other. The incremental shift for the methine protons is about +0.8 p.p.m. in the μ-oxo Sc complex (83)[220], [as compared to Sc(OEP)[185]], 1.3 p.p.m. in the layered Hg complex (82)[219], and 2.1 p.p.m. [as compared to $Ru^{II}(OEP)$[193]] in the dimeric ruthenium complex (84)[193]. The latter value serves as an additional argument for the proposed direct metal—metal bond. Like the stacked phthalocyanines[18], complexes with stacked porphyrins may serve as a useful probe for the ring current effect in the spatial region above the conjugated system, which is otherwise only accessible with axial ligands, porphyrin cyclophanes[221], or structures like the fused cyclopropano-chlorins (46) and (47).

10.2.8.4. Peripheral complexes

While in all metalloporphyrins so far discussed the metal is bound to the inner nitrogen atom, two new types of metal complexes have been recently characterized in which the metal is bound to peripheral substituents of the macrocyclic system. Logan et al.[222] investigated π complexes of $Cr(CO)_3$ with one or more of the phenyl rings in $H_2(TPP)$. Hyperfine interactions are essentially confined to the ring(s) to which the chromium is bound, with the chemical shifts in the latter comparable in magnitude to those observed in the chromium carbonyl—benzene complex.

A second group of complexes related to the chlorophylls, but with the metal bound at the periphery, was recently investigated by Scheer et al.[80]. In these compounds, the metal ion is chelated by the β-keto-ester function present in ring E. The ^1Hmr spectrum of the peripheral Mg complex of methylpheophorbide-a indicates a uniformly reduced aromatic ring current, presumably arising from the electron-withdrawing effect of the chelate. Most signals are shifted to higher fields, with $\Delta\delta$ values similar for protons in similar environment. However, the 8-CH_3 doublet, the 7-H multiplet, as well as the 10b-CH_3 singlet, are deshielded.

These deshielding effects in the vicinity of the isocyclic ring E can be rationalized in terms of conformational changes. In the chelate, the β-keto-ester system is essentially coplanar with the macrocycle. This brings the 10b-CH_3 protons into a more deshielding region of the ring current field

---

* In the indor double resonance technique, n.m.r. transitions of a heteronucleus are scanned with a strong RF field, while one line of a coupled (usually proton) multiplet is monitored[1].

References, p. 514

TABLE 24

[1]Hmr chemical shifts (δ[p.p.m.] from TMS) of methyl pheophorbide-*a* (87) and its peripheral Mg complex (88), and incremental shifts ($\Delta\delta$ [p.p.m.]) of (88) vs. (87).

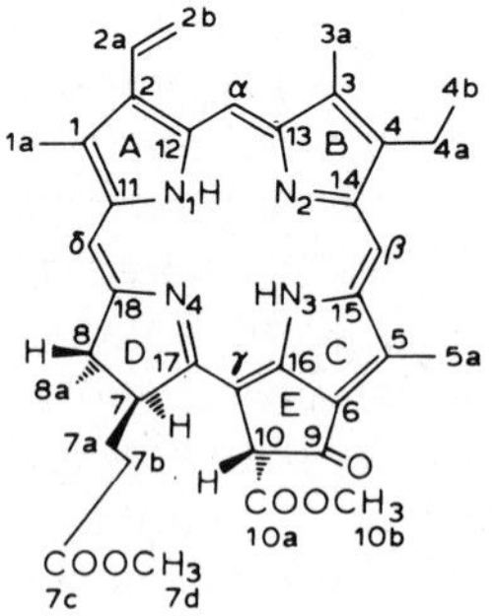

(87) 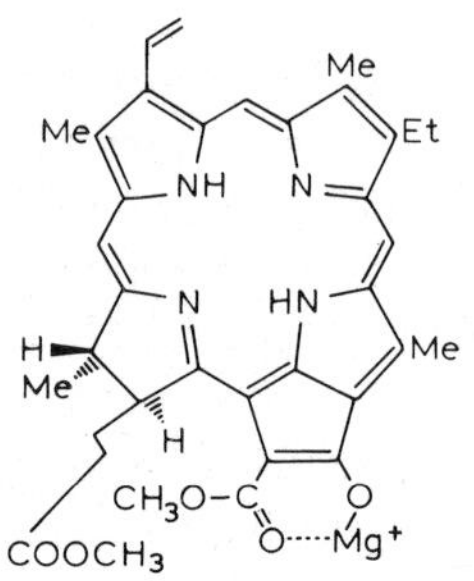   (88)

	Methyl pheophorbide (87)	Peripheral Mg Complex (88)	$\Delta\delta$	Multiplicity
β-H	9.75	9.01	+0.74	*s*
α-H	9.57	8.83	+0.74	*s*
δ-H	8.71	8.00	+0.71	*s*
H_X	8.08	7.77	+0.31	*dd, J*=11,17
Vin H_A	6.23	6.06	+0.17	*dd, J*=2,17
H_B	6.05	5.87	+0.18	*dd, J*=2,11
10-H	6.61	—	—	
7-H	4.29	4.65	−0.36	*m*
8-H	4.42	4.10	+0.32	"*q*"
10b-CH$_3$	3.76	3.83	−0.07	*s*
7d-CH$_3$	3.52	3.38	+0.14	*s*
5a-CH$_3$	3.42	3.11	+0.31	*s*
1a-CH$_3$	3.21	2.95	+0.26	*s*
3a-CH$_3$	3.08	2.83	+0.25	*s*
8-CH$_3$	1.66	1.73	−0.07	*d, J*=7
4-CH$_2$	3.54	3.29	+0.25	*q, J*=7
4-CH$_3$	1.53	1.39	+0.14	*t, J*=7
N—H	+0.74	2.44	−1.70	*s*, broad
	−1.48	2.04	−3.52	*s*, broad

2×10^{-3} M in pyridine-[2]H$_5$, and pyridine-[2]H$_5$ saturated with anhydrous Mg(ClO$_4$)$_2$; 30°C; Ref. 80a.

(Fig. 1), and the signal at $\delta = 3.83$ p.p.m. is therefore assigned to this group. In addition, the increased steric hindrance of the 10-substituent with the substituents at C-7 induces a conformational change in ring D by which both the 7-H and the 8-CH$_3$ group are forced into a more deshielding region. This effect is well established in δ-substituted chlorins[124], where incremental shifts of the same magnitude are observed.

Peripheral metal chlorin complexes are unstable to water and competitive Mg^{2+} chelating agents such as acetylacetone or 2-carbethoxy-cyclopentanone. As metal ion exchange in these complexes is slow on the n.m.r. time scale, two distinct sets of resonances are observed during titrations, one set of which corresponds to the free methyl-pheophorbide and the other to its peripheral complex. If part of the complex is destroyed by addition of water, temperature-dependent equilibrium for the reaction of water with the peripheral complex can be determined by n.m.r. Complex formation is favored by higher temperatures. For a 27-(39)fold molar excess of water, the net reaction enthalpy is 5 (9.1) kcal/mole, and equal amounts of free methyl pheophorbide and its Mg^{2+} complex are present at 30°C (90°C), respectively.

10.2.8.5. Paramagnetic metal complexes

The salient features of ^{1}Hmr spectra of compounds with unpaired spins are determined by hyperfine electron-nuclear interactions and by relaxation processes[224]. Although broad n.m.r. lines are observed in many cases, many paramagnetic metalloporphyrins have electron spin relaxation times fast enough to result in sufficiently sharp lines under high resolution conditions[22]. In complexes of porphyrins with paramagnetic metal ions, the large chemical shifts generated by the macrocyclic ring current are often small in comparison to the hyperfine shifts resulting from interactions with the unpaired spins, and in these cases the ^{1}Hmr spectrum can extend over more than 50 p.p.m. The hyperfine shifts in paramagnetic metalloporphyrins leads to a considerably enhanced resolution of signals in very similar chemical environment (as compared to the diamagnetic porphyrins). This fact renders the spectra extremely sensitive to structural and electronic changes, and Fe porphyrins are now widely used as a sensitive n.m.r. probe in hemoproteins. (For an application in the analysis of porphyrin isomers, see Ref. 225.)

The hyperfine interactions can be split into two major components. The first is the contact shift, which results from the leakage of unpaired spin to the nucleus under observation by n.m.r., and the second is the pseudocontact shift, which results from dipole—dipole couplings in molecules with anisotropic g-tensors and/or zero field splitting (ZFS)*. (For a detailed discussion and leading references, see Refs. 224 and 226.)

Pseudocontact interactions occur through space, while contact interactions occur through chemical bonds. The contact term thus allows a detailed

---

* The g-tensor characterizes the spatial distribution of the electron g-factor, and the ZFS parameters D and E characterize the coupling of unpaired electrons in systems with more than one free electron or hole[38,224].

References, p. 514

TABLE 25

Incremental hyperfine shifts (Δ [p.p.m.] relative to the respective diamagnetic reference compound) of paramagnetic Fe-porphyrins and of their μ-oxo dimers.

Complex (Reference compound)	Axial Ligand (s)	Hyperfine Shifts (Pseudocontact contribution)	Conditions	Ref.
Low Spin Fe^{III}				
$\mathbf{Fe^{III}}$ (P) [Zn(P)]	$(CN)_2$	Methine—H: +9 β-H: +24	Pyridine—$^2H_5/^2H_2O$, 46°C	22
Fe^{III} (TPP) [Ni(TPP)]	im_2 [a]	β-H: +25.3 (+5.8) o-H: +3.09 (+3.09) m-H: +1.49 (+1.44) p-H: +1.37 (+1.27)	C^2HCl_3, 29°C	27
Fe^{III}(T-n-PrP) [b] [Ni(T-n-PrP)]	im_2 [a]	β-H: +21.0 (+5.8) $meso$-CH$_2$: −0.6 (+4.5) CH$_2$: +0.5 CH$_3$: +1.3	C^2HCl_3, 29°C	27
Fe^{III} (OEP) [Ni(OEP)]	im_2 [a]	Methine—H: +7.0 (+9.3) CH$_3$: −1.97 (+3.2) CH$_3$: +1.60	$C^2H_2Cl_2$	27
High Spin Fe^{III}				
Fe^{III}(TPP) [Ni(TPP)]	Cl	β-H: −70.2 (−9.6) o-H: +1.7 (−6.3, −3.2) m-H: −4.50, −5.62 (−2.6, −2.0) p-H: +1.45 (−2.1)	C^2HCl_3, 29°C	
Fe^{III} (T-n-PrP) [b] [Ni(T-n-PrP)]	Cl	β-H: −76.8 (−9.6) $meso$-CH$_2$: −57.2 CH$_2$: 0 CH$_3$: −1.3	C^2HCl_3, 29°C	29
Fe^{III} (OEP) [Ni(OEP)]	Cl	Methine—H: +65 CH$_2$: −35.4, −39.0 CH$_3$: −4.7	C^2HCl_3, 29°C	29

Fe^{III} Dimers

	Structure			Solvent	Ref.
[Fe^{III}(TPP)]$_2$O {[Sc(TPP)]$_2$O}	—O—	β-H:	−5.02	C^{2}HCl$_3$, 29°C	29
		o,m,p-H:	~+0.05		
[Fe^{III}(T-n-PrP)]$_2$O {[Sc(TPP)]$_2$O and T-n-PrP ^{b}}	—O—	β-H:	−6.3	C^{2}HCl$_3$, 29°C	29
		meso-CH$_2$:	~−1.3		
		CH$_2$:	~+0.5		
		CH$_3$:	~0		
[Fe^{III}(OEP)]$_2$O {[Sc(OEP)]$_2$O}	—O—	Methine-H:	~+3.9	C^{2}HCl$_3$, 29°C	29
		CH$_2$:	−2.26, −1.30		
		CH$_3$:	−0.19		

Co^{II} (Low Spin)

Co(TPP) [Ni(TPP)]	Solvent	β-H:	−7.0 (−9.4)	C^{2}HCl$_3$, 35°C	28
		o-H:	−5.0 (−5.0)		
		m-H:	−2.15 (−2.3)		
		p-H:	−2.03 (−2.03)		
Co(OEP) [Ni(OEP)]	Solvent	Methine-H:	−19.0 (−15.0)	C^{2}HCl$_3$, 35°C	28
		CH$_2$:	−3.55 (−5.2)		
		CH$_3$:	−4.05		

^{a} im = imidazole

^{b} T-n-PrP = *meso*-tetra-*n*-propyl-porphyrinate

References, p. 514

insight into the electronic structure of the molecule, especially the spin density distributions and the spin transfer mechanism, while the pseudocontact term can give valuable information on the magnetic anisotropy and the zero field splitting parameters. The magnitude of the pseudo-contact shift induced by paramagnetic shift reagents in favorable circumstances yield information on the solution structure and conformation of a molecule. Next to the assignment of the resonances, the separation of these contributions to the spectrum is therefore of considerable interest in most investigations.

Since the first detection of sharp hyperfine-shifted lines in the ^{1}Hmr spectrum of cytochrome-c[20], a great deal of ^{1}Hmr work has been done on heme-proteins and related structures primarily directed to the structures in solution and structure-function relationships. The reader is referred to two excellent reviews[21,22] for a description of these studies. A recent series of publications by Perutz et al.[227–229] shows the potentialities and limitations of the n.m.r. method in hemoprotein studies when used in conjunction with other methods. Here we propose to focus on some of the basic investigations which have been carried out on iron porphyrins, and on some porphyrin complexes containing other paramagnetic metals.

10.2.8.5.1. Fe-complexes

Several types of iron complexes are observed in porphyrins, depending on oxidation state and ligand field: (a) low-spin complexes in which the ligand field splits the energy levels of the d-orbitals sufficiently far apart that a maximal number of the d-electrons are paired, resulting in a net spin of $S = 0$ for Fe^{II} and $S = 1/2$ for Fe^{III}; (b) high-spin complexes with a net spin of $S = 4/2$ for Fe^{II} and $S = 5/2$ for Fe^{III}; (c) Fe^{IV} complexes with $S = 2$; and (d) Fe^I complexes*. Complexes containing low-spin Fe^{II} are diamagnetic (see Section 10.2.8.1), all the others are paramagnetic. In addition to these 1 : 1 complexes, some μ-oxo Fe^{III} dimers are known to show antiferromagnetic coupling ($S_0 = 0$) of the two Fe^{III} atoms.

(a) *Low-spin Fe^{III}*: The ^{1}Hmr spectrum of low-spin Fe^{III} protoporphyrin-IX dimethyl ester dicyanide $[Fe^{III}(\text{Proto-IX-DME})(CN)_2]$ was first investigated by Wüthrich et al.[22,230] together with some related low spin Fe^{III} porphyrins. Groups of resonances were originally assigned by intercomparison and by the relative intensities of the resonances, but the assignment of all four β-pyrrole CH_3 signals and some methine proton resonances was recently achieved by total synthesis of selectively deuterated compounds[231,232]. From the chemical shift of the ester protons, which experience only pseudo-contact contributions, small pseudo-contact contributions to the shifts of the hyperfine-shifted protons were originally estimat-

---

* An intermediate spin state ($S = 3/2$) was recently reported for a Fe^{III} porphyrin[229a].

TABLE 26

^{1}Hmr chemical shifts (δ[p.p.m.] from TMS) of miscellaneous paramagnetic metalloporphyrins

Complex	Chemical Shift (δ[p.p.m.])		Conditions, Remarks	Ref.
$Fe^{IV}(TPP)Cl^+$	β-H:	+68.6	$C^2H_2Cl_2$, 90% Fe^{IV}	246
	Phenyl-H:	+12.3, +5.8		
$[Fe(TPP)]_2O^+ClO_4^-$	β-H:	12.2	C^2HCl_3, 40 C	246
	o,p-H:	11.4		
	m-H:	3.4		
$meso$-tetra-CH_2NO_2-Fe^{III}-$(OEP)^+Cl^-$	$meso$-CH_2:	−1.4(br), +0.2, +1.3, +3.2(br)	C^2HCl_3	245
	CH_2Me:	−40.1, −37.5, −35.2, −33.2		
$Cr^{III}(TPP)^+ X^-$ $X=Cl^-, I^-, N_3^-$	Broad peak at δ = −7.5—8 p.p.m.		C^2HCl_3, 35°C	26
$Cr^{III}(TPP)^+ Cl^-$	Aromatic—H:	~−7.5 (br)	C^2HCl_3, 35°C	26
	CH_3:	−2.35(br)		
Mn^{III} (Etio-I)$^+$ Cl^-	CH_3:	−35.3	C^2HCl_3, 35°C	24
	CH_2CH_3:	−22.6, −2.6		
	Methine—H:	+10.5		
$Eu^{III}(p$-$CH_3 \cdot TPP)$ ^{a}	o-H:	−13.31, −8.13	C^2HCl_3, −21°C	259
	m-H:	−9.33, −8.13		
	p-CH_3:	−3.44		
	Acac ^{a}:	0.88		

^{a} Acetylacetonate (Acac) as fifth and sixth ligand.

ed[22,230,233]. Assuming that the hyperfine shifts arise from contact interactions, a high spin density at the β-positions and a much smaller one (perhaps one-third as large) for the *meso* protons were inferred. A spin transfer mechanism mediated predominantly by the π-system was inferred by Shulman et al.[25], Kurland et al.[23], and Hill et al.[234]. The latter investigators correlated increasing hyperfine shifts in dipyridinates of Fe^{III}(Proto-IX-DME) with decreasing basicity of the (suitably substituted) pyridine ligand in the 5th and 6th axial coordination positions. As the electron density at the coordinated Fe^{III} decreases with decreased basicity of the ligand, it was concluded that a spin transfer by charge transfer from the ligand to the metal occurs.

La Mar et al.[27] have investigated the Fe^{III} low-spin complexes of the three key porphyrins H_2(TPP), H_2(OEP), and *meso*-tetra n-propylporphyrin. All of these compounds are highly symmetric and this enhances the sensitivity of the ^{1}Hmr data acquisition and facilitates assignments. In addition, this series of compounds makes it possible to compare hyperfine shifts for protons and the $-CH_2$-methylene group in both the *meso*- and the β-position. Under the assumption that the porphyrin frontier orbitals are identical in all three compounds, the following conclusions are arrived at for the complexes of low spin Fe^{III}: As suggested by other spin-transfer studies[23,25,230,234], the spin-in these compounds is transferred to the π-system of the ligand by charge transfer to the metal, and the spin resides primarily in the highest occupied molecular orbital that has high-spin density at the β-positions and low-spin density at the *meso*-positions. Because the spin transfer from the *meso*-positions to the phenyl ring is hindered, the phenyl protons are expected to experience only pseudo-contact shifts, and therefore their chemical shifts were used to separate the pseudo- from the true contact contribution. In contrast to earlier results[22,230−233,234], both contributions to the chemical shifts are found to be of the same order of magnitude. The dipolar shift is positive throughout, the contact shift can be either positive or negative. Both show Curie $(1/T)$ behavior, and deviations observed in OEP were explained by hindered rotation of the ethyl substituents. At ambient temperatures, the effective[25] g-tensor is axial, with the axis perpendicular to the macrocycle plane, a conclusion that follows because only one set of signals for each set of equivalent substituents is observed. In spite of their $1/T$ behavior, the hyperfine shifts usually do not extrapolate to zero at $T = \infty$[22,27,235−237], which is discussed in terms of second order Zeemann effects and mixing in of excited states into the ground state. Both contributions in low-spin Fe^{III} complexes have recently been critically investigated by Horrocks[237].

(b) *High-spin* Fe^{III}: The lines in high-spin Fe^{III} complexes are spread over more than 80 p.p.m. and are generally[26] considerably broadened. The first n.m.r. spectrum of a Fe^{III} (high spin) porphyrin complex, Fe^{III} (TPP)Cl, was published by Eaton et al.[238], and a series of high-spin hemins was studied by Kurland et al.[23]. Broadened lines were observed, and some signals were assigned by their relative intensity and by intercomparison with each other. The assignment of particular resonances in high-spin Fe^{III} porphyrins relative to that of the respective low-spin complexes was investigated by Gupta and Redfield[239,240] by an elegant cross-relaxation double resonance method.

The ligand effect on the hyperfine shift in deuterohemins was studied by Caughey et al.[241], who found a correspondence of the magnitude of the hyperfine shifts to the D value of the zero field splitting parameters, suggesting a pseudo-contact contribution to the observed shifts[226]. In contrast to the low-spin complexes, a significant σ-spin transfer is generally[238] observed in high-spin Fe^{III} complexes[23,29,241,242]. La Mar et al.[29] studied a series of high-spin Fe^{III} complexes of symmetric porphyrins in detail. These investigators proved[238] that the spin transfer occurs from the central metal to the ligand, with about equal spin density at the β- and the methine-positions of the macrocycle. A spin transfer from the metal to the ligand is further supported by the preferred stabilization of the high-spin form in π-complexes of aromatic acceptors with Fe^{III} porphyrins[243]. Although the g-tensor in high-spin Fe^{III} complexes is essentially anisotropic, an appreciable pseudo-contact contribution to the shift was found, the magnitude of which could be evaluated via the phenyl proton shift as described above for the low-spin series[27]. From a $1/T^2$ term in the (non-Curie) temperature dependence of the chemical shifts, the authors[29] concluded that the pseudo-contact shift arises from the zero field splitting as discussed earlier[23,226,241], and the D value calculated from the chemical shift data was in good agreement with the value previously deduced from IR measurements for high-spin Fe^{III} porphyrins[244]. The recently reported[245] spectrum of a fully substituted porphyrin Fe^{III} complex, the *meso*-tetra-CH_2-NO_2 substituted $Fe^{III}(OEP)^+$ shows, surprisingly, four $-CH_2-$ resonances, which might be explained by a ruffling of the molecule caused by steric hindrance and an out-of-plane position of the central metal ion.

Whereas either high- or low-spin complexes are observed in most instances, deviations from the $1/T$ law and the wide range of shifts in some hemin azides has been attributed to a mixture of both forms[246a] in the same compound[22]. The exchange between high- and low-spin hemin was studied together with the ligand exchange (pyridine, water) by Degani and Fiat[233] by relaxation measurements (see Section 10.4.1).

Fe^{IV} complexes are postulated by Felton et al.[246]. The 1 : 1 complex $Fe^{IV}(TPP)Cl^+$ has a net spin of $S = 2$, and shows a broad signal at $\delta = 68.6$ p.p.m. assigned to the resonances of the β-pyrrole protons. A more stable Fe^{IV} compound, the μ-oxo dimer of $Fe(TPP)$ has formally one Fe^{III} and one Fe^{IV} atom, and the removal of one electron from a Fe^{III}-O-Fe^{III} orbital is discussed*. The alternative formulation as a π-cation radical of $Fe^{III}(TPP)$ was discussed critically by Fuhrhop[248] on the basis of the uv-vis spectrum and redox potentials.

(c) Di-nuclear Fe^{III} complexes: μ-Oxo-bridged dinuclear Fe^{III} porphyrins[249] have been the object of attention by several groups[29,246,250—252]. The two iron atoms are anti-ferromagnetically coupled[253], which results in a diamagnetic ground state (S_0) and paramagnetic excited states ($S_1, S_2 ...$) the spacing of which is characterized by the exchange parameter J. Boltzmann population of the paramagnetic levels ($S_1, S_2 ...$) leads to hyperfine shifts for the protons. Assuming only contact contributions and identical electron-proton coupling constants A_n for all excited states, Boyd et al.[251] determined J from the temperature dependence of the hyperfine shifts. However, Wicholas et al.[252] determined that about 80% of the shift should be attributed to the first excited state, S_1, and 20% to the S_2 state, although the

---

* The unusual redox properties of $[Fe^{III}(TPP)]_2O$ are further illustrated by the report of a Fe^{I} species obtained by reduction with Na/Hg[247].

References, p. 514

latter is populated only to the extent of about 3%. These investigators were also able to show that the coupling constants A_1 and A_2 corresponding to S_1 and S_2, are unequal with $A_1 > A_2$. This relative order of coupling constant magnitude has been proved by La Mar et al.[29], who again investigated a series of highly symmetric model compounds. These authors[29] take into account the ring current shifts due to the neighboring ring by using the diamagnetic μ-oxo-scandium complexes for comparison, and evaluated the possibility of dipolar contributions that were neglected in earlier publications[150,246,251,252]. A μ-oxo dimer of hemin-a is described by Caughey[250], and the n.m.r. behavior of a heterometallic Fe^{III}-Cu^{II} dimer is reported by Bayne et al.[254].

10.2.8.5.2. Metals other than Fe

The hyperfine shifts of low-spin Co^{II} porphyrins were shown[28,256] to be dominated by the pseudo-contact term[255] (for the application of Co^{II} porphyrins as shift reagents in n.m.r., see Section 10.4.1.2). This term does not follow Curie ($1/T$) behavior, a circumstance that was shown to arise not from zero field splitting, but rather from a temperature (and more important, solvation) dependence of the g-tensor.

Several Mn^{III} porphyrins, including the Mn^{III} complex of a pheophorbide, were studied by Janson et al.[24]. The increased shifts upon increase of the porphyrin donor strength suggest a charge transfer from the ligand to the metal, and spin transfer through the π-system is invoked.

Abraham et al.[10] investigated the ligation of Ni^{II}(Meso-IX-DME). While square planar Ni^{II} porphyrins are usually diamagnetic, paramagnetic complexes are formed upon addition of a fifth ligand[256]. Strong shifts in resonances are observed for the methine protons, much smaller ones ($\sim$1/6) for the signals of the protons in the β substituents, and a spin transfer via the π-system is advanced to account for the spectra.

The nuclear spin relaxation mechanism was studied for Cr^{III}, Mn^{III}, and high-spin Fe^{III} complexes of $meso$-tetra-p-tolyl-porphyrin by a linewidth analysis[26]. For the Fe^{III} complexes, the linewidth is proportional to the electron spin relaxation time, T_{1e}, which is determined by the modulation of the zero field splitting parameter, D, by the molecular tumbling. It can be varied considerably by the axial ligand. This relaxation mechanism is less important for Mn^{III}, and Cr^{III} shows relaxation times corresponding to the tumbling correlation time. A dependence on D (or rather D^2) could be demonstrated for Fe^{III} and Mn^{III} complexes by variation of the fifth (axial) ligand[26]. Thus, a suitable choice of solvent can aid considerably in the resolution of the ^{1}Hmr spectra. Most of the investigations described here focus on resonances arising from the porphyrins. The broadening of resonance lines of axial ligands in paramagnetic porphyrin complexes, which has been used by several authors[233,257,258] as a probe for the mechanism of ligand exchange is discussed in Section 10.4.1.2.

A Eu^{III} complex of (TPP) was recently reported by Wong et al.[259]. As evidenced by the non-equivalence of the o- and m-phenyl proton resonances (Sections 10.2.8.1. and 10.4.2.2), the metal ion is considerably out-of-plane. Assuming only pseudo-contact shifts, a considerably larger Eu—N distance than in other Eu complexes was estimated.

10.3. Nuclei other than ^1H

The extensive use of ^1H as the basic n.m.r. probe for large organic molecules such as porphyrins to the exclusion of other nuclei was originally dictated by the high sensitivity of ^1Hmr as compared to that of the other elements present (carbon, nitrogen, and oxygen), which constitute the structural backbone. A combination of one or more of the following properties of a given nucleus (Table 27) is responsible for the problems involved in the n.m.r. spectroscopy of nuclei other than ^1H: (a) low inherent sensitivity at a given magnetic field, which depends on the third power of the gyromagnetic ratio; (b) low natural abundance; (c) spin $S \neq 1/2$, which either completely precludes n.m.r. spectroscopy ($S = 0$), or renders interpretation of spectra difficult because of complex coupling patterns produced when $S \geqslant 1$; (d)

TABLE 27

Nmr characteristics for some nuclei important in porphyrins and metalloporphyrins

Nucleus	Inherent Sensitivity [a]	Natural Abundance (%)	S	T_1 [b]
^1H	100.0	99.9	1/2	++
^2H	0.36	0.015	1	++/+++
^{12}C	0	98.9	0	—
^{13}C	1.6	1.1	1/2	+/++
^{14}N	0.04	99.63	1	+/+++
^{15}N	0.1	0.38	1/2	+
^{16}O	0	99.76	0	—
^{17}O	0.25	0.04	5/2	+++
^{18}O	0	0.20	0	—
^{199}Hg	0.57	16.8	1/2	+/++
^{203}Tl	18.7	29.5	1/2	+/++
^{205}Tl	19.2	70.5	1/2	+/++

Only crude approximations for the spin lattice relaxation time T_1 are given.
[a] As compared to ^1H at the same magnetic field strength.
[b] +: $T_1 > 10^2$ sec; ++: $10^0 < T_1 < 10^2$ sec; +++: $T_1 < 10^0$ sec.

quadrupole-induced line broadening; and (e) long inherent spin lattice relaxation times, T_1, of a nucleus, which easily leads to saturation of the n.m.r. resonances.

Early attempts to increase the spectral sensitivity of exotic nuclei involved enrichment of the nucleus of interest above its natural abundance level, S/N enhancement by signal averaging techniques, methods of circumventing the relaxation problem by adding small amounts of paramagnetic compounds (such as Cr acac$_3$) to facilitate relaxation, the use of cross relaxation techniques (nuclear Overhauser enhancement), and flow methods. The principal technical advance, however, in the n.m.r. spectroscopy of formerly exotic nuclei was made possible by the development of pulse Fourier transform (PFT) n.m.r. spectroscopy in recent years[260], a technique that can be further combined with some of the above-mentioned procedures. In conventional continuous wave (cw) mode for recording n.m.r. spectra, only one frequency at a time is observed, and most of the time spent in recording the spectrum is lost in collecting noise instead of signal. In the PFT mode, all nuclei are excited simultaneously by a strong radio frequency pulse, and the decay of the thus induced magnetization (free induction decay, FID) is observed in the time domain. If the spin—spin relaxation time, T_2, (or better T_2*) is comparable to the spin lattice relaxation time, T_1, a signal is collected over most of the measurement time, which is thus used much more efficiently. The Fourier transformation from this time domain signal, (i.e., the FID) into the frequency domain (i.e., the usual spectrum), and any additional necessary processing of the spectrum is then done by a digital computer. The sensitivity enhancement by PFT as compared to cw is usually from one to two orders of magnitude. The sensitivity of PFT can be even further increased by some of the above-cited techniques, particularly the nuclear Overhauser enhancement that results from a simultaneous irradiation of coupled protons while ^{13}C spectra are recorded*. For a detailed discussion of pulse FT spectroscopy[260] and its applications to ^{13}Cmr[30], the reader is referred to recent monographs.

10.3.1. ^{13}Cmr of porphyrins
10.3.1.1. ^{13}Cmr of diamagnetic porphyrins

The majority of n.m.r. studies on nuclei other than ^{1}H have been made on

---

* Although pulse Fourier transform techniques are usually employed in heteronuclear (i.e., nuclei other than ^{1}H) n.m.r., we want to emphasize the advantages of PFT for ^{1}Hmr as well. This is especially true for porphyrins, as the greatly increased sensitivity overcomes their sometimes poor solubility and their pronounced aggregation. For example, the spectrum of porphin shown in Fig. 3a was obtained in our laboratory in about 15 min in a solution estimated to be 0.0005 M. PFT n.m.r. has also been applied[173] for the study of photoreactions directly in n.m.r. sample tubes, which require low concentrations because of the high light absorption of porphyrins.

^{13}C, not only because carbon is a universal component of organic structures (^{13}C is present to the extent of 1.1% of the carbon present), but also because ^{13}C possesses a comparatively high sensitivity (Table 27) and a comparatively low price in high isotopic purity, which makes the synthesis and biosynthesis of ^{13}C enriched compounds a practical proposition.

Only two ^{13}C studies of porphyrins have appeared in which at least some of the spectra were recorded in the cw mode[36,37]. The compounds employed were enriched in ^{13}C to 15 and 95%, respectively, but nonetheless, long signal averaging times were still necessary to obtain reasonable signal-to-noise in the spectra.

By contrast, and illustrating the great technical advances embodied in modern spectroscopic equipment, a wide variety of porphyrins at natural abundance or only moderate ^{13}C enrichment, used at concentrations of less than 0.1 M, have now been investigated by ^{13}C PFT-n.m.r. spectroscopy. The technique of moderate (biosynthetic) enrichment to about 15% ^{13}C has proven especially useful[33,37]. ^{13}C—^{13}C Spin—spin coupling in compounds at this enrichment is still negligible, while the gain in sensitivity over natural abundance is considerable. Enrichments higher than 20% are desireable primarily for studies of ^{13}C—^{13}C spin—spin interactions[36], and, to some extent, for selective labeling experiments[35,261,262].

(a) *Assignments*: Due to the favorably spaced and well-assigned ^{1}Hmr spectra of porphyrins, carbon atoms bearing hydrogen atoms can be directly assigned from multiplicity of their resonance, and the assignment can be assisted by either single frequency off-resonance decoupling[34], or (more unambiguously) by single frequency on-resonance decoupling[37]. The very closely spaced methine ^{13}C resonances in H$_2$(Proto-IX-DME) and (2,4-diacetyl-Deut-IX-DME) were assigned by (synthetic) selective enrichment at these specific positions with ^{13}C[261,262], and Katz et al.[37] and Lincoln et al.[263] used stepwise chemical modifications to clarify questionable assignments.

The assignment of the quaternary carbons in large molecules presented a major challenge, especially as the chemical shifts of these carbons may be expected to yield valuable information otherwise unavailable from ^{1}Hmr. All, or almost all, of the expected quaternary carbon atom resonances are usually observed as resolved singlets, which are easily differentiated by their multiplicity in the undecoupled spectrum and by their relatively low intensity as compared to the proton-bearing carbons in the broad-band ^{1}H decoupled spectrum. The latter effect is due to longer relaxation times and the small nuclear Overhauser enhancement of the quaternary carbons.

Matwiyoff et al.[36] and Abraham et al.[32] discuss the (exchange) broadening of the α-pyrrole ^{13}C signals in free base porphyrins[9,261], by tautomeric N—H exchange[32], as a way to differentiate between α- and β-pyrrole carbon resonances. The latter authors also discuss the effect of metalation on the absolute chemical shifts and spacing of the α- and β-pyrrolic resonances as a

References, p. 514

potential aid in assigning these resonances (see also Ref. 33). Lincoln et al.[263] use gradual structural changes to assign some of the quaternary ^{13}C resonances, and ^{13}C—^{13}C couplings in highly enriched porphyrins can establish some assignments in the vicinity of *meso*-carbon atoms[36]. The most direct approach to assignment of the quaternary carbon atoms is the modified indor (heteronuclear double resonance) techniques used by Boxer et al.[33] to assign all the macrocyclic quaternary carbons in chlorophyll-*a* and some of its derivatives. In conventional indor spectroscopy, the absorption level or intensity of a satellite line of a (usually proton) multiplet is monitored, while a second radio frequency is swept through the absorption range of the respective nucleus coupled to the proton. In 'center line indor', the absorption intensity at the center of the resonance line is monitored instead. Although no distinct multiplets are observed, the center lines of the proton resonances of the β-pyrrole substituents are broadened by long range spin—spin couplings with the quaternary carbon atoms, and the absorption is increased by collapse of the unresolved proton multiplet when a transition of the coupled quaternary carbon atom is induced. Obviously, the utility of this technique is again enhanced by the use of compounds at moderate ^{13}C enrichment. The assignment of the ^{13}C resonances of chlorophyll-*a* obtained[33] in this way are not only self-consistent, but also confirm the earlier assignment of the ^{1}Hmr spectrum[13] in all respects.

(b) *^{13}C Chemical shifts*: Although the ^{13}C chemical shift can be broken down into the same components as discussed for protons, the relative importance of their contributions is different. The ^{13}C nucleus is comparatively shielded from the surrounding environment, but carbon nuclei have a multitude of accessible hybridization states that affect the ^{13}C shifts strongly and spread ^{13}C resonances over a range of several hundred p.p.m. Thus, ring current effects on the chemical shifts of protons are roughly ±5—10 p.p.m. and this strongly determines the appearance of a proton spectrum that in most instances extends over only a slightly larger range. Although ring current effects are of the same absolute magnitude for ^{1}H and ^{13}C, they contribute less than 10% to ^{13}C shifts[31], which are spread over about 200 p.p.m. Reliable ring current increments to the chemical shift of carbon atoms in the conjugation pathway of porphyrin macrocycles would be a valuable probe of the magnetic anisotropy in these regions of the molecule that are not accessible to ^{1}Hmr. So far, the ambiguities in the interpretation of ^{13}C shifts do not permit the separation of these terms. Ring current contributions may be responsible, however, for the same order of chemical shifts for both the ^{1}H and ^{13}C signals of sets of β-pyrrole CH_3 groups.

The ^{13}Cmr spectra (insofar as available) of the same selected archetypical porphyrins discussed at the beginning of this chapter (Section 10.2.1) are summarized in Table 28. Rather than discussing each spectrum in detail we propose to outline such fundamental and general features of porphyrin ^{13}C n.m.r. as can be drawn from the yet very incomplete work. The ^{13}C spectra

TABLE 28

^{13}Cmr chemical shifts (δ [p.p.m.] from TMS) of some principal porphyrins, and ^{15}N spectrum (δ [p.p.m.] rel. to external $^{15}NH_4Cl$) of methyl pheophorbide-*a*-$^{15}N_4$.

				Solvent	Ref.
(11)	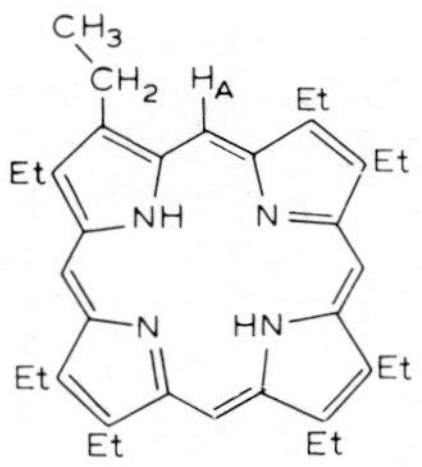	Methine—C:	96.8	C^2HCl_3/TFA	32
		$\alpha_{pyrrole}$—C:	142.2		
		$\beta_{pyrrole}$—C:	141.1		
		CH_2:	19.8		
		CH_3:	18.0		
(13)	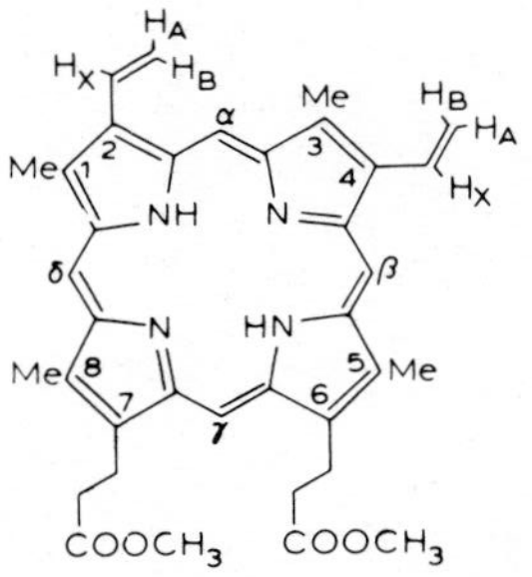	Methine—C (α,β):	97.5, 97.0	C^2HCl_3	35
		Methine—C (γ,δ):	95.6, 96.6		
		CH_3:	11.5—12.7		
		CH_2CH_2:	21.8, 36.8		
		$COOCH_3$:	172.8, 51.6		
		$CHCH_2$:	129.8, 120.2		

C	δ^c	$^1J_{^{13}C-^{13}C}$ [d]	C	δ^e	$^1J_{^1H-^{13}C}$ [f]
1	131.6	44	1a	11.6	129
2	136.3		2a	128.4	155
3	135.9	45	2b	121.9	160, 68[h]

References, p. 514

TABLE 28 (continued)

(87)

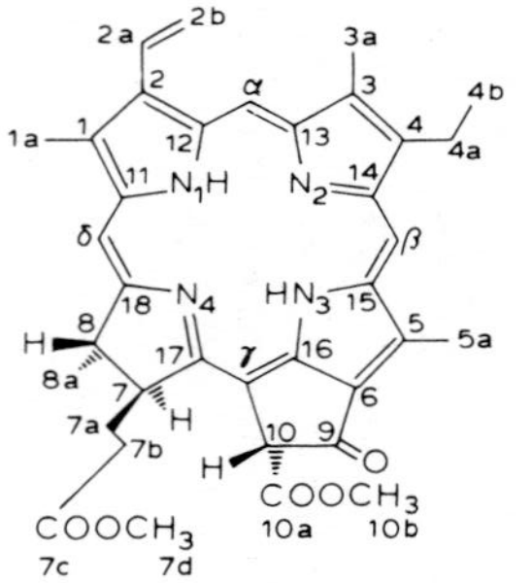

C	δ^c	$^1J_{^{13}C-^{13}C}$ [d]	C	δ^e	$^1J_{^1H-^{13}C}$ [f]
4	144.9	42	3a	10.4	126
5	128.8	44	4a	18.7	125
6	161.2		4b	16.9	160
7	51.0 [a]	129 [b]	5a	11.6	129
8	49.9 [a]	46, 130 [b]	10a	169.2	58 [i]
9	189.0 [a]		10b	52.5	148
10	64.6 [a]	136 [b]	7a	30.9	130
11	141.9		7b	29.7	126
12	135.9		7c	172.9	
13	155.3		8a	22.8	125
14	150.7		α	96.8	155, 70 [g]
15	137.8		β	103.7	145, 70 [g]
16	149.6		γ	104.9	
17	173.3		δ	92.7	157, 70 [g]
18	172.0				

N	δ^j	J_{N-N} [k]		J_{N-H} [k]
1	102.5	$^2J_{12} = 2.0$, $^2J_{14} = 2.5$		$^1J_{N-H} = 98$
2	219			
3	110.9	$^2J_{23} = 5.7$, $^2J_{34} = 1.4$		$^3J_{N-H} = 3$ [l]
4	272.8			

[a] From Ref. 37.

[b] $^1J_{^1H-^{13}C}$ from Ref. 37.

[c] From Ref. 33 in δ[p.p.m.] relative to internal TMS.

[d] Coupling constant for the β-pyrrolic C and the adjacent substituent C in Hz, from Ref. 36.

[e] In δ[p.p.m.] relative to internal TMS, from Ref. 37.

[f] In Hz, from Ref. 37.

[g] $^1J_{^{13}C-^{13}C}$ in Hz between the methine-C and the adjacent α-pyrrolic C, from Ref. 36.

[h] $^1J_{^{13}C_{2a}-^{13}C_{2b}}$ in Hz, from Ref. 36.

[i] $^1J_{^{13}C_{10}-^{13}C_{10a}}$ in Hz, from Ref. 33.

[j] In δ[p.p.m.] relative to external $^{15}NH_4Cl$, from Ref. 33.

[k] In Hz, from Ref. 33.

[l] Three bond $trans$-coupling between ^{15}N and the methine proton.

of porphyrins can be more or less arbitrarily subdivided into four regions: the aliphatic carbon region with chemical shifts in the range ~10—70 p.p.m.;* the methine carbon region (~90—100 p.p.m.); the aromatic and olefinic carbon region (130—170 p.p.m.); and the carbonyl region in the most strongly deshielded portion of the spectrum (170—190 p.p.m.). Although there may be some overlap in chemical shifts, especially in the low-field regions, such a situation is usually readily resolvable from the number and multiplicity of the resonances.

The signals of all proton-bearing sp^3 hybridized carbon atoms are observed in the high field region between 0 and 70 p.p.m. These chemical shifts fall well within the usual ^{13}C expectation ranges[30]. The resonances of the carbon atoms in the aliphatic side-chains occur in the range of $\delta = 10$—40 p.p.m., and for similar substituents the same order of chemical shifts in both the ^1H and the ^{13}C spectrum is found[33,37]. The chemical shifts for several important β-pyrrole substituents are listed in Table 29, and these shifts seem to be fairly constant in various porphyrins. The 7 and 8 carbons in chlorins come into resonance at about 50 p.p.m.[33,213,263], and the carbon atoms adjacent to ester or carbonyl functions have resonances in the range of 50—70 p.p.m.

The chemical shift range from 90—100 p.p.m. contains the resonances of the methine carbons. These signals are closely spaced in protoporphyrin-IX[35] and related alkyl- or vinyl-substituted porphyrins[32,34,263] but are spread out by β-substitution with other groups[35,263] and in chlorins[33,213,263]. Alkyl-substitution of a β-pyrrole position shifts the neighboring methine carbon resonances upfield by about 3.5 p.p.m.[32], as compared to the β-unsubstituted compounds. In chlorins, the signal(s) of the methine carbons next to the reduced ring occur (as in the proton spectrum) as separate resonances at higher fields, probably because of the high electron density at these sites. However, in pheophorbides the quaternary γ-C methine resonance occurs in the region of the α- and β-methine carbon atom signals. The above-mentioned shielding clearly seems to be compensated for by quaternization and bond angle deformations[128].

In the region between 130 and 170 p.p.m. the resonances of the quaternary α- and β-pyrrole carbons occur, with the latter at higher field and usually without overlap of the α-carbon atoms at lower field. Again, these two sets of α- and β-pyrrole carbon resonances are closely spaced in alkyl-substituted porphyrins[9,32,35], while the resonance peaks are spread out in

---

* All ^{13}C chemical shifts are given in δ [p.p.m.], down-field relative to internal $Si(^{13}CH_3)_4$ (TMS). For conversion of chemical shifts from other internal standards sometimes used in the literature, the following reference values relative to TMS were used here: $^{13}CS_2$: $\delta = 193.7$ p.p.m.; benzene: $\delta = 128.5$ p.p.m.[30]. TMS has the considerable convenience in that both ^1H and ^{13}C chemical shifts are referred to the same internal standard compound.

TABLE 29

Expectation ranges (δ[p.p.m.] from TMS) of the ^{13}Cmr resonances of β-pyrrole substituents in porphyrins

Substituent	Chemical Shift δ[p.p.m]	Ref., remarks
β-pyrrole		
CH_3	10.4—15.7 (22.8—23.1 = 8 CH_3 in chlorins)	9,32,33,34,35,37, 261,263
CH_2CH_3	18.7—22.5/16.9—18.5 (12.8[35])	32,34,36,37,261, 263
$CH=CH_2$	128.8—131.6/120.1—122.6	34,35,36,37,261, 263
$CH_2—CH_2Br$	18.3	263
$CHOH—CH_3$	65.3/26.0	34
$CH_2CH_2COOCH_3$	21.6—23.5/35.8—39.2/172.7—174.7/51.4—53.0	9,32,34,35,37,
in chlorins	30.9—31.1/29.6—29.9/172.9—173.5/51.6—51.7	261,263
CHO	187.4	36,263
$COCH_3$	33.1—34.3	34,35
CO—R	189—196.1	9-CO in chlorins, 37,263
$COOCH_3$	169.4—173.0/52.0—53.1	263
γ-Methine		
CH_2COOCH_3	38.3—38.6/169.4—173.0/52.0—53.1	263

If not otherwise indicated, the chemical shifts are listed according to the carbon atom sequence in the substituent formula (from left to right).

the less symmetrically substituted porphyrins[9,32,34,35,263] and in the chlorins[33,213,263]. (For a recent discussion of the ^{13}C resonances in non-alternant hydrocarbons, see Ref. 264.) Abraham et al.[32] observed a distinct incremental shift of about 2 p.p.m. for β-pyrrole carbon atoms next to a carbomethoxy-ethyl substituent as compared to a methyl substituent. In pheophorbides[33], the α-pyrrole carbon atoms in ring D are more deshielded by almost 20 p.p.m. than those in the remaining pyrrole rings, thus indicating a more pyridine-like character for the pyrroline ring (see below). A similar low-field shift is also observed for carbon 6 in ring C, which not only is subject to shielding by the adjacent 9-carbonyl group, but also has distorted bond angles[128] that change its hybridization.

The unsubstituted β-pyrrole C-2 and C-4 atoms in H_2(Deut-IX) occur on the high-field side of the quaternary carbon resonance region[32,34,35,263] (120—130 p.p.m.), along with the resonances of the carbon atoms in olefinic substituents[33,34,263]. This similarity in chemical shift was considered by Doddrell and Caughey[34] to be a strong hint for the presence of an inner (16-annulene di-anion) conjugation pathway (see formula 4, 5, Section

10.1.2), which makes the peripheral double bonds essentially olefinic. This interpretation was rejected by Abraham et al.[9,32], partly[32] upon the observation of similar chemical shifts for the α- and β-pyrrole carbon atoms, both of which come into resonance at considerably lower field than does the methine carbon atom. Ambiguities in the interpretation of the ^{13}C shifts still remaining do not permit a definitive decision on this point by [13]Cmr.

Abraham et al.[32] were able to show that the resonances of the α-pyrrole carbon atoms in the coproporphyrin isomers are close to coalescence at room temperature with respect to N—H tautomerism. This N—H exchange is slower in chlorins (see Section 10.4.2.1) and the more localized N—H protons generate two distinct types of rings as far as the [13]C (and ^{15}N) spectrum is concerned. Rings A and C are pyrrole-like, ring B and especially ring D resemble pyridine[33]. Upon metalation, the differences between the chemical shifts of the carbon and the nitrogen atoms in the different pyrrole rings becomes less pronounced, and at the same time the average of the α- and β-pyrrole ^{13}C resonances is shifted to lower field[32,33]. These effects are discussed by Boxer et al.[33] in terms of a redistribution of charge densities upon metalation within the macrocyclic system and a change in its absolute value, which results in part from a leveling effect on the non-bonding nitrogen orbital energies, and a simultaneous increase in their average value. The effect of protonation in [13]Cmr spectra has been investigated by Abraham et al.[32]. Upon initial addition of TFA, the α-pyrrole carbons are shielded, and the β-pyrrole (and methine) carbons are deshielded, as in the case of other N-heterocyclic compounds. At higher TFA concentrations, all signals are deshielded due to solvent effects and/or further protonation. Shifts of the α and β meso carbon lines in chlorin spectra upon addition of TFA have been used[213] to identify these carbons as well as clarify the site of meso methylation in the Chlorobium chlorophylls '660'.

The signals of the carbonyl carbons are observed in the region from 170 to 190 p.p.m. Katz et al.[37,265] found a pronounced downfield shift of the C-9 carbonyl resonance in chlorophyll-chlorophyll self-aggregates ($\Delta = -2.4$ p.p.m.). An even stronger downfield shift is observed in the resonances of the carbonyl carbon atoms of 2,4-pentanediones upon coordination with Mg^{2+} [37]. Obviously, any shielding from the ring current of the adjacent macrocycle in the chlorophyll dimer is more than compensated for by the deshielding effect of the coordination interactions of the non-bonding C = O orbital with the metal ion. Similar downfield shifts due at least in part to hydrogen bonding are reported for the ester carbonyl carbons in the coproporphyrin isomers[32] in strong acids.

(c) [13]C Spin—spin coupling: [1]H—[13]C — Natural abundance ^{13}C spectra are usually recorded under partial (single frequency off-resonance) or full (broad band) proton decoupling to obtain sensitivity enhancement from the nuclear Overhauser effect. Under these conditions, however, the [1]H—[13]C couplings are reduced or removed, respectively. Although gated decoupling is expected to yield correct coupling constant values, $^1H-^{13}C$ coupling constants are reported to our knowledge only for ^{13}C enriched compounds in spectra recorded without double irradiation. $^1H_{1H-13C}$ of about 150 and 130 Hz are observed for aromatic and olefinic, and for benzylic carbon atoms, respectively, in chlorins[36,37] and porphyrins (methine carbon atoms only)[35]; these values fall within the usual range expected for these groups[30]. While these coupling constants are fairly consistent, the two non-benzylic methyl groups in methyl pheophorbide-*a* at 4b and 8a, show a marked difference in their coupling constants, viz. 160 and 125 Hz, respectively, which probably reflects differences in steric hindrance in these groups. No long range $^1H-^{13}C$ couplings have thus far been reported; such couplings are usually unresolved and result only in line broadening[32,33]. A small ($J < 2$ Hz) coupling of the α-pyrrole carbons (probably) with the methine protons is observed in the OEP dication[135].

$^{13}C-^{13}C$ — While the enrichment ($\leqslant 20\%$ ^{13}C) optimal for ^{13}Cmr data collection effectively suppresses carbon—carbon couplings, some $^1J_{13C-13C}$ values are reported by Matwiyoff et al.[36] in highly ^{13}C-enriched chlorophyll-*a* and -*b*. Although the aromatic part of the spectrum is obscured by the various extensive couplings, it was possible to extract some $^{13}C-^{13}C$ coupling constants from the multiplets of the β-pyrrole substituents and the methine carbon atoms. The following values for the 1 bond $^{13}C-^{13}C$ constants are reported[36] (Table 28): 44 ± 2 Hz for the coupling of benzylic carbon atoms of the aliphatic side-chain to the ring carbon atoms; 34 Hz for J_{4a-4b}; 50 Hz J_{3-3a} in Chl-*b*; 68 Hz for J_{2a-2b}; 58 Hz for J_{10-11}; and 70 Hz for the coupling constant of the methine carbon atoms with both α-pyrrole neighbors. The values of $^1J_{13C-13C}$ are mainly dependent on the hybridization states of the coupled carbons (for a recent review, see Refs. 30 and 266). The only unusual coupling value, as judged from similarly substituted benzenes[30], is the (equal) coupling constant of 70 Hz observed between the methine carbon atoms and their α-pyrrole carbon neighbors, a value which is closer to that of ethylene (68 Hz) than to that of the benzene carbon atoms (~ 60 Hz).

$Tl-^{13}C$ — Abraham et al.[9,32] discuss in some detail the (long-range) $Tl-^{13}C$ couplings in Tl^{III}-porphyrins. No difference was noticeable for the two magnetically very similar isotopes ^{203}Tl and ^{205}Tl. Two to four bond couplings are observed in these cases, which are (in accordance with earlier Tl studies)[267] about 60 times larger but in the same relative order as the $^1H-^1H$ couplings, indicative of dominant contact couplings via spin transfer through the σ bond system.

10.3.1.2. ^{13}C of paramagnetic metalloporphyrins

In addition to the ^1Hmr spectra of Fe^{III} porphyrins, Wüthrich et al. recently studied the ^{13}Cmr spectra of some low-spin Fe^{III} porphyrins[268,269], as well as the Fe^{III} complexes of Proto-IX[268,270] and Deut-

IX^{270}. The chemical shift assignment of carbon atoms bearing protons was achieved by single frequency off-resonance, and, in some cases, by single frequency (on-resonance) decoupling, and the quaternary carbons were assigned by intercomparisons. The treatment of the ^{13}C data is, as in the case of the ^{1}Hmr data, directed to the separation of the hyperfine shifts from all other contributions present in the diamagnetic porphyrins. Furthermore, the separation of the hyperfine shifts into contact and pseudo-contact contributions (Section 10.2.8.5), and the further subdivision of both hyperfine terms into contributions from the σ and π electron framework and the metal ion was attempted. As in the ^{1}Hmr investigations, the Zn complexes were again used as reference compounds to evaluate the diamagnetic shifts* (average value of sets of ^{13}C resonances from chemically equivalent nuclei). It should be noted, however, that the pronounced differences in the ^{13}C chemical shifts of diamagnetic porphyrins observed for different metals[32] make these reference shifts of lesser value than in the case of the ^{1}H spectra, because they are sometimes of similar magnitude as the hyperfine shifts.

A semi-quantitative treatment provides limits for the pseudo-contact shifts, which were calculated from the g-factor anisotropy (see Section 10.2.8.5) in frozen solution (an upper limit), and from values estimated in earlier ^{1}Hmr work (a lower limit).

The results from ^{13}Cmr are in many points complementary to the conclusions derived from ^{1}H hyperfine shifts. The ^{13}Cmr results provide a distinct refinement, as they indicate pronounced differences in the (hyper-conjugation) coupling parameters Q for different substituents[270]. These differences can be interpreted in geometrical terms and in principle yield a better insight into the conformation of the various substituents in solution. It has been further shown that, in contrast to ^{1}H shifts, the ^{13}C pseudo-contact shift contribution from spin transferred to the ligand is no longer negligible.

10.3.2. ^{15}Nmr of porphyrins

^{15}N in magnetic resonance spectroscopy may have such very long relaxation times that acquisition of ^{15}Nmr data may be prevented. The ^{15}N spectrum of highly (95%) enriched ^{15}N pheophytin-a (87b) (Table 28) has been recorded by the use of the very long pulse interval of 60 sec[33], and by addition of $Cr^{III}acac_3$ **[271] to induce more rapid relaxation. The ^{15}N shifts of the related Mg-complex (chlorophyll-a) were obtained indirectly by

---

* In contrast to the usual n.m.r. spectroscopic definition, the diamagnetic shift refers here to all terms (diamagnetic as well as paramagnetic) *except* the hyperfine terms arising from the nuclear interaction of the unpaired spin.
** By comparison with the values given by Boxer et al.[33] the ^{15}N-spectrum of the algal pigment mixture[271] is obviously that of pheophytin-a and -b (probably demetalated by $Cr^{III}acac_3$) rather than that of the chlorophylls.

References, p. 514

heteronuclear (^{1}H–^{15}N) double resonance experiments[33]. For both chloro-phyll-a and pheophytin-a, the ^{15}N resonances were assigned by single fre-quency decoupling of the methine proton resonances[33]. The ^{15}N spectrum in the free base pheophytin-a shows two sets of resonances, with chemical shifts characteristic for pyrroles (ring A, C) and pyridines (ring D, and some-what intermediate, ring B). The order of the chemical shifts was interpreted to reflect the relative order of the energy levels of the non-bonding nitrogen orbitals, with ring D > B $\geqslant$ C ~ A[33]. This order is in agreement with ESCA data on porphyrins[272]. Various ^{15}N–^{15}N couplings via the inner hydrogen atoms are observed, and the data allow an estimate of the sharing of the inner hydrogen atoms between the nitrogen atoms (Table 28, see also Sec-tion 10.4.2.1)[33]. The ^{15}N–^{1}H coupling constant with the inner hydrogen atoms is $^1J_{^{15}N-^1H}$ = 98 Hz. In addition, long-range (*trans*) coupling of $^3J_{^{15}N-^1H}$ = 3 Hz, with the methine protons (used to assign the ^{15}N reso-nances) in chlorophyll-a, was observed, and five bond couplings ($^5J_{^{15}N-^1H}$) with the methine protons via the central Mg are discussed[33].

10.3.3. Magnetic resonance of central metals in metalloporphyrins

In spite of the wide variety of porphyrin metal complexes available for study, only a very few investigations deal with the n.m.r. spectroscopy of these metals and their (nuclear) spin-spin interactions with the porphyrin ligand. Abraham et al. investigated in some detail long-range spin–spin couplings between Tl and ^{13}C[9,32] and ^{1}H[8,11]. ^{203}Tl and ^{205}Tl are mag-netically very similar and their proton coupling constants are identical, in agreement with results of Maher and Evans on Tl organic compounds[267]. The ^{13}Cmr results show a predominant contact coupling mechanism with spin transfer through the σ-system. A similar mechanism was advanced for the long-range ^{1}H–Tl couplings with the protons of the β-pyrrole side-chains, whereas the methine protons appeared to be coupled *via* the π-sys-tem.

The μ-diporphinato–trimercury complex (82) was studied by (^{1}H-{^{199}Hg})indor spectroscopy (see Table 23)[219,223]. The observation of two ^{199}Hg in indor lines at 17.8866 and 17.8854 MHz in an intensity ratio of 2 : 1, is one of the basic results acquired to establish the stacked structure proposed for this compound.

10.3.4. ^{2}Hmr of porphyrins

^{2}Hmr is of low sensitivity, and the resonance lines are broadened because of quadrupolar relaxation, (for a recent review see Ref. 273). The few ^{2}Hmr studies available, however, indicate certain technical advantages to this spec-troscopic technique. For example, the ^{2}Hmr spectra are first order (because of the higher Δδ/J ratio), and resolution is much better in the case of (paramagnetically) broadened lines[274]. Isotope effects on chemical shifts and coupling constants are furthermore of considerable theoretical interest.

In evaluating the potentialities and advantages of bio-molecules of unnatural isotopic composition, the Argonne group reported the ^{2}Hmr spectrum of methyl pheophorbide-a-^{2}H$_{35}$ (7d-CH$_3$) and chlorophyll-a-^{2}H$_{72}$[293]. Although the ^{2}H lines are broadened by quadrupolar relaxation, the resolution is good for the former compound (2—7 Hz line widths) and sufficient in Chl-a for the identification of the major resonances. The ^{2}H chemical shifts of the porphyrin moiety of the molecules are very similar to the ^{1}H shifts, with generally positive (shielding) isotope effects of less than +0.05 p.p.m. The ^{2}H-phytyl side-chain, however, is strongly shielded. An isotope effect of 0.63 p.p.m. is observed in the ^{2}H-chlorophyll, which is probably accounted for by the integrated (shielding) isotope effect in the aliphatic chain, although aggregation shifts involving ring current effects cannot be completely excluded.

10.4. Introduction to applications section

In this section, the applications of n.m.r. spectroscopy to three major areas are discussed: the aggregation of porphyrins (including ligand exchange processes) is mainly studied by using ring current induced shifts (RIS) as a probe for molecular interaction; dynamic processes involving tautomerism and rotation of substituents; and the stereochemistry of porphyrins. To complete this somewhat arbitrary selection, pertinent applications outside this scope are listed under miscellaneous without further discussion.

10.4.1. Aggregation

The very early ^{1}Hmr studies of porphyrins revealed a remarkable solvent and concentration dependence of the chemical shifts of the solute (the porphyrins), the solvent, and co-solutes. These effects arise from self-aggregation of the porphyrins, or are the result of more or less specific interactions with nucleophiles that may be present. The reason for aggregation shifts in the porphyrins, in a general way, lies in the combination of the strong magnetic anisotropy of the porphyrins with strong, and often specific, molecular interactions of the porphyrins with each other or with other species (nucleophiles) present in solution. The study of the chemical shift consequences of porphyrin molecular interactions thus provides detailed insight into both self-aggregation (endogamous aggregation)[59] and porphyrin—ligand interactions (exogamous interactions)[59].

Porphyrin aggregation can involve either or both π—π and metal—ligand interactions. The π—π forces are relatively weak (for an exception, see Ref. 276), fairly insensitive to solvent*, and generally produce only upfield shifts

---

* Disaggregation in TFA is a result of dication formation rather than a result of a solvent effect.

of the species ligated to the porphyrins. The sources of these upfield shifts are obvious from the magnetic anisotropy of the macrocycle (Fig. 1), and result from the positioning of protons above or below the plane of another macrocycle. The surface defining zero shielding is not perpendicular to the macrocycle plane, and only associated molecules which are substantially larger in area than the porphyrin itself, can protrude into the deshielding areas. π—π Interactions are the main aggregation forces* in free base porphyrins. In metalloporphyrins, the π—π forces are often outranged in magnitude by metal—ligand coordination interactions, which are strongly solvent dependent and may result (especially in the case of porphyrin self-interactions) in both low and high-field shifts. In diamagnetic metalloporphyrins, aggregation shifts are ring current induced and can amount to as much as 2 p.p.m. or more for proton chemical shifts. In paramagnetic complexes, the chemical shifts in aggregates are dominated by nuclear hyperfine interactions with the unpaired spin(s), and thus can be an order of magnitude larger (see Section 10.2.8.5).

In addition to self-aggregated species we also include in this section a discussion of covalently bound axial ligands involving the central metal ion of metalloporphyrins. Molecular aggregates of this kind show exactly the same incremental shifts, and [1]Hmr data have proven extremely useful in evaluating and comparing the magnetic anisotropy of the porphyrin macrocycle.

10.4.1.1. Porphyrin self-aggregation

The two self-aggregation extremes studied by [1]Hmr are the coproporphyrin tetramethyl esters[6,9] and the chlorophylls[37]. The concentration dependence of the proton chemical shifts in Copro was first noted by Abraham et al.[7] and later investigated in detail by [1]Hmr[6] and [13]Cmr[9]. As a by-product these studies made possible a useful analytical technique for distinguishing-III and IV isomers of coproporphyrins, thus solving an old problem. From symmetry considerations, both of these isomers are expected to yield similar multiplicity patterns for sets of chemically equivalent protons. This is indeed observed. For example, the methine resonances in the monomers present in TFA solution show three signals with intensities in the ratio 1 : 2 : 1. However, the significantly different geometry of the porphyrin aggregates in C^2HCl_3 solution leads to a partial collapse of the methine resonances to two signals (intensities in the ratio 2 : 2) in the III isomer[6]. In other cases, distinctions between isomers in which not all of the expected resonances are visible can be made from the fine structure in the spectra of the aggre-

---

* This is true only if no strongly aggregating substituents are present. Thus. for example, the recently studied 2a-hydroxypheophorbides show aggregation via H-bonding[132,213,277].

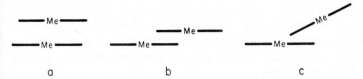

Fig. 5. Endogamous (self)-aggregation in porphyrins. (a) π—π charge transfer interaction, plane-to-plane distance 8—10 A, macrocycles laterally displaced. (From Ref. 68; see also Refs. 6,8,282.); (b) strong π—π interactions between ring A and D in Fe^{III}(Proto-IX)dicyanide (see Ref. 276); (c) strong n.m.r. averaged metal-ligand interaction between the central magnesium atom and the ring E keto carbonyl group in chlorophylls (see Ref. 37).

gates[92]. A quantitative analysis of the concentration dependence of the ^{1}H chemical shifts[6] indicated that: (a) only a monomer—dimer equilibrium in porphyrin π—π aggregation has to be considered in the concentration range up to about 0.2 m; (b) in the porphyrin dimer, the two porphyrin molecules are parallel to each other and on the average about 8 Å apart; and (c) the dimer components are persuaded by steric interactions of the bulky (propionic ester) side-chains to adopt an orientation such that the side-chains are staggered. Thus, the macrocycles are laterally displaced by about 2 Å (Fig. 5a). Of particular interest, Abraham[6] remarks that the separation distance observed in these dimers is more typical of charge transfer complexes rather than entities generated by generalized π—π interactions. A similar conclusion was reached by Ogoshi et al.[180] from infrared studies of the aggregation of porphyrin di-acids, which were formulated by these workers as cation—anion complexes, and the aggregation of N-methyl OEP with its mono-cation[142] provides further support for this view. On the other hand, a separation of only 4.5 Å characteristic for π—π aggregation was recently reported[276] for a low-spin protohemin (Fig. 5b).

A similar aggregation behavior was demonstrated for Meso-, Proto- and Deuteroporphyrins-IX in a high magnetic field study[68]. In these cases, the aggregates show a 10 Å separation of the macrocycle planes. Moreover, it was shown that at 220 MHz all magnetically non-equivalent nuclei were resolved[68]. The self-aggregation of pheophorbides-a and -b was studied by Closs et al.[13] and that of pyropheophorbide-a by Pennington et al.[214]. Because all of the ring proton signals in these compounds are assigned, aggregation in the dimers can be mapped in detail. The dimer structure in these compounds is similar to that of the porphyrins, but here steric interactions in the chlorin rings causes a pronounced lateral displacement of the macrocycles, with ring B showing the strongest overlap in the dimer.

Both the endogamous and exogamous interactions of the chlorophylls, which represent the other extreme of strong (axial) ligand—metal interactions, have been studied in detail by the Argonne group[13,214]. The central magnesium atom in the chlorophylls can be considered to be coordinatively unsaturated, which leads in the absence of extraneous ligands (nucleophiles,

Lewis bases) to pronounced chlorophyll—chlorophyll aggregation. In aliphatic hydrocarbon solvents, large chlorophyll—chlorophyll aggregates with aggregation numbers higher than 20 are observed in concentrated chlorophyll solutions (0.1 M)[188,278]. These large oligomers, with molecular weights in excess of 20,000 appear to have linear polymeric structures in which the central Mg atom in one chlorophyll molecule is ligated (principally) to the 9-keto carbonyl group of the next molecule (Fig. 5c). The large aggregates that may be present in aliphatic hydrocarbon solutions not only lead to complex n.m.r. spectra, because of ring current effects, but the resonance lines are broadened by the longer correlation times of the aggregates and by a slow exchange of the subunits. The ^{1}Hmr spectra of these oligomers are of such ill-defined nature as to prohibit detailed analysis. The basic chlorophyll—chlorophyll interaction can be studied, however, in 'soft' non-polar solvents such as chloroform, benzene or carbon tetrachloride. In these less hostile solvents, the chlorophylls are present largely as dimers[279] or small oligomers undergoing fairly fast exchange. Participation of the 9-keto C = O group in chlorophyll-a aggregation is demonstrated in the ^{1}Hmr spectra by the strong high-field shifts of all ^{1}H resonances in the vicinity of the isocyclic ring E (see aggregation map, Fig. 6), by the similar aggregation behavior of pyrochlorophylls lacking the 10-carbomethoxy substituent[214], and by the upfield shift of the 9-CO ^{13}C resonance[37] upon disaggregation (see Section 3.1). The dimer and oligomer structure defined by n.m.r. is the weighted average of all the conformers that are present, as the exchange of chlorophyll molecules between the species present is fast on the n.m.r. time scale. Thus, only one set of lines is visible at room temperature*. A structure (Fig. 5c), in which the macrocycle planes form an angle with each other was inferred from (approximate) ring current induced shift (RIS) calculations (for a related CD study, see Houssier and Sauer[280]). These conclusions have received additional support recently from an analysis of the lanthanide induced shifts (LIS) in a Chl-a dimer—Eu(fod)$_3$ complex[279]. Although the LIS reagent changes the chlorophyll dimer structure to some extent, small solvent-dependent differences for the average dimer conformation in benzene and carbon tetrachloride were nevertheless inferred.

The ^{1}Hmr spectra of chlorophyll dimers in non-polar solvents are essentially concentration independent, unlike the case for π—π aggregates, whose spectra are strongly concentration dependent. The strong keto C=O···Mg interactions that form aggregates can be disrupted by the addition of bases that compete with the 9-CO group as the fifth (or sixth) axial ligand for the central Mg atom[59]. Disaggregation of chlorophyll oligomers and dimers by titration with base can easily be followed by ^{1}Hmr because of the pronounced proton chemical shift changes caused by disaggregation[13,188,214].

---

* Preliminary results obtained on pyrochlorophyll-a indicate the presence of at least three distinct conformers below —45°C, for which both high and low field shifts are observed[80].

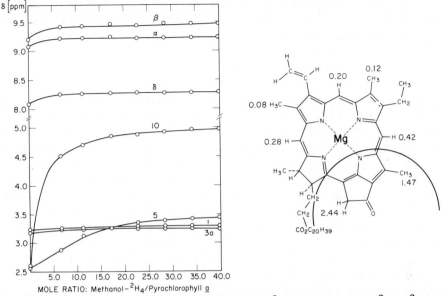

Fig. 6. (a) Titration of pyrochlorophyll-a (5×10^{-2} M in CCl_4) with $C^2H_5O^2H$; and (b) aggregation map of pyrochlorophyll-a. The chemical shift values indicated in the structural formula refer to the incremental shifts observed upon complete disaggregation of the dimer in CCl_4 solutions by titration with $C^2H_3O^2H$ (from Refs. 80, 214).

The order of base-strength for coordination to Mg relative to the 9-CO function for a variety of bases has been established by both [1]Hmr and by uv-vis and infrared measurements[405].

Between the extremes of $\pi-\pi$ interactions in the free base porphyrins on the one hand, and the metal coordination interactions in the chlorophylls on the other, a gradual mixing of both effects can be observed in porphyrin complexes with other central metals. A gradual increase in self-aggregation involving ligand—metal interactions was shown to occur in the chlorophylls for the series free base, Ni^{II}, Cu^{II} (by infrared only), Zn^{II}, and Mg^{II} by both [1]Hmr and infrared spectra measurements[281]. The presence of both types of interactions in Ni(Meso-IX-DME) was also inferred from n.m.r. aggregation studies by Doughty and Dwiggins[282]. Because of the increased strength of interaction in the metalloporphyrin, separation of the macrocycle planes in the dimer is decreased from 10 Å in the free base dimer[68] to 7.9 Å in the Ni^{II} complex dimer[282]. The four isomeric Tl^{III}-Copro's were studied by Abraham et al.[11]. The geometry of the (weak) metalloporphyrin aggregates is very similar to that of the free bases, but the somewhat increased molecular interaction as compared to the free bases is evidenced by larger equilibrium constants (4.11 vs. 3.55 l/mole^{-1}). A unique type of strong $\pi-\pi$ interactions in low-spin Fe^{III} porphyrins was recently reported by LaMar et al.[276]. Quantitative analysis of the selective paramagnetic broadening of ring A and ring D substituents yielded a structure in which ring A and D interact strongly and specifically, and in which the macrocycles are only about 4.5 Å apart (Fig. 5b).

10.4.1.2. Hetero- or exogamous aggregation

As in the preceding section, we again classify the interaction products of porphyrins with nucleophiles by the type of interactions, i.e., metal atom coordination interactions involving the central metal, $\pi-\pi$ interactions with the aromatic system, and interactions via certain substituents involving hydrogen-bonding. Only little information is available for specific aggregation of free base porphyrins with non-porphyrin molecules. As a characteristic feature of porphyrin solutions, concentration-dependent chemical shifts are not only observed for the porphyrin themselves, but also for the solvent and the resonances of the protons of the internal standard. Abraham et al.[6] found incremental shifts of −0.65 Hz for the protons in tetramethylsilane (TMS) and about −12 Hz* for chloroform upon dilution of concentrated coproporphyrin solutions (from 100 mg/ml to 34 mg/ml) and extrapolated to infinite dilution. The TMS shift can be accounted for by bulk susceptibility changes with concentration, the $CHCl_3$ shifts by the randomly averaged effect of the ring current on the solvent. Obviously, the latter effect cannot be neglected in quantitative studies[6,11], but it is due to a random process rather than to specific solvent—porphyrin interactions that generate complexes. Assuming fast exchange, an upper incremental shift limit of 1 Hz for the $CHCl_3$ resonance was estimated, which is negligible in most situations. As an example of specific $\pi-\pi$ interactions, the aggregation of the strong acceptor 1,3,5-trinitrobenzene with metalloporphyrins has been reported, and a model has been advanced in which pyrrole ring B acts preferentially as the donor[243,255,283].

Specific aggregation interactions are also observed for free base porphyrins with other porphyrins (viz. pheophytins with chlorophylls, J.J. Katz, unpublished results). As an interesting application, Wolf and Scheer[126] determined the enantiomeric purity of chiral pheoporphyrins by adding an excess of another chiral enantiomeric pure porphyrin, in this case pyromethylpheophorbide-a, to the solution. The diastereomeric collision complexes that are formed show sufficiently large chemical shift differences in the methine chemical shift region to be of analytical value.

The main interest in porphyrin heteroaggregation with exogamous nucleophiles has focussed on metalloporphyrin interactions because of the stronger and more specific interactions characteristic of such systems. Porphyrin-induced shift (PIS) reagents as a complement and alternative to lanthanide-induced shift (LIS) reagents have been studied by Storm et al.[17], Hill et al.[14,255], and extensively by Kenney, Maskasky, Janson and co-workers

---

* At 100 MHz.

(Ref. 19 and citations therein). Compounds suitable for shift reagent use, which show both ring current and additional pseudo-contact shifts, have been developed and critically examined[19]. Pronounced upfield shifts for methanol CH_3-protons bound to the central Mg in chlorophylls by coordination through oxygen were first reported by Closs et al.[13], and the corresponding incremental shifts of pyridine[16] and other ligands[51] in Mg-porphyrins were used to map the magnetic domain of space above and below the macrocycle plane. From a refined analysis of these data, the displacement of the central metal in some metalloporphyrins in solution from the plane of the macrocycle could be inferred[17], with conclusions as to the position of the metal ion in good agreement with conclusions derived from crystallographic data[128a,284].

The use of group IV metalloporphyrins and phthalocyanines and of the corresponding Fe^{II} and Ru^{II} complexes as shift reagents has recently been summarized by Maskasky and Kenney[19]. The PIS reagents are generally inferior to the LIS reagents as far as the magnitude of the chemical shift is concerned (the maximal shifts are about 8 p.p.m.[15]), but the PIS reagents are more stable and selective. Fe^{II} phthalocyanine[18,19] and its Ru^{II} analog[19] have been shown to interact very selectively with amines and to have ligand exchange kinetics optimal for recording spectra. Although the parent compounds are paramagnetic, the amine complexes are diamagnetic and the complexes show pure ring current shifts. Of the group IV metalloporphyrins, the germanium compound is specially valuable, for it forms covalent bonds with ligands and the products are well-defined compounds that can be purified and crystallized. Compounds of this type can be very valuable for combined X-ray/n.m.r. investigation to obtain the conformation of the adducts in both the crystal and in solution. The magnetic anisotropy is best mapped for the phthalocyanines[40] (see Section 10.1.2), but for better solubility the porphyrin complexes are recommended. The (TPP) complexes are easily accessible, but quantitative interpretation is not only difficult because the ring current properties of this porphyrin is less well known, but is also difficult because the phenyl rings cause steric interactions and additional (benzene) ring current shifts. Germanium porphin is recommended as a shift reagent but its use is restricted by its high price.

The use of Co^{II} porphyrins as pseudo-contact shift reagents has been studied by Hill et al.[14,255]. In the 1 : 1 complex of trinitrobenzene with Co^{II}(Meso-IX), the benzene ring is very probably situated above one of the pyrrole rings[255], indicating the operation of substantial π—π interactions. The hyperfine shifts are interpreted as arising only from pseudo-contact contributions, which add to the smaller ring current contribution. On the basis of these results, complexes of some steroids with Co^{II} porphyrins were studied, and in one case, that of the steroid cortisone, the solution structure was successfully determined by a quantitative analysis of the induced shifts[14].

References, p. 514

Studies of porphyrin or phorbin model systems of biological importance often exhibit complicated sets of overlapping [1]Hmr spectra, which may be difficult to analyze. Katz et al.[285] circumvented these problems by the use of mixtures of compounds in which one of the partners is extensively or fully deuterated and thus invisible in [1]Hmr, allowing detailed observation of the other component[51,285,286]. The aggregation interaction of lutein, a xanthophyll important in photosynthesis, with chlorophyll-a, is an example[285]. Adducts form via ligation of the hydroxyl group of lutein with the central magnesium atom, positioning a portion of the lutein molecule above the macrocycle, with effects clearly visible from the RIS of the lutein proton resonances. This aggregation complex was studied with fully deuterated chlorophyll-a, and lutein of normal isotopic composition, while in an inverse isotope experiment, in which chlorophyll interaction with sulfolipids were studied, the [1]Hmr resonances of the latter were deleted by the use of fully deuterated sulfolipid obtained by biosynthesis. In the latter type of experiment, however, no specific binding site can be established because the resonances of the ligand are absent in the [1]Hmr spectrum.

Ligand—metal interactions and ligation kinetics of Fe-porphyrins and related compounds have received considerable attention because of their biochemical importance in hemoproteins[21,22,227−229]. As in the discussion of the paramagnetic metal complexes (Section 10.2.8.5), we wish to discuss here only some of the principal model systems that have been studied.

The spin state of Fe-porphyrins is determined by the ligand field, which reflects and is determined to a great extent by the ligands present in the axial positions. The effect of axial ligands in complexes of both Fe^{II} and Fe^{III} porphyrins[287] have been studied[199] by [1]Hmr and infrared spectroscopy and these studies have been reviewed by Caughey et al.[250]. The results are interpreted in terms of the relative strength of the bonding of the central metal to the porphyrin and to the ligands in the fifth and sixth axial position. For various axial ligands, a gradual increase in the hyperfine shifts is observed from the low-spin complexes with two identical axial ligands to the high-spin complexes to an extent that depends on the binding of the axial ligands.

In addition to equilibrium studies, the ligation kinetics of paramagnetic metalloporphyrins have been investigated in some recent publications. The complexes formed are usually orders of magnitudes less stable than the complexes of the same ligand with the same metal ion not coordinated to a porphyrin[194,257]. Complexes of metal porphyrins with nitrogen bases have been most extensively studied because of interest in these complexes as models for heme—ligand interactions. For a series of substituted pyridines, the stability of the complex increased with increasing pK_a of the amine, but was decreased by steric repulsions[194]. For nitrogen ligands, S_N1 type ligand exchange reactions have been observed in which dissociation is the rate determining step[194,288,289]. For a high-spin chlorohemin, an $S_N 2$ type exchange reaction is reported[290], which has a tetragonal—bipyramidal transition state in which the metal ion moves through the macrocyclic plane. A third process, an intramolecular ligand exchange was proposed by Tsutsui et al.[291] in which the binding site in cyclic diamines is changed. Although this mechanism was shown to be incorrect by cross-relaxation experiments[289] for the case of imidazole as ligand[291], the intramolecular exchange suggested was shown[194] to occur with another ligand, pyridazine. The solvation of high-spin metalloporphyrins was studied by several groups who made use of the paramagnetic contributions to the linewidth of the ligands as the probe[233,257,258,289].

10.4.2. Dynamic processes
10.4.2.1. N—H Tautomerism
Several tautomers involving N—H exchanges can be formulated for the free base porphyrins (Fig. 7), and additional structures are possible in which the protons are shared by two (or more) ring N-atoms*. The tautomerism is generally fast on the n.m.r. time scale**. This phenomenon was first dis-

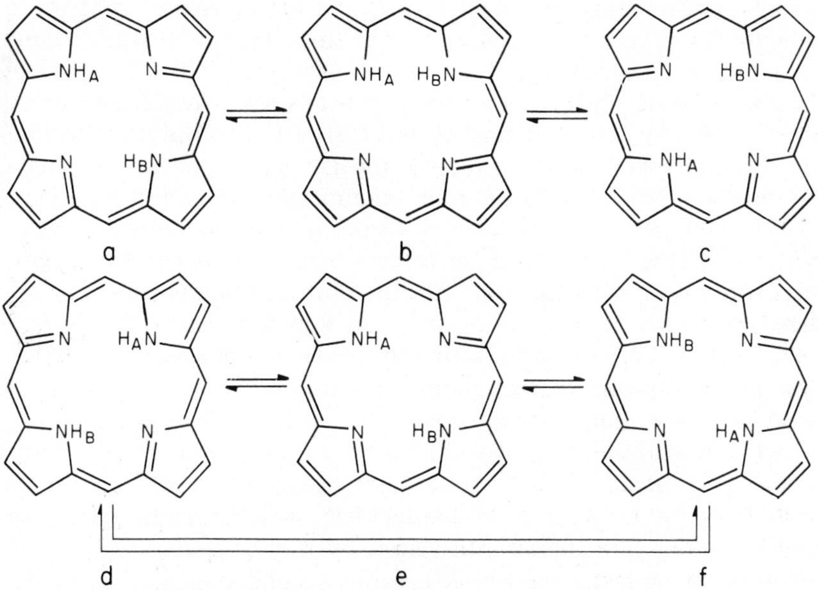

Fig. 7. N—H tautomeric equilibria in porphyrins. Non-concerted mechanism (ab,bc) with both N—H protons exchanging independently, and concerted mechanism with N—H exchanging simultaneously between neighboring (de,df), or, opposite nitrogen atoms (ef).

---

* In addition to intramolecular N—H exchange, Tsutsui[292] demonstrated recently fluxional behavior in Re^I and Tc^I complexes, in which the interchange of N—H, N—Re tautomers can be observed.
** For a relevant discussion of X-ray results, see Ref. 296.

References, p. 514

cussed by Becker et al.[293] who attributed the magnetic equivalence of the methyl groups in H_2(Copro-I) to tautomerism even at low temperatures, and the same explanation was used to explain the methine signals in H_2(Copro-III) studied by Abraham[6].

The non-equivalence of neighboring pyrrole rings due to slow N—H exchange was first observed by Storm et al.[74], who found two resolved lines for the β-protons in H_2(TPP) and deuteroporphyrin-IX dimethyl ester at low temperatures. The signals for H_2(TPP) coalesce at $-53°$C, and the tautomerism shows an extremely high kinetic isotope effect[74,75] when the inner protons are replaced by deuterium. The tautomerism was explained by a concerted mechanism (Fig. 7b). The much smaller isotope effect for H_2(Deut-IX-DME) was attributed to the decreased symmetry in the latter, which biases the different tautomer equilibria. This problem was critically reinvestigated[61,76], and Abraham et al.[76] attributed the enormous isotope effect observed by Storm to a neglect of the activation entropy. From the [13]Cmr coalescence of two different carbon atoms at the same temperature in TPP (N—[1]H) and TPP (N—[2]H), respectively, the isotope effect k_{1H}/k_{2H} on the tautomerization was determined to be 12.1 at $35°$C. This value is well within the expectation range for such an isotope effect and is compatible with an independent exchange mechanism for the two N-hydrogen atoms (Fig. 7a).

In the less symmetric chlorins, the N—H protons are considerably more localized on the nitrogen atoms of rings A and C adjacent to the reduced ring rather than on the nitrogen atoms of rings B and D[74,75,294,295]. The single broad N—H resonances at $\delta = -1.38$ p.p.m. in chlorin-e_6 trimethyl ester (14) splits into two peaks at $\delta = -1.35$ and -1.42 p.p.m. upon cooling[75], and at high magnetic field (Fig. 3b[80]). Similar effects have been noted for several other 7,8-chlorins with γ-substituents[74]. In the phorbins bearing an isocyclic five-membered ring, which may be regarded as substitution at the 6- and γ-positions, (Section 10.2.3), two separate N—H resonances about 1–2 p.p.m. apart are already observed at room temperature[80,127,138]. One N—H resonance occurs in the range usually observed for the N—H signals in chlorins ($\delta \approx -1.5$ p.p.m.) and bacteriochlorins ($\delta \approx -1$ p.p.m.), and is thus assigned to the N_1 proton. The other resonance assigned to the N_3 proton is considerably shifted by 1–2 p.p.m. to higher field, which must be related to the steric deformations introduced into ring C by ring E formation[128]. The implication of more or less localized N-protons in phorbins at N_1 and N_3 was recently proved by [15]Nmr data on pheophorbides of the a series[33] (Section 10.3.2).

From the [15]N—[15]N coupling constants (via the inner hydrogen atoms) and the ^{15}N chemical shifts, a decrease in the tendency to protonation in the order $N_1 \geqslant N_3 > N_2 > N_4$ was inferred[33], which corresponds well to X-ray crystal structure data for methyl pheophorbide-a[128]. Tautomeric structures similar to those in porphyrins have been advanced on the basis of

[1] Hmr data, with the participation of the lone pair on the nitrogen atom of ring D in the conjugated system in structures protonated at N_4[74]. ^{15}N—^{15}N coupling constants (via the inner hydrogen, Section 10.3.2) provide a direct measure for the importance of this mechanism[33]. Sharing of the proton is most prominent between N_2—N_3, it is less for N_1—N_4, even less for N_1—N_2 and N_3—N_4, and negligible for N_1—N_3 and N_2—N_4. The pronounced differences strongly favor a non-converted mechanism (Fig. 7a) and argue especially against hydrogen exchange between opposite N-atoms in the pheophorbides studied.

10.4.2.2. Conformation of β-pyrrole substituents

For certain metal complexes of OEP ($Tl^{III,11}$; Pb^{II}, $Sn^{II,190}$; Al^{III}, Ga^{III}, In^{III}, Ge^{IV}, $Sn^{IV,169}$; $Fe^{III,29}$) the methylene protons of the ethyl substituents give rise to a complex signal instead of the usually observed quadruplet. The complex pattern observed for the CH_2-signal in Tl^{III}(OEP) was analyzed by Abraham[186] as an ABR_3X, or better, an ABC_3X spectrum[11] and two different coupling constants of the A and B methylene protons to the central Tl^{III} ion (6.1 and 18.1 Hz) have been determined. Two effects are invoked in the interpretation. The rotation of the ethyl side-chains is slow ($\Delta H \sim 20$ kcal) and probably correlated with its next neighbor, and the central metal ion is in an out-of-plane position[296]. While the out-of-plane metal ion is not expected to affect the rotation to any considerable extent, it does increase the magnetic anisotropy of the (diastereotopic) methylene protons sufficiently for differentiation by [1] Hmr. Splitting of β-pyrrole methylene groups in (OEP) complexes is, therfore, an indication of an out-of-plane central metal ion, and/or of asymmetric ligation.

Hindered rotation between two distinct conformers was also advanced to account for split signals in some pheoporphyrins and pheophorbides with 2-CHOH—CH_3 substituents[132]. According to recent results of Brockmann and Trowitzsch[277], aggregation plays an additional role in this phenomenon as to enhance the magnetic anisotropy. These authors studied the effect in some detail for compounds related to the *Chlorobium* chlorophylls, which bear a 1-hydroxyethyl substituent at the 2 position (Section 10.2.8.2). Only one set of signals was found for the pure 2′ epimer, while the racemic mixture gave two sets of signals which merge upon dilution. The splitting can, therefore, be attributed to slow rotation of the side-chain combined with formation of diastereomeric aggregates. Recently, the slow rotation of a methyl group in ferric myoglobin was studied by n.m.r.[296a].

Only few n.m.r. data are available on the preferred conformation of β-pyrrole substituents in solution. From comparison of acyclic and cyclic conjugated substituents, a weaker interaction with the aromatic system was inferred for the former from [1] Hmr data (see Section 10.2.2), indicating the presence of non-coplanar conformers. The non-equivalence of ethyl CH_2-protons discussed above was interpreted to arise from preferred confor-

References, p. 514

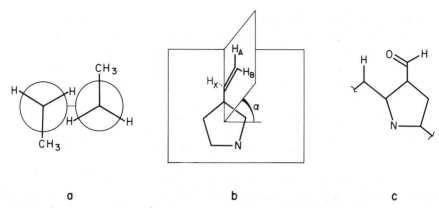

a   b   c

Fig. 8. Conformation of β-pyrrole substituents; only partial structures are shown. (a) Ethyl side-chains in H_2(OEP), schematic representation viewed parallel to the plane of the macrocycle, (from Ref. 296); (b) vinyl group in chlorophylls and (Proto-IX) derivatives, α = dihedral angle between the plane of the macrocycle and the plane of the vinyl group; and (c) 3-CHO in chlorophyll-b derivatives (from Ref. 263).

mations with the CH_3-groups of neighboring ethyl substituents out-of-plane and transoid to each other (Fig. 8a). Different conformations of vinyl substituents in heme and chlorophyll derivatives, respectively, can be inferred from [1]Hmr data. In chlorin-e_6 trimethyl ester (14), for example, the H_B resonance (see Fig. 8b) occurs at lower field than the H_A resonance, while the opposite is true for H_2(Proto-IX-DME) (13, see Section 10.2.1). H_B is closer to the macrocycle* than H_A; in case of a planar substituent it is therefore expected to be more strongly deshielded by the ring current, as observed in the chlorophyll derivatives. If the vinyl group is rotated, H_A remains almost co-planar with the macrocycle, while H_B is forced out-of-plane (Fig. 8b), thus occupying a less deshielding region, which accounts for the observed high field shift of H_B (relative to H_A) in H_2(Proto-IX) and related porphyrins. The conformation of the 2-formyl substituent in chlorophyll-b derivatives was investigated by using the pronounced magnetic anisotropy of the carbonyl group as a probe[73,263]. The data support a coplanar conformation with the aldehyde C = O oxygen atom oriented towards the α-methine position (Fig. 8c).

10.4.2.3. Conformation of meso-substituents
   On the basis of X-ray structures[48,112,297,298], [1]Hmr long-range shielding effects (see Section 10.2.3) and space-filling models, nonlinear substituents have been shown to assume a conformation in which the planes of the macrocycle and the substituent are nearly perpendicular to each other. The

---

* This is true both for the S-cis and the S-trans conformation of the vinyl group with respect to the α-H.

rotation of *meso*-phenyl substituents has received considerable attention. Atropisomers of *o*-substituted $H_2(TPP)$ derivatives[299,300−302] have been shown to be stable on the [1]Hmr time scale up to 198°C Ni(o-Me-TPP)[248], and a lower limit of 26 kcal has been estimated for the activation enthalpy of rotation. With bulky substituents in the *o*-phenyl positions, the atropisomers are stable even on prolonged refluxing in THF[300].

In unsubstituted $H_2(TPP)$, the two *o*-(as well as the two *m*-)protons are enantiotopic. This magnetic equivalency is removed, however, in metal complexes with an out-of-plane metal and/or with different axial ligands[29,194,201,303]. For $Ru^{II}(TPP)CO$, in which the non-equivalency is due to asymmetric ligation, an activation enthalpy of 18 kcal/mole has been determined from the coalescence of the *o*-phenyl proton resonances[183].

10.4.3. Stereochemistry

10.4.3.1. The macrocycle

The stereochemistry of the porphyrin macrocycle has been studied extensively by X-ray diffraction. [See Hoard[298] (Chapter 8) and Fleischer[48] for reviews.] These crystal structure studies show the macrocycle system to be fairly flexible. Planarity of the macrocycle is rather an exception and its shape has been described as domed, ruffled or roof shaped[169,298]. Large deviations from planarity are especially observed in $H_2(TPP)$ derivatives[48]. The pyrrole rings are maximally twisted about 28° out-of-plane defined by the four central nitrogen atoms in the TPP dication[181].

Although about 100 porphyrin X-ray crystal structures have been published, relatively little is known about the stereochemistry and the rigidity of the macrocycle in solution. A first attempt to determine the solution structure of chlorophyll directly by the use of LIS shift reagents[279] gave results consistent with the X-ray parameters, but obviously the accuracy of the method is limited and serves to detect only relatively large deviations from planarity. If it is assumed that the crystal structure represents a conformation that is easily accessible in solution, deviations from planarity in the solid may primarily reflect the response to crystal packing of the macrocycle rather than its actual solution conformation. This is indicated, for example, by the different conformations assumed by $H_2(TPP)$ in the triclinic[183] and tetragonal[297] crystal forms, as well as by the very anisotropic thermal ellipsoids deduced from X-ray diffraction, which show a pronounced out-of-plane mobility of most atoms in the crystal[48,297]. Ring current calculations usually assume the aromatic system to be planar. The consequences of pronounced deviations from planarity have been discussed only for $H_2(TPP)$ in solution[69], which is well-known for its atypical behavior as compared to naturally-occurring porphyrins and the usual model compounds with β-substituents.

The strong incremental ring current shifts observed as a result of structural modifications that increase steric hindrance in the molecule may be taken as

References, p. 514

an indication that the assumption of an essentially planar macrocyclic conjugation system in solution for most porphyrins is justified. Many of the
examples in the foregoing sections show, moreover, that the conjugation
pathway tends to make adjustments that serve to circumvent steric obstacles
effectively. While the conformational analysis and assignment of the basic
macrocycle chosen as the reference is therefore somewhat ambiguous, some
reasonably successful attempts have been made to correlate incremental
n.m.r. shifts with conformational changes of the macrocycle in cases where
marked deviations in conformation have been observed.

The [1]Hmr spectra of N-mono-substituted porphyrins have been interpreted in terms of the effects of steric hindrance[140]. A solution structure
was inferred in which the N-substituted pyrrole-ring is twisted considerably
out-of-plane; the neighboring rings are twisted to a lesser extent in the
opposite direction; and, the opposite ring remains essentially in-plane. The
X-ray structure of N-ethoxycarbonyl-OEP[141] provides convincing support
for the first two conclusions. The inclination of 11.7° observed for the
opposite pyrrole ring, which is intermediate between the N-substituted
(19.1°) and its neighboring rings (4.6°, 2.2°) contradicts the third conclusion, but this deviation may arise at least in part from packing[141]. Similar
sterically induced conformation changes have been invoked in the interpretation of the n.m.r. spectra of mono-, di- and tri-N-alkyl porphyrins and their
cations, and the results have been interpreted in terms of the conformation
and the rigidity of the ring system[142]. In mono-meso-substituted porphyrins, the pronounced decrease of the ring-current and the deshielding of the
methine proton opposite to the substituent can be rationalized on the basis
of a structure folded like a peaked roof along the axis connecting these two
positions (see Section 10.2.3). A recent X-ray analysis[112] of meso-benzoy-
loxy-octaethylporphyrin provided some support for a folded structure, for
the macrocycle is considerably folded at the substituted methine position,
and, to a lesser extent, at the opposite one. In addition, the entire macrocycle is stretched along the fold, and the inner cavity deformed into a rectan-
gle. A similar roof-shaped structure proposed[169,170] mainly on n.m.r. arguments for α,γ-dimethyl-β,γ-porphodimethenes (63) in solution was supported as well by a crystal structure determination[171]. The conformational
changes resulting from δ-substitution are especially pronounced in the more
flexible reduced ring D of 7,8 chlorins, which provides independent proof of
the 7,8-trans configuration of the hydrogen atoms of the pyrroline
ring D[124].

10.4.3.2. Metalloporphyrins

Two other types of stereochemical effects become important in metalloporphyrins. The metal ion can be out-of-plane, and it can be ligated in
various distinct ways (Fig. 9). The position of the central metal ion with
respect to the macrocycle is determined by the ionic radius of the metal ion

(a)

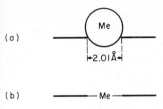

(b)

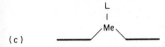

(c)

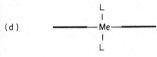

(d)

Fig. 9. Stereochemistry of metalloporphyrins. (a) Metal ion considerably larger than 2.01 Å diameter, e.g., Pb^{II} (Refs. 109,191); (b) metal ion ⩽2.01 Å diameter, no axial ligands; (c) metal ion ⩽2.01 Å diameter, one axial ligand; and (d), metal ion ⩽2.01 Å diameter, 2 axial ligands. (See also Tables 21 and 25.) Schematic representation, viewed parallel to the plane of the macrocycle.

and the mode of ligation. The diameter of the central cavity of the macrocycle is fairly restricted[48,298] and the porphyrin cannot easily accomodate metal ions with an ionic radius significantly larger than 2.01 Å (Fig. 9a). Below this critical ionic radius, the type of coordination essentially determines whether the metal is in-plane or out-of-plane. Square planar and tetragonal bipyramidal configuration have the metal ion in-plane (Fig. 9b,d), while in the tetragonal pyramidal configuration (Fig. 9c) the axial ligand forces the metal out-of-plane. (See, for example, Ref. 17 and citations therein.) A qualitative indication for an out-of-plane metal ion is the splitting observed for the o-phenyl proton signals in (TPP) metal complexes, and for the multiplicity of the methylene proton signals in the ^{1}Hmr spectra of H$_2$(OEP) complexes (Section 10.4.2); in H$_2$(TPP), the splitting is due to hindered rotation[201,302]. Although the methylene protons of the (OEP) complexes are diastereotopic per se, their magnetic non-equivalence is enhanced by the out-of-plane position of the metal and by a correlated rotation of neighboring ethyl groups[186]. A quantitative estimate of the extent of metal ion displacement is possible from the ring-current induced ^{1}Hmr chemical shifts of ligand protons bound in the metal axial positions[16,17,51,52]. Assuming pyridine to be ligated at right angles to the macrocycle in metalloporphyrin-pyridinates, the incremental shift of the 4-pyridine proton can be used to estimate the distance of the ligand from the porphyrin plane. With the metal-N distance and the pyridine geometry well established from known pyridine compounds, the apparent N-to-metal ion distance deduced from ^{1}Hmr makes an estimate of the metal displacement

References, p. 514

from the macrocycle possible[17]. In Co^{III} (Meso-IX-DME), the metal ion is essentially in plane, while 'Zn(TPP) and Mg(TPP) have metal ions that are out-of-plane by 0.3—0.5 Å and 0.7—0.8 Å, respectively. Ring current arguments can also be used to estimate the plane-to-plane distances in layered structures with two or more metalloporphyrins parallel to each other (see Section 10.2.8.3).

With metal ions that can assume a paramagnetic state, the ligand field may determine the spin state, which thus can serve as a probe for the metal coordination type (see Section 10.2.8.5 for leading references). Axial ligation of square-planar, diamagnetic Ni-etioporphyrin leads to a paramagnetic tetragonal pyramidal complex[10], and similar behavior is found for Fe^{II} porphyrin complexes when one of the two axial ligands of low-spin Fe^{II} complexes is removed. In Fe^{III} complexes, both the five and six-coordinate states are paramagnetic, but they show differences characteristic of their ligand field. In most cases, even with different axial substituents, a clearly defined spin state is present, but a high-spin, low-spin mixture is often observed for azidohemins[22,246].

10.4.3.3. Non-centrosymmetric stereoisomerism

An interesting new case of stereoisomerism in porphyrins was recently observed by Hudson et al.[219]. The diporphinato-trimercury complex (82) of Etio-I was shown to exist in two diastereomeric forms. In the racemic forms, the two porphyrins are ligated 'face to face', in the meso form they are ligated 'face to back' (Fig. 10). Obviously, isomerism of this kind is not dependent on hindered rotation of the porphyrins around their common axis, and similar isomers are possible for all porphyrins that do not possess a

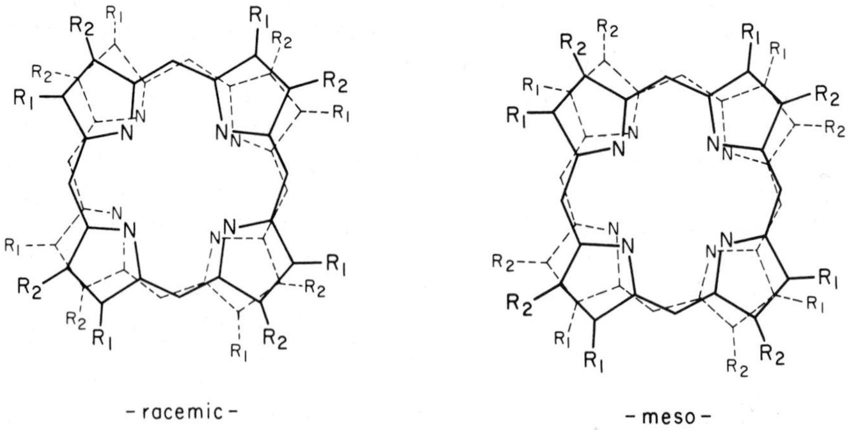

- racemic -   - meso -

Fig. 10. Stereoisomerism in porphyrin dimers with structures similar to (82). For details, see text.

σ_ν plane, as for instance, Copro-I and -III, but not Copro-II or -IV. This type of isomerism is generally possible in all diporphinato complexes and could be valuable in studying porphyrin—ligand exchange reactions in these molecules. Such isomerism is likewise a possibility in porphyrin π—π dimers present in concentrated solutions, and may be an alternative explanation (in addition to lateral displacement)[6] for the additional fine structure observed in the [1]Hmr spectra of some of porphyrin self-aggregates.

In phenyl-substituted H_2(TPP) derivatives, the hindered rotation around the methine—phenyl bond leads to the possibility of atropisomerism[302]. meso-Tetra-o-tolyl—porphyrin shows a complex pattern for the o-CH_3 resonances, which was attributed by Walker[299] to originate from a statistical mixture of the four possible atropisomers. Recently, the four meso-tetra-o-aminophenyl—porphyrins have been separated and characterized by [1]Hmr[300]. Both the tolyl and o-amino H_2(TPP) isomers are stable at room temperature, and even at 180°C no indication of line broadening is observed[299] (see Section 10.4.2).

10.4.3.4. Asymmetric carbon atoms

In fully unsaturated porphyrins all macrocyclic carbon atoms are sp^2 hybridized, but asymmetric C-atoms can result from introduction of substituents, or formed by reduction of the macrocycle system to the chlorin or porphodimethene states among others. Pheoporphyrins with one asymmetric C-atom of defined configuration were first described by Wolf et al.[126,138]. The low signal-to-noise ratio of their ORD spectra made the determination of enantiomeric purity difficult, but this problem can be overcome by an absolute determination of enantiomeric purity of the compounds by [1]Hmr[126]. In a chiral enantiomeric environment, previously enantiotopic (i.e., indistinguishable by [1]Hmr), protons become diastereotopic and thus, in principle, distinguishable by n.m.r.[304a]. This effect was studied in a variety of compounds in chiral aromatic solvents (for leading references, see Ref. 304). The pronounced aggregation exhibited by porphyrins made it possible to use the conventional achiral solvent ($CDCl_3$) and adding a chiral porphyrin as co-solute, which form diastereotopic aggregates with the porphyrin enantiomers. The n.m.r. spectra of the chiral pheoporphyrin showed sufficient differences in the methine region in the presence of excess pyromethylpheophorbide-a (with the natural 7S, 8S configuration) to distinguish between the R and S form of the pheoporphyrins, and thus to determine the enantiomeric purity by integration of the appropriate signals[126]. In a similar case, the pronounced aggregation of 2-(α-hydroxyethyl)pheoporphyrins caused by hydrogen-bonding was shown[213] to be responsible for the splitting observed for most of the [1]Hmr signals[132]. This splitting occurs only for 2a-epimeric mixtures, but not for the pure epimers[213]. Splitting thus reflects diastereomeric aggregates, in which the chemical shift differences are enhanced by specific aggregation of the systems and by hindered rotation at the 2—2a bond (see Section 10.4.2.2).

The stereochemistry of peripherally-reduced porphyrins has been studied extensively because of their relevance to the chlorophylls. The 7,8-*trans* figuration of the chlorophylls suggested by the racemization experiments of Fischer and Gibian[305] was proved by chromic acid degradation[306,307], and by the small *transoid* vicinal coupling constant ($J \leqslant 2$ Hz) of the 7,8 protons[13]. In the model compound H_2(OEC) the 'extra' hydrogen atoms in the reduced pyrroline ring are essentially equivalent and thus show no coupling, and in some other pheophorbides with unnatural configuration, the 7,8 coupling constants were not observable. In these cases, however, the *cis* and *trans* isomers can be differentiated by other differences in their n.m.r. spectra[126]. In the *cis*-H_2(OEC), the ring current is somewhat reduced by steric hindrance in the *cis* ethyl groups as compared to the *trans* epimer[65,308], and the altered coupling pattern of the extra hydrogen atoms with the neighboring methylene protons indicates a change in the conformation of the reduced ring[126]. The differences between the epimers are much more pronounced in 7,8-*cis* as compared to 7,8-*trans*-mesopyromethylpheophorbide[126]. In the 7,8-*cis* compound, all proton signals, especially those from protons in the vicinity of the reduced ring D, are shifted to higher field. An additional characteristic difference in the pyropheophorbide compounds is the increased anisochrony of the 10-methylene* protons in the 7,8-*cis* ($\Delta(H_\beta - H_\alpha)$ = 0.38 p.p.m.) as compared to the 7,8-*trans* epimer ($\Delta(H_\beta - H_\alpha)$ = 0.11 p.p.m.)[126]. Obviously, the shielding effect of the *transoid* 7,8-alkyl side-chains in the *trans*-pheophorbide are partially compensated. An indirect approach to the relative configuration of chlorins is possible by analysis of the induced conformational changes that result from introduction of a large substituent at a neighboring *meso*-position[124]. Because of the steric interaction of the δ-substituent with the 8-CH_3 group in δ-Cl-chlorin-e_6 trimethyl ester, ring D is tilted out of the macrocycle plane, and the high field shift of both the 8-CH_3 and the 7-H signals thus proves their *cisoid* relationship. Similar changes have recently been observed by perturbation of the γ-position in peripheral complexes of pheophorbides[80] (see Section 10.2.8.4).

In cyclopropano-chlorins and -bacteriochlorins[66] the chlorin substituents are necessarily in the *cis* configuration. The endo- and exo-positions of the cyclopropane substituents show characteristic differences in chemical shift arising from the anisotropy of the ring current shift. Exo-substituents are essentially in the plane of the macrocycle and are thus strongly deshielded, while the endo-substituents are above the plane and resonate at substantially higher field. In the case of protons, a further additional assignment is possible on the basis of coupling constants with the chlorin protons[66].

---

* α and β are used as in terpenes, α designating a substituent below, β a substituent above the plane in the structure shown the conventional way.

Most chlorophylls possess an additional asymmetric center at C-10 in the isocyclic ring E, whose configuration can be linked to that at C-7 and C-8 by means of incremental substituent shifts[138]. For chlorophyll-*a*, the *transoid* configuration of the 7 and 10 substituents was suggested on the basis of the high field shift of the 7-H signal (relative to that of the 8-H) due to the shielding by carbonyl groups in the 10-COOCH$_3$ substituent. This interpretation is somewhat ambiguous because the incremental shift of the anisotropic ester group is difficult to estimate, and because only one of the epimers is readily available in pure form. The *transoid* configuration was proved by a combined ^{1}Hmr and spectropolarimetric[307a] investigation of diastereomeric 10-alkoxy pheophorbides[79,138]. In these compounds, the conformation of the C-10 atom is stabilized, and the two C-10 epimers can thus be studied separately.

With the assumption that incremental shifts are more pronounced for *cisoid* than for *transoid* substituents, the relative position of two substituents can then be determined by n.m.r. as shown in the following example (Fig.

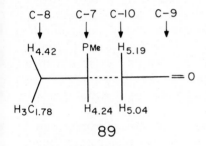

89

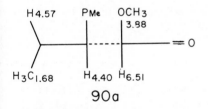

90a

90b

Fig. 11. ^{1}Hmr chemical shifts (δ[p.p.m.] from TMS) of pyromethylpheophorbide-*a* (89), and its enantiomeric (R,S)-methoxy derivatives (90a,b). Schematic representation of the periphery of rings D and E, viewed parallel to the macrocycle (see Ref. 79).

References, p. 514

11). Substitution of one of the 10-H protons in pyromethylpheophorbide-a (89) by —OCH$_3$ causes shifts of the 7-H proton signal by —0.16 (—0.35) p.p.m., the 8-H quadruplet by —0.15 (—0.12) p.p.m., and the 8-CH$_3$ doublet by +0.10 (—0.13) p.p.m. in the two epimers (90a) and (90b). Obviously, the 10-OCH$_3$ substituent has a pronounced deshielding effect on the neighboring ring protons. This effect is expected to be more pronounced for *cisoid* than for *transoid* substituents, and from the relative magnitude of the shifts *cisoid* configuration to the 8-H can be inferred in the epimer (90a), *cisoid* configuration to the 7-H and 8-CH$_3$ in the epimer (90b). The different incremental shifts observed for both positions of the 10-OCH$_3$ groups (i.e., the positive increment for the 8-CH$_3$ resonance in 90a), indicate that the substituent shifts are accompanied by pronounced shifts due to other (conformational) reasons, clearly demonstrating that an unambiguous analysis requires data on both epimers. In the same way as the configuration at C-10 affects the chemical shifts of the substituents at C-7 and C-8, the configuration at C-7 affects (but to a lesser extent) the chemical shift of the substituents at C-10. Thus, in the compounds cited above the 10-OCH$_3$ is more strongly deshielded when *cisoid* to the propionic ester side-chain than when it is *transoid* to it, and the stereochemical assignment of the non-equivalent 10-methylene protons in pyropheophorbides[79] and in the chlorophyll 'prime' epimers (i.e., Chl-a', Section 10.2.8.2)[215] can be carried out in the same way. Similar relationships have been observed for the pheophorbide-b series as well[309], and the effects are characteristic of a wide variety of 10-alkoxy pheophorbides[130,138].

The configuration of C-9 in 9-desoxo-9-hydroxy-pheophorbides can be linked to that at C-10 in an analogous manner, and the reciprocal influence can again be detected by a careful analysis of the n.m.r. signals[125,130]. In the case of the diastereomeric 9-desoxo-9-hydroxy-methylpheophorbides-a (91a—c) (Table 12) the relative configuration at C-9 and C-10 is also deducible from the 9-H, 10-H coupling constants (7 Hz for 9,10-*cis* as compared to ≤2 Hz for 9,10-*trans*)[125].

In a different approach, configuration correlations can be made by studying specific intramolecular interactions. Hydrogen-bonding of a 9-OH group with substituents at C-10 leads to a broadening of the 9-H signals from residual HCOH coupling, which is removed upon deuterium exchange of the 9-OH group[125,130]. This broadening is obviously characteristic of a hydrogen bond of intermediate strength, for both weak and strong hydrogen bonds are expected to yield sharp signals, with the latter showing a distinct coupling constant[310]. Gradual changes in the strength of the hydrogen bonds as explified in the [1]Hmr spectra can be expected to be correlated with the O—H stretch frequency in the infrared spectra[125,130]. In favorable cases, hydrogen-bonding is not only observed to occur between the 9-OH group and the neighboring groups, but also with the carbonyl function in the 7-propionic side-chain (91b)[125,136].

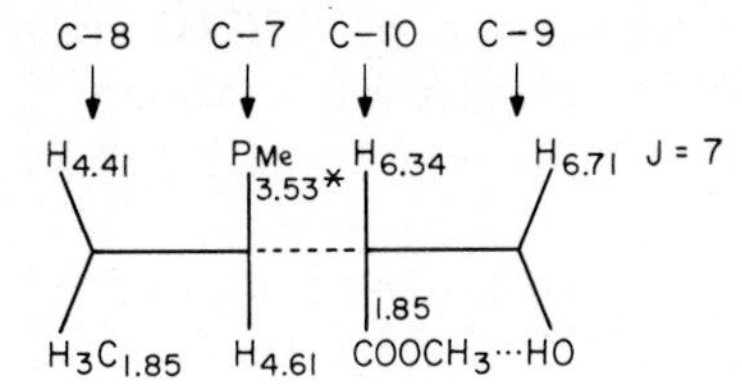

Fig. 12. [1]Hmr chemical shifts (δ[p.p.m.] from TMS) of the diastereomeric 9-desoxo-9(R,S)-hydroxy-10(R,S)-methylpheophorbides-a (91a,b,c). Schematic representation as in Fig. 11. (See Ref. 125.)

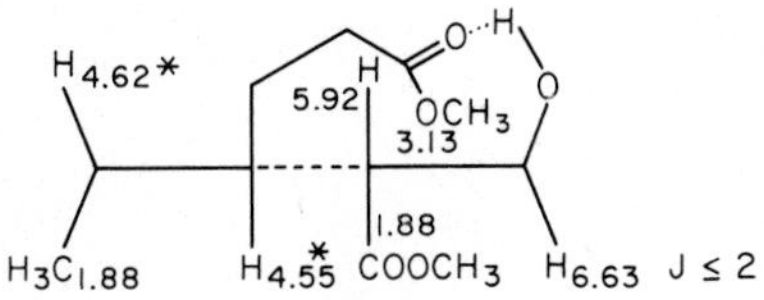

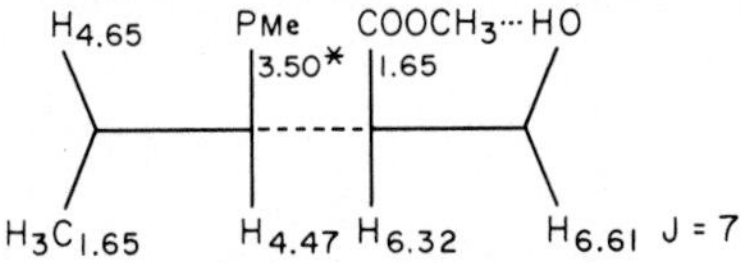

10.4.4. Miscellaneous

[1]Hmr has been widely used in mechanistic studies by monitoring the ^{1}H/^{2}H exchange with the medium. Pertinent examples include the nucleophilic exchange of the methine protons in acidic media[56,78,144,163,207,209] and during and after metalation[135,311], the exchange of benzylic protons[79,127,135,138,215] and substituents[130], the phlorin-chlorin and related equilibria[57,93,103,121,123,164], the photochemical[90,126,173,311a] and electrochemical[312] reduction of porphyrins and redox processes in hemins[313].

[1]Hmr has often been used as a decisive tool to distinguish between isomers. For the application of symmetry arguments, the reader is referred to standard monographs[1] as well as to examples cited in Section 10.2.2, 10.2.3, 10.2.8, and 10.4.3.

The application of stable isotopes (^{2}H, ^{13}C, ^{15}N) in conjuction with n.m.r. measurements is of increasing interest in biosynthetic studies. The method is superior to radioactive labeling because the position and the amount of the label can be measured directly, provided the n.m.r. resonances are assigned. In a pair of mirror image experiments, [1]H$_2$O plus succinic

References, p. 514

acid-$^2H_4$ vs. ^2H_2O + succinic acid-$^1H_4$, the incorporation of $^1H(^2H)$ from the succinic acid and the medium into bacteriochlorophyll-a was studied[314], and recently the ^{13}C pathway in Proto-IX biosynthesis was investigated by ^{13}Cmr[35,261,262] (see also Chapter 3).

The photo-oxidation of chlorophyll with quinones has been studied[315] by chemical induced dynamic nuclear polarization[316] of the quinone resonances, and selective ^1Hmr line-broadening as a result of triplet energy transfer in photo-excited methyl pyropheophorbide-a makes possible the assignment of hyperfine coupling constants in the triplet[317].

Acknowledgments

H.S. acknowledges the grant of a Research Stipendiate from the Deutsche Forschungsgemeinschaft, Bonn-Bad Godesberg. We thank Dr. Thomas R. Janson for helpful discussions in preparing this manuscript. We are indebted to all who have communicated unpublished work.

References

1. L.M. Jackmann, 'Applications of NMR Spectroscopy in Organic Chemistry', 2nd ed., Pergamon, New York (1969); J.A. Pople, W.G. Schneider, and H.J. Bernstein, 'High Resolution NMR', McGraw-Hill, New York (1959); E.D. Becker, 'High Resolution NMR, Theory and Chemical Applications', Academic Press, New York (1969); F.A. Bovey, 'NMR Spectroscopy', Academic Press, New York (1969); J.W. Emsley, J. Feeney, and L.H. Sutcliffe, 'High Resolution NMR Spectroscopy', Pergamon, New York (1965—66).

2. (a) E.D. Becker and R.B. Bradley, J. Chem. Phys., 31, 1413 (1959).

2. (b) J. Ellis, A.H. Jackson, G.W. Kenner, and J. Lee, Tetrahedron Lett., 23 (1960).

3. R.J. Abraham, Mol. Phys., 4, 145 (1961).

4. W.S. Caughey and W.S. Koski, Biochemistry, 1, 923 (1962).

5. K.M. Smith, Quart. Rev., 25, 31 (1971).

6. R.J. Abraham, P.A. Burbidge, A.H. Jackson, and D.B. Macdonald, J. Chem. Soc., 620 (1966).

7. R.J. Abraham, P.A. Burbridge, A.H. Jackson, and G.W. Kenner, Proc. Chem. Soc., London, 134 (1963).

8. R.J. Abraham, G.H. Barnett, E.S. Bretschneider, and K.M. Smith, Tetrahedron, 29, 553 (1973).

8. (a) R.J. Abraham, A.H. Jackson, G.W. Kenner, and D. Warburton, J. Chem. Soc., 853 (1963).

8. (b) R.J. Abraham, A.H. Jackson, and G.W. Kenner, J. Chem. Soc., 3468 (1961).

9. R.J. Abraham, G.E. Hawkes, and K.M. Smith, J. Chem. Soc., Chem. Commun., 401 (1973).

10. R.J. Abraham and P.F. Swinton, J. Chem. Soc. C, 903 (1969).

11. R.J. Abraham, G.H. Barnett, and K.M. Smith, J. Chem. Soc., Perkin Trans. 1, 2142 (1973).

12. H.H. Inhoffen, J.W. Buchler, and P. Jäger, Fortschr. Chem. Org. Naturst., **26**, 785 (1968).
13. G.L. Closs, J.J. Katz, F.C. Pennington, M.R. Thomas, and H.H. Strain, J. Am. Chem. Soc., **85**, 3809 (1963).
14. H.A.O. Hill, P.J. Sadler, R.J.P. Williams, and C.D. Barry, Ann. N.Y. Acad. Sci., **206**, 247 (1973).
15. J.E. Maskasky and M.E. Kenney, J. Am. Chem. Soc., **93**, 2060 (1971).
16. C.B. Storm and A.H. Corwin, J. Org. Chem., **29**, 3700 (1964).
17. C.B. Storm, J. Am. Chem. Soc., **92**, 1423 (1970).
18. J.E. Maskasky, J.R. Mooney, and M.E. Kenney, J. Am. Chem. Soc., **94**, 2132 (1972).
19. J.E. Maskasky and M.E. Kenney, J. Am. Chem. Soc., **95**, 1443 (1973).
20. A. Kowalsky, Biochemistry, **4**, 2382 (1965); A. Kowalsky, 'Hemes and Hemoproteins,' 529 (1966).
21. W.D. Philips, in 'NMR of Paramagnetic Molecules', Eds. G.N. LaMar, W. DeW. Horrocks Jr., and R.H. Holm, Academic Press, New York (1973), p. 422.
22. K. Wüthrich, Struct. Bonding Berlin, **8**, 53 (1970).
23. R.J. Kurland, R.G. Little, D.G. Davis, and C. Ho, Biochemistry, **10**, 2237 (1971).
24. T.R. Janson, L.J. Boucher, and J.J. Katz, J. Am. Chem. Soc., **92**, 2725 (1970).
25. R.G. Shulman, S.H. Glarum, and M. Karplus, J. Mol. Biol., **57**, 93 (1971).
26. G.N. LaMar and F.A. Walker, J. Am. Chem. Soc., **95**, 6950 (1973).
27. G.N. LaMar and F.A. Walker, J. Am. Chem. Soc., **95**, 1782 (1973).
28. G.N. LaMar and F.A. Walker, J. Am. Chem. Soc., **95**, 1790 (1973).
29. G.N. LaMar, G.R. Eaton, R.H. Holm, and F.A. Walker, J. Am. Chem. Soc., **95**, 63 (1973).
30. G.C. Levy and G.L. Nelson, '^{13}C Nuclear Magnetic Resonance for Organic Chemists', Wiley, New York (1972); J.B. Stothers, '^{13}C NMR Spectroscopy', Academic Press, New York (1972).
31. H. Günther, H. Schmickler, H. Königshofen, K. Recker, and E. Vogel, Angew. Chem., **85**, 261 (1973); R. du Vernet and V. Boekelheide, Proc. Nat. Acad. Sci. U.S.A., **71**, 2961 (1974).
32. R.J. Abraham, G.E. Hawkes, and K.M. Smith, J. Chem. Soc., Perkin Trans. 2, 627 (1974).
33. S.G. Boxer, G.L. Closs, and J.J. Katz, J. Am. Chem. Soc., **96**, 7058 (1974).
34. D. Doddrell and W.S. Caughey, J. Am. Chem. Soc., **94**, 2510 (1972).
35. A.R. Battersby, G.L. Hodgson, M. Ihara, E. McDonald, and J. Saunders, J. Chem. Soc., Perkin Trans. 1, 2923 (1973).
36. N.A. Matwiyoff and B.F. Burnham, Ann. N.Y. Acad. Sci., **206**, 365 (1973); C.E. Strouse, V.H. Kollman, and N.A. Matwiyoff, Biochem. Biophys. Res. Commun., **46**, 328 (1972).
37. J.J. Katz and T.R. Janson, Ann. N.Y. Acad. Sci., **206**, 579 (1973).
38. A. Carrington and A.D. McLachlan, 'Introduction to Magnetic Resonance', Harper and Row, New York (1967).
39. L. Pauling, J. Chem. Phys., **4**, 673 (1936).
40. T.R. Janson, 'The Ring-Current Effect of the Phthalocyanine Ring', Dissertation, Case Western Reserve University, Univ. Microfilms, Ann Arbor, 72-51, (1971).
41. R.C. Haddon, V.R. Haddon, and L.M. Jackman, Fortschr. Chem. Forsch., **16**, 103 (1971).
42. J.A. Pople, J. Chem. Phys., **24**, 1111 (1956).
43. J.S. Waugh and R.W. Fessenden, J. Am. Chem. Soc., **79**, 846 (1957).
44. C.E. Johnson, Jr. and F.A. Bovey, J. Chem. Phys., **29**, 1012 (1958).
45. F. London, J. Phys. Radium, **8**, 397 (1937).
46. J.A. Pople, Mol. Phys., **1**, 175 (1958).
47. R. McWeeny, Mol. Phys., **1**, 311 (1958).

48. E.B. Fleischer, Acc. Chem. Res., **3**, 105 (1970).

48. (a) G.G. Hall and A. Hardisson, Proc. Roy. Soc. London, Ser. A, **268**, 328 (1962).

48. (b) V.M. Mamaev, G.V. Ponomarev, S.V. Zenin and R.P. Evstigneeva, Teor. Eksp. Khim. **6**, 60 (1970).

49. J.H. Fuhrhop, Angew. Chem., **86**, 363 (1974).

50. J.F.M. Oth, H. Baumann, J.M. Gilles, and G. Schröder, J. Am. Chem. Soc., **94**, 3498 (1972).

51. J.J. Katz, H.H. Strain, D.L. Leussing, and R.C. Dougherty, J. Am. Chem. Soc., **90**, 784 (1968).

52. T.R. Janson, A.R. Kane, J.F. Sullivan, K. Knox, and M.E. Kenney, J. Am. Chem. Soc., **91**, 5210 (1969).

53. F. Sondheimer, Proc. Roy. Soc. London, Ser. A, **297**, 173 (1967).

54. I.C. Calder, P.J. Garratt, and F. Sondheimer, Chem. Commun., 41 (1967).

55. A.M. Shulga, Biofizika, **18**, 32 (1973); Transl. **18**, 30 (1973).

56. M.J. Broadhurst, R. Grigg, and A.W. Johnson, Chem. Commun., 1480 (1969).

57. M.J. Broadhurst, R. Grigg, and A.W. Johnson, Chem. Commun., 807 (1970).

58. C.O. Bender, R. Bonnett, and R.G. Smith, J. Chem. Soc. D, 345 (1969).

59. J.J. Katz, in 'Inorganic Biochemistry,' Ed. G. Eichhorn, Elsevier (1973), p. 1022.

60. K.N. Solov'ev, V.A. Mashenkov, A.T. Gradyushko, A.E. Turkova, and V.P. Lezina, J. Appl. Spectrosc. USSR, **13**, 1106 (1970).

61. A.M. Ponte Goncalves and R.A. Gottscho, J. Am. Chem. Soc., in press.

62. P.S. Clezy, V. Diakiw, and A.J. Liepa, Aust. J. Chem., **25**, 201 (1972).

63. T.S. Srivastava, C.-P. Hrung, and M. Tsutsui, J. Chem. Soc., Chem. Commun., 447 (1974).

64. D. Dolphin, R.H. Felton, D.C. Borg, and J. Fajer, J. Am. Chem. Soc., **92**, 743 (1970).

65. H.W. Whitlock, Jr., R. Hanauer, M.Y. Oester, and B.K. Bower, J. Am. Chem. Soc., **91**, 7485 (1969).

66. H.J. Callot, Tetrahedron Lett., 1011 (1972).

67. T.R. Janson, H. Scheer, and J.J. Katz, unpublished results.

68. T.R. Janson and J.J. Katz, J. Magn. Reson., **6**, 209 (1972).

69. R.J. Abraham, G.E. Hawkes, and K.M. Smith, Tetrahedron Lett., 71 (1974).

70. A.D. Adler, F.R. Longo, J.D. Finarelli, J. Goldmacher, J. Assour, and L. Korsakoff, J. Org. Chem., **32**, 476 (1967).

71. H.H. Inhoffen, J.H. Fuhrhop, H. Voight, and H. Brockmann Jr., Ann. Chem., **695**, 133 (1966).

72. H.H. Inhoffen, G. Klotmann, and G. Jeckel, Ann. Chem., **695**, 112 (1966).

73. G. Jeckel, Dissertation, Technische Hochschule Braunschweig, West Germany (1967).

74. C.B. Storm, Y. Teklu, and E.A. Sokolski, Ann. N.Y. Acad. Sci., **206**, 631 (1973).

75. C. Storm and Y. Teklu, J. Am. Chem. Soc., **94**, 1745 (1972).

76. R.J. Abraham, G.E. Hawkes, and K.M. Smith, Tetrahedron Lett., 1483 (1974).

77. R.B. Woodward, Pure Appl. Chem., **2**, 383 (1961).

78. R.B. Woodward and Škarić, J. Am. Chem. Soc., **83**, 4676 (1961).

79. H. Wolf, H. Brockmann Jr., H. Biere, and H.H. Inhoffen, Ann. Chem., **704**, 208 (1967).

80. H. Scheer and J.J. Katz, unpublished results.

80. (a) H. Scheer and J.J. Katz, J. Am. Chem. Soc., in press.

81. P.S. Clezy and A.J. Liepa, Aust. J. Chem., **24**, 1027 (1971).

82. R. Chong and P.S. Clezy, Aust. J. Chem., **20**, 951 (1967).

83. P.S. Clezy and N.W. Webb, Aust. J. Chem., **25**, 2217 (1972).

84. P.S. Clezy, A.J. Liepa, and N.W. Webb, Aust. J. Chem., **25**, 2687 (1972).

85. P.S. Clezy, A.J. Liepa, and N.W. Webb, Aust. J. Chem., **25**, 1991 (1972).

86. H.H. Inhoffen, H. Brockmann Jr., and K.M. Bliesener, Ann. Chem., **730**, 173 (1969).

87. H. Fischer, and A. Kirstahler, Hoppe-Seyler's Z. Physiol. Chem., **198**, 43 (1931); H. Fischer and G. Wecker, Hoppe-Seyler's Z. Physiol. Chem., **272**, 1 (1941).

88. P.S. Clezy and V. Diakiw, J. Chem. Soc., Chem. Commun., 453 (1973).

89. M.T. Cox, A.H. Jackson, G.W. Kenner, S.W. McCombie, and K.M. Smith, J. Chem. Soc., Perkin Trans. 1, 516 (1974).

90. H. Scheer and H. Wolf, Ann. Chem., 1741 (1973).

91. G.L. Collier, A.H. Jackson, and G.W. Kenner, J. Chem. Soc. C, 66 (1967).

92. A.H. Jackson, G.W. Kenner, and G.S. Sach, J. Chem. Soc. C, 2045 (1967).

93. A.H. Jackson, G.W. Kenner, and K.M. Smith, J. Chem. Soc. C, 302 (1968).

94. G.W. Kenner, S.W. McCombie, and K.M. Smith, Ann. Chem., 1329 (1973).

95. J.A.S. Cavaleiro, G.W. Kenner, and K.M. Smith, J. Chem. Soc., Perkin Trans. 1, 2478 (1973).

96. A.H. Jackson, G.W. Kenner, and J. Wass, J. Chem. Soc. Perkin Trans. 1, 480 (1974).

97. T.T. Howarth, A.H. Jackson, and G.W. Kenner, J. Chem. Soc., Perkin Trans. 1, 502 (1974).

98. G.W. Kenner, S.W. McCombie, and K.M. Smith, J. Chem. Soc., Perkin Trans. 1, 527 (1974).

98. (a) M.T. Cox, T.T. Howarth, A.H. Jackson, and G.W. Kenner, J. Chem. Soc., Perkin Trans. 1, 512 (1974).

99. R.V.H. Jones, G.W. Kenner, and K.M. Smith, J. Chem. Soc., Perkin Trans. 1, 531 (1974).

100. H.J. Callot, Tetrahedron, **29**, 899 (1973).

101. H.J. Callot, Bull. Soc. Chim. Fr., 1492 (1974).

102. P. Burbidge, G.L. Collier, A.H. Jackson, and G.W. Kenner, J. Chem. Soc. C, 930 (1967).

103. A.W. Johnson and D. Oldfield, Tetrahedron Lett., 1549 (1964).

104. R. Bonnett and G.F. Stephenson, J. Org. Chem., **30**, 2791 (1965).

105. H.H. Inhoffen and A. Gossauer, Ann. Chem., **723**, 135 (1969).

106. P.S. Clezy and V. Diakiw, Aust. J. Chem., **24**, 2665 (1971).

106. (a) R. Bonnett, M.J. Dimsdale, and G.F. Stephenson, J. Chem. Soc. C, 564 (1969).

107. P.S. Clezy and A.J. Liepa, Aust. J. Chem., **23**, 2477 (1970).

108. P.S. Clezy and A.J. Liepa, Aust. J. Chem., **23**, 2461 (1970).

109. P.S. Clezy, A.J. Liepa, and G.A. Smythe, Aust. J. Chem., **23**, 603 (1970).

110. R. Bonnett, M.J. Dimsdale, and K.D. Sales, J. Chem. Soc., Chem. Commun., 962 (1970).

111. J.-H. Fuhrhop, S. Besecke, and J. Subramaniam, J. Chem. Soc., Chem. Commun., 1 (1973).

112. M.B. Hursthouse and S. Neidle, J. Chem. Soc., Chem. Commun., 449 (1972).

113. J.-H. Fuhrhop, Dissertation, Technische Hochschule Braunschweig, West Germany (1966).

114. R.Bonnett, I.A.D. Gale, and G.F. Stephenson, J. Chem. Soc. C, 1600 (1966).

115. R. Grigg, Chem. Commun., 1238 (1967).

116. G.H. Barnett, M.F. Hudson, S.W. McCombie, and K.M. Smith, J. Chem. Soc., Perkin Trans. 1, 691 (1973).

117. R. Bonnett and M.J. Dimsdale, J. Chem. Soc., Perkin Trans. 1, 2540 (1972).

118. M.T. Cox, A.H. Jackson, and G.W. Kenner, J. Chem. Soc. C, 1974 (1971).

119. J.W. Buchler and L. Puppe, Ann. Chem., **740**, 142 (1970).

120. D. Dolphin, J. Heterocycl. Chem., **7**, 275 (1970).

121. P.S. Clezy and C.J.R. Fookes, J. Chem. Soc., Chem. Commun., 1268 (1971).

122. O. Somaya, Dissertation, Technische Hochschule Braunschweig, West Germany, 1967.

123. P.S. Clezy and G.A. Smythe, Chem. Commun., 127 (1968).

124. G. Brockmann and H. Brockmann jr., IIT NMR Newsletter, 117-62 (1968).

125. H. Wolf and H. Scheer, Ann. Chem., **745**, 87 (1971).
126. H. Wolf and H. Scheer, Ann. Chem., 1710 (1973).
127. C.D. Mengler, Dissertation, Technische Hochschule Braunschweig, West Germany (1966).
128. M.S. Fischer, Ph.D. Dissertation, University of California, Berkeley (1969), (UCLR-19524); J. Gassmann, I. Strell, F. Brandl, M. Sturm, and W. Hoppe, Tetrahedron Lett., 4609 (1971); R.C. Petersen and L.E. Alexander, J. Am. Chem. Soc., **90**, 3873 (1968); R.C. Petersen, J. Am. Chem. Soc., **93**, 5629 (1971).
129. F.C. Pennington, H.H. Strain, W.A. Svec, and J.J. Katz, J. Am. Chem. Soc., **89**, 3875 (1967).
130. H. Wolf and H. Scheer, Tetrahedron, **28**, 5839 (1972).
131. H. Brockmann, K.-M. Bliesener, and H.H. Inhoffen, Ann. Chem., **718**, 148 (1968).
132. H. Scheer, Dissertation, Technische Universität Braunschweig, West Germany (1970).
133. J. Bode, Dissertation, Technische Universität Braunschweig, West Germany (1972).
134. I. Richter, Dissertation, Technische Universität Braunschweig, West Germany, (1969).
135. H. Scheer, unpublished results.
136. H. Brockmann jr. and J. Bode, Ann. Chem., 1017 (1974).
137. G.W. Kenner, S.W. McCombie, and K.M. Smith, J. Chem. Soc., Perkin Trans. 1, 2517 (1973).
138. H. Wolf, H. Brockmann jr., I. Richter, C.-D. Mengler, and H.H. Inhoffen, Ann. Chem., **718**, 162 (1968).
139. R.C. Dougherty, H.H. Strain, W.A. Svec, R.A. Uphaus, and J.J. Katz, J. Am. Chem. Soc., **92**, 2826 (1970).
140. W.S. Caughey and P.K. Ibers, J. Org. Chem., **28**, 269 (1963).
141. G.M. McLaughlin, J. Chem. Soc., Perkin Trans. 2, 136 (1974).
142. A.H. Jackson and G.R. Dearden, Ann. N.Y. Acad. Sci., **206**, 151 (1973).
143. H.H. Inhoffen and W. Nolte, Tetrahedron Lett., 2185 (1967); H.H. Inhoffen and W. Nolte, Ann. Chem., **725**, 167 (1969).
143. (a) R. Grigg, G. Shelton, A. Sweeney, and A.W. Johnson, J. Chem. Soc., Perkin Trans. 1, 1789 (1972).
143. (b) P. Batten, A. Hamilton, A.W. Johnson, G. Shelton, and D. Ward, J. Chem. Soc., Chem. Commun., 550 (1974).
143. (c) D.K. Lavallee and A.E. Gebala, Inorganic Chem., **13**, 2004 (1974).
144. R. Bonnett, I.A.D. Gale, and G.F. Stephenson, J. Chem. Soc. C, 1168 (1969).
145. J.-H. Fuhrhop and T. Lumbantobing, Tetrahedron Lett., 2815 (1970).
146. H.H. Inhoffen, J.W. Buchler, and R. Thomas, Tetrahedron Lett., 1141 (1969).
147. R. Thomas, Dissertation, Technische Hochschule Braunschweig, West Germany (1967).
148. D.G. Whitten and J.C.N. Yau, Tetrahedron Lett., 3077 (1969).
149. P.S. Clezy and D.B. Morell, Aust. J. Chem., **23**, 1491 (1970).
150. Y. Chang, P.S. Clezy, and D.B. Morell, Aust. J. Chem., **20**, 959 (1967).
151. R.B. Woodward, Ind. Chim. Belge, 1293 (1962).
152. A.E. Pullman, J. Am. Chem. Soc., **85**, 366 (1963).
153. H.H. Inhoffen, P. Jager, R. Mahlop, and C.D. Mengler, Ann. Chem., **704**, 188 (1967).
154. H. Brockmann jr., Habilitation, Technische Hochschule Braunschweig, West Germany (1969).
155. H. Scheer, W.A. Svec, B.T. Cope, M.H. Studier, R.G. Scott, and J.J. Katz, J. Am. Chem. Soc., **96**, 3714 (1974).
156. J.J. Katz, R.C. Dougherty, and L.J. Boucher, in 'The Chlorophylls', Eds. L.P. Vernon and G.R. Seely, Academic Press, New York (1966).
157. H.J. Callot and A.W. Johnson, Chem. Commun., 749 (1969).

158. H.J. Callot, A.W. Johnson, and A. Sweeney, J. Chem. Soc., Chem. Commun., 1424 (1973).

158. (a) J.-H. Fuhrhop, T. Lumbantobing, and J. Ulrich, Tetrahedron Lett., 3771 (1970).

158. (b) G.N. Sinyakov, V.P. Suboch, A.M. Shul'ga and G.P. Gurinovich, Dokl. Akad. Nauk Beloruss. SSR, 17, 660 (1973).

159. A. Eschenmoser, R. Scheffold, E. Bertele, M. Pesaro, and H.G. Schwend, Proc. Roy. Soc. London, Ser. A, 288, 306 (1965); R. Bonnett and D.G. Redman, Proc. Roy. Soc. London, Ser. A, 288, 342 (1965); H.A.O. Hill, J.M. Pratt, and R.P.J. Williams, J. Chem. Soc. C, 2859 (1965); H.A.O. Hill, B.E. Mann, J.M. Pratt, and R.J.P. Williams, J. Chem. Soc. C, 564 (1968); J.D. Brodie and M. Poe, Biochemistry, 10, 914 (1971); J.D. Brodie and M. Poe, Biochemistry, 11, 2534 (1972); S.A. Cockle, H.A.O. Hill, R.J.P. Williams, B.E. Mann, and J.M. Pratt, Biochim. Biophys. Acta, 215, 415 (1970); C.E. Brown, D. Shemin, and J.J. Katz, J. Biol. Chem., 248, 8015 (1973); A.R. Battersby, M. Ihara, E. McDonald, J.R. Stephenson, and B.T. Golding, J. Chem. Soc., Chem. Commun., 404 (1973).

160. R. Bonnett and A.F. McDonagh, J. Chem. Soc., Perkin Trans. 1, 881 (1973); J.-H. Fuhrhop, P.K.W. Wasser, J. Subramanian, and U. Schrader, Ann. Chem., 1450 (1974); A. Gossauer and W. Hirsch, Ann. Chem., 1496 (1974).

161. R. Grigg, J. Chem. Soc. C, 3664 (1971).

162. S. Besecke, and J.-H. Fuhrhop, Angew. Chem., 86, 125 (1974); S. Besecke and J.-H. Fuhrhop, Angew. Chem., Int. Ed. Engl., 13, 150 (1974).

162. (a) J.N. Esposito, L.E. Sutton, and M.E. Kenney, Inorg. Chem., 6, 1116 (1967).

163. R. Grigg, A. Sweeney, and A.W. Johnson, Chem. Commun., 1237 (1970).

164. H.W. Whitlock and M.Y. Oester, J. Am. Chem. Soc., 95, 5738 (1973).

165. G.L. Closs and L.E. Closs, J. Am. Chem. Soc., 85, 818 (1963).

166. R. Mählhop, Dissertation, Technische Hochschule Braunschweig, West Germany (1966); H.H. Inhoffen and R. Mählhop, Tetrahedron Lett., 4283 (1966).

166. (a) V.P. Suboch, A.M. Shul'ga, G.P. Gurinovich, Yu.V. Glazkov, A.G. Zhuravlev and A.N. Sevchenko, Dokl. Akad. Nauk SSSR, 204, 404 (1972).

167. H.H. Inhoffen, J.-H. Fuhrhop, and F. von der Haar, Ann. Chem., 700, 92 (1966).

168. D.A. Savel'ev, A.N. Sidorov, R.P. Evstigneeva, and G.V. Ponomarev, Dokl. Akad. Nauk. SSSR, 167, 135 (1966).

169. J.W. Buchler, L. Puppe, K. Rohbock, and H.H. Schneehage, Ann. N.Y. Acad. Sci., 206, 116 (1973).

170. J.W. Buchler and L. Puppe, Ann. Chem., 1046 (1974).

171. P.N. Dwyer, J.W. Buchler, and W.R. Scheidt, J. Am. Chem. Soc., 96, 2789 (1974).

172. A.M. Shul'ga, G.N. Sinyakov, V.P. Suboch, G.P. Gurinovich, Yu.V. Glazkov, A.G. Zhuravlev, and A.N. Sevchenko, Dokl. Akad. Nauk. SSSR, 207, 457 (1972).

173. H. Scheer and J.J. Katz, Proc. Natl. Acad. Sci. U.S.A., 71, 1626 (1974).

174. K.M. Smith, J. Chem. Soc., Chem. Commun., 540 (1971).

174. (a) A.M. Shul'ga, I.F. Gurinovich, G.P. Gurinovich, Yu.V. Glazkov and A.G. Zhuravlev, Zh. Prikl. Spektr., 15, 106 (1971).

175. J.-H. Fuhrhop, J. Chem. Soc., Chem. Commun., 781 (1970).

176. H. Fischer and A. Treibs, Ann. Chem., 457, 209 (1927).

177. H.H. Inhoffen and N. Müller, Tetrahedron Lett., 3209 (1969).

178. P.M. Müller, S. Farooq, B. Hardegger, W.S. Salmond, and A. Eschenmoser, Angew. Chem., 85, 954 (1973).

179. A. Neuberger and J.J. Scott, Proc. Roy. Soc. London, Ser. A, 213, 307 (1952).

180. H. Ogoshi, E. Watanabe, and Z. Yoshida, Tetrahedron, 29, 3241 (1973).

181. A. Stone and E.B. Fleischer, J. Am. Chem. Soc., 90, 2735 (1968).

182. S.J. Silvers and A. Tulinsky, J. Am. Chem. Soc., 89, 3331 (1967).

183. S.J. Silvers and A. Tulinsky, J. Am. Chem. Soc., 86, 927 (1964).

184. M.J. Hamor, T.A. Hamor, and J.L. Hoard, J. Am. Chem. Soc., **86**, 1938 (1964).
185. J.W. Buchler, G. Eikelmann, L. Puppe, K. Rohbock, H.H. Schneehage, and D. Weck, Ann. Chem., **745**, 135 (1971).
186. R.J. Abraham and K.M. Smith, Tetrahedron Lett., 3335 (1971).
187. J.-H. Fuhrhop, Z. Naturforsch., **25b**, 255 (1970).
188. J.J. Katz, R.C. Dougherty, and L.J. Boucher, 'The Chlorophylls', Eds. L.P. Vernon and G.R. Seely, Academic Press, New York (1966).
189. J.W. Buchler, L. Puppe, K. Rohbock, and H.H. Schneehage, Chem. Ber., **106**, 2710 (1973).
190. D.G. Whitten, J.C. Yau, and F.A. Carrol, J. Am. Chem. Soc., **93**, 2291 (1971).
191. D.G. Whitten, T.J. Meyer, F.R. Hopf, J.A. Ferguson, and G. Brown, Ann. N.Y. Acad. Sci., **206**, 516 (1973).
192. J.-H. Fuhrhop, Tetrahedron Lett., 3205 (1969).
193. G.W. Sovocool, F.R. Hopf, and D.G. Whitten, J. Am. Chem. Soc., **94**, 4350 (1972).
194. S.S. Eaton, G.R. Eaton, and R.H. Holm, J. Organomet. Chem., **39**, 179 (1972).
195. Z. Yoshida, H. Ogoshi, T. Omura, E. Watanabe, and T. Kurosaki, Tetrahedron Lett., 1077 (1972).
196. H. Ogoshi, T. Omura, and Z. Yoshida, J. Am. Chem. Soc., **95**, 1666 (1973).
197. J.W. Buchler and P.D. Smith, Angew. Chem., **86**, 378 (1974); J.W. Buchler and P.D. Smith, Angew. Chem. Int. Ed. Engl., **13**, 341 (1974).
198. H. Ogoshi, E. Watanabe, N. Koketzu, and Z. Yoshida, J. Chem. Soc., Chem. Commun., 943 (1974).
199. W.S. Caughey, C.H. Barlow, D.H. O'Keeffe, and M.C. O'Toole, Ann. N.Y. Acad. Sci., **206**, 296 (1973).
199. (a) D.A. Clarke, R. Grigg, and A.W. Johnson, Chem. Commun., 208 (1966).
200. G.A. Smythe and W.S. Caughey, Chem. Commun., 809 (1970).
200. (a) J.-H. Fuhrhop, K.M. Kadish, and D.G. Davis, J. Am. Chem. Soc., **95**, 5140 (1973).
200. (b) D.A. Clarke, R. Grigg, A.W. Johnson, and H.A. Pinnock, Chem. Commun., 309 (1967).
201. S.S. Eaton and G.R. Eaton, J. Chem. Soc., Chem. Commun., 576 (1974).
202. H.W. Whitlock, Jr. and B.K. Bower, Tetrahedron Lett., 4827 (1965).
203. H.H. Strain, B.T. Cope, G.N. McDonald, W.A. Svec, and J.J. Katz, Phytochemistry, **10**, 1109 (1971).
204. J.J. Katz, H.H. Strain, A.L. Harkness, M.H. Studier, W.A. Svec, T.R. Janson, and B.T. Cope, J. Am. Chem. Soc., **94**, 7938 (1972).
205. A.S. Holt, in 'The Chlorophylls', Eds. L.P. Vernon and G.R. Seely, Academic Press, New York, (1966), Chap. 4.
206. H. Brockmann jr., personal communication (1974); W. Trowitzsch, Dissertation, Technische Hochschule Braunschweig, West Germany (1974).
207. J.W. Mathewson, W.R. Richards, and H. Rapoport, J. Am. Chem. Soc., **85**, 364 (1963).
208. H. Rapoport and H.P. Hamlow, Biochem. Biophys. Res. Commun., **6**, 134 (1961).
209. J.H. Mathewson, W.R. Richards, and H. Rapoport, Biochem. Biophys. Res. Commun., **13**, 1 (1963).
210. A.S. Holt, D.W. Hughes, H.J. Kende, and J.W. Purdie, J. Am. Chem. Soc., **84**, 2835 (1962).
211. K.E. Eimhjellen, O. Aasmundrud, and A. Jensen, Biochem. Biophys. Res. Commun., **10**, 232 (1963); K.E. Eimhjellen, H. Steensland, and J. Traetteberg, Arch. Mikrobiol., **59**, 82 (1967); T. Meyer, S.E. Kennel, S.M. Tedro, and M.D. Kamen, Biochim. Biophys. Acta, **292**, 634 (1974).
212. F.C. Pennington, S.D. Boyd, H. Horton, S.W. Taylor, D.G. Wulf, J.J. Katz, and H.H. Strain, J. Am. Chem. Soc., **89**, 387 (1967).

213. K.M. Smith and J.F. Unsworth, Tetrahedron, **31**, 367 (1975).
214. F.C. Pennington, H.H. Strain, W.A. Svec, and J.J. Katz, J. Am. Chem. Soc., **86**, 1418 (1964).
215. J.J. Katz, G.D. Norman, W.A. Svec, and H.H. Strain, J. Am. Chem. Soc., **90**, 6841 (1968).
216. H.H. Strain and W.M. Manning, J. Biol. Chem., **146**, 275 (1942).
217. A.A. Krasnovksii, Dokl. Akad. Nauk. SSSR, **60**, 421 (1948).
218. G.R. Seely, in 'The Chlorophylls', Eds. L.P. Vernon and G.R. Seely, Academic Press, New York (1966), p. 523; A.N. Sidorov, in "Elementary Photoprocesses in Molecules', Ed. B.S. Neporent, Plenum Press, New York (1968); A.A. Krasnovskii, M.I. Bystrova, and F. Lang, Dokl. Akad. Nauk. SSSR, **194**, 308 (1970).
219. M.F. Hudson and K.M. Smith, J. Chem. Soc., Chem. Commun., 515 (1973).
220. J.W. Buchler and H.H. Schneehage, Z. Naturforsch., **28b**, 433 (1973).
221. H. Diekman, C.K. Chang, and T.G. Traylor, J. Am. Chem. Soc., **93**, 4068 (1971).
222. N.J. Logan and Z.U. Siddiqui, Can. J. Chem., **50**, 720 (1972).
223. M.F. Hudson and K.M. Smith, Tetrahedron Lett., 2223 (1974).
224. G.N. La Mar, W.DeW. Horrocks, Jr., and R.H. Holm (Eds.), 'NMR of Paramagnetic Molecules', Academic Press, New York (1973).
225. A.M. d'A. Rocha Gonsalves, Tetrahedron Lett., 3711 (1974).
226. R.J. Kurland and B.R. McGarvey, J. Magn. Reson., **2**, 286 (1970).
227. M.F. Perutz, J.E. Ladner, S.R. Simon, and C. Ho, Biochemistry, **13**, 2163 (1974).
228. M.F. Perutz, A.R. Fersht, S.R. Simon, and G.C.K. Roberts, Biochemistry, **13**, 2174 (1974).
229. M.F. Perutz, E.J. Heidner, J.E. Ladner, J.G. Beetlestone, C. Ho, and E.F. Slade, Biochemstry, **13**, 2187 (1974).
229. (a) M. Maltempo, J. Chem. Phys., **61**, 2540 (1974).
230. K. Wüthrich, R.G. Shulman, B.J. Wyluda, and W.S. Caughey, Proc. Nat. Acad. Sci. U.S.A., **62**, 636 (1969).
231. J.A.S. Cavaleiro, A.M.d'A. Rocha Gonsalves, G.W. Kenner, K.M. Smith, R.G. Shulman, A. Mayer, and T. Yamane, J. Chem. Soc., Chem. Commun., 392 (1974).
232. J.A.S. Cavaleiro, A.M.d'A. Rocha Gonsalves, G.W. Kenner, and K.M. Smith, J. Chem. Soc., Perkin Trans. 1, 1771 (1974); G.W. Kenner and K.M. Smith, Ann. N.Y. Acad. Sci., **206**, 138 (1973).
233. H.A. Degani and D. Fiat, J. Am. Chem. Soc., **93**, 4281 (1971).
234. H.A.O. Hill and K.G. Morallee, Chem. Commun., 266 (1970).
235. K. Wüthrich, R.G. Shulman, and T. Yamane, Proc. Nat. Acad. Sci. U.S.A., **61**, 1199 (1968).
236. K. Wüthrich, R.G. Shulman, and J. Peisach, Proc. Nat. Acad. Sci. U.S.A., **60**, 373 (1968).
237. W.DeW. Horrocks, Jr. and E.S. Greenberg, Mol. Phys., **27**, 993 (1974).
238. D.R. Eaton and E.A. LaLancette, J. Chem. Phys., **41**, 3534 (1964).
239. R.K. Gupta and A.G. Redfield, Biochem. Biophys. Res. Commun., **41**, 273 (1970).
240. R.K. Gupta and A.G. Redfield, Science, **169**, 1204 (1970).
241. W.S. Caughey and L.F. Johnson, Chem. Commun., 1362 (1969).
242. R.J. Kurland, D.G. Davis, and C. Ho, J. Am. Chem. Soc., **90**, 2700 (1968).
243. G.N. La Mar, J.D. Satterlee, and R.V. Snyder, J. Am. Chem. Soc., **96**, 7137 (1974).
244. G.C. Brackett, P.L. Richards, and W.S. Caughey, J. Chem. Phys., **54**, 4383 (1971).
245. J.C. Fanning, T.L. Gray, and N. Datta-Gupta, J. Chem. Soc., Chem. Commun., 23 (1974).
246. R.H. Felton, G.S. Owen, D. Dolphin, and J. Fajer, J. Am. Chem. Soc., **93**, 6332 (1971).
246. (a) J. Beetlestone and P. George, Biochemistry, **3**, 707 (1964).
247. I.A. Cohen, D. Ostfeld, and B. Lichtenstein, J. Am. Chem. Soc., **94**, 4522 (1972).

248. J.-H. Fuhrhop, Struct. Bonding Berlin, **18**, 1 (1974).
249. N. Sadasivan, H.I. Eberspaecher, W.H. Fuchsman, and W.S. Caughey, Biochemistry, **8**, 534 (1969).
250. W.S. Caughey, J.O. Alben, W.Y. Fujimoto, and J.L. York, J. Org. Chem., **31**, 2631 (1966).
250. (a) W.S. Caughey, Adv. Chem. Ser., **100**, 248 (1971).
251. P.D.W. Boyd and T.D. Smith, Inorganic Chem., **10**, 2041 (1971).
252. M. Wicholas, R. Mustacich, and D. Jayne, J. Am. Chem. Soc., **94**, 4518 (1972).
253. E.B. Fleischer and T.S. Srivastava, J. Am. Chem. Soc., **91**, 2403 (1969).
254. R.A. Bayne, G.A. Smythe, and W.S. Caughey, in 'Probes for Membrane Structure and Function', Eds. B. Chance, M. Cohn, C.P. Lee, and T. Yonetani, Academic Press, New York (1971).
255. H.A.O. Hill, B.E. Mann, and R.J.P. Williams, J. Chem. Soc., Chem. Commun., 906 (1967).
256. B.D. McLees and W.S. Caughey, Biochemistry, **7**, 642 (1967).
257. J. Hodgkinson and R.B. Jordan, J. Am. Chem. Soc., **95**, 763 (1973).
258. L. Rusnak and R.B. Jordan, Inorganic Chem., **11**, 196 (1972).
259. C.-P. Wong, R.F. Venteicher, and W.DeW. Horrocks, Jr., J. Am. Chem. Soc., **96**, 7149 (1974).
260. T.C. Farrar and E.D. Becker, 'Pulse and Fourier Transform NMR', Academic Press, New York (1971).
261. A.R. Battersby, J. Moron, E. McDonald, and J. Feeney, J. Chem. Soc., Chem. Commun., 920 (1972).
262. A.R. Battersby, E. Hunt, E. McDonald, and J. Moron, J. Chem. Soc., Perkin Trans. 1, 2917 (1973).
263. D.N. Lincoln, V. Wray, H. Brockmann, Jr., and W. Trowitzsch, to be published.
264. A.J. Jones, T.D. Alger, D.M. Grant, and W.M. Litchman, J. Am. Chem. Soc., **92**, 2386 (1970).
265. J.J. Katz, T.R. Janson, A.G. Kostka, R.A. Uphaus, and G.L. Closs, J. Am. Chem. Soc., **94**, 2883 (1972).
266. G.E. Maciel, in "Nuclear Magnetic Resonance Spectroscopy of Nuclei Other Than Protons", Eds. T. Axenrod and G.A. Webb, John Wiley and Sons, New York (1974), Chap. 13, p. 187.
267. J.P. Maher, M. Evans, and M. Harrison, J. Chem. Soc., Dalton Trans., 188 (1972).
268. K. Wüthrich and R. Baumann, Ann. N.Y. Acad. Sci., **222**, 709 (1973).
269. K. Wüthrich and R. Baumann, Helv. Chim. Acta, **56**, 585 (1973).
270. K. Wüthrich and R. Baumann, Helv. Chim. Acta, **57**, 336 (1974).
271. A. Lapidot, C.S. Irving, and Z. Malik, Proceedings of the First International Conference on Stable Isotopes in Chemistry, Biology, and Medicine, Argonne, CONF-730525, U.S. Atomic Energy Commission (1973), p. 127.
272. D. Karweik, N. Winograd, D.G. Davis, and K.M. Kadish, J. Am. Chem. Soc., **96**, 591 (1974). H. Falk, O. Hofer and H. Lehner, Monatsh. Chem. **105**, 366 (1974).
273. P. Diehl, in 'Nuclear Magnetic Resonance Spectroscopy of Nuclei Other than Protons', Eds. T. Axenrod and G.A. Webb, John Wiley and Sons, New York (1974), Chap. 18, p. 275.
274. J. Reuben and D. Fiat, J. Am. Chem. Soc., **91**, 1242 (1969); A. Johnson and G.W. Everett, Jr., J. Am. Chem. Soc., **92**, 6705 (1970).
275. R.C. Dougherty, G.D. Norman, and J.J. Katz, J. Am. Chem. Soc., **87**, 5801 (1965).
276. G.N. La Mar and D.B. Viscio, J. Am. Chem. Soc., **96**, 7354 (1974).
277. H. Brockmann, Jr. and W. Trowitzsch, personal communication, (1974).
278. K. Ballschmiter, K. Truesdell, and J.J. Katz, Biochim. Biophys. Acta, **184**, 604 (1969).

279. A.D. Trifunac and J.J. Katz, J. Am. Chem. Soc., **96**, 5233 (1974).

280. C. Houssier and K. Sauer, J. Am. Chem. Soc., **92**, 779 (1970).

281. L.J. Boucher and J.J. Katz, J. Am. Chem. Soc., **89**, 4703 (1967).

282. D.A. Doughty and C.W. Dwiggins, Jr., J. Phys. Chem., **73**, 423 (1969).

283. J.R. Larry and Q. Van Winkle, J. Phys. Chem., **73**, 570 (1969).

284. C.E. Strouse, Proc. Nat. Acad. Sci. U.S.A., **71**, 325 (1974).

285. J.J. Katz, Dev. Appl. Spectrosc., **6**, 201 (1968).

286. J.J. Katz and H.L. Crespi, Pure Appl. Chem., **32**, 221 (1972).

287. C.E. Castro, Bioinorg. Chem., **4**, 45 (1974).

288. G.N. La Mar and F.A. Walker, J. Am. Chem. Soc., **94**, 8607 (1972).

289. J.W. Faller and J.W. Sibert, J. Organometal. Chem., **31**, C5 (1971).

290. G.N. La Mar, J. Am. Chem. Soc., **95**, 1662 (1973).

291. M. Tsutsui, D. Ostfeld, and L.M. Hoffman, J. Am. Chem. Soc., **93**, 1820 (1971).

292. M. Tsutsui and C.P. Hrung, J. Am. Chem. Soc., **96**, 2638 (1974).

293. E.D. Becker, R.B. Bradley, and C.J. Watson, J. Am. Chem. Soc., **83**, 3743 (1961).

294. L.E. Webb and E.B. Fleischer, J. Am. Chem. Soc., **87**, 667 (1965).

295. K.M. Smith, private communication, 1974.

296. J.W. Lauher and J.A. Ibers, J. Am. Chem. Soc., **95**, 5148 (1973); P.W. Codding and A. Tulinsky, J. Am. Chem. Soc., **94**, 4151 (1972).

296. (a) I. Morishima and T. Iizuka, J. Am. Chem. Soc., **96**, 7365 (1974).

297. M.J. Hamor, T.A. Hanmor and J.L. Hoard, J. Am. Chem. Soc., **86**, 1938 (1964).

298. J.L. Hoard, Science, **174**, 1295 (1971).

299. F.A. Walker, Tetrahedron Lett., 4949 (1971).

300. J.P. Collman, R.R. Gagne, T.R. Halbert, J.-C. Marchon and C.A. Reed, J. Am. Chem. Soc., **95**, 7868 (1973).

301. J.P. Collman, R.R. Gagne, H.B. Gray, and J.W. Hare, J. Am. Chem. Soc., **96**, 6522 (1974).

302. L.K. Gottwald and E.F. Ullman, Tetrahedron Lett., 3071 (1969).

303. W. Bhatti, M. Bhatti, S.S. Eaton, and G.R. Eaton, J. Pharm. Sci., **62**, 1574 (1973).

304. M. Raban and K. Mislow, in 'Topics in Stereochemistry', Eds. N.L. Allinger and E.L. Eliel, Vol. 2, Interscience, New York (1967), p. 199; W.H. Pirkle and S.D. Beare, J. Am. Chem. Soc., **91**, 5150 (1969); W.H. Pirkle, S.D. Beare, and R.L. Muntz, Tetrahedron Lett., 2295 (1974).

305. H. Fischer and H. Gibian, Ann. Chem., **550**, 208 (1942); H. Fischer and H. Gibian, Ann. Chem., **552**, 153 (1942).

306. G.E. Ficken, R.B. Johns, and R.P. Linstead, J. Chem. Soc., 2272 (1956).

307. H. Brockmann, Jr., Ann. Chem., **754**, 139 (1971).

307. (a) H. Wolf and H. Scheer, Ann. N.Y. Acad. Sci., **206**, 549 (1973).

308. H.H. Inhoffen, J.W. Buchler, and R. Thomas, Tetrahedron Lett., 1145 (1969).

309. H. Wolf, I. Richter, and H.H. Inhoffen, Ann. Chem., **725**, 177 (1969).

310. W.B. Moniz, C.F. Poranski, Jr., and T.N. Hall, J. Am. Chem. Soc., **88**, 190 (1966) and citations therein.

311. A.M.d'A. Rocha Gonsalves, G.W. Kenner, and K.M. Smith, J. Chem. Soc. D, 1304 (1971); G.W. Kenner, K.M. Smith, and M.J. Sutton, Tetrahedron Lett., 1303 (1973).

311. (a) A.M. Shul'ga, I.F. Gurinovich, Yu.V. Glazkov and G.P. Gurinovich, Zh. Prikl. Spektr., **15**, 671 (1971), A.M. Shul'ga, G.P. Gurinovich and I.F. Gurinovich, Biofizika **18**, 32 (1973).

312. H.H. Inhoffen and P. Jäger, Tetrahedron Lett., 1317 (1964).

313. C.E. Castro and H.F. Davis, J. Am. Chem. Soc., **91**, 5405 (1969).

314. R.C. Dougherty, H.L. Crespi, H.H. Strain, and J.J. Katz, J. Am. Chem. Soc., **88**, 2854 (1966); J.J. Katz, R.C. Dougherty, H.L. Crespi, and H.H. Strain, J. Am. Chem. Soc., **88**, 2856 (1966); J.J. Katz and H.L. Crespi, in 'Recent Advances

Phytochemistry', Eds. M.K. Seikel and V.C. Runeckles, Vol. II, Appleton-Century-Crofts, New York (1969).
315. M. Tomkiewicz and M.P. Klein, Proc. Nat. Acad. Sci., U.S.A., **70**, 143 (1973).
316. A.R. Lepley and G.L. Closs (Eds.), 'Chemically Induced Dynamic Nuclear Polarization', Wiley, New York (1973).
317. S. Boxer and G.L. Closs, private communication (1974).

VIBRATIONAL SPECTROSCOPY OF PORPHYRINS AND METALLOPORPHYRINS

HANS BÜRGER

Institut für Anorganische Chemie der Technischen Universität, D-33 Braunschweig, West Germany

11.1. Introduction

For many years the application of infrared spectroscopy has been limited to the characterization and recognition of side-chains in porphyrins and metalloporphyrins. It was not possible to obtain Raman spectra, and the observed infrared absorptions could not be associated with distinct vibrational motions. Only in exceptional cases could structural implications be deduced from the spectra.

The information gained from infrared spectra up to the late sixties has been reviewed[1,2]; restriction of space does not allow for any repetition or detailed discussion of this material in the following chapter.

New methods and a more sophisticated application of classical methods have since accomplished considerable progress in the study of porphyrin spectra. The most pronounced features are as follows:

(a) Raman spectra are beginning to provide the hitherto lacking information concerning the symmetrical, infrared forbidden vibrations. Both normal Raman spectra[3,4] and resonance Raman spectra (see below), with excitation within an electronic transition, have been reported.

(b) In order to detect metal-nitrogen (MN) and metal-axial ligand (ML) vibrations, metalloporphyrins with different pure metal isotopes were studied over a large frequency range[5-8].

(c) The far infrared region, which apparently provides valuable information on MN vibrations, became experimentally accessible and has been frequently investigated[5-11].

(d) The assignment and the mutual influence (mixture) of skeletal vibrations of metalloporphyrins has been proved by a normal coordinate analysis[7].

(e) $^{16}O/^{18}O$ substitution assisted to establish that protoheme dimerizes to form a Fe-O-Fe bridge[12].

Porphyrins and Metalloporphyrins, ed. Kevin M. Smith
© 1975, Elsevier Scientific Publishing Company, Amsterdam, The Netherlands

(f) The vibrations involving the *meso*-CH groups were recognized by the ease with which they undergo deuteration[13].

(g) The appearance of anomalously (inverse) polarized resonance Raman lines ($3/4 < \rho < \infty$), e.g. close to 1590 cm^{-1}, upon excitation of the resonance Raman spectrum in the region of the α,β electronic transition indicates a planar N_4M skeleton of a (low spin) metalloporphyrin[14-16]. For pyramidal (high spin) complexes, or upon excitation remote from the α,β band (π-π^* transition) e.g. in the Soret band, only polarized and depolarized resonance Raman lines are observed. Inverse polarized resonance Raman lines require a_{2g} symmetry of the appropriate vibrations and hence planar (D_{4h}) symmetry of the total molecule.

11.2. Vibrations of the porphin macrocycles

Today, even the vibrational spectrum of the simplest example to deal with, namely porphin itself, has not yet been completely recorded or fully understood. Nevertheless a comparatively reliable assignment of the infrared active vibrations has been achieved by means of a normal coordinate analysis[7]. Though unsubstituted porphins play only a minor role in porphyrin chemistry, the vibrational behaviour of a simple 37 atom metalloporphyrin with D_{4h} symmetry is typical for the porphin part of more complex molecules. Its 105 vibrations can be classified[7] as follows:

in-plane vibrations		out-of-plane vibrations	
a_{1g} (Raman)	9	a_{1u} —	3
a_{2g} —	8	a_{2u} (infrared)	6
b_{1g} (Raman)	9	b_{1u} —	5
b_{2g} (Raman)	9	b_{2u} —	4
e_u (infrared)	18	e_g (Raman)	8
	71		34

Consequently a maximum of 35 Raman lines and 24 infrared absorptions is likely to be observed; substituents increase the number of observable vibrations both by the larger number of atoms and possibly by lowering of symmetry. Peripheral substituents are of considerable influence on the vibrations of the porphin macrocycle, but not in an additive manner: the Raman spectrum of etioporphyrin is *not* a superposition of the spectra of octamethyl- and octaethyl-porphyrins[4]. As an example the infrared spectrum of Ni(OEP) (from Ref. 3) is reproduced in Fig. 1. An assignment of the inner vibrations of the porphin macrocycle is, as far as apparently trustworthy, given in Table 1. If not quoted otherwise, all frequencies refer to infrared spectra. The resonance Raman spectra mentioned in Section 11.1 with normal and inverse depolarization ratios provide information concerning the

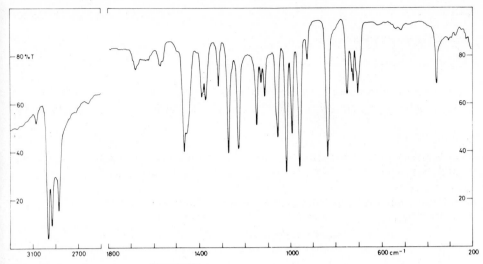

Fig. 1. Infrared spectrum of Ni(OEP).

Raman active and inactive symmetrical vibrations. It is, however, not yet possible to associate them with specific motions, though the inverse polarized ones apparently involve an a_{2g} type vibration.

The spectra of free porphyrin bases and metalloporphyrins of the same ligand differ considerably, the latter generally exhibiting less and sharper infrared absorptions due to their higher symmetry. The differences are more pronounced for the vibrations of the porphin skeleton than for the inner vibrations of the peripheral substituents. They are most obvious in the far infrared where the MN vibrations of metalloporphins are located. These cannot be considered to be isolated (see below), since they are coupled with skeletal bending modes. In cases where the frequency is apparently dependent on the metal, it is marked in Table 1 with 'M-sens'.

A detailed discussion of the vibrations of the porphin macrocycle is given in Refs. 2,9, and 17; for Raman data see Ref. 18. The assignment of strong infrared absorptions of metalloporphins (M = Zn, Cu, Ni) according to a normal coordinate analysis[7] is displayed in Table 2.

11.3. Vibrations of non-porphyrin macrocycles

Porphyrins exhibit only weak infrared absorptions between 1470 and ~1700 cm^{-1} while the strongest Raman and resonance Raman lines are observed in this region. The low infrared intensity of these vibrations is obviously due to the selection rules associated with the highly symmetrical macrocycle. Hydrogenation both of the *meso* (→ phlorins etc.) and/or the pyrrole ring(s) (→ chlorins etc.) causes the appearance of a strong infrared

References, p. 534

TABLE 1

Porphin skeletal vibrations

			Ref.
ν NH	3300—3360	generally	(2)
	3320	porphin	(6)
	3355	H$_2$(OEC)*	(6)
	3315	H$_2$(Proto-IX-DME)	
δ NH in-plane	1110	H$_2$(Proto-IX-DME)	} (9)
δ NH out-of-plane	638, 739	H$_2$(Proto-IX-DME)	
Skeletal stretch, anomalously polarized, of planar MN$_4$	} 1582—1609 resonance Raman	Cu(OEP), Co(OEP), Ni(Etio-I)	(14)
	most pronounced: 1585, 1313 resonance Raman	ferrocytochrome-*c*	
	1589, 1342, 1305 resonance Raman	oxyhemoglobin	} (16)
Skeletal stretch $3/4 > \rho > 0$	1583—1592 resonance Raman	Fe^{III} (Proto-IX-DME), [Fe(OEP)]$_2$O V(O)(Etio-I), Zn(OEP), Fe(OEP)Cl Mg(OEP)	} (14)
Skeletal stretch	1569, 1641 Raman	Ni(OEP)	} (3)
	1566, 1635 Raman	Pd(OEP)	
ν CH (*meso*)	3110	Zn(OEP)	} (13)
ν CD (*meso*)	2260	Zn(OEP)-d$_4$	
δ CH (*meso*) in plane	1217...1229	M(OEP), M=Mg, Zn, Cu, Co, Ni, Pd	} (13)
δ CD (*meso*) in plane	942	Zn(OEP)-d$_4$	
δ CH (*meso*) out-of-plane	799 ± 2	M(Methylpheophorbide-*a,b*)	(19)
	831...839	M(OEP), M=Mg, Hg, Pd	(6)
	834...839	M(OEP), M=Mg, Zn, Cu, Co, Ni, Pd	} (13)
δ CD (*meso*) out-of-plane	638	Zn(OEP)-d$_4$	
δ Porphin	920...980 M-sens.	H$_2$(Hemato-IX-DME) M(Hemato-IX-DME)	} (9)
δ Porphin	493...529 M-sens.	H$_2$(Proto-IX-DME) M(Proto-IX-DME)	

* Octaethylchlorin

absorption close to 1600 cm^{-1} (methene, chlorin band). These absorptions are apparently not associated with the newly formed methylene or pyrroline groups, but arise from the activation of a porphin vibration which is forbid-

TABLE 2

Assignment of Medium and Strong Infrared Absorptions >600 cm^{-1} of Metalloporphins According to a Normal Coordinate Analysis[7]

Intensity**	M=Zn	Cu	Ni	Approximate description*
w—m	1550	1567	1547	$\nu\ C_p{-}C_p$
m	1520	1534	1547	$\nu\ C'_p{-}N$, $\nu\ C_p{-}C_p$
s	1389	1387	1398	$\nu\ C_p{-}C'_p$, $\delta\ C_p{-}H$
s	1302	1310	1320	$\delta\ C_m{-}H$, $\delta\ C_p{-}H$
s	1155	1151	1151	$\delta\ C_p{-}H$, $\delta\ C_m{-}H$
s	1057	1057	1068/62	δ CCN, $\nu\ C'_p{-}N$, $\nu\ C'_p{-}C_m$, $\delta\ C_p{-}H$, $\nu\ C_p{-}C_p$
s—vs	993	998	994	$\nu\ C_p{-}C'_p$, $\nu\ C'_p{-}N$
s	860/48	861/48	856/46	δ CH out-of-plane
s	763	774	769	δ ring
m—s	740/03	745/02	744	δ ring out-of-plane
s	699	698	700	δ CCN, δ ring

* C_m = C *meso*, C_p = C_{1-8}, C'_p = C linked to N, C_p and C_m
** S = strong; m = medium; w = weak.

TABLE 3

Pyrromethene and chlorin bands

		Ref.
Methene bands		
1590—1620 (vs) and additionally: ~1220 ~1000 ~850	M[α,γ-$(Me)_2$-α,γ$(H)_2$-OEP]	(20,21)
1620	Ni[α,γ-$(Me)_2$-α,γ-$(H)_2$—OEP]	(3)
Chlorin bands		
1650 (infrared vs)	Ni(OEC)	(3)
1639	Cu(OEC)	(6)
1627	Zn(OEC)	(6)
1612	H_2(OEC)	(6,22)
(1608 *cis*, 1605 *trans* } 1612	H_2(3,4,7,8-$(H)_4$-OEP)	(23)
1613...1616	several chlorins and *meso*chlorins	(24)
1504 ± 3 (resonance Raman) (corrin ring breathing)	vitamin B_{12} derivatives	(25)
1500 (resonance Raman)		(26)

References, p. 534

den or weak in the infrared under D_{4h} symmetry. Pyrromethene and chlorin bands are of particular analytical value. Table 3 summarizes some experimental data.

11.4. Vibrations of peripheral groups

The strongest absorptions in the infrared spectra of porphyrins are frequently associated with peripheral substituents of the macrocycle[9]. If these absorptions appear in characteristic regions of the spectrum, e.g. ν OH, ν CH (vinyl), ν C = O (keto, ester, acid), they may be used diagnostically to prove the presence of the appropriate groups. The region between $\sim$700 and $\sim$1400 cm^{-1} is, however, crowded with vibrations of the porphin skeleton itself. In this region inner vibrations of the peripheral substituents should be assigned with great care unless they are of extraordinary intensity. Table 4 shows some vibrations which can be recognized and assigned with certainty.

TABLE 4

Infrared Absorptions of peripheral substituents

			Ref.
ν CH (vinyl)	3010; 3090	H_2(Proto-IX-DME) M(Proto-IX-DME)	} (9)
	3077—3106 2976—3012	Proto-IX derivatives	(27)
ν C=C (vinyl)	1613 or 1626		
δ CH in-plane (vinyl)	1295		
δ CH out-of-plane (vinyl)	986	H_2(Proto-IX-DME)	(27)
δ CH_2 out-of-plane (vinyl)	900		
γ CH (*meso*-phenyl)	$\sim$700	H_2(TPP), M(TPP)	(2)
ρ CH_3 (ethyl)	912, 955	M(OEP)	
δ_{as} CH_2 (ethyl)	1450, 1470		(11)
δ CH_3 (ethyl)	1370, 1380	M=Mg, Zn, Cu, Co, Ni, Pd	
ν C=O (ester)	1739 ± 2	H_2(Proto-IX-DME)	} (2,9,2
ν C—O (ester)	1170	M(Proto-IX-DME) etc.	
ν C—O (ester)	1156...1175	H_2(Proto-IX-DME) and H_2(Deut-IX-DME) derivatives	}(27)
ν C=O (ethylester)	1745	Rh(Meso-IX-DEE) Ir(Meso-IX-DEE)(CO)	}(18)
ν C=O (ketone) (C—9)	1703 ± 10 (free) 1650 (associated)	M(Methylpheophorbide-*a,b*) M=Mg, Zn, Cu, Ni	}(19)
ν C=O (β-keto)	1660—1668	H_2[2,4-di-formyl/acetyl/ propionyl-(Deut-IX-DME)]	}(27)

Additionally, less characteristic vibrations (e.g. ν C = C, δ HCH, phenyl vibrations) may be assigned according to references 1 and 2 and standard textbooks dealing with the application of group frequencies.

11.5. Metal-nitrogen vibrations

Besides metal-sensitive vibrations of the porphyrin macrocycle, the MN_4 unit of metalloporphyrins gives rise to 4 MN stretching and 5 NMN bending vibrations. According to the structure they obey the following selection rules:

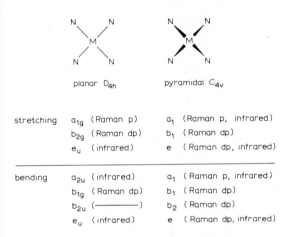

	planar D_{4h}	pyramidal C_{4v}
stretching	a_{1g} (Raman p)	a_1 (Raman p, infrared)
	b_{2g} (Raman dp)	b_1 (Raman dp)
	e_u (infrared)	e (Raman dp, infrared)
bending	a_{2u} (infrared)	a_1 (Raman p, infrared)
	b_{1g} (Raman dp)	b_1 (Raman dp)
	b_{2u} (————)	b_2 (Raman dp)
	e_u (infrared)	e (Raman dp, infrared)

Among the stretching vibrations those of a_{1g} and b_{2g} symmetry (D_{4h}) do not involve any motion of the M atom and are hence insensitive to the mass of M and to isotopic effects; the others can, in principle, be recognized from their isotopic shifts. It is not yet clear whether distorted (e.g. roof-like) molecules obey D_{4h}/C_{4v} or their appropriate selection rules (e.g. C_{2v}).

Isotopic substitution enables one to recognize, first of all, the degenerate $(e_u, e)MN_4$ stretching vibration in the infrared. Experimental results[5-7] and a normal coordinate analysis[8] have, however, shown that the observed isotopic shift is smaller than expected. This suggests that the degenerate MN stretching vibration is mixed with bending vibrations (e.g. δ CCN) of the macrocycle.

According to the most reliable investigations, the MN stretching coordinate predominates[5,7,11] in a vibration between 200 and 300 cm^{-1}, but may additionally contribute to a smaller extent to vibrations between 300 and 400 cm^{-1}. Metallophthalocyanines exhibit HMN stretching vibrations at even lower wavenumbers, ranging from 98 (M = Zn) to 185 cm^{-1} (M = Ni)[10].

Comparative investigations prove that the ν MN frequency varies sys-

References, p. 534

TABLE 5

Metal-ligand vibrations

			Ref.
Metal-Oxygen			
ν Ge—OMe	560	Ge(OEP)(OMe)$_2$	(37)
ν Sn—OMe	490	Sn(OEP)(OMe)$_2$	(37)
ν Sc—O—Sc	710, 500	[Sc(OEP)]$_2$ O	(21)
ν TiO	960	Ti(O)(OEP)	(21)
ν VO	990	V(O)(OEP)	(21)
	1003 free, 986 ass.	V(O)(Meso-IX-DME)	(38)
ν MoO	952	Mo(O)(OEP)	(21)
	938	Mo(O)(OEP)F	(21)
	896	Mo(O)(OEP)OMe)	(37)
ν Mo—OMe	442	Mo(O)(OEP)(OMe)	(37)
ν WO	944	W(O)(OEP)F	(21)
	901	W(O)(OEP)(OMe)	(37)
ν W—OMe	450	W(O)(OEP)(OMe)	(37)
ν ReO	953	Re(O)(OEP)F	(21)
ν Fe—O—Fe	840/780($^{16/18}$O)	[Fe(Deut-IX-DME)]$_2$O	(29)
	903/858($^{16/18}$O)	[Fe(Proto-IX)]$_2$O	(12,28)
	870/790	[Fe(OEP)]$_2$O	(21)
ν Fe—OMe	530	Fe(OEP)(OMe)	(37)
	540	Fe(Deut-IX-DME)(OMe)	(29)
ν_{as}/ν_s OsO$_2$	825/875	Os(O)$_2$(OEP)	(39)
Metal-Nitrogen			
ν FeN	315	Fe(OEP)NCS	(8)
	421	Fe(OEP)N$_3$	(8)
	373	[Fe(OEP)(γ-Picolin)$_2$]ClO$_4$	(8)
	377	[Fe(OEP)(Imidazole)$_2$]ClO$_4$	(8)
	333	[Fe(OEP)(Benzimidazole)$_2$]ClO$_4$	(8)
Metal-Halogen			
ν MoF	485	Mo(O)(OEP)F	(21)
ν WF	490	W(O)(OEP)F	(21)
ν ReF	520	Re(O)(OEP)F	(21)
ν FeF	606	Fe(OEP)F	(8)
	570	Fe(Deut-IX-DME)F	(29)
ν GeCl	360	Ge(OEP)Cl$_2$	(11)
ν MnCl	260...280	various Chloro(aquo-porphyrin-Mn(III)Complexes	(40)
ν FeCl	357	Fe(OEP)Cl	(8)
	330	Fe(Deut-IX-DME)Cl	(29)
ν FeBr	270	Fe(OEP)Br	(8)
ν FeI	246	Fe(OEP)I	(8)

tematically with the metal M. The following order has been observed for a series of M(OEP) complexes[5-8,11].

Zn(OEP)	Mg(OEP)	Cu(OEP)	Co(OEP)	Fe(OEP)L	Pd(OEP)	Ni(OEP)
203	214	234	264	263...280	275	287 cm^{-1}

So far no information concerning the symmetrical MN vibrations (a_{1g}, b_{2g}/a_1, b_1) has become available.

11.6. Vibrations of axial ligands

Axially substituted metalloporphyrins have been investigated by infrared spectroscopy with respect to both characteristic, definitely assignable inner-ligand vibrations and metal-ligand vibrations, the assignment of which, in part, was simplified by the use of pure metal isotopes[8]. Typical examples are quoted in Tables 5 and 6.

TABLE 6

Inner vibrations of axial ligands

			Ref.
ν C≡O	1891...1926 according to L	Os(OEP)(CO)L	(30)
	2100	Rh(TPP)(CO)Cl	(31)
	2060	Ir(Hemato-IX-DEE)CO	(18)
		Ir(Meso-IX-DEE)CO	(18)
	1955	Fe(TpivPP)CO(THF)$_2$*	(32)
ν C≡N	2082...2130 according to L	(Co^{III}N$_4$)(CN)L (Vitamin B$_{12}$)	(33)
ν NO	1700	Co(TPP)NO	(34)
ν CO (OCOCF$_3$)	1665, 790	Tl(OEP)OCOCF$_3$	(35)
ν CO (OCOMe)	1565, 680	Tl(OEP)OCOMe	(35)
	1560...1665	M(OEP)(OCOMe)$_n$ M=Ge, Sn, Sc, Zr, Hf, Mn, MoO, WO, Fe	(21)
	1594, 1338	Mn(Etio-I)OCOMe	(36)
ν CO(O—Me)	1070—1100	M(OEP)(OMe) M=Al, Si, Ge, Sn, Mo, W, Fe	(37)
ν O$_2$	1385	Fe(TPivPP)O$_2$(1 MeIm)*	(41)
	1385	Fe(TPivPP)O$_2$(N-n-BuIm)*	(41)
ν_{as}NCS	2042	Fe(OEP)NCS	(8)
ν_{as}NNN	2055	Fe(Deut-IX-DME)	(29)

* TPivPP = meso-tetra($\alpha,\alpha,\alpha,\alpha$-o-pivalamidephenyl)porphyrinato

By means of $^{16/18}$O substitution infrared spectroscopy has proved that heme-type Fe(III) porphyrins dimerize forming (nearly) linear Fe-O-Fe bridges[12,21,28,29]. Similarly the (temperature dependent) association of vanadyl(Meso-IX-DME) *via* V = O...V = O bridges has been detected by an infrared investigation of ν V = O. It should be pointed out that one cannot deduce unequivocally whether an infrared absorption near 900 cm^{-1} belongs to ν M = O of a monomeric or ν_{as} M—O—M of a bridged species.

C $\equiv$ O and C $\equiv$ N stretching vibrations of free CO and CN ligands are associated with strong infrared absorptions at about 2000 cm^{-1}. ν CO drops below 1800 cm^{-1} if CO behaves as a bridging group.

Very recently, O$_2$ complexes of *meso*-tetra(α,α,α,α-O-pivalamidephenyl)-porphyrin derivatives have been shown[41] to exhibit a strong infrared absorption at 1385 cm^{-1} at $-175°$C in KBr discs. This absorption is attributed to the OO stretching vibration of axially coordinated dioxygen, this assignment being supported by previous X-ray work.

Acknowledgement

Valuable discussions with Dr. J.W. Buchler are gratefully acknowledged.

References

1. J.E. Falk, 'Porphyrins and Metalloporphyrins', Elsevier, Amsterdam (1964).
2. G.P. Gurinovich, A.N. Sevchenko, and K.N. Solovev, "Infrared Spectra of Chlorophyll and Related Compounds", in: "Spectroscopy of Chlorophyll and Related Compounds", Izdatel'stvo Nauka, Tekhnika (1968), Translation Series United States Atomic Energy Commission (1971).
3. H. Bürger, K. Burczyk, J.W. Buchler, J.-H. Fuhrhop, F. Höfler, and B. Schrader, Inorg. Nucl. Chem. Lett., 6, 171 (1970).
4. K.N. Solovyov, N.M. Ksenofontova, S.F. Shkirman, and T.F. Kachura, Spectrosc. Lett., 6, 455 (1973).
5. H. Bürger, K. Burczyk, and J.-H. Fuhrhop, Tetrahedron, 27, 3257 (1971).
6. K. Burczyk, Thesis, Braunschweig (1970).
7. H. Ogoshi, Y. Saito, and K. Nakamoto, J. Chem. Phys., 57, 4194 (1972).
8. H. Ogoshi, E. Watanabe, Z. Yoshida, J. Kincaid, and K. Nakamoto, J. Am. Chem. Soc., 95, 2845 (1973).
9. L.J. Boucher and J.J. Katz, J. Am. Chem. Soc., 89, 1340 (1967).
10. T. Kobayashi, Spectrochim. Acta, 26A, 1313 (1970).
11. H. Ogoshi, N. Masai, Z. Yoshida, J. Takemoto, and K. Nakamoto, Bull. Chem. Soc. Jpn., 44, 49 (1971).
12. S.B. Brown, P. Jones, and I.R. Lantzke, Nature London, 223, 960 (1969).
13. H. Ogoshi and Z. Yoshida, Bull. Chem. Soc. Jpn., 44, 1722 (1971).
14. R.H. Felton, N.T. Yu, D.C. O'Shea, and J.A. Shelnutt, J. Am. Chem. Soc., 96, 3675 (1974).
15. L.A. Nafie, M. Pezolet, and W.L. Peticolas, Chem. Phys. Lett., 20, 563 (1973).

16. T.G. Spiro and T.C. Strekas, Proc. Nat. Acad. Sci. U.S.A., **69**, 2622 (1972).
17. S.F. Mason, J. Chem. Soc., 976 (1958).
18. N. Sadasivan and E.B. Fleischer, J. Inorg. Nucl. Chem., **30**, 591 (1968).
19. L.J. Boucher and J.J. Katz, J. Am. Chem. Soc., **89**, 4703 (1967).
20. J.W. Buchler and L. Puppe, Ann. Chem., **740**, 142 (1970).
21. J.W. Buchler, L. Puppe, K. Rohbock, and H.H. Schneehage, Ann. N.Y. Acad. Sci., **206**, 116 (1973).
22. H.H. Inhoffen, J.W. Buchler, and R. Thomas, Tetrahedron Lett., 1145 (1969).
23. H.H. Inhoffen, J.W. Buchler, and R. Thomas, Tetrahedron Lett., 1141 (1969).
24. H.R. Wetherell, M.J. Hendrickson, and A.R. McIntyre, J. Am. Chem. Soc. **81**, 4517 (1959).
25. F. Mayer, D.J. Gardiner, and R.E. Hester, J. Chem. Soc., Faraday Trans. 2, **69**, 1350 (1973).
26. W.T. Wozniak and T.G. Spiro, J. Am. Chem. Soc., **95**, 3402 (1973).
27. W.S. Caughey, J.O. Alben, W.Y. Fujimoto, and J.L. York, J. Org. Chem., **31**, 2631 (1966).
28. I.A. Cohen, J. Am. Chem. Soc., **91**, 1980 (1969).
29. N. Sadasivan, H.I. Eberspaecher, W.F. Fuchsmann, and W.S. Caughey, Biochemistry, **8**, 534 (1969).
30. J.W. Buchler and K. Rohbock, J. Organometal. Chem., **65**, 223 (1974).
31. I.A. Cohen and B.C. Chow, Inorg. Chem., **13**, 488 (1974).
32. J.P. Collman, R.R. Gagué, and C.A. Reed, J. Am. Chem. Soc., **96**, 2629 (1974).
33. R.A. Firth, H.A.O. Hill, J.M. Pratt, R.G. Thorp, and R.J.P. Williams, Chem. Commun., 400 (1967).
34. B.B. Wayland, J.V. Minkiewicz, and M.E. Abd-Elmageed, J. Am. Chem. Soc., **96**, 2795 (1974).
35. R.J. Abraham, G.H. Barnett, and K.M. Smith, J. Chem. Soc., Perkin Trans. 1, 2173 (1973).
36. A. Yamamoto, L.K. Phillips, and M. Calvin, Inorg. Chem., **7**, 847 (1968).
37. J.W. Buchler, L. Puppe, K. Rohbock, and H.H. Schneehage, Chem. Ber., **106**, 2710 (1973).
38. F.E. Dickson and L. Petrakis, J. Phys. Chem., **74**, 2850 (1970).
39. J.W. Buchler and P.D. Smith, Angew. Chem., Int. Ed. Engl., **6**, 341 (1974).
40. L.J. Boucher, J. Am. Chem. Soc., **92**, 2725 (1970).
41. J.P. Collman, R.R. Gagne, H.B. Gray, and J.W. Hare, J. Am. Chem. Soc., **96**, 6522 (1974).

ELUCIDATION OF ELECTRONIC STRUCTURE

MÖSSBAUER SPECTROSCOPY AND MAGNETOCHEMISTRY OF METALLOPORPHYRINS

PETER HAMBRIGHT

Department of Chemistry, Howard University, Washington, D.C. 20059, USA

and

ALAN J. BEARDEN

Donner Laboratory, University of California, Berkeley, California 94720, USA

12.1. Mössbauer Spectroscopy
12.1.1. Introduction

Mössbauer spectroscopy, in contrast to electron paramagnetic resonance or magnetic susceptibility measurements, can give detailed information about electronic configurations of diamagnetic as well as paramagnetic states[1-3]. A serious limitation of the method is that a nuclide having a low-lying isomeric nuclear energy must be employed as an absorbing nucleus and the excited state of that same nuclear isomer must be populated as a result of a nuclear reaction or decay. ^{57}Fe, ^{119}Sn and ^{129}I Mössbauer studies have been used to probe the electronic environment in the vicinity of the coordinated metal ion or axial ligand in metalloporphyrin systems.

For ^{57}Fe Mössbauer spectroscopy, ^{57}Co sources with activities of 1 to 50 mCi are commercially available. The spectroscopic arrangement then consists of a source, an absorber, a detector of the transmitted 14.4 keV gamma radiation (a proportional gas counter or a scintillation counter), and some means of providing a relative velocity between the source and absorber (to vary the energy of the gamma ray). Both the source and absorber must be in solid form to insure that the recoil free gamma transition will be favored. The minimum in the transmission of 14.4 keV gamma rays will only occur at zero relative velocity when both the source and absorber are in the same identical chemical state. If however a 'standard' source material is used with a variety of absorbers, chemical information about the absorber can be obtained.

Porphyrins and Metalloporphyrins, ed. Kevin M. Smith
© 1975, Elsevier Scientific Publishing Company, Amsterdam, The Netherlands

Since the discovery of Mössbauer spectroscopy (MS) in 1958, several thousand research papers have appeared, most dealing with applications to solid-state physics, chemistry, geochemistry, and biochemistry. Introductory descriptions of Mössbauer spectroscopy are to be found in books by Frauenfelder[4], Danon[5] and Wertheim[6]; applications to chemistry are stressed in books by Goldanskii and Herber[7], and by Greenwood and Gibb[8]. Monographs on specific applications of Mössbauer spectroscopy are found in the series edited by Gruverman[9]. A MS data index[10] for the years 1958—1965 and an updated listing of biological MS publications[11] are available.

Information about the chemical surroundings of the Mössbauer nuclide is contained in the fact that small changes in the local symmetry and magnitude of electronic wave functions affect to an observable degree the energies of the nuclear energy levels connected by the Mössbauer absorption (or emission) line(s). The observation of these small energy shifts or splittings is made possible by the high monochromaticity of Mössbauer emission lines, typically $1 : 10^{13}$ for ^{57}Fe. Although the data are best discussed in terms of an appropriate 'spin-only' Hamiltonian, it is useful to describe Mössbauer spectra in terms of four distinct types of interactions. These are: the electric monopole interaction (isomer shift or chemical shift); the electric quadrupole interaction; the nuclear Zeeman interaction; and the nuclear hyperfine interaction.

The nuclear monopole interaction which gives rise to the isomer shift (δ_0) has its origin in the Coulombic interaction between electronic charge density at the Mössbauer nuclide position and the intranuclear charge distribution. As only s-orbitals have electron density at the nucleus, the relative isomer shift will be dependent on the extent of s-character of the electronic wave function. This is related to the shielding of s electrons by p and d electrons, and the bonding properties of ligands surrounding the nuclide. Both free atom[12] and covalency[13] calculations have been made for ^{57}Fe.

The nuclear quadrupole interaction in ^{57}Fe occurs between the nuclear quadrupole moment of the $I = 3/2$ excited state (^{57m}Fe, 14.4 keV) and nonzero electric field gradients present at the nuclear position due to either atomic charge configurations, ligand charge configurations, or both[14]. In the absence of other interactions, materials displaying a nuclear quadrupole interaction show two Mössbauer absorption lines. The separation between lines, the quadrupole splitting (ΔE), is a measure of the energy difference between the $m_I = \pm 1/2$ and the $m_I = \pm 3/2$ nuclear energy states. The nuclear quadrupolar interaction is also described by the asymmetry parameter, η, which is defined as, $\eta = (V_{xx} - V_{yy})/V_{zz}$ where V_{xx}, V_{yy}, and V_{zz} represent the second derivatives of the electric potential along orthogonal crystal axes with the origin at the Mössbauer nuclide. The sign of η can be determined by the application of an external magnetic field of sufficient intensity (30 to 50 kG) to produce observable differences in the nuclear Zeeman interaction of the $m_I = \pm 1/2$ and $m_I = \pm 3/2$ states[15].

The nuclear Zeeman interaction is the Mössbauer analog of the optical Zeeman Effect in which degenerate nuclear paramagnetic states show splittings in the presence of an applied magnetic field[7]. An additional magnetic interaction, the magnetic hyperfine interaction, occurs when there is unpaired spin density near the Mössbauer nuclide. The interaction is of the form $aI \cdot S$ where I is the nuclear spin (1/2 and 3/2 for the ground and first excited state of ^{57}Fe), and S is the unpaired electron spin. The nuclear hyperfine coupling constant, a, is a measure of the strength of the interaction[7]. Observation of the nuclear hyperfine interaction is facilitated by imposing a small external magnetic field either parallel or perpendicular to the direction of the Mössbauer gamma radiation; this magnetic field produces a polarization of electronic magnetic moments dependent on H/T. The external field strength can usually be chosen such that it does not produce an observable direct nuclear Zeeman splitting.

In general all four interactions may be present in a Mössbauer spectrum; judicious use of external parameters such as sample temperature and applied magnetic field are often necessary to sort out the interactions and to determine which electronic configurations might give rise to the observed spectrum. Calculation of expected Mössbauer spectra is in many cases essential to an understanding of the experimental data. There may be important time dependencies in the effects of these interactions, particularly for the nuclear hyperfine component. It is well to remember that Mössbauer spectroscopy senses the energy of nuclear energy levels for times of the order of 10^{-9} to 10^{-6} sec. ^{57}Fe Mössbauer spectroscopy has a characteristic time of 10^{-7} sec, the lifetime of the ^{57}Fe excited state. Time dependencies such as those due to electronic-spin relaxation can often be seen by working in the liquid helium range.

With ^{57}Fe, spectra taken in zero magnetic field can only sort out high-spin Fe(II) from the other spin states of Fe(II) and Fe(III), as the quadrupole splittings and isomer shifts for the other three common spin states lie too close to each other and can be identical for many combinations of ligand symmetry and ligand-field strength. In the case of high-spin Fe(III)porphyrins, the spin state ($^6S_{5/2}$) is well-established by other techniques, particularly the observation of electron paramagnetic resonance spectra[16] with g-values of 6,6 and 2 and low temperature susceptibility data[7]. Therefore the important role of Mössbauer spectroscopy of high-spin iron-porphyrins is to supply precise values of the nuclear isomer shift and the nuclear quadrupole interaction coupling constant for comparison with theoretical values calculated on the basis of modern quantum mechanical methods for the electronic configuration of the iron-porphyrin structure[18-21].

12.1.2. High-spin ferric porphyrins

Mössbauer spectra of protohemin chloride have been obtained almost from the beginning of the chemical application of this spectroscopy[22-30].

References, p. 551

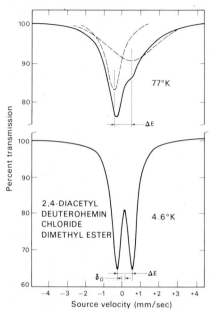

Fig. 1. Typical Mössbauer spectra of a high-spin five coordinated ferric porphyrin. The dashed lines show the computer resolution of the unresolved quadrupole doublet at $77°K$. ΔE is the quadrupole splitting and δ_0 is the isomer shift, relative to sodium nitroprusside (δ_0 = 0.00 mm/sec). See Ref. 27.

The most striking feature of the spectra is that a quadrupole doublet is only observed at temperatures below about $10°K$. Above this temperature, the lines have equal absorption areas, and widths approach the ratio $3:1$ (Fig. 1). Blume[31] has explained this phenomenon as arising from electron spin-spin relaxation as a consequence of the large zero field splitting (D) for such ferric porphyrins[32]. ($2D$ is ~ 10 cm^{-1} rather than ~ 0.1 cm^{-1} as is usually found for high-spin ferric complexes). The crystal field produces three doublets; the ground $S_z = \pm 1/2$ state, an intermediate $S_z = \pm 3/2$ level and the upper $S_z = \pm 5/2$ state, which are separated by $2D$ and $4D$, respectively. D can be obtained very accurately in the far infrared[32], and approximate values are found by Mössbauer[30] and PMR measurements. At helium temperatures, the spin relaxation is fast in the $\pm 1/2$ state, and the effective field at the nucleus averages to zero. Thus no magnetic hyperfine spectra are expected, and a well resolved quadrupole doublet is obtained. As the temperature increases, the $\pm 3/2$ and $\pm 5/2$ states become occupied. These states have slower relaxation rates than the $\pm 1/2$ level, and produce asymmetric spectra. Thus as the temperature increases, states that produce asymmetric behavior increase, and the quadrupole lines broaden[31]. The sign of the electric field gradient is positive since the highest energy peak displays the broadening at $300°K$.

Johnson[28] has demonstrated the existence of an internal field in proto-hemin chloride. For Mössbauer measurements at high H/T (30 kG, 1.6–4.2°K), the doublet splits into six hyperfine components. 10^{-3} M protohemin chloride in tetrahydrofuran at 4.2°K is magnetically dilute, and shows a hyperfine spectrum which was better resolved in an applied magnetic field[33].

Other high-spin ferric porphyrins behave in essentially the same manner as protohemin chloride (Table 2). Work has been done with TPP[34−36], (p-Cl)[37], (p-F)[37], (p-NO$_2$)[34] and (p-OMe)TPP[37], various substituted deuterohemins[27,30,38] and meso-tetrapyridylporphyrin[29] coordinated by dif-ferent axial ligands. At about 5°K, the isomer shifts (with respect to sodium nitroprusside) range from 0.58–0.77 mm/sec, and average[37] to ~0.67 mm/sec. This is typical for high-spin ionic Fe(III) in a five coordinate envi-ronment. With a given porphyrin type, δ_0 increases in the order $I^- > CH_3-CO_2^- > N_3^- \sim NCS^- \sim Br^- > CF_3CO_2^- \sim Cl^- > F^-$, which parallels their pi-bonding interactions with iron[37].

In general, the quadrupole splitting with hemins increases slightly with decreasing temperature, and values at 4°K are about 10–15% greater than at 300°K. For chloro, bromo and iodo complexes, ΔE increases[37], as (p-OMe)TPP > (p-F)TPP > (p-Cl)TPP > TPP. The axial ligand also affects ΔE, and the trend with (p-OMe)TPP[37] is $I^- > CF_3CO_2^- > Br^- > Cl^- > CH_3CO_2^- > \sim N_3^- > F$. Similar results[30,38] have been found with Deutero-IX-DME. The order follows the directly observed[32] zero-field splitting parameters found for the deuterohemin complex: I^- ($D = 16.4$ cm^{-1}) > Br^- (11.8) > Cl^- (8.95) > N_3^- (7.32) > F^- (5.55). Such D values can also be determined by computer simulation of the temperature dependence of the spectrum, which was assumed to be composed of three pairs of Lorentzian lines representing thermal populations of the three zero-field states[30]. The origin of the frequently observed reversal in intensities of the quadrupole lines as the temperature decreases has been attributed to a thermally induced change in the magnetic field direction[35,37].

MS studies on Fe(III)Hemato-IX and a hematoheme—internal histidine complex with the axial ligand $^{129}I^-$ indicated that the Fe—I bond is extremely covalent in both derivatives[39]. While the net charge on F^- in the ionic (P)Fe—F is ~$0.95\epsilon^-$, that of I^- in (P)Fe-I is ~$0.1\epsilon^-$. This charge is localized in the porphyrin ring, not on the iron. The electron capture decay of ^{125}I was used to produce a ^{125}Te axial ligand in hematoheme.

Low spin bis-amine hemichromes show quadrupole-split doublets at 300°K, which increase slightly in width at lower temperatures[36,40,41]. The asymmetric d^5 S = 1/2 configuration produces ΔE values in the range 1.9 to 2.5 mm/sec, much larger than found with the symmetric S = 5/2 ferric porphyrins (0.4–1.3 mm/sec). The pyridine hemichromes have a smaller ΔE than the imidazoles, which are better pi donor ligands and ΔE is insensitive to the counter ion[40]. The isomer shifts range from 0.37 to 0.47 mm/sec

References, p. 551

TABLE 1

Mössbauer parameters for selected iron porphyrins

Spin State	Porphyrin	Isomer shift (δ_0, mm/sec) [a]	Quadrupole splitting (ΔE, mm/sec) [a]	Temp. [b]	Ref.
Fe(III), S = 5/2	Fe(Proto-IX)Cl	0.61	0.83	H	30
	Fe(Meso-IX)Cl	0.58	0.93	H	30
	Fe(TPP)I	0.72	0.75	H	37
	Fe(p-Me · TPP)I	0.76	1.33	H	37
	Fe(p-Me · TPP)—CF_3CO_2	0.65	1.10	H	37
	Fe(p-Me · TPP)—Br	0.65	1.07	H	37
	Fe(p-Me · TPP)—Cl	0.64	1.03	H	37
	Fe(p-Me · TPP)—CH_3CO_2	0.73	0.92	H	37
	Fe(p-Me · TPP)—N_3	0.71	0.67	H	37
	Fe(p-Me · TPP)—NCS	0.69	0.63	H	37
Fe(III), S = 1/2	(p—OMe · TPP)Fe(Im)$_2$ [c], Cl	0.44	2.06	A	40
	(p—OMe · TPP)Fe(Im)$_2$,Br	0.43	2.05	A	40
	(TPP)Fe(Im)$_2$,Cl	0.40	2.11	A	40
	(p—Cl · TPP)Fe(Im)$_2$,Cl	0.42	2.11	A	40
Fe(II), S = 0	(p—OMe · TPP)Fe—(Pip)$_2$ [c]	0.69	1.49	A	40
	(p—Me · TPP)Fe—(Pip)$_2$	0.69	1.53	A	40
	(Pip)$_2$—Fe(Proto—IX)	0.67	1.47	A	40
	(Py)$_2$—Fe(Proto—IX)	0.63	1.27	A	40
	(Eim)$_2$—Fe(Proto—IX)	0.64	1.03	A	40
	(Morp)$_2$—Fe(Proto—IX)	0.68	1.52	A	40
Fe(III), S = 5/2	[(p—OMe · TPP)Fe]$_2$—O	0.68	0.63	N	46
	[(o—OMe · TPP)Fe]$_2$—O	0.59	0.65	N	46
	[2-PyridylPFe]$_2$—O [d]	0.70	0.64	N	46
	[3-PyridylPFe]$_2$—O	0.69	0.63	N	46
	[4-PyridylPFe]$_2$—O	0.69	0.67	N	46

[a] Reported with respect to sodium nitroprusside, δ_0 = 0.0 mm/sec.
[b] H, 4.2—6°K; A, 295—300°K; N, 77—80°K.
[c] Im, imidazole; Pip, piperidine; Py, pyridine; Eim, ethylenime; Morp, morpholine.
[d] *meso*-tetrakis(2-Pyridyl)porphyrin.

(versus nitroprusside), and are consistent with hexacoordination. δ_0 is rather insensitive to porphyrin basicity. The pi-bonding characteristics of low spin ferric porphyrins, where the iron is in the porphyrin plane, might suggest that the iron delocalizes added charge into both the porphyrin and ligand orbitals, and maintains an effective electroneutrality.

The spectra of protohemin chloride have been studied in applied pressures up to 170 k bar[42]. Both δ_0 and ΔE increase with pressure, indicating reduction of the initial Fe(III) to intermediate or mixed spin states of Fe(II). At high temperatures and low pressures, the bis-imidazole(Proto-IX)chloro complex is 30% reduced to low spin Fe(II), while high pressures favor low and intermediate Fe(II) states. The complex itself is in a mixed spin state[41] at 300°K.

12.1.3. Ferrous porphyrins

Low spin bis-amine hemochromes are readily prepared due to the spontaneous reduction of the Fe(III) form in many amine solvents. Well defined quadrupole doublets are obtained at 300°K, and ΔE is fairly insensitive to temperature[36,40]. For a number of porphyrin types and axial substituents, δ_0 (300°K) is in the narrow range 0.64—0.72 mm/sec; an indication that the Pauling electroneutrality principle outlined for the hemichromes may also apply to the d^6 hemochromes. With a given porphyrin, ΔE depends somewhat on the coordinated ligand, and values from 1.2 to 1.6 mm/sec are common for related nitrogenous bases[40]. Little variation of ΔE is found for the same ligand with various (p-X)TPP complexes. Some work has been done on S = 2 Fe(TPP) and Fe(TPP)(THF)$_2$ complexes[43] and with mono-amine ferrous porphyrins that carry molecular oxygen[44].

12.1.4. Oxy-bridged ferric porphyrin dimers

The MS parameters[38,45,46] of these S = 5/2 antiferromagnetically coupled μ-oxy dimers were essentially the same for seven different meso-tetra-arylporphyrin types[45]. δ_0 = 0.70 ± 0.02 mm/sec (nitroprusside) and ΔE = ~0.67 ± 0.05 mm/sec at both 77° and 4.2°K.

Twelve Sn(IV) derivatives in the meso-tetra-arylporphyrin series with bis-chloro, fluoro or hydroxy ligands were studied by [119]Sn spectroscopy[47]. Only in the case of Sn(p-Me)TPP(OH)$_2$ was a quadrupole splitting observed. The isomer shifts varied with the electronegativity of the axial group, and the corresponding phthalocyanines had larger quadrupole interactions. Other tin porphyrin work has been reported in relation to the sitting-a-top complex problem[48]. Doublets have been found in the Sn(p-X)TPP(Cl)$_2$ series[115].

12.2. Magnetic susceptibility
12.2.1. Introduction

Magnetic susceptibility measurements are routinely used by inorganic chemists to probe the stereochemistry and electronic state of metal ions in coordination compounds. Many excellent monographs have been devoted to this subject[49-53]. The susceptibility of solids or solutions can be measured on large (>100 mg) samples by the Gouy method. The Faraday technique, however, is more useful for porphyrin studies, where only a limited amount of solid material (<5 mg) is available. Balances for measurements from ambient temperatures to 77°K can be inexpensively constructed[54,55]. Either method requires a standard of known susceptibility and Hg[Co(SCN)$_4$][56]

has been most widely used. Several groups[54,57] have shown that the temperature dependence of the susceptibility of this complex when measured by the Faraday method differs from that obtained by the Gouy technique.

Many workers determine metalloporphyrin susceptibilities in solution by the Evans[58] PMR method[59]. The shift of a proton resonance line of an inert reference substance due to a paramagnetic species is given by $(\Delta H/H) = (2\pi/3)\Delta\kappa$, where $\Delta\kappa$ is the change in volume susceptibility. The same precision[60] in the measurement of the paramagnetic susceptibility can be obtained by a PMR study on 50 mg of material at 60 MHz as with the Gouy method on 400 mg. At 220 MHz only 7 mg is required under favorable conditions.

When the observed susceptibility is corrected for ferromagnetic impurities, temperature independent paramagnetism and diamagnetic contributions, the temperature dependence of the molar paramagnetic susceptibility (χ_p) usually follows the Curie law, $\chi_p = C/T$ or the Curie—Weiss law, $\chi_p = C/(T + \theta)$. C is the Curie constant. The Weiss constant, θ, does not often have a simple interpretation[52]. The effective magnetic moment (μ_{eff}) in units of Bohr Magnetons (BM) is given by the expression

$$\mu_{eff} = 2.83 \, (\chi_p T)^{\frac{1}{2}} \tag{1}$$

and results should be reported in this form[52]. The aim of most metalloporphyrin susceptibility measurements is to confirm or establish the oxidation state of the metal ion, and to find if any peculiar magnetic interactions are present. For this purpose, μ_{eff} is compared with the 'spin only' value computed from the equation $\mu_{so} = (N(N + 2))^{\frac{1}{2}}$, where N is the number of unpaired electrons. Many tabulations show the range of μ_{eff} values to be expected from a particular magnetically dilute metal ion in a given environment[50]. More information can of course be obtained from susceptibility studies at low temperatures and high field strengths, using single crystals. In the following sections, we will briefly describe magnetic studies on porphyrin compounds under the general categories (A) oxidation state determinations, (B) testing theoretical predictions, and (C) spin-state equilibria and magnetic exchange phenomena.

12.2.2. Oxidation states

Most metalloporphyrins[61] obey the Curie law to 77°K, and the resulting magnetic moments (Table 2) are fairly unambiguous indications of the metal ions' oxidation state. Earlier work in this area has been reviewed[62,63]. Phthalocyanines[64] have more complicated magnetic properties. Difficulties arise when the purity[65] or form of the complex is in question. The low moments for protohemin both in basic solution and as solids isolated under basic conditions[66] could arise from aggregation and oxy-bridge dimerization[45,67]. Decarboxylation of protohemin chloride at high temperatures[17]

TABLE 2

Representative magnetic susceptibility results for metalloporphyrins

Con-figura-tion	Complex [a]	Coor-dina-tion num-ber	Unpair-ed elec-trons	μ (spin only)	μ (obs.)	Temp.[b]	Ref.
d^9	Cu(II)Hemato-IX-DME	4	1	1.73	1.93	N	61
	Ag(II)Hemato-IX-DME	4	1	1.73	1.94	N	61
d^8	Ni(II)Hemato-IX-DME	4	0	0	diam	N	61
	Pd(II)Hemato-IX-DME	4	0	0	diam	N	61
	Pt(II)Etio—I	4	0	0	diam	N	61
	Na[Co(I)TPP]	4	0	0	diam	R	101
	Ag(III)OEP	4	0	0	diam	R	102
	H[Rh(I)TPP]	4	0	0	diam	R	103
	Ni(II)TPP—Pip	5	0	0	diam	R	114
	Ni(II)TMPyP—(Im)$_2$	6	2	2.84	3.3	R	73
d^7	Co(II)Meso-IX-DME	4	1	1.73	2.89	N	61
	Na[Fe(I)TPP]—(THF)$_2$	6	3	3.87	c	N	104
	Fe(I)TPP	4	1	1.73	1.8	R	104
	Rh(II)TPP	4	1	1.73	1.2	R	103
d^6	Fe(II)TPP	4	4	4.89	4.85	R	105
	Fe(II)TPP(Py)$_2$	6	0	0	diam	R	106
	Fe(II)TPP(THF)$_2$	6	4	4.89	5.02	R	75
	Fe(II)TPP(NO)	5	4	4.89	4.75	R	107
	Rh(III)TPP(CO,Cl)	6	0	0	diam	R	108
	Ir(III)Hemato-IX-DME-(Cl,CO)	6	0	0	0.4	R	109
	Rh(III)Meso-IX-DME(Cl)	5	0	0	0.4	R	109
	Co(III)Hemato-IX(Cl,H$_2$O)	6	0	0	0.23	N	61
d^5	Fe(III)Proto-IX—Cl	5	5	5.92	5.89	N	17
	Fe(III)TPP—(Im)$_2$	6	1	1.94	2.36	H	36
	Mn(II)TPP—(Py)$_2$	6	5	5.92	6.02	R	106
	Rh(II)TPP—(C$_6$H$_5$,Cl)	6	1	1.94	1.95	R	110
	K[Ru(III)TPP](CN)$_2$	6	1	1.94	1.9	N	65
d^4	Mn(III)Hemato-IX-Cl	5	4	4.89	4.88	N	61
	Fe(IV)TPP$^+$	?	4	4.89	c	R	111
	Cr(II)TPP	4	4	4.89	4.9	R	70
d^3	Cr(III)TPP—OCH$_3$	5	3	3.87	3.61	N	112
	Mn(IV)Hemato-IX—(OH)$_2$	6	1	1.73	2.0	R	113
d^1	VO-Hemato-IX-DME	5	1	1.73	1.79	N	61
	O=MoTPP—Cl	6	1	1.73	1.71	N	113

[a] Im, imidazole; Py, pyridine; THF, tetrahydrofuran.
[b] N, 77—300°K; R, 298°K; H, 4.2—50°K.
[c] Observed magnetic moment not reported.

References, p. 551

could account for the unusual magnetic behavior found in this range[68]. The low moment of 2.84 BM shown by Cr(II)Meso-IX-DME in the solid state[69] indicates magnetic interactions that are absent in solution, where μ_{eff} = 5.19 BM. This interaction is not found with solid Cr(II)TPP (μ_{eff} = 4.9 BM) where the phenyl rings may inhibit close contact between the metal centers[70,71]. Magnetic measurements have elucidated the equilibria in nickel-porphyrin solutions due to complexation by pyridine[72] or water[73] molecules (see Chapter 6).

The high spin-low spin transition and the resulting movement of the iron atom upon oxygen addition to ferrous hemoglobin may account for the cooperative subunit interactions[74]. Both diamagnetic[44] (O_2—Fe(II)P—(1-MeIm)) and paramagnetic[75] (O_2—Fe(II)P—THF) dioxygen complexes have been synthesized. In the coboglobin system where only a formal spin change is realized upon dioxygen formation (low spin $d^7 \to d^6$), the doming of the entire porphyrin framework has been suggested as the stereochemical trigger.

The following molar diamagnetic susceptibilities have been reported. $H_2(TPP)^{65}$ = -386×10^{-6} cgs units, $H_2(Proto\text{-}IX\text{-}DME)^{76}$ = 585×10^{-6} and $H_2(Meso\text{-}IX\text{-}DME)^{76}$ = 595×10^{-6} cgs units.

12.2.3. Theoretical predictions

Figure 2 shows that protohemin chloride deviates markedly from Curie behavior below 40°K, and the Curie—Weiss law with C = 7.09×10^{-3} and θ = 3.1 is a good representation of the data[17]. Such deviations may be a reflection of differing populations of the $^6A_1$, zero-field states with temperature, and the Kotani[77]—Griffith[16] equation shows a relationship between the observed moments (μ^2) and the zero-field splitting parameter D, where x = D/kT.

$$\mu^2 = \frac{19 + [6x^{-1} \exp(-2x)(19-11x^{-1})] + \exp(-6x)(25-5x^{-1})}{1 + \exp(-2x) + \exp(-6x)} \tag{2}$$

At high temperatures where $kT \gg 2D$, the moment will reach the spin-only value of 5.92 BM, and at low temperature where only the ±1/2 state is populated ($2D \gg kT$), μ should approach 4.36 BM. Studies of protohemin chloride by several groups[17,78] show that μ = 5.92 or 5.89 BM at 293°, 5.86 BM at 77°K, and 4.58 BM at 4°K. The chloride, bromide and SCN salts of Fe(III)TPP behave in a similar fashion[78]. With the exception of possibly Fe(TPP)Br, the observed results are not well represented by the Kotani—Griffith equation for any value of D, and similar deviations at low temperature are found for other hemiproteins[79]. The origin of these deviations has been discussed[17,78]. The temperature dependence of the moments are in fair agreement with the Harris—Loew calculations[18]. At high values of H/T (22 kG, 2°K), 90% magnetic saturation was demonstrated for protohemin chloride[18].

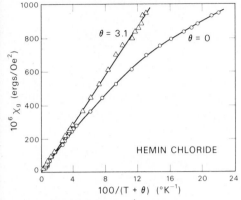

Fig. 2. Temperature dependence of the magnetic susceptibility of Cl—Fe(III)—Protopor-phyrin-IX, (hemin chloride) from 4.5 to 300°K. The results are expressed in terms of the Curie—Weiss law, $\chi = C/(T + \theta)$. $\theta = 0$ is the observed data and $C = 7.09 \times 10^{-3}$, $\theta = 3.1$ are the best-fit parameters to the equation. See Ref. 17.

12.2.4. Spin state equilibria and magnetic exchange

The magnetic susceptibility[80] of the protohemin dimer [μ-oxy-bis(Proto-IX-DME)Fe(III)] was studied from 1.5 to 300°K at field strengths from 2 to 18 kG. Antiferromagnetism was indicated, as the moments increased with increasing temperature ($\mu^2 = 2.4$ BM at 296°K, 1.5 at 213°, and 0.09 at 78°K). By using the Heisenberg $2JS_1 S_2$ model, a coupling constant 2J of 380° was found for the interaction between two high spin (S = 5/2) ferric ions in the dimer. The temperature dependence of the Knight shifts of the pyrrole protons of ferric porphyrin dimers[81] was used to determine isotopic exchange parameters ($-J$) of 309 cm^{-1} for O-[Fe(III)TPP]$_2$ and 335 cm^{-1} for O-[Fe(III)(p-Me · TPP)]$_2$. However, PMR results on related porphy-rin[82] dimers indicate that dipolar interactions and zero field splitting param-eters are not negligible, and thus the TPP dimer results[81] are probably inaccurate.

Mn(III)[83], Mo and W[84] porphyrins also form oxybridged adducts. The nature of the interactions have not been studied in detail. Several solid vanadyl porphyrins have low room temperature moments, which have been attributed to antiferromagnetic interactions[85].

The magnetic properties of an aligned crystal[86] of Cu(TPP) has been determined at low temperature. The hyperfine interactions between the nuclear and electronic spins of copper affect the perpendicular susceptibility, and the parallel susceptibility obeyed the Curie—Weiss law above 20°K with $\theta = 3°$.

The spin-state equilibrium behavior found in many heme proteins[87] and inorganic complexes[88] has also been observed with iron porphyrins. PMR measurements have demonstrated that bis-pyridine Fe(III)Proto-IX may be

References, p. 551

an (S = 1/2 → S = 3/2) system[89], with the singlet state predominant at low temperature. Similar results[90] are found with substituted pyridines in methanol solutions. Low spin bis-pyridine Fe(II) and Fe(III) complexes exchange electrons rapidly[91], and the addition of CO to form (CO—Fe(II)P—Py) blocks the exchange. The higher spin ferric forms do not exchange electrons with the low spin complexes[91,92]. The [H_2O—Fe(III)-P—Py]Proto—IX derivative[93] is about 20% in the S = 5/2 form at 300°K, and the monoimidazole complex of Fe(III)TPP may be high spin[94]. The nature of the species present in pyridine solutions of ferric porphyrins are not well understood. The high spin complex [Py—Fe(III)P—Cl] is present at 300°K in pure pyridine[38], in agreement with monopyridinate stability constant results[95]. The transformation of metmyoglobin-hydroxide between its high and low spin states was found to be a very rapid process[97]. Therefore the much slower rates that have been attributed to spin state equilibrium in porphyrin systems[91,93] may in fact involve axial ligand substitution at the iron center.

In the solid state, monoamine adducts of Fe(III)OEP—ClO_4 have been prepared[98]. The monopyridinate is low-spin between 77° and 240°K, while the 4-CHO and 4-CN pyridine derivatives have moments higher (4.3 and 3.3 BM) than the low spin value of 2.2 BM at 77°K. The perchlorate complex itself had a moment of 4.8 BM, and ClO_4^- was probably coordinated to the iron. The solid bis-imidazole [Fe(III)Proto-IX] chloride[41] is low spin at 77°K, and 23% in the assumed S = 5/2 spin state at 300°K. Upon heating to 530°K, only one imidazole is lost, and the resulting complex was shown to be high spin. In contrast, the bisimidazole (Meso-IX) and (Deut-IX) derivatives are low spin at 300°K.

In a novel approach, 2 moles of Fe(III)TPP—N_3 and 1 mole of Fe(II)TPP were reacted to form 2 : 1 complexes that could be represented[99] as either [—Fe(III)—NNN—Fe(II)—NNN—Fe(III)—], [N_3—Fe(III)—Fe(II)—Fe(III)—N_3] [Fe(III)—N_3—Fe(II)—N_3—Fe(III)] (—N—configuration). The magnetic moment was unchanged from 77° to 300°K, with μ_{eff} = 5.0 BM. The interpretation was that Fe(II) is diamagnetic, and each Fe(III) has a normal spin-only moment of four unpaired electrons (S = 2). The fifth electron from each ferric ion was paired in an orbital that extended across the entire system. The polymer [—Fe(TPP)—(imidazolate)—]$_n$ has been shown by magnetic techniques to involve spin coupling through the imidazolate ion[100].

Acknowledgements

We thank the following agencies for their support. NSF GB-37881 (AJB) and USPHS, Research Career Development Award (AJB), K 04 GM 24494 GMS, and USAEC, AT-(40-1)-4047 (PH).

References

1. A.J. Bearden and W.R. Dunham, Struct. Bonding (Berlin), 8, 1 (1970).
2. G. Lang, Quart. Rev. Biophys., 3, 1 (1970).
3. A.J. Bearden and T.H. Moss, in "Magnetic Resonance in Biological Systems", Eds. A. Ehrenberg, B.G. Malmström, and T. Vänngärd, Pergamon Press, Oxford (1967), p. 391.
4. H. Frauenfelder, "The Mössbauer Effect", Benjamin, New York (1962).
5. J. Danon, "Lectures on the Mössbauer Effect", Gordon and Breach, New York (1968).
6. G.K. Wertheim, "Mössbauer Effect: Principles and Applications", Academic Press, New York (1964).
7. V.J. Goldanskii and R.H. Herber, "Chemical Applications of Mössbauer Spectroscopy", Academic Press, New York (1968).
8. N.N. Greenwood and T.C. Gibb, "Mössbauer Spectroscopy", Chapman and Hall, London (1971).
9. "Mössbauer Effect Methodology", A series edited by I. Gruverman, Plenum Press, New York.
10. A.H. Muir, Jr., K.J. Ando, and H.M. Coogan, "Mössbauer Effect Data Index 1958-1965", Interscience, New York (1966).
11. Professor L. May, Catholic University of America, Washington, D.C. 20001.
12. L.R. Walker, G.K. Wertheim, and V. Jaccarino, Phys. Rev. Lett., 6, 98 (1961).
13. V.I. Goldanskii, E.F. Makarov, and R.A. Studan, Teoriya i Eksperim. Khim. Akad. Nauk Ukr. SSR. 2, 504 (1966).
14. O.C. Kistner and A.W. Sunyar, Phys. Rev. Lett., 4, 412 (1960).
15. R.L. Collins, J. Chem. Phys., 42, 1072 (1965).
16. J.S. Griffith, Biopolymers Symp., 1, 35 (1964).
17. S. Sullivan, P. Hambright, B.J. Evans, A. Thorpe, and J. Weaver, Arch. Biochem. Biophys., 131, 51 (1970).
18. G.H. Lowe and R.L. Ake, J. Chem. Phys. 51, 3143 (1969); G. Harris ibid, 48, 2129 (1968).
19. M. Zerner, M. Gouterman, and H. Kobayashi, Theor. Chim. Acta, 6, 366 (1967).
20. P. Moutsos, J.G. Adams, and R.R. Sharma, J. Chem. Phys., 60, 1447 (1974).
21. P.S. Han, T.P. Das, and M.F. Rettig, J. Chem. Phys., 56, 3861 (1972).
22. P.G. Reizenstein and J.B. Swan, Int. Biophysics Conf., 147 (1961).
23. L.M. Epstein, J. Chem. Phys., 36, 2731 (1962).
24. U. Gonser, J. Phys. Chem., 66, 564 (1962).
25. W. Karger, Ber. Bunsenges. Phys. Chem., 68, 793 (1964).
26. R.G. Shulman and G.K. Wertheim, Rev. Mod. Phys., 36, 459 (1964).
27. A.J. Bearden, T.H. Moss, W.S. Caughey, and C.A. Beaudreau, Proc. Nat. Acad. Sci. U.S.A., 53, 1246 (1965).
28. C.E. Johnson, Phys. Lett., 21, 491 (1966).
29. C. Wynter, P. Hambright, C. Cheek, and J.J. Spijkerman, Nature London, 216, 1105 (1967).
30. T.H. Moss, A.J. Bearden, and W.S. Caughey, J. Chem. Phys., 51, 2624 (1969).
31. M. Blume, Phys. Rev. Lett., 18, 305 (1967); 14, 96 (1965).
32. G.C. Brackett, P.L. Richards, and W.S. Caughey, J. Chem. Phys., 54, 4383 (1971).
33. (a) G. Lang, T. Asakura, and T. Yonetani, Phys. Rev. Lett., 24, 981 (1970);
    (b) A. Amusa, P. Debrunne, H. Frauenfelder, E. Münck, and G. Depasqua, J. Phys. C., 7, 1881 (1974);
    (c) J.C. Chang, T.P. Das, and K.J. Duff, Bul. Am. Phys. Soc., 20, 482 (1975);
    (d) R. Sharma and P. Moutsos, Phys. Rev. B, 11, 1840 (1975).

552   PETER HAMBRIGHT and ALA J. BEARDEN

34. A.A. Owusu, Proc. Symp. Peaceful Uses of Atomic Energy Afr., 445 (1969).
35. C. Maricondi, D.K. Straub, and L.M. Epstein, J. Am. Chem. Soc., **94**, 4157 (1972).
36. L.M. Epstein, D.K. Straub, and C. Maricondi, Inorg. Chem., **6**, 1720 (1967).
37. M.A. Torrens, D.K. Straub, and L.M. Epstein, J. Am. Chem. Soc., **94**, 4162 (1972).
38. W.S. Caughey, in "Inorganic Biochemistry", Ed. G. Eichhorn, Elsevier, Amsterdam (1973), Chap. 24.
39. M. Pasternack, P.G. Debrunner, G. DePasquali, L.P. Hager, and L. Yeoman, Proc. Nat. Acad. Sci. U.S.A., **66**, 1142 (1970).
40. D.K. Straub and W.M. Conner, Ann. N.Y. Acad. Sci., **206**, 383 (1973).
41. L. Bullard, R. Panayappan, A. Thorpe, P. Hambright, and G. Ng, Bioinorg. Chem., **3**, 41 (1973).
42. D.C. Grenoble, C.W. Frank, C.B. Bargeron, and H.C. Drickamer, J. Chem. Phys. **55**, 1633 (1971).
43. H. Kobayashi, Y. Maeda, and Y. Yanagawa, Bull. Chem. Soc. Jpn, **43**, 2342 (1970).
44. (a) J.P. Collman, R. Gagne, C.A. Reed, T.R. Halbert, G. Lang, and W. Robinson, J. Am. Chem. Soc., **97**, 1427 (1975).
    (b) G.C. Wagner and R.J. Kassner, J. Am. Chem. Soc., **96**, 5593 (1974).
45. I.A. Cohen, J. Am. Chem. Soc., **91**, 1980 (1969).
46. M.A. Torrens, D.K. Straub, and L.M. Epstein, J. Am. Chem. Soc., **94**, 4160 (1972).
47. M. O'Rourke and C. Curran, J. Am. Chem. Soc., **92**, 1501 (1970).
48. B.F. Brunham and J.J. Zuckerman, J. Am. Chem. Soc., **90**, 1547 (1970).
49. A. Earnshaw, "Introduction to Magnetochemistry", Academic Press, London (1968).
50. B.N. Figgis and J. Lewis in "Modern Coordination Chemistry", Ed. J. Lewis and R.J. Wilkins, Interscience, New York (1960), Chap. 6.
51. B.N. Figgis, "Introduction to Ligand Fields", Interscience, New York (1966).
52. F.E. Mabbs and D.J. Machin, "Magnetism and Transition Metal Complexes", Chapman and Hall, London (1973).
53. L.N. Mulay and I. Mulay, Anal. Chem., **42**, 325R (1970).
54. S. Sullivan, A.N. Thorpe, and P. Hambright, J. Chem. Educ., **48**, 345 (1971).
55. L.F. Lindoy, V. Katovic, and D. Busch, J. Chem. Educ., **49**, 117 (1972).
56. B.N. Figgis and R.S. Nyholm, J. Chem. Soc., 4190 (1958); 338 (1959).
57. G.A. Candela and R.E. Munday, IRE Trans. Instrum., 106 (1962).
58. D.F. Evans, J. Chem. Soc., 2003 (1959).
59. K. Bartle, D.W. Jones, and S. Maricic, Croat. Chem. Acta, **40**, 227 (1968).
60. K. Bartle, B.J. Dale, D.W. Jones, and S. Maricic, J. Magn. Reson., **12**, 286 (1973).
61. P. Hambright, A. Thorpe, and C. Alexander, J. Inorg. Nucl. Chem., **30**, 3139 (1968).
62. F.E. Senftle and P. Hambright, in "Biological Effects of Magnetic Fields", Ed. M. Barnothy, Vol. 2. (1969) p. 261.
63. R. Havemann, W. Haberditzl, and P. Grzegorzewski, Z. Phys. Chem., **218**, 71 (1961).
64. A.B.P. Lever, J. Chem. Soc., 1821 (1965).
65. B.C. Chow and I.A. Cohen, Bioinorg. Chem., **1**, 57 (1971).
66. W.A. Rawlinson and P.B. Scott, Aust. J. Sci. Res., **5A**, 173 (1952).
67. P. Jones, K. Prudhoe, and S.B. Brown, J. Chem. Soc., Dalton Trans., 912 (1974).
68. G. Schoffa and W. Scheler, Naturwissenschaften, **44**, 464 (1957).
68a. D. Brault and M. Rougee, Biochemistry, **13**, 4598 (1974).
69. M. Tsutsui, M. Ichikawa, F. Vohwinkel, and K. Suzuki, J. Am. Chem. Soc., **88**, 854 (1966).
70. N.J. Gogan and Z.U. Siddiqui, Can. J. Chem., **50**, 720 (1972).
71. D. Ostfeld and M. Tsutsui, Acc. Chem. Res., **7**, 52 (1974).
72. B.D. McLees and W.A. Caughey, Biochemistry, **7**, 642 (1968).
73. R.F. Pasternack, E.G. Spiro, and M. Teach, J. Inorg. Nucl. Chem., **36**, 599 (1974).
74. J.L. Hoard and W.R. Scheidt, Proc. Nat. Acad. Sci. U.S.A., **70**, 3919 (1973).
75. J.P. Collman, R.R. Gange, and C.A. Reed, J. Am. Chem. Soc., **96**, 2629 (1974).

76. R. Havemann and W. Haberditzl, Z. Phys. Chem., **217**, 91 (1961).
77. M. Kotani, Prog. Theor. Phys., Suppl., **17**, 4 (1961).
78. C. Maricondi, W. Swift, and D.K. Straub, J. Am. Chem. Soc., **91**, 5205 (1969).
79. A. Akira, J. Ostuka, and M. Kotani, Biochim. Biophys. Acta, **140**, 284 (1967).
80. T.H. Moss, H.R. Lilienthal, G. Moleski, G.A. Smythe, M.C. McDaniel, and W.S. Caughey, J. Chem. Soc., Chem. Commun., 263 (1972).
81. P.D.W. Boyd and T.D. Smith, Inorg. Chem., **10**, 2041 (1971).
82. G.N. LaMar, G.E. Eaton, R.H. Holm, and F.A. Walker, J. Am. Chem. Soc., **95**, 63 (1973).
83. E.B. Fleischer, J.M. Palmer, T.S. Srivastava, and A. Chatterjee, J. Am. Chem. Soc., **93**, 3162 (1971).
84. J.W. Buchler and K. Rohbock, Inorg. Nucl. Chem. Lett., **8**, 1073 (1972).
85. E. Higginbotham and P. Hambright, Inorg. Nucl. Chem. Lett., **8**, 747 (1972).
86. J.L. Imes, G.L. Neiheisel, and W.R. Pratt, Jr., Amer. Phys. Soc. Bull., **19**, 49 (1974).
87. P. George, Biochemistry, **3**, 707 (1964).
88. E.K. Barefield, D.H. Busch, and S.M. Nelson, Quart. Rev. (London), **22**, 457 (1968).
89. H.A.O. Hill and K.G. Morallee, J. Am. Chem. Soc., **94**, 731 (1972).
90. H.A.O. Hill and K.G. Morallee, J. Chem. Soc., Chem. Commun., 266 (1970).
91. J.A. Weightman, N.J. Hoyle, and R.J.P. Williams, Biochim. Biophys. Acta, **244**, 567 (1971).
92. K. Yamamoto and T. Kwan, Bull. Chem. Soc. Jpn, **45**, 664 (1972).
93. H.A. Degani and D. Fiat, J. Am. Chem. Soc., **93**, 4281 (1971).
94. G.N. LaMar and F.A. Walker, J. Am. Chem. Soc., **94**, 8607 (1972).
95. P. Hambright, J. Chem. Soc., Chem. Commun., 470 (1967).
96. J.K. Beattie and R.J. West, J. Am. Chem. Soc., **96**, 1933 (1974).
97. J.K. Beattie, N. Sutin, D.H. Turner, and G.W. Flynn, J. Am. Chem. Soc., **95**, 2052 (1973).
98. H. Ogoshi, E. Watanabe, and Z. Yoshida, Chem. Lett., 989 (1973).
99. I.A. Cohen, Ann. N.Y. Acad. Sci., **206**, 453 (1973).
100. I.A. Cohen and D. Ostfeld, Adv. Chem. Series, **5**, 221 (1974).
101. H. Kobayashi, T. Hara, and Y. Kaizu, Bull. Chem. Soc. Jpn., **45**, 2148 (1972).
102. K. Kadish and D.G. Davis, Angew. Chem., Int. Ed. Engl., **11**, 1014 (1972).
103. B.R. James and D.V. Stynes, J. Am. Chem. Soc., **94**, 6225 (1972).
104. I.A. Cohen, D. Ostfeld, and B. Lichtenstein, J. Am. Chem. Soc., **94**, 4522 (1972).
105. S.M. Husain and J.G. Jones, Inorg. Nucl. Chem. Lett., **10**, 105 (1974).
106. H. Kobayashi and Y. Yanagawa, Bull. Chem. Soc. Jpn., **45** 450 (1972).
107. L. Vaska and H. Nakai, J. Am. Chem. Soc., **95**, 5431 (1973).
108. E.B. Fleischer, R. Thorp, and D. Venerable, J. Chem. Soc., Chem. Commun., 475 (1969).
109. N. Sadasivan and E.B. Fleischer, J. Inorg. Nucl. Chem., **30**, 591 (1968).
110. E.B. Fleischer and D. Lavallee, J. Am. Chem. Soc., **89**, 7132 (1967).
111. R.H. Felton, G.S. Owen, D. Dolphin, and J. Fajer, J. Am. Chem. Soc., **93**, 6332 (1971).
112. E.B. Fleischer and T.S. Srivastava, Inorg. Chim. Acta, **5**, 151 (1971).
113. P.A. Loach and M. Calvin, Biochemistry, **2**, 361 (1963).
114. F.A. Walker, E. Hui, and J.M. Walker, J. Am. Chem. Soc., **97**, 2390 (1975).
115. N. Debye and A.D. Adler, Inorg. Chem., **13**, 3037 (1974).

ELECTRON PARAMAGNETIC RESONANCE SPECTROSCOPY OF PORPHYRINS AND METALLOPORPHYRINS

J. SUBRAMANIAN

Gesellschaft für Molekularbiologische Forschung mbH, D-3301 Stöckheim bei Braunschweig, Mascheroder Weg 1, West Germany

13.1. General background

13.1.1. Introduction

Electron paramagnetic resonance (EPR) has now become a popular and powerful tool in the hands of chemists and biologically oriented scientists. A large number of investigators working on porphyrins and related systems have utilized this method in probing into the structural and dynamic aspects of porphyrins as well as their role in biological systems. The large number of monographs and review articles on various aspects of EPR[1-20] obviate, in the present article, any detailed treatment of the basic theory and techniques of EPR. Only the specific aspects of EPR that have been helpful in the field of porphyrins will be discussed in brief. Also, in view of the voluminous literature on the EPR of porphyrins, an exhaustive discussion of the available data is not attempted here. Rather a selective treatment of different aspects (transition metal complexes, free radicals, triplet states) of this field will be given; for further details the reader is referred to the cited literature.

The EPR technique enables one to detect and in many cases to characterize systems with one or more unpaired electrons. In the case of paramagnetic porphyrin systems, one or more unpaired electrons may reside either on the π-ligand system or in the central metal atom, or in both. The basic features of EPR spectra are strongly dependent on the number as well as the location of the unpaired electrons in a system, thus making the technique very useful in porphyrin-radical chemistry.

13.1.2. Basic principles and information obtainable from EPR spectroscopy

13.1.2.1. g-values and symmetry[1-7,10-13,19,20]

The magnetic moment $\hat{\mu}$ of an unpaired electron is given by

$$\hat{\mu} = g\beta_e \hat{S} \ , \tag{1}$$

Porphyrins and Metalloporphyrins, ed. Kevin M. Smith
© 1975, Elsevier Scientific Publishing Company, Amsterdam, The Netherlands

where g is the ratio of the magnetic moment to the total angular momentum, $\hat{S}$ is the spin angular momentum in units of $h/2\pi$, and β_e is the Bohr magneton. In the presence of an external magnetic field $\hat{H}$, the component of $\hat{S}$ in the direction of $\hat{H}$ is quantized and can have two values $M_s = \pm\frac{1}{2}$. Thus we have two energy levels $\pm(\frac{1}{2})g\beta_e H$, and the frequency of electromagnetic radiation required to induce transitions between the two states is $\nu = (g\beta_e/h)H$. For a magnetic field of 3500 G, the absorption frequency lies in the X-band (3 cm) microwave region. The selection rule for an EPR transition is $\Delta M_s = \pm 1$, with the oscillating magnetic field of the radiation being perpendicular to the external magnetic field, H.

The g-value is a dimensionless parameter which can be determined from the experimental EPR spectrum. It is determined solely by the spin and orbital angular momenta of the unpaired electron. In a spherically symmetric environment (S-state) the orbital angular momentum is zero, and the g has the free-spin value 2.0023. Thus, in an organic free radical, the electron is highly delocalized, leading to very little orbital angular momentum and consequently only small deviations from the free spin g-value. For transition metal ions, the molecular or electric fields arising from ligand atoms or from neigboring ions quench the orbital motion of the electrons partly or completely. One therefore observes deviations from the free spin g-value, which can be correlated with the effect of spin—orbit interactions. With solids, the g-value depends on the orientation of the magnetic field to the symmetry axes of the electric field about the ion, leading to anisotropy in the g-values. If the ligand field has an axial symmetry, EPR is observed with polycrystalline samples at two g-values: a major absorption if the external magnetic field is perpendicular to the symmetry axis (say z), ($g_x = g_y \equiv g_\perp$) and a minor absorption when the applied magnetic field is parallel to the symmetry axis ($g_z \equiv g_\parallel$). This situation is observed in most of the metalloporphyrins since the square-planar ligand field possesses an axial symmetry. When the environment is orthorhombic, one observes three different g-values in the solid state. With solutions where the tumbling motions of molecules produce an average electric field (if the tumbling rate is faster than the g-anisotropy), only a single average g-value, i.e. $g_{av} = 1/3\,(g_x + g_y + g_z)$ or $g_{av} = 1/3\,(2g_\perp + g_\parallel)$ is observed. The g-value is characteristic of the metal ion, its oxidation state and its molecular environment.

13.1.2.2. Fine structure and zero field splitting[1−7,10−13,19,20]

If there are two or more unpaired electrons in systems ($S > 1$), the individual magnetic moments interact with the magnetic fields generated by other electrons. The interactions among the electrons may be of purely magnetic dipolar nature (triplet states of organic molecules) or may arise through the effect of electric or ligand fields in the molecule (transition metal ion). In both cases the spin multiplets are not degenerate even in the absence of an external magnetic field, leading to a zero-field splitting. For

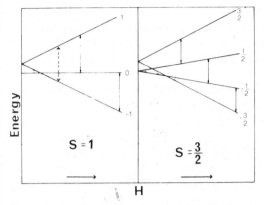

Fig. 1. Energy levels for $S = 1$ and $S = 3/2$ as a function of magnetic field for an axially symmetric system. The magnetic field is along the main symmetry axis. The dotted lines indicate the $\Delta M_s = \pm 2$ transition.

transition metal ions, zero-field splitting is observed whenever there is a deviation from cubic symmetry. The EPR spectra are very much dependent on the relative magnitudes of the zero-field splitting energy and the microwave frequency of the spectrometer. Figure 1 depicts the schematic energy levels for $S = 1$ and $S = 3/2$ systems. From EPR measurements of the zero-field splittings for organic triplet-state molecules, one can obtain the average distance between two unpaired electrons. From studies on transition metal ions one can obtain information on the symmetry of the ligand field.

13.1.2.3. Nuclear hyperfine interaction[1−7,10,13,19,20]

The interaction of the magnetic moment of an unpaired electron with the moments of the magnetic nuclei present in a system manifests itself in the appearance of hyperfine structure in the EPR spectra. The magnetic interactions between electrons and nuclei fall into two categories: Fermi contact interaction and dipolar interaction. Fermi contact interaction arises due to the presence of a fraction of the unpaired electron (spin density) in the s-orbital of the atom concerned. This interaction is isotropic and is observed even with solutions. In an organic π-radical the unpaired electron in a p-π-orbital yields hyperfine structure by interacting with σ-bonded protons and/or other nuclei. Though the π-orbital has a node in the plane of the π-bonds, spin density at the 1s orbital of an H atom arises by a σ-π-spin polarization mechanism. A similar contact interaction occurs between a magnetic nucleus of a transition metal ion and an unpaired electron in the d-orbital. The spin density in an s-orbital can arise in the following way. A transition metal ion with the configuration $1s^2\ 2s^2\ 2p^6\ ...\ 3d^n$ can have an admixture of excited states like $1s^2\ 2s^1\ 2p^6\ ...\ 3d^n\ 4s^1$, thus inducing a net spin density in the s-orbital. More recently Abragam[1] has proposed a core polarization mechanism: each of the paired electrons in a s-orbital (α- and β-spins) interact differently with an unpaired electron (say α-spin) in a d-orbital, leading to a net spin density in the s-orbital. The hyperfine couplings observed for transition metal ions are much larger than those for organic free radicals.

The second type of interaction between electrons and nuclei is the classical magnetic dipolar interaction and is observed only in solids. It depends on the orientation of the electron and nuclear moments with respect to the magnetic field and hence is anisotropic. This interaction vanishes in solution due to molecular tumbling motions. Additional complications arise for transition metal ions, since both the spin and orbital angular moments of the electron have dipolar interaction with the nuclear moments, which again depends on the relative orientation of the molecular symmetry axes with respect to the magnetic field. The anisotropic hyperfine couplings give valuable information regarding the ground-state wave-function for the transition metal ion. In favorable cases like Cu, Ag and VO porphyrins one also observes — in addition to the metal hyperfine coupling — couplings from the ligand nitrogen nuclei. From these data one can estimate the nature of charge transfer from ligand to metal and vice versa very precisely. The usefulness of the metal and ligand hyperfine couplings for the understanding of the porphyrin/metal ion bonding will be discussed in detail in Section 13.2.

13.1.2.4. The concept of a spin Hamiltonian[1−7,10−13,19,20]

We have seen earlier that, in the case of transition metal ions, contributions to the EPR spectra arise from the interactions of electron spin moments ($\hat{S}$) and orbital angular moments ($\hat{L}$) with an external magnetic field ($\hat{S} \cdot \hat{H}$ and $\hat{L} \cdot \hat{H}$) and also with magnetic nuclei ($\hat{S} \cdot \hat{I}$ and $\hat{L} \cdot \hat{I}$). In addition one must also consider ($\hat{S} \cdot \hat{L}$) spin orbit interactions and spin—spin ($\hat{S}_1 \cdot \hat{S}_2$) interactions. It is difficult to interpret the experimental EPR spectra in terms of these interactions in a straightforward way. One fits the experimental data into a 'spin Hamiltonian' containing only the spin operators $\hat{S}$ and $\hat{I} \cdot \hat{S}$ is an effective spin operator. The orbital parts of the interactions are merged with the g-values g_x, g_y, g_z and the hyperfine couplings A_x, A_y, A_z. One can correlate the experimental data with theory by starting with a suitable wave function and computing the spin Hamiltonian parameters. Thus the spin Hamiltonian bridges the gap between theory and experiment. Examples of spin Hamiltonians will be encountered in Section 13.2. The spin Hamiltonian must be defined with respect to a coordinate system in which the g and the A tensors are diagonal.

13.1.2.5. Spin relaxation and linewidths[14−18,20]

The requirements for an EPR absorption, in addition to the resonance condition, are the occurrence of a higher population in the low-energy spin state at thermal equilibrium (Boltzmann distribution) and an efficient mechanism for spin relaxation. The relaxation of spins involves the dissipation of energy of the excited state, i.e. without emission of radiation. This occurs by two mechanisms: spin—lattice relaxation and spin—spin relaxation. The former mechanism proceeds through spin—orbit interaction. The efficiency of this mechanism thus depends on spin—orbit coupling. The linewidths in the EPR spectra are related to the inverse of the relaxation times, T_1 and T_2. Strong spin—orbit coupling leads to a short spin—lattice relaxation time (T_1) and consequently to broad lines. In such cases, as it happens in Fe(III) systems, the EPR lines are too broad for detection at ambient temperature. At low temperature T_1 increases, and once can observe an EPR spectrum at liquid-nitrogen or liquid-helium temperatures. Also the population at the lower energy state increases on cooling, favoring improved EPR absorption. In free radicals, spin—orbit interactions are negligible and T_1 is long, leading to narrow lines even at room temperature. Spin—spin relaxation occurs due to the intermolecular interactions between the spin systems. Neighboring spin momenta cause static and oscillating magnetic fields at the site of a given spin. These fields produce fluctuations in the actual external magnetic field ($H_0 + H_{local}$) experienced by the spin and also induce transitions within the spin system. The spin—spin relaxation time T_2 decreases with increasing concentration. Narrow-line EPR spectra can be obtained by dilution of the sample. Apart from the dipolar broadening discussed above, exchange interactions between spins also contribute to T_2. Both the relaxation times T_1 and T_2 are quite sensitive to molecular dynamics. Dynamic processes which modulate the g and hyperfine tensors will affect the linewidth and hence the latter gives valuable information regarding the nature of molecular motions[17,18].

13.1.3. Experimental aspects[4,6,8,9]

As a number of monographs are now available on the instrumentation in EPR, this aspect need not be discussed here. However, details useful for those working in the porphyrin field will be presented. For EPR spectroscopy of paramagnetic metalloporphyrins in solutions, chloroform, dichloromethane or admixtures with methanol are used as the common solvents. One-electron reductions and oxidations are usually done electrolytically under controlled potentials in an inert atmosphere or under vacuum. Chemical oxidation and reductions are also possible. In any case, one should always check the identity of the species formed after oxidation reduction by spectrophotometric measurements, before and after EPR measurements are taken. A number of designs of electrolysis cells are available which can be used for both EPR and optical measurements[8].

The EPR spectra are usually obtained as derivatives of the absorption spectra because of the high-frequency modulation (around 100 kHz) and phase-sensitive detection used in EPR instrumentation. It is common practice to work at low microwave power levels to avoid saturation. This is especially important when measurements are taken at liquid-helium temperatures (hemes and heme proteins). At such low temperatures, low-frequency modulation (around 400—1000 Hz) and superheterodyne detection are used to improve the sensitivity.

For special techniques of handling biological materials the reader is referred to a recent book by Schwartz, Bolton and Borg[4].

13.2. Porphyrins which contain paramagnetic metals

Porphyrins form highly symmetric (D_{4h} or C_4) square-planar metal complexes. A large number of transition metal complexes of porphyrins have been investigated by EPR in order to correlate the spin Hamiltonian parameters with the geometry of the ligand field and also to estimate the covalent character of the metal-bond. The molecular orbital (MO) theory of metal ligand binding in porphyrins has been treated in detail by various groups[33,80,126]. Only the usefulness of ESR data for obtaining MO coefficients will be considered here. The schematic one-electron energy levels of the five 3d-orbitals of a transition metal ion in octahedral, square-pyramidal and square-planar ligand fields are represented in Fig. 2a. In the molecular orbital scheme, the unpaired electrons present in those d-orbitals which are directly involved in bonding, are promoted to the antibonding orbitals with predominantly metal character. From the EPR data one obtains the coefficients of the orbitals of the metal and the ligand in the antibonding level[12].

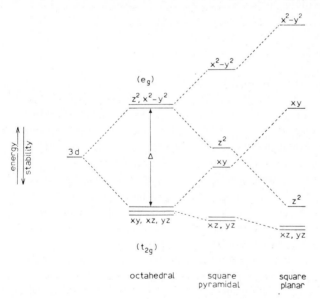

Fig. 2a. One-electron energy levels of metal d-orbitals in various ligand fields.

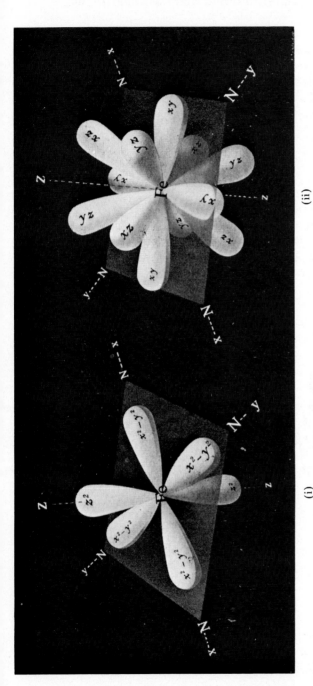

Fig. 2b. Spatial diagram of metal d orbitals shown in respect to the four N-atoms of an iron porphyrin chelate. (i) The bonding l_g orbitals: a component of a d_{z^2} orbital in the plane of the porphyrin ring is omitted for simplicity. (ii) The bonding t_{2g} orbitals.

(ii)

(i)

References, p. 586

By interchanging these coefficients one obtains the wave function corresponding to the bonding orbital for the metal and the ligand atoms.

13.2.1.1. Copper and silver porphyrins[21-34]

A number of EPR studies have been reported on Cu(II), Ag(II) and VO(IV) porphyrins. All these systems contain a single unpaired electron in the metal. Measurements have been made on dilute single crystals, frozen solutions and in fluid media[21-30]. The spin Hamiltonian in the fluid media is quite simple because only the average values of g and hyperfine couplings are observed (Fig. 3).

$$\mathcal{H}_{\text{isotr}} = g\beta_e H_z S_z + a_M \hat{I}_M \cdot \hat{S} + \hat{S} \cdot \sum_i a_{1i} \hat{I}_{1i} \tag{2}$$

where g, a_M and a_{1i} are the measured isotropic g-value, the metal hyperfine coupling and the ligand hyperfine coupling respectively. I_M and I_{1i} are the nuclear spins of the metal and the ligand nuclei respectively. With fluid media one observes only the isotropic hyperfine couplings arising from Fermi contact interaction. With porphyrins, ligand hyperfine couplings from four equivalent nitrogen nuclei are usually observed.

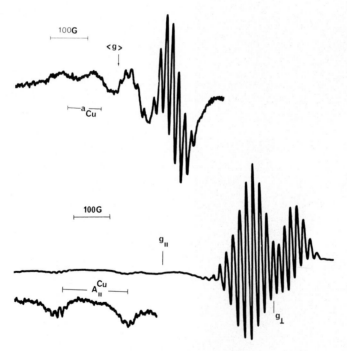

Fig. 3. X-band EPR spectrum of Cu(II)OEP (a) in chloroform at 290°K (b) in chloroform—EBBA liquid crystal at 100°K.

From the ESR spectra of solids (frozen solutions or single crystals) one gets information about the components of g and hyperfine tensors. The spin Hamiltonian for a single unpaired electron in an axially symmetric crystal field is (including hyperfine interaction with the metal nucleus)

$$\mathcal{H} = \beta_e [g_{\parallel} H_z S_z + g_{\perp}(H_x S_x + H_y S_y)] + A_{\parallel} S_z I_z + A_{\perp}(S_x I_x + S_y I_y). \tag{3}$$

For a nucleus with spin $I \geqslant 1$ one must also consider the quadrupolar interactions but we shall neglect the refinements for the present discussion. The operator $\hat{S}$ is a fictitious spin operator which actually involves both the orbital and spin angular momenta. For transition metal ions with unpaired electrons in the d-orbitals, the effect of spin—orbit coupling is treated as a perturbation over the crystal field. The contributions from spin orbit interactions ($\lambda \hat{L} \cdot \hat{S}$) and other interactions of the type $\hat{L} \cdot H$ and $\hat{L} \cdot \hat{I}$ enter as second-order corrections to the g- and A-values. Let us consider an unpaired electron in a d^9 system in a square-planar ligand field (e.g. Cu(II) porphyrin). If the x and y axes are chosen along the M—N bonds, then the antibonding orbitals are[25]:

$$|x^2 - y^2\rangle (B_{1g}) = \alpha d_{x^2 - y^2} + (\alpha'/2)(-\sigma_x^a + \sigma_y^b + \sigma_x^c - \sigma_y^d) \tag{4}$$

$$|xy\rangle (B_{2g}) = \beta_1 d_{xy} - (\beta_1'/2)(p_y^a + p_x^b - b_y^c - p_x^d) \tag{5}$$

$$|3z^2 - r^2\rangle (A_{1g}) = \alpha_1 d_{3z^2 - r^2} - (\alpha_1'/2)(\sigma_x^a + \sigma_y^b - \sigma_x^c - \sigma_y^d) \tag{6}$$

$$\begin{array}{l} |xz\rangle \\ \quad (E_g) \\ |yz\rangle \end{array} \left\{ \begin{array}{l} \beta d_{xz} - \beta'(p_z^{(1)} - p_z^{(3)})/\sqrt{2} \\ \\ \beta d_{yz} - \beta'(p_z^{(2)} - p_z^{(4)})/\sqrt{2} \end{array} \right. \tag{7}$$

The orbitals σ_x, σ_y etc. are the sp^2 hypridized σ lone pairs of the nitrogen atoms, which are bonded to the metal ion. Normalization of the B_{1g}-orbital yields $\alpha^2 + \alpha'^2 - 2\alpha\alpha'S = 1$, where S is the overlap integral between the metal and the ligand orbitals. The coefficients α and α' from the ESR data may be obtained in the following way.

$$g_{\parallel} = 2.0023 - 8\rho[\alpha\beta_1 - \alpha'\beta_1 S - \alpha'(1 - \beta_1^2)^{1/2}T(n)/2] \tag{8}$$

$$g_{\perp} = 2.0023 - 2\mu[\alpha\beta - \alpha'\beta S - \alpha'(1 - \beta^2)^{1/2}T(n)/\sqrt{2})] \tag{9}$$

$$A_{\parallel} = P[-\alpha^2(\tfrac{4}{7} + \kappa_0) + (g_{\perp} - 2) + \tfrac{3}{7}(g_{\parallel} - 2)$$

$$+ \text{ additional terms involving } T(n)] \tag{10}$$

$$A_{\perp} = P[\alpha^2(\tfrac{2}{7} - \kappa_0) + \tfrac{11}{14}(g_{\perp} - 2) + \text{ terms involving } T(n)] \tag{11}$$

where λ_0 is the spin orbit coupling constant for the free metal ion and

$$\rho = \lambda_0 \alpha \beta_1 / (E_{xy} - E_{x^2 - y^2})$$

$$\mu = \lambda_0 \alpha \beta / (E_{xz} - E_{x^2 - y^2})$$

$$P = 2.0023 \, g_N \beta_N \beta_e \langle r^{-3} \rangle$$

and κ_0 is the Fermi-contact term for the free ion. $T(n)$ are correction terms involving Slater orbital exponents. One requires a knowledge of the quantities λ_0, κ_0, the d—d transitions energies $(E_{xy} - E_{x^2 - y^2})$ etc. as well as the value of $\langle r^{-3} \rangle$ for the d-orbitals. The magnitudes of the d—d transitions are usually obtained from the optical spectra, λ_0 and κ from the data for the free ion and the value of $\langle r^{-3} \rangle$ from theoretical calculations. For porphyrin complexes the d—d transitions have to be obtained by indirect methods. Hence it is possible to obtain the value of α^2 which is the coefficient of the metal d orbital in the MO for the unpaired electron. The coefficient α' can be obtained through the normalization constant for the orbital B_{1g}. One can also obtain the coefficient α' directly from the ligand superhyperfine structure, which will be discussed now.

The superhyperfine structure arising from the ligand nuclei gives direct information on the charge transfer in the metal—ligand bond. With porphyrins, one observes the superhyperfine couplings due to four equivalent nitrogens. The superhyperfine part of the Hamiltonian is

$$\mathcal{H}_{SUHP} = \hat{S} \cdot \sum_{n=1}^{4} A_n \hat{I}_n \tag{12}$$

where n is the index for the nitrogen atoms. When all the four nitrogen atoms are magnetically equivalent (D_{4h} symmetry), the superhyperfine part can be written as

$$\mathcal{H}_{SUHP} = A_{\parallel}^N S_z I^N + A_{\perp}^N (S_x I^N + S_y I_y^N) \tag{13}$$

where I_x^N, I_z^N and I_z^N are the sums of the nuclear spin components of the ligand nitrogen along the three coordinates. We use the same coordinate system as that employed for metal hyperfine structure. For a number of metalloporphyrins the superhyperfine tensors have been found to be predominantly isotropic (i.e. $A_{\parallel}^N \cong A_{\perp}^N$) The isotropic and anisotropic parts of $A_{\parallel}^N$ and $A_{\perp}^N$ can be expressed in terms of contributions from the spin densities at the nitrogen 2s- and 2p-orbitals[24].

$$A_{\parallel}^N = g_e g_N \beta_E \beta_N \left(\frac{\alpha'}{2}\right)^2 [\gamma^2 (8\pi/3)|s(0)|^2 + \tfrac{4}{5}(1 - \gamma^2)\langle r^{-3}\rangle 2p] \tag{14}$$

$$A_{\perp}^N = g_e g_N \beta_e \beta_N \left(\frac{\alpha'}{2}\right)^2 [\gamma^2 (8\pi/3)|s(0)|^2 - \tfrac{2}{5}(1 - \gamma^2)\langle r^{-3}\rangle 2p] \tag{15}$$

The first term inside the square brackets is the isotropic part of the hyperfine coupling arising from the spin density at the 2s-orbital of the nitrogen atom. The second term is the dipolar part arising from the presence of spin density in the p_x- and p_y-orbitals of the ligand atom. The ligand σ-orbital is described as the hybridized function of 2s- and 2p-orbitals of the nitrogen. $|\sigma> = \gamma|2s> + (1 - \gamma^2)^{1/2} |p_\sigma>$. From the magnitudes of $A_{\parallel}^N$ and $A_{\perp}^N$ once can obtain the values of γ^2, the hybridization coefficient and α' in the wave function (4).

Thus we get independently the coefficients α and α' from the metal and ligand hyperfine couplings. The other coefficients β_1 and α_1 can also be obtained by using all the experiment data, that is, from the $g_\perp$, $g_{\parallel}$, $A_\perp$ and $A_{\parallel}$ values. So far we have discussed the evaluation of the MO coefficients without reference to any specific metalloporphyrin. We shall now consider the different d^9-systems and the complementary d^1-systems as well.

Copper porphyrins and the related copper phthalocyanins have been extensively studied by many groups. The g-values, hyperfine and superhyperfine tensors have been obtained both from the measurements in frozen solutions and in dilute single crystals (see Table 1 for the EPR data). The coefficients α, α', β and β_1 in equations (4)—(7) have been obtained from the EPR data by the procedure described above. The following features emerge from the EPR data for Cu(II) *meso*-tetraphenylporphyrin, Cu(TPP)[24]. The in-plane σ-bond between Cu and N is quite strong with α = 0.85—0.88 and α' = 0.53—0.55. The in-plane π-bonding is negligible (β_1 = 1.0) and the out-of-plane π-bonding between the metal d-orbitals and the ligand π-orbitals is very small (β = 0.92—0.95). The above values of the bonding coefficients are independent of the nature of the substituents in the porphyrin periphery. In the case of silver (TPP), the metal—ligand σ-bond is tighter than that observed for the copper complex (α = 0.82 for Ag(TPP)). Both the in-plane π-bonding (β_1 = 0.61) and the out-of-plane π-bonding (β = 0.67) are significantly higher in Ag(TPP) than in Cu(TPP). These results are quite in agreement with the general trend that ions of the second transition series form stronger covalent bonds with ligands than the corresponding ions of the first transition series. The EPR results for the copper porphyrins are in reasonable agreement with the corresponding parameters calculated using MO wave functions[33].

Solvent effects and the influence of axial ligands on the EPR parameters for the Cu porphyrins have been reported[21,22]. An interesting feature had been noticed in the EPR spectra of concentrated ($>10^{-3}$ M) solutions of Cu porphyrins[22,34]. Hyperfine couplings due to two Cu nuclei had been observed, and the EPR spectra could be interpreted as arising from a system containing two unpaired electrons and two Cu nuclei. This situation arises from the formation of dimers in concentrated solutions. The interaction between the two unpaired electrons in the dimer leads to the observation of zero-field splitting and a forbidden transition around g = 4. With the assumption of dominant dipolar interaction between the unpaired electrons, one can estimate the distance between the unpaired electrons or in other words, the distance between the two Cu nuclei. This had been found to lie in the

References, p. 586

TABLE 1

Selected EPR data for Cu(II), Ag(II), VO(II) and MoO(II) porphyrins

Compound	Medium	$T°K$	$g_\parallel$	$g_\perp$	$A_\parallel^M \times 10^{-4}\,cm^{-1}$	$A_\perp^M \times 10^{-4}\,cm^{-1}$	$A_{SUHP} \times 10^{-4}\,cm^{-1}$	Reference
Cu(II)Etio-II	Benzene	RT	\multicolumn{2}{}{$\langle g \rangle = 1.097$}	$\langle A \rangle = 89.28$		$\langle A^N \rangle = 14.3$	23
Cu(II)Etio-II	Castor oil	RT	2.169	2.061	188.0	39.0		23
Cu(II)TPP	Chloroform	RT	\multicolumn{2}{}{$\langle g \rangle = 2.107$}	$\langle A \rangle = 97.7$		$\langle A^N \rangle = 15.9$	27
Cu(II)TPP	H_2(TPP) single crystal	77	2.190	2.045	210.9	33.03	$A_\perp^N = 14.56$ $A_\parallel^N = 16.14$	24
Cu(II)OEP	Chloroform–EBBA mixture	100	2.173	2.078	186.6	41.11	$A_\parallel^N = 15.0$ $A_\perp^N = 18.33$	31
Ag(II)(Deut-IX-DME)	Castor oil	RT	2.104	2.029	71.0	31.0	$A_\parallel^N = 20.0$ $A_\perp^N = 23.0$	29
Ag(II)TPP	H_2(TPP) single crystal	77	2.108	2.037	56.62	28.28	$A_\parallel^N = 20.9$ $A_\perp^N = 24.05$	24
VO(IV)TPP	Chloroform	RT	\multicolumn{2}{}{$\langle g \rangle = 1.980$}	$\langle A \rangle = 89.4$			
VO(IV)TPP	Chloroform	77	1.961	1.989	161.2	88.7	$A_\parallel^N = 2.9$ $A_\perp^N = 2.8$	27
VO(IV)P	Triphenylene single crystal	77	1.964	1.985	174.0	59.0		36
MoO(TPP)OH	Benzene	RT	\multicolumn{2}{}{$\langle g \rangle = 1.983$}	$\langle A \rangle = 46.28$			37

Abbreviations and symbols used: H_2(TPP) = meso-tetraphenylporphyrin; H_2(OEP) = octaethylporphyrin; H_2(Etio-II) = Etioporphyrin-II; H_2(Deut-IX-DME) = Deuteroporphyrin-IX dimethyl ester; H_2(P) = porphin; A^M = hyperfine coupling for the metal nucleus; A_{SUHP} = superhyperfine couplings from the ligand; $\langle \rangle$ indicates isotropic value; RT = unspecified room temperature; EBBA = N-(p-ethoxybenzilidene)-p-n-butyl aniline liquid crystal.

range 3—3.5 Å for a number of Cu porphyrins. With the axial symmetry indicated by the EPR data, structures have been suggested for the dimer in which the two porphyrin planes are parallel[34]. EPR studies of the related systems Cu(II) chlorins and Cu(II) biliverdins have also been reported[31], the EPR parameters of Cu porphyrins and chlorins being almost identical. From the analysis of EPR data, it had been suggested that the ligand in Cu(II) formylbiliverdin has a helical structure with only moderate deviations from planarity. Preliminary X-ray studies have substantiated this conclusion[31a].

13.2.1.2. Vanadyl and molybdyl porphyrins[27,35,37]

Vanadyl porphyrins contain vanadium in the +4 state with a single unpaired electron in the d-orbital. Since sharp-line EPR spectra are observed at room temperature, the unpaired electron is in a non-degenerate orbital. On this basis the unpaired electron is located in the d_{xy}-orbital. This orbital is not involved in the σ-bonding with the ligand nitrogens. In addition to the bonding with the porphyrin, vanadium forms a σ- and a π-bond with the oxygen as the fifth ligand. One can write down the wave functions as was done for the d^9-case and obtain MO coefficients from the EPR data (Table 1).

The significant feature in the EPR spectra of vanadyl porphyrins[35] is that the superhyperfine coupling from the ^{14}N ligand nitrogen is very small (2.8 G) as compared with the large ^{14}N couplings observed with Cu and Ag porphyrins (14—15 G). The MO coefficients for vanadyl porphyrins indicate that the covalency of the V—N σ-bond is of the order of that of the Cu—N bond in Cu porphyrins. Again the in-plane π-bonding is negligible leading to almost complete localization of the unpaired electron on the metal d_{xy}-orbital. Significant π-bonding exists between vanadium and oxygen. The small superhyperfine coupling had been explained in terms of spin polarization of the nitrogen core by the unpaired electron on the vanadium.

As with the Cu porphyrins, evidence for dimer formation had been obtained from the ESR spectra of strong solutions of vanadyl porphyrins[34]. The distance between the two vanadium nuclei in the dimer is estimated to be in the range of 4—4.5 Å. The behavior of monomeric MoO porphyrins had been found to be similar to that of vanadyl porphyrin monomers (see Table 1)[37].

13.2.2. d^2- and d^8-Systems

EPR of metalloporphyrins with d^2-configuration has not been reported ported so far. Since the ligand field is square-planar, the d^8-system of Ni(II) and Ag(III) porphyrins are diamagnetic. In the presence of strong axial ligands, Ni porphyrin assumes a distorted octahedral configuration, leading to paramagnetism. Only the magnetic susceptibility and NMR contact shifts have been measured under these conditions[37a]. No EPR data are available for paramagnetic Ni(II) porphyrins at present.

References, p. 586

13.2.3. d^3- and d^7-systems: Co(II) porphyrins and related systems[38−65]

Co(II) porphyrins have been extensively studied by EPR because of their importance in oxygen adsorption. The d^7-configuration is in the low-spin state, leading to a $S = \frac{1}{2}$ system. In square-planar or square-pyramidal ligand fields, low-spin Co(II) has a configuration $d_{xz}^2\, d_{yz}^2\, d_{xy}^2\, d_{z^2}^1\, d_{x^2-y^2}$. With the unpaired electron in the d_{z^2}-orbital, one expects the following relationships between the EPR parameters and the electronic structure (to a first order)[44,45]

$$g_\| = 2.002 \tag{16}$$

$$g_\perp = 2.002 - 6\lambda/(E_{xz,yz} - E_{z^2}) \tag{17}$$

$$A_\|(^{59}\mathrm{Co}) = P[-\kappa + (\tfrac{4}{7}) - (\tfrac{1}{7})(g_\| - 2.002)] \tag{18}$$

$$A_\perp(^{59}\mathrm{Co}) = P[-\kappa - (\tfrac{2}{7}) - (\tfrac{45}{42})(g_\perp - 2.002)] \tag{19}$$

where A, γ, ρ and κ have the same meaning as in equations (10) and (11). Normally, P and κ values are estimated using the experimental data. The P and κ values can then be compared to the free-ion values P_0 and κ_0 to estimate the extent of covalent bonding between the metal and the ligand.

$E_{xz,yz} - E_{z^2}$ is the energy separation between the d-orbitals concerned. This energy gap is highly sensitive to the nature of the axial ligand, and hence the changes in the EPR parameters can be used to identify adducts of Co(II) porphyrins with π- or σ-donors or acceptors. The high sensitivity of the EPR parameters to solvent or adducts is in itself evidence for the presence of the unpaired electron in an A_{1g}-orbital (d_{z^2}) which is directly involved in axial interactions (see Table 2 for selected EPR data on Co(II) porphyrins).

Three distinct types of EPR spectra of Co(II) porphyrins have been reported so far[42]. In the first type obtained from the adducts of Co(II) porphyrins with π-acceptors, one finds the following typical EPR parameters. $g_\|$ = 1.7—1.9, $g_\perp$ = 2.9—3.2, $A_\|$ = 0.0130—0.0180 cm^{-1} and $A_\perp$ = 0.0250—0.0350 cm^{-1}. The characteristic features of this type of species are: $g_\|$ is less than 2.002 and $A_\|$ and $A_\perp$ are large and of comparable magnitude. In the second type of EPR spectra obtained from Co(II) porphyrins complexed with Lewis bases (generally a five coordinate complex) the following EPR parameters are found. $g_\|$ = 2.002, $g_\perp$ = 2.3, $A_\|$ = 0.0080—0.0090 cm^{-1} and $A_\perp$ = 0.0030—0.0040 cm^{-1}. One also observes superhyperfine coupling from the axial ligand nuclei like ^{14}N, ^{31}P (Figs. 4 and 5).

A third type of EPR spectrum is obtained from the molecular oxygen adducts of Co(II) porphyrins. The EPR spectra of these adducts are characterized by small g-anisotropy and very small values of the hyperfine couplings for the ^{59}Co nucleus and — more important — much longer electron spin relaxation times (smaller linewidths) than with the other two types of

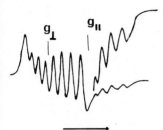

Fig. 4. X-band EPR spectrum of polycrystalline Co(II)(p-OCH$_3$)TPP. $g_\perp$ = 3.285, $g_\parallel$ = 1.79, $A_\perp$ = 0.039 cm^{-1} and $A_\parallel$ = 0.015 cm^{-1}. (Type I spectrum, see text).

Co complexes ($g_\parallel$ = 2.07, $g_\perp (g_x, g_y)$ = 1.99, $A_\parallel$ = 0.0015—0.0020 cm^{-1}, $A_\perp$ = 0.0009—0.0012 cm^{-1}) (Fig. 6).

The first category of EPR spectra has been reported for the polycrystalline free base doped with Co(II)TPP[27], for the complexes of Co(II)TPP with a number of π-acids[40—42] and also for β-Co(II) phthalocyanine[38]. Though it is clear that the unpaired electron is in a d$_{z^2}$-orbital, the large values for $A_\parallel$ and $A_\perp$ are not adequately accounted for by the existing theories. Hill et al. have studied NMR and EPR spectra of a number of complexes of Co(II) porphyrins with π-acceptors, steroids etc. and have proposed detailed models for these complexes[40,41,43].

In the second type of EPR spectra, axial ligand superhyperfine couplings have been obtained with imidazole type bases and ligands like CO, Et$_3$P and PF$_3$[39,42,44—50]. The bonding characteristics of the axial ligand with Co had been estimated from the ligand hyperfine couplings. It has been sug-

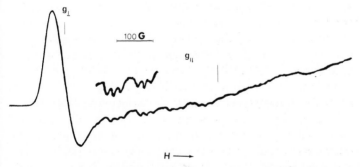

Fig. 5. X-band EPR spectrum of Co(II)OEP—imidazole (1 : 1) complex in chloroform—methanol at 100°K. The triplet splitting of the parallel lines is due to the interaction with a single ^{14}N nitrogen of imidazole. (Type II spectrum, see text). $g_\perp$ = 2.30, $g_\parallel$ = 2.02, $A_\parallel^{Co}$ = 0.0081 cm^{-1}, $A_\parallel^{N}$ = 0.0016 cm^{-1}.

References, p. 586

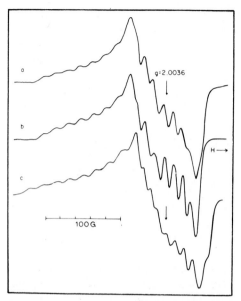

Fig. 6. X-band EPR spectra of superoxocobalt(III) complexes: (a) cobalamine, (b) *bis*-dimethylglyoximate cobalt, (c) cobalt mesoporphyrin-IX (ref. 125).

gested that CO is linearly complexed with the Co atom, the unpaired electron being predominantly (87%) on the Co atom[45].

Oxygen interacts with Co(II) porphyrin without axial ligand (base-off) and axially ligated Co(II) porphyrins (base-on) [L · Co(II) · porphyrin] to form 1 : 1 dioxygen complexes. An important feature of these complexes is that their EPR spectra can be observed at relatively higher temperatures than are normally required to obtain the spectra from oxygen-free Co(II) complexes[44−51]. This indicates that the unpaired electron in the dioxygen complex has very little orbital momentum. One normally observes a rhombic g-tensor (g_x, g_y, g_z) rather than an axial one. This is taken as evidence for a bent Co—O_2 unit. The hyperfine couplings from ^{59}Co are very small in the dioxygen complex (Table 2). Also, the axial ligand hyperfine coupling is considerably reduced in the L · Co(II)TPP · O_2 complexes ($<a>$ for ^{31}P in Co(II)TPP · Bu_3P is 0.0196 cm^{-1} while in Co(II)TPP · Bu_3P · O_2 is a_p = 0.0017 cm^{-1}). Recent EPR measurements on an ^{17}O-enriched Co(II)TPP · O_2 complex showed a large hyperfine coupling for ^{17}O (21 G) and also indicated the equivalence of the two oxygen atoms in the complex[52]. All the above-mentioned data, particularly the low values for the hyperfine couplings of ^{59}Co and the axial ligand suggest that the unpaired electron spends much less time on the Co atom than on the π^*-levels of O_2 molecules. It has also been observed that stronger σ-donating axial ligands lead to smaller ^{59}Co hyperfine couplings thus promoting the localization of the π-electrons on the oxygen. The hyperfine coupling of ^{59}Co arises mainly from the polarization of the σ-bond Co—O_2 by the unpaired electron in the π_g-orbital formed by the overlap of d-orbitals of Co and the π^*-orbital of O_2 molecules. The structure of the dioxygen adduct of Co(II) porphyrin is represented by Co(III)P · B · O_2^-, where P stands for porphyrin and B for axial base. From an EPR study of the oxygen adduct of a Schiff base Co(II) complex, Tovrog and Drago[52a] have suggested that the assumption of the configuration with Co(III) in the dioxygen complexes is untenable. But a recent ab initio calculation of the electronic energies of dioxygen adducts of Co(II) Schiff base complexes with different axial base ligands has indicated that both the structures Co(III)O_2^- and Co(II)O_2 are of comparable energy and the former is more stable[53]. The contribution of the former structure increases with stronger σ-donating ligands in the axial position. These results support the generalization based on experimental data that ligands which stabilize Co(III) relative to Co(II) would give systems with the highest affinity for oxygen[50].

EPR studies on single crystals of Co(II) myoglobin and cobalt oxymyoglobin have been reported[51]. A number of studies have also been reported on the related cobalamines, the Co(II) complexes of corrins, tetradehydrocorrins and corroles[54—57]. Of these, the Co(II) complexes of corrins and tetradehydrocorrins have their ground states similar to those of Co(II) porphyrins. In the Co(II) complexes of corroles the ground state has been found to be $x^2 - y^2$ (x and y axes bisect the N—Co—N bond angle)[57]. Other d^7-systems include Ni(III)TPP$^+$, obtained by electro-oxidation of Ni(II)TPP for which an axially symmetric g-tensor ($g_\perp$ = 2.295, $g_{||}$ = 2.116) has been reported. A d^7-Fe(I)TPP$^-$ has been identified[59—61]. It has been obtained by chemical as well as electro-chemical reduction of μ-oxo-bis-meso-tetraphenylporphyrin Fe(III). The compound Fe(I)TPP$^-$ gives an EPR spectrum indicating axial symmetry ($g_\perp$ = 2.30, $g_{||}$ = 1.93). Since $g_\perp > g_{||}$, the unpaired electron is assigned to the d_{z^2}-orbital.

13.2.4. d^4- and d^6-systems: Fe(II) and Mn(III) porphyrins[62—65]

The high-spin Fe(II) porphyrins (d^6, $S = 2$) and Mn(III) porphyrins (d^4, $S = 2$) do not give any EPR spectra at the spectrometer frequencies available. Very large zero-field splittings and short spin—lattice relaxation times are considered as reasons for the absence of EPR signals. Fe(II) porphyrins form paramagnetic adducts with nitric oxide, of the type Fe(II)P · NO. The EPR spectrum of Fe(II)P · NO in solution at room temperature consists of 3 lines with a separation of 17 G. The Fe atom is in a low-spin state with the unpaired electron predominantly being in the π^*-level of nitric oxide. For frozen solutions, rhombic g- and hyperfine tensors are reported. The g-values agree with a $d_{z^2}^1$-configuration arising from the overlap of a d_{z^2}-orbital with the orbitals of nitric oxide. The rhombic g-tensor indicates the presence of a bent Fe—NO unit. Crystal structure studies also support this view. Analogous

TABLE 2

Typical EPR data for Co(II) porphyrins and related systems and their adducts with π-donors and acceptors, Lewis bases and diatomic molecules

Compound	Medium	$T°K$	$g_\parallel$	$g_\perp$	$A_\parallel^{Co}$ $\times 10^{-4} cm^{-1}$	$A_\perp^{Co}$ $\times 10^{-4} cm^{-1}$	axial A_{SUHP} $\times 10^{-4} cm^{-1}$	Reference
Co(TPP)	H₂(TPP)	77	1.798	3.322	197	315	—	27
	Toluene	77	1.966	2.848	138	272	—	27
Co(p-CH₃)TPP + 1,3,5-trinitro benzene	Toluene	77	1.894	3.051	133	301		42
Co(II)(Meso-IX-DME) + deoxycorticosterone (1:1)	Chloroform	123	2.020	3.310	292	227		40
Co(p-OCH₃)TPP with pyridine (1:1)	Toluene	100	2.025	2.327	76.4	12	15.6 ($A_\parallel^N$)	42
Co(p-OCH₃)TPP with imidazole (1:1)	Toluene	100	2.028	2.308	77.1	12	16.2 ($A_\parallel^N$)	42
Co(TPP) with methyl-isonitrile (CH₃NC) (1:1)	Toluene	123	2.025	2.247	67.0	37	— $A_\parallel(^{13}C) = 59.7$	44
Co(TPP) with carbon-monoxide	Toluene	123	2.017	2.217	73.6	34	$A_\perp(^{13}C) = 53.5$	44
Co(TPP) with trimethoxy-phosphine	Toluene	123	2.024	2.244	69.2	28	$A_\parallel^P = 295$ $A_\perp^P = 264$	45

Compound	Solvent	Temp	g	$\langle A \rangle (^{59}Co)$	A	other	Ref
Co(p-OCH$_3$)TPP · O$_2$	Toluene	100	2.083 $g_x = 1.996$ $g_y = 2.006$	29.2	$A_x = 17.3$ $A_y = 18.7$		42
Co(p-OCH$_3$)TPP · O$_2$ with trinitro benzene	Toluene	100	b $g_x = 1.986$ $g_y = 1.973$	b	$A_x = 34.2$ $A_y = 34.2$		42
Co(p-OCH$_3$)TPP · O$_2$ with pyridine	Toluene	100	2.077 2.002	· 16.6	10.7		42
Co(p-CH$_3$)TPP · O$_2$ with imidazole	Toluene	100	2.077 2.003	17.7	10.6		
Co(TPP) · O$_2$ with trimethoxy phosphine	Toluene	133	$\langle g \rangle = 2.022$	$\langle A \rangle \langle^{59}Co\rangle = 7.8$		$\langle A \rangle \langle^{31}P\rangle = 14.8$	45
α-Co(II)Pc	ZnPc powder	77	2.007 2.422	116	66	—	38
β-Co(II)Pc	ZnPc powder	77	1.910 $g_x = 2.92$ $g_y = 2.89$	160.0 136.7	$A_x = 270.0$ $A_y = 260.0$	—	38
Co(II)Cn$^+$ClO$_4^-$	THF	77	1.993 2.374	136.7	88.7	$A_{\parallel}^N = 18.8$	55
Co(II)Cn$^+$ClO$_4^-$	Pyridine	77	1.998 2.180	98.9	33.6		55
Co(II)Cobyr I	Toluene	77	2.003 2.263	112.0	0.0	$A_{\parallel}^I = 140.0$	56
Co(III)Cobyr$^+$I$^-$ · O$_2^-$	Toluene	77	2.060 2.001	19.2	11.0		56
Co(II)Tdh$^+$ClO$_4^-$	THF	77	1.991 2.374	136.0	89.0	—	57
Co(II)Tdh$^+$ClO$_4^-$	Pyridine	77	1.998 2.180	99.0	34.0	$A_{\parallel}^N = 18.0$	57
Co(II)Corrole$^-$	THF	77	1.966 $g_x = 2.776$ $g_y = 2.321$	47.0	$A_x = 140.0$ $A_y = 102.0$		57

Abbreviations and symbols: H$_2$(TPP) = *meso*-tetraphenylporphyrin; H$_2$(Meso-IX-DME) = mesoporphyrin-IX dimethyl ester; Pc = phthalocyanine; Cn = 1,19-diethoxycarbonyltetradehydrocorrin; Cobyr = cobyric acid heptamethyl ester; A_{SUHP}^{axial} = superhyperfine coupling from the axial ligand; THF = tetrahydrofuran; b = unresolved.

References, p. 586

complexes of NO with Fe(II) myoglobin and hemoglobins have been reported. Fe(II) porphyrin · NO systems serve as a model for the understanding
of the nature of hemoprotein complexes with diatomic molecules like CO
and O_2. The complexes of the latter two systems with Fe(II) porphyrins are
diamagnetic[66].

13.2.5. d^5-systems: Mn(II) and Fe(III) porphyrins[66−78]

Considerable interest has been shown in the EPR studies of Fe(III) porphyrins in recent years. Certain aspects of the physicochemical studies on
biologically important Fe porphyrins are discussed in Ref. 127. A large
number of review articles have also appeared on the correlation of electronic
structure with EPR studies on Fe porphyrins. Hence only a brief treatment
of this subject will be presented here.

In the high-spin state of Fe(III) and Mn(II), all the five d-orbitals are
singly occupied, and the system is orbitally non-degenerative ($^6S_{5/2}$). The
6-fold degeneracy of the spins ($|\pm 1/2>$, $|\pm 3/2>$, $|\pm 5/2>$) is removed
even in nearly cubic ligand fields due to the higher-order spin—spin interactions
among the electrons. The zero-field splitting arising from these interactions is
small, and under these conditions EPR spectra show five transitions with
$\Delta M_s = \pm1$. The separation between the transitions gives the value of the
zero-field splitting energy. In strong crystal fields of tetragonal symmetry
(like that of porphyrins) the zero-field splitting is very large, and the lowest
state $|\pm 1/2>$ is separated by about 5—10 cm^{-1} from the other two states
$|\pm 3/2>$ and $|\pm 5/2>$. For protohemin the interaction energy is about
7 cm^{-1}. Since this quantity is much larger than the spectrometer frequency
(~ 3 cm^{-1}), one observes only the transition $\pm 1/2$ for the $S = 5/2$ system.
This transition is highly anisotropic with $g_\perp \cong 6.0$ and $g_\parallel \cong 2.0$. These
g-values are considered as fingerprints for high-spin d^5 in strong tetragonal
fields with axial symmetry. Distortions from axial symmetry cause the line
at $g = 6$ to split. In the high-spin configuration, Fe in protohemin is slightly
above the plane of the ligand. When strong axial ligands like N_3^-, CN^- or
SH^- are introduced, the high-spin form is converted into the low-spin form
with $S = 1/2$. Then one observes axial ($g_\perp$, $g_\parallel$) or rhombic ($g_x \neq g_y \neq g_z$)
g-values around 2.0 (e.g. for the protohemin—imidazole complex, $g_1 = 2.78$,
$g_2 = 2.26$, $g_3 = 1.72$)[72,73,75]. Fe(III) porphyrins with weak axial ligands
like Cl^- are typical of high-spin systems. Those with strong axial ligands
always exist in the low-spin state with the Fe atom in the plane of the
porphyrin (e.g. hemichromes) (Fig. 7). EPR studies on single crystals of
perylene doped with protohemin have been reported. The small superhyperfine couplings (4 G) from the porphyrin nitrogens were found to be of predominantly isotropic character.

The rhombic nature of the g-values in the low-spin form arises due to the
distortion of axial symmetry by the fifth and sixth ligands. In hemoglobin
this form of distortion is due to different conformations of the protein

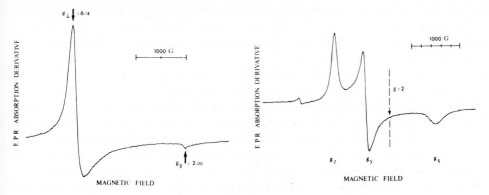

Fig. 7. X-band EPR spectrum of (a) chlorodeuterohemin dimethyl ester in dimethyl formamide at 77°K, (b) di-(imidazole) complex of chlorodeuterohemin dimethyl ester in dimethyl formamide (ref. 74).

moiety around the protohemin fragment[68-71]. When Fe(III) hemoglobin is denatured[73], the final EPR absorption lies in the range from $g = 1$ to $g = 3$. Thus distortions of varying degrees are present in this system. Model studies have been carried out on the effect of different axial ligands on g-values with the low-spin form of Fe(III) porphyrins and Fe(III) chlorins[73].

An important characteristic of Fe(III) porphyrins is that they do not give any EPR spectra at room temperature because the spin—lattice relaxation times are very short. One obtains reasonably good EPR spectra at liquid-nitrogen temperature (linewidths of the order of 100 G) and sharper and more intense lines at liquid-helium temperatures.

Mn(II) porphyrins, though unstable, can be obtained in oxygen-free solutions by the reduction of Mn(III) porphyrins. In frozen solutions porphyrins do not yield any EPR spectra, not even at liquid-helium temperature, as on freezing strong aggregation and solvent-solute segregation broaden the EPR lines considerably. An ingenious method was developed by Yonetani et al.[78] who replaced the natural heme in hemoglobins and cytochrome c by synthetic porphyrins. Thus, in a hemoprotein containing Mn(III) porphyrin, Mn(III) can be reduced to Mn(II), and EPR spectroscopy becomes feasible. The spectra obtained in this way are quite well resolved and show hyperfine structure of 6 lines ($I_{Mn} = 5/2$) from the Mn nucleus. Mn(II) is in the high-spin configuration and yields two g-values ($g_\perp = 5.9$ and $g_\parallel = 2.0$). By studying the variation of the absorption at $g_\perp = 5.9$ at several microwave frequencies, the zero-field splitting parameter was estimated to be 0.5 cm^{-1}. Small rhombic distortions were also observed through these measurements.

13.3. Porphyrins with unpaired electrons on the ligand
13.3.1. Radical cations of free-base and metalloporphyrins
13.3.1.1. A brief review of the electronic structure of the porphyrin ligand[33,79,80]

The porphyrin macrocycle with 26 π-electrons is highly conjugated. In many metalloporphyrins, the π-system undergoes facile reversible one and two-electron oxidation and reduction. The free base porphyrin is of D_{2h} symmetry and the outermost bonding orbitals are non-degenerate. Many metalloporphyrins with a divalent central metal belong to the point group D_{4h}, and the highest bonding orbitals are nearly degenerate. The lowest unoccupied orbitals have E_g symmetry. This situation prevails with metalloporphyrins with or without alkyl substituents in the β-pyrrolic positions and no substituent at the *meso* positions. The charge distribution in the two highest occupied MO's are complementary. In one of them (A_{1u}) the charge densities at the *meso* positions and at the nitrogen atoms are very small[79,88] (see also Chapter 15). In the other MO with A_{2u} symmetry the *meso* positions have the highest charge density, and the orbital coefficients for the nitrogen atoms are also appreciable. The energy difference between the two MO's is very small (of the order of 0.1 eV) and hence is very sensitive to even small perturbations. When all the *meso* positions are substituted by phenyl or alkyl groups, the A_{2u} level is raised considerably so that the energy difference between A_{1u} and A_{2u} becomes appreciable. The E_g levels are truly degenerate in both these situations (i.e. A_{1u} and A_{2u} ground states).

Cation radicals are formed by one-electron oxidation of porphyrins either chemically or by electrolysis[81-90]. The radical cations from the free-base porphyrins are unstable; but those from metalloporphyrins with closed-shell metal ions are quite stable, and the oxidation potentials are dependent on the nature of the central metal ion[87a]. The unpaired electron is in an A_{1g} or A_{2g} level, depending on the nature of the porphyrin and also the environment in solution.

The EPR spectra of the free radical system in solution are explained by means of the isotropic spin Hamiltonian

$$\mathcal{H} = g\beta H S_z + S_z \sum_i a_i M_i \tag{20}$$

where i is the index for a group of magnetically equivalent nuclei and M_i is the nuclear magnetic quantum number for the group i. If there are n equivalent nuclei of spin I in a group, the total spin I_n is equal to nI. Such a group of nuclei will give rise to $2nI + 1$ hyperfine lines of spacing a. When there is a large number of different groups of nuclei in a system, the EPR spectrum becomes complicated and to extract the hyperfine coupling constants, one must resort to computer simulation.

The g-values of π-radical systems are very close to that of the free-electron spin value, since in a delocalized system the orbital angular momenta are completely quenched. Consequently one observes EPR spectra of such radical systems in solutions even at room temperatures and above. Very small linewidths ($\leqslant 100$ mG) are realized in these spectra. The assignment of hyperfine couplings to a particular center in a molecule cannot be done just from the EPR spectrum. On the basis of symmetry (No. of equivalent nuclei) this is sometimes possible. But generally the assignment is done by isotopic substitution (^{2}H for ^{1}H, or ^{15}N for ^{14}N, etc.). Deuterium (^{2}H) has a spin of $I = 1$ and a magnetic moment six times smaller than that of a proton. Hence the number and the spacing of the hyperfine lines will change as a consequence of isotopic substitution and thus allow the assignment of hyperfine couplings.

13.3.1.2. Radical cations from meso-tetraphenyl- and octaethyl-porphyrin (OEP) systems

The EPR spectra obtained from metalloporphyrin radical cations can be classified into two types. The unsubstituted porphyrins, exemplified by OEP's with A_{1u} ground state, yield EPR spectra with small hyperfine couplings from the four equivalent *meso* protons[83,86,88] ($a_H^{meso} = 1.57$ G). Hyperfine coupling from the ^{14}N's are estimated to be less than 0.3 G. The density of the unpaired electrons at a carbon π-orbital can be estimated from the proton hyperfine couplings using the McConnell relation[91]

$$a_H = Q_{C-H} \rho_\pi^C \tag{21}$$

where Q is a measure of σ-π spin polarization ($Q \cong -25$ G). One can also calculate the spin densities by the MO method. The results confirm the presence of the unpaired electron in an A_{1u} orbital. Radical cations of *meso*-tetraphenyl- and *meso*-tetraalkyl-porphyrins yield an entirely different type of EPR spectrum[88]. (Fig. 8). The lines are very narrow (linewidth = 100 mG); large spin densities are indicated at the *meso* positions. Hyperfine couplings from the α-CH$_2$ protons in the *meso*-tetraalkyl-substituted porphyrins are of the order of 3—4 G, and the nitrogen hyperfine couplings 1.5—2 G. The EPR data undoubtedly show the presence of the unpaired electron in the A_{2u} state.

The proton and nitrogen hyperfine couplings are not very sensitive to the nature of the central metal ion[88]. Hyperfine couplings[88] from the central metal atoms have also been observed for ^{67}Zn ($I = 1/2$), ^{59}Co ($I = 7/2$) and ^{205}Tl ($I = 1/2$)[88]. Except for ^{205}Tl, other metal ions yield only small hyperfine couplings. The quantities a_{Zn} and a_{Co} decrease with a decrease in the nitrogen spin densities. [$a_{Co} = 1.2$ G and 5.6 G respectively in Co(III)OEP$^{++\cdot}$ and Co(III)TPP$^{++\cdot}$]. In the former case the unpaired electron is in an A_{1u} orbital with low spin density at the nitrogen atoms. In (TPP), with an A_{2u} ground state, the nitrogen spin density is appreciable. These observations, combined with the fact that the optical spectra of these cation

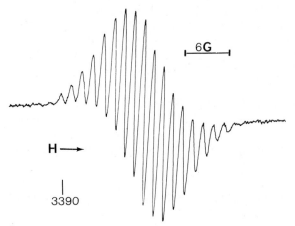

Fig. 8. X-Band EPR spectrum of the cation radical of *meso*(undecyl)porphyrin, obtained by oxidation with iodine. $a_H^{(\alpha-CH_2)} = 3.00$ G. $a_N = 1.5$ G. $<g> = 2.0030$.

radicals and the parent compounds are insensitive to the nature of the metal ions, point to negligibly small interactions between the ligand π-orbitals and the corresponding metal d-π orbitals. The hyperfine couplings of the metal ions are thus explainable solely in terms of the spin-polarization of the N-metal σ-bond by the unpaired electron on the p-π orbital of the N atom.

The situation in the radical cations of Tl(III) porphyrins (Fig. 9), however, is different[90]. Large Tl-hyperfine couplings were found for these radical cations ($a_{Tl} = 12$–60 G), since the optical hyperfine coupling constant of ^{205}Tl is very large (~7000 G). π-Interactions of even very small magnitude are thus detectable in Tl-containing radical ions. The magnitude of a_{Tl} decreases when going from (OEP) ($a_{Tl} = 60$ G) to (TPP) ($a_{Tl} = 12$ G). Also, CN^- as an axial ligand increases the Tl-hyperfine coupling from 12 G to 58 G in the radical cation of Tl(III)TPP$^{+\cdot}$. On the other hand, a_{Tl} decreases from 60 G to 19 G in Tl(III)OEP$^{+\cdot}$ under the same conditions. These observations have been explained by taking into account the σ-π spin polarization of

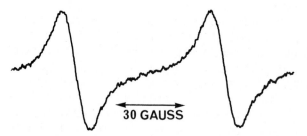

Fig. 9. EPR spectrum of the cation radical of Tl(III)OEP$^+$ ClO$_4^-$ obtained by electro-oxidation in dichloromethane. $a_{Tl} = 65$ G. $<g> = 2.0029$.

the Tl—N bond and direct π-interactions of Tl orbitals with the ligand π-orbitals. The contributions from these two types of mechanisms to the Tl-hyperfine coupling are of opposite signs. Direct π-interactions have also been suggested by NMR studies on Tl—H spin couplings in the Tl(III) porphyrins (see p. 492)[93]. Although the EPR and NMR results indicate the presence of direct π-interactions between the metal and the ligand in the Tl(III) porphyrins, optical spectra and redox potentials do not show any abnormalities. The out-of-plane π-bonding is not significant enough to affect the visible spectra and redox potentials. On the other hand, these effects are reflected by the EPR and NMR spectra owing to the very large optical hyperfine coupling of the Tl nucleus; hence a_{Tl} is quite sensitive to even small variations of spin densities. From the NMR data it had been suggested that the Tl atom is not in the plane of the porphyrin ring[93], and this has been confirmed in an X-ray investigation (p. 347).

What is the implication of the above EPR studies on porphyrin radical cations? First of all, the ease of formation and the stability of the radical cations are of basic importance in their role in photosynthesis. Second, the mapping of spin densities by the EPR method helps us to rationalize the regioselectivity found in electrophilic and radical reactions of these systems.

An important aspect of porphyrin-radical chemistry is their ability to form stable dimers (see p. 618). When the radical cations of Mg and Zn (OEP) are cooled to $-30°C$ and below, the EPR signal diminishes in intensity and disappears completely around $-60°C$. This process is reversible and has been explained as due to the formation of diamagnetic dimers (see p. 619 for a detailed discussion of these dimers)[94].

13.3.1.3. Chlorins and chlorophylls[90,94a−97]

Chlorins and bacteriochlorins are more easily oxidized than the porphyrins. These two systems form the basic chromophore of chlorophyll a and bacteriochlorophyll respectively. EPR spectra of the radical cations of Mg and Zn octaethylchlorins (OEC) consist of three lines with intensity $1:2:1$ and a separation of 4.8 G. EPR studies of the radical cation of partly deuterated (meso)-$\gamma\delta$-d$_2$-Zn(II)OEC indicated that the hyperfine coupling may arise either from the equivalent meso protons (α,β) or from the two β-protons in the partly reduced pyrrole ring[90]. The EPR spectrum of the radical cation of Zn(II) 1,1,3,4,5,6,7,8-octaethyl-2-oxo-chlorin (Fig. 10) consists only of one narrow single line, indicating that the hyperfine coupling of 5.8 G can be assigned to the two protons in the reduced pyrrole ring[116]. This assignment was further confirmed by a recent ENDOR study of chlorophyll[96].

In ENDOR, a double-resonance technique, the electron-spin levels ($M_s \pm 1/2$) are first saturated by strong microwave power. Thus the EPR signal becomes very broad and of low intensity. Then a radiofrequency signal is applied to the sample to polarize the spin levels. Due to cross-relaxation, an

References, p. 586

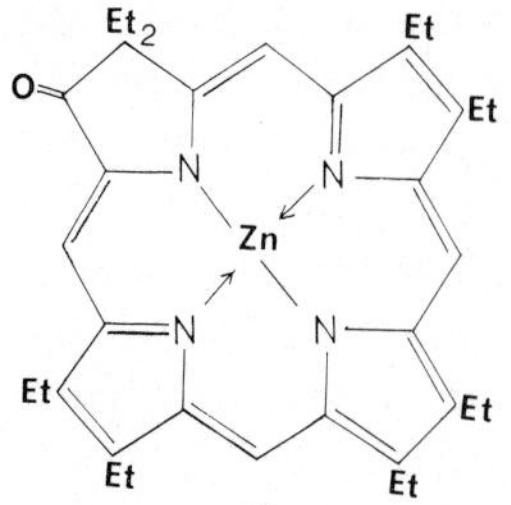

Fig. 10. 1,1,3,4,5,6,7,8-Octaethyl-2-oxochlorin.

EPR signal will appear when the radiofrequency equals that corresponding to the hyperfine coupling energy. This method has been used extensively in obtaining the proton hyperfine couplings in chlorophyll systems[96,97]. In the normal EPR mode, these systems yield only broad, unresolvable single-line EPR spectra of a width of 5—10 G.

13.3.2. Radical anions from porphyrins and related systems[98—101]

13.3.2.1. Metalloporphyrins[98,99]

As mentioned earlier, the lowest empty orbitals of a metalloporphyrin (with D_{4h} or C_4 symmetry) are truly degenerate. When a radical anion is formed, the unpaired electron thus goes into a degenerate orbital and consequently has a large orbital momentum. According to the Jahn—Teller theorem, this degeneracy may be removed by static or dynamic distortion of the molecular system. Nevertheless, very low-lying excited states in such a system lead to a significant amount of orbital momentum which is not completely quenched. This is reflected by the appearance of broad, unresolved EPR lines for the radical anions of metalloporphyrins[98].

13.3.2.2. Radical anions of bacteriochlorins

A well-resolved EPR spectrum has been reported for the radical anion obtained by electroreduction of meso-tetraphenylbacteriochlorophyll[100]. The corresponding anion of bacteriochlorophyll yields an only partially resolved EPR spectrum[100].

13.3.2.3. Phlorins

Reductions of Zn complexes of etioporphyrin and tetrabenzoporphyrin as well as the free base etioporphyrin in ether by alkali metals first leads to the formation of the respective radical anions[101]. The latter may disproportionate in the following way:

$$2ZnP^- + H^+ \rightleftharpoons ZnP + ZnPH^- \tag{22}$$

The diamagnetic phlorin anion is further reduced to the radical dianion which gives a resolved EPR spectrum. SCF-π calculations indicate with certainty that there is no orbital degeneracy in the lowest unoccupied levels of phlorins. The phlorin—porphyrin equilibrium is discussed in Chapters 14—17.

13.3.3. Triplet states of porphyrins and metalloporphyrins[103−115]

The formation and chemistry of triplet states of porphyrins is discussed in Chapter 16. A triplet state has two unpaired electrons ($S = 1$). The relaxation from a triplet- to the ground-state (singlet) is symmetry-forbidden and hence is slow. Thus the triplet state has a longer lifetime (from milliseconds to seconds) than the singlet excited state. However, molecular collisions and other dynamic processes usually reduce the lifetime of a triplet. Also, in fluid media the anisotropic interactions in a triplet are averaged to zero, and no magnetic information is obtainable. Hence, EPR spectroscopy of photo-excited triplets has been carried out in rigid glasses or oriented organic single crystals[102].

In the triplet state of a conjugated system, the unpaired electrons move in delocalized π-orbitals. Consequently they have negligible orbital angular momenta. The magnetic interactions (zero-field splittings) between the electrons are mainly of dipolar origin and hence very small (10^{-2} cm^{-1}) compared to the large zero-field splittings ($\leqslant 1$ cm^{-1}) observed for transition metal ions. In the latter systems, it must be remembered, the ZFS is strongly coupled to the spin-orbit interactions.

The spin Hamiltonian for a triplet state of a π-system is

$$\mathcal{H} = g\beta \hat{H} \cdot \hat{S} + D(S_z^2 - \tfrac{1}{3}\hat{S}^2) + E(S_x^2 - S_y^2) + \hat{S} \cdot \underset{\sim}{A} \cdot \hat{I} \tag{23}$$

where D and E are the ZFS parameters. E indicates the deviation of the molecular system from axial symmetry. Since $S = 1$, there are two allowed transitions with $\Delta M_s = \pm 1$, the oscillating field being *perpendicular* to H_z. This selection rule holds only at high magnetic fields, that is, where the total electron spin S is quantized. At lower magnetic fields, however, S_z is not a good quantum number, and a new transition ($\Delta M_s = \pm 2$) is allowed when the oscillating field is *parallel* to the component H_z of the external magnetic field H_0. This transition gives rise to the 'half-field line' ($g = 4.0$) which is isotropic and more intense than $\Delta M_s = \pm 1$ transitions. In the earlier studies on triplet states of porphyrins only the $\Delta M_s = \pm 2$ lines were detected. Improvement of instrumentation in recent years made it possible to detect $\Delta M_s = \pm 1$ lines[105−107]. A new technique, called optically detected magnetic resonance (ODMR)[111], allows very accurate determinations of the parameters D and E[111−115]. In the triplet states of conjugated systems, D is a measure of the delocalization of the π-electrons. The value of D decreases with increasing electron delocalization. The hyperfine couplings, if observed, yield information on the charge distribution in the triplet state.

References, p. 586

From the EPR studies of the triplet states in porphyrins and chlorophylls, the following results have emerged[105,108,109,115].

(i) The lowest triplet arises from a π-π^* exitation and not from a n-π^* transition.

(ii) The free-base porphyrins have higher ZFS's ($D = 0.046$ cm^{-1}) than the metalloporphyrins with D_{4h} symmetry ($D = 0.034$ cm^{-1}), indicating larger delocalization in the latter.

(iii) In contrast to the porphyrins, in chlorophyll b (Mg-complex) ($D = 0.030$ cm^{-1}) and pheophytin (free-base chlorophyll) ($D = 0.033$ cm^{-1}) the ZFS's are very similar. E-values are also appreciable ($E = 0.003$ cm^{-1})

(iv) Electron delocalization, as estimated from the D-values, is larger in Mg phthalocyanine than that in Mg porphyrin.

Two near-degenerate triplet states are possible for the porphyrins, namely $^3|A_{1u} - E_g|$, $^3|A_{2u} - E_g|$. This near-degeneracy opens up the possibility of spin—orbit interaction contributing to the ZFS. This effect is offset by the crystal-field perturbations and vibronic interactions arising from the Jahn—Teller effect. As a result the ZFS's are accounted for predominantly by spin—dipolar interactions, including only second-order contributions from orbital degeneracy. In etio- and octaethyl-porphyrins, the triplet state has been indentified as $^3|A_{1u} - E_g|$, and in $meso$-tetraphenylporphyrins as $^3|A_{2u} - E_g|$[109].

13.4. Porphyrins with unpaired electrons in the metal as well as in the ligand
13.4.1. Oxidation products of Cu porphyrins and chlorins

One-electron oxidation and reduction of Cu $meso$-tetraphenylporphyrin and phthalocyanines have indicated the presence of two unpaired electrons in the molecular system as shown by the measurements of their magnetic susceptibilities at room temperature[84,117]. The visible spectra of these systems also resemble those of the radical cations of metalloporphyrins. On the other hand neither at room temperature nor around 80° K could EPR spectra be observed.

The absence of EPR signals at room temperature and the paramagnetism of the compound point to a high population of triplet states at these temperatures; their absence at 80° K may be either due to a very large ZFS or to the antiferromagnetic coupling of the two unpaired electrons. To establish the electronic state of these systems, more experimental data are required. The one-electron oxidation product of Cu(II)OEP also behaves in a similar way[82].

Copper octaethylchlorin, Cu(II)OEC, is easily oxidized by iodine to a radical type species, as indicated by the visible spectrum[116]. This radical, Cu(II)OEC$^{+\cdot}$, does not give any EPR signals at room temperature; but frozen solutions at around 100° K do yield a spectrum characteristic of a triplet state ($D = 0.02$, $E = 0$). Both $\Delta M_s = \pm 1$ and $\Delta M_s = \pm 2$ transitions were observed, the intensity of the latter transition being much weaker than that of the former. Hyperfine couplings from two equivalent $I = 3/2$ nuclei, ascribed to Cu, were also noticed. The hyperfine coupling ($A^{Cu} = 0.009$ cm^{-1}) is half of the corresponding value for the monomer Cu(II) chlorin ($A^{Cu} = 0.018$ cm^{-1}). The species at low temperature is thus identified as the dimer of the radical Cu(II)OEC$^{+\cdot}$. The unpaired electrons on the ligand are antiferromagnetically coupled, leaving the two unpaired electrons on the copper nuclei. Exchange interactions ($J \gg a_{Cu}$) between the electrons at the Cu-orbitals perhaps reduce the Cu-hyperfine coupling to one half of its original value. A structure has been suggested for this dimer in which an axial ligand, possibly I$^-$, is sandwiched between the porphyrin planes[116].

13.4.2. Oxidation products of iron porphyrins

The Fe-porphyrins Fe(III)OEP$^+$, Fe(III)TPP$^+$ and [Fe(III)TPP]$_2$O undergo reversible one-electron oxidations[117a]. At low temperatures the oxidized products of [Fe(III)TPP]$_2$O and [Fe(III)OEP]$_2$O yield single-line EPR spectra ($g = 1.99$), but the oxidation products of Fe(III)OEP$^+$ and Fe(III)TPP$^+$ yield none. A magnetic susceptibility of 5.1 BM had been obtained for the latter two systems. An oxidation state of +4 ($S = 2$) has been proposed for the Fe atom in these systems; whether they contain one unpaired electron on the ligand (π-cation) with the Fe atom in the +3 oxidation state or a closed-shell π-system with Fe(IV), is still an open question. Visible spectra of these systems do not yield unambiguous indications.

A paramagnetic dioxygen complex of a Fe(II) porphyrin has recently been reported[120]. Apart from the characteristic signals of a high-spin Fe(III), a free-radical type signal was also noticed for this species. It has been suggested that Fe(II) is oxidized to Fe(III) and that the additional electron goes to the E_g level of the porphyrin, forming a radical anion.

13.4.3. Oxidation products of vanadyl porphyrins

The EPR spectra of solutions of VO porphyrins at room temperature consist of eight lines (^{51}V has $I = 7/2$). Although the absorption intensities for all the lines are equal, their widths are different; thus the variation in linewidths leads to a variation in the amplitudes (or heights) of the derivative lines (Fig. 11a). The linewidth variations have been explained as due to the modulation of the g- and hyperfine tensors by molecular tumbling (see Section 13.1.2.5.). As a result, the linewidths δ depend on the nuclear quantum numbers and take the form

$$\delta = A + Bm_I + Cm_I^2 + Em_I^3 \tag{24}$$

where A, B, C and E represent the contributions from different modulating mechanisms. Normally one observes a preferential broadening of the outer lines ($C > 0$) with VO systems. One-electron oxidation of VO(OEP) forms a

References, p. 586

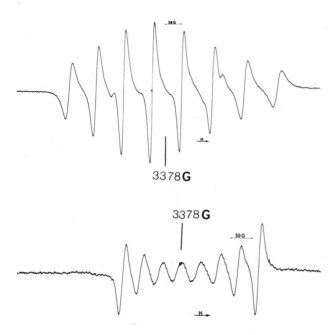

Fig. 11. X-band EPR spectrum of (a) VO(OEP) in chloroform at 290°K, $<g>$ = 1.9799. (b) Cation radical of VO(OEP) obtained by electro-oxidation in dichloromethane at 290°K, $<g>$ = 1.9905.

species whose EPR spectrum is presented in Fig. 11b[119]. Its dominant feature is the broadening of the inner lines (C < O). According to the existing theories, this phenomenon can be accounted for only by considering the presence of more than one unpaired electron[124]; indeed, magnetic susceptibility measurements indicate the presence of two unpaired electrons. The EPR spectra are also independent of concentration in the range from 10^{-3} to 10^{-5} M, and the linewidth effects persist even at temperatures around 80°C. Based on these observations, it had been proposed that the species under investigation is a π-radical cation of VO(OEP); the hyperfine coupling of ^{51}V is one-half of the value found with solutions of unoxidized VO(OEP). The decrease of hyperfine coupling is explainable in terms of exchange interactions between the two unpaired electrons.

Frozen solutions of the radical cation yield EPR signals similar to those obtained for the radical cations of Cu(II)OEC. The ZFS parameters for the dimer of the radical cation of VO(OEP) are quite similar to those of the Cu(II)OEC analog[116,119].

13.5. Biological applications

The applications of EPR to biological systems containing the porphyrin chromophore are mainly based on the following:

(i)   Identification of paramagnetic centers.

(ii) Distinctions between a free-radical type system and a transition metal ion and also identification of more than one paramagnetic center (through g-values, linewidths and temperature dependence).

(iii) Identification of the electronic ground state (through hyperfine couplings in a free radical, ZFS in triplet states of π-systems and ZFS or the g- and the A-tensors for transition metal ions).

(iv) Characterization of the symmetry of the ligand field and the nature of the metal–ligand bond (through g-values, the A-tensor and the ZFS parameters).

(v)   Estimation of the concentration of the paramagnetic center (through integration of the signal and comparison with a standard).

Examples of free radicals and triplet states are found in the EPR studies of chlorophylls and photosynthetic systems. Identification and characterization of the radical species involved have been the main basis for the postulation of mechanisms of the primary process in the photosynthesis[121-123].

EPR has become an indispensable tool for the identification of the oxidation states as well as the spin states (i.e. high-spin or low-spin) of Fe in hemoproteins[123]. The dependence of linewidths on molecular motions has been exploited in the 'spin-labeling' technique[125]. The motions of the protein moiety around the heme has also been studied by this method. Suggestions for the electronic configuration of Fe in oxyhemoglobin have been made on the basis of EPR studies on related Co(II) porphyrin-O_2 complexes and oxygen-containing heme enzymes. By analogy with the Co complex, Fe in oxyhemoglobin is considered to be in the +3 oxidation state, with a low-spin ($S = 1/2$) configuration. The unpaired electron in the O_2 molecule is antiferromagnetically coupled with the unpaired electron in the Fe atom, leading to a diamagnetic Fe(III) $\cdot$ O_2^- system[123].

The closely related techniques, namely magnetic susceptibility measurements (Chapter 12) and Mössbauer spectroscopy (Chapter 12) have also been used in many cases, along with EPR spectroscopy, to extract the maximum amount of information on the electronic structure of Fe in hemoproteins. Fe(II) systems which do not yield any EPR signals had been investigated by the Mössbauer technique.

Acknowledgement

The author is very much indebted to Dr. J.-H. Fuhrhop and Dr. E. Lustig who carefully read the entire manuscript and made innumerable valuable suggestions.

This work was supported by the Ministry of Research and Technology (BMFT) of the Federal Republic of Germany within its Technology program and by grants of the Deutsche Forschungsgemeinschaft.

References, p. 586

References

1. A. Abragam and B. Bleaney, "Paramagnetic Resonance of Transition Metal Ions", Oxford University Press (1970).
2. J. Wertz and J.R. Bolton, "Electron Spin Resonance — Elementary Theory and Applications", McGraw Hill, New York (1972).
3. A. Carrington and A.D. McLachlan, "Introduction to Magnetic Resonance", Harper and Row, New York (1967).
4. H.M. Schwartz, J.R. Bolton and D.C. Borg (Eds.), "Biological Applications of Electron Spin Resonance", Wiley—Interscience, New York (1972).
5. G. Feher, "EPR Applications to Selected Problems in Biology" Gordon and Breach, New York (1970).
6. D.J.E. Ingram, "Biological Applications of Electron Spin Resonance", Adam Hilger, London (1969).
7. H. Beinert and G. Palmer in "Advances in Enzymology", F.F. Nord (Ed.), Vol 27, Interscience, New York (1965), p. 105.
8. C.P. Poole, "Electron Spin Resonance — A Comprehensive Treatise on Experimental Techniques", Wiley—Interscience, New York (1967).
9. R. Alger, "Electron Paramagnetic Resonance — Techniques and Applications", Wiley—Interscience, New York (1968).
10. A. Ehrenberg, B. Malström and T. Vanngård (Eds.), "Magnetic Resonance in Biological Systems", Pergamon, New York (1967).
11. C. Frankoni (Ed.), "Magnetic Resonances in Biological Research", Gordon and Breach, New York (1971).
12. B.R. McGarvey in "Transition Metal Chemistry", Vol. 3, R.L. Carlin (Ed.), Marcel Dekker, New York (1966), p. 89.
13. T. Fu Yen, (Ed.), "Electron Spin Resonance of Metal Complexes, Plenum, New York (1969).
14. L.T. Muus and P.W. Atkins (Eds.), "Electron Spin Relaxation in Liquids", Plenum, London (1972).
15. S. Geschwind (Ed.), "Electron Paramagnetic Resonance", Plenum, New York (1972).
16. C.P. Poole and H.A. Farach, "Relaxation in Magnetic Resonance" Academic Press, New York (1971).
17. J.H. Freed and G.K. Fraenkel, J. Chem. Phys. **39**, 326 (1963).
18. A. Hudson and G.R. Luckhurst, Chem. Rev. **69**, 191 (1969).
19. H.A. Kuska and Max. T. Rogers, "Radical Ions", E.T. Kaiser and L. Kevan (Eds.), John Wiley, New York (1968).
20. G.E. Pake and T.S. Estle, "The Physical Principles of Electron Paramagnetic Resonance", W.A. Benjamin, New York (1973).
21. A. McCragh, C.B. Storm and W.S. Koski, J. Am. Chem. Soc. **87**, 1470 (1965).
22. W.E. Blumberg, J. Peisach, B.A. Wittenberg and J.B. Wittenberg, J. Biol. Chem. **243**, 1854 (1968).
23. E.M. Roberts and W.S. Koski, J. Am. Chem. Soc. **82**, 3006 (1960).
24. P.T. Manoharan and Max. T. Rogers, "Electron Spin Resonance of Metal Complexes", Teh Fu Yen (Ed.), Plenum, New York (1969), p. 143.
25. D. Kivelson and P. Neiman, J. Chem. Phys. **35**, 149 (1961).
26. D.J.E. Ingram, J.E. Bennet, P. George and J.M. Goldstein, J. Am. Chem. Soc. **78**, 3541 (1956).
27. J.M. Assour, J. Chem. Phys. **43**, 2477 (1965).
28. J.S. Griffith, Discussion Farad. Soc. **26**, 81 (1958).
29. F.K. Kneubühl, W.S. Koski and W. Caughey, J. Am. Chem. Soc. **83**, 1607 (1961).
30. E.M. Roberts, W.S. Koski and W.S. Caughey, J. Chem. Phys. **34**, 591 (1961).

31. J. Subramanian, J.-H. Fuhrhop, A. Salek, and A. Gossauer, J. Mag. Res. 15, 19 (1974).
31a. G. Struckmeyer, U. Thewald and J.-H. Fuhrhop (to be published).
32. M. Abkowitz, I. Chen and J.H. Sharp, J. Chem. Phys. 48, 4561 (1968).
33. B. Roos and M. Sundbom, J. Mol. Spectr. 36, 8 (1970).
34. P.D.W. Boyd, T.D. Smith, J.H. Price and J.R. Pilbrow, J. Chem. Phys. 52, 1253 (1972).
35. D. Kivelson and S.K. Lee, J. Chem. Phys. 41, 1896 (1964).
36. J. Bohandy, B.F. Kim, and C.K. Jen, J. Mag. Res. 15, 420 (1974).
37. E.B. Fleischer and T. Srivastava, J. Am. Chem. Soc. 92, 5518 (1970).
37a. R.J. Abraham and P.F. Swinton, J. Chem. Soc. London B, 903 (1969).
38. J.M. Assour and W.K. Kahn, J. Am. Chem. Soc. 87, 207 (1965).
39. F.A. Walker, J. Am. Chem. Soc. 92, 4235 (1970).
40. H.A.O. Hill, P.J. Sadler, R.J.P. Williams, and C.D. Barry, Ann. N.Y. Acad. Sci. 206, 247 (1973).
41. C.D. Barry, H.A.O. Hill, P.J. Sadler and R.J.P. Williams, Proc. Roy. Soc. (London) A 334, 493 (1973).
42. F.A. Walker, J. Mag. Res. 15, 201 (1974).
43. H.A.O. Hill, P.J. Sadler, and R.J.P. Williams, J. Chem. Soc., Dalton Trans., 1663 (1974).
44. B.B. Wayland, J.V. Minkiewicz, and M.E. Abd-Elmageed, J. Am. Chem. Soc. 96, 2795 (1974).
45. B.B. Wayland and Abd-Elmageed, J. Am. Chem. Soc. 96, 4809 (1974).
46. S.F. Ginsberg and N.B. Ol'shanskaya, Zh. Prikl. Spectrosk. 20, 250 (1974).
47. G.N. La Mar and F.A. Walker, J. Am. Chem. Soc. 95, 1790 (1973).
48. S.F. Ginsberg, L.N. Butseva, V.V. Kharpov, and V.I. Stanko, Teor. Eksp. Khim. 9, 841 (1973).
49. J.H. Bayston, N.K. King, F.D. Looney, and M.E. Winfield, J. Am. Chem. Soc., 91, 2775 (1969).
50. H.C. Stynes and J.A. Ibers, J. Am. Chem. Soc. 94, 5125 (1972).
50a. W.R. Scheidt and J.L. Hoard, J. Am. Chem. Soc. 95, 8281 (1973).
51. J.C.W. Chien and C. Dickenson, Proc. Natl. Acad. Sci. USA 69, 2783 (1972).
52. E. Melamud, B.L. Silver and Z. Dori, J. Am. Chem. Soc. 96, 4698 (1974).
52a. B.S. Tovrog and R.S. Drago, J. Am. Chem. Soc. 96, 6765 (1974).
53. A. Dediew and A. Veillard, Theor. Chim. Acta (Berlin) 36, 231 (1975).
54. J.R. Pilbrow and M.E. Winfield, Mol. Phys. 25, 1073 (1973).
55. N.S. Hush and I.S. Woolsey, J. Am. Chem. Soc. 94, 4107 (1972).
56. V.A. Zelewsky, Helv. Chim. Acta, 55, 2941 (1972).
57. N.S. Hush and I.S. Woolsey, J. Chem. Soc., Dalton Trans. 25 (1974).
58. A. Wolberg and J. Manassen, Inorg. Chem. 9, 2365 (1970).
59. I.A. Cohen, D. Ostfeld and B. Lichtenstein, J. Am. Chem. Soc. 94, 4522 (1972).
60. D. Lexa, M. Momenteau, J. Mispelter, and J.-M. Lhoste, Bioelect. Bioenerg. 1, 108 (1974).
61. K.M. Kadish, G. Larson, D. Lexa, and M. Momenteau, J. Am. Chem. Soc. 97, 282 (1975).
62. B.B. Wayland and L.W. Olson, J. Am. Chem. Soc. 96, 6037 (1974).
63. J.C.W. Chien, J. Chem. Phys. 51, 4220 (1969).
64. H. Kon, J. Biol. Chem. 243, 4350 (1968).
65. T. Yonetani, H. Yamamoto, J.E. Erman, J.S. Leigh, and G.H. Reed, J. Biol. Chem. 247, 2447 (1972).
66. M. Weissbluth, "Hemoglobin" Molecular Biology, Biochemistry and Biophysics, Vol. 15, Springer Verlag, Berlin (1974).
67. J.S. Griffith, Proc. Roy. Soc. (London) A 235, 23 (1956).

68. J. Peisach, W.E. Blumberg, S. Ogawa, E.A. Rachmilewitz, and R. Oltzick, J. Biol. Chem. **246**, 3342 (1971).
69. J. Peisach and W.E. Blumberg in "Probes of Structure and Function of Macromolecules and Membranes", Vol 2, B. Chance, T. Yonetani and A.S. Mildvan (Eds.), Academic Press, New York (1971), p. 231.
70. D.J.E. Ingram in "Magnetic Resonances in Biological Research" ed. C. Franconi, Gordon and Breach, New York (1971), p. 41.
71. W.E. Blumberg and J. Peisach, in "Magnetic Resonances in Biological Research", C. Franconi, Gordon and Breach (Eds.), New York (1971), p. 67.
72. A. Röder and E. Bayer, European J. Biochem. **11**, 89 (1969).
73. J. Peisach, W.E. Blumberg and A. Adler, Ann. N.Y. Acad. Sci. **206**, 310 (1973) and the references therein.
74. M. Momenteau, Biochim. Biophys. Acta **304**, 814 (1973).
75. M. Momenteau, J. Mispelter and D. Lexa, Biochim. Biophys. Acta **320**, 652 (1973).
76. M. Momenteau and B. Loock, Biochim. Biophys. Acta **343**, 535 (1974).
77. C.P. Scholes, J. Chem. Phys. **52**, 4890 (1970).
78. T. Yonetani, H.R. Drott, J.S. Leigh Jr., G.H. Reed, M.R. Waterman and T. Asakura, J. Biol. Chem. **245**, 2998 (1970).
79. G. Weiss, H. Kobayashi and M. Gouterman, J. Molec. Spectr. **16**, 415 (1965).
80. G.M. Maggiora, J. Am. Chem. Soc. **95**, 6555 (1973).
81. J.-H. Fuhrhop and D. Mauzerall, J. Am. Chem. Soc. **90**, 3875 (1968).
82. J.-H. Fuhrhop and D. Mauzerall, J. Am. Chem. Soc. **91**, 4174 (1969).
83. R.H. Felton, D. Dolphin, D.C. Borg, and J. Fajer, J. Am. Chem. Soc. **91**, 196 (1969).
84. A. Wolberg and J. Manassen, J. Am. Chem. Soc. **92**, 2983 (1970).
85. D. Dolphin, A. Forman, D.C. Borg, J. Fajer, and R.H. Felton, Proc. Natl. Acad. Sci. USA **68**, 614 (1971).
86. R.H. Felton, G.S. Owen, D. Dolphin, and J. Fajer, J. Am. Chem. Soc. **93**, 6332 (1971).
87. A. Forman, D.C. Borg, R.H. Felton, and J. Fajer, J. Am. Chem. Soc. **93**, 2790 (1971).
87a. J.-H. Fuhrhop, K. Kadish, and D. Davis, J. Am. Chem. Soc. **95**, 5140 (1973).
88. J. Fajer, D.C. Borg, A. Forman, R.H. Felton, L. Vegh and D. Dolphin, Ann. N.Y. Acad. Sci. **206**, 349 (1973).
89. Yu V. Glazkov, A.G. Zhuravlev, P.V. Kuzovkov and A.M. Shul'ga, Zh. Prikl. Spektrosk. **18**, 117 (1973).
90. C. Mengersen, J. Subramanian, J.-H. Fuhrhop and K.M. Smith, Z. Naturforsch. **29a**, 1827 (1974).
91. H.M. McConnell and D.B. Chesnut, J. Chem. Phys. **28**, 107 (1958).
92. P. Kusch and H. Taub in "Handbuch der Physik", **37/1**, 103 (1959).
93. R.J. Abraham, G.H. Barnett and K.M. Smith, J. Chem. Soc, Perkins Trans. I, 2141 (1973).
94. J.-H. Fuhrhop, P.K.W. Wasser, D. Riesner, and D. Mauzerall, J. Am. Chem. Soc. **94**, 7996 (1972).
95. M.E. Druyan, J.R. Norris and J.J. Katz, J. Am. Chem. Soc. **95**, 1682 (1973).
96. J.R. Norris, H. Scheer, M. Druyan, and J.J. Katz, Proc. Natl. Acad. Sci. USA **71**, 4897 (1974).
97. J.J. Katz in "Inorganic Biochemistry", G.L. Eichhorn, Ed., Elsevier, Amsterdam (1973), p. 1022.
98. R.H. Felton and H. Linschitz, J. Am. Chem. Soc. **88**, 1113 (1966).
99. A.P. Bobrovski and V.E. Kholmogrov, Biofizika **19**, 50 (1974).
100. J. Fajer, D.C. Borg, A. Forman, D. Dolphin and R.H. Felton, J. Am. Chem. Soc. **95**, 2739 (1973).

101. N.S. Hush and J.R. Rowlands, J. Am. Chem. Soc. **89**, 2976 (1967).
102. C.A. Hutchison and B.W. Mangam, J. Chem. Phys. **34**, 908 (1961).
103. G.T. Rikhireva, Z.P. Gribova, L.P. Kayushin, A.V. Umrikhina, and A.A. Krasnovskii, Dokl. Akad. Nauk. SSSR **159**, 196 (1964); **181**, 1485 (1968).
104. J.-M. Lhoste, Studia Biophys. (Berlin) **12**, 135 (1968).
105. J.-M. Lhoste, C. Helene, and M. Ptak in "The Triplet State", A.B. Zahlan (Ed.), Cambridge University Press, London (1967), p. 479.
105a. A.M.P. Goncalves and R.P. Burger, J. Chem. Phys. **61**, 2975 (1974).
106. J.-M. Lhoste, C.R. Acad. Sci. Paris **D266**, 1059 (1968).
107. H. Levanon and A. Wolberg, Chem. Phys. Lett. **24**, 96 (1974).
108. M. Gouterman, B.S. Yamanashi and A.L. Kwiram, J. Chem. Phys. **56**, 4073 (1972).
109. S.R. Langhoff, E.R. Davidson, M. Gouterman, W.R. Leenstra and A.L. Kwiram, J. Chem. Phys. **62**, 169 (1975).
110. H. Levanon and S. Vega, J. Chem. Phys. **61**, 2265 (1974).
111. M.S. de Groot, I.A.M. Hesselmann and J.H. van der Waals, Mol. Phys. **10**, 241 (1966).
112. R.H. Clarke and R.H. Hofeldt, J. Am. Chem. Soc. **96**, 3005 (1974).
113. G.W. Canters, J. Van Egmond, T.J. Schaafsma and J.H. Van der Waals, Mol. Phys. **24**, 1203 (1972).
114. I.Y. Chan, W.G. van Dorp, T.J. Schaafsma and J.H. van der Waals Mol. Phys. **22**, 741, 753 (1971).
115. G.W. Canters, J. Van Egmond, T.J. Schaafsma, I.Y. Chan, W.G. van Dorp and J.H. van der Waals, Ann. N.Y. Acad. Sci. **206**, 711 (1973).
116. C. Mengersen, J. Subramanian and J.-H. Fuhrhop (to be published).
117. L.d. Rollman and R.T. Iwomoto, J. Am. Chem. Soc. **90**, 1455 (1968).
117a. R.H. Felton, G.S. Owen, D. Dolphin, A. Forman, D.C. Borg and J. Fajer, Ann. N.Y. Acad. Sci. USA **206**, 504 (1973).
118. W.H. Fuchsman, C.H. Barlow, W.J. Wallace and W. Caughey, Biochem. Biophys. Res. Commun. **61**, 635 (1974).
119. J. Subramanian, C. Mengersen and J.-H. Fuhrhop (Submitted for publication).
120. D.H. Kohl in "Biological Applications of Electron Spin Resonance", H.M. Schwartz, J.R. Bolton and D.C. Borg (Eds.), Wiley—Interscience, New York (1972), p. 213.
121. E.C. Weaver and H.E. Weaver in "Photophysiology", A.C. Giese (Ed.), Vol 7, Academic Press (1972), p. 1.
122. A.J. Bearden and R. Malkin, Quart. Rev. Biophys. **7**, 131 (1974).
123. J.B. Wittenberg, B.A. Wittenberg, J. Peisach and W.E. Blumberg, Proc. Natl. Acad. Sci. U.S.A. **67**, 1846 (1970).
124. G.R. Luckhurst and G.F. Pedulli, Chem. Phys. Lett. **7**, 49 (1970).
125. H.A.O. Hill in "Inorganic Biochemistry", G.L. Eichhorn (Ed.), Elsevier, Amsterdam (1973), p. 1067.
126. M. Zerner and M. Gouterman, Theor. Chim. Acta, **4**, 44 (1963).
127. 'Hemes and Hemoproteins', eds. B. Chance, R.W. Estabrook, and T. Yonetani, Academic Press, New York, (1966); R. Lemberg and J. Barrett, "Cytochromes", Academic Press, London, (1973).

CHEMICAL REACTIVITY

REVERSIBLE REACTIONS OF PORPHYRINS
AND METALLOPORPHYRINS AND ELECTROCHEMISTRY

JÜRGEN-HINRICH FUHRHOP

Gesellschaft für Molekularbiologische Forschung,3301 Stöckheim/Braunschweig, West Germany and *Institut für Organische Chemie der T.U. Braunschweig, 3300 Braunschweig, Schleinitzstr., West Germany*

14.1. Introduction

Perhaps the most common reversible reaction encountered in the porphyrin area is that of protonation; since this is one of the basic properties of the porphyrin nucleus, and it is discussed in detail in Chapters 1 and 6, it will not be further exemplified here. The transfer of electrons through the π-system of porphyrin type compounds $[H_2(P)]$ plays an important role in photosynthesis[1] and possibly in other biological redox chains involving metalloporphyrins (MP)[2,3]. Two types of reversible in vitro reactions of metalloporphyrins are of interest in this connection and will be the main subject of this chapter: the removal (or addition) of electrons from (or to) the π-electron core of the porphyrin ligand and the formation of π-complexes with aromatic compounds. Redox reactions of the central metal ions are discussed in Chapter 6, but the problem of differentiation of these from those on the porphyrin periphery will be discussed in this chapter.

Sometimes the reversible removal of an electron is accompanied by the reversible addition of a nucleophile (e.g. OMe), leading to substituted phlorins; on the other hand, and more important, porphyrin-anions, formed by electrochemical or chemical reduction, add protons to the methine bridges. Both addition reactions thus lead to 'phlorin' derivatives, which are also the most important primary products of irreversible porphyrin reactions which will be discussed in Chapter 15.

All the reversible electrochemical reactions discussed in this chapter are so rapid that equilibrium is achieved at the working electrode and the Nernst equation (1) can be used in the calculation of electrode potentials:

Porphyrins and Metalloporphyrins, ed. Kevin M. Smith
© 1975, Elsevier Scientific Publishing Company, Amsterdam, The Netherlands

$$E = E_0 + \frac{RT}{2.3\,nF} \log \frac{[MP_{Ox}]}{[MP_{Red}]} \tag{1}$$

R is the Universal Gas Constant, T is the absolute Temperature, n the number of electrons exchanged, and F the Faraday. A more stable oxidized state of the molar concentration $[MP_{Ox}]$ corresponds to a more negative standard oxidation potential E_0, which, in aqueous systems is normally expressed with reference to the standard hydrogen electrode. The reference electrode to which all potentials in this chapter will be correlated is the saturated calomel electrode (SCE), which is approx. 240 mV less than the hydrogen electrode. Metalloporphyrin potentials, however, are usually measured in organic solvents, where reference to an absolute standard becomes difficult and comparison of metalloporphyrins is mostly based on the relative order of the numerical potential values determined under identical conditions. $[MP_{Ox}]$ may relate either to a cation radical (e.g. Mg $P^{+\cdot}$) or an oxidized central ion [e.g. Fe(III)]; $[MP_{Red}]$ then expresses the concentration of the reduced state in equilibrium [e.g. the neutral porphyrin; Fe(II)]. The usual analytical tools like potentiometry, polarography, and cyclic voltammetry have been successfully applied to the study of porphyrin redox reactions in various solvents and even in the biological environment. The relative redox potentials of the porphyrin ligands in their various metal complexes obtained with these methods can be used to predict the general reactivity of the porphyrin π-system, so that optimal complexes can be selected for each reaction type. Electrochemical regularities also give insight into the character of metal—nitrogen bonds and their influence on the porphyrin system and thereby enable one to check on the results of model calculations. Hence, the numerical values of redox potentials are of general interest and the more common techniques to obtain them will be discussed briefly. Finally, the redox steps of transition metal ions [e.g. Fe(II)] have to be identified and separated from reactions of the porphyrin ligands before an interpretation of electrochemical data becomes possible. This aspect will also be taken into account.

14.2. Chemical oxidation and reduction potentiometry

Rabinowitch[4], and later Goedheer[5], have shown that chlorophyll-a and magnesium and zinc porphyrins react with Fe(III) salts and other oxidants to give products which do not have sharp absorption bands in their visible spectra. These bands could, however, be restored more or less quantitatively by the addition of reductants to the solution.

Fuhrhop and Mauzerall[6,7] titrated various metallo-octaethylporphyrins and -chlorins (M = Mg, Zn, Cu, Ni, Pd) in chloroform/methanol solutions with chemical oxidants [I_2,Br_2,NBS,DCCB, and Fe(III) salts] and determined the midpoint potentials of the underlying redox reactions:

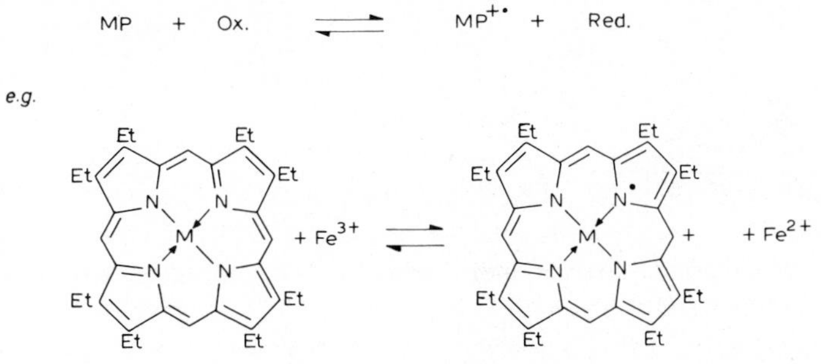

$$MP \; + \; Ox. \; \rightleftharpoons \; MP^{+\cdot} \; + \; Red.$$

e.g.

(1)   (2)

M = Mg, Zn, Cu, Ni, Pd

by using potentiometry and simultaneous spectrophotometry. At equilibrium, the potentials of the systems Ox/Red and $MP^{+\cdot}$/MP are equal to each other and to a first approximation, an equilibrium constant can be defined from eq. 3:

$$\log K = 2.3 \; n.F \; \frac{E_{1/2}^{Ox/Red} - E_{1/2}^{MP^{+\cdot}/MP}}{RT} \tag{2}$$

where $\quad k = \dfrac{[Red][MP^{+\cdot}]}{[Ox][MP]} \tag{3}$

As an example, oxidation of Mg(OEP) with bromine will be considered. The midpoint potential of the $Br_2/2Br^-$ pair in water is 830 mV vs. SCE, the corresponding potential of $Mg(OEP)^{+\cdot}/Mg(OEP)$ in methanol is 420 mV vs. SCE; therefore

$$\log k = \frac{0.41}{0.059} \cong 7; \qquad k = 10^7 \;\; \text{and} \;\; MP^{+\cdot} = 10^7 \times \frac{[Ox][MP]}{[Red]}$$

which would mean that oxidation of Mg(OEP) should be complete with equimolar amounts of oxidants added. It has been found experimentally that such titrations in organic solvents with no excess of oxidant are only possible for complexes with midpoint potentials below 300 mV [e.g. Mg(OEC), Zn(OEC), BChl-*a*] and iodine as oxidant[6,7]. In all other cases varying excesses of oxidants are needed to complete the oxidation of the metalloporphyrins, probably because the organic solvents, or reducing impurities contained in them, consume oxidant at potentials above 500 mV. Another, as yet unexplained, observation has to be mentioned here: the potential of an oxidized metalloporphyrin solution is usually constant within a few millivolts over minutes after the addition of oxidants, whereas the potentials of

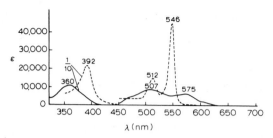

Fig. 1. Electronic spectra of palladium octaethylporphyrin (- - - - - -) and its one-electron oxidation product (————) (from Ref. 6).

the inorganic Ox/Red pairs in organic solvents [e.g. Fe(III)/Fe(II)] decay quickly.

Thus it is usually not possible to determine the concentration of metalloporphyrins or their π-cation radicals by direct potentiometric titrations, but the potentials measured must be correlated with other data which are proportional to the concentration of the porphyrin species. The most convenient method is electronic absorption spectroscopy at a wavelength where the spectra of the metalloporphyrin and its oxidation product have largely different extinction coefficients. A typical pair of spectra are given in Fig. 1;

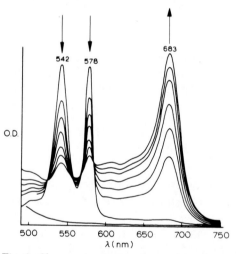

Fig. 2. Changes in electronic spectra of magnesium octaethylporphyrin on titration with oxidants.

an example of a set of absorption curves obtained by photometric titration is in Fig. 2. The ratio $MP^{+\cdot}/MP$, which is needed for a Nernstian plot is then obtained from:

$$\frac{[MP^{+}\cdot]}{[MP]} = \frac{O.D._{MP}-O.D.}{O.D.-O.D._{MP^{+}\cdot}} \tag{4}$$

where $O.D._{MP}$ and $O.D._{MP+}$. are the optical densities at zero or 100% oxidation respectively and O.D. indicates the optical density at the actual titration point. Another possibility is the ESR-spectroscopic titration, where the signal obtained with large excess of oxidant is set equal to 100% and intermediate concentrations are directly indicated by the relative intensity of the observed signals. Some typical Nernstian plots obtained by such titrations are given in Fig. 3.

Oxidometric titrations offer the advantage over the more common technique of polarography and voltammetry, that only a pH-meter, a microburet, and a spectrophotometer are needed as equipment and, more importantly, that the spectra, (which give an indication of the chemistry taking place) and the potentials, can be conveniently obtained simultaneously. The most serious limitation of this method is that it can only be used in general for oxidations in the range from about zero to 800 mV vs. SCE. The use of xenon difluoride in butyronitrile pushes the upper limit to about 1 V[8].

Some midpoint potentials obtained by this method are summarized in Table 1. The results will be discussed in conjunction with the more extensive voltammetric data. Here it should only be pointed out that metallochlorins

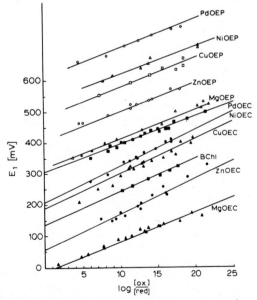

Fig. 3. Plot of E_1 (first porphyrin oxidation) vs. log [Ox]/[Red] of some metalloporphyrins, metallochlorins, and bacteriochlorophyll.

References, p. 620

TABLE 1

Midpoint potentials of metallo-porphyrins and -chlorins from potentiometric titrations[6,7]

ligand central ion	octaethyl-porphyrin	octaethyl-chlorin
Mg	427	107
Zn	525	197
Cu	601	331
Ni	636	356
Pd	726	422
Sn(IV)-(OH)$_2$ (or Oxo dimer)	≥1100 (irrev)	600 (irrev)
Fe(III)-OH (or Oxo-dimer)	~850 (irrev)	—

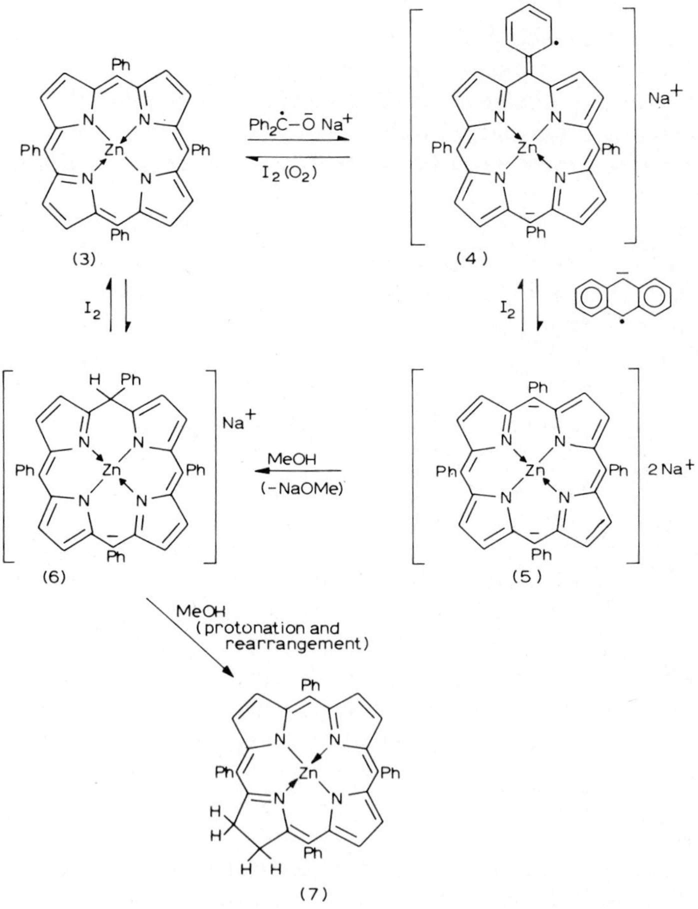

are oxidized at potentials approx. 300 mV lower than the corresponding porphyrins[7].

In a pioneering study, Closs and Closs[9] prepared the π-monoanion (4) of Zn(TPP) (3) by reduction with sodium benzophenone ketyl in tetrahydrofuran, and the π-dianion (5) with anthracene negative ion. Both reactions are fully reversible on addition of iodine or air. Protonation of dianion (5) with methanol produces the phlorin zinc complex salt (6), which with excess methanol rearranged slowly to the chlorin (7). Approximate midpoint potentials of the redox pairs (3 ⇌ 4) and (4 ⇌ 5) have been estimated to be approx. −1.1. V and −1.6 V by utilizing different sodium ketyls and sodium aromatics. The anionic species (4) and (5) were obtained in crystalline form and were also titrated quantitatively with iodine. Later, Hush[10,11] reduced a variety of porphyrins and closely related compounds with metallic sodium in ether solvents (e.g. tetrahydrofuran) and reported clean electronic spectra of mono- and dianions[10] as well as well-resolved ESR spectra of the π-anion radicals[11].

14.3. Polarography and voltammetry with a rotating platinum electrode

Voltammetry deals with the effect of the variation of potential of a polarized electrode in an electrolysis cell on the current that flows through it. Working electrodes are usually either dropping mercury (in polarography[12,18]), rotating platinum disk (in voltammetry with stirred solutions[13−16]), stationary platinum disk[14,16], button[23] or hanging mercury drop[12,19,20] electrodes (in cyclic voltammetry). The techniques where the electrode or solution is in motion are the more simple and are usually the methods of choice for measurement of redox potentials and concentrations; the rapid stationary electrode techniques have to be applied if the kinetics of an electron transfer are of interest. Here, the results from the application of 'slow' techniques in the porphyrin field will be summarized.

The high overpotential for the reduction of hydrogen ions or water on a mercury surface allows one to investigate processes that can occur only at very low potentials, e.g. the formation of porphyrin-anions which typically necessitate potentials below −1.5 V vs. SCE. A dropping electrode is applied to achieve constant renewal of the electrode surface. The solvent most often employed in polarography of metalloporphyrins is DMSO. On the other hand, mercury is quite easily oxidized, so that potentials above approx. 200 mV vs. SCE cannot be secured; here a platinum wire, or better, a disk electrode rotated at constant high speed to avoid complications by slow natural convection processes, is used. The solvent of choice is usually n-butyronitrile and the range of potentials which can be covered without discharge of hydrogen ions or oxidation of solvent is about from zero to +1.5 V vs. SCE. If ions are present in the solution or are formed during electrolysis, then they will move through the solution when a potential is applied and lead to a 'migration current'. All practical voltammetric work is done under conditions which render this current negligible, namely by addition to the solution of at least a 50 times excess of supporting electrolyte relative to the electroactive porphyrin. The most common salt is tetra-n-butyl-ammonium perchlorate, which is soluble in organic solvents and redox inert. Commercial recording potentiometers with circuit modules providing linear compensation of the large iR drops (i = current, R = resistance) found in organic solvents have mostly been used. This compensation of iR losses necessitates the introduction of a third 'auxiliary' electrode besides the working and reference electrodes[12−14].

A typical voltammogram is shown in Fig. 4. The height of the wave is determined by the rate of diffusion of molecules from the solution into the double layer on the electrode surface. The resulting diffusion current, I, is proportional to the concentration of the electroactive species and for reversible electron transfers the potential is given by

$$E = E_{1/2} - \frac{RT}{nF} \ln \left(\frac{i_g - i}{i} \right) \tag{5}$$

for a reduction wave. For oxidation reactions, the sign of the second term is positive. Semiquantitative equations for i_g, which contain standard physical data such as area and type of the electrode, diffusion coefficients of the ions, viscosity of the solvent, and so on, are available[12−14], but are rarely applied in experiments with metalloporphyrins. Here the system is generally standardized with a porphyrin solution of known midpoint potential and concentration and with the porphyrin under investigation a few experiments at

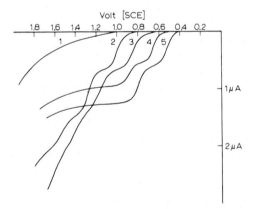

Fig. 4. Typical voltammograms obtained in 0.1 M $LiClO_4$ solutions in propionitrile with a rotating platinum electrode: 1) 0.1 M $LiClO_4/C_2H_5ĊN$; 2) Pheophytin-b; 3) Pheophytin-a; 4) Chlorophyll-b; 5) Chlorophyll-a (from Ref. 15).

TABLE 2

Anodic half-wave potentials of metalloporphyrins from Voltammetry with a rotating platinum electrode

Porphyrin	$E\frac{1}{2}$ (1)		$E\frac{1}{2}$ (2)	References
Ag-Tetraphenyl	0.54		—	15
Ba-Tetraphenyl	0.46		0.75	15
Cd-Tetraphenyl	0.63		0.93	15
Co-Tetraphenyl	0.32	1.06	1.26	15
Cu-Tetraphenyl	0.90		1.16	15
Deuteroporphyrin-IX dimethyl ester	0.60		(1.04)	15
Hematoporphyrin-IX dimethyl ester	0.77		—	15
Mesoporphyrin-IX dimethyl ester	0.78		—	15
meso-Tetraphenyl	0.97		(1.12)	15
Mg-Etio	0.40		0.77	15
Mg-Tetraphenyl	0.54		0.86	15
Ni-Tetraphenyl	0.95		—	15
Mg-Mesoporphyrin-IX dimethyl ester	0.54		0.94	15
Pb-Tetraphenyl	0.63		0.96	15
Pt-Etio	0.75		—	15
Proto-IX dimethyl ester	0.83		—	15
Zn-Tetraphenyl	0.71		1.03	15
Zn-Deuteroporphyrin-IX dimethyl ester	0.60		(1.04)	15
Zn-Mesoporphyrin-IX dimethyl ester	0.50		0.97	15
Zn-*N*-Methyl etio	1.09		—	15
Zn-Protoporphyrin-IX di-*n*-amyl ester	0.61		—	15
Zn-Hematoporphyrin-IX dimethyl ester	0.53		1.00	15
Chlorophyll-*a*	0.52		0.77	16
Chlorophyll-*b*	0.65		0.87	16
Cu-Porphin	0.93		—	63
meso-Amino-octaethyl	0.34		—	60
meso-Nitro-octaethyl	0.78		—	60
meso-Oxy-octaethyl	0.35		—	60
meso-Oxy-octaethyl anion	−0.05			
meso-Thio-octaethyl	0.79		—	60
Mg-Porphin	0.60		—	60
Ni-Porphin	0.97		—	63
Porphin	0.91		—	63
Zn-Amino-octaethyl	0.25		—	60
Zn-α,γ-Dioxo-octaethylporphodimethene	0.84		—	60
Zn-Oxa-octaethyl	0.34		—	60
Zn-Oxy-octaethyl	0.20		—	60
Zn-Porphin	0.72		—	63
Zn-Thio-octaethyl	0.52		—	60

References, p. 620

various concentrations are made, to check whether wave height and concentration depend linearly on each other and whether a shift in midpoint potentials is observed.

Stanienda[15,16] described in some detail the experimental procedure to obtain oxidation potentials (Table 2) of porphyrin solutions in butyronitrile with the aid of a rotating platinum disk electrode as well as on the stationary electrode. Furthermore, these early papers contain linear plots of concentration against the limiting current, which offer evidence for predominant diffusion control of the electron transfers.

Hush[17] was the first to report extended polarographic data on metalloporphyrins with redox inert central ions, which was followed by related work from various laboratories[18-22]. Representative reduction potentials are summarized in Table 3. The small 'prewaves' often observed in the polarography of metalloporphyrins and chlorophylls presumably correspond to the reduction of the aggregates, formed on the surface of the electrode,

TABLE 3

Cathodic halfwave potentials of metalloporphyrins from polarography

Porphyrin	$-E_{\frac{1}{2}}$ (1)	$-E_{\frac{1}{2}}$ (2)	$-E_{\frac{1}{2}}$ (3)	References
Bacteriochlorophyll	(0.70)1.05	1.55	—	28,62
Cd-Tetraphenyl	(0.67)1.25	1.70	—	18
Chlorophyll-*a*	(0.85)1.12	1.54	—	22,62
Chlorophyll-*b*	(0.85)1.05	1.46	—	62
Co-Tetraphenyl	0.82	1.87	—	18
Cu(Etio-IV)	1.48	1.99	—	17
Cu-Tetraphenyl	(0.72) 1.20	1.68	—	18
Deutero-IX dimethyl ester	1.29	2.53	—	19
H_2(Etio-I)	(0.92)1.37	1.80	—	17,18
meso-Amino-octaethyl	1.43	—	—	60
Meso-IX dimethyl ester	1.34	1.73	2.57	19
meso-Tetraphenyl-bacteriochlorin	1.10	1.55	—	19
meso-Tetraphenyl	(0.70)1.08	1.45	2.38	18,18,19
meso-Thio-octaethyl	1.68	—	—	60
Mg-Octaphenyltetraaza	0.68	1.11	1.81	17
Na_2-Tetraphenyl	—	—	—	18
Ni-Tetraphenyl	1.18	1.75	—	18
Mg-Tetraphenyl	(0.67)1.35	1.80	—	18
Sn^{IV}-Tetraphenyl-diacetate	(0.3) 0.81	1.26	2.14	18
Pb-Tetraphenyl	(0.83)1.10	1.52	—	18
Zn-Etio-I	(0.89)1.62	2.00	2.77	17,18
Zn-*meso*-oxa-octaethyl	1.00	—	—	60
Zn-Tetraphenylchlorin	2.07	—	—	18
Zn-Tetraphenyl	(0.71)1.31	1.72	2.45	17,18

and do not usually constitute more than 10—15% of the total number of porphyrin molecules present in solution over the concentration range from 10^{-5}—5×10^{-4} M[18,22]. Felton and Linschitz[18] also estimated an approximate diffusion constant and diffusion current for metalloporphyrins in DMSO from standard expressions, and found that the calculated diffusion current ($I_d = 0.86\ n$) compared nicely with the experimental constants (I_d - 0.89; 0.90) for the first and second waves of Zn(TPP) reduction.

14.4. Cyclic voltammetry

The polarographic and voltammetric techniques discussed so far do not yield detailed information on the kinetics of electron transfer reactions and no clear cut criteria for the reversibility of a redox reaction can be derived from these current-potential curves[13,14]. Rapid cyclic voltammetry provides the means to investigate the 'pathway' of an electrochemical reaction in greater detail, and because of its relative experimental simplicity and speed it is often used instead of the conventional slow techniques, even for the simple determination of redox potentials.

A stationary working electrode, ususally a platinum disk, in an unstirred solution is employed, and the potential is varied at a finite rate as a linear function of time. Typical sweep rates are in the range of 0.5—100 V/sec, and the current potential curves no longer have the S-shape known from techniques with infinitesimally small sweep rates, but exhibit peaks. If a rapid *triangular sweep* is applied, and an oscilloscope or a fast *x-y*-recorder is used a *cyclic voltammogram* is obtained (Fig. 5). If, for example, the sweep

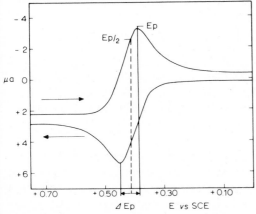

Fig. 5. Example for a cyclic voltammogram. The ΔE_p given is 60 mV and therefore corresponds to a reversible reaction.

References, p. 620

towards lower potentials produces reduced molecules in the vicinity of the electrode, then they are reoxidized during the reverse sweep to the same concentration which was initially present in the solution, provided that the rate at which the potential is taken back to its initial value is more rapid than the diffusion processes needed to establish equilibrium with the bulk of the solution. If the reduced species undergoes irreversible secondary reactions in a shorter period than that of triangular sweep, then new waves may be found in the reverse sweep and the height of the wave corresponding to the forward reaction is reduced.

The maximum potential E_P in the current-potential curve in the linear scan voltammetric experiment occurs at:

$$E_p = E_{1/2} - \frac{28.5}{n} \text{ mV at } 25°\text{C} \tag{5}$$

The half-peak potential $E_{P/2}$, which is often easier to measure, is at:

$$E_{P/2} = E_{1/2} + \frac{28.0}{n} \text{ mV at } 25°\text{C} \tag{6}$$

where $E_{1/2}$ is the polarographic midpoint potential, and n the number of electrons transfered. These equations are for reductions; for oxidations, the sign of the numerical term is reversed.

On the reverse scan, the position of the reoxidation peak is not identical with the potential of the forward scan. It depends on vt_s, the switching potential, if the reverse sweep starts at less than $100/n$ mV cathodic of the reduction peak. If the switching potential, however, is set further apart, then the separation of the two peaks will be $59/n$ mV and is independent of the scan rate of the potential scan (Fig. 5). These two criteria, together with the equal height of the steps in the forward and reverse reactions, are commonly taken as diagnostic for a *reversible, purely diffusion controlled charge transfer*, and cyclic voltammetry has been used simply to obtain potentials which could also have been measured by slow techniques, and at the same time to demonstrate reversibility of the electron transfers. It is also noteworthy that cyclic voltammograms down to potentials of −1.4 V can be obtained with the same platinum button electrode used for the oxidation experiments. Table 4 lists redox potentials of metal complexes of octaethylporphyrins obtained using this technique[23].

If the scan rate in cyclic voltammetry is increased to values greater than 0.1 V/sec, then metalloporphyrin redox couples do not behave like ideal reversible systems because electron transfer rates are not infinitely large and the current is controlled by a mixture of diffusion *and* charge transfer kinetics. This is called the *quasi-reversible case* and cyclic voltammetry at varying scan rates can be used to measure electron transfer rates. These are hetero-

TABLE 4

Peak potentials of metalloporphyrins from cyclic voltammetry with a platinum button electrode

	Ligand	Oxida-tion	Metal	Ligand	Reduc-tion
	$E_{\frac{1}{2}}(2)$	$E_{\frac{1}{2}}(1)$	$E_{\frac{1}{2}}$	$E_{\frac{1}{2}}(1)$	$E_{\frac{1}{2}}(2)$
$H_2(OEP)$	1.30	0.81		−1.46	−1.86
$H_4(OEP)^{2+}$		1.65			
Mono-N-Me(OEP)		1.37		−0.73	−1.17
Mono-N-Me(OEP)		0.86		−1.37	
Ag(OEP)		1.10	0.44	−1.29	
			III ⇌ II		
Al(OEP)(OH)	1.28	0.95		−1.31	
Ca(OEP)	0.86	0.50		−1.68	
Cd(OEP)	1.04	0.55		−1.52	
Co(OEP)		1.00	Irreversible		
			III ⇌ II		
			−1.05		
			II ⇌ I		
Cr(OEP)(OH)	1.22	0.99	0.79	−1.35	
			IV ⇌ III		
			−1.14		
			III ⇌ II		
Cu(OEP)	1.19	0.79		−1.46	
Fe(OEP)(OH)*	1.24	1.00	−0.24	−1.33	
			III ⇌ II		
Ga(OEP)(OH)	1.32	1.01		−1.34	−1.80
Ge(OEP)(OH)$_2$	1.36	1.09		−1.31	
In(OEP)(OH)	1.36	1.08		−1.19	−1.59
Mg(OEP)	0.77	0.54		−1.68	
Mn(OEP)(OH)	1.4	1.12	−0.42	−1.61	
			III ⇌ II		
MoO(OEP)(OH)		1.43	−0.21	−1.30	−1.72
			V ⇌ IV		
Ni(OEP)		0.73		−1.5	
Pb(OEP)		0.65	0.9	−1.30	
			II ⇌ IV		
Pd(OEP)		0.82		−1.53	
Sb(OEP)(OH)		1.4		−1.07	
Sc(OEP)(OH)	1.03	0.70		−1.54	
Si(OEP)(OH)$_2$	1.19	0.92		−1.35	
Sn(OEP)(OH)$_2$*		1.4		−0.90	−1.30
TiO(OEP)	1.32	1.03		−1.21	−1.69
Tl(OEP)(OH)	1.31	1.00		−1.24	
VO(OEP)	1.25	0.96		−1.25	−1.72
Zn(OEP)	1.02	0.63		−1.61	

* Oxo-dimer.

References, p. 620

genous rates of the electrode reaction in cm sec^{-1} (which should not be confused with homogenous rates which are given in mol l^{-1} sec^{-1} etc.): at low scan rates, the difference in peak potentials E_P for the forward and reverse reactions are close to the theoretical $59/n$ (mV), but when the scan rate is increased, the Nernstian equilibrium corresponding to the electrode potential is not reached due to the slowness of electron transfer and the two peaks get further apart.

The peak separations, E_P, are related to the charge transfer rate constant, K_s, by the equation:

$$\psi = \frac{K_s}{\left(\dfrac{\pi\,nFD}{RT}\,v\right)^{\frac{1}{2}}} \tag{7}$$

where the dimensionless kinetic function ψ can be directly converted into K_s by use of working curves or tables from the literature, v is the scan rate in volts per second, D is the diffusion constant, and the other constants have their usual significance. Experimentally, a scan rate is chosen which yields peak separations somewhere between $120/n$ and $60/n$ mV because certain simplifications concerning D are then possible and the appropriate charge transfer coefficients can be evaluated by matching various theoretical curves with the experimental one[13,14].

A summary of redox potentials obtained by cyclic voltammetry is given in Table 4, and will be discussed in section 14.6.

Some electron transfer rate constants for metalloporphyrin redox reactions from the work of Kadish and Davis[24] are given in Tables 5 and 6. The rates for the oxidation of the porphyrin ring are faster for bivalent metalloporphyrins [e.g. Mg(OEP)], than for complexes containing metals in higher oxidation states [e.g. Si^{IV}(OEP)(OH)$_2$]. This has been rationalized with the assumption of a more highly ordered solvation sphere around positively

TABLE 5

Heterogeneous rate constant determination for the reactions Mg(OEP) $\rightleftharpoons$ Mg(OEP)†

	Scan Rate, v, V/sec	$n\Delta E$p		k^0 cm/sec
Mg(OEP)	4	72	2.0	8.6 × 10^{-1}
	10	80	3.2	8.2 × 10^{-1}
	15	84	3.9	8.3 × 10^{-1}
	30	90	5.5	9.0 × 10^{-1}
			Average 8.5 ± 0.3 × 10^{-1}	

TABLE 6

Heterogeneous rate constant for oxidation of various octaethylporphyrin metal complexes

Compound	k^0 (cm/sec)
First Ligand Oxidation	
Mg(OEP)	0.1
Pd(OEP)	0.1
Cr(OEP)(OH)	4×10^{-2}
VO(OEP)	4×10^{-2}
Si(OEP)	5×10^{-2}

charged porphyrin complexes. The addition of an electron to the porphyrin π-system occurs with uniform, low transfer rates, which suggests similar mechanisms for transfers of electrons from the platinum electrode in all cases. Rate constants for metal redox reactions are discussed on page 255.

14.5. Preparative electrolysis of metalloporphyrins

Electrochemical synthesis[25] is governed by four interdependent variables: current, potential, concentration, and time. If one wants to prepare a porphyrin in any oxidation state different from the one at hand, then it has to be oxidized (or reduced) at a potential slightly above (or below) the potential of the voltammetric wave. This limits the voltage to be applied to values between around ± 1 V. Because of the usually low solubility of most metalloporphyrins in water, electrolytes of low conductivity have to be used, which cause a large iR drop and indefinitely low turnover rates at the potentials applied. Therefore similarly to voltammetry, an auxiliary electrode is introduced, which allows a much higher voltage to be applied while keeping the potential between working and reference electrodes at the desired value. With the electrolysis going on, the resistance of the solution goes up and the current and voltage fall. Therefore one has either to regulate the voltage of the power supply manually to keep the voltage in the electrolysis cell constant (constant voltage electrolysis) or one couples the cell to a commercial polarography apparatus with electronic potentiostats, which then automatically holds the working electrode at a constant potential (potentiostatic electrolysis). The limiting factors in electrochemical synthesis are then the relative low concentrations ($\sim 10^{-3}$ M) obtainable with porphyrin solutions and, sometimes, the limited lifetime of the π-cations and π-anions, which do not allow extended electrolysis times.

Felton and coworkers[8] describe a three electrode electrolysis cell with a

References, p. 620

platinum gauze basket as working electrode, which has been used to prepare various crystalline metalloporphyrin cation radicals[8,56−58]. Such cells can be easily adapted to the measurements of electronic and/or ESR spectra during electrolysis. For the production of porphyrin π-anions either a platinum or mercury electrode can be used[27]. For bulk electrolysis, a mercury pool[27−31] or gold minigrid[19−21] is usually preferred, whereas in ESR work a platinum electrode[8,27] is easier to handle. Inhoffen and co-workers[29−31] prepared a variety of chlorin-phlorins from chlorophyll derivatives on a large scale, using a 300 ml electrolysis cell with a mercury pool electrode. Up to 100 mg of a chlorin could be reduced within a few hours at potentials around −0.6 V and currents in the order of a few milliamperes. At the lower extreme of the preparative scale, Wilson et al.[19−21] applied the technique of thin-layer electrochemistry[32], where an optically transparent gold minigrid served as the electrode (transparency about 45%) and a cell of about 0.1 nm path length was used. Here quantitative electronic spectra of various oxidation states were obtained within a minute; an example is given in Fig. 6.

Thus, reversible redox reactions of the porphyrin ligand, sometimes followed by reversible phlorin formation (see section 14.9), can be achieved by chemical (see 14.2) photochemical (see page 678) or electrochemical methods. A sequence of electrochemical, photochemical, and chemical reactions, leading in vitro from chlorophyll-*a* to chlorophyll-*b*, is given on page 615.

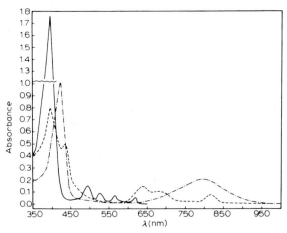

Fig. 6. Electronic spectra of reduction products of deuteroporphyrin-IX dimethyl ester obtained with the gold minigrid electrode cell. (———) porphyrin; (- - - - - -) anion radical; (— · — ·) phlorin anion (from Ref. 19).

14.6. Differentiation between reactions of central ions and the porphyrin periphery

From voltammetry one obtains information on potentials, number of electrons, reversibility, and kinetics involved in an electron transfer reaction, but no direct evidence for the location (metal or porphyrin periphery) of the reaction can be extracted. The following four criteria have been used[3,23] to answer this crucial question:

1) An *electronic spectrum* with two sharp absorption bands in the visible and a narrow Soret band indicates a metalloporphyrin (MP), where the porphyrin ligand is in its zero oxidation state. If this spectrum is found in an oxidation or reduction product from a metalloporphyrin, then only the metal has changed its oxidation state (e.g. $Ag^{II}P \rightleftharpoons Ag^{III}P$; see Fig. 7)[34]. If, however, after oxidation the electronic spectrum broadens out considerably, then the porphyrin ligand might have reacted (e.g. $Cu^{II}P \rightleftharpoons Cu^{II}P^+$, Fig. 8)[6]. New and intense bands above 700 nm in reduction products often indicate formation of a phlorin or a porphyrin π-anion. Complications in the application of these simple criteria have been discussed[3].

2) If only one unpaired electron is found in the product, then its location is easily determined by examination of linewidth, *g*-values, hyperfine structure, etc. in the *ESR spectrum* (see p. 576).

3) When both the metal and the porphyrin ligand contain unpaired electrons, then no ESR signal is usually detectable. This is true, for example, for the Cu(II) porphyrin π-cation radicals[6]. Here, measurements of *bulk susceptibilities* using a Gouy balance[34] or Evans' NMR shift method[36−38] help to locate the site of a reaction [e.g. $Ag^{III}(OEP)$ is diamagnetic; $Cu^{II}(TPP)^{+\cdot}$ has a molar susceptibility of 2.88 BM].

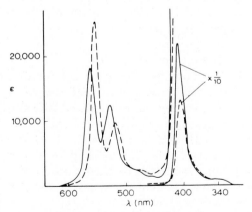

Fig. 7. Electronic spectra of silver[II] octaethylporphyrin (———) and its one-electron oxidation product $Ag^{III}(OEP)$ (------) (from Ref. 34).

References, p. 620

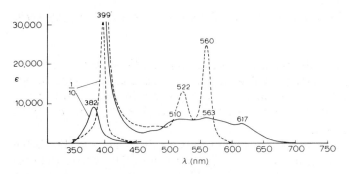

Fig. 8. Electronic spectra of copper octaethylporphyrin (- - - - - -) and its one-electron oxidation product $Cu^{II}(OEP)^+$ (————) (from Ref. 6).

4) From Table 4, two useful *electrochemical rules* can be deduced: the normal difference between the first one-electron oxidation and reduction potentials of the porphyrin ligand is 2.25 ± 0.15 Volts and the second removal or addition of an electron follows after an interval of approx. 0.3 V. Voltammetric waves corresponding to potentials which are far away from these patterns are then tentatively associated with reactions of the central metal ions.

14.7. Redox potentials and chemical reactivity of the porphyrin ligand

The numerical values of potentials of reversible one-electron oxidations and reductions of alternant aromatic hydrocarbons have been found to increase linearly with rising gaps between the nonbonding energy level and the calculated energies of the outer orbitals [highest occupied molecular orbital, (HOMO); lowest unoccupied molecular orbital, (LUMO)], from which an electron is removed or to which an electron is added[39]. Two important effects influence the energies of the "frontier" orbitals, and therefore the redox properties and chemical reactivity of a molecule:

1) The more extended the conjugated system is, the closer the frontier orbitals approach to each other and to the nonbonding level, and the easier the compound is to oxidize and reduce. The energy gap between HOMO and LUMO decreases with rising delocalization energy (conjugative effect).

2) Electron donating substituents (e.g. CH_3) facilitate oxidation of the molecule, by raising the HOMO energy towards the nonbonding level, and render the same molecule more stable against reduction by lifting the LUMO energy. The gap between HOMO and LUMO is roughly constant (inductive effect).

The effect of various metal ions on the energy of the outer orbitals of porphyrin ligands is largely inductive. The criterion of constant energy gaps

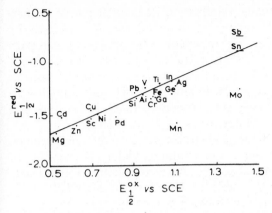

Fig. 9. A plot of $E_{1/2}$ for the first ring oxidation vs. $E_{1/2}$ for the first ring reduction (see text) (from Ref. 23).

between the outer orbitals corresponds with the experimental finding that electronic excitation energies of the porphyrin π-system are almost independent of the central metal ion[40,41] (see Chapter 16) and that the differences in potential between the first reduction and oxidation of the porphyrin ring are, with two exceptions (Mn, Mo), constant (2.25 ± 0.15 V, Fig. 9)[42]. A scheme of the outer orbitals (neglecting degeneracies) for a metalloporphyrin with central ions of different oxidation state (Mg^{II}, Pd^{II}, Sn^{IV}) is given in Fig. 10. Energy differences as high as one Volt are found, and have been traced back to the electrostatic properties of the metals, which lead to more or less negative net charge on the porphyrin ring (e.g. $Mg^{0.6+}$, $P^{0.6-}$, $Pd^{0.2+}$, $P^{0.2-}$)[23,24]. An inductive parameter 'h', analogous with Pauling's electronegativities, has been proposed to describe the strong influence of metals in high oxidation states[23].

The variations of electrochemical behavior of the porphyrin ligand in

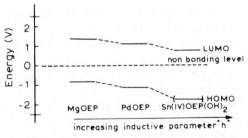

Fig. 10. A schematic representation of relative energies of outer orbitals of the porphyrin π-system in different metal complexes as deduced from electrochemical data (from Ref. 23).

References, p. 620

different metal complexes find many parallels in its general reactivity pattern: tin(IV) porphyrins are stable against bromine oxidation[43], and are the only metalloporphyrins which can be quantitatively photoreduced with amines to phlorins and chlorins[43,44]. Only metalloporphyrins (Mg, Cd) with midpoint potentials below 430 mV could be photo-oxygenated[45] (see p. 814). Electrophilic Vilsmeier formylation is only possible with metalloporphyrins where the porphyrin periphery is negatively charged, e.g. copper and nickel porphyrins, but not with Sn(IV) porphyrins[46] (page 649).

A comparison of midpoint oxidation potentials of some different magnesium porphyrins, which are taken here as an example, yields information on the effects of substituents at the periphery. Alkyl or phenyl substituents induce only minor changes (from Table 2; Mg(P) 0.60, Mg(OEP) 0.53, Mg(TPP) 0.54 V), whereas electron-withdrawing groups may raise the potentials considerably (from Table 1; Mg(OEC) 0.11, Chl-a 0.55 V). It is also found that reduction of peripheral double bonds lowers the oxidation potential by about 300 mV [from Table 1: Mg(OEP) 0.43, Mg(OEC) 0.11; Chl-a 0.55, BChl 0.27 V].

Voltammetric oxidation potentials of *meso*-hydroxy and amino porphyrins (Table 2), which can be converted into oxo- or imino-phlorins respectively (see page 630), are also much lower than those of unsubstituted porphyrins. All of these results are in qualitative agreement with Gouterman's four orbital model.

14.8. Properties of porphyrin π-radicals

Solutions of metalloporphyrin π-radicals, as well as of the diamagnetic π-dianions and π-dications, are often stable for many hours, and can therefore be studied at leisure with normal techniques. The main decomposition pathway is usually the slow formation of the starting metalloporphyrin. Many of the radicals have been crystallized and kept over periods of months under nitrogen without appreciable decomposition being observed. Felton and coworkers achieved the first X-ray analysis of a porphyrin radical, the perchlorate of zinc tetraphenylporphyrin cation, and reported the shortest zinc-to-oxygen bond yet observed, showing that the perchlorate anion is bound covalently to the metal[88]. Table 7 lists some representative publications which contain detailed information on the more stable species. There is, however, one porphyrin ligand which shows voltammetric waves in quite normal potential ranges, but where the radicals formed immediately polymerize to black, graphite-like precipitates, such that no ESR spectra can be observed. This is the case with unsubstituted porphin[63] (Table 2) and shows that the stability of unsubstituted porphyrin radicals is largely due to steric hindrance to the formation of porphyrin polymers. Another remark is in place here: the radicals of free base porphyrins are much less stable than

TABLE 7

Guide to publications which contain detailed analytical data (quantitative electronic spectra, ESR-spectra with hfs, gravimetric analysis, etc.) of oxidized and reduced porphyrin ligands

	References			
	−e	−2e	+e	+2e
Bacteriochlorophyll	5,6	—	28,47	—
Chlorophyll-*a*	5,6	—	—	—
Co^{III} Octaethylporphyrin	57,90	—	—	—
Co^{III} Tetraphenylporphyrin	57,90	90	—	—
CuOctaethylporphyrin	6,37	—	—	—
Cu^{II} Tetraphenylporphyrin	57	—	—	—
Deuteroporphyrin-IX dimethyl ester	—	—	19	19
Fe^{III} Tetraphenylporphyrin	26,56	—	—	—
Fe^{III} Octaethylporphyrin	26,56	—	—	—
Mesoporphyrin-IX dimethyl ester	—	—	19,21	19,21
meso-Tetraphenylbacteriochlorin	—	—	20,28	20
meso-Tetraphenylchlorin	—	—	20	20
meso-Tetraphenylporphyrin	—	—	20,21	20,21
meso-Oxy-octaethylporphyrin	55,60	—	—	—
Mg-Octaethylporphyrin	6,8,57	8,57	—	—
Mg-Octaphenyltetra-azaporphyrin	—	—	10	10
Mg-Tetrabenzporphyrin	5	—	—	—
Mg-Tetraphenylporphyrin	8,57,58	57	—	—
Ni-Octaethylporphyrin	6	—	—	—
Ni^{II} Tetraphenylporphyrin	50	50	—	—
Pd-Octaethylporphyrin	6	—	—	—
Pb^{II}-Octaethylporphyrin	52	—	—	—
Ru^{II} Octaethylporphyrin	53	—	—	—
Zn-*meso*-Oxyoctaethylporphyrin	55,60	—	—	—
Zn-Octaethylporphyrin	6,51	57,58,6	—	—
Zn-Octaethylchlorin	7	—	—	—
Zn-Tetrabenzporphyrin	—	—	10	10
Zn-Tetraphenylporphyrin	8,57,58	57	9	9
Zn-Tetramethylporphyrin (and other *meso*-tetraalkyls)	59	—	—	—

those of the metal complexes and have not yet been obtained in solid form.

Some examples of electronic spectra have been given in previous sections and have been discussed in recent reviews[2,3]. Chapter 13 deals with the ESR data.

Radicals obtained by the oxidation of the porphyrin π-systems play an important role in photosynthesis[1,3] (see also Chapters 16 and 17), react with oxygen (see Chapters 15 and 16), add and decompose peroxides,

References, p. 620

undergo intramolecular cyclizations (see Chapter 15), form stable diamagnetic $\pi-\pi'$-dimers and other π-complexes (see section 14.10), and are active in charge transportation through hydrophobic membranes (see Chapter 17). One can be certain that many more known porphyrin reactions with oxidants or electrophiles involve primary radical formation and that many more reactions will be discovered in the future.

The chemistry of reduced porphyrin π-systems, e.g. the π-anion radicals or π-dianions, is less well known. Crystalline material has been reported only once[9]. The only well-defined reactions are the addition of protons to the methine bridges (see Section 14.10), dimethylation with methyl iodide (see Chapter 15), disproportonation of phlorin radicals (see Chapter 16) and, of course, the oxidation to the starting material.

14.9. Phlorins, porphodimethenes and porphyrinogens

Like the electron transfers from and to the porphyrin periphery, hydrogenation of a bridge carbon atom is a fully reversible reaction which is of fundamental importance in porphyrin chemistry. The conversion of a methine into a methylene bridge can be achieved either by addition of protic solvents to porphyrin anions, hydride transfer to a neutral porphyrin (which is often a photo reaction), or by catalytic hydrogenation of a neutral porphyrin. In bivalent metalloporphyrins four different states of hydrogenation (8—11) are possible. In metal-free derivatives, the negative charges are neutralized by the addition of protons.

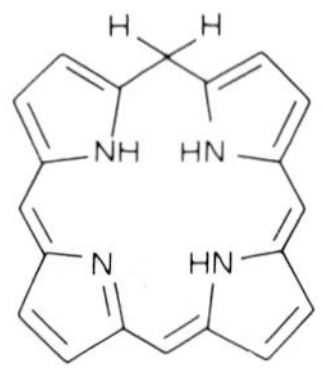

(8)
Phlorin

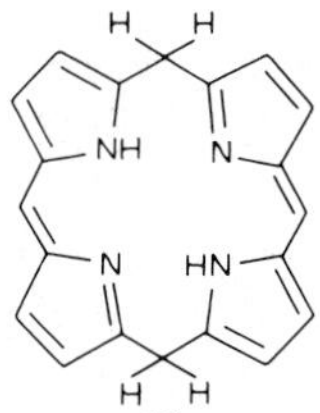

(9)
Porphodimethene

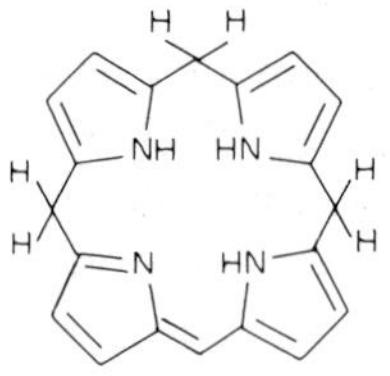

(10)
Porphomethene

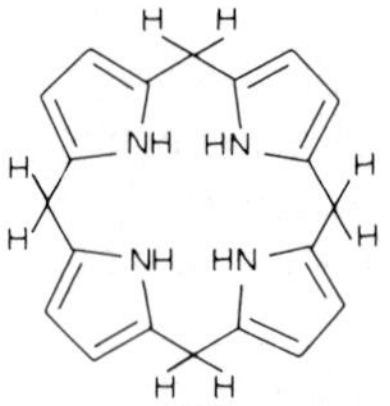

(11)
Porphyrinogen
(metal complexes are
unstable)

The most interesting reduced state of a porphyrin is undoubtedly the phlorin (8) which was discovered independently by Mauzerall[64-66] and Woodward[67,68]. Phlorins are very stable compounds under non-oxidizing conditions; they do, for example, withstand the action of concentrated sulfuric acid[68]. They are comparable in energy content to the completely conjugated cyclic electronic system of the porphyrin, and therefore equilibria between phlorins and porphyrins can be maintained under appropriate conditions.

In porphyrins with substituents on the methine bridge and on both of the adjacent β-pyrrolic carbons, phlorin formation relieves the strain induced by steric overcrowding of the macrocyclic plane, and is therefore particularly favored. Two reactions found during Woodward's ingenious chlorophyll synthesis illustrate this principle: a γ-propionic acid (12) isomerizes on simple heating in acetic acid in a reversible reaction to a phlorin, but when thiolacetic acid is used, the reversible nucleophilic addition of one solvent molecule is observed[68]. The photoreduction of porphyrins to other phlorins and chlorins, originated by Mauzerall, is described in Chapter 16. Later, Closs and Closs[9] produced zinc tetraphenyl phlorin anion (6) by protonation of the dianion (5) and observed facile conversion into the chlorin (7). Buchler[69] investigated the latter rearrangement with various metal complexes of octaethylporphyrin, and found that acid labile chelates (e.g. Mg, Zn) gave the most satisfactory results. Finally, Whitlock[70] reported that the reaction (6) → (7) can be almost completely reversed by heating the chlorin (7) in tetrahydrofuran—potassium tert.-butoxide in tert. butyl alcohol for 37 h at 130°C. Pulse radiolytic studies of the reduction of hematoporphyrin and its zinc complex by anion radicals of various carbonyl compounds [e.g. $(CH_3)_2CO^-$] also revealed that the primary product is always of the phlorin type[89].

The most surprising synthetic application of phlorins is certainly the successful conversion of a chlorophyll-a derivative, chlorin-e_6 trimethyl ester (12) into a chlorophyll-b derivative, rhodin-g_7 trimethyl ester (19), achieved by Inhoffen's group[33]. The chlorin-β-phlorin (13) was formed in almost 100% yield by electrolysis (see p. 814); the overall yield of (19) was about 1%.

The electronic spectra of phlorin salts (e.g. Refs. 9,19,65,68) and metal complexes typically produce two bands of comparable oscillator strength around 440 and 800 nm (Fig. 11) which are almost identical with the spectra of ring-opened metal bilitrienes[71]. Metal-free, neutral phlorins are monoacid bases of pK 9, and absorb around 620 nm in the visible range[68]. All phlorins are readily converted into porphyrins by various oxidizing agents (e.g. oxygen, iodine, quinones) and consume precisely the amount of oxidant needed to remove two electrons.

Porphodimethenes (9) are obtained in the Krasnovskii photoreduction procedure of porphyrins in acid media (see Chapter 16) or by direct acidifi-

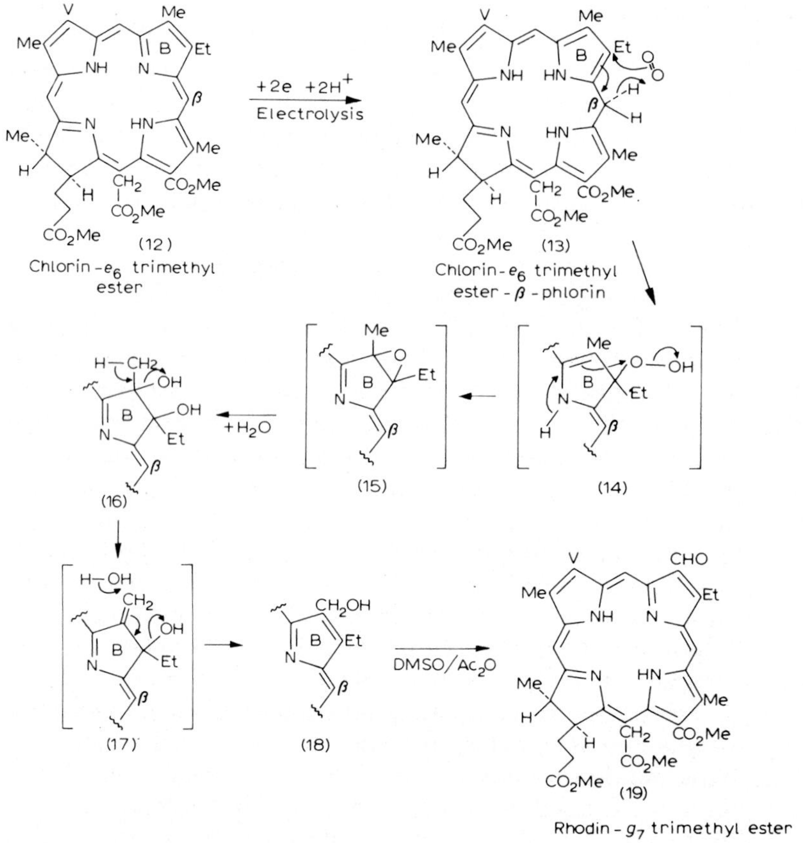

(12)
Chlorin-e_6 trimethyl ester

$+2e\ +2H^+$
Electrolysis

(13)
Chlorin-e_6 trimethyl ester-β-phlorin

(16)

$+H_2O$

(15)

(14)

(17)

(18)

DMSO/Ac$_2$O

(19)
Rhodin-g_7 trimethyl ester

cation of phlorins[68]. Only the α,γ (or β,δ)-hydrogenated isomers are formed, because in phlorins the methine bridge opposite to the methylene bridge has the highest electron density[68,72]. Porphodimethenes are readily oxidized to porphyrins, but rather air stable derivatives, the α,γ-dimethyl porphodimethenes (20) can be obtained, when porphyrin π-dianions react with methyl iodide[69]. Some typical spectra of these compounds are shown in Fig. 12.

Porphomethenes (10) have been prepared by the photoreduction of por-

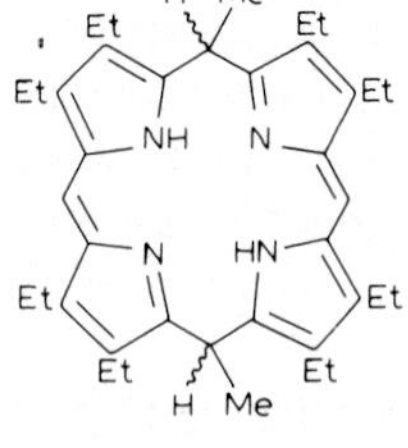

(20)

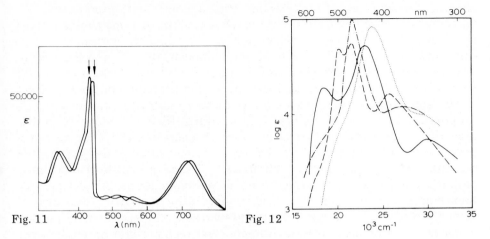

Fig. 11. Visible spectra of two phlorin cations (from Ref. 68).

Fig. 12. Electronic spectra of α,γ-dihydro-octaethylporphodimethene (from Ref. 69). (. . . . . .) Free base; (— · — ·) zinc complex; (- - - - - -) copper complex; (————) nickel complex.

phyrins with amines[65] (see Chapter 16) and afford electronic spectra and chemical properties similar to those of the porphodimethenes.

The porphyrinogens (11) are colorless hexahydroporphyrins (λ_{max} 200 nm)[65] and are the primary macrocyclic tetra-pyrroles found in the biosynthetic pathway of the hemes, chlorophylls, and vitamin B_{12} (see Chapter 3). It is at the porphyrinogen stage that all of the transformations of the acetic and propionic acid side-chains incorporated from porphobilinogen are carried out, resulting in those substituents finally appearing in protoporphyrin-IX (methyl, propionic acid, and vinyl)[73]. Uroporphyrinogen-III has also been identified[74] as a precursor of vitamin B_{12}, and it is therefore hardly surprising that uroporphyrinogens have been frequently employed in biosynthetic studies: the methods of choice are enzymic preparation of uroporphyrinogen-I or -III from porphobilinogen[74,75], chemical synthesis of a statistical isomer mixture from porphobilinogen in acidic solution[74,76] or reduction of uroporphyrin-III with sodium amalgam in water[64,77,78]. Porphyrinogens are rapidly oxidized to porphyrins by air, even in the solid state, and therefore have to be prepared under strictly anaerobic conditions shortly before they are used.

The rather unique fact that the macrocycle with twenty carbon atoms and four nitrogens of porphyrinogens is formed spontaneously in acidic solution in almost quantitative yield[76] from monomeric porphobilinogen has been explained with a single statistical argument. First, linear self-condensations take place until a tetrapyrrane (bilane) is formed. This is the first linear oligomer, which can without strain form a bond from head-to-tail. The maximum distance between the ends is 18 Å (for the all *trans* configuration), the mean distance can be calculated as 9 Å[76]. This corresponds to a concentration of the bilane and porphobilinogen, of 4 M in a bimolecular reaction, and easily explains why intramolecular cyclizations are highly favored over intermolecular additions of further porphobilinogen molecules.

The condensation of porphobilinogen under neutral conditions gives the uroporphyrinogen-I isomer expected from straightforward condensation reactions of the amino methyl grouping of porphobilinogen with the free α-positions[76]. In acidic solutions, however, rapid isomerization of the pyrromethanes and uroporphyrinogens is observed[66,76] and a statistical mixture of uroporphyrinogens I-IV is formed from porphobilinogen as well as from pure isomer uroporphyrinogen-I. The acid lability of the bond from the α-pyrrolic carbon atoms to the bridge carbons and its reversible cleavage in uroporphyrinogens is a most characteristic difference between porphyrinogens and porphyrins, the latter being stable in acidic solutions. The mechanism of specific formation of uroporphyrinogen-III formation in biological systems is discussed in Chapter 3. Basically similar isomerizations have been observed in acidic solutions of bilirubin[79], and the main provision for this reaction to occur seems to be the mobility of the pyrrole units which bear a tetrahedral carbon atom in the protonated form. In porphyrins a protonated α-pyrrolic position leads to a considerable strain in the macrocycle; in protonated phlorins this strain would be much less, and a helical type of configuration could be assumed, whereas in porphyrinogens, or ring-opened tetrapyrroles, no sterical inhibition of addition reactions at α-pyrrole carbons, which are inherent to pyrrole chemistry[80], is expected.

14.10. *Aggregation through interactions of the porphyrin π-electron core with other π-electron systems*

R. Willstätter and M. Fischer were the first to prepare crystalline 'Molekülverbindungen' from etioporphyrin with styphnic and picric acids[82] in chloroform. H. Fischer and his collaborators extended this work to a large variety of natural porphyrin derivatives and gave a systematic survey of their work in Fischer's book[83]. The main results were that porphyrins form mostly 1 : 1 or 1 : 2 complexes with styphnic, picric, picrolonic and flavianic acids, with typical electronic spectra; these addition compounds were often much more soluble in ether than the pure porphyrins and they often yielded excellent crystals and dissociated when alcohols were added to the chloroform or ether solutions.

Mauzerall studied the spectra of uroporphyrin molecular complexes with pyridine derivatives in aqueous solution with regard to biological implications[84]. He observed the formation of 1 : 1 and 1 : 2 complexes which gave small shifts in the visible spectra, a drop in the intensity of the Soret band and a decrease in the total intensity of emission spectra by a factor of between two and five. A similar complex between chlorophyll-a and flavin mononucleotide in acetone/water produces a new long wavelength absorption at 705 nm and is photochemically active[85]. This aggregate as well as those with quinones therefore are described in detail in Chapter 16. The most dramatic spectroscopic effects, however, are observed in self-aggregates of porphyrins, which deserve interest because they presumably play a role in the photosynthetic and respiratory redox chains. Only two typical examples of well-defined porphyrin aggregates will be discussed here; chlorophyll aggregates are referred to in Chapter 16.

When protochlorophyllide is dissolved in solvents of low dielectric constant (benzene, chloroform, ether) the spectrum 1 of Fig. 13 slowly converts to spectrum 2[86]. This can be fully reversed by addition of a few drops of methanol. The 650 nm compound does not fluoresce and its spectrum resembles that of photoactive protochlorophyll in plants, whereas the 632 nm pigment is like the protochlorophyllide in etiolated plants, which is not converted into chlorophylls by light. A simple correlation of these observations is that only protochlorophyllide molecules in a non-polar matrix, e.g. a lipoprotein, can be photo-reduced and that the molecule in such an environment is always in its aggregated state.

The molecular complexes described so far are formed by the usual weak interactions of organic molecules (ionic bonds, van-der-Waals interactions, etc.) and the binding energies are presumably all of the order of 2—3 Kcal/mol or less. Much stronger binding forces are found between some metalloporphyrin π-cation radicals. Zinc octaethylporphyrin radical, which has been studied in detail[51], is taken as an example here. Crystals of this compound, obtained by chemical oxidation of zinc octaethylporphyrin, are diamagnetic (Gouy balance experiments) and show intense absorption in the near infrared ($\sim$930 nm). The same strong band (oscillator strength 0.75) is observed in an acetonitrile solution at room temperature and in methanol—chloroform 10 : 1 at $-50°$C (Fig. 14). If one warms up the latter solution to $+50°$C, the 930 nm band disappears completely and a normal π-cation radical spectrum is found (Fig. 14). These spectral changes are fully reversible and it has also been shown that the increase of the band in the near infrared parallels the decrease in intensity of the ESR signal of the radical. From quantitative measurements of the temperature dependence of electronic and ESR absorption peaks, a ΔH of 17.5 Kcal/mole^{-1} has been determined, which is five times more than with diamagnetic dimers of smaller π-radicals, e.g. Wurster's blue[87]. Temperature jump experiments in methanol—chloroform proved that the reaction was second order with respect to the zinc porphyrin radical and showed that the reaction rate was close to diffusion controlled ($K_{12} = 1.3 \times 10^8$ l. mol^{-1} sec^{-1}). The dimer has also been characterized by NMR and mass spectroscopy. In chloroform solution the

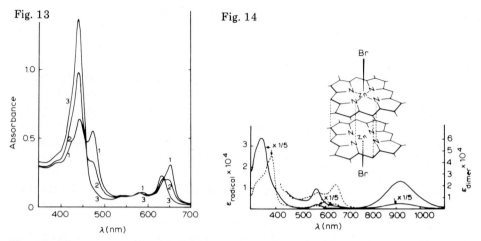

Fig. 13. Electronic spectra of protochlorophyllide monomer (1) and dimer (3). [(2) corresponds to an intermediate stage in the dimerization process.] (from Ref. 86).

Fig. 14. Electronic spectra of zinc octaethylporphyrin π-radical (- - - - -) and its diamagnetic π—π' dimer (———). (From Ref. 51).

radical did not dimerize at room temperature. The stability of the dimer and the strong charge transfer band have been attributed to a novel type of π-π'-bond formed by overlap of the half-filled π-orbitals in the face-to-face dimer of porphyrin rings. Similar absorptions have been observed in cooled solutions of magnesium octaethylporphyrin[8]. Zinc hematoporphyrin-IX radicals in water methanol 10 : 1 dimerize quantitatively at room temperature[51].

To summarize the interactions of porphyrin π-electrons with other aromatic systems, one might say that the porphyrin macrocycle forms well-defined, stable molecular complexes in various solvents and that electron transfer from one partner to the other, presumably perpendicular to the porphyrin plane, is very efficient; this is especially important in their photochemical behavior (see Chapter 17). Aggregates which are formed by interactions of the central metal ions are discussed in Chapters 5, 6, and 7.

References

1. D.H. Kohl in "Biological Applications of Electron Spin Resonance", Eds. H.M. Swartz, J.R. Bolton, and D.C. Borg, Wiley — Interscience, New York (1972), p. 213.
2. D.C. Borg in ref. 1, p. 265.
3. J.-H. Fuhrhop, Struct. Bonding Berlin, 18, 1 (1974).
4. E. Rabinowitch and J. Weiss, Proc. Soc., London, Ser. A 162, 251 (1937).

5. J.C. Goedheer, in "The Chlorophylls" Eds. L.P. Vernon and G.R. Seely, Academic Press, New York (1966), p. 164.
6. J.-H. Fuhrhop and D. Mauzerall, J. Am. Chem. Soc., 90, 3875 (1968); 91, 4174 (1969).
7. J.-H. Fuhrhop, Z. Naturforsch., Teil B, 25, 255 (1970).
8. J. Fajer, D.C. Borg, A. Forman, D. Dolphin, and R.H. Felton, J. Am. Chem. Soc., 92, 3451 (1970).
9. G.L. Closs and L.E. Closs, J. Am. Chem. Soc., 85, 818 (1963).
10. N.S. Hush and J.W. Dodd, J. Chem. Soc., 4607 (1964).
11. N.S. Hush and J.R. Rowlands, J. Am. Chem. Soc., 89, 2976 (1967).
12. L. Meites, "Polarographic Techniques", Interscience, New York (1965).
13. S. Piekarsky and R.N. Adams in "Physical Methods of Chemistry", Vol. IIa, Eds. A. Weissberger and B.W. Rossiter, Wiley - Interscience, New York (1971), p. 566.
14. R.N. Adams, "Electrochemistry at Solid Electrodes", Marcel Dekker, New York (1969).
15. A. Stanienda, Z. Phys. Chem., 229, 259 (1964).
16. A. Stanienda, Z. Phys. Chem., Neue Folge, 52, 254 (1967).
17. D.W. Clack and N.S. Hush, J. Am. Chem. Soc., 87, 4238 (1965).
18. R.H. Felton and H. Linschitz, J. Am. Chem. Soc., 88, 1113 (1966).
19. G. Peychal-Heiling and G.S. Wilson, Anal. Chem., 43, 545 (1971).
20. G. Peychal-Heiling and G.S. Wilson, Anal. Chem., 43, 550 (1971).
21. G.S. Wilson and B.P. Neri, Ann. N.Y. Acad. Sci., 206, 568 (1973).
22. B.A. Kizelev, Y.N. Kozlov, and V.B. Yevstigneeva, Biofizika, 15, 594 (1970).
23. J.-H. Fuhrhop, K. Kadish, and D.G. Davis, J. Am. Chem. Soc., 95, 5140 (1973).
24. K. Kadish and D.G. Davis, Ann. N.Y. Acad. Sci., 206, 495 (1973).
25. J. Chang, R.F. Large, and G. Popp, in ref. 13, Vol. IIb, p. 2.
26. R.H. Felton, G.S. Owen, D. Dolphin, and J. Fajer, J. Am. Chem. Soc., 93, 6332 (1971).
27. L.O. Rollman and R.T. Iwamoto, J. Am. Chem. Soc., 90, 1455 (1968).
28. J. Fajer, D.C. Borg, A. Forman, D. Dolphin, and R.H. Felton, J. Am. Chem. Soc., 95, 2739 (1973).
29. H.H. Inhoffen and P. Jäger, Tetrahedron Lett., 3387 (1965).
30. H.H. Inhoffen, P. Jäger, R. Mählhop, and C.-D. Mengler, Ann. Chem., 704, 188 (1967).
31. P. Jäger, Dissertation Braunschweig, 1965.
32. W.R. Heinemann, J.N. Burnett, and R.W. Murray, Anal. Chem. 40, 1794 (1968), and references therein.
33. H.H. Inhoffen, P. Jäger and R. Mählhop, Ann. Chem., 749, 109 (1971).
34. K. Kadish, D.G. Davis, and J.-H. Fuhrhop, Angew. Chem., 84, 1072 (1972).
35. R. Havemann, W. Haberditzl, and K.H. Mader, Z. Phys. Chem., 218, 71 (1961).
36. D.F. Evans, J. Chem. Soc., 2003 (1959).
37. A. Wolberg and J. Manassen, J. Am. Chem. Soc., 92, 2982 (1970).
38. J.W. Buchler, G. Eikelmann, L. Puppe, K. Rohbock, H.H. Schneehage, and D. Weck, Ann. Chem., 745, 135 (1971).
39. E. Heilbronner and H. Bock, "Das HMO Modell und seine Anwendung", Chaps. 7 and 12, Verlag Chemie (1968).
40. M. Gouterman, J. Chem. Phys., 30, 1139 (1959).
41. M. Gouterman, G.H. Wagnière, and L.C. Snyder, J. Mol. Spectrosc., 11, 108 (1963).
42. M. Zerner and M. Gouterman, Theor. Chim. Acta, 4, 44 (1966).
43. J.-H. Fuhrhop and T. Lumbantobing, Tetrahedron Lett., 2815 (1970).
44. A.H. Corwin and O.D. Collins, J. Org. Chem., 27, 3060 (1962).
45. P.K.W. Wasser and J.-H. Fuhrhop, Ann. N.Y. Acad. Sci., 206, 533 (1973).

46. H.H. Inhoffen, J.-H. Fuhrhop, H. Voigt, and H. Brockmann Jr., Ann. Chem., **695**, 133 (1966).
47. D. Dolphin and R.H. Felton, Acc. Chem. Res., **6**, 1 (1974).
48. D.H. Kohl, J. Townsend, G. Commoner, H.L. Crespi, R.C. Dougherty, and J.J. Katz, Nature London, **206**, 1105 (1965).
49. J.D. McElroy, G. Feher, and D.C. Mauzerall, Biochim. Biophys. Acta, **267**, 363 (1972).
50. A. Wolberg and J. Manassen, Inorg. Chem., **9**, 2365 (1970).
51. J.-H. Fuhrhop, P. Wasser, D. Riesner, and D. Mauzerall, J. Am. Chem. Soc., **94**, 7996 (1972).
52. J.A. Ferguson, T.J. Meyer, and D.G. Whitten, Inorg. Chem., **11**, 2766 (1971).
53. G.W. Sovocool, F.R. Hopf, and D.G. Whitten, J. Am. Chem. Soc., **94**, 4350 (1972).
54. G.M. Brown, F.R. Hopf, J.A. Ferguson, T.J. Meyer, and D.G. Whitten, in press.
55. J.-H. Fuhrhop, S. Besecke, and J. Subramanian, J. Chem. Soc., Chem. Commun., 1 (1973).
56. R.H. Felton, G.S. Owen, D. Dolphin, A. Forman, D.C. Borg, and J. Fajer, Ann. N.Y. Acad. Sci., **206**, 504 (1973).
57. D. Dolphin, Z. Muljiani, K. Rousseau, D.C. Borg, and R.H. Felton, Ann. N.Y. Acad. Sci., **206**, 177 (1973).
58. J. Fajer, D.C. Borg, A. Forman, R.H. Felton, L. Vegh, and D. Dolphin, Ann. N.Y. Acad. Sci., **206**, 349 (1973).
59. J. Fajer, D.C. Borg, A. Forman, A.D. Adler, and V. Varadi, J. Am. Chem. Soc., **96**, 1238 (1974).
60. J.-H. Fuhrhop, S. Besecke, J. Subramanian, Ch. Mengersen, and D. Riesner, in preparation.
61. R.H. Felton, G.M. Sherman, and H. Linschitz, Nature London, **203**, 637 (1964).
62. H. Berg and K. Kramarczyk, Biochim. Biophys. Acta, **131**, 141 (1967).
63. J.-H. Fuhrhop and R. Schlözer, in preparation.
64. D. Mauzerall and S. Granick, J. Biol. Chem., **232**, 1141 (1958).
65. D. Mauzerall, J. Am. Chem. Soc., **84**, 2437 (1960).
66. D. Mauzerall and G. Feher, Biochim. Biophys. Acta, **84**, 2437 (1972).
67. R.B. Woodward, Angew. Chem., **72**, 651 (1960).
68. R.B. Woodward, Ind. Chim. Belge, 1293 (1962).
69. J.W. Buchler and L. Puppe, Ann. Chem., **740**, 142 (1970).
70. H.W. Whitlock and M.Y. Oester, J. Am. Chem. Soc., **95**, 5738 (1973).
71. J.-H. Fuhrhop, A. Salek, J. Subramanian, and Ch. Mengersen, Ann. Chem., in press.
72. J.V. Knop and J.-H. Fuhrhop, Z. Naturforsch., Teil B, **25**, 729 (1970).
73. S. Granick and D. Mauzerall, in "Metabolic Pathways", Ed. D.M. Greenberg, Vol. II, Academic Press, New York (1961), p. 526.
74. A.I. Scott, C.A. Townsend, K. Okada, M. Kajiwara, and R.J. Cushley, J. Am. Chem. Soc., **94**, 8269 (1972).
75. L. Bogorad, Methods Enzymol., **5**, 891 (1962).
76. D. Mauzerall, J. Am. Chem. Soc., **82**, 2601, 2605 (1960).
77. B. Franck, D. Gantz, F.-P. Montforts, and F. Schmidtchen, Angew. Chem., **84**, 433 (1972).
78. H. Fischer and A. Stern, "Die Chemie des Pyrrols", Vol. II, Part 2, Akademische Verlagsgesellschaft, Leipzig (1940), p. 420.
79. A.F. McDonagh and R. Bonnett, Chem. Commun., 238 (1970).
80. A. Gossauer, "Die Chemie der Pyrrole", Springer, 1974 Chap. 3.
81. J.A. Guzinski and R.H. Felton, Chem. Commun., 715 (1973).
82. R. Willstätter and M. Fischer, Ann. Chem., **400**, 192 (1913).
83. H. Fischer and H. Orth, "Die Chemie des Pyrrols", Vol. II, part 1, Akademische Verlagsgesellschaft, Leipzig (1937), p. 612.

84. D. Mauzerall, Biochemistry, **4**, 1801 (1965).
85. S.J. Tu and J.H. Wang, Biochem. Biophys. Res. Commun., **36**, 79 (1969); S.J. Tu, Y.J. Tan, and J.H. Wang Bioinorg. Chem., **1**, 79 (1971).
86. B. Ke and C.J. Seliskar, Biochim. Biophys. Acta, **153**, 685 (1968).
87. K.H. Hausser and N.N. Murrell, J. Chem. Phys., **27**, 500 (1957).
88. L.D. Spaulding, P.G. Eller, J.A. Bertrand, and R.H. Felton, J. Am. Chem. Soc., **96**, 982 (1974).
89. Y. Harel and D. Meyerstein, J. Am. Chem. Soc., **96**, 2720 (1974).
90. D.C. Borg, J. Fajer, R.H. Felton, and D. Dolphin, Proc. Nat. Acad. Sci., U.S.A., **67**, 813 (1970).

IRREVERSIBLE REACTIONS AT THE PORPHYRIN PERIPHERY (EXCLUDING PHOTOCHEMISTRY)

JÜRGEN-HINRICH FUHRHOP

Gesellschaft für Molekularbiologische Forschung 3301 Stöckheim/Braunschweig, West Germany and *Institut für Organische Chemie der T. U. Braunschweig 3300 Braunschweig, Schleinitzstr. West Germany*

15.1. General aspects of reactivity at the porphyrin periphery

15.1.1. Electronic reactivity parameters of reaction sites

The α- and β-pyrrolic positions and the methine bridges constitute the reactive sites of the porphyrin (1) periphery. A simple theory of the reactiv-

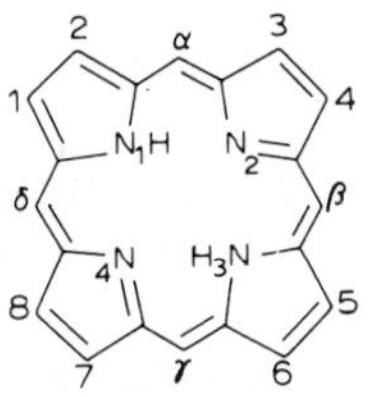

(1)

ity of conjugated systems, namely Fukui's frontier model[1], states that the relative reactivities of reaction sites are determined by the magnitude of the square of the coefficient of the atomic orbital in the highest occupied (HOMO) and the lowest unoccupied (LUMO) molecular orbitals, at this site. A relatively high electron density in the HOMO at a particular atom means that an electrophile is most readily added to this site; a large atomic coefficient in the LUMO indicates high reactivity of this center towards nucleophiles. The reactivity towards electroneutral radicals can be obtained either by averaging electron densities of the HOMO and the LUMO, or by independent McLachlan calculation[2] of unpaired electron densities. Table 1 lists results of PPP-SCF calculations of Fukui's reactivity parameters[3]. The main prediction from this table is that *all* reactions of the porphyrin periphery (namely electrophilic, nucleophilic, and radical additions) should occur preferentially on the methine bridges. Reactivity towards radicals (Table 2)[3,4] is highest on the α-pyrrole carbons, and therefore suggests a special vulnerability

Porphyrins and Metalloporphyrins, ed. Kevin M. Smith

TABLE 1

Electron densities and reactivity parameters of free base porphyrin (from Ref. 3)

Position	Electron density	Electrophilic	Nucleophilic	Radical
α	0.961	0.278	0.121	0.200
1	1.031	0.006	0.106	0.056
3	1.024	0.046	0.032	0.039
$N_{1,3}$	1.324	0.140	0.125	0.133
$N_{2,4}$	1.607	0.157	0.000	0.078

of this site in the reactions of porphyrin radicals. The reactivity parameters of the chlorin macrocycle (2) (Table 3) indicate that the γ,δ-methine bridges

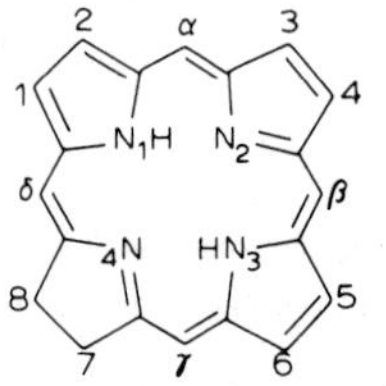

(2)

adjacent to the hydrogenated pyrrole ring should be more nucleophilic than the α,β-carbon atoms, and that peripheral double bonds of the pyrrole units adjacent to the pyrroline ring should be more reactive towards both nucleophilic and electrophilic addition reactions than is the opposite ring.

The physical background of the frontier orbital model is mainly intuitive and it does not take into account any effect of solvation, sterical hindrance,

TABLE 2

Electron densities and reactivity parameters of free base chlorin (from Ref. 3)

Position	Electron density	Electrophilic	Nucleophilic	Radical
α	1.067	0.311	0.167	0.239
β	0.924	0.235	0.265	0.250
3	1.049	0.050	0.104	0.077
4	1.013	0.033	0.130	0.081
5	1.039	0.005	0.010	0.007
N_1	1.258	0.239	0.158	0.198
$N_{2,4}$	1.595	0.142	·0.003	0.072
N_3	1.310	0.130	0.145	0.137

TABLE 3

Electron densities and reactivity parameters of free base phlorin (from Ref. 3)

Position	Electron density	Electrophilic	Nucleophilic	Radical
β	0.99	0.25	0.30	0.28
γ	1.09	0.36	0.08	0.22
δ	0.86	0.01	0.26	0.13
1	1.03	0.03	0.09	0.06
2	1.08	0.04	0.05	0.04
3	1.07	0.02	0.01	0.01
4	1.04	0.11	0.10	0.11
5	1.01	0.05	0.16	0.11
6	1.04	0.09	0.09	0.08
7	1.07	0.14	0.01	0.08
8	1.01	0.00	0.09	0.04

weak interactions of reactants and so on. Therefore, calculated electron densities can only be taken as a quantitative evaluation of one parameter out of many others, which may or may not reflect experimental reactivity patterns of a molecule. However, as will be shown in the descriptions of individual reactions at the porphyrin periphery, general rules derived from MO theory fit experimental findings surprisingly well. R.B. Woodward[5] has arrived at some of these generalizations by descriptive valence bond arguments, and introduced the perceptual idea that the four pyrrole units in porphyrins tend to reach individual electron sextets, therefore withdrawing electron density from the methine bridges. This argument rationalizes in a simple way the surprisingly strong electrophilic character of the *meso* carbons in the electron-rich porphyrin macrocycle and the loss of this feature in methine bridges which are connected with hydrogenated pyrrole units.

15.1.2. Stereochemical aspects

On his way to a synthesis of chlorophyll-a[5,62], R.B. Woodward demonstrated that bulky substituents on the peripheral positions, together with substituents on the methine bridges 'overload' the porphyrin plane (Fig. 1). The steric strain of such fully peripherally substituted structures is relieved when either the methine bridge carbon atom (phlorin (3) formation) or the peripheral carbons (chlorin (2) formation) become tetrahedral. This trend is fully borne out by many in vitro experiments (see Section 15.3.2) and can also be verified by observations on natural tetrapyrrole pigments; in pigments with methine bridge substituents, at least one of the neighboring pyrrole units is saturated in the peripheral positions (chlorophyll-a: isocyclic ring; vitamin B_{12}: methyl).

References, p. 662

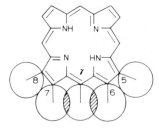

(3)

A second steric effect of peripheral substituents is their shielding of reactions on the methine bridges and pyrrolic carbon atoms. It is probably for reasons of steric hindrance that the usual Friedel-Crafts type reactions cannot be carried out on the methine bridges of fully peripherally substituted metalloporphyrins. Attack at the α-pyrrolic positions has so far only been possible with molecular oxygen and hydrogen peroxide, which are small reactants presumably without stable solvation spheres.

Peripheral substituents also prevent radical polymerization of the porphyrin macrocycles. Metal complexes of porphin, which are free from such restrictions, do not form stable radicals, but after oxidation[6] aggregate immediately and irreversibly to afford black precipitates.

Finally it should be pointed out that addition reactions to α-pyrrolic carbon atoms in porphyrins are probably always followed by immediate and irreversible opening of the macrocycle. Reports in the literature, especially abundant in papers on porphyrin disruption to bile pigments[7,8], which formulated porphyrins with substituents on α-pyrrolic carbons have either been conclusively shown to be incorrect[9-11] or are very probably so. The reason for the apparent instability of the porphyrin macrocycle after saturation of an α-pyrrolic carbon atom may be the steric strain imposed on the whole conjugated system, but this is by no means obvious because a slight deformation of the cycle towards a helical structure should take care of this strain. It is also well established that the tetradehydrocorrins, which have a similar macrocyclic structure, may bear one or two substituents on α-pyrrolic positions, without being disrupted[12,13]. One might therefore expect that porphyrin compounds with α-pyrrolic substituents will be prepared in

Fig. 1. Steric crowding in porphyrins with substituents in *meso* and peripheral positions (see text; from Ref. 45).

the future, and their properties revealed, but so far they seem to be exceedingly unstable.

15.1.3. Influence of central metal ions

This has been discussed in Chapter 14 and will be only shortly expanded here. Magnesium or zinc complexes should be the ideal compounds for reactions with oxidants and electrophiles because they induce the highest negative charge into the porphyrin periphery. These complexes are, however, so acid labile that they do not survive the conditions of most electrophilic reactions. Under such conditions nickel or copper porphyrins are suitable compromises[14,15]. On the other hand, tin(IV) complexes are ideally suited for hydrogenations or nucleophilic substitutions of the porphyrin π-system[16]. These complexes, however, can only be demetalated under quite rigorous conditions, so that yields of metal-free products are usually low. Here the application of acid labile antimony complexes might be indicated, but no pertinent experiments have been carried out with them so far.

In general, the activation of the porphyrin periphery by chelation with appropriate metal ions is preferable to the activation by protonation with acids (as in the Krasnovskii reduction procedure[17]) or deprotonation with strong bases (as in Fischer's photo-oxygenation[18]), because the primary products of porphyrin reactions (e.g. phlorins, formylbiliverdins) undergo a variety of irreversible secondary reactions with acids and bases. Reactions with metalloporphyrins on the other hand can often be carried out in inert, anhydrous, and aprotic solvents.

15.2. Oxidation

15.2.1. Introduction

One of the main functions of heme in nature is the activation of molecular oxygen by the donation of electrons from the iron σ-orbitals into the antibonding π-orbitals or oxygen. The *superoxide anion* is thought to be formed by the heme enzyme tryptophane dioxygenase (= pyrrolase) which catalyzes the *addition* of an oxygen molecule to a double bond of tryptophane [(4) → (6)], whereas the multienzyme complex containing cytochrome P450 pro-

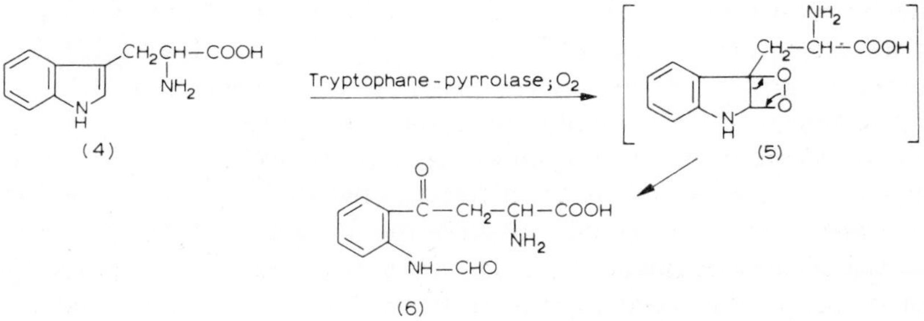

References, p. 662

duces an *'oxenoid'* oxygen which *inserts* an oxygen atom into an unactivated C—H bond, e.g. into the 11-position of a steroid [(7) → (8)][19,20] (see also

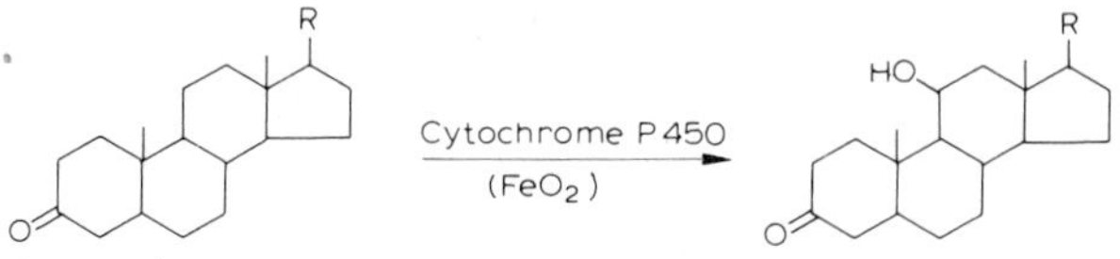

Chapter 16). Both types of reaction are also encountered in the biological breakdown of heme itself, yielding biliverdin (see Chapter 4), and it has been found that even in vitro the porphyrin ligand is very reactive towards molecular oxygen. Because of the biological significance of these reactions, they have been investigated ever since Warburg[21] obtained a 'green hemin' by coupled oxidation of heme in pyridine, whereby the biological degradation of the porphyrin macrocycle can be simulated, in vitro, with high yields under appropriate conditions. Porphyrin oxidations using singlet oxygen are discussed in Chapter 16; chemical oxidations and the properties of products are described in the following paragraphs.

15.2.2. Oxyporphyrins, oxophlorins, and their π-radicals
15.2.2.1. Structure

meso-Oxyporphyrins (9) have been obtained by total synthesis (see Chapter 2), by coupled oxidation[9,21,28], by reduction of xanthoporphyrinogens[22,24], and by oxidation of porphyrins with benzoyl peroxide[25]. By far the most convenient and efficient procedure, however, is the oxidation of zinc porphyrin complexes with thallium(III) trifluoroacetate, reported by Smith[26]. The tautomeric equilibrium of metal-free oxyporphyrins (9) is exclusively on the side of the keto-form, called oxophlorin (10), which, with its mesomeric dipolar forms, produces electronic absorption spectra intermediate between a porphyrin and a phlorin (Fig. 2, spectrum (a) 1)[9]. On titration with strong acids a monocation (11) is formed by protonation on nitrogen, and the long wavelength band shifts towards 700 nm giving an absorption spectrum very similar to that of 'real' phlorins with interrupted conjugation in the macrocycle (Fig. 2, spectrum (a) 6, and Fig. 11 in Chapter 14). Further addition of acid to a chloroform solution of (11) gives the oxyporphyrin dication (12) with a fairly typical (see Chapter 1) two-banded visible spectrum and a strong Soret band (Fig. 2, spectrum (b) 7). These spectral changes provide qualitative evidence for three important properties of the oxyporphyrins: 1) the unprotonated nitrogens are more basic than the carbonyl bridge, 2) the energy difference between the porphyrin and phlorin macrocycles has to be small, and 3) in the neutral oxyporphyrin (i.e. oxophlorin) the electronic spectrum in chloroform points to a distribution of the electrons between oxygen and the bridge carbon which is intermediate

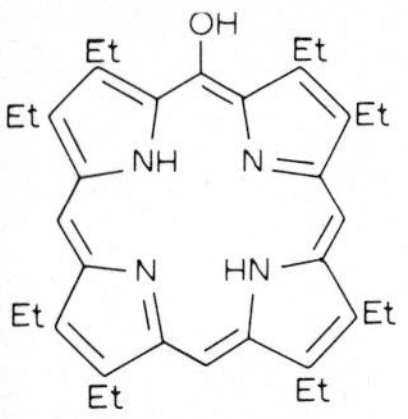

(9)

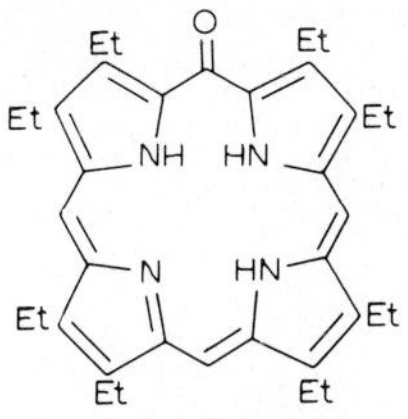

(10)

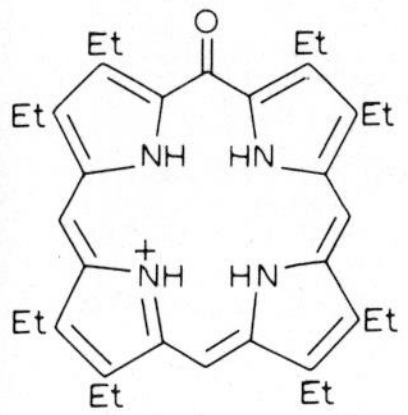

(11)

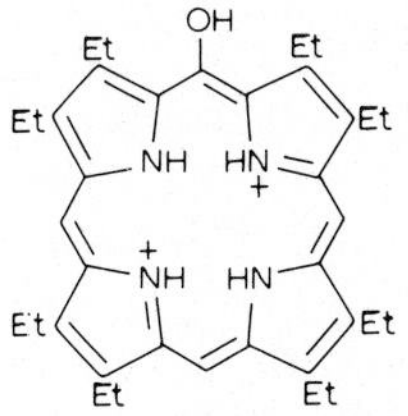

(12)

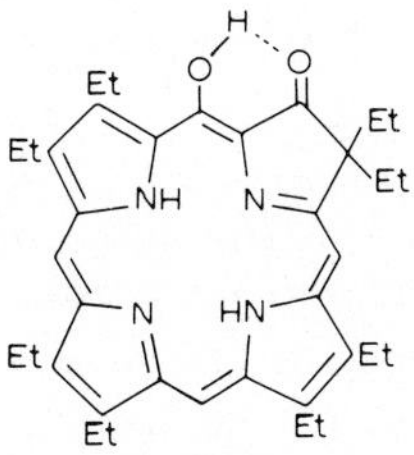

(13)

between a single and double bond. Similar spectral changes are found when the oxophlorins are titrated with strong bases, leading to deprotonated anions[9,31]. A pure *meso*-hydroxy structure can be obtained in neutral solution with divalent metal complexes of oxophlorins, or when a peri keto group [such as in (13)] stabilizes the hydroxyl group by intramolecular hydrogen bonding[27].

Oxophlorins and oxyporphyrins yield *meso*-acetoxyporphyrins in quantitative yield when treated with acetic anhydride in pyridine[9]; the benzoate is also known, but no keto derivatives (e.g. an oxime or hydrazone) have been obtained so far. A similar situation exists with the dipyrrolic pyrroketone counterparts which behave more like vinylogous amides than ketones. In fact, metal-free oxophlorins react[9] with triethyloxonium tetrafluoroborate to give *meso*-ethoxyporphyrins.

Metal complexes of oxyporphyrins always produce porphyrin type electronic spectra[9,28], and are easily oxidized in quantitative yield to π-radicals[29,31,37]; this is also true for the metal-free oxophlorins[29]. The electronic spectra of these radicals (Fig. 3) produce a broad absorption band covering the whole visible range and sometimes show weak bands in the near

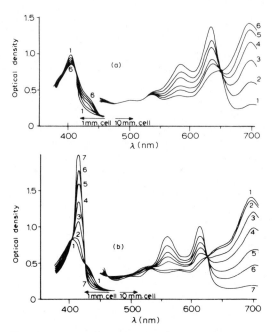

Fig. 2. Spectrophotometric titration of oxophlorin e.g. (10) in chloroform with trifluoro-acetic acid (TFA). a) Free base (10) → monocation (11). b) Monocation (11) → dication (12) (from Ref. 9).

infrared. ESR spectra of these radicals have four partly-resolved lines arising from the coupling of the unpaired electron with the methine hydrogens, and the signal reversibly disappears on cooling a chloroform solution to $-60°C$[29,31]. Fast temperature-jump kinetics of this phenomenon showed that a dimerization of the radical species was responsible for this behavior, which is accompanied by typical changes in the electronic absorptions[31]. The π-radicals of oxophlorins can be quantitatively reduced to diamagnetic oxophlorins by the addition of mild reductants e.g. triethylamine. Thus, the

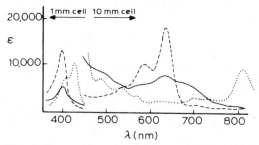

Fig. 3. Electronic spectra of oxophlorin (10) (- - - - - -), its π-radical (14a) (———), and zinc complex (14b) (. . . . . .) (from Ref. 29).

oxophlorin-radicals behave very similarly to unsubstituted metalloporphyrin π-radicals (see Chapters 13 and 14). There is however, one important difference: the oxophlorin-radicals are produced at much lower potentials than the corresponding metalloporphyrins; (10) produces voltammetric waves at −50 and +350 mV[31], whereas the unsubstituted octaethylporphyrin is oxidized at 810 mV. Such lowering of the oxidation potential of the porphyrin ligand has also been observed with *meso*-amino-substituted porphyrins. Oxy- and amino-porphyrins are also the only porphyrins which, depending on solvent and pH, produce both porphyrin and phlorin type spectra. These substituents have been called 'type II'.

Their exceptional influence on the reactivity of the whole porphyrin π-system presumably relates to their ability to form exo-macrocyclic double bonds, e.g. (15), thereby interrupting the aromatic inner conjugation path.

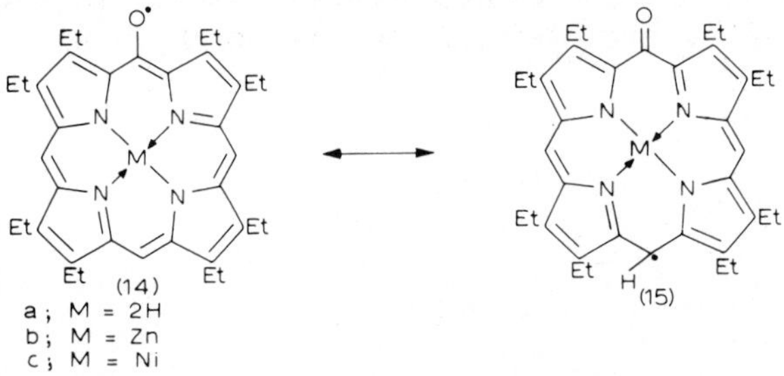

(14)
a ; M = 2H
b ; M = Zn
c ; M = Ni

(15)

This assumption is supported by the infrared spectra of oxophlorins and their protonated cations which possess an intense absorption around 1600 cm^{-1}. This band is absent in metal oxyporphyrins, but reappears in their π-radicals and anions. A final indication of the interruption of the π-system in neutral oxyporphyrins (= oxophlorins) is the large downfield shift of the methine proton signals in the p.m.r. spectrum when the exocyclic oxygen is protonated with deuterotrifluoroacetic acid. The resulting diprotonated species has a porphyrin type ring current (chemical shift of methine proton $\tau \sim 0$), whereas the nonprotonated species does not show low field signals (methine protons at $\tau \sim 3$). The latter value had to be extrapolated from acid titration curves, because neutral oxophlorins always contain small amounts of their π-radical, which broaden out their p.m.r. peaks[9].

15.2.2.2. Reactivity

Oxophlorins can be reconverted in 60—80% yield into porphyrins by reduction with sodium amalgam in methanol—acetic acid, hydrogenation and diborane[9], or lithium aluminum hydride in tetrahydrofuran[22−24].

Oxidation of oxophlorins with molecular oxygen first leads to π-radicals and thereafter oxygenation to α,γ-dioxoporphodimethenes (16) is observed. Both oxidations are accelerated by light and adsorption on alumina[29,31].

The proton of the hydroxy group has been substituted by acetyl[9,23]

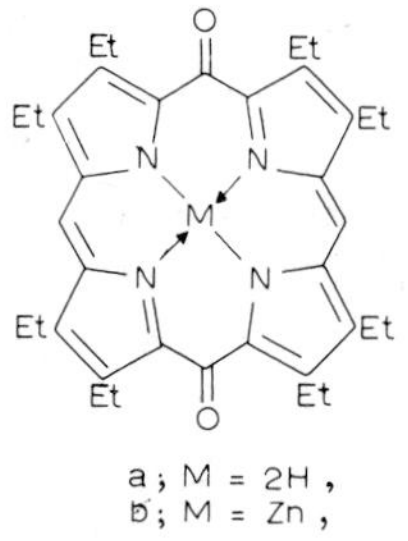

a ; M = 2H ,
b ; M = Zn ,

(16)

(17a), methyl (17b), or ethyl (17c) groups[9]. The resulting derivatives do not show any phlorin properties but behave like normal (= type I) porphyrins. The acyl derivatives can be reconverted into oxophlorins by short treatment with base, or into porphyrins by hydrogenation to give porphyrinogens, followed by re-oxidation with DDQ.

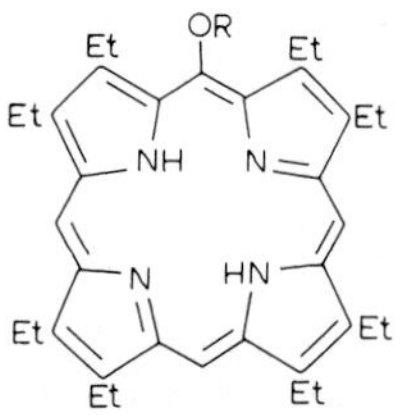

(17)

a ; R = COCH₃
b ; R = Me
c ; R = Et

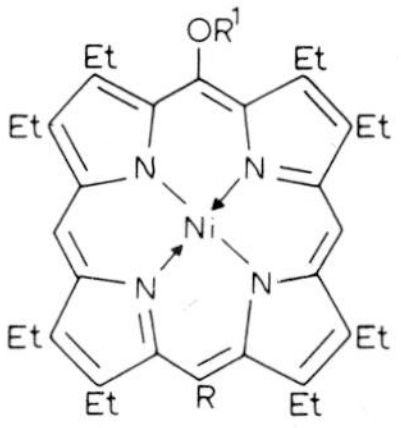

(18)

a ; R = CHO , R¹ = H
b ; R = CHNOH , R¹ = H
c ; R = CN , R¹ = COCH₃

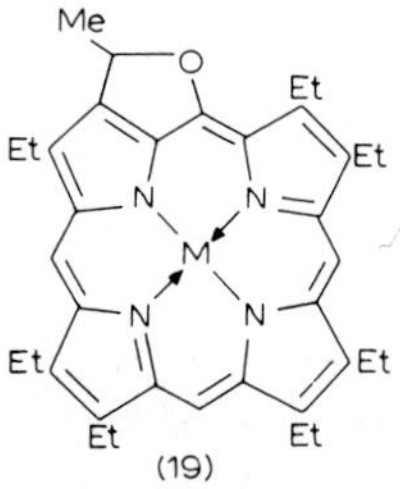

(19)

a ; M = 2H
b ; M = Zn
c ; M = Ni

Vilsmeier formylation of nickel oxyoctaethylporphyrin leads in 60% yield to the γ-formyl compound (18a)[31] and to about 10% of a β,γ-diformyl product[31]. (18a) is paramagnetic due to admixture with some free radicals which could not be removed by reductants. By standard procedures the diamagnetic oxime (18b) and nitrile (18c) have been prepared from (18a).

Irradiation of zinc or nickel oxyoctaethylporphyrin radicals (14b,c) in apolar solvents under nitrogen affords the oxole (19). This heterocyclic derivative of a porphyrin is stable as a nickel complex but decomposes quite rapidly in air when in the form of a zinc or metal free derivative (19a,b).

Photo-oxygenation of zinc oxyoctaethylporphyrin radical yields the zinc oxaporphyrin (20a) (see p. 638), which is opened by base to give the biliverdin analogue (21); the same type of oxygen-containing macrocycle is found as the major product (22) when heme is oxidized with hydrogen peroxide or ascorbic acid—oxygen in pyridine[9,21,28,32]. A likely precursor is the α-pyrrolic peroxide of type (23), which could rearrange either to ring-closed

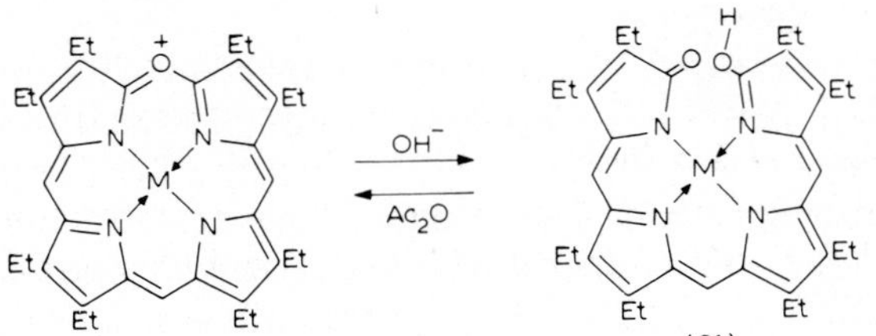

(20)

(21)

a; M = 2 H
b; M = Zn

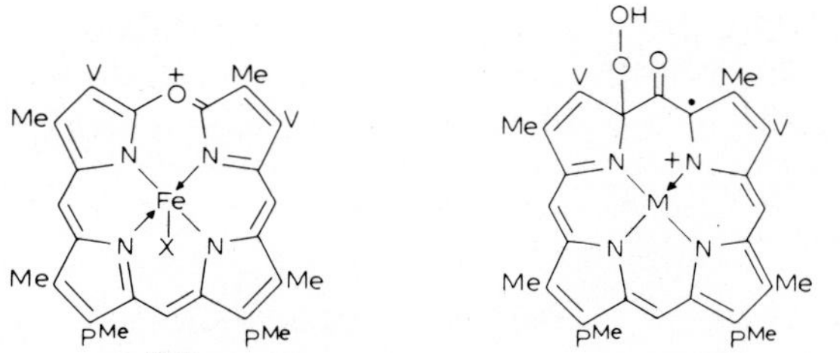

(22)

(23)

V = CH=CH$_2$
pMe = CH$_2$CH$_2$CO$_2$Me

oxoporphyrins by splitting off hydrogen peroxide (from two molecules of the peroxide) and carbon monoxide, or to ring-opened biliverdins by intramolecular migration of a hydroxyl radical again with loss of carbon monoxide[31]. The first pathway would correlate with the one observed in in vitro oxidation of oxophlorins by molecular oxygen or peroxides, whereas the latter pathway is presumably followed in natural heme degradation which does not involve hydrolytic reactions like (20) → (21)[35] (see p. 143). Oxidation of octaethylchlorin or its zinc complex with excess of thallium(III) trifluoroacetate yields directly the hydrogenated biliverdin (24), without the primary formation of an oxachlorin[34]. The reasons why in some cases only oxaporphyrins, and in other cases only biliverdins are formed (both often in quite high yield), are not known.

References, p. 662

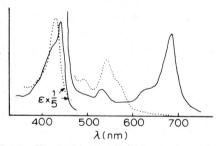

(24)

15.2.2.3. Metal complexes of oxyporphyrins

Electronic spectra of metallo-oxyporphyrins (9b) are extraordinarily sensitive to changes in solvent and pH[28]. The hydroxy group is sometimes easily deprotonated by weak bases (e.g. pyridine) and the porphyrin visible spectrum changes into a phlorin spectrum with bands close to 700 nm (Fig. 4)[28]. It is difficult to obtain p.m.r. spectra of metallo-oxyporphyrins because they often contain some quantities of the corresponding π-radical. Reasonably well-resolved spectra can be obtained at low temperatures where the radicals form diamagnetic dimers[31]. It is sometimes difficult to differentiate between metal complexes of oxyporphyrins, oxaporphyrins, biliverdins, and related products, which all possess electronic spectra with broad bands around 700 and 400 nm; they are often unstable, cannot be demetalated without decomposition and do not produce well-resolved p.m.r. spectra. Oxyporphyrins can be identified here by their reduction with sodium amalgam or reaction with acetic anhydride to yield porphyrins or *meso*-acetoxyporphyrins in high yield.

15.2.3. Dioxo-porphodimethenes

α,γ-Dioxo-porphodimethenes such as (16a) have been obtained by chemical or photochemical oxygenation of *meso*-oxyporphyrins (9)[31] and *meso*-aminoporphyrins[36]. They are stable towards further oxidation by oxygen and can be isolated in crystalline form either as free base or metal complexes[36]. When the xanthoporphinogen (25) is reduced with sodium

Fig. 4. Electronic spectrum of dipyridine cobalt(III) octaethyloxophlorin in ether (————) and after treatment with HCl (. . . . . . .) (from Ref. 28).

borohydride the dihydro-dioxo-compound (26a) is the first isolable compound[24]. It is surprisingly stable and only very slowly oxidizes to (16) with
air. Reduction of (25) with zinc in acetic acid leads to the tetrahydro compound (27), which is, in contrast to (26a), unstable in air. 3% Sodium amalgam in ethanol and subsequent air oxidation converts the dioxo compound
(16) into octaethylporphyrin[36].

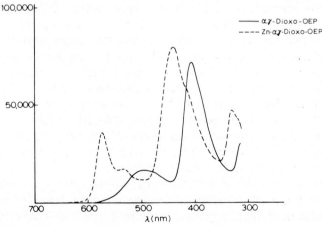

a; R = H
b; R = Me
c; R = Et

(25)   (26)   (27)

The zinc complex (16b) reacts with nucleophiles (e.g. MeLi, EtLi) to form
the β-substituted derivatives (26b,c) in good yield[106].

The electronic spectrum of dioxo-porphodimethene (16a) (Fig. 5) resembles pyrromethene spectra, the methine proton p.m.r. signals appear at τ =
3.36, and the mass spectrum usually exhibits a strong M + 2 peak[36].

It has been found that the diketone (16a) and its zinc complex (16b)
resclve into two main fractions on thin layer chromatography. Both metal-
free products have identical electronic, infrared, p.m.r. and mass spectra, but
the two zinc complexes behave chemically different: the complex of low R_f
value is a hydrogen bonded solvated dimer (X-ray)[161] and forms stable green
and violet π-radicals on nucleophilic attack or reduction (g = 2.003, linewidth 2 G), whereas the product with higher R_f does not form stable radicals.

Fig. 5. Electronic spectra of α,γ-dioxo-octaethylporphodimethene (16a) (———) and its
zinc complex (16b) (- - - - - -).

15.2.4. Xanthoporphyrinogens (meso-tetraoxoporphyrinogens)

By analogy with a classical reaction in the indigo series, H. Fischer tried to dehydrogenate porphyrins by the action of lead dioxide in chloroform—acetic acid to give dehydro porphyrins. Surprisingly, stable yellow products were isolated from a variety of porphyrins[38], all of which contained four more oxygen atoms than the starting porphyrin and could be reconverted into porphyrins with sodium amalgam. Fischer considered the right structure (25) for these xanthoporphyrinogens (xanthos = yellow; porphyrinogen = procreator of porphyrins). He finally rejected it because reduction with zinc in acetic acid removed only two oxygens, which seemed to be incompatible with the formulation of four identical oxygen functions in the oxidized porphyrin. The structural problem has been solved by modern spectroscopic methods[24]; the pathway of the xanthoporphyrinogen reductions, described in the preceding paragraph, remains unclarified, although a tentative explanation has been offered[39]. Porphyrins with oxidation sensitive side-chains (e.g. vinyl) break down to undefined brown pigments of low molecular weight (<400) when treated with lead dioxide.

Octaethyl-xanthoporphyrinogen produces an ultraviolet absorption peak at 340 nm ($\epsilon = 5 \times 10^4$), holds two molecules of water of crystallization, has a strong infrared band at 1602 cm^{-1} and an NH p.m.r. signal at $\tau = 3.0$. The most interesting aspect of its analytical properties lies in the very typical mass spectroscopic fragmentation patterns, which can be used to establish the sequence of pyrrole rings in porphyrins (see Chapter 9). The X-ray[162] shows (25) to be highly distorted with two H_2O molecules hydrogen bonded to the central NH's.

15.2.5. Oxaporphyrins

Oxidation of Fe(III) oxyporphyrins with oxygen in pyridine in the dark forms 'verdohemin' in 50% yield[9], which in fact has the structure of an oxaporphyrin (22). The methine carbon is lost as carbon monoxide, and this has been shown qualitatively in this system[155]. Photo-oxygenation of zinc octaethyloxophlorin radical in methylene chloride-benzene yields the oxaporphyrin (28) in about 80% yield under appropriate conditions and CO evolution has been semi-quantitatively determined[31,37]. Oxaporphyrin (20) is opened by trace amounts of base and acids of medium strength to biliverdin (21), this reaction being fully reversible with acetic anhydride. The ring closure only proceeds in good yields with metal complexes of biliverdins and metal free oxaporphyrins are therefore not easily available.

The electronic spectrum of (20) is a typical bilatriene type spectrum (Fig. 6), the p.m.r. spectrum points to a strong ring current [methine proton signals at $\tau = 0.73$ (1) and 0.82 (2)], and in the infrared, two weak absorptions at 1600 and 1750 cm^{-1} were observed. Many metallo-oxaporphyrins yield intense molecular ions in mass spectroscopy, which makes them par-

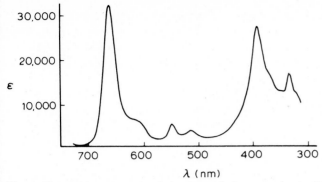

Fig. 6. Electronic spectrum of zinc octaethyloxaporphyrin (20b) (from Ref. 31).

ticularly useful derivatives of metallo biliverdinates; these cannot usually be analyzed directly by mass or p.m.r. spectroscopy.

15.2.6. β,β'-Dihydroxychlorins and β-oxochlorins

Whereas porphyrins and metalloporphyrins in neutral or basic solution add oxygen or hydrogen peroxide mainly to the inner 'aromatic' 16-membered macrocycle, (as has been demonstrated in the previous sections) protonated porphyrins are oxidized by hydrogen peroxide exclusively in the peripheral positions. The primary product is usually a dihydroxychlorin (28) which undergoes a pinacol rearrangement in concentrated sulfuric acid to yield chlorin type ketones (29)[25,41−45]. When peripheral hydroxylation with hydrogen peroxide is unsuccessful, then often osmium tetroxide was applied. In the hands of Inhoffen and coworkers these reactions have, sometimes followed by further reactions of the keto group, yielded a variety of interesting chlorin, corphin, and tetradehydrocorrin derivatives[43−45], some of which (29−34) are given below [see also structure (13)]. Such compounds are of interest because they either are, or can be converted into, water soluble artificial corphin or corrole precursors, which might form vitamin B_{12} antagonists in microorganisms.

The spectroscopic data of these peripherally oxygenated derivatives correspond closely to the related hydrogenated porphyrins, and are therefore not discussed.

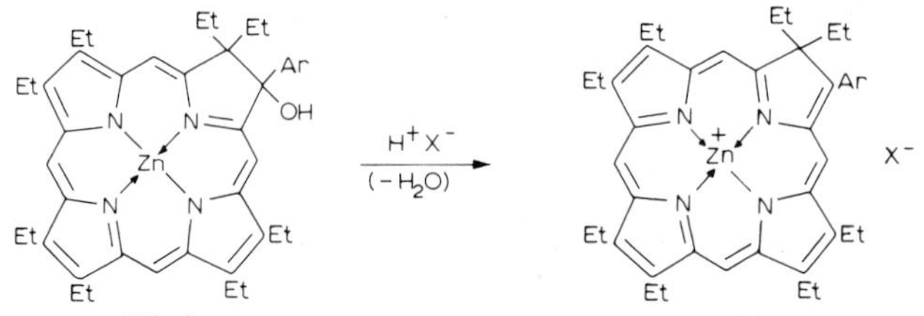

(30 a) $\xrightarrow[(-H_2O)]{H^+X^-}$ (30 b)

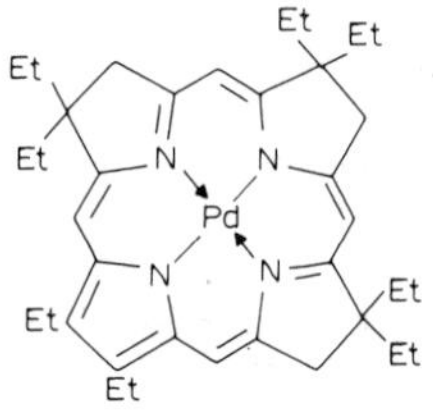

(31)

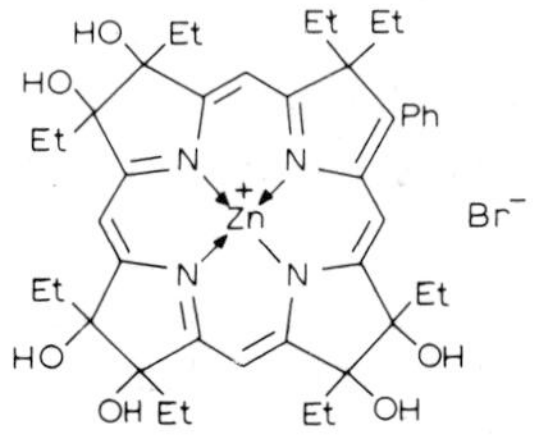

(32)

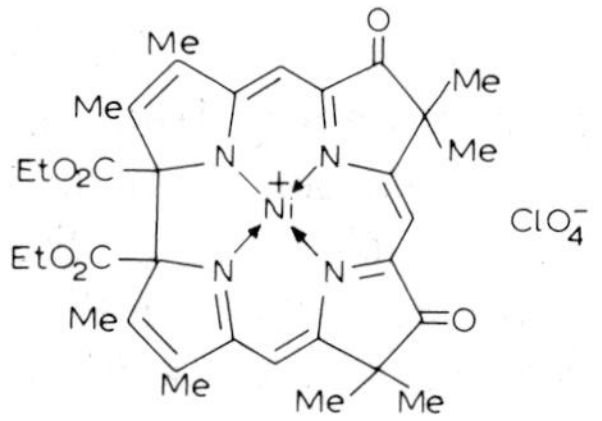

(33)

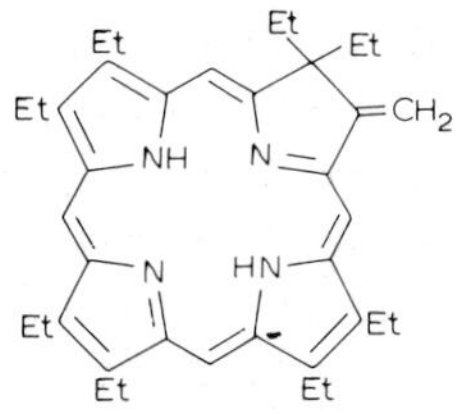

(34)

15.3. Peripheral (β,β')hydrogenation
15.3.1. Introduction

Addition of hydrogen to the methine bridges, leading to phlorins, is the primary step. This is a reaction which is fully reversed by the action of oxygen, and which was therefore discussed in Chapter 14. Hydrogenation of the peripheral positions leads to more stable products, chlorins in the first instance, and will therefore be discussed under the heading of 'irreversible reactions'. This choice is somewhat arbitrary because all peripherally hydrogenated porphyrins can be almost quantitatively reconverted into porphyrins by treatment with quinones, the vinylogs of oxygen. The general experience is, however, that chlorins are stable under most reaction conditions, whereas phlorins are not. Five types of peripherally hydrogenated porphyrins are possible: chlorins (35), a-tetrahydroporphyrins (36), b-tetrahydroporphyrins (bacteriochlorins) (37), hexahydroporphyrins (38), and corphins (39). All of these conjugated systems have been synthesized and, except for (38), fully characterized. Moreover, chlorins and bacteriochlorins are natural products, and it was hoped for a long time that corphin might be the elusive interme-

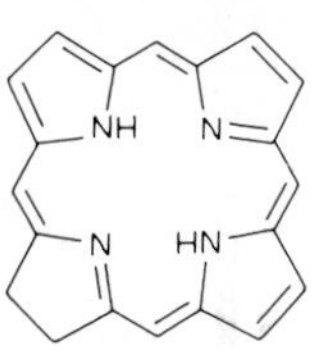

(35)

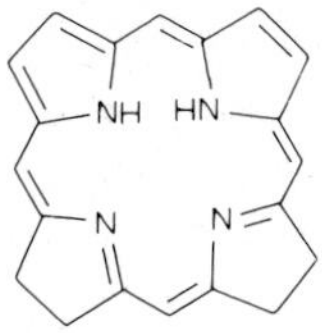

(36)

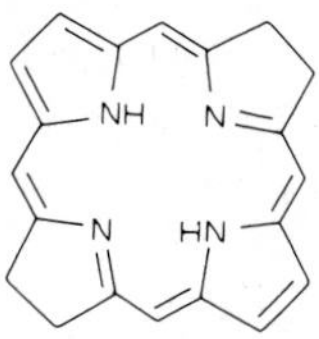

(37)

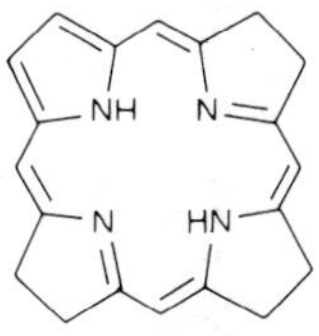

(38)

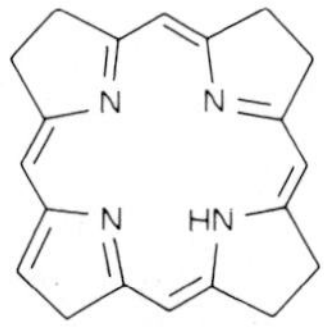

(39)

diate between the porphyrin and corrin biosynthetic pathways. If hydrogenation is achieved at both the peripheral positions and the methine bridges then thirty isomeric dihydroporphyrins are possible[46], but so far the only 'mixed' hydrogenation products reported are the chlorin-phlorins (see p. 615); for this reason these products will not be discussed further.

15.3.2. Chlorins

H. Fischer's procedure[47,48], later adopted by Eisner[49] for the reduction of iron porphyrins with sodium in boiling isoamyl alcohol is still the method of choice for the preparation of trans-chlorins in good yield (provided no labile side-chains are present) (see p. 815). Other methods with similar limitations include the chemical[50,51] or photochemical[16,52] reduction of tin(IV) porphyrins (where the removal of the central metal ion is a particular problem)[16], isomerization of phlorins in strongly basic solution[53], reduction of various metalloporphyrins with sodium anthracenide in tetrahydrofuran and subsequent protolysis[54], heating with sodium ethoxide[55], and photochemical reductions of zinc porphyrins with ascorbic acid and amines[56]. cis-Chlorins have been obtained by porphyrin reduction with di-imide[51,57] (see p. 815) (which gives the best yields), hydrazine[45,58], ascorbic acid and DABCO[59], and diborane[60].

None of the above procedures can be applied to porphyrins without protection of reducible side-chains, e.g. vinyl groups, and none is selective for differently substituted pyrrole units in the porphyrin macrocycle. Such selective hydrogenation is possible for porphyrins with large substituents on adjacent methine bridges, rendering the neighboring pyrrole units highly susceptible towards addition reactions. This principle (see p. 628) was first applied by H. Fischer in hydrogenations of chlorophyll derivatives[61] (40) → (41), later by Corwin[50], and was used in a most extraordinary manner by Woodward[62] in one of the final stages (42) → (46) of his chlorophyll synthe-

References, p. 662

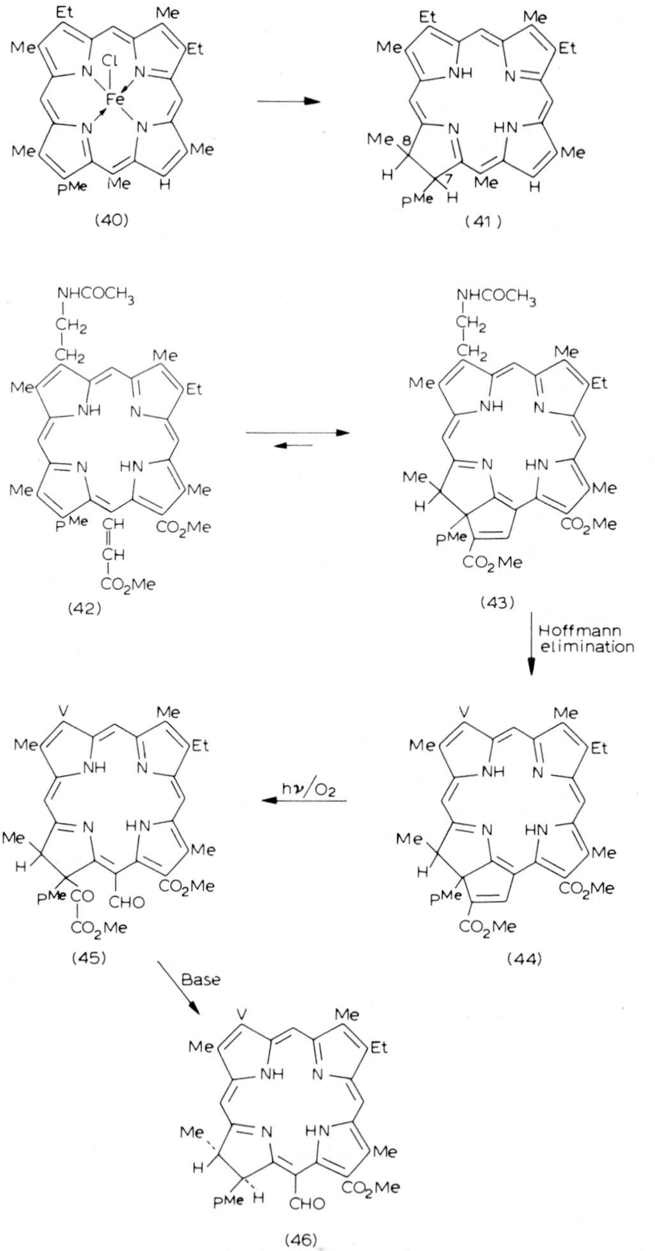

(40)

(41)

(42)

(43)

Hoffmann
elimination

(45)

hν/O₂

(44)

Base

(46)

sis. This pathway to chlorins does not include any hydrogenation and can therefore be applied to porphyrins with vinyl groups. Only the two pyrrole units neighboring the methine substituent are attacked, and selectivity between these two rings can be achieved when one of the peripheral positions, [where the cyclization, e.g. (42) → (43) takes place] is deactivated by a carboxyl group, which might be removed at a later stage.

Electrochemical reduction of metalloporphyrins usually leads to phlorins, but unexpected rearrangements sometimes occur, which in one case has been used to synthesize either chlorophyll-*b* or bacteriochlorophyll derivatives from chlorin-e_6 trimethyl ester (see p. 615).

Metallo-chlorins form stable π-radical cations at lower potentials than do metalloporphyrins (see p. 598). Chlorins and their metal complexes can be dehydrogenated to porphyrins by air[63,64] and, more effectively, by quinones[65,66]. Their physical properties and specific reactivity are always discussed together with the related phenomena on porphyrins in this book.

15.3.3. a- and b-Tetrahydroporphyrins

The two isomeric octaethyltetrahydroporphyrins of type a (47) and b (48) were first prepared by Eisner[49] by the reduction of iron octaethylporphyrin with sodium in isopentyl alcohol. The b-isomer, however, is usually

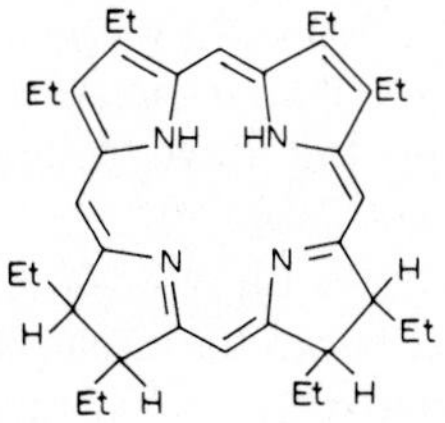

(47)

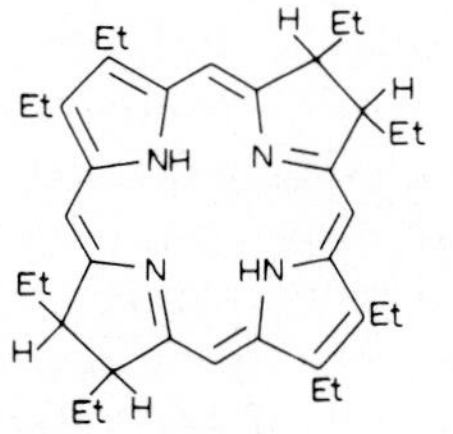

(48)

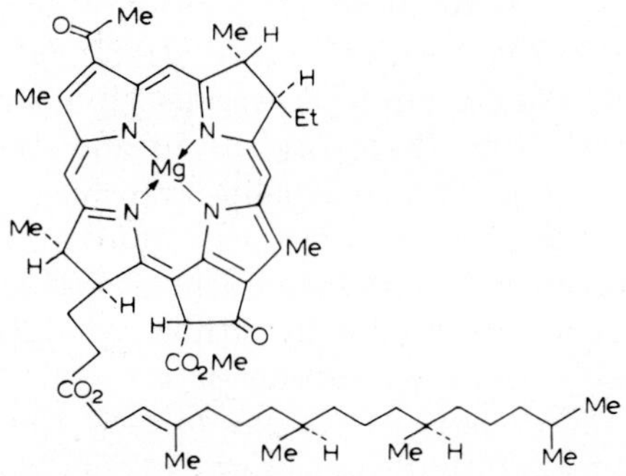

(49)

only formed in trace amounts, whereas the a-isomer can be produced in good yield. The same difficulty to produce b-tetrahydroporphyrins from *meso*-unsubstituted porphyrins has been found in other chemical[51,57−60,67] and photochemical[16] reduction procedures, which all yield preferentially the a-isomer. Reduction of *meso*-tetraphenylchlorin, however, with Raney nickel/hydrogen[68] or of *meso*-tetraphenylporphyrin with di-imide[57] yields the b-isomers. This finding points to the steric strain in *meso*-tetrasubstituted porphyrins, which presumably is less when all four methine bridges are

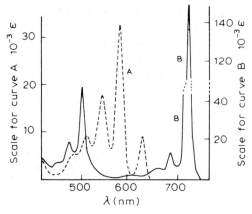

Fig. 7. Electronic spectra a-octaethyltetrahydroporphyrin (47, spectrum A) and b-octa-ethyltetrahydroporphyrin (48, spectrum B), (from Ref. 49).

flanked by a reduced pyrrole unit. The finding that in general, when steric interactions are of no importance, chlorins are reduced to a-tetrahydropor-phyrins is in good agreement with results of model calculations: the periphe-ral positions of pyrrole units neighboring the pyrroline ring have much higher reactivity indices than those of the opposite ring[3] (see Table 2).

The most prominent b-tetrahydroporphyrin is the magnesium complex, bacteriochlorophyll (49), which can be reversibly oxidized to its π-radical at low potential (see p. 612). All of the tetrahydroporphyrins are, however, easily dehydrogenated to chlorins by almost any oxidant, and care has to be taken to preclude extended contact with air in isolation procedures. Heating of degassed alkaline solutions of tetrahydrotetraphenylporphyrins also leads to dehydrogenation[53]. Apart from these oxidations and dehydrogenations the chemistry of tetrahydroporphyrins is virtually unknown.

The p.m.r. spectra of the tetrahydroporphyrins point to ring current ef-fects similar to those observed in chlorins (see Chapter 10); fragmentation patterns observed in mass spectra resemble also those of chlorins (see Chap-ter 9) and infrared spectra usually contain a skeletal vibration band ('chlorin band') between 1550 and 1660 cm^{-1} (see Chapter 11). The visible absorption spectra of a-tetrahydroporphyrins (Fig. 7) are comparable to those of porphyrin monocations and monoanions, whereas spectra of b-tetrahydroporphyrins[69] are of the chlorin-type [Soret bands: (47) λ_{max} 370 nm (103,500), (48) λ_{max} 374 (180,700)[49]]. Metalation does not signifi-cantly change the number and relative intensities of bands[51,69].

15.3.4. Hexahydroporphyrins and corphins

Octaethyl-hexahydroporphyrin has been prepared by Eisner[49] who re-ported the visible spectrum (λ_{max} 500 nm) and the almost quantitative conversion into octaethylporphyrin by three moles of dichlorodicyanoben-

zoquinone. Whether this product is of type (38), however, or is partly hydrogenated at methine bridge positions remains to be established. The inner
conjugation pathway of the porphyrins is only interrupted at one pyrrole
ring in (38), and one would therefore expect a bilatriene-type absorption
spectrum with bands above 600 nm.

On the other hand, Eschenmoser has synthesized a new ligand system (50)
which also has the oxidation state of a hexahydroporphyrin and no reduced
α-pyrrolic or methine bridge positions[70]. Compound (50) is a structural
hybrid of corrins and porphyrins and has therefore been named
corphin[69,70]. It is first protonated on a nitrogen atom to give (51) whereas
the dication (52) in trifluoroacetic acid has an extra proton on a peripheral

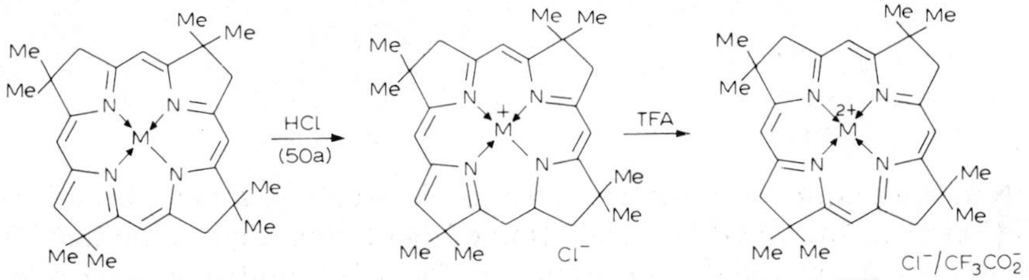

a ; M = Zn$^+$
b ; M = Ni$^+$
c ; M = Pd$^+$
d ; M = Co(CN)$_2^-$
e ; M = H$_2^+$

carbon atom. The p.m.r. methine proton signals in these compounds are
always close to $\tau = 4$, which clearly indicates the expected lack of ring current. The palladium complex (50c) (λ_{max}^{EtOH} nm, 298 (23,000), 336 (33,000),
400 (5,100), 488 (12,300), sh. 552 (3,500) is also reversibly protonated by
acids ($\lambda_{max}^{CF_3COOH}$ end abs. 270 nm, 342 (89,000), 400 (22,000), 426
(78,000). A water soluble hexahydroxycorphin (32) has been described by
Inhoffen (see p. 639).

15.4. Electrophilic substitution and addition reactions
15.4.1. General aspects

Electrophilic substitution on porphyrins might either occur on unsubstituted methine bridges or peripheral positions. Electronegative porphyrin
ligands are obtained by introduction of divalent central ions (Mg > Zn > Cu
> Ni > Pd) and these metal complexes are usually substituted on the
methine bridges. Magnesium porphin (53b), for example, is brominated by
N-bromosuccinimide exclusively in the meso-positions, to give (54)[71], and
Vilsmeier formylation on the copper complex (53c) shows the same preference and reacts to form (55)[71]. Electrophilic metals and protons in the
porphyrin center, on the other hand tend to deactivate the methine bridges

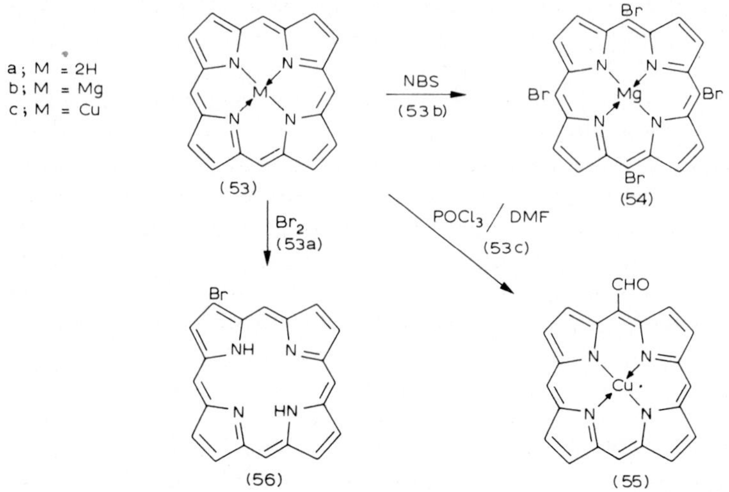

and activate the peripheral positions towards electrophilic attack. Bromination of (53a) to afford (56)[72] and addition of dichloromethyl to the tin(IV) complex (57) to form (58)[73] are examples for this tentative general rule, which is, however, not strictly followed in all known cases.

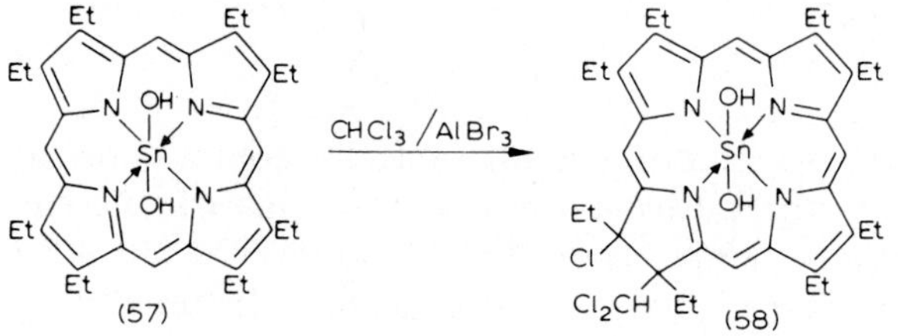

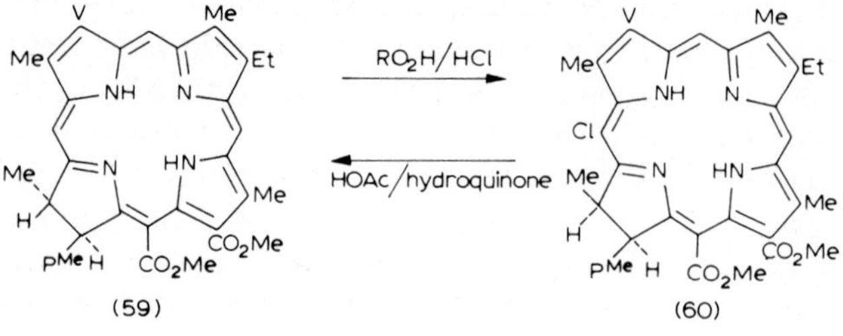

Another important generalization, first stated by Woodward, says that methine bridges of chlorins neighboring reduced pyrrole rings are more susceptible to electrophilic attack than those between non reduced pyrrolic units. This is exemplified by the chlorination of chlorin-p_6 trimethyl ester (59) to give (60). This particular reaction caused the famous mistake of H. Fischer, who formulated a substitution of the two pyrroline hydrogen atoms in the chlorophyll derivative by hydroxyl groups. This error was then corrected 24 years later by R.B. Woodward[74−76].

15.4.2. Deuteration

Deuterated porphyrins are useful objects in the study of their biosynthesis (see Chapter 3) and biological environments in hemoproteins (pp. 473—480) and molecular complexes (see Chapter 14). They may be obtained by total synthesis (see Chapter 3) or by feeding of deuterated precursors in the biosynthetic pathway of chlorophylls to appropriate microorganisms (see Chapter 10). The methine protons of porphyrins, hower, can also be exchanged for deuterium by simple treatment with deuterated acids.

R.B. Woodward found in the course of his chlorophyll synthesis[76] that rhodochlorin dimethyl ester exchanged the methine protons adjacent to the pyrroline ring within two hours when heated to 80°C in deuterated acetic acid. The other methine bridge protons were not exchanged under these conditions. It was mainly Bonnett who investigated the *meso* hydrogen/deuterium exchange in acid media further and his results could be rationalized by the plausible assumption, that the rate constants of deuteration are fastest on the most highly electronegative methine bridges[77]. This generalization was later nicely exemplified in Kenner's group, where it was shown that methine protons of porphyrins can be exchanged even in pyridine solution by deuterated methanol, when the porphyrin ligand is activated by the formation of a magnesium complex[78] (see Chapter 14). Johnson[79], and Dolphin[80] extended the exchange studies to metalloporphyrins with central ions of oxidation state +3 (Co, Fe) and found complicated kinetic correlations. This might be related to activation of the weak acids used by the positive charge of the central ions, which would lead to a high local concentration of protons. Smith used a similar mechanism (61) → (62) to explain the proton splitting from methanol in his exchange experiments in pyridine[78].

The general mechanism of deuterium exchange in metalloporphyrins (63) clearly involves the formation of isoporphyrin cations (64), which are reversibly deprotonated to (65). Isoporphyrins like (64) cannot be detected in appreciable concentration under normal acid conditions, unless they are stabilized by a large *meso*-substituent (see p. 449). It was also found that deuterium is incorporated into the *meso* positions during the reduction of iron(III) porphyrins to iron(II) complexes by iron powder in neutral deuterated media[81]. This was taken as direct evidence for attack on the porphyrin periphery during the reduction of the central ion[81], but this conclusion has been questioned[80]. Effects of substituent on deuteration of porphyrins have also been reported. In aminoporphyrins[88], and oxophlorins[9,25] the γ-bridge proton is exchanged rapidly, which is also explained by the high electron density at this position[31]. Steric effects may also be important: in α,γ-dimethyl octaethylporphyrin the β,δ-protons are much more amenable to ex-

References, p. 662

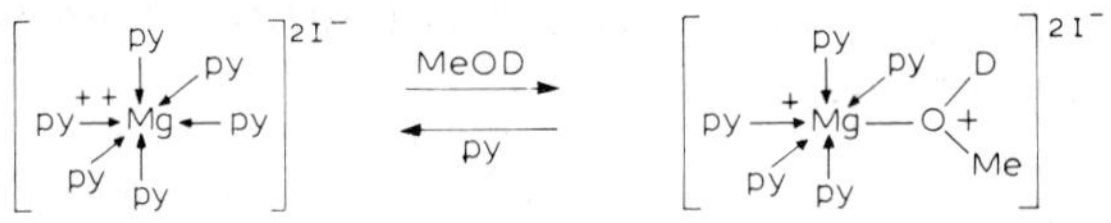

(61)   (62)

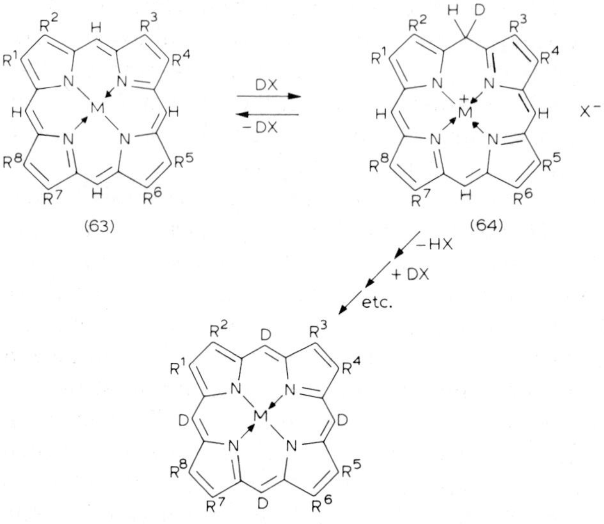

(63)   (64)

(65)

change than in the parent porphyrin[89]. The disturbed planarity of the porphyrin macrocycle may favor the formation of an isoporphyrin (64). Electron-
withdrawing substituents (e.g. acetyl), on the other hand, retard *meso*-deu-
teration[89].

In the chlorophyll series, Katz et al. found that not only the methine
bridge protons are exchanged in deuterated acids but also some of the substituent protons[82,83], which should be remembered in work with complicated porphyrins.

15.4.3. Acylation and methylation

The acetylation of deuterohemin (66) with acetic anhydride and tin(IV)
chloride as Lewis acid, to form (67), was one of the final steps in the total
synthesis of hemin by H. Fischer; this provides the classical example for

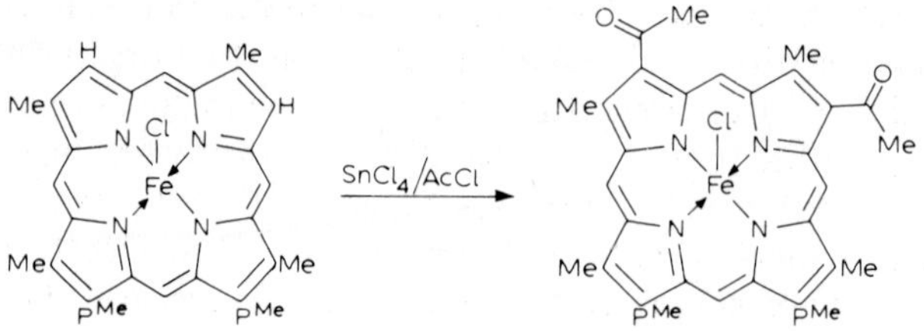

(66)   (67)

Friedel-Crafts type reactions on the porphyrin periphery. It was then already known that metal complexes (e.g. Cu, Ni) were much more reactive than free base porphyrins[84,85]. When the copper complex of deuteroporphyrin-XI dimethyl ester was treated with the same reagents under milder conditions the two monoacetylated derivatives could also be isolated[86]. For copper complexes of 2-desvinyl chlorophyll derivatives, tin(II) chloride was found to be a strong enough Lewis acid to accomplish the acetylation of the 2-position in 94% yield[87]. Analogous formylation of unsubstituted peripheral positions was achieved by H. Fischer with (dichloromethyl methyl ether (68) → (69).

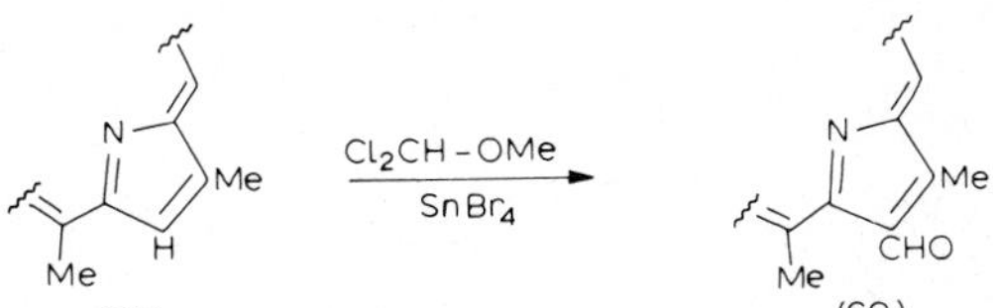

(68)   (69)

None of the above and many other Friedel-Crafts type reactions were useful for introduction of carbon substituents into the methine bridges. Presumably the complexes between Lewis acids and acylium ions are too bulky for attack at the *meso* positions. In 1966 Inhoffen introduced the Vilsmeier formylation into porphyrin chemistry[14], which is the only reaction so far known to introduce functional carbon groups into the *meso* positions in high yield. Copper porphyrins are generally mono-formylated on a methine bridge[14,71] [e.g. (53c) → (55)], but with copper deuteroporphyrin-IX dimethyl ester, 2- and 4-formylation has also been observed[86]. Assignments of structures to these products have been corrected[90]. Protohemin has also been subjected to the Vilsmeier procedure and the relative reactivities appeared to be vinyl (see p. 656) > *meso* > peripheral[91].

meso-Methylation of palladium octaethylporphyrin to give (70) has been achieved with methyl fluorosulphonate in 36% yield[79]. An electrophilic addition reaction has been observed between tin(IV) octaethylporphyrin (57) and chloroform tin(IV) bromide. The resulting dichloromethylchlorin (58) could be reduced and demetalated to the methyl chlorin (71)[73].

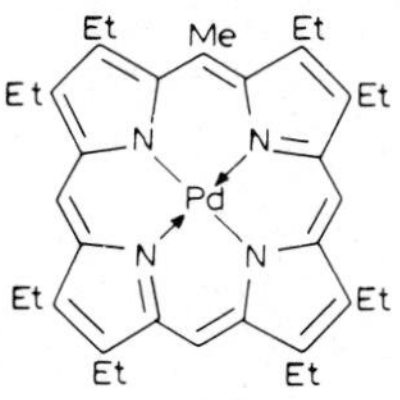

(70) 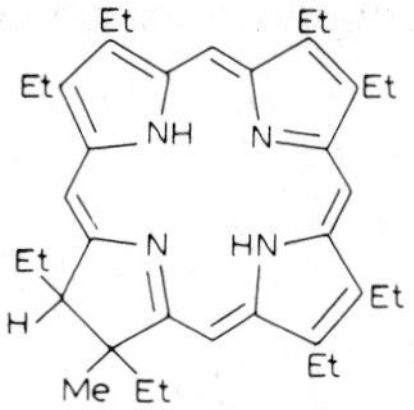   (71)

Further reactions of the functional side-chains of porphyrins are discussed in Section 15.6, but it should be mentioned here that acetyl and formyl groups, especially in the *meso* position, are easily cleaved at high temperatures or in strong acid or basic media[45,87].

15.4.4. Reactions with carbenes, nitrenes, and nitrogen tetroxide/dichloromethane

Copper octaethylporphyrin (72) was heated with ethyl diazoacetate in benzene solution and was converted in good yield to the isomeric cyclopropane chlorins (73)[92]. Minor products were the *meso*-ethoxycarbonylporphyrin (74) and chlorin (75)[92]. The stereochemistry of *meso*-tetraphenyl

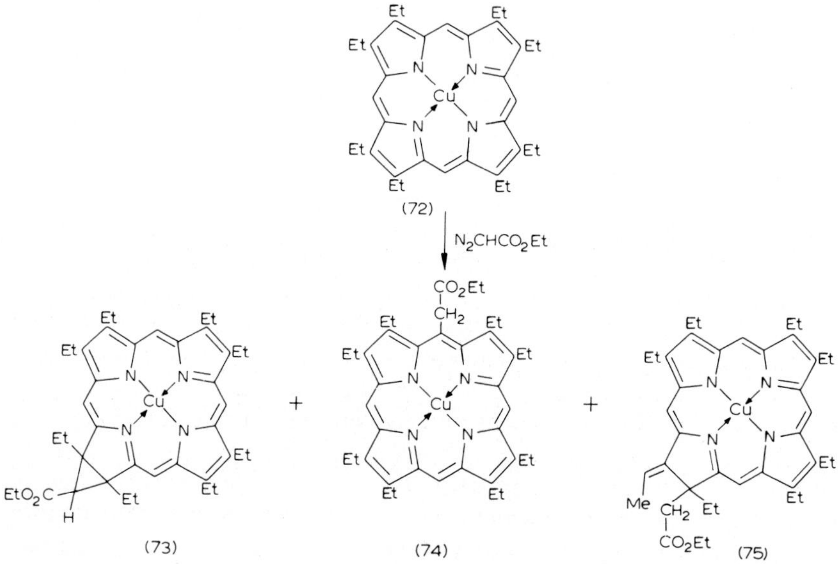

analogs of (73) has been evaluated[93]. Ethoxycarbonylnitrene, another highly electron deficient reagent, reacted with (72) in a completely different manner: it was inserted in high yield between the methine bridge and peripheral carbon atoms and the only isolable product was the *meso*-ethoxycarbonylaminoporphyrin (77). The same reaction with metal free octaethylporphyrin led first to the isolable homo-azaporphyrin (76), which readialy isomerized to (77) on heating or chelation with various metals[94].

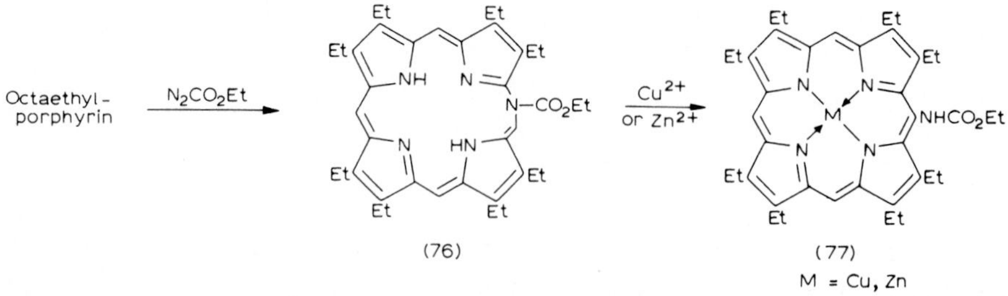

Another promising reaction of this type, which might involve electrophilic or radical attack (or both), has recently been found with octaethylhemin. This reacts rapidly with nitrogen tetroxide in methylene chloride to form the *meso*-tetranitromethyl derivative (78) in 50% yield[156].

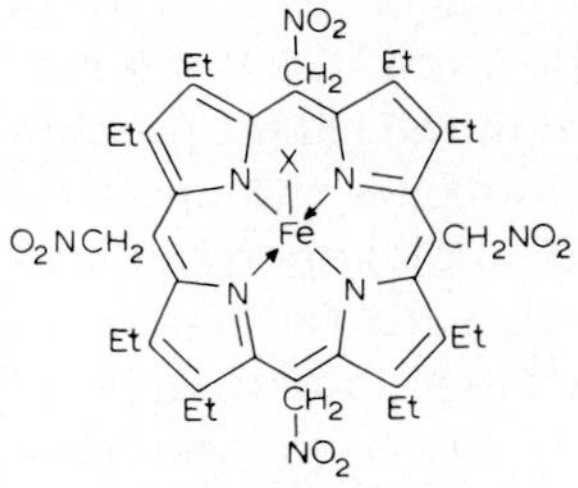

(78)

15.4.5. Nitration

Nitration of the methine bridges is one of the most gratifying ventures in preparative porphyrin chemistry. Yields are usually high, up to three *meso*-positions can be substituted sequentially, and the materials obtained often crystallize very well. The method of choice[95] for porphyrins is usually concentrated nitric acid in acetic acid at 0°C, whereas oxidation-sensitive chlorins are nitrated with nitronium tetrafluoroborate in sulfolane[95]. Treatment of π-cation radicals of magnesium porphyrins with sodium nitrite also affords *meso*-mononitroporphyrins in good yields[160]. Surprisingly, deuteroporphyrin is not nitrated in the 2,4-positions (e.g. to give (81)), but only on the methine bridges[96,97] (e.g. to give (80)). This is in clear contrast to all

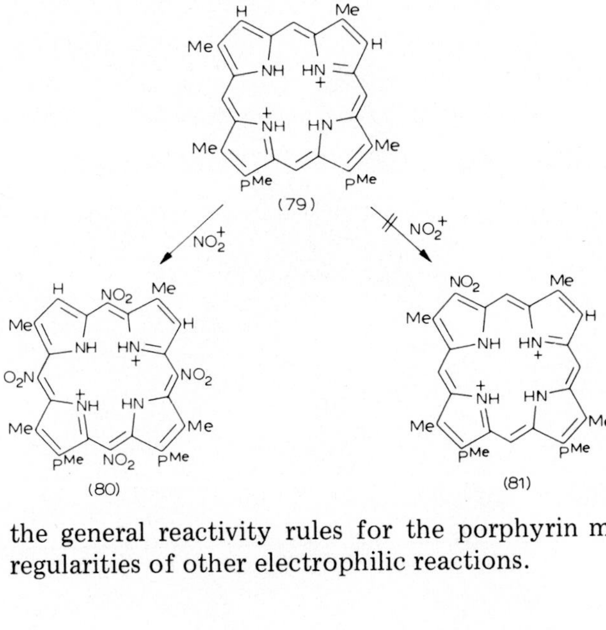

(79)

(80)

(81)

the general reactivity rules for the porphyrin macrocycle as deduced from regularities of other electrophilic reactions.

References, p. 662

The nitro groups can be reduced to the amino oxidation state by tin(II) chloride in conc. hydrochloric acid[95] or, in better yield, with sodium borohydride and palladium charcoal in methanol[97]. The amino substituent dramatically changes the chemical properties of the porphyrin macrocycle[36]: phlorin type electronic spectra are observed in acid media and an imino proton corresponding to structure (82) can be traced in the p.m.r. spectra; the oxidation potentials are much lower than in the parent porphyrins (see Table 2 p. 601) and π-radicals of metal complexes are unstable; aminoporphyrins are readily oxidized and hydrolyzed to dioxoporphodimethenes (e.g. (16)); the γ-methine proton is readily exchanged for deuterium in deuterated trifluoroacetic acid[88]. All of these facts can be rationalized with the assumption that a *meso*-amino substituent forces phlorin type behavior upon the porphyrin macrocycle, and this compares well with the similar effect of the hydroxyl group (see p. 630).

Nitrosation of aminoporphyrins with sodium nitrite in tetrafluoroboric acid at −5°C yields the corresponding diazonium fluoroborate (83) in 90% yield[97].

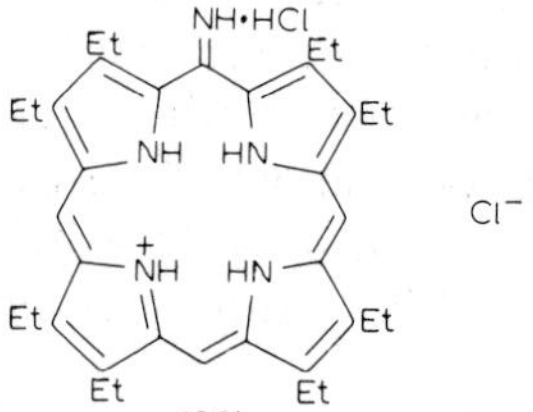

(82)

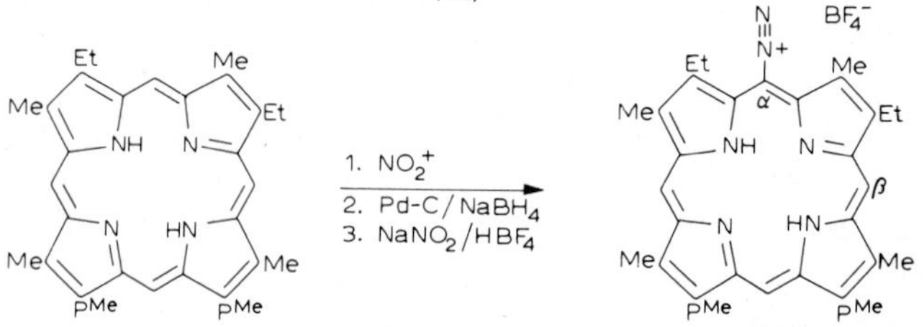

1. NO_2^+
2. Pd-C / NaBH$_4$
3. NaNO$_2$/HBF$_4$

(83)
+ β isomer

160°
Vacuo

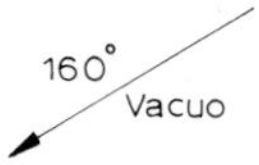

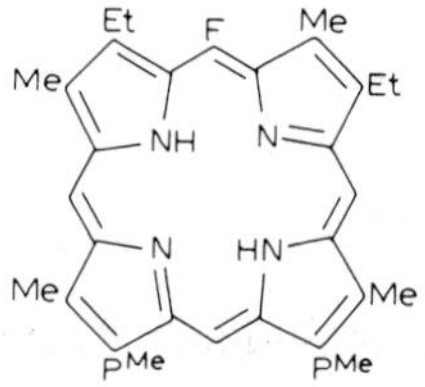

(84)

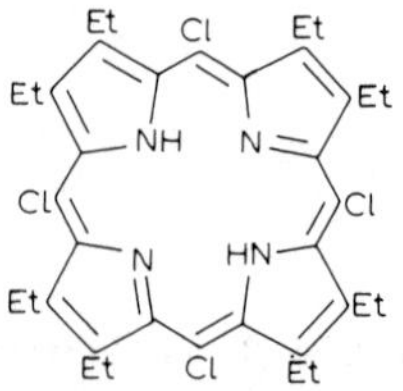

(85)

15.4.6. Halogenation

Balz-Schiemann type decomposition[98] (heating to $160°C$) of the diazonium tetrafluoroborate (83) leads to the expected *meso*-fluoroporphyrin (84) in low yield[97]. The ^{19}F n.m.r. of an $\alpha(\beta)$-monofluorinated isomeric mixture of mesoporphyrin-IX dimethyl ester in trifluoroacetic acid showed two resonances at 42.57 and 43.77 p.p.m. upfield from the solvent absorption[97].

Chlorination of free base porphyrins either with hydrogen peroxide—hydrochloric acid mixtures in tetrahydrofuran or acetic acid[99] or with sulfuryl chloride in chloroform[100] leads in good yield to the *meso*-tetrachlorinated products e.g. (85). Preparation of mono- or di-chlorinated porphyrins in large quantities is difficult and is usually performed in carefully controlled two phase reactions where the porphyrin is dissolved in the organic layer and the chlorine evolving system in the water phase[99]. Johnson reported that the attempted Vilsmeier formylation of etioporphyrin with phosphorous oxychloride and dimethyl formamide did not give the expected aldehyde, but the *meso*-monochloro product in 16% yield[15]. This is another example of the general finding that porphyrins tend to be substituted by chlorine when chloride anions and oxidants e.g. air, are present (see also p. 646). Chlorins are chlorinated exclusively in the methine bridges adjacent to the pyrroline ring[99].

The electronic spectrum of $\alpha,\beta,\gamma,\delta$-tetrachloro octaethylporphyrin displays an α-band above 700 nm. Its diprotonated dication has a phlorin type spectrum with an intense band at 690 nm[99].

Mg Porphin reacts with N-bromosuccinimide to give $\alpha,\beta,\gamma,\delta$-tetrabromoporphin (54) and no peripheral substitution is observed[71]. Free base porphin, on the other hand, was monosubstituted on a peripheral carbon atom to give (56)[72]. Deuteroporphyrin-IX dimethyl ester (79) can be converted into the 2,4-dibromo derivative with pyridinium bromide perbromide[101]. The peripheries of fully β-substituted porphyrins do not react with brominating agents, but *meso* brominated derivatives can be obtained by bromination of the corresponding chlorins and subsequent dehydrogenation[99]. A substitution of a vinyl group by bromine has been reported[102a].

Porphyrins with an iodinated periphery do not appear to have been prepared.

15.4.7. Thiocyanation

β-Sulfonation of porphyrins has been reported by Treibs[102b] but the properties of the sulfonated products are not known. Thiocyanation of copper porphyrins on a *meso* position is a very effective reaction[103]. Hydrolysis of the thiocyano group with concentrated sulfuric acid yields the mercapto derivative[103], which, in contrast to oxy and amino porphyrins is said not to form a double bond to the methine carbon (see pp. 630 and 652).

15.5. Nucleophilic substitution and addition reactions

The electron deficiency of the porphyrin methine bridges (see p. 626) should render these reaction sites very susceptible to nucelophilic attack. Apart from the easy formation of phlorins and porphyrinogens with hydrogen (see pp. 615 and 617) this theoretical finding is not reflected in the experimental literature, where electrophilic attack is by far more important. One exception is the reversible addition of thioacetic acid to a sterically crowded *meso* position[5]:

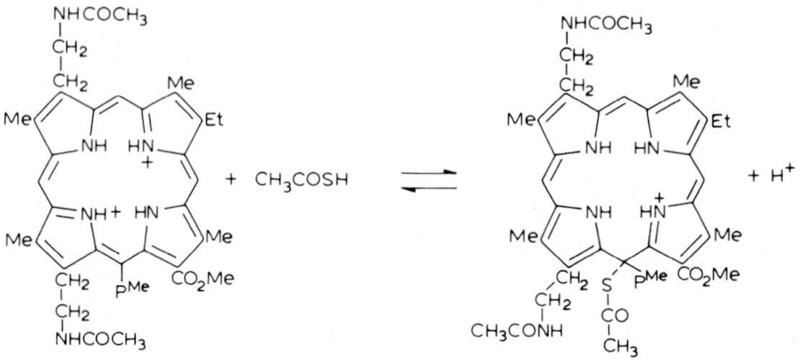

(86)   (87)

15.6. Reactions of carbon substituents

Natural porphyrins bear functional groups which are important for the attachment of the porphyrin redox system to their complex biological environment: the vinyl group of heme is attached via a thioether linkage to the protein in cytochrome-c and via long chain alcohols in cytochromes-a and -d[104]; a propionic acid of the chlorophylls is often esterified with a long chain alcohol and their isocyclic keto group forms a weak bond to the magnesium central ion of another chlorophyll molecule[105]. Covalent bonding of side-chains to proteins, their weak interactions with lipid membranes, and their participation in the formation of molecular complexes are certainly of primary importance to the biological functions of the porphyrins, and have been studied in in vitro models. Apart from these aspects, the reactions of porphyrin substituents are sometimes of chemical interest in their own right, because the properties of the porphyrin macrocycle might be altered in subtle but well defined ways by changes in the substituents. Here only the reactions of carbon substituents will be discussed. Reactions of hetero atom substituents have been given in the previous sections.

15.6.1. Ethyl groups

An ethyl group of octaethylporphyrin (88) has been chlorinated with tert. butyl hypochlorite and 2,2'-azobisisobutyronitrile in chloroform to give

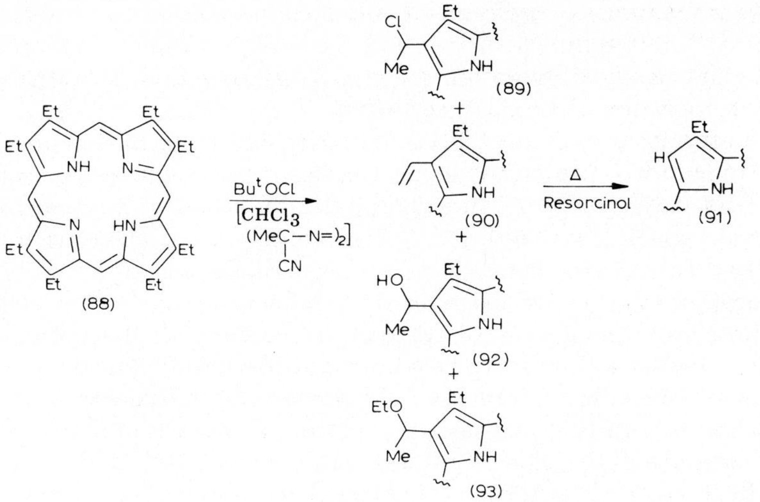

1'-chloro-octaethylporphyrin (89) in 7% yield. From the reaction mixture 37% of monovinylheptaethylporphyrin (90), 4% of 1'-hydroxy-octaethylporphyrin (92), and 7% of 1'-ethoxy-octaethylporphyrin (93) were also isolated. Heating of the vinyl compound (90) with resorcinol yielded heptaethylporphyrin (91)[106].

Another radical reaction of an ethyl group in octaethylporphyrin has been described on p. 634.

15.6.2. Vinyl groups

The vinyl groups of protoporphyrin can be efficiently removed by heating in a resorcinol melt according to Schumm[107]. The adduct (94) has been isolated in Kenner's group and the devinylation to deuterohemin (95) was rationalized as follows[157]:

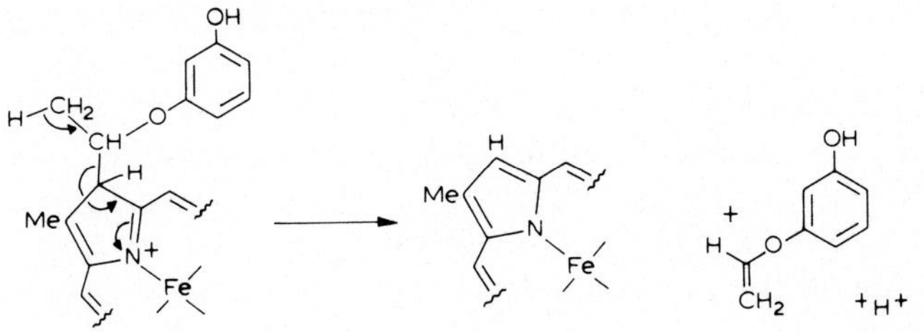

Hydrogenation of the vinyl group, leading into the 'meso' series, has been achieved with diborane[108], hydrazine[109], hydroiodic acid in acetic acid, and by heterogenous catalytic reduction[110].

Markownikoff hydration of the vinyl group is achieved with hydrobromic—acetic acid mixtures and subsequent hydrolysis[112]. The reverse

References, p. 662

reaction occurs when hematoporphyrin-IX is heated in vacuo at $135-150°C$ for 5–10 min[113] or is refluxed in benzene containing p-toluene sulfonic acid[112]. Anti-Markownikoff hydration (to give 2-hydroxyethyl derivatives) occurs with hydroboration and oxidative workup[108].

Oxidation to formyl groups may be achieved by refluxing the vinylporphyrin in acetone with five molar equivalents of potassium permanganate[112,114]. The vinyl groups of protoporphyrin-IX are also oxidized by osmium tetroxide which leads to the 2',2'',4',4''-tetrahydroxy mesoporphyrin-IX derivative. This can then be cleaved with periodic acid to 2,4-diformyl-deuteroporphyrin and converted into the 2,4-diacrylic acid[115a], which has been hydrated to the 2,4-bis (β-hydroxy-propionic acid, the porphyrinogen of which was once thought to be a possible biosynthetic precursor of protoporphyrin-IX. β-Hydroxypropionic acid deuteroporphyrin was found to be quite stable in both acid and neutral conditions, whereas in the corresponding porphyrinogen the side-chain was easily dehydrated and decarboxylated in high yield[115b]. Acrylic acids have been obtained from $meso$-formylporphyrins[116].

The photo-oxygenation of protoporphyrin-IX to $Spirographis$ porphyrin derivatives is described in Chapter 16.

Addition of diazoacetic acid to the vinyl groups in ether solution leads to cyclopropane carboxylic esters (97), which survive the oxidative degradation to imides. Therefore these derivatives have often been used in degradative structural studies[117,118].

Vilsmeier reactions on the vinyl groups of heme resulted in the formation of acrolein derivatives (98)[91].

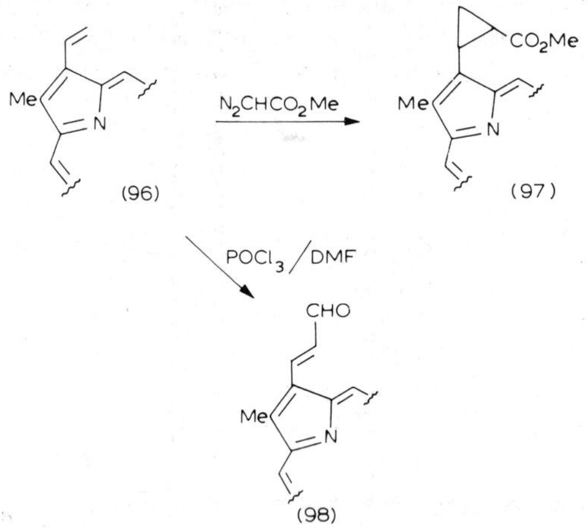

15.6.3. Carboxylic and propionic acids

The methyl esters are usually obtained by treatment with 5% sulfuric acid in dry methanol at low temperatures[119]. Hematoporphyrins are partially dehydrated by acid treatment, and diazomethane is preferable. Hemins, however, should not be treated with diazomethane[120]. Long chain alcohols, e.g. phytol, have usually been esterified in pyridine solution with the aid of phosgene[121]. Similar techniques have been used to obtain amides[122-124] e.g. the 'myoglobin model' (99)[124]. Sulfuric anhydrides have also been pre-

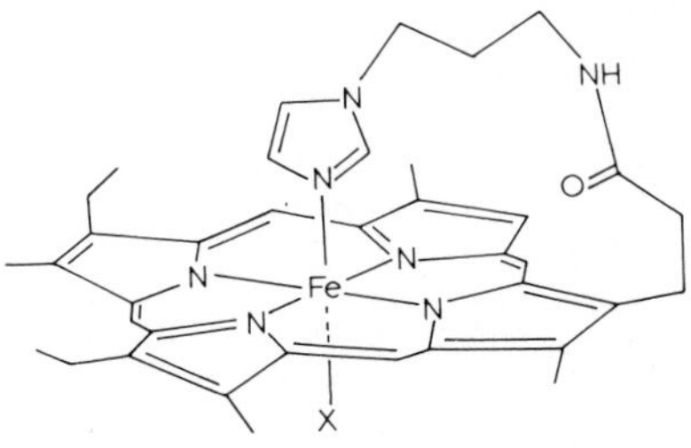

(99)

pared and were converted into amides[125]. Well-defined bis-porphyrins (100a,b) were obtained by reaction of the monomeric porphyrin acid chloride with p-phenylene-diamine or ethylene diamine[126].

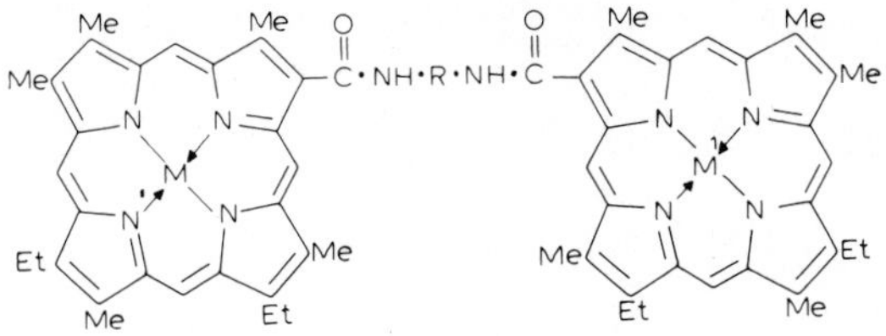

(100)

a ; R = —CH$_2$·CH$_2$—

b ; R =

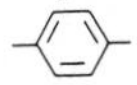

Reduction of mesoporphyrin-IX dimethyl ester to the 6,7-bis-(γ-hydroxypropyl)porphyrin was achieved with lithium aluminum hydride in tetrahydrofuran[14]. Mesylation of this alcohol and further reduction with the same hydride yielded the corresponding propyl compound[14]. Pyrolytic decarboxylations of propionic and selectively, acetic acid[158,159], side-chains have also been performed. Treatment of porphyrin propionic acids with oleum yields 'rhodins', e.g. (101) from mesoporphyrin-IX[127]. The Curtius degradation of propionic acid side-chains produces vinyl groups; coproporphyrin-I, for example, has been converted into the tetravinylporphyrin (102)[128].

The synthetic porphyrin carboxylic acid (103) was converted via (104) and (105) into the β-keto-ester (106). When (106) was chelated with magnesium and oxidised to the π-cation radical with iodine under basic conditions, a 7% yield of the isocyclic ring containing compound (107a) was pro-

(101a) (101b)

(102) (103) (104)

(105) (106) (107)

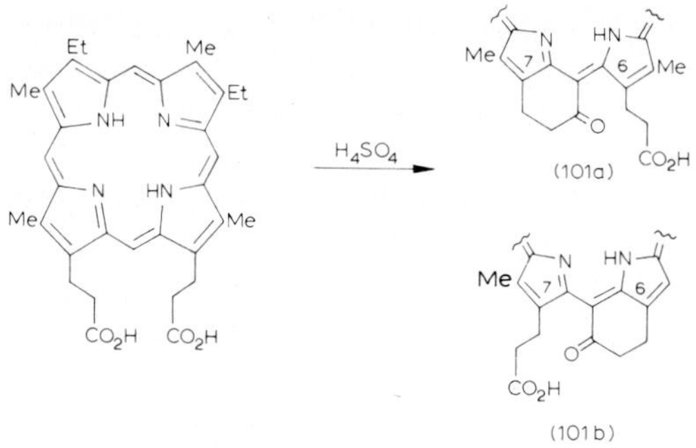

duced[129a]. An analogous photochemical cyclization of a bis thallium derivative of (106) to give (107b) after a suitable work-up has been reported[129b,c] along with more efficient methods for the synthesis of porphyrin β-keto-

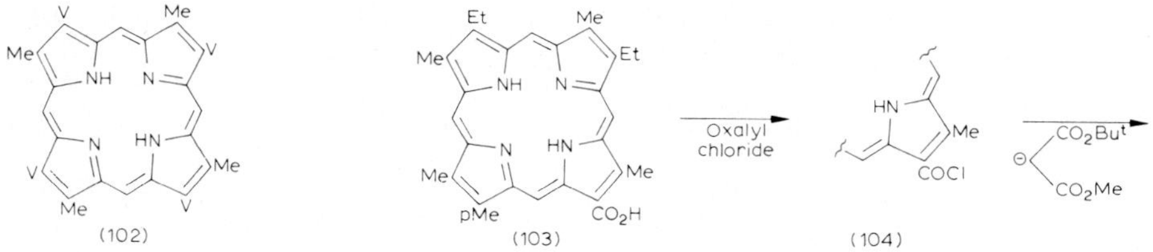

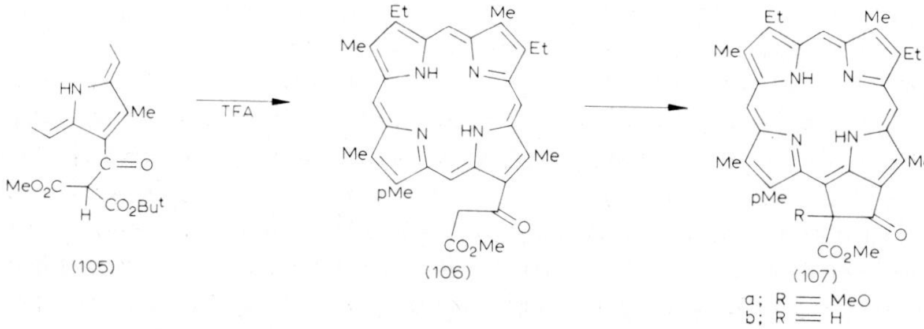

esters by treatment[129c,d] of porphyrin imidazolides with the magnesium complex of methyl hydrogen malonate.

Woodward's purpurin reaction is discussed in Section 15.3.2.

15.6.4. Miscellaneous

The complicated reactions of the isocyclic ring belong to the realms of chlorophyll chemistry and will not be discussed here. Pertinent references are Treibs' book (see p. 662) for a review of H. Fischer's contributions and Refs. 130—135 for more recent work.

The synthetic acetamidoporphyrin (108) undergoes an intramolecular Vilsmeier reaction and rearrangement to form the spirocyclic chlorin (109) when treated with phosphoryl chloride in pyridine[136].

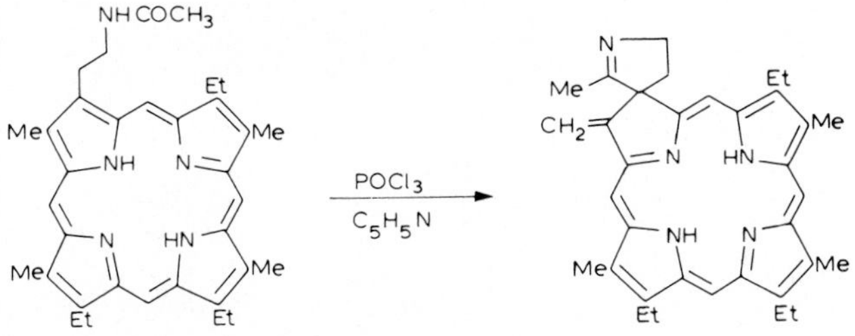

(108) (109)

15.7. Degradation of the porphyrin nucleus

Degradation of tetrapyrroles to monopyrroles or other units which can be recognized has played an important part in establishment of the structure of both porphyrins and chlorophylls. Three types of degradative reaction are available: oxidative degradation with chromic acid, oxidative degradation with $KMnO_4$, and reductive degradation with hydriodic acid. The main products obtained when mesoporphyrin-IX is subjected to each of these reactions are shown in Table 4.

In the reductive degradation with HI, each pyrrole ring of mesoporphyrin can give rise to all four of the substituted monopyrroles shown in Table 4[137]. The usefulness of this method for determination of the nature and relative positions of side-chains in porphyrin is very limited. Mac Donald, however, has pointed out that methine substituents can be best investigated by this method, where they appear as α-CH_2R group in the resulting pyrroles. Reductive in situ alkylation (Scheme 1) produces more easily identifiable pyrroles[138] which can be identified by glpc.

CrO_3 oxidation[139,140] has been used extensively, both for structure determination and for biosynthetic studies in which the origin of the N and C atoms of protoheme was traced using ^{15}N or [14]C. Protoheme is converted into mesoporphyrin-IX before the oxidation, which is carried out at low

TABLE 4
Degradation products from mesoporphyrin-IX

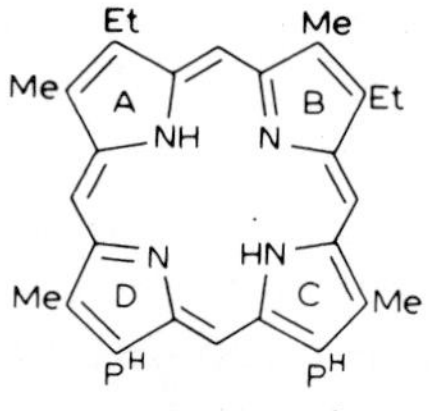

Mesoporphyrin-IX

Origin of product	Product	Formula
CrO$_3$ Oxidation		
Rings A, B	Methylethylmaleimide	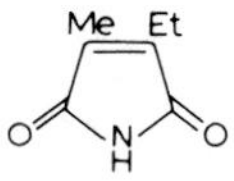
Rings C, D	Hematinic acid	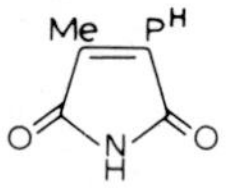
meso-Carbons	Carbon dioxide	CO_2
KMnO$_4$ Oxidation		
Rings A, B	3-Ethyl-4-methylpyrrole-2,5-dicarboxylic acid	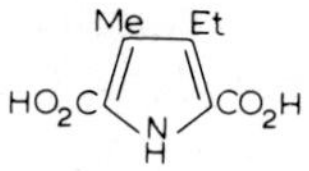
Rings C, D	3-Carboxyethyl-4-methyl-pyrrole-2,5-dicarboxylic acid	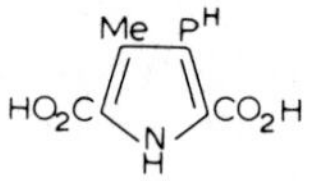
HI Reduction		
Rings A, B, C, D	'Opsopyrrole'	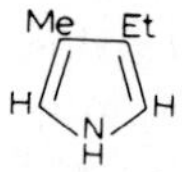
	'Kryptopyrrole'	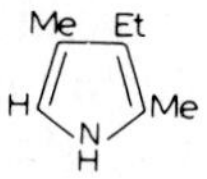
	'Hemopyrrole'	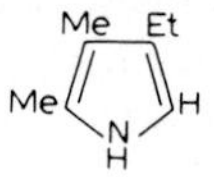
	'Phyllopyrrole'	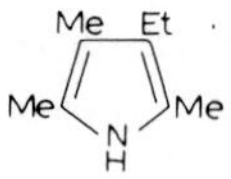

Scheme 1 Pyrroles expected from one pyrrole sub-unit in a porphyrin: Column A, after reduction with HI/HOAc Column B, after "reductive C-alkylation" with HI/HOAc/R^3-CHO (*NB*. Though the product mixture from just one of the four pyrrole sub-units in the porphyrin appears complex, considerable simplification can usually be expected. For example, in the very common case where $R^1 = R^2 = R^3 = H$, *one pyrrole only* is produced from this sub-unit).

Scheme 1

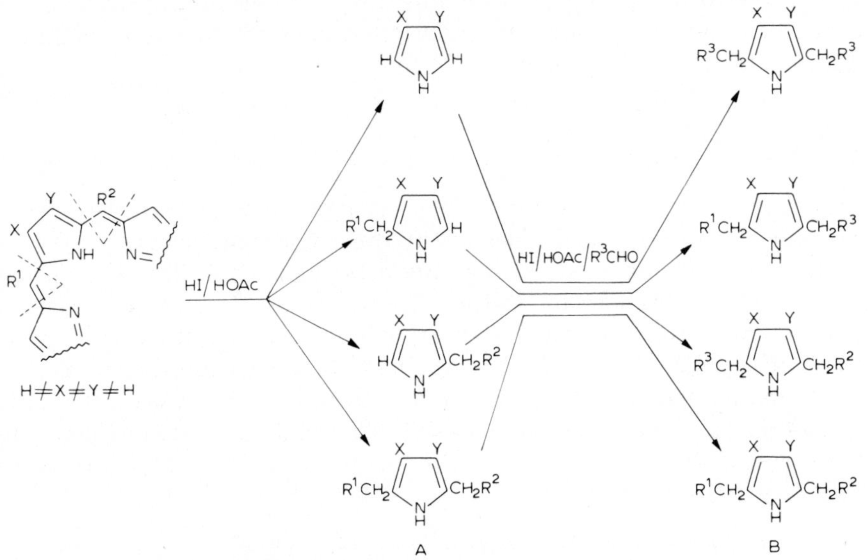

temperature. The porphyrin is dissolved in 50% (v/v) H_2SO_4, and a solution of CrO_3 added slowly, the methylethylmaleimide and hematinic acid are isolated from the reaction mixture and crystallized[141-145]. During the oxidation, the methene bridge carbon atoms are evolved as CO_2, and by collection of this CO_2 from the ^{14}C-labeled porphyrin it has been shown conclusively that these carbon atoms arise from α-carbon atoms of glycine[140,143,145,146]. By conversion of the isotopically labeled protoheme into mesoporphyrin-IX, oxidation of this to methylethylmaleimide and hematinic acid and stepwise degradation of these imides, the biosynthetic origin of every atom in rings (I and II) and (III and IV) has been traced[144,147,148]. An example of the use of this method for structure determination is provided by Morley and Holt[142,149,150], who have developed gas chromatographic methods for isolation of the maleimides obtained by CrO_3 oxidation of *Chlorobium* chlorophyll, as well as for the monopyrroles obtained by HI reduction (cf. Table 4).

Nicolaus[151] has developed a procedure for identification, by paper chromatography, of the products of oxidation of porphyrins by $KMnO_4$. The porphyrin side chains —Me, —Et, —$CH_2CH_2CO_2H$, —CO_2H, —$COCH_3$, and —$CH(OH)CH_3$ persist in the degradation products, but —CH = CH_2 and —CHO side-chains are both oxidized to —CO_2H. The vinyl group is readily

References, p. 662

identified, however, by comparing the products of oxidation obtained before and after its reduction on the original porphyrin to ethyl or conversion to cyclopropyl carboxylic acid (see p. 656), and the formyl group may be identified by conversion to the nitrile. As has been mentioned, the identification by Nicolaus of 3-cyanopyrrole-4-propionic acid-2,5-dicarboxylic acid after $KMnO_4$ oxidation of porphyrin-a played a very important part in proving that the formyl group occupies position 8 on this porphyrin.

Recent examples of oxidative degradation studies are the stereochemical investigations of H. Brockmann Jr.[152-154].

References

The experimental contributions of Hans Fischer to the chemistry of porphyrins, and his more general ideas about this subject are summarized in a unique, logical, truly devoted, and enthusiastic manner in 'Das Leben und Wirken von Hans Fischer' by Alfred Treibs (Hans Fischer Gesellschaft, 1971). This "historically invaluable effort which succeeds in conveying the magnitude of Fischer's work and the qualities of this extraordinary man ... makes fascinating and nostalgic reading', [K. Bloch, J. Am. Chem. Soc., 96, 2658 (1974)]. The recent general availability of this book has enabled concise treatment of the classical work in the foregoing chapter in favor of more contemporary investigations.

1. K. Fukui, T. Yonezawa, and H. Shingu, J. Chem. Phys., 20, 722 (1952); K. Fukui, T. Yonezawa, C. Nagata, and H. Shingu, J. Chem. Phys., 22, 1433 (1954).
2. A.D. McLachlan, Mol. Phys., 2, 233 (1960).
3. J.V. Knop and J.-H. Fuhrhop, Z. Naturforsch., Teil B, 25, 729 (1970).
4. J.-H. Fuhrhop, A. Salek, J. Subramanian, Chr. Mengersen, and S. Besecke, Ann. Chem., in press.
5. R.B. Woodward, Ind. Chim. Belge, 1293 (1962).
6. R. Schlözer, Diplomarbeit Braunschweig, (1974).
7. R. Lemberg, Rev. Pure. Appl. Chem., 6, 1 (1956).
8. J.-H. Fuhrhop and D. Mauzerall, Photochem. Photobiol., 13, 453 (1971).
9. A.H. Jackson, G.W. Kenner, and K.M. Smith, J. Chem. Soc. C, 302 (1968).
10. P.K.W. Wasser and J.-H. Fuhrhop, Ann. N.Y. Acad. Sci., 206, 533 (1973).
11. J.-H. Fuhrhop, P.K.W. Wasser, J. Subramanian, and U. Schrader, Ann. Chem., 1450 (1974).
12. D. Dolphin, R.L.N. Harris, J.L. Huppatz, A.W. Johnson, and I.T. Kay, J. Chem. Soc. C, 30 (1966).
13. H.H. Inhoffen, N. Schwarz, and K.-P. Heise, Ann. Chem., 146 (1973).
14. H.H. Inhoffen, J.-H. Fuhrhop, H. Voigt, and H. Brockmann Jr., Ann. Chem., 695, 133 (1966).
15. A.W. Johnson and D. Oldfield, J. Chem. Soc. C, 794 (1966).
16. J.-H. Fuhrhop and T. Lumbantobing, Tetrahedron Lett., 2815 (1970).
17. A.V. Umrikhina, G.A. Yusupova, and A.A. Krasnovskii, Dokl. Akad. Nauk SSSR, 175, 1400 (1967).
18. H. Fischer and K. Herrle, Hoppe-Seyler's Z. Physiol. Chem., 251, 85 (1938).
19. V. Ullrich, Angew. Chem., 84, 689 (1972).

20. J.-H. Fuhrhop, Struct. Bonding Berlin, 18, 1 (1974).
21. O. Warburg and E. Negelein, Chem. Ber., 63, 1816 (1930).
22. H. Fischer and A. Treibs, Ann. Chem., 457, 209 (1927).
23. E. Stier, Hoppe-Seyler's Z. Physiol. Chem., 272, 239 (1942).
24. H.H. Inhoffen, J.-H. Fuhrhop, and F.v.d. Haar, Ann. Chem., 700, 92 (1966).
25. R. Bonnett, M.J. Dimsdale, and G.F. Stephenson, J. Chem. Soc. C, 564 (1969).
26. K.M. Smith, G.H. Barnett, M.F. Hudson, and S.W. McCombie, J. Chem. Soc., Perkin Trans. 1, 691 (1973).
27. H.H. Inhoffen and A. Gossauer, Ann. Chem., 723, 135 (1969).
28. R. Bonnett and M.J. Dimsdale, J. Chem. Soc., Perkin Trans. 1, 2540 (1972).
29. J.-H. Fuhrhop, S. Besecke, and J. Subramanian, Chem. Commun., 1 (1973).
30. R. Bonnett, M.J. Dimsdale, and K.D. Sales, Chem. Commun., 962 (1970).
31. J.-H. Fuhrhop, S. Besecke, J. Subramanian, Ch. Mengersen, and D. Riesner, in press.
32. H. Libowitzky and H. Fischer, Hoppe Seyler's Z. Physiol. Chem., 255, 209 (1938).
33. C.R.E. Jefcoate, J.R.L. Smith, and R.O.C. Norman, J. Chem. Soc. B, 1013 (1969).
34. J.A.S. Cavaleiro and K.M. Smith, J. Chem. Soc., Perkin Trans. 1, 2149 (1973).
35. R. Tenhunen, H. Marver, N.R. Pimstone, W.F. Trager, D.J. Cooper, and R. Schmid, Biochemistry, 11, 1716 (1972).
36. J.-H. Fuhrhop, Chem. Commun., 781 (1970).
37. S. Besecke and J.-H. Fuhrhop, Angew. Chem., 86, 125 (1974); Angew. Chem., Int. Ed. Engl., 13, 150 (1974).
38. (a) H. Fischer and A. Treibs, Ann. Chem., 451, 209 (1927). (b) H. Fischer and H. Orth, 'Die Chemie des Pyrrols' Vol. II, Part 2, Akademische Verlagsgesellschaft, Leipzig (1940), p. 423.
39. J.-H. Fuhrhop, Dissertation Braunschweig (1966).
40. (a) H. Fischer and H. Eckholdt, Ann. Chem., 544, 138 (1940); (b) H. Fischer and H. Orth, 'Die Chemie des Pyrrols', Vol. II, part 1, Akademische Verlagsgesellschaft, Leipzig (1937), p. 269.
41. R. Bonnett, D. Dolphin, A.W. Johnson, D. Oldfield, and G.F. Stephenson, Proc. Chem. Soc., London, 371 (1964).
42. H.H. Inhoffen and W. Nolte, Tetrahedron Lett., 2185 (1967); W. Nolte, Dissertation Braunschweig, (1967).
43. H.H. Inhoffen, J. Ullrich, H.A. Hoffmann, and G. Klinzmann, Tetrahedron Lett., 613 (1969).
44. H.H. Inhoffen and N. Müller., Tetrahedron Lett., 3209 (1969).
45. H.H. Inhoffen, J.W. Buchler, and P. Jäger, Prog. Chem. Org. Nat. Prod., 26, 284 (1968).
46. D. Mauzerall, J. Am. Chem. Soc., 84, 2437 (1962).
47. H. Fischer, K. Platz, J.H. Helberger, and H. Niemer, Ann. Chem., 479, 41 (1930).
48. H. Fischer, R. Lambrecht, and K. Mittenzwei, Hoppe-Seyler's Z. Physiol. Chem., 253, 1 (1938).
49. U. Eisner, J. Chem. Soc., 3461 (1957).
50. A.H. Corwin and O.D. Collins, J. Org. Chem., 27, 3060 (1962).
51. (a) H.H. Inhoffen, H. Parnemann, and R.G. Foster, Illinois Inst. Techn. NMR Newsletter, 65, 45 (1964). (b) H. Parnemann, Dissertation Braunschweig (1964).
52. D.G. Whitten, J.C. Yau, and F.A. Carroll, J. Am. Chem. Soc., 93, 2291 (1971).
53. H.W. Whitlock and M.Y. Oester, J. Am. Chem. Soc., 95, 5738 (1973).
54. J.W. Buchler and H.H. Schneehage, Tetrahedron Lett., 3803 (1972).
55. A. Treibs and E. Widemann, Ann. Chem., 471, 146 (1929).
56. G.R. Seely and K. Talmadge, Photochem. Photobiol., 3, 195 (1964).
57. H.W. Whitlock, R. Hanauer, M.Y. Oester, and B.K. Bower, J. Am. Chem. Soc., 91, 7485 (1969).
58. A.N. Sidorov, Russ. Chem. Rev., 35, 153 (1966).

59. H. Wolf and H. Scheer, Tetrahedron Lett., 1115 (1972).
60. H.H. Inhoffen, J.W. Buchler, and R. Thomas, Tetrahedron Lett., 1145 (1969).
61. Ref. 38 b, p. 144.
62. R.B. Woodward, Angew. Chem., 72, 651 (1960).
63. Ref. 45, p. 320.
64. H. Fischer and K. Herrle, Ann. Chem., 527, 138 (1937).
65. U. Eisner and R.P. Linstead, J. Chem. Soc., 3749 (1955).
66. H.H. Inhoffen and H. Biere, Tetrahedron Lett., 5145 (1966).
67. H.H. Inhoffen, J.W. Buchler, and R. Thomas, Tetrahedron Lett., 1141 (1969).
68. G.D. Dorough and J.R. Miller, J. Am. Chem. Soc., 74, 6106 (1952).
69. A.P. Johnson, P. Wehrli, R. Fletcher, and A. Eschenmoser, Angew. Chem., 80, 622 (1968); P.S. Müller, S. Farooq, B. Hardegger, W.S. Salmond, and A. Eschenmoser, ibid., 85, 954 (1973).
70. P.M. Müller, Dissertation, Elektrotechnische Hochschule Zürich (1973).
71. J.-H. Fuhrhop and R. Schlözer, Angew. Chem., in press.
72. E. Samuels, R. Shuttleworth, and T.S. Stevens, J. Chem. Soc. C, 145 (1968).
73. J.-H. Fuhrhop, T. Lumbantobing, and J. Ullrich, Tetrahedron Lett., 3771 (1970).
74. H. Fischer and W. Lautsch, Ann. Chem., 528, 247 (1937).
75. H. Fischer and K. Kahr, Ann. Chem., 531, 209 (1937).
76. R.B. Woodward and V. Skarić, J. Am. Chem. Soc., 83, 4674 (1961).
77. R. Bonnett, I.A.D. Gale, and G.F. Stephenson, J. Chem. Soc. C, 1168 (1967).
78. G.W. Kenner, K.M. Smith, and M.J. Sutton, Tetrahedron Lett., 1303 (1973).
79. R. Grigg, A. Sweeney, and A.W. Johnson, Chem. Commun., 1237 (1970).
80. J.B. Paine and D. Dolphin, J. Am. Chem. Soc., 93, 4080 (1971).
81. C.E. Castro and H.F. Davis, J. Am. Chem. Soc., 91, 5405 (1969).
82. J.J. Katz, R.C. Dougherty, and L.J. Buocher in "The Chlorophylls", Eds. L.P. Vernon, G.R. Seely, Academic Press, New York (1966), p. 245.
83. H. Scheer, W.A. Svec, B.T. Cope, M.H. Studier, R.G. Scott, and J.J. Katz, J. Am. Chem. Soc., 96, 3714 (1974).
84. H. Fischer and K. Zeile, Ann. Chem., 468, 98 (1929).
85. H. Fischer and A. Treibs, Ann. Chem., 466, 188 (1928).
86. H. Brockmann Jr., K.M. Bliesener, and H.H. Inhoffen, Ann. Chem., 718, 148 (1968).
87. H.H. Inhoffen, G. Klotmann, and G. Jeckel, Ann. Chem., 695, 112 (1966).
88. A.W. Johnson and D. Oldfield, J. Chem. Soc., 4303 (1965).
89. Ref. 45, p. 327.
90. P.S. Clezy and V. Diakiw, J. Chem. Soc., Chem. Commun., 453 (1973).
91. A.W. Nichol, J. Chem. Soc. C, 903 (1970).
92. H.J. Callot, A.W. Johnson, and A. Sweeney, J. Chem. Soc. Perkin Trans. 1, 1424 (1973).
93. H.J. Callot, Tetrahedron Lett., 1011 (1972).
94. R. Grigg, J. Chem. Soc. C, 3664 (1971).
95. R. Bonnett and G.F. Stephenson, J. Org. Chem., 30, 2791 (1965).
96. H. Fischer and W. Klendauer, Ann. Chem., 547, 123 (1941).
97. M.J. Billig and E.W. Baker, Chem. Ind., 654 (1969).
98. (a) G. Balz and G. Schiemann, Chem. Ber., 60, 1186 (1927); (b) H. Suschitzky, Adv. Fluorine Chem., 4, 1 (1965).
99. R. Bonnett, I.A.D. Gale, and G.F. Stephenson, J. Chem. Soc. C, 1600 (1966).
100. H. Voigt, Dissertation, Braunschweig (1964).
101. W.S. Caughey, J.O. Alben, W.Y. Fujimoto, and J.L. York, J. Org. Chem., 31, 2631 (1966).
102. (a) Y. Chang, P.S. Clezy, and D.B. Morrell, Aust. J. Chem., 20, 959 (1967). (b) A. Treibs, Ann. Chem., 506, 196 (1933).

103. P.S. Clezy and C.J.R. Fookes, Chem. Commun., 1971, 1268.
104. R. Lemberg and J. Barrett, in "Cytochromes", Academic Press, London (1973), p. 8.
105. Ref. 82, p. 235.
106. J. Dörffel, Dissertation Braunschweig (1969).
107. O. Schumm, Hoppe-Seyler's Z. Physiol. Chem., 178, 11 (1928).
108. R. Thomas, Dissertation Braunschweig (1966).
109. H. Fischer and H. Gibian, Ann. Chem., 548, 183 (1941).
110. S. Schwartz, M.H. Berg, I. Bossenmaier, and H. Dinsmore, in 'Methods of Biochemical Analysis', Ed. D. Glick, Vol. 8, Interscience, New York (1960), p. 221.
111. H. Fischer, K. Herrle, and H. Kellermann, Ann. Chem., 524, 222 (1936).
112. P.S. Clezy and J. Barrett, Biochem. J., 78, 798 (1961).
113. H. Fischer and R. Müller, Hoppe-Seyler's Z. Physiol. Chem., 142, 120, 155 (1925).
114. R. Lemberg and J. Parker, Aust. J. Exp. Biol. Med. Sci., 30, 163 (1952).
115. (a) F. Sparatore and D. Mauzerall, J. Org. Chem., 25, 1073 (1960); (b) S. Sano, J. Biol. Chem., 241, 5276 (1966).
116. J.-H. Fuhrhop and L. Witte, in preparation.
117. H. Fischer and H. Medick, Ann. Chem., 517, 245 (1935).
118. M.J. Parker, Biochim. Biophys. Acta, 35, 496 (1959).
119. J.E. Falk, E.I.B. Dresel, A. Benson, and B.C. Knight, Biochem. J., 63, 87 (1956).
120. P.S. Clezy and D.B. Morrell, Biochim. Biophys. Acta, 71, 150 (1965).
121. H. Fischer and W. Schmidt, Ann. Chem., 519, 244 (1935).
122. H. Fischer, E. Haarer, and F. Stadler, Hoppe-Seyler's Z. Physiol. Chem., 241, 201 (1936).
123. A.E. Vasilev, A.A. Khachatur'yan, B.W. Borisov, M.M. Vshakova, and G. Ya. Rozenberg, Zh. Org. Khim., 8, 1973 (1972).
124. C.K. Chang and T.G. Traylor, Proc. Nat. Acad. Sci. U.S.A., 70, 2647 (1973); J. Am. Chem. Soc., 95, 5810 (1973).
125. P.K. Warme and L.P. Hager, Biochemistry, 9, 1599 1606, 4244 (1970).
126. F.P. Schwarz, M. Gouterman, Z. Muljiani, and D. Dolphin, Bioinorg. Chem., 2, 1 (1972).
127. H. Fischer and C.G. Schroeder, Ann. Chem., 537, 250; 541, 196 (1939).
128. H. Fischer, E. Haarer, and F. Stadler, Hoppe-Seyler's Z. Physiol. Chem., 241, 201 (1936).
129. (a) M.T. Cox, T.T. Howarth, A.H. Jackson, and G.W. Kenner, J. Am. Chem. Soc., 91, 1232 (1969); J. Chem. Soc., Perkin Trans. 1, 512 (1974); (b) G.W. Kenner, S.W. McCombie, and K.M. Smith, J. Chem. Soc., Perkin Trans. 1, 527 (1974); (c) ibid., J. Chem. Soc., Chem. Commun., 844 (1972); (d) M.T. Cox, A.H. Jackson, G.W. Kenner, S.W. McCombie, and K.M. Smith, J. Chem. Soc., Perkin Trans. 1, 516 (1974).
130. F.C. Pennington, H.H. Strain, W.A. Svec, and J.J. Katz, J. Am. Chem. Soc., 86, 1418 (1964).
131. G.R. Seely, in "The Chlorophylls", Eds. L.P. Vernon and G.R. Seely, Academic Press, New York (1966), p. 67.
132. F.C. Pennington, H.H. Strain, W.A. Svec, and J.J. Katz, J. Am. Chem. Soc., 89, 3875 (1967).
133. H. Wolf, H. Brockmann Jr., H. Biere, and H.H. Inhoffen, Ann. Chem., 704, 208 (1967).
134. H. Wolf and H. Scheer, Tetrahedron Lett., 1111 (1972).
135. G.W. Kenner, S.W. McCombie, and K.M. Smith, J. Chem. Soc., Perkin Trans. 1, 2517 (1973).
136. G.L. Collier, A.H. Jackson, and G.W. Kenner, J. Chem. Soc. C, 66 (1967).

137. Ref. 38b, part 1, p. 48.
138. R.A. Chapman, M.W. Roomie, T.C. Morton, P.T. Krajcarski, and S.F. MacDonald, Can. J. Chem., **49**, 3544 (1971).
139. H. Fischer and H. Wenderoth, Ann. Chem., **537**, 170 (1939).
140. H. Muir and A. Neuberger, Biochem. J., **45**, 163 (1949).
141. G.E. Ficken, R.B. Johns, and R.P. Linstead, J. Chem. Soc., 2272 (1956).
142. H.V. Morley and A.S. Holt, Can. J. Chem., **39**, 755 (1961).
143. H. Muir and A. Neuberger, Biochem. J., **47**, 97 (1950).
144. D. Shemin, Methods Enzymol., **4**, 643 (1957).
145. J. Wittenberg and D. Shemin, Fed. Proc. Fed. Am. Soc. Exp. Biol., **9**, 247 (1950).
146. H. Muir and A. Neuberger, Biochem. J., **45**, XXXIV (1939).
147. D. Shemin and S. Kumin, J. Biol. Chem., **198**, 827 (1952).
148. J. Wittenberg and D. Shemin, J. Biol. Chem., **185**, 103 (1950).
149. A.S. Holt, D.W. Hughes, H.J. Kende, and J.W. Purdie, J. Am. Chem. Soc., **84**, 2835 (1962).
150. D.W. Hughes and A.S. Holt, Can. J. Chem., **40**, 171 (1962).
151. R.A. Nicolaus, Rass. Med. Sper., (Suppl. 2), **7**, 1 (1960).
152. H. Brockmann Jr., Ann. Chem., **754**, 139 (1971).
153. H. Brockmann Jr. and J. Bode, Ann. Chem., **748**, 20 (1971).
154. H. Brockmann Jr. and G. Knobloch, Chem. Ber., **106**, 803 (1973).
155. T. Sjoestrand, Acta Physiol. Scand., **26**, 328, 334, 338 (1952); G.D. Ludwig, W.S. Blakemore, and D.L. Drabkin, Biochem. J., **66**, 38 P (1957).
156. J.C. Fanning and T.L. Gray, J. Chem. Soc., Chem. Commun., 23 (1974).
157. P.A. Burbidge, G.L. Collier, A.H. Jackson, and G.W. Kenner, J. Chem. Soc. B, 930 (1967).
158. D. Mauzerall, J. Am. Chem. Soc., **82**, 2601 (1960).
159. P.S. Clezy and J. Barrett, Biochem. J., **78**, 798 (1961).
160. G.H. Barnett and K.M. Smith, J. Chem. Soc., Chem. Commun., 772 (1974).
161. W.S. Sheldrick, personal communication.
162. W.S. Sheldrick and J.-H. Fuhrhop, Angew. Chem., in press.

PHOTOCHEMISTRY OF PORPHYRINS AND METALLOPORPHYRINS

FREDERICK R. HOPF and DAVID G. WHITTEN

Department of Chemistry, University of North Carolina, Chapel Hill, North Carolina 27514, U.S.A.

16.1 Introduction
16.1.1. Excited states of porphyrins and metalloporphyrins

A discussion of the photochemistry of porphyrins and metalloporphyrins is probably best begun with a consideration of the properties of the excited states involved. Although there exist still some controversies over the exact assignments of various transitions in certain compounds, particularly for the biologically important iron complexes and some heavy metal complexes[1-6], it has been generally accepted that the prominent electronic transition of porphyrins and their metal complexes are $\pi \rightarrow \pi^*$ transitions associated with the porphyrin ring. Most of the luminescence and photochemistry observed from these compounds is also associated with the porphyrin π,π^* states even though, as will be developed, the lifetimes and reactivities of these states depend strongly on the metal ion incorporated.

16.1.1.1. Luminescence of porphyrins and metalloporphyrins
16.1.1.1.1. General

Although luminescence was early recognized as characteristic of several porphyrins and their metal complexes[7] and even used in many cases as an analytical technique, the first systematic study of the influence of different metals on porphyrin fluorescence and phosphorescence was by Becker and Kasha in 1955[8]. In this study the luminescence of free base and metal complexes [Zn(II), Cu(II), Ni(II), and Mg(II)] of etioporphyrin-II, phthalocyanine, pheophorbide and the chlorophylls was investigated in EPA glass at 77° K. This study was followed by more extensive investigations of different metal complexes of mesoporphyrin-IX dimethyl ester under a variety of temperature-solvent conditions[9-12]. Although subsequent investigations by a number of different workers have amplified and modified these findings[13-23], the work by Becker and Allison remains a basic reference in this field. Tables 1 and 2 summarize the emission properties of free base and

Porphyrins and Metalloporphyrins, ed. Kevin M. Smith
© 1975, Elsevier Scientific Publishing Company, Amsterdam, The Netherlands

TABLE 1

Luminescence properties of porphyrins in room temperature ($300°$K) solution

Non-luminescent	Fluorescence only	Phosphorescence only	Fluorescence and phosphorescence
Ni(II), VO, Sn(II)	Zn(II), Mg(II)	Co(III)	Pd(II)
Ru(II)L$_2$, Ru(III)	Sn(IV), Pb(II)	Rh(III)	Pt(II)
Cu(II), Ag,	Al, Cd, FB,	Ir	Ru(II)CO
Co(II)	Si(IV), Ge(IV),		
	Ba, Sr, Be,		
	Sc(III), Ti(IV),		
	Zr(IV), Hf(IV),		
	Nb(V), Ta(V)		

metalloporphyrins reported to date at room teperature in fluid medium and in liquid nitrogen temperature ($77°$K) glasses respectively. In general most free base porphyrins, chlorins and related compounds show strong fluorescence at room temperature and both fluorescence and phosphorescence in rigid glasses. Metalloporphyrin luminescence falls into several categories, dependent largely on the electronic structure of the metal (vide infra). In general free base octa-alkyl porphyrins show fluorescence near 618—620 nm[9,10]. From the onset of this emission (relaxed) singlet energies of approx. 47 kcal/mol can be assigned. Phosphorescence of octa-alkyl (free base) porphyrins generally is observed near 750 nm and from the phosphorescence onset a triplet energy of approx. 40 kcal/mol can be assigned[10]. For simple $\alpha\beta\gamma\delta$-tetraphenylporphyrin fluorescence λ max and singlet energies are 660 nm and 45 kcal/mol respectively.

For octa-alkylporphyrin metal complexes, phosphorescence and fluores-

TABLE 2

Luminescence properties of porphyrin at low temperature in rigid medium

Phosphorescence only	Phosphorescence and fluorescence	Non-luminescent
Hg, Pb.,	Pb, Pt, Ru,	Ag
VO, Mn(II)	Mg, Zn, FB,	
Fe(II), Fe(III)	Be, Ca, Sr,	
Co(II), Co(III)	Cd, Sn(IV),	
Ni, Cu	Ba, Sc(III),	
Ir	Ti(IV), Zr(IV),	
	Hf(IV), Nb(V),	
	Ta(V)	

TABLE 3

Triplet energies for metalloporphyrins estimated from onset of phosphorescence

Complex	E_t(kcal/ mol)	Medium	Temp. ($^\circ$K)	Ref.
Fe(III)(Meso-IX-DME)(OAc)	43	EPA	77	10
Fe(II)(Meso-IX-DME)	43	EPA	77	10
Co(II)(Meso-IX-DME)	43.5	EPA	77	[10]
Co(III)(Meso-IX-DME)(OAc)	43	EPA	77	[10]
Mn(II)(Meso-IX-DME)	39.5	EPAF	77	10
Mn(III)(Meso-IX-DME)(OAc)	44.5	EPAF	77	10
Mg(II)(Etio-II)	42.2	EPA	77	9
Mg(II)(Etio-II)	40.5	3MeP	77	10
Ba(II)(Meso-IX-DME)	38.9	EPAF	77	10
Ca(II)(Meso-IX-DME)	40.9	EPAF	77	10
Zn(II)(Meso-IX-DME)	41.5	EPA	77	10
Sr(II)(Meso-IX-DME)	39.7	EPAF	77	10
Cd(II)(Meso-IX-DME)	40	EPA	77	10
Sn(IV)(Meso-IX-DME)(OAc)$_2$	41	EPA	77	10
Pb(II)(Meso-IX-DME)	41.5	3MeP	77	10
Hg(II)(Meso-IX-DME)	38	EPAF	77	10
Cu(II)(Meso-IX-DME)	44	EPAF	77	9
Pd(II)(Meso-IX-DME)	44.6	EPAF	77	9
Ni(II)(Meso-IX-DME)	43	EPAF	77	[9]
Pt(II)(Etio-I)	45.3	EPAF	77	13
Pt(II)(Etio-I)	45	decane	81	13
Pd(II)(Etio-I)	44	nonane	81	13
Pd(II)(Etio-I)	44.3	EPAF	77	13
Pd(II)(TPP)	42.6	EPAF	77	13
Pt(II)(TPP)	44.6	EPAF	77	13
Pd(II)(OEP)	44.8	Benzene	300	
Ru(II)(OEP)(CO)	46.1	Benzene	300	
Rh(III)(Etio-I)(Cl)	44.3	2-MTHF	77	17
Si(IV)(OEP)(Cl)$_2$	41	2-MTHF	77	18
Pb(IV)(OEP)	41	MeOH	77	18
Ge(IV)(OEP)(Cl)$_2$	41.4	2-MTHF	77	18
Sn(IV)(OEP)(Cl)$_2$	41.4	2-MTHF	77	18
Sn(IV)(TPP)(Cl)$_2$	41	2-MTHF	77	18
VO(Etio-I)	41	PMA	74	19
VO(TPP)	39.7	PMA	10	19

Entries in brackets, [ ], refer to compounds whose luminescence properties have been questioned.

cence generally occur at shorter wavelengths than the corresponding free bases. Consequently triplet and singlet state energies for metal complexes are somewhat higher than for the free bases. In general both absorption and luminescence spectra of many metal complexes are strongly solvent sensitive[24-26]; therefore, excited state energies are solvent-dependent. Table 3 lists representative triplet energies determined for several metalloporphyrins from phosphorescence onsets.

16.1.1.1.2. Luminescence of specific types of metalloporphyrin complexes
(a) *Closed-shell metal complexes.* As reported initially by Becker[11] and later amplified by Gouterman and others[1], the closed shell metals having empty or full d shells form complexes with octa-alkyl- and tetraphenylporphyrins that generally show only fluorescence at room temperature in fluid media but both fluorescence and phosphorescence in rigid media at liquid nitrogen temperature. The fluorescence yield tends to decrease with increasing atomic number of the metal as does the phosphorescence lifetime. Both absorption and emission spectra are shifted (usually to longer wavelengths) as the square-planar metalloporphyrin accepts axial ligands in the fifth and sixth coordination positions[25] and it has been shown that exchange of these ligands can occur for zinc and magnesium complexes during the excited singlet lifetime[25].

Although most metalloporphyrins show no important wavelength effects in their luminescence spectra, it has recently been found[27] that zinc tetra-benzporphyrin gives fairly strong $S_2 \rightarrow S_0$ emission ($\phi \sim 0.016$) upon activation in the Soret band in Argon matrices at $20°$K. Similar upper state emission has been reported for certain chlorophylls[28]; phthalocyanines, including the magnesium complex, have been found to emit from upper vibrational levels of the first excited singlet at $77°$K and at $300°$K in α-chloronapththalene[29,30]. Several groups have recently investigated the appearance of narrowband fine structure (quasilines) in low temperature absorption and luminescence (fluorescence and phosphorescence) of porphyrins and metalloporphyrins (primarily zinc and magnesium complexes)[27,31-35]. At present the origin of this fine structure must probably be regarded as somewhat uncertain, although a variety of explanations have been advanced.

(b) *Open-shell diamagnetic metal complexes.* The original study by Becker and Allison[10] indicated only phosphorescence for porphyrin complexes with diamagnetic transition metal ions having partially filled d shells. In general phosphorescent lifetimes were relatively short with these metals (Mn, Fe, Co and Ni) and emission was generally observed only in rigid media at $77°$K. Subsequent investigations have yielded somewhat different results in a few cases. Re-examination of several Ni and Co complexes suggests that these complexes do not emit[13,19] or that at least the phosphorescent yield is very low ($\phi_p < 0.0005$)[13]. It has been suggested that low lying d-d excited states in Ni(II) complexes are responsible for the lack of emission[19]. The Pd and Pt complexes (d^8) phosphoresce very strongly, both in rigid media at $77°$K and at room temperature in fluid (degassed) solution. In the case of Pd and Pt porphyrins, fluorescence is also observed at the usual wavelength[13,15]; it has been shown that this fluorescence consists of both prompt and delayed components in the case of Pd and only delayed fluorescence in the case of Pt. The delayed fluorescence results not from triplet—triplet annihilation but rather from thermal equilibrium (repopulation, 'E type') between singlet and triplet states[15].

Although the energy gap between the lowest singlet and triplet is fairly large in these compounds (~8 kcal/mol), the vast difference in radiative decay rates between singlet and triplet states allows fast emission from the energetically disfavored singlet to compete with the much slower emission from the triplet. Although examples of this phenomenon are relatively rare, a prominent case previously reported is that of benzophenone[36]. The small singlet—triplet separation, absorption spectral characteristics, and room-temperature solution phosphorescence of the Pd, Pt and Ru porphyrin make them extremely attractive sensitizers for use in studying photosensitized reactions and energy transfer processes. A rhodium(III) (d^6) complex has also been shown to fluoresce at room temperature[17]. This emission is apparently prompt fluorescence. Carbonyl complexes of ruthenium(II) Etio-I (d^6) and OEP show phosphorescence and fluorescence both at low temperature and at room-temperature in fluid solution[37,39]. The fluorescence observed is evidently entirely delayed (Type E) fluorescence. Interestingly it has been found that for ruthenium(II) only the CO complexes emit strongly; the corresponding dipyridinate complex and other species do not give any detectable emission[38]. Apparently in the complexes described above, the open shell but diamagnetic metal greatly enhances the forbidden radiative $T^* \to S_0$ transition while not greatly effecting radiationless decay. Thus the non-emitting Zn(OEP) triplet has a lifetime of approx. 400 μsec in benzene while the strongly emitting Pd(OEP) triplet has a lifetime of >500 μsec[38].

(c) *Paramagnetic metal complexes.* In metalloporphyrin complexes where the central metal is paramagnetic, studies to date have revealed only phosphorescence with relatively short lifetimes. In the case of copper porphyrins, certain complexes show as many as three distinct luminescent lifetimes suggesting the possibility of several close-lying but non-equilibrating excited states. It was reported that vanadyl etioporphyrin-I gives prominent fluorescence in room temperature and low-temperature pyridine solutions[39]; the authors were able to rationalize the contrast in behavior between the vanadyl complex and the copper(II) and cobalt(II) complexes, which also contain an unpaired electron but show no fluorescence, on the basis of the extended Hückel calculations which indicated more complete electron localization in d-orbitals in the vanadyl complex when compared to copper and cobalt. However, subsequent investigations by the same authors[40] revealed that (1) the emission was due to an impurity and (2) that more thorough calculations reveal extensive electron delocalization in the vanadyl complex comparable to that in copper porphyrins.

References, p. 695

(d) *Luminescence of modified porphyrin derivatives.* The luminescence of chlorophyll and derivatives has been widely investigated in works too numerous to cite[41]. However, relatively few systematic studies of these or of related compounds, such as the chlorins have been reported to the present time. Recently it was reported[42] that several metal complexes of tetraphenylchlorins (including Cd, Cu and Pd) show luminescence behavior similar to the corresponding porphyrins; however at $77°K$ the luminescence from these complexes was much weaker than for the corresponding porphyrins.

Structural modification of the porphyrin macrocycle by an isocyclic cyclopentanone or cyclopentane ring as in phylloerythrin and desoxophylloerythrin has been reported to cause major changes in the spectroluminescent properties of free base, Pd and Zn complexes[43]. Both radiative and non-radiative rates are strongly affected by the ring.

Quantum yields of luminescence have been measured by several groups as a function of temperature, central metal, substituents, and other factors[16,44−50]. In most cases it has been found that $S_1 \rightarrow S_0$ radiationless decay is unimportant and that inefficiency in luminescence arises from the radiationless $T_1 \rightarrow S_0$ process. Exceptions to this are the Cu(II) porphyrins[50] and the Ag(II), Ni(II) and Co(III) porphyrins[48]. For the latter compounds it has been proposed that an intramolecular electron transfer process (ligand to metal) is responsible for deactivation of excited singlets.

The general decrease in fluorescence and increase in phosphorescence with heavy central metals has been partially attributed to the 'heavy atom effect' which is generally believed to arise from spin orbit coupling. Internal substitution of porphyrins by halogens (Br or I) increases rates of intersystem crossing[49] with the anticipated correlation between the intersystem crossing probability and the square of the spin—orbit interaction constant for these atoms. External heavy atom effects have also been observed[43,45] using alkyl iodides and KI; this has proved a useful technique in the measurement of quantum yields.

16.1.1.2. Porphyrin excited states as studied by flash photolysis

Although triplet states of free base porphyrins and closed-shell metalloporphyrins do not emit in fluid media at room temperature, they exhibit prominent triplet—triplet absorption of duration long enough for convenient study with microsecond flash techniques. Some of the earliest studies were performed on chlorophyll and related magnesium complexes[51−55]. Extensions[56−60] showed that similar patterns were observed for most free base and metalloporphyrin complexes. An intense transient t—t absorption is generally observed slightly to the red of the Soret band. If the flash is intense the decay consists of several components which can be resolved by eq (1):

$$\frac{-dC^*}{dt} = k_1 C^* + k_2 (C^*)^2 + k_3 (C^*)(Cg) \tag{1}$$

where C* and Cg are concentrations of excited and ground state forms of the porphyrin[58]. Typical first-order decay constants are in the range $10^3 - 10^4$ sec^{-1} for zinc, magnesium and free-base porphyrins at $300°K$. Triplet states of porphyrins containing paramagnetic metal ions are generally too short-lived to be studied in room temperature solution by conventional flash (microsecond) techniques. However, triplets of porphyrins containing open shell diamagnetic metals such as Pd, Pt and Ru have long lifetimes in room temperature solutions, ($\sim 0.1 - 1$ msec)[13,16]; for Pt and Pd complexes, non-radiative rates are relatively slow compared to Zn and Mg complexes since prominent $T^* \rightarrow S^0$ luminescence is observed for the former complexes.

The variation of excited state lifetime of porphyrins with metal and ligand substitution remains an intriguing but rather poorly understood phenomenon. With metal complexes it would be expected that lifetimes should be affected by paramagnetic effects due to unpaired metal d electrons and to spin—orbit coupling. The former effect has been observed, as mentioned previously. However the effect of spin—orbit coupling has not been clearly shown and for non-paramagnetic metals other factors are evidently important. However, as mentioned in the preceding section, the effect of spin—orbit coupling of substituent halogen atoms on porphyrin excited state lifetimes appears to be 'normal'[49].

Free base porphyrins have triplet lifetimes relatively shorter than the corresponding zinc and magnesium complexes and it has been suggested[61-65] that this is due to involvement of N—H tautomerism in radiationless $T^* \rightarrow S^0$ deactivation. Recently[63] it has been shown that deuteration of the N—H hydrogens produces large increases (approx. $2\times$) in non-radiative lifetimes of several free-base porphyrins at $77°K$. A significant deuterium isotope effect has also been observed on the rate of tautomerism of several free-base porphyrins[66,67]. The effect of deuteration on non-radiative decay vanishes as the temperature is lowered to $4.7°K$; the authors have developed a dynamic model in which the interconversion between tautomeric forms brings about a relaxation not observed in the metal complexes[64,65].

As will be developed in subsequent sections, the use of the flash spectroscopic technique to investigate quenching processes and transients produced in these interactions has provided vital information in unravelling the mechanism of many porphyrin photoreactions.

16.1.2. Porphyrin photochemistry — summary of emerging reaction patterns

In the most general terms, it could be anticipated that excited states of porphyrins and their metal complexes might participate in three possible primary processes. (1) Unimolecular reactions (i.e., isomerization or fragmentation), (2) bimolecular reactions (including complex formation, electron transfer and addition reactions), and/or (3) energy transfer. There appears to be little indication that the first of these three reactions occur prominently since, in an 'inert' environment, most porphyrins appear to be almost indefinitely photostable. Possible exceptions to this include metastable species such as the tin(II) porphyrins, which undergo a photochemical intramolecular electron transfer process to yield the more stable tin(IV) porphyrin dianions[68], and porphyrins containing ligands such as CO which may be lost unimolecularly on photolysis[37,69]. The latter two reactions occur prominently although there is frequently some confusion as to which of the two occur as the primary process in complex photoreactions. A fourth possible process (4) which appears important in several photobiological processes involves reactions of photoexcited porphyrins with some reactive molecule complexed with the porphyrin prior to light absorption.

Probably the simplest and most straight-forward method for demonstrating the occurrence of processes 2 or 3 and investigating mechanistic details is a kinetic investigation of excited state quenching[68]. The possible involvement of an excited singlet state is indicated by fluorescence quenching which follows the Stern—Volmer relationship for intensities (eq. 2) or lifetimes (eq. 3), where k_q^s refers to the bimolecular rate constant for the primary process 2 or 3 (eq. 4) and ϕ^0 and ϕ_f refer to quantum efficiencies

$$\phi_f^0/\phi_f = 1 + k_q^s \tau_f^0 [Q] \tag{2}$$

$$\tau_f^0/\tau_f = 1 + k_q^s \tau_f^0 [Q] \tag{3}$$

$$P^{*1} + Q \xrightarrow{k_q^s} \text{products (2)} \; or \; P^0 + Q^{*1} \; (3) \tag{4}$$

for fluorescence in the absence and presence of quencher at concentration [Q] respectively and τ_f^0 and τ_f refer to singlet lifetimes for unquenched porphyrin respectively. If products are observed as a consequence of the reaction, a second Stern—Volmer relationship for the quantum efficiency of

$$\frac{1}{\phi_{Prod}} = \frac{1}{a} \left(1 + \frac{1}{k_q^s \tau_f^0 [Q]} \right) \tag{5}$$

product formation is observed (eq. 5). where a is a constant and ϕ_{prod} is the quantum efficiency for product formation. The intercept/slope from the plot of eq. (5) should equal the slope of the plot of eq. (2) or (3) to confirm singlet involvement in the process.

Triplet state involvement in process 2 or 3 can be determined by a flash photolysis study; observation of a Stern—Volmer lifetime relationship analogous to eq. (3) would provide confirmation. Similarly, for the several metalloporphyrins which phosphoresce in room temperature solutions, a Stern—Volmer relationship for phosphorescence intensity analogous to eq. (2) would reveal triplet participation. Numerous such quenching studies have been used to determine the mechanisms of photoreactions involving porphyrins and metalloporphyrins. Several specific cases will be discussed in later sections of this chapter.

16.2. Energy transfer, complex formation, quenching phenomena: porphyrins as photosensitizers

16.2.1. General, quenching phenomena and intermolecular energy transfer

There are numerous reports of excited state quenching phenomena where no significant porphyrin decomposition occurs. Many of these processes involve energy transfer although, as will be developed subsequently, in certain cases complex formation or other events can provide channels for non-radiative decay. Singlet—singlet energy transfer involving chlorophyll has been well-established as a key step in photosynthesis and will not be dealt with here in detail. Triplet—triplet energy transfer involving porphyrins as donors has also been observed in several cases. In many ways the porphyrins and their metal complexes are nearly ideal photosensitizers. The combination of relatively small singlet—triplet splitting, high intersystem crossing yield[71,72] and long triplet lifetimes in certain cases in combination with intense long wavelength transitions affords the opportunity for selective population of porphyrin triplets followed by efficient energy transfer to acceptors having lower than 40 kcal/mol triplet energies. The use of porphyrins to sensitize reactions such as azobenzene isomerization[73] and thioindigo cis—trans isomerization[74] has been reported in addition to other processes.

16.2.2. Intramolecular energy transfer

Triplet—triplet energy transfer involving porphyrins as donors has also been observed in intramolecular cases involving coupled bichromophoric systems[78−81]. In systems involving different metalloporphyrins linked by ethylene or p-phenylene bridges, it has been found that efficient intramolecular energy transfer of triplet excitation occurs from one porphyrin to the other (Zn → Cu; Zn → Co) even though singlet—singlet energy transfer does not occur[78]. An exchange mechanism is suggested for the ethylene bridged Zn—Cu system while a mechanism involving charge—transfer complex formation has been proposed for the Zn—Co system[78]. It has very recently been reported that efficient singlet—singlet intramolecular energy transfer occurs between porphyrin units in the different chains of the hemoglobin macromolecule where individual heme units are replaced with protoporphyrin or zinc protoporphyrin[81]. Transfer between unlike chains occurs with that from α to β favored over the reverse. Changes in the rate on converting a heme to oxyhemoglobin suggest changes in the relative heme—heme orientation on oxygenation. For Zn, Mg and Co porphyrins complexed to the isomerizable olefinic ligands 4-stilbazole and 1-(α-napththyl)-2-(4-pyridyl)-ethylene (NPE) reversible transfer of triplet excitation from porphyrin to olefinic ligand has been observed[79,80]. In the case of NPE complexes with Zn and Mg porphyrins the combination of energy transfer and rapid ligand exchange coupled with the long lifetime of the metalloporphyrin triplet allows a quantum chain process of cis → trans isomerization of NPE to occur[79,80].

References, p. 695

16.2.3. Sensitization of singlet oxygen and photodynamic deactivation

One of the most prominent energy transfer processes involving porphyrins as energy donors involves the sensitization of singlet oxygen by porphyrin triplets (eq. 6).

$$P^3* + {}^3O_2 \rightarrow P^0 + {}^1O_2^*$$ (6)

It has been known that various dyes, including porphyrins, can mediate the photo-oxidation of a variety of organic substrates[82-84]. More recent investigations have established that the intermediate in these reactions is usually singlet oxygen produced by energy transfer from the dye (porphyrin) triplet in eq. (6)[85-87]. It has subsequently been well-established for porphyrins that oxygen efficiently quenches the triplet and that singlet oxygen is indeed a product of the quenching process[85,88-90]. Several instances of light-mediated damage to biological systems probably involve the sequence: light absorption by porphyrins or their metal complexes, energy transfer to oxygen (eq. 6), and subsequent attack of singlet oxygen on unsaturated molecules. A case where the evidence is fairly clear on this point involves photohemolysis of red cell membranes in the genetic disorder erythropoietic protoporphyria. It has relatively long been known that patients with this disorder have unusually large concentrations of free protoporphyrin[91]. Further photohemolysis occurs with visible light and the action spectrum has its λ_{max} near 400 nm. Molecular oxygen is required for photohemolysis and the net result is destruction of the red-cell membrane. In a series of recent, elegant experiments it has been rather convincingly demonstrated that the protoporphyrin triplet sensitizes singlet oxygen which probably subsequently attacks the lipid components of the red cell membrane[92-94]. One of the chief reactivity membrane components appears to be cholesterol which undergoes the 'ene' reaction to yield 3β-hydroxy-5α-hydroperoxy-Δ6-cholestene (eq. 7).

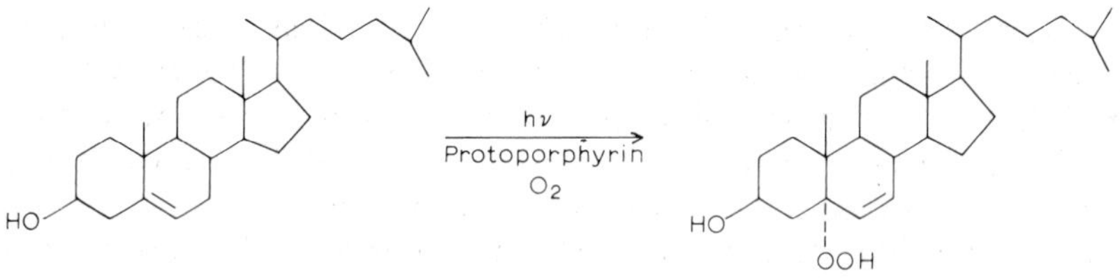

(7)

Other unsaturated lipid components can react with singlet oxygen themselves or be decomposed by radical chain oxidation following decomposition of the cholesterol hydroperoxide[94]. Additional evidence supporting the above sequence as the molecular mechanism for erythropoietic porphyria includes the observation that massive doses of β-carotene and α-tocopherol (here there is an optimum level at moderate dosage) protect patients from photohemolysis[95]. Both of these substances have been found to be efficient quenchers of singlet oxygen; the former has been recognized for some time as a quencher[86,87] while it has been more recently demonstrated that α-tocopherol is also an effective quencher of singlet oxygen[94].

Related events at the molecular level are probably involved in several cases involving porphyrin-mediated 'photodynamic deactivation'[96−103]. Photo-oxidative damage to amino acids especially methionine, histidine, trypto-phane and tyrosine, in the vicinity of heme groups has been shown to occur[97,103,104]. It has been found that substitution of free base hemato-porphyrin or Mg complexes greatly accelerates the photo-oxidation process[96,99].

Photo-oxygenation of the porphyrin-related pigment bilirubin (which may be involved in jaundice phototherapy[105,106]) may also involve sensitization of singlet oxygen since bilirubin sensitizes formation of $^1O_2$ efficiently[107] and itself reacts rapidly and efficiently with $^1O_2$[108] to give several products.

16.2.4. Electron transfer and complex formation in the excited state

One of the key roles of electronically excited porphyrins and related molecules — indeed perhaps the cornerstone of photosynthesis — is their dual function as both electron donors and electron acceptors. It has been well-established that both singlet and triplet states can function as electron donors and electron acceptors under a variety of conditions[109−120]; several of these reactions will be discussed in a later section, particularly those leading to modification of the porphyrin ring. However there have also been several reported cases of quenching of porphyrin excited states by electron donors and electron acceptors under conditions where electron-transfer products cannot or have not been detected[121−124]. This is particularly true where excited porphyrins are quenched by electron donors or acceptors in non-polar solvents under conditions where ion formation would be dis-favored. In recent studies it has been shown that the quenching is due to formation of excited state complexes (exciplexes) from singlets and triplets of the porphyrins and metalloporphyrins[125−128]. These complexes fre-quently live long enough to be detected by flash spectroscopy. Although in the cases studied to date the usual fate of the complex is simply return to the ground state of porphyrin and quencher, it appears reasonable that long-lived exciplexes may well be intermediate in other porphyrin photoreactions.

16.3. Photoreduction reactions

A qualitative treatment of electron distribution in the porphyrin macro-cycle is of value in reviewing photoreduction reactions of free base porphy-rins and metalloporphyrins. A useful model is that proposed by Wood-ward[129]. Each of the pyrrolenine units of the porphyrin macrocycle (1)

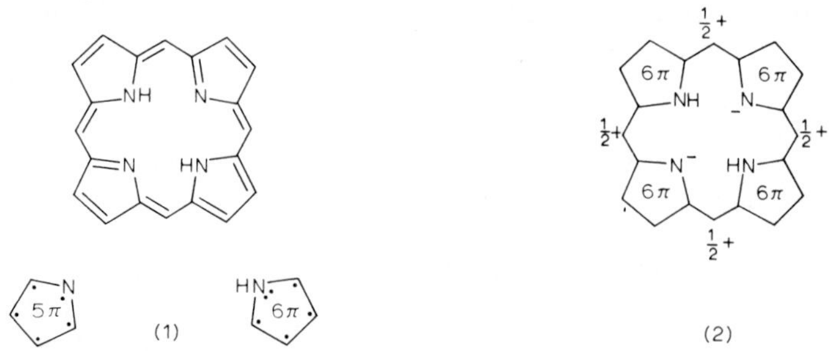

Fig. 1.

contains 5π electrons and should be expected to remove electron density from its macrocyclic surrounding to achieve an aromatic sextet. Within the porphyrin ring system such electron withdrawal can only take place from the methine bridge positions, since the pyrrole rings already contain an aromatic sextet and will resist such withdrawal. Structure (2) represents the extreme resonance contributor (Fig. 1).

In the neutral porphyrin ligand, i.e., formally when no electrons are transferred from the metal to the porphyrin and no redox reaction has taken place on the porphyrin, the methine bridges carry a partial positive charge and nucleophilic additions should occur readily. However, if electrons are introduced from the central metal ion of metalloporphyrins or by reduction of the ligand, this should have little effect on the aromatic system of the pyrrole rings but should have a pronounced effect on the electron density at the methine bridge positions.

If this electron density distribution is considered qualitatively correct, the reactivities of the pyrrole carbon atoms should be similar to those in pyrrole, i.e., they should all be accessible to electrophilic substitutions. In contrast, the reactivity of the methine bridges in porphyrin complexes with different metals should reflect the diversity that can be expected from these complexes according to their variable redox behavior[130,131]. There is qualitative agreement between theory and experiment, inasmuch as most nucleophilic, electrophilic and free radical addition reactions of the porphyrin ligand begin at the methine bridges.

16.3.1. Free base porphyrins

The first reversible photochemical reduction of porphyrins was reported by Krasnovskii[132]. Since then a large number of papers by Krasnovskii and others[133-135] have dealt with many aspects of photoreduction under a variety of reaction conditions[136-139]. Mauzerall et al.[140,141] reported a study of the photochemical reduction of free base uroporphyrin esters under mild conditions with a variety of reducing agents including ascorbic acid,

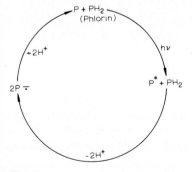

(8)

glutathione and ethylenediamine tetraacetic acid and ethyl acetoacetate. The products of these reductions were suggested to be the di-, tetra- and hexahydroporphyrins (3—5) (eq. 8). Kinetic and spectral evidence as well as comparison with products formed by non-photochemical reductants was utilized in determination of the intermediate photoreduction product structures. Further evidence in support of the phlorin structure (3) as the initial dihydrogenation product is provided by thermodynamic studies[142]. The photoreversibility of phlorin formation has been demonstrated[143]. A mixture of phlorin and porphyrin in glycol undergoes conproportionation on irradiation (Fig. 2) to form two porphyrin radical anions, which are converted back to starting materials by a relatively slow dark reaction.

Fig. 2.

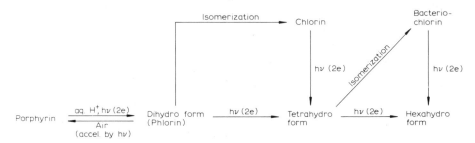

Fig. 3.

Gurinovich et al.[144] have shown that photoreduction of pheophytin and its derivatives leads to results similar to those for simple porphyrins when the porphyrin does not contain a cyclopentanone ring. The presence of a carbocyclic ring caused more complex product mixtures.

Sidorov[145,146] has investigated the reversible photoreduction of pyridine solutions of *meso*-tetraphenylporphyrin in the presence of hydrazine or hydrogen sulfide as reductants. From visible, i.r. and kinetic evidence it was proposed that two pathways were operative under these conditions. One led to phlorin type products; the other resulted in disruption of the macrocyclic ring.

Later work[147] by Sidorov on the hydrazine/pyridine photoreduction of *meso*-tetraphenylchlorin indicates that photoaddition of two atoms of hydrogen occurs only at the two central nitrogen atoms and not at the bridge positions. *meso*-Tetraphenylbacteriochlorin, under the same conditions, was reduced to a colorless product which reversibly reverted to the starting material upon exposure to air. No evidence was given to indicate whether hydrogenation occurred at the bridge positions or only at the central nitrogen atoms.

There are numerous reports of photoreductions in strongly acidic media (e.g., aqueous acid or alcohol-acid). The results obtained are apparently similar to those in basic media. Krasnovskii et al.[148] have done extensive work on photoreduction of free base porphyrins in acidic solutions. A scheme summarizing their results is given in Fig. 3. Electron spin resonance evidence implicates the presence of radicals as short-lived intermediates[149] in the overall two-electron processes.

16.3.2. Metalloporphyrins
16.3.2.1. Reduction of the porphyrin ligand

Seely and Calvin[150] demonstrated the reversible photochemical reduction of zinc *meso*-tetraphenylporphyrin, using benzoin or dihydroxyacetone as reductants. Their results are summarized in Fig. 4 [Zn(TPP) is simply represented as porphyrin].

A subsequent paper by Seely and Talmadge[151] reports on the photo-

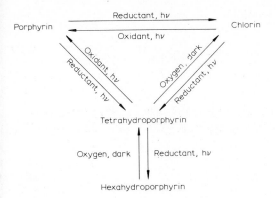

Fig. 4.

chemical reduction of zinc porphin by ascorbic acid in degassed solutions. Extensive kinetic and spectral evidence is presented which supports the authors' formulation of the initial dihydro-derivative as one having hydrogen atoms added to one bridge carbon and one peripheral carbon. The steps in the proposed mechanism are outlined in eq. (8—14) (P represents zinc porphin and QH_2 represents reductant).

$$P \xrightarrow{h\nu} P^{*1} \xrightarrow{i.s.c.} P^{*3} \tag{9a}$$

$$P^* + QH_2 \rightarrow PH^\cdot + QH^\cdot \tag{9b}$$

$$PH^\cdot + QH^\cdot \rightarrow P + QH_2 \tag{10}$$

$$2\,PH^\cdot \rightarrow P + PH_2 \tag{11}$$

$$2\,QH^\cdot \rightarrow Q + QH_2 \tag{12}$$

$$PH^\cdot + PH_2 \rightarrow P + PH_3^\cdot \tag{13}$$

$$PH_3^\cdot + QH^\cdot \;(\text{or } Q) \rightarrow \text{chlorin} + QH_2 \;(\text{or } QH^\cdot) \tag{14}$$

Likewise the further photoreduction of chlorin was studied, and the steps for its reduction to a tetrahydrochlorin are given in eq. (15—20).

$$\text{Chlorin} \xrightarrow{h\nu} \text{chlorin}^{*1} \rightarrow \text{chlorin}^{*3} \tag{15}$$

$$\text{Chlorin}^{*3} + QH_2 \rightarrow \text{chlorin}\,H^\cdot + QH^\cdot \tag{16}$$

$$\text{Chlorin}\,H^\cdot + QH^\cdot \rightarrow \text{chlorin} + QH_2 \tag{17}$$

2 Chlorin H$\cdot$ $\rightarrow$ tetrahydrochlorin + chlorin $\hspace{3em}$ (18)

Chlorin H$\cdot$ + P $\rightarrow$ chlorin + PH$\cdot$ $\hspace{3em}$ (19)

Chlorin H$\cdot$ + PH$_2$ $\rightarrow$ Chlorin + PH$_3^\cdot$ $\hspace{3em}$ (20)

The back oxidation of PH$_2$ to P is first order and very rapid in the presence of partly oxidized ascorbate, probably proceeding via eq. (21–22), with (21) being rate limiting.

Q + QH$_2$ $\rightarrow$ 2 QH$\cdot$ $\hspace{3em}$ (21)

QH$\cdot$ + PH$_2$ $\rightarrow$ QH$_2$ + PH$\cdot$ $\hspace{3em}$ (22)

The variation in quantum yields observed by the authors can be explained if steps (23) and (24) are considered as competing quenching reactions for steps (10) and (16).

P* + P $\rightarrow$ 2 P $\hspace{3em}$ (23)

Chlorin* + chlorin $\rightarrow$ 2 chlorin. $\hspace{3em}$ (24)

The conversion of PH$_2$ into chlorin does not occur in the dark, but requires some substance produced only by light. Since excited porphyrins do not react with PH$_2$ directly (except possibly by quenching), the reduced porphyrin radical PH$\cdot$ was implicated. However if this is so, and if the dark oxidation of PH$_2$ proceeds via radicals, it is difficult to explain why no chlorin is formed in the dark. A possible explanation for this discrepancy is proposed: the radical formed by photoreduction of porphyrin differs from and is a stronger reducing agent than the radical formed by oxidation of PH$_2$. If the former radical is designated PH$_{p.r.}$, then instead of eq. (9b):

P* + QH$_2$ $\rightarrow$ PH$_{p.r.}^\cdot$ + QH$\cdot$ $\hspace{3em}$ (25)

and step (13) becomes

PH$_{p.r.}^\cdot$ + PH$_2$ $\rightarrow$ P + PH$_3^\cdot$ $\hspace{3em}$ (26)

while all other reactions remain as outlined. The structures proposed for the intermediates are outlined in Fig. 5. Similar reactions have subsequently been observed by Sidorov for magnesium and cadmium porphyrins[152]. However, the authors propose a different set of steps for the photoreduction sequence, i.e., metallochlorins are formed directly from metalloporphyrins while metal-PH$_2$ is a by-product[153,154].

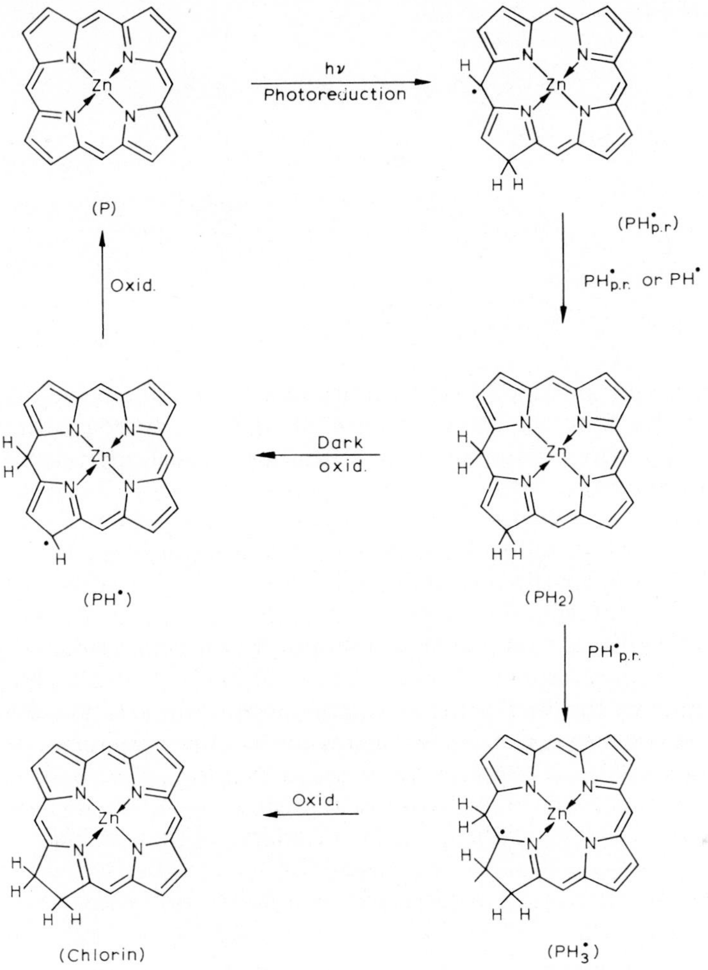

Fig. 5.

Shul'ga et al.[155], have attempted to determine interconnections between photoreductions of free base porphyrins and their metal complexes. Some of their results can be interpreted in terms of a common photoreduction mechanism for free base porphyrins and metalloporphyrins.

Suboch et al.[156,157] report the photoreduction (degassed ascorbic acid—propanol—pyridine) of protochlorophyll and a series of its derivatives. Conditions were found for considerably increasing the yield of chlorin derivatives[158,159]. Suboch et al.[156], have isolated and identified the products of photoreduction of a series of metalloporphyrins. Differences (electronic, i.r. and p.m.r.) were discovered between photochemically prepared chlorins and natural chlorins. These result from the fact that photochemically prepared

References, p. 695

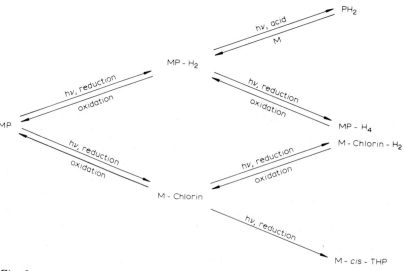

chlorins have a *cis* arrangement of hydrogen atoms on the 7,8-linkage (7), whereas natural chlorins have the *trans* arrangement (6). *Cis* hydrogenation also occurred upon photoreduction of simple, synthetic metalloporphyrins such as zinc etioporphyrin (8).

For explanation of this selective *cis*-photohydrogenation, Suboch proposes direct formation of metallochlorins from metalloporphyrins. At present the best explanation for the diverse experimental results requires dual reaction pathways. Depending upon the reaction conditions, photoreduction follows two parallel paths: (i) reversible metalloporphyrin photoreduction and (ii) metalloporphyrin photoreduction to the corresponding metallochlorin (Fig. 6). According to the first pathway, metalloporphyrin (M-P) photoreductions proceeds via a number of intermediates to the dihydrometallopor-

Fig. 6.

phyrin (M-PH$_2$) which can either oxidize to starting metalloporphyrin or, by subsequent irradiation, be converted via a number of intermediates into tetrahydrometalloporphyrin (M-PH$_4$) or (in some cases) demetalate to dihydroporphyrin (PH$_2$)[160].

A parallel pathway effects direct photoreduction of the metalloporphyrin to the metallochlorin (M-Chlorin). The metallochlorin may react by a different pathway to reversible metallochlorin photoreduction product (M-Chlorin H$_2$), or by direct photoreduction to a metallo-*cis*-tetrahydroporphyrin (M-*cis*-THP).

Numerous investigators have done work which has led to the present understanding of photoreduction. Visible spectral investigation[161], specific e.s.r. studies[162-166] and luminescence studies[167-169] have all had an important role in mechanism elucidation.

Fuhrhop[170] reports a photoreduction unique to Sn(IV)- and Ge(IV)octaethylporphyrins. The irradiation of Sn(IV)P or Ge(IV)P leads first to metallophlorins, which upon further irradiation rapidly and nearly quantitatively rearrange to metallochlorins without the intermediate formation of metalloporphodimethenes (Fig. 7). Evidently the strongly electron-attracting Sn(IV) and Ge(IV) polarize the porphyrin ligand so greatly that even the β-pyrrole carbon atoms acquire a partial positive charge and become easier to reduce.

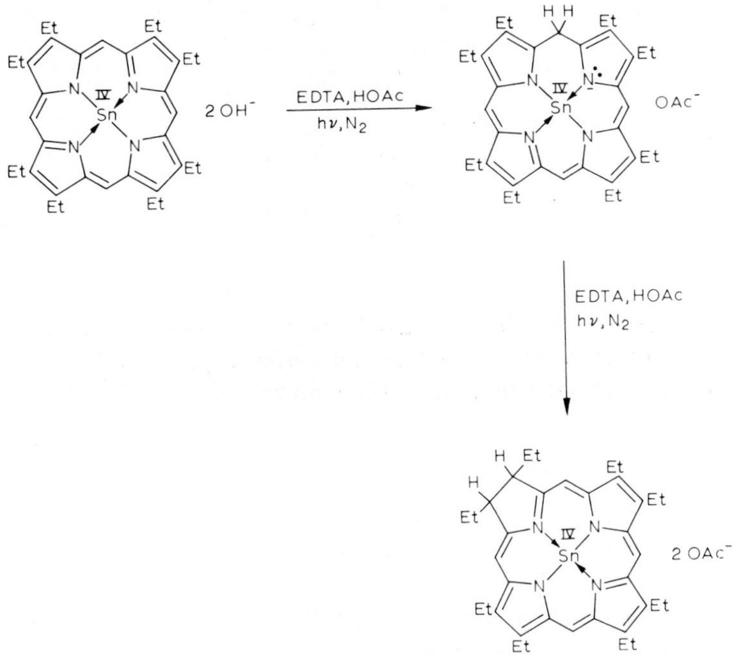

Fig. 7.

This argument is supported by recent X-ray investigations[171] which show a lengthening of the β,β'-pyrrole bond in Sn(IV) porphyrins, compared to other metalloporphyrins.

Similar mono- and dianions[172] of a series of metalloporphyrins and metallophthalocyanines have been reported as intermediates in hydrazine photoreductions[173,174]. Whitten et al.[68,119], have reported the photochemical reduction of Sn(IV) porphyrins to the Sn(IV) chlorin, with further photoreduction leading to the *vic*-tetrahydroporphyrintin(IV) (eq. 27).

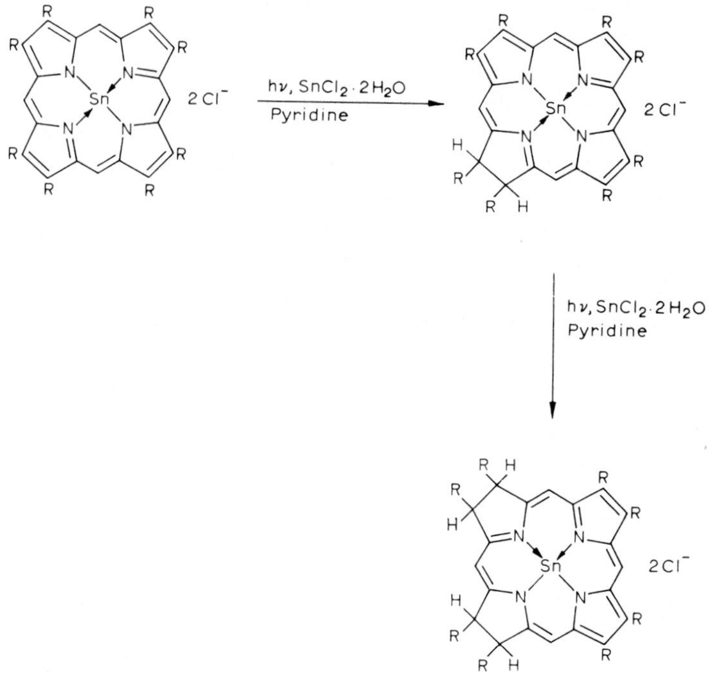

(27)

Under the reaction conditions (SnCl$_2$ · 2H$_2$O/pyridine) electron transfer from the SnCl$_2$ to the porphyrin excited triplet evidently occurs first, followed by protonation of the resulting Sn(IV) porphyrin dianion (9)[175,176].

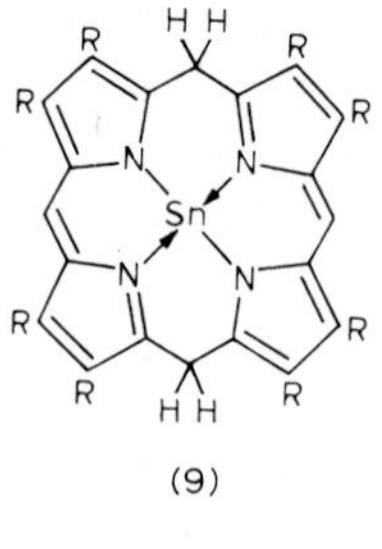

(9)

Preparation of the highly reactive Sn(II) porphyrins is also described by Whitten. Activation of the Sn(II)-porphyrins by heat or light leads to ring-reduced Sn(IV) porphyrin species, evidently via intramolecular electron transfer.

16.3.2.2. Photoreduction of the central metal

Calvin et al.[177,178] report the reversible photoredox or 'breathing' of manganese phthalocyanines in pyridine solutions. These reactions will be considered in the next section of this chapter (photo-oxidation). Whitten[68] has reported the photoreduction of Co(III) porphyrins to the corresponding Co(II) porphyrins in degassed $SnCl_2 \cdot 2H_2O$/pyridine. Whitten[179] has demonstrated that the 1-(1-naphthyl)-2-(4-pyridyl)ethylene complexes of Co(III)etioporphyrin-I are photoreduced in polar solvents (Fig. 8). This photoreduction presumably occurs via the intermediate excited triplet state of $Co(III)(NPE)_2$ Etio-I which undergoes intramolecular electron transfer to form Co(II)(NPE)Etio-I and $NPE^+\cdot$.

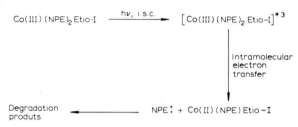

Fig. 8.

16.4. Photo-oxidation

16.4.1. Free-base porphyrins

Mauzerall et al.[180] report the photo-oxidation of the octamethyl esters of uroporphyrinogen (10) to the corresponding uroporphyrin (13) (Fig. 9) in pH < 8.5 solutions containing oxygen. Visible spectral evidence indicated that the oxidation proceeded through the porphomethene (11) and porphodimethene (12) intermediates. White light was used. This reaction exhibited a marked induction period. Control experiments showed that addition of a small percentage of uroporphyrin caused a marked shortening of the induction period. Attempts to determine an action spectrum by visible spectrometry were unsuccessful, and no direct proof of the proposed porphomethene intermediates was obtained. Rigorously degassed solutions of uroporphyrinogen are only slightly oxidized even when exposed to intense white light for long periods of time. Various oxidation inhibitors were shown to retard the photo-oxidation. The rate of photo-oxidation increased with decreasing pH. Yields of porphyrin also increased as the pH was lowered to an optimum in 1 M HCl. Clearly both oxygen and porphyrin products are important in this photoreaction. A plausible explanation may be: (eq. 28—30)

$$\text{Porphyrin} \xrightarrow{h\nu} \text{porphyrin}^{*1} \xrightarrow{i.s.c.} \text{porphyrin}^{*3} \tag{28}$$

$$\text{Porphyrin}^{*3} + O_2{}^1 \rightarrow \text{porphyrin} + O_2{}^{*1} \tag{29}$$

$$O_2{}^{*1} + \text{porphyrinogen} \rightarrow \rightarrow \rightarrow \text{porphyrin} \tag{30}$$

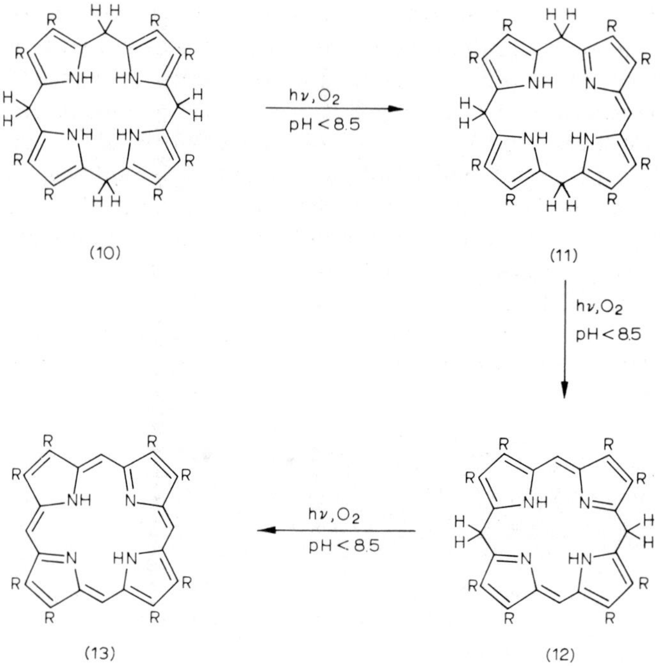

Fig. 9.

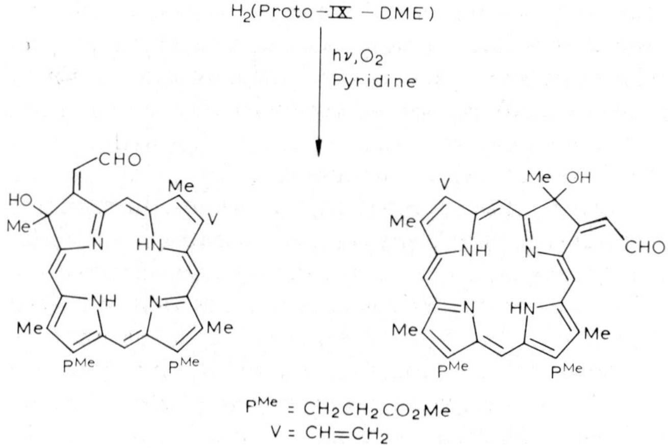

$$P^{Me} = CH_2CH_2CO_2Me$$
$$V = CH=CH_2$$

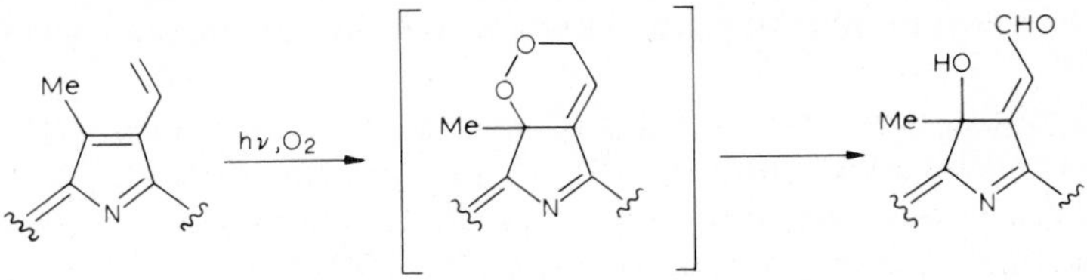

Another photo-oxidation of a hydroporphyrin involves the use of 1,2-naphthoquinone to convert tetrahydro-*meso*-tetraphenylporphyrin into *meso*-tetraphenylchlorin[181].

Recent work[189] on heme-derived porphyrins has shown that unsaturated side-chains are photolabile and can be selectively oxidized by molecular oxygen in organic solvents such as pyridine or dichloromethane. Protoporphyrin-IX dimethyl ester, when irradiated into any of the visible bands or the Soret band, is converted to two products (eq. 31). The reaction occurs via 1,4-addition of oxygen to the vinyl-substituted pyrrolenine rings (eq. 32). With subsequent decomposition of the peroxide. The absolute quantum yields for air oxidation of protoporphyrin were determined in various solvents, e.g., $\phi = 0.033$ in chloroform and 0.006 in benzene[190].

Among the earliest porphyrin photo-oxidation reactions are those reported by Fischer[191]. The reactions were run in very basic media on the disodium salts of porphyrins. The products appear to be the non-metallated analogs of those reported above by Fuhrhop[186,187] (formyl-biliverdins). As a result of the basic conditions employed, other nucleophilic additions occurred immediately after formation of the formyl-biliverdin which complicated structure determination.

16.4.2. Metalloporphyrins

Some of the earliest reports of photo-oxidation involved the conversion of metallochlorins into metalloporphyrins. Calvin et al.[7,182,183] report photo-oxidation of Zn- and Mg-*meso*-tetraphenylchlorins by a series of quinones in benzene solution. Kinetic data point to the intermediacy of the metallochlorin triplet state in the reaction. Quantum yields (~0.001 to 0.008) were measured for several series of quinones. Straight line plots of quantum yield vs oxidation potential of the quinones were obtained. These authors also reported[184] that Zn- and Mg-*meso*-tetraphenylchlorins could be photo-oxidized by molecular oxygen in a manner analogous to that for ortho- and para-quinones. However the H_2O_2 produced underwent a secondary reaction with the porphyrin to yield a product similar to that obtained by bleaching chlorophyll in the presence of oxygen.

Gurinovich et al.[185], reported on the air photo-oxidation of a series of porphyrins and their zinc derivatives in organic solvents. They suggest that irreversible addition of oxygen occurs with destruction of the conjugated bond system. Partial reduction of the products was effected by photochemical reduction with ascorbic acid/pyridine, but little quantitative data were supplied.

A clear case of photo-oxidation leading to disruption of the porphyrin macrocycle is reported by Fuhrhop et al.[186,187]. Magnesium octaethylporphyrin is photo-oxygenated with molecular oxygen to the magnesium formyl-biliverdin according to the following sequence (eq. 33).

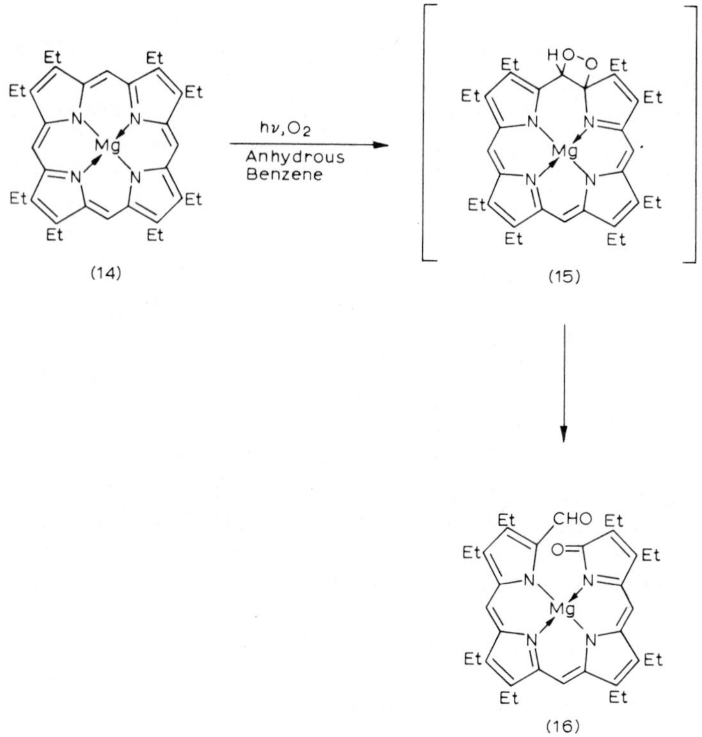

$$(33)$$

Extensive p.m.r., i.r., visible and mass spectral data are given to substantiate the structure of the photo-oxidation product.

Rate inhibition studies utilizing β-carotene at very low concentration (3×10^{-5} M) indicate that the β-carotene is reacting with an intermediate of lifetime <3 μsec. This follows from the maximum rate constant for a diffusion limited reaction in benzene solution of $\sim 10^{10}$ M^{-1} sec^{-1}, and the failure to observe molecular complexes at these concentrations. However, all triplets studied thus far are quenched by oxygen at rates close to the diffusion controlled limit. Considering the solubility of oxygen in benzene, this indicates that the lifetime of the triplet state of Mg(OEP) cannot be more than 0.03 μsec. This clearly suggests that photo-oxygenation occurs via the excited singlet state of oxygen, probably formed on quenching the triplet state of Mg(OEP) according to the following scheme (eq. 34—36):

$$Mg(OEP) \xrightarrow{h\nu} [Mg(OEP)]^{*1} \to [Mg(OEP)]^{*3} \tag{34}$$

$$[Mg(OEP)]^{*3} + (O_2)^3 \to Mg(OEP) + (O_2)^{*1} \tag{35}$$

$$(O_2)^{*3} + Mg(OEP) \to \text{oxygen adduct} \tag{36}$$

Other metalloporphyrins do not photo-oxygenate in this manner. The authors suggest that Mg(OEP) uniquely reacts in this way because it has a sufficiently low oxidation potential. By utilizing the fact that metallochlorins have a first oxidation potential ~300 mV lower than the corresponding metalloporphyrin, Fuhrhop[187] successfully photo-oxygenated zinc octaethylchlorin to give products analogous to those obtained from Mg(OEP). Now however two isomers are observed, as indicated below,

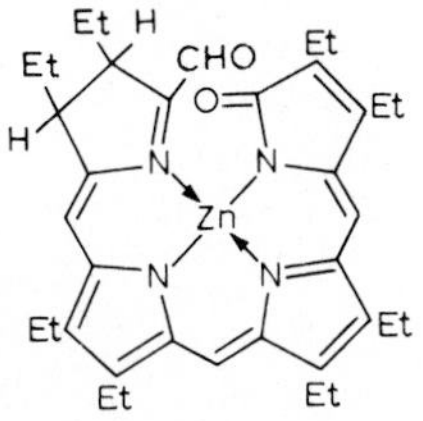

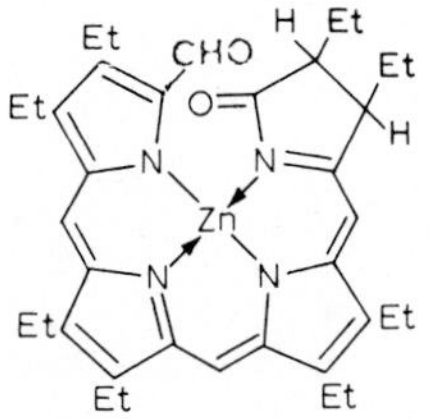

1'-desoxo-1'-formyl-1,2-dihydrooctaethylbiliverdin zinc complex (17) and 1'-desoxo-1'-formyl-7,8-dihydrobiliverdin zinc complex (18).

Yet a different degradation sequence occurs[188] when a bridge carbon is substituted as in octaethyl-α-hydroxyporphyrinatozinc(II). The reaction (Fig. 10) is formally an oxidative decarbonylation which transforms the compound into a biliverdin residue (25). The corresponding nickel complex undergoes a similar initial reaction to form the nickel analog of (21); however, further irradiation leads to the cyclic ether (26). Further irradiation of (21) in the absence of oxygen also leads to the zinc-containing analog of (26).

A final type of photo-oxidation reaction involves change in oxidation state of the central metal. Although such metal redox reactions are not published for metalloporphyrins, an example is provided by the related manganous phthalocyanines (Fig. 11). The complex (27) is assumed to be the species present when manganous phthalocyanine is dissolved in oxygen-free pyridine; it cannot be isolated. When air is introduced, oxidation to the trivalent species (28) occurs. This complex also could not be isolated. The complex is photosensitive, intense white light causing reduction to (27), whereas in the dark disproportionation to (27) and (29) occurs. In the presence of oxygen (28) slowly oxidizes to (29) in the dark, a process which is accelerated by light. In the absence of oxygen, (29) is reduced to (27)

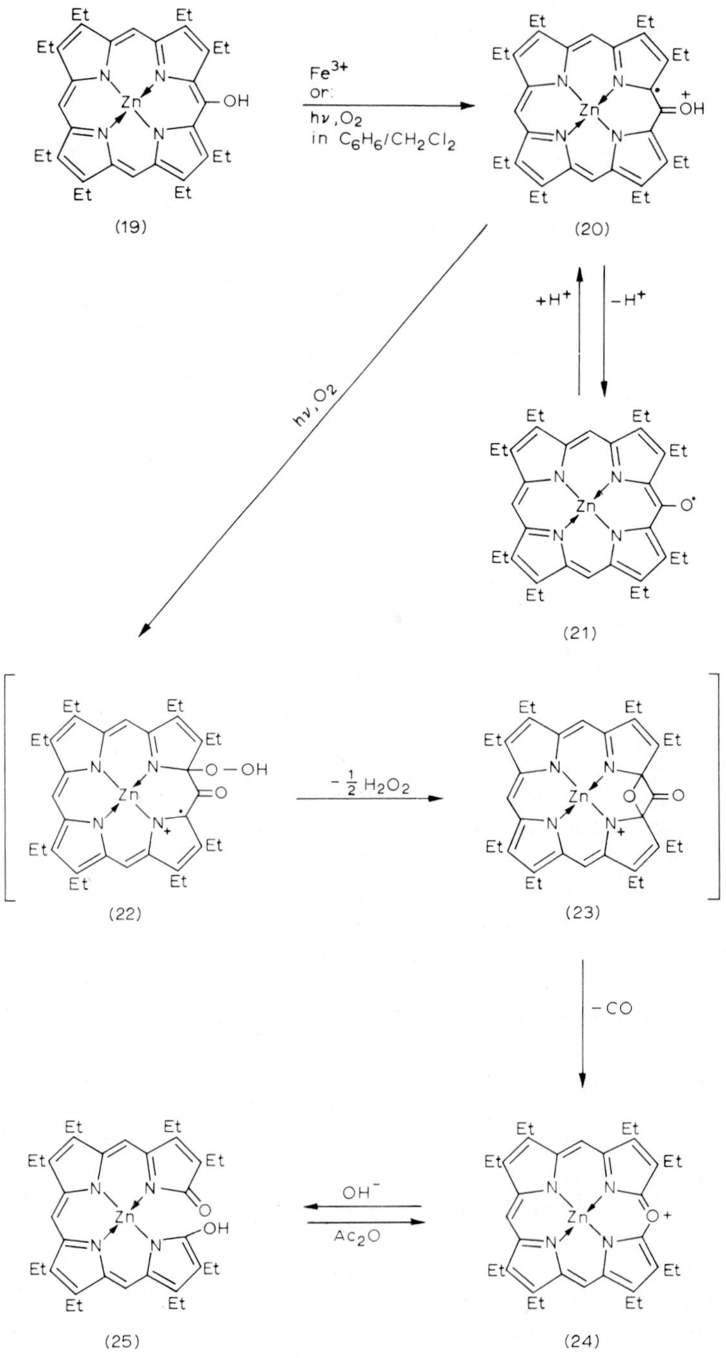

Fig. 10.

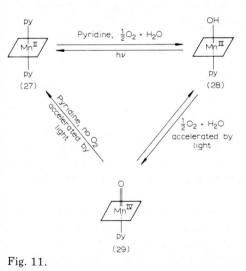

(26)

apparently without the intermediate formation of (28). This process is rapid in sunlight and slow in the dark.

Although many authors have speculated about the mechanistic details of porphyrin photo-oxidation reactions, little definitive evidence is available. Much of the mechanistic work done in this area has been done with chlorophyll derivatives rather than simple porphyrins. Changes in ground state u.v.-visible spectra have been utilized to obtain information about relatively long-lived intermediate species[193–195].

Flash spectroscopic techniques have been used to investigate reversible photoreactions of chlorophyll at low temperatures[196].

A promising development has been the coupling of flash photolysis and e.s.r. techniques. With this technique, light-induced changes in spin properties can be studied. By use of this technique it has been shown that the light-induced radical in *Rhodospirillum rubrum* arises from a radical of bacteriochlorophyll[197]. Other applications of this technique have been in the study of chlorophyll-*a* photo-oxidation by molecular oxygen[198] and the

Fig. 11.

References, p. 695

reversible photo-oxidation observed in a chlorophyll-benzoquinone system[199].

Recently chemically induced dynamic electron polarization (CIDEP)[200] has been used to study the light-induced reaction between bacteriochlorophyll (BChl) and benzoquinone (BQ) at low temperatures. The results of these experiments provide good evidence that an initial, direct light-induced electron transfer occurred between bacteriochlorophyll and benzoquinone, leading to formation of $BChl^{+\cdot}$ and $BQ^{-\cdot}$. Other e.s.r. work by the same authors[201], utilizing continuous illumination of chlorophyll or bacteriochlorophyll in degassed ethanol, suggests that photo-oxidation occurs via the lowest excited singlet states to produce cation radicals. Chibisov[202] has studied the effect of water on the yield of chlorophyll cation radical during photo-oxidation. Numerous other papers deal with photo-induced effects on chlorophyll as well as the effects of. light on various in vivo systems related to photosynthesis. However they are of limited value in further understanding the mechanisms of simple photo-oxidation. Chlorophyll systems will be more fully covered in another chapter of this text[203].

16.5. Ligand photoejection

At present very few examples of ligand photoejections which lead to permanent chemical alteration of metalloporphyrin structure are known. It is anticipated that as more metal—carbonyl—porphyrin complexes are prepared, they will prove to be a rich source of interesting photochemical reactions. Photochemical ejection of CO from Ru(II)(CO)(py)TPP to form Ru(II)(py)$_2$TPP has been reported by Chow and Cohen[69]. Whitten et al.[37] have studied the photoejection of CO (with subsequent ligand substitution) from a number of ruthenium (II) carbonyl porphyrins (eq. 37).

$$Ru(II)(CO)(L)P \xrightarrow[\text{solution of donor ligand, L}]{h\nu, \text{ degassed}} Ru(II)(L)_2 P + CO \qquad (37)$$

The substituted ligands included pyridine, dimethylsulfoxide, tetrahydrofuran and several aliphatic and aromatic amines[204]. The porphyrin moieties studied included (OEP), (Etio-I), and (Meso-IX-dioctadecyl ester)[38]. Similarly, photoejection of CO from degassed solutions of Os(II)(CO)(py)OEP to form Os(II)(py)$_2$OEP has been observed[205].

Preliminary results[206] indicate that degassed solutions of ruthenium nitrosyl porphyrins (formulated as Ru(II)(NO)$_2$P by Tsutsui[207]) undergo a slow NO displacement in the dark. For solutions of Ru(II)(NO)$_2$OEP this process is accelerated by irradiation into either the visible or Soret bands.

References

1. M. Gouterman, in 'Excited States of Matter', Graduate Studies Texas Tech. U., No. 2, Ed. C.W. Shoppee, Texas Tech. Univ. Press. Lubbock, Texas (1973), p. 63.
2. C. Weiss, H. Kobayashi, and M. Gouterman, J. Mol. Spectrosc., **16**, 415 (1965) and references therein.
3. A.H. Corwin, Ann. N.Y. Acad. Sci., **206**, 201 (1973).
4. S.R. Platt, in "Radiation Biology", Ed. A. Hollander, Vol. III, McGraw-Hill, New York, (1956), Chap. 2.
5. A.H. Corwin, A.B. Chivvis, R.W. Poor, D.G. Whitten, and E.W. Baker, J. Am. Chem. Soc., **90**, 6577 (1968).
6. L.V. Iogansen, Dokl. Akad. Nauk SSSR, **205**, 390 (1972).
7. See for example, M. Calvin and G.D. Dorough, J. Am. Chem. Soc., **70**, 699 (1948).
8. R.S. Becker and M. Kasha, J. Am. Chem. Soc., **77**, 3669 (1955).
9. J.B. Allison and R.S. Becker, J. Chem. Phys., **32**, 1410 (1960).
10. R.S. Becker and J.B. Allison, J. Phys. Chem., **67**, 2662 (1963).
11. R.S. Becker and J.B. Allison, J. Phys. Chem., **67**, 2669 (1963).
12. J.B. Allison and R.S. Becker, J. Phys. Chem., **67**, 2675 (1963).
13. D. Eastwood and M. Gouterman, J. Mol. Spectrosc., **35**, 359 (1970).
14. G.D. Dorough, J.R. Miller, and F.M. Huennekens, J. Am. Chem. Soc., **73**, 4315 (1951).
15. J.B. Callis, M. Gouterman, Y.M. Jones, and B.H. Henderson, J. Mol. Spectrosc., **39**, 410 (1971).
16. P.G. Seybold and M. Gouterman, J. Mol. Spectrosc., **31**, 1 (1969).
17. L.K. Hanson, M. Gouterman, and J.C. Hanson, J. Am. Chem. Soc., **95**, 4822 (1973).
18. M. Gouterman, F.P. Schwarz, P.D. Smith, and D. Dolphin, J. Chem. Phys., **59**, 676 (1973).
19. M.P. Tsvirko, K.N. Solov'ev, and V.V. Sapunov, Opt. Spektrosk, **36**, 335 (1974).
20. K.N. Solov'ev, Vestsi Akad. Navuk B. SSR, Ser. Fiz. -Tekh. Navuk., **3**, 27 (1962). see Chem. Abstr., **58**, 3013e (1963).
21. D. Djuric, Ark. Farm. (Belgrade), **11**, 1 (1961).
22. M.P. Tsvirko and V.V. Sapunov, Opt. Spektrosk., **34**, 1094 (1973).
23. For a review of spectroscopy of porphyrins see G.P. Gurinovich, A.N. Sevchenko and K.N. Solov'ev, "Spectroscopy of Chlorophyll and Related Compounds", Science Press, Minsk, 1968, English Translation issued 1971: Nat'l. Technical Information Service, U.S. Dept. of Commerce, Springfield, Va. 22151.
24. S.B. Broyde and S.S. Brody, J. Chem. Phys., **46**, 3334 (1967).
25. D.G. Whitten, I.G. Lopp and P.D. Wildes, J. Am. Chem. Soc., **90**, 7196 (1968).
26. D. Mauzerall, Biochemistry, **4**, 1801 (1965).
27. L. Bajema, M. Gouterman and C.B. Rose, J. Mol. Spectrosc., **39**, 421 (1971).
28. G.I. Kobyshev and A.N. Terenin, "Proc. Int. Conf. Luminescence 1966", Ed. G. Szigeti, Akad. Kiado., Budapest, Hungary (1968).
29. K.E. Rieckhoff, E.R. Menzel and E.M. Voight, Phys. Rev. Lett., **28**, 261 (1972).
30. E.R. Menzel, K.E. Rieckhoff and E.M. Voight, Chem. Phys. Lett., **13**, 604 (1972).
31. A.T. Gradyushko, V.A. Mashnekov and K.N. Solov'ev, Biofizika, **14**, 827 (1969).
32. A.N. Sevchenko, K.N. Solov'ev, A.T. Gradyushko and S.F. Shkirman, Dokl. Sov. Phys., **11**, 587 (1967).
33. S.F. Shkirman and K.N. Solov'ev, Izv. Akad. Nauk SSSR, Ser. Fiz., **29**, 1378 (1965).
34. K.N. Solov'ev, N.M. Ksenofontova, S.F. Shkirman and J.F. Kachura, Spectrosc. Lett., **6**, 455 (1973).
35. G.W. Canters, J. Van Egmond, T.J. Schaafsma and J.H. Van der Waals, Mol. Phys., **24**, 1203 (1972).

36. J. Saltiel, H.C. Curtis, L. Metts, J.W. Miley, J. Winterle, and M. Wrighton, J. Am. Chem. Soc., **92**, 410 (1970).
37. G.W. Sovocool, F.R. Hopf, and D.G. Whitten, J. Am. Chem. Soc., **94**, 4350 (1972).
38. F.R. Hopf and D.G. Whitte, unpublished results.
39. D. Eastwood and M. Gouterman, J. Mol. Spectrosc., **23**, 210 (1967).
40. D. Eastwood and M. Gouterman, J. Mol. Spectrosc., **25**, 547 (1968).
41. D.T. Holmes and R. Livingstone, Photochem. Photobiol., **4**, 629 (1965).
42. G.D. Egorova, V.A. Mashenkov, K.N. Solov'ev, and N.A. Yusnkevich, Zh. Prikl. Spektrosk., **19**, 838 (1973).
43. G.D. Egorova, V.A. Mashenkov, and K.N. Solov'ev, Biofizika, **18**, 40 (1973).
44. A.T. Gradyushko, V.A. Mashenkov, A.N. Sevchenko, K.N. Solov'ev, and M.P. Tsvirko, Dokl. Akad. Nauk SSSR, **182**, 64 (1968).
45. A.T. Gradyushko, A.N. Sevchenko, K.N. Solov'ev, and M.P. Tsvirko, Izv. Akad. Nauk SSSR, Ser. Fiz., **34**, 636 (1970).
46. A.T. Gradyushko, V.A. Mashenkov, K.N. Solov'ev, and M.P. Tsvirko, Zh. Prikl. Spektrosk., **9**, 514 (1968).
47. B. Dzhagarov and G.P. Gurinovich, Opt. Spektrosk., **30**, 425 (1971).
48. G.P. Gurinovich and B.M. Dzhagarov, Izv. Akad. Nauk SSSR, Ser. Fiz., **37**, 383 (1973).
49. K.N. Solov'ev, A.T. Gradyushko, and M.P. Tsvirko, Izv. Akad. Nauk SSSR, Ser. Fiz., **36**, 1107 (1972).
50. B.M. Dzhagarov, Izv. Akad. Nauk SSSR, Ser. Fiz., **36**, 1093 (1972).
51. H. Linschitz and K. Sarkanen, J. Am. Chem. Soc., **80**, 4826 (1958).
52. R. Livingstone and V.A. Ryan, J. Am. Chem. Soc., **75**, 2176 (1953).
53. G.R. Seely, in "The Chlorophylls", Eds. L.P. Vernon and G.R. Seely, Academic Press, New York (1966), p. 523.
54. -P.A. Shakhverdov, Elem. Fotopro Sessy Mol., Akad. Nauk SSSR, 283 (1966).
55. A.K. Chibisov, Photochem. Photobiol., **10**, 331 (1969) and references therein.
56. P.A. Shakhverdov and A.N. Terenin, Dokl. Akad. Nauk SSSR, **150**, 1311 (1963).
57. G.P. Hurinovich, A.N. Sevchenko and K.N. Solov'ev, UFN, **79**, 196 (1963).
58. L. Pekkarinen and H. Linschitz, J. Am. Chem. Soc., **82**, 2407 (1960).
59. P.J. McCartin, Trans. Faraday Soc., **60**, 1694 (1964).
60. R. Livingstone and P.J. McCartin, J. Phys. Chem., **67**, 2511 (1963).
61. H. Linschitz, C. Steel, and J.A. Bell, J. Phys. Chem., **66**, 2574 (1962).
62. J.S. Connolly, D.S. Gorman, and G.R. Seely, Ann. N.Y. Acad. Sci., **206**, 649 (1973).
63. A.T. Gradyushko and M.P. Tsvirko, Opt. Spektrosk., **31**, 291 (1971).
64. R.P. Burgner and A.M. Ponte Goncalves, J. Chem. Phys., **60**, 2942 (1974).
65. R.P. Burgner and A.M. Ponte Goncalves, Abstracts, Int. Conf. Photochem., Nashville, Tenn. (1974).
66. C.B. Storm and Y. Teklu, J. Am. Chem. Soc., **94**, 1745 (1972).
67. C.B. Storm, Y. Teklu, and E.A. Sokoloski, Ann. N.Y. Acad. Sci., **206**, 631 (1973); R.J. Abraham, G.E. Hawkes, and K.M. Smith, Tetrahedron Lett., 1483 (1974).
68. D.G. Whitten, J.C. Yau, and F.A. Carroll, J. Am. Chem. Soc., **93**, 2291 (1971).
69. B.C. Chow and I.A. Cohen, Bioinorg. Chem., **1**, 57 (1971).
70. For reviews of photochemical kinetics and their use in the study of mechanisms of photochemical reactions see refs. 69 and 70.
71. A.A. Lamola and N.J. Turro, in 'Energy Transfer and Organic Photochemistry' Tech. Org. Chem. Ser., Vol. XIV, Eds. P.A. Leermakers and A. Weissberger, Wiley-Interscience, New York (1969).
72. N.J. Turro, "Molecular Photochemistry", W.A. Benjamin, Inc., New York (1967).
73. G.P. Gurinovich, A.I. Patsko, and A.N. Sevchenko, Dokl. Phys. Chem., **174**, 402 (1967).

74. B. Dzhagarov, Opt. Spektrosk., **28**, 66 (1970).
75. P.D. Wildes, J.G. Pacifici, G. Irick, and D.G. Whitten, J. Am. Chem. Soc., **93**, 2004 (1971).
74. G.M. Wyman, B.M. Zarnegar, and D.G. Whitten, J. Phys. Chem., **77**, 2584 (1973).
75. E. Fujimori and R. Livingstone, Nature London, **180**, 1036 (1957).
76. R. Raman and G. Tollin, Photochem. Photobiol., **13**, 15 (1971).
77. A.K. Chibisov, A.V. Karayakin, and M.E. Zubrilina, Dokl. Akad. Nauk SSSR, **177**, 226 (1967).
78. F.P. Schwarz, M. Gouterman, Z. Muljiani, and D. Dolphin, Bioinorg. Chem., **2**, 1 (1972).
79. P.D. Wildes and D.G. Whitten, J. Am. Chem. Soc., **92**, 7609 (1970).
80. D.G. Whitten, P.D. Wildes, and C.A. DeRosier, J. Am. Chem. Soc., **94**, 7811 (1972).
81. J.J. Leonard, T. Yonetani, and J.B. Callis, Biochemistry, **13**, 1460 (1974).
82. K. Gollnick and G.O. Schenck, Pure Appl. Chem., **9**, 507 (1964).
83. G.O. Schenk, Angew. Chem., **69**, 579 (1957).
84. A. Nickon and W.L. Mendelson, J. Am. Chem. Soc., **87**, 3921 (1965).
85. C.S. Foote, Acc. Chem. Res., **1**, 104 (1968) and references therein.
86. C.S. Foote, Y.C. Chang, and R.W. Denny, J. Am. Chem. Soc., **92**, 5216 (1970).
87. C.S. Foote, Y.C. Chang, and R.W. Denny, J. Am. Chem. Soc., **92**, 5218 (1970).
88. L. Sibel'dina, Z.P. Gribova, and L.P. Kayushin, Biofizika, **15**, 816 (1970).
89. A.P. Bobrovskii and V.E. Kholmogorov, Khim. Vys. Energ., **6**, 125 (1972).
90. L.A. Sibel'dina, Z.P. Gribova, L.P. Kayushin, and B.S. Marinov, Dokl. Akad. Nauk SSSR, **181**, 482 (1968).
91. H. Langhof, H. Müller, and L. Rietschel, Arch. Klin. Exp. Dermatol., **212**, 506 (1961); I.A. Magnus, A. Jarrett, T.A. Prankerd, and C. Rimington, Lancet, 1961-II, 448 (1961).
92. A.A. Lamola, T. Yamane, and A.M. Trozzolo, Science, **179**, 1131 (1973).
93. F.H. Doleiden, S.R. Fahrenholtz, A.A. Lamola, and A.M. Trozzolo, Photochem. Photobiol. (1974), in press; S.R. Fahrenholtz, F.H. Doleiden, A.M. Trozzolo, and A.A. Lamola, Photochem. Photobiol. (1974), in press.
94. A.M. Trozzolo, F.H. Doleiden, S.R. Fahrenholtz, A.A. Lamola, A.M. Mattucci, and T. Yamane, Abstracts, 2nd Ann. Mtng., Amer. Soc. Photobiol. (1974), p. 95.
95. A.A. Schothorst, J. Van Steveninck, L.N. Went, and D. Suurmond, Clin. Chim. Acta, **28**, 41 (1970).
96. G. Jori, G. Galiazzo, and E. Scoffone, Experientia, **27**, 379 (1971); Biochemistry, **8**, 2868 (1964).
97. G. Jori, G. Gennari, G. Galiazzo, and E. Scoffone, FEBS Lett. **6**, 267 (1970).
98. G. Galiazzo, A.M. Tamburro, and G. Jori, Eur. J. Biochem., **12**, 362 (1970).
99. G. Jori, G. Galiazzo, A.M. Tamburo, and E. Scoffone, J. Biol. Chem., **245**, 3375 (1970).
100. L.C. Harber, J. Hsu, H. Hsu, and B.D. Goldstein, J. Invest. Dermatol., **58**, 373 (1972).
101. A.A. Schothorst, J. Van Steveninck, L.N. Went, and D. Suurmond, Clin. Chim. Acta, **39**, 161 (1972).
102. J.S. Beilin and G. Oster, Proc. Intern. Congr. Photobiol., 3rd, Copenhagen 1960, 254 (1961).
103. M.R. Mauk and A.W. Girotti, Biochemistry, **13**, 1757 (1974).
104. G. Jori, G. Gennari, M. Folin, and G. Galiazzo, Biochim. Biophys. Acta, **229**, 525 (1971); M. Folin, A. Azzi, A.M. Tamburo, and G. Jori, ibid., **285**, 337 (1972).
105. J.D. Ostrow, Prog. Liver Dis., **4**, 447 (1972).
106. J.D. Ostrow, Semin. Hematol., **9**, 113 (1972).
107. R. Bonnett and J.C.M. Stewart, Biochem. J., **130**, 985 (1972).

108. J.J. Lee, B.C. Mathewson, J.E. Wamplen, R.D. Etheridge, and N.U. Curry, Fed. Proc. Biochemistry II, Abstr., 2522 (1972).
109. H. Linschitz and J. Rennert, Nature London, **169**, 193 (1952).
110. I.N. Chernyuk and I.I. Dilung, Dokl. Akad. Nauk SSSR, **165**, 1350 (1965).
111. G.R. Seely, J. Phys. Chem., **73**, 117 (1969).
112. J.R. Harbour and G. Tollin, Photochem. Photobiol., **19**, 163 (1974).
113. D. Djuric, Arh. Farm. (Belgrade), **12**, 263 (1962); **12**, 19 (1962).
114. L.A. Sibel'dina, Eur. Biophys. Congr., Proc., 1st 1971, **4**, 249 (1971).
115. K.P. Quinlaw and E. Fujimori, J. Phys. Chem., **71**, 4154 (1967).
116. G. Tollin and G. Green, Biochim. Biophys. Acta, **60**, 524 (1962); G. Tollin, K.K. Chattergee, and G. Green, Photochem. Photobiol., **4**, 592 (1965).
117. G.R. Seely, J. Phys. Chem., **69**, 2779 (1965).
118. A.K. Bannerjee and G. Tollin, Photochem. Photobiol., **5**, 315 (1966).
119. D.G. Whitten and J.C.N. Yau, Tetrahedron Lett., 3077 (1969).
120. J.K. Roy and D.G. Whitten, J. Am. Chem. Soc., **94**, 7162 (1972).
121. R. Livingstone, L. Thompson, and M.V. Ramarao, J. Am. Chem. Soc., **74**, 1073 (1952).
122. R. Livingstone and C.L. Ke, J. Am. Chem. Soc., **72**, 909 (1950).
123. R. Livingstone, Quart. Rev. (London), **14**, 174 (1960).
124. S.L. Bondarev, G.P. Burinovich, and V.S. Chernikov, Izv. Akad. Nauk SSSR, Ser. Fiz., **34**, 641 (1970).
125. I.G. Lopp, R.W. Hendren, P.D. Wildes, and D.G. Whitten, J. Am. Chem. Soc., **92**, 6440 (1970).
126. J.K. Roy and D.G. Whitten, J. Am. Chem. Soc., **93**, 7093 (1971).
127. J.K. Roy, F.A. Carroll, and D.G. Whitten, J. Am. Chem. Soc., **96**, in press (1974).
128. D.G. Whitten, J.K. Roy, and F.A. Carroll, Proc. Intern. Exciplex Conf., 1974, in press.
129. R.B. Woodward, Ind. Chim. Belge., 1293 (1962).
130. J.-H. Fuhrhop, K. Kadish, and D.G. Davis, J. Am. Chem. Soc., **95**, 5140 (1973).
131. J.-H. Fuhrhop, Struct. and Bonding (Berlin), **18**, 1 (1974).
132. A.A. Krasnovskii and K.K. Voinovskaya, Dokl. Akad. Nauk SSSR, **96**, 1209 (1954).
133. A.A. Krasnovskii and E.V. Pakshina, Dokl. Akad. Nauk SSSR, **120**, 581 (1958).
134. A.A. Krasnovskii and A.V. Umrikhima, Dokl. Akad. Nauk SSSR, **122**, 1061 (1958).
135. (a) A.A. Krasnovskii, J. Chim. Phys., **55**, 968 (1958);
     (b) A.A. Krasnovskii, Ann. Rev. Plant Physiol., **11**, 363 (1960).
136. A.N. Sidorov and D.A. Savel'ev, Biofizika, **13**, 933 (1968).
137. A.M. Shul'ga and G.P. Gurinovich, Biofizika, **13**, 42 (1968).
138. L.V. Slopolyanskaya, I.M. Byteve, and G.P. Gurinovich, Dokl. Akad. Nauk. B. SSR, **16**, 1048 (1972).
139. G.P. Gurinovich, A.I. Patsko, A.M. Shul'ga, and A.N. Sevchenko, Dokl. Akad. Nauk SSSR, **156**, 125 (1964).
140. D. Mauzerall, J. Am. Chem. Soc., **82**, 1832 (1960).
141. D. Mauzerall, J. Am. Chem. Soc., **84**, 2437 (1962).
142. D. Mauzerall, J. Am. Chem. Soc., **82**, 2601 (1960).
143. D. Mauzerall and G. Feher, Biochim. Biophys. Acta, **88**, 658 (1964).
144. V.P. Subach and G.P. Gurinovich, Dokl. Akad. Nauk B. SSR, **15**, 365 (1971).
145. A.N. Sidorov and A.N. Terenin, Dokl. Akad. Nauk SSSR, **145**, 1092 (1962).
146. A.N. Sidorov, V.G. Vorob'ev, and A.N. Terenin, Dokl. Akad. Nauk SSSR, **152**, 919 (1963).
147. A.N. Sidorov, Dokl. Akad. Nauk SSSR, **161**, 128 (1965).
148. A.V. Umrekhina, G.A. Yusupova, and A.A. Krasnovskii, Dokl. Akad. Nauk SSSR, **175**, 1400 (1967).

149. G.T. Rikhireva, A.V. Umrikhina, L.P. Kayushin, and A.A. Krasnovskii, Dokl. Akad. Nauk SSSR, **163**, 491 (1965).
150. G.R. Seely and M. Calvin, J. Chem. Phys., **23**, 1068 (1955).
151. G.R. Seely and K. Talmadge, Photochem. Photobiol., **3**, 195 (1964).
152. A.N. Sidorov, in 'Elementary Photoprocesses in Molecules', Ed. B.S. Neporent, Consultants Bureau, Plenum Press, New York (1968), p. 201.
153. A.N. Sidorov, Dokl. Akad. Nauk SSSR, **158**, 937 (1964).
154. D.A. Soveljev, A.N. Sidorov, R.P. Eustigneeva, and G.V. Ponomarjov, Dokl. Akad. Nauk SSSR, **167**, 135 (1966).
155. A.M. Shul'ga, G.P. Gurinovich, and A.N. Sevchenka, Dokl. Akad. Nauk SSSR, **169**, 1206 (1966).
156. W.P. Suboch, A.P. Losev, and G.P. Gurinovich, Photochem. Photobiol., **20**, 183 (1974).
157. G.P. Gurinovich, A.P. Losev, and V.P. Suboch, Proc. Int. Congr. Photosynth. Res., 2nd, (published 1972), **1**, 299 (1971).
158. V.P. Suboch, A.P. Losev, G.P. Gurinovich, and A.N. Sevchenko, Dokl. Akad. Nauk SSSR, **194**, 721 (1970).
159. A.A. Krasnovskii, M.I. Bystrova, and F. Lang, Dokl. Akad. Nauk SSSR, **194**, 1441 (1970).
160. T.T. Bonnister, Plant. Physiol., **34**, 246 (1959).
161. A.N. Sidrov, Dokl. Akad. Nauk SSSR, **158**, 973 (1964).
162. V.E. Kholmogorov, Biofizika, **16**, 378 (1971).
163. Z.B. Gribova, V.A. Umrikhina, and L.P. Kayushin, Biofizika, **11**, 353 (1966).
164. S.B. Gribova, Abh. Dtsch. Akad. Wiss. Berlin, Kl. Med., 331 (1965).
165. A.V. Umrikhina, N.V. Bublichenko, and A.A. Krasnovskii, Biofizika, **18**, 565 (1973).
166. Z.P. Gribova, Ultrafiolet Izluck., No. 4, 41 (1966).
167. G.P. Gurinovich, M.V. Poteeva, and A.M. Shul'ga, Izv. Akad. Nauk SSSR, Ser. Fiz., **27**, 777 (1963).
168. G.P. Gurinovich, I.F. Gurinovich, and A.M. Shul'ga, Dokl. Akad. Nauk B. SSR, **8**, 292 (1964).
169. G.P. Gurinovich and G.N. Sinyakov, Biofizika, **10**, 946 (1965).
170. J.-H. Fuhrhop and T. Lumbantobing, Tetrahedron Lett., 2815 (1970).
171. J.L. Hoard, Science, **174**, 1295 (1971).
172. G.L. Closs and L.E. Closs, J. Am. Chem. Soc., **85**, 818 (1963).
173. V.G. Moslov and A.N. Sidorov, Teor. Eksp. Khim., **7**, 832 (1971).
174. R.P. Evstigneeva, V.G. Moslov, A.F. Mironov, and A.N. Sidorov, Biofizika, 999 (1971).
175. H.H. Inhoffen, Pure Appl. Chem., **17**, 443 (1968).
176. J.W. Buchler and L. Puppe, Ann. Chem., **740**, 142 (1970).
177. M. Calvin, P.A. Loach, and A. Yamamoto, in: 'Theory and Structure of Complex Compounds', Ed. B. Jezorskar-Tryebistawska, Macmillan (1964) p. 13.
178. G. Englesma, A. Yamamoto, E. Markham, and M. Calvin, J. Phys. Chem., **66**, 2517 (1962).
179. D.G. Whitten and C.A. DeRosier, unpublished data.
180. D. Mauzerall and S. Granick, J. Biol. Chem., **232**, 1141 (1958).
181. G.D. Dorough and J.R. Miller, J. Am. Chem. Soc., **74**, 6106 (1952).
182. G.D. Dorough and M. Calvin, Science, **105**, 433 (1947).
183. F.M. Huennekens and M. Calvin, J. Am. Chem. Soc., **71**, 4024 (1949).
184. F.M. Huennekens and M. Calvin, J. Am. Chem. Soc., **71**, 4031 (1949).
185. I.F. Gurinovich, G.P. Gurinovich, and A.N. Sevchenko, Dokl. Akad. Nauk SSSR, **164**, 201 (1965).

186. J.-H. Fuhrhop and D. Mauzerall, Photochem. Photobiol., **13**, 453 (1971).
187. P.K.-W. Wasser and J.-H. Fuhrhop, Ann. N.Y. Acad. Sci., **206**, 533 (1973).
188. (a) S. Besecke and J.-H. Fuhrhop, Angew. Chem., **86**, 125 (1974); (b) S. Besecke and J.-H. Fuhrhop, Angew. Chem., Int. Ed., Engl., **13**, 150 (1974).
189. H.H. Inhoffen, H. Brockman and K.-M. Bliesnerv, Ann. Chem., **730**, 173 (1969).
190. I.F. Gurinovich, I.M. Byteva, V.S. Chernikov, and O.M. Petsol'd, Zh. Org. Khim., **8**, 842 (1972).
191. (a) H. Fischer and M. Dürr, Ann. Chem., **501**, 112 (1933); (b) H. Fischer and K. Herrle, Hoppe-Seyler's Z. Physiol. Chem., 251 85 (1938).
192. J.A. Elvidge and A.B.P. Lever, Proc. Chem. Soc., 195 (1959).
193. K.P. Quinlan, Biochim. Biophys. Acta, **267**, 493 (1972).
194. I.F. Gurinovich, G.P. Gurinovich, and A.N. Sevchenko, Dokl. Akad. Nauk SSSR, **164**, 201 (1965).
195. Axel Madsen, Proc. Int. Congr. Photobiol., 3rd, Copenhagen, 1960, p. 567 (published 1961).
196. I. Dilung and I. Chernyuk, Abh. Dtsch. Akad. Wiss. Berlin, Kl. Med., 325 (1966).
197. J.D. McElroy, G. Faher, and D. Mauzerall, Biochim. Biophys. Acta, **267**, 363 (1972).
198. V.B. Eustigneev, N.A. Sadovnikova, A.P. Kostikov, and L.P. Koyushin, Dokl. Akad. Nauk SSSR, **203**, 1343 (1972).
199. B.J. Hales and J.R. Bolton, J. Am. Chem. Soc., **94**, 3314 (1972).
200. S.K. Wong and J.K.S. Wan, J. Am. Chem. Soc., **94**, 7197 (1972); S.K. Wong, D.A. Hutchinson, and J.K.S. Wan, J. Am. Chem. Soc., **95**, 622 (1973).
201. J.R. Harbour and G. Tollin, Photochem. Photobiol., **19**, 69 (1974).
202. V.M. Kutyurin, T.D. Slavnova, and A.K. Chibisov, Biofizika, **18**, 1004 (1973).
203. See chapter 17.
204. F.R. Hopf, T.P. O'Brien, W.R. Scheidt, and D.G. Whitten, submitted for publication.
205. F.R. Hopf, D.G. Whitten, and J.W. Buchler, unpublished results.
206. D.G. Whitten and F.R. Hopf, unpublished results.
207. T.S. Srivastava, L. Hoffman, and M. Tsutsui, J. Am. Chem. Soc., **94**, 1385 (1972).

PHOTOCHEMISTRY OF PORPHYRINS IN MEMBRANES AND PHOTOSYNTHESIS

DAVID MAUZERALL and FELIX T. HONG

The Rockefeller University New York, N.Y. 10021, U.S.A.

17.1. Introduction

The aim of this chapter is to show a relation between electron transfer processes in photoexcited porphyrins and that in photosynthesis. To do this we will briefly summarize the known photochemistry of photosynthesis. We will then present what we believe to be relevant photochemistry of the porphyrins, first in solution, then in monolayers and bilayer lipid membranes. We will show that progress has been made towards obtaining efficient, reproducible photoreactions which transiently store appreciable fractions of the photon energy in reactive intermediates. The goal now is two fold: to understand these systems in all physical detail and to couple out this energy into more long term storage, as is done in photosynthesis.

The basic photochemistry of porphyrins is reviewed by Whitten in Chapter 16 of this book, and also in Ref. (1). Electron transfer of porphyrins and metalloporphyrins is reviewed in Ref. (2). The oxidized (cations) and reduced (anions) species of the porphyrins are reviewed by Fuhrhop in Chapters 14 and 15 and by Subramanian in Chapter 13. Electronic absorption spectra of porphyrins and their derivatives are discussed in Chapter 1. The particular properties of chlorophyll have been well covered in the book of that name[3]. For introductions to photosynthesis the works by Kamen[4], Clayton[5], Rabinowitch and Govindjee[6], and Gregory[7] are available.

17.2. Electron transfer
17.2.1. Photosynthesis

The standard equation of photosynthesis: $CO_2 + H_2O + h\nu \rightarrow (CH_2O) + O_2$ shows stoichiometry, not mechanism. The basic photochemical mechanism of photosynthesis appears to be a charge separation, followed by utili-

Porphyrins and Metalloporphyrins, ed. Kevin M. Smith
© 1975, Elsevier Scientific Publishing Company, Amsterdam, The Netherlands

zation of the free energy stored in these charges through biochemical cycles leading to the formation of ATP and the final products of photosynthesis. This mechanism is known to be true for photosynthetic bacteria. It is most likely true for the very similar photosystem I in plants and algae, and is assumed true for photosystem II which makes oxygen. The isolation of the essence of bacterial photosynthesis, the reaction centers, has greatly clarified this field. Basic to this work is the concept of the photosynthetic unit, wherein large numbers of chlorophyll and other accessory pigments capture photons and transfer the excitation energy to a special or trap chlorophyll(s) where the photochemistry occurs. The intriguing possibility of actually isolating the trap, or reaction center, was made a reality by Reed and Clayton[8] in 1968. These centers have been brought to chemical purity[9] and we will list their remarkable properties. The center is a highly hydrophobic protein isolated from a caroteneless mutant of *Rhodopseudomonas spheroides*. It is made up of three peptide chains of mol. wt. about 25,000, and contains almost 70% nonpolar amino acids[10]. It contains not one or two but six pigment molecules; four bacteriochlorophylls and two bacteriopheophytins[11]. It also contains one ubiquinone molecule, one iron atom and five or six half-cystines. On photoexcitation an electron is transferred from a pair of bacteriochlorophylls[12-14] to the acceptor. This leaves the cation $(Bchl)_2^+$ and most probably reduces the iron[15,16] or iron—quinone complex. If the iron is removed, the electron ends on the quinone, forming the semiquinone radical anion[17-19]. At room temperature, and when the reaction center is coupled to the remainder of the photosynthetic apparatus, electron transfer to secondary acceptors and donors occurs. In the reaction center, or in chromatophores or intact bacteria at cryogenic temperatures, further electron transfer is not possible and the electron returns with a time constant of 30 msec to form $(Bchl)_2$ and the oxidized acceptor[20]. This return rate is completely independent of temperature from 1.5 to 80°K[20] and thus most likely occurs through quantum mechanical tunneling. We believe this barrier of 30 msec is the basic reason that photosynthesis works. This barrier to the electron return provides ample time, on the molecular scale, for the secondary acceptors and donors to take up the separated charges and convert their free energy into products useful to the cell. Consistent with this picture, it is known that transfer to the secondary acceptor in *Chromatium* occurs in about 70 μsec[21] and the transfer from the secondary donors occurs in 1 μsec to 1 msec depending on the bacteria[22].

17.2.2. Theory

We have developed a simple theory of electron transfer reactions based on quantum mechanical tunneling[2,23]. The most important parameters are the energy levels of the molecules involved and their distance of separation. The distance is a generalized coordinate and includes the orientation of the molecules. Once the molecules are determined, e.g., chlorophyll (because of the energy distribution of sunlight) and Fe—quinone (because of the need to store a reasonable fraction ($\sim 50\%$) of the photon energy), it is the distance parameter that determines the efficiency of the process. The electron tunneling makes the probability of transfer a powerful function of the distance, and the requirement of efficient trapping creates a sharp maximum in the probability versus distance curve. Too far, and the probability of transfer is too low during an excited state lifetime. Too close, and the back transfer to the ground state becomes too fast. We believe Nature has optimized this parameter in the design of a reaction center and we are now obtaining evidence for these concepts by studying in vitro or model systems. As will be shown below, we have determined that electron transfer can occur over a large distance, 20 Å, from excited states of porphyrins, and that the charge transfer across a lipid bilayer—water interface is also very fast.

The role of modern photosynthesis is to supply the free energy gradient for all living matter on this earth. It does this by producing oxygen and reduced organic matter. This thermodynamic gradient is used by all cells in respiration. However, life originated under reducing conditions. The present atmosphere of oxygen is largely the product of modern photosynthesis. Therefore to be useful in the evolutionary sense, photosynthesis at that time would have evolved hydrogen and oxidized organic compounds[24]. That is, photosynthesis is useful if it opposes or replenishes the favorable, downhill, thermodynamic gradient of the era. Now, the favored photochemical reaction of metalloporphyrins, and chlorophyll, is to reduce organic compounds, while porphyrins themselves photo-oxidize organic compounds in the absence of oxygen. Thus chlorophyll does modern photosynthesis and porphyrins may have done primitive photosynthesis. The biosynthetic pathway, which may be a window looking back into evolution, first forms porphyrins then metalloporphyrins and chlorophylls (see Chapter 3). Moreover, the thrust of evolution has clearly been to form efficient, organized structures in cells, always associated with lipid in membranes. This is clearly seen in the steadily decreasing charge on the intermediates of the biosynthetic pathway to chlorophyll thus localizing it in such hydrophobic areas. Chlorophyll is the latest stage of that evolution. But life originated in the sea, so the first photosynthetic porphyrin should be very water soluble. The name of the first porphyrin formed in biosynthesis, uroporphyrin, attests to its excellent water solubility and uroporphyrin does oxidize organic compounds under anaerobic conditions. From the photochemical viewpoint, the great gain of efficiency on forming highly organized structures may be traced to the epoch-making discovery of a photochemical path to oxygen. Previous to this era photoreactions in solution could proceed efficiently by the long lived triplet states. As we will see, even 10^{-6} M concentration of certain ions can completely trap the triplet state of zinc uroporphyrin. However, oxygen quenches (and occasionally destroys) such triplet states at encounter limited rates ($k_2 \sim 10^{10}$ M^{-1} sec^{-1}). Therefore an organized, almost solid-state, structure of donor—acceptor pairs was required to allow the charge transfer act to occur in sub-nanosecond times and so compete very well with oxygen. Such a structure is the modern photosynthetic reaction center. But there is a further advantage to being ordered. A photoreaction in an ordered membrane system allows the formation of an electric field across the membrane. The photosynthetic apparatus synergistically makes use of this field to make ATP (see Section 17.7.6.).

17.2.3. Reduction

The photoreduction of chlorophyll was observed by Krasnovskii some twenty five years ago. He has reviewed this early work[25]. The photoreduction of porphyrins was also observed[26]. The structure of these reduced porphyrins was established more recently[27]. The photoreduction of porphyrins in the presence of mild reducing agents usually proceeds through the one electron reduced porphyrin free radicals[28,29] which disproportionate to porphyrin and dihydroporphyrin. The dihydroporphyrin or phlorin has the hydrogen on a single methine carbon (Fig. 1). Of the 30 possible dihydroporphyrins this was the only structure compatible with a monoprotonic pK of 9[27]. The structure of phlorins was also determined by NMR[30]. It is interest-

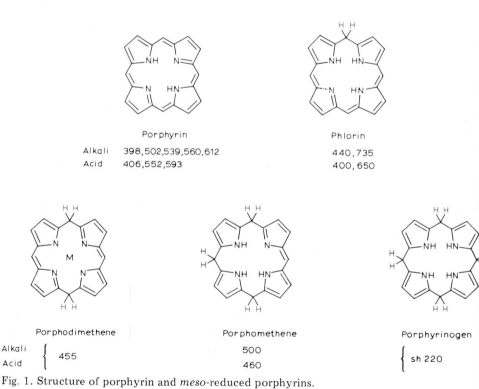

	Porphyrin	Phlorin
Alkali	398,502,539,560,612	440,735
Acid	406,552,593	400,650

	Porphodimethene	Porphomethene	Porphyrinogen
Alkali	455	500	sh 220
Acid		460	

Fig. 1. Structure of porphyrin and *meso*-reduced porphyrins.

ing that under normal conditions the phlorins are not the thermodynamically stable products. Kinetic control is often the rule in these photochemical reactions, quite possibly because of the excess energy present in the excited state and the consequent rapidity of the reactions. In water solution, simple phlorins disproportionate slowly ($t_{1/2} \sim 1$ h at $100°$C, 1 day at $25°$C) (Ref. 27) to porphyrins and tetrahydroporphyrins (porphomethenes Fig. 1), while in aprotic solvents and with hindered phlorins, isomerization to chlorins occurs[31]. The equilibrium in this case may be delicate since the reverse isomerization is also seen[32]. The photoreduction of Sn(IV) porphyrins also yields chlorins possibly because of the disfavoring of the pyramidal structure by Sn. Whereas the protoreduction of porphyrins occurs with ease with very mild reducing agents, the reduction of metalloporphyrins is more difficult. For example the photoreduction of uroporphyrin by a bis tertiary amine, EDTA, at pH 7 has a quantum yield of at least 0.4[33], close to the maximum 0.5 expected from disproportionation of the photoformed free radicals. By contrast even with reducing agents such as ascorbic acid, the quantum yield of photoreduction of zinc porphyrin is about 0.03[34]. The structure of the zinc dihydroporphyrin has been shown by NMR to be a porphodimethene (β,δ-dihydro, Fig. 1) for the case of zinc α,γ-dimethyloctamethylporphyrin[35]. This method has also been used to show that the highly unstable Krasnovskii reduced chlorophyll has the β,δ-dihydro structure[36]. All of the reduced metalloporphyrins with a strong ($\epsilon > 10^5$) absorption near 460 nm and a weak band near 500 nm can be argued to have this structure since it is just what is expected for the metalloporphodimethene. The closely held parallel transition moments of the pyrromethene chromophore will interact to yield a shorter wavelength allowed transition having about twice the transition moment of the monomer, and a much weaker band to longer wavelengths. The energy of interaction indicates a distance of about 5 Å. Some evidence for the intermediate free radicals in the formation of reduced porphyrins also exists[37,38].

17.2.4. Oxidation

In contrast to the ease of photoreduction of porphyrins, the porphyrins chelating a closed shell ion, e.g., Mg(II), Zn(II), are very readily photooxidized to cation radicals[39,40]. Many early observations on greening color changes and light sensitivity of zinc or magnesium porphyrins or chlorins can now be readily explained. This difference between porphyrins and their metal chelates reflects the chemical redox properties of these pigments as shown by their redox potentials. The reduction potentials of porphyrins are less in magnitude than those of the metalloporphyrins and the reverse holds for the oxidation potentials[41]. One explains this difference by simply noting that the closed shell metalloporphyrins are essentially porphyrin dianions enclosing a dipositive metal. Thus the porphyrin ring will readily lose an electron, but not so readily gain one because of the coulomb repulsion

References, p. 722

energy. The porphyrin dianions prepared with alkali metal monocations are very readily oxidized. The redox potentials of the metalloporphyrins[42] and chlorins[43] increase with the increasing electronegativity of the metal ion. The tri-positive metal and particularly the Sn(IV) porphyrins are very difficult to oxidize. Vice versa, in the free base porphyrins, the inner protons are covalently bonded to the nitrogens, the residual charge arising from other valence resonance forms is distributed around the ring, and the system is difficult to oxidize. The low-lying energy levels of the conjugated system can now accept an electron without excess repulsion, and easy reduction occurs. In protonic solvents, hydrogen adds to the *meso* position to give the stable phlorins after disproportionation.

17.3. Cyclic reactions in solution

Since our aim is to show a relation between the photochemistry of porphyrins and photosynthetis, the importance of cyclic reactions must be stressed. A cyclic reaction is one in which following photoexcitation and reaction to form products, the reactants are regenerated in the dark. These reactions are of inherent interest since, by definition, they store more or less of the photon energy in the photoproducts. In principle, it is possible to tap off this energy by further cyclic reactions as is done in photosynthesis.

17.3.1. Porphyrin—phlorin

The simplest example of a photoredox cycle is that of the reaction of an excited porphyrin with a phlorin to form a pair of porphyrin radicals (Fig. 2). These radicals disproportionate in the dark via clean second order kinetics to give porphyrin and phlorin. Both ESR and optical evidence for this cycle was obtained[2,29]. The spectral changes between the porphyrin and the radical: broadening of the allowed Soret band ($20 \to 35$ nm), red shift of the weak visible bands ($500-600 \to 600-700$ nm), and weak transitions in

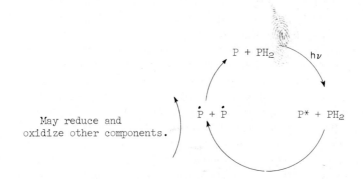

Fig. 2. Cyclic photoreaction of porphyrin and phlorin.

the near infrared (970 and 1150 nm) are as expected on adding one electron to an aromatic system[44]. This cyclic reaction combines the storage of considerable free energy in the porphyrin free radicals for extended lengths of time (1 sec to 1 day). The latter property depends on the high charge of the porphyrin radicals, which adds an electrostatic repulsion to the encounter limited rate of disproportionation (see below).

17.3.2. Metalloporphyrin—acceptor

Electrostatic repulsion was used to determine the distance of electron transfer between an excited metalloporphyrin and an electron acceptor[24,40]. The triplet state of zinc uroporphyrin was chosen for its charge (−8) and convenient lifetime (10 msec). The acceptors ranged from N-benzyl-nicotinamide (charge +1) through NADP and ferricyanide (charge −3). The pseudo-first order rate constant of quenching of the triplet state at various concentrations of acceptors was measured by flash photolysis in well de-oxygenated solutions. This allowed determination of the second order rate constant for the reaction of triplet and acceptor. The total electrostatic contribution to this rate constant was isolated by repeating the measurements at various ionic strengths and extrapolating to zero ionic strength to remove ionic shielding effects. This result was checked by extrapolation to infinite ionic strength which gave a constant value, equal to that measured with an uncharged acceptor: 2×10^8 M^{-1} sec^{-1}. The fact that NADP and ferricyanide react at the same rate whereas their one electron redox potentials are so different (−0.7 vs. +0.4 V) shows that the rate is completely uncorrelated with redox potential. These large rate constants also make the reaction very efficient. Micromolar solutions of some donors are sufficient to trap the porphyrin triplet state.

The slope of the plot of log quenching constant versus charge product for the full coulombic effect yields the distance of electron transfer reaction: 22 ± 4 Å. This is the average distance, assuming it is constant for all members of the class. Calculation of the distance for each individual reaction requires specific assumptions, and yields a range of 15 to 30 Å. Further confirmation of this large distance of electron transfer arises from the self-consistency of the data. The ionic strength dependence of the reaction rate constant for NADP shows unequivocally that the acceptor is negatively charged, with magnitude about 3. Yet the actual acceptor portion of the molecule is the pyridinium ring, with a charge of +1. The zinc uroporphyrin triplet state is seeing the overall or net charge of the molecule, not that of the acceptor portion. Thus the porphyrin—acceptor distance must be at least several times the internal charge separation distance in NADP (~5 Å) to feel the net charge.

Two important points emerge from this data. First, the distance of electron transfer from excited states can be far greater than the usually assumed nearest neighbor. This is consistent with the results on bacterial reaction centers and is explainable with our hypothesis of electron tunneling. Second, the efficiency of such a transfer can be very high, that is, the lifetime of the intermediate free radicals is far longer than the formation time under conditions that all excited states are trapped. Following the tunneling hypothesis, this is expected since the probability of electron transfer at a given distance to the ground state is smaller because of the lowering of the energy of the electron in the relaxed free radical species. The molecules must diffuse closer together to back react and thus their probability to escape, particularly in the face of a repulsive potential, is enhanced. This concept is further supported by two further independent arguments. The 3-carbamido-pyridinium salts used as electron acceptors form molecular complexes with porphyrins[45]. Flashing the preformed porphyrin—acceptor complex gives no detectable transient, or radical of lifetime greater than 10 μsec[40]. Thus at the distance of nearest neighbors, the reverse electron transfer must be very fast. Not only is the triplet state quenched, but the yield of fluorescence is also decreased 50%[45]. One can also apply the electrostatic repulsion argument to explain the slow, charge dependent rate of disproportionation of the reduced uroporphyrin radicals (see above). The radius necessary to explain the data using the Debye treatment[46] of encounter limits in the presence of an electrostatic potential is about 10 Å, i.e., about the sum of the molecular radii. Thus in the ground states, a closer approach appears to be necessary. In passing it must be pointed out that some confusion exists on the 'radius' used in calculations on encounter limited reactions. By assuming the hydrodynamic radius equals the 'physical' radius or radius of reaction, and by use of the Stokes—Einstein equation, the limiting rate constant depending only on the viscosity of the solvent (η) is obtained: $k = RT/\eta$. If the radius of reaction is larger than the hydrodynamic radius as it is for excited states, this result must be multiplied by the ratio of these radii. Moreover, for charged ions, the 'radius' of the ion enters in the treatment for ionic strength effects[47]. These three physically distinct radii must be carefully distinguished.

17.4. Sensitization

The porphyrin pigments have been widely used as sensitizers for photochemical reactions. Most of these are thermodynamically downhill reactions, i.e., the photoexcited pigment only supplies a mechanism for an otherwise allowed reaction. The photo-oxidations in air mostly proceed through the $'\Delta_g$ state of O_2 formed by energy transfer from the dye triplet state[48]. The photo-reductions and direct photo-oxidation in the absence of air very likely occur through the free radicals discussed above. A review of the Russian work is available[49]. The number of cases where free energy is stored in the products, is still quite small. The original study by Vernon[50] of the porphyrin and chlorophyll sensitized formation of NADH from ascorbic acid with use of enzymes is a clear example. The enzyme preparation, NAD reductase, is necessary to trap the substituted nicotinamide radicals formed by the reaction outlined above. Eisenstein and Wang[51] have given spectral evidence for the reduction of ferredoxin by glutathione, sensitized by hematoporphyrin.

17.5. Colloidal systems

Since chlorophyll is insoluble in water and colloidal preparations of chlorophyll in water have spectra vaguely resembling that of chlorophyll in vivo, there have been many reports on their activity. Rabinovitch[52] is a good source for this early work, and for more recent work see Krasnovskii's review[53]. Bannister and Bernardini[54] have prepared fluorescent colloidal chlorophyll in the presence of a detergent that sensitizes the autoxidation of paratoluenediamine, or undergoes the Krasnovskii reaction. Most of the work involves aggregates of poorly defined nature, but the recurring report that oxygen does not inhibit reactions photosensitized by these preparations (see, e.g., Ref. 55) as it does those in solution may be significant. The work of the Katz group[56] on the structure of such aggregates, determined by NMR, has greatly clarified this field. They have shown the critical importance of water or other polar ligands on the structure and size of the aggregates. The photochemistry of these known aggregates has not been much studied aside from observations of photoinduced, very narrow ESR signals[57]. They may arise from a highly delocalized electron (or hole) in the aggregate. We[58] have observed an ESR signal from bacteriochlorophyll as narrow as 0.54 Gauss, g = 2.00248 in dry CCl_4. If the narrowing is caused only by delocalization over equivalent sites, the aggregate has ~500 units.

Bacteriochlorophyll in 20% v/v acetone—water, and ~1 mole of lauryldimethylamine oxide forms aggregates absorbing at 790 and 850 nm, not unlike the absorption in the reaction centers[58]. The form absorbing at 850 nm is preferentially oxidized by I_2. Bacteriopheophytin forms a more unique aggregate absorbing at 835 nm, and is not oxidized by I_2. Mixtures of bacteriochlorophyll and bacteriopheophytin form aggregates whose spectra are strictly non-additive mixtures of the component spectra. The electronic interaction between the molecules is thus quite strong.

17.6. Monolayers
17.6.1. Air—water

The concept of monomolecular layers of chlorophyll being an active component of photosynthetic system is as old as the first observation of such layers by Blodget and Langmuir[59]. The spectral properties of such films have been well described in the classical work of Trurnit and Colmano[60] and of Bellamy, Gaines, and Tweet[61]. A review[62] and extension to multi-layers[63] is available. In pure chlorophyll monolayers, the angle of the plane of the macrocycle with the water is thought to vary with the surface pressure, and this angle for a complete monolayer may be about 60°. In dilute monolayers (e.g., with oleyl alcohol) if the macrocycle is assumed to be at about this angle, the angle of the emission moment to the water[64] is about 20°.

Because of technical difficulties, little photochemical work has been done on these monolayers. Compressed films of chlorophyll are photo-oxidized (determined by an increase in molecular area) with a quantum yield many times higher than in solution[65].

17.6.2. Solid—water

The physical advantage of monolayers can be combined with the ease of photochemistry in solution by absorbing the pigment on to very small poly-styrene spheres and suspending these in water[66]. The emulsifier must be first removed from the latex. The pigment could be absorbed by various methods, the most useful being dilution of an acetone solution with water in the presence of the spheres while stirring rapidly. When about 10^{-3} of the surface of the sphere is covered by the pigment pheophytin-a, its absorption and emission spectra are identical to that in solution. The wavelength maxima are slightly shifted from that in toluene because of the mixed refractive index at the polystyrene—water interface. Further, the photochemistry of the isolated pigment on this surface is the same as in solution. This was determined by using the pheophytin to photosensitize the reduction of the azo dye amido-naphthol red by ascorbate or N-benzyl-dihydronicotinamide. This particular azo dye was chosen to avoid the catalytic effects of the phenylenediamine product of other azo dyes.

A simple method using an integrating cylinder (orange juice can) to measure absorption changes in the highly scattering suspension was developed. The measurements were linearized by working at nearly constant dye absorption: bleached dye was replaced by addition from a microburette. The quantum yield of pheophytin-sensitized dye reduction was the same in solution (with uncoated polystyrene spheres added for constant scatter) in 90% methanol as when absorbed to the spheres in water. As the surface coverage was increased to a complete monolayer, the quantum yield of fluorescence decreased rapidly after 0.5% coverage, and the absorption spectra broadened considerably at 10% coverage, becoming similar to that seen in pheophytin mono-layers at the air water interface. The quantum yield of sensitized dye reduction decreased at high coverage also, but was still 65% of the low coverage when the fluorescence yield had fallen to less than 5% of the original value (Fig. 3). A simple model for linear aggregates was used, based on a one-dimensional Ising model. The only adjustable parameter in this model is the effective nearest neighbor interaction energy, and the broadening of the

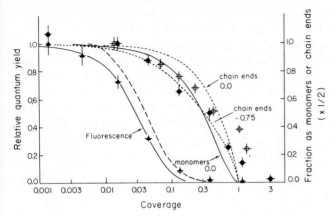

Fig. 3. The relative quantum yield of dye reduction (○—○—○) and (●—●—●) and of fluorescence (△—△—△) of pheophytin-a on polystyrene spheres is plotted as a function of surface coverage. The solid and dotted curves are theoretical and refer to calculations where only monomers or chain ends were considered to be photoactive. The appended numbers refer to the neighbor binding energy in kcal/mol. The dashed curve is the relative fluorescence yield of chlorophyll-oleyl alcohol monolayers. For details, see Ref. 66.

absorption spectra was fit by a value of zero to −0.75 kcal/mol. This small energy is expected because the London dispersion interaction between pheophytins is not much greater than that between pheophytin and polystyrene. The one dimensional model is adequate since the aggregates of pheophytin, or of any planar aromatic molecule, will form card-pack-type aggregates because of the anisotropic polarizability, the source of the 'dispersion' force. The out of phase component of the polarizability determines the observed spectra of the aggregate. This one dimensional model was later applied to the aggregation of chlorophyll on polyvinylpyridine by Seely[67]. With this model for the aggregates and the theory of energy transfer in two dimensions, a quantitative fit of the fluorescence quenching curve on the sphere was obtained. That is, energy transfer occurs efficiently near 1% surface coverage and the small number of aggregates quench fluorescence well before they are detected by absorption changes. The data clearly show that energy transfer to some aggregates, at least dimers, still results in photochemical reaction. This seems to be the only case where objective evidence exists for energy transfer to photoreactive aggregates, as is observed in the photosynthetic unit.

The above analysis of pigment aggregation suggests that on a polar surface, the interpigment attraction would exceed the pigment surface attraction, and the pigment would aggregate at low surface coverage[66]. In fact the absorption spectrum of pheophytin on the aluminosilicate surface is broadened even below 1% surface coverage. At high surface coverage, water appears to act as a surface barrier, increasing the aggregation, since on drying the particles, the spectra narrow considerably. The fluorescence spectra also mimic the absorption changes.

We have observed an interesting effect of the degree of surface coverage of polystyrene spheres when bacteriochlorophyll is the pigment[68]. At low coverage the pigment is rapidly photo-oxidized on illumination in air, while at high coverage it is much more resistant. A similar effect is seen in solution. Thus the high concentration of pigment in bacterial chromatophores or chloroplasts may already prevent some photo-oxidative degradations. In photosynthetic systems the carotenes are usually assumed to provide this protection.

17.7. Bilayer lipid membranes*

The technique of forming bilayer lipid membranes (BLM) in vitro[69] provides an opportunity to study photoelectrical phenomena in an environment close to that present in photosynthetic systems. Tien[70,71] reported in 1968 photoelectrical effects with a BLM made from a chloroplast extract. This subject has been reviewed by Tien[72]. The general subject of BLM has also been summarized by Jain[73]. We will restrict our discussion to chlorophyll or porphyrin-containing BLM. We will first discuss structural aspects, then summarize published work and present our work in greater detail.

17.7.1. Structure
Because of the anisotropic structure of BLM, the pigments contained therein may not be randomly oriented. Several attempts have been made to determine the orientation of the porphyrin ring with respect to the BLM by means of dichroic measurements[74-76]. The values reported are about 45°. Both Hoff[77] and Steinemann et al.[75] have pointed out that when the transition moment makes an angle of 35°16' with respect to the plane of the BLM, dichroism vanishes even for a well-oriented system. The reported values of the transition moments are close to this critical value, and the dichroic ratio is a weak function of the tilt-angle. Therefore, the determined value of the tilt-angle of the porphyrin ring with respect to the BLM is subject to large error. Hoff[77] has studied oriented chlorophyll and bacteriochlorophyll lecithin multilayers and the precision of his measurements permits him to conclude that the pigment molecules are oriented in the multilayers, and that the tilt-angle is about 35°.

The phytol chain of the chlorophylls most likely lies parallel with the hydrocarbon chains of the phospholipid in BLM and is therefore perpendicu-

---

* Bilayer lipid membrane is synonymous with bimolecular lipid membrane, black lipid membrane, and lipid bilayer, and is abbreviated BLM.

lar to the plane of BLM. The above-mentioned measurements are based on the assumption that the porphyrin ring maintains a fixed angle with the phytol chain. It was also assumed that all chlorophylls are homogeneous in this orientation. These two assumptions are questionable. An NMR study with ^{13}C or ^{2}H labeled pigment would be revealing.

There appears to be no preferred azimuthal orientation of the pigment in BLM. Attempts to generate induced dichroism by 'bleaching' some pigment with an intense polarized laser beam show that no induced dichroism can be detected after 10 μsec[78]. Presumably, the pigment rotates about the azimuthal angle in a time short compared to 10 μsec.

The location of the porphyrin ring in a BLM is important to our understanding of its photoreactions. Because the fluorescence of a chlorophyll-BLM is reduced by half when $K_2S_2O_8$ is present in *one* aqueous phase and is diminished to zero when $K_2S_2O_8$ is present in *both* aqueous phases, Steinemann et al.[79] concluded that the porphyrin ring extends into the aqueous phase. Huebner[80] pointed out that $K_2S_2O_8$ may quench the chlorophyll fluorescence by making the electron transfer from the excited state very probable. Cherry et al.[81] instead concluded that the porphyrin ring lies in the polar head region of the BLM, because of the following observations. The expected change of capacitance due to chlorophyll in the hydrocarbon region has not been observed. Therefore, chlorophyll is not located in the hydrocarbon region. The expected BLM thickness change due to chlorophyll extending into the aqueous phases has also not been observed. Therefore, chlorophyll does not extend into the aqueous phase. Because of the small size of the predicted effects, both conclusions are very weak. In reality, the interface is not a mathematical plane and is, instead, a region of finite thickness. It is chemically reasonable that the porphyrin ring lies in the lipid polar head region. Further investigation is required to elucidate this important question.

Both Steinemann et al.[79] and Cherry et al.[82] reported a surface concentration of chlorophyll of the order of 10^{13} molecules per cm^2 (or 10^5 molecules per μ^2) corresponding to a mean distance of approximately 30 Å. The proportion of chlorophyll in the BLM is consistently one order of magnitude smaller than that in the membrane-forming solution. Steinemann et al.[79] observed that the absorption in the BLM is linear up to chlorophyll : lecithin ratio of 0.45, while the corresponding fluorescence signal goes through a maximum of a molar ratio of 0.3. Presumably, this is due to quenching by energy transfer to aggregates at higher concentration. This is quite analogous to that observed in monolayers[64] or on the surface of polystyrene spheres[66].

17.7.2. Photoelectrical effects

Photoelectrical effects in a pigment-containing BLM were first reported by Tien[70,71]. Tien used a chloroplast extract to form a BLM which separates symmetrical aqueous salt solutions. Photovoltages of a few millivolts were observed under illumination with continuous light. The photovoltage increases approximately linearly with light intensity. The action spectrum of this photovoltage roughly corresponds to the absorption spectrum of the photopigment. The polarity of the photovoltage depends on the direction of the incident light; the illuminated side being negative. This implies that the cause is a gradient of light intensity across the annulus of the BLM.

It soon became apparent that larger photovoltages could be achieved in an asymmetrical system, i.e., where different electron acceptors and donors are present in the two aqueous phases[83]. The polarity of the photovoltage now depends solely on the aqueous redox composition and not on the incident direction of the exciting light. Similar photoeffects were observed in BLM which contain retinal[84,85], cyanine dyes[86], magnesium porphyrins[87,88], and flavins[89]. Photoeffects of chlorophyll-containing BLM sensitized by various dyes present in the aqueous phases have also been reported[90].

Most of the early reports on photoeffects of pigmented BLM concern continuous light responses. However, steady state measurements seldom shed light on the underlying molecular mechanism. Therefore, excitation by pulsed light and analysis of the time course of the photoresponses have been attempted by a number of laboratories[85,86,88,89,91−93]. As we shall see later, the measured responses now depend critically on details of the electrical measurement apparatus. Using the open circuit method and excitation with flash light of 3 μsec in duration, Tien[91] observed a transient photovoltage with risetime in the range of the light flash, followed by two slower components (risetime 20 msec and 1 sec, respectively) and a very slow component with the electrical relaxation time of the membrane. The complex dependence of the photovoltage on pH gradient, redox gradient of the aqueous phases and applied voltage gradient were measured. Trissl and Läuger[92] investigated chlorophyll-containing BLM under two different conditions. In a chlorophyll-BLM coupled to TMPD in one aqueous phase and sodium ascorbate in the other aqueous phase, steady state photocurrent and photovoltage were observed under continuous illumination. In a chlorophyll-BLM coupled to cytochrome-c in one aqueous phase, a transient photocurrent rises with the instrumental response time of 60 msec and decays with a 3 sec time constant under step function illumination. The authors suggest light-induced continuous mobile charge production in the first case and generation of an electrical double layer in the second case. Hong and Mauzerall[88] studied a BLM system which contains lipid soluble Mg(Meso-IX) diesters or Mg(OEP) and separates two aqueous phases of ferricyanide and ferrocyanide. By using a short excitation of 0.3 μsec and an ultrafast feedback amplifier, photocurrent can be measured with an instrumental time resolution of 150 nsec. The photocurrent has two components of opposite polarity. Analysis of the time course of the photocurrent gives direct information of the interfacial chemical kinetics. The observation is consistent with the picture of two interfacial redox reactions coupled by diffusion of the membrane-bound pigment molecules, both charged and uncharged[9 3].

17.7.3. Methodology

As indicated in Section 17.2, the efficient modern photosynthetic apparatus is contained in a highly organized structure in membranes. In order to function continuously as a light energy transducer, the photoreaction and the related dark reaction must be quantitatively cyclic (Section 17.3). As a first step in the study of the BLM model system, we chose the magnesium porphyrins ferricyanide—ferrocyanide system. This is the simplest of its kind, since no intermediate electron acceptor—donor is involved in the system. The scheme of photo and dark reactions is depicted in Fig. 4. Since electrons can be transferred only over limited range, and the pigment and the inorganic electron acceptors and donors are separated by the membrane—water interface, the electron transfer reaction must be essentially interfacial. Small pigments such as Mg(OEP) and Mg(Meso-IX-DAE)* are mobile in the liquid crystal-like environment of BLM. Therefore, we have a

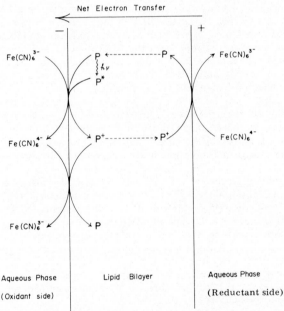

Fig. 4. The photochemical reaction cycle in the BLM-water system. The ground state magnesium porphyrin P and the electron acceptor $Fe(CN)_6^{3-}$ undergo reversible one-electron oxidation-reduction to form the stable monocation P^+ and $Fe(CN)_6^{4-}$. Oxidation is much enhanced with the excited state P^*. The pigment molecules and the inorganic electron acceptor—donor are localized in the membrane phase and in the aqueous phase, respectively. The electron transfer reactions at both interfaces are coupled by diffusion of the pigment molecules, both charged and uncharged.

---

* *Abbreviation:* Mg(Meso-IX-DAE) = Magnesium mesoporphyrin-IX di-*n*-amyl ester.

system of two interfacial photoreactions coupled by diffusion of the pigment across the bilayer. If the redox composition of the two aqueous phases are identical, then the extent of the electron transfer reactions at the two interfaces will be equal but opposite in direction. No net dark or photovoltage can be observed. However, an increased conductance is expected, as is indeed observed. On the other hand, if the aqueous redox composition is asymmetrical in the two sides, a net dark voltage and a net photovoltage across the BLM will be observed; the acceptor-rich side is negative with respect to the donor-rich side.

The reverse electron transfer reaction:

$$P^+_{(BLM)} + Fe(CN)_6^{4-}{}_{(aq.)} \rightarrow P_{(BLM)} + Fe(CN)_6^{3-}{}_{(aq.)} \tag{1}$$

can take place at both interfaces. Therefore, two parallel and competing processes are going on both in the dark and in the light. The pigment monocation can be discharged at the same interface where it was formed, or diffuses across the BLM and becomes discharged at the opposite interface. Only the latter event is observed in a steady state continuous light illumination. On the other hand, if the exciting light has sufficiently fast rise and fall time, the former event will dominate the transient photoresponse.

In general, there are two types of electrical measurements in the studies of BLM systems. 1) Open circuit method: the voltage across the BLM is measured via a pair of electrodes by means of a high input impedance device such as an electrometer in voltage mode. The measuring device draws negligible current from the BLM system. 2) Voltage clamp method: the voltage across the membrane is controlled at a fixed value and the current flowing through the BLM system is monitored. This method is widely used in neurophysiological studies[94] and is best implemented by a feedback circuit.

In the study of photoeffects in BLM, the latter method is preferred, because it provides the kinetic data specific to the photoreaction and is therefore easier to analyze in terms of molecular kinetic processes. We have measured the rate constant of interfacial charge recombination by means of a tunable voltage clamp method[93] which is a modification of the conventional voltage clamp method specifically adapted to this type of measurement.

Methods of preparing lipid—solvent mixtures for forming a BLM have been described in detail[69,95]. The nature of the phospholipid used greatly affects the mechanical stability of the BLM. The presence of lipid peroxide is detrimental to the stability[96]. The following method is used to restore a sample of phospholipid contaminated with peroxide. Ten times the weight of sodium ascorbate is added to an ethanol solution of the peroxide-containing phospholipid sample. The mixture is allowed to stand in the dark for 3 h, and then evaporated to dryness under a stream of dry nitrogen gas. The phospholipid is ready to be used or can be stored in liquid nitrogen in sealed ampules. Samples of pigment-containing membrane-forming solution can be stored in liquid nitrogen for many months without significant loss of photoreactivity.

The starting material for making Mg(Meso-IX-diester) is the free base, dicarboxylic acid of mesoporphyrin-IX, which is prepared from its dimethyl ester after purification of the latter by alumina column chromatography. Fifty mg of H_2(Meso-IX) is ground in a mortar and pestle, and then transferred to a glass-stoppered test tube. Two ml of thionyl chloride (distilled) are added in the hood. The tube is sealed under nitrogen gas and let stand at room temperature in the dark. The completion of the reaction is tested by adding a drop of the mixture to methanol and separating ester and acid by TLC. When the reaction is complete the test tube is connected to a well-trapped vacuum line and evaporated to dryness at about 10 μ Hg vacuum. One ml of liquid long-chain alcohol (e.g., octanol) is premixed with an equal volume of methylene chloride and added to the dried mesoporphyrin acid chloride. The mixture is stirred with a dry glass rod to ensure thorough dissolution. It is neutralized with 1 M sodium acetate and washed with distilled

water. The excess alcohol is removed by vacuum distillation in a sublimation apparatus at
1 μ Hg. The final trace of residual alcohol and free acid mesoporphyrin is then removed
by alumina column chromatography. The purified mesoporphyrin ester is then used for
magnesium chelation according to the standard procedure[97]. The overall yield is at least
90%. The absorption spectrum in ether is: λ_{max} 408 · 5 (ϵ 450,000), 545 (16,800), and
581 nm (17,000).

17.7.4. Results

The BLM is surrounded by a thick annulus (Plateau—Gibbs border). In
photoexperiments, 10^5 times more photons are absorbed in the annulus than
in the thin bilayer region. The optically thick annulus therefore introduces a
gradient of light — a problem not existing in the bilayer region (estimated
absorbance at 590 nm $<10^{-4}$). A scanning experiment[88] was carried out to
distinguish the photoresponses from the annular region and from the bilayer
region (Fig. 5). The transient pulsed response originates exclusively from the
thin bilayer region while the steady state continuous response is more prominent at the annular region. In view of this fact, data obtained without
focusing the light source should be interpreted with caution.

The dark and the light reaction of the pigment cause the membrane conductance to increase 10 to 100 times. This increase in conductance is caused
by creation of a new conductance channel in parallel with the ionic leakage
conductance always present. The new current is carried exclusively by the
pigment monocation P^+ [87].

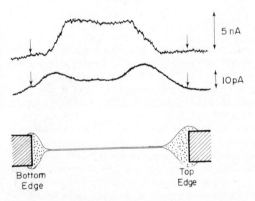

Fig. 5. The photocurrent response to a focused light beam versus the distance across the
membrane with the membrane voltage clamped at zero. The peak response to a 50 μsec
argon ion laser pulse (50 μJ; mainly 514.5 nm) was measured by means of a boxcar
integrator with aperture time of 25 μsec (upper trace), and the steady state response to a
filtered xenon arc light (60 μW) was recorded directly with a strip chart recorder. The
beam was focused to one-sixth of the membrane diameter (1.96 mm), and was scanned
vertically across the membrane diameter. The arrows indicate the points where the light
beam reached the membrane annulus or departed completely from it. The membrane-
forming aperture together with a membrane are schematically below the two traces.
(Reproduced from Nature, New Biol., 240, No. 100, pp. 154—155, November 29, 1972.)

The steady state photoresponses to continuous light depend on the redox composition of the two aqueous phases. In the chloroplast extract-BLM[70,71], an emf was observed which is that expected of a Nernstian concentration cell. In the magnesium porphyrin/ferricyanide—ferrocyanide system, polarization occurs and the emf is less than the expected Nernstian

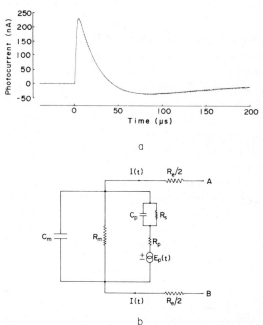

a

b

Fig. 6. (a) Superposition of measured (dotted curve) and computed (smooth curve) photoresponses to a dye laser pulse. The measured curve is taken by the tunable voltage clamp method at V = 0 with an access resistance R_e = 5.1 K Ω, from a BLM which contains Mg(Meso-IX-DAE) and separates two aqueous phases with 20 mM $K_3Fe(CN)_6$ and 0.5 mM $K_4Fe(CN)_6$ on the acceptor-rich side, and 20 mM $K_4Fe(CN)_6$ on the donor-rich side. Both aqueous phases also contain 1 M NaCl and 10 mM phosphate buffer at pH = 7.2. The temperature is held constant at 26.0 ± 0.2° C. The exciting laser beam is focused to illuminate 19% of the thin bilayer of area 1.7 mm². The instrumental time constant is 1.5 μsec. The average of 16 measurements from the same membrane is obtained with a signal averager. The input data for the computation are all obtained from experimental measurements: τ_s = 29 μsec, τ_1 = 63 μsec, R_m = 1.5 × 10^9 Ω, C_m = 8.2 nF, R_s = 10^9 Ω, and $E_p(t)$ is taken to have the same time course of an equilateral triangular pulse with a half width of 0.3 μsec, i.e., the same as the exciting light pulse. The low pass filtering effect of the feedback RC loop of time constant 1.5 μsec is also taken into account. The computed responses are normalized with respect to the peak. The calculated parameters are τ_p = 44 μsec, R_p = 33 K Ω, C_p = 1.3 nF. The value of all RC parameters refer to the entire area of the thin bilayer. (b) The equivalent circuit of the pigmented BLM system. $E_p(t)$ is the photo-emf. R_m and C_m are the ordinary membrane resistance and capacitance, respectively. R_p and C_p are the chemical resistance and capacitance, respectively. R_s is the transmembrane resistance. R_e is the access resistance. I(t) is the photocurrent. The negative feedback amplifier maintains points A (donor-rich side) and B (acceptor-rich side) at equipotential.

potential. In the latter case, the observed signal is markedly influenced by stirring of the aqueous phases, showing that polarization gradients extend into the aqueous phases. In contrast, the transient photoeffect to a short laser pulse (~ 1 μsec) is remarkably insensitive to stirring. Furthermore, it can be made specific to a single interface[88]. The magnitude and the time course of the transient photocurrent is strongly dependent on the ferrocyanide concentration of the acceptor-rich side. Increased ferrocyanide concentration causes a decrease of both amplitude and relaxation time of the photocurrent.

These observations can be quantitatively accounted for by the electrical equivalent circuit shown in Fig. 6b. Figure 6a shows the photocurrent measured by the tunable voltage clamp method and the photocurrent predicted by the equivalent circuit[93].

The equivalent circuit includes a novel feature. A chemical capacitance (C_p) in series with the photo-emf is also present in addition to the ordinary membrane capacitance (C_m).

The parameter τ_p is remarkably constant from one pigment to another in the series of long-chain esters of Mg(Meso-IX). Apparently the interfacial reaction does not depend on the molecular size of the pigment. It is thus worthwhile to obtain independent kinetic information on the pigment molecule itself. This is achieved by using the double pulse technique, a standard approach in photobiology[98]. A saturating exciting light pulse is first delivered to the pigmented BLM such that the photoeffect cannot be enhanced by further increase of exciting light intensity. After a variable delay, a second weaker pulse is delivered to test the recovery of the photosystem. At short delays, the test response is diminished in amplitude but its kinetics are unchanged. By repeating the experiment with various delays, the recovery of the pigment system is determined. The results are plotted as log fraction of inhibition versus time of delay (Fig. 7). The data for four different pigments are shown. The dependence of the recovery time constants on the pigment structure is evident. Notice that the inhibition at time zero is not 100%, indicating a fraction having very rapid recovery. This recovery could be due to molecular rotation. As mentioned above dichroic measurement with time resolution of 10 μsec fails to detect any induced dichroism, consistent with this hypothesis. The results of both the steady-state continuous light experiments and the transient pulsed light experiments are consistent with the working hypothesis proposed in Fig. 4.

17.7.5. Discussion

The dark-emf and the photo-emf have their origin at the interfaces, being the tendency of the pigment to donate an electron to the acceptor in the aqueous phases and of the pigment cation to accept an electron from the donor. In the range of linear dependence on light, the photo-emf has the same time course as that of the exciting light pulse. This is physically reason-

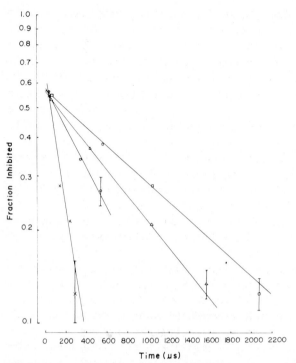

Fig. 7. Results of double pulse experiments. The fraction of the photoresponse inhibited on a logarithmic scale is plotted against the time delay between the saturating and test laser pulse. Mg(Meso-IX-diester); diphytyl ($\square$—$\square$—$\square$), diamyl ($\triangle$—$\triangle$—$\triangle$), dimethyl ($\bigcirc$—$\bigcirc$—$\bigcirc$). Mg(OEP) ($\times$—$\times$—$\times$). Temperature is maintained at 25 ± 0.2°C.

able, because the electron transfer from the excited state must be extremely fast. The singlet state of the magnesium porphyrin lives about 1 nsec. The triplet state lives less than 100 nsec under aerobic conditions. The photo-emf will be appreciable only if the acceptor concentration considerably exceeds the donor concentration, and therefore, this emf is associated mainly with the acceptor-rich side of the two interfaces.

The rate constant $1/\tau_p$ is interpreted as the pseudo-first-order rate constant of the reverse electron transfer reaction (eq. 1). The slope $(3.5 \pm 0.4) \times 10^7$ M^{-1} sec^{-1} of the linear plot of $1/\tau_p$ versus ferrocyanide concentration is interpreted as the second order rate constant of the reaction.

The transmembrane resistance R_s, essentially that due to pigment resistance, is of the order of 10^7 Ω/cm^2, and is dependent on the pigment structure, indicating that diffusion of the pigment is involved. In contrast, the chemical resistance R_p is independent of pigment structure but is approximately linearly related to the ferrocyanide concentration of the acceptor-rich side. Its magnitude is about three orders of magnitude lower than that of the transmembrane resistance R_s. The chemical resistance R_p is interpreted as the resistance encountered by the reverse electron transfer at the acceptor-rich interface.

In the transient response to pulsed light excitation, the photocurrent flows mainly through C_p and bypasses R_s, thus allowing R_p and C_p to be measured. In contrast, measurements of the pigment conductance in continuous light or in the dark give essentially R_s. This explains why the pulsed response is independent of the composition of the donor-rich side while the continuous response depends on both aqueous phases. The results of the scanning experiment are also explained. The photocurrent is higher in the thin bilayer region where C_p is higher, given the same photo-emf.

The chemical capacitance C_p is, however, physically distinct from the ordinary membrane capacitance C_m. The light pulse induces a charge separation at the acceptor-rich interface forming P^+ and reduced acceptor A^- (the ferrocyanide ion). The excess charge on the reduced acceptor becomes rapidly screened by excess ions in the aqueous phases, while P^+ in the bilayer is only partially screened. Thus, the photo-generated P^+ forms a layer of effectively fixed surface charge in the μsec time range. The partially screened charge P^+ can thus polarize the dielectric of the bilayer and an accumulation of net negative charges will appear at the donor-rich interface. Extra photo-generated P^+ disappears through charge recombination at both interfaces, but chiefly at the acceptor-rich interface. The polarization of the bilayer and the electrical double layers disappears. This is tantamount to the charging and discharging of a capacitance (the chemical capacitance). Photogenerated charges can be stored and retrieved in this capacitance. This interpretation has been substantiated by a theoretical calculation based on Gouy—Chapman double layer theory[99].

The transient responses as observed by Ullrich and Kuhn[86] and Trissl and Läuger[92] can be readily understood in terms of the present model.

17.7.6. Relation to photosynthesis

The relation between the primary events in photosynthesis and electron transfer reactions was discussed in the introduction. Here we wish to discuss the special effect of an ordered system on these photoreactions. Witt[100], Jackson and Crofts[101] and others have presented evidence that in plants and photosynthetic bacteria, respectively, light induced electric fields form across the photosynthetic membranes. It has been suggested that the electric field serves to drive a proton flux, which is coupled at ATP synthesis in accordance with Mitchell's hypothesis[102].

Our system of pigmented BLM is a drastically simplified photosystem in which no intermediate or secondary electron acceptor—donor is present. The primary step is a charge transfer reaction but is, nevertheless, an interfacial one. Addition of membrane-bound intermediate acceptor—donor would render it closer to the natural photosystem. In fact, the addition of plastoquinones to this BLM does increase the conductance by one order of magnitude[103].

Because of the very small size of the photosynthetic apparatus, direct electrical measurement with intracellular electrode has so far not been realized. However, Fowler and Kok[104] claimed to have measured this potential directly in a suspension of chloroplast fragments in the presence of a gradient of light. Understandably, the measurement is of the open circuit type. A comparison of electric fields in vivo and in the BLM system is of great interest. The risetime of the electric field formation in algae e.g. *Chlorella*, is extremely fast, 20 nsec[102]. The relaxation of this electric field is slower and is influenced by the coupling to ATP formation[102]. The results of the present investigation on pigmented BLM suggest that each and every interfacial reaction will have a (chemical) capacitance associated with it. The emf's generated by the photoreaction are specific to each capacitance. The overall electrical response to a pulse of light will decay with more than one time constant. The observation of at least two relaxation times in the optical measurement of chloroplasts[105] could well be due to the presence of chemical capacitance. Further studies in a more realistic pigmented BLM system would contribute to and supplement our understanding of the natural process.

The pigmented BLM system offers a straight-forward way of precise and highly informative measurement on a model which can be made asympotically similar to that in photosynthesis.

Acknowledgement

We thank Dr. J.-H. Fuhrhop for a gift of Mg(OEP) and Drs. A. Finkelstein and A. Mauro for helpful discussions concerning membrane phenomena. This work was supported by NIH Grant GM-20729.

References

1. D.G. Whitten, in "The Porphyrins", Ed. D. Dolphin, Academic Press, New York, in press.
2. D. Mauzerall, in "The Porphyrins", Ed. D. Dolphin, Academic Press, New York, in press.
3. "The Chlorophylls" Ed. L.P. Vernon and G.R. Seely, Academic Press, New York (1966).
4. M. Kamen, "Primary Processes in Photosynthesis", Academic Press, New York (1963).
5. R.K. Clayton, "Molecular Physics of Photosynthesis", Blaisdell Publ. Co., New York (1965).
6. E. Rabinowitch and Govindjee, "Photosynthesis", John Wiley and Sons, Inc., New York (1969).

7. R.P.F. Gregory, "Biochemistry of Photosynthesis", Wiley-Interscience, London (1971).
8. D.W. Reed and R.K. Clayton, Biochem. Biophys. Res. Commun., 30, 471 (1968).
9. G. Feher, Photochem. Photobiol., 14, 373 (1971).
10. M.Y. Okamura, L.A. Steiner, and G. Feher, Biochemistry, 13, 1395 (1974).
11. S. Strayley, R. Clayton, W. Parsons, and D. Mauzerall, Biochim. Biophys. Acta, 305, 597 (1973).
12. J.D. McElroy, G. Feher, and D. Mauzerall, Biochim. Biophys. Acta, 267, 363 (1972).
13. J.R. Norris, M.E. Druyan, and J.J. Katz, J. Am. Chem. Soc., 95, 1680 (1973).
14. G. Feher, A.J. Hoff, R.A. Isaacson, and J.D. McElroy, Biophys. J. 13, 61a (1973).
15. J. McElroy, G. Feher, and D. Mauzerall, Biophys. J., 10, 204a (1970).
16. J.S. Leigh Jr. and P.L. Dutton, Biochem. Biophys. Res. Commun., 46, 414 (1972).
17. G. Feher, M.Y. Okamura, and J.D. McElroy, Biochim. Biophys. Acta, 267, 222 (1972).
18. R.L. Hall, M.C. Kung, M. Fu, B.J. Hales, and P.A. Loach, Photochem. Photobiol., 18, 505 (1973).
19. L. Slooten, Biochim. Biophys. Acta, 275, 208 (1972).
20. J.D. McElroy, D.C. Mauzerall, and G. Feher. Biochim. Biophys. Acta, 333, 261 (1974).
21. W.W. Parsons, Biochim. Biophys. Acta, 189, 384 (1969).
22. W.W. Parsons, Biochim. Biophys. Acta, 153, 248 (1968).
23. D. Mauzerall, unpublished work.
24. D. Mauzerall, Ann. N.Y. Acad. Sci., 206, 483 (1973).
25. A.A. Krasnovskii, J. Chim. Phys., 55, 968 (1958).
26. A.A. Krasnovskii and K.K. Voinovshaya, Dokl. Akad. Nauk SSSR, 96, 1209 (1954); Chem. Abstr., 48, 13441b.
27. D. Mauzerall, J. Am. Chem. Soc., 84, 2437 (1962).
28. D. Mauzerall and G. Feher, Biochim. Biophys. Acta, 79, 430 (1964).
29. D. Mauzerall and G. Feher, Biochim. Biophys. Acta, 88, 658 (1964).
30. R.B. Woodward, Ind. Chim. Belge., 1293 (1962).
31. G.L. Closs and L.E. Closs, J. Am. Chem. Soc., 85, 818 (1963).
32. H.W. Whitlock and M.Y. Oester, J. Am. Chem. Soc., 95, 5738 (1973).
33. D. Mauzerall, J. Phys. Chem., 66, 2531 (1962).
34. G.R. Seely and K. Talmadge, Photochem. Photobiol., 3, 195 (1964).
35. A.M. Shul'ga, G.N. Sinyakov, V.P. Suboch, G.P. Gurinovich, Yu.V. Glazkov, A.G. Zhuravlev, and A.N. Sevchenko, Dokl. Akad. Nauk SSSR, 207, 457 (1972); Chem. Abstr. 78, 69294.
36. H. Scheer and J.J. Katz, Proc. Nat. Acad. Sci. U.S.A. 71, 1626 (1974).
37. R.P. Evstigneeva, V.G. Maslov, A.F. Mironov, and A.N. Sidorov, Biofizika, 6, 999 (1971); Chem. Abstr., 76, 80438f.
38. Y. Harel and D. Meyerstein, J. Am. Chem. Soc., 96, 2720 (1974).
39. A.N. Sidorov, V.E. Kholmogorov, R.P. Evstigneeva, and G.N. Kal'tsova, Biofizika, 13, 143 (1968); Chem. Abstr., 68, 84419h.
40. P. Carapellucci and D. Mauzerall, Ann. N.Y. Acad. Sci., 244, 214 (1975).
41. J.-H. Fuhrhop, Angew. Chem., Int. Ed. Engl., 13, 321 (1974).
42. J.-H. Fuhrhop and D. Mauzerall, J. Am. Chem. Soc., 91, 4174 (1969).
43. J.-H. Fuhrhop, Z. Naturforsch., B, 25, 255 (1970).
44. G.J. Höytnik, N.H. Vetthorst, and P.J. Zanstara, Mol. Phys., 3, 533 (1960).
45. D. Mauzerall, Biochemistry, 4, 1801 (1965).
46. P.W. Debye, Trans. Electrochem. Soc., 82, 265 (1942).
47. S.R. Logan, Trans. Faraday Soc., 62, 3416 (1966).
48. D.R. Kearns, Chem. Rev., 71, 395 (1971).
49. V.B. Evstigneev., Photochem. Photobiol., 4, 171 (1965).

50. L.P. Vernon, Acta Chem. Scand., **15**, 1651 (1961).
51. K.K. Eisenstein and J.H. Wang. J. Biol. Chem., **244**, 1720 (1969).
52. E. Rabinowitch, "Photosynthesis and Related Processes," Interscience Publ. N.Y. Vol. I, (1945) p. 67—69 and 483—525; Vol. II (1951) pt. 1. p. 603—691 and 740—804; Vol. II pt. 2, p. 1487—1525.
53. A.A. Krasnovskii, Ann. Rev. Plant Physiol., **11**, 363 (1960).
54. T.T. Bannister and J.E. Bernardini, Photochem. Photobiol., **2**, 535 (1963).
55. V.B. Evstigneev and V.A. Gavrilova, Dokl. Akad. Nauk SSSR, **126**, 410 (1959) p. 136; Engl. Transl. Am. Inst. Biol. Sci., Chem. Abstr., **54**, 1665f (1960).
56. J.J. Katz, R.C. Dougherty, and L.J. Boucher, in "The Chlorophylls", Eds. L.P. Vernon and G.R. Seely, Academic Press, N.Y. (1966), p. 186.
57. J.J. Katz, K. Ballschmiter, M. Garcia-Morin, H.H. Strain, and R.H. Uphaus. Proc. Nat. Acad. Sci. U.S.A., **60**, 100 (1968).
58. D. Mauzerall and G. Feher, unpublished observations, University of California, San Diego, (1971).
59. I. Langmuir and V.J. Shaefer, J. Am. Chem. Soc., **59**, 2075 (1937).
60. H.J. Trurnit and G. Colmano, Biochim. Biophys. Acta, **31**, 434 (1959).
61. W.D. Bellamy, G.L. Gaines, and A.G. Tweet, J. Chem. Phys., **39**, 2528 (1963).
62. B. Ke, in "The Chlorophylls", Ed. L.P. Vernon and G.R. Seely, Academic Press, New York (1966), p. 253.
63. W. Sperling and B. Ke, Photochem. Photobiol., **5**, 857 (1966).
64. T. Trosper, R.B. Park, and K. Sauer, Photochem. Photobiol., **7**, 451 (1968).
65. J. Aghion, S.B. Broyde, and S.S. Brody, Biochemistry, **8**, 3120 (1969).
66. R.A. Cellarius and D. Mauzerall, Biochim. Biophys. Acta, **112**, 235 (1966).
67. G.R. Seely, J. Phys. Chem., **71**, 2091 (1967).
68. J. Sharp and D. Mauzerall, unpublished observations, University of California, San Diego (1966).
69. P. Mueller, D.O. Rudin, H.T. Tien, and W.C. Wescott, Nature London, **194**, 979 (1962).
70. H.T. Tien, Nature London, **219**, 272 (1968).
71. H.T. Tien, J. Phys. Chem., **72**, 4512 (1968).
72. H.T. Tien, "Bilayer Lipid Membranes (BLM); Theory and Practice", Marcel Dekker, New York (1974).
73. M.K. Jain, "The Bimolecular Lipid Membrane: A system," Van Nostrand Reinhold, New York (1972).
74. R.J. Cherry, K. Hsu, and D. Chapman, Biochim. Biophys. Acta, **267**, 512 (1972).
75. A. Steinemann, G. Stark, and P. Läuger, J. Membrane Biol., **9**, 177 (1972).
76. H. Weller, Jr., Biophys. J., **13**, 254a (1973).
77. A.J. Hoff, Photochem. Photobiol., **19**, 51 (1974).
78. F. Hong, and D. Mauzerall, unpublished observation, Rockefeller University (1973).
79. A. Steinemann, N. Alamuti, W. Brodmann, O. Marschall, and P. Läuger, J. Membrane Biol., **4**, 284 (1971).
80. J.S. Huebner, J. Membrane Biol., **8**, 403 (1972).
81. R.J. Cherry, K. Hsu, and D. Chapman, Biochim. Biophys. Acta, **288**, 12 (1972).
82. R.J. Cherry, K. Hsu, and D. Chapman, Biochem. Biophys. Res. Commun., **43**, 351 (1971).
83. H.T. Tien, and S.P. Verma, Nature London, **227**, 1232 (1970).
84. N. Kobamoto and H.T. Tien, Biochim. Biophys. Acta, **241**, 129 (1971).
85. M. Schadt, Biochim. Biophys. Acta, **323**, 351 (1973).
86. H.-M. Ullrich and H. Kuhn, Biochim. Biophys. Acta, **266**, 584 (1972).
87. F. Hong and D. Mauzerall, Biochim. Biophys. Acta, **275**, 479 (1972).
88. F. Hong and D. Mauzerall, Nature New Biol., **240**, 154 (1972).
89. O. Froehlich and B. Diehn, Nature London, **248**, 802 (1974).

90. H.T. Tien, Photochem. Photobiol., 16, 271 (1972).

91. J.S. Huebner and H.T. Tien, Biochim. Biophys. Acta, **256**, 300 (1972).

92. H.-W. Trissl and P. Läuger, Biochim. Biophys. Acta, **282**, 40 (1972).

93. F. Hong and D. Mauzerall, Proc. Nat. Acad. Sci. U.S.A., **71**, 1564 (1974).

94. J.W. Moore and K.S. Cole, in "Physical Techniques in Biological Research", Vol. VI, Part B, Ed. W. Nastuk, Academic Press, New York (1963), p. 263.

95. T. Hainai, D.A. Haydon, and J. Taylor, Proc. Royal Soc. London, Ser. A, **281**, 377 (1964).

96. H. Van Zutphen and D.G. Cornwell, J. Membrane Biol., **13**, 79 (1973).

97. S.J. Baum, B.F. Burham, and R.A. Plane, Proc. Nat. Acad. Sci. U.S.A., **52**, 1439 (1964).

98. D. Mauzerall, Proc. Nat. Acad. Sci. U.S.A., **69**, 1358 (1972).

99. F. Hong, Fed. Proc., **33**, 1268 Abstr. No. 249 (1974).

100. H.T. Witt, in "Nobel Symp. V", Ed. S. Claesson, Almqvist and Wiksell, Stockholm; Interscience, New York, London, Sydney (1967), p. 261.

101. J.B. Jackson and A.R. Crofts, FEBS Lett., **4**, 185 (1969).

102. H.T. Witt, Q. Rev. Biophys., **4**, 365 (1971).

103. B. Masters and D. Mauzerall, unpublished observation, Rockefeller University, (1974).

104. C.F. Fowler and B. Kok, Biochim. Biophys. Acta, **357**, 308 (1974).

105. P. Joliot and R. Delosme, Biochim. Biophys. Acta, **357**, 267 (1974).

STRUCTURAL ANALOGS OF PORPHYRINS

18

STRUCTURAL ANALOGS OF PORPHYRINS

A.W. JOHNSON

School of Molecular Sciences, University of Sussex, Falmer, Brighton, U.K.

Nomenclature

The I.U.P.A.C. recommended nomenclature (1) will be used throughout this chapter since it will be concerned with derivatives of corrin as well as of porphyrin.

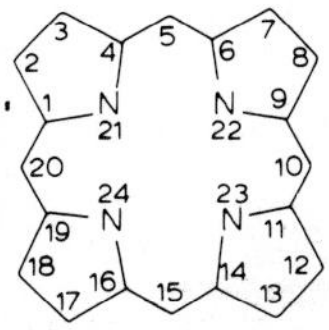

(1)

18.1. Porphyrin analogs with mixed hetero-atoms

Of the available methods for the synthesis of the porphyrin ring one of the most widely used is the so-called 2 + 2 type synthesis whereby two pyrromethanes or pyrromethenes are condensed to give an open-chain tetrapyrrole, which is then either isolated and cyclized to a porphyrin in a separate step, or it undergoes direct cyclization to the porphyrin. This method was extended to a preparation of a 21,22-dioxaporphyrin (3) by condensation of a 5,5′-diformyldifurylmethane (2) with a pyrromethane-5,5′-dicar-

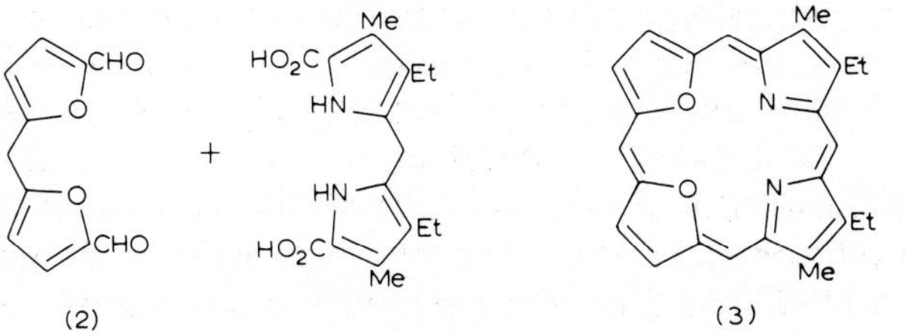

(2)   (3)

Porphyrins and Metalloporphyrins, ed. Kevin M. Smith
© 1975, Elsevier Scientific Publishing Company, Amsterdam, The Netherlands

boxylic acid[1]. Preliminary studies directed towards the synthesis of dioxapor-phyrins[2] were based on a 2 + 1 + 1 approach, the condensation of 2-formyl-furans with a pyrromethane-5,5'-dicarboxylic acid, but the required macro-cycles were not obtained. However a general method for the preparation of porphyrins containing one or two non-adjacent furan and/or thiophene rings based on a 3 + 1 approach has been developed:

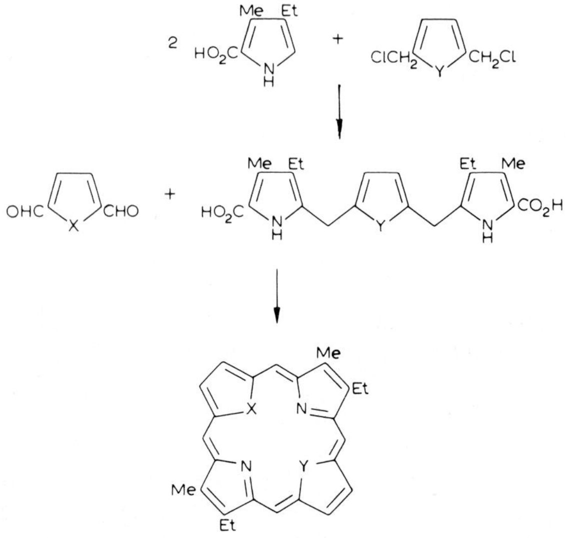

(4)

By this method the 21-oxa-((4); X = O; Y = NH), 21-thia((4); X = S; Y = NH), 21,23-dioxa-((4); X = Y = O), 21,23-dithia-((4); X = Y = S) and 21-oxa-23-thia-porphyrins ((4); X = O; Y = S) were obtained.

The properties of the new ring systems reflected their aromatic character and their overall resemblance to the porphyrins, although some interesting spectral variations were observed. Thus, in trifluoroacetic acid solution, where all the products were present as dications, the Soret band was shifted to shorter wavelengths in the oxa compounds, but to longer wavelengths in the thia derivatives and the intensities of absorptions were much lower than those observed for porphyrins (Table 1).

Another interesting effect was noted in the position of the quartet of so-called Q bands (480—680 nm; Table 2) of the visible spectra of the compounds in pyridine solution. Separation of the (I and II) and (III and IV) bands remained constant while the (II) and (III) band separation increased markedly with the number of nitrogen atoms replaced (Table 3). Also the intensity of band (IV) relative to those of the other bands increased similarly (Table 4). No marked variations of the proton chemical shifts were observed relative to those of etioporphyrin-I, but the unusually high basicity of the dioxa-porphyrins was unexpected and the free bases were not obtained in

TABLE 1

Soret bands (B-bands) in TFA (nm)

Compound	$\lambda_{max}(\epsilon)$	
21,23-Dioxaporphyrin ((4); X = Y = O)	369.5 (sh.), (196,000),	376.5 (248,000)
21,22-Dioxaporphyrin (3)	375.5, (212,000),	385.5 (229,000)
21-Oxaporphyrin ((4); X = O; Y = NH)	385.5 (sh.), (220,000),	389.5 (23,000)
Etioporphyrin-I	395.5 (467,500)	
21-Oxa-23-thiaporphyrin ((4); X = O, Y = S)	398 (205,000)	
21-Thiaporphyrin ((4); X = S; Y = NH)	410 (175,000)	
21,23-Dithiaporphyrin ((4); X = Y = S)	412 (156,200)	

the pure state. The mass spectra revealed that the most intense ion in the molecular ion region ($M + 2$ for the oxa compounds with significant peaks for $M + 4$) did not necessarily represent the molecular weight, although the relative intensities varied considerably with experimental conditions. Finally it was apparent that, of the new macrocycles, only the 21-monoxa derivative

TABLE 2

Q Bands in pyridine (nm)

Compound	Band IV	Band III	Band II		Band I
21-Oxaporphyrin	491 ϵ, 16,020	523 9320	590.5, 599 4160, 3550		651 1880
21-Oxa-23-thiaporphyrin	488.5 ϵ, 15,900	518.5 6750	616.5, 627 1830, 1720	649.5 445	680.5 440
Etioporphyrin-I	500 ϵ, 11,890	533 8340	571 5460		627 4360
21-Thiaporphyrin	500 ϵ, 17,820	528.5 11,340	596, 603 4320, 3440	625 818	656 1605
21,23-Diathiaporphyrin	494 ϵ, 11,030	521 4140	613, 621 1140, 1006	645 252	674.5 640

TABLE 3

Approximate* band separations (nm) in pyridine

Compound	(I-II)	(III-IV)	(II-III)
21-Oxaporphyrin	56	32	71
21-Oxa-23-thiaporphyrin	58	30	103
Etioporphyrin-I	56	33	38
21-Thiaporphyrin	56	28.5	71.5
21,23-Dithiaporphyrin	56.5	27	95

* Measured from centre of doublet of band II

formed stable metal complexes; although a zinc derivative was obtained from the 21-monothia compound, it was stable only in the presence of excess zinc ions.

The preparation of certain 5-thiaporphyrins (5) by condensation of 5,5'-diformyl-dipyrrylsulfides and pyrromethane-5,5'-dicarboxylic acids is referred to in Section 18.2.1 in connection with corrole synthesis. Such compounds readily lose sulfur to form corroles.

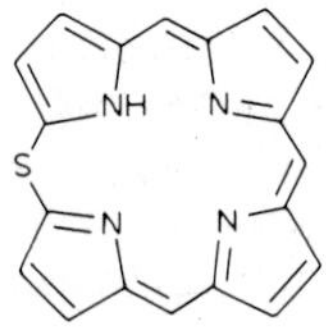

(5)

18.2. Corroles and tetradehydrocorrins
18.2.1. Synthesis

The three ring systems to be considered are corrole ((6), including metal complexes), 1-methyltetradehydrocorrin [as its nickel complex, (7)] and

TABLE 4

Relative intensities of Q bands in pyridine

Compound	Band IV	Band III	Band II	Band I
21-Oxaporphyrin	8.55	4.95	2.2	1
21-Oxa-23-thiaporphyrin	35.2	15.3	4.1	1
Etioporphyrin-I	2.7	1.9	1.25	1
21-Thiaporphyrin	11.0	7.1	2.4	1
21,23-Dithiaporphyrin	17.25	6.45	1.7	1

1,19-dimethyltetradehydrocorrin [as its metal complexes (8)]. Early attempts to prepare corrole comprised the attempted cyclization of open-chain tetra-pyrrolic compounds already containing a direct linkage between two of the

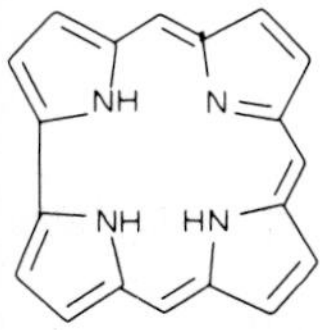

(6)

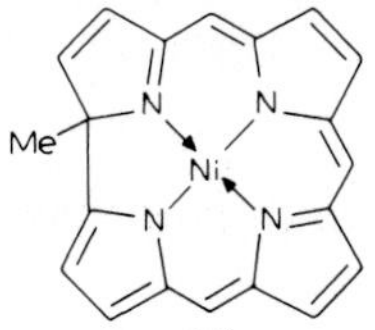

(7)

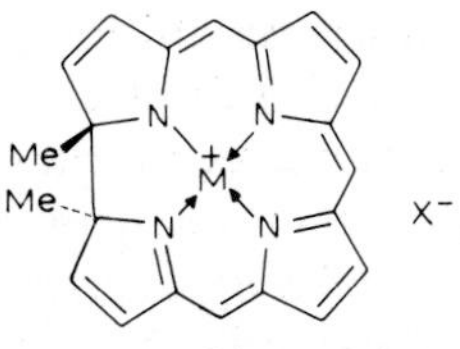

(8)

rings e.g. (9)[3]. However, the introduction of a C_1 bridge to complete the macrocycle from (9), or a variety of related compounds[4], was not achieved. The first successful corrole synthesis[5] resulted from a brief irradiation with visible light of 1,19-diunsubstituted biladienes-*ac* (10) which were themselves prepared by condensation of two equivalents of a 2-formylpyrrole with a pyrromethane-5,5'-dicarboxylic acid or, alternatively, two equivalents of an α-unsubstituted pyrrole with a 5,5'-diformylpyrromethane (11)[cf.6]. In the presence of metal ions, the cyclization of the biladienes-*ac* gave the corre-sponding metal corroles.

Although 2 + 2 type syntheses of corroles by condensation of 5,5'-difor-myl-2,2'-bipyrroles (12) and pyrromethane-5,5'-dicarboxylic acids were unsuccessful, the addition of cobalt salts and triphenylphosphine caused the formation of the corresponding cobalt(III) complex[7]:

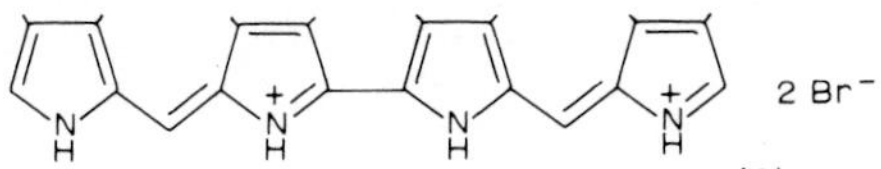

(9)

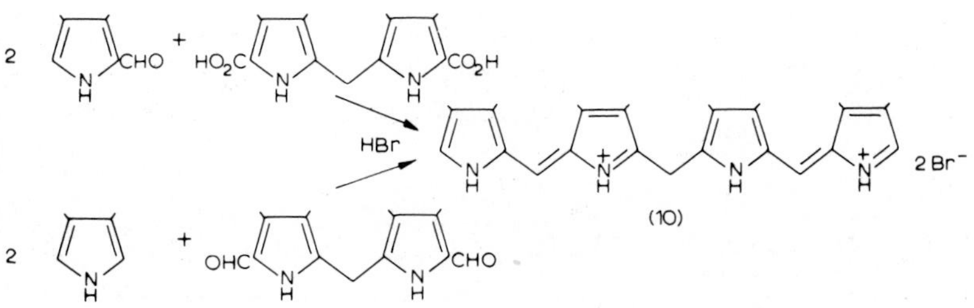

(10)

(11)

References, p. 752

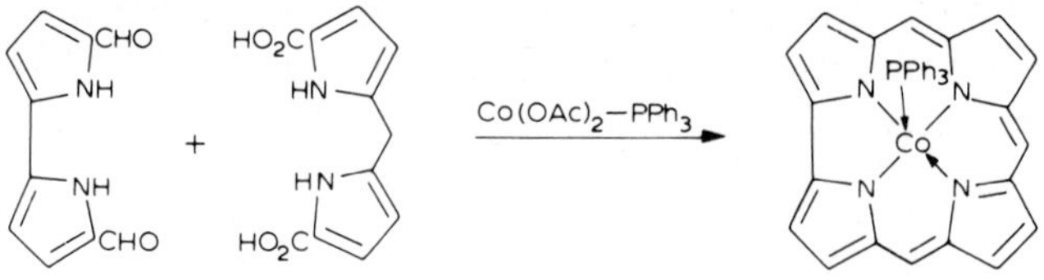

(12)

This method was used for the preparation of the triphenylphosphinatocobalt(III) complex of corrole itself[7].

A third method for corrole synthesis introduced a new principle in that a sulfur atom was introduced into the macrocycle and was subsequently removed by a chelotropic elimination[8]. The required *meso*-thia intermediates were obtained by condensation of a 5,5'-diformyldipyrryl sulfide (13; R = Me) with a pyrromethane-5,5'-dicarboxylic acid at $-10°$C in presence of hydrogen chloride. In an all-alkyl series, the purple-red non-aromatic product (10—20%) was formulated as (14), on the basis of its n.m.r. spectrum, and it was accompanied by an unstable green product, isolated as a stable charged zinc complex (5; R = Me), *i.e.* a derivative of a thiaporphyrin. Treatment of the purple-red product with zinc acetate slowly gave the zinc thiaporphyrin, and when it was heated with triphenylphosphine in boiling *o*-dichlorobenzene it gave the corresponding corrole (15; R = Me, 15%). In another series, the 5,5'-diformyldipyrryl sulfide (13; R = CO_2Et) was condensed with the pyrromethane-5,5'-dicarboxylic acid as before, when a blue air-stable crystalline product (52%) was obtained which was formulated as the *meso*-thiaphlorin (16) on the basis of spectral and chemical evidence. Oxidation of (16) with 2,3-dichloro-5,6-dicyano-1,4-benzoquinone gave the unstable *meso*-thia-

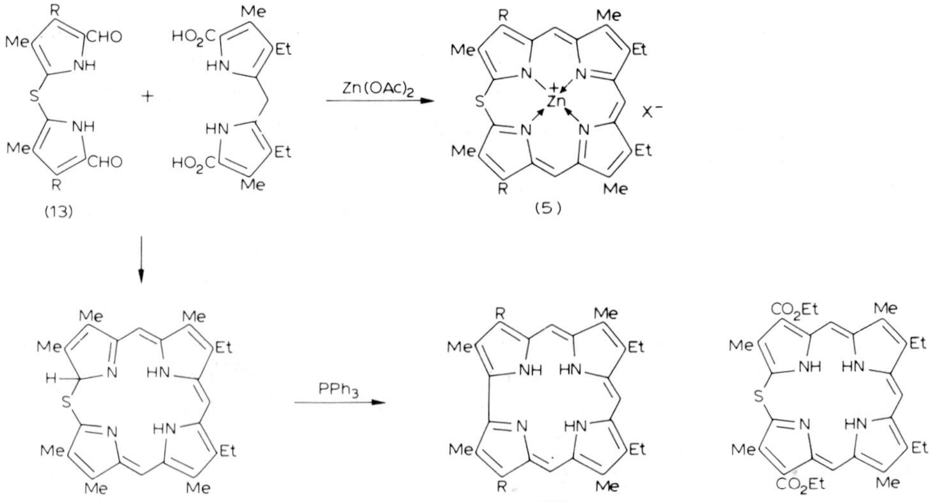

(13)   (5)

(14)   (15)   (16)

porphyrin, readily isolated as the zinc complex (5; R = CO_2 Et). In this series also, sulfur was extruded from the *meso*-thiaphlorin (16) in boiling *o*-dichlorobenzene yielding the corrole (15; R = CO_2 Et; 35—40%) but the yield was increased to 60% when the desulfurization was carried out in presence of triphenylphosphine.

Nickel 1-methyltetradehydrocorrins (7) were obtained from 1-bromo-19-methylbiladienes-*ac* (17) by cyclization in presence of base and nickel ions in an oxygen-free atmosphere[9]. It will be recalled that the same intermediates yield porphyrins when they are heated in solution[10]. The formation of the nickel 1-methyltetradehydrocorrins (7) involves a loss of hydrogen bromide and the cyclization has been formulated as follows:

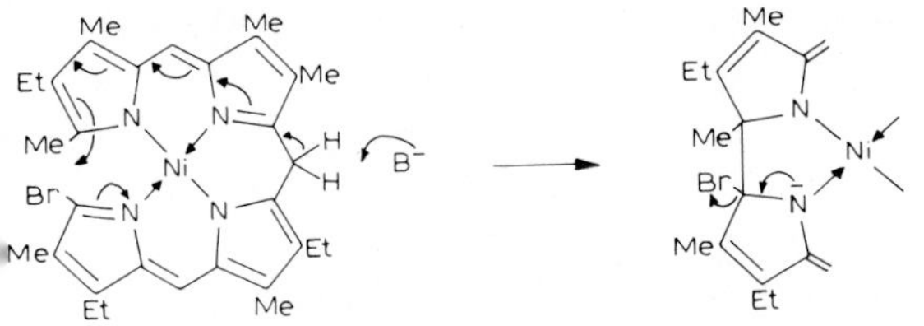

(17)

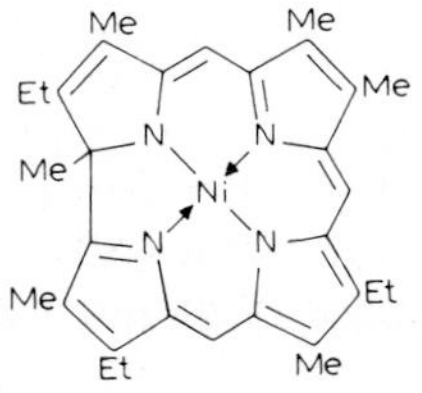

(7)

Only the nickel(II) complexes have been obtained as stable crystalline derivatives and the corresponding cobalt(II) complexes have been observed only in solution. Cyclization of the biladienes in presence of copper salts yielded porphyrins. Examination of other structural variants showed that, unlike bromine, neither 1-alkoxy[11] nor 1-ethoxycarbonyl substituents[9] were extruded although the latter could be eliminated in a subsequent reaction and therefore can be used as a protecting group. Examples were also provided of the cyclization of 1-bromo-19-ethyl and 1-bromo-19-isopropyl-biladienes-*ac*.

The remaining group in this series, the cationic 1,19-dimethyltetradehydrocorrins (8) have been obtained in the form of their nickel(II) and cobalt-(II)[12] or palladium(II) salts[13] by cyclizations of the corresponding 1,19-dimethylbiladienes-*ac* (18) in presence of the appropriate metal acetate, base, and air. An earlier attempt to effect this cyclization in presence of copper(II) salts led to the formation of porphyrins with the elimination of one carbon atom[14], a method which has since been used extensively for porphyrin

References, p. 752

synthesis using variously dimethylformamide[15], pyridine[16], and methanol—acetic acid[17] as solvent. Many 1,19-dimethyltetradehydrocorrin metal salts have been prepared by the above cyclization procedure and the methyl groups shown to occupy the *trans* configuration by resolution of the D-camphorsulfonate of a nickel complex[18]. Consideration of the mechanism of the electrocyclic cyclization to give such a product shows that it must be a conrotatory process, but this is forbidden for the corresponding anion, a $4n + 2$ π-electron system, which is obtained initially by the action of alkali on the nickel complex of the biladiene-*ac*. However if the anion is oxidized to the cation (19), the cyclization then becomes an allowed process. This view of the reaction pathway was confirmed by treatment of the nickel biladiene-*ac* complex with trityl perchlorate, i.e. hydride abstraction, in absence of oxygen when the cyclized product was obtained in good yield.

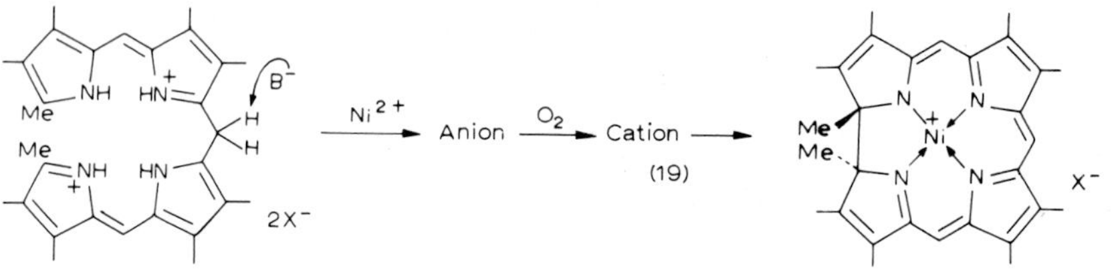

Variations on the metal 1,19-disubstituted tetradehydrocorrin salt structure include 1,19-dialkoxy[19], 1-alkoxycarbonyl-19-alkyl[12] and 1,19-dialkoxycarbonyl[12] derivatives as well as examples containing β-acetic[20], β-propionic ester[15,20] and β-hydroxy[21] substituents.

A useful method for the preparation of 1-methyl-19-alkyltetradehydrocorrin nickel salts, i.e. containing different quaternary substituents, is the alkylation of the nickel 1-methyltetradehydrocorrins with alkyl halides[22] $(7 \rightarrow 8)$ when the additional alkyl group is introduced exclusively at the 19-position. Although this is not an obvious point of reactivity towards electrophilic reagents, α-substitution of pyrroles is usually more favored than β-substitution, and of all the pyrrolic α-positions in (7), substitution at the 19-position causes the minimum interference with the conjugated system in the transition state. When a solution of a cobalt(II) 1-methyltetradehydrocorrin was prepared in degassed solvents in absence of oxygen and was treated with methyl iodide, a similar methylation occurred.

18.2.2. Physical properties of corroles and metal tetradehydrocorrins

Corrole contains an aromatic 18 π-electron chromophore and the spectroscopic properties of 8,12-diethyl-2,3,7,13,17,18-hexamethylcorrole[5] reflect this, e.g. the *meso*-proton n.m.r. signals at τ 0.6—1.2 and imino proton signals near τ 13.5, and the molecular ion being the base peak in the mass

TABLE 5

Visible absorption spectra of nickel(II) macrocyclic complexes, λ_{max} nm (ϵ).

Corrole[5]	1-Methyltetradehydrocorrin[9]	1,19-Dimethyltetradehydrocorrin (Chloride)[12]
349 (45,100)	294 (14,500)	273 (30,550)
652 (9200)	321 (16,000)	353 (32,600)
	349 (14,800)	408 (5570)
	417 (30,400)	456 (4500)
	794 (8360)	558 (13,980)

spectrum[23]. The visible spectrum showed a strong Soret band at 396 nm (ϵ, 123,100) with other maxima at 536, 550 and 593 nm (ϵ, 18,100, 18,100 and 21,450). Nevertheless an X-ray examination of the corrole showed appreciable deviations from the plane[24].

All three ring systems (6), (7) and (8) form metal complexes and the above table summarizes the maxima (ϵ) in the visible spectra of the nickel(II) complexes of β-alkyl derivatives.

Probably because of distortions from the plane, the nickel corroles are paramagnetic[5], but the nickel 1-methyltetrahydrocorrins show n.m.r. signals corresponding to the *meso*-protons at τ 2.8—4.0 and the nickel 1,19-dimethyltetradehydrocorrin salts at τ 2.3—2.5 reflecting the loss of aromatic character consequent on the break in conjugation. The chemical shifts of the *meso*-protons in 5-substituted nickel tetradehydrocorrin salts (see below) have been correlated with the ρ-constants of the substituents[25].

18.2.3. Chemical properties of corroles and their metal complexes

Corroles (6) contain an 18 π-electron system and are aromatic. They can form anions (20) and cations (21) and both of these are aromatic also, the charge being localized over the ring[26]. In strong acids (H_2SO_4, FSO_3H, etc.) a non-aromatic diprotonated species is obtained, formulated as (22) on the basis of n.m.r. spectra.

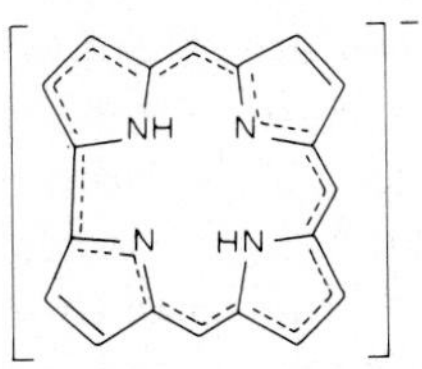

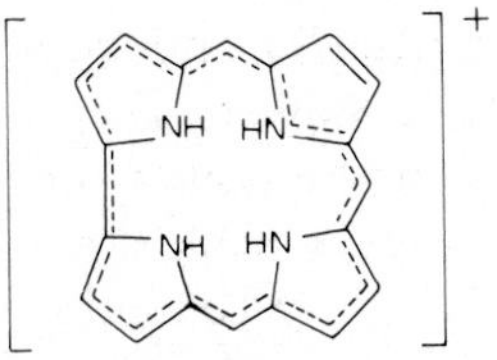

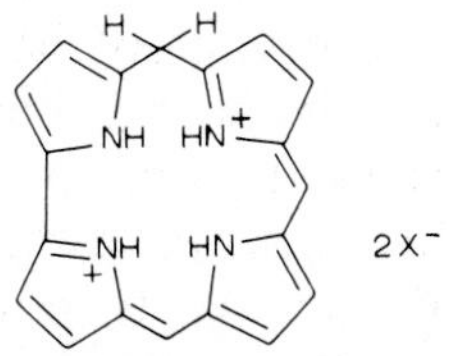

Reaction of the corrole anions with alkyl (methyl, ethyl, allyl, and 3,3-dimethylallyl) halides causes formation of a separable mixture of the N(21) (23) and N(22)-alkylcorroles (24), of which the latter appeared to possess a higher degree of aromatic character (possibly a measure of deviation from the planar structure) as judged by n.m.r. and visible spectra. With acetyl chloride only the $N(21)$-acetylcorrole was characterized. Further methylation of the $N(21)$-methylcorrole gave the $N,N(21,22)$-dimethylcorrole salt which was thermally unstable and reverted to the $N(21)$-monomethyl derivative, the more stable isomer, on heating. This conclusion was reached also through a study of $N(22)$-allylcorrole which, in refluxing toluene, formed 24% of the $N(21)$-isomer and 29% of the parent corrole. This rearrangement was shown to be intermolecular as a similar reaction of the $N(22)$-3,3-dimethylallyl derivative proceeded without inversion, and the yield of the rearranged product was reduced appreciably in the presence of cumene, a radical trapper[26].

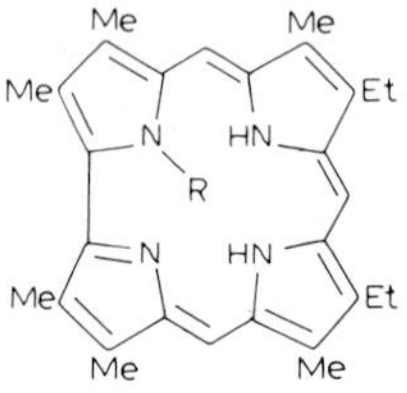

Like the porphyrins, the corroles readily form metal derivatives, but in the case of the nickel(II) and copper(II) corroles, only two of the three imino hydrogens are replaced and the location of the third is still uncertain, although in the case of cobalt, it is the cobalt(III) derivative which is obtained where all three imino hydrogens are replaced by the metal. These cobalt(III) complexes can exist as square planar, pyramidal or octahedral complexes depending on the number of axial ligands[7]. It is the pyramidal form (25; $R = Py$, PPh_3, p-CN $\cdot$ C$_6$H$_4$ $\cdot$ CH$_3$) which is obtained normally, but the pyridino complex can lose pyridine to form the paramagnetic square planar cobalt(III) complex when it is heated in chloroform solution, and all of the pyramidal complexes in presence of excess pyridine tend to form the octahedral dipyridinocobalt(III) species. The structures of these cobalt complexes were supported by electrochemical studies. An iron(III) corrole was also described[7].

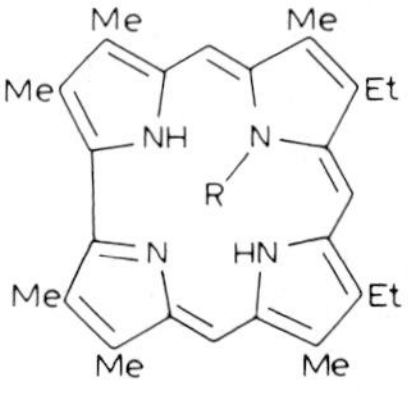

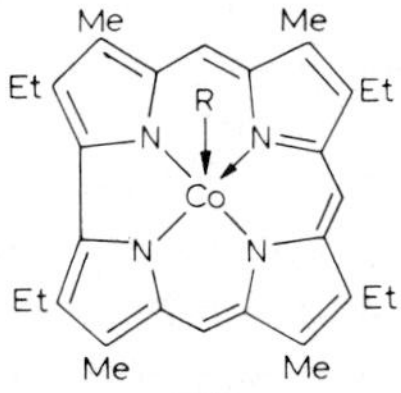

Unfortunately both the nickel(II) and copper(II) corroles are paramagnetic and n.m.r. studies with a view to locating the 'extra' hydrogen have been of no avail, although from the visible spectra, the nickel and palladium complexes may differ in this respect from the copper complex[27]. The 'extra' hydrogen can be removed readily by the action of base with the formation of aromatic metal corrole anions, and palladium corrole anions have been isolated as crystalline dihydrates of the pyridinium salts.

Alkylation of the nickel corrole anions also gave aromatic alkyl derivatives where the new alkyl group is appreciably shielded (τ 12.6 for the methyl singlet). The nickel $N(21)$-methyl corrole structure was proposed after an

X-ray examination of the analogous copper derivative[28] and a later X-ray study of the nickel corrole ethylation product (26) confirmed this[29]. Moreover the nickel $N(21)$-methylcorroles could also be formed by reaction of hexakis(methyl cyanide) nickel(II) perchlorate with $N(21)$-methylcorroles[30]. Palladium and copper can be introduced directly into $N(21)$-methylcorroles likewise and a copper $N(22)$-methylcorrole was also prepared. As the visible spectra of copper corrole and copper $N(21)$-methylcorrole were similar, it is possible that the 'extra' hydrogen of copper corrole (but not nickel corrole) is located at N-21[26].

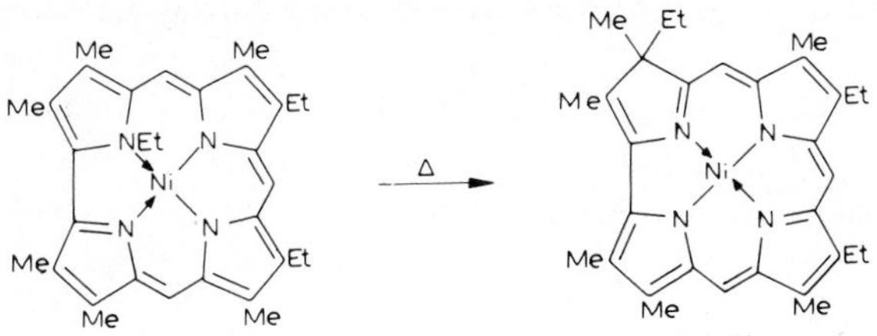

(26)   (27)

When the nickel $N(21)$-methylcorroles were heated in boiling chlorobenzene they rearranged to isomeric products [e.g. (27)] containing gem-dialkyl groups at C-3[30]. This was proved by the use of ethyl marker groups and by n.m.r. studies. The corresponding $N(21)$-ethyl and -propyl derivatives rearranged even in refluxing benzene and a reaction of the nickel corrole anion with allyl bromide gave the nickel 3-alkyl-3-allylcorrole directly together with several nickel *meso*-allylcorroles. The size of the alkyl group is an important feature in these alkylation-rearrangement reactions for similar reaction of the nickel corrole anion with isobutyl iodide in refluxing acetone gave the 3,3-disubstituted product (58%) with none of the $N(21)$-isobutyl derivative. The n.m.r. *meso*-proton signals rise from τ 0.6 in the $N(21)$-alkyl derivatives to 2.43—3.32 in the rearranged products, and the driving force to the rearrangement clearly does not involve a gain in aromatic character and must involve factors such as the formation of a strong carbon—carbon σ-bond from a weak carbon—nitrogen σ-bond, relief of steric strain and the gain in ligand field stabilization energy when nickel returns to the more ideal square planar geometry assumed in the product. The rearrangement is regarded as two consecutive [1,5]-sigmatropic shifts of the alkyl group across the face of the molecule and the absence of 'crossed' products in the thermolysis of mixtures is confirmation of the intramolecular nature of the reaction as also was the insensitivity of the reaction to the presence of radical trapping agents[30].

Methylation of the palladium corrole anion caused the formation of the $N(21)$-methyl derivative (6%) together with the palladium 3,3-dialkylcorrole (24%) and both products were shown to be stable under the reaction condi-

References, p. 752

tions. In refluxing o-dichlorobenzene the palladium N(21)-methyl derivatives underwent rearrangement but on the other hand, thermolysis of the copper N(21)-methyl corroles caused loss of the methyl groups and regeneration of the parent copper corrole[26].

Treatment of the square planar cobalt(III) corrole in tetrahydrofuran with phenyllithium or with phenylmagnesium bromide gave a green product formulated as the cobalt(II) N(21)-phenylcorrole (28) but attempts to prepare the analogous N(21)-alkyl derivatives have been unsuccessful[7].

Chemical oxidation of triphenylphosphinecobalt(III) 2,8,12,18-tetra-ethyl-3,7,13,17-tetramethylcorrole with an excess of 2,3-dichloro-5,6-dicyano-1,4-benzoquinone gave, as the major product, the corresponding 3,17-di-formyl derivative (29; 45%).

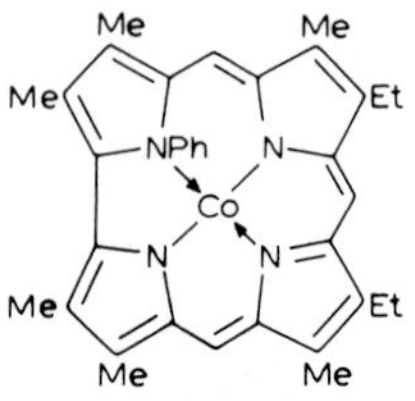

(28) 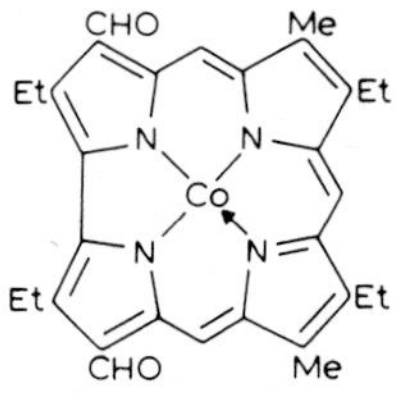   (29)

18.2.4. Chemical properties of nickel 1-methyltetradehydrocorrins

The n.m.r. spectra (above) of these compounds (7) suggested that they have little aromatic character and this is substantiated by the chemical behavior. Thus, reaction with dimethyl acetylenedicarboxylate occurs in Diels—Alder fashion to yield (30; isolated as the corresponding salts)[31] by a [16 + 2] type thermal addition. When the nickel 1-alkyltetradehydrocorrins [e.g. (31)] were heated, the products were the same nickel 3,3-dialkylcorroles (27) as were formed from the thermolytic rearrangements of the nickel N(21)-alkylcorroles [e.g. (27)][32]. By the use of ethyl groups as markers, and taking into account the greater migratory aptitude of ethyl compared with methyl, it was shown that the C-1 substituent migrated first to C-2 and then to C-3 by a concerted mechanism. In extensions of these observations, it was shown that C-1 allyl groups underwent rearrangement to C-3 particularly easily, and that 3,3-dimethylallyl groups were not inverted in the course of the rearrangement. Moreover, ethoxycarbonyl angular substituents could also be rearranged from C-1 to C-3, and these novel migrations also proceeded under experimental conditions milder than those required for the rearrangement of alkyl groups.

As discussed above, both protonation and alkylation of the nickel 1-methyltetradehydrocorrins occurs at C-19 and yields the corresponding nickel 1,19-disubstituted tetradehydrocorrin salts.

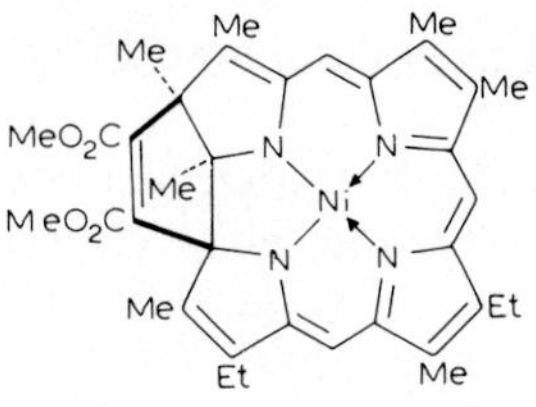

(30)

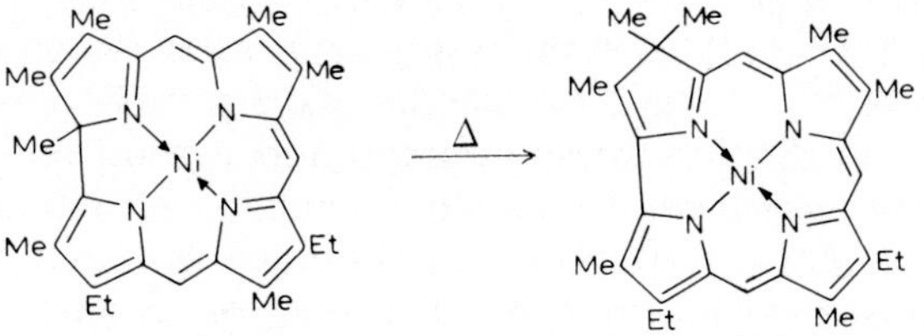

(31)   Δ →   (27)

18.2.5. Chemical properties of metal 1,19-disubstituted tetradehydrocorrin salts

18.2.5.1. Protonation

The protonation of the nickel and cobalt tetradehydrocorrin salts by strong acids has been discussed[33].

18.2.5.2. Removal of angular ester groups

Angular ester substituents in the metal tetradehydrocorrin salts can be readily eliminated by hydrolysis and decarboxylation. The esters may therefore serve a useful protecting function in synthesis, the nickel 19-alkoxycarbonyl-1-methyl derivatives yielding the nickel 1-methyltetradehydrocorrins

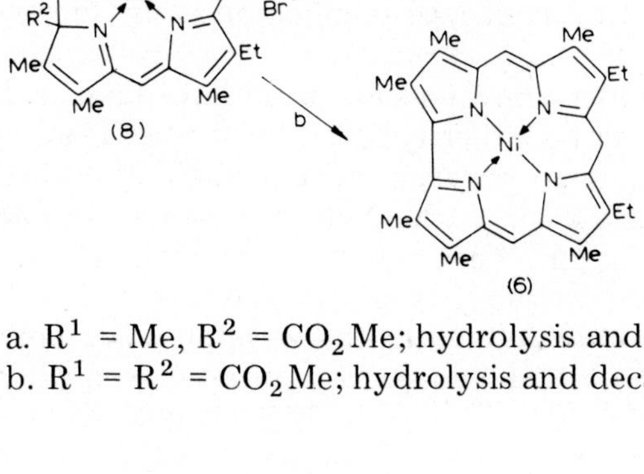

a. R^1 = Me, R^2 = CO_2Me; hydrolysis and decarboxylation
b. R^1 = R^2 = CO_2Me; hydrolysis and decarboxylation

and the nickel 1,19-dialkoxycarbonyl derivatives yielding the nickel corroles after removal of the ester groups[12].

In view of the difficulty of preparing the cobalt 1-methyltetradehydrocorrins, the use of the 19-ethoxycarbonyl protecting group could have value in the preparation of ring systems related to vitamin B_{12}[34].

18.2.5.3. Thermolysis

In contrast to the thermal rearrangement of the nickel 1-alkyltetradehydrocorrins (above), the nickel 1,19-dialkyltetradehydrocorrin salts (8) give porphyrins when they are heated[35]. Of the two angular alkyl groups, one, usually methyl, is the source of the porphyrin fourth *meso* carbon and, depending on the nature of the anion, the other alkyl tends to be either retained as a *meso* substituent [e.g. (32)] (anion = perchlorate) or expelled (anion = bromide). In the case of the chlorides, oxidation accompanies the rearrangement and it is mainly the *meso*-substituted chlorin oxide (33) which is obtained[35,36] although, as with the other anions, mixtures of porphyrin derivatives are obtained. The first stage in this interesting rearrangement is the opening of the macrocycle by a Hofmann-type elimination to yield (34) and in certain cases, e.g. nickel 1-benzyl-19-methyltetradehydrocorrins these intermediates were isolated. The fission of the 1,19-bond of the tetradehydrocorrin could also occur by a valence isomerization to the 1,19-dimethylbilatriene-*abc* (35) which by elimination of HX would give the same intermediate (34). The valence isomerization step should be largely independent of the nature of the anion whereas the Hofmann-type elimination where the anion acts as the base, would be sensitive to variation of the anion. The reversible operation of either ring fission process should result in racemization of optically active starting material since both (34) and (35) are capable on inversion of configuration by passage of the terminal substituents past each other. Various optically active salts of the nickel 1,19-dimethyltetradehydrocorrin were therefore examined[18], the temperature utilized for each racemization being such that the rotation dropped to half its original value to 1—2 h. The iodide (α_D + 990°) and bromide (α_D + 985°) achieved this rate of racemization at ca. 80°, but the camphorsulfonate (α_D + 940°) and perchlorate (α_D + 1150°) required 110° and 189° respectively to achieve a similar rate. Hence the Hofmann-type elimination is the favored mechanism for the ring-opening step.

Cyclization of (34) in the manner shown yields an intermediate from which the angular methyl group can be either expelled by further reaction with the bromide anion or can undergo rearrangement to the *meso*-substituted porphyrin when the anion is a weak nucleophile such as perchlorate. Similar rearrangements to porphyrins have been reported for cobalt(II) tetradehydrocorrin salts[35].

An interesting side reaction was observed during the thermolysis of nickel 1,2,3,7,8,12,13,17,18,19-decamethyltetradehydrocorrin nitrate, when a β-

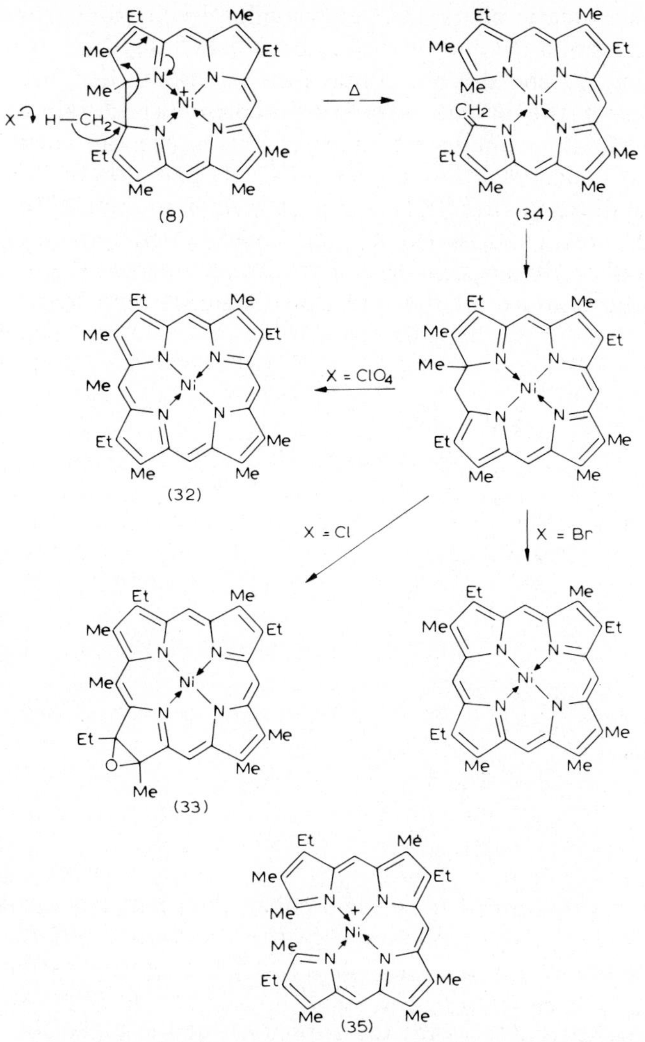

methyl group, possibly at C-3, was converted into a nitrile presumably by the route $CH_3 \rightarrow CH = NOH \rightarrow CN$, the nitrosation of the methyl being caused by the nitrous fumes liberated by decomposition of the nitrate anion[35].

18.2.5.4. Hydrogenation

In the absence of β-substituents the hydrogenation of the parent nickel(II) and cobalt(II) 1,19-dimethyltetradehydrocorrin salts (36) gave the corresponding crystalline metal corrin salts at room temperature and pressure and in fact the complexed metal could act as catalyst for the reaction[37]. The nickel corrin perchlorate (37; M = Ni) showed a n.m.r. spectrum containing singlets at τ 3.68 (1 H) and 4.0 (2 H) corresponding to the *meso*-protons indicating an absence of aromatic character. A crystalline by-product from the hydrogenation was also obtained and proved to be a nickel dihydrocorrin salt although the position of the double bonds was not established with certainty. Hydrogenation of the cobalt tetradehydrocorrin perchlorate at room temperature and 25 atmospheres of hydrogen also gave the corresponding corrin derivative isolated either as the cobalt(II) perchlorate (37; M = Co) or as the neutral dicyanocobalt(III) corrin. A smaller amount of the dicyanocobalt(III) dihydrocorrin was also identified. It was shown that under acid conditions all of these products exchanged the *meso*-protons very rapidly for deuterons. This total synthesis of the corrin involves only three stages.

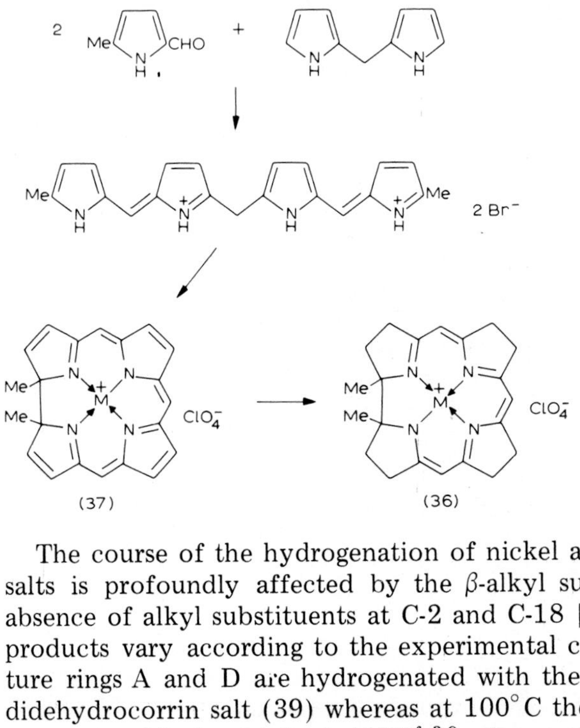

(37)   (36)

The course of the hydrogenation of nickel and cobalt tetradehydrocorrin salts is profoundly affected by the β-alkyl substitution pattern[38]. In the absence of alkyl substituents at C-2 and C-18 [e.g. (38)] the hydrogenation products vary according to the experimental conditions. At room temperature rings A and D are hydrogenated with the formation of the nickel BC-didehydrocorrin salt (39) whereas at 100°C the isomeric nickel AD-didehydrocorrin salt is obtained (40)[cf.39]. Base-catalyzed dehydrogenations of rings B and C (but not A and D) occurred very easily in presence of atmospheric oxygen, and the same pattern of dehydrogenation was observed in the mass spectrometer. Further hydrogenation of these products gave the amorphous nickel corrin salts often accompanied by the corresponding nickel dihydrocorrin salts, some of which were obtained crystalline.

Hydrogenation of nickel tetradehydrocorrin salts (8) bearing alkyl groups at positions 1, 2, 18 and 19 also involves the $\beta\beta$-double bonds of rings B and C in the first instance[39], but vigorous hydrogenation yields either the crystalline nickel isocorrin perchlorate (41) or the monodehydro-isocorrin

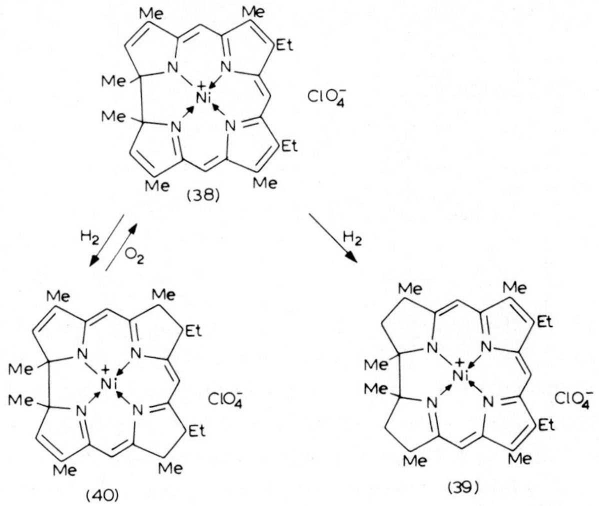

(38)

(40)   (39)

derivative (42) and their formation is a consequence of the steric crowding of the alkyl substituents around the 1,19-linkage[38]. However, hydroxylation of the tetradehydrocorrins appears to involve all four $\beta\beta$-linkages in (8) (see below) and the presence of methyl substituents at the *meso* positions, C-5 and C-15, may also have an effect, so that the conversion of the fully substituted nickel tetradehydrocorrin salts to the corresponding corrins remains a possibility.

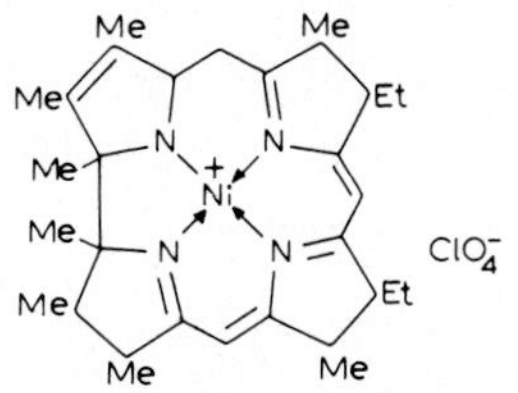

(41) 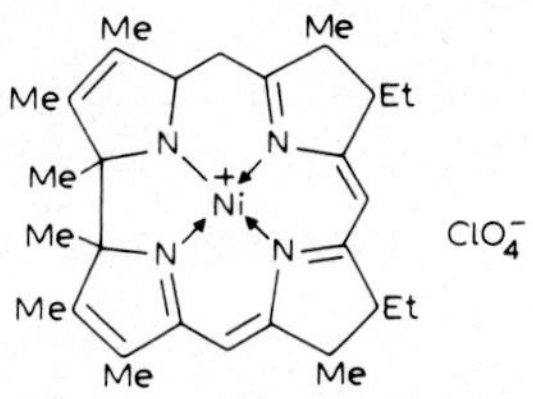   (42)

When the nickel AD-didehydrocorrin perchlorate (40) was treated with base and methyl iodide in an atmosphere of nitrogen the color of the solution changed from orange to deep green and the methylation product (43) was obtained in which methylation of the enamine ring B has occurred as well as dehydrogenation of ring C[39]. The salt of the product was formulated as (44) on the basis of its n.m.r. and mass spectra. Such structures are of interest because of their relation to the vitamin B_{12} chromophore which contains a β-gem dialkyl group in each of the four rings.

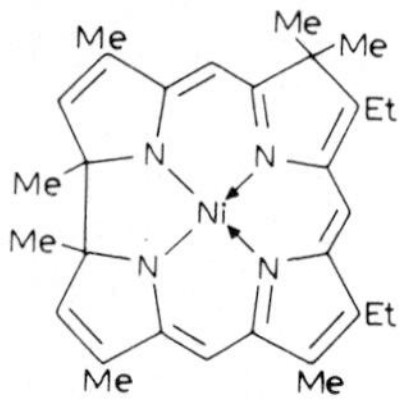

(43)

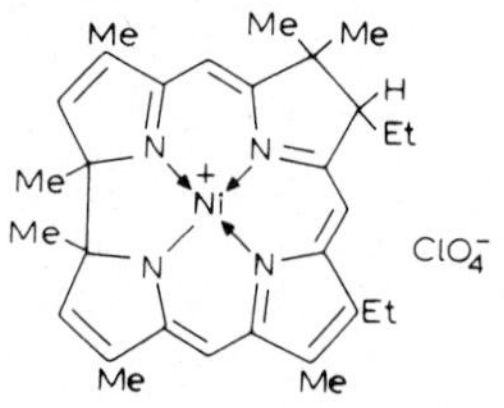

(44)

18.2.5.5. Hydroxylation

As in the porphyrin series[40], hydroxylation of the nickel and cobalt tetradehydrocorrin salts involves the $\beta\beta$-double bonds and as with hydrogenation, rings B and then C are involved in the first place, to yield diols (45) and tetra-ols (46)[41]. Further hydroxylations involve rings A and D to give hexa-ols and octa-ols. It had been shown with the porphyrins[40] that pinacol rearrangement of the 1,2-diols yielded ketones containing gem dimethyl groups, and in the present series similar rearrangements were also brought about by acid treatment, but because of the reduced symmetry of the molecule the diol (45) gave a mixture of two monoketones [part formulae (47) and (48)] and the tetra-ol (46) gave two diketones (49) and (50) which were separated. Other examples of these reactions have been given[20]. Hydrogenolysis of the two diketones gave the corresponding didehydrocorrins (51) and (52).

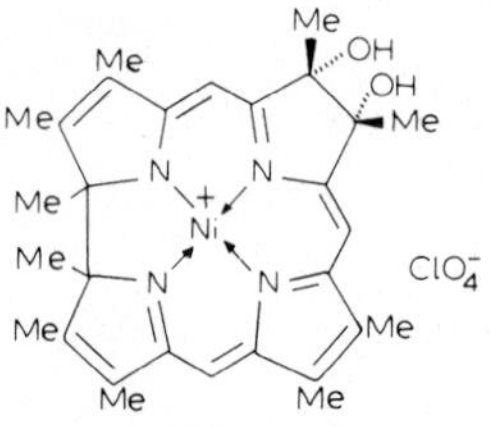

(45)

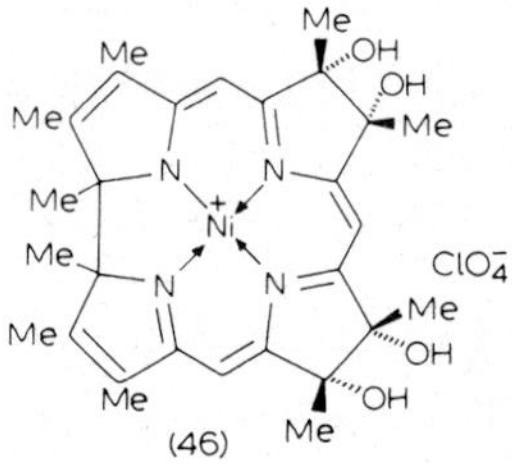

(46)

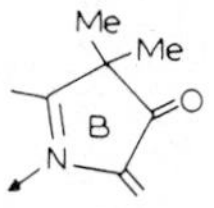

(47)

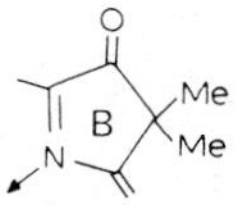

(48)

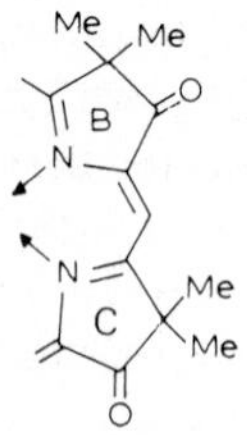

(49)

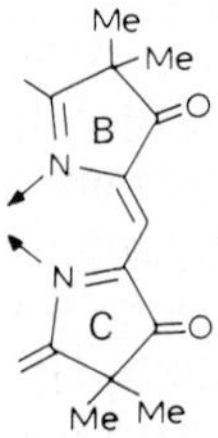

(50)

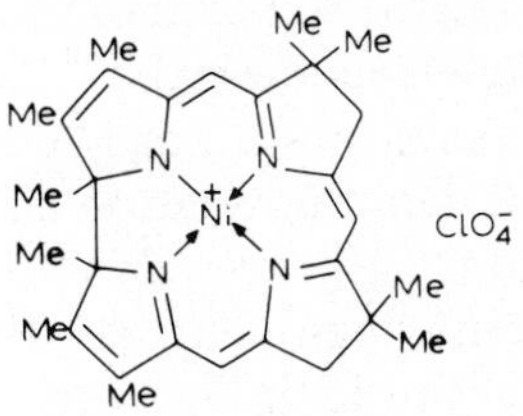

(51) 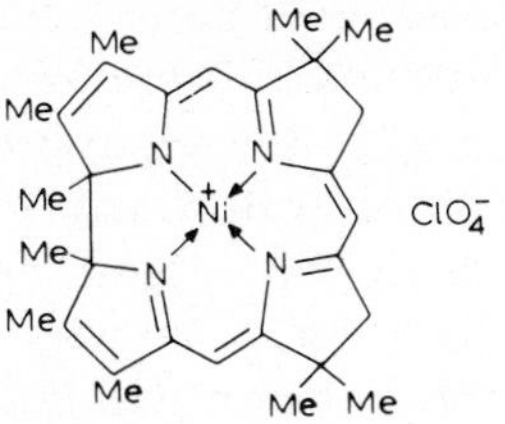   (52)

In a related series where the nickel tetradehydrocorrins contained 1,19-di-ethoxycarbonyl groupings, Inhoffen[42] has shown that the nickel can be removed from the diol, tetraol and rearranged ketones [cf. (45) to (50)] by reduction, e.g. with sodium hydrosulfite, which also opens the macrocycle, and that the resulting bilatriene can be recyclized with cobalt salts, to the cobalt macrocycles.

Compounds related to the above ketones can be obtained by the incorporation of 3,3-dimethylpyrrolidine-2,4-dione (53) as one of the terminal rings of the open-chain tetrapyrrolic intermediate[11] prior to the cyclization. Related condensations to incorporate the unit (53) at both ends of the linear tetracyclic system have been described by Inhoffen[43].

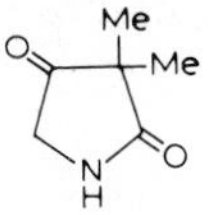

(53)

18.2.5.6. Substitution and redox reactions of nickel salts

The action of alkali on the nickel 1,19-dimethyltetradehydrocorrin salts yields the neutral nickel 5-keto compounds (54; R = H) and a variety of other reactions give the 15-substitution products of the 5-keto derivative (54)[36], e.g. the 15-formyl derivative from the action of chloroform and alkali, the 15-cyano derivative from the action of cyanide, the 15-bromo derivative by oxidation of a monobromo substitution product[44], and the 15-nitro- and 10,15-dinitro derivatives from nitrations[36]. In the cyanation reaction stable mono- and di-cyano radicals were obtained which were converted to 5-mono- and 5,15-di-cyanotetradehydrocorrin salts (55; M = Ni) with acids. Earlier work[9] had indicated that the meso-positions of the nickel tetradehydrocorrin salts would undergo ready deuterium exchange, and halogenation also causes substitution at these positions[35,44]. A dibromo derivative could be isolated as its unstable salts and these, as well as the unstable chlorination product could be converted to the 5,15-dimethyl derivative by reaction with lithium copper methyl[35]. Some disparity exists regarding the nature of the nitration products[35,44,45] but aminomethylation, hydroxymethylation, and chloromethylation reactions yield the 5-substituted tetradehydrocorrin salts and transformations of these products have been described[46]. Vilsmeier formylation of a nickel tetradehydrocorrin perchlorate has been claimed to yield the 10-(5?) formyl derivative[47]. Reductive methylation of nickel decamethyltetradehydrocorrin iodide with sodium anthracenide and methyl iodide caused methylation at C-10 with concurrent ring-opening, but recyclization gave the 10-methyl derivative of the nickel tetradehydrocorrin salt[48].

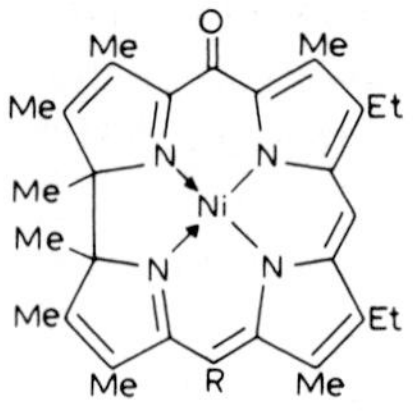

Reduced products of a nickel(II) 1,19-diethoxycarbonyltetradehydrocorrin nitrate have been generated by sodium film reduction in tetrahydrofuran under high vacuum[49]. Both one- and two-electron reduction products were observed, the former being a stable free radical (cf. the cyanation reaction described above) and the latter a nickel(II) complex with both extra electrons in the ligand π-orbitals.

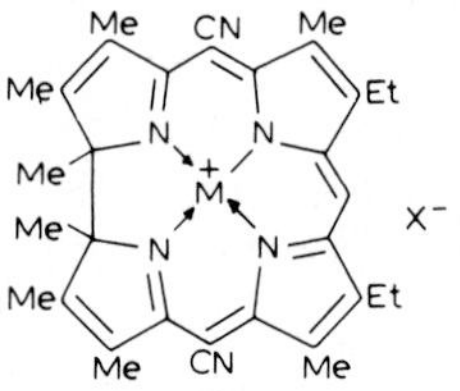

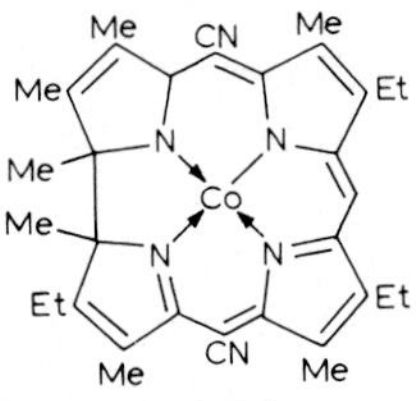

18.2.5.7. Substitution and redox reactions of cobalt salts

In contrast to the nickel salts where reactions involve the ligand, the cobalt salts usually undergo reactions at the metal resulting in a change in oxidation state[50]. Thus reaction of the cobalt(II) salt with cyanide, azide, or nitrite ions yields neutral cobalt(III) derivatives. In contrast to the thermolysis of cobalt(II) salts which gave porphyrins (Section 18.2.5.3), thermolysis of the neutral dicyanocobalt(III) derivatives gave a mixture of the stable neutral 5-cyano- and 5,15-dicyano tetradehydrocorrins. The cobalt(I) state was stabilized by the *meso*-electron-attracting cyano groups and the dicyano derivative (55a) could be isolated as a stable solid. The structures were confirmed by electrochemical studies and the neutral cobalt(I) derivatives were shown to be converted reversibly to the corresponding cobalt(II) salts (e.g. 55; M = Co) by the action of acids. 5,15-Dicyano-10-nitro- and 5,10,15-trinitro-cobalt(I) species were also described.

The reduction products of a cobalt(II) 1,19-diethoxycarbonyltetradehydrocorrin iodide formed by sodium film reduction in tetrahydrofuran have been studied. The one-electron reduction product is a stable cobalt(I) species but the two-electron product is a cobalt(II) complex with both electrons located in the ligand[49].

18.3. Corroles with mixed hetero-atoms

Structural variations on corrole include the palladium complexes of the ring systems (56; X = O, S, NMe, NH or tautomer) prepared by cyclization of the palladium derivative of bis(5-bromo-2,2'-dipyrromethenyls) (57) in presence of hydrochloric acid, sodium sulfide, methylamine or ammonia respectively[51]. The macrocycle (56; X = O) was historically the first example containing the direct linkage between two of the pyrrole rings[52]. Examples of the sulfur-containing macrocycle (56; X = S) were also obtained by desulfurization of the dithia-macrocycle (58) in presence of triphenylphosphine when a 1 : 1 mixture (42%) of the two possible monothiacorroles was obtained in a slow reaction which was accelerated by use of the corresponding zinc compounds. Removal of the second sulfur atom did not occur[8].

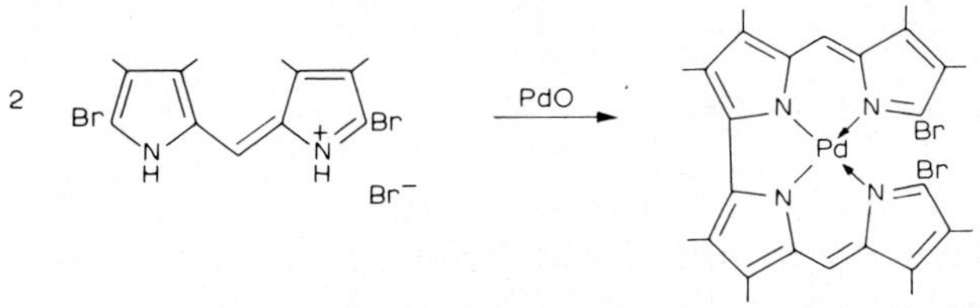

(57)

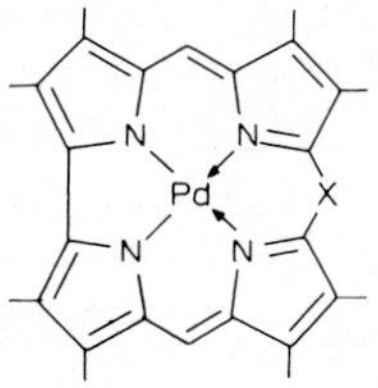

(56)

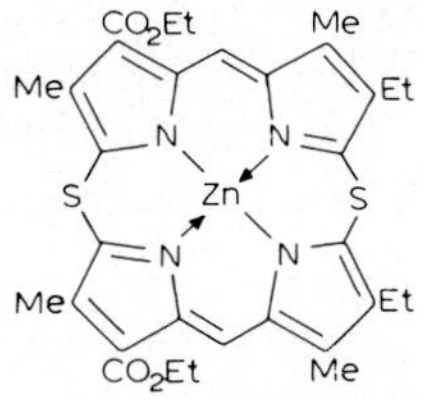

(58)

In another variant on the corrole structure, two adjacent pyrrole rings were replaced by furans in the 21,24-dioxacorroles (59). This ring system was synthesized by application of the sulfur extrusion principle (Section 18.2.1) whereby bis(5-formylfuryl) sulfide (60) was condensed with a pyrromethane-5,5'-dicarboxylic acid in presence of hydrogen bromide and gave the dioxacorrole (59) directly (27—30%)[8].

Reaction of the dioxacorrole with methyl iodide at 100°C in presence of NN-diisopropylethylamine gave a mixture of the N-methyl- ((61); R^1 = Me, R^2 = H) and NN'-dimethyl ((61); R^1 = R^2 = Me) derivatives. Evidence for the trans-disposition of the N-ethyl groups in ((61); R^1 = R^2 = Et) was obtained by partial resolution of the D-camphorsulfonate, and the enhanced basicity of the N-monoalkyl derivatives compared with the parent system (cf. porphyrins and corroles) was marked. Treatment of the dioxacorrole with acetyl chloride and aluminum chloride gave a mixture of the 5-monoacetyl derivative and a nuclear substitution product, possibly a 2,18- or a 3,17-diacetyl derivative[26].

References, p. 752

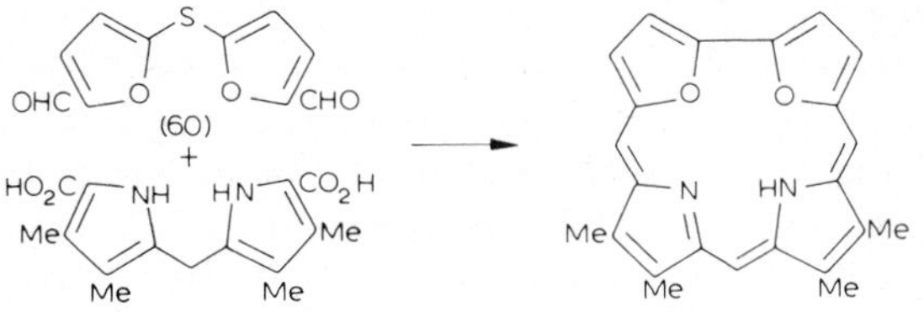

(60)
(59)
(61) 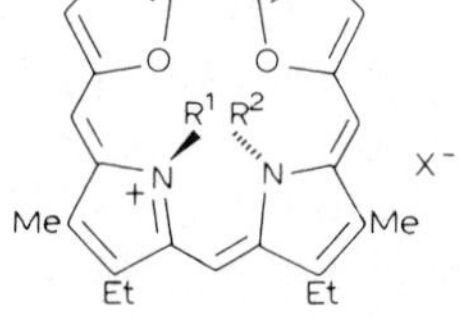

18.4. Sapphyrins and related ring systems

An attempted synthesis of the 21,24-dioxacorrole ring system (above) by condensation of a 5,5'-diformyl-2,2'-bifuryl with a pyrromethane-5,5'-dicarboxylic acid gave a by-product which was identified (n.m.r. and mass spectra) as a dioxasapphyrin (62), a 22 π-electron macrocycle[52]. The all-pyrrole analog (63), sapphyrin, had been referred to previously[53] although experimental details are not available. The synthesis of these 22 π systems has been rationalized by a 2 + 3 approach, e.g. by condensation of a 5,5'-diformyl-2,2'-bipyrrole (64) with a tripyrranedicarboxylic acid (65) to give the sapphyrin (46%), which on n.m.r. evidence was aromatic (τ, meso-H, -0.3 to -0.5; NH, 13.9) and which showed a Soret band at 450 nm (ϵ, 530,200).

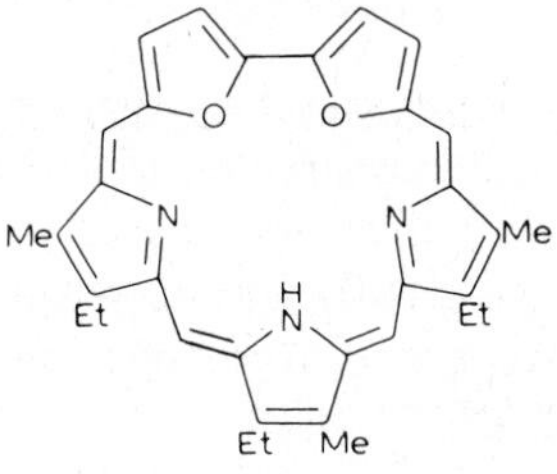

(62)

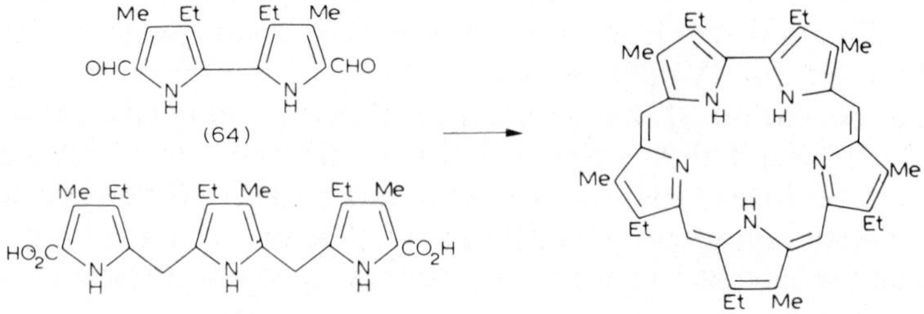

(64)

(65)

(63)

Related syntheses were used for preparations of the dioxasapphyrin (62), where a sulfur extrusion type synthesis from bis(5-formylfuryl) sulfide (60) was also employed successfully, a thiasapphyrin (66), a dioxanorsapphyrin (67), the first macrocycle containing two direct linkages between the five-membered rings, and spectral evidence was obtained for the existence of a norsapphyrin (68).

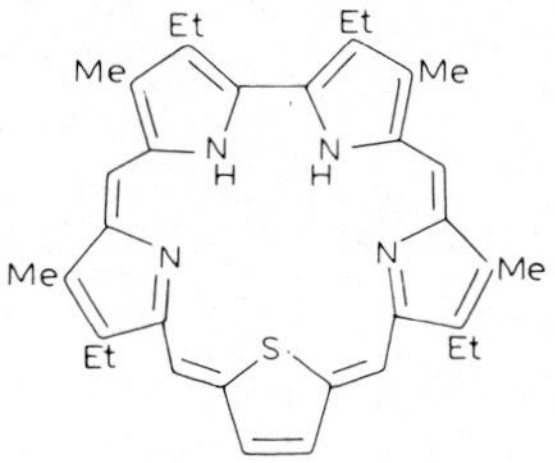

(66)

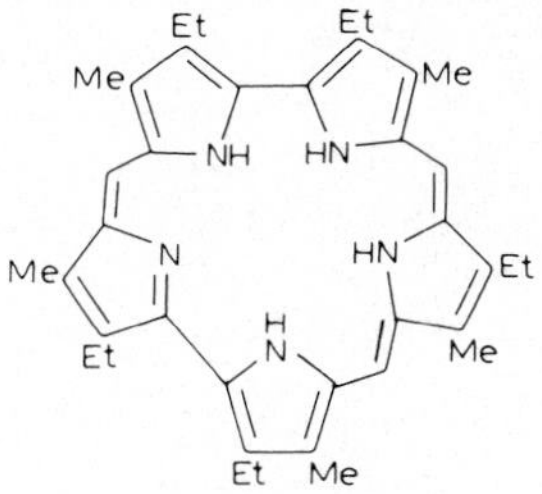

(67)

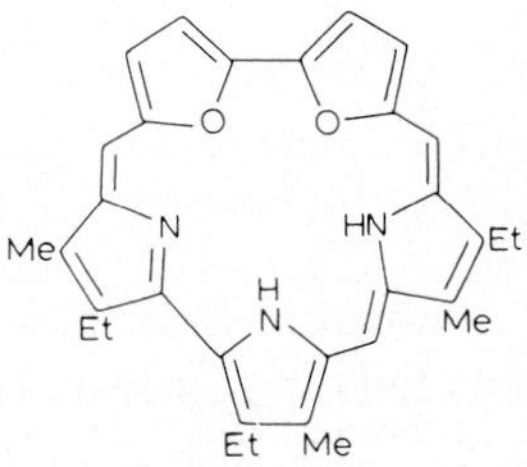

(68)

The mass spectra of these compound showed strong (M + 2) peaks especially the dioxasapphyrin which exhibits strong basicity like the dioxaporphyrins[2] and certain dioxacorrole derivatives[26]. The 22 π-electron macrocycles are of interest in view of the prediction that [4n + 2] annulenes should be aromatic up to and including $n = 5$, but that [26] annulene should be non-aromatic[54]. Experimental evidence[55] however, proved not to be in agreement with the theoretical prediction.

18.5. Corphins

The ring system (69), isomeric with hexahydroporphyrin, has been named corphin[56]. The first example was prepared from the bicyclic lactam (70) by two successive cyclizations of the palladium (but not nickel or cobalt) complex in presence of triethyloxonium fluoroborate (Meerwein reagent) and base. In presence of trifluoroacetic acid the palladium corphin rearranged to the fully conjugated form (71).

References, p. 752

The so-called porphyrin gemini-ketones [e.g. (72)] formed from porphyrins by hydroxylation in one, two, or three rings, with osmium tetroxide and subsequent acid rearrangement[40], can be regarded as dehydro-corphin derivatives (Section 18.2.5.5).

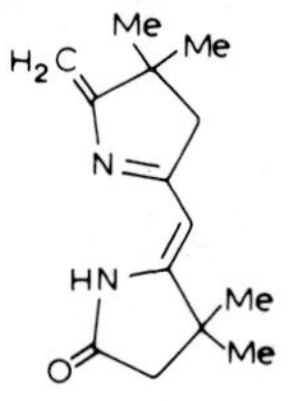

(70) 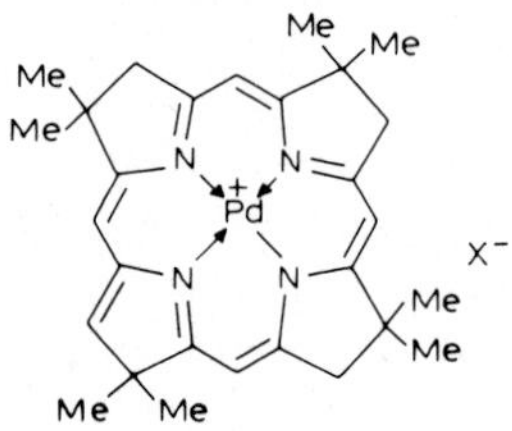   (69)

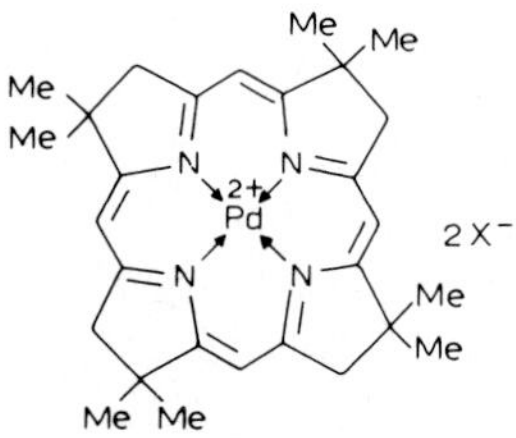

(71) 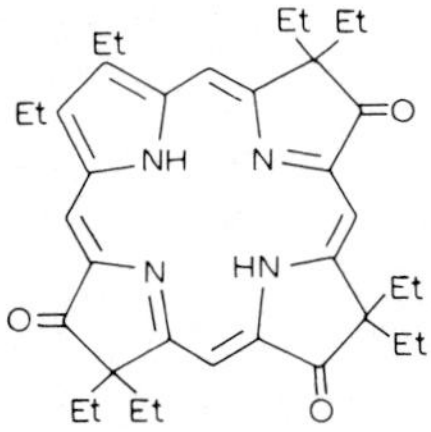   (72)

As with corrins, no method exists for removal of transition metals such as Ni, Co, Pd, or Pt from the robust corphin chelating system; however, metal-free corphins have recently become available[57] by transmetalation of a Ni or Pd precorphinoid (by treatment with cyanide and then zinc perchlorate) to give the zinc precorphin, and cyclization. The zinc atom is easily removed from the resultant zinc complex, giving the metal-free corphin, isolated as the crystalline chloride, bromide or perchlorate salt (73).

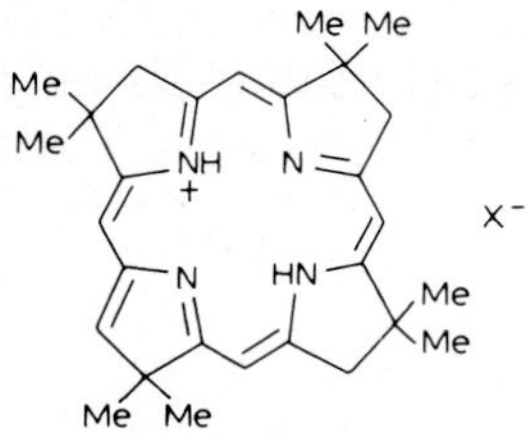

(73)

$X = Cl, Br, ClO_4$

References

1. Professor T.J. King, University of Nottingham, private communication.
2. M.J. Broadhurst, R. Grigg, and A.W. Johnson, J. Chem. Soc. C, 3681 (1971).

3. H. Fischer and A. Stächel, Hoppe-Seyler's Z. Physiol. Chem., **258**, 121 (1939).
4. D. Dolphin, R.L.N. Harris, J.L. Huppatz, A.W. Johnson, I.T. Kay, and J. Leng, J. Chem. Soc. C, 98 (1966).
5. A.W. Johnson and I.T. Kay, J. Chem. Soc., 1620 (1965).
6. A.R. Battersby, G.L. Hodgson, M. Ihara, E. McDonald, and J. Saunders, J. Chem. Soc., Perkin Trans. 1, 2923 (1973).
7. J.M. Conlon, A.W. Johnson, W.R. Overend, D. Rajapaksa, and C.M. Elson, J. Chem. Soc., Perkin Trans. 1, 2281 (1973).
8. M.J. Broadhurst, R. Grigg, and A.W. Johnson, J. Chem. Soc., Perkin Trans. 1, 1124 (1972).
9. D.A. Clarke, R. Grigg, R.L.N. Harris, A.W. Johnson, I.T. Kay, and K.W. Shelton, J. Chem. Soc. C, 1648 (1967).
10. R.L.N. Harris, A.W. Johnson, and I.T. Kay, J. Chem. Soc. C, 22 (1966).
11. J.H. Atkinson, A.W. Johnson, and W. Raudenbusch, J. Chem. Soc. C, 1155 (1966).
12. D. Dolphin, R.L.N. Harris, J.L. Huppatz, A.W. Johnson, and I.T. Kay, J. Chem. Soc. C, 30 (1966).
13. A. Gossauer, H. Maschler, and H.H. Inhoffen, Tetrahedron Lett., 1277 (1974).
14. A.W. Johnson and I.T. Kay, J. Chem. Soc., 2418 (1961).
15. R. Grigg, A.W. Johnson, R. Kenyon, V.B. Math, and K. Richardson, J. Chem. Soc. C, 176 (1969).
16. P.S. Clezy and A.J. Liepa, Aust. J. Chem., **24**, 1027 (1971).
17. A.F. Mironov, V.D. Rumyantseva, B.V. Rumyantseva, and R.P. Evstigneeva, Zh. Org. Khim., **7**, 165 (1971); Chem. Abstr., **74**, 112026 (1971).
18. R. Grigg, A.P. Johnson, A.W. Johnson, and M.J. Smith, J. Chem. Soc. C, 2457 (1971).
19. H.H. Inhoffen, H. Maschler, and A. Gossauer, Ann. Chem., 141 (1973).
20. H.H. Inhoffen, F. Fattinger, and N. Schwarz, Ann. Chem., 412 (1974).
21. H.H. Inhoffen, N. Schwarz, and K.-P. Heise, Ann. Chem., 146 (1973).
22. R. Grigg, A.W. Johnson, and K.W. Shelton, J. Chem. Soc. C, 1291 (1968).
23. A.H. Jackson, G.W. Kenner, K.M. Smith, R.T. Aplin, H. Budzikiewicz, and C. Djerassi, Tetrahedron, **21**, 2913 (1965).
24. D.M. Hodgkin, H.R. Harrison, and O.J.R. Hodder, J. Chem. Soc. B, 640 (1971).
25. T.A. Melent'eva, N.D. Pekel, N.S. Genokhova, and V.M. Berezovskii, Zh. Obshch. Khim, **43**, 1337 (1973); Chem. Abs., **79**, 91211 (1973).
26. M.J. Broadhurst, R. Grigg, G.S. Shelton, and A.W. Johnson, J. Chem. Soc., Perkin Trans. 1, 143 (1972).
27. R. Grigg, A.W. Johnson, and G. Shelton, J. Chem. Soc., Perkin Trans. 1, 2287 (1971).
28. R. Grigg, T.J. King, and G. Shelton, Chem. Commun., 56 (1970).
29. G. Shelton, Ph.D. Thesis, University of Nottingham (1970).
30. R. Grigg, A.W. Johnson, and G. Shelton, Ann. Chem., **746**, 32 (1971).
31. A.P. Johnson and M.J. Smith, unpublished results, University of Nottingham.
32. R. Grigg, A.W. Johnson, K. Richardson, and M.J. Smith, J. Chem. Soc. C, 1289 (1970).
33. T.A. Melent'eva, N.D. Pekel, and V.B. Berezovskii, Zh. Obshch. Khim., **42**, 183 (1972); Chem. Abs., **77**, 19623 (1972).
34. J.M. Conlon, J.A. Elix, G.I. Feutrill, A.W. Johnson, M.W. Roomi, and J. Whelan, J. Chem. Soc., Perkin Trans. 1, 713 (1974).
35. R. Grigg, A.W. Johnson, K. Richardson, and K.W. Shelton, J. Chem. Soc. C, 655 (1969).
36. A. Hamilton and A.W. Johnson, J. Chem. Soc. C, 3879 (1971).
37. A.W. Johnson and W.R. Overend, J. Chem. Soc., Perkin Trans. 1, 2681 (1972).

38. A.W. Johnson, W.R. Overend, and A. Hamilton, J. Chem. Soc., Perkin Trans. 1, 991 (1973).
39. I.D. Dicker, R. Grigg, A.W. Johnson, H. Pinnock, and P. van den Broek, J. Chem. Soc. C, 536 (1971).
40. A.W. Johnson and D. Oldfield, J. Chem. Soc., 4303 (1965); H.H. Inhoffen and D. Nolte, Ann. Chem., 725, 167 (1969); R. Bonnett, M.J. Dimsdale, and G.F. Stephenson, J. Chem. Soc. C, 564 (1969).
41. H.H. Inhoffen, J. Ulrich, H.A. Hoffmann, G. Klinzman, and R. Scheu, Ann. Chem., 738, 1 (1970).
42. A. Gossauer and H.H. Inhoffen, Ann. Chem., 738, 18 (1970); A. Gossauer, D. Miehe, and H.H. Inhoffen, ibid., p. 31.
43. H.H. Inhoffen and H. Maschler, Ann. Chem., 1269 (1974).
44. T.A. Melent'eva, N.D. Pekel, and V.M. Berezovskii, Zh. Obshch. Khim., 44, 939 (1974); Chem. Abs., 81, 9223 (1974).
45. T.A. Melent'eva, N.D. Pekel, N.S. Genokhova, and V.M. Berezovskii, Doklady Akad Nauk SSSR, 194, 591 (1970); Chem. Abs., 74, 87104 (1971).
46. T.A. Melent'eva, N.S. Genokhova, and V.B. Berezovskii, Zh. Obshch. Khim., 44, 934 (1974); Chem. Abs., 81, 9222 (1974); Idem, Doklady Akad Nauk SSSR, 194, 591 (1970); Chem. Abs., 74, 87104 (1971).
47. T.A. Melent'eva, N.D. Pekel, and V.M. Berezovskii, Zh. Obshch. Khim. 42, 180 (1972); Chem. Abs., 77, 19622 (1972).
48. H.H. Inhoffen, J.W. Buchler, L. Puppe, and K. Rohbock, Ann. Chem., 747, 133 (1971).
49. N.S. Hush and I.S. Woolsey, J. Am. Chem. Soc., 94, 4107 (1972).
50. C.M. Elson, A. Hamilton, and A.W. Johnson, J. Chem. Soc., Perkin Trans. 1, 775 (1973).
51. A.W. Johnson, I.T. Kay, and R. Rodrigo, J. Chem. Soc., 2336 (1963).
52. A.W. Johnson and R. Price, J. Chem. Soc., 1649 (1960); A.W. Johnson and I.T. Kay, Proc. Chem. Soc., London, 168 (1961).
53. Professor R.B. Woodward, Aromaticity Conference, Sheffield (1966).
54. M.J.S. Dewar and G.T. Gleicher, J. Am. Chem. Soc., 87, 685 (1965).
55. B.W. Metcalf and F. Sondheimer, J. Am. Chem. Soc., 93, 5271 (1971).
56. A.P. Johnson, P. Wehrli, R. Fletcher, and A. Eschenmoser, Angew. Chem., Int. Ed. Eng., 7, 623 (1968).
57. P.M. Müller, S. Farooq, B. Hardegger, W.S. Salmond, and A. Eschenmoser, Angew. Chem., Int. Ed. Engl., 12, 914 (1973).

LABORATORY METHODS

LABORATORY METHODS

JÜRGEN-HINRICH FUHRHOP

Gesellschaft für Molekularbiologische Forschung, D-3301 Stöckheim über Braunschweig, and Institut für Organische Chemie der T. U. Braunschweig, D-3300 Braunschweig, West Germany

and

KEVIN M. SMITH

The Robert Robinson Laboratories, University of Liverpool, Liverpool, U.K.

19.1. Introduction

One of the most attractive features of Falk's 'Porphyrins and Metalloporphyrins' was the inclusion of a 'laboratory methods' section which made the book virtually unique. In this chapter it is our intention to provide a fairly wide-ranging compilation of laboratory methods similar to that in the original[1]. In some cases there has been virtually no improvement in procedures since 1964, and no textual changes are made from the account written by Falk. In other areas, and particularly in synthesis, completely new sections are added. The chapter finishes with an appendix containing a comprehensive list of electronic absorption spectra of porphyrins and metalloporphyrins, measured in various solvents.

19.2. Typical synthetic procedures

Choice of procedures for inclusion in this section is necessarily difficult. Invariably there are several modern approaches to each of the compounds chosen; that used under normal circumstances will usually depend on whether or not an isotopically labeled product is required, and if so, upon the precise position of the label. In making the difficult decision regarding which synthetic approach to describe, we have tended to rely upon the most efficient and/or convenient rather than most versatile synthesis. In most cases, experience of the reactions has been gained at first hand.

Porphyrins and Metalloporphyrins, ed. Kevin M. Smith
© 1975, Elsevier Scientific Publishing Company, Amsterdam, The Netherlands

Scheme 1

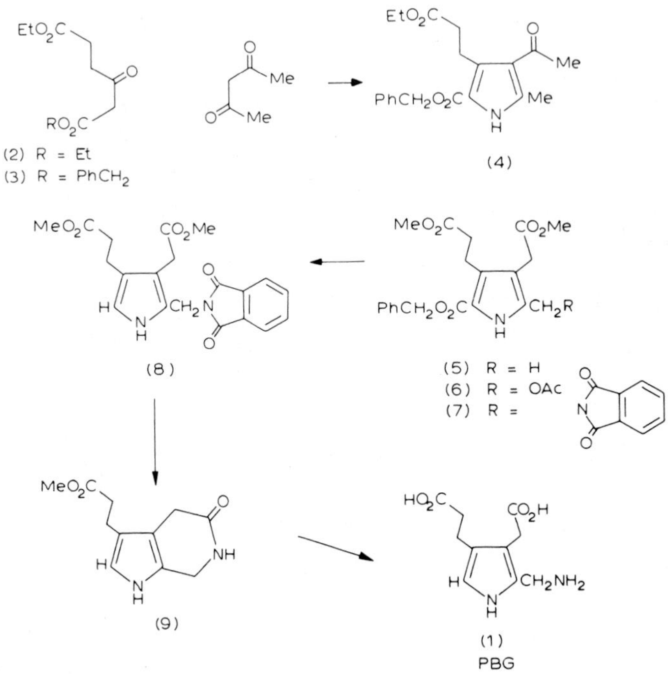

(2) R = Et
(3) R = PhCH₂

(4)

(8)

(5) R = H
(6) R = OAc
(7) R =

(9)

(1)
PBG

19.2.1. Porphobilinogen (1) (Scheme 1)[2,3]

19.2.1.1. Benzyl ethyl β-keto-adipate (3)

Diethyl β-keto-adipate (2) (Note 1) (44.4 g)[4] and benzyl alcohol (100 ml) are mixed and then stirred at 140°C (oil bath) under nitrogen with an air condenser fitted for 3 h (to remove ethanol); the reaction can be followed by n.m.r. [Ph·CH₂·OH (τ 5.4), Ph·CH₂·OCO (τ 4.8)]. Excess benzyl alcohol and unreacted β-keto-adipate are distilled off (0.5 mm Hg, ≤160°C) and the residue is used directly without further purification. Yield 45 g (80%)

Notes

(1) Diethyl β-keto-adipate, before use, should be taken up in ether, washed with saturated NaHCO₃, saturated NaCl, then dried (Na₂SO₄) and the ether evaporated. Similarly, the benzyl alcohol should be distilled from calcium oxide, taken into ether, washed as for the β-keto-adipate, and then the ether removed.

(2) Distillation of the benzyl ethyl β-keto-adipate results in some decomposition; the crude material is efficient in pyrrole syntheses.

19.2.1.2. Benzyl 4-acetyl-3-(2-ethoxycarbonylethyl)-5-methylpyrrole-2-car-boxylate (4)

Freshly distilled amyl nitrite (22.3 ml) is added during 1 h to a stirred solution of conc. HCl (0.5 ml) in benzyl ethyl β-keto-adipate (44.6 g) cooled in an ice-salt bath. After standing at room temperature overnight, the solution is added during 30 min to a stirred suspension of acetylacetone (16.5 ml) in acetic acid (150 ml) and zinc powder (25 g); the reaction mixture is kept at 65°C throughout the addition, and a mixture of zinc powder (55 g) and ammonium acetate (50 g) is added in portions so that zinc is always present in excess. After standing for 15 min, the mixture is heated on a boiling water bath with efficient stirring during 2 h. The hot solution is then decanted from excess zinc into iced water (2 l) from which the product is collected by filtration. Recrystallization from di-isopropyl ether gives 23.8 g (42%) of white crystals, m.p. 90—91°C.

Note

Sodium nitrite and sodium acetate are often used in pyrrole syntheses in place of amyl nitrite and ammonium acetate (see p. 761).

19.2.1.3. Benzyl 3-(2-methoxycarbonylethyl)-4-methoxycarbonylmethyl-5-methylpyrrole-2-carboxylate (5)

The foregoing pyrrole (20 g) in anhydrous methanol (150 ml) is treated with a solution of thallium(III) nitrate trihydrate (24.9 g) in methanol (150 ml) and conc. HNO_3 (2 ml) (Note 1). After stirring during 48 h (Note 2) at 25°C the precipitated thallium(I) nitrate is filtered off and the filtrate diluted with water and methylene chloride. Drying, evaporation, and crystallization from methylene chloride—*n*-hexane gives 17.3 g (83%) of white crystals, m.p. 78—9°C.

Notes

(1) Perchloric acid can be used in place of nitric.

(2) Thallium(I) nitrate is precipitated almost immediately; the reaction is run over 48 h to ensure complete transesterification of the ethyl ester.

19.2.1.4. Benzyl 5-acetoxymethyl-3-(2-methoxycarbonylethyl)-4-methoxy-carbonylmethylpyrrole-2-carboxylate (6)

Lead tetra-acetate (6 g) is added to a solution of the foregoing pyrrole (5 g) and acetic anhydride (2 ml) in acetic acid (100 ml). The mixture is warmed on a steam bath for 10 min to dissolve the lead tetra-acetate and then left stirring at room temperature overnight. The resultant solution is added dropwise over 30 min to well stirred water (300 ml) and the precipitated solid is filtered, washed with water, dissolved in methylene chloride, and dried. The slightly cloudy solution is filtered through Celite. Evaporation and crystallization from methylene chloride—*n*-hexane gives a white solid (4.9 g; 85%), m.p. 106—7°C.

References, p. 861

Note

In larger scale reactions it is often advisable to add the lead tetra-acetate in batches over a period of a few hours.

19.2.1.5. Benzyl 3-(2-methoxycarbonylethyl)-4-methoxycarbonylmethyl-5-phthalimidomethylpyrrole-2-carboxylate (7)

Potassium phthalimide (4.4 g) in dry dimethylsulfoxide (150 ml) is mixed with a solution of the foregoing pyrrole (8 g) in dimethylsulfoxide (50 ml). After stirring for 2 h at room temperature a white precipitate of potassium acetate is apparent. Evaporation (1 mm Hg) gives a solid which is dissolved in methylene chloride, washed with water, dried, evaporated, and crystallized by trituration with ether to give white crystals (8.5 g; 89%), m.p. 132—3°C.

19.2.1.6. 3-(2-Methoxycarbonylethyl)-4-methoxycarbonylmethyl-5-phthalimidomethylpyrrole (8)

To a rapidly stirred solution of the foregoing pyrrole (2 g) in anisole (200 ml) (Note 1) is added, all at once, 10% v/v conc. H_2SO_4 in CF_3CO_2H (200 ml). After stirring at room temperature for 2 h the solution is diluted with methylene chloride, and the organic layer is washed several times with water, then saturated NaCl, and finally water again. After drying, evaporation, and crystallization from ethanol, a white solid (0.93 g; 63%), m.p. 127—8°C is obtained.

Notes

(1) Anisole is used as a benzyl carbonium ion scavenger; in its absence a by-product, presumably the 2-benzylpyrrole, is obtained.

(2) If a sodium bicarbonate washing is included in the work-up, losses are encountered.

19.2.1.7. Porphobilinogen lactam methyl ester (9)

To a solution of hydroxylamine hydrochloride (640 mg; 3 equiv.) in methanol (30 ml) is added 4 N sodium methoxide in methanol (4.5 ml; 7 equiv.). A fine precipitate of sodium chloride forms and this suspension is added to a well-stirred solution of the foregoing pyrrole (1 g; 1 equiv.) in tetrahydrofuran (30 ml). After stirring at room temperature for 20 min the orange emulsion is poured into water (20 ml) and extracted with chloroform until the organic layer no longer gives a positive Ehrlich test (approx. 6 × 200 ml). The organic extracts are combined, dried, evaporated, and crystallized from methanol to give a white solid (356 mg; 62%), m.p. 245—6°C.

19.2.1.8. Porphobilinogen (1)[3,5]

The foregoing pyrrole (475 mg) is dissolved in 2 N KOH solution (2 ml) by heating on a steam bath for 10 min. The pale yellow solution is then stirred at room temperature for 3 days, filtered through a sintered glass

funnel, washing the sinter with distilled water (1 ml). The pH of the filtrate
is adjusted to 7 with 7 N acetic acid (pH meter). Crystallization then occurs
at room temperature but the solution is left at 0°C for 2 h before filtering
off the white solid, which is washed with distilled water, methanol, and
ether, dried and stored at −70°C. Yield 350 mg (67%).

Note

Purity of the product is best estimated by paper chromatography (see p.
791).

19.2.2. Coproporphyrin-III tetramethyl ester (10) (Scheme 2)
19.2.2.1. Benzyl 4-(2-methoxycarbonylethyl)-3,5-dimethylpyrrole-2-carbox-ylate (11)[6]

A well stirred mixture of benzyl acetoacetate (134 g) in acetic acid (160
ml) is treated with a solution of sodium nitrite (51.5 g) in water (100 ml) at
a rate that allows the temperature to be kept below 10°C (ice bath). After
3 h further stirring, the solution is stored overnight in a refrigerator. This
solution (380 ml) is then added during 2 h to a solution of methyl 4-acetyl-
5-oxohexanoate (76 g) in acetic acid (260 ml) kept at 65°C and simultaneously, an intimate mixture of zinc dust (76 g) and sodium acetate (76 g) is
added at a rate such that zinc is always in excess. The reaction mixture is
then kept at 65°C for a further 1.5 h and then poured into water (10 l). The
product is filtered off, recrystallized from methanol—water, and dried to give
white crystals (50—75 g; 40—60%), m.p. 100—101°C.

19.2.2.2.   2-Benzyloxycarbonyl-4-(2-methoxycarbonylethyl)-3-methyl-pyrrole-5-carboxylic acid (12)[7]

The foregoing pyrrole (42.1 g) is suspended in carbon tetrachloride (400
ml) (Note 1) and treated, with stirring, with freshly distilled sulfuryl chloride
(34 ml; 3.2 equiv.) dropwise during 7 h. After being left at room temperature overnight the solvent is evaporated and dioxan (1.5 l) and water (1 l)
containing sodium acetate (216 g) are added (Note 2). The mixture is refluxed on a boiling water-bath for 2 h and then left overnight at room
temperature. Ether (500 ml) is added and the solution is extracted with
aqueous $NaHCO_3$, and then aqueous Na_2CO_3. The combined aqueous layers
are flushed with a rapid stream of air (to remove ether) and then sulfur
dioxide to pH approx. 6. The precipitated carboxylic acid is washed with
water and dried (33 g; 71%) and is normally used without further purification.

Notes

(1) Using carbon tetrachloride as solvent allows the reaction to be monitored by n.m.r. spectroscopy, observing firstly the disappearance of the
5-methyl resonance, and then of the 5-chloromethyl and 5-dichloromethyl
resonances.

Scheme 2

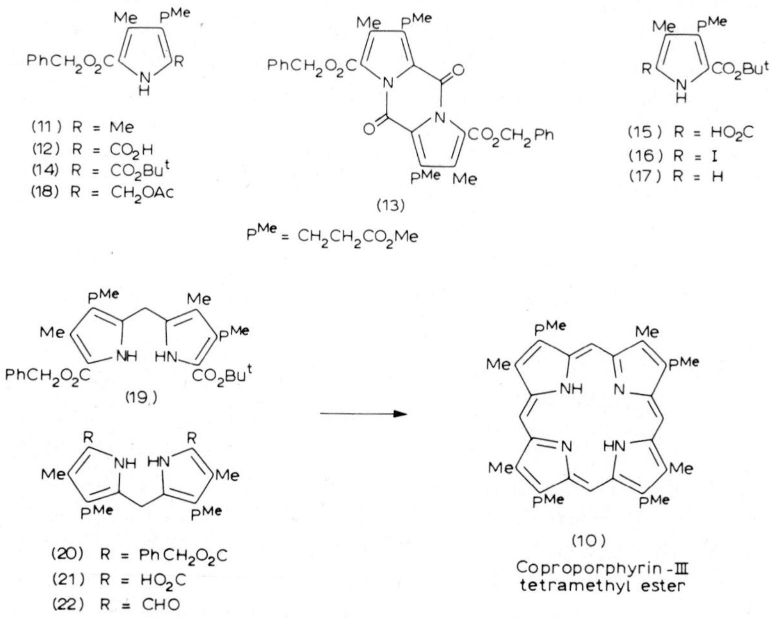

(11) R = Me
(12) R = CO₂H
(14) R = CO₂Buᵗ
(18) R = CH₂OAc

(13)

pMe = CH₂CH₂CO₂Me

(15) R = HO₂C
(16) R = I
(17) R = H

(19.)

(20) R = PhCH₂O₂C
(21) R = HO₂C
(22) R = CHO

(10)

Coproporphyrin - III
tetramethyl ester

(2) The hydrolysis of the trichloromethylpyrrole is carried out at high dilution in order to avoid formation of the highly insoluble pyrrocoll (13).

19.2.2.3. Benzyl 5-t-butoxycarbonyl-4-(2-methoxycarbonylethyl)-3-methyl-pyrrole-2-carboxylate (14)[8]

The foregoing pyrrole acid (20.8 g) in chloroform (400 ml) is cooled in an ice-bath and treated with isobutene (200 ml) and conc. H_2SO_4 (3 ml) (Note 1). The mixture is stirred overnight at room temperature in a flask with a crude pressure release valve (a ground glass stopper with rubber tubing wrapped over it and secured with a clamp and boss at the neck of the flask). An aqueous solution of $NaHCO_3$ is added to make the mixture alkaline and the mixture is stirred in a hood for a further 2 h without a stopper. The organic phase is washed with water, dried, evaporated, and crystallized from light petroleum to give white crystals (20 g; 83%), m.p. 91—92°C (Note 2).

Notes

(1) This method seems more efficient than that involving treatment of the corresponding acid chloride with t-butyl alcohol and N,N-dimethylaniline[9].

(2) If the product is impure [due to by-products carried forward from the preparation of (12)], the crude material can be chromatographed on a short column of alumina (grade III) eluting with toluene or benzene.

19.2.2.4. 5-t-Butoxycarbonyl-4-(2-methoxycarbonylethyl)-3-methylpyrrole-2-carboxylic acid (15)[9]

The foregoing pyrrole (20 g) in tetrahydrofuran (125 ml) and triethylamine (0.2 ml) containing 10% palladized charcoal (2 g) is hydrogenated at room temperature and atmospheric pressure until uptake of hydrogen has ceased (45 min; 1345 ml). The catalyst is filtered off on a bed of Celite, washed well with methylene chloride, and the combined filtrates are evaporated to give an oil which is crystallized from methylene chloride-light petroleum, giving white needles (15.2 g; 98%), m.p. 142—3°C.

19.2.2.5. t-Butyl 2-iodo-4-(2-methoxycarbonylethyl)-3-methylpyrrole-5-carboxylate (16)[9]

The foregoing pyrrole (4 g) in methanol (30 ml) is treated with sodium bicarbonate (3.1 g) in water (30 ml) and warmed to 60°C before addition of a solution of iodine (3.15 g) and potassium iodide (5.0 g) in methanol (40 ml) and water (10 ml) with stirring and at such a rate that there is no permanent coloration. Water (40 ml) is added dropwise and after keeping the solution at 60°C for a further 1 h it is cooled to 20°C. The precipitated solid is filtered off, dried at 50°C, and recrystallized from ether-light petroleum to give white needles (3.7 g; 74%), m.p. 133—4°C.

19.2.2.6. t-Butyl 4-(2-methoxycarbonylethyl)-3-methylpyrrole-5-carboxylate (17)[10]

The foregoing pyrrole (14.4 g) in methanol (300 ml) containing NaOAc trihydrate (14 g) is hydrogenated at room temperature and atmospheric pressure over Adams catalyst (1 g) until hydrogen uptake (820 ml) has ceased. The catalyst is filtered off through a bed of Celite and the solution is concentrated to approx. 100 ml before dilution with water (200 ml) and methylene chloride. After several extractions with methylene chloride, the organic phases are dried, evaporated to dryness, and the residual oil is crystallized under high vacuum. Yield 9.8 g (100%), m.p. 63—4°C.

19.2.2.7. Benzyl 5-acetoxymethyl-4-(2-methoxycarbonylethyl)-3-methylpyrrole-2-carboxylate (18)

Benzyl 4-(2-methoxycarbonylethyl)-3,5-dimethylpyrrole-2-carboxylate (11, 35.2 g) in acetic acid (400 ml) and acetic acid anhydride (8 ml) is treated with portions of lead tetra-acetate (total 51.9 g) over a period of 2.5 h with stirring at room temperature. After stirring overnight the solution is added dropwise to water (500 ml) with vigorous stirring. The product is filtered off, dried, to give white crystals (40 g; 95%) (Notes 1 and 2).

Notes

(1) The product is usually pure enough for subsequent reactions. If not, it can be recrystallized from hexane.

References, p. 861

(2) In small scale reactions (p. 759) the lead tetra-acetate can be added in one portion.

19.2.2.8. Benzyl 5'-t-butoxycarbonyl-3,4'-di(2-methoxycarbonylethyl)-3',4-dimethylpyrromethane-5-carboxylate (19)[10,11]

The foregoing pyrrole (2.1 g) is suspended in methanol (30 ml) containing t-butyl 4-(2-methoxycarbonylethyl)-3-methylpyrrole-5-carboxylate [1.5 g; (17)] and toluene p-sulfonic acid hydrate (60 mg) and then heated under nitrogen at $40°C$ during 4 h. 10% Aqueous $NaHCO_3$ solution (3 ml) is added slowly and the product is collected by filtration. It is recrystallized from hot methanol to give a white solid (2.05 g; 64%), m.p. $150—153°C$.

Note
On occasions, acetic acid can be used as solvent in place of methanol[11]. The latter has the disadvantage that it often gives rise to 5-methoxymethyl-pyrrole by-products, but the acetic acid method requires an organic extraction work-up.

19.2.2.9. Dibenzyl 3,3'-di(2-methoxycarbonylethyl)-4,4'-dimethylpyrromethane-5,5'-dicarboxylate (20)

Benzyl 4-(2-methoxycarbonylethyl)-3,5-dimethylpyrrole-2-carboxylate [(11), 35.5 g] in ether (1 l) (slight heating required) is treated with stirring with bromine (7 ml) in ether (500 ml) during 3 min. The mixture is then stirred for 1.5 h and evaporated (HBr!) to give a pale pink, fluffy solid. This is taken up in methanol (135 ml) which is refluxed for 4 h and then set aside at room temperature overnight. Filtration, washing with cold methanol, and then recrystallization from hot methanol, gives white crystals (26.5 g; 74%), m.p. $102—3°C$.

19.2.2.10. 5,5'-Diformyl-3,3'-di(2-methoxycarbonylethyl)-4,4'-dimethylpyrromethane (22)[12]

The foregoing pyrromethane (20 g) in tetrahydrofuran (600 ml) (Note 1) and triethylamine (0.5 ml) is hydrogenated at room temperature and atmospheric pressure over 10% palladized charcoal (2 g) until uptake (1500 ml) has ceased (1 h). The catalyst is filtered off on Celite and evaporation of the filtrate gives a white solid [(21); 13 g, 93%].

The pyrromethane-5,5'-dicarboxylic acid [(21); 5 g] is heated under reflux in dimethylformamide (30 ml) for 30 min. After cooling to $0°C$, benzene (11 ml) and benzoyl chloride (9.65 ml) (Note 2) are added, keeping the temperature below $5°C$ with an ice-bath (20 min for addition). The mixture is stirred for 20 min and allowed to come to room temperature before addition of benzene (26 ml). The precipitate is collected after 1 h (5.5 g, imine salt), dissolved in methanol (100 ml), treated with saturated aqueous sodium acetate (100 ml), and stirred overnight at room temperature. Filtration and drying gives the product (2.3 g; 50%) which is used without further purification.

Notes

(1) The hydrogenation is best carried out at fairly high dilution, otherwise the di-acid may precipitate and it is then difficult to remove from the catalyst after filtration.

(2) Dimethylformamide and benzoyl chloride should be distilled before use.

19.2.2.11. *Coproporphyrin-III tetramethyl ester* (10)[10]

Benzyl   5'-*t*-butoxycarbonyl-3,4'-di(2-methoxycarbonylethyl)-3',4-dimethylpyrromethane-5-carboxylate [(19); 456 mg] in tetrahydrofuran (100 ml) and triethylamine (2 drops) is hydrogenated at room temperature and atmospheric pressure over 10% palladized charcoal (46 mg) until uptake of hydrogen has ceased (1 h). The catalyst is filtered off on Celite and the filtrate evaporated to give a residue which is taken up in trifluoroacetic acid (25 ml) and kept under N_2 for 45 min before evaporation. Methylene chloride and water are added and the organic phase is washed with $NaHCO_3$ solution, then water, dried and made up to a vol of 150 ml with dry methylene chloride. This solution is added to 5,5'-diformyl-3,3'-di(2-methoxycarbonylethyl)-4,4'-dimethylpyrromethane [(22); 250 mg] in methylene chloride (100 ml) in a darkened flask (aluminum foil) (Note 1) and then treated with a solution of toluene *p*-sulfonic acid hydrate (900 mg) in methanol (12.5 ml). After stirring for 6 h in the dark a saturated solution of zinc(II) acetate in methanol (12.5 ml) is added (Note 2) and the solution is set aside overnight. It is washed with water, $NaHCO_3$ solution, and then water again, dried, evaporated to dryness, taken up in 5% v/v H_2SO_4 in methanol (100 ml) and set aside overnight in the dark. The mixture is then poured into water and methylene chloride, washed with $NaHCO_3$ solution, water, then dried and evaporated. The residue is chromatographed on alumina (grade III) eluting with methylene chloride. Evaporation of the red eluates and crystallization from methylene chloride—methanol gives the porphyrin (180 mg; 41%), m.p. 150—3°C with remelting at 179—180°C.

Notes

(1) In the early stages, the cyclization mixture should be kept completely in the dark.

(2) Addition of zinc(II) acetate is believed to prevent large scale linear polymerization by templation.

19.2.3. *Etioporphyrin-I* (23) (Scheme 3)[13]

A solution of bromine (31.2 ml) in acetic acid (140 ml) is added to a stirred solution of *t*-butyl 4-ethyl-3,5-dimethylpyrrole-2-carboxylate (24;

Scheme 3

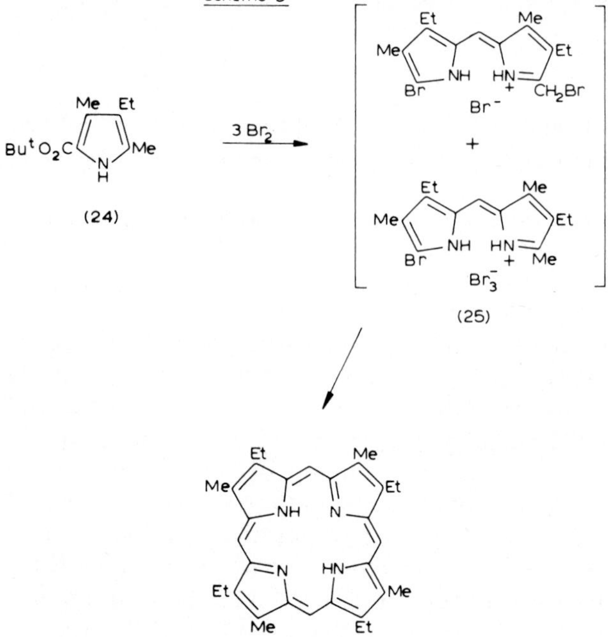

(24)

(25)

(23)

Etioporphyrin-I

44.6 g)[14] in acetic acid (150 ml) and then stirred overnight. The pyrrome-thene product is filtered off, washed with petroleum ether, dried (41 g) and then heated under reflux in formic acid (100 ml) for 3 h. Excess formic acid is distilled off and the residue is dissolved in chloroform (hot), washed with water, NaHCO$_3$ solution, water, then dried and evaporated to give a brown residue. Methylene chloride (50 ml) is added and the mixture is filtered. The residue is washed with methanol (200 ml) and then a little ether, dried, then recrystallized from chloroform—methanol to give the required porphyrin (4.96 g; 21%) as purple microneedles.

Note

The filtrates from the crystallizations can be evaporated and chromato-graphed to afford a further small quantity of porphyrin.

19.2.4. Octaethylporphyrin (26) (Scheme 4)[15], (Scheme 5)[16]
19.2.4.1. Ethyl 4-acetyl-3-ethyl-5-methylpyrrole-2-carboxylate (27)[17]

A mixture of ethyl 3-oxopentanoate (118 g) and acetic acid (350 ml) is treated with sodium nitrite (70 g) in water (170 ml) keeping the temperature below 5°C, and the mixture is kept at 0°C for 4 h and then at room tempera-ture overnight. This solution is added to a well-stirred mixture of acetyl-

Schemes 4 and 5

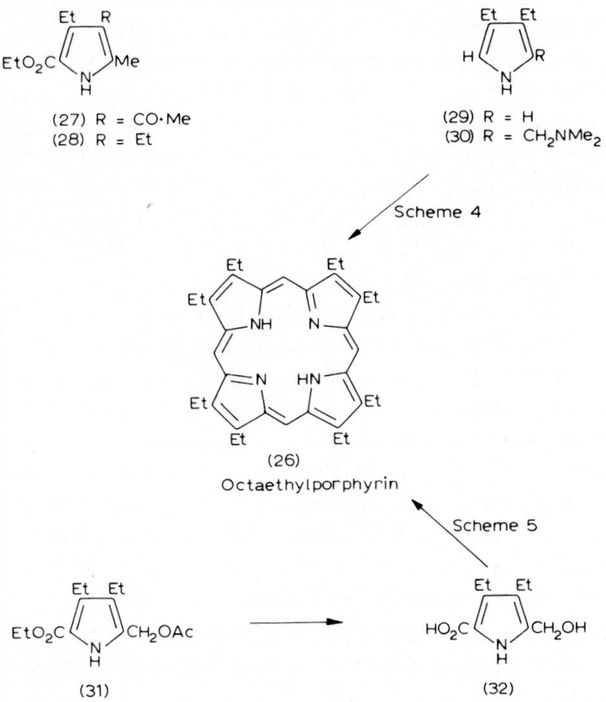

(27) R = CO·Me
(28) R = Et

(29) R = H
(30) R = CH$_2$NMe$_2$

Scheme 4

(26)
Octaethylporphyrin

Scheme 5

(31)

(32)

acetone (96 g), zinc dust (135 g), and sodium acetate (100 g) at a rate such that the mixture refluxes gently. The mixture is heated under reflux for 4 h and then poured into ice water (15 l). The product is filtered off, washed with water, dried, and recrystallized from ethanol, giving white crystals (98 g; 55%), m.p. 118.5°C.

19.2.4.2. Ethyl 3,4-diethyl-5-methylpyrrole-2-carboxylate (28)[15]

The foregoing pyrrole (100 g) in pure tetrahydrofuran (1 l) under nitrogen is stirred until completely dissolved and then cooled to 5°C. Sodium borohydride (39 g) is then added and after this is completely dispersed boron trifluoride etherate (200 g) (freshly distilled) is added dropwise at such a rate (1.5 h) that the temperature is maintained near 10°C; after complete addition the temperature is raised to 25°C and the mixture is stirred for 1 h. The mixture is cooled to below 15°C and 5% HCl (1 l) is added cautiously dropwise so that the temperature remains below 35°C. Water is added and the solution is extracted with ether, which is washed with saturated NaCl, dried, and evaporated to give pink crystals (95 g; 100%), m.p. 75°C.

19.2.4.3. 3,4-Diethylpyrrole (29)

The foregoing pyrrole (95.2 g) in dry ether (800 ml) is cooled under nitrogen in an ice-salt bath before addition of freshly distilled sulfuryl chloride (111 ml) dropwise so that the temperature remains below 0°C. The solution is then kept at 0°C for 36 h before addition of ice-water (125 ml) and evaporation of the ether. To the oily residue is added a hot solution of sodium acetate (75 g) in water (1.5 l) and the mixture is refluxed with stirring until a brown precipitate appears (30 min). This is filtered off, dissolved in 10% aqueous sodium hydroxide (570 ml) which is then extracted with ether and heated under reflux for 2 h before cooling to 15°C and acidification with stirring to congo red with hydrochloric acid, giving the 2,5-dicarboxylic acid (89.75 g; 94%), as a purple solid. This is added in small portions to quinoline (150 ml) and barium promoted copper chromite catalyst[18] (1 g) under nitrogen at 190°C and the mixture is stirred at this temperature until gas evolution has ceased. The temperature is then raised and the product is distilled (head temperature approx. 240°C), the distillate being taken into cold 5% HCl (1.4 l) and then extracted with ether. The combined organic extracts are dried, evaporated, and the residue is distilled under nitrogen, to give an oil (41.7 g; 80%), b.p. 202—5°C at 740 mm, m.p. 13°C.

19.2.4.4. 3,4-Diethyl-2-dimethylaminomethylpyrrole (30)

The foregoing pyrrole (41.7 g) in methanol (325 ml) under nitrogen at −15°C (acetone—dry ice) is treated with a solution of dimethylamine hydrochloride (28.1 g), potassium acetate (33.8 g), and 37% aqueous formaldehyde (27.8 g) in water (130 ml) dropwise during 1.75 h keeping the temperature between −15 and −10°C. The mixture is then kept at 0°C for 12 h before adding cold 5% HCl (400 ml) slowly with stirring. The cold solution is extracted with ether and the aqueous layer is made basic by slow addition of 2 N NaOH (500 ml) with stirring and cooling in an ice-bath. This solution is extracted with ether which is dried and evaporated to give the crude product (61 g; 100%) as a brown oil.

Note

The product can be distilled for purification and this then affords a higher yield in the polymerization step; however, this is not normally considered necessary.

19.2.4.5. Octaethylporphyrin (26)[15]

The above mentioned Mannich base (30) (61.0 g) in purified (KMnO$_4$) acetic acid (500 ml) is heated under reflux for 1 h with stirring and passage of a rapid stream of oxygen through the solution. The solution is cooled, acetic acid is evaporated in vacuo, and methanol (500 ml) is added to the residue. The product is filtered off and washed with methanol until the filtrate is no longer brown. After drying, 24.1 g of crude product is obtained which is recrystallized from toluene to give octaethylporphyrin (23.51 g; 52%), m.p. 324—5°C.

19.2.4.6. Ethyl 5-acetoxymethyl-3,4-diethylpyrrole-2-carboxylate (31)[16]

Ethyl 3,4-diethyl-5-methylpyrrole-2-carboxylate [(28); 17.7 g] in acetic acid (490 ml) is stirred during addition of lead tetra-acetate (37.5 g) in portions. The mixture is stirred for 2 h, evaporated, and the residue is treated with chloroform, washed with water, dried, evaporated, and crystallized from methanol to give white crystals (15.2 g; 68%), m.p. 91—2°C. Another 2.8 g is obtained by further treatment of the mother liquors.

19.2.4.7. 3,4-Diethyl-5-hydroxymethylpyrrole-2-carboxylic acid (32)

The foregoing pyrrole (15.2 g) and KOH (30 g) are heated under reflux in methanol (210 ml) during 4 h. The solution is evaporated before addition of water (150 ml), extraction with ether, and then acidification of the aqueous phase with stirring and cooling, dropwise with dil. HCl (350 ml). The product is filtered off, washed with aqueous sodium acetate and water, and then dried to give 10.7 g (96%) of the required pyrrole carboxylic acid, m.p. 115—20°C.

19.2.4.8. Octaethylporphyrin (26)[16]

The foregoing pyrrole (10 g) in acetic acid (40 ml) containing potassium ferricyanide (1 g) is heated at 100°C with stirring for 1 h. The mixture after standing overnight at room temperature affords octaethylporphyrin (2 g) which is filtered off. Further treatment of the filtrate, followed by chromatography of the porphyrinic fraction on alumina (grade III), gives a second crop (1 g; total yield 44%).

19.2.5. meso-Tetraphenylporphyrin (33) (Note 1)
19.2.5.1. Crude material[19]

Benzaldehyde (66.5 ml) and pyrrole (46.5 ml) are added simultaneously to refluxing propionic acid (2.5 l) (Note 2) and the mixture is refluxed for 30 min before being allowed to cool and stand at room temperature overnight. The product is filtered off and washed with water and methanol to give glistening purple crystals (20.4 g; 19.8%). Concentration of the propionic acid filtrate to approx. 500 ml gives a second crop (6.0 g; total yield 26%). The combined material contains[20] approx. 5% meso-tetraphenylchlorin impurity.

Notes

(1) The meso-tetraphenylporphyrin, though of very high macrocyclic purity, is invariably contaminated with between 2 and 10% of the corresponding chlorin (for purification, see below). The extent of contamination is best judged by measurement of the visible absorption spectrum in ben-

zene. Impurity in the H_2TPP is apparent from the enhanced absorption around 650 nm due to superposition of the long-wavelength H_2TPC band upon band I of the porphyrin. Using the facts that pure H_2TPP has the spectrum λ_{max} 485 (ϵ 3400), 515 (18,700), 592 (5300), and 647 nm (3400), and pure H_2TPC, λ_{max} 518 (ϵ 15,000), 543 (10,800), 600 (5800), and 654 nm (41,700), it is possible to estimate[21] the percentage contamination.

(2) Propionic acid is superior to acetic acid, mainly because H_2TPP is less soluble in the former. It is, however, more expensive than acetic acid but can be recovered by concentration of the filtrate. The recovered propionic acid should be refluxed for approx. 3 h with potassium dichromate (30 g) and then distilled before re-use (particularly if a different tetra-arylporphyrin is to be prepared).

19.2.5.2. 'Chlorin-free' meso-tetraphenylporphyrin (33)[20,22]

Crude H_2TPP (20 g) in refluxing ethanol-free chloroform (2.5 l) is treated with a solution of 2,3-dichloro-5,6-dicyanobenzoquinone (5 g) in dry benzene (150 ml) and the mixture is refluxed for 3 h (Note 1). The hot, yellow tinged solution is filtered under suction through a sintered glass funnel (2.5″ diam. × 8″ approx.) containing alumina (300 g; grade I) which has a filter paper placed on top of it to avoid disturbance of the bed during pouring. The alumina is washed with methylene chloride (200 ml) and the combined filtrates are concentrated to approx. 200 ml before addition of methanol (200 ml). Filtration affords the product (19.2 g; 96%) as glistening purple crystals (Note 2).

Notes

(1) In smaller scale reactions (approx. 1 g) only about 30 min refluxing is required[22].

(2) The overall recovery from the DDQ procedure depends largely upon the purity of the crude H_2TPP. That obtained from the procedure in Section 19.2.5.1. is invariably of high purity.

(3) Economies in the amount of DDQ used are possible; however, this again depends upon the purity of the crude material.

(4) This procedure is also applicable to purification of other tetra-arylporphyrins[e.g. 23].

19.3. Porphyrins derived from protoheme
19.3.1. Protoporphyrin-IX

The preparation of small samples of protoporphyrin for use, for example, as a marker for paper chromatography or as a spectroscopic reference substance, is best done by treatment of hemin with ferrous sulfate at room temperature (see p. 800). Crystalline hemin may be used; it is often convenient to use small samples of hemin extracted from blood with acetone—HCl and transferred to ether.

All other methods (removal of iron from hemes, p. 800) require heating in acid media, and yield impure products. For work on the scale of several grams the Grinstein method (p. 802) is usually used. This yields protoporphyrin ester in a crystalline, but very impure form. It may be purified by column chromatography and is best converted to the free porphyrin by hydrolysis with alcoholic KOH to avoid the hydration of the vinyl groups which occurs in aqueous acids. Further purification of the free porphyrin may be achieved by countercurrent distribution or, less efficiently, by ether—HCl fractionation in separating funnels.

The potassium salt of protoporphyrin is insoluble, and may be crystallized; the sodium salt is slightly more soluble. Protoporphyrin dihydrochloride is fairly soluble in chloroform, into which it may be extracted from solutions in 1.5 N HCl.

The ester may be prepared from the free porphyrin by esterification (p. 834), or directly from hemin by the Grinstein method (p. 802). It crystallizes well from chloroform—methanol. An efficient route to the dimethyl ester from commercially available hematoporphyrin-IX di-hydrochloride is as follows[24]: A solution of hematoporphyrin-IX di-hydrochloride (2.5 g) and toluene p-sulfonic acid hydrate (7.8 g) in o-dichlorobenzene (220 ml) is heated at $140°C$ (oil-bath) for 35 min with agitation of the solution with a stream of nitrogen. The cooled mixture is diluted with methylene chloride (1 l) and extracted with water. The organic phase is dried and evaporated to dryness, the o-dichlorobenzene being removed at 0.1 mm Hg. A solution of the residue in methanol (250 ml) and conc. H_2SO_4 (11 ml) is kept at $0°C$ for 2 days in the dark before being diluted with water and methylene chloride. The organic phase is washed with aqueous $NaHCO_3$, dried and evaporated to give a residue which is chromatographed on alumina (grade III) eluting with methylene chloride. After evaporation of the red eluates the residue is crystallized from methylene chloride—methanol, giving purple crystals (1.56 g; 71%), m.p. 226—8°C.

Note

Further elution of the column with $CHCl_3$ containing 1% methanol gives a small quantity of hematoporphyrin-IX dimethyl ester.

19.3.2. Hematoporphyrin-IX and derivatives
19.3.2.1. Hematoporphyrin-IX[25]

Hemin (3 g) is shaken for 24 h (1 h for samples of a few mg) with 75 g of HBr-acetic acid (sp.gr. 1.41), the mixture poured into excess water, and after allowing a few minutes for hydrolysis of the HBr-adduct, is neutralized to congo paper with NaOH. The precipitate is filtered, washed, dissolved in dilute NaOH, the solution filtered if necessary, and the sodium salt of hematoporphyrin precipitated by adding 33% (w/v) NaOH. For small samples it is more convenient to take the hematoporphyrin from the acidic reaction mixture to ether, in which it may be washed and esterified directly. A yield of 35% has been reported by Granick et al.[26], who purified the dihydrochloride by dissolving it in 1 N HCl, filtering from some insoluble material and precipitating the porphyrin with sodium acetate. The neutral porphyrin was filtered, dried, dissolved in ether, filtered, extracted into 0.025 N HCl and the dihydrochloride again crystallized by concentration in a desiccator over H_2SO_4.

19.3.2.2. Hematoporphyrin-IX dimethyl ester

Esterification of the free porphyrin, best by diazomethane, and crystallization from chloroform-methanol are carried out as described above. Even after purification by countercurrent distribution, the ester was found by paper chromatography to contain two minor porphyrin contaminants[26]. J. Barrett (personal communication) purifies the ester by chromatography on alumina washed with water. The purified material began to melt at 217°C, and birefringence disappeared completely at 223°C.

19.3.2.3. Hematoporphyrin-IX dimethyl ether[27]

Hemin (2.5 g) is shaken for 4 days with 6.5 g of HBr-acetic acid (sp.gr. 1.41). The mixture is filtered and evaporated to dryness in vacuo at 50°C, and the residue warmed for 24 h with 30 ml of methanol at 50°C. The product is recovered by extraction into ether, extraction from this with aqueous HCl (10% w/v) (HBr was used by Fischer) and precipitation with sodium acetate. After hot extraction of the powdered, dry porphyrin from a thimble with ether, the porphyrin crystallized from the ether on long standing.

19.3.2.4. Hematoporphyrin-IX dimethyl ester dimethyl ether

Two methods are usually used for the preparation of this compound. (1) The ester is prepared by esterification of the free porphyrin dimethyl ether with HCl-methanol or diazomethane. It crystallizes well, but Fischer and Orth[27] reported four crystalline forms; needles of m.p. 185°C, double pyramids of m.p. 178.5°C, cubes of m.p. 140°C and needles of m.p. 110°C. (2) Granick et al.[26] refluxed protohemin with H_2SO_4-methanol as described by Fischer and Lindner[28], and report that after recrystallization of the resulting hematohemin, paper chromatography revealed one spot only. After removal of iron and esterification by the Grinstein method (see p. 802) the porphyrin ester was purified by countercurrent distribution, and finally crystallized from benzene-petroleum ether. The material melted over the range 178—182°C, birefringence disappearing at 187°C. The melting behavior is complicated, and the papers of Granick et al.[26] and Fischer and Müller[29] should be consulted. Paper chromatography of this material revealed two extra spots, one of which appeared to be 'trimethylhematoporphyrin', and some of this was removed by a further crystallization from CCl_4-petroleum ether.

19.3.3. Mesoporphyrin-IX

The classical method of preparation involves treatment of protohemin with HI-acetic acid[29], which should have a sp.gr. of 1.9 for the best results; ascorbic acid is sometimes added[30]. The method has been applied to crude hemin obtained by evaporation of acetone-oxalic acid extracts of blood[31].

The method of choice, however, is catalytic hydrogenation over Pd in formic acid[32]. Either protoporphyrin (or its ester) or protohemin may be used, iron being removed from the latter during the reaction. Mesohemin is best prepared by introduction of iron (see p. 803) into mesoporphyrin. In the method described below, protoporphyrin dimethyl ester is converted to mesoporphyrin dimethyl ester.

The dimethyl ester is prepared as follows[33]: In 120 ml of anhydrous formic acid (Note 1) 1.3 g of methylmethacrylate (Perspex turnings) are dissolved, and 185 mg (75 mg Pd per 1 g protoporphyrin) of 40% Pd-charcoal are added. Protoporphyrin dimethyl ester (1 g) is now added, the flask placed in a bath at 50°C, and H_2 passed until reduction is complete (approx. 45 min) (Note 2). The mixture is then poured into washed ether, the solution decanted from the precipitated methacrylate and the solid washed several times with ether. The mesoporphyrin is now extracted from the ether with 2.5% HCl, the hydrochloride extracted from the HCl into $CHCl_3$, and the $CHCl_3$ solution washed (water, 2 N NH_4OH, water) and evaporated to dryness. The mesoporphyrin ester is crystallized from $CHCl_3$-methanol. Yield, about 95%.

Notes

(1) Anhydrous formic acid is prepared by refluxing formic acid over anhydrous $CuSO_4$, and distilling under reduced pressure.

(2) The course of the reduction is followed (hand spectroscope) by the disappearance of the neutral protoporphyrin absorption band at 631 nm and the appearance of the mesoporphyrin band at 620 nm. Small samples are taken from the reaction mixture, chloroform added, and the formic acid neutralized by shaking with aqueous sodium acetate. The appearance of a band at 650 nm (chlorin?) indicates over-reduction, but this does not occur in less than 60-min hydrogenation.

19.3.4. Deuteroporphyrin-IX[34]

Protohemin (1 g) is ground with 3 g of resorcinol, and then heated under an air-condenser in an oil-bath at 150—160°C (Schumm used 190—200°C) for 45 min. After cooling, the dark brown solid is ground and extracted with washed ether or ethyl acetate until the extracts are almost colorless. The solid residue of crude deuterohemin (about 1 g) is dried, and may be purified by recrystallization or converted to deuteroporphyrin or its ester. In a modification[35], the crude reaction mixture was treated directly with iron powder, acetic acid and HCl, and the resulting deuteroporphyrin purified by extraction into ethyl acetate. The same authors recommend chromatography on a $CaCO_3$ column to purify the porphyrin ester.

For small quantities of hemins (1—10 mg) heating for about 15 min is sufficient; the hemin may then be taken to ether and impurities removed by washing with 0.15 N HCl.

19.3.5. Deuteroporphyrin-IX 2,4-disulfonic acid
19.3.5.1. Dimethyl ester[36]

Deuteroporphyrin dimethyl ester (100 mg) is mixed thoroughly in a mortar with 1.5 g of N-pyridinium sulfonic acid (Note). The mixture is fused for 10 min at 185—190°C, cooled and dissolved in a small volume of water. On standing, the deuteroporphyrin dimethyl ester disulfonic acid crystallizes, and may be recrystallized from water; yield, 82.6 mg. The free porphyrin is prepared similarly from free deuteroporphyrin.

Note

N-pyridinium sulfonic acid is prepared as follows: 16.2 ml (15.8 g) of pyridine are dissolved in 48 ml of chloroform and the solution stirred mechanically in an ice-salt bath. Chlorosulfonic acid (6.2 ml, 11.7 g) is added dropwise in a fume cupboard (HCl is evolved), and the thick white precipitate of pyridinium sulfonic acid is filtered, washed with water at 0°C, and dried over P_2O_5. Yield, 4.5 g.

19.3.5.2. Tetramethyl ester

This is one of the most water-soluble porphyrin esters available, and for this reason has often been used for physicochemical studies. An interesting sulfonated product of unknown structure has been obtained by the action of sodium bisulfite on protohemin, but its elemental analysis shows that it is not deuteroporphyrin disulfonic acid; cysteine- and other thiol-adducts of protohemin are discussed elsewhere. The tetra-methyl ester may be prepared[37] by treatment of the dimethyl ester, or of deuteroporphyrin disulfonic acid itself, with diazomethane.

19.4. Compounds derived from chlorophylls (Structures, pp. 52—55)
19.4.1. Chlorophylls from plant tissue

Procedures employed for extraction of chlorophyll pigments from fresh plant material are variable, depending upon the type and form of tissue used, the scale of the extraction, and the chlorophylls required. All of the variables have been examined and a thorough review is available[38]. No particular procedure is applicable to all circumstances, but typically[38], fresh spinach leaves (200 g) free of midribs are dropped into boiling water (2 l) and after 1 or 2 min the water is cooled by dilution with excess cold water. After decantation, the leaves are washed once with cold water and then squeezed between cloth and paper towels (under foot). The leaves (approx. 100 g) are then placed separately in methanol (500 ml) plus petroleum ether (125 ml). After a few min the deep green extract is filtered through a pad of cotton wool (the leaves are extracted twice more with methanol—petroleum ether) and the extracts (separately) are diluted with saturated brine (1 l), thereby transferring the pigments to the petroleum ether layer. These layers are then further treated, depending upon the type of pigments required[38,39], since they contain a wide variety of compounds apart from the required chlorophylls-*a* and -*b*. Chromatography on sugar columns serves both to remove the by-products and separate the two major chlorophylls.

19.4.2. Pheophytins

Treatment of chlorophylls with acidic solutions[40] causes demetalation, affording grey-brown pheophytins-*a* and -*b*.

19.4.2.1. Separation of pheophytins-a and -b

The scale of operation for separation of chlorophyll pigments by sugar chromatography is strictly limited owing to the high adsorbent to pigment ratio. A report[41] of the use of Girard 'T' reagent for some separations in this field has resulted in the development of a useful, large-scale separation procedure[42]: pheophytin mixture (14.5 g) in chloroform (400 ml) is heated under reflux for 1 h with Girard's reagent 'T' (6 g) in methanol (500 ml) and acetic acid (70 ml). The solution is evaporated in vacuo, the residue taken up in methylene chloride (approx. 75 ml, warm) which is filtered through glass wool and then chromatographed on alumina (grade III, 500 g), eluting with methylene chloride. Evaporation of the eluates gives a black, sticky solid which is recrystallized from methylene chloride—methanol to give pheophytin-*a* (8.8 g; 89%) (yield assumes approx. 3 : 1 ratio). Elution of the column with chloroform—methanol (20 : 1) gives the series *b* derivative after evaporation of the eluates, and the residue is stirred overnight with a mixture of methanol (400 ml), acetone (20 ml), and conc. H_2SO_4 (13 ml) before dilution with chloroform and washing with water. The organic layer is evaporated and the residue is chromatographed on alumina (grade III, 300 g) eluting with methylene chloride. Evaporation of the eluates and recrystallization from methylene chloride-methanol gives (Note) methyl pheophorbide-*b* (2.5 g; 68%).

Note

The pheophytin-*b* is unable to survive the conditions for decomposition of the Girard derivative, and this results in the production of the corresponding pheophorbide.

References, p. 861

19.4.3. Pheophorbides

Methanolysis of the pheophytin mixture affords the corresponding pheophorbides[42]: pheophytin mixture (5 g) is stirred for 15 h in the dark with 5% v/v H_2SO_4 in methanol (400 ml). The blue-green mixture is diluted with chloroform (400 ml), washed with water (2 × 1 l), dried, evaporated to dryness, and crystallized from methylene chloride—methanol to give the methyl pheophorbide mixture (2.85 g; 83%), normally used without further purification. This mixture can then be separated as described above for the pheophytin mixture, or alternatively, the pure pheophorbides can be obtained directly by methanolysis of pheophytin-*a* and from the separation described above for methyl pheophorbide-*b*.

19.4.4. Purpurin-7 trimethyl ester[42]

Pheophytin-*a* (2.65 g) in warm pyridine (40 ml) is diluted with ether (1.25 l) before passage of a stream of air through the mixture during addition of KOH (20 g) in *n*-propanol (70 ml). The bright green mixture is stirred and aerated for a further 30 min and then extracted with water (2 × 500 ml). The ethereal solution is discarded and the aqueous extract is acidified with conc. H_2SO_4 (20 ml) in water (100 ml) and then extracted with methylene chloride. The extracts are washed with water and then immediately treated with an excess of ethereal diazomethane. After standing at room temperature for 10 min the mixture is evaporated and the residue is chromatographed on alumina. Evaporation of the appropriate eluates and recrystallization from methylene chloride—*n*-hexane gives the purpurin triester (1.31 g; 67%).

Note

If the pheophytin-*a* is hydrogenated [(2.0 g) in acetone (400 ml) over palladized charcoal (10%, 200 mg)] and then subjected to the KOH reaction, mesopurpurin-7 trimethyl ester (0.78 g; 52%) is obtained.

19.4.5. Purpurin-18 methyl ester[42]

Pheophytin-*a* (0.5 g) is oxidized in alkaline solution as described above. The methylene chloride extract (prior to CH_2N_2 treatment) containing the 'unstable chlorin' is subjected to repeated evaporation and redissolution in methylene chloride—benzene (Note 1) before treatment with excess ethereal diazomethane. Chromatography on a short column of alumina and recrystallization from the appropriate evaporated eluates furnishes the purpurin (220 mg; 66%).

Notes

(1) This is carried out until spectrophotometry shows no further increase in absorption at 695 nm.

(2) Prior hydrogenation and then treatment as described here results in the formation of mesopurpurin-18 methyl ester.

19.4.6. 2-Vinylrhodoporphyrin-XV dimethyl ester[42]

Purpurin-7 trimethyl ester (320 mg) in 2,4,6-collidine (40 ml) is heated at 190—200°C during 2 h. After cooling, the solvent is evaporated off (0.1 mm Hg) and the residue recrystallized from methylene chloride—methanol to give the rhodoporphyrin (207 mg; 76%).

19.4.7. Rhodoporphyrin-XV dimethyl ester[42]

Similar treatment of mesopurpurin-7 trimethyl ester (325 mg) gives rhodoporphyrin-XV dimethyl ester (228 mg; 81%).

19.4.8. Methyl pyropheophorbide-a (Note)[42]

Methyl pheophorbide-a (100 mg) is heated under reflux in collidine (25 ml) for 90 min during slow passage of a stream of nitrogen. The solution is evaporated (0.1 mm Hg) and the residue recrystallized from methylene chloride—methanol. Yield 82 mg; 91%.

Note

Application of this procedure to methyl pheophorbide-b and the corresponding mesopheophorbides-a and -b affords the analogous pyro-derivatives also.

19.4.9. Chlorin-e₆ trimethyl ester and rhodin-g₇ trimethyl ester[43]

Chlorin-e_6 trimethyl ester and rhodin-g_7 trimethyl ester

Pheophytin mixture (5.0 g) in dry tetrahydrofuran (200 ml) under nitrogen is stirred and treated with a solution of sodium (1.0 g) in dry methanol (200 ml). The mixture is stirred for 1 h, then treated with acetic acid (3 ml) and evaporated to dryness. The residue was chromatographed (carefully) on alumina (grade III, 1 kg) eluting slowly with methylene chloride—benzene (2 : 1). The chlorin-e_6 formed a faster moving grey band, and this was followed by the red-brown rhodin. Both chlorins were recrystallized from methylene chloride—n-hexane. Yields, chlorin-e_6 trimethyl ester (1.3 g) and rhodin-g_7 trimethyl ester (0.3 g).

Note

The same reaction can be carried out with separated pheophytins or pheophorbides, thereby avoiding the chromatographic separation.

19.4.10. Phylloerythrin methyl ester[42]

Methyl mesopyropheophorbide-a (100 mg) and 2,3-dichloro-5,6-dicyanobenzoquinone (DDQ, 70 mg) are refluxed during 1 h in acetone (20 ml) and benzene (20 ml). The mixture is evaporated and the residue is chromatographed on alumina, eluting with methylene chloride. The mobile purple-grey band affords the product (35 mg; 35%) after crystallization from methylene chloride—methanol.

References, p. 861

19.4.11. 2-Vinylpheoporphyrin-a₅ dimethyl ester[44]

Methyl pheophorbide-*a* (500 mg) in dry acetone (50 ml) is heated under reflux (nitrogen) during dropwise addition of a solution of DDQ (500 mg) in benzene (50 ml) over 30 min. The mixture is then refluxed for 30 min, evaporated, chromatographed on alumina (grade V) eluting with methylene chloride, and evaporation of the appropriate eluates gives a residue which is stirred during 30 min (Note) with hexane (20 ml) and ether (100 ml). The porphyrin is filtered off and recrystallized from methylene chloride—methanol. Yield 140 mg, 28%.

Note

The product is invariably contaminated with a little chlorin material (spectrophotometry); this is removed in this extraction.

19.4.12. Pheoporphyrin-a₅ dimethyl ester[44]

Similar treatment of methyl mesopheophorbide-*a* affords the corresponding pheoporphyrin. A direct approach from methyl pheophorbide-*a*: methyl pheophorbide-*a* (600 mg) in dry tetrahydrofuran (300 ml) and triethylamine (1 drop) containing 10% palladized charcoal (300 mg) is hydrogenated at room temperature and atmospheric pressure until the blue-black color has disappeared. The solution is filtered through Celite and then treated with DDQ (700 mg) in benzene (100 ml). After 5 min the reddish purple solution is evaporated and the residue chromatographed on alumina (grade III). Elution with methylene chloride gives the porphyrin (295 mg; 49%) recrystallized from methylene chloride—methanol.

19.5. Porphyrins in natural materials

19.5.1. Coproporphyrins

The best biological source of isomer I is meconium (Ref. 25, pp. 479, 490), and isomer III is best obtained from bacterial cultures[45]. Coproporphyrins are extracted easily from biological materials by treatment with glacial acetic acid, followed by addition of ether. They are extracted from ether by very weak aqueous acids, and while washing ether solutions free of acid, the aqueous phase must be kept above a pH of about 4.5 to prevent loss of porphyrin; dilute $NaHCO_3$ is useful for this purpose. The alkali salts of coproporphyrins are relatively water-soluble.

19.5.2. Uroporphyrins

Isomers I and III only are known to occur in nature, and mixtures of these are often found. They are the most water-soluble of the naturally occurring porphyrins, and in fact cannot be brought into solution in ether. The isomer analysis, and separations of these mixtures are difficult. The best natural source of isomer I is the urine of human subjects or cattle suffering from congenital porphyria, and it may be prepared from porphobilinogen by treatment with alkali[46] or enzymically[47]. Pure uroporphyrin III has been obtained from the feathers of *Turacus sp.*, where it occurs in very high concentration as its Cu-chelate[48], though in some samples isomers I and III have both been reported[49]. Porphobilinogen may be converted to uroporphyrin III in acid conditions or enzymically[50]; non-enzymic formation from excreted porphobilinogen is the origin[51] of most or all the uroporphyrin III (and lesser amounts of isomer I) found in the urine of acute porphyria patients. Statistical mixtures of all four uroporphyrins can be prepared[52] from porphobilinogen.

Pseudouroporphyrin ('phyriaporphyrin') is a hepta-carboxylic porphyrin which is discussed elsewhere (pp. 97, 848).

19.5.3. Detection in biological materials

Porphyrins may be detected directly by their characteristic orange to red fluorescence in many natural materials such as the bones, teeth and tissues of porphyric animals, egg shells, and the root nodules of legumes; moistening the material with HCl (about 2 N) intensifies the fluorescence, probably by increasing the solubility of the porphyrin in a surface layer. Alkali was necessary to reveal the fluorescence of porphyrins bound covalently to protein[53], no fluorescence being seen in acid suspensions, and it appears that the soluble porphyrin—protein complex was the fluorescent species.

When present in relatively high concentration as in porphyric urines, the characteristic absorption bands of the porphyrin or, if present, its Zn- or Cu-chelate may be observed with a hand spectroscope. At lower concentrations, the sensitivity is greatly increased by shaking the urine with ethyl acetate—acetic acid (9 : 1, v/v) and observing the fluorescence of the porphyrin in the ethyl acetate layer.

Porphyrinogens in urines are detected similarly after their oxidation to porphyrins. Certain urines (acute porphyria) contain, in addition to porphyrins, considerable amounts of porphobilinogen, and to prevent the conversion of this to porphyrin the samples should be stored frozen. Heating converts porphobilinogen and porphyrinogens to porphyrins. An excellent discussion of the analysis of urine for porphyrins, porphyrinogens and porphobilinogen has been provided by Schwartz et al.[54].

19.5.3.1. Determination of protoporphyrin and uro (or copro)porphyrin in whole blood[311]

Blood (2 μl) is transferred to a small test tube which serves as a fluorescence cuvette. Ethyl acetate—acetic acid (2 : 1, 300 μl) is rapidly added with a Schwarz—Mann 1 μl biopipette, followed by 0.5 N HCl (300 μl). After 3 sec on a vibrator mixer, the solution is left to stand for 10 min (or centrifuged), and, without separation of the layers, directly examined in a fluorescence spectrophotometer. The ratio of the two fluorescence peaks

(605/655 nm) distinguishes between protoporphyrin and uro- or copropor-phyrin.

19.5.4. Concentration from biological materials

It is often necessary to concentrate porphyrins from large volumes of solution (e.g. urines). Many adsorbents have been used, including talc[55], calcium phosphate[56], kieselguhr[57], lead acetate[58], alumina[59], magnesium oxide[60] and infusorial earth[61]. The adsorbed porphyrins may be eluted with ammonia, sulfuric acid, or other solvents, or eluted and esterified in the one treatment, preparatory to column chromatography, with methanol contain-ing H_2SO_4 or HCl. These methods are perhaps more useful for the recovery of large amounts of material than for quantitative analysis of the porphyrins, but have been used for both purposes. Some of the methods are described briefly here as examples.

19.5.4.1. The calcium phosphate method of Sveinsson et al.[56b] (Note 1)

To 1 ml of urine, 1 ml of 3% (w/v) $CaCl_2$ is added; urinary phosphates are precipiated as calcium phosphate; excess calcium ions are now precipitated by addition of 2 ml of N NaOH, and the combined precipitate with adsorbed porphyrins is washed at the centrifuge with 0.1 N NaOH and water, and is then dissolved in 10 ml of 0.5 N HCl. The HCl solution is filtered, and the porphyrin(s) (Note 2) determined spectrophotometrically.

Notes

(1) Since this method was introduced by Garrod[62] in 1894, many work-ers have precipitated urinary porphyrins on calcium phosphate. The method given here is the result of a careful study[56b] of the conditions for quantita-tive recoveries.

(2) Uroporphyrins-I and -III, as well as coproporphyrin, are adsorbed quantitatively on the calcium phosphate, but most other urinary pigments including porphyrinogens, porphobilinogen, urobilin, and non-pyrrole pig-ments, are largely excluded. When the method was used[63] to determine the low uroporphyrin content of normal urines, however, other pigments were adsorbed sufficiently from the large volumes of urine used to interfere with spectrophotometric determinations. They were removed by esterifying the uroporphyrin and extracting the ester from ether with 3 N HCl.

19.5.4.2. The Kieselguhr method[57]

The urine is filtered, acidified with glacial acetic acid (8 ml/l) and left at 5°C overnight. The precipitate is collected at the centrifuge, the porphyrins eluted by stirring with dilute ammonia, and precipitated by addition of dilute acetic acid. Coproporphyrin may be removed from the precipitated uroporphyrin by boiling with glacial acetic acid. The porphyrins were esteri-fied and the esters purified by column chromatography.

19.5.4.3. The talc method[55]

Porphyria urine (5 l) containing 20—40 mg of porphyrins per l, is passed through a Buchner filter of 25 cm diameter, containing a layer of talc 2—3 mm thick. Foaming is controlled by addition of ethanol. The fluorescence of adsorbed porphyrins is readily seen on talc (but not on Kieselguhr), and the filtrate is passed through layers of talc until all the porphyrin is adsorbed. Batchwise addition of talc to the urine can be used also. The porphyrins are eluted from the moist talc with N HCl—acetone (1 : 9, v/v), or may be eluted and esterified in the one step with H_2SO_4—methanol.

19.5.5. Analysis by solvent extraction

For analysis of urines, and any other tissue fluids (blood, tissue breis) it is convenient to extract the porphyrins into a water-immiscible solvent; a number of methods of this kind are described in the following pages. Treatment of the tissue directly with methanol—sulfuric acid mixtures as used for esterification of porphyrins has been suggested for the extraction of uro- and copro-porphyrins[64].

19.5.5.1. Copro- and proto-porphyrins and porphyrinogens[1,54,65]

If intact erythrocytes are to be analyzed, they must be spun down, hemolyzed with water, and recombined with the supernatant before extraction. The procedure for urines is summarized in Note 1. A mixture of ethyl acetate and glacial acetic acid (3 : 1 v/v) is then added slowly with vigorous stirring and the mixture allowed to stand for at least several hours. The precipitated protein is removed at a sintered-glass Büchner funnel, and washed well by stirring with ethyl acetate—acetic acid. The filtrate is transferred to a separating funnel, and washed twice with about 0.5 vol of saturated aqueous sodium acetate. The washings are extracted with fresh ethyl acetate which is added to the main ethyl acetate layer and the combined ethyl acetate layers are washed once with 0.1 vol of 3% (w/v) sodium acetate.

The protein residue may still contain considerable amounts of uroporphyrin[65]. This is extracted to completion with 10% NH_4OH, the filtrate added to the combined sodium acetate layers, and the whole set aside for determination of uroporphyrin.

The copro- and proto-porphyrins are now extracted completely from the ethyl acetate with 15% (w/v) HCl, and if porphyrinogens are present, the ethyl acetate shaken with iodine and extracted again with acid as described on p. 794. The porphyrin-free ethyl acetate layer now contains all the heme which was present, and may be set aside for recovery of this.

The porphyrins are now transferred quantitatively from the 15% HCl to peroxide-free ether. The ether solution is washed twice with saturated and once with 3% (w/v) sodium acetate, and the aqueous layers combined and extracted with fresh ether. The combined ether layers are washed with dilute

NaHCO$_3$, avoiding excess (Note 2). Exhaustive extraction with 0.36% (w/v) HCl now removes coproporphyrin; protoporphyrin is then removed by extraction with 10% (w/v) HCl.

Notes

(1) In the case of urine, 100 ml ethyl acetate is added to 5 ml of the urine, and the mixture shaken well with 5 ml of a mixture of glacial acetic acid and saturated sodium acetate (4 : 1 v/v). The ethyl acetate layer is washed with 1% (w/v) sodium acetate until no more red fluorescent material (uroporphyrin) appears in the washings. The coproporphyrin is then extracted and purified as described above. Schwartz et al.[54] recommend extraction of the coproporphyrin from the ethyl acetate with 1.5 N HCl, followed by fluorimetric determination (see p. 783) in the HCl solution.

(2) Washing with water is likely to cause loss of coproporphyrin or other porphyrins with low HCl numbers since traces of acetic acid remaining in the ether layer are extracted, forming an acid aqueous layer.

(3) A small proportion of the protoporphyrin is extracted by the 0.36% HCl, and when relatively large amounts of protoporphyrin are present, the fluorescence is not an adequate indicator of the complete extraction of coproporphyrin. When the extraction is nearing completion, samples of the 0.36% HCl extracts should be examined at the spectrophotometer, and the extraction stopped when the Soret maximum has moved to 404—406 nm. The whole of the coproporphyrin fraction should then be transferred to fresh ether and re-fractionated with 0.36% and 10% HCl, to remove and recover traces of protoporphyrin.

Schwartz et al.[54] determine the coproporphyrin in a 0.36% HCl extract of the original ethyl acetate solution, and the protoporphyrin in a subsequent 5% HCl extract.

The copro- and protoporphyrin fractions described here contain, usually as minor fractions, porphyrins with intermediate numbers of carboxyl groups if present.

19.5.5.2. Zinc(II) coproporphyrin

Free coproporphyrin may be extracted from ether or ethyl acetate by weak HCl (0.5%), without extracting its Zn-chelate; on shaking now with 5% HCl, the Zn is liberated and the resulting free coproporphyrin is extracted[66].

19.5.5.3. Spectrophotometric determinations

After separations and purifications as described above, the porphyrins are obtained in aqueous HCl solutions; it is thus convenient to determine them in this medium, which is particularly suitable since porphyrin dications have extinction coefficients two to three times greater than the neutral molecules.

Porphyrin solutions, e.g. following extraction from tissues, often contain impurities which absorb in a more or less linear manner over the visible wavelength scale. To correct the observed density readings for the absorption of such impurities, the expression:

$$D_{corr} = \frac{2D_{max} - (D_{430} + D_{380})}{k}$$

is used, where k is a factor determined empirically by measurements on the pure porphyrin[67]. The expression was introduced for uro- and copro-porphyrins[67b], and has since been applied to other porphyrins[67a] using the values for k given in Table 1a.

A method for the determination of copro- and protoporphyrins in solutions containing both is described on p. 794.

19.5.5.4. Fluorimetric determination of uro-, copro-, and protoporphyrins

Using the isolated 405 nm line from a mercury lamp, prophyrin fluorescence may be measured with a degree of specificity and with much higher sensitivity than is possible by spectrophotometry[68]. A linear relationship has been found[54,69] between porphyrin concentration and fluorescence intensity up to concentrations of 5 μg/100 ml, when the cuvettes were of 1—2 cm path length and the incident light was at 90° to the detector. The linear part of the curve may be extended by using a smaller angle of incidence and by reducing the path-length of the cuvette. Schwartz et al.[54] use solutions of coproporphyrin in 1.5 N HCl as standards for spectrofluorimetric determinations of uro- and proto-porphyrins, and give the correction factors 0.75 and 1.25 respectively for these. The many factors which cause quenching of porphyrin fluorescence are discussed by Schwartz et al.[54].

19.5.5.5. Uroporphyrins[54,65]

Three main methods are used to recover the uroporphyrins from the above aqueous (sodium acetate) fractions: extraction with cyclohexanone or ethyl acetate of the aqueous solution brought to pH 1.5 or 3.2 respectively, or adsorption from the aqueous layer onto an alumina column, followed by elution with HCl.

19.5.5.5.1. The cyclohexanone method[65,70]

The combined aqueous extracts (sodium acetate layers and NH_4OH extract, p. 781) are brought to pH 1.5—1.8 with concentrated HCl, and shaken with two lots of cyclohexanone. The cyclohexanone is mixed with an equal volume of ether, and the uroporphyrin extracted into 5% (w/v) HCl. Transference to the cyclohexanone is usually complete after two extractions, but further extractions should be made until no further porphyrin is extractable by 5% HCl from the cyclohexanone layer mixed with ether; this test is necessary because the fluorescence of low concentrations of porphyrins is not detectable in cyclohexanone. The pH must be checked, and adjusted if necessary, between cyclohexanone extractions.

References, p. 861

TABLE 1

Porphyrin dications in aqueous HCl

Porphyrin	Solvent[b]	λ_{max}(nm) ϵ_{mM}	Soret	II	I	Ref.
Uro-I	0.5 N HCl	λ	405.5	552	594	
		ϵ	541	18.3	6.52	67a[a]
Copro-III	0.1 N HCl	λ	399.5	548	590	
		ϵ	489	16.8	6.65	67a[a]
Proto-IX	2.7 N HCl	λ	408	554	598	
		ϵ	262	13.5	5.78	67a[a]
Deutero-IX	0.1 N HCl	λ	398	548	588	
		ϵ	433	13.7	5.34	67a[a]
Meso-IX	0.1 N HCl	λ	399	547	590	
		ϵ	445	15.1	5.55	a
Diacetyl-deutero-IX	2.7 N HCl	λ	420	562	608	
		ϵ	244	13.6	6.68	a
Deutero-2,4-disulfonic acid dimethyl ester	pH 0.8 (HCl)	λ Ratios		555 1.0	600 0.329	c

[a] J.E. Falk and A. Johnson, unpublished. For the first four compounds, the extinction values for the Soret band (Rimington[67a]) were used for determination of the ϵ_{mM} of the visible bands.
[b] It should be noted that the absorption maxima and values of porphyrin cations vary somewhat with the HCl concentration.
[c] A. Neuberger and J.J. Scott, Proc. R. Soc. (London), Ser. A, 213, 307 (1952).

TABLE 1a

Porphyrin dications in aqueous HCl; correction factors[67a]

Porphyrin	HCl concentration(N)	Soret max. (nm)	Correction maxima (nm)	k[a]
Uro	0.5	405.5	380, 430	1.844
Copro	0.1	399.5	380, 430	1.835
Deutero	0.1	398	375, 425	1.868
Proto	1.37	408	380, 430	1.668
Phylloerythrin	3.0	421	395, 445	1.753

[a] For use with the correction formulae given on p. 783. For uroporphyrin methyl ester[94] in chloroform solution $k = 1.56$; for uroporphyrin in extracts from blood a modified formula has been used[65].

19.5.5.5.2. The ethyl acetate method[54,71] (Note)

The aqueous solution containing uroporphyrin is brought to pH 3.0—3.2 with concentrated HCl, and extracted with successive portions of ethyl acetate until upon extraction with 2% (w/v) HCl no fluorescence can be observed in the aqueous phase. The pH must be checked, and adjusted if necessary, between ethyl acetate extractions. The porphyrin is then extracted from the combined ethyl acetate layers with 2% HCl.

Note

It was long thought[72] that while uorporphyrin-III was extractable by ethyl acetate from aqueous solutions of pH 3.1, uroporphyrin-I did not pass to ethyl acetate at any pH value. Experiments with pure samples of the two isomers gave[70] recoveries between 90 and 100% of both uroporphyrins I and III.

19.5.5.5.3. The alumina column method[54,59,73]

A column of Al_2O_3 (washed with 3% acetic acid) is prepared, and the aqueous uroporphyrin solution passed through it. The column is washed with 3% acetic acid, and the uroporphyrin is then eluted with 5% HCl. Batch adsorption on Al_2O_3 may be used, followed by washing at the centrifuge with half-saturated sodium acetate and elution with HCl[59].

19.5.5.5.4. Determination

The uroporphyrin in aqueous HCl solution obtained by the above methods may be determined spectrophotometrically as described on p. 782 (Table 1).

The ϵ_{mM} for uroporphyrins[67b] I and III in 2% (w/v) HCl is 541; if the measurements are made on 5% HCl solutions, the corrected extinction must be multiplied by 1.05.

19.5.6. Porphyrins in petroleum and shale

A variety of chlorophyll degradation products, and their complexes with metals such as Cu, Ni and V, are found in shales and petroleum oils. The methods will not be described in detail here, but some of the more important references[74] are noted.

19.6. Porphyrin precursors
19.6.1. δ-Aminolevulinic acid (ALA)
19.6.1.1. Determination in biological materials

The methods for determination of aminoketones depend upon a reaction with picrate in NaOH, yielding a red-brown substance of unknown constitution[75], or upon the conversion, by a Knorr condensation, of the aminoketone to a pyrrole which is then determined by the Ehrlich reaction[75,76].

References, p. 861

Of methods of the second type, Mauzerall and Granick's[76a] procedure in which the aminoketone is condensed with acetylacetone appears to be the most satisfactory. This method is also very sensitive, and is probably the method of choice, but the picrate method, especially as modified by Elliott[77], is useful where other Ehrlich-reacting compounds (e.g. PBG) are present. If the pyrrole method of assay is to be used, PBG must be removed by column chromatography.

19.6.1.1.1. Elliott's picric acid method[77]

To 0.25 ml of a solution containing 0.05—0.8 μmol ALA, or 0.05—0.5 μmol of aminoacetone, 0.25 ml of a solution of picric acid saturated at room temperature and cooled to 0°C in an ice-bath is added. Sodium hydroxide (0.5 ml of 5 N at 0°C) is added, the solution mixed and allowed to stand at 0°C for 14 min (ALA) or 6 min (aminoacetone). Hydrochloric acid (1.75 ml of 4 N) is then added, the solution mixed and brought to room temperature in a water-bath. After a further 10 min the absorption is read at 495 nm (Notes 1, 2 and 3).

Notes

(1) Keto-acids cause negligible interference in most instances, but should be removed if present in large amounts; Elliott describes a chromatographic method (Dowex-50) which removes keto-acids with quantitative recoveries of the aminoketones.

(2) Porphobilinogen unless present in great excess does not interfere[75], as, of course, it does in the acetoacetate method in which a color is finally developed with Ehrlich's reagent.

(3) Reading at 495 nm instead of 450 nm as recommended by Shuster, avoids any absorption due to picric acid, and renders its removal unnecessary[78]. At 495 nm, the absorption is proportional to ALA concentration up to 0.8 μmol and to aminoacetone concentration up to 0.5 μmol. The color with both aminoketones fades at the rate of 2% per 10 min at room temperature. Recovery experiments (rat liver homogenate) showed that internal standards should be included in these determinations.

19.6.1.1.2. Mauzerall and Granick's acetylacetone method[76a] (Note 1)

To the ALA solution, 0.2 ml of acetylacetone is added and the volume brought to 10 ml with acetate buffer pH 4.6 (Note 2). The stoppered vessel is heated for 10 min in a boiling water-bath and cooled to room temperature. To 2 ml of this solution, 2 ml of modified Ehrlich reagent are added, and after 15 min (Note 3) the density is read at 553 nm against a reagent blank. The apparent ϵ_{mM} is 2.86. The optical density and ALA concentration are linearly related for density values up to 0.8, and the limit of detection is about 0.003 μmol ALA per ml. Recovery of known amounts of ALA, from 0.032—0.260 μmol, was approximately 95%.

Notes

(1) The acetylacetone method is less sensitive than the acetoacetate method to color interference in the presence of high concentrations of amino acids, ammonia or glucosamine[76a].

(2) The acetate buffer is prepared by adding 1 mol (57 ml) of glacial acetic acid to 1 mol (136 g) of sodium acetate trihydrate, and diluting to 1 l.

(3) The color is stable for at least a further 15 min[76a].

19.6.1.1.3. Mauzerall and Granick's ethylacetoacetate method[76a,(75)]

To the ALA solution, 0.2 ml of ethylacetoacetate is added, and the volume made up to 10 ml with 0.5 M phosphate buffer, pH 6.8. The solution is heated in a boiling water-bath for 10 min and cooled. An aliquot is mixed with an equal volume of Mauzerall and Granick's modified Ehrlich reagent and after standing for 5 min the density is read at 553 nm. Extinction values up to 0.6 are linearly related to ALA concentration, and in this range correspond to an apparent molar extinction of $7.2 \cdot 10^4$.

19.6.1.2. Chromatographic separation of ALA and PBG

The following method, described by Mauzerall and Granick[76a], permits the separation and the quantitative determination of both ALA and PBG in 1-ml samples of urines, and may, of course, be adapted to other fluids, as well as to the isolation of larger amounts of the precursors.

19.6.1.2.1. Separation of PBG and ALA on the Dowex-2 column

Dowex-2-X8 is converted to the acetate form by washing on a column with 3 N sodium acetate until the washings are chloride-free, and then with water until the eluate is free of sodium acetate. Columns of resin 2 cm high in tubes of diameter 0.7 cm, and flow rates of about 3 ml/min are used.

A 1.0-ml sample (urine at pH 5—7) is placed on the Dowex-2 column, and two portions of 2 ml of water are added. The eluate, which contains ALA, is set aside and the PBG eluted from the column by 2 ml of 1 N acetic acid followed, after this has drained thoroughly, by 2 ml of 0.2 N acetic acid. The combined eluates are diluted quantitatively to 10 ml and PBG determined as described below.

19.6.1.2.2. Concentration of ALA by a Dowex-50 column

Dowex-50 is allowed to stand overnight with 2 N NaOH, washed until neutral, and converted to the acid form by treatment with 1 column-volume of 4 N HCl, 6 vol of 2 N HCl, 6 vol of N HCl and then 6 vol of water. Columns of resin 2 cm high in tubes of diameter 0.7 cm and flow rates of about 3 ml/min, are used.

The water eluate from the Dowex-2 column, containing ALA, is transferred quantitatively to the Dowex-50 column, and the column washed with water (16 ml) to remove urea (Note 1). Sodium acetate (3 ml of 0.5 M) is then added (Note 2) and, after draining, the ALA eluted by addition of 7 ml of 0.5 M sodium acetate. This eluate is made up to 10 ml and ALA determined as above.

Notes

(1) Removal of urea may be followed by the yellow color it gives with Ehrlich's reagent. Losses in the water washings of up to 25% of the ALA have been found by De Matteis and Prior[79], who recommend use of the entire washing and eluate for the determination. They did not encounter interference by urea.

(2) The color of the column should become lighter three-fourths, but not all the way down.

(3) In a method for the separation and determination of PBG, ALA and aminoacetone, Urata and Granick[80] use Dowex-1 instead of Dowex-2 to adsorb the PBG, and effect separation of the ALA and aminoacetone in the filtrate by passing this through a column of Amberlite IRC-50, pH 5, ALA being eluted by water and aminoacetone by 2 N HCl[77].

19.6.1.3. Paper chromatography of ALA

ALA may be chromatographed with the solvents usually used for amino acids. Elliott[77] has given an R_F value of 0.15 for ALA using the solvent system ethanol—1.0 N acetic acid—pyridine—water (95 : 10 : 3 : 3 by vol) and has discussed the identification of aminoacetone in the presence of ALA. With ninhydrin, ALA gives a yellow spot on paper chromatograms[81], and may also be detected by its pale white fluorescence under ultraviolet light or by spraying with saturated picric acid and then with 5 N NaOH followed by 5 N HCl, when a red-orange spot, fading to light-brown, is formed[75].

19.6.2. Porphobilinogen (PBG)

The preparation by Westall[82] of the first pure crystalline sample of PBG led to the elucidation of its structure[46,83] and the proof of its role as a porphyrin precursor[84].

19.6.2.1. Isolation from urine

The column method described on p. 787 for the separation of ALA and PBG can, of course, be adapted for preparation of the compound in bulk. Some loss and decomposition of PBG upon columns (alumina, Dowex-2, Deacidite FF) has been reported[46].

The following method (Cookson and Rimington[46]) avoids the use of columns. The urine is brought to pH 4 with acetic acid. Porphyrins may be removed if desired by shaking with talc and filtering. A solution of mercuric acetate (15% w/v) is now added until the supernatant is free of PBG, as shown by Ehrlich's reagent, and the precipitate is collected by filtration and washed with 1% mercuric acetate.

The precipitate is suspended in a minimum of water, and decomposed by passing H_2S. The HgS formed is removed by filtration or centrifugation, and a stream of air is passed through the filtrate to remove excess H_2S.

A solution of lead acetate (10% w/v) is now added until no further precipitate is formed. This precipitate is removed, washed well with 1% lead acetate, and discarded. Much pigment is removed by this treatment.

To the combined filtrate and washings a solution of $AgNO_3$ (20% w/v) is added until precipitation is complete. The precipitate is removed by filtration, washed well with 1% $AgNO_3$, and discarded. In this step much material with a white-blue fluorescence under ultraviolet light is removed.

To the combined filtrate and washings a slight excess of 15% mercuric acetate solution is added. The precipitate, now almost white, is collected by centrifuging and decomposed without addition of water by passing H_2S into the paste, which is stirred with the tip of the gassing tube. The HgS formed is removed by centrifuging. If the resultant supernatant still has considerable white-blue fluorescence, the $AgNO_3$ treatment and the subsequent steps are repeated.

A stream of air is passed through the final supernatant to remove H_2S and the mixture brought to pH 4 by addition of concentrated NH_4OH in microdrops, when PBG crystallizes as the monohydrate[46]. After standing in a refrigerator overnight, the porphobilinogen may be recovered by filtration and washed with a minimum of ice-cold 1% (w/v) acetic acid followed by acetone.

19.6.2.2. Preparation of PBG hydrochloride

A sintered-glass funnel of fine porosity is provided with a water-jacket and set in a Büchner flask. A warm (45°C) solution of PBG in 2 N HCl is filtered in this by suction; the funnel is replaced by a fresh one, the filtrate transferred to this, and ice or an ice—salt mixture placed in the jacket. On cooling, the PBG hydrochloride crystallizes; the mother liquor is removed by suction and the crystals washed with ice-cold 2 N HCl, sucked as dry as possible, and then washed with ice-cold acetone. The hydrochloride is very soluble at room temperature, but in this way yields up to about 80% are obtained.

19.6.2.3. Microbial preparation of PBG from ALA on a large scale[312,313]

Wet cells (2 kg) of *Chromatium vinosum* and ALA · HCl (30 g) are suspended in 0.05 M Tris—HCl buffer (60 l) of pH 8.6 at 30°C. A nitrogen stream is passed through the solution and after 2 h the cells are centrifuged at 0°C, washed twice with Tris-HCl buffer (5 l) and all buffer solutions (70 l) are combined. The pH is lowered to 4 with 10% acetic acid, and 20% mercuric acetate solution (200 ml) is added. The precipitate is centrifuged,

References, p. 861

suspended in distilled water (30 l) and decomposed with hydrogen sulfide. After further centrifugation and degassing of the solution with a rapid stream of nitrogen, uroporphyrin may be precipitated at this stage by addition of lead acetate and centrifugation. Further contaminants are removed by addition of 20% silver acetate (300 ml) and centrifugation. One more precipitation with mercuric acetate and re-precipitation from 0.5 N ammonia with acetic acid (to pH 4) yields 12.5 g (60%) of crystalline PBG.

19.6.2.4. Qualitative identification

The conversion of porphobilinogen to uroporphyrin by heating in acid (0.03 N HCl or 5% acetic acid) provides unequivocal proof of its presence (see Section 21.6.2.6). This conversion can occur in urines on standing, and, of course, can lead to false-negative PBG tests; to delay the conversion, urine should be stored frozen.

Attention must be paid to the differentiation of PBG from other compounds, particularly in urines, which give colored products with Ehrlich's reagent[54,73,85]. When large amounts of PBG are present, as in some porphyric urines, the interference is not important[86], but when low concentrations are to be determined the following tests are useful. After addition of the Ehrlich's reagent, colored products other than PBG pass to the organic phase on shaking with a mixture of amyl and benzyl alcohols (3 : 1 v/v)[86] or with n-butanol[54]; chloroform[87] is less effective. PBG may be adsorbed on Al_2O_3 from a solution in 50% acetone just acidified with acetic acid. Any urobilinogen adsorbed is removed by washing with water, and PBG may then be eluted with 3% acetic acid, leaving adsorbed most other Ehrlich-reacting materials, the elution of which requires much stronger acid[54]. The observation[54] that PBG develops its maximal color on addition of Ehrlich's reagent alone, while other compounds in urine develop color mainly after the addition of excess sodium acetate, may also be helpful in the differentiation.

19.6.2.4.1. The Ehrlich reaction

The identification of PBG in biological materials depends primarily upon the colored complex it forms with Ehrlich's reagent[88,89]. The chemistry of the reaction of Ehrlich's reagent with pyrroles has been elucidated by Treibs and Herrmann[90], who found that contrary to what had been thought, a free α(2)-position in the pyrrole is not obligatory for formation of the colored product. Electron-donating (alkyl) substituents, as well as a free α-position in the pyrrole and also excess of the aldehyde reagent were found to favor the reaction, and electronegative substituents or excess of the pyrrole to hinder it. These and other aspects of the reaction have been discussed[49,54,76a,89,91] in relation to the specificity and the sensitivity of the Ehrlich color reaction. Thiol compounds interfere with the color development, and $CuSO_4$ has been used to overcome this[92].

19.6.2.5. Quantitative determination by Ehrlich's reagent

For the determination of PBG, Ehrlich's reagent[89] usually consists of 2% (w/v) *p*-dimethylaminobenzaldehyde in 5 N HCl[83,86]. The aldehyde is purified by recrystallization from aqueous methanol; it is most convenient to dissolve it in 10 N HCl and then to dilute to 5 N. Equal volumes of this reagent and a PBG solution are mixed, and the extinction is read at 555 nm in a 1-cm cuvette against a blank prepared by mixing equal volumes of the PBG solution and 5 N HCl. Full color development takes a little time, and the reading of maximum intensity is taken. Under these conditions the apparent ϵ_{mM} of PBG is 34.6.

This method is suitable when appreciable amounts of PBG are present, and has been used widely for determinations of PBG in urine, in trichloroacetic acid extracts of liver[93] and of blood cells[94], and in saline extracts of liver purified by chromatography on Al_2O_3[95].

The modified reagent of Mauzerall and Granick[76a] is less convenient for general use, since it must be made freshly each day. It is more sensitive, however, and is useful when smaller amounts of PBG are to be determined. It is made as follows. In 30 ml of glacial acetic acid, 1 g of *p*-dimethylaminobenzaldehyde is dissolved; 8 ml of 70% perchloric acid are added, and the volume made up to 50 ml with glacial acetic acid. This solution is 2 N in respect to perchloric acid. Color development is faster if the perchloric acid concentration is increased, but it should never exceed 4 N because of the possible danger of spontaneous decomposition. The reagent rapidly turns brown, and must be used on the day it is made.

Equal volumes of this reagent and the PBG solution are mixed, and after 15 min the extinction is read at 555 nm, against a blank made from equal volumes of the reagent and water. The color is stable for about 10 min longer. Under these conditions, PBG has an apparent ϵ_{mM} of 61, and extinction values up to about 0.8 are directly proportional to the PBG concentration.

19.6.2.6. Paper chromatography

A number of incidental reports have been made of the paper chromatography of PBG, but Westall's[82] original study still provides the most definitive data. Westall found that spots containing as little as 0.25 μg of PBG could be detected by spraying with Ehrlich's reagent. In addition, on heating the paper to 100°C for 15 min in an atmosphere containing acetic acid vapor, PBG was converted to uroporphyrin, which of course was detectable by its pink fluorescence under ultraviolet light filtered through Woods glass. This conversion to porphyrin provides unequivocal identification of PBG, and is more sensitive than the Ehrlich reaction. Westall found absolute correspondence between the values as revealed by Ehrlich's reagent and by fluorescence after conversion to uroporphyrin. The single-phase solvent system, *n*-butanol—acetic acid—water (63 : 11 :26, v/v) is recommended by Rimmington.

Heikel[96] has studied the paper electrophoresis of porphobilinogen.

19.6.3. Porphyrinogens

It is now recognized that porphyrinogens are intermediates in the formation of protoporphyrin and protoheme, and since the demonstration in 1956 that uroporphyrinogen is enzymically converted to protoheme[97], the preparation in the laboratory of different porphyrinogens for use as substrates in enzymic experiments has become important. The small amount of coproporphyrin normally found in human urines is excreted mainly in the porphyrinogen form, and porphyrinogens occur in certain pathological urines in considerable quantities.

Some reactions which reduce porphyrins to di- and tetra-hydroporphyrins have been discussed earlier.

19.6.3.1. Preparation with sodium amalgam[53,98—101]

The porphyrin (400 nmol) is dissolved in 4 ml of 0.05 N NH_4OH, and 1.0 ml of M potassium phosphate buffer of pH 7.4 is added (Note 1). Nitrogen gas is passed, excess sodium amalgam (Note 2) is added, the flask is stoppered, wrapped in light-proof foil (Note 3) and placed on a shaker at room temperature until the porphyrin fluorescence has disappeared; the time required varies with different porphyrins, from about 2—60 min (Note 7). The flask is now opened, and the porphyrinogen solution is quickly transferred, by means of a Pasteur pipette plugged with glass wool to form a filter, to a 10-ml volumetric vessel wrapped in light-proof foil while nitrogen gas is passed, and the vessel cooled in an ice-bath. The solid residues from the amalgam are washed with distilled water and the washings added to the volumetric flask. The pH of the solution is now brought carefully to 7.4 with acetic acid (Note 4) and the volume made up to 10 ml with O_2-free distilled water.

To determine the porphyrinogen content of the solution, two 0.5-ml aliquots are taken. One is diluted with 3 ml of HCl (to give a final HCl concentration at which the extinction coefficient of the porphyrin is known, see Table 1) and the absorption at the Soret peak is determined immediately (the acid should be added at the spectrophotometer). To the second aliquot an aqueous solution of I_2 (0.05 N) is added dropwise until a faint straw-color persists. When all the porphyrinogen is oxidized to porphyrin (Note 5) a small crystal of cysteine is added to decolorize the excess iodine (Note 6) and the extinction of this solution is read at the Soret maximum. From the difference between the readings on the two solutions the percentage reduction (Note 7) of porphyrinogen in the preparation is readily calculated[53,98].

Notes

(1) Most dicarboxylic porphyrins have more soluble potassium salts than sodium salts.

(2) Sodium amalgam, containing 3% Na in Hg, should be stored in vacuo over solid NaOH, and used no more than a week after its preparation.

(3) The autoxidation of porphyrinogens is slow in neutral or alkaline conditions[98,99], but is strongly catalyzed by light.

(4) Autoxidation of porphyrinogens is rapid in acid conditions, and care must be taken not to add excess acetic acid during the neutralization.

(5) Iodine has been reported to attack the vinyl groups of protoporphyrinogen[100], apparently in ethyl acetate—acetic acid solution, but virtually complete recoveries of spectroscopically pure protoporphyrin have been found when this oxidation was carried out[53,98] at pH 7.4.

(6) Thiol compounds condense with the vinyl groups of protoporphyrinogen[100], but much less readily with the porphyrin.

(7) Falk[1] has found 95—98% reduction of meso-, uro- and copro-porphyrins and other porphyrins with saturated side-chains quite regularly. The vinyl side-chains of protoporphyrin may be reduced by sodium amalgam; to minimize this side-effect, Sano and Granick[100,101] carry out the reaction at 80°C for short times.

19.6.3.2. *Preparation with sodium borohydride*[99,102,103]

In ether or in methanol[102], $NaBH_4$ reduces keto or aldehyde side-chains on porphyrins and chlorophylls, but vinyl groups are not reduced, nor are porphyrinogens formed. In aqueous conditions, however, $NaBH_4$ reduces porphyrins to porphyrinogens. Aqueous KBH_4 has been used for the preparation of protoporphyrinogen[103], but spectroscopic evidence has revealed considerable reduction of the vinyl side-chains under these conditions. It is probably safer to restrict the use of this method to porphyrins without unsaturated side-chains.

19.6.3.3. *Detection of enzymically formed porphyrinogens*

Porphyrinogens have negligible absorption in the Soret region but uroporphyrinogen, formed in enzymic experiments, can be detected by following the increase of absorption at the Soret maximum during its autoxidation to uroporphyrin.

A similar method has been used[53,98] to detect enzymically formed protoporphyrinogen in the presence of its precursor, coproporphyrinogen. At the end of a 2-ml incubation, 0.6 ml of conc. HCl was added rapidly, followed by 16 ml of 5% (w/v) HCl. The mixture was quickly transferred to a closable centrifuge tube (Spinco, 30 rotor), O_2-free nitrogen gas was passed, and the tube sealed. The protein precipitate was centrifuged down (10,000 rev/min for 10 min), and the supernatant drawn into an evacuated spectrophotometer cuvette through an N_2-filled line. Extinctions were determined at

References, p. 861

401.5 nm and 408 nm, and the cuvette was then opened, and exposed to air and sunlight. The increases in extinction were then recorded as the photo-chemical oxidation of the porphyrinogens proceeded, and the amount of each porphyrin present was calculated as described below. By extrapolation back of the porphyrin formation curves, the approximate amounts present as porphyrinogens at the time of stopping the enzymic incubation could be calculated.

Copro- and proto-porphyrins were determined together[98] in 5% (w/v) HCl by solving the following simultaneous equations, where the constants are the ϵ_{mM} of the two porphyrins in this solvent at both 401.5 and 408 nm.

$$E_{1\ cm}^{401.5} = 408[copro] + 191[proto]$$

$$E_{1\ cm}^{408} = 235[copro] + 275[proto]$$

19.6.3.4. Analysis of urinary porphyrinogens

Recognition of the occurrence of porphyrinogens in urines stems from Saillets' studies in 1896 of a compound he called 'urospectrin'[104], but it was not until about 1950 that much attention was paid to these compounds[56a,68,105]; it gradually became clear that the main preformed chromogen was coproporphyrinogen or a closely related compound. Uroporphyrinogen may be formed in quantity, of course, from non-enzymic condensation of porphobilinogen in stored urines, but does not appear to be excreted preformed in more than trace amounts.

The methods for analysis of porphyrinogens in urines depend upon the differential adsorption of porphyrins, but not the chromogens[107] (including porphobilinogen) onto calcium phosphate, or upon the transference of all the pigments to ethyl acetate, extraction of preformed coproporphyrin by aqueous HCl, oxidation of the porphyrinogen remaining in the ethyl acetate layer, and extraction and determination of the porphyrin so formed in aqueous HCl. The method is not precise, because some oxidation occurs during the extraction.

The method of Watson et al.[68]

To the urine an equal volume of a mixture of glacial acetic acid and saturated aqueous sodium acetate (4 : 1, v/v) is added. The mixture is extracted once with 10 vol of ethyl acetate, and the ethyl acetate layer washed with aqueous sodium acetate (1% w/v). Preformed coproporphyrin is extracted completely from the ethyl acetate with 1.5 N HCl (until no further fluorescence in HCl extracts), and is determined. The ethyl acetate residue is now shaken with 0.05 vol of 0.005% (w/v) aqueous I_2, and the porphyrin arising from oxidation of the porphyrinogen extracted as above with 1.5 N HCl and determined. DDQ efficiently accomplishes[e.g.7] transformation of porphyrinogens into porphyrins.

Note

Watson et al. found that when the original ethyl acetate extract was treated at once with I_2, and the total porphyrin (preformed, and arising from oxidation of the porphyrinogen) was then extracted by HCl, somewhat higher values were obtained than by the differential extraction method.

19.7. Preparation of metalloporphyrins from porphyrins and isolation from natural sources

For a thorough discussion, with selected examples, of the general problem of metal insertion into the porphyrin system, see Chapter 5.

19.7.1. Insertion of metal ions other than iron

A general discussion of the metalation of free base porphyrins, together with a detailed appraisal of all known metalloporphyrins, has been given in Chapters 5 and 6. Here, a few typical experimental procedures, mainly concerning main group and first transition row metals, will be described in detail. They exemplify the following general methods which are thought to be the most convenient and efficient.

(1) Heating of the porphyrin with the alcoholate of alkali earth ions under anhydrous conditions (e.g. Li, Na, K).

(2) Refluxing a pyridine solution of the porphyrin with a metal perchlorate (e.g. Mg, Ca, Cd).

(3) Heating the porphyrin with a metal salt in solvents such as chloroform-methanol, carbon disulfide, acetic acid, pyridine, dioxane, and ethylene glycol. This is the general method for most divalent metal ions, especially transition metals.

(4) Reaction of porphyrins with metal acetylacetonates in phenol or in melts of nitrogenous bases (e.g. imidazole). This is the method of choice for trivalent ions, where carbonyl compounds of low oxidation state are not easily accessible (e.g. Sc, Al).

(5) Decomposing metal carbonyls or other heat labile metal compounds at high temperatures in the presence of porphyrins in solvents such as diethylene glycol or decalin. This method allows the introduction of transition metal ions, when the oxidation number of a metal is +3 or higher in its stable salts.

19.7.1.1. Li, Na, K (Ba, Rb, Cs)

Dorough et al.[107] describe the following procedure. Of an approx. 10^{-5} M solution of *meso*-tetraphenylporphyrin in anhydrous pyridine, 3.5 ml are placed in a spectrophotometer cuvette. A 10^{-1} M solution of the alkali metal hydroxide in absolute methanol (0.1 ml) is added and the cuvette is sealed. In the case of Na, K, Li, the resulting green solutions of the salts are stable for many hours. Benzene[108] or dimethylsufoxide[109] can be used instead of pyridine. Benzene solutions of porphyrin dianions often turn turbid after some time, but after precipitation of excess base by centrifugation, they are stable. Similar procedures have been used with ions such as Ba, Rb, and Cs[107].

19.7.1.2. Mg, Ca, Cd

Fischer and Dürr's[110] method as modified by Granick[111] uses Grignard reagents and is nowadays only recommended for heat sensitive porphyrins. In general the magnesium perchlorate—pyridine method of Baum, Burnham, and Plane[112] is less time consuming and gives better yields.

19.7.1.2.1. Magnesium protoporphyrin-IX dimethyl ester[113]

In a 500 ml, 3 necked flask (reflux condenser with KOH drying tube, gas inlet, stopper) protoporphyrin-IX dimethyl ester (1.2 g) and magnesium perchlorate (8.0 g) are heated to boiling under nitrogen in dry pyridine (200 ml, distilled over KOH) in dim light (aluminum foil). Refluxing is continued for 3—4 h. After 2—3 h a sample is taken, diluted with ether and examined with a hand spectroscope or spectrophotometer. No band at 630 nm should be present. The cooled solution is filtered, the dark residues washed with ether until colorless and the pyridine is evaporated to a vol of approx. 25 ml. This solution is poured into dry peroxide-free ether (1.2 l), extracted with water, the aqueous extracts being back extracted with ether, and the water is then discarded. Solid material which precipitates at the interface between ether and water is always incorporated into the ether phase. After evaporation of the ether the residue is dried and crystallized from benzene—petrol ether, to give 0.95—1.0 g (75—80%).

19.7.1.2.2. Dipotassium salt of magnesium protoporphyrin-IX[113]

The foregoing diester (800 mg, crude is satisfactory) in 30% KOH in absolute methanol (100 ml) is kept at 40—45°C for 30 min before decantation cautiously from some red-black residue into water (300 ml). This is kept at 5°C for 30 min and then centrifuged. The precipitate is dissolved in methanol (100 ml), 10 ml of a solution made up by diluting the methanolic KOH 1 : 10 with water is added and the resulting solution is slowly cooled to 5°C and then to —20°C overnight. Crystals are filtered off and washed with dry methanol. The mother liquors are evaporated to approx. 40 ml and again placed at —20°C overnight. The total yield is about 800 mg (approx. 95%).

Often, magnesium porphyrins crystallize with two molecules of pyridine which cannot be removed by simple drying in vacuo. The following method for pyridine-free magnesium porphyrins has been applied to magnesium octaethylporphyrin[114]. The dipyridinate (60 mg) in peroxide-free ether (300 ml) is washed with about 200 ml (×6) of 0.1 N HCl, then with saturated

NaHCO$_3$ (100 ml) and water (3 × 200 ml). The ether is evaporated and the residues are crystallized from benzene—petroleum ether.

The same procedure has been used for other highly acid labile metalloporphyrins (e.g. Ca, Pb, Cd) and chlorins[114–116].

19.7.1.3. Method (3)
19.7.1.3.1. Al(III)

To a suspension of octaethylporphyrin (300 mg) in carbon disulfide (200 ml) three additions (each of 1 g) of solid aluminum tribromide are made within 10 min. The mixture is then stirred in a dry atmosphere at room temperature. Chromatography on a dry column (CHCl$_3$) and crystallization from methanol gives Al(OEP)OH (60%)[116].

19.7.1.2.3. Si(IV)[117], [Ge(IV), Ga(III), and In(III)]

Octaethylporphyrin (100 mg) and silicon tetrachloride (0.5 ml) are mixed with pyridine (10 ml) and heated in a sealed glass tube at 180—190°C for 4—6 h. After cooling to room temperature, the visible spectrum is examined through a hand spectroscope (a safety shield should be placed between the pressure tube and the spectroscope); no bands at 590 nm (dication) or 620 nm should be apparent. If they are, the reaction should be continued. The tube is then opened and the solution diluted first with a mixture of ethanol (20 ml) and conc. HCl (2 ml) and then with chloroform (300 ml). This is washed with NaHCO$_3$ solution and water, evaporated to dryness, and then dry chromatographed (Woelm) with chloroform. Crystallization from chloroform—methanol yields Si(OEP)(OH)$_2$ (80 mg).

Similar procedures have been used to obtain Ge(IV), Ga(III), and In(III) porphyrins[116].

19.7.1.3.3. Sn(IV), [Mn(III), Pd(II), and Pt(II)]

Octaethylporphyrin (100 mg), tin(II) chloride (1 g) and sodium acetate (1 g) are heated under reflux in acetic acid (100 ml) for 30 min. A sample is then taken, diluted with chloroform and checked with a hand spectroscope for the presence of a 620 nm band. If the test is negative, the acetic acid solution is evaporated to a volume of approx. 30—40 ml, refluxed again for 10 min and then left at room temperature overnight. The crystals of Sn(IV)OEP(OAc)$_2$ are filtered off and recrystallized from acetic acid. Yield, approx. 90%[118].

Similar procedures are successful in the preparation of Mn(III) (from MnCl$_2$), Pd(II) (PdCl$_2$), and Pt(II) (K$_2$PtCl$_4$) porphyrins, but chromatographic purification (dry column, Woelm alumina) is required in these cases. Platinum and palladium tend to oxidize chlorins to porphyrins. Addition of a 3-fold excess of acetylacetone (with respect to metal salt) has been found[115] to prevent this; another possibility is to replace the acetic acid solvent with dioxane, or in some cases to use only freshly distilled acetic acid[119].

References, p. 861

19.7.1.3.4. Zn(II), [Co(II), Ni(II), and Cu(II)]

To the porphyrin (250 mg) in boiling chloroform (10—100 ml depending upon solubility) is added a saturated solution of zinc acetate in methanol (1 ml). After a few min refluxing and checking by spectrophotometry, the mixture is concentrated, diluted with a little methanol, and after cooling the zinc complex is filtered off in virtually quantitative yield.

Cu(II), Co(II), and Ni(II) porphyrins can be obtained by the same procedure. The last of these requires longer reaction times.

19.7.1.3.5. Vanadyl

The porphyrin is heated in a sealed glass tube with vanadium tetrachloride at 165°C for 4 h. The resulting vanadyl complex (Note) is chromatographed (dry column, Woelm alumina, $CHCl_3$) and crystallized from hot benzene—methanol[120,121]. Yield approx. 80%.

Note

If a porphyrin with carboxylic side-chains is used (e.g. mesoporphyrin-IX) then an esterification step (HCl—methanol, or H_2SO_4—methanol) is required before chromatography.

19.7.1.3.6. Ag(II), Pb(II), Hg(II)

The porphyrin and metal salt [e.g. $AgNO_3$, $Pb(OAc)_2$, $Hg(OAc)_2$] are refluxed in pyridine until complete metalation has taken place (hand spectroscope or spectrophotometer)[107,122]. The acid labile lead and mercury complexes are, after dilution with ether and then water, extracted into and crystallized from ether (Note).

Note

The mercury complexes tend to bind molecules of pyridine; this can be removed by heating in vacuo.

19.7.1.4. Metal diketones

Metal acetylacetonates are often commercially available. Otherwise they may be prepared following the procedure of Fackler[123].

19.7.1.4.1. Sc(III), Al(III)

Octaethylporphyrin (105 mg), scandium acetylacetonate (210 mg), and imidazole (1.5 g) are heated together at 235°C during 20 min. The mixture is then sublimed, firstly at 90—170°C (imidazole), and then at 190—225°C to afford Sc(OEP)OAc (60 mg)[124]. The Al(III) complex of octaethylporphyrin is obtained in 85% yield by heating octaethylporphyrin (276 mg), aluminum acetylacetonate (1.05 g) and phenol (2.87 g) at 230°C for 2 h. The cooled melt is extracted several times with boiling 2 N NaOH, the residue filtered off, washed with water to neutrality, and then dissolved in methylene chloride. Chromatography on alumina (grade IV) and crystallization from methanol—water gives 257 mg of red-violet platelets[124].

19.7.1.4.2. Lanthanides, e.g. Eu(III)

meso-Tetraphenylporphyrin (1 mmol) and hydrated tris(2,4-pentanedionato)europium(III) (2 mmol) are refluxed in 1,2,4-trichlorobenzene (214°C) for 3—4 h; the solvent is then removed under reduced pressure and the product is purified by column chromatography. This reaction is general to the whole lanthanide series and a number of meso-tetra-arylporphyrins have been metalated with several different β-diketone complexes of the lanthanides[125].

19.7.1.5. Method (5)

Many metal carbonyls or other heat-labile metal compounds can be purchased at the large suppliers of chemicals. More exotic metal carbonyls are available from Alpha Inorganics, Hercules Inc., Strem, and Ventron etc.

19.7.1.5.1. O=Mo(V), Cr(II), O=Ti(IV)

O=Mo(V)[TPP]OH

H_2(TPP) (1 g) and Mo(CO)$_5$ are suspended in decalin (400 ml) and refluxed for 5—6 h in a nitrogen atmosphere. The solution, after cooling, is passed through a dry column (alumina, Woelm), eluting firstly with benzene (to remove decalin) and then with chloroform. The green fraction is collected and rechromatographed. Crystallization from chloroform—methanol yields 25—40% of O=Mo(V)[TPP]OH[126].

Cr(II) mesoporphyrin-IX dimethyl ester

Mesoporphyrin-IX dimethyl ester (200 mg) and Cr(CO)$_6$ (1.0 g) are heated in decalin at 170°C under nitrogen for 1.5 h. Solvent and unreacted Cr(CO)$_6$ are removed under reduced pressure and the residue is dissolved in toluene which is concentrated and then the product is precipitated by addition of n-pentane. One or two reprecipitations yield about 40% of pure material[127].

O=Ti(IV)[OEP]

Octaethylporphyrin (200 mg) is suspended in diethylene glycol (100 ml) with dicyclopentadienyltitanium dichloride (1.0 g) and then refluxed for 2 h. Completion of the reaction is checked by spectrophotometry. The solution is cooled, mixed with chloroform, washed with dilute HCl, then NaHCO$_3$ solution and water, and evaporated to dryness. The residue is chromatographed on a dry column of alumina (Woelm) and eluted with chloroform. Evaporation and crystallization from chloroform—methanol gives 160 mg (75%) of large rhombic crystals. A further 50 mg of fine needles is obtained from the mother liquors.

References, p. 861

19.7.2. Insertion and removal of iron

Preparative methods for insertion and removal of iron in porphyrins depend upon the fact that in these compounds the ferrous ion is very much less firmly held than is the ferric ion. Porphyrins coordinate readily with ferrous ions, but not with ferric, in hot acetic acid; the resulting ferrous chelate is rapidly autoxidized to the stable ferric state. To remove iron from a hemin, this is first reduced to the ferrous state (heme); if a strong acid is then added, protons are able to displace the coordinated ferrous ion with formation of the porphyrin dication. Fischer and Zerweck[128] have reported that pyridinium hemochromes, upon treatment with aqueous HCl are converted into porphyrins. A similar reaction has been reported by Clezy and Morell[129]; following the reaction of diazomethane with a hemin, a diazomethane—heme complex is formed which on treatment with aqueous HCl yields the porphyrin related to the original hemin. These reactions provide examples of the removal of iron in the presence of water, which is usually inhibitory[130].

19.7.2.1. Removal of iron from hemins

Reducing agents which have been used include iron powder[131,132], ferrous acetate[133], ferrous sulfate[130,134], palladium and hydrogen[131,135], stanous chloride[136], sodium amalgam[137], pyruvic acid[138] and hydrogen bromide in glacial acetic acid[139]. The last reagent cannot be used for hemins with vinyl side-chains, the HBr adducts being formed (see hematoporphyrin preparation, p. 771); it does not remove iron from lactoperoxidase hemin, probably because of the presence of electrophilic side-chains (D.B. Morell, private communication). Sodium amalgam, iron powder, and catalytic palladium can cause reduction of unsaturated side-chains as well as reduction of the nucleus under some conditions (see p. 792), though quantitative reoxidation is usually possible. While Pt-black in alkali or acetic acid[140], or PdO in formic acid[141] both reduce vinyl side-chains, PtO in boiling formic acid is reported to reduce protohemin to the ferrous form, with concurrent removal of the iron, without reducing the vinyl groups. The acids which have been used to displace the ferrous ion include formic, oxalic, acetic and hydrochloric acids. Concentrated sulfuric acid removes iron and other metals from metalloporphyrins; the porphyrin nucleus is stable to this reagent at room temperature, but side-chain modifications can occur; thus protohemin is converted to hematoporphyrin.

19.7.2.1.1. The ferrous sulfate method (Lemberg et al.[134], Morell et al.[130])

This is the mildest and the most convenient method, and the aqueous conditions permit close control of the reaction as is desirable with rather labile compounds such as hemin-a_2[136], hemin-a[134] and lactoperoxidase hemin (D.B. Morell, private communication). It is desirable to use this method also for the preparation of protoporphyrin for use as a chromatographic marker, less by-products being formed than by most other methods. The factors affecting the rate of the reaction have been examined[130]. The scale of the method is similar to that of the ferrous acetate method (Note 1).

The hemin (1.0 mg) is dissolved in a minimum of pyridine (Note 2) and the solution diluted to 10 ml with glacial acetic acid (Note 3). If the reaction is to be performed at room temperature, it is carried out in an atmosphere of N_2. A fresh solution of $FeSO_4$ (Note 4) in concentrated HCl (0.4 ml) is added, and the passage of N_2 over the solution is continued for 5 min; the reaction should then be complete. The solution is poured into a mixture of ether and aqueous sodium acetate in a separating funnel, and the porphyrin recovered.

Notes

(1) For maximum scale reactions with hemins stable to acid, it may be desirable to heat as in the ferrous acetate method.

(2) If much pyridine is present, the amount of concentrated HCl added must be increased accordingly; for the room-temperature reaction the total water present (including the HCl) should not exceed 5% (v/v).

(3) For hemins with formyl side-chains it was found[134] that the acetic acid should be refluxed for 1 h over $FeSO_4$, and then distilled from Cu-free apparatus, to prevent the Cu-catalyzed oxidation of formyl to carboxylic acid groups.

(4) For hemins without labile side-chains (e.g. formyl) a near-saturated solution of $FeSO_4$ in conc. HCl is best used. For labile hemins, 40 mg of $FeSO_4$ in 0.4 ml of conc. HCl is a convenient concentration. Ferrous ions appear to catalyze the oxidation of formyl and possibly other side-chains (D.B. Morell, private communication).

19.7.2.1.2. The ferrous acetate—acetic acid method (Warburg and Nege-lein[133])

The hemin is dissolved in glacial acetic acid with the aid of a small volume of pyridine. A solution of ferrous acetate in acetic acid is prepared by refluxing iron powder with glacial acetic acid under a stream of N_2 or CO_2. For small-scale preparations the heat source is removed, the mixture allowed to settle, and a small volume of the water-clear supernatant solution of ferrous acetate in acetic acid transferred, with a Pasteur pipette, to a solution of the gently refluxing hemin solution (Note). For 10 ml of this solution, approx. 1 ml of concentrated hydrochloric acid is added immediately after the ferrous acetate, when conversion of the hemin to the porphyrin is virtually instantaneous. This may be verified by the absence of the broad hemin band at approx. 630 nm together with the appearance of the porphyrin dication spectrum (see Fig. 2a, p. 24). The porphyrin may be recovered by flocculation at pH approx. 4, or better by transference to ether, followed by extraction of the porphyrin with aqueous HCl, which leaves unchanged hemin in the ether.

References, p. 861

Note

For large-scale work a closed system allowing transfer of the ferrous acetate without access to air must be used; the transfer line should include a sintered-glass filter to hold back unreacted iron powder, which of course can cause production of hydrogen with consequent reduction of unsaturated side-chains in the porphyrin.

19.7.2.1.3. The iron-powder method (Corwin and Krieble[142])

This method is useful for work on the 1-g scale with porphyrins which have no unsaturated side-chains.

The hemin (1 g) is refluxed in 500 ml glacial acetic acid containing 5 ml of 35% (w/v) HCl. Iron powder (100 mg) is added in small amounts over 10 min; the color changes from brown to violet, and the reaction is continued until no unchanged hemin remains. The porphyrin may be precipitated from the mixture with NaOH.

19.7.2.1.4. The formic acid method[132]

In a 3-l round-bottom, 3-neck flask, fitted with a mechanical stirrer and a reflux condenser, 15.5 g of hemin and 776.5 g (630 ml) of formic acid (98—100%) are heated to boiling and stirred vigorously. Iron powder ('reduced iron', 15.6 g) is added in 2.6-g portions at intervals of 5 min, and refluxing is continued for 25 min, after the last addition. The mixture is cooled and filtered, the precipitate washed with a small amount of formic acid, and the filtrate poured into 2.4 litres of water. Protoporphyrin is precipitated by addition to this aqueous solution of about 300 g of solid ammonium acetate (until fluorescence in the supernatant is minimal). After standing overnight, the porphyrin is recovered by filtration and dried at $50°C$ over P_2O_5 in vacuo. Yield 11.7 g (87.5%).

19.7.2.1.5. Removal of iron with concurrent esterification (Grinstein[143])

The method of Grinstein may be used on any scale. It is particularly convenient, and has been used widely for larger-scale preparations; in two or three days some 5 g of crystalline protoporphyrin ester may be prepared. The product requires purification by column chromatography but may be used for larger-scale preparative work if a subsequent product may conveniently be purified.

A solution of the hemin is prepared in methanol (Note 1) containing 10% (w/v) of oxalic acid. For the preparation of protoporphyrin ester, protohemin is obtained conveniently on the large scale by extraction of an acetone powder of blood or, better, of packed washed red cells (Note 2) with oxalic acid—methanol. To 200 ml of the filtered hemin solution in a 250-ml conical flask, about 2 g of powdered $FeSO_4$ is added; gaseous HCl is passed rapidly into the solution (Note 3) until the absorption band due to hemin (about 630 nm) is replaced by those of the protoporphyrin dication (601,

585 nm). The reaction mixture is cooled, excess water added, and the mixture extracted with three successive amounts of chloroform. The chloroform solution is washed twice, with two volumes of water each time; the washing should be carried out as soon and as quickly as possible, to prevent hydrolysis of the ester by aqueous HCl.

The chloroform extracts from a number of 200-ml reactions are collected together, and washed once with about an equal volume of 2 N NH_4OH. A large white precipitate of ammonium oxalate is removed by a Büchner filter, the precipitate washed with chloroform, and the chloroform layer from the filtrate washed three times with water. The chloroform solution is dried conveniently by filtering by gravity through a triple thickness of folded Whatman No. 1 paper wet with freshly-washed and dried chloroform, and is then evaporated to dryness. The residue may be crystallized directly from chloroform—methanol but should be purified by column chromatography.

Notes

(1) Methanol used for the extraction should have been distilled, and any chloroform used after the ammonia wash must be freshly washed and dried over $CaCl_2$.

(2) Acetone powders are prepared by stirring at least 3 vol of acetone into the blood or cell suspension, and allowing the mixture to stand for at least 1 h; standing overnight is better. After filtering the precipitated protein as dry as possible at the pump, the precipitate is quickly washed with ether. A friable, easily-handled mass suitable for the extraction of hemin then results.

(3) The reaction is best carried out in batches not larger than 200 ml. For successful, complete conversion of hemin to porphyrin ester it is essential to pass the HCl gas in a very fast stream.

(4) The method may be modified in various ways; thus the oxalic acid serves no purpose except that of splitting the hemin from protein and bringing it into solution. Any hemin may be dissolved in a small volume of aqueous NaOH or pyridine, methanol and $FeSO_4$ added, and HCl gas then passed to bring about the conversion to porphyrin ester. There is then usually no need to filter after washing the chloroform with NH_4OH since ammonium oxalate is not formed.

19.7.2.2. Insertion of iron
19.7.2.2.1. The ferrous sulfate method

This adaptation of the method of Lemberg, Morell et al.[130,134], for removal of iron (see above), was developed as a very mild method suitable for the insertion of iron into porphyrin-*a* and other porphyrins labile to hot acid.

The porphyrin is dissolved in 1 ml pyridine and 50 ml of pure acetic acid, and immediately afterwards 1 ml of a strong solution of $FeSO_4$ in water are added. A stream of CO_2 or N_2 is then passed while the solution is kept at

80°C for 10 min. Conversion to the hemin is about 90—95% complete. The mixture is allowed to stand in air for 10 min to allow autoxidation to the ferric complex to occur, and then the pigments are taken to ether and residual porphyrin is removed by extraction with 25% (w/v) HCl. The addition of pyridine decreases the proton concentration and alters the equilibrium in favor of hemin formation.

19.7.2.2.2. The ferrous acetate—acetic acid method (Warburg and Negelein[133])

This method is not safe for use with labile hemins; even with protohemin some alteration of side-chains occurs.

A solution of ferrous acetate in acetic acid is prepared as described above (Warburg and Negelein method for removal of iron) and transferred as described there to a solution of the porphyrin refluxing gently in glacial acetic acid. Conversion to the heme is virtually instantaneous, and autoxidation to the stable (ferric) hemin is complete after a few minutes further refluxing. If chloride ions have been supplied (by adding solid NaCl, or $SrCl_2$, or a very little concentrated HCl) the chlorohemin usually crystallizes on cooling. If desired, the pigments may be extracted at pH 4 into ether or ethyl acetate, and any residual free porphyrin removed by extraction with aqueous HCl.

19.7.3. Hemes
19.7.3.1. Determination as pyridine hemochromes

The hemochromes are complexes formed by the coordination of two molecules of a base to a ferroporphyrin (heme). As Lemberg and Legge[144] point out, hemochromes had been prepared in crystalline form long before their relationship to hemoglobin was correctly understood. Thus Van Zeyneck[145] in 1898 prepared ammonia hemochrome in the solid state by reduction of 'hematin' in alcoholic solution with hydrazine hydrate. Crystalline pyridine hemochromes also have been prepared[146].

The sharp and characteristic α-bands of pyridine hemochromes are often used for their identification and determination. The extinction coefficients (ε) of a number of pyridine hemochromes are given in Table 2, in which the absorption coefficient (β), often used in European publications, is also defined, and a factor given for its conversion to ε values.

Pyridine hemochrome spectra are measured in aqueous alkaline pyridine solutions after reduction with sodium dithionite. Paul et al.[147] have found that the spectrum of protohemochrome is not changed when the alkali concentration is varied from 0.02—0.5 N NaOH, at a pyridine concentration of 25% (v/v) (ca. 3 M), or when the pyridine concentration is varied from 1.2—6.2 M at an NaOH concentration of 0.1 N. Paul et al. found that the pyridine to heme ratio should be at least 150,000 : 1.

Paul et al.[147] recommend final concentration of 0.075 N NaOH and 2.1 M pyridine, and these conditions have been found very satisfactory.

TABLE 2

Pyridine hemochromes

(a) Absolute spectra

Heme	λ_{max}(nm) $\epsilon_{mM}{}^b$	Soret	Ref.	β max	min	α max	Ref.
Copro	λ			516		545	1
Proto-IX	λ	418.5	a	526	540	557	
	ϵ	191.5	a	17.5	9.9	34.4	147
	λ	420	129				
Deutero-IX	λ	406	129	515	530.5	545	
	ϵ	97.3	a			24.0	227[a]
Meso-IX	λ	407	a	518	531	547	227
	ϵ	140.4	a			33.2	1
	λ	412-13	129				
Hemato-IX	λ			519	535	549	227
2(4)-Vinyl-4(2)-hydroxyethyl-deutero-IX	λ			520	536	552	227
2(4)-Hydroxy-methyldeutero-IX	λ			515	534	546	227
2(4)-Formyl-deutero-IX	λ	428	129	532	553	581	129, 227
2,4-Diformyl-deutero-IX	λ	450	129	549.7		584.3	129, 222
2(4)-Acetyl-deutero-IX	λ			530		571	222
	λ	422-3	129	526		570	129
2,4-Diacetyl-deutero-IX	λ			539.7		575.2	222
	λ	439	129	540		573	129
2-Formyl-4-vinyl-deutero-IX (chlorocruoro)	λ			545.1		583.1	222
	λ			543	562	583.5	227
	λ	434	129	532-4		580.5	129
Heme-a	λ	427	129			585	129
	λ					587	222
Heme-c	λ			522		551	
	ϵ			18.6		29.1	1

[a] R.J. Porra, unpublished; the ϵ values for deuterohemochrome are unusually low; the sample was recrystallized to constant ϵ.

[b] The absorption coefficient β is often used in European publications, β is equal to $1/c(\ln I^\circ/I)$ for a 1-cm path length, with c in moles per ml; $\epsilon_{mM} = \beta \cdot 10^{-6}/2.303$.

References, p. 861

Pyridine Hemochromes

(b) Reduced minus oxidized spectra[227]

Heme	λ_{max}(nm) $\Delta\epsilon = \epsilon_{mM}(\alpha - min)$	β max	min	α max
Proto-IX	λ $\Delta\epsilon$	526	541	557 20.7
Deutero-IX	λ $\Delta\epsilon$	515	530.5	545 15.3
Meso-IX	λ $\Delta\epsilon$	518	531	547 21.7
2(4)-Formyldeutero-IX	λ	532	553	581
2-Formyl-4-vinyl-deutero-IX (chloro-cruoro)	λ	543	562	583.5

A number of factors can lead to inaccuracy in pyridine hemochrome determinations. The degradation of 'hematins' in aqueous alkali can be very rapid; even on storage overnight at 4°C in 0.075 N NaOH significant alterations to the spectrum of some preparations has been observed[147]. It is best to dissolve the hemin or hemoprotein in a mixture of pyridine and alkali to give the correct final concentration, and to reduce and measure the spectrum at once. In any case, the alkali should be added as late as possible. The most sensitive index of degradation and also of contamination by impurities with non-specific absorption, is the ratio of intensity at the α-maximum to the minimum between the α- and β-band (see Table 2). For recrystallized proto-hemin Paul et al. found this ratio to be 3.46 and for recrystallized[148] myoglobin 3.47.

Care should be taken also during the reduction. Just before reading at the spectrophotometer, a few crystals of sodium dithionite (not more than 3 mg per ml)[149] are added and dissolved by rotating the cuvette gently to avoid aerating the solution. If desired, the mixture may be overlayered with liquid paraffin to help exclude oxygen, and so minimize autoxidation to the ferric (hemichrome) state. After adding the dithionite, readings are taken as quickly as possible every 2 nm over the expected α-maximum, then at the maximum; a little more dithionite is added, and if there is an increase in extinction at the maximum the process is rapidly repeated until no further increase shows that the compound has been reduced completely.

Commercial samples of dithionite are very variable; a free-flowing sample with no odor of SO_2 should be chosen and transferred at once to small, tightly-closed containers, which, once opened, are only used for a few days each. If an odor of SO_2 develops, or the particles begin to stick together, the material should be discarded[149].

Hemochromes have sharp, strong Soret bands (Table 2); these may of course be used for their determination, but in tissue extracts are subject to much more interference by other compounds than are the α-bands.

It has been found[150] that reduced minus oxidized hemochrome spectra permit the determination of heme in the presence of considerable amounts of porphyrins, as well as decreasing the errors due to non-specific absorption. The accuracy is increased further by using the difference millimolar extinction coefficient ($\Delta\epsilon$), which is defined as the difference in millimolar extinction between the maximum of the α-band and the minimum which occurs between the α- and β-bands. Values for $\Delta\epsilon$ for several hemochromes are given in Table 2b.

In practice, to 4.2 ml of a solution of the hemin, 1 ml of pyridine and then 0.5 ml of N NaOH are added. In each of two cuvettes, 3 ml of this solution are placed; to one, 0.05 ml of $3 \cdot 10^{-3}$ M $K_3 Fe(CN)_6$ is added, and to the other 2 mg of sodium dithionite. The reduced minus oxidized hemochrome spectrum is then recorded, and the $\Delta\epsilon$ is determined.

19.7.3.2. Solvent extraction from tissues

Protoheme, and other hemes (chlorocruoroheme, heme-a, heme-a_2) which are not covalently bound to protein may be extracted from biological materials by acidified organic solvents. In this way the prosthetic groups of hemoglobins, myoglobins, erythrocruorins, catalases, some peroxidases, and cytochromes-b, -a_3 and -a_2 are readily split, and their proteins precipitated. Heme-c, milk peroxidase heme and erythrocyte peroxidase heme are covalently bound to protein, and are not split off by these reagents.

19.7.3.2.1. Ether—acetic acid or ethyl acetate—acetic acid

These are the most commonly used solvent mixtures for extraction of porphyrins and hemes; the procedure is described in detail on p. 781. Ether should be washed free of peroxides but is preferable to ethyl acetate because of its much greater ease of removal by evaporation, during which the solutes are subjected to much less heating.

19.7.3.2.2. Acid—acetone

Acetone containing 1—5% by vol of conc. HCl is a more vigorous extraction mixture than ethyl acetate—acetic acid. The tissue material to be extracted should not contain too much water; after filtration from the protein precipitate, the filtrate is mixed with ether, neutralized with sodium acetate, and the ether solution of the hemins washed with water. The acid—acetone method has been used widely[151] for the preparation of native globin from hemoglobin. A detailed description of the isolation of hemin-a from bovine heart has been given by Lynen et al.[152]. This has since been simplified by

Caughey et al.[153] by partly replacing the acetone with chloroform and development of a Celite column chromatographic procedure for heme-*a*. It should be noted that in the presence of HCl, acetone can condense slowly with formyl side-chains on hemins.

19.7.3.2.3. Methylethylketone

Unlike acetone, this solvent is immiscible with water; in mildly acidic conditions at low temperature the hemin may be extracted directly into the organic phase, leaving the undenatured protein in the aqueous phase[154]. The salt-free hemoprotein solution at $0°C$ is brought to pH 2 with 0.1 N HCl, and shaken with an equal volume of ketone. The method has been found (unpublished observations) to extract porphyrins and also plant pigments rather efficiently.

The extraction of hemes from yeast cells is difficult, and Barrett[155] has obtained good extraction by grinding dried yeast cells with No. 12 or 14 ballottini beads in a mixture of methylethylketone, acetonitrile and 3 N HCl (6 : 4 : 1 by vol). For moist yeast, 3 parts by volume of 11 N HCl are used.

Alcoholic phosphotungstic acid also has been used for hemin extractions[156].

19.7.3.3. Protohemin

19.7.3.3.1. Preparation of crystalline hemin

Several methods are available, and the choice of method depends upon various considerations. The classical method, which appears to have been introduced in 1885 by Schalfejeff[157] involves[158] slow addition of blood to glacial acetic acid saturated with NaCl, at $100°C$. The reaction may be carried out with a drop of blood on a microscope slide, the formation of characteristic rhomb-shaped ('Teichmann') crystals of hemin providing a microtest for blood[139b]. Contamination of the crystals with protein is often troublesome during their filtration in this procedure. Protein contamination can be avoided, at the loss of some yield, by filtering off the hemin crystals before the mixture has cooled completely. For bulk preparations, however, the modified method of Labbe and Nishida[159] is preferred. These methods are described in detail below.

In isotopic-tracer experiments where ^{14}C-labeled hemin is to be counted, it should be remembered that porphyrins adsorbed on the hemin crystals prepared by the above methods can be an important source of error, and should be removed before crystallization.

For preparation of solid samples for counting at infinite thickness, it was found convenient[65] both for purification to constant radioactivity and for easier planchette packing, to convert the crude hemin recovered from the ethyl acetate to protoporphyrin ester and thence to its copper chelate.

Small samples of hemin for use as a chromatographic marker and other purposes are quickly obtained by extraction of blood with acetone—HCl followed by transference of the hemin to ether.

(a) The glacial acetic acid method[158]

Citrated, heparinized or defibrinated blood (100 ml) is strained through gauze cloth, and run in a thin stream, which is not allowed to touch the stirrer or the sides of the beaker, into 300 ml of glacial acetic acid at 100—102°C saturated with solid NaCl and stirred mechanically. The temperature must not exceed 103°C, nor drop below 90°C during the addition after which it is maintained at 100°C for 15 min. Hemin crystals separate on cooling, and should be filtered off quickly when the solution has cooled to about 60°C, and washed at the centrifuge or on the filter with 50% acetic acid, water, alcohol and ether. Yield about 300 mg. If cooling is allowed to proceed too far gelatinous material which cannot be washed away from the crystals sometimes appears.

(b) The acetone-acetic acid $SrCl_2$ method (Labbe and Nishida[159,160])

A stock solution of 2% (w/v) $SrCl_2 \cdot 6H_2O$ in glacial acetic acid is made. Just before use (Note 1), the extraction solvent is made by mixing 1 vol of the $SrCl_2$ solution with 3 vol of acetone. To 12 vol (Note 2) of this solvent, 1 vol of blood is added with stirring, and the mixture allowed to stand for 30 min with occasional stirring. The mixture may be heated briefly to boiling during this period to improve the precipitation of protein (Note 3). After cooling, the mixture is filtered through a sintered glass Büchner, the precipitate washed with two portions of the extraction solvent, and the filtrate transferred to a beaker. The acetone is now evaporated (Note 3) by heating to 100°C, but not more than 102°C. Hemin crystallization proceeds as the solution becomes concentrated, and is completed on cooling. The crystals are washed at the centrifuge or on a filter with 50% acetic acid, water, alcohol and ether. Yield, 75—80%. A micro-scale adaptation of this method permits crystallization of hemin in 50% yield from as little as 0.1 ml blood[161].

Notes

(1) $SrCl_2$ crystallizes if the mixture is stored.

(2) It has been found (J.E. Falk and A. Johnson, unpublished) that 7 volumes of acetone are sufficient, but 6 vol insufficient, for complete precipitation of the protein when whole blood is used.

(3) Porous pot, or boiling stones, and stirring are necessary to control bumping.

(c) Recrystallization[158]

There is no evidence that well-prepared hemin crystals are improved by recrystallization. In experiments with ^{59}Fe, however, it provides a useful criterion of radioactive purity.

To 5 mg hemin (Note 1) in a 100-ml flask, 25 ml pyridine are added; when the hemin has dissolved, 40 ml of chloroform are added and the mixture shaken for 15 min. The solution is filtered through a small fast fluted paper, and the paper washed with 15 ml of chloroform. Glacial acetic acid (350 ml) is heated to boiling, 5 ml saturated aqueous NaCl (Note 2) and 4 ml conc. HCl are added, the heat-source removed, and the hemin solution added in a steady stream with mechanical stirring. After standing at room temperature for 12 h the hemin crystals are recovered at a centrifuge or a filter, and washed with 50% acetic acid, water, alcohol and ether. Yield, 75—85%.

Notes

(1) A similar method is used to prepare crystalline hemin from the material recovered from ethyl acetate after porphyrins have been extracted (p. 782); after evaporation of the ethyl acetate to dryness, the residue is dissolved in pyridine and treated as described here. It has been found important to recover and crystallize the hemin from ethyl acetate solutions as soon as possible, and not to store the solutions.

The method has been adapted to the small scale[162]. For the hemin residue from the ethyl acetate—acetic acid extraction of 2 ml blood, 0.15 ml pyridine and 1.5 ml chloroform are used plus 1 ml further chloroform to wash the filter; the solution is added to 3.75 ml of glacial acetic acid saturated with NaCl, the mixture heated rapidly to 108°C on a sand bath, removed at once from the bath, and one drop of 2 N HCl added. Hemin crystallizes on standing at room temperature overnight. The ethyl acetate should be washed with $FeSO_4$ and water before use, and the crude hemin solutions should not be stored at any stage.

(2) Labbe and Nishida[159] use instead of NaCl, 15 ml of 2% (w/v) $SrCl_2$ per 100 mg of hemin in an otherwise similar procedure.

19.7.3.3.2. Formiatohemin and other hemin—anion complexes
(a) Preparation of formiatohemin[163]

Defibrinated blood (100 ml) is hemolyzed by the addition of 200 ml water, mixed with 100 ml of 85% formic acid, heated gradually on a water-bath, with stirring, to 75°C, and maintained at that temperature for 10 min. After cooling, the crystalline formiatohemin is recovered at the centrifuge, washed three times with 1% formic acid, then water, and dried. Yield, 570 mg.

For recrystallization, 1 g of the crystals is dissolved in 100 ml methanol containing 5 g of KOH, and the solution poured into 25 ml of 85% formic acid. The crystals are recovered and washed as before. Yield, 0.6 g.

(b) Preparation of other hemin—anion complexes

A solution of formiatohemin in KOH—methanol, as used for the recrystallization above, is poured into hot glacial acetic acid containing NaCl, when chlorohemin crystallizes on cooling. Other halide complexes could obviously be prepared in an ánalogous way. For hypophosphitohemin, the KOH—methanol solution was made strongly acidic with oxalic acid, filtered, and sodium hypophosphite added.

Chloro-, bromo-, iodo- and fluoro-hemin esters have been prepared[164] by introducing iron into a porphyrin ester by the ferrous acetate—acetic acid method in the presence of the respective sodium halides. For recrystallization the following mixtures were used; chloroform and carbon tetrachloride (chloro-); ethylene bromide and hexane (bromo-); ethyl iodide and hexane (iodo-); the fluorohemin was dissolved in hot methanol and precipitated with a saturated solution of sodium fluoride in methanol. The fluoro compound is unstable.

19.7.3.4. Splitting the heme from cytochrome-c
The silver sulfate method of Paul[165]

To 3 ml of cytochrome-c solution (containing about $7 \cdot 10^{-4}$ gramatoms of nitrogen), 0.6 ml of glacial acetic acid and 3 ml of $AgSO_4$ solution (800 mg/ml) are added and the mixture is heated at $60°C$ for 80 min (Note).

After cooling, the hemin is extracted with three portions of ether containing 25% (v/v) acetic acid and the ether extract washed three times with 5% (w/v) NaCl or sodium acetate[166]. The hemin may then be extracted from the ether with KH_2PO_4 (0.5% w/v, pH 7.0)[167], or the ether may be evaporated to dryness and the hemin dissolved in aqueous NaOH[168].

Note

$AgNO_3$ has been used[166] instead of $AgSO_4$, but it may be safer to avoid the nitrate in view of the possibility of oxidation by nitric acid[164]. Silver acetate has been used by Margoliash et al.[169]. J. Barrett[170] has reinvestigated the use of mercury salts, tested but not favored by Paul[165], and found better and more reproducible yields of hemin from cytochromes-c from mammalian tissue, yeast, *Rhodospirillum rubrum* and *Chromatium* than were given by the $AgSO_4$ method. To 2 mg of the cytochrome-c, 4.5 ml of a solution of mercurous sulfate is added, the mixture left overnight at $30°C$, and the hemin extracted with ether—acetic acid as above. The mercurous sulfate solution is made by warming 64 mg of this compound in 100 ml of distilled water with 0.4 ml of glacial acetic acid.

19.8. Reactions at the meso and peripheral positions of porphyrins and metalloporphyrins

Only a few typical reactions of the porphyrin macrocycle will be described in experimental detail. In order to avoid detailed description of separation of isomers or re-esterification of carboxylic functions, the typical porphyrin will usually be octaethylporphyrin. Most of the reactions can be

References, p. 861

performed in comparable yields with other porphyrins; vinyl groups, however, usually need to be reduced to ethyl (p. 773) or else protected (p. 823).

19.8.1. Oxidation
19.8.1.1. Metalloporphyrin π-cation radicals

Up to 500 mg of metallo-octaethylporphyrin (M = e.g. Mg, Zn, Cu, Ni, Pd) is dissolved in a minimum volume of methylene chloride and then mixed rapidly with a slight excess of bromine in methylene chloride. Spectrophotometry should attest the complete absence of visible absorption bands. The solvent and excess bromine are immediately removed on a flash evaporator without heating. The residue can be crystallized from glyme containing about 10% bromine relative to the amount of porphyrin present[171,172]. For typical electronic absorption spectra see Chapter 14.

An electrochemical oxidation procedure in methylene chloride containing 0.1 N tetrapropylammonium perchlorate at a potential slightly above the mid-point potential has been described by Fajer et al.[171]. The solvent is then evaporated, the residue re-dissolved in benzene and filtered to remove the electrolyte, and the filtrate is then evaporated to dryness and the radical crystallized from methylene chloride. Purification in the absence of oxidants usually leads to some reduction of the radical to metalloporphyrin.

19.8.1.2. Octaethyloxophlorin[173]

Zinc(II) octaethylporphyrin (415 mg) in tetrahydrofuran (30 ml) and methylene chloride (100 ml) is treated with a solution of dry thallium(III) trifluoroacetate (480 mg) in tetrahydrofuran (20 ml) and then stirred for 1 min. Water (0.25 ml) in tetrahydrofuran (10 ml) is added and after stirring for a further 10 min the solution is treated briefly with sulfur dioxide gas. Concentrated HCl (2 ml) is then added and the solution is stirred for 5 min before being poured into water (250 ml) and extracted with methylene chloride (250 ml). The organic phase is washed with water, dried, and evaporated to dryness to give a residue which is chromatographed on alumina (200 g; grade III) with methylene chloride as eluant. The blue eluates are collected (other minor bands are present) (Note 1) and evaporated to dryness. Crystallization from methylene chloride—methanol gives deep blue needles (302 mg; 79%) m.p. 254—256°C.

Notes

(1) A fore-run containing[174,175] αγ-dioxo-octaethylporphodimethene (p. 636) and αβ disubstituted compounds is invariably present.

(2) The chromatography must be carried out as rapidly as possible; otherwise, the oxophlorin is converted into its π-radical[174,176] (see next Section).

19.8.1.3. Octaethyloxophlorin π-radical[174,176]

Slow chromatography (time >3 h) of octaethyloxophlorin on alumina (dry column, Woelm) yields more than 90% of brownish green crystals of the corresponding π-radical (from methylene chloride—methanol); this compound was the reason for difficulties in measurement of n.m.r. spectra of free-base oxophlorins owing to peak broadening. If the radical is left for much longer periods on the column (e.g. overnight) then it is converted into the brown αγ-dioxoporphodimethene in yields over 60%.

19.8.1.4. Octaethylxanthoporphyrinogen[177]

Octaethylporphyrin (2 g) in chloroform (300 ml) and acetic acid (60 ml) is treated with PbO_2 (12 g) and then stirred for 3—4 h. The PbO_2 is filtered off and the filtrate is evaporated completely to dryness (acetic acid is removed by repeated evaporation with benzene). The residue is dissolved in a minimum volume of hot acetone, filtered hot, and then diluted with an equal volume of methanol until large crystals appear. After standing overnight at −10° C the large rhombic yellow crystals (2.7 g) are collected. Recrystallization can be carried out from pure ethanol; the crystals are occasionally brownish and this can be overcome by chromatography on silicagel (in benzene—acetone, 4 : 1), m.p. 272—276° C.

The same procedure can be applied to many other porphyrins. Protoporphyrin is degraded quantitatively to undefined products under these conditions.

19.8.1.5. 1,2,3,4,5,6,8,8,-Octaethyl-7-oxochlorin[178]

Octaethylporphyrin (150 mg) in conc. H_2SO_4 (75 ml) is treated for 6 min with 20% aqueous hydrogen peroxide (10 ml) at 10—15° C (ice bath). The mixture is then poured into ice water (600 ml) and extracted with ether which is then washed with aqueous $NaHCO_3$, and water. The ether is evaporated and the residue chromatographed on alumina (Spence H) eluting firstly with petroleum ether (for removal of over-oxidized materials) and then petroleum ether—benzene (1 : 1) which elutes the 7-oxochlorin. Crystallization from methylene chloride yields 31 mg (20%) of green needles.

19.8.1.6. Conversion of an oxyhemin into a biliverdin derivative[179,180]

β-Hydroxymesoporphyrin-IX chlorohemin dimethyl ester (78 mg; Note 1) in pyridine (15 ml) is flushed with oxygen for 20 h (Note 2) in the dark. The pyridine is evaporated, 2% methanolic hydrogen chloride (10 ml) is added and then removed under reduced pressure after 1 min. The residual green oil is taken up in chloroform, washed with water, and crystallized twice from methylene chloride—ether (1 : 1) to yield 46 mg (49%) of green needles of β-oxa-mesoporphyrin-IX chlorohemin dimethyl ester ('verdohemin'[179]), λ_{max} ($CHCl_3$), 372 nm (ε 38,000), 634 (12,500).

The oxaporphyrin iron complex (32 mg) is treated for 1 min with 2% KOH in methanol (10 ml), diluted to 5 ml with 25% methanolic HCl and then poured into methylene chloride. Work-up, chromatography on alumina, and crystallization from methylene chloride—n-hexane gives 19 mg (71%) of blue rods of mesobiliverdin-IXβ dimethyl ester, m.p. 220—221°C, λ_{max} (CHCl$_3$), 365 nm (ϵ 50,000), 643 (14,000).

Notes

(1) This chlorohemin was obtained by heating a "paste" of the synthetic β-oxymesoporphyrin-IX dimethyl ester (25 mg) with sodium acetate (1 g), NaCl (0.5 g), and ferrous sulfate (0.2 g) in acetic acid (2 ml) at 100°C for 15 min. After extracting the hemin into ether and methylene chloride 26.5 mg (92%) of the above iron complex was isolated.

(2) Such a long reaction time has been shown[181] to be unnecessary.

19.8.1.7. Zinc octaethyl-1'-formylbiliverdin[182,183]

Magnesium octaethylporphyrin (30 mg) in dry benzene (1 l) in a round bottomed flask (1 l capacity) is treated with 3 Å molecular sieves (50 g) and the solution is stirred vigorously with a magnetic bar. The flask is then irradiated with a 1250 W movie light from a distance of 35 cm until the visible absorption bands of MgOEP have completely disappeared (40—60 min). The benzene solution is diluted with dry ether (1 l), filtered from the molecular sieves, and evaporated to a total volume of approx. 300 ml. A solution of zinc acetate (10 mg) in methanol is added and the reaction mixture is refluxed for a few min. The solution is evaporated, the residue re-dissolved in chloroform (200 ml), washed several times with water, dried, and evaporated to dryness. Crystallization from ether or methylene chloride—methanol gives 22 mg of long brown prisms, λ_{max} (CHCl$_3$), 408 nm (ϵ 32,000), 830 (12,000). Sometimes a green product is formed in the course of crystallization; this is the bis-helical dimer, which absorbs at 740 nm[183].

The same procedure was successful for the oxygenation of magnesium protoporphyrin-IX dimethyl ester but methylene chloride instead of benzene was used as solvent for the irradiations owing to solubility difficulties.

19.8.2. Hydrogenations and reductions
19.8.2.1. β-Phlorin of chlorin-e$_6$ trimethyl ester[184]

This is a relatively stable phlorin and can therefore be prepared in the solid form without great difficulty. Less stable phlorins can be prepared in solution following the same procedure but isolation requires rigorous exclusion of air.

The chlorin (500 mg) in benzene (150 ml) and methanol (150 ml) containing lithium chloride (0.6 g) and boric acid (0.5 g) is filtered and then electrolyzed at a silver wire anode at a potential of −1.0 V and a current of approx. 10 mA. When, after 8—10 h, the current has dropped to less than 1 mA the electrolysis is stopped. Chloroform (300 ml) is added and the solution is washed several times with water and then evaporated to dryness. Chromatography on alumina (grade IV) eluting with chloroform under nitrogen yields, after evaporation, the red chlorin—phlorin (411 mg), pure by n.m.r. standards; $\lambda_{max} \sim 540$ nm (broad).

19.8.2.2. trans-Octaethylchlorin[185—187]

Iron(III) octaethylporphyrin (1 g) is refluxed under nitrogen in iso-amyl alcohol (100 ml) for 10 min before addition of sodium (10 g). After 20—30 min the solution has turned green; it is cooled to 0°C and methanol (25 ml) is added slowly. The resulting solid mass is broken up with a spatula and water (100 ml) is added to redissolve the solid material. The mixture is now extracted with benzene (300 ml) which is washed with water (200 ml), conc. HCl (150 ml) and then several times with water. The water phase is re-extracted with benzene before being discarded. The combined benzene phases are concentrated to about 75 ml and then mixed with acetic acid (500 ml). Under nitrogen, saturated ferrous sulfate solution in conc. HCl (50 ml) is added with rigorous stirring. After 2 min, saturated NaOAc solution (300 ml) is added and then benzene (200 ml). The benzene layer is washed with water and evaporated to dryness. The residue is dissolved in benzene and extracted with 50% phosphoric acid; this removes octaethylporphyrin. When the 50% phosphoric acid remained colorless it is substituted with 75% phosphoric acid which extracts the required chlorin. The combined 75% acid extracts are diluted with an equal volume of water and the chlorin is extracted into benzene which is washed neutral with water, dried, and evaporated to dryness. The residue is crystallized twice from ethanol. Yield, 404—600 mg (47—70%) of green needles of H_2(OEC), m.p. 232°C. Whitlock[186] reports a yield of 75%.

19.8.2.3. cis-Octaethylchlorin[186]

A mixture of octaethylporphyrin (0.5 g), anhydrous potassium carbonate (3.8 g) and freshly distilled β-picoline (25 ml) is refluxed under nitrogen and a solution of p-toluene sulfonylhydrazine (5.0 g) in picoline (15 ml) is added dropwise over a period of 2.5 h. The mixture is refluxed for a further 2 h, cooled, and extracted with benzene, which is washed with cold dil. HCl and extracted with three 35 ml portions of 85% phosphoric acid. The combined phosphoric acid extracts are diluted to 60% and extracted with benzene. The benzene extracts are then washed with 60% phosphoric acid and water and evaporated to dryness. Crystallization from chloroform—methanol yields 56 mg (11%) of the cis-chlorin, m.p. 216—217°C.

Whitlock's paper[186] also describes in detail the preparation of a- and b-tetrahydro-octaethylporphyrins and the corresponding meso-tetraphenylporphyrin derivatives.

19.8.2.4. Uroporphyrinogen-III[99]

Freshly ground 3% sodium amalgam (2 g) is added to 4 ml of a 4×10^{-4} M solution of uroporphyrin-III in very dilute KOH. The flask is

References, p. 861

flushed with nitrogen, stoppered, and shaken vigorously keeping the solution in the dark (aluminum foil). After fluorescence has disappeared the solution is rapidly filtered through a fine sintered glass disk under suction, the filtrate is brought to pH 6.8 with 40% H_3PO_4 and stored in the dark at dry ice temperatures.

Uroporphyrinogen could be extracted with ethyl acetate from the above solution when this is acidified to pH 3. Uroporphyrin is not extracted under these conditions.

A wide variety of other methods have been employed for the preparation of porphyrinogens (see Section 19.6.3).

19.8.2.5. Conversion of an oxophlorin into a porphyrin[7,188]

Crude oxophlorin (approx. 300 mg) in pyridine (30 ml) and acetic anhydride (8 ml) is stirred at room temperature for approx. 30 min before evaporation of the deep red solution to dryness. Chromatography on alumina (grade III) eluting with methylene chloride and evaporation of the red eluates gives 90—95% yield of the corresponding meso-acetoxyporphyrin after crystallization from methylene chloride—methanol. The meso-acetoxyporphyrin in tetrahydrofuran (75 ml) and triethylamine (0.2 ml) is hydrogenated at room temperature and atmospheric pressure over 10% palladized charcoal (150 mg) until the supernatant solution is colorless (approx. 6 h). The solution is filtered through Celite which is washed with a little more tetrahydrofuran. The combined tetrahydrofuran filtrate is treated immediately with 2,3-dichloro-5,6-dicyanobenzoquinone (approx. 350 mg) in dry benzene (5 ml) before quickly evaporating the solution to dryness and immediate chromatography on alumina (grade III) eluting with methylene chloride. Evaporation of the red eluates and crystallization from methylene chloride—methanol gives 80—90% of the porphyrin.

Oxophlorins can also be transformed into porphyrins by reduction with sodium amalgam[188] or by hydrogenation followed by diborane reduction and re-oxidation with air or DDQ[188].

19.8.3. Reactions at the meso positions of porphyrins and metalloporphyrins

The substitution reactions described in the following section occur at the methine bridges when the periphery is fully substituted. In porphyrins with unsubstituted β-pyrrolic positions substitution at these carbons is also possible and the relative yields of peripherally and meso substituted products is highly dependent upon the choice of reaction conditions (review[189]) (see Chapter 15).

19.8.3.1. Deuteration
19.8.3.1.1. meso-Tetradeutero-octaethylporphyrin[190]

Octaethylporphyrin (80 mg) is dissolved in 2 ml of D_2SO_4—D_2O (9 : 1, w/v) and kept at room temperature for 18 h. The solution is poured into ice-water (100 ml) and extracted with chloroform (100 ml) which is washed with aqueous $NaHCO_3$, water, and then dried. Evaporation and crystallization from methylene chloride—methanol gives 60 mg of the tetradeuterated porphyrin.

19.8.3.1.2. meso-Tetradeutero-coproporphyrin-I tetramethyl ester[191]

A milder procedure for *meso* deuteration has been described by Smith et al.: coproporphyrin-I tetramethyl ester (123 mg) in dry pyridine (3 ml) is heated under reflux in an atmosphere of nitrogen during 4 h with 20 ml of hexapyridyl magnesium di-iodide solution (Note 1). The cooled solution is diluted with chloroform and poured into 3% citric acid in water. The chloroform layer is washed with water, dried, evaporated, and the residue is re-esterified in 5% (v/v) H_2SO_4 in methanol. The solution is poured into water, extracted with methylene chloride which is then washed with water, dried and evaporated to dryness, (Note 2). Crystallization from methylene chloride—methanol gives 90% of the tetradeuterated porphyrin (Note 3).

Notes

(1) Magnesium (800 mg), iodine (1.5 g) and ether (30 ml) are refluxed under nitrogen until colorless (20 min). The yellowish solution is filtered, evaporated to dryness, and then the residue is dissolved in pyridine (50 ml) containing deuteromethanol (2 ml) (Note 4).

(2) The material can be chromatographed if required (and particularly if *meso*-tritiated material is prepared), using alumina (grade III) and eluting with methylene chloride.

(3) Exchange occurs upon the magnesium complex of the porphyrin. If the reaction is carried out on other metalloporphyrins, the rate of deuteration is dependent upon the chelating metal ion.

(4) D_2O can be used in place of MeOD, however, the rate of magnesiation is somewhat slower. If a little tritiated water is used instead of the deuterium source then highly tritiated products are obtained[7]. The reaction is applicable to a wide variety of porphyrins and metalloporphyrins.

19.8.3.1.3. γ,δ-Dideutero-trans-octaethylchlorin[190,192,193]

trans-Octaethylchlorin (120 mg) in dioxane (5 ml) and deuteroacetic acid (15 ml) is heated under nitrogen for 6 h at 95°C. Methylene chloride (50 ml) is added and the solution is washed with water (100 ml) containing sodium acetate (15 g) before being dried and evaporated to dryness. The residue can be chromatographed on alumina (grade III) eluting with methylene chloride. Evaporation of the green eluates and crystallization from methylene chloride—benzene gives 113 mg (94%) of the dideuterochlorin[193]. N.m.r. spectroscopy indicates more than 95% deuteration at the γ and δ positions.

References, p. 861

19.8.3.2. Vilsmeier formylation of copper octaethylporphyrin[194,195] and Knoevenagel condensation[310]

To a mixture of dimethylformamide (10 g) and phosphorus oxychloride (20.3 g) a solution of copper octaethylporphyrin (1.0 g) in ethylene dichloride (900 ml) is added dropwise over a period of 20—30 min at 50—60°C. After 20 min further standing at this temperature the mixture is cooled to room temperature and treated with saturated aqueous sodium acetate (1 l) before stirring for 2 h at 60°C. The cooled solution is diluted with chloroform (500 ml) which is washed several times with water, dried, and evaporated to dryness. The residue is dissolved in conc. H_2SO_4, kept at 50°C for 1 h and poured into ice-water (1 l) saturated with ammonium acetate. The porphyrins are extracted with chloroform, washed with water, dried, and evaporated. Preparative thick layer chromatography (4 plates 100 × 20 cm, 2 mm thick silicagel, chloroform elution) and crystallization from chloroform—methanol yields 700 mg of α-formyl-octaethylporphyrin, m.p. 254—256°C.

Titanium tetrachloride (11 ml) in carbon tetrachloride (25 ml) are dropped into dry tetrahydrofuran (200 ml) which is cooled to 0°C. A yellow precipitate forms and to this mixture are added α-formyloctaethylporphyrin (100 mg) in tetrahydrofuran (25 ml) and malonic acid (5 g) at ice temperature. With continued cooling, pyridine (16 ml) in tetrahydrofuran (30 ml) are added dropwise over 3—4 h (Note), and the solution is stirred another 10 h at room temperature. Water is added, the dicarboxylic acid is extracted into ether, which is washed with saturated brine and evaporated. Preparative thick-layer chromatography (1 plate, 100 × 20 cm × 2 mm silicagel, chloroform/methanol 10 : 1 eluent) and recrystallization from chloroform—methanol gives 80 mg (75%) of the meso-(vinyl-β,β-dicarboxylic acid)porphyrin[310].

Note

Faster addition leads to greasy precipitates.

19.8.3.3. Mono-nitration of octaethylporphyrin[196]

Octaethylporphyrin (200 mg) is treated with an ice-cold mixture of fuming nitric acid (d = 1.5) and acetic acid (1 : 1, 32 ml). The mixture is shaken for 1.5 min without further cooling, poured into ice-water, extracted with ether, and worked up as in the preceding section. Chromatography on dry column silicagel (Woelm, chloroform elution) and crystallization from benzene yields 160 mg of red-brown needles, m.p. 250—252°C.

19.8.3.4. α-Nitro-etioporphyrin-I[197]

Magnesium etioporphyrin-I (25 mg) in chloroform (20 ml) is treated with iodine (26.7 mg) in chloroform (15 ml) and to this green solution is added sodium nitrite (140 mg) in acetonitrile (5 ml) and methanol (35 ml) (warm to dissolve nitrite and then cool). After a few minutes stirring, trifluoroacetic acid (2 ml) is added and the solution is poured immediately into water, washed with methylene chloride, which is in turn washed several times with water. The methylene chloride is dried, evaporated, and the residue is chromatographed on alumina (grade III) eluting with benzene. Evaporation of the eluates and crystallization from methylene chloride—methanol gives 22 mg (85%) of mono-nitro-etioporphyrin-I.

19.8.3.5. Mono- and di-chlorination of octaethylporphyrin[198]

A solution of octaethylporphyrin (100 mg) in tetrahydrofuran (160 ml) is refluxed with a mixture of 3% hydrogen peroxide (10 ml) and 0.5 N HCl (110 ml) for 30 min. The cooled mixture is extracted with ether which is then washed with water, dried, and evaporated. The residue is chromatographed twice on alumina (grade III), eluting with benzene—light petroleum (1 : 1). The more mobile band is collected, evaporated and the residue crystallized from chloroform—methanol to give 22 mg (20%) of α,γ-dichloro-octaethylporphyrin, m.p. 252—253°C. The second band is similarly treated and affords 25 mg (24%) of α-chloro-octaethylporphyrin, m.p. 270—272°C. Solutions of these chlorinated porphyrins in neutral solvents do not fluoresce.

19.8.3.6. Mono-thiolation of copper octaethylporphyrin[199]

Thiocyanogen (3 mmol) in acetic acid (Note) and copper octaethylporphyrin (200 mg; 0.3 mmol) in acetic acid (100 ml) are stirred at room temperature for 10 min, (color change red to violet). The solvent is evaporated and the residue dissolved in ice-cold 90% sulfuric acid (150 ml) to give a dark green solution. After 5 min the acid is neutralized with sodium hydroxide in the cold and extracted with chloroform which is washed with water, dried, and evaporated to dryness. The residue is chromatographed on dry column silicagel (Woelm) eluting with benzene—chloroform (1 : 1). Crystallization from chloroform—methanol yields 110 mg (63%) of α-mercapto-octaethylporphyrin, m.p. >300°C.

Note

This reagent is prepared as follows: ammonium thiocyanate (532 mg) is dissolved in a minimum volume of acetic acid and bromine is added dropwise with vigorous stirring at room temperature. A yellow precipitate is formed and addition of bromine is stopped when the solution turns orange and the precipitate becomes grey. After filtration, the filtrate contains approx. 3 mmol of thiocyanogen which should be used immediately.

19.9. Modification and identification of porphyrin side-chains

Microchemical methods applicable directly to the intact porphyrin or hemin molecule are described here, together with some larger scale reactions, interconversions, and side-chain protections of porphyrins of natural origin. In studies of natural hemes it is important, if possible, to compare reactions of the pyridinium hemochrome of the isolated heme with the pyridinium hemochrome prepared from the intact hemoprotein in order to guard against unsuspected side-chain alterations which may have occurred during the isolation.

19.9.1. Methoxyl groups

Granick[200] has adapted the method of Boos[201] for use with small amounts of porphyrins (1 mg or less), 100 μg or less of methanol being determined accurately. About 1 mg of porphyrin ester is weighed on a platinum boat, and transferred to bulb A of the vessel illustrated in Fig. 1. Chloroform is added, and the dissolved ester distributed in a thin film inside the bulb and the solvent removed by evaporation. A mixture is made of 3 ml of 5% H_3PO_4, 5 ml of 5% $KMnO_4$ and 8 ml of water; 0.8 ml of this mixture is placed in bulb B. The bulbs are now cooled to $-40°C$, 2 ml of 6 N H_2SO_4 are added to bulb A, and after freezing the vessel is evacuated and sealed. Bulb A is then warmed to room temperature while bulb B is kept in an ice-bath. After 2 h hydrolysis of the porphyrin ester in bulb A is complete, and this bulb is then warmed so as to distil 0.5 ml of liquid into bulb B, which is then raised to room temperature. After 15 min at room temperature oxidation of the methanol to formaldehyde, in bulb B, is complete; longer times should be avoided.

The vessel is opened and just enough saturated $NaHSO_3$ added to bulb B to decolorize the solution. Bulb B is again cooled to $-40°$, and 4 ml of concentrated sulfuric acid and 0.2 ml of 2% chromotropic acid added. The vessel is stoppered, and bulb B heated to 60° for 15 min, cooled, and diluted to the 10-ml mark with water. The absorption of the colored solution is read at 580 nm against a reagent blank in a 1-cm cuvette, and the methanol

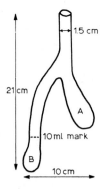

Fig. 1. Apparatus for micro-methoxyl determination.

content of the porphyrin determined from a curve obtained by analyzing known amounts of methanol in the same way. The standard curve found by Granick was a straight line with a slope of 0.0049 optical-density units (log I/I_0; 1 cm), with a deviation of $\pm 5\%$ in the range 10–130 μg. Determinations on protoporphyrin mono- and di-methyl esters agreed well with the calculated values.

19.9.2. Active hydrogen atoms

The Zerewitinoff method for the determination of potentially ionizable hydrogen atoms (e.g. —SH, —OH, —CONH$_2$ and imide groups) is well-known. Among pyrroles, bile pigments and porphyrins, reasonably sharp results may be obtained and the analysis has proved of some value, particularly in studies of the structure of chlorophyll derivatives[202,203].

19.9.3. Vinyl side-chains
19.9.3.1. Reduction to ethyl

In addition to catalytic hydrogenation, this reduction may be carried out by treatment of the vinyl-porphyrin with HI in acetic acid, as described for the preparation of mesoporphyrin.

19.9.3.2. Reaction with diazoacetic ester[204]

This reagent, which is conveniently prepared from glycine ethyl ester, forms an adduct with vinyl groups, saturating the double bond and removing the electrophilic effects of the vinyl group on the spectrum.

Parker[205] has described the following procedure. To an ether solution of the porphyrin, diazoacetic ester in ether is added; while passing a stream of nitrogen, the vessel is heated to 60°C and the ether allowed to evaporate. The vessel is now stoppered and heated at 60°C for 20 h. After cooling, the residue is dissolved in ether and the porphyrin—diazoacetic ester adduct extracted from the ether into 10% (w/v) HCl. It may then be transferred to fresh ether or another suitable solvent for examination of its spectrum.

Notes

(1) Barrett has found that the reaction yields cleaner products if carried out at room temperature.

(2) If carbonyl groups are present, they should be protected; thus Parker[205] protected formyl groups as their oximes.

19.9.3.3. Hydration by HBr—acetic acid

This process has been used widely for the preparation of hematoporphyrin from protoporphyrin.

It is convenient[206] to dissolve the porphyrin in 50% (w/v) HBr in acetic acid in a stoppered tube and to leave it at room temperature overnight; for small preparations, 4 h standing is sufficient, and it is useful to pass N$_2$

References, p. 861

through the solution to remove any free Br_2. Then excess 5% HCl is added and the solution left to stand at $0°C$ in an atmosphere of N_2 for 24 h (a few hours for small preparations) to complete hydrolysis of the bromoethyl groups. The resulting α-hydroxyethyl porphyrin is then extracted into ether.

19.9.3.4. Removal in the resorcinol melt

Vinyl side-chains of porphyrins and hemins are replaced by hydrogen atoms when the compound is fused with resorcinol and kept at $150-160°C$ for 15 min.

As well as vinyl side-chains, α-hydroxyethyl, hydroxymethyl, and formyl groups are removed by this treatment, but α-ketonyl, β-hydroxyethyl, and carboxylic acid groups attached directly to the porphyrin, are not removed unless[206] the temperature exceeds $200°C$.

19.9.3.5. Oxidation of vinyl to formyl groups

This may be achieved[206,207] by refluxing the porphyrin in acetone with five equivalents of $KMnO_4$, added in small portions cver several minutes. Vinyl oxidation is inhibited by other electrophilic substituents on the porphyrin, and a pre-existing formyl (as in porphyrin-a) must first be reduced to hydroxymethyl. Hydroxymethyl groups are little affected by the oxidation and may be protected fully by acetylation. Lemberg and Parker obtained[207] only 5% of 2,4-diformyldeuteroporphyrin-IX from protoporphyrin-IX using permanganate. A better route has been reported by Sparatore and Mauzerall[208] who improved one of Fischer's reactions[209] using osmium tetroxide.

To protoporphyrin-IX dimethyl ester (550 mg) in dry dioxane (200 ml) is added osmium tetroxide (500 mg) in ether (25 ml) followed by pyridine (0.35 ml). After 24 h standing in the dark under nitrogen the ether is evaporated and sodium sulfite (1.1 g) in water (20 ml) is added and the mixture is heated at $100°C$ for 40 min. After filtering off the black osmium precipitates and the usual work-up procedure 45% of 2,4-di(α,β-dihydroxyethyl)deuteroporphyrin-IX dimethyl ester is isolated. This diol (900 mg) is cleaved with sodium periodate (1.15 g) in dioxane-water to give a mixture of 2(4)monoformyl-4(2)monoglycol- and 2,4-diformylporphyrins, which are readily separated by fractional extraction and crystallization. About 30% of pure diformyl product is obtained.

19.9.3.6. Conversion of vinyl into β-hydroxypropionic acid[208,210]

2,4-Diformyldeuteroporphyrin-IX dimethyl ester is prepared[208] as described above. To a solution of this porphyrin (150 mg) in pyridine (100 ml) containing piperidine (0.03 ml) is added malonic acid (6 g) and piperidine (0.1 ml) in pyridine (60 ml) dropwise at $95°C$ during 6 h, and then again, malonic acid (2 g) in pyridine (20 ml) containing piperidine (0.02 ml) during 1 h at $118°C$. Precipitation in the cold with petroleum ether yields the raw product, the 2,4-diacrylic acid of deuteroporphyrin-IX dimethyl ester[208].

The above diacrylic acid (130 mg) is treated with 40% HBr-acetic acid (50 ml) at 40°C for 4 h before being poured into water (250 ml) and heated at 100°C for 30 min. Fractionated extractions, column chromatography, and crystallization from methanol—benzene (1 : 1) yields 45% of the 2,4-bis(β-hydroxypropionic acid) of deuteroporphyrin-IX dimethyl ester[210].

19.9.3.7. Protection of vinyl groups

Markownikoff hydration of vinyl groups (e.g. protoporphyrin → hematoporphyrin) offers some protection to vinyl groups, for example against photo-oxidation. However, on account of the ease with which hematoporphyrin tends to dehydrate under the mildest of conditions, an anti-Markownikoff hydration route has been developed[211].

Protoporphyrin-IX dimethyl ester (118 mg) in methylene chloride (30 ml) and methanol (5 ml) is treated with thallium(III) nitrate trihydrate (310 mg; approx. 3 equiv.) at 40°C for 10 min. Sulfur dioxide gas is then briefly passed through the solution, conc. HCl (0.5 ml) is added, and the porphyrin is extracted into methylene chloride. The organic phase is washed with water, dried, evaporated to dryness, and the residue is chromatographed on alumina (grade V) eluting with methylene chloride. The red eluates are evaporated and the residue is crystallized from methylene chloride—n-hexane to give 131 mg (92%) of the 2,4-bis(2,2-dimethoxyethyl)deuteroporphyrin-IX dimethyl ester, m.p. approx. 230°C with decomposition.

The foregoing porphyrin bis-acetal (300 mg) in tetrahydrofuran (100 ml) and water (3 ml) is heated under reflux with conc. HCl (1 ml) for 5 min. After cooling, methylene chloride and water are added and the porphyrin is extracted into the organic phase, dried, and evaporated to give a residue which is dissolved in methylene chloride (100 ml) and treated at 0°C with NaBH$_4$ (1 g) in ice-cold methanol (25 ml). After stirring at room temperature for 10 min, acetic acid (2 ml) is added and the porphyrin is worked-up as before. After re-esterification (5% v/v H$_2$SO$_4$ in methanol) and chromatography (alumina, grade V; elution with chloroform containing 1% methanol), the 2,4-bis(2-hydroxyethyl)deuteroporphyrin-IX dimethyl ester is obtained from chloroform—methanol. Yield, 203 mg (70%), m.p. 225—226°C.

The anti-Markownikoff protected product is re-converted into protoporphyrin as follows[7]: The bis(2-hydroxyethyl)porphyrin (176 mg) in chloroform (80 ml) and dimethylformamide (15 ml) containing K$_2$CO$_3$ (5.5 g) is treated with thionyl chloride (5 ml). After stirring for 6 h the mixture is poured cautiously into water; the organic phase is separated, washed with water, dried, and evaporated to dryness (last traces of dimethylformamide being removed under high vacuum). After optional chromatography (alumina, grade III) the bis(2-chloroethyl)porphyrin (149 mg; 80%) is obtained from methylene chloride—n-hexane. The foregoing porphyrin (140 mg) is then converted into its zinc chelate (methylene chloride solution treated with saturated zinc acetate in methanol, aqueous work-up) which is taken up in tetrahydrofuran (25 ml) and treated in the dark at room temperature during 3 days with M potassium t-butoxide in t-butyl alcohol (50 ml). The mixture is then treated with acetic acid (2 ml) and worked up with chloroform and aqueous washings to give a residue which is re-esterified (5% v/v H_2SO_4 in methanol) and then chromatographed (alumina, grade III; elution with methylene chloride). Evaporation of the red eluates gives 105 mg (84%) of the divinylporphyrin (from methylene chloride—methanol).

Dehydration of hematoporphyrin-IX dimethyl ester to give protoporphyrin-IX dimethyl ester is discussed on p. 771.

19.9.4. Hydroxyethyl groups in side-chains

No natural and few synthetic porphyrins are known to have hydroxyl groups directly substituted on the pyrrole rings, and the methods described here have been developed particularly for studies of porphyrin-a. Like vinyl groups, α-hydroxyethyl groups are removed by fusion with resorcinol (see above).

19.9.4.1. Acetylation

The porphyrin ester is allowed to stand overnight in a mixture of 1 volume of acetic anhydride and 10 volumes of pyridine. The product is purified via an ether solution. Acetylation of a hydroxyl group causes an increase in R_F in suitable paper chromatographic analyses (see Table 5, p. 846).

19.9.4.2. Oxidation to a keto group

Oxidation of an α-hydroxyethyl group to an acetyl group leads to a carbonyl group in conjugation with the porphyrin nucleus; this modification causes a characteristic spectroscopic change which can be reversed by ketone reagents.

The porphyrin is dissolved in acetone, a stream of nitrogen is passed and after 5 min a few drops of chromium trioxide reagent (Note 1) are added. After 1 min (Note 2) the nitrogen source is removed, the porphyrin extracted into ether and the solution washed with aqueous $FeSO_4$ and then water, and small amounts of impurities are removed[206] by extraction with 1% (w/v) Na_2CO_3.

Notes

(1) The oxidation reagent[212] is prepared by dissolving 26.72 g of chromium trioxide in 23 ml of concentrated sulfuric acid, and diluting with water to 100 ml.

(2) Heating with this reagent leads to degradation of the porphyrin nucleus[213].

(3) When this method was applied to porphyrin-a with its α-hydroxyethyl

group protected by acetylation, the formyl side-chain was oxidized to a carboxyl group[206] (see below).

(4) A dilute solution of CrO_3 in benzene has been used to oxidize a CHOH group in the isocyclic ring of a chlorophyll derivative[102].

The equilibrium between α-hydroxyethyl and vinyl side-chains on porphyrins in aqueous HCl is discussed elsewhere. Complete dehydration of α-hydroxyethyl groups has been achieved[214] by heating the porphyrin in vacuo at 135—150°C for 5—10 min. The method has been adapted to comparisons of rates of dehydration, followed spectrophotometrically, as a test of the identity of hematohemin derived from cytochrome-*c* and synthetic hematohemin[165].

Because this procedure was not suitable for use with porphyrin-*a* Clezy and Barrett[206] developed the following very convenient method: the porphyrin ester (1 mg) is heated for 6 h in benzene with 2 mg of toluene-*p*-sulfonic acid. Dehydration of the hydroxyethyl side-chains of hematoporphyrin was complete in this time. Among a series of porphyrins with electrophilic side-chains, the rate of dehydration of hydroxyethyl side-chains decreased as the electrophilic power of the other side-chains increased. The toluene-*p*-sulfonic acid method has been modified for large-scale preparation of protoporphyrin-IX from hematoporphyrin-IX (p. 771).

19.9.5. Carboxylic acid side-chains

Like insulated carboxyl groups, such as those in propionic acid side-chains, carboxyl groups directly substituted to the porphyrin nucleus may be esterified, and their presence revealed by the effect of esterification upon the paper chromatographic behavior of the porphyrin. Unlike propionic acid groups, carboxyls in conjugation with the porphyrin nucleus have a marked effect upon the absorption spectrum. They may be replaced by hydrogen atoms by fusion with resorcinol above 200°C, with a consequent change in the spectrum. The electrophilic effect of the carboxyl group upon the spectrum disappears if the carboxyl group is ionized, and Clezy[215] has made use of this fact in developing a specific spectroscopic test. He found in a series of carboxyl-substituted hemins a consistent difference between the hemochrome α-bands of the hemins and their methyl esters, the hemochromes of the esters absorbing consistently at wavelengths 11—14 nm longer than those of the acids, which were dissociated in the alkaline medium.

19.9.5.1. Decarboxylation of acetic acid side-chains

This process is particularly important for the conversion of uroporphyrins to coproporphyrins. Fischer and Zerweck[216] carried out the decarboxylation by heating the uroporphyrin in 0.3 N HCl in a sealed tube at 180°C; simple modifications have been introduced by various authors[217—219] and yields approaching 100% of coproporphyrin may be obtained by the following procedure.

References, p. 861

Uroporphyrin (5—100 μg) is dissolved in 1 ml of 0.3 N HCl in a Carius tube of total volume 3—15 ml (Note 1). The tube is evacuated to a pressure of 1.5 mm of Hg for 10 min, and then sealed at this pressure. The temperature is raised to 180°C over a period of about 20 min, and maintained at 180°C for 3 h. After cooling, the solution is brought to pH 4 with sodium acetate, and the coproporphyrin extracted into ether. Yield[218], 100% (Note 2).

A perhaps more reliable method is that of Mauzerall[52]. The solution of uroporphyrin in 1 M hydrochloric acid is thoroughly de-aerated by freeze—thaw cycling, and then sealed under vacuum and heated at 180°C for 4 h.

Notes

(1) If uroporphyrin ester is used, it is transferred to the Carius tube in chloroform solution, the chloroform is removed completely, the porphyrin dissolved in a small volume of 7.5 N HCl and water added to dilute the HCl to 0.3 N.

(2) In air the yield[216,218,219] is about 45% and at a pressure[54] of 10 mm Hg, about 95%. By heating for shorter times at 140°C instead of 180°C, partial decarboxylation occurs and products with 5, 6 and 7 carboxyl groups may be obtained[219].

19.9.5.2. Decarboxylation of propionic acid side-chains

Propionic acid side-chains may be decarboxylated by heating the solid porphyrin in vacuum at about 350°. In this way uro-, copro- and meso-porphyrins are converted to etioporphyrin. The yields reported do not exceed[214] about 20%.

19.9.5.3. Formation of amide bonds in propionic side-chains

Examples of amides prepared from hemin have been described[220]. One preparation of potential importance for ligand binding studies of iron porphyrins is given here:

Bis-histidine mesohemin[221]

Pure mesohemin (Fe^{III} mesoporphyrin-IX chloride) (100 mg) in N LiOH (6 ml) is diluted with water (300 ml) and precipitated with N HCl (6 ml). The mesohemin lithium salt is centrifuged, washed with water, recentrifuged and then dried over P_2O_5. This product is dissolved, under nitrogen, in M SO_3-dimethylformamide solution (2 ml), triethylamine (0.3 ml) is added and the mixture is left for 30 min at room temperature. A mixture of mono- and bis-sulfuric anhydrides is formed under these conditions; to form the histidine amide from these activated acid derivatives, a solution of 0.32 M histidine buffer (20 ml; pH 9.0) is added at 0°C and reacted for 2 h at room temperature. Chromatography on de-aerated *Unisil* silicic acid (p. 856) with *t*-butyl alcohol—acetic acid—water (4 : 1 : 1) and precipitation of the three separated fractions from N KOH with N HCl yields 20% of the bis-histidine mesohemin and 50% of the monohistidine derivatives. The yield of bis-histidine mesohemin can be improved by doubling the amount of histidine buffer used. On amino acid analysis it shows 1.94 moles histidine per mesohemin.

19.9.6. Keto side-chains

Carbonyl groups conjugated with the porphyrin nucleus are readily identified even in the presence of other electrophilic side-chains, by the reversal of their electrophilic effects upon spectra by oxime formation (for conditions, see formyl groups, below), and the identification of keto groups then becomes a matter of differentiation from formyl groups.

Unlike the formyl group, keto groups such as acetyl are not removed in the resorcinol melt at temperatures below 200°C; unlike formyl groups they do not give the bisulfite or hydrazine reactions, nor do they form stable acetals (see below).

Keto groups are, however, reduced to alcohols by $NaBH_4$ under conditions similar to those used for reduction of formyl groups, Low concentrations of $NaBH_4$ in methanol reduced the 3-formyl group of chlorophyll-b, without reduction of the carbonyl group (C_9) in the isocyclic ring. The latter was reduced, in chlorophyll-b, pheoporphyrin-a_5 and phylloerythrin, when higher concentrations of $NaBH_4$ were used[102].

19.9.7. Formyl side-chains
19.9.7.1. Oxime formation

The porphyrin ester is refluxed for 15 min in pyridine with addition of hydroxylamine hydrochloride. Addition of Na_2CO_3 is not usually essential, the pyridine sufficing to neutralize the HCl liberated. The oxime of the porphyrin ester is transferred to ether, and after washing with water until free of pyridine, the ether solution is dried over anhydrous Na_2SO_4 and evaporated to dryness. Oximes of porphyrin esters may be crystallized from chloroform—methanol; in the case of diformyldeuteroporphyrin ester dioxime, chloroform was a better crystallizing solvent[222]. The esters have characteristic melting points.

The reaction with hydroxylamine is given by formyl but not by acetyl (keto) side-chains on hemes under the following conditions[222]: to a solution of the hemin in 0.5 ml pyridine, 1.5 ml of water and then 3 mg of the neutral hydroxylamine mixture (above) are added. After dissolving the solids by gentle rotation of the tube, the mixture is allowed to stand at 20°C for 30 min. A little sodium dithionite is added, and the absorption spectrum measured. Formyl hemes treated in this way gave shifts of 17—20 nm to shorter wavelengths; hemes with acetyl side-chains, or with the ketonyl isocyclic ring (phylloerythrin hemin, pheoheme-a_5) gave little or no shift.

Oximes of $meso$-formylporphyrins have also been prepared[195,223] and these have been dehydrated to give the corresponding $meso$-cyanoporphyrins.

References, p. 861

19.9.7.2. Hydrazone formation

Under the same conditions as the hydroxylamine reaction for hemes, but using hydrazine hydrate instead of hydroxylamine, formyl hemes gave shifts as large as 30 nm to shorter wavelength, and again the keto compounds did not react[222].

19.9.7.3. Reaction with bisulfite[205,206]

The porphyrin is dissolved in a mixture of 4 volumes pyridine with 1 volume water, a few crystals of $NaHSO_3$ added and the solution allowed to stand at room temperature for 30 min. The sulfite complex formed with the formyl group removes its electrophilic effect on the spectrum. The complex is dissociated on heating.

19.9.7.4. Acetal formation

The porphyrin (1 mg) is dissolved in 10 ml methanol containing 6—8 drops of conc. HCl, and the solution heated in a boiling water-bath for 5 min. The porphyrin is then taken to ether, and the solution washed free of acetic acid with dilute $NaHCO_3$; if neutralization is not complete some splitting occurs. The spectrum is at shorter wavelengths than that of the formyl porphyrin, the electrophilic effect of the formyl group being removed. The acetals may be hydrolyzed by 5% HCl, and can be used to protect formyl groups in reactions carried out at neutrality. The reaction is not given by acetyl groups or by the isocyclic ring keto-group.

The acetal from neopentyl glycol has been found to be a useful formyl protecting group[224]: The formylporphyrin (30 mg) in a minimal volume of methylene chloride is treated with neopentyl glycol (50 mg) in benzene (30 ml) containing a crystal of toluene p-sulfonic acid hydrate. The methylene chloride and about 10 ml of benzene are distilled off and the remaining solution is refluxed for 2 h before addition of saturated aqueous sodium acetate and methylene chloride. After extraction, the organic phase is washed with Na_2CO_3 solution, water, then dried and evaporated to dryness. Crystallization from methylene chloride—n-hexane gives approx. 90% yield of the required neopentyl acetal.

Removal of the acetal is accomplished by treatment of the protected porphyrin under esterifying conditions (5% v/v H_2SO_4 in methanol).

Formyl side-chains on porphyrins condense with acetone in the presence of HCl[225], and also with primary amino-groups to form Schiffs bases which are, however, unstable in acid conditions. The instability to acid provides a valuable test for spectroscopic changes due to Schiffs base formation between formyl side-chains on natural hemes and amino-groups in alkali-denaturated proteins[226].

19.9.7.5. Reduction to hydroxymethyl groups[206]

The formyl porphyrin (1 mg) is dissolved in ether, and 20 mg $NaBH_4$ added, forming a suspension; this is allowed to stand overnight at room temperature, and then 0.5% (w/v) HCl is added to decompose excess $NaBH_4$, the hydroxymethyl porphyrin remaining in the ether.

19.9.7.6. Reduction to a methyl group[206] (Woff-Kishner reduction)

The formylporphyrin (20 mg) is dissolved in 50 ml of ethylene glycol, and 200 mg of hydrazine sulphate and 400 mg of KOH dissolved in 3 ml of water are added. The mixture is heated on an oil-bath under a water reflux condenser for 1 h, then water is removed from the condenser and the temperature in the reaction flask raised to $190-200°C$ for 1 h. The mixture is cooled, poured into water, acidified with HCl and the porphyrin extracted into ether.

19.9.7.7. Oxidation to a carboxyl group[206]

Treatment with the chromium trioxide reagent in acetone, under the conditions described above for oxidation of hydroxyethyl to acetyl groups, converts formyl side-chains on porphyrins to carboxyl groups, this oxidation proceeding more slowly than that of the hydroxyethyl group.

19.9.7.8. Conversion into ethylene-epoxide groups

It has been observed (P. Clezy, personal communication) that if a formyl porphyrin in ether solution is allowed to stand in the dark for several days in the presence of diazomethane, the formyl group is converted to an ethylene-epoxide group, which may be hydrolyzed to a 1,2-glycol.

19.9.7.9. Reaction with Girard's reagent

'Girard T' (carboxymethyltrimethylammonium chloride hydrazide) reacts much more readily with formyl than with keto groups; it has been used in a very convenient bulk separation of chlorophylls-a and -b[42] (see p. 775).

19.10. Special techniques
19.10.1. Solubilities

Most natural porphyrins and many synthetic ones have carboxylic acid side-chains, and all have basic nitrogen atoms. They are thus ampholytes, and have some solubility in both aqueous acids and aqueous alkalies. If the carboxylic groups are masked, as by esterification, alkali-solubility is lost, and if the nitrogen atoms are coordinated to a metal ion the solubility in aqueous acids disappears. The esters of porphyrins and their metal chelates are, of course, much more soluble in organic solvents than the porphyrin free acids.

19.10.1.1. Aqueous acids

Like most organic bases, solid or crystalline porphyrins and porphyrin esters are not easy to dissolve in dilute mineral acids; they usually dissolve

References, p. 861

readily in concentrated acids (e.g. 10 N HCl), and then remain in solution after dilution. They are readily soluble in glacial acetic acid, formic acid and concentrated sulfuric acid.

Porphyrins with free carboxylic acid groups may be precipitated from solutions in aqueous acid or alkali by bringing to the isoelectric point. The dissociation constants vary according to the nature of the side-chains but in practice most porphyrins precipitate in the region of pH 4. The isoelectric point is indicated conveniently by bromphenolblue paper (neutral, grey color) or by loss of fluorescence (ultraviolet light filtered by Woods glass). With dilute solutions, colloidal precipitates are often formed first; high salt concentration can hasten flocculation. The resulting amorphous precipitates are often difficult to filter, and are best recovered and washed at the centrifuge.

The hydrochlorides of some porphyrins may be crystallized from strong hydrochloric acid solutions, and some carboxylic porphyrins may be crystallized as their sodium salts. Certain porphyrins may be extracted into chloroform from aqueous hydrochloric acid solutions, there being critical acid concentrations for particular porphyrins.

Uroporphyrins are the most water-soluble of the common naturally-occurring porphyrins and are not soluble in ether. They may, however, be transferred from aqueous solution to certain organic solvents. Their esters behave like those of other porphyrins.

19.10.1.2. Transference from aqueous acid to ether

Porphyrins are not readily dissolved in ether or ethyl acetate from the solid state, but may be transferred quantitatively to these solvents from their solutions in aqueous acids. The acid solution is overlayered, in a separating funnel, with a relatively large volume of ether, the pH adjusted to about 4 (bromphenolblue paper is very useful) and the mixture shaken vigorously before precipitation occurs. Sodium acetate is useful for adjustment of the pH, since it forms a buffer system in the desired region, but titration to the isoelectric point with other alkalies (ammonia, NaOH, carbonates) is sometimes used. On washing the ether solution, free acetic acid from the ether layer can cause extraction into the aqueous layer of porphyrins of low HCl number (e.g. coproporphyrins, cf. Table 3, p. 14). Complete neutralization of acetic acid in the ether may be achieved by shaking the solution with dilute aqueous sodium carbonate, avoiding addition of excess; any porphyrin remaining in the aqueous layer is re-extracted into fresh ether. On the other hand, the maintenance in solution of porphyrins in ether often depends upon the presence of some acetic acid; hematoporphyrin can even be brought into ether from dilute aqueous HCl solution by the addition of a little acetic acid. Similar considerations apply to ethyl acetate solutions.

19.10.1.3. Aqueous alkalies

Porphyrins with free carboxylic acid side-chains and their metal complexes may be dissolved in aqueous alkalies; solubility is lower than in acid media, and colloidal systems are often obtained. Electron-attracting side-chains on simple porphyrins tend to decrease alkali-solubility, but other factors can play a part; thus Lemberg (unpublished results) has observed that porphyrin-*a*, which has rather large alkyl substituents as well as a formyl group, is extracted from ether more readily by highly dilute NaOH than by stronger solutions.

Comparison of the extinction in the Soret region with that of a solution of the same concentration in pyridine is a useful test for solution in the monomeric state. Detergents can assist the formation of monomeric solutions, and Porra and Jones[227] have found that 1% of Emasol 4130 (Tween 80) is sufficient to give monomeric solutions of meso- and proto-porphyrins at concentrations of 3.10^{-5} M at pH 7—10. Solid or crystalline hemins are often difficult to bring into solution in aqueous alkali; it is usually convenient to wet the solid material thoroughly with 1 N NaOH, and then to stir with successive portions of water until all is in solution. True solutions do not result; in aqueous alkali iron-porphyrins form dimers or further polymers. In addition, changes occur on standing in alkaline media, and for most physicochemical and biological purposes such "solutions" should not be stored for more than a few hours; storage at low temperatures, in the dark, minimizes the changes.

19.10.1.4. Solubilization in aqueous detergents

The biological environment of the porphyrin and heme pigments is essentially aqueous, and it is desirable to be able to study their physicochemical properties in such media. Measurements of such properties as their ionization and coordination have been limited by their very low solubility in water. Some measurements have been carried out in non-aqueous and mixed-solvent media, but such systems are unsuitable for electrochemical studies because of unknown ionic activity effects.

Phillips[228] introduced the important technique of solubilization of porphyrins and metalloporphyrins in aqueous detergent solutions. Final pigment concentrations from 2.10^{-6} M to 5.10^{-5} M are convenient for most physicochemical studies. In these systems porphyrins and their esters, ferro- and ferri-porphyrins and their esters, and other metalloporphyrins, are monomerically dispersed, and it has been possible to study quantitatively a number of aspects of their behavior such as the ionization of porphyrins, the formation and stability constants of porphyrin metal chelates, and stability constants for the further coordination of iron-porphyrins with extra ligands. Anionic (e.g. sodium dodecyl sulfate), cationic (e.g. cetyltrimethyl ammonium salts) or non-ionic detergents may be used for particular purposes, and neutral detergents are, of course, convenient for enzymic studies[227].

Solutions of porphyrin esters in aqueous detergents may be prepared by addition of a solution in acetone to the detergent solution. The acetone may then be removed by drawing N_2 or air through the warmed solution for a few hours. As an example, meso- and proto-porphyrin dimethyl esters may be solubilized monomolecularly, at concentrations of $4 \cdot 10^{-5}$ M and $3 \cdot 10^{-5}$ M respectively, in 4.5% (w/v) aqueous sodium dodecyl sulfate (SDS), cetyltrimethylammonium bromide (CTAB) or Tween 20. Another method[227] is to extract the porphyrin from an ether solution into 2 N NH_4OH containing 1% Emasol 4130, the aqueous layer being freed of ether by evacuation. In this way solutions up to $3 \cdot 10^{-5}$ M may be obtained. Hemes may be brought into solution in detergents from their solutions in aqueous alkali, or may be first dissolved in very small volumes of dioxane or bases such as pyridine or ethanolamine. More recently the properties of hemin monomers in SLS micelles have been evaluated[314]. For details of porphyrin visible absorption spectra in aqueous sodium dodecyl sulfate, see the Appendix to this Chapter, p. 888.

19.10.1.5. Organic solvents

Porphyrins and their metal chelates are soluble in pyridine, dioxan, acetone and related solvents. They are not easily dissolved in ether or ethyl acetate, but relatively concentrated solutions may be obtained by mixing these solvents with strong solutions of the porphyrins in acetone—HCl, pyridine etc., which may then be washed away with dilute aqueous acid followed by water. Precautions must be taken if porphyrins with low HCl numbers are present, since they may be lost in acid washes. Porphyrins may also be transferred to ether or ethyl acetate from solutions in aqueous acids by bringing to the isoelectric point as described above.

Unlike the free carboxylic acids, porphyrin esters and many of their metal chelates are soluble in chloroform, carbon tetrachloride, benzene, carbon bisulphide, ethyl acetate, as well as pyridine, dioxan, ether, etc. They are relatively insoluble in paraffins and alcohols, and methanol—chloroform or petroleum ether—chloroform mixtures are often used for their crystallization. Porphyrins without polar side-chains, e.g. etioporphyrins, have similar solubilities to the esters of carboxylic porphyrins. The ether solubility of porphyrin esters is decreased by formyl substituents.

The porphyrins are conveniently separated from porphyrin esters by extracting them from an ether solution with NH_4OH. Porphyrin esters can be extracted into chloroform from their solutions in aqueous mineral acids. This procedure allows separation from the free porphyrin carboxylic acids, which remain in the aqueous phase if the acid concentration is reasonably high. Some free porphyrin carboxylic acid hydrochlorides, however, may be extracted into chloroform from aqueous HCl at specific proton concentrations.

Chloroform which has not been adequately washed often contains enough protons to convert porphyrin esters dissolved in it to their dications; the resulting spectrum has occasionally been mistaken for that of a porphyrin metal chelate. Shaking with a little solid sodium carbonate converts the cation to the free base. Solid or crystalline hemins are readily dissolved in organic solvents such as pyridine, dioxan and related compounds and in acidified ether, ethyl acetate or acetone (see below).

Solutions of pyridine hemochromes in benzene or ether can be obtained either by dissolving the hemin ester in pyridine, reducing with a minimum of sodium dithionite in water, and extracting into the organic solvent, or by mixing a concentrated solution of the hemin ester in pyridine with benzene, and reducing by shaking with aqueous sodium dithionite[229].

19.10.1.6. Acidified organic solvents

Hemins and porphyrins are readily dissolved in acidified organic solvents, such as acetone—HCl, ethyl acetate—acetic acid, ether—acetic acid and oxalic acid—methanol. These solvents are of course protein precipitants, and are used widely for the extraction of the tetrapyrroles from biological materials. Ethyl acetate solutions so obtained may be washed free of acid, and the porphyrins extracted with aqueous mineral acid which does not extract the metal chelates. Even more than with porphyrins, maintenance of hemins in solution in ethyl acetate or ether depends upon the presence of some acetic acid. The splitting of hemin from hemoglobin by shaking a solution brought to about pH 2 with methylethylketone is described on p. 888; the method is useful for extraction of porphyrins and chlorophylls also.

19.10.2. Separations by solvent-partition
19.10.2.1. Partition between ether and water

The ether phase is much more strongly favored than neutral aqueous phases by most dicarboxylic porphyrins, partition coefficients of the order of $10^4 - 10^7$ being found[230]. Of the natural porphyrins, the octacarboxylic uroporphyrins alone favor the aqueous phase, and indeed are virtually insoluble in ether. Other 'water-soluble' porphyrins, dissolving to a certain extent in water at neutral pH, include porphyrins such as porphyrin-c, its synthetic analogs, and other synthetic compounds such as deuteroporphyrin-2,4-disulfonic acid and meso-tetra(p-sulfonatophenyl)porphyrin[231].

19.10.2.2. Partition between ether and aqueous HCl

The partitioning of porphyrins and chlorophylls between ether and aqueous hydrochloric acid solutions forms the basis of an important technique for their separation and purification, of which an example is given on p. 330. Partition chromatography between ether and HCl on celite columns has also been used[232].

The concept of acid (HCl) number, as introduced by Willstätter[233] and used extensively by Fischer[234] has already been discussed (p. 15).

19.10.2.3. Countercurrent distribution

The separation of porphyrins by virtue of their different HCl numbers is achieved much more efficiently by the use of countercurrent distribution[235—237].

The distribution coefficient of a compound is calculated by the use of the formula:

$$N = (n\text{-}1)\,\frac{1}{K+1}$$

where n is the total number of tubes, and N is the number of that tube containing the highest concentration of the compound. Granick and Bogorad[237] introduced a number of modifications to the standard technique. A spectrophotometer cell-holder was designed to allow rapid, direct determinations of the porphyrin in the upper and lower phases. It was found that to overcome the rather low solubility of porphyrins, the initial solution could be spread over the first 5 or 6 tubes without changing significantly the final distribution in 100-tube runs; tetrahydrofuran was found useful to increase solubility, but had the disadvantage that peroxides, which lead to decomposition of porphyrins were formed during the experiments. Acetone may be used in the same way[236], but diminishes the separation between some porphyrins[237]; dioxan has been found satisfactory for proto- and hemato-porphyrins[236]. More recently, the system isobutyl-methyl-ketone and t-butyl alcohol (1 : 1) (top phase) and dilute sulfuric acid (0.5—2 M, depending on the porphyrins) (lower phase) has been used[238] with some success. The organic phase does not suffer the disadvantages of peroxide formation and the lower phase lacks the potentially troublesome chloride ion. It is significant that no separation of porphyrin type-isomers has yet been achieved using countercurrent distribution.

19.10.2.4. Partition between ether and aqueous buffers

The pH number of a porphyrin has been defined[239] as the pH of a buffer solution which extracts half of the porphyrin from four volumes of a solution of it in ether.

19.10.3. Esterification and hydrolysis of esters

The most convenient method for esterification of porphyrins is a simple treatment with an alcohol—mineral acid mixture, e.g. methanol—H_2SO_4. This method may be used on the microgram or the gram scale. Some porphyrins are labile to mineral acids, however, and for these the diazomethane method is to be preferred; the procedure described below is quick, relatively safe, and convenient.

19.10.3.1. Esterification of porphyrins using diazomethane

Nitrosomethylurea is unstable, and may decompose explosively. It should be stored at $0°C$. Diazomethane is also readily obtained[240] from more stable precursors, such as 'Diazald' (p-toluenesulfonylmethylnitrosamide), which is commerically available (Aldrich) and is greatly preferable to nitrosomethylurea. Diazomethane is highly toxic, and must be used in an efficient fume-cupboard.

Method
The porphyrin in ether solution is washed with water, and the ether solution (it is not necessary to dry) is placed in an ice-bath in a fume-cupboard. In the ice-bath a solution of diazomethane in ether is also prepared, by cautious addition of nitrosomethylurea (1 g) to ether underlayered with 20% (w/v) KOH (7 ml). The yellow solution of diazomethane in ether is separated from the aqueous KOH in a separating funnel in the fume-cupboard, and washed once with distilled water. Excess of the diazomethane solution is added to the porphyrin solution with gentle swirling, and after 5 min the excess diazomethane is destroyed by addition of dilute acetic acid. The porphyrin solution is washed repeatedly with water to remove the acetic acid, once with dilute NH_4OH to remove any unesterified porphyrin, and finally with water.

Notes
(1) Hemins should not be esterified with diazomethane.
(2) The presence of even traces of unesterified porphyrin is readily revealed by applying a spot of the mixture to Whatman No. 1 paper and chromatographing for a few minutes with chloroform; unesterified material does not move away from the origin.
(3) The method can be used for porphyrins with formyl side-chains, but see p. 829.

19.10.3.2. Esterification of porphyrins using alcohols with mineral acid
The carboxylic side-chains of porphyrins may be esterified at room temperature or below, with an alcohol (usually methanol, Note 1) saturated with gaseous HCl or containing concentrated sulfuric acid (usually 5% w/v).

20 h at room temperature in the dark (Note 2) is sufficient for full esterification of even the octacarboxylic uroporphyrins by HCl—methanol, but after 20 h at $-10°C$ esterification is incomplete[236]. Protoporphyrin, on the other hand, is fully esterified in either methanol—HCl or methanol—H_2SO_4 after 18 h at $-10°C$ (Note 3). Treatment with 5% H_2SO_4—methanol for 48 h at $0°C$ has been found suitable for esterification of porphyrin-a.

After esterification, the porphyrin esters are conveniently extracted into ether or chloroform, and the solution washed with 2 N Na_2CO_3 or NH_4OH, followed by water, and evaporated to dryness. The esters may then be crystallized.

References, p. 861

Notes

(1) Pure, dry methanol must be used.

(2) Porphyrins, especially those with unsaturated side-chains, are light-sensitive.

(3) In addition to its photosensitivity, protoporphyrin is hydrated to a degree in aqueous HCl, its vinyl side-chains being converted to the α-hydroxyethyl groups of hematoporphyrin (cf. p. 821) but it was shown[236] by countercurrent distribution studies that full esterification of protoporphyrin without hydration of vinyl groups occurred in 18 h at −10°C with 5% sulfuric acid—methanol. The purification of protoporphyrin is discussed on p. 771. Hematoporphyrin may be esterified in dry methanol containing 15% (w/v) of anhydrous HCl[26], but both this, and the treatment with methanol—sulfuric acid causes some dehydration of the hydroxyethyl groups to vinyl groups, and diazomethane has been found the most suitable esterifying agent.

19.10.3.3. Esterification of hemins

Like the porphyrins, the hemins may be esterified in 5% (w/v) H_2SO_4—methanol. This method has been used successfully even for the unstable hemin-*a*.

Hemin esters may be prepared also by insertion of iron into the porphyrin ester (p. 803) and in particular circumstances this may be the method of choice. For example, the esters of porphyrin-*a* and hemin-*a* cannot easily be separated by ether—HCl fractionation, and in the particular case where it is wished to prepare hemin-*a* ester from the porphyrin, hemin-*a* must first be prepared and freed of residual porphyrin as below, and then esterified with methanol—sulfuric acid. Hemins react with diazomethane, which is not useful for their esterification[241]. A method has been described[229], which involves refluxing the hemin in the alcohol in the presence of trifluoroacetic acid.

19.10.3.4. Hydrolysis of esters
19.10.3.4.1. Aqueous HCl

The most common method is simply to dissolve the porphyrin ester in 25% (w/v) HCl and allow the solution to stand at room temperature, in the dark, for sufficient time to effect complete hydrolysis. For most purposes, 48 h is a convenient time, and is necessary in some cases (e.g. uroporphyrin octamethyl esters). Shorter periods are, of course, desirable for compounds like protoporphyrin, and Grinstein[143] has found that the dimethyl ester of this porphyrin is completely hydrolyzed in 5 h at room temperature. It has been shown, however[236], by countercurrent distribution analysis, that it is not possible to hydrolyze protoporphyrin methyl ester with aqueous HCl of any strength without some hydration of the vinyl groups occurring. Esters of carboxyl groups directly attached to the porphyrin nucleus are extremely difficult to hydrolyze in aqueous HCl, and KOH—methanol (below) is better in this case.

Protoporphyrin monomethylester has been prepared by partial hydrolysis, with aqueous HCl, of the dimethyl ester[242].

After hydrolysis, the porphyrin may be recovered by standing the micro-beaker containing the solution over KOH in a vacuum desiccator, or by bringing the acid solution to pH 4 and transferring the porphyrin to ether.

19.10.3.4.2. KOH-methanol

Both porphyrin esters and hemin esters may be hydrolyzed by dissolving in methanol containing 1% KOH, and either refluxing or allowing to stand at room temperature. Some water must, of course, be present.

19.10.3.4.3. KOH-water/tetrahydrofuran[238b]

The porphyrin ester is dissolved in tetrahydrofuran and treated with an equal volume of 2 N aqueous KOH. The two phase solution is shaken or stirred overnight in the dark, after which time all of the porphyrin has transferred to the aqueous phase. Disposal of the colorless tetrahydrofuran phase and acidification with aqueous HCl precipitates the porphyrin car-boxylic acid.

19.10.4. Crystallization and melting points
19.10.4.1. Crystallization
19.10.4.1.1. Porphyrin salts

The di-hydrochlorides of porphyrins may usually be precipitated, some-times in crystalline form, from concentrated solutions in mineral acids, and the sodium salts of carboxylic porphyrins from strong alkaline solutions; these precipitations are of some preparative value.

19.10.4.1.2. Porphyrins

The porphyrins in their undissociated forms may be crystallized by allow-ing solutions in organic solvents, usually ether, to evaporate slowly. They may be crystallized from pyridine also, chloroform being added to permit filtration and then removed in vacuo. Well-formed micro-crystals may be obtained in this way.

19.10.4.1.3. Porphyrin esters

Unlike the porphyrin carboxylic acids, the porphyrin esters have melting points of some usefulness, and they are also, of course, much more soluble in organic solvents and thus more readily purified, particularly by chromato-graphy.

The methyl esters are most commonly used, though esters with many other alcohols have been made. Though the methyl esters of mono- and di-formyldeuteroporphyrins may be safely prepared, methanol can cause

References, p. 861

acetal formation with very active formyl side-chains as in photoprotoporphyrin (cf. pp. 688—689); if diazomethane is used, the time of exposure to this
reagent should be kept to a minimum.

The general method for the crystallization of porphyrin methyl esters, and
their chelates with divalent metal ions, is as follows. The porphyrin ester is
dissolved in a small volume (1 ml for 20—30 mg ester) of freshly washed and
dried chloroform. To the boiling solution, an equal volume of boiling, absolute dry methanol is added. As boiling continues, chloroform distils off and
the methanol concentration increases; this process is allowed to continue
until crystals form on cooling. If too much chloroform is removed, an
amorphous precipitate of the ester, which is practically insoluble in methanol, is obtained.

The crystals may be recovered by centrifuging or on a small Büchner or
Hirsch filter, and washed with cold methanol. Washing with boiling water is
often useful to remove inorganic material when crystalline metal chelates are
prepared. Elementary analysis of porphyrin esters and their metal chelates
crystallized in this way has occasionally revealed the presence of chloroform
of crystallization, and for the metal chelates, at least, crystallization from
benzene is preferable.

Other pairs of solvents have been used, e.g. mixtures of chloroform or
benzene with ether or petroleum ether; chloroform—ether mixtures are particularly useful for porphyrin esters with formyl side-chains. For the crystallization of Mg(II) protoporphyrin dimethyl ester[111], Granick added petroleum ether to an ether—xylene solution of the compound. Chloroform—
acetone has been used for some uroporphyrin esters[243] and occasionally
acetone itself for recrystallization by extraction from a thimble. Benzene has
been found to be of general usefulness for hot extraction at the thimble and
crystallization of chelates of porphyrin esters with divalent cations. Porphyrin esters may be crystallized from ether at −16°.

19.10.4.2. Crystal form and melting points

Porphyrin esters crystallize in a variety of forms, from plates and prisms
to the tangled hair-like forms often found with uroporphyrin esters. The
crystals are generally small and dark, with a metallic sheen. The high melting
points of this group of compounds, usually in the range 200—300°C, are best
determined on a Kofler block, or by the use of a hot-stage microscope, and
the disappearance of birefringence, observed with crossed Nicol prisms is the
usual criterion of melting. Double melting points, due to crystals of higher
melting point forming in the first melt or on cooling it, are common with
some uroporphyrin and coproporphyrin isomers, and polymorphism also is
common (see below).

The methyl esters of deutero-, meso- and proto-porphyrins have fairly
reproducible and reasonably sharp melting points (at 218—220°C,
213—216°C and 222—223°C, with a few degrees variation between different

authors), and their characterization by mixed melting point determinations with known material is useful technique[111].

The copro- and uro-porphyrin methyl esters on the other hand are known to be polymorphic[243,244] and, as has long been known, form mixed crystals (solid solutions). This is most unfortunate, since distinction between the position-isomers (-I, -II, -III, and -IV) of copro- and uro-porphyrins is often important, and melting points of the esters clearly do not provide suitable criteria.

The mixed melting point curves for coproporphyrin-I and -III methyl esters[245] and for uroporphyrin-I and -III methyl esters[246] are thus of limited usefulness for the determination of the proportions of these isomers in mixtures.

The Cu(II)-chelates of coproporphyrin esters are less subject to variations in crystal form, and their melting points are useful (isomer III, sintering about 213°C, m.p. 216—219°C; isomer IV, m.p. 230—233°C)[244] and have been used as criteria of the identity with reference materials of coproporphyrins obtained by decarboxylation of uroporphyrins and a heptacarboxyl porphyrin isolated from natural materials[247]. The melting point of optically active hematoporphyrin dimethyl ester prepared from cytochrome-c was depressed on mixing with synthetic, optically inactive hematoporphyrin ester[165].

More recently, shift differences in the n.m.r. spectra of porphyrins in the presence of lanthanide shift reagents[248] have proved to be a useful guide to identity between two porphyrins when melting point and mixed m.p. are unhelpful[249].

19.11. Chromatography
19.11.1. Paper chromatography of porphyrins and hemins
19.11.1.1. Introduction
19.11.1.1.1. Solvents

Peroxides must be rigorously excluded from all solvents used for chromatography of porphyrins.

19.11.1.1.2. Detection of bands and spots

On column chromatograms, the bands of porphyrins and metalloporphyrins can usually be observed by their color. The fluorescence of porphyrins, and of some metalloporphyrins (Zn, Cd, Mg) under ultraviolet light provides a much more sensitive means of detection.

The fluorescence on paper chromatograms of porphyrins and their esters is intensified greatly by spraying with iso-octane[250], 0.005 μg being observable. Kerosene gives a similar intensification.

Spots of non-fluorescent porphyrin metal complexes are best observed by spraying the paper with a solution of fluoranthene in n-pentane, and then

References, p. 861

illuminating with light at 366 nm; the metal complexes show up as dark spots against a fluorescent background, 0.04 µg being observable[250].

Spots containing as little as $3 \cdot 10^{-4}$ µg of hemin may be observed by treatment with the following spray[251], made up not more than 3 h before use: 25 ml of absolute methanol are shaken for 1 min with excess of benzidine hydrochloride. The solution is decanted, and to it is added 12.5 ml of water, 5.0 ml of glacial acetic acid, 2.5 ml of 3% H_2O_2 and 0.5 ml of pyridine. Benzidine is carcinogenic, however, and o-tolidine, or o-dianisidine[252] which is more sensitive than benzidine, should be used.

19.11.1.1.3. Choice of method

Of the various methods described below, that of choice for a first examination of unknown materials is the chromatography of porphyrin free acids with water—lutidine mixtures. This method separates porphyrins[51,236,253,254] and hemins[160] approximately in accordance with the number of carboxylic acid groups present, and under some conditions can be used to separate isomeric coproporphyrins.

Other methods, in which the porphyrin esters are used, are available for the identification of dicarboxylic porphyrins, of isomeric coproporphyrins and uroporphyrins, of porphyrins with hydroxyl groups in their side-chains, and of hemins.

19.11.1.1.4. Choice of paper

The paper commonly used is Whatman No. 1 for chromatography. Schleicher and Schüll paper 2043b has been found to behave similarly in most of the methods described, in which neither washing nor drying of the paper prior to use is necessary. The solvent is usually run in the direction of the machine grain of the paper.

19.11.1.1.5. Application to paper

Spots of 0.1—3.0 µg are applied to the paper by micropipette, usually from solutions in dilute NH_4OH or organic solvents. The spots are dried before development, a warm air-blower (hair-dryer) being convenient. Spots are best applied in a series of small applications, with drying between to minimize spread; their diameter should not exceed 1 cm. Overloading is the most common cause of tailing, streaking and lack of resolution.

19.11.1.1.6. Interference and decomposition

High salt concentrations interfere with the separations; salts should be removed by washing with water a solution of the porphyrins in an organic solvent.

In materials extracted from tissues, lipids sometimes follow the porphyrins closely through extraction procedures, and interfere on paper chromatograms. They may be removed by washing the material on a column chromatogram with ether or petroleum ether, or sometimes by a preliminary development of the paper chromatogram with petroleum ether.

Porphyrins, and particularly minute amounts adsorbed on paper or on columns, should be protected from strong light.

19.11.1.1.7. R_F values

While R_F values are used below as a convenient indication of the extent of separations, they are not criteria of identity for porphyrins since they often vary with temperature and other conditions. Correspondence with known (marker) compounds, chromatographed alongside the unknown, and co-chromatography with a known compound, are more reliable criteria.

19.11.1.1.8. Types of separation on paper

The lutidine—water method (see below) is clearly an example of partition chromatography, as is the reversed-phase development on silicone-treated paper. The various separations of porphyrin esters with non-aqueous organic solvents are probably mainly adsorption separations, though some partitioning may be involved.

19.11.1.2. Porphyrin free acids (lutidine method)

The method was introduced in 1949 by Nicholas and Rimington[255] who used as solvent system the organic phase obtained by saturating a mixture of 2,4- and 2,5-lutidines with water at 21°C. 2,6-Lutidine, unlike the other lutidines, is miscible with water up to 40°C; it was first used for porphyrin chromatography by Kehl and Stich[254] and its use has now become general. Any temperature around 20° is suitable, as long as it remains constant during the development, and as long as known markers are run alongside the unknown substances. Some typical R_F values are shown in Table 3.

Method[236]

The porphyrins (about 3 μg) are spotted onto Whatman No. 1 (or Schleicher and Schüll 2043b[254]) paper, the solvent dried off, and descending chromatograms are run overnight at 21°C, in the direction of the grain of the paper. The developing solvent is a mixture of 2,6-lutidine and water (5 : 3.5 v/v) and separate beakers of 2,6-lutidine, water, and 7 N NH$_4$OH are placed in the bottom of the tank. These are renewed daily. After the development the paper is dried, and examined under ultraviolet light as described above.

Notes

(1) The isomer separation which occurs with all lutidine methods (see below), of course, precludes strict interpolation of R_F's of unknown porphyrins in the curve of R_F against number of carboxyl groups. Happily, the isomers of uroporphyrins do not separate under these conditions; isomers of

TABLE 3

R_F Values of porphyrins in 2,6-lutidine-water systems

Porphyrin	R_F^a	R_F^b
Uroporphyrin	0.26	0.06
Pentacarboxylic porphyrin		0.42
Coproporphyrin	0.54	0.56
Hemin	0.7	
Tricarboxylic porphyrin		0.68
Protoporphyrin	0.84	0.83
Mesoporphyrin	0.86	
Hematoporphyrin	0.87	
Deuteroporphyrin	0.88	
Etioporphyrin	1.0	
Porphyrin esters	1.0	

[a] Kehl and Stich[254], Schleicher and Schüll paper 2043b, 25°C, ascending, NH_3 vapor.
[b] Eriksen[253], Whatman No. 1 paper, 20°C, ascending, NH_3 vapor.

dicarboxylic (proto-, deutero-)porphyrins are not known in nature, and the existence of tricarboxylic porphyrins of isomeric types other than that derivable from coproporphyrin-III is most unlikely. Thus the region in which the chromatograms must be interpreted with caution is that where the R_F values correspond to penta-, hexa- or hepta-carboxyl porphyrins. The occurrence of isomers among such porphyrins has been studied by Chu and Chu[256]. In comparisons of the behavior of polycarboxyl porphyrins on paper electrophoresis and on 2,6-lutidine paper chromatography, the non-linearity of the R_F—carboxyl group relationship, particularly with higher numbers (8, 7, 6) of carboxyl groups has been demonstrated by Lockwood and Davies[64]. Apart from questions of isomerism, the R_F of some porphyrins (e.g. porphyrin-a) is influenced by side-chains other than carboxyl groups[225].

(2) Adaptations of this method have been made for rapid, routine examinations of urinary porphyrins[257], for radial (circular) chromatography[258] for the examination of porphyrins occuring in petroleum aggregates[259], and for the chromatography of a number of chlorophyll derivatives[258]. An apparatus has been described in which a number of paper strips may be run simultaneously with observation of the position of spots under ultraviolet light[260].

(3) With[261] has described an ascending development with aqueous lithium chloride in an atmosphere of NH_4OH. The separation (according to the number of carboxyl groups) is in the reverse order to that found with

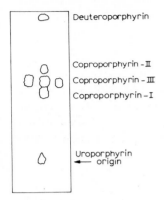

Fig. 2. Separation of porphyrins, including coproporphyrin isomers-I, -II and -III, by 2,6-lutidine.

lutidine development, uroporphyrins moving fastest, and the method is rapid, full development taking about 1 h only.

19.11.1.3. Coproporphyrin isomers (lutidine method)

It was observed by Benson and Falk in 1954[51] that the 2,4-, 2,5-lutidine and the 2,6-lutidine developments cause separations of coproporphyrin isomers. The 2,6-lutidine method described above separates coproporphyrins-I, -III plus -IV, and -II, in increasing order of R_F[51,236]. Smaller spots (about 1 μg) are applied for this purpose. Typical separations are shown in Fig. 2. No method has been found for the separation of coproporphyrins-III and -IV, which have identical R_F values on these chromatograms; high pressure liquid chromatography has potential here[262].

The modification by Eriksen[263] gives excellent and reproducible separations of the coproporphyrin isomers, though porphyrins with higher numbers of carboxyl groups are not well differentiated from each other.

It is a matter of choice, depending in part upon whether the esters or the free acids are to hand, whether the lutidine or the propanol method (below) is used. The lutidine method has the advantage of separating three out of the four isomers, while in the propanol method isomers-I and -II, and isomers-III and -IV move in pairs. This is perhaps not of great moment, since the occurrence of isomers-II and -IV in nature is very unlikely.

Separations of isomers on paper chromatograms developed with methanol[264], and separations of coproporphyrins-I and -III on Al_2O_3 columns developed with ether—acetic acid[265] have not been confirmed[54]. A separation of the coproporphyrins by elution from Al_2O_3 with 35% acetone (isomer III) followed by 100% acetone (isomer I)[266] has been found in the same laboratory to give inconsistent results[54]. A method depending upon the quenching of fluorescence of mixtures of the methyl esters of coproporphyrins-I and -III in an acetone—buffer mixture is described fully by Schwartz et al.[54]

Method[263]

The porphyrins (about 1 μg) are spotted as described above onto Whatman No. 1 paper. A petri dish containing a mixture of 2,6-lutidine and water (5 : 2) is placed in the bottom of a glass cylinder, and a small bottle containing NH_4OH (s.g. 0.880) is stood in the dish. The paper is rolled into a cylinder, stood upright in the petri dish, and ascending development carried out overnight.

Notes

(1) It is important to allow the atmosphere to become saturated with the solvent vapors before beginning the development, and to seal the lid well with petroleum jelly or another suitable compound.

(2) The Eriksen method has been used[267] for the separation of mesoporphyrins-I and -IX. Spots of about 0.3 μg are applied from a fresh solution in pyridine to Whatman No. 1 paper, and ascending development with 2,6-lutidine-water (5 : 1) in an atmosphere of ammonia is carried out as described above.

(3) The method has been adapted very successfully to use in thin-layer chromatography[268].

19.11.1.4. Coproporphyrin isomers (n-propanol method)

In this method, the methyl esters of the porphyrins are used. It was introduced by Chu *et al.*[269] and provided the first paper chromatographic separation of isomeric porphyrins. It is useful for several other purposes also (see Notes 1 and 2), and furnished the basis of a method for the separation of the isomeric uroporphyrins-I and -III (see below).

Method[269,270]

Whatman No. 1 paper is cut into sheets 12.5 × 14.5 cm; the porphyrin esters (0.5 μg) in an organic solvent are spotted on a base-line drawn across the grain of the paper, 2 cm from the edge, and the spots (diameter not more than 2 mm) are dried. *First development*: Kerosene is added to the tank shortly before the development to saturate the atmosphere. The paper is then subjected to ascending development at about 22° for about 20 min (solvent front about 75 mm from baseline), with a mixture of kerosene and chloroform (1 : 1, v/v). The paper is dried at 105—110°. *Second development*: In an atmosphere saturated with kerosene the paper is developed in the same direction for about 1 h with a mixture of kerosene and n-propanol (5 : 0.9, v/v), and removed from the tank and dried. R_F values are calculated from the position of the solvent front in this development. A typical chromatogram is shown in Fig. 3, and some R_F values in Table 4.

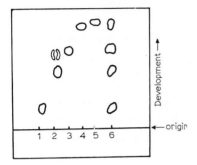

Fig. 3. Paper chromatogram of methyl esters of porphyrins[269] (see text). The porphyrin methyl esters are: 1, uro-I; 2, copro-I; 3, copro-III; 4, proto-IX; 5, meso-IX; 6, a mixture of all five compounds.

Notes

(1) Blumer[250] has modified the method for use with porphyrin compounds occurring in petroleum and sedimentary rocks. He found that with this solvent mixture the preliminary development used by Chu et al.[269] was not necessary. Porphyrins with free carboxylic acid groups (e.g. proto-, meso-) remained at the origin; they could be esterified on the paper with diazomethane and run with the same solvent mixture in the second direction. The maximum amount of porphyrin or metal complex that could be chromatographed without tailing was 0.5 μg. It was found that with care, determinations by measuring the area of the spots could be made to an accuracy of $\pm 10\%$; a calibration curve for the determination of mesoporphyrin in this way was given.

TABLE 4

R_F Values of porphyrin methyl esters[a]

	Second run, n-propanol with:				
	kerosene[b]	n-decane[c]	n-do-decane[c]	n-tetra decane[c]	n-hexane[c]
Uroporphyrin-I	0.17	0.14	0.20	0.13	0.15
Coproporphyrin-I	0.47	0.42	0.52	0.45	0.47
Coproporphyrin-III	0.67	0.70	0.76	0.74	0.66
Protoporphyrin-IX	0.84	0.86	0.92	0.92	0.89
Mesoporphyrin-IX	0.89	0.92	0.96	0.95	0.93

[a] Chu et al.[269]
[b] First run, chloroform-kerosene, 2.6 : 4.0.
[c] First run, chloroform with the same alkane as used in the second run. Propanol-alkane, 1 : 5.

References, p. 861

TABLE 5

R_F values of hydroxylated porphyrins and chlorins and their acetylated products[a]

	Chloroform-kerosene, 2.6:4		Propanol-kerosene, 1:5	
	Alcohol	Acetate	Alcohol	Acetate
Monohydroxyethyl deuteroporphyrin	0.34	0.64	0.38	0.68
Monovinyl-monohydroxyethyl deuteroporphyrin	0.29	0.60	0.34	0.66
Hematoporphyrin	0.03	0.56	0.18	0.57
Monohydroxymethyl deuteroporphyrin	0.19	0.54	0.24	0.57
Monohydroxymethyl-monovinyl deuteroporphyrin	0.22	0.54	0.31	0.56
Dihydroxymethyl deuteroporphyrin	0.01	0.56	0.14	0.56
2-Formyl-4-hydroxyethyl deuteroporphyrin	0.16	0.58	0.26	0.64
Porphyrin-a	0.10	0.56	0.26	0.62
2-Ethyleneglycol deuteroporphyrin	0.30	0.64	0.32	0.61
2-α-Hydroxymesorhodochlorin	0.31	0.80	0.34	0.78
2-α-Hydroxymesochlorin-p_6	0.33	0.82	0.35	0.80
2-α-Hydroxymesopheophorbide-a	0.10	0.78	0.34	0.72
Chlorin-a_2	0.30	0.65	0.40	0.60

[a] Barrett[271], chromatography by the method of Chu et al.[269].

(2) The method of Chu et al.[269] has been adapted by Barrett[271] for the separation of the methyl esters of porphyrins with hydroxyl groups in their side-chains, use being made of the differences in R_F before and after acetylation (Table 5).

19.11.1.5. Uroporphyrin isomers (dioxan method)

Falk and Benson[236,272] found that the methyl esters of uroporphyrins-I and -III may be separated by the use of dioxan, in a method similar to that of Chu et al.[269] (above) for the coproporphyrin isomers. Isomer II runs identically to isomer I, and isomer III identically to isomer IV[236]. Other complications of the method are discussed below (Note 2).

Method[218,272]

To a base line, 2 cm from the edge of a 21-cm square of Whatman No. 1 paper, about 1 μg of the porphyrin ester is applied from a solution in freshly washed and dried chloroform. The solvent is dried off, and the two ascending developments described below are carried out at 22°C in the direction of the grain of the paper. From 15 to 30 min are allowed for equilibration of the atmosphere before inserting the development mixture and the paper. *First development:* The developing solvent is a mixture of 6 volumes of chloroform containing 1% by volume of absolute ethanol with 4 volumes of kerosene. Vapor phase: chloroform containing 1% ethanol. The function of this development is simply to move the porphyrin esters away from the origin, leaving behind impurities and unesterified porphyrins. When the esters have moved about 4 cm from the base line (about 20 min), the paper is dried, and then cut off about 0.5 cm below the spots, and subjected to the second development. *Second development:* The developing solvent is a mixture of 4 volumes of kerosene with 1 volume of dioxan. Vapor phase: dioxan. The chromatogram is developed in the same direction as before, until the solvent front has just reached the top of the paper.

Typical results of the use of this method are shown in Fig. 4.

Notes

(1) The first development, with kerosene-chloroform, serves as a quick and convenient test for the presence of unesterified material in all preparations of porphyrin esters.

(2) Assessment of the proportions of uroporphyrin esters-I and -III in mixtures chromatographed by this method is made uncertain by the entrainment of isomer I with isomer III when considerable proportions of the latter are present[272]. There has long been evidence[55,246] for some kind of "molecular compound" formation between porphyrins of this type, and Bogorad

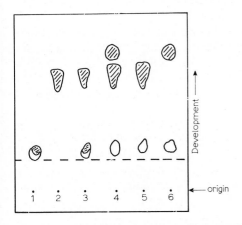

Fig. 4. Paper chromatography of uroporphyrins[273]. Open spots show positions reached after preliminary development, after which the paper was cut off at the broken line. Hatched spots show positions of porphyrins after final development. 1, Uro-I; 2, Uro-III; 3, Uro-I and Uro-III, 50-50 mixture; 4, Uro-III and pseudouroporphyrin (also known as 'phyriaporphyrin'); 5, Uro-III after separation of (4) on a CaCO$_3$ column; 6, Pseudouroporphyrin after separation of (4) on CaCO$_3$ column.

References, p. 861

and Marks[274], using isotopically labeled uroporphyrin-I and -III, have found that when mixtures of them are chromatographed by this method not only is I entrained with III, but III may be held with I, though when chromatographed singly the isomers take up their proper R_F positions. While entrainment in paper chromatograms of a slower-moving substance in the spot of a faster-moving one is not uncommon, the virtually equal partition between the two positions found by Bogorad and Marks appears to be a new phenomenon.

Cornford and Benson[218] have studied the quantitative aspects of the separation of uroporphyrins-I and -III by eluting the 'I' and 'III' spots, decarboxylating to the coproporphyrins, separating the coproporphyrins on paper chromatograms by the 2,6-lutidine method (above), and finally eluting and determining the resulting coproporphyrin spots spectrophotometrically. They have confirmed the findings of Bogorad and Marks, and have found that under strictly defined conditions the correct proportions of uroporphyrins-I and -III in a mixture can be determined from the results of dioxan chromatograms. After the development, the spots are eluted with chloroform, the uroporphyrin content of each eluate determined spectrophotometrically, and the proportions of each isomer present estimated from a correction table which they provide.

This method is the only one available for the separation of isomeric uroporphyrins. It has been used widely and is still useful, but there is a real need for a method which does not suffer from such complications. There is need, in addition, for a method capable of separating all four isomeric uroporphyrins though it is unlikely that uroporphyrins-II and -IV occur naturally.

(3) Pseudouroporphyrin, a heptacarboxylic porphyrin, which may be an important compound for the elucidation of the biosynthetic pathway, was so named because of its chemical similarity to a uroporphyrin. It has also been named 'phyriaporphyrin'[247]. On decarboxylation it yields coproporphyrin-III, but it differs chromatographically, in both the lutidine and the dioxan methods described above, from all four isomeric uroporphyrins[236] (cf. Fig. 4).

A porphyrin behaving similarly has been isolated from certain porphyric urines by Canivet and Rimington[275]. Grinstein et al.[55,247,276] have isolated another porphyrin from porphyric urines, the methyl ester of which has a melting point of 208°C; chromatography by the dioxan method has shown that it differs from pseudouroporphyrin. The structure of pseudouroporphyrin has recently been defined[249] (see p. 97) as uroporphyrin-III with the 8-acetic acid (ring D) decarboxylated to a methyl group.

19.11.1.6. Dicarboxylic porphyrins

The methods described above do not effect useful separations from each other of porphyrins with two carboxylic groups, for example hemato-, deutero-, meso- and proto-porphyrins. The following method serves this purpose.

Method[277]

Whatman No. 1 paper is cut into sheets 20 × 14 cm with the short end parallel to the grain of the paper. The porphyrin esters are spotted, from an organic solvent, as shown in Fig. 5. *First development*: Ascending development is carried out in a 2-liter tank, in the long direction of the paper, with a mixture of kerosene (5 ml), tetrahydropyran (1.4 ml) and methyl benzoate (0.35 ml) at about 20°, until the solvent front reaches the top of the paper (about 3 h).

The atmosphere is saturated with kerosene (15 ml) by damping a paper liner in the tank, and with tetrahydropyran (0.4 ml) placed in the bottom of the tank. The volume of tetrahydropyran used here is rather critical.

The paper is removed and dried at 105—110° for 10 min. The developed spots are marked under u.v. light and a new base-line drawn through their centers. *Second development*: The paper is trimmed, dipped in a 12.5% (w/v) solution of Dow-Corning silicone No. 550 fluid in petroleum ether (b.p. 65—110°), and dried at 105—110° for 3 min. Wetting the paper with petroleum ether just before dipping in the silicone solution gives a more even coating.

Ascending development is carried out in a 1-liter tank, in the short direction of the paper, in an atmosphere saturated with water vapor, with a mixture of water (3.8 ml), acetonitrile (1 ml), *n*-propanol (2 ml) and pyridine (0.5 ml), until the solvent front has moved about 8 cm (about 1.5 h).

Notes

(1) The second development is an example of reversed-phase partition chromatography.

(2) Distilled fractions of kerosene, or pure *n*-decane, *n*-dodecane or *n*-tetradecane gave no better results than crude kerosene.

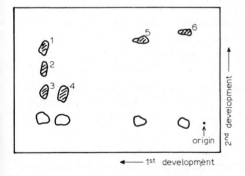

Fig. 5. Paper chromatogram of dimethyl esters of dicarboxylic porphyrins[277]. 1, Hemato dimethyl ether; 2, Deutero-IX; 3, Meso-IX; 4, Proto-IX; 5, Monovinylmonohydroxyethyl-deutero-IX; 6, Hemato-IX. Open spots, first development, hatched spots, second development (see text).

References, p. 861

(3) Tetrahydrofuran may be used instead of tetrahydropyran.

(4) Either development may be used singly for the particular separations it effects. For example, the separation of a mixture containing hemato-, deutero- and proto- (or meso-)porphyrins, is conveniently achieved on silicone-treated paper by a single development for 50 min in a 1-liter tank with a mixture of water (2.3 ml), acetonitrile (2.8 ml) and dioxan (0.8 ml).

(5) Despite some false reports[278], no isomer of protoporphyrin other than -IX has been detected in natural materials. Protoporphyrin is easily reduced to mesoporphyrin, and the method[267] for separation of mesoporphyrins-I and -IX would now permit the detection of small amounts of protoporphyrin-I. It is now clear, however[98], that this isomer of protoporphyrin cannot be formed enzymically. Protoporphyrin-I has recently been synthesized[7] and its physical constants reported. The dimethyl ester was inseparable from that of protoporphyrin-IX by t.l.c., but separation was readily accomplished using high pressure liquid chromatography (p. 859).

19.11.1.7. Hemins

Chu and Chu[160] found that a 2,6-lutidine—water system separates hemins, as it does porphyrins, in accordance with the number of carboxyl side-chains they contain. A similar separation may be achieved[160] by chromatographing the hemin esters by the n-propanol method[269] used for the separation of isomeric coproporphyrins.

A more versatile method, however, separates uro- and copro-hemins, and the dicarboxylic hemato-, deutero-, meso- and proto-hemins, as the free acids, in a single development[160].

Method

Hemins are spotted on a base-line 1.2 cm from the edge of a 14 × 14 cm · sheet of Whatman No. 1 paper which has been treated with silicone as described on p. 849. The paper is placed in a tank lined with filter paper wet with water, and containing 0.4 ml pyridine per 1-liter tank-volume in the bottom of the tank. Ascending development is then carried out at about 20° until the solvent front has moved about 12 cm (about 1 h), with a mixture of water, n-propanol and pyridine 5.5 : 0.1 : 0.4 by volume. The paper is dried at 105—110°C and the hemins are located as dark spots under ultraviolet light, or by the o-dianisidine spray (see p. 840). A typical chromatogram is shown in Fig. 6.

Notes

(1) For application to the chromatograms, hemins should be dissolved when possible in organic solvents such as ethyl acetate or acetone. Solutions in pyridine[279] or in aqueous alkali[280] must be freshly prepared. The hematins have the same R_F values as the corresponding hemins.

(2) The pyridine concentration in the developing solvent is critical; with

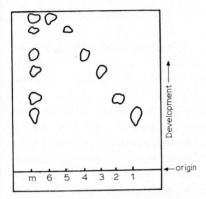

Fig. 6. Paper chromatogram of hemins (see text). 1, Proto; 2, Meso; 3, Deutero; 4, Hemato; 5, Copro-I or -III; 6, Uro-I; m, a mixture of all six hemins.

too little, protohemin streaks; with higher concentrations, the R_F values are increased and faster-moving hemins are less well separated. If fast-moving compounds (*e.g.* uro-, copro-, hemato-hemins) are the only compounds present, separation is improved if the pyridine concentration is reduced (water—*n*-propanol—pyridine, 6 : 0.1 : 0.1).

(3) Morrison and Stotz[281] and Connelly *et al.*[251] have used silicone-treated paper, with several different solvent mixtures, for paper chromatographic studies of heme-*a* and heme-*c* and other compounds.

19.11.2. Column chromatography

19.11.2.1. Introduction

Talc columns were used for the purification of porphyrin free acids by Fischer and Hofmann[282] as early as 1937, and later by Comfort[283]. For separations of porphyrin esters, columns of Al_2O_3 and $CaCO_3$ have been used, and a systematic study of these adsorbents as well as MgO and $MgCO_3$ was made by Nicholas[284] in 1951.

When the adsorptive power of the materials and the conditions of development were standardized, reproducible separations of porphyrin esters were obtained[284]. As in paper chromatography, the porphyrins separate on these columns largely according to the number of carboxylic side-chains they possess; thus on columns of Al_2O_3 or $CaCO_3$, mobility increases from uro- to copro- to proto-porphyrin, as on most paper chromatograms. On MgO and on $MgCO_3$, however, the order is reversed, as it is on silicone-treated paper chromatograms. It is very convenient in practice to have the choice of reversal of order of elution; thus with a mixture of porpyrins, it is convenient to choose for a first separation a column from which the porphyrin(s) required pure are eluted easily, and for further purification of these fractions to use a column on which they are more strongly adsorbed.

References, p. 861

Mixtures of dicarboxylic porphyrins also (as their esters) may be separated on columns of MgO or $CaCO_3$.

In the column chromatography of porphyrin esters, development is usually carried out with gradually increasing concentrations of an eluting solvent (methanol > chloroform > benzene) diluted by a relatively inert solvent such as petroleum ether.

The columns are by far the most useful and most widely used for the routine purification of porphyrin esters. For various special purposes, some other excellent methods which are available are described. Synthetic resin columns have been little used for porphyrins, but Heikel[285] has described the quantitative separation of urinary porphyrins on Dowex-2.

The separation, by gas chromatography and paper chromatography, of monopyrrolic degradation products made during the determination of structure of porphyrins is discussed on p. 659.

19.11.2.2. Porphyrin esters
19.11.2.2.1. Al_2O_3, MgO, $MgCO_3$, $CaCO_3$

The procedure is illustrated for an Al_2O_3 column; similar considerations apply to all the methods.

Method[284]: A column is prepared by pouring a slurry, in A.R. benzene, of Al_2O_3 grade IV into a chromatographic tube. The column, which must never be allowed to run dry, is settled by vibration. The top of the adsorbent may be protected from disturbance during addition of solutions by the addition of a thin layer of acid-washed sand or by inserting a polythene disk. For 1 mg of porphyrin esters, a column of approximately 2 × 20 cm of adsorbent is required.

The porphyrin ester is added to the column in as small a volume of benzene as is practicable; if necessary, the ester is brought into solution with a few drops of chloroform, and excess of benzene then added. Development is carried out first with chloroform—benzene (1 : 10, v/v), when proto-, meso-, deutero- and other dicarboxylic porphyrins should be eluted. Increasing the proportion of chloroform to benzene to 1 : 1 elutes tetracarboxylic (e.g. copro-) porphyrins, and a mixture of chloroform—methanol (100 : 1) is required to elute octacarboxylic porphyrins. These separations were found with pure porphyrins; the developing solvents have to be adjusted empirically for the development of fractions with intermediate numbers of carboxyl groups, or when impurities which change the chromatographic conditions are present. Preliminary developments with petroleum ether or benzene are often useful to remove lipids from materials of biological origin.

Notes

(1) Porphyrin free acids present in preparations of the esters remain as an immobile band at the top of these columns. They may, of course, be recovered and re-esterified.

(2) For preliminary separations, and especially for work on a large scale, Al_2O_3 is the most convenient adsorbent.

(3) Metal chelates of porphyrin esters are purified conveniently on columns of Al_2O_3 or of $MgCO_3$, on which they are readily separated from the metal-free compounds. Certain metal chelates, however (e.g. Hg and Cd) may lose their metal on Al_2O_3 columns.

(4) Protoporphyrin ester, prepared by the method of Grinstein[143] which is convenient for the large scale, often contains considerable amounts of an impurity absorbing at about 660 nm. This is easily removed by chromatography on columns of Al_2O_3 grade V developed with benzene.

(5) Lemberg and Parker[207] have reported the separation, on columns of BDH 'Alumina for chromatographic separation', of the methyl esters of protoporphyrin and 2,4-divinyl-, 2-formyl-4-vinyl-(chlorocruoro-) and 2,4-diformyldeutero-porphyrins, and of the closely related 2- and 4-monoformyldeuteroporphyrins. The columns were packed as a slurry in a mixture of equal volumes of chloroform and ether, and the same mixture used for development.

(6) Columns of $CaCO_3$ have a particular usefulness for separations of porphyrins with many carboxyl groups, such as uroporphyrins, the porphyrin of m.p. 208°C described by Watson and collaborators[286], pseudouroporphyrin (p. 848), and other compounds with similar behavior. These columns are, however, often difficult to manage. They are packed by tamping small portions at a time of the dry material into the column; samples which pack without 'springing' should be selected. In addition, $CaCO_3$ is difficult to inactivate, and a relatively inactive sample should be chosen so that it can be activated to the desired grade by heating (100°C or even higher for about 12 h may be required). After packing dry, slight suction is applied at the bottom of the column, and benzene—chloroform, 10 : 1 (v/v) is added slowly until the column is uniformly wet; with care, this procedure largely eliminates formation of cracks and air-pockets. The solution of porphyrin esters is then added and development carried out, the flow-rate being controlled by adjustment of the suction.

19.11.2.2.2. Celite (Hyflo Supercel)

Chu and Chu[270,287] have found that incomplete resolution of isomeric octacarboxylic (uro-) and heptacarboxylic porphyrins may be achieved on celite (Hyflo Supercel) columns, which in addition give the usual separation of porphyrins according to the number of carboxyl groups present. The fewer carboxyl groups, the faster the migration (see also Ref. 232).

Method[270,287]. *Separation according to carboxyl groups*: A column of celite 1.8 × 14 cm is packed dry by tamping. The porphyrin esters in an organic solvent (*e.g.* chloroform) are mixed with a little celite; this is dried, and packed by tamping on the top of the column. Development is carried out, with suction, with a mixture of chloroform and petroleum ether (b.p.

References, p. 861

30—60°), 1 : 2 by volume. The bands are not eluted, but are removed mechanically after drying the column. The porphyrins may be eluted from the
celite by chloroform, to which about 1% of methanol may be added to elute
very strongly adsorbed materials. An example of use of the method is given
in Fig. 7.

Separation of isomeric hepta- and octa-carboxylic porphyrins: A column
of celite 3 X 45 cm is packed, and a mixture of 2-3 mg of the porphyrins to
be analyzed is applied as described above. The column is developed with
benzene, then with chloroform-benzene (1 : 2 by volume) with increasing
additions of ethanol, until the porphyrin front almost reaches the bottom of
the column. The column is dried, and the celite removed in arbitrary fractions which are analyzed by paper chromatography.

Notes

(1) When a mixture of uroporphyrins-I and -III was chromatographed in
this way, isomer III was found near the top of the fluorescent zone and
isomer I near the bottom, with graded mixtures between.

(2) As has long been evident (for literature, see Chu and Chu[287]), porphyrins with 3, 5, 6 and 7 carboxyl groups occur in various natural materials.
Chu and Chu[287,288] have prepared, by stepwise decarboxylation of uroporphyrins I and III, hepta-, hexa- and penta-carboxylic porphyrins related to
both these isomeric types. Porphyrins isolated from porphyric urines have
been compared with these reference substances in respect to their behavior
on the celite columns, paper chromatography, melting points and infrared

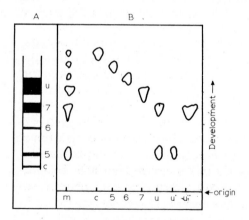

Fig. 7. Column chromatography of methyl esters of urinary porphyrins[287]. A: Celite
column chromatogram; B: Paper chromatogram, using kerosene-chloroform and then
kerosene-dioxan as solvent systems, m, the original mixture; c, coproporphyrin; 5, 6 and
7, penta-, hexa- and hepta-carboxylic porphyrins; u, uroporphyrin mixture from column;
u', u'', Uroporphyrin-I and -III markers.

spectra. The evidence confirms the existence in the natural materials of hepta-, hexa- and penta-carboxylic porphyrins of both isomeric forms.

(3) By chromatography on celite columns of the porphyrins from the erythrocytes of porphyric cattle, Chu and Chu[289] have isolated a porphyrin which appears to be monocarboxylic.

19.11.2.2.3. Silicic acid

Separations on the above columns are due to adsorptive forces. A method of partition chromatography, in which the porphyrin esters were chromatographed on columns of hydrated silica gel, was introduced by Lucas and Orten[290]. More recently (J.M. Orten and E.G. Larsen, personal communication[1]), hydrated silicic acid has been used, and this method is described below.

The order of elution is again according to the number of carboxyl groups, materials with fewer carboxyl groups having greater mobility. In addition, dicarboxylic porphyrins may be separated. Unesterified porphyrins remain at the top of the column. These columns have much greater capacity than most others, 1 g (dry wt.) of adsorbent being sufficient for 1 mg of a mixture of porphyrin esters.

Method (Orten and Larsen): Silicic acid powder (Mallinckrodt, analytical reagent grade) is sieved. The portion of particle size between 100 and 150 mesh per inch is dried at 150° to constant weight, and 4 g are mixed thoroughly with 2 ml water. The mixture is covered with *n*-hexane (b.p. approximately 69°C), and, care being taken to keep the mixture covered with hexane to prevent drying out, it is triturated (a stainless steel spatula is convenient) until it is homogeneous. The slurry is then poured, about one third at a time, into a column (10 × 300 mm), each portion being packed under pressure (15—20 cm Hg) until no further settling occurs.

The porphyrin esters are dissolved in a minimal volume of chloroform and an equal volume of *n*-hexane is added. The air pressure is released from the column and this solution added, and a smooth horizontal front is established by stirring the top 1—2 mm of the column. *Development*: A flow rate of 8—12 drops per minute is maintained by a pressure of 5—10 cm Hg. Protoporphyrin is mobilized by a mixture of chloroform (30 volumes) and *n*-hexane (70 volumes) and is eluted when the chloroform concentration is increased to 40%. The same solvent mixture mobilizes coproporphyrins I or III, and these are eluted when the chloroform is increased to 50%. Uroporphyrins migrate with 50% chloroform, and are eluted with 60% chloroform in *n*-hexane. Tri-, penta-, hexa- and hepta-carboxyl porphyrins are eluted in the expected intermediate positions.

19.11.2.3. Porphyrin free acids

Cellulose powder: In all the above methods the porphyrin esters are used. The lutidine method of paper chromatography using the porphyrin free acids has been adapted by Eriksen[291] for large-scale separations on columns. The R_F values after development of the columns with 2,6-lutidine—water are similar to those found in the usual paper chromatography (p. 841) with this solvent system. Attempts at isomer-separations on such columns have not yet been reported.

Method[291]: A 15-cm column of cellulose powder (Whatman, 'for chromatography') is packed by tamping. A mixture of the porphyrins to be separated, dissolved in a little acetone-containing NH_4OH, is mixed with a little of the same cellulose powder, the material dried in a desiccator, and packed on top of the column by tamping.

Development with 2,6-lutidine—water (6 : 2, v/v) separates the porphyrins; the approximate R_F values of uro-, copro- and meso-porphyrins are 0, 0.53 and 0.95 respectively.

Further development with the same solvent mixture, under slight positive pressure, caused elution of the mesoporphyrin. On development with 2,6-lutidine—water (6 : 4), the coproporphyrin was eluted, and by the addition of a drop of concentrated NH_4OH to 25 ml of the same solvent mixture, the uroporphyrin was eluted.

Note: Cellulose columns developed with ether have been used successfully for the purification of porphyrin-a[292].

19.11.2.4. Hemins

Column chromatography has been applied to hemins mainly for studies of the prosthetic groups of cytochromes-a, -a_2 and -c. Kiese and Kurz[293] used columns of Al_2O_3 to separate the hemin-a of heart muscle from protohemin, and Morrison and Stotz[294] achieved further separations on silicic acid columns. For the isolation of hemin-a_2, Barrett[271] used silica gel columns. The hemins from ox heart have been separated also on celite columns[295], and hemin-c has been chromatographed on silicone-treated cellulose powder[296]. Hemin esters may be purified on columns of Al_2O_3 or $MgCO_3$ using similar organic solvents to those used for porphyrin esters.

19.11.2.4.1. Silicic acid

Method[294]: Silicic acid (Merck or Mallinckrodt) 500 g, is mixed with 800 ml of ethanol containing concentrated HCl (ethanol—HCl, 5 : 1, v/v). The mixture is filtered on a Büchner funnel and washed three times, or until colorless, with the same solvent mixture. After three further washings with 700 ml of 95% ethanol, and three more with 800 ml of ether, the silicic acid is air-dried in thin layers for 24—48 h and then sieved (80 mesh per inch) and dried in a vacuum desiccator.

The silicic acid (50 g) is mixed well with 20 ml of 0.15 N HCl, and the resulting powder mixed well with 80 ml of chloroform to give a fine slurry. The slurry is poured into a chromatographic tube to make a column of the adsorbent 3.5 × 11 cm.

To the column wet with chloroform, 10 ml *n*-hexane are added, and with 2—5 ml of hexane lying above the column, the hemins dissolved in a little chloroform are carefully added. The chloroform solution settles below the hexane layer and precipitation can be avoided. The column is washed with 100 ml of chloroform—hexane 1 : 1, to remove lipids, and the elution of heart-muscle hemins is achieved by development with chloroform.

19.11.2.4.2. Silicagel

Method[271]: Silica gel (B.D.H., 100 mesh/inch) is washed twice with 95% ethanol and twice with water and dried at 22°. A finely dispersed preparation, essential for smoothly working columns, is obtained by shaking mechanically, under N_2, 9 g of the silica gel with 4 ml of methanol—water (7 : 3, v/v). The particles are then suspended in 60 ml of petroleum ether (b.p. 68°C) for preparation of the chromatographic column.

The hemins are applied to the column in benzene solution, lipids eluted with petroleum ether (b.p. 68°C), and development follows with petroleum ether equilibrated with methanol—water (7 : 3, v/v) and finally with wet benzene with increasing additions of methanol. Heme-*a* is eluted when the benzene contains 1% of methanol.

19.11.2.5. Dry column chromatography[315]

This relatively new method in many cases gives superior separations than those which can be achieved by normal 'wet' column chromatography; these separations, however, are not so good that one could expect pure type-isomers to be obtained from a mixture. Several companies now sell silica and alumina supports specially prepared for dry chromatography; these do tend to be expensive and in some cases normal chromatographic or thin layer grade materials can be used.

Method: Dry column or ordinary silicagel or alumina is dropped gently into a glass column (Note 1) or plastic tube (Note 2). After tapping and settling, the porphyrin in a minimum volume of methylene chloride or chloroform is placed on top of the column (Note 3). The column is then eluted with the solvent or solvents of choice (Note 3), and the bands (in the case of a glass column) are collected. If a commerical flexible plastic tube is used, the column is eluted until the bands are seen to be separated from each other. The column is then dissected with a sharp knife and the contents of each section are eluted with solvent slightly more polar than the mixture being used for elution of the intact column.

Notes

(1) A glass funnel with the narrow end drawn out to leave a small hole is suitable. The weighed amount of adsorbent is then placed into the glass funnel resting on top of the column, and it is automatically packed slowly

References, p. 861

and uniformly; the bottom end of the column must, of course, be partially blocked with a plug of glass or cotton wool. If a nylon tube is used, pin holes should be pierced at the bottom of the tube. Best results are obtained if the column is tamped two or three times during packing.

(2) Flexible, solvent resistant plastic tubes are now available commercially. The lower end of the tube is restricted with a narrow piece of glass tube secured with an elastic band; a plug of glass or cotton wool is added to prevent the adsorbent being lost. These plastic tubes have the advantage that they can be cut. Complete elution of the bands off the end of the column is therefore unnecessary, and large amounts of solvent are thereby saved.

(3) Suction from a water pump can be applied at this stage if the column is particularly slow running.

19.11.3. Thin-layer chromatography[297]

Thin-layer chromatography is now a common technique used in all areas of chemistry. Excellent separations are obtained on analytical thin plates of either silicagel or alumina; porphyrins tend to streak on the latter. Common solvents for development of plates are chloroform, methylene chloride, or mixtures of these with hexane, or petroleum ether. In addition, mixtures of acetone and hexane, or ethyl acetate and benzene, have been used. The chromatograms should always be run in the dark[298]. With the characteristic colors of porphyrins, measurement of R_F is usually easy. However, methods for the quantitative analysis of porphyrins in the sub-microgram range have been developed[299].

Thin plates can be prepared in the normal way, using a commercially available spreader[297]. It is often convenient, however, to make analytical plates on microscope slides for use in monitoring reactions. This can be done very easily by preparing a suspension of silicagel in stabilized chloroform (the relative quantities of each should be determined by trial and error — the less adsorbent present in the chloroform, then the thinner are the plates) in a bottle with a screw-on cap. The suspension is shaken vigorously, the cap taken off, and a microscope slide held with a conventional spring clothes peg is immersed in the suspension. It is then carefully withdrawn from the bottle (replace cap) and dried for a few minutes in an oven at about 110°C. The spots are placed on the plate and it is developed in a small beaker containing the appropriate solvent, a filter paper around the inside (to help saturation of the air space with solvent), and with a watch glass or Petri dish as a lid. R_F's determined by this method are unreliable, but it is useful for monitoring reactions wherein comparison spots can be run on the same plate.

Of equal use in porphyrin chemistry is thick layer chromatography. Here, 'large' quantities of porphyrins can be separated. Plates are best prepared using a commercial spreader, though a method using masking tape placed around the edge of a glass plate upon which is poured a suspension of adsorbent, has its uses. Typically, thicknesses of 1—2 mm of adsorbent are used: For 1.5 mm plates, water (310 ml) and silicagel (150 g) [or aluminum oxide (300 g)] are placed in a 1 liter flask which is stoppered and the suspension is shaken vigorously for 1 min until completely homogeneous. It is allowed to stand for 15 min, is shaken briefly once more and then placed in the spreader (gap 2 mm) and laid onto the glass plates slowly. In this method, 5 20 × 20 cm or 1 100 × 20 cm plates of 1.5 mm thickness are obtained.

Ready coated thick layer plates are commercially available; as might be expected, these tend to be comparatively expensive, though they are highly efficient and provide reproducible separations.

After development of the chromatogram, and this can be done several times using the same plate, the adsorbent is removed from the plate and extracted with a solvent mixture slightly more polar than that used for development óf the chromatogram. Repeated development of the chromatogram tends to give better separations than a single development using a solvent mixture of sufficiently high polarity to move the bands concerned to the same R_F value.

19.11.4. High-pressure liquid chromatography[300]

This technique has only relatively recently become generally available. In the organic chemistry sense, it gained its most dramatic foothold as a result of its use in the total synthesis of vitamin B_{12}[301]. At the present time, practitioners are largely dependent upon commercially available pumping and detection systems, but this might not always be the case. A variety of silica and alumina supports are available.

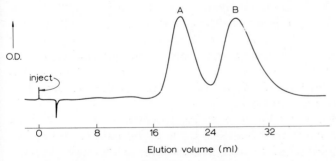

Fig. 8. High-pressure liquid chromatographic separation of harderoporphyrin trimethyl ester [B, 4,6,7-tri(2-methoxycarbonylethyl)-1,3,5,8-tetramethyl-2-vinyl-porphin)] from isoharderoporphyrin trimethyl ester [A, 2,6,7-tri-(2-methoxycarbonylethyl)-1,3,5,8-tetra-methyl-4-vinylporphin)]. A Waters Associates ALC 202-401 instrument was used[238b] with the 6000 pumping system and a 254 nm UV detector. Samples were introduced with a syringe onto two 2 ft × 1/8 in (OD) columns containing Corasil-II (Waters) and elution was carried out at a rate of 2 ml/min with chloroform (stabilized with 0.2% alcohol) containing small amounts of cyclohexane.

References, p. 861

The most useful detector for porphyrin work is undoubtedly the UV type (254 or 280 nm commonly), and, with the relatively small amount of work which has so far been reported, it appears that porphyrin methyl esters (rather than free acids) on normal (rather than reverse) phase column supports provide the best separations.

Figure 8 shows the separation of harderoporphyrin trimethyl ester [4,6,7-tri(2-methoxycarbonylethyl)-1,3,5,8-tetramethyl-2-vinylporphin] from its 2(propionic ester)-4-vinyl isomer[238b]. In this instance, all other chromatographic techniques, as well as countercurrent distribution, had failed to give a good separation of these two porphyrin isomers of great potential biosynthetic significance[238b]. Other porphyrin type-isomers have been separated, e.g. protoporphyrin-I and protoporphyrin-IX dimethyl esters[7]. The tetramethyl esters of some of the coproporphyrin isomers have also been separated[302] but the efficient separations of all four coproporphyrin isomers from each other has yet to be disclosed.

19.11.5. Electrophoresis

Electrophoresis has been applied to porphyrins on paper[64,303—307] and on agar gel[308,309]. Other solid media, including starch paste, starch gel, paper powder and glass powder were found to be inferior to agar gel[308]. The methods have a particular usefulness for polycarboxylic porphyrins[64], but dicarboxylic porphyrins do not move away from the origin on paper electrophoresis. Separations of isomers have not yet been achieved by this method. Some representative methods are described briefly here.

19.11.5.1. Paper electrophoresis

Method: Lockwood and Davies[64] used 0.04 M Na_2CO_3 with 10^{-4} M EDTA on Whatman 3 MM paper, freely suspended horizontally between glass supports at the ends, 20 cm apart. The porphyrins were applied from solutions in concentrated NH_4OH at the cathode end of the wet paper, in streaks containing 0.5—50 μg of porphyrin per cm; after equilibration for 15 min, the papers were subjected to a current of 10 V/cm for 3 h. The relative mobilities of porphyrins with 8, 7, 6, 5 and 4 carboxyl groups were 1.0, 0.93, 0.75, 0.42 and 0.15 respectively.

Porphyrins were subjected by With[306] to electrophoresis at 7.5—8 V/cm for 1—3 h, on Whatman No. 1 paper, with 0.05 M barbiturate buffer of pH 8.6. The distances moved from the origin (near the anode) were 10—15, 0.5—2.0 and 0—0.3 cm for uro-, copro- and proto-porphyrins respectively.

Notes

(1) It has been shown[307] that while bilirubin added to normal human serum is immediately attached to albumin, proto- and hemato-porphyrins, and uro- and copro-porphyrins-I and -III move independently of the serum proteins.

(2) The porphyrin precursors δ-aminolevulinic acid and porphobilinogen have been separated from porphyrins by paper electrophoresis[303].

(3) Quite good recoveries of porphyrins have been achieved after paper electrophoresis[304].

19.11.5.2. Agar gel electrophoresis

Method[308]: Troughs, 50 × 18 × 1.5 cm, containing 0.5 or 1.0% agar gel prepared in veronal buffer of pH 8.6 and ionic strength 0.05, were used. The porphyrin free acids were applied in small holes in the gel, and the electrophoresis was carried out at 4°C. For a film thickness of 2—5 mm, 6 V/cm were applied, and separation was complete after 10—12 h. Excellent separation was found of octa-, hexa-, penta-, tetra- and tri-carboxylic porphyrins.

References

1. J.E. Falk, 'Porphyrins and Metalloporphyrins', Elsevier, Amsterdam (1964).
2. G.W. Kenner, K.M. Smith, and J.F. Unsworth, J. Chem. Soc., Chem. Commun., 43 (1973).
3. G.W. Kenner, K.M. Smith, and J.F. Unsworth, unpublished results.
4. E.C. Taylor and A. McKillop, Tetrahedron, 23, 897 (1967).
5. B. Frydman, S. Reil, M.E. Despuy, and H. Rapoport, J. Am. Chem. Soc., 91, 2338 (1969).
6. A.W. Johnson, E. Markham, R. Price, and K.B. Shaw, J. Chem. Soc., 4254 (1958).
7. J.A.S. Cavaleiro, G.W. Kenner, and K.M. Smith, J. Chem. Soc., Perkin Trans. 1, 2478 (1973).
8. J.A.S. Cavaleiro, G.W. Kenner, and K.M. Smith, unpublished results.
9. A.H. Jackson, G.W. Kenner, and K.M. Smith, J. Chem. Soc. C, 502 (1971).
10. R.J. Abraham, G.H. Barnett, E.S. Bretschneider, and K.M. Smith, Tetrahedron, 29, 553 (1973).
11. J.A.S. Cavaleiro, A.M.d'A. Rocha Gonsalves, G.W. Kenner, and K.M. Smith, J. Chem. Soc., Perkin Trans. 1, 2471 (1973).
12. R. Chong, P.S. Clezy, A.J. Liepa, and A.W. Nichol, Aust. J. Chem., 22, 229 (1969).
13. G.H. Barnett and K.M. Smith, unpublished results.
14. A.W. Johnson, I.T. Kay, E. Markham, R. Price, and K.B. Shaw, J. Chem. Soc., 3416 (1959).
15. H.W. Whitlock and R. Hanauer, J. Org. Chem., 33, 2169 (1968).
16. H.H. Inhoffen, J.-H. Fuhrhop, H. Voigt, and H. Brockmann Jr., Ann. Chem., 695, 133 (1966).
17. U. Eisner, A. Lichtarowicz, and R.P. Linstead, J. Chem. Soc., 733 (1957).
18. W.A. Lazier and H.R. Arnold, Org. Synth., Coll. Vol. II, Ed. A.H. Blatt, Wiley, New York, (1943), p. 142.
19. A.D. Adler, F.R. Longo, J.D. Finarelli, J. Goldmacher, J. Assour, and L. Korsakoff, J. Org. Chem., 32, 476 (1967).
20. G.H. Barnett, M.F. Hudson, and K.M. Smith, J. Chem. Soc., Perkin Trans. 1, in press.
21. G.M. Badger, R.A. Jones, and R.L. Laslett, Aust. J. Chem., 17, 1028 (1964).
22. G.H. Barnett, M.F. Hudson, and K.M. Smith, Tetrahedron Lett., 2887 (1973).
23. R.J. Abraham, G.E. Hawkes, M.F. Hudson, and K.M. Smith, J. Chem. Soc., Perkin Trans. 2, 204 (1975).

24. G.W. Kenner, S.W. McCombie, and K.M. Smith, Ann. Chem., 1329 (1973).
25. H. Fischer and H. Orth, 'Die Chemie des Pyrrols', Akademische Verlagsgesellschaft, Leipzig Vol. II, part 1, (1937), p. 421.
26. S. Granick, L. Bogorad, and H. Jaffe, J. Biol. Chem., **202**, 801 (1953).
27. Ref. 25, p. 425.
28. H. Fischer and F. Lindner, Hoppe-Seyler's Z. Physiol. Chem., **168**, 152 (1927).
29. H. Fischer and R. Müller, Hoppe-Seyler's Z. Physiol. Chem., **142**, 120, 155 (1925).
30. M. Grinstein and C.J. Watson, J. Biol. Chem., **147**, 671 (1943); M.O. Schultze, ibid., **142**, 89 (1942).
31. E.J. Chu, J. Biol. Chem., **166**, 463 (1946).
32. H. Fischer and B. Pützer, Hoppe-Seyler's Z. Physiol. Chem., **154**, 39 (1926); S. Granick, J. Biol. Chem., **172**, 717 (1948); H. Muir and A. Neuberger, Biochem. J., **45**, 163 (1949); J.F. Taylor, J. Biol. Chem., **135**, 569 (1940); J. Wittenberg and D. Shemin, ibid., **185**, 103 (1950).
33. H. Muir and A. Neuberger, Biochem. J., **45**, 163 (1949).
34. Ref. 25, p. 414; O. Schumm, Hoppe-Seyler's Z. Physiol. Chem., **178**, 1 (1928).
35. T.C. Chu and E.J. Chu, J. Am. Chem. Soc., **74**, 6276 (1952).
36. P. Baumgarten, Chem. Ber., **59**, 1166 (1926); J.B. Neilands, J. Biol. Chem., **190**, 763 (1951).
37. Ref. 25, p. 566.
38. H.H. Strain and W.A. Svec, in "The Chlorophylls", Eds. L.P. Vernon and G.R. Seely, Academic Press, New York (1966), p. 21.
39. H. Fischer and A. Stern, "Die Chemie des Pyrrols", Vol. 2, Akademische Verlagsgesellschaft, Leipzig (1940), p. 48.
40. Ref. 39, p. 56.
41. H.R. Wetherell and M.J. Hendrickson, J. Org. Chem., **24**, 710 (1959).
42. G.W. Kenner, S.W. McCombie, and K.M. Smith, J. Chem. Soc., Perkin Trans. 1, 2517 (1973).
43. G.W. Kenner, S.W. McCombie, and K.M. Smith, unpublished results.
44. G.W. Kenner, S.W. McCombie, and K.M. Smith, J. Chem. Soc., Perkin Trans. 1, 527 (1974).
45. J.B. Neilands and J.A. Garibaldi, Biochem. Prep., **7**, 36 (1960).
46. G.H. Cookson and C. Rimington, Biochem. J., **57**, 476 (1954).
47. L. Bogorad, J. Biol. Chem., **233**, 501, 510, 516 (1958); Methods Enzymol., **5**, 855 (1962).
48. R.E.H. Nicholas and C. Rimington, Biochem. J., **50**, 194 (1951).
49. C.J. Watson and M. Berg, J. Biol. Chem., **214**, 537 (1955).
50. J.E. Falk and E.I.B. Dresel, Biochim. Biophys. Acta, **39**, 458 (1960); J.E. Falk, E.I.B. Dresel, and C. Rimington, Nature London, **172**, 292 (1953); R.G. Westall, ibid., **170**, 614 (1952).
51. J.E. Falk, Biochem. Soc. Symp. (Cambridge, England), **12**, 17 (1954).
52. D. Mauzerall, J. Am. Chem. Soc., **82**, 2601 (1960).
53. R.J. Porra and J.E. Falk, Biochem. Biophys. Res. Comm., **5**, 179 (1961).
54. S. Schwartz, M.H. Berg, I. Bossenmaier, and H. Dinsmore, in "Methods of Biochemical Analysis", Ed. D. Glick, Vol. 8, Interscience, New York (1960), p. 221.
55. M. Grinstein, S. Schwartz, and C.J. Watson, J. Biol. Chem., **157**, 323 (1945); A.G. MacGregor, R.E.H. Nicholas, and C. Rimington, Arch. Intern. Med., **90**, 483 (1952); T.K. With, Biochem. J., **68**, 717 (1958).
56. (a) L. Eriksen, Scand. J. Clin. Lab. Invest., **3**, 121, 128, 135 (1951); A.E. Garrod, J. Physiol., **13**, 598 (1892); (b) S.L. Sveinsson, C. Rimington, and H.D. Barnes, Scand. J. Clin. Lab. Invest., **1**, 2 (1949).
57. C. Rimington, Biochem. J., **37**, 443 (1943); C. Rimington and P.A. Miles, ibid., **50**, 202 (1951).

58. L.M. Corwin and J.M. Orten, Anal. Chem., **26**, 608 (1954).
59. E.L. Talman in "Standard Methods of Clinical Chemistry", Ed. O. Seligson, Vol. 2, Academic Press, New York (1958), p. 137.
60. F.E. De Salamanca, F.P. Mesorana, and J.O. de la Gandara, Arch. Med. Exp. (Madrid), **12**, 25 (1949).
61. W. Grotepass, Hoppe-Seyler's Z. Physiol. Chem., **253**, 276 (1938).
62. A.E. Garrod, J. Physiol., **17**, 349 (1894).
63. W.H. Lockwood, Aust. J. Exp. Biol. Med. Sci., **31**, 453 (1953).
64. W.H. Lockwood and J.L. Davies, Clin. Chim. Acta, **7**, 301 (1962).
65. E.I.B. Dresel and J.E. Falk, Biochem. J., **63**, 72 (1956).
66. E.M. Kapp and A.F. Coburn, Brit. J. Exp. Pathol., **17**, 255 (1936).
67. (a) C. Rimington, Biochem. J., **75**, 620 (1960); (b) C. Rimington and S.L. Sveinsson, Scand. J. Clin. Lab. Invest., **2**, 209 (1950); (c) T.K. With, ibid., **7**, 193 (1955); H.A. Zondag and E.J. van Kampen, Clin. Chim. Acta, **1**, 127 (1956).
68. C.J. Watson, R.P. de Mello, S. Schwartz, V. Hawkinson, and I. Bossenmaier, J. Lab. Clin. Med., **37**, 831 (1951).
69. S. Schwartz, L. Zieve, and C.J. Watson, J. Lab. Clin. Med., **37**, 843 (1951).
70. E.I.B. Dresel, C. Rimington, and B.E. Tooth, Scand. J. Clin. Lab. Invest., **8**, 73 (1956); G.Y. Kennedy, ibid., **8**, 79 (1956).
71. E.I.B. Dresel and B.E. Tooth, Nature London, **174**, 271 (1954).
72. J. Waldenström, H. Fink, and W. Hoerburger, Hoppe-Seyler's Z. Physiol. Chem., **233**, 1 (1935).
73. S. Schwartz, M. Keprios, and R. Schmid, Proc. Soc. Exp. Biol. Med., **79**, 463 (1952).
74. S. Groennings, Anal. Chem., **25**, 938 (1953); J. Sanik, Anal. Chim. Acta, **21**, 572 (1959); J.M. Sugihara and L.R. McGee, J. Org. Chem., **22**, 795 (1957); J.R. Vallentyne, Can. J. Bot., **33**, 304 (1955); G.W. Hodgson, Ann. N.Y. Acad. Sci., **206**, 670 (1973) and refs. therein.
75. L. Shuster, Biochem. J., **64**, 101 (1956).
76. (a) D. Mauzerall and S. Granick, J. Biol. Chem., **219**, 435 (1956); (b) S. Granick and H.G. Vanden Schrieck, Proc. Soc. Exp. Biol. N.Y., **88**, 270 (1955).
77. W.H. Elliott, Biochem. J., **74**, 90 (1960).
78. W.G. Laver, A. Neuberger, and S. Udenfriend, Biochem. J., **70**, 4 (1958).
79. F. de Matteis and B.E. Prior, Biochem. J., **83**, 1 (1962).
80. G. Urata and S. Granick, J. Biol. Chem., **238**, 811 (1963).
81. N.I. Berlin, A. Neuberger, and J.J. Scott, Biochem. J., **64**, 80 (1956).
82. R.G. Westall, Nature London, **170**, 614 (1952).
83. G.H. Cookson, Nature London, **172**, 457 (1953); G.H. Cookson and C. Rimington, ibid., **171**, 875 (1953).
84. J.E. Falk, E.I.B. Dresel, and C. Rimington, Nature London, **172**, 292 (1953).
85. J.B. Jepson, in "Chromatographic Techniques: Clinical and Biochemical Applications", Ed. I. Smith, Heinemann, London, (1958); M. Markovitz, J. Lab. Clin. Med., **50**, 367 (1957); B. Vahlquist, Hoppe-Seyler's Z. Physiol. Chem., **259**, 213 (1939).
86. C. Rimington, Assoc. Clin. Path., Broadsheet 20 (1958), 36 (1961).
87. C.J. Watson and S. Schwartz, Proc. Soc. Exp. Biol. Med., **47**, 393 (1941).
88. A. Rossi-Fanelli, E. Antonini, and A. Caputo, Biochim. Biophys. Acta, **30**, 608 (1958); C.J. Watson and S. Schwartz, Proc. Soc. Exp. Biol. Med., **47**, 393 (1941).
89. B. Vahlquist, Hoppe-Seyler's Z. Physiol. Chem., **259**, 213 (1939).
90. A. Treibs and E. Herrmann, Hoppe-Seyler's Z. Physiol. Chem., **299**, 168 (1955).
91. O.H. Gibson and D.C. Harrison, Biochem. J., **46**, 154 (1950); F.T.G. Prunty, ibid., **39**, 446 (1945); C. Rimington, S. Krol, and B. Tooth, Scand. J. Clin. Lab. Invest., **8**, 251 (1956).
92. K.D. Gibson, A. Neuberger, and J.J. Scott, Biochem. J., **61**, 618 (1955).

93. A. Merchante, B.L. Wajchenberg, and S. Schwartz, Proc. Soc. Exp. Biol. Med., **95**, 221 (1957).
94. E.I.B. Dresel and J.E. Falk, Biochem. J., **63**, 388 (1956).
95. F.T.G. Prunty, Biochem. J., **39**, 446 (1945).
96. T. Heikel, Scand. J. Clin. Lab. Invest., **8**, 172 (1956).
97. R.A. Nevé, R.F. Labbe, and R.A. Aldrich, J. Am. Chem. Soc., **78**, 691 (1956).
98. R.J. Porra and J.E. Falk, Biochem. J., **90**, 69 (1963).
99. D. Mauzerall and S. Granick, J. Biol. Chem., **232**, 1141 (1958).
100. S. Sano and S. Granick, J. Biol. Chem., **236**, 1173 (1961).
101. S. Sano and C. Rimington, Biochem. J., **86**, 203 (1963).
102. A.S. Holt, Plant Physiol., **34**, 310 (1959).
103. G. Nishida and R.F. Labbe, Biochim. Biophys. Acta, **31**, 519 (1959).
104. Saillet, Rev. Med. Paris, **16**, 542 (1896).
105. D.N. Raine, Biochem. J., **47**, xiv (1950).
106. L. Eriksen, Nature London, **167**, 691 (1951).
107. G.D. Dorough, J.R. Miller, and F.M. Huennekens, J. Am. Chem. Soc., **73**, 4315 (1951).
108. H. Fischer and K. Herrle, Hoppe-Seyler's Z. Physiol. Chem., **251**, 85 (1938).
109. A. Salek, Diplomarbeit, Braunschweig, (1970).
110. H. Fischer and M. Dürr, Ann. Chem., **501**, 107 (1933).
111. S. Granick, J. Biol. Chem., **175**, 333 (1948).
112. S.J. Baum, B.F. Burnham, and R.A. Plane, Proc. Nat. Acad. Sci. U.S.A., **52**, 1439 (1964).
113. J.-H. Fuhrhop and S. Granick, Biochem. Prep., **13**, 55 (1971).
114. J.-H. Fuhrhop and D. Mauzerall, J. Am. Chem. Soc., **91**, 4174 (1969).
115. J.-H. Fuhrhop, Z. Naturforsch., B, **25**, 255 (1970).
116. J.-H. Fuhrhop, K.M. Kadish, and D.G. Davis, J. Am. Chem. Soc., **95**, 5140 (1973).
117. D.B. Boylan and M. Calvin, J. Am. Chem. Soc., **89**, 5472 (1967).
118. H. Parnemann, Dissertation, Braunschweig, (1964).
119. G.D. Dorough and F.M. Huennekens, J. Am. Chem. Soc., **74**, 3974 (1952).
120. A. Treibs, Ann. Chem., **509**, 103 (1934).
121. J.G. Erdman, V.G. Ramsey, N.W. Kalenda, and W.E. Hanson, J. Am. Chem. Soc., **78**, 5844 (1956).
122. H. Fischer and F.W. Neumann, Ann. Chem., **494**, 225 (1932).
123. J.P. Fackler Jr., Prog. Inorg. Chem., **7**, 361 (1966).
124. J.W. Büchler, G. Eikelmann, L. Puppe, K. Rohbock, H.H. Schneehage, and D. Weck, Ann. Chem., **745**, 135 (1971).
125. C.-P. Wong, R.F. Venteicher, and W. DeW. Horrocks Jr., J. Am. Chem. Soc., **96**, 7149 (1974).
126. T.S. Srivastava and E.B. Fleischer, J. Am. Chem. Soc., **92**, 5518 (1970); Inorg. Chem. Acta, **5**, 151 (1971).
127. M. Tsutsui, R.A. Velapoldi, K. Suzuki, F. Vohwinkel, M. Ichikawa, and T. Koyano, J. Am. Chem. Soc., **91**, 6262 (1969).
128. H. Fischer and W. Zerweck, Hoppe-Seyler's Z. Physiol. Chem., **137**, 176 (1924).
129. P.S. Clezy and D.B. Morell, Biochim. Biophys. Acta, **71**, 150, 165 (1963).
130. D.B. Morell and M. Stewart, Aust. J. Exp. Biol. Med. Sci., **34**, 211 (1956).
131. H. Fischer and B. Pützer, Hoppe-Seyler's Z. Physiol. Chem., **154**, 39 (1926).
132. V.G. Ramsey, Biochem. Prep., **3**, 39 (1953).
133. O. Warburg and E. Negelein, Biochem. Z., **244**, 9 (1932).
134. R. Lemberg, B. Bloomfield, P. Caiger, and W. Lockwood, Aust. J. Exp. Biol., **33**, 435 (1955); D.B. Morell, J. Barrett, and P.S. Clezy, Biochem. J., **78**, 793 (1961).
135. S. Granick, J. Biol. Chem., **172**, 717 (1948).

136. M. Grinstein and F.B. Camponovo, Rev. Soc. Argent. Biol., **21**, 301 (1940); A. Hamsik, Hoppe-Seyler's Z. Physiol. Chem., **196**, 195 (1931).
137. A. Papendiek, Hoppe-Seyler's Z. Physiol. Chem., **152**, 215 (1926); O. Schumm, ibid., **139**, 219 (1924).
138. K.G. Paul, Acta Chem. Scand., **4**, 1221 (1950); O. Schumm and E. Mertens, Hoppe-Seyler's Z. Physiol. Chem., **177**, 15 (1928).
139. (a) H. Fischer and J. Hasenkamp, Ann. Chem., **519**, 42 (1935); (b) M. Nencki and J. Zalesky, Hoppe-Seyler's Z. Physiol. Chem., **30**, 384 (1900).
140. O. Warburg and H.S. Gewitz, Hoppe-Seyler's Z. Physiol. Chem., **288**, 1 (1951).
141. A.H. Corwin and J.G. Erdman, J. Am. Chem. Soc., **68**, 2473 (1946); J.G. Erdman, V.G. Ramsey, N.W. Kalenda, and W.E. Hanson, ibid., **78**, 5844 (1956); J.F. Taylor, J. Biol. Chem., **135**, 569 (1940).
142. A.H. Corwin and R.H. Krieble, J. Am. Chem. Soc., **63**, 1829 (1941).
143. M. Grinstein, J. Biol. Chem., **167**, 515 (1947).
144. R. Lemberg and J.W. Legge, "Haematin Compounds and Bile Pigments", Interscience, New York (1949).
145. R. v. Zeyneck, Hoppe-Seyler's Z. Physiol. Chem., **25**, 492 (1898).
146. R.C. Alcock, Biochem. J., **27**, 754 (1933); H. Fischer, A. Treibs, and K. Zeile, Hoppe-Seyler's Z. Physiol. Chem., **195**, 1 (1931); R. Hill, Proc. R. Soc. London, Ser. B, **100**, 419 (1926); E. Kalmus, Hoppe-Seyler's Z. Physiol. Chem., **70**, 217 (1910); R. v. Zeyneck, ibid., **70**, 224 (1911); A.H. Corwin and S.D. Bruck, J. Am. Chem. Soc., **80**, 4736 (1958).
147. K.-G. Paul, H. Theorell, and Å. Åkeson, Acta Chem. Scand., **7**, 1284 (1953).
148. C.A. Appleby and R.K. Morton, Biochem. J., **73**, 539 (1959).
149. R. Lemberg, D.B. Morell, N. Newton, and J.E. O'Hagan, Proc. R. Soc. London, Ser. B, **155**, 339 (1961).
150. R.J. Porra and O.T.G. Jones, Biochem. J., **87**, 181 (1963).
151. M.L. Anson and A.E. Mirsky, J. Gen. Physiol., **13**, 469 (1930); D.L. Drabkin, J. Biol. Chem., **158**, 721 (1945); U.J. Lewis, ibid., **206**, 109 (1954); J.B. Sumner and A.L. Dounce, ibid., 127, 439 (1938); A. Rossi-Fanelli, E. Antonini, and A. Caputo, Biochim. Biophys. Acta, **30**, 608 (1958); H. Theorell, Ark. Kemi, Mineral. Geol., **14B**, No. 20, 1 (1940).
152. M. Grassl, G. Augsburg, U. Coy, and F. Lynen, Biochem. Z., **337**, 35 (1963); M. Grassl, U. Coy, R. Seyffert, and F. Lynen, ibid., **338**, 771 (1963).
153. J.L. York, S. McCoy, D.N. Taylor, and W.S. Caughey, J. Biol. Chem., **242**, 908 (1967).
154. F.W.J. Teale, Biochim. Biophys. Acta, **35**, 543 (1959).
155. J. Barrett, Biochim. Biophys. Acta, **54**, 580 (1961).
156. A.C. Crooke and C.J.O. Morris, J. Physiol., **101**, 217 (1942).
157. M. Schalfejeff, Chem. Ber., **18**, 232 (1885).
158. H. Fischer, Org. Syn., **3**, 442 (1955).
159. R.F. Labbe and G. Nishida, Biochim. Biophys. Acta, **26**, 437 (1957).
160. T.C. Chu and E.J. Chu, J. Biol. Chem., **212**, 1 (1955).
161. P. Formijne and N.J. Poulie, Proc. K. Ned. Akad. Wet., Ser. C, **57**, 438 (1954).
162. J. Wildy, A. Nizet, and A. Benson, Biochim. Biophys. Acta, **54**, 414 (1961).
163. A. Hamsik and M. Hofman, Hoppe-Seyler's Z. Physiol. Chem., **305**, 143 (1956).
164. J.G. Erdman and A.H. Corwin, J. Am. Chem. Soc., **69**, 750 (1947).
165. K.-G. Paul, Acta Chem. Scand., **4**, 239 (1950); **5**, 389 (1951).
166. M. Morrison and E. Stotz, J. Biol. Chem., **213**, 373 (1955).
167. J.R. Postgate, J. Gen. Microbiol., **14**, 545 (1956).
168. I. Sekuzu, Y. Orii, and K. Okunuki, J. Biochem. Tokyo, **48**, 214 (1960).
169. E. Margoliash, N. Frohwirt, and E. Wiener, Biochem. J., **71**, 559 (1959).

170. J. Barrett and M.D. Kamen, Biochim. Biophys. Acta, **50**, 573 (1961).
171. J. Fajer, D.C. Borg, A. Forman, D. Dolphin, and R.H. Felton, J. Am. Chem. Soc., **92**, 3451 (1970).
172. P.K.W. Wasser, Diplomarbeit, Braunschweig, (1971).
173. G.H. Barnett, M.F. Hudson, S.W. McCombie, and K.M. Smith, J. Chem. Soc., Perkin Trans. 1, 691 (1973).
174. J.-H. Fuhrhop, S. Besecke, J. Subramanian, Chr. Mengersen, and D. Riesner, in preparation.
175. G.H. Barnett, B. Evans, and K.M. Smith, Tetrahedron, in press.
176. J.-H. Fuhrhop, Chem. Commun., 781 (1970); J.-H. Fuhrhop, S. Besecke, and J. Subramanian, ibid., 1 (1973).
177. H.H. Inhoffen, J.-H. Fuhrhop, and F. v.d. Haar, Ann. Chem., **700**, 92 (1966); H. Fischer and A. Treibs, ibid., **451**, 209 (1927).
178. R. Bonnett, M.J. Dimsdale, and G.F. Stephenson, J. Chem. Soc. C, 564 (1969).
179. H. Fischer and H. Libowitzky, Hoppe-Seyler's Z. Physiol. Chem., **255**, 209 (1938); **251**, 198 (1938); H. Libowitzky, ibid., **265**, 191 (1940).
180. A.H. Jackson, G.W. Kenner, and K.M. Smith, J. Am. Chem. Soc., **88**, 4539 (1966); J. Chem. Soc. C, 302 (1968).
181. R. Bonnett and A.F. McDonagh, J. Chem. Soc., Perkin Trans. 1, 881 (1973).
182. J.-H. Fuhrhop and D. Mauzerall, Photochem. Photobiol., **13**, 453 (1971).
183. J.-H. Fuhrhop, P.K.W. Wasser, J. Subramanian, and U. Schrader, Ann. Chem., 1450 (1974); G. Struckmeier, unpublished results.
184. R. Mählhop, Disseration, Braunschweig, (1966); H.H. Inhoffen, P. Jäger, and R. Mählhop, Ann. Chem., **749**, 109 (1971).
185. U. Eisner, A. Lichtarowicz, and R.P. Linstead, J. Chem. Soc., 733 (1957).
186. H.W. Whitlock, R. Hanauer, M.Y. Oester, and B.K. Bower, J. Am. Chem. Soc., **91**, 7485 (1969).
187. P.K.W. Wasser, Dissertation, Braunschweig, (1973).
188. A.H. Jackson, G.W. Kenner, G. McGillivray, and K.M. Smith, J. Chem. Soc. C, 294 (1968).
189. H.H. Inhoffen, J.W. Büchler, and P. Jäger, Fortschr. Chem. Org. Naturst., **26**, 284 (1968).
190. R. Bonnett, I.A.D. Gale, and G.F. Stephenson, J. Chem. Soc. C, 1169 (1967).
191. G.W. Kenner, K.M. Smith, and M.J. Sutton, Tetrahedron Lett., 1303 (1973).
192. R.B. Woodward and V. Škarić, J. Am. Chem. Soc., **83**, 4676 (1961).
193. J.A.S. Cavaleiro and K.M. Smith, J. Chem. Soc., Perkin Trans., 1, 2149 (1973).
194. H. Voigt, Dissertation, Braunschweig, (1964).
195. H.H. Inhoffen, J.-H. Fuhrhop, H. Voigt, and H. Brockmann, Jr., Ann. Chem., **695**, 133 (1966).
196. R. Bonnett and G.F. Stephenson, J. Org. Chem., **30**, 2791 (1965).
197. G.H. Barnett and K.M. Smith, J. Chem. Soc., Chem. Commun., 772 (1974).
198. R. Bonnett, I.A.D. Gale, and G.F. Stephenson, J. Chem. Soc. C, 1000 (1966).
199. P.S. Clezy and G.A. Smythe, Chem. Commun., 127 (1968).
200. S. Granick, J. Biol. Chem., **232**, 1101 (1958).
201. R.N. Boos, Anal. Chem., **20**, 964 (1948).
202. H. Fischer and H. Orth, 'Die Chemie des Pyrrols', Akademische Verlagsgesellschaft, Leipzig Vol. I, (1934).
203. H. Fischer and J.J. Postowsky, Hoppe-Seyler's Z. Physiol. Chem., **152**, 300 (1926); H. Fischer and P. Rothemund, Chem. Ber., **61**, 1268 (1928).
204. Ref. 39, p. 17.
205. M.J. Parker, Biochim. Biophys. Acta, **35**, 496 (1959).
206. P.S. Clezy and J. Barrett, Biochem. J., **78**, 798 (1961).

207. R. Lemberg and J. Parker, Aust. J. Exp. Biol., Med. Sci.,30, 163 (1952).
208. F. Sparatore and D. Mauzerall, J. Org. Chem., 25, 1073 (1960).
209. H. Fischer and K. Deilmann, Hoppe-Seyler's Z. Physiol. Chem., 280, 186 (1944).
210. S. Sano, J. Biol. Chem., 241, 5276 (1966).
211. G.W. Kenner, S.W. McCombie, and K.M. Smith, Ann. Chem., 1329 (1973).
212. C. Djerassi, R.R. Engle, and A. Bowers, J. Org. Chem., 21, 1547 (1956).
213. H. Fischer, A. Treibs, and G. Hummel, Hoppe-Seyler's Z. Physiol. Chem., 185, 33 (1929).
214. H. Fischer and R. Müller, Hoppe-Seyler's Z. Physiol. Chem., 142, 120, 155 (1925).
215. P.S. Clezy, Nature London, 192, 750 (1961).
216. H. Fischer and W. Zerweck, Hoppe-Seyler's Z. Physiol. Chem., 137, 242 (1924).
217. G.P. Arsenault, E. Bullock, and S.F. MacDonald, J. Am. Chem. Soc., 82, 4384 (1960).
218. P.A.D. Cornford and A. Benson, J. Chromatogr., 10, 141 (1963).
219. P.R. Edmondson and S. Schwartz, J. Biol. Chem., 205, 605 (1953).
220. A.E. Vasil'ev, A.A. Khachatur'yan, B.N. Borisov, M.M. Ushakova, and G.Ya. Rozenberg, Zh. Org. Khim., 8, 1973 (1972).
221. P.K. Warme and L.P. Hager, Biochemistry, 9, 1599 (1970).
222. R. Lemberg and J.E. Falk, Biochem. J., 49, 674 (1951).
223. A.W. Johnson and D. Oldfield, J. Chem. Soc. C, 794 (1966).
224. A.H. Jackson, G.W. Kenner, and J. Wass, J. Chem. Soc., Perkin Trans. 1, 480 (1970).
225. R. Lemberg, P.S. Clezy, and J. Barrett, in 'Haematin Enzymes', Eds. J.E. Falk, R. Lemberg, and R.K. Morton, Pergamon Press, London (1961), p. 344.
226. R. Lemberg, Nature London, 193, 373 (1962); R. Lemberg and R. Newton, Proc. R. Soc. London, Ser. B, 155, 364 (1961).
227. R.J. Porra and O.T.G. Jones, Biochem. J., 87, 181 (1963).
228. J.N. Phillips, Rev. Pure Appl. Chem., 10, 35 (1960).
229. J.H. Wang, A. Nakahara, and E.B. Fleischer, J. Am. Chem. Soc., 80, 1109 (1958).
230. K. Zeile and B. Rau, Hoppe-Seyler's Z. Physiol. Chem., 250, 197 (1937).
231. E.B. Fleischer, J.M. Palmer, T.S. Srivastava, and A. Chatterjee, J. Am. Chem. Soc., 93, 3162 (1971).
232. D.W. Hughes and A.S. Holt, Can. J. Chem., 40, 171 (1962).
233. R. Willstätter and W. Mieg, Ann. Chem., 350, 1 (1906).
234. H. Fischer and H. Orth, 'Die Chemie des Pyrrols', Akademische Verlagsgesellschaft, Leipzig, Vol. IIi (1937), and Vol. IIii (1940).
235. L.C. Craig, C. Golumbic, H. Mighton, and E. Titus, J. Biol. Chem., 161, 321 (1945); K.-G. Paul, Scand. J. Clin. Lab. Invest., 5, 212 (1953).
236. J.E. Falk, E.I.B. Dresel, A. Benson, and B.C. Knight, Biochem. J., 63, 87 (1956).
237. S. Granick and L. Bogorad, J. Biol. Chem., 202, 781 (1953).
238. (a) A.H. Jackson, G.W. Kenner, and J. Wass, J. Chem. Soc., Perkin Trans. 1, 1475 (1972); G.Y. Kennedy, A.H. Jackson, G.W. Kenner, and C.J. Suckling, FEBS Lett., 6, 9 (1970); 7, 205 (1970); (b) J.A.S. Cavaleiro, G.W. Kenner, and K.M. Smith, J. Chem. Soc., Perkin Trans. 1, 1188 (1974).
239. A. Treibs and E. Wiedemann, Ann. Chem., 471, 146 (1929).
240. L.F. Fieser and M. Fieser, 'Reagents for Organic Synthesis', Vol. 1, Wiley, New York (1967), p. 191.
241. P.S. Clezy and D.B. Morell, Biochim. Biophys. Acta, 71, 150 (1963).
242. S. Granick, J. Biol. Chem., 236, 1168 (1961).
243. S.F. MacDonald and K.H. Michl, Can. J. Chem., 34, 1768 (1956).
244. F. Morsingh and S.F. MacDonald. J. Am. Chem. Soc., 82, 4377 (1960).
245. E.M. Jope and J.R.P. O'Brien, Biochem. J., 39, 239 (1945).
246. R.E.H. Nicholas and C. Rimington, Biochem. J., 55, 109 (1953).

247. A.M. del C. Batlle and M. Grinstein, Biochim. Biophys. Acta, **57**, 191 (1962).
248. M.S. Stoll, G.H. Elder, D.E. Games, P. O'Hanlon, D.S. Millington, and A.H. Jackson, Biochem. J., **131**, 429 (1973).
249. A.R. Battersby, E. Hunt, M. Ihara, E. McDonald, J.B. Paine III, F. Satoh, and J. Saunders, J. Chem. Soc., Chem. Commun., 994 (1974).
250. M. Blumer, Anal. Chem., **28**, 1640 (1956).
251. J.L. Connelly, M. Morrison, and E. Stotz, J. Biol. Chem., **233**, 743 (1958).
252. J.A. Owen, H.J. Silberman, and C. Got, Nature London, **182**, 1373 (1958).
253. L. Eriksen, Scand. J. Clin. Lab. Invest., **5**, 155 (1953).
254. R. Kehl and W. Stich, Hoppe-Seyler's Z. Physiol. Chem., **289**, 6 (1951).
255. R.E.H. Nicholas and C. Rimington, Scand. J. Clin. Lab. Invest., **1**, 12 (1949); Biochem. J., **48**, 306 (1951).
256. T.C. Chu and E.J. Chu, J. Biol. Chem., **234**, 2747 (1959).
257. A.H. Corwin and Z. Reyes, J. Am. Chem. Soc., **78**, 2437 (1956).
258. D.A. Rappoport, C.R. Calvert, R.K. Loeffler, and J.G. Gast, Anal. Chem., **27**, 820 (1955).
259. H.N. Dunning and J.K. Carlton, Anal. Chem., **28**, 1362 (1956).
260. G.Y. Kennedy, Scand. J. Clin. Lab. Invest., **5**, 281 (1953).
261. T.K. With, Scand. J. Clin. Lab. Invest., **9**, 395 (1957).
262. A.R. Battersby, K.H. Gibson, E. McDonald, L.N. Mander, J. Moron, and L.M. Nixon, J. Chem. Soc., Chem. Commun., 768 (1973).
263. L. Eriksen, Scand. J. Clin. Lab. Invest., **10**, 319 (1958).
264. S. Nishikawa, J. Jpn. Biochem. Soc., **24**, 52 (1952).
265. I. Sumegi, Acta Morphol. Acad. Sci. Hung., **1**, 459 (1951).
266. C.J. Watson and S. Schwartz, Proc. Soc. Exp. Biol. Med., **44**, 7 (1940).
267. C. Rimington and A. Benson, J. Chromatogr., **6**, 350 (1961).
268. J. Jensen, J. Chromatogr., **10**, 236 (1963).
269. T.C. Chu, A.A. Green, and E.J. Chu, J. Biol. Chem., **190**, 643 (1951).
270. T.C. Chu and E.J. Chu, J. Biol. Chem., **227**, 505 (1957).
271. J. Barrett, Biochem. J., **64**, 626 (1956); Nature London, **183**, 1185 (1959).
272. J.E. Falk and A. Benson, Biochem. J., **55**, 101 (1953).
273. J.E. Falk, in "Porphyrin Biosynthesis and Metabolism", Eds. G.E. Wolstenholme and E.C.P. Millar, Churchill, London (1955), p. 63.
274. L. Bogorad and G.S. Marks, Biochim. Biophys. Acta, **41**, 356 (1960).
275. J. Canivet and C. Rimington, Biochem. J., **55**, 867 (1953).
276. J.E. Falk and A. Benson, Arch. Biochem. Biophys., **51**, 528 (1954).
277. T.C. Chu and E.J. Chu, J. Biol. Chem., **208**, 537 (1954).
278. H. Fischer, Hoppe-Seyler's Z. Physiol. Chem., **259**, 1 (1939); H. Fischer and C.G. Schroeder, Ann. Chem., **541**, 196 (1939).
279. A. Treibs, Hoppe-Seyler's Z. Physiol. Chem., **168**, 68 (1927).
280. J. Shack and W.M. Clark, J. Biol. Chem., **171**, 143 (1947).
281. M. Morrison and E. Stotz, J. Biol. Chem., **228**, 123 (1957).
282. H. Fischer and H. Hofmann, Hoppe-Seyler's Z. Physiol. Chem., **246**, 15 (1937).
283. A. Comfort, Biochem. J., **44**, 111 (1949).
284. R.E.H. Nicholas, Biochem. J., **48**, 309 (1951).
285. T. Heikel, Scand. J. Clin. Lab. Invest., **10**, 193 (1958).
286. C.J. Watson, S. Schwartz, and V. Hawkinson, J. Biol. Chem., **157**, 345 (1945); R. Hill, Biochem. J., **19**, 341 (1925).
287. T.C. Chu and E.J. Chu, Anal. Chem., **30**, 1678 (1958); J. Biol. Chem., **234**, 2741 (1959).
288. T.C. Chu and E.J. Chu, J. Biol. Chem., **234**, 2751 (1959).
289. T.C. Chu and E.J. Chu, Biochem. J., **83**, 318 (1962).

290. J. Lucas and M. Orten, J. Biol. Chem., **191**, 287 (1951).
291. L. Eriksen, Scand. J. Clin. Lab. Invest., **9**, 97 (1957).
292. R. Lemberg and M. Stewart, Aust. J. Exp. Biol. Med. Sci., **33**, 451 (1955); D.B. Morell, J. Barrett, and P.S. Clezy, Biochem. J., **78**, 793 (1961).
293. M. Kiese and H. Kurz, Biochem. Z., **325**, 299 (1954).
294. M. Morrison and E. Stotz, J. Biol. Chem., **213**, 373 (1955).
295. W.S. Caughey and J.L. York, J. Biol. Chem., **237**, P.C. 2414 (1962).
296. F.H. Hulcher and W. Vishniac, Brookhaven Symp. Biol., **11**, 348 (1939).
297. J.G. Kirchner, "Technique of Organic Chemistry", Volume XII, Eds. E.S. Perry and A. Weissberger, Interscience, New York (1967).
298. M. Doss and B. Ulshöfer, Biochim. Biophys. Acta, **237**, 356 (1971).
299. M. Doss, S. Afr. J. Lab. Clin. Med., Special Issue, **45**, 221 (1971); Z. Anal. Chem., **252**, 104 (1970).
300. L.R. Snyder and J.J. Kirkland, 'Introduction to Modern Liquid Chromatography', Wiley, New York (1974).
301. J. Schreiber, Chimia, **25**, 405 (1971); R.B. Woodward, Pure Appl. Chem., **33**, 145 (1973); A. Eschenmoser, "23rd International Congress of Pure and Applied Chemistry", Boston, July 1971, Vol. 2, Butterworths, London (1972).
302. K.M. Smith, unpublished results.
303. T. Heikel, Scand. J. Clin. Lab. Invest., **7**, 347 (1955); E.G. Larsen, I. Melcer, and J.M. Orten, Fed. Proc., Fed. Am. Soc. Exp. Biol., **14**, 440 (1955).
304. R.E. Sterling and A.G. Redeker, Scand. J. Clin. Lab. Invest., **9**, 407 (1957).
305. N. Verghese, J. Clin. Pathol., **11**, 191 (1958).
306. T.K. With, Scand. J. Clin. Lab. Invest., **8**, 113 (1956).
307. T.K. With, Scand. J. Clin. Lab. Invest., **10**, 186 (1958).
308. L. Eriksen, Scand. J. Clin. Lab. Invest., **10**, 39 (1958).
309. J.E. Kench and S.C. Papastamatis, Nature London, **170**, 33 (1952).
310. L. Witte and J.-H. Fuhrhop, Angew. Chem., in press.
311. S. Granick, S. Sassa, L. Granick, R.D. Levere, and A. Kappas, Proc. Nat. Acad. Sci. U.S.A., **69**, 2381 (1972).
312. H. Vogelmann, unpublished results.
313. G. Müller, Z. Naturforsch., B, **27**, 473 (1972).
314. J. Simplicio, Biochemistry, **11**, 2526 (1972).
315. B. Loev and M.M. Goodman, Progr. Separ. Purif., **3**, 73 (1970).

APPENDIX: ELECTRONIC ABSORPTION SPECTRA

In the following Tables, values for absorption maxima and extinction coefficients where available are given for the visible and Soret bands of porphyrins, metalloporphyrins, and chlorins in organic solvents, as well as values for porphyrin free bases, mono-cations, and di-cations in aqueous sodium dodecyl sulfate. Tables of absorption spectra of porphyrin dications in aqueous HCl (p. 784) and of pyridine hemochromes (pp. 805, 806) can be found in the text.

A comprehensive listing of absorption maxima can also be found in Treibs' book, 'Das Leben und Wirken von Hans Fischer', Hans Fischer Gesellschaft, Munich (1971), pp. 462—503.

Porphyrins and Metalloporphyrins, ed. Kevin M. Smith
© 1975, Elsevier Scientific Publishing Company, Amsterdam, The Netherlands

TABLE 1

Electronic absorption spectra of porphyrins in organic solvents

Porphyrins	Solvent	λ_{max} (nm) ϵ_{mM}	Soret	IV	III	II	Ia	I	Spectral Type	Ref.
Uro-I and -III octamethyl ester	CHCl$_3$	λ	406	502	536	572		627	Etio	1
		ϵ	215	15.8	9.35	6.85		4.18		
	Dioxan	λ		499	531	569	596	624		2
		ϵ		15.25	9.26	6.99	1.39	3.93		
Copro-I and -III tetramethyl ester	CHCl$_3$	λ	400	498	532	566	594	621	Etio	3
		ϵ	180	14.34	9.92	7.13	1.48	5.0		
	Ether	λ	397	498	527	568	596	623		3
		ϵ	172	15.12	10.58	7.27	1.32	6.69		
	Pyridine	λ	401.5	498	531	567	595-6	622		3
		ϵ	173	15.05	10.06	7.28	1.72	5.15		
	Dioxan	λ	398	497	530	566	595	621		3
		ϵ	177	15.35	10.3	7.24	1.4	5.41		
Proto-IX di-methyl ester	CHCl$_3$	λ	407	505	541	575	603	630	Etio	3
		ϵ	171	14.15	11.6	7.44	2.03	5.38		
	Ether	λ	404	503	536	576	605	633		3
		ϵ	158	14.8	11.86	6.63	1.54	6.57		
	Pyridine	λ	409	506	541	576	605	631		3
		ϵ	163	14.89	11.87	7.48	2.0	5.54		
	Dioxan	λ	406	504	538	575	603	631		3
		ϵ	164	14.7	11.59	6.86	1.41	5.60		
Deutero-IX di-methyl ester	CHCl$_3$	λ	399.5	497	530	566	593	621	Etio	3
		ϵ	175	13.36	10.1	8.21	2.21	4.95		
	Ether	λ	395	492	524	566	595	621		3
		ϵ	170	14.04	8.61	6.19	1.24	5.18		
	Pyridine	λ	400	497	529	566	593	620		3
		ϵ	175	14.5	7.84	6.32	1.32	3.85		
	Dioxan	λ	397	495	525	565	593	618		4
		ϵ	170	15.95	8.59	6.8	1.29	4.33		

Meso-IX di-	$CHCl_3$	λ	400	499	533	567	594	621	Etio	3
methyl ester		ε	166	13.56	9.62	6.48	1.69	4.87		
	Ether	λ	395.5	497	526	567	596	623		3
		ε	158	13.81	10.21	6.64	1.35	6.64		
	Pyridine	λ	401	498	532	567	595	621		3
		ε	160	14.37	10.0	6.92	1.68	5.2		
	Dioxan	λ	397	497	529	567	595	621		3
		ε	166	14.29	9.67	6.56	1.36	5.36		
Hemato-IX di-	Pyridine	λ	402	499.5	532	569.2	596	623	Etio	5
methyl ester		ε	175.5	14.7	9.04	6.57	1.26	4.35		
Etio-I-IV	$CHCl_3$	λ	399.5						Etio	6
		ε	160							
	Dioxan	λ		496	528	566	595	621		7
		ε		13.6	9.5	5.95	1.36	5.18		
2(4)-Monofor-	Ether	λ		510.5	550.5	578		640.5		8
myldeutero-IX	$CHCl_3$	λ		515	555	580		641		9
dimethyl ester		Ratios		1.0	1.79	1.037		0.175		
deuter-IX	Dioxan	λ		512	551	578.5		640		9
		Ratios		1.0	1.43	0.85		0.223	Rhodo	
2,4-Diformyl-	Ether	λ		521	557.5	593		648	Etio	8
deutero-IX		λ	435	526	562.5	595		651		
dimethyl ester	$CHCl_3$	ε	137.5	12.6	7.70	6.48		3.48		10
		Ratios	10.9	1.0	0.61	0.51		0.28		
	$CHCl_3$	λ	437	527	563	596		650.5		11
		Ratios	10.12	1.0	0.58	0.47		0.265		
2(4)-Monoace-	Ether	λ		505.5	543	576		634.5		8
tyldeutero-IX	$CHCl_3$	λ		511	548	578		634		
dimethyl ester		Ratios		1.0	1.145	0.735		0.193	Rhodo	9
2,4-Diacetyl-	Ether	λ		512.5	546.5	586		639		8
deutero-IX	$CHCl_3$	λ		518	552	588		640	Etio	9
dimethyl ester		Ratios		1.0	0.545	0.46		0.258		
	Dioxan	λ		513	548	584		635		
		ε		14.11	7.79	6.56		3.58		
	Dioxan	λ		516	552	588		639		9
		Ratios		1.0	0.525	0.425		0.246		

TABLE 1 (continued)

Porphyrins	Solvent	λ_{max} (nm) ϵ_{mM}	Soret	IV	III	II	Ia	I	Spectral Type	Ref.
Deutero-IX-2,4-diacrylic acid	Pyridine	λ	428	517	554	586		641	Etio	11
		ϵ	151	14.9	13.6	8.05		5.77		
Deutero-IX-2,4-disulfonic acid dimethyl ester	Water	λ		510	543	572		625	Phyllo	12
		Ratios		1.0	0.622	0.67		0.288		
2-Vinyl-4-formyldeutero-IX (chlorocruoro) dimethyl ester	Ether	λ		514.5	555	583		642		8
	Dioxan	λ		514	553	581		639	Rhodo	13
		ϵ		18.23	21.98	14.02		4.06		
	Dioxan	λ		515	554	583		641		9
		Ratios		1.0	1.21	0.728		0.213		
	CHCl$_3$	λ		518.5	558.5	584		644		9
		Ratios		1.0	1.4	0.917		0.209		
	CHCl$_3$	λ	421	520	560	585		644.5		10
		ϵ	161	10.6	14.8	9.33		2.29		
		Ratios	15.2	1.0	1.4	0.885		0.217		
Porphyrin-a	Ether	λ		518	558	582		647		14
		ϵ								
dimethyl ester	CHCl$_3$	λ	418.5	520	563.5	584.5		646	Oxorhodo	15
		Ratios	18.0	1.0	2.11	1.4		0.26		
Cryptoporphyrin-a	Ether	λ		515	555	584		642		8
dimethyl ester	CHCl$_3$	λ		519	559	584		642.5	Rhodo	16
		Ratios		1.0	1.32	0.86		0.21		
2(4)-Vinyl-4(2)-hydroxyethyl-deutero-IX	Ether	λ	401	502	533	574		628		17
Deutero-IX-4-propionic acid	Ether	λ	399	499	534	569		623		18

Compound	Solvent								Type	Ref
Deutero-IX-4-acrylic acid	Ether	λ	412	505	544.5	575		637		18
Pheoporphyrin-a_5 monomethyl ester	Dioxan	λ	417	521	562	583		634	Oxorhodo	2,19
		ϵ	193	9.39	16.06	12.61		1.88		
Phylloerythrin monomethyl ester	Dioxan	λ		517.5	557.5	581.5		634	Rhodo/ Oxorhodo	2
		ϵ		10.51	15.94	10.51		2.3		
Desoxophylloerythrin monomethyl ester	Dioxan	λ		496	530	564	589	615	Phyllo	2
		ϵ		16.67	3.56	6.29	1.31	6.54		
Phylloporphyrin-XV monomethyl ester	Dioxan	λ		502	533.5	573.5		627	Phyllo	2
		ϵ		15.98	5.06	6.07		1.53		
2-Vinylpheoporphyrin-a_5 monomethyl ester	Dioxan	λ	419	525	567	588		638	Oxorhodo	19
		ϵ	193	8.25	17.7	14.0		2.15		
Chloroporphyrin-e_6 trimethyl ester	Dioxan	λ		506	543	576		629	Etio	2
		ϵ		12.63	8.09	6.73		1.7		
Pemptoporphyrin dimethyl ester	CH$_2$Cl$_2$	λ	402.5	501	534	571	600	625	Etio	20
		ϵ	141	11.3	8.13	5.2	1.1	3.2		
Harderoporphyrin trimethyl ester	CH$_2$Cl$_2$	λ	405	507	540	574		630	Etio	21
		ϵ	178	17.4	13.0	10.2		5.37		
Rhodoporphyrin-XV dimethyl ester	CHCl$_3$	λ	404	508	547	574		632	Rhodo	22
		ϵ	179	10.0	13.4	7.78		2.00		
2-Vinylrhodoporphyrin-XV dimethyl ester	CHCl$_3$	λ	408	514	554	579		635	Oxorhodo	22
		ϵ	190	9.6	17.9	11.3		1.92		
2,4-Divinyl-rhodoporphyrin-XV dimethyl ester	CHCl$_3$	λ	413	515	556	583		637	Rhodo	23
		ϵ	177	11.0	13.7	8.1		1.8		

TABLE 1 (continued)

Porphyrins	Solvent	λ_{max} (nm) ϵ_{mM}	Soret	IV	III	II	Ia	I	Spectral Type	Ref.
6-β-Keto-ester from Rhodo-XV dimethyl ester	CHCl$_3$	λ	410	509	544	573		630	Rhodo	24
		ϵ	222	12.1	15.9	10.7		2.07		
	O·IM NaOMe in MeOH	λ	396	497	533	568		620	Etio	24
		ϵ	172	12.9	9.65	6.35		3.8		
6-β-Keto-ester from 2-Vinyl-rhodo-XV dimethyl ester	CH$_2$Cl$_2$	λ	409	512.5	553	574		635	Oxorhodo	23
		ϵ	176	7.2	15.4	9.9		1.3		
	O·IM NaOMe in MeOH	λ	401	504.5	542	572		625	Rhodo	23
		ϵ	154	10.5	12.4	6.9		2.2		
Octaethylporphyrin	Benzene	λ	400	498	532	568	596	622	Etio	25
		ϵ	159	14.5	10.8	6.8	1.5	5.8		
meso-Tetraphenylporphyrin	Benzene	λ	419	515	548	592		647	Etio	26
		ϵ	470	18.7	8.1	5.3		3.4		
Porphin	Benzene	λ	395	490	520	563	569	616		27
		ϵ	261	16	3.0	5.2	4.4	0.89		

References

1. D. Mauzerall, J. Amer. Chem. Soc., **82**, 2601 (1960).
2. A. Stern and H. Wenderlein, Hoppe-Seyler's Z. Physiol. Chem., **174**, 81 (1935).
3. J.E. Falk, 'Porphyrins and Metalloporphyrins', Elsevier, Amsterdam, (1964).
4. A. Stern and H. Wenderlein, Hoppe-Seyler's Z. Physiol. Chem., **175**, 405 (1936).
5. S. Granick, L. Bogorad, and H. Jaffe, J. Biol. Chem., **202**, 801 (1953).
6. C. Rimington, Biochem. J., **75**, 620 (1960).
7. A. Stern and H. Wenderlein, Hoppe-Seyler's Z. Physiol. Chem., **170**, 337 (1934).
8. P.S. Clezy and J. Barrett, Biochem. J., **78**, 798 (1961).
9. R. Lemberg and J.E. Falk, Biochem. J., **49**, 674 (1951).
10. R. Lemberg and J. Parker, Aust. J. Expt. Biol., **30**, 163 (1952).
11. F. Sparatore and D. Mauzerall, J. Org. Chem., **25**, 1073 (1960).
12. A. Neuberger and J.J. Scott, Proc. Roy. Soc. (London) Ser. A, **213**, 307 (1952).

13. A. Stern and H. Molvig, Z. Phys. Chem., **177**, 365 (1936).
14. D.B. Morell, J. Barrett, and P.S. Clezy, Biochem. J., **78**, 793 (1961).
15. R. Lemberg and M. Stewart, Aust. J. Expt. Biol., **33**, 451 (1955).
16. M.J. Parker, Biochim. Biophys. Acta, **35**, 496 (1959).
17. R.J. Porra and J.E. Falk, Biochem. J., **90**, 69 (1963).
18. R.J. Porra, Ph.D. Thesis, Canberra, (1962).
19. S. Granick, J. Biol. Chem., **183**, 713 (1950).
20. A.H. Jackson, G.W. Kenner, and J. Wass, J. Chem. Soc., Perkin Trans. I, 480 (1974).
21. J.A.S. Cavaleiro, G.W. Kenner, and K.M. Smith, J. Chem. Soc., Perkin Trans. I, 1188 (1974).
22. T.T. Howarth, A.H. Jackson, and G.W. Kenner, J. Chem. Soc., Perkin Trans. I, 502 (1974).
23. M.T. Cox, A.H. Jackson, G.W. Kenner, S.W. McCombie, and K.M. Smith, J. Chem. Soc., Perkin Trans. I, 516 (1974).
24. M.T. Cox, T.T. Howarth, A.H. Jackson, and G.W. Kenner, J. Chem. Soc., Perkin Trans. I, 512 (1974).
25. U. Eisner, A. Lichtarowicz, and R.P. Linstead, J. Chem. Soc., 733 (1957).
26. G.M. Badger, R.A. Jones, and R.L. Laslett, Aust. J. Chem., **17**, 1028 (1964).
27. U. Eisner and R.P. Linstead, J. Chem. Soc., 3749 (1955).

TABLE 2

Electronic absorption spectra of *meso*-substituted porphyrins (in chloroform unless otherwise stated)

meso-Substituent (parent porphyrin)	λ_{max} nm (ϵ_{mM})					Ref.
	Soret	Band IV	Band III	Band II	Band I	
Formyl-(Etio-I)	407(142)	509(9.7)	538(6.9)	577(6.6)	634(4.2)	1
Oximinoformyl-(Etio-I)	401(127)	500(10.4)	535(6.1)	569(4.4)	619(2.8)	1
Cyano-(Etio-I)	405.5(145)	515(9.96)	553(13.4)	587(5.76)	641(13.3)	1
Carboxy-(Etio-I)	402(154)	500(12.8)	535(7.5)	572(5.8)	620(3.9)	1
Hydroxymethyl-(Etio-I)	402(160)	504(13.3)	537(8.8)	570(6.8)	623(4.1)	1
Methyl-(Etio-I) in CH_2Cl_2	406(169.4)	506(14.0)	539(5.6)	576(5.7)	627(1.3)	2
Hydrazone of formyl (Etio-I)	407(148)	505(11.2)	537(6.3)	574(5.05)	628(2.8)	1
Nitro-(Octaethyl)	400(128)	504(12.5)	538(8.4)	571(6.3)	623(5.3)	3
$\alpha\beta$-Dinitro-(Octaethyl)	382(82.6), 396(80.8)	508(10.1)	538(5.6)	578(4.7)	628(3.0)	3
$\alpha\beta\gamma$-Trinitro-(Octaethyl)	387(76.4), 405(71.6)	514(10.3)	537(5.4)	590(4.0)	638(1.9)	3
Amino-(Etio-I)	416(158)	519(11.75)	553(5.4)	586.5(3.6)	646(6.4)	1
Benzylideneamino-(Etio-I)	414(144)	512(13.0)	543(3.04)	580(4.64)	634(0.76)	1
Acetamido-(Etio-I)	406(158)	505(13.9)	537(8.2)	575(6.3)	633(4.4)	4
Chloro-(Octaethyl)	406(161)	507(14.8)	540(5.3)	578(5.6)	628(1.6)	5
$\alpha\gamma$-Dichloro-(Octaethyl)	411(185)	514(12.7)	548(3.2)	586(4.1)	637(0.6)	5
$\alpha\beta\gamma\delta$-Tetrachloro-(Octaethyl)	446(188)	522(4.4 inf), 550(10.6)	597(9.5)	634(3.9 inf)	712(4.6)	5
Bromo-(Octaethyl)	409(184)	510(13.9)	543(6.0)	580(5.7)	630(1.8)	5
Octaethyloxophlorin	404.5(144)		547(4.8 inf)	586(9.2)	632(16.3)	6
β-Mercapto-(Copro-II-TME)	402(140)	507(9.1)	550(8.2)	574(6.6)	635(5.7)	7
β-Methoxy-(Meso-IX dimethyl ester)	404(246)	504(18.2)	535(5.1)		627(1.3)	8
β-Ethoxy-(Meso-IX dimethyl ester)	405(199.5)	504(15.9)	537(4.6)	574(5.9)	628(1.14)	8
β-OCO$_2$Me-(Meso-IX dimethyl ester)	399(224)	499(17.0)	531(7.1)	571(6.9)	624(2.34)	8

Acetoxy-(Etio-I) in CH$_2$Cl$_2$	402(160)	500(14.1)	532(5.15)	572(5.05)	625(1.55)	9
Benzoyloxy-(Octaethyl)	402(174)	501(16.1)	533(6.54)	571(6.2)	623.5(1.9)	6
Trifluoroacetoxy-(Octaethyl) in CH$_2$Cl$_2$	399(156)	500(14.3)	533(7.4)	570(5.7)	625(2.6)	10
αβ-Diacetoxy-(Etio-I) in CH$_2$Cl$_2$	404(168)	501(14.9)	532(4.4)	575(5.3)	624(0.76)	2
αβ-Dibenzoyloxy-(Etio-I) in CH$_2$Cl$_2$	405(200)	502(18.8)	532(3.7)	575(5.3)	6.25(0.59)	2
α-Benzoyloxy-γ-chloro-(Etio-I) in CH$_2$Cl$_2$	407(202)	503(17.5)	534(3.9)	573(6.0)	625(0.97)	2

References

1. A.W. Johnson and D. Oldfield, J. Chem. Soc. C, 794 (1966).
2. B. Evans, (Liverpool), unpublished results.
3. R. Bonnett and G.F. Stephenson, J. Org. Chem., **30**, 2791 (1965).
4. A.W. Johnson and D. Oldfield, J. Chem. Soc., 4303 (1965).
5. R. Bonnett, I.A.D. Gale, and G.F. Stephenson, J. Chem. Soc. C, 1600 (1966).
6. R. Bonnett, M.J. Dimsdale, and G.F. Stephenson, J. Chem. Soc. C, 564 (1969).
7. P.S. Clezy and C.J.R. Fookes, Chem. Commun., 1268 (1971).
8. A.H. Jackson, G.W. Kenner, and K.M. Smith, J. Chem. Soc. C, 302 (1968).
9. G.H. Barnett, (Liverpool), unpublished results.
10. G.H. Barnett, M.F. Hudson, S.W. McCombie, and K.M. Smith, J. Chem. Soc., Perkin Trans. I, 691 (1973).

TABLE 3

Electronic absorption spectra of chlorins in organic solvents

Compound	Solvent	λ_{max} (nm) α^* and ϵ_{mM}	350—380	380—400	400—420	420—45(
Chlorophyll-*a*	Ether	λ			410	430
		α			85.2	131.5
	Acetone	λ				433
	(80%)	α				101.5
Chlorophyll-*b*	Ether	λ				430
		α				62.7
	Acetone	λ				
	(80%)	α				
Chlorophyll-*c*	Ether	λ				447
		α^{**}				227.0
	Acetone	λ				¦42
Chlorophyll-*d*	Ether	λ		392		¦47
		α^{**}		58.9		97.8
Bacteriochlorophyll	Ether	λ	358.5	391.5		
		α	80.5	52.8		
	Methanol	λ	365			
		α	59.2			
Pheophytin-*a*	Ether	λ			408.5	
		α			132.0	
	Dioxan	λ				
		ϵ_{mM}				
	Acetone	λ			409	
	(80%)	α			130.9	
Methyl pheophor-bide-*a*	Dioxan	λ				
		ϵ_{mM}				
Chlorin-*e*$_6$ methyl ester	Dioxan	λ				
		ϵ_{mM}				
Pheophytin-*b*	Ether	λ			412.5	434
		α			83.0	216.0
	Acetone	λ				436
	(80%)	α				181
Methyl pheophor-bide-*b*	Dioxan	λ				
		ϵ_{mM}				
Rhodin-*g*$_7$ trimethyl ester	Dioxan	λ				
		ϵ_{mM}				
Bacteriopheophytin	Ether	λ	357.5	384		
		α	127.9	70.6		
	Chloroform	λ	363	390		
		α	111.9	59.4		
trans-Octaethylchlorin (OEC)	Benzene	λ		391		
		ϵ_{mM}		188.5		

0—500	500—520	520—550	550—580	580—600	600—650	650—700	700—800	Ref.
		533.5	578		615	662		1,2
		4.22	9.27		16.3	100.9		
		536		582	618	665		3
		4.78		11.6	19.6	90.8		
5		549		595	644			2
4.8		7.07		12.7	62.0			
0		(536)	(558)	600	648.9	(665)		3
8.0		(6.37)	(7.91)	14.3	52.5	(10.8)		
			579.5		628			2
			20.6		22.0			
			580		628			4
	512	548.5		595	643	688		2
	1.98	4.03		9.47	14.3	110.4		
		(530)	577			(697)	773	2
		(3.0)	22.9			(10)	100	
					608	(685)	772	2
					16.9	(9.5)	46.1	
1	505	534	560		609.5	667		2
5.1	14.6	12.6	3.6		9.8	63.7		
	506	535	559		609	667		5
	10.1	8.65	3.2		6.98	43.0		
2	505	536	(558)		610	666—667		3
6.3	15.0	13.12	(5.23)		11.9	56.6		
	506	535	560		610	666		6
	11.04	9.27	2.84		7.77	52.76		
	500	529	559		610	664		6
	13.2	5.21	2.0		4.09	53.25		
		525.5	555	599	655			2
		14.2	8.7	9.5	42.1			
		527	(558)		600	655	(666)	3
		14.9	(9.48)		10.7	35.7	(10.4)	
		525	555		600	652.5		7
		11.8	7.72		7.8	30.64		
		521	557	596	640			7
		11.56	7.48	5.81	27.06			
		525.5			625	680	749	2
		31.9			4.1	12.0	76.0	
		533			630	687	757.5	2
		29.5			4.4	13.0	71.4	
7		520		593	617			8
2.5		4.1		4.0	4.5			
6		544			647			
3.4		1.6			73.2			

TABLE 3 (continued)

Compound	Solvent	λ_{max} (nm) α^* and ϵ_{mM}	350—380	380—400	400—420	420—45•
cis-Octaethylchlorin	Benzene	λ		393		
		ϵ_{mM}		183		
γ-Chloro-OEC	Chloroform	λ		400		
		ϵ_{mM}		196		
$\gamma\delta$-Dichloro-OEC	Chloroform	λ			405	
		ϵ_{mM}			210	
γ-Bromo-OEC	Chloroform	λ		400		
		ϵ_{mM}		209		
$\gamma\delta$-Dibromo-OEC	Chloroform	λ			409	
		ϵ_{mM}			194	
γ-Nitro-OEC	Chloroform	λ		392		
		ϵ_{mM}		90		
$\gamma\delta$-Dinitro-OEC	Chloroform	λ		394		
		ϵ_{mM}		86		
meso-Tetraphenyl-chlorin	Benzene	λ				
		ϵ_{mM}				
Octaethyl-7-oxochlo-rin	Chloroform	λ			408	
		ϵ_{mM}			156	

Notes

* α = specific absorption coefficient = D/lc expressed in $l \cdot$ g/cm, where D = optical density, cell length, c = concentration of pigment in g/l.

** These figures approximate only.

Values in parentheses are for wavelengths other than absorption maxima, or for weak bands.

References

1. A.F.H. Anderson and M. Calvin, Nature, **194**, 285 (1962).
2. J.H.C. Smith and A. Benitez, in 'Modern Methods of Plant Analysis', Eds. K. Paech and M. Tracey, Vol. 4, Springer, Berlin, (1955), p. 142.

50—500	500—520	520—550	550—580	580—600	600—650	650—700	700—800	Ref.
89		522		596	620	651		8
12.6		3.5		4.1	4.4	69.1		
97		548						
12.6		1.6						
	501	548		595	649			9
	15.5	2.0		4.6	45			
	508	534	556		607	661		9
	14.1	5.7	1.9		3.9	47		
	501	527	550	597	620	650		9
	16	2.3	1.9	4.6	3.8	54		
	510	538			612	664		9
	13.6	5.8			3.7	46		
97		536			605	650		10
10.8		2.0			3.4	32.8		
					621			
					3.7			
99		530	560		617	662		10
12.2		2.8	1.3		3.9	43		
					626			
					4.1			
	518	543			600	654		11
	15	10.8			5.8	41.7		
	512	550		588	615	642		12
	7.9	10.9		5.4	1.8	32.9		

3. L.P. Vernon, Anal. Chem., 32, 1144 (1960).
4. S.W. Jeffrey, Nature, 194, 600 (1962).
5. A. Stern and H. Wenderlein, Hoppe-Seyler's Z. Physiol. Chem., 175, 405 (1936).
6. A. Stern and H. Wenderlein, Hoppe-Seyler's Z. Physiol. Chem., 174, 81 (1935).
7. A. Stern and H. Wenderlein, Hoppe-Seyler's Z. Physiol. Chem., 174, 321 (1935).
8. H.W. Whitlock, R. Hanauer, M.Y. Oester, and B.K. Bower, J. Amer. Chem. Soc., 91, 7485 (1969).
9. R. Bonnett, I.A.D. Gale, and G.F. Stephenson, J. Chem. Soc. C, 1600 (1966).
10. R. Bonnett and G.F. Stephenson, J. Org. Chem., 30, 2791 (1965).
11. G.M. Badger, R.A. Jones, and R.L. Laslett, Aust. J. Chem., 17, 1028 (1964).
12. R. Bonnett, M.J. Dimsdale, and G.F. Stephenson, J. Chem. Soc. C, 564 (1969).

TABLE 4

Electronic absorption spectra of representative monometallic metallo-octa-alkylporphyrins in organic solvents (compiled by P.D. Smith)

In this table the compounds are listed according to increasing atomic number of the central metal ion. Octaethylporphyrin derivatives are normally given, and only in those cases where the M(OEP) case is insufficiently characterized the corresponding mesoporphyrin-IX dimethyl ester complex is listed; the low extinction coefficients found for some of these are not characteristic for M(Meso-IX-DME) complexes.

Porphyrin	Solvent	Soret		β Band		α Band		Ref.
		$\lambda_{max}(nm)$	ϵ_{mM}	$\lambda_{max}(nm)$	ϵ_{mM}	$\lambda_{max}(nm)$	ϵ_{mM}	
Mg(OEP)Py$_2$	Benzene	411	468	546	20.0	582	19.1	1
Al(OEP)OPh	Benzene	399	288	533	11.2	572	22.4	2
Si(OEP)(OMe)$_2$	Benzene	408	282	538	12.6	573	14.5	3
Sc(OEP)OAc	Benzene	406	331	538	12.3	574	16.6	4
TiO(OEP)	Benzene	406	363	536	17.4	574	33.9	5,6
VO(OEP)	CH$_2$Cl$_2$	407	331	533	16.2	572	31.6	2
Cr(Meso-IX-DME)	CHCl$_3$	415	47	536	1.3	572	4.4	7
Cr(OEP)OPh(PhOH)*	Benzene	425	158	544	11.2	576	9.5	8
Mn(Meso-IX-DME)Py$_2$*	Pyridine/H$_2$O	426	158	549	16.5	581	8.3	9
Mn(OEP)Br(Py)*	Benzene	437	42	462	20.4	557	8.9	2
		355	49					
Fe(OEP)Py$_2$	Pyridine	400	125	518	14.2	548	19.9	3
Fe(OEP)OMe*	CH$_2$Cl$_2$/MeOH	396	105	477	11.8	592	8.5	4
		356	55					
Co(OEP)	Benzene	394	229	519	11.7	554	27.5	2
Co(OEP)Br(Py)	Benzene	427	95	537	13.2	568	11.7	2
Ni(OEP)	Dioxan	391	219	516	11.0	551	33.1	10
Cu(OEP)	CHCl$_3$/MeOH	399	305	522	13.0	560	24.5	11
Zn(OEP)	Dioxan	407	417	536	22.9	572	24.5	10
Ga(OEP)OPh	Benzene	404	288	537	16.2	574	21.9	3
Ge(OEP)(OMe)$_2$	Benzene	410	380	538	18.2	574	18.6	3
As(OEP)Cl	CH$_2$Cl$_2$	403	195	535	11.0	574	11.2	12

Compound	Solvent	λ	ε	λ	ε	λ	ε	Ref.
Zr(OEP)(OAc)$_2$	CH$_2$Cl$_2$	400	380	529	12.0	567	31.6	2
Nb(OEP)F$_3$	CH$_2$Cl$_2$	403	186	532	8.3	569	24.0	8
MoO(OEP)	CH$_2$Cl$_2$	412	182	540	12.0	578	20.9	5
MoO(OEP)OMe	CH$_2$Cl$_2$	443	87	562	15.1	595	10.0	2
		342	54					
Ru(OEP)CO(Py)	CH$_2$Cl$_2$	396	235	518	16.0	549	24.5	13
Rh(OEP)Me	CHCl$_3$	396	129	512	20.0	544	32.4	14
Pd(OEP)	Benzene	395	191	515	14.5	548	58.9	15
Ag(OEP)	Pyridine	410	269	527	13.9	561	20.8	3
[Ag(OEP)]ClO$_4$	CHCl$_3$/MeOH	404	131	516	9.8	522	25.2	16
Cd(OEP)Py	Pyridine	421	288	551	21.9	586	12.9	3
In(OEP)OPh	Benzene	411	347	543	16.6	579	17.4	2
Sn(OEP)(OMe)$_2$	Benzene	410	316	542	20.4	578	18.2	17
Sb(OEP)Cl	CH$_2$Cl$_2$	400	219	534	14.5	572	12.9	12
SbO(OEP)Cl	CH$_2$Cl$_2$	461	72	572	11.7	598	4.5	12
		379	87			sh		
Hf(OEP)(OAc)$_2$	Benzene	401	417	530	14.1	569	42.7	8
Ta(OEP)F$_3$	CH$_2$Cl$_2$	399	174	528	8.3	567	26.9	8
WO(OEP)OPh	Benzene	433	191	562	12.3	598	7.9	8
		356	18					
ReO(OEP)OPh	Benzene	454	62	578	9.5	596	8.5	8
		337	65					
Os(OEP)CO(Py)	CH$_2$Cl$_2$	394	302	510	12.9	540	20.4	8
Os(OEP)(OMe)$_2$	CH$_2$Cl$_2$/MeOH	370	123	497	9.4	530	7.9	13
OsO$_2$(OEP)	CH$_2$Cl$_2$	428	18	470	15.1	578	8.9	18
		378	123					
Ir(Meso-IX-DME)CO/Cl	CHCl$_3$	394	42	512	4.9	546	6.3	19
Pt(OEP)	Benzene	382	282	503	12.6	536	56.2	15
Hg(OEP)	Benzene	411	270	534	11.8	565	12.6	3
Tl(OEP)OH · H$_2$O	CH$_2$Cl$_2$	415	327	543	18.2	581	14.2	20
Pb(OEP)	Benzene	462	138	540	3.2	582	14.5	21
		365	65					
BiO(OEP)NO$_2$	CH$_2$Cl$_2$	460	72	572	13.2	606	4.4	12
		374	51			sh		

Footnotes

sh indicates a shoulder

* Indicates an additional long wavelength or near infrared absorption reported in the literature.

TABLE 4 (continued)

References

1. J.E. Falk and R.S. Nyholm, in 'Current Trends in Heterocyclic Chemistry', Eds. A. Albert, G.M. Badger, and C.W. Shoppee, Butterworths, London, (1958), p. 130.
2. J.W. Buchler, G. Eikelmann, L. Puppe, K. Rohbock, H.H. Schneehage, and D. Weck, Ann. Chem., **745**, 135 (1971).
3. J.W. Buchler, K.L. Lay, L. Puppe, and H. Stoppa, unpublished results.
4. J.W. Buchler and H.H. Schneehage, Z. Naturforsch., Teil B, **28**, 433 (1973).
5. J.W. Buchler and K. Rohbock, unpublished results; K. Rohbock, Dissertation, Technische Hochschule, Aachen, (1972).
6. J.-H. Fuhrhop, Tetrahedron Lett., 3205 (1969).
7. M. Tsutsui, R.A. Velapoldi, K. Suzuki, F. Vohwinkel, M. Ichakawa, and T. Koyano, J. Am. Chem. Soc., **91**, 6262 (1969).
8. J.W. Buchler and K. Rohbock, Inorg. Nucl. Chem. Lett., **8**, 1073 (1972).
9. L.J. Boucher and H.K. Garber, Inorg. Chem., **9**, 2644 (1970).
10. J.W. Buchler and L. Puppe, Ann. Chem., **740**, 142 (1970).
11. J.-H. Fuhrhop and D. Mauzerall, J. Am. Chem. Soc., **91**, 4174 (1969).
12. J.W. Buchler and K.L. Lay, Inorg. Nucl. Chem. Lett., **10**, 297 (1974).
13. J.W. Buchler and P.D. Smith, unpublished results.
14. H. Ogoshi, T. Omura, and Z. Yoshida, J. Am. Chem. Soc., **95**, 1666 (1973).
15. J.W. Buchler and L. Puppe, Ann. Chem., 1046 (1974).
16. K. Kadish, D.G. Davis, and J.-H. Fuhrhop, Angew. Chem., **84**, 1072 (1972).
17. J.A. Milroy, J. Physiol. London, **38**, 384 (1909).
18. J.W. Buchler and P.D. Smith, Angew. Chem., **86**, 378 (1974); Angew. Chem. Int. Ed. Engl., **13**, 341 (1974).
19. N. Sadasivan and E.B. Fleischer, J. Inorg. Nucl. Chem., **30**, 591 (1968).
20. R.J. Abraham, G.H. Barnett, and K.M. Smith, J. Chem. Soc., Perkin Trans. 1, 2142 (1973).
21. J.W. Buchler and L. Puppe, unpublished results; L. Puppe, Dissertation, Technische Hochschule, Aachen, (1972).

TABLE 5

Electronic absorption spectra of Hemins and Hematins in ether[1]

Heme	Hemin absorption maxima (nm)							Hematin absorption maxima (nm)		
	S'*	S*	IV	III	II	I	S/S'	S	II	I
Meso-IX	370	397	506—508	534	585	632—634	0.84	392	564	590
Deutero-IX	370	399	505	530	570	626—628	0.96	391	560	585
Proto-IX	381—382	407	512	539—540	585	638	0.97	399	570	595—600
2(4)-Acetyldeutero-IX	365—375	406	507	540	585	633	1.29	401	574	610
4-Formyldeutero-IX	365—370	411	507	547	600	650	1.32	404	573	616
2-Formyl-4-vinyl-deutero-IX (chlorocruoro)	370—380	415	510	548	605—610	650	1.24	406	577	620
Heme-a	370—380	416	505	550	600	660	1.26	410	573	621
2,4-Diacetyldeutero-IX	360—365	417	515	544	590	640	1.54	411	580	610—615
2,4-Diformyldeutero-IX	360	425	517	550	590	645	1.51	425	587	630

* The two components of the Soret band.
[1] P.S. Clezy and D.B. Morell, Biochim. Biophys. Acta, 71, 165 (1963).

TABLE 6

Electronic absorption spectra of porphyrins in aqueous sodium dodecyl sulfate[1]

Porphyrin	λ_{max} (nm) ϵ_{mM}	Soret	IV	III	II	I
Meso-IX dimethyl ester	λ	399	498	529	569	624
	ϵ	208	11.9	9.2	7.1	5.32
Deutero-IX dimethyl ester	λ	398	497	526	567	623
	ϵ	195	15.2	9.3	8.1	5.25
Copro-III tetramethyl ester	λ	399	498.5	533.5	562.5	615
	ϵ	207	11.1	10.3	8.56	3.36
Proto-IX dimethyl ester	λ	408	505	540.5	578	633
	ϵ	151	14.1	11.7	6.76	6.4
2,4-Diacetyldeutero-IX di-methyl ester	λ	423.5	517	552	584.5	639
	ϵ	100	9.2	6.0	5.0	1.8
Pyrro-XV monomethyl ester	λ	399	497	528	568	623
	ϵ	176	9.8	7.6	6.0	3.2
Phylloerythrin monomethyl ester	λ	422.5	524.5	569.5	594.5	644
	ϵ	137	6.6	12.2	9.0	2.8
Pheophorbide-a monomethyl ester	λ	408.5	510	544	614	673.5
	ϵ	60	6.8	7.4	7.2	40.0

[1] B. Dempsey, M.Sc. Thesis, University of Sydney, (1961); J.E. Falk, 'Porphyrins and Metalloporphyrins', Elsevier, Amsterdam, (1964), p. 237.

TABLE 7

Electronic absorption spectra of porphyrin monocations in aqueous sodium dodecyl sulfate[1]

Porphyrin	λ_{max} (nm) ϵ_{mM}	Soret	IV	III	II	I
Meso-IX dimethyl ester	λ	389		523-9	559	601.5
	ϵ	183		9.6	14.0	4.12
Deutero-IX dimethyl ester	λ	390.5		525	554.5	598.5
	ϵ	168		10.0	13.8	5.45
Copro-III tetramethyl ester	λ	390		524-30	559.5	601.5
	ϵ	178		9.52	13.9	4.2
Proto-IX dimethyl ester	λ	398.5		533-8	568	609.5
	ϵ	187		9.64	14.1	4.92
2,4-Diacetyldeutero-IX dimethyl ester	λ	414.5		538	570	616
	ϵ	108		8.4	10.8	4.4
Pyrro-XV monomethyl ester	λ	391-2		526	556	599.5
	ϵ	144		7.4	10.8	2.8
Phylloerythrin monomethyl ester	λ	415.5		521.5	559	592.5
	ϵ	198		3.6	9.6	12.0
Pheophorbide-a mono-methyl ester	λ	419	530.5	567	604.5	656
	ϵ	121	5.8	5.6	6.4	38.0

[1] B. Dempsey, M.Sc. Thesis, University of Sydney, (1961); J.E. Falk, 'Porphyrins and Metalloporphyrins', Elsevier, Amsterdam, (1964), p. 238.

TABLE 8

Electronic absorption spectra of porphyrin dications in aqueous sodium dodecyl sulfate[1]

Porphyrin	λ_{max}(nm) ϵ_{mM}	Soret	IV	III	II	I
Meso-IX dimethyl ester	λ	404			549	592
	ϵ	406			15.2	6.56
Deutero-IX dimethyl ester	λ	402			549	589.5
	ϵ	394			14.7	5.7
Copro-III tetramethyl ester	λ	404			549.5	592
	ϵ	414			16.4	6.56
Proto-IX dimethyl ester	λ	412			557	602
	ϵ	282			15.1	6.84
2,4-Diacetyldeutero-IX dimethyl ester	λ	421.5			565	610
	ϵ	192			11.4	4.6
Pyrro-XV monomethyl ester	λ	404			549	591.5
	ϵ	352			13.0	5.2
Phylloerythrin monomethyl ester	λ	418			565	592.5
	ϵ	214			9.8	7.8
Pheophorbide-*a* monomethyl ester	λ	418	531.5— 537.5	581.5	619— 621.5	666
	ϵ	109	3.6	6.4	8.4	44.0

[1] B. Dempsey, M. Sc. Thesis, University of Sydney, (1961); J.E. Falk, 'Porphyrins and Metalloporphyrins', Elsevier, Amsterdam, (1964), p. 239.

SUBJECT INDEX

Absorption spectra, in acidic solvents, 24, table 784
— —, in alkaline solvents, 24
— —, chlorins, 25, table 880—883
— —, in detergents, tables 888—889
— —, etio type, 20
— —, hemochromes, table 805
— —, *meso*-substituted porphyrins, 23, table 878—879
— —, metalloporphyrins, 25, 187—191, table 884—886
— — —, hyper type, 190
— — —, hypso type, 190
— — —, normal type, 189
— — —, in organic solvents, table 872—877
— —, oxophlorins, 26
— —, phlorins, 25
— —, phyllo type, 23
— —, porphyrins in aqueous acid, table 784
— —, porphyrins with isocyclic ring, 23
— —, porphyrin β-keto-esters, 23
— —, porphyrin monocations, 24
— —, rhodo type, 21
— —, substituent effects, 22
— —, tables 871—889
'Accidental' heme cleavage, 146—150
Accumulation of biosynthetic intermediates with ammonium ions or hydroxylamine, 79—82
Acetals from formylporphyrins, 828
Acetate method for metal insertion, 179
meso-Acetoxyporphyrins from oxophlorins, 41, 631, 816
Acetyl acetonate method for metal insertion, 182, 286, preparative method 798
Acetylation of hydroxyl groups, preparative method, 824
—, of oxophlorins, 41, 631, 816
—, step in aminolevulinic acid synthesis, 70

Acetylporphyrins, from α-hydroxyethylporphyrins, preparative method 824
Acid/base properties, 11—15, 234—238
Acid catalyzed conversion of porphobilinogen into uroporphyrinogens, 75, 82
— —, solvolysis of metalloporphyrins, 246
Acid number, 14, 15, 833
— —, table 14
Acid solvolysis reactions of metalloporphyrins, 243—247
Activated oxygen, 134, 135, 142, 629, 676
— —, in heme cleavage, 134, 135, 142
Active H atoms, determination, 821
Acylation, 648
Agar gel electrophoresis, 861
π-π Aggregation, 294, 493—501, 618
Algal bile pigments, origin, 125, 144
Alkali metal insertion, 795
Alkylation, reductive, 659—661
Amide bonds in side-chains, preparative method, 826
Aminolevulinic acid, binding to dehydratase, 73
— —, detection in natural materials, 785—787
— —, dehydratase, 72
— — —, Sepharose-bound, 72
— —, formation, 62—74
— — —, in plants, 66
— —, synthesis 62—74
— — —, acylation step, 70
— — —, decarboxylation step, 70, 71
— —, synthetase, 66
— — —, activity in plants, 66
— — —, properties, 67
— — —, inhibition, 68, 69, 72
— —, paper chromatography, 788
meso-Aminoporphyrins, 636
—, reactivity, 652
Ammonium ions, use for accumulation of biosynthetic intermediates, 79—82